P9-DEC-184

FREE VIBRATION ANALYSIS OF BEAMS AND SHAFTS

FREE VIBRATION ANALYSIS OF BEAMS AND SHAFTS

DANIEL J. GORMAN

Associate Professor of Mechanical Engineering
The University of Ottawa

A WILEY-INTERSCIENCE PUBLICATION

JOHN WILEY & SONS,

New York · London · Sydney · Toronto

Library of Congress Cataloging in Publication Data

Gorman, Daniel J. 1930–
Free vibration analysis of beams and shafts.

"A Wiley-Interscience publication."
Includes bibliographical references and index.
1. Girders–Vibration. 2. Shafting–Vibration.

I. Title.

TA660.B4G64 1975 624'.1772 74-20504
ISBN 0-471-311770-5

Printed in the United States of America

10 9 8 7 6 5 4 3 2 1

PREFACE

The objective of this book is to greatly simplify the general problem of free vibration analysis of beams and shafts for both the vibration specialist and the design engineer. For the specialist, it will obviate the need for much of the time-consuming work of formulating the general solution, enforcing the boundary conditions, and programming the computer. For the design engineer who is not a vibration specialist it will make possible, in many cases for the first time, the free vibration analysis of beams and shafts. It is also hoped that this book will form a welcome addition to the personal library of the student of mechanical vibration. With this view in mind, care has been taken to include all the theoretical considerations underlying each of the systems considered. Students who have mastered the usual undergraduate introduction to vibration of mechanical systems should have no difficulty in comprehending all the material presented in this book and thereby gain a more thorough understanding of the free vibration of beams and shafts.

The subject matter deals exclusively with free lateral and torsional vibration. In the first chapter the general theory underlying lateral vibration is reviewed, and a technique is introduced for the concise orderly storage of tabulated data for the vibration analysis of virtually any beam with any type of end support. In subsequent chapters the vibration of numerous kinds of beam systems is discussed, and the data required for their vibration analysis are tabulated. These include beams with local and distributed spring support, beams with concentrated masses, multispan (carry through) beams, and so on.

In Chapter 7 the problem of torsional vibration of shafts is considered for numerous systems. These include shafts with concentrated rotational masses and shafts with torsion spring couplings.

Lateral vibration of beams with step discontinuities in stiffness and mass distribution is considered in Chapter 8. Analysis is provided for beams with single discontinuities and beams with combinations of two discontinuities symmetrically distributed about the center of the beam.

Several miscellaneous problems in lateral beam vibration are considered in Chapter 9. These include beams with axial loading, tapered beams, and beams on elastic foundations.

Throughout the book illustrative problems are worked out in detail for the benefit of students and designers.

Many people have participated in various ways during the preparation of this book. The author is extremely indebted to Mrs. L. Sprouse for typing the manuscript and for the encouragement she provided throughout. He is also grateful to Mrs. D. Champion-Demers and Mr. K. C. Nguyen for the many hours they spent typing the tables.

Preparation of this book would have been impossible, of course, without the participation of students at the University of Ottawa who helped set up the governing equations and programmed the digital computer. This latter work was carried out by Mr. R. K. Sharma, Mr. D. S. Nguyen, and Mr. E. O. Ong. Mr. Sharma also prepared the tables for photoreduction, assisted on many occasions by Mr. D. Seaman, the departmental administrative assistant. All design graphs contained herein were prepared by Mr. K. G. Gupta.

Finally, the author would like to express his deep appreciation to the staff of the University Computing Center and the Communications Center where all tables underwent photoreduction. These people provided excellent assistance at all times. He would like to further acknowledge that this work was made possible in large part by the National Research Council of Canada which provided support in many forms, including student financial support as well as helping to make the computing facilities available.

DANIEL J. GORMAN

Ottawa, Ontario
August 1974

CONTENTS

SYMBOLS

Lateral Beam Vibration

A	Beam cross-sectional area
E	Young's modulus for beam material
f	Frequency of beam vibration
f_n	Frequency of beam vibration in nth mode
f^*	Ordered frequencies of double span beams
g	Gravitational acceleration
I	Area moment of inertia of beam cross section
I_0	Rotational moment of inertia of mass attached at beam boundary
I^*	$= \rho AL^3/I_0$
K	Spring constant for linear coil spring
K^*	Dimensionless linear spring constant $= KL^3/EI$
L	Beam overall length
M	Concentrated mass attached to beam
N	Number of spans in multispan beam
P	Axial compressive force in beam
PL^2/π^2EI	Ratio of axial compression force P to Euler buckling load for beam with simple-simple support
Q	Measure of how far beam frequency lies between limiting frequencies
$r(x)$	Beam modal shape as a function of x
$r(\xi)$	Beam modal shape as a function of ξ
t	Time
T	Spring constant for torsional spring (moment per radian deflection)
T^*	Dimensionless torsion spring constant = TL/EI
$v(x, t)$	Beam lateral displacement
x	Distance measured along beam
ξ	$= x/L$

η	$= \rho AL/M$
μ	Ratio of span length to overall beam length
ρ	Density of beam material
ω	Circular frequency of vibration of beam
β	Eigenvalue appearing in beam vibration problem

Torsional Vibration of Shafts

Symbols are identical to those exployed for lateral vibration problems except as described below.

G	Shear modulus of elasticity of shaft material $= E/2(1+\nu)$
I_0	Moment of inertia of rotational mass attached to shaft
I_0^*	$= I_p\rho L/I_0$
I_p	Polar moment of inertia of shaft cross section
K	Torsion spring constant
K^*	Dimensionless torsion spring constant KL/GI_p
T	Torque in shaft
$\theta(x)$	Modal shape of torsional vibration as a function of x
$\theta(\xi)$	Modal shape of torsional vibration as a function of ξ
$\phi(x, t)$	Angular rotation of shaft as a function of x and t
ν	Poisson's ratio for shaft material

Subscripts

SS	Simple-simple
SF	Simple-free
CC	Clamped-clamped
CS	Clamped-simple
CF	Clamped-free
FF	Free-free

CHAPTER ONE

THE FUNDAMENTAL THEORY OF LATERAL VIBRATION OF BEAMS

The lateral, torsional, and longitudinal vibration of beams has long been a subject of great interest in the industrialized world. It is now recognized that virtually no high speed equipment can be properly designed without obtaining solutions to what are essentially lateral or torsional vibration problems. The same is true in many static design problems in which time-varying forces act on various members. Examples of the former are torsional vibrations in gear train and motor-pump shafts. Examples of the latter are flow-induced vibration of heat exchanger tubes and nuclear reactor fuel rods. The fundamentals of the subject of lateral vibration of beams are discussed in this chapter, while the fundamentals of torsional vibration are discussed in Chapter 7. The less frequently encountered problems of longitudinal vibration are not discussed in this book.

1. The Free Vibration of Beams

There are essentially two kinds of lateral vibrations that a beam can undergo. These are free vibration and forced vibration. In free vibration a beam undergoes oscillatory motion while free of any external forces, whereas in forced vibration the beam responds to a system of time-varying external forces. An understanding of the free vibration of any beam is virtually a prerequisite to the understanding of its response in forced vibration. Furthermore, it is found that in the majority of design problems the need for solving the more complicated problem of forced vibration response is obviated, once a solution for the free vibration problem is obtained. This is accomplished through the "tuning out" of resonances, as discussed later.

It is known that lateral free vibration of beams can occur in an infinite number of modal shapes, referred to as modes, and that each modal shape has a discrete frequency associated with it. The first or fundamental mode is the mode associated with the lowest frequency. The second mode is that associated with the second lowest frequency, the third mode is associated with the third lowest frequency, and so on. Fortunately, the second, third, and higher modes of vibration can often be neglected by the analyst. This is primarily because the number of nodal points (points on the beam having zero displacement) increases directly with the mode number, first, second, third, and so on, and much more energy is required to excite the higher modes to appreciable amplitude. There may be cases, however, in which external driving forces are available at a frequency close to the higher mode frequencies and knowledge of these frequencies becomes essential.

2. Utilization of Free Vibration Analysis

Both experience and theoretical analysis have shown that even small time-varying concentrated or distributed external forces can excite large amplitude vibration of beams if certain conditions are fulfilled. The most important condition is that the time-varying forces contain frequencies close to the free vibration frequencies of the beam. In the specific case of concentrated forces, the force must be located at points other than nodal points for the associated free vibration frequency and modal shape. These forces may be periodic with time, such as the forces exerted on the base of a reciprocating engine, or they may vary randomly with time, such as the wall pressure forces exerted by turbulent fluid flow. The situation prevailing when the driving force varies sinusoidally with time and has a frequency coinciding with one of the beam free vibration frequencies is known as resonance. It is also possible to have periodic or random driving forces whose bandwidth contains frequencies which coincide with one or more of the beam natural (free vibration) frequencies. Frequencies that coincide are said to be in resonance.

In many if not most design problems, there is little that can be done to change the nature of the driving forces. It is obvious therefore that in order to minimize dynamic vibration response the responsibility of the vibration analyst is to alter the configuration of the beam system, where possible, so that its free vibration frequencies are removed from those available in the driving forces. In the technical world this process is often referred to as tuning out. In fact, this is the principal function of the vibration analyst. It might be added that the main objective of this book is to permit the analyst to evaluate readily the free vibration modal shapes and frequencies for almost any beam system of practical interest.

There are other less obvious reasons for requiring a knowledge of free vibration frequencies and modal shapes. In beams excited by concentrated forces, it is sometimes possible to move the point of application of these forces to nodal points for the modal shape they are about to excite. This virtually eliminates their ability to impart energy to the beam and thus keeps the dynamic response within tolerable limits. It is obvious that knowledge of modal shapes as well as free vibration frequencies is required in order to achieve this end.

It is also true that in many development problems it is desired to measure dynamic response by means of strain gauges, accelerometers, and other devices. Again, a knowledge of the modal shapes is required so that the measuring devices can be located where they will give maximum response.

While the subject of dynamic response of beams, that is, forced vibration response, is not considered in this book, it is recognized that free vibration modal shapes and frequencies are utilized by analysts in solving for dynamic response. It is expected therefore that the need for much of the work associated with solution for beam dynamic response will be obviated by means of the material contained herein.

3. *The Governing Differential Equation of Free Vibration*

There are two commonly utilized methods for obtaining solutions to the problem of free vibration of beams. The method most frequently used, where possible, is to solve the beam differential equation that expresses equilibrium between inertia forces and elastic restoring forces, subject to prescribed boundary conditions. The second method is an energy method which consists essentially of utilizing the fact that in free vibration the sum of the beam potential energy, due to its departure from a static equilibrium configuration, and the kinetic energy due to the motion of its particles is constant. Exact solutions are obtained to all problems discussed in this book by means of the former method. A short reference list indicating some of the many excellent books that deal with the subject of mechanical vibration is provided at the end of the book.

The differential equation governing the free vibration of uniform beams is discussed in most textbooks on vibration. We focus our attention on the small beam element of length dx, as shown in Fig. 1.1. Bending moments and shear forces M and V, respectively, act on the ends of the element.

The transverse shear force in the beam is given by the expression

$$V = EI\frac{\partial^3 v(x,t)}{\partial x^3}$$

Considering the beam displacements and associated slopes to be suffi-

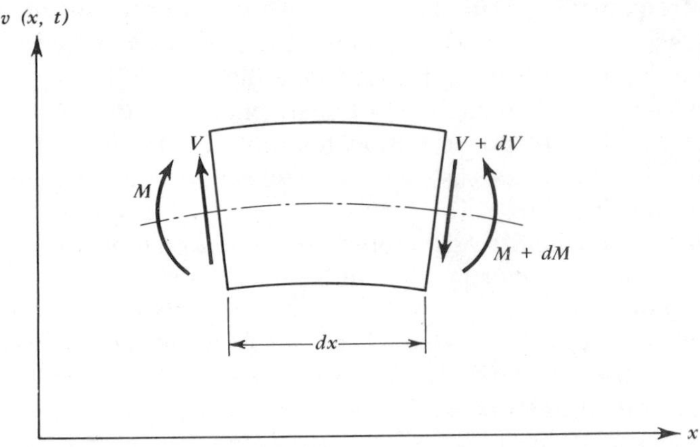

Fig. 1.1

ciently small, and setting the net transverse force acting on the element equal to its mass times acceleration, we obtain

$$\frac{-\partial^2 v(x,t)}{\partial t^2} = \frac{EI}{\rho A}\frac{\partial^4 v(x,t)}{\partial x^4} \tag{1.1}$$

It should be noted that the effects of shear strain and rotary inertia are neglected in this equation. In special cases in which these assumptions are not permitted, the analyst should consult the literature on this subject. It should also be noted that for Eq. 1.1 to be valid the distance between nodal points must be large in comparison to beam transverse dimensions, so that the effects of shear strain may be neglected. It is found that Eq. 1.1 is readily solved by means of separation of variables, and we proceed therefore to express the displacement $v(x,t)$ as follows.

$$v(x,t) = p(t)r(x) \tag{1.2}$$

Substituting Eq. 1.2 into Eq. 1.1 we obtain

$$-r(x)\frac{d^2p(t)}{dt^2} = \frac{EI}{\rho A}p(t)\frac{d^4r(x)}{dx^4} \tag{1.3}$$

It is noted that Eq. 1.3 contains total derivatives only, and it may be rearranged so that the left-hand side is a function of time only, while the right-hand side is a function of distance x:

$$-\frac{1}{p(t)}\frac{d^2p(t)}{dt^2} = \frac{EI}{\rho A}\frac{1}{r(x)}\frac{d^4r(x)}{dx^4} \tag{1.4}$$

In order for the left-hand side to equal the right-hand side, they both must be equal to a constant. It can be shown that for a solution to exist this constant must be positive, and it is denoted here as ω^2. We therefore obtain two ordinary homogeneous differential equations,

$$\frac{d^2p(t)}{dt^2} + \omega^2 p(t) = 0 \tag{1.5}$$

and

$$\frac{d^4r(x)}{dx^4} - \beta^4 r(x) = 0 \tag{1.6}$$

where

$$\beta^4 = \frac{\rho A \omega^2}{EI} \tag{1.7}$$

The solutions for these equations are

$$r(x) = A \sin \beta x + B \cos \beta x + C \sinh \beta x + D \cosh \beta x \tag{1.8}$$

and

$$p(t) = \cos(\omega t - \alpha) \tag{1.9}$$

where α is a phase angle depending on starting conditions. It is obvious that for free vibration to occur the quantity β must take on values that permit a nontrivial solution for at least one of the constants of Eq. 1.8. Having obtained these values for β we obtain associated values for frequency of vibration from Eq. 1.7.

It is the determination of the values of β in Eq. 1.8 that permit nontrivial solutions for the constants A, B, C, and D that constitutes the subject of free vibration analysis of beams. Equation 1.6 is a fourth order differential equation, and its solution is subjected to four boundary conditions. Utilizing these boundary conditions and the solution for $r(x)$ in Eq. 1.8, four simultaneous homogeneous equations can be written which relate the constants of Eq. 1.8. In order for a nontrivial solution to exist, the determinant of the coefficient matrix of the constants appearing in the four simultaneous equations must be equal to zero.

For illustrative purposes a simple problem is solved. Consider a uniform beam with simple (pinned) support at each end. At x equals zero we have zero displacement and zero bending moment. We may therefore write,

$$r(x) = \frac{d^2r(x)}{dx^2} = 0 \bigg|_{x=0}$$

and substituting in Eq. 1.8 obtain

$$B + D = 0 \tag{1.10}$$

and

$$-B + D = 0 \tag{1.11}$$

Therefore

$$B = D = 0$$

Similar conditions prevail at x equals L, therefore utilizing the results of Eqs. 1.10 and 1.11 we write,

$$A \sin \beta L + C \sinh \beta L = 0 \tag{1.12}$$

and

$$-A \sin \beta L + C \sinh \beta L = 0 \tag{1.13}$$

For a nontrivial solution for A and C to exist we have the following requirement.

$$\sin \beta L \sinh \beta L = 0 \tag{1.14}$$

Equation 1.14 is satisfied for values of βL equal to $n\pi$, where n is equal to any positive integer. Solution for n equal zero is of no interest, since no vibration occurs, and negative values of n are of no interest, since they lead to redundancy. It is now possible to obtain values for the first and all higher frequencies of vibration of the beam from Eq. 1.7. The shape of the associated modes can be obtained by means of Eq. 1.12 or 1.13. It is evident from Eq. 1.12 that C equals zero for $\beta L = n\pi$, hence $r(x) = A \sin \beta x$. Since we are interested in modal shapes only, we are at liberty to set A equal to 1 and write $r(x) = \sin \beta x$.

In the work to be undertaken in this book it is highly advantageous to work with nondimensionalized variables of space and displacement. Accordingly, we introduce the variable $\xi = x/L$, and the symbol r denotes lateral displacement divided by L when r denotes a function of ξ.

Equation 1.6 may be written

$$\frac{d^4 r(x/L)}{L^3 d(x/L)^4} - \beta^4 L r(x/L) = 0 \tag{1.15}$$

or

$$\frac{d^4 r(\xi)}{d\xi^4} - \beta^4 r(\xi) = 0 \tag{1.16}$$

where now

$$\beta^4 = \frac{\rho A \omega^2 L^4}{EI} \tag{1.17}$$

and

$$f = \frac{\omega}{2\pi} = \frac{\beta^2}{2\pi L^2} \sqrt{\frac{EI}{\rho A}} \tag{1.18}$$

The solution to Eq. 1.16 is then

$$r(\xi) = A \sin \beta\xi + B \cos \beta\xi + C \sinh \beta\xi + D \cosh \beta\xi \tag{1.19}$$

where $0 \leqslant \xi \leqslant 1$.

The solution to all lateral vibration problems of uniform beams presented in this book is obtained by subjecting the general solution (Eq. 1.19) to the appropriate boundary conditions.

4. *Boundary Conditions for Beams in Free Vibration*

The two classes of boundary conditions to which beams may be subjected are denoted classical and nonclassical boundary conditions and are discussed in that order.

Classical Boundary Conditions. These are boundary conditions that have been widely utilized in the classical literature and involve only the shape of the beam deflection curve at its boundaries. Their mathematical formulation is presented here along with the designation used to specify the beams to which they pertain. Expressions for the modal shapes are included for completeness.

1. Free-free beams.

$$\left.\frac{d^2r(\xi)}{d\xi^2} = \frac{d^3r(\xi)}{d(\xi)^3} = 0\right|_{\substack{\xi=0\\ \xi=1}}$$

$$r(\xi) = \sin \beta\xi + \sinh \beta\xi + \gamma(\cos \beta\xi + \cosh \beta\xi)$$

where

$$\gamma = \frac{\sin \beta - \sinh \beta}{\cosh \beta - \cos \beta}$$

2. Clamped-free beams.

$$\left.r(\xi) = \frac{dr(\xi)}{d\xi} = 0\right|_{\xi=0} \qquad \left.\frac{d^2r(\xi)}{d\xi^2} = \frac{d^3r(\xi)}{d\xi^3} = 0\right|_{\xi=1}$$

$$r(\xi) = \sin \beta\xi - \sinh \beta\xi - \gamma(\cos \beta\xi - \cosh \beta\xi)$$

where

$$\gamma = \frac{\sin \beta + \sinh \beta}{\cos \beta + \cosh \beta}$$

3. Clamped-clamped beams.

$$\left.r(\xi) = \frac{dr(\xi)}{d\xi} = 0\right|_{\substack{\xi=0\\ \xi=1}}$$

$$r(\xi) = \sinh \beta\xi - \sin \beta\xi + \gamma(\cosh \beta\xi - \cos \beta\xi)$$

where

$$\gamma = \frac{\sinh \beta - \sin \beta}{\cos \beta - \cosh \beta}$$

4. Simple-simple beams.

$$r(\xi) = \frac{d^2r(\xi)}{d\xi^2} = 0\bigg|_{\substack{\xi=0\\ \xi=1}}$$

$$r(\xi) = \sin \beta\xi$$

5. Clamped-simple beams.

$$r(\xi) = \frac{dr(\xi)}{d\xi} = 0\bigg|_{\xi=0} \qquad r(\xi) = \frac{d^2r(\xi)}{d\xi^2} = 0\bigg|_{\xi=1}$$

$$r(\xi) = \sinh \beta\xi - \sin \beta\xi + \gamma(\cosh \beta\xi - \cos \beta\xi)$$

where

$$\gamma = \frac{\sinh \beta - \sin \beta}{\cos \beta - \cosh \beta}$$

6. Simple-free beams.

$$r(\xi) = \frac{d^2r(\xi)}{d\xi^2} = 0\bigg|_{\xi=0} \qquad \frac{d^2r(\xi)}{d\xi^2} = \frac{d^3r(\xi)}{d\xi^3} = 0\bigg|_{\xi=1}$$

$$r(\xi) = \sin \beta\xi + \gamma \sinh \beta\xi$$

where

$$\gamma = \frac{\sin \beta}{\sinh \beta}$$

The values of β, denoted herein as eigenvalues, are presented for the first 10 modes of vibration of beams with classical boundary conditions in Table 1.1. By utilizing these values and Eq. 1.18, the frequencies of vibration can be obtained for any of these beams. The associated modal shapes can be obtained by using the expressions provided. They can be expressed as functions of x by simply replacing the variable ξ with x/L.

TABLE 1.1. EIGENVALUES FOR BEAMS WITH CLASSICAL BOUNDARY CONDITIONS

Mode	Clamped-clamped	Clamped-simple	Clamped-free	Simple-simple	Simple-free	Free-free
1	4.730	3.927	1.875	3.142	0.0	0.0
2	7.853	7.069	4.694	6.283	3.927	4.730
3	10.996	10.210	7.855	9.425	7.069	7.853
4	14.137	13.352	10.996	12.566	10.210	10.996
5	17.274	16.493	14.137	15.708	13.352	14.173
6	20.420	19.635	17.279	18.850	16.493	17.274
7	23.562	22.777	20.420	21.991	19.635	20.420
8	26.703	25.918	23.562	25.133	22.777	23.562
9	29.845	29.060	26.704	28.274	25.918	26.703
10	32.987	32.201	29.845	31.416	29.060	29.845

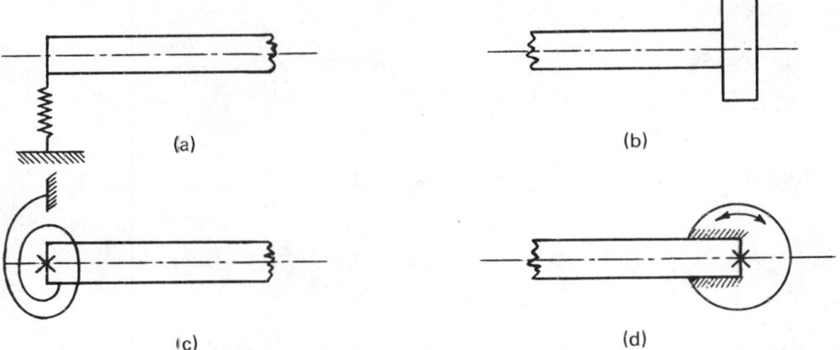

Fig. 1.2

Nonclassical Boundary Conditions. There are four kinds of nonclassical boundary conditions considered herein. They are illustrated in Fig. 1.2. It is expected that they contain most of the conditions commonly encountered. The first is presented in terms of both conventional coordinates and nondimensionalized coordinates for illustrative purposes. The remainder are presented in nondimensional form, since this is the only form utilized in the work contained in this book.

1. Beams with Lateral Coil Spring Support at One End. Consider first a beam with a coil spring at the end $x = 0$. Setting the shear force on the beam equal to the negative of the spring restoring force we obtain

$$EI\frac{d^3r(x)}{dx^3} = -Kr(x)\bigg|_{x=0}$$

or

$$\frac{EI}{L^2}\frac{d^3r(x/L)}{d(x/L)^3} = -KLr(x/L)\bigg|_{x=0}$$

and finally,

$$\frac{d^3r(\xi)}{d\xi^3} = -\frac{KL^3}{EI}r(\xi)\bigg|_{\xi=0}$$

With the coil spring located on the other end of the beam, the boundary condition becomes

$$\frac{d^3r(\xi)}{d\xi^3} = \frac{KL^3}{EI}r(\xi)\bigg|_{\xi=1}$$

2. Beams with Torsional Spring Support at One End.

$$\frac{d^2r(\xi)}{d\xi^2} = \frac{TL}{EI}\frac{dr(\xi)}{d\xi}\bigg|_{\xi=0} \quad \text{or} \quad \frac{d^2r(\xi)}{d\xi^2} = \frac{-TL}{EI}\frac{dr(\xi)}{d\xi}\bigg|_{\xi=1}$$

3. Beams with Concentrated Mass M at One End.

$$\frac{d^3r(\xi)}{d\xi^3}=\frac{M}{\rho AL}\beta^4 r(\xi)\bigg|_{\xi=0} \quad \text{or} \quad \frac{d^3r(\xi)}{d\xi^3}=\frac{-M}{\rho AL}\beta^4 r(\xi)\bigg|_{\xi=1}$$

4. Beams with Rotary Inertial Mass at One End (see Fig. 1.2d).

$$\frac{d^2r(\xi)}{d\xi^2}=\frac{-I_0}{\rho AL^3}\beta^4\frac{dr(\xi)}{d\xi}\bigg|_{\xi=0} \quad \text{or} \quad \frac{d^2r(\xi)}{d\xi^2}=\frac{I_0}{\rho AL^3}\beta^4\frac{dr(\xi)}{d\xi}\bigg|_{\xi=1}$$

Note that in each case terms involving time are common to both sides, hence are eliminated.

5. *Formulation of the Mathematical Model*

In order to facilitate presentation of the background underlying preparation of the free vibration tables, it is advantageous to discuss a particular illustrative case in detail. We therefore look at the problem of the vibration of a cantilever beam supported at its outer end with a transverse linear coil spring (Fig. 1.3). The general solution is given by Eq. 1.19. There are three classical boundary conditions to be imposed, which are evidently

$$\frac{dr(\xi)}{d\xi}=0\bigg|_{\xi=0}; \qquad \frac{d^2r(\xi)}{d\xi^2}=0\bigg|_{\xi=1}; \qquad \text{and } r(\xi)=0\bigg|_{\xi=0}$$

The fourth boundary condition is a nonclassical one and is expressed as

$$\frac{d^3r(\xi)}{d\xi^3}=K^*r(\xi)\bigg|_{\xi=1}$$

It is readily shown that the enforcing of these boundary conditions requires that $A=-C$, $B=-D$, and solution for the vibration frequencies

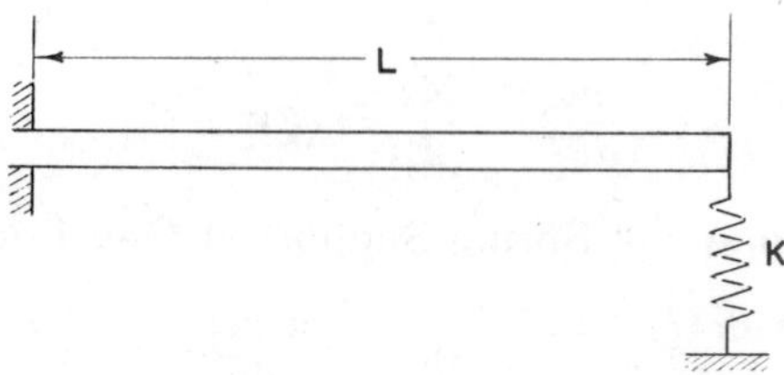

Fig. 1.3

requires finding those values of β that satisfy the determinant equation

$$\begin{vmatrix} \cos\beta + \cosh\beta + \dfrac{K^*}{\beta^3}(\sin\beta - \sinh\beta) & \sinh\beta - \sin\beta + \dfrac{K^*}{\beta^3}(\cos\beta - \cosh\beta) \\ \sin\beta + \sinh\beta & \cos\beta + \cosh\beta \end{vmatrix} = 0 \tag{1.20}$$

It is found, as would be expected, that as K^* approaches zero this equation approaches that governing the vibration of a clamped-free beam. As K^* approaches infinity the equation approaches that of a clamped-simply supported beam. Solutions for both of these classical cases are easily obtained with the aid of Table 1.1. The frequencies for any actual beam of the type under consideration will take on values somewhere between these two well-defined limits. Up to this point, however, it is not immediately obvious what these values will be.

We choose at this time to introduce a new parameter for the beam-spring system. The new parameter is called a Q factor and will be designated by the letter Q. The Q factor is a measure of how far the actual beam frequencies lie between those associated with the two limiting cases. Concentrating for the moment on the fundamental frequency, we may write for the beam frequency

$$f = f_{cF} + Q(f_{cs} - f_{cF}) \tag{1.21}$$

It is obvious that if the relationship between Q and K^* is known the frequency of vibration for any beam of the type under consideration is easily established. We now obtain this relationship by means of relationships between Q and β and between β and K^*.

It follows from the definition of β (Eq. 1.17) that we may write

$$\frac{\beta}{\beta_{CF}} = \sqrt{f/f_{cF}}, \qquad \text{or } \frac{\beta}{\beta_{cs}} = \sqrt{f/f_{cs}}$$

and utilizing Eq. 1.21

$$\beta = \beta_{cF}\sqrt{f/f_{cF}} = \beta_{cF}\sqrt{1 + Q\left(\frac{f_{cs}}{f_{cF}} - 1\right)}$$

$$= \beta_{cF}\sqrt{1 + Q\left(\frac{\beta_{cs}^2}{\beta_{cf}^2} - 1\right)} \tag{1.22}$$

With Q selected we therefore obtain a corresponding value of β. Simplifying Eq. 1.20 we obtain

$$K^* = \frac{\beta^3(1 - \cos\beta\cosh\beta)}{\sinh\beta\cos\beta - \sin\beta\cosh\beta} \tag{1.23}$$

With β selected we obtain a unique value for K^* from Eq. 1.23.

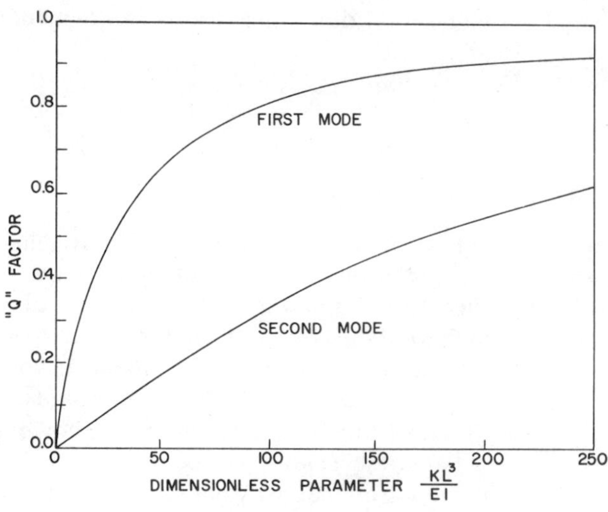

Fig. 1.4

A plot of Q versus K^* for the first and second modes of vibration is presented in Fig. 1.4. While this graph is useful for illustrative purposes, it has one obvious disadvantage in practice. It is not possible for the curves to cover the full range of Q, that is, $0 < Q < 1$, since the upper limit of K^* is infinity. Furthermore, the curves move further and further out in the direction of the positive K^* axis as we go to higher modes. A possible alternative would be to choose a logarithmic scale for the K^* axis, however, in the interest of minimizing space requirements it is preferable to present the data in tabulated form. This has been done for the first two modes of the illustrative problem under consideration, in Table 1.2. The format and frequency of points given in Table 1.2 has been selected with a view to establishing frequencies for any beam with an error not greater than 1%. In general, the properties of beams and supports are not known to this degree of accuracy. For values of Q equal to or less than 0.8,

TABLE 1.2 K^* AS FUNCTION OF Q FACTOR AND MODE NUMBER FOR CANTILEVER BEAM WITH LATERAL SPRING SUPPORT AT OUTER END

	Q									
Mode	0.05	0.10	0.20	0.30	0.40	0.50	0.60	0.70	0.80	0.90
1	1.148	2.509	5.948	10.54	16.64	24.93	36.77	55.29	89.96	188.0
2	14.55	28.22	55.75	86.25	122.5	168.2	230.0	323.0	492.2	961.8

intermediate values are obtained through linear interpolation. In order to satisfy accuracy requirements, a smaller interval between values of Q is utilized near the origin. It is observed that the curves of Fig. 1.3 approach the upper limit ($Q = 1.0$) in an asymptotic fashion. For this reason, the following relationship between Q and K^* is utilized for intermediate values of Q greater than 0.8.

$$Q = 1 - .2\frac{K^*_{Q=.8}}{K^*} \tag{1.24}$$

where $K^*_{Q=.8}$ is the value of K^* at $Q = .8$. An analysis of the tabulated data indicates that utilization of Eq. 1.24 also gives frequencies with error not greater than 1 percent.

It is advantageous at this time to demonstrate the use of Table 1.2 in determining the fundamental frequency of an actual beam of this type.

ILLUSTRATIVE PROBLEM 1.1

Consider an aluminum beam with the following geometric properties:

$$\text{Width} = 1\text{ in.} \qquad \text{depth} = \tfrac{1}{2}\text{ in.} \qquad \text{length} = 30\text{ in.}$$

We take the modulus of elasticity as being equal to 10.0×10^6 psi and the density of aluminum as 9.75×10^{-2} lb mass/in^3. Consider the beam to be clamped at one end and given lateral support at its other end by a linear coil spring with stiffness coefficient $K = 80.0$ lb/in.

(a) Determine the fundamental frequency.
Utilizing Eq. 1.18 and introducing the gravitational constant g to obtain compatible units we proceed as follows.

1. $f_{CF} = \dfrac{\sqrt{EIg/\rho A}}{2\pi L^2}\beta_{CF}^2 = 5.07 \times 1.875^2 = 17.8$ Hz

2. $f_{CS} = 5.07\beta_{CS}^2 = 5.07 \times 3.927^2 = 78.2$ Hz

3. $K^* = \dfrac{KL^3}{EI} = 20.8$

4. From Table 1.2 (interpolating),

$$Q = 0.4 + \frac{20.8 - 16.6}{24.9 - 16.6} \times 0.1 = .451$$

5. From Eq. 1.21

$$f = 17.8 + .451(78.2 - 17.8) = 45.0 \text{ Hz}$$

(b) Consider next the same beam with $K = 2000$ lb/in. Let us compute the second mode frequency.

1. $f_{CF} = 5.07 \times 4.694^2 = 112$ Hz
2. $f_{CS} = 5.07 \times 7.069^2 = 254$ Hz
3. $K^* = \dfrac{KL^3}{EI} = 520$
4. From Table 1.2 (note that $Q > 0.8$)

$$Q = 1 - .2 \times \frac{492.2}{520} = .811$$

5. $f = 112 + .811(254 - 112) = 227$ Hz

Summary

In this chapter we have looked at the basic differential equation that governs the free lateral vibration of uniform beams, and at its solution under the commonly encountered classical and nonclassical boundary conditions that must be enforced. Both the solution and boundary conditions have been expressed in dimensionless form. There were two important developments in this chapter: (1) the observation that beams with nonclassical boundary conditions become, in their limiting cases, beams with classical boundary conditions, and (2) the demonstration that by introducing a parameter, designated a Q factor, it is possible to present an orderly, concise array of tabulated data which in turn permits establishment of how far the actual beam frequency lies between the two well-defined limits. The technique has been illustrated with a specific example.

It will be seen in the chapters to follow that, utilizing this basic technique, an orderly storage of data is presented which facilitates the rapid establishment of frequencies and modal shapes of virtually any beam and beam-support system of interest where nonclassical boundary conditions are involved. Illustrative problems are worked out from time to time for the benefit of the reader.

CHAPTER TWO

FREE VIBRATION OF BEAMS WITH NONCLASSICAL BOUNDARY CONDITIONS

In this chapter, a study is made of beams with single nonclassical boundary conditions and combinations of two nonclassical boundary conditions. Tabulated data are provided for the establishment of the first 10 lateral vibration frequencies and modal shapes. It is believed that virtually all beams of this type likely to be encountered in practice are discussed. All vibration tables can be found at the end of the chapter.

Part 1. Beams with a Single Nonclassical Boundary Condition

Each beam is given a reference number. The associated table for Q versus mode number has the same reference number. Immediately following the reference number the reader will find a verbal description of the beam, the determinant equation, the expression for its modal shape, and the formula for obtaining the frequency for the nth mode. For illustrative problems the 1 in. wide by $\frac{1}{2}$ in. deep aluminum beam of 30 in. in length (see Chapter 1) will be utilized.

CASE 2.1. Beam pinned with torsion spring at one end and free at the other.

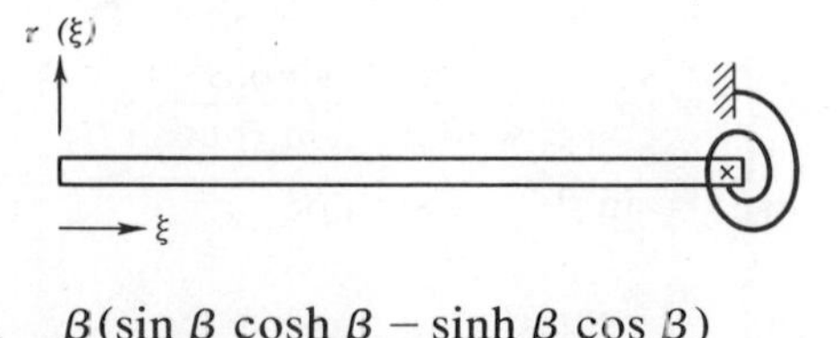

$$T^* = \frac{\beta(\sin\beta\cosh\beta - \sinh\beta\cos\beta)}{1+\cos\beta\cosh\beta}$$

$$r(\xi) = -(\sin\beta\xi + \sinh\beta\xi) + \gamma(\cos\beta\xi + \cosh\beta\xi)$$

where

$$\gamma = \frac{\sin\beta + \sinh\beta}{\cos\beta + \cosh\beta}$$

$$f_n = f_{SFn} + Q(f_{CFn} - f_{SFn})$$

ILLUSTRATIVE PROBLEM 2.1

The reference aluminum beam is pinned with torsion spring support at one end and free at the other. The torsion spring constant equals 2000 in. lb/radian angular deflection. Determine the first mode frequency and shape.

1. $f_{SF} = 0$ (Table 1.1)
2. $f_{CF} = 17.8$ Hz (Illustrative Problem 1.1)
3. $T^* = (TL/EI) = (2000 \times 30)/(10.0 \times 10^6 \times 1.04 \times 10^{-2}) = .576$
4. $Q = .300 + .100(.576 - .407)/(.781 - .407) = .345$
5. $f = 0 + .345 \times 17.8 = 6.15$ Hz

Referring to the development of Eq. 1.22, it is seen that we may write

$$\beta = \beta_{CF}\sqrt{\frac{f}{f_{CF}}} = 1.875\sqrt{\frac{6.15}{17.8}} = 1.10$$

that is, β is proportional to $\sqrt{f}$.

$$\therefore r(\xi) = \sin 1.10\xi + \sinh 1.10\xi + \gamma(\cos 1.10\xi + \cosh 1.10\xi)$$

where

$$\gamma = \frac{\sin 1.10 + \sinh 1.10}{\cos 1.10 + \cosh 1.10} = 1.05$$

CASE 2.2. Beam pinned with torsion spring at one end and simply supported at the other.

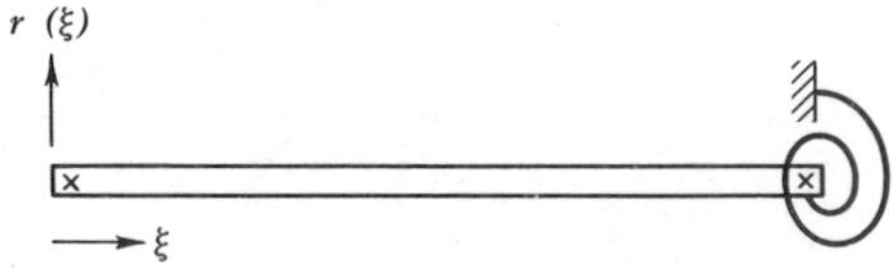

$$T^* = \frac{2\beta \sin\beta \sinh\beta}{\cos\beta \sinh\beta - \sin\beta \cosh\beta}$$

$$r(\xi) = \sin\beta\xi + \gamma \sinh\beta\xi$$

where

$$\gamma = \frac{-\sin\beta}{\sinh\beta}$$

$$f_n = f_{SSn} + Q(f_{CSn} - f_{SSn})$$

CASE 2.3. Beam pinned with torsion spring at one end and clamped at the other.

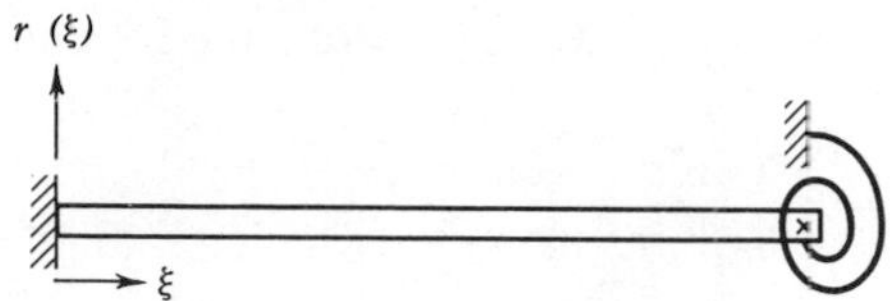

$$T^* = \frac{\beta(\sin\beta\cosh\beta - \sinh\beta\cos\beta)}{1 - \cos\beta\cosh\beta}$$

$$r(\xi) = \sinh\beta\xi - \sin\beta\xi + \gamma(\cosh\beta\xi - \cos\beta\xi)$$

where

$$\gamma = \frac{\sin\beta - \sinh\beta}{\cosh\beta - \cos\beta}$$

$$f_n = f_{CSn} + Q(f_{CCn} - f_{CSn})$$

CASE 2.4. Beam pinned at one end, and with lateral spring support at the other.

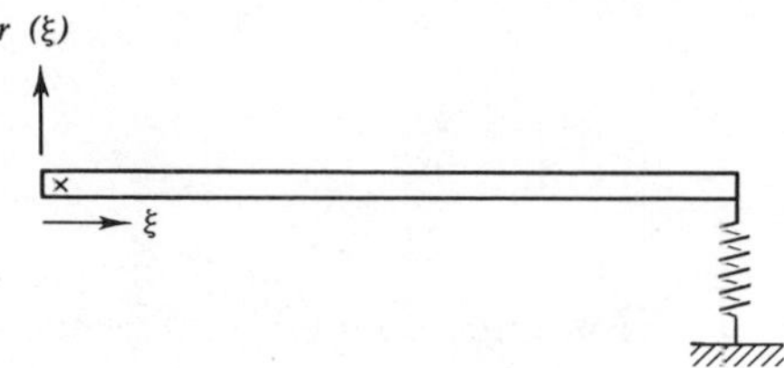

$$K^* = \frac{\beta^3(\sin\beta\cosh\beta - \cos\beta\sinh\beta)}{2\sin\beta\sinh\beta}$$

$$r(\xi) = \sin\beta\xi + \gamma\sinh\beta\xi$$

where

$$\gamma = \frac{\sin\beta}{\sinh\beta}$$

$$f_n = f_{SFn} + Q(f_{SSn} - f_{SFn})$$

CASE 2.5. Beam clamped at one end, and with lateral spring support at the other.

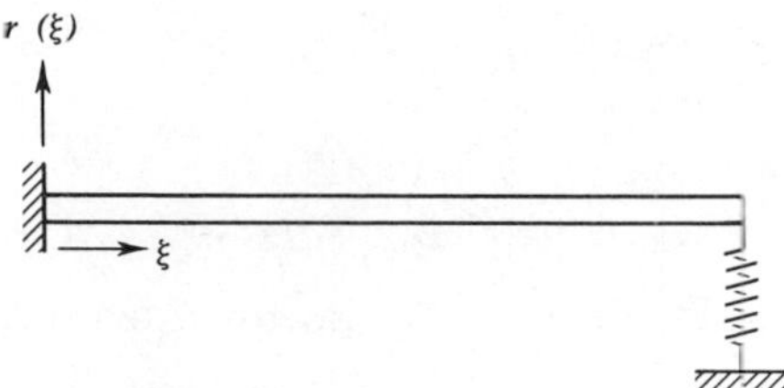

$$K^* = \frac{\beta^3(1 + \cos\beta \cosh\beta)}{\sinh\beta \cos\beta - \sin\beta \cosh\beta}$$

$$r(\xi) = \sinh\beta\xi - \sin\beta\xi + \gamma(\cosh\beta\xi - \cos\beta\xi)$$

where

$$\gamma = \frac{-(\sin\beta + \sinh\beta)}{\cosh\beta + \cos\beta}$$

$$f_n = f_{CFn} + Q(f_{CSn} - f_{CFn})$$

CASE 2.6. Beam pinned at one end, and with concentrated mass at the other.

r (ξ)

ξ

$$\eta = \frac{2\beta \sin\beta \sinh\beta}{\cos\beta \sinh\beta - \sin\beta \cosh\beta}$$

$$r(\xi) = \sin\beta\xi + \gamma \sinh\beta\xi$$

where

$$\gamma = \frac{\sin\beta}{\sinh\beta}$$

$$f_n = f_{SSn} + Q(f_{SF(n+1)} - f_{SSn})$$

CASE 2.7. Beam clamped at one end, and with concentrated mass at the other.

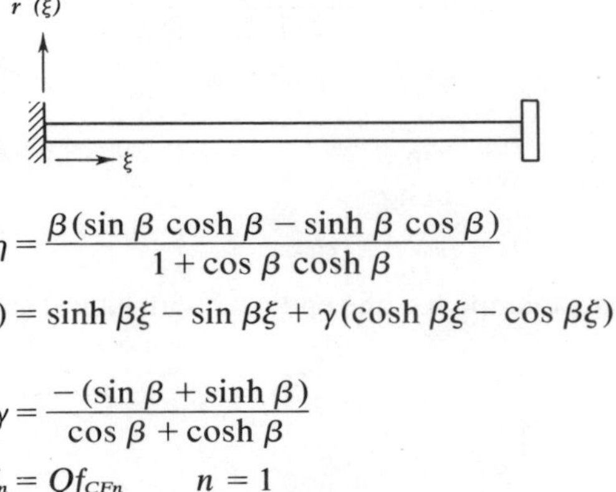

$$\eta = \frac{\beta(\sin\beta \cosh\beta - \sinh\beta \cos\beta)}{1 + \cos\beta \cosh\beta}$$

$$r(\xi) = \sinh\beta\xi - \sin\beta\xi + \gamma(\cosh\beta\xi - \cos\beta\xi)$$

where

$$\gamma = \frac{-(\sin\beta + \sinh\beta)}{\cos\beta + \cosh\beta}$$

$$f_n = Qf_{CFn} \qquad n = 1$$
$$= f_{CS(n-1)} + Q(f_{CFn} - f_{CS(n-1)}) \qquad n > 1$$

CASE 2.8. Beam free at one end and pinned with a rotational mass at the other.

$$I^* = \frac{\beta^3(1 + \cos\beta \cosh\beta)}{\sinh\beta \cos\beta - \sin\beta \cosh\beta}$$

$$r(\xi) = \sin\beta\xi + \sinh\beta\xi + \gamma(\cos\beta\xi + \cosh\beta\xi)$$

where

$$\gamma = \frac{-(\sin\beta + \sinh\beta)}{\cos\beta + \cosh\beta}$$

$$f_n = f_{CFn} + Q(f_{SF(n+1)} - f_{CFn})$$

CASE 2.9. Beam pinned at one end and pinned with a rotational mass at the other.

$$I^* = \frac{\beta^3(\sin\beta \cosh\beta - \cos\beta \sinh\beta)}{2 \sin\beta \sinh\beta}$$

$$r(\xi) = \sin\beta\xi + \gamma \sinh\beta\xi$$

where

$$\gamma = \frac{-\sin\beta}{\sinh\beta}$$

$$f_n = Qf_{SSn} \qquad n = 1$$

$$= f_{CS(n-1)} + Q(f_{SSn} - f_{CS(n-1)}) \qquad n > 1$$

CASE 2.10. Beam clamped at one end and pinned with a rotational mass at the other.

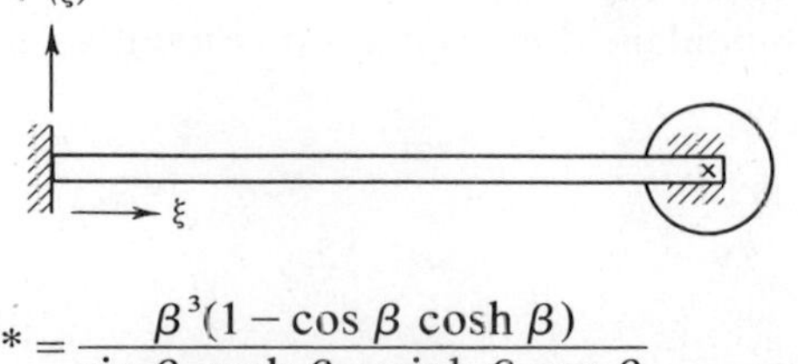

$$I^* = \frac{\beta^3(1 - \cos\beta \cosh\beta)}{\sin\beta \cosh\beta - \sinh\beta \cos\beta}$$

$$r(\xi) = \sin\beta\xi - \sinh\beta\xi + \gamma(\cos\beta\xi - \cosh\beta\xi)$$

where

$$\gamma = \frac{\sinh\beta - \sin\beta}{\cos\beta - \cosh\beta}$$

$$\begin{aligned} f_n &= Qf_{CSn} \qquad n = 1 \\ &= f_{CC(n-1)} + Q(f_{CSn} - f_{CC(n-1)}) \qquad n > 1 \end{aligned}$$

CASE 2.11. Beam free at one end, and with a concentrated mass at the other.

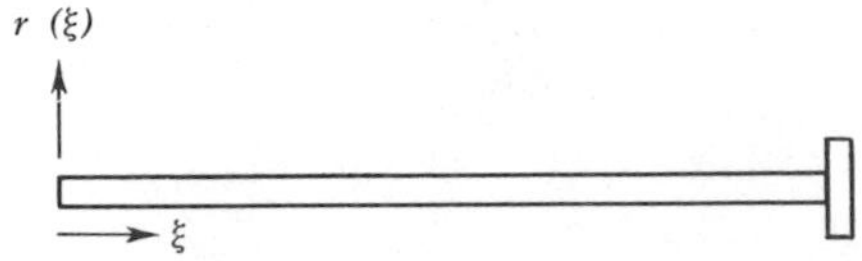

$$\eta = \frac{\beta(\sinh\beta\cos\beta - \cosh\beta\sin\beta)}{1 - \cos\beta\cosh\beta}$$

$$r(\xi) = \sin\beta\xi + \sinh\beta\xi + \gamma(\cos\beta\xi + \cosh\beta\xi)$$

where

$$\gamma = \frac{\sin\beta - \sinh\beta}{\cosh\beta - \cos\beta}$$

$$f_n = f_{SF(n+1)} + Q(f_{FF(n+1)} - f_{SF(n+1)})$$

Part 2. Beams with Combinations of Two Nonclassical Boundary Conditions

The preparation of tables for free vibration analysis of these beams is accomplished utilizing a procedure very similar to that employed for the beams in Part 1. There are, however, two main variables of interest now, and this makes the procedure slightly more involved. In order to clarify the procedure and to facilitate use of the tables, a typical representative problem is worked out in detail.

Consider the problem of free vibration analysis of a uniform beam pinned with torsion spring support at each end (Case 2.22).
The appropriate boundary conditions are evidently

1. $r(0) = 0$
2. $\left.\dfrac{d^2r(\xi)}{d\xi^2} = T_1^* \dfrac{dr(\xi)}{d\xi}\right|_{\xi=0}$
3. $r(1) = 0$
4. $\left.\dfrac{d^2r(\xi)}{d\xi^2} = -T_2^* \dfrac{dr(\xi)}{d\xi}\right|_{\xi=1}$

Utilizing the general solution (Eq. 1.19) and enforcing the boundary conditions 1 through 4 we obtain algebraic equations relating the constants A, B, C, and D as follows.

$$B + D = 0 \tag{2.1}$$

$$-B + D - \frac{T_1^*}{\beta}(A + C) = 0 \tag{2.2}$$

$$A \sin \beta + B \cos \beta + C \sinh \beta + D \cosh \beta = 0 \tag{2.3}$$

$$-A \sin \beta - B \cos \beta + C \sinh \beta + D \cosh \beta$$
$$= \frac{-T_2^*}{\beta}\{A \cos \beta - B \sin \beta + C \cosh \beta + D \sinh \beta\} \tag{2.4}$$

The constants C and D may be eliminated from Eqs. 2.3 and 2.4 by means of Eqs. 2.1 and 2.2, and the determinant equation of the resulting coefficient matrix is expressed as

$$T_1^{*2} + T_1^* \frac{\beta(1+\alpha)(\sin \beta \cosh \beta - \cos \beta \sinh \beta)}{\alpha(1 - \cos \beta \cosh \beta)} + \frac{2\beta^2 \sin \beta \sinh \beta}{\alpha(1 - \cos \beta \cosh \beta)} = 0 \tag{2.5}$$

where

$$\alpha = \frac{T_2^*}{T_1^*}$$

Since T_2^* can be chosen so that it is always equal or less than T_1^*, the only range of α which is of interest is given by

$$0 < \alpha \leqslant 1$$

For a prescribed value of α, Eq. 2.5 is quadratic in T_1^*, and the mathematical values for T_1^* are readily obtained. It is necessary therefore to select values of β that are of interest, and these are obtained as in Part 1. It is evident that the lower limiting cases for the modes of vibration of this beam are simple-simple cases, while the upper limits are clamped-clamped, regardless of the value of α. It is found, as would be expected, that only one positive root is obtained for T_1^* when β is allowed to vary within these limits, and no positive root is found outside the limits. Tables similar to those in Part 1 are presented for values of α between .1 and 1, with α varying in increments of .1. In problems that lack the symmetry of this one, an additional set of tables with the inverse of α varying in a similar manner between .1 and .9 is also presented.

The range for the parameter α was selected after giving some consideration to the dimensionless parameters involved. If we analyze the response of a cantilever beam with a static transverse concentrated load applied at the free end, we find that the load per unit deflection equals $3EI/L^3$. For the parameter $K^* = 1$, the spring constant K equals EI/L^3. Therefore a value of $K^* = 1$ corresponds to a spring whose stiffness is of the same order as the transverse stiffness of the outer end of a cantilever beam and is probably in the range of spring stiffness of frequent interest. Examining the other dimensionless spring parameter T^*, we find that a value of one is likely to be representative of values in the range of frequent interest. For this reason it is expected that the range of the parameter α discussed above is sufficient for most applications. In analyzing problems in which the ratio α takes on intermediate values, it is in general necessary to determine the frequencies associated with adjacent values of α and then use interpolation. It is believed that the degree of accuracy that can thereby be obtained for virtually any problem of interest will be at least compatible with the accuracy to which the properties of the system are known. Should the analyst decide in a special case that greater accuracy is appropriate, he may utilize the quadratic equations provided. Situations may arise when either α or $1/\alpha$ lies between 0.0 and 0.1. In situations like this it is suggested that backward interpolation be utilized (for example, projecting linearly from known values associated with $\alpha = .2$ and $\alpha = .1$). It is recommended that the validity of the results obtained in such cases be checked by utilizing the eigenvalue equation provided, and that the results be corrected by iteration using this equation, if necessary.

All the beam systems in Part 2 are assigned a reference number; they are given a brief description, and tabulated data are provided in a manner similar to that of Part 1.

There are a limited number of cases in which the upper and lower limits for adjacent modes are found to overlap. Typical is that of a beam with lateral spring support at each end (Case 2.20). We consider the situation in which the springs are of equal stiffness. The upper and lower limits are evidently simple-simple and free-free, respectively. As expected, the determinant equation is quadratic in the system parameter K^*. Solution for K^* in terms of the eigenvalue β indicates that there is not always only one positive root. This might seem perplexing at first, but further study indicates that this is to be expected, and the reasons are explained with the aid of Fig. 2.1. It is known that the lowest eigenvalue for a free-free beam is zero and represents a limiting case for first mode vibration of the beam under study (K^* approaching zero). It should be recognized that there are two modes of vibration associated with this zero value, namely, rigid body translation and rotation, both of infinite period. Translations represent a symmetric mode and, as K^* increases, the limiting case is first mode simple-simple vibration with $\beta = \pi$. However, rotation represents an antisymmetric

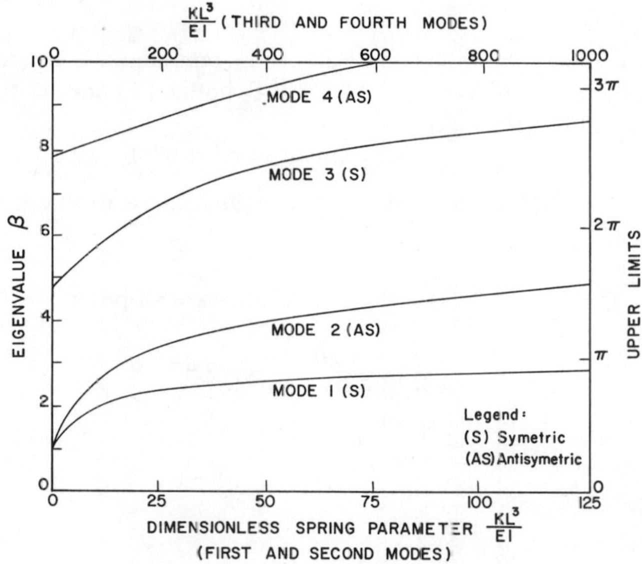

Fig. 2.1

mode and, as K^* increases, the limiting case is second mode simple-simple vibration with $\beta = 2\pi$ and a nodal point midway between the ends of the beam. A plot of β versus K^* is presented in Fig. 2.1 for both symmetric and antisymmetric modes. It is seen that the eigenvalues for every mode begin at those pertaining to free-free beams, including the zero eigenvalue. As K^* is increased for symmetric modes, the next higher simple-simple symmetric mode limit is approached. An analogous rule applies for antisymmetric modes. The lowest free-free eigenvalue, zero, is unique in that it corresponds to a starting point for both symmetric and antisymmetric modes. Even with springs that are not equal in stiffness the limits remain unchanged. Designers should be aware of the implications of Fig. 2.1. It is evident, for example, that if a resonant condition exists due to a driving force of a particular frequency, altering K^* to move one mode out of resonance may move another in, if the alteration is not judiciously chosen.

The problem of vibration of a beam with rotational masses at each end is similar in nature. The lower frequency limit for first mode vibration of this beam is also zero (as the rotational masses approach infinity). In the lower limit the symmetric mode corresponds to counterrotation of the masses, while the antisymmetric mode corresponds to the situation in which the masses have angular rotation in the same directions.

ILLUSTRATIVE PROBLEM 2.2

The reference aluminum beam (Part 1) is pinned with torsion spring support at each end (Case 2.22). T_1 and T_2 equal 150 and 86.5 in.-lb per degree of rotation, respectively.

(a) Determine the first mode frequency and modal shape.

$$f_{SS} = 5.07 \times 3.14^2 = 50.0 \text{ Hz} \qquad \text{(Illustrative Problem 1.1)}$$

$$f_{CC} = 5.07 \times 4.73^2 = 113.4 \text{ Hz}$$

$$T_1 = 150 \times \frac{180}{3.14} = 8594 \text{ in.-lb/unit beam slope}$$

$$T_1^* = \frac{T_1 L}{EI} = \frac{8594 \times 30}{10 \times 10^6 \times 1.04 \times 10^{-2}} = 2.47$$

$$\alpha = \frac{T_2^*}{T_1^*} = \frac{T_2}{T_1} = \frac{86.5}{150} = 0.577$$

For $\alpha = 0.50$,

$$Q = .20 + \frac{2.47 - 2.23}{3.94 - 2.23} \times 0.1 = .214$$

For $\alpha = 0.60$,

$$Q = .20 + \frac{2.47 - 2.07}{3.63 - 2.07} \times 0.1 = .226$$

For $\alpha = 0.577$ (interpolating),

$$Q = .214 + .77(.226 - .214) = .223$$

$$\therefore f = f_{SS} + Q(f_{CC} - f_{SS}) = 50.0 + .223(113.4 - 50) = 64.1 \text{ Hz}$$

$$\beta = \beta_{CC}\sqrt{\frac{f}{f_{CC}}} = 4.73\sqrt{\frac{64.1}{113.4}} = 3.56 \qquad \text{(Eq. 1.22)}$$

$$\therefore r(\xi) = -(\sin 3.56\xi + \sinh 3.56\xi) + \gamma(\cos 3.56\xi + \cosh 3.56\xi)$$

where

$$\gamma = \frac{\sin 3.56 + \sinh 3.56}{\cos 3.56 + \cosh 3.56} = 1.03$$

(b) It is desired to bring the first mode frequency up to 105 Hz using torsion springs of equal stiffness at each end of the beam. Determine the spring stiffness required.

$$f = 50.0 + Q(113.4 - 50.0) = 105$$

$$\therefore Q = \frac{55.0}{113.4 - 50.0} = .867$$

But

$$Q = 1 - \frac{.2T^*_{1(Q=.8)}}{T^*_1} \qquad \text{(Eq. 1.24)}$$

Rearranging and referring to Table 2.12 ($\alpha = 1$),

$$T^*_1 = \frac{.2 \times 28.18}{.133} = 42.4$$

$$\therefore T_1 = T_2 = \frac{42.4EI}{L} = 147 \times 10^3 \text{ in.-lb/unit deflection}$$

$$= \frac{147 \times 10^3 \times 3.14}{180} = 2.57 \times 10^3 \text{ in.-lb/degree deflection}$$

1. Two Coincident Nonclassical Boundary Conditions

CASE 2.12. Beam clamped at one end, and with lateral spring support and concentrated mass at the other.

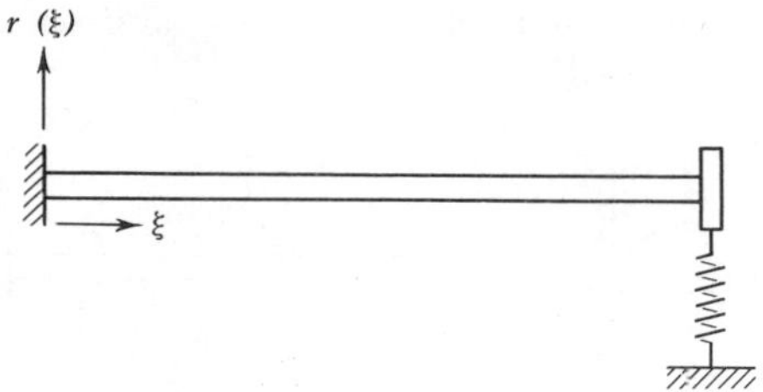

$$K^{*2} + K^* \frac{\beta^3(1 + \cos\beta \cosh\beta)}{\sin\beta \cosh\beta - \sinh\beta \cos\beta} - \frac{\beta^4}{\alpha} = 0$$

$$\alpha = \frac{\eta}{K^*}$$

$$r(\xi) = \sin\beta\xi - \sinh\beta\xi + \gamma(\cos\beta\xi - \cosh\beta\xi)$$

where

$$\gamma = \frac{-(\sin\beta + \sinh\beta)}{\cos\beta + \cosh\beta}$$

$$f_n = Qf_{CSn} \qquad n = 1$$

$$= f_{CS(n-1)} + Q(f_{CSn} - f_{CS(n-1)}) \qquad n > 1$$

CASE 2.13. Beam simply supported at one end, and with lateral spring support and concentrated mass at the other.

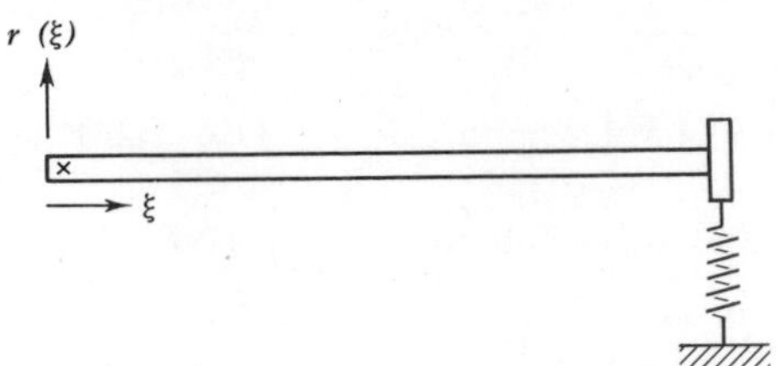

$$K^{*2} + K^* \frac{\beta^3(\cos\beta \sinh\beta - \sin\beta \cosh\beta)}{2\sin\beta \sinh\beta} - \frac{\beta^4}{\alpha} = 0$$

$$\alpha = \frac{\eta}{K^*}$$

$$r(\xi) = \sin\beta\xi + \gamma \sinh\beta\xi$$

where

$$\gamma = \frac{\sin\beta}{\sinh\beta}$$

$$f_n = Qf_{SSn} \qquad n = 1$$
$$= f_{SS(n-1)} + Q(f_{SSn} - f_{SS(n-1)}) \qquad n > 1$$

CASE 2.14. **Beam pinned with torsion spring and rotational mass at one end and clamped at the other.**

$$T^{*2} + T^* \frac{\beta(\sin\beta \cosh\beta - \cos\beta \sinh\beta)}{1 - \cos\beta \cosh\beta} - \frac{\beta^4}{\alpha} = 0$$

$$\alpha = \frac{I^*}{T^*}$$

$$r(\xi) = \sin\beta\xi - \frac{\sin\beta}{\sinh\beta}\sinh\beta\xi + \gamma\left(\cos\beta\xi - \cosh\beta\xi + \frac{\cosh\beta - \cos\beta}{\sinh\beta}\sinh\beta\xi\right)$$

where

$$\gamma = \frac{\cos\beta \sinh\beta - \sin\beta \cosh\beta}{\sin\beta \sinh\beta + \cos\beta \cosh\beta + \sinh^2\beta - \cosh^2\beta}$$

$$f_n = Qf_{CCn}, \qquad n = 1$$
$$= f_{CC(n-1)} + Q(f_{CCn} - f_{CC(n-1)}), \qquad n > 1$$

CASE 2.15. Beam pinned with torsion spring and rotational mass at one end and simply supported at the other.

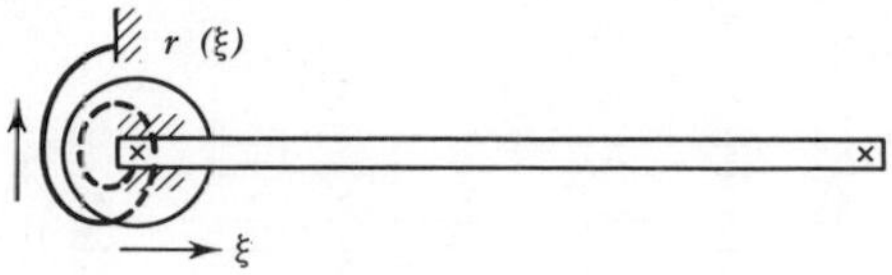

$r\ (\xi)$

ξ

$$T^{*2} + T^* \frac{(2\beta \sin \beta \sinh \beta)}{(\sinh \beta \cosh \beta - \sinh \beta \cos \beta)} - \frac{\beta^4}{\alpha} = 0$$

$$\alpha = \frac{I^*}{T^*}$$

$$r(\xi) = \sin \beta\xi - \sinh \beta\xi + \gamma \left\{ \cos \beta\xi - \cosh \beta\xi \frac{-2\beta \sinh \beta\xi}{[T^* - (\beta^4/I^*)]} \right\}$$

where

$$\gamma = \frac{\sinh \beta - \sin \beta}{\cos \beta - \cosh \beta - \dfrac{2\beta \sinh \beta}{[T^* - (\beta^4/I^*)]}}$$

$$f_n = Qf_{CSn}, \qquad n = 1$$
$$= f_{CS(n-1)} + Q(f_{CSn} - f_{CS(n-1)}), \qquad n > 1$$

CASE 2.16. Beam pinned with torsion spring and rotational mass at one end and free at the other.

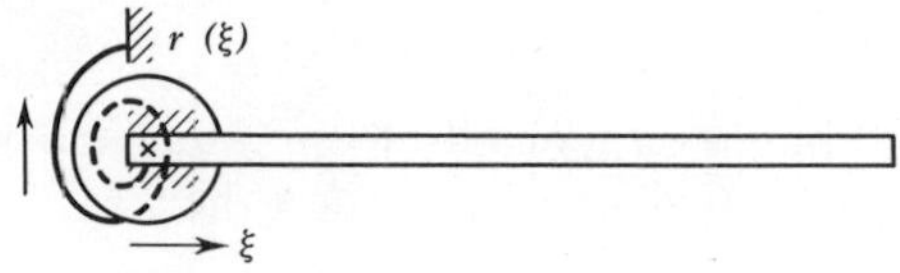

$r\ (\xi)$

ξ

$$T^{*2} + T^* \frac{\beta(\cos \beta \sinh \beta - \sin \beta \cosh \beta)}{1 + \cos \beta \cosh \beta} - \frac{\beta^4}{\alpha} = 0$$

$$\alpha = \frac{I^*}{T^*}$$

$$r(\xi) = \sin \beta\xi - \sinh \beta\xi - \gamma \left\{ \cos\ \beta\xi - \cosh\ \beta\xi \frac{-2\beta}{[T^* - (\beta^4/I^*)]} \sinh\ \beta\xi \right\}$$

where

$$\gamma = (\sin \beta + \sinh \beta) \Big/ \left\{ \cos \beta + \cosh \beta \frac{+2\beta \sinh \beta}{[T^* - (\beta^4/I^*)]} \right\}$$

$$f_n = Qf_{CFn}, \qquad n = 1$$
$$= f_{CF(n-1)} + Q(f_{CFn} - f_{CF(n-1)}), \qquad n > 1$$

CASE 2.17. Beam clamped at one end and with lateral spring support and torsion spring support at the other.

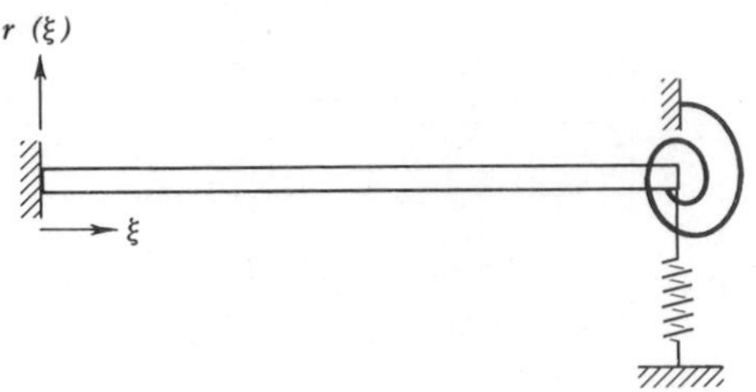

$$K^{*2} + K^* \times \frac{[(-\beta/\alpha)(\sinh\beta\cos\beta - \sin\beta\cosh\beta) + \beta^3(\cos\beta\sinh\beta + \cosh\beta\sin\beta)]}{1 - \cos\beta\cosh\beta} + \frac{\beta^4}{\alpha}\frac{1 + \cos\beta\cosh\beta}{1 - \cos\beta\cosh\beta} = 0$$

$$\alpha = \frac{T^*}{K^*}$$

$$r(\xi) = \sin\beta\xi - \sinh\beta\xi + \gamma(\cos\beta\xi - \cosh\beta\xi)$$

where

$$\gamma = \frac{\cos\beta + \cosh\beta + (K^*/\beta^3)(\sin\beta - \sinh\beta)}{\sin\beta - \sinh\beta + (K^*/\beta^3)(\cosh\beta - \cos\beta)}$$

$$f_n = f_{CFn} + Q(f_{CCn} - f_{CFn})$$

CASE 2.18. Beam simply supported at one end, and with lateral spring support and torsion spring support at the other.

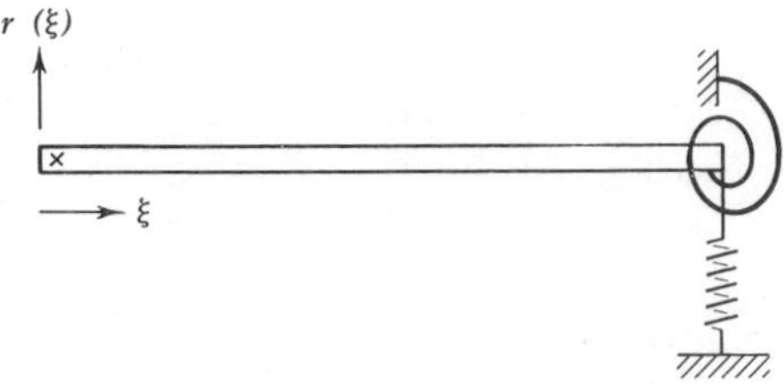

$$K^{*2} + K^* \frac{-(2\beta/\alpha)(\sin\beta\sinh\beta) - 2\beta^3(\cos\beta\cosh\beta)}{\cos\beta\sinh\beta - \sin\beta\cosh\beta} - \frac{\beta^4}{\alpha} = 0$$

$$\alpha = \frac{T^*}{K^*}$$

$$r(\xi) = \sin\beta\xi + \gamma\sinh\beta\xi$$

where

$$\gamma = \frac{\cos\beta + (K^*/\beta^3)\sin\beta}{\cosh\beta - (K^*/\beta^3)\sinh\beta}$$

$$f_n = f_{SFn} + Q(f_{CSn} - f_{SFn})$$

CASE 2.19. Beam free at one end, and with lateral spring support and torsion spring support at the other.

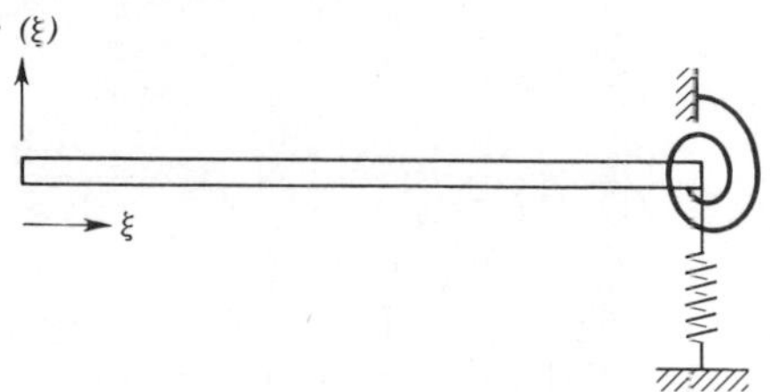

$$K^{*2} + K^* \times \frac{[(\beta/\alpha)(\sinh\beta\cos\beta - \cosh\beta\sin\beta) - \beta^3(\sin\beta\cosh\beta + \cos\beta\sinh\beta)]}{1+\cos\beta\cosh\beta} + \frac{\beta^4}{\alpha}\frac{1-\cos\beta\cosh\beta}{1-\cos\beta\cosh\beta} = 0$$

$$\alpha = \frac{T^*}{K^*}$$

$$r(\xi) = \sin\beta\xi + \sinh\beta\xi + \gamma(\cos\beta\xi + \cosh\beta\xi)$$

where

$$\gamma = \frac{\cos\beta - \cosh\beta + (K^*/\beta^3)(\sin\beta + \sinh\beta)}{\sin\beta + \sinh\beta - (K^*/\beta^3)(\cos\beta + \cosh\beta)}$$

$$\begin{aligned} f_n &= Qf_{CFn} \qquad n=1 \\ &= Qf_{CFn} \qquad n=2 \\ &= f_{FF(n-1)} + Q(f_{CFn} - f_{FF(n-1)}) \qquad n>2 \end{aligned}$$

2. Beams with Two Noncoincident Nonclassical Boundary Conditions

CASE 2.20. Beam with lateral spring support at each end.

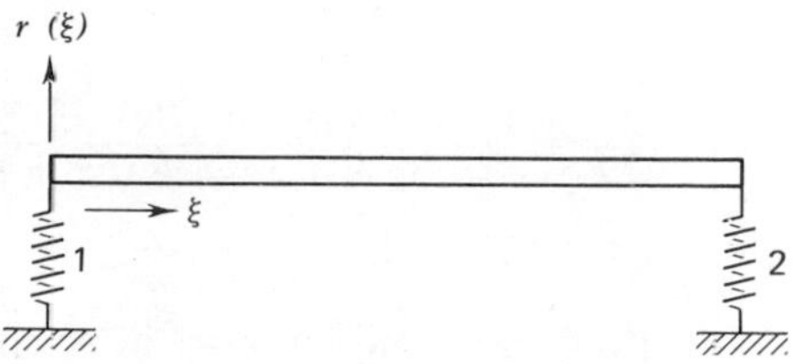

$$K_1^{*2} + K_1^* \frac{\beta^3(1+\alpha)(\sinh\beta\cos\beta - \sin\beta\cosh\beta)}{2\alpha\sin\beta\sinh\beta} + \frac{\beta^6(1-\cos\beta\cosh\beta)}{2\alpha\sin\beta\sinh\beta} = 0$$

$$\alpha = \frac{K_2^*}{K_1^*}$$

$$r(\xi) = \sin\beta\xi + \frac{\sin\beta}{\sinh\beta}\sinh\beta\xi + \gamma\left[\cos\beta\xi + \cosh\beta\xi \frac{-(\cosh\beta - \cos\beta)}{\sinh\beta}\sinh\beta\xi\right]$$

where

$$\gamma = \frac{\sinh\beta - \sin\beta}{2(K_1^*/\beta^3)\sinh\beta + \cos\beta - \cosh\beta}$$

$$f_n = Qf_{SSn}, \qquad n = 1, \quad n = 2$$
$$= f_{FF(n-1)} + Q(f_{SSn} - f_{FF(n-1)}), \qquad n > 2$$

CASE 2.21. Beam pinned with torsion spring support at one end and with lateral spring support at the other.

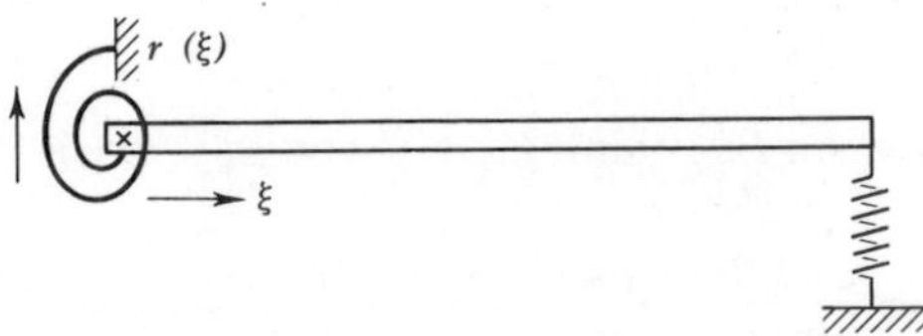

$$T^{*2} + T^* \frac{-(\beta^3/\alpha)(1+\cos\beta\cosh\beta) - 2\beta\sin\beta\sinh\beta}{\cos\beta\sinh\beta - \sin\beta\cosh\beta} - \frac{\beta^4}{\alpha} = 0$$

$$\alpha = \frac{K^*}{T^*}$$

$$r(\xi) = \sin\beta\xi + \frac{\sin\beta}{\sinh\beta}\sinh\beta\xi + \gamma\left(\cos\beta\xi - \cosh\beta\xi + \frac{\cos\beta + \cosh\beta}{\sinh\beta}\sinh\beta\xi\right)$$

where

$$\gamma = \frac{-(1 + (\sin\beta/\sinh\beta))}{(2\beta/T^*) + ((\cos\beta + \cosh\beta)/\sinh\beta)}$$

$$f_n = f_{SFn} + Q(f_{CSn} - f_{SFn})$$

CASE 2.22. Beam with torsion spring support at each end.

$$T_1^{*2} + T_1^* \frac{\beta(1+\alpha)(\sin\beta\cosh\beta - \cos\beta\sinh\beta)}{\alpha(1-\cos\beta\cosh\beta)} + \frac{2\beta^2 \sin\beta\sinh\beta}{\alpha(1-\cos\beta\cosh\beta)} = 0$$

$$\alpha = \frac{T_2^*}{T_1^*}$$

$$r(\xi) = \sin\beta\xi - \sinh\beta\xi + \gamma[\cos\beta\xi - \cosh\beta\xi - (2\beta/T_1^*)\sinh\beta\xi]$$

where

$$\gamma = \frac{\sinh\beta - \sin\beta}{\cos\beta - \cosh\beta - (2\beta/T_1^*)\sinh\beta}$$

$$f_n = f_{SSn} + Q(f_{CCn} - f_{SSn})$$

CASE 2.23. Beam with lateral spring support at one end and with rotational mass at the other.

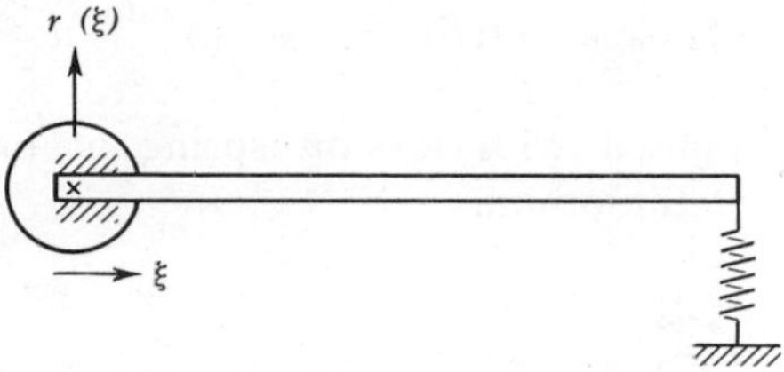

$$K^{*2} + K^* \frac{\beta^3(1+\alpha)(\cos\beta\sinh\beta - \sin\beta\cosh\beta)}{2\alpha\sin\beta\sinh\beta} - \frac{\beta^6(1+\cos\beta\cosh\beta)}{2\alpha\sin\beta\sinh\beta} = 0$$

$$\alpha = \frac{I^*}{K^*}$$

$$r(\xi) = \sin\beta\xi + \frac{\sin\beta}{\sinh\beta}\sinh\beta\xi + \gamma\left(\cos\beta\xi - \cosh\beta\xi + \frac{\cos\beta + \cosh\beta}{\sinh\beta}\sinh\beta\xi\right)$$

where

$$\gamma = \frac{\sin\beta + \sinh\beta}{(2I^*/\beta^3)\sinh\beta - \cos\beta - \cosh\beta}$$

$$f_n = Qf_{SSn}, \qquad n = 1$$

$$= f_{CF(n-1)} + Q(f_{SSn} - f_{CF(n-1)}), \qquad n > 1$$

CASE 2.24. Beam pinned with rotational mass at one end and pinned with torsion spring at the other.

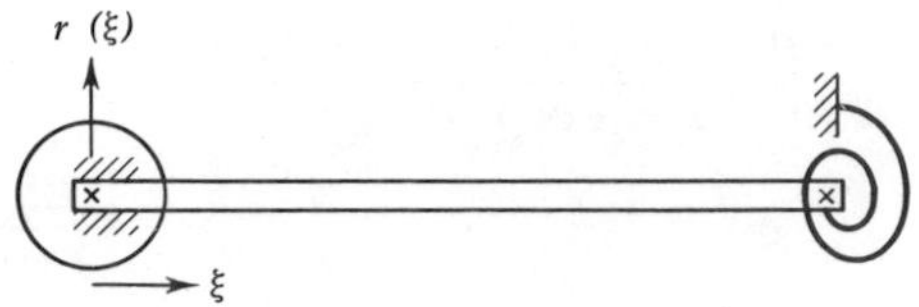

$$T^{*2} + T^* \frac{2\beta \sin\beta \sinh\beta + (\beta^3/\alpha)(\cos\beta \cosh\beta) - \beta^3/\alpha}{\sin\beta \cosh\beta - \sinh\beta \cos\beta} - \frac{\beta^4}{\alpha} = 0$$

$$\alpha = \frac{I^*}{T^*}$$

$$r(\xi) = \sin\beta\xi - \sinh\beta\xi + \gamma[\cos\beta\xi - \cosh\beta\xi + (2I^*/\beta^3)\sinh\beta\xi]$$

where

$$\gamma = \frac{\sinh\beta - \sin\beta}{\cos\beta - \cosh\beta + (2I^*/\beta^3)\sinh\beta}$$

$$f_n = Qf_{CSn}, \qquad n = 1$$
$$= f_{CS(n-1)} + Q(f_{CSn} - f_{CS(n-1)}), \qquad n > 1$$

CASE 2.25. Beam pinned with torsion spring at one end and with concentrated mass at the other.

$$T^{*2} + T^* \frac{\beta(1+\alpha)(\cos\beta \sinh\beta - \sin\beta \cosh\beta)}{\alpha(1+\cos\beta\cosh\beta)} - \frac{2\beta^2 \sin\beta \sinh\beta}{\alpha(1+\cos\beta\cosh\beta)} = 0$$

$$\alpha = \frac{\eta}{T^*}$$

$$r(\xi) = \sin\beta\xi - \sinh\beta\xi + \gamma[\cos\beta\xi - \cosh\beta\xi - (2\beta/T^*)\sinh\beta\xi]$$

where

$$\gamma = \frac{-(\sin\beta + \sinh\beta)}{\cos\beta + \cosh\beta + (2\beta/T^*)\sinh\beta}$$

$$f_n = Qf_{CFn}, \qquad n = 1$$
$$= f_{SS(n-1)} + Q(f_{CFn} - f_{SS(n-1)}), \qquad n > 1$$

CASE 2.26. Beam pinned with rotational mass at each end.

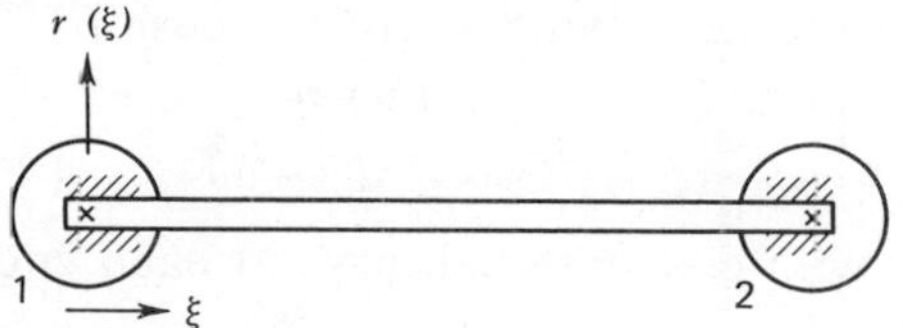

$$I_1^{*2} + I_1^* \frac{\beta^3(1+1/\alpha)(\cos\beta \sinh\beta - \sin\beta \cosh\beta)}{2\sin\beta \sinh\beta} + \frac{\beta^6(1-\cos\beta\cosh\beta)}{2\alpha \sin\beta \sinh\beta} = 0$$

$$\alpha = \frac{I_2^*}{I_1^*}$$

$$r(\xi) = \sin\beta\xi - \frac{\sin\beta}{\sinh\beta}\sinh\beta\xi + \gamma\left[\cos\beta\xi - \cosh\beta\xi \frac{+(\cosh\beta - \cos\beta)}{\sinh\beta}\sinh\beta\xi\right]$$

where

$$\gamma = \frac{(\sin\beta/\sinh\beta) - 1}{[(\cosh\beta - \cos\beta)/\sinh\beta] - (2I_1^*/\beta^3)}$$

$$f_n = Qf_{SSn}, \qquad n = 1, 2$$
$$= f_{CC(n-2)} + Q(f_{SSn} - f_{CC(n-2)}), \qquad n > 2$$

CASE 2.27. Beam pinned with rotational mass at one end, and with concentrated mass at the other.

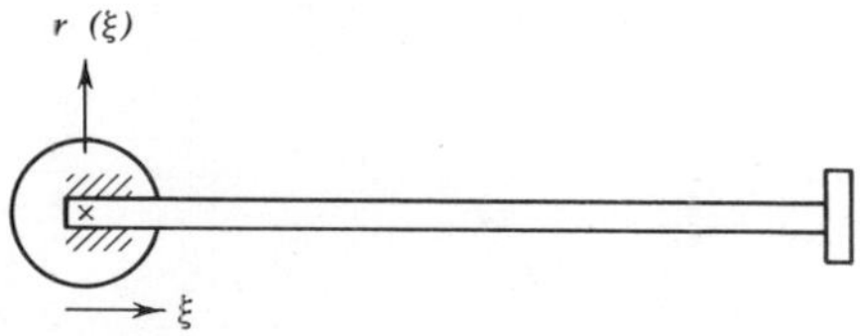

$$\eta^2 + \eta \frac{1 + \cos\beta \cosh\beta + (2\alpha/\beta^2)\sin\beta \sinh\beta}{(\alpha/\beta^3)(\cosh\beta \sin\beta - \sinh\beta \cos\beta)} - \frac{\beta^4}{\alpha} = 0$$

$$\alpha = \frac{I^*}{\eta}$$

$$r(\xi) = \sin\beta\xi - \sinh\beta\xi + \gamma\left(\cos\beta\xi - \cosh\beta\xi + \frac{2I^*}{\beta^3}\sinh\beta\xi\right)$$

where

$$\gamma = \frac{\sin\beta + \sinh\beta}{(2I^*/\beta^3)\sinh\beta - (\cos\beta + \cosh\beta)}$$

$$f_n = Qf_{SF(n+1)}, \qquad n = 1$$
$$= f_{CS(n-1)} + Q(f_{SF(n+1)} - f_{CS(n-1)}), \qquad n > 1$$

CASE 2.28. Beam with concentrated mass at each end.

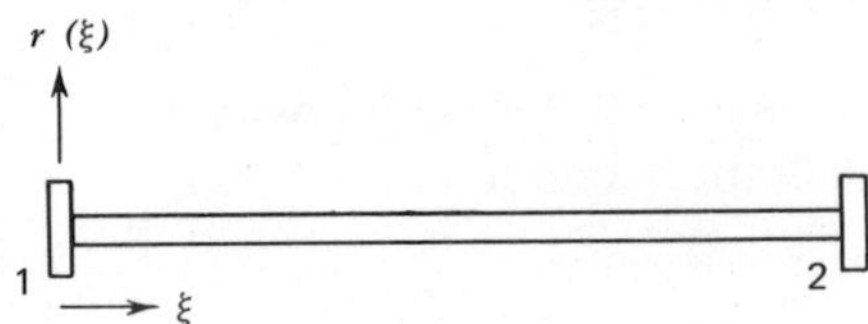

$$\eta_1^2 + \eta_1 \frac{\beta(1+\alpha)(\sin\beta\cosh\beta - \cos\beta\sinh\beta)}{\alpha(1-\cos\beta\cosh\beta)} + \frac{2\beta^2\sin\beta\sinh\beta}{\alpha(1-\cos\beta\cosh\beta)} = 0$$

$$\alpha = \frac{\eta_2}{\eta_1}$$

$$r(\xi) = \sin\beta\xi + \sinh\beta\xi + \gamma\left(\cos\beta\xi + \cosh\beta\xi + 2\frac{\beta}{\eta_1}\sinh\beta\xi\right)$$

where

$$\gamma = \frac{\sin\beta - \sinh\beta}{\cosh\beta - \cos\beta + 2(\beta/\eta_1)\sinh\beta}$$

$$f_n = f_{SSn} + Q(f_{FF(n+1)} - f_{SSn})$$

TABLE 2.1: T*

Q: MODE:	0.05	0.10	0.20	0.30	0.40	0.50	0.60	0.70	0.80	0.90
(1)	.0103	.0416	.1715	.4065	.7813	1.364	2.294	3.903	7.189	17.15
(2)	.3673	.7735	1.732	2.956	4.577	6.834	10.20	15.79	26.93	60.28
(3)	.6125	1.286	2.863	4.856	7.475	11.10	16.47	25.35	43.00	95.70
(4)	.8702	1.824	4.048	6.844	10.50	15.54	23.00	35.30	59.69	132.5
(5)	1.127	2.361	5.229	8.827	13.52	19.98	29.52	45.23	76.36	169.2
(6)	1.383	2.896	6.410	10.81	16.54	24.41	36.04	55.16	93.02	205.8
(7)	1.640	3.432	7.591	12.79	19.57	28.86	42.56	65.09	109.7	242.6
(8)	1.896	3.970	8.773	14.78	22.59	33.29	49.08	75.03	126.4	279.3
(9)	2.153	4.506	9.955	16.76	25.61	37.73	55.61	84.97	143.1	316.0
(10)	2.410	5.041	11.14	18.74	28.64	42.18	62.13	94.91	159.8	352.8

TABLE 2.2 : T*

Q: MODE:	0.05	0.10	0.20	0.30	0.40	0.50	0.60	0.70	0.80	0.90
(1)	.2924	.6180	1.393	2.392	3.723	5.587	8.380	13.03	22.31	50.12
(2)	.5492	1.154	2.571	4.366	6.729	9.999	14.86	22.90	38.88	86.62
(3)	.8060	1.690	3.752	6.348	9.746	14.43	21.31	32.81	55.52	123.3
(4)	1.063	2.226	4.933	8.331	12.77	18.87	27.89	42.74	72.19	160.0
(5)	1.320	2.763	6.115	10.31	15.79	23.30	34.40	52.67	88.84	196.6
(6)	1.576	3.299	7.296	12.30	18.81	27.75	40.93	62.61	105.5	233.4
(7)	1.832	3.835	8.477	14.28	21.83	32.19	47.45	72.54	122.2	270.1
(8)	2.089	4.372	9.660	16.27	24.86	36.63	53.97	82.47	138.9	306.9
(9)	2.346	4.907	10.84	18.25	27.88	41.06	60.50	92.42	155.6	343.5
(10)	2.602	5.443	12.02	20.23	30.90	45.50	67.02	102.4	172.2	380.2

TABLE 2.3 : T*

Q: MODE:	0.05	0.10	0.20	0.30	0.40	0.50	0.60	0.70	0.80	0.90
(1)	.3462	.7306	1.643	2.813	4.369	6.541	9.787	15.18	25.94	58.14
(2)	.6142	1.289	2.870	4.867	7.491	11.12	16.50	25.40	43.07	95.85
(3)	.8702	1.824	4.047	6.843	10.50	15.54	23.00	35.29	59.69	132.5
(4)	1.127	2.361	5.229	8.827	13.52	19.98	29.52	45.23	76.36	169.2
(5)	1.383	2.896	6.410	10.81	16.54	24.41	36.04	55.16	93.02	205.8
(6)	1.640	3.432	7.591	12.79	19.57	28.86	42.56	65.09	109.7	242.6
(7)	1.896	3.970	8.773	14.78	22.59	33.29	49.08	75.03	126.4	279.3
(8)	2.153	4.506	9.955	16.76	25.61	37.73	55.61	84.97	143.1	316.0
(9)	2.410	5.041	11.14	18.74	28.64	42.18	62.13	94.91	159.8	352.8
(10)	2.665	5.577	12.32	20.73	31.66	46.61	68.65	104.9	176.5	389.6

TABLE 2.4 : K*

Q: MODE:	0.05	0.10	0.20	0.30	0.40	0.50	0.60	0.70	0.80	0.90
(1)	.0813	.3267	1.332	3.101	5.806	9.784	15.73	25.29	43.38	94.70
(2)	8.912	17.55	35.59	56.30	81.49	113.7	157.8	224.6	346.6	685.9
(3)	45.05	85.59	162.1	240.9	329.9	438.2	581.4	793.9	1178	2242
(4)	128.3	240.8	444.1	643.4	859.9	1116	1449	1937	2816	5248
(5)	278.5	519.1	944.4	1350	1779	2279	2920	3857	5537	10190
(6)	515.0	956.5	1726	2445	3195	4058	5157	6754	9615	17530
(7)	857.8	1589	2851	4015	5216	6585	8318	10830	15330	27770
(8)	1326	2453	4383	6146	7950	9992	12570	16290	22950	41380
(9)	1941	3584	6385	8923	11500	14410	18060	23330	32750	58850
(10)	2720	5018	8919	12430	15990	19970	24960	32160	45010	80650

TABLE 2.5 : K*

Q: MODE:	0.05	0.10	0.20	0.30	0.40	0.50	0.60	0.70	0.80	0.90
(1)	1.148	2.509	5.948	10.54	16.64	24.93	36.77	55.29	89.96	188.0
(2)	14.55	28.22	55.75	86.25	122.5	168.2	230.0	323.0	492.2	961.8
(3)	60.69	114.8	215.6	317.8	431.9	569.5	750.4	1018	1501	2839
(4)	158.8	297.4	546.2	788.1	1049	1356	1754	2336	3384	6289
(5)	328.8	612.1	1111	1584	2082	2661	3401	4481	6418	11780
(6)	590.0	1095	1973	2790	3640	4614	5854	7654	10880	19800
(7)	962.4	1782	3194	4492	5829	7349	9272	12060	17040	30840
(8)	1466	2709	4837	6777	8757	11000	13820	17890	25180	45370
(9)	2119	3912	6966	9729	12530	15690	19650	25360	35580	63870
(10)	2943	5427	9643	13430	17260	21550	26930	34670	48500	86820

TABLE 2.6 : η

Q: MODE:	0.05	0.10	0.20	0.30	0.40	0.50	0.60	0.70	0.80	0.90
(1)	.2924	.6180	1.393	2.392	3.724	5.587	8.380	13.03	22.31	50.12
(2)	.5492	1.154	2.571	4.366	6.729	9.999	14.86	22.90	38.88	86.62
(3)	.8060	1.690	3.752	6.348	9.746	14.43	21.37	32.81	55.53	123.3
(4)	1.063	2.226	4.934	8.331	12.77	18.87	27.89	42.74	72.19	160.0
(5)	1.320	2.763	6.115	10.31	15.79	23.30	34.41	52.67	88.85	196.6
(6)	1.576	3.299	7.296	12.30	18.81	27.75	40.93	62.61	105.5	233.4
(7)	1.832	3.835	8.479	14.28	21.83	32.19	47.45	72.54	122.2	270.1
(8)	2.089	4.372	9.662	16.27	24.86	36.63	53.97	82.49	138.9	306.9
(9)	2.346	4.909	10.84	18.25	27.88	41.06	60.50	92.42	155.6	343.6
(10)	2.603	5.447	12.03	20.24	30.90	45.51	67.02	102.4	172.3	380.2

TABLE 2.7 : η

Q: MODE:	0.05	0.10	0.20	0.30	0.40	0.50	0.60	0.70	0.80	0.90
(1)	.0103	.0416	.1715	.4065	.7813	1.364	2.294	3.903	7.190	17.15
(2)	.3673	.7735	1.732	2.956	4.577	6.834	10.20	15.79	26.93	60.28
(3)	.6125	1.286	2.863	4.856	7.475	11.10	16.47	25.35	42.99	95.69
(4)	.8702	1.824	4.047	6.844	10.50	15.54	23.00	35.30	59.69	132.5
(5)	1.127	2.361	5.229	8.827	13.52	19.98	29.52	45.23	76.36	169.2
(6)	1.382	2.896	6.409	10.81	16.54	24.41	36.03	55.15	93.01	205.8
(7)	1.639	3.432	7.591	12.79	19.57	28.86	42.56	65.09	109.7	242.6
(8)	1.896	3.968	8.773	14.78	22.59	33.29	49.08	75.02	126.4	297.3
(9)	2.153	4.505	9.955	16.76	25.61	37.73	55.61	84.97	143.1	316.0
(10)	2.410	5.041	11.14	18.74	28.63	42.18	62.13	94.90	159.7	352.8

TABLE 2.8 : I*

Q: MODE:	0.05	0.10	0.20	0.30	0.40	0.50	0.60	0.70	0.80	0.90
(1)	1.148	2.509	5.948	10.54	16.64	24.93	36.77	55.29	89.96	188.0
(2)	14.55	28.22	55.75	86.25	122.5	168.2	230.0	323.0	492.2	961.8
(3)	60.69	114.8	215.6	317.8	431.9	569.5	750.4	1018	1501	2839
(4)	158.8	297.4	546.2	788.1	1049	1356	1754	2336	3384	6283
(5)	328.8	612.1	1111	1584	2082	2661	3401	4481	6418	11780
(6)	590.0	1095	1973	2970	3640	4614	5854	7654	10880	19800
(7)	962.4	1782	3194	4492	5829	7349	9272	12060	17040	30840
(8)	1466	2709	4837	6777	8757	11000	13820	17890	25180	45370
(9)	2119	3912	6966	9729	12530	15690	19650	25360	35580	63870
(10)	2943	5427	9643	13430	17260	21550	26930	34670	48500	86820

TABLE 2.9 : I*

Q: MODE:	0.05	0.10	0.20	0.30	0.40	0.50	0.60	0.70	0.80	0.90
(1)	.0813	.3267	1.332	3.101	5.806	9.784	15.73	25.29	43.38	94.70
(2)	8.912	17.55	35.59	56.30	81.49	113.7	157.8	224.6	346.6	685.9
(3)	45.05	85.59	162.1	240.9	329.9	438.2	581.4	793.9	1178	2242
(4)	128.3	240.8	444.1	643.4	859.9	1116	1449	1937	2816	5248
(5)	278.5	519.1	944.4	1350	1779	2279	2920	3857	5537	10190
(6)	515.0	956.4	1726	2445	3195	4058	5156	6754	9614	17530
(7)	857.8	1589	2851	4015	5216	6585	8318	10830	15330	27770
(8)	1326	2453	4383	6146	7950	9992	12570	16290	22950	41380
(9)	1940	3583	6385	8924	11500	14410	18060	23330	32750	58850
(10)	2720	5018	8919	12430	15980	19970	24960	32160	45010	80640

TABLE 2.10 : I*

Q: MODE:	0.05	0.10	0.20	0.30	0.40	0.50	0.60	0.70	0.80	0.90
(1)	.1488	.5977	2.433	5.647	10.53	17.64	28.14	44.73	75.58	161.5
(2)	14.66	28.50	56.36	87.17	123.7	169.6	231.7	325.3	495.5	968.3
(3)	60.69	114.8	215.5	317.7	431.7	569.3	750.2	1018	1501	2838
(4)	158.8	297.4	546.3	788.2	1049	1356	1754	2337	3384	6283
(5)	328.8	612.1	1111	1584	2082	2661	3401	4481	6418	11780
(6)	590.0	1095	1973	2790	3640	4614	5854	7654	10880	19800
(7)	962.4	1782	3194	4492	5829	7349	9272	12060	17040	30840
(8)	1466	2709	4837	6777	8757	11000	13820	17890	25180	45370
(9)	2119	3912	6966	9729	12530	15690	19650	25360	35580	63870
(10)	2943	5427	9643	13430	17260	21550	26930	34670	48500	86820

TABLE 2.11 : η

Q: MODE:	0.05	0.10	0.20	0.30	0.40	0.50	0.60	0.70	0.80	0.90
(1)	.3462	.7307	1.643	2.813	4.369	6.541	9.787	15.18	25.94	58.15
(2)	.6142	1.290	2.870	4.867	7.491	11.12	16.50	25.40	43.07	95.87
(3)	.8702	1.824	4.047	6.843	10.50	15.54	23.00	35.30	59.69	132.5
(4)	1.127	2.361	5.229	8.827	13.52	19.98	29.52	45.23	76.37	169.2
(5)	1.384	2.898	6.411	10.81	16.55	24.42	36.04	55.17	93.05	206.0
(6)	1.641	3.434	7.594	12.80	19.57	28.86	42.57	65.11	109.7	242.7
(7)	1.898	3.971	8.776	14.78	22.59	33.30	49.09	75.05	126.4	297.5
(8)	2.155	4.508	9.958	16.77	25.62	37.74	55.62	84.99	143.1	316.3
(9)	2.412	5.044	11.14	18.75	28.64	42.19	62.14	94.93	159.8	353.0
(10)	2.669	5.581	12.32	20.73	31.66	46.63	68.67	104.9	176.5	389.8

TABLE 2.12 : K* ($\alpha = 0.1$)

Q: MODE:	0.05	0.10	0.20	0.30	0.40	0.50	0.60	0.70	0.80	0.90
(1)	1.396	3.807	9.404	15.83	23.24	32.05	43.08	58.39	84.50	156.2
(2)	20.81	40.40	72.21	102.4	135.8	176.0	228.2	304.1	439.0	811.6
(3)	33.00	77.26	178.8	279.2	384.0	504.5	657.4	876.8	1262	2312
(4)	45.31	112.5	321.2	569.6	825.3	1107	1454	1940	2784	5055
(5)	57.30	145.6	486.7	991.1	1515	2073	2738	3654	5223	9418
(6)	69.19	177.5	667.1	1563	2513	3493	4633	6178	8801	15781
(7)	81.07	208.9	856.4	2309	3881	5460	7261	9674	13738	24522
(8)	92.93	239.9	1050	3249	5680	8068	10747	14301	20257	36020
(9)	104.8	270.7	1246	4406	7973	11408	15213	20220	28578	50656
(10)	116.6	301.3	1441	5804	10821	15574	20782	27592	38921	68803

TABLE 2.12 : K* ($\alpha = 0.2$)

Q: MODE:	0.05	0.10	0.20	0.30	0.40	0.50	0.60	0.70	0.80	0.90
(1)	.8098	2.438	6.550	11.56	17.65	25.27	35.33	50.08	76.44	150.2
(2)	11.17	24.61	51.54	79.36	111.3	150.7	203.1	280.5	418.9	798.5
(3)	16.82	42.35	122.7	221.3	328.4	452.4	609.7	834.9	1228	2291
(4)	22.85	59.13	212.8	462.0	730.8	1024	1381	1880	2737	5027
(5)	28.79	74.98	311.7	820.8	1377	1958	2641	3574	5162	9381
(6)	34.70	90.49	414.0	1319	2329	3345	4511	6080	8727	15737
(7)	40.62	105.9	516.7	1982	3649	5280	7115	9557	13651	24470
(8)	46.54	121.1	618.1	2831	5399	7854	10577	14165	20157	35960
(9)	52.45	136.4	717.8	3891	7644	11162	15018	20066	28464	50589
(10)	58.36	151.6	815.5	5186	10443	15295	20562	27420	38793	68728

TABLE 2.12 : K* ($\alpha = 0.3$)

Q: MODE:	0.05	0.10	0.20	0.30	0.40	0.50	0.60	0.70	0.80	0.90
(1)	.5765	1.849	5.287	9.679	15.21	22.36	32.10	46.75	73.40	148.2
(2)	7.651	18.04	42.39	69.40	101.1	140.6	193.4	271.8	411.8	794.0
(3)	11.29	29.29	97.86	196.5	306.1	432.5	592.1	820.1	1217	2284
(4)	15.28	40.14	165.4	416.8	694.3	994.2	1356	1858	2720	5017
(5)	19.22	50.51	236.4	750.5	1325	1916	2607	3547	5142	9369
(6)	23.16	60.74	307.3	1220	2261	3293	4469	6047	8702	15722
(7)	27.10	70.91	376.5	1851	3565	5217	7065	9517	13621	24453
(8)	31.04	81.04	443.8	2666	5300	7781	10518	14119	20123	35940
(9)	34.98	91.16	509.1	3691	7528	11077	14952	20014	28426	50566
(10)	38.92	101.3	572.7	4949	10311	15200	20488	27362	38750	68703

TABLE 2.12 : K* ($\alpha = 0.4$)

Q: MODE:	0.05	0.10	0.20	0.30	0.40	0.50	0.60	0.70	0.80	0.90
(1)	.4490	1.507	4.534	8.560	13.78	20.69	30.28	44.93	71.79	147.1
(2)	5.822	14.34	36.93	63.60	95.29	135.0	188.3	267.3	408.1	791.8
(3)	8.498	22.41	83.12	182.3	293.9	421.9	583.0	812.4	1211	2280
(4)	11.48	30.39	137.5	391.1	674.6	978.3	1343	1848	2712	5012
(5)	14.43	38.09	192.7	711.1	1297	1895	2590	3533	5131	9363
(6)	17.38	45.71	246.4	1165	2226	3266	4448	6030	8689	15715
(7)	20.33	53.31	298.0	1779	3522	5185	7040	9497	13607	24444
(8)	23.29	60.89	347.6	2577	5248	7743	10489	14096	20106	35930
(9)	26.24	68.46	395.6	3583	7468	11035	14918	19988	28406	50555
(10)	29.19	76.03	442.3	4823	10244	15152	20451	27333	38729	68690

TABLE 2.12 : K* ($\alpha = 0.5$)

Q: MODE:	0.05	0.10	0.20	0.30	0.40	0.50	0.60	0.70	0.80	0.90
(1)	.3682	1.279	4.021	7.798	12.82	19.58	29.10	43.77	70.79	146.5
(2)	4.699	11.94	33.21	59.73	91.54	131.5	185.1	264.4	405.9	790.4
(3)	6.813	18.15	73.09	172.9	286.1	415.3	577.4	807.8	1207	2278
(4)	9.190	24.45	118.7	374.3	662.3	968.5	1335	1841	2707	5009
(5)	11.55	30.57	163.6	685.6	1280	1882	2580	3525	5125	9359
(6)	13.91	36.64	206.4	1130	2204	3250	4435	6020	8682	15710
(7)	16.27	42.71	247.2	1734	3495	5165	7025	9485	13598	24439
(8)	18.63	48.76	286.1	2521	5217	7721	10471	14082	20095	35924
(9)	20.99	54.81	323.9	3516	7432	11009	14898	19972	28395	50548
(10)	23.36	60.86	360.5	4744	10203	15123	20428	27315	38716	68683

TABLE 2.12 : K* ($\alpha = 0.6$)

Q: MODE:	0.05	0.10	0.20	0.30	0.40	0.50	0.60	0.70	0.80	0.90
(1)	.3123	1.115	3.643	7.238	12.11	18.78	28.26	42.97	70.11	146.0
(2)	3.940	10.24	30.46	56.94	88.89	129.0	182.8	262.5	404.4	789.5
(3)	5.685	15.26	65.70	166.1	280.7	410.8	573.6	804.6	1205	2277
(4)	7.663	20.46	105.0	362.4	653.9	961.9	1329	1837	2704	5007
(5)	9.628	25.53	142.6	667.6	1269	1873	2573	3519	5121	9357
(6)	11.59	30.58	178.0	1106	2189	3239	4427	6013	8677	15707
(7)	13.56	35.62	211.5	1702	3477	5152	7014	9477	13592	24435
(8)	15.53	40.66	243.4	2482	5196	7705	10460	14073	20089	35920
(9)	17.50	45.70	274.3	3469	7408	10991	14885	19962	28387	50544
(10)	19.46	50.74	304.4	4690	10176	15104	20414	27303	38707	68678

TABLE 2.12 : K* ($\alpha = 0.7$)

Q: MODE:	0.05	0.10	0.20	0.30	0.40	0.50	0.60	0.70	0.80	0.90
(1)	.2712	.9908	3.348	6.804	11.58	18.18	27.64	42.37	69.62	145.7
(2)	3.392	8.971	28.32	54.80	86.91	127.2	181.2	261.2	403.4	788.9
(3)	4.878	13.16	59.97	161.0	276.7	407.5	570.8	802.4	1203	2276
(4)	6.571	17.59	94.39	353.5	647.7	957.1	1325	1834	2702	5006
(5)	8.255	21.92	126.7	654.3	1260	1867	2568	3515	5118	9355
(6)	9.937	26.23	156.7	1087	2178	3231	4421	6008	8673	15705
(7)	11.62	30.55	184.9	1679	3464	5143	7007	9471	13588	24433
(8)	13.31	34.87	211.8	2453	5181	7695	10451	14067	20084	35917
(9)	15.00	39.18	238.0	3436	7391	10979	14875	19954	28382	50541
(10)	16.68	43.50	263.4	4651	10156	15090	20403	27295	38701	68674

TABLE 2.12 : K* (α = 0.8)

Q: MODE:	0.05	0.10	0.20	0.30	0.40	0.50	0.60	0.70	0.80	0.90
(1)	.2397	.8926	3.112	6.455	11.15	17.70	27.16	41.92	69.24	145.5
(2)	2.978	7.988	26.60	53.12	85.38	125.9	180.0	260.1	402.6	788.4
(3)	4.271	11.57	55.37	157.0	273.7	405.0	568.8	800.7	1202	2275
(4)	5.752	15.42	85.97	346.6	643.1	953.5	1322	1831	2700	5005
(5)	7.224	19.20	114.1	644.0	1254	1862	2564	3512	5116	9354
(6)	8.696	22.97	140.1	1074	2170	3226	4416	6004	8670	15703
(7)	10.17	26.75	164.4	1661	3455	5136	7002	9467	13584	24431
(8)	11.65	30.52	187.6	2432	5169	7686	10445	14062	20080	35915
(9)	13.12	34.30	210.2	3410	7377	10970	14868	19949	28378	50538
(10)	14.60	38.07	232.2	4621	10141	15079	20395	27289	38697	68671

TABLE 2.12 : K* (α = 0.9)

Q: MODE:	0.05	0.10	0.20	0.30	0.40	0.50	0.60	0.70	0.80	0.90
(1)	.2148	.8129	2.915	6.166	10.80	17.32	26.77	41.56	68.95	145.3
(2)	2.654	7.202	25.18	51.75	84.15	124.8	179.0	259.3	401.9	788.0
(3)	3.799	10.32	51.57	153.8	271.2	403.0	567.1	799.4	1201	2274
(4)	5.114	13.73	79.06	341.0	639.4	950.6	1320	1830	2699	5004
(5)	6.422	17.08	104.0	635.7	1249	1859	2561	3510	5114	9353
(6)	7.731	20.43	126.7	1062	2164	3221	4412	6001	8668	15702
(7)	9.043	23.79	148.0	1647	3447	5130	6997	9464	13582	24429
(8)	10.35	27.14	168.4	2414	5160	7680	10440	14058	20077	35913
(9)	11.67	30.49	188.2	3390	7367	10963	14863	19944	28374	50536
(10)	12.98	33.85	207.6	4598	10130	15071	20388	27284	38693	68669

TABLE 2.12 : K* (α = 1.0)

Q: MODE:	0.05	0.10	0.20	0.30	0.40	0.50	0.60	0.70	0.80	0.90
(1)	.1946	.7468	2.750	5.923	10.50	17.00	26.45	41.27	68.71	145.2
(2)	2.394	6.558	23.98	50.61	83.14	123.9	178.3	258.6	401.4	787.7
(3)	3.421	9.320	48.36	151.1	269.3	401.5	565.8	798.3	1200	2274
(4)	4.604	12.37	73.28	336.4	636.4	948.3	1318	1828	2698	5004
(5)	5.781	15.39	95.50	629.0	1245	1856	2558	3508	5113	9352
(6)	6.958	18.40	115.7	1053	2159	3217	4409	5999	8667	15701
(7)	8.139	21.41	134.6	1636	3441	5126	6994	9461	13580	24428
(8)	9.319	24.43	152.7	2400	5153	7675	10436	14055	20075	35912
(9)	10.50	27.45	170.4	3373	7359	10957	14858	19941	28372	50535
(10)	11.68	30.47	187.7	4579	10120	15065	20383	27280	38690	68668

TABLE 2.12 . K* (1/α = 0.1)

Q: MODE:	0.05	0.10	0.20	0.30	0.40	0.50	0.60	0.70	0.80	0.90
(1)	.0206	.0939	.6833	3.020	7.395	13.93	23.58	38.75	66.72	144.0
(2)	.2436	.7346	8.829	39.17	74.02	116.2	171.7	253.2	397.3	785.2
(3)	.3434	.9563	9.760	125.6	252.2	388.2	555.0	789.6	1194	2270
(4)	.4611	1.251	10.82	294.3	611.2	929.4	1303	1816	2689	4998
(5)	.5786	1.548	12.02	568.8	1211	1831	2539	3493	5101	9345
(6)	.6962	1.847	13.39	974.6	2117	3187	4386	5981	8653	15693
(7)	.8142	2.147	14.87	1537	3391	5090	6966	9439	13564	24419
(8)	.9322	2.448	16.40	2282	5095	7633	10404	14030	20057	35901
(9)	1.050	2.749	17.98	3234	7292	10910	14822	19913	28351	50523
(10)	1.168	3.050	19.57	4420	10045	15012	20343	27248	38667	68654

TABLE 2.12 : K* (1/α = 0.2)

Q: MODE:	0.05	0.10	0.20	0.30	0.40	0.50	0.60	0.70	0.80	0.90
(1)	.0410	.1816	1.070	3.325	7.850	14.33	23.93	39.04	66.95	144.1
(2)	.4863	1.448	11.70	40.74	75.14	117.1	172.4	253.8	397.7	785.5
(3)	.6864	1.907	16.55	128.9	254.2	389.7	556.2	790.6	1194	2270
(4)	.9220	2.499	20.20	299.6	614.1	931.6	1305	1818	2690	4999
(5)	1.157	3.094	23.23	576.1	1215	1834	2541	3495	5103	9346
(6)	1.392	3.692	26.27	984.0	2122	3190	4388	5983	8654	15694
(7)	1.628	4.293	29.37	1548	3396	5094	6969	9442	13566	24420
(8)	1.864	4.894	32.52	2295	5101	7638	10407	14032	20059	35902
(9)	2.100	5.496	35.73	3250	7300	10915	14826	19916	28353	50524
(10)	2.336	6.099	38.94	4438	10053	15018	20347	27252	38670	68656

TABLE 2.12 : K* (1/α = 0.3)

Q: MODE:	0.05	0.10	0.20	0.30	0.40	0.50	0.60	0.70	0.80	0.90
(1)	.0610	.2641	1.372	3.940	8.262	14.72	24.27	39.33	67.17	144.2
(2)	.7280	2.142	13.92	42.20	76.23	118.0	173.2	254.4	398.2	785.8
(3)	1.029	2.852	22.09	132.0	256.2	391.2	557.4	791.5	1195	2271
(4)	1.383	3.744	28.59	304.6	616.9	933.7	1307	1819	2691	4999
(5)	1.735	4.638	33.78	583.2	1219	1837	2543	3496	5104	9347
(6)	2.088	5.536	38.68	993.3	2127	3194	4391	5985	8656	15695
(7)	2.442	6.437	43.52	1560	3402	5098	6972	9444	13568	24421
(8)	2.796	7.340	48.37	2309	5108	7643	10411	14035	20061	35904
(9)	3.150	8.243	53.26	3266	7307	10920	14830	19919	28356	50525
(10)	3.504	9.148	58.14	4456	10062	15024	20352	27255	38672	68657

TABLE 2.12 : K* (1/α = 0.4)

Q: MODE:	0.05	0.10	0.20	0.30	0.40	0.50	0.60	0.70	0.80	0.90
(1)	.0808	.3422	1.628	4.301	8.642	15.08	24.61	39.62	67.40	144.4
(2)	.9687	2.817	15.80	43.58	77.29	118.8	173.9	255.0	398.7	786.1
(3)	1.372	3.792	26.89	135.1	258.1	392.7	558.7	792.5	1196	2271
(4)	1.843	4.986	36.25	309.6	619.8	935.8	1308	1820	2692	5000
(5)	2.314	6.179	43.77	590.2	1223	1839	2546	3498	5105	9348
(6)	2.784	7.378	50.67	1002	2131	3197	4394	5987	8657	15696
(7)	3.256	8.580	57.36	1571	3408	5102	6975	9446	13569	24422
(8)	3.728	9.784	63.97	2323	5114	7647	10415	14038	20063	35905
(9)	4.200	10.99	70.58	3282	7315	10925	14834	19922	28358	50527
(10)	4.673	12.20	77.15	4474	10070	15029	20356	27259	38675	68659

TABLE 2.12 : K* (1/α = 0.5)

Q: MODE:	0.05	0.10	0.20	0.30	0.40	0.50	0.60	0.70	0.80	0.90
(1)	.1004	.4166	1.855	4.624	8.995	15.43	24.93	39.91	67.62	144.5
(2)	1.209	3.477	17.46	44.88	78.32	119.7	174.7	255.7	399.1	786.3
(3)	1.714	4.726	31.18	138.0	260.1	394.2	559.9	793.5	1196	2272
(4)	2.304	6.225	43.35	314.3	622.6	937.9	1310	1822	2693	5001
(5)	2.892	7.719	53.29	597.0	1227	1842	2548	3500	5106	9348
(6)	3.480	9.218	62.28	1011	2136	3201	4396	5989	8659	15697
(7)	4.070	10.72	70.89	1582	3413	5106	6979	9449	13571	24423
(8)	4.660	12.23	79.31	2336	5121	7652	10418	14041	20065	35906
(9)	5.250	13.73	87.69	3297	7322	10931	14838	19925	28360	50528
(10)	5.841	15.24	95.99	4492	10079	15035	20361	27262	38677	68660

TABLE 2.12 . K* (1/α = 0.6)

Q: MODE:	0.05	0.10	0.20	0.30	0.40	0.50	0.60	0.70	0.80	0.90
(1)	.1197	.4877	2.061	4.920	9.328	15.77	25.25	40.19	67.84	144.6
(2)	1.447	4.120	18.96	46.12	79.33	120.6	175.4	256.3	399.6	786.6
(3)	2.056	5.655	35.10	140.8	262.0	395.7	561.1	794.4	1197	2272
(4)	2.764	7.461	49.99	319.0	625.4	940.0	1312	1823	2694	5001
(5)	3.470	9.257	62.38	603.6	1230	1845	2550	3501	5108	9349
(6)	4.176	11.06	73.55	1020	2141	3204	4399	5991	8661	15698
(7)	4.884	12.86	84.13	1593	3419	5110	6982	9451	13573	24424
(8)	5.592	14.67	94.42	2349	5127	7656	10422	14043	20067	35907
(9)	6.300	16.48	104.6	3313	7329	10936	14842	19928	28363	50529
(10)	7.009	18.29	114.7	4509	10087	15041	20365	27266	38680	68662

TABLE 2.12 : K* (1/α = 0.7)

Q: MODE:	0.05	0.10	0.20	0.30	0.40	0.50	0.60	0.70	0.80	0.90
(1)	.1387	.5560	2.250	5.194	9.642	16.09	25.56	40.46	68.06	144.8
(2)	1.685	4.749	20.34	47.31	80.31	121.4	176.1	256.9	400.0	786.9
(3)	2.397	6.579	38.73	143.5	263.8	397.1	562.3	795.4	1198	2272
(4)	3.224	8.694	56.25	323.5	628.2	942.1	1313	1824	2695	5002
(5)	4.048	10.79	71.12	610.2	1234	1847	2552	3503	5109	9350
(6)	4.872	12.89	84.50	1028	2145	3207	4402	5993	8662	15699
(7)	5.698	15.00	97.11	1604	3424	5114	6985	9454	13575	24425
(8)	6.524	17.11	109.3	2362	5134	7661	10425	14046	20069	35908
(9)	7.350	19.22	121.3	3328	7337	10941	14846	19931	28365	50531
(10)	8.177	21.33	133.2	4527	10095	15047	20370	27269	38682	68663

TABLE 2.12 : K* (1/α = 0.8)

Q: MODE:	0.05	0.10	0.20	0.30	0.40	0.50	0.60	0.70	0.80	0.90
(1)	.1576	.6218	2.427	5.451	9.942	16.40	25.86	40.73	68.28	144.9
(2)	1.922	5.365	21.62	48.45	81.28	122.2	176.8	257.5	400.5	787.2
(3)	2.739	7.497	42.12	146.1	265.7	398.6	563.5	796.4	1199	2273
(4)	3.684	9.923	62.19	327.9	630.9	944.2	1315	1825	2696	5002
(5)	4.625	12.33	79.53	616.6	1238	1850	2554	3505	5110	9350
(6)	5.567	14.73	95.16	1037	2150	3211	4404	5995	8664	15699
(7)	6.512	17.14	109.8	1615	3430	5118	6988	9456	13576	24426
(8)	7.456	19.55	124.0	2375	5140	7666	10429	14049	20071	35910
(9)	8.400	21.96	137.9	3343	7344	10946	14850	19935	28367	50532
(10)	9.344	24.38	151.5	4544	10104	15053	20374	27273	38685	68665

TABLE 2.12 : K* (1/α = 0.9)

Q: MODE:	0.05	0.10	0.20	0.30	0.40	0.50	0.60	0.70	0.80	0.90
(1)	.1762	.6853	2.593	5.693	10.23	16.71	26.16	41.00	68.49	145.0
(2)	2.158	5.968	22.83	49.55	82.22	123.1	177.6	258.1	401.0	787.4
(3)	3.080	8.411	45.32	148.6	267.5	400.0	564.6	797.3	1199	2273
(4)	4.144	11.15	67.86	332.2	633.7	946.3	1317	1827	2697	5003
(5)	5.203	13.86	87.65	622.8	1241	1853	2556	3507	5111	9351
(6)	6.263	16.56	105.6	1045	2155	3214	4407	5997	8665	15700
(7)	7.325	19.28	122.3	1625	3435	5122	6991	9459	13578	24427
(8)	8.387	21.99	138.5	2388	5147	7670	10432	14052	20073	35911
(9)	9.449	24.71	154.2	3358	7352	10952	14854	19938	28369	50533
(10)	10.51	27.42	169.7	4562	10112	15059	20379	27276	38688	68666

TABLE 2.13 : K* (α = 0.1)

Q: MODE:	0.05	0.10	0.20	0.30	0.40	0.50	0.60	0.70	0.80	0.90
(1)	1.602	3.289	6.944	11.04	15.72	21.25	28.18	37.89	54.77	102.4
(2)	16.18	29.69	51.37	72.68	96.69	125.7	163.4	218.0	315.0	583.6
(3)	29.93	68.14	148.3	225.1	306.0	400.1	520.3	693.8	999.7	1835
(4)	42.28	104.0	283.0	485.5	694.0	926.3	1213	1620	2326	4232
(5)	54.31	137.5	443.6	872.4	1316	1792	2364	3156	4515	8156
(6)	66.23	169.6	621.0	1405	2231	3090	4095	5463	7788	13982
(7)	78.11	201.1	808.5	2105	3501	4912	6529	8700	12365	22093
(8)	89.97	232.2	1001	2995	5187	7351	9788	13029	18468	32869
(9)	101.8	263.0	1197	4095	7350	10499	13998	18610	26317	46683
(10)	113.7	293.7	1393	5431	10054	14450	19280	25604	36135	63917

TABLE 2.13 : K* (α = 0.2)

Q: MODE:	0.05	0.10	0.20	0.30	0.40	0.50	0.60	0.70	0.80	0.90
(1)	1.145	2.376	5.130	8.350	12.20	16.96	23.27	32.61	49.66	98.70
(2)	8.901	18.57	36.75	55.81	78.11	106.0	143.3	198.7	298.1	572.4
(3)	15.35	38.03	102.7	177.3	259.1	355.1	478.5	656.6	969.5	1816
(4)	21.36	55.06	189.2	391.6	609.7	851.3	1148	1564	2282	4206
(5)	27.31	71.06	286.5	718.8	1189	1685	2274	3081	4458	8121
(6)	33.23	86.64	388.3	1180	2059	2951	3980	5369	7717	13940
(7)	39.15	102.0	491.1	1800	3281	4740	6389	8588	12281	22043
(8)	45.06	117.3	592.9	2600	4918	7146	9624	12898	18370	32811
(9)	50.98	132.6	693.1	3605	7033	10261	13809	18461	26206	46618
(10)	56.89	147.8	791.3	4839	9688	14179	19066	25436	36010	63844

TABLE 2.13 : K* (α = 0.3)

Q: MODE:	0.05	0.10	0.20	0.30	0.40	0.50	0.60	0.70	0.80	0.90
(1)	.9425	1.973	4.331	7.174	10.67	15.14	21.24	30.50	47.74	97.40
(2)	6.167	13.86	30.28	48.48	70.29	97.97	135.4	191.4	292.1	568.6
(3)	10.32	26.53	82.51	156.9	240.1	337.8	463.0	643.2	959.0	1809
(4)	14.29	37.50	148.0	352.0	576.9	823.5	1124	1544	2267	4197
(5)	18.24	47.94	218.6	655.2	1141	1647	2242	3055	4438	8109
(6)	22.18	58.19	289.7	1089	1995	2901	3940	5337	7693	13926
(7)	26.12	68.37	359.4	1677	3201	4680	6341	8550	12252	22026
(8)	30.06	78.52	427.1	2444	4822	7075	9568	12854	18338	32791
(9)	34.00	88.64	493.0	3414	6921	10179	13745	18410	26169	46596
(10)	37.94	98.75	557.0	4612	9560	14087	18994	25380	35969	63820

TABLE 2.13 : K* (α = 0.4)

Q: MODE:	0.05	0.10	0.20	0.30	0.40	0.50	0.60	0.70	0.80	0.90
(1)	.8220	1.732	3.857	6.482	9.787	14.10	20.09	29.35	46.72	96.74
(2)	4.723	11.15	26.42	44.20	65.83	93.53	131.2	187.6	289.0	566.7
(3)	7.777	20.40	70.49	145.0	229.6	328.5	454.9	636.4	953.7	1806
(4)	10.74	28.44	123.7	329.4	559.1	809.0	1112	1534	2260	4192
(5)	13.69	36.17	179.0	619.4	1115	1627	2226	3042	4429	8104
(6)	16.64	43.81	233.2	1038	1962	2876	3920	5321	7681	13919
(7)	19.60	51.41	285.3	1610	3160	4649	6316	8531	12238	22018
(8)	22.55	59.00	335.3	2359	4773	7038	9540	12832	18321	32782
(9)	25.50	66.57	383.7	3311	6863	10138	13712	18385	26150	46585
(10)	28.46	74.14	430.7	4490	9494	14040	18958	25351	35948	63807

TABLE 2.13 : K* (α = 0.5)

Q: MODE:	0.05	0.10	0.20	0.30	0.40	0.50	0.60	0.70	0.80	0.90
(1)	.7397	1.569	3.536	6.015	9.195	13.41	19.36	28.62	46.09	96.34
(2)	3.829	9.374	23.79	41.32	62.92	90.68	128.5	185.2	287.1	565.5
(3)	6.238	16.58	62.30	137.2	222.9	322.7	449.8	632.2	950.4	1804
(4)	8.600	22.91	107.3	314.6	548.0	800.0	1104	1528	2255	4190
(5)	10.96	29.05	152.5	596.2	1100	1615	2216	3034	4423	8100
(6)	13.32	35.13	196.0	1005	1941	2861	3907	5311	7674	13915
(7)	15.68	41.19	237.2	1567	3135	4630	6302	8519	12230	22013
(8)	18.04	47.25	276.5	2305	4743	7017	9523	12819	18311	32776
(9)	20.41	53.30	314.5	3246	6829	10113	13693	18370	26139	46579
(10)	22.77	59.35	351.4	4414	9455	14012	18936	25334	35935	63800

TABLE 2.13 : K* (α = 0.6)

Q: MODE:	0.05	0.10	0.20	0.30	0.40	0.50	0.60	0.70	0.80	0.90
(1)	.6790	1.448	3.300	5.675	8.768	12.92	18.83	28.12	45.66	96.07
(2)	3.220	8.105	21.85	39.24	60.85	88.70	126.7	183.6	285.8	564.7
(3)	5.208	13.97	56.26	131.6	218.2	318.7	446.4	629.4	948.3	1803
(4)	7.171	19.18	95.21	304.1	540.3	793.9	1099	1524	2252	4188
(5)	9.136	24.27	133.4	579.8	1089	1606	2209	3029	4419	8098
(6)	11.10	29.32	169.4	982.7	1927	2850	3899	5305	7669	13912
(7)	13.07	34.36	203.3	1537	3117	4618	6292	8511	12224	22009
(8)	15.04	39.41	235.5	2268	4722	7002	9511	12810	18305	32772
(9)	17.01	44.44	266.6	3202	6805	10096	13680	18360	26132	46574
(10)	18.98	49.48	296.9	4362	9428	13993	18922	25323	35927	63795

TABLE 2.13 : K* (α = 0.7)

Q: MODE:	0.05	0.10	0.20	0.30	0.40	0.50	0.60	0.70	0.80	0.90
(1)	.6319	1.354	3.118	5.414	8.443	12.55	18.45	27.75	45.35	95.87
(2)	2.778	7.149	20.34	37.65	59.29	87.23	125.3	182.4	284.9	564.2
(3)	4.470	12.07	51.57	127.3	214.7	315.8	444.0	627.3	946.7	1802
(4)	6.150	16.50	85.93	296.2	534.7	789.5	1096	1521	2250	4186
(5)	7.833	20.84	118.9	567.6	1081	1600	2205	3025	4416	8096
(6)	9.518	25.16	149.4	965.8	1917	2843	3893	5300	7666	13910
(7)	11.20	29.48	178.0	1515	3105	4609	6285	8506	12220	22007
(8)	12.89	33.79	205.2	2241	4708	6992	9503	12803	18300	32769
(9)	14.58	38.11	231.5	3169	6788	10084	13671	18352	26126	46571
(10)	16.27	42.43	257.1	4325	9409	13980	18911	25315	35921	63792

TABLE 2.13 : K* (α = 0.8)

Q: MODE:	0.05	0.10	0.20	0.30	0.40	0.50	0.60	0.70	0.80	0.90
(1)	.5939	1.279	2.971	5.206	8.186	12.27	18.15	27.46	45.11	95.73
(2)	2.443	6.401	19.13	36.39	58.08	86.09	124.3	181.6	284.2	563.8
(3)	3.915	10.63	47.80	124.0	212.1	313.6	442.1	625.8	945.6	1801
(4)	5.383	14.47	78.52	290.0	530.5	786.1	1093	1519	2248	4185
(5)	6.855	18.26	107.3	558.2	1075	1596	2201	3023	4414	8095
(6)	8.329	22.03	133.8	952.7	1909	2837	3889	5297	7663	13908
(7)	9.804	25.81	158.5	1498	3096	4602	6280	8502	12217	22005
(8)	11.28	29.58	181.9	2221	4697	6984	9497	12798	18297	32767
(9)	12.76	33.36	204.6	3145	6776	10075	13664	18347	26122	46569
(10)	14.23	37.13	226.7	4296	9395	13969	18903	25309	35916	63789

TABLE 2.13 : K* (α = 0.9)

Q: MODE:	0.05	0.10	0.20	0.30	0.40	0.50	0.60	0.70	0.80	0.90
(1)	.5624	1.216	2.851	5.035	7.977	12.03	17.91	27.23	44.93	95.61
(2)	2.181	5.798	18.12	35.36	57.10	85.20	123.5	180.9	283.7	563.4
(3)	3.483	9.496	44.68	121.3	209.9	311.9	440.7	624.6	944.6	1801
(4)	4.787	12.89	72.42	285.1	527.1	783.5	1091	1517	2247	4185
(5)	6.095	16.25	97.96	550.6	1070	1593	2198	3020	4412	8094
(6)	7.404	19.60	121.2	942.4	1903	2833	3885	5294	7661	13907
(7)	8.716	22.95	142.8	1485	3088	4597	6276	8498	12214	22004
(8)	10.03	26.30	163.3	2204	4688	6978	9492	12795	18294	32766
(9)	11.34	29.66	183.3	3125	6766	10068	13658	18343	26119	46567
(10)	12.65	33.01	202.8	4273	9384	13962	18897	25304	35913	63787

TABLE 2.13 : K* (α = 1.0)

Q: MODE:	0.05	0.10	0.20	0.30	0.40	0.50	0.60	0.70	0.80	0.90
(1)	.5358	1.164	2.749	4.893	7.803	11.84	17.71	27.05	44.78	95.52
(2)	1.969	5.302	17.27	34.50	56.30	84.46	122.8	180.3	283.2	563.2
(3)	3.136	8.580	42.04	119.0	208.2	310.5	439.5	623.6	943.9	1800
(4)	4.309	11.62	67.31	281.0	524.4	781.4	1089	1516	2246	4184
(5)	5.486	14.63	90.17	544.4	1067	1590	2196	3019	4411	8093
(6)	6.665	17.65	110.8	933.9	1899	2829	3883	5292	7659	13906
(7)	7.844	20.66	130.0	1474	3083	4592	6272	8496	12213	22003
(8)	9.024	23.68	148.3	2191	4681	6973	9488	12792	18292	32764
(9)	10.21	26.70	166.0	3109	6758	10062	13654	18339	26117	46565
(10)	11.39	29.72	183.4	4255	9375	13955	18892	25300	35910	63785

TABLE 2.13 : K* (1/α = 0.1)

Q: MODE:	0.05	0.10	0.20	0.30	0.40	0.50	0.60	0.70	0.80	0.90
(1)	.2019	.5156	1.579	3.362	6.063	10.03	15.95	25.47	43.53	94.78
(2)	.2023	.6242	6.607	25.67	48.90	78.04	117.3	175.6	279.7	561.0
(3)	.3152	.8877	9.569	97.29	193.2	298.6	429.7	615.7	938.0	1797
(4)	.4318	1.177	10.56	243.2	501.2	763.9	1075	1505	2238	4179
(5)	.5492	1.474	11.70	488.8	1035	1567	2178	3004	4400	8087
(6)	.6669	1.772	13.04	859.5	1859	2800	3860	5274	7646	13898
(7)	.7848	2.072	14.50	1380	3034	4558	6246	8475	12197	21994
(8)	.9027	2.373	16.02	2077	4625	6933	9457	12767	18274	32754
(9)	1.021	2.674	17.58	2975	6693	10017	13619	18312	26097	46554
(10)	1.139	2.975	19.17	4101	9301	13904	18853	25269	35888	63772

TABLE 2.13 : K* (1/α = 0.2)

Q: MODE:	0.05	0.10	0.20	0.30	0.40	0.50	0.60	0.70	0.80	0.90
(1)	.2651	.6340	1.772	3.589	6.300	10.26	16.17	25.66	43.67	94.86
(2)	.4034	1.221	8.619	26.92	49.82	78.80	117.9	176.2	280.1	561.3
(3)	.6301	1.768	15.51	100.1	195.0	299.9	430.8	616.6	938.6	1797
(4)	.8633	2.351	19.40	248.0	503.9	765.9	1077	1506	2238	4180
(5)	1.098	2.945	22.48	495.6	1039	1569	2180	3006	4401	8088
(6)	1.334	3.543	25.51	868.4	1863	2804	3863	5276	7648	13899
(7)	1.569	4.143	28.60	1391	3040	4562	6249	8477	12199	21995
(8)	1.805	4.744	31.73	2090	4631	6937	9461	12770	18276	32755
(9)	2.042	5.346	34.92	2991	6700	10022	13622	18315	26099	46555
(10)	2.278	5.949	38.14	4118	9309	13910	18857	25273	35890	63774

TABLE 2.13 : K* (1/α = 0.3)

Q: MODE:	0.05	0.10	0.20	0.30	0.40	0.50	0.60	0.70	0.80	0.90
(1)	.3140	.7281	1.936	3.794	6.522	10.48	16.37	25.84	43.81	94.95
(2)	.6032	1.793	10.18	28.06	50.72	79.55	118.6	176.7	280.5	561.5
(3)	.9446	2.642	20.23	102.9	196.7	301.3	431.9	617.5	939.3	1798
(4)	1.295	3.522	27.16	252.6	506.5	767.9	1078	1507	2239	4180
(5)	1.647	4.414	32.53	502.2	1042	1572	2182	3007	4403	8088
(6)	2.000	5.311	37.47	877.1	1868	2807	3865	5278	7649	13900
(7)	2.354	6.212	42.32	1402	3045	4566	6251	8480	12201	21996
(8)	2.708	7.115	47.16	2103	4638	6941	9464	12773	18278	32756
(9)	3.062	8.018	52.03	3006	6707	10027	13626	18318	26101	46556
(10)	3.416	8.923	56.92	4136	9318	13915	18862	25276	35893	63775

TABLE 2.13 : K* (1/α = 0.4)

Q: MODE:	0.05	0.10	0.20	0.30	0.40	0.50	0.60	0.70	0.80	0.90
(1)	.3554	.8086	2.081	3.982	6.732	10.69	16.58	26.02	43.95	95.03
(2)	.8018	2.344	11.50	29.13	51.59	80.29	119.2	177.2	280.9	561.7
(3)	1.259	3.509	24.26	105.4	198.5	302.6	433.0	618.4	940.0	1798
(4)	1.726	4.688	34.16	257.0	509.2	769.9	1080	1508	2240	4181
(5)	2.196	5.881	41.98	508.7	1046	1574	2184	3009	4404	8089
(6)	2.667	7.079	48.98	885.6	1872	2810	3868	5280	7651	13901
(7)	3.139	8.280	55.70	1413	3051	4570	6254	8482	12202	21997
(8)	3.610	9.484	62.32	2116	4644	6946	9468	12776	18280	32757
(9)	4.083	10.69	68.92	3021	6715	10032	13630	18321	26103	46558
(10)	4.555	11.90	75.51	4153	9326	13921	18866	25280	35895	63777

TABLE 2.13 : K* (1/α = 0.5)

Q: MODE:	0.05	0.10	0.20	0.30	0.40	0.50	0.60	0.70	0.80	0.90
(1)	.3920	.8801	2.213	4.156	6.930	10.90	16.78	26.20	44.09	95.11
(2)	.9990	2.876	12.67	30.14	52.43	81.01	119.8	177.7	281.3	562.0
(3)	1.573	4.370	27.85	107.9	200.1	304.0	434.1	619.3	940.6	1798
(4)	2.157	5.852	40.58	261.3	511.8	771.8	1081	1510	2241	4181
(5)	2.745	7.345	50.92	515.0	1049	1577	2186	3011	4405	8090
(6)	3.333	8.844	60.09	894.0	1877	2813	3870	5282	7652	13902
(7)	3.923	10.35	68.77	1423	3056	4573	6257	8484	12204	21998
(8)	4.513	11.85	77.22	2129	4650	6950	9471	12778	18282	32759
(9)	5.104	13.36	85.60	3036	6722	10037	13634	18324	26106	46559
(10)	5.694	14.87	93.92	4171	9334	13927	18870	25283	35898	63778

TABLE 2.13 : K^* ($1/\alpha = 0.6$)

Q: MODE:	0.05	0.10	0.20	0.30	0.40	0.50	0.60	0.70	0.80	0.90
(1)	.4251	.9451	2.334	4.319	7.119	11.10	16.97	26.37	44.23	95.20
(2)	1.195	3.390	13.73	31.09	53.24	81.73	120.4	178.3	281.6	562.2
(3)	1.886	5.224	31.10	110.3	201.8	305.3	435.2	620.1	941.3	1799
(4)	2.588	7.012	46.56	265.4	514.3	773.7	1083	1511	2242	4182
(5)	3.293	8.807	59.44	521.1	1053	1580	2188	3012	4406	8090
(6)	4.000	10.61	70.84	902.3	1881	2816	3873	5284	7654	13903
(7)	4.708	12.41	81.53	1434	3061	4577	6260	8487	12206	21999
(8)	5.415	14.22	91.87	2142	4656	6955	9474	12781	18284	32760
(9)	6.124	16.03	102.1	3051	6729	10042	13638	18327	26108	46560
(10)	6.832	17.84	112.2	4188	9342	13932	18875	25287	35900	63779

TABLE 2.13 : K^* ($1/\alpha = 0.7$)

Q: MODE:	0.05	0.10	0.20	0.30	0.40	0.50	0.60	0.70	0.80	0.90
(1)	.4555	1.005	2.447	4.473	7.300	11.29	17.16	26.55	44.37	95.28
(2)	1.390	3.888	14.70	32.00	54.04	82.43	121.0	178.8	282.0	562.5
(3)	2.199	6.072	34.10	112.6	203.5	306.6	436.3	621.0	941.9	1799
(4)	3.018	8.168	52.16	269.5	516.9	775.7	1085	1512	2243	4182
(5)	3.841	10.27	67.58	527.1	1056	1582	2190	3014	4407	8091
(6)	4.666	12.37	81.25	910.4	1885	2820	3875	5286	7655	13903
(7)	5.492	14.48	94.02	1444	3067	4581	6263	8489	12207	22000
(8)	6.317	16.59	106.3	2154	4663	6959	9478	12784	18286	32761
(9)	7.145	18.70	118.3	3066	6736	10047	13642	18330	26110	46561
(10)	7.971	20.81	130.2	4205	9350	13938	18879	25290	35903	63781

TABLE 2.13 : K^* ($1/\alpha = 0.8$)

Q: MODE:	0.05	0.10	0.20	0.30	0.40	0.50	0.60	0.70	0.80	0.90
(1)	.4839	1.061	2.553	4.619	7.474	11.48	17.35	26.72	44.51	95.36
(2)	1.584	4.372	15.61	32.86	54.81	83.12	121.6	179.3	282.4	562.7
(3)	2.512	6.914	36.91	114.8	205.1	307.9	437.4	621.9	942.6	1799
(4)	3.449	9.322	57.46	273.4	519.4	777.6	1086	1513	2244	4183
(5)	4.390	11.72	75.39	533.0	1060	1585	2192	3015	4408	8092
(6)	5.332	14.13	91.37	918.3	1890	2823	3878	5288	7656	13904
(7)	6.276	16.54	106.3	1454	3072	4585	6266	8491	12209	22001
(8)	7.220	18.95	120.5	2167	4669	6964	9481	12786	18288	32762
(9)	8.165	21.36	134.4	3081	6744	10052	13646	18333	26112	46563
(10)	9.110	23.78	148.1	4221	9358	13944	18884	25293	35905	63782

TABLE 2.13 : K^* ($1/\alpha = 0.9$)

Q: MODE:	0.05	0.10	0.20	0.30	0.40	0.50	0.60	0.70	0.80	0.90
(1)	.5106	1.114	2.654	4.759	7.641	11.66	17.53	26.89	44.64	95.44
(2)	1.777	4.843	16.46	33.70	55.57	83.79	122.2	179.8	282.8	562.9
(3)	2.824	7.750	39.54	116.9	206.7	309.2	438.4	622.8	943.3	1800
(4)	3.879	10.47	62.50	277.3	521.9	779.5	1088	1515	2245	4184
(5)	4.938	13.18	82.91	538.8	1063	1587	2194	3017	4410	8092
(6)	5.999	15.89	101.2	926.2	1894	2826	3880	5290	7658	13905
(7)	7.060	18.60	118.2	1464	3077	4589	6269	8493	12211	22002
(8)	8.122	21.32	134.5	2179	4675	6968	9485	12789	18290	32763
(9)	9.185	24.03	150.3	3095	6751	10057	13650	18336	26115	46564
(10)	10.25	26.75	165.9	4238	9366	13950	18888	25297	35908	63784

TABLE 2.14 : T^* ($\alpha = 0.1$)

Q: MODE:	0.05	0.10	0.20	0.30	0.40	0.50	0.60	0.70	0.80	0.90
(1)	2.067	5.370	12.38	19.53	26.79	34.16	41.72	49.64	58.45	70.91
(2)	63.91	75.01	90.21	103.8	117.1	130.3	143.8	157.7	173.0	193.7
(3)	180.6	199.5	223.5	244.4	264.5	284.6	304.8	325.6	348.1	377.7
(4)	359.9	386.8	419.6	447.7	474.7	501.5	528.5	556.1	585.9	624.3
(5)	601.4	636.4	678.1	713.4	747.3	780.8	814.5	849.1	886.1	933.3
(6)	905.3	948.4	999.0	1041	1082	1122	1163	1204	1248	1304
(7)	1271	1322	1382	1432	1479	1526	1573	1622	1673	1738
(8)	1700	1759	1827	1885	1939	1993	2047	2102	2161	2235
(9)	2191	2258	2336	2400	2461	2522	2583	2645	2711	2793
(10)	2744	2820	2906	2978	3046	3113	3181	3250	3323	3415

TABLE 2.14 : T^* ($\alpha = 0.2$)

Q: MODE:	0.05	0.10	0.20	0.30	0.40	0.50	0.60	0.70	0.80	0.90
(1)	1.205	3.403	8.283	13.34	18.52	23.81	29.29	35.14	41.88	52.41
(2)	41.90	50.83	62.27	72.19	81.80	91.41	101.2	111.5	123.1	140.4
(3)	121.4	137.1	155.5	170.8	185.5	200.0	214.7	230.1	247.1	271.3
(4)	244.9	267.7	293.1	313.8	333.5	353.0	372.6	393.0	415.4	446.6
(5)	412.4	442.4	474.8	500.9	525.7	550.0	574.6	600.0	627.8	666.0
(6)	624.0	661.1	700.6	732.2	761.9	791.2	820.7	851.1	884.4	929.6
(7)	879.7	924.0	970.5	1007	1042	1076	1110	1146	1185	1237
(8)	1179	1231	1284	1327	1366	1406	1445	1485	1529	1589
(9)	1523	1582	1642	1690	1735	1779	1823	1869	1918	1985
(10)	1911	1977	2045	2098	2148	2197	2246	2297	2352	2425

TABLE 2.14 : T^* ($\alpha = 0.3$)

Q: MODE:	0.05	0.10	0.20	0.30	0.40	0.50	0.60	0.70	0.80	0.90
(1)	.8604	2.562	6.487	10.61	14.86	19.23	23.79	28.71	34.55	44.26
(2)	32.32	40.17	49.92	58.20	66.19	74.17	82.35	91.03	101.1	116.8
(3)	95.35	109.5	125.4	138.2	150.5	162.6	174.9	187.8	202.4	224.2
(4)	194.2	215.0	237.1	254.5	271.0	287.2	303.6	320.7	339.9	367.9
(5)	329.0	356.5	384.7	406.8	427.5	447.8	468.3	489.6	513.4	547.6
(6)	499.7	533.9	568.4	595.1	620.0	644.4	669.0	694.6	723.0	763.5
(7)	706.3	747.4	788.2	819.5	848.6	877.1	905.8	935.6	968.7	1015
(8)	949.0	996.9	1043	1079	1113	1145	1178	1212	1250	1303
(9)	1227	1282	1335	1376	1413	1450	1487	1525	1568	1627
(10)	1542	1604	1663	1708	1750	1791	1832	1874	1921	1987

TABLE 2.14 : T^* ($\alpha = 0.4$)

Q: MODE:	0.05	0.10	0.20	0.30	0.40	0.50	0.60	0.70	0.80	0.90
(1)	.6714	2.076	5.425	8.991	12.68	16.50	20.51	24.88	30.18	39.43
(2)	26.69	33.86	42.57	49.87	56.88	63.90	71.11	78.83	87.90	102.7
(3)	79.95	93.13	107.4	118.8	129.6	140.2	151.1	162.5	175.7	196.1
(4)	164.1	183.7	203.7	219.2	233.7	247.9	262.4	277.6	294.9	321.0
(5)	279.4	305.3	331.1	350.7	368.9	386.8	404.9	423.8	445.2	477.1
(6)	425.7	458.2	489.7	513.4	535.4	556.9	578.6	601.3	626.8	664.4
(7)	603.1	642.2	679.5	707.4	733.1	758.2	783.6	810.0	839.6	883.0
(8)	811.8	857.4	900.5	932.5	962.0	990.7	1019	1049	1083	1132
(9)	1051	1103	1152	1188	1222	1254	1287	1320	1358	1413
(10)	1322	1381	1436	1476	1513	1549	1585	1623	1665	1726

TABLE 2.14 : T^* ($\alpha = 0.5$)

Q: MODE:	0.05	0.10	0.20	0.30	0.40	0.50	0.60	0.70	0.80	0.90
(1)	.5514	1.754	4.706	7.888	11.19	14.63	18.27	22.27	27.20	36.16
(2)	22.92	29.57	37.56	44.19	50.54	56.89	63.45	70.50	78.92	93.19
(3)	69.51	81.97	95.21	105.6	115.4	125.0	134.9	145.3	157.5	177.0
(4)	143.7	162.3	180.9	195.1	208.2	221.2	234.3	248.2	264.1	289.0
(5)	245.6	270.5	294.5	312.4	329.0	345.2	361.7	378.9	398.7	428.9
(6)	375.3	406.5	435.9	457.7	477.7	497.2	516.9	537.6	561.2	596.8
(7)	532.8	570.4	605.3	630.9	654.3	677.1	700.1	724.2	751.6	792.7
(8)	718.2	762.3	802.6	831.9	858.8	884.9	911.2	938.7	969.9	1016
(9)	931.5	982.0	1027	1060	1091	1120	1150	1181	1216	1268
(10)	1172	1229	1280	1317	1351	1384	1417	1451	1490	1547

TABLE 2.14 : T^* ($\alpha = 0.6$)

Q: MODE:	0.05	0.10	0.20	0.30	0.40	0.50	0.60	0.70	0.80	0.90
(1)	.4681	1.524	4.180	7.076	10.10	13.26	16.61	20.34	25.00	33.75
(2)	20.17	26.42	33.87	40.00	45.85	51.72	57.78	64.36	72.30	86.14
(3)	61.86	73.74	86.19	95.84	104.9	113.8	122.9	132.6	144.1	162.9
(4)	128.7	146.5	164.1	177.3	189.5	201.4	213.6	226.5	241.5	265.4
(5)	220.7	244.7	267.4	284.2	299.5	314.5	329.7	345.8	364.3	393.4
(6)	338.1	368.4	396.3	416.5	435.1	453.1	471.4	490.6	512.7	547.0
(7)	481.0	517.5	550.5	574.4	596.1	617.2	638.5	660.9	686.6	726.0
(8)	649.3	692.0	730.3	757.7	782.6	806.8	831.1	856.7	885.9	930.6
(9)	843.0	892.1	935.5	966.5	994.6	1021	1049	1077	1110	1160
(10)	1062	1117	1166	1200	1232	1262	1292	1324	1361	1416

TABLE 2.14 : T^* ($\alpha = 0.7$)

Q: MODE:	0.05	0.10	0.20	0.30	0.40	0.50	0.60	0.70	0.80	0.90
(1)	.4068	1.349	3.775	6.447	9.252	12.19	15.33	18.84	23.29	31.89
(2)	18.06	23.98	31.00	36.74	42.21	47.70	53.39	59.58	67.15	80.67
(3)	55.96	67.37	79.18	88.25	96.69	105.0	113.6	122.7	133.6	151.9
(4)	117.0	134.3	151.0	163.5	174.9	186.1	197.5	209.6	223.8	247.0
(5)	201.4	224.8	246.4	262.2	276.6	290.7	304.9	320.0	337.6	365.8
(6)	309.3	338.8	365.4	384.6	401.9	418.9	436.0	454.1	475.1	508.2
(7)	440.7	476.3	508.0	530.5	550.9	570.7	590.6	611.7	636.1	674.2
(8)	595.7	637.5	674.2	700.0	723.4	746.0	768.9	793.0	820.7	863.8
(9)	774.2	822.2	863.9	893.2	919.5	945.0	970.7	997.8	1028	1077
(10)	976.3	1030	1077	1109	1139	1167	1196	1226	1260	1313

TABLE 2.14 : T^* ($\alpha = 0.8$)

Q: MODE:	0.05	0.10	0.20	0.30	0.40	0.50	0.60	0.70	0.80	0.90
(1)	.3598	1.212	3.451	5.942	8.569	11.33	14.29	17.63	21.91	30.40
(2)	16.39	22.03	28.70	34.12	39.28	44.46	49.84	55.73	63.00	76.27
(3)	51.23	62.24	73.54	82.13	90.11	97.99	106.1	114.8	125.2	143.1
(4)	107.7	124.5	140.5	152.3	163.1	173.7	184.5	196.0	209.6	232.3
(5)	186.0	208.7	229.5	244.5	258.2	271.4	284.9	299.3	316.1	343.6
(6)	286.2	314.9	340.6	358.8	375.3	391.3	407.5	424.6	444.7	477.0
(7)	408.4	443.2	473.7	495.1	514.4	533.1	552.1	572.1	595.3	632.5
(8)	552.6	593.5	628.9	653.5	675.7	697.1	718.7	741.6	768.1	810.0
(9)	718.9	765.9	806.1	834.0	859.0	883.1	907.5	933.1	962.9	1009
(10)	907.2	960.4	1005	1036	1064	1091	1118	1146	1179	1231

TABLE 2.14 : T^* ($\alpha = 0.9$)

Q: MODE:	0.05	0.10	0.20	0.30	0.40	0.50	0.60	0.70	0.80	0.90
(1)	.3226	1.101	3.184	5.525	8.004	10.62	13.44	16.63	20.77	29.17
(2)	15.02	20.42	26.79	31.95	36.86	41.78	46.90	52.54	59.56	72.63
(3)	47.35	58.00	68.87	77.07	84.66	92.16	99.86	108.2	118.2	135.8
(4)	99.98	116.3	131.8	143.1	153.4	163.4	173.7	184.7	197.9	220.0
(5)	173.2	195.4	215.5	229.9	242.9	255.5	268.4	282.1	298.3	325.1
(6)	267.0	295.2	320.0	337.5	353.1	368.4	383.8	400.3	419.5	451.1
(7)	381.6	415.7	445.3	465.8	484.2	502.1	520.1	539.2	561.6	597.9
(8)	516.9	557.1	591.4	615.0	636.1	656.6	677.2	699.0	724.5	765.5
(9)	673.0	719.3	758.3	785.0	808.8	831.8	855.1	879.6	908.2	953.9
(10)	849.9	902.2	946.0	975.8	1002	1027	1053	1080	1112	1163

TABLE 2.14 : T^* ($\alpha = 1.0$)

Q: MODE:	0.05	0.10	0.20	0.30	0.40	0.50	0.60	0.70	0.80	0.90
(1)	.2924	1.009	2.961	5.173	7.527	10.02	12.71	15.78	19.81	28.14
(2)	13.86	19.06	25.18	30.12	34.00	39.51	44.42	49.04	56.66	69.56
(3)	44.08	54.43	64.92	72.79	80.06	87.23	94.61	102.6	112.4	129.6
(4)	93.49	109.4	124.5	135.3	145.1	154.8	164.6	175.2	187.9	209.7
(5)	162.4	184.1	203.7	217.5	229.9	242.0	254.4	267.5	283.2	309.6
(6)	250.8	278.5	302.6	319.4	334.4	349.0	363.8	379.6	398.3	429.2
(7)	358.9	392.5	421.3	441.1	458.7	475.8	493.1	511.4	533.1	568.6
(8)	486.8	526.3	559.7	582.5	602.7	622.3	642.0	663.0	687.6	727.8
(9)	634.3	679.8	717.9	743.6	766.4	788.5	810.7	834.3	861.9	906.7
(10)	801.5	853.1	895.7	924.5	949.9	974.4	999.2	1025	1055	1105

TABLE 2.14 : T^* ($1/\alpha = 0.1$)

Q: MODE:	0.05	0.10	0.20	0.30	0.40	0.50	0.60	0.70	0.80	0.90
(1)	.0311	.1228	.4688	.9946	1.673	2.512	3.576	5.073	7.762	16.23
(2)	1.929	3.360	5.500	7.315	9.004	10.95	13.05	15.73	20.01	32.03
(3)	7.215	11.22	15.82	19.11	22.06	25.01	28.21	32.09	37.89	52.70
(4)	17.23	24.79	32.25	37.10	41.27	45.33	49.64	54.75	62.13	80.01
(5)	32.61	44.36	54.85	61.28	66.69	71.87	77.31	83.65	92.64	113.7
(6)	53.73	70.04	83.64	91.69	98.34	104.6	111.2	118.8	129.4	153.6
(7)	80.78	101.9	118.7	128.3	136.2	143.7	151.3	160.2	172.4	199.9
(8)	113.9	139.9	159.9	171.2	180.3	188.9	197.7	207.8	221.7	252.4
(9)	153.1	184.2	207.4	220.3	230.7	240.4	250.4	261.7	277.2	311.2
(10)	198.4	234.7	261.1	275.7	287.3	298.1	309.2	321.8	339.0	376.2

TABLE 2.14 : T^* ($1/\alpha = 0.2$)

Q: MODE:	0.05	0.10	0.20	0.30	0.40	0.50	0.60	0.70	0.80	0.90
(1)	.0618	.2389	.8590	1.716	2.738	3.918	5.316	7.124	10.03	18.19
(2)	3.652	5.965	9.060	11.56	13.94	16.38	19.05	22.27	27.00	38.81
(3)	13.10	18.88	24.97	29.24	33.08	36.88	40.91	45.60	52.13	67.05
(4)	30.15	40.27	49.62	55.74	61.07	66.24	71.65	77.82	86.21	104.5
(5)	55.41	70.31	83.05	91.04	97.86	104.4	111.2	118.9	129.1	150.8
(6)	89.16	109.1	125.3	135.1	143.5	151.4	159.6	168.8	180.9	206.1
(7)	131.5	156.7	176.3	188.1	197.9	207.2	216.8	227.5	241.5	270.2
(8)	182.6	213.0	236.2	249.8	261.1	271.9	282.8	295.0	310.9	343.1
(9)	242.5	278.2	304.8	320.4	333.2	345.3	357.7	371.4	389.2	424.9
(10)	311.1	352.2	382.3	399.8	414.1	427.6	441.4	456.6	476.3	515.5

TABLE 2.14 : T^* ($1/\alpha = 0.3$)

Q: MODE:	0.05	0.10	0.20	0.30	0.40	0.50	0.60	0.70	0.80	0.90
(1)	.0920	.3491	1.201	2.312	3.586	5.015	6.657	8.698	11.79	19.85
(2)	5.225	8.170	11.91	14.88	17.70	20.57	23.66	27.28	32.38	44.24
(3)	18.19	25.10	32.13	37.11	41.59	46.02	50.67	55.96	63.07	78.28
(4)	40.95	52.56	63.12	70.15	76.32	82.32	88.54	95.53	104.7	123.5
(5)	74.01	90.70	104.9	114.0	121.8	129.4	137.2	145.9	157.1	179.6
(6)	117.6	139.6	157.4	168.6	178.2	187.3	196.7	207.1	220.4	246.5
(7)	171.9	199.2	220.7	234.0	245.3	256.0	267.0	279.1	294.5	324.3
(8)	236.9	269.6	294.9	310.2	323.2	335.6	348.1	361.9	379.4	413.0
(9)	312.6	350.9	379.8	397.3	412.0	425.9	440.0	455.5	475.1	512.4
(10)	399.2	442.9	475.6	495.1	511.5	527.0	542.8	560.0	581.7	622.7

TABLE 2.14 : T^* ($1/\alpha = 0.4$)

Q: MODE:	0.05	0.10	0.20	0.30	0.40	0.50	0.60	0.70	0.80	0.90
(1)	.1218	.4544	1.508	2.831	4.314	5.947	7.790	10.02	13.28	21.33
(2)	6.681	10.12	14.35	17.71	20.89	24.11	27.54	31.51	36.92	48.90
(3)	22.75	30.47	38.23	43.77	48.78	53.73	58.90	64.70	72.30	87.83
(4)	50.42	63.08	74.56	82.32	89.19	95.88	102.8	110.5	120.3	139.6
(5)	90.14	108.0	123.3	133.3	142.1	150.5	159.2	168.7	180.0	203.9
(6)	142.1	165.4	184.6	196.8	207.4	217.6	228.0	239.4	253.7	280.7
(7)	206.5	235.3	258.3	272.8	285.2	297.2	309.4	322.6	339.2	370.0
(8)	283.2	317.6	344.4	361.2	375.6	389.3	403.2	418.3	437.2	471.9
(9)	372.4	412.3	443.1	462.1	478.4	493.8	509.5	526.5	547.6	586.2
(10)	474.0	519.6	554.2	575.5	593.6	610.8	628.3	647.2	670.5	713.1

TABLE 2.14 : T^* ($1/\alpha = 0.5$)

Q: MODE:	0.05	0.10	0.20	0.30	0.40	0.50	0.60	0.70	0.80	0.90
(1)	.1512	.5553	1.790	3.297	4.960	6.771	8.789	11.19	14.60	22.67
(2)	8.043	11.88	16.53	20.22	23.71	27.23	30.97	35.23	40.92	53.05
(3)	26.92	35.26	43.62	49.65	55.13	60.53	66.15	72.40	80.43	96.28
(4)	58.95	72.42	84.66	93.06	100.5	107.8	115.3	123.6	134.0	153.8
(5)	104.6	123.4	139.6	150.4	159.9	169.1	178.5	188.8	201.6	225.3
(6)	163.9	188.3	208.5	221.7	233.2	244.3	255.6	267.9	283.1	310.8
(7)	237.2	267.1	291.3	306.9	320.5	333.5	346.7	361.0	378.6	410.3
(8)	324.3	359.9	388.1	406.1	421.7	436.6	451.7	468.0	488.0	523.8
(9)	425.3	466.6	498.9	519.3	536.9	553.7	570.7	589.0	611.4	651.3
(10)	540.2	587.2	623.6	646.4	666.0	684.7	703.6	724.0	748.8	792.7

TABLE 2.14 : T^* ($1/\alpha = 0.6$)

Q: MODE:	0.05	0.10	0.20	0.30	0.40	0.50	0.60	0.70	0.80	0.90
(1)	.1801	.6523	2.052	3.724	5.548	7.518	9.692	12.25	15.80	23.91
(2)	9.326	13.50	18.51	22.49	26.26	30.06	34.07	38.60	44.55	56.83
(3)	30.77	39.64	48.51	54.97	60.87	66.68	72.71	79.36	87.78	103.9
(4)	66.79	80.91	93.82	102.8	110.8	118.7	126.7	135.5	146.4	166.6
(5)	117.7	137.3	154.4	165.9	176.0	185.9	196.0	206.9	220.4	244.7
(6)	183.8	209.0	230.2	244.2	256.5	268.4	280.5	293.6	309.6	338.1
(7)	265.1	296.0	321.3	337.8	352.3	366.3	380.4	395.6	414.2	446.8
(8)	361.6	398.2	427.6	446.7	463.4	479.4	495.5	512.9	534.0	570.8
(9)	473.3	515.7	549.3	571.0	589.8	607.8	626.0	645.5	669.2	710.1
(10)	600.3	648.4	686.3	710.4	731.4	751.4	771.7	793.4	819.6	864.7

TABLE 2.14 : T^* ($1/\alpha = 0.7$)

Q: MODE:	0.05	0.10	0.20	0.30	0.40	0.50	0.60	0.70	0.80	0.90
(1)	.2087	.7459	2.298	4.120	6.091	8.206	10.52	13.22	16.90	25.06
(2)	10.54	15.02	20.34	24.59	28.61	32.66	36.92	41.69	47.88	60.32
(3)	34.38	43.69	53.02	59.87	66.15	72.34	78.74	85.76	94.55	111.0
(4)	74.07	88.74	102.2	111.7	120.3	128,6	137.1	146.4	157.8	178.5
(5)	129.9	150.2	167.9	180.1	190.9	201.4	212.1	223.6	237.7	262.5
(6)	202.2	228.1	250.1	264.9	278.0	290.6	303.5	317.3	334.0	363.2
(7)	290.8	322.5	348.8	366.3	381.6	396.4	411.4	427.5	446.9	480.3
(8)	395.9	433.4	464.0	484.1	501.7	518.7	535.9	554.3	576.3	614.0
(9)	517.5	560.8	595.7	618.5	638.4	657.5	676.8	697.5	722.2	764.2
(10)	655.6	704.8	743.9	769.4	791.6	812.8	834.3	857.3	884.7	931.0

TABLE 2.14 : T^* ($1/\alpha = 0.8$)

Q: MODE:	0.05	0.10	0.20	0.30	0.40	0.50	0.60	0.70	0.80	0.90
(1)	.2370	.8363	2.530	4.491	6.598	8.847	11.30	14.13	17.93	26.14
(2)	11.70	16.44	22.05	26.54	30.80	35.08	39.57	44.58	50.98	63.58
(3)	37.78	47.47	57.23	64.44	71.06	77.60	84.35	91.72	100.8	117.6
(4)	80.90	96.05	110.1	120.1	129.1	137.9	146.9	156.6	168.5	189.5
(5)	141.4	162.2	180.6	193.3	204.7	215.7	227.0	239.1	253.8	279.2
(6)	219.3	245.9	268.7	284.2	297.9	311.3	324.8	339.3	356.8	386.5
(7)	314.9	347.3	374.4	392.7	408.9	424.5	440.3	457.2	477.4	511.6
(8)	428.0	466.3	497.8	518.9	537.4	555.3	573.4	592.7	615.7	654.3
(9)	558.8	602.9	638.9	662.7	683.7	703.8	724.2	745.9	771.6	814.6
(10)	707.2	757.2	797.6	824.2	847.6	870.0	892.6	916.7	945.2	992.6

TABLE 2.14 : T^* ($1/\alpha = 0.9$)

Q: MODE:	0.05	0.10	0.20	0.30	0.40	0.50	0.60	0.70	0.80	0.90
(1)	.2648	.9240	2.750	4.841	7.075	9.449	12.02	14.98	18.89	27.16
(2)	12.82	17.78	23.66	28.38	32.86	37.36	42.06	47.28	53.90	66.65
(3)	41.00	51.04	61.18	68.73	75.68	82.55	89.62	97.32	106.8	123.8
(4)	87.35	102.9	117.5	127.9	137.3	146.5	156.0	166.2	178.5	199.9
(5)	152.1	173.4	192.4	205.7	217.7	229.3	241.1	253.7	268.9	294.8
(6)	235.5	262.6	286.1	302.3	316.7	330.7	344,9	360.0	378.1	408.5
(7)	337.5	370.5	398.5	417.6	434.5	450.8	467.4	485.1	506.0	540.9
(8)	458.2	497.1	529.6	551.6	571.0	589.7	608.7	628.8	652.7	692.1
(9)	597.5	642.4	679.5	704.3	726.2	747.3	768.7	791.3	818.0	862.0
(10)	755.6	806.5	848.0	875.7	900.2	923.7	947.4	972.5	1002	1050

TABLE 2.15 : T^* ($\alpha = 0.1$)

Q: MODE:	0.05	0.10	0.20	0.30	0.40	0.50	0.60	0.70	0.80	0.90
(1)	1.365	3.617	8.446	13.40	18.46	23.62	28.98	34.69	41.30	51.73
(2)	43.85	53.03	66.09	77.94	89.55	101.2	113.0	125.3	138.8	157.4
(3)	145.6	162.5	184.3	203.3	221.7	240.1	258.6	277.8	298.5	325.8
(4)	309.2	334.1	364.7	391.0	416.3	441.4	466.7	492.7	520.6	556.8
(5)	535.2	568.1	607.6	641.1	673.3	705.1	737.2	770.0	805.2	850.2
(6)	823.4	864.5	912.9	953.7	992.7	1031	1070	1109	1152	1206
(7)	1174	1223	1280	1328	1374	1419	1465	1511	1561	1624
(8)	1587	1644	1710	1765	1818	1870	1923	1976	2033	2105
(9)	2062	2128	2203	2265	2325	2384	2443	2503	2567	2648
(10)	2600	2674	2758	2827	2894	2960	3025	3092	3164	3253

TABLE 2.15 : T^* ($\alpha = 0.2$)

Q: MODE:	0.05	0.10	0.20	0.30	0.40	0.50	0.60	0.70	0.80	0.90
(1)	.7870	2.274	5.631	9.143	12.76	16.49	20.41	24.70	29.90	39.12
(2)	28.46	35.72	45.48	54.09	62.49	70.91	79.53	88.61	98.94	114.5
(3)	97.38	111.4	128.0	142.0	155.4	168.7	182.2	196.3	212.0	234.4
(4)	209.9	230.9	254.6	273.9	292.4	310.6	329.0	348.1	369.2	398.6
(5)	366.4	394.6	425.2	450.0	473.5	496.6	519.9	544.1	570.6	607.0
(6)	567.0	602.3	640.0	670.2	698.7	726.8	755.0	784.2	816.1	859.6
(7)	811.6	854.2	898.9	934.6	968.1	1001	1034	1068	1105	1156
(8)	1100	1150	1201	1243	1281	1319	1357	1396	1439	1497
(9)	1433	1490	1549	1595	1639	1682	1725	1769	1817	1882
(10)	1810	1874	1940	1992	2041	2088	2136	2186	2239	2311

TABLE 2.15 : T^* ($\alpha = 0.3$)

Q: MODE:	0.05	0.10	0.20	0.30	0.40	0.50	0.60	0.70	0.80	0.90
(1)	.5585	1.702	4.397	7.263	10.23	13.33	16.61	20.27	24.86	33.61
(2)	21.79	28.11	36.37	43.54	50.51	57.51	64.70	72.36	81.28	95.52
(3)	76.23	88.79	103.1	114.8	126.0	137.1	148.3	160.2	173.7	193.9
(4)	166.2	185.3	205.8	222.1	237.5	252.6	268.0	284.1	302.1	328.6
(5)	291.9	317.8	344.4	365.4	385.0	404.2	423.7	444.0	466.7	499.3
(6)	453.6	486.2	519.1	544.7	568.5	591.9	615.5	640.0	667.2	706.1
(7)	651.3	690.7	729.8	760.0	788.1	815.6	843.2	872.0	903.9	949.0
(8)	885.0	931.2	976.6	1011	1043	1075	1107	1140	1176	1227
(9)	1154	1207	1259	1298	1335	1371	1406	1444	1485	1542
(10)	1460	1520	1578	1622	1662	1702	1742	1784	1830	1893

TABLE 2.15 : T^* ($\alpha = 0.4$)

Q: MODE:	0.05	0.10	0.20	0.30	0.40	0.50	0.60	0.70	0.80	0.90
(1)	.4341	1.373	3.668	6.146	8.734	11.45	14.35	17.63	21.86	30.36
(2)	17.89	23.60	30.95	37.26	43.37	49.52	55.86	62.68	70.75	84.23
(3)	63.74	75.37	88.27	98.65	108.5	118.2	128.1	138.7	150.8	169.8
(4)	140.2	158.1	176.7	191.2	204.7	218.1	231.6	245.9	262.1	286.8
(5)	247.6	272.0	296.3	314.9	332.2	349.2	366.3	384.3	404.7	435.1
(6)	386.2	417.0	447.1	469.8	490.9	511.5	532.3	554.0	578.5	614.7
(7)	555.9	593.3	629.1	656.0	680.7	705.0	729.4	754.9	783.5	825.4
(8)	756.7	800.7	842.3	873.3	901.8	929.7	957.7	987.0	1019	1067
(9)	988.7	1039	1086	1121	1154	1185	1217	1250	1287	1340
(10)	1251	1309	1362	1401	1437	1472	1508	1544	1585	1645

TABLE 2.15 : T^* ($\alpha = 0.5$)

Q: MODE:	0.05	0.10	0.20	0.30	0.40	0.50	0.60	0.70	0.80	0.90
(1)	.3554	1.157	3.176	5.387	7.712	10.16	12.81	15.83	19.82	28.18
(2)	15.28	20.54	27.25	32.97	38.50	44.07	49.83	56.07	63.57	76.55
(3)	55.28	66.24	78.15	87.61	96.52	105.3	114.4	124.0	135.2	153.4
(4)	122.6	139.6	156.9	170.1	182.4	194.5	206.8	219.8	234.9	258.4
(5)	217.5	240.8	263.5	280.5	296.2	311.6	327.2	343.6	362.4	391.3
(6)	340.2	369.9	397.9	418.8	437.9	456.6	475.5	495.3	517.9	552.2
(7)	490.8	526.8	560.3	584.9	607.5	629.5	651.7	675.0	701.4	741.1
(8)	669.3	711.7	750.6	779.1	805.0	830.3	855.8	882.5	912.7	957.9
(9)	875.6	924.5	968.8	1001	1030	1059	1087	1117	1151	1202
(10)	1109	1165	1214	1250	1283	1315	1347	1381	1419	1475

TABLE 2.15 : T^* ($\alpha = 0.6$)

Q: MODE:	0.05	0.10	0.20	0.30	0.40	0.50	0.60	0.70	0.80	0.90
(1)	.3011	1.002	2.816	4.828	6.958	9.216	11.67	14.50	18.31	26.58
(2)	13.39	18.30	24.53	29.81	34.91	40.04	45.38	51.19	58.27	70.89
(3)	49.10	59.52	70.70	79.47	87.70	95.85	104.2	113.1	123.7	141.3
(4)	109.6	126.0	142.2	154.5	165.9	177.1	188.5	200.6	214.7	237.4
(5)	195.3	217.8	239.2	255.1	269.6	283.9	298.3	313.6	331.2	359.0
(6)	306.4	335.1	361.7	381.1	398.8	416.1	433.6	452.0	473.2	506.2
(7)	442.9	477.8	509.6	532.5	553.4	573.8	594.3	616.0	640.7	678.9
(8)	604.8	646.0	683.0	709.5	733.6	757.0	780.6	805.4	833.7	877.0
(9)	792.2	839.7	881.8	911.9	939.2	965.7	992.3	1020	1052	1100
(10)	1005	1058	1106	1139	1170	1199	1229	1260	1296	1349

TABLE 2.15 : T^* ($\alpha = 0.7$)

Q: MODE:	0.05	0.10	0.20	0.30	0.40	0.50	0.60	0.70	0.80	0.90
(1)	.2612	.8846	2.539	4.396	6.373	8.480	10.78	13.47	17.14	25.36
(2)	11.95	16.57	22.43	27.36	32.12	36.92	41.92	47.39	54.16	66.50
(3)	44.33	54.31	64.91	73.14	80.84	88.47	96.30	104.7	114.8	131.9
(4)	99.58	115.4	130.9	142.4	153.1	163.6	174.3	185.7	199.1	221.0
(5)	178.1	199.9	220.4	235.3	249.0	262.3	275.9	290.2	307.0	333.9
(6)	280.2	308.1	333.5	351.8	368.4	384.6	401.0	418.4	438.5	470.4
(7)	405.7	439.7	470.1	491.8	511.4	530.5	549.8	570.1	593.6	630.5
(8)	554.8	595.0	630.4	655.4	678.0	700.0	722.1	745.4	772.3	814.2
(9)	727.4	773.8	814.3	842.7	868.3	893.1	918.1	944.4	974.6	1021
(10)	923.6	976.2	1021	1053	1082	1109	1137	1166	1200	1252

TABLE 2.15 : T^* ($\alpha = 0.8$)

Q: MODE:	0.05	0.10	0.20	0.30	0.40	0.50	0.60	0.70	0.80	0.90
(1)	.2308	.7930	2.317	4.049	5.902	7.887	10.07	12.64	16.20	24.38
(2)	10.81	15.18	20.73	25.39	29.87	34.40	39.13	44.34	50.84	62.97
(3)	40.51	50.12	60.24	68.04	75.32	82.53	89.93	97.93	107.6	124.3
(4)	91.53	106.9	121.7	132.7	142.8	152.7	162.8	173.6	186.5	207.9
(5)	164.3	185.6	205.2	219.4	232.3	244.9	257.8	271.4	287.4	313.7
(6)	259.0	286.3	310.8	328.2	343.9	359.2	374.8	391.2	410.5	441.6
(7)	375.8	409.1	438.4	459.0	477.6	495.6	513.8	533.1	555.6	591.5
(8)	514.5	553.9	588.0	611.9	633.3	654.0	675.0	697.1	722.8	763.6
(9)	675.2	720.7	759.8	786.8	811.1	834.5	858.2	883.2	912.1	957.7
(10)	858.0	909.7	953.6	983.8	1010	1037	1063	1091	1123	1173

TABLE 2.15 : T^* ($\alpha = 0.9$)

Q: MODE:	0.05	0.10	0.20	0.30	0.40	0.50	0.60	0.70	0.80	0.90
(1)	.2067	.7191	2.135	3.762	5.513	7.396	9.478	11.95	15.42	23.58
(2)	9.879	14.04	19.33	23.75	28.01	32.31	36.82	41.80	48.09	60.05
(3)	37.38	46.66	56.39	63.82	70.75	77.60	84.66	92.31	101.6	118.0
(4)	84.90	99.81	114.1	124.6	134.2	143.7	153.3	163.6	176.0	197.0
(5)	152.9	173.7	192.7	206.2	218.5	230.5	242.8	255.8	271.2	296.9
(6)	241.6	268.3	291.9	308.6	323.6	338.2	353.0	368.8	387.3	417.7
(7)	351.0	383.6	412.0	431.8	449.5	466.7	484.1	502.5	524.1	559.2
(8)	481.1	519.8	552.9	575.8	596.2	616.0	636.0	657.1	681.8	721.6
(9)	632.0	676.8	714.6	740.6	763.7	786.1	808.6	832.5	860.3	904.9
(10)	803.8	854.6	897.1	926.2	952.0	977.0	1002	1028	1059	1108

TABLE 2.15 : T^* ($\alpha = 1.0$)

Q: MODE:	0.05	0.10	0.20	0.30	0.40	0.50	0.60	0.70	0.80	0.90
(1)	.1871	.6581	1.983	3.521	5.184	6.981	8.978	11.36	14.77	22.91
(2)	9.104	13.09	18.14	22.37	26.44	30.55	34.86	39.66	45.76	57.59
(3)	34.75	43.74	53.12	60.25	66.88	73.43	80.20	87.55	96.56	112.7
(4)	79.32	93.86	107.7	117.8	127.0	136.0	145.3	155.2	167.2	187.8
(5)	143.3	163.6	182.0	195.1	206.9	218.4	230.1	242.6	257.5	282.7
(6)	226.9	253.0	276.0	292.1	306.5	320.4	334.6	349.7	367.7	397.5
(7)	330.1	362.2	389.8	408.8	425.8	442.2	458.9	476.6	497.5	531.9
(8)	453.0	491.0	523.3	545.3	564.9	583.8	602.9	623.3	647.1	686.2
(9)	595.6	639.6	676.5	701.5	723.7	745.1	766.7	789.6	816.5	860.1
(10)	757.9	807.9	849.4	877.4	902.2	926.1	950.2	975.7	1005	1053

TABLE 2.15 : T^* ($1/\alpha = 0.1$)

Q: MODE:	0.05	0.10	0.20	0.30	0.40	0.50	0.60	0.70	0.80	0.90
(1)	.0198	.0784	.3050	.6643	1.151	1.789	2.661	4.014	6.727	15.73
(2)	1.205	2.183	3.798	5.269	6.755	8.354	10.20	12.58	16.49	27.81
(3)	5.485	8.733	12.68	15.58	18.23	20.91	23.83	27.41	32.81	46.89
(4)	14.24	20.84	27.57	32.02	35.88	39.66	43.70	48.50	55.48	72.58
(5)	28.24	38.90	48.62	54.65	59.75	64.65	69.80	75.84	84.43	104.7
(6)	47.90	63.04	75.86	83.51	89.84	95.86	102.1	109.4	119.6	143.1
(7)	73.46	93.34	109.3	118.6	126.2	133.3	140.7	149.3	161.1	187.7
(8)	105.0	129.8	149.0	159.9	168.7	177.0	185.6	195.3	208.8	238.7
(9)	142.7	172.5	194.9	207.5	217.5	226.9	236.6	247.6	262.7	295.9
(10)	186.5	221.5	247.1	261.3	272.6	283.1	293.9	306.2	322.9	359.3

TABLE 2.15 : T^* ($1/\alpha = 0.2$)

Q: MODE:	0.05	0.10	0.20	0.30	0.40	0.50	0.60	0.70	0.80	0.90
(1)	.0393	.1530	.5634	1.153	1.884	2.766	3.869	5.414	8.184	16.80
(2)	2.306	3.941	6.356	8.428	10.45	12.57	14.91	17.77	22.06	33.10
(3)	10.06	14.87	20.17	24.00	27.47	30.92	34.61	38.93	45.00	59.11
(4)	25.13	34.12	42.64	48.30	53.25	58.08	63.14	68.94	76.86	94.29
(5)	48.30	61.99	73.87	81.39	87.83	94.04	100.5	107.8	117.6	138.4
(6)	79.91	98.58	113.9	123.3	131.2	138.8	146.7	155.5	167.1	191.5
(7)	120.1	143.9	162.7	174.0	183.5	192.4	201.7	212.0	225.5	253.3
(8)	169.0	198.1	220.4	233.5	244.5	254.9	265.5	277.3	292.8	324.1
(9)	226.7	261.1	286.9	301.9	314.4	326.1	338.1	351.5	368.8	403.7
(10)	293.1	332.8	362.1	379.1	393.1	406.2	419.6	434.5	453.7	492.1

TABLE 2.15 : T^* ($1/\alpha = 0.3$)

Q: MODE:	0.05	0.10	0.20	0.30	0.40	0.50	0.60	0.70	0.80	0.90
(1)	.0585	.2244	.7916	1.559	2.469	3.525	4.797	6.493	9.348	17.76
(2)	3.326	5.455	8.424	10.92	13.33	15.82	18.53	21.75	26.36	37.39
(3)	14.07	19.89	26.07	30.54	34.60	38.64	42.89	47.77	54.36	68.70
(4)	34.31	44.70	54.37	60.88	66.63	72.23	78.06	84.62	93.27	111.2
(5)	64.77	80.16	93.42	102.0	109.4	116.6	124.0	132.3	143.0	164.5
(6)	105.7	126.3	143.3	153.9	163.1	171.8	180.8	190.8	203.6	228.8
(7)	157.3	183.3	203.9	216.6	227.5	237.8	248.4	260.1	275.0	303.9
(8)	219.7	251.0	275.3	290.2	302.7	314.7	326.8	340.2	357.2	389.8
(9)	292.7	329.6	357.6	374.5	388.8	402.3	416.0	431.1	450.2	486.5
(10)	376.5	418.9	450.6	469.6	485.6	500.7	516.1	532.9	554.0	594.1

TABLE 2.15 : T^* ($1/\alpha$ — 0.4)

Q: MODE:	0.05	0.10	0.20	0.30	0.40	0.50	0.60	0.70	0.80	0.90
(1)	.0775	.2928	.9982	1.914	2.970	4.169	5.580	7.404	10.35	18.64
(2)	4.281	6.804	10.21	13.04	15.78	18.57	21.59	25.11	29.98	41.09
(3)	17.68	24.24	31.09	36.08	40.63	45.15	49.88	55.22	62.26	76.88
(4)	42.38	53.77	64.31	71.52	77.92	84.18	90.65	97.85	107.1	125.5
(5)	79.06	95.64	110.0	119.4	127.7	135.7	143.9	153.0	164.5	186.6
(6)	128.0	149.9	168.1	179.8	189.9	199.7	209.6	220.6	234.3	260.3
(7)	189.2	216.6	238.7	252.6	264.6	276.1	287.9	300.7	316.7	346.5
(8)	262.9	295.8	321.7	337.9	351.8	365.1	378.6	393.2	411.5	445.3
(9)	348.9	387.5	417.3	435.7	451.5	466.5	481.7	498.3	518.8	556.5
(10)	447.4	491.6	525.3	546.0	563.6	580.4	597.4	615.8	638.6	680.2

TABLE 2.15 : T^* ($1/\alpha$ — 0.5)

Q: MODE:	0.05	0.10	0.20	0.30	0.40	0.50	0.60	0.70	0.80	0.90
(1)	.0963	.3588	1.188	2.233	3.415	4.739	6.270	8.208	11.24	19.46
(2)	5.181	8.033	11.80	14.92	17.93	21.00	24.29	28.06	33.19	44.39
(3)	20.99	28.13	35.54	40.97	45.95	50.89	56.04	61.79	69.23	84.11
(4)	49.68	61.84	73.10	80.90	87.89	94.71	101.7	109.5	119.3	138.1
(5)	91.88	109.4	124.6	134.8	143.8	152.5	161.4	171.2	183.4	206.1
(6)	147.8	170.8	190.0	202.6	213.6	224.2	235.0	246.8	261.4	288.1
(7)	217.6	246.1	269.3	284.3	297.3	309.9	322.6	336.4	353.4	384.2
(8)	301.2	335.4	362.6	380.0	395.1	409.5	424.1	439.9	459.4	494.2
(9)	398.7	438.6	469.9	489.7	506.8	523.1	539.6	557.5	579.3	618.1
(10)	510.2	555.8	591.1	613.3	632.4	650.6	669.1	688.9	713.2	756.0

TABLE 2.15 : T^* ($1/\alpha$ — 0.6)

Q: MODE:	0.05	0.10	0.20	0.30	0.40	0.50	0.60	0.70	0.80	0.90
(1)	.1149	.4224	1.366	2.526	3.821	5.254	6.894	8.934	12.04	20.22
(2)	6.036	9.170	13.25	16.63	19.89	23.20	26.72	30.74	36.08	47.41
(3)	24.07	31.69	39.57	45.41	50.77	56.08	61.60	67.72	75.52	90.68
(4)	56.39	69.17	81.06	89.40	96.90	104.2	111.8	120.0	130.3	149.5
(5)	103.6	121.8	137.8	148.7	158.3	167.7	177.2	187.6	200.5	223.7
(6)	165.9	189.7	209.8	223.2	235.0	246.4	258.0	270.5	285.9	313.3
(7)	243.3	272.8	297.1	313.0	326.9	340.4	354.0	368.7	386.6	418.2
(8)	336.0	371.2	399.6	418.1	434.2	449.6	465.3	482.2	502.6	538.4
(9)	443.9	484.9	517.5	538.5	556.7	574.2	591.9	611.0	633.9	673.9
(10)	567.1	613.8	650.6	674.1	694.6	714.1	733.8	755.0	780.5	824.6

TABLE 2.15 : T^* ($1/\alpha$ — 0.7)

Q: MODE:	0.05	0.10	0.20	0.30	0.40	0.50	0.60	0.70	0.80	0.90
(1)	.1333	.4840	1.532	2.797	4.195	5.729	7.467	9.603	12.79	20.94
(2)	6.851	10.23	14.59	18.21	21.69	25.22	28.96	33.19	38.75	50.20
(3)	26.96	34.99	43.30	49.49	55.20	60.86	66.72	73.18	81.31	96.73
(4)	62.63	75.94	88.40	97.22	105.2	113.0	121.0	129.7	140.5	160.1
(5)	114.4	133.3	150.0	161.4	171.7	181.6	191.8	202.8	216.2	240.0
(6)	182.6	207.1	228.0	242.1	254.7	266.8	279.1	292.3	308.4	336.5
(7)	267.1	297.4	322.6	339.4	354.2	368.4	382.9	398.4	417.1	449.5
(8)	368.1	404.2	433.7	453.1	470.2	486.6	503.2	521.0	542.4	579.1
(9)	485.6	527.4	561.2	583.3	602.7	621.3	640.1	660.1	684.2	725.1
(10)	619.5	667.2	705.3	730.1	751.7	772.5	793.4	815.8	842.5	887.8

TABLE 2.15 : T^* ($1/\alpha$ — .80)

Q: MODE:	0.05	0.10	0.20	0.30	0.40	0.50	0.60	0.70	0.80	0.90
(1)	.1514	.5437	1.690	3.052	4.544	6.172	8.002	10.23	13.49	21.63
(2)	7.631	11.23	15.84	19.68	23.37	27.10	31.05	35.48	41.23	52.80
(3)	29.68	38.07	46.77	53.30	59.33	65.30	71.49	78.27	86.70	102.4
(4)	68.50	82.27	95.23	104.5	112.9	121.1	129.6	138.7	149.9	169.9
(5)	124.6	144.0	161.3	173.3	184.1	194.6	205.3	216.9	230.8	255.1
(6)	198.2	223.3	245.0	259.8	273.0	285.7	298.7	312.6	329.4	358.0
(7)	289.3	320.3	346.3	363.9	379.5	394.5	409.8	426.1	445.6	478.6
(8)	398.1	434.9	465.3	485.7	503.7	521.0	538.5	557.2	579.5	616.9
(9)	524.4	567.1	602.0	625.1	645.5	665.0	684.8	705.9	731.0	772.9
(10)	668.4	717.0	756.3	782.2	804.9	826.8	848.9	872.3	900.2	946.5

TABLE 2.15 : T^* ($1/\alpha$ — .90)

Q: MODE:	0.05	0.10	0.20	0.30	0.40	0.50	0.60	0.70	0.80	0.90
(1)	.1694	.6017	1.840	3.292	4.873	6.587	8.503	10.81	14.15	22.28
(2)	8.381	12.18	17.02	21.06	24.94	28.87	33.01	37.63	43.56	55.26
(3)	32.27	40.98	50.03	56.87	63.21	69.48	75.96	83.04	91.77	107.7
(4)	74.04	88.22	101.6	111.3	120.2	128.8	137.6	147.2	158.8	179.1
(5)	134.2	154.1	171.9	184.5	195.8	206.8	218.0	230.1	244.5	269.3
(6)	212.9	238.6	260.9	276.4	290.2	303.6	317.2	331.7	349.0	378.3
(7)	310.3	341.8	368.7	387.0	403.3	419.0	435.0	452.1	472.3	506.0
(8)	426.3	463.7	495.1	516.3	535.1	553.2	571.6	591.1	614.2	652.5
(9)	561.0	604.4	640.3	664.3	685.6	706.2	726.9	748.9	774.9	817.7
(10)	714.4	763.7	804.1	831.1	854.9	877.8	900.9	925.5	954.3	1001

TABLE 2.16 : T^* (α — 0.1)

Q: MODE:	0.05	0.10	0.20	0.30	0.40	0.50	0.60	0.70	0.80	0.90
(1)	.5611	1.133	2.311	3.545	4.855	6.283	7.916	9.975	13.19	21.75
(2)	9.742	13.95	20.61	26.87	33.09	39.39	45.88	52.78	60.72	73.11
(3)	63.09	74.13	89.39	103.1	116.4	129.8	143.4	157.4	172.7	193.6
(4)	180.7	199.6	223.6	244.4	264.6	284.6	304.8	325.6	348.1	377.7
(5)	359.9	386.8	419.6	447.7	474.7	501.5	528.4	556.1	585.9	624.3
(6)	601.4	636.4	678.1	713.4	747.3	780.8	814.5	849.1	886.1	933.3
(7)	905.3	948.4	999.0	1041	1082	1122	1163	1204	1248	1304
(8)	1271	1322	1382	1432	1479	1526	1573	1622	1673	1738
(9)	1700	1759	1827	1885	1939	1993	2047	2102	2161	2235
(10)	2191	2258	2336	2400	2461	2522	2583	2645	2711	2793

TABLE 2.16 : T^* (α — 0.2)

Q: MODE:	0.05	0.10	0.20	0.30	0.40	0.50	0.60	0.70	0.80	0.90
(1)	.3983	.8073	1.660	2.571	3.560	4.672	6.002	7.791	10.84	19.69
(2)	5.982	9.108	13.97	18.49	22.98	27.56	32.33	37.52	43.77	54.66
(3)	41.37	50.23	61.70	71.68	81.36	91.04	100.9	111.3	123.0	140.2
(4)	121.4	137.2	155.5	170.9	185.5	200.0	214.8	230.1	247.1	271.3
(5)	244.9	267.7	293.1	313.8	333.5	353.0	372.6	393.0	415.4	446.6
(6)	412.4	442.4	474.8	500.9	525.7	550.0	574.6	600.0	627.8	666.0
(7)	624.0	661.1	700.6	732.2	761.9	791.2	820.7	851.1	884.4	929.6
(8)	879.7	924.0	970.5	1007	1042	1076	1110	1146	1185	1237
(9)	1179	1231	1284	1327	1366	1406	1445	1485	1529	1589
(10)	1523	1582	1642	1690	1735	1779	1823	1869	1918	1985

TABLE 2.16 : T^* (α — 0.3)

Q: MODE:	0.05	0.10	0.20	0.30	0.40	0.50	0.60	0.70	0.80	0.90
(1)	.3262	.6631	1.372	2.140	2.988	3.963	5.166	6.850	9.863	18.92
(2)	4.408	7.004	11.05	14.79	18.51	22.32	26.33	30.76	36.27	46.56
(3)	31.91	39.70	49.45	57.78	65.82	73.87	82.10	90.84	100.9	116.6
(4)	95.38	109.6	125.4	138.3	150.5	162.6	174.9	187.8	202.4	224.2
(5)	194.2	215.0	237.1	254.5	271.0	287.2	303.6	320.7	339.9	367.9
(6)	329.0	356.5	384.7	406.8	427.5	447.8	468.3	489.6	513.4	547.6
(7)	499.7	533.9	568.4	595.1	620.0	644.4	669.0	694.6	723.0	763.5
(8)	706.3	747.4	788.2	819.5	848.6	877.1	905.8	935.6	968.7	1015
(9)	949.0	996.9	1043	1079	1113	1145	1178	1212	1250	1303
(10)	1227	1282	1335	1376	1413	1450	1487	1525	1568	1627

TABLE 2.16 : T^* ($\alpha = 0.4$)

Q: MODE:	0.05	0.10	0.20	0.30	0.40	0.50	0.60	0.70	0.80	0.90
(1)	.2832	.5771	1.201	1.883	2.648	3.544	4.674	6.305	9.313	18.50
(2)	3.516	5.772	9.312	12.58	15.85	19.20	22.75	26.73	31.80	41.77
(3)	26.36	33.45	42.17	49.51	56.57	63.63	70.89	78.66	87.76	102.6
(4)	79.97	93.16	107.5	118.9	129.6	140.3	151.1	162.6	175.7	196.1
(5)	164.1	183.7	203.7	219.2	233.7	247.9	262.4	277.6	294.9	321.0
(6)	279.4	305.3	331.1	350.7	368.9	386.8	404.9	423.8	445.2	477.1
(7)	425.7	458.2	489.7	513.4	535.4	556.9	578.6	601.3	626.8	664.4
(8)	603.1	642.2	679.5	707.4	733.1	758.2	783.6	810.0	839.6	883.0
(9)	811.8	857.4	900.5	932.5	962.0	990.7	1019	1049	1083	1132
(10)	1051	1103	1152	1188	1222	1254	1287	1320	1358	1413

TABLE 2.16 : T^* ($\alpha = 0.5$)

Q: MODE:	0.05	0.10	0.20	0.30	0.40	0.50	0.60	0.70	0.80	0.90
(1)	.2538	.5185	1.084	1.709	2.418	3.260	4.343	5.942	8.956	18.25
(2)	2.935	4.945	8.134	11.08	14.03	17.07	20.31	23.98	28.76	38.54
(3)	22.63	29.21	37.20	43.86	50.25	56.65	63.25	70.34	78.79	93.02
(4)	69.53	81.99	95.23	105.6	115.4	125.0	134.9	145.3	157.5	177.0
(5)	143.7	162.3	180.9	195.1	208.2	221.2	234.3	248.2	264.1	289.0
(6)	245.6	270.5	294.5	312.4	329.0	345.2	361.7	378.9	398.7	428.9
(7)	375.3	406.5	435.9	457.7	477.7	497.2	516.9	537.6	561.2	596.8
(8)	532.8	570.4	605.3	630.9	654.3	677.1	700.1	724.2	751.6	792.7
(9)	718.2	762.3	802.6	831.9	858.8	884.9	911.2	938.7	969.9	1016
(10)	931.5	982.0	1027	1060	1091	1120	1150	1181	1216	1260

TABLE 2.16 : T^* ($\alpha = 0.6$)

Q: MODE:	0.05	0.10	0.20	0.30	0.40	0.50	0.60	0.70	0.80	0.90
(1)	.2322	.4752	.9976	1.580	2.248	3.052	4.102	5.680	8.704	18.08
(2)	2.523	4.345	7.269	9.980	12.69	15.50	18.51	21.95	26.51	36.17
(3)	19.92	26.10	33.54	39.70	45.59	51.50	57.60	64.21	72.17	85.97
(4)	61.88	73.77	86.21	95.86	104.9	113.8	122.9	132.6	144.1	162.9
(5)	128.7	146.5	164.1	177.3	189.5	201.4	213.6	226.5	241.5	265.4
(6)	220.7	244.7	267.4	284.2	299.5	314.5	329.7	345.8	364.3	393.4
(7)	338.1	368.4	396.3	416.5	435.1	453.1	471.4	490.6	512.7	547.0
(8)	481.0	517.5	550.5	574.4	596.1	617.2	638.5	660.9	686.6	726.0
(9)	649.3	692.0	730.3	757.7	782.6	806.8	831.1	856.7	885.9	930.6
(10)	843.0	892.1	935.5	966.5	994.6	1021	1049	1077	1110	1160

TABLE 2.16 : T^* ($\alpha = 0.7$)

Q: MODE:	0.05	0.10	0.20	0.30	0.40	0.50	0.60	0.70	0.80	0.90
(1)	.2153	.4416	.9306	1.480	2.116	2.891	3.917	5.482	8.517	17.95
(2)	2.215	3.005	6.600	9.123	11.65	14.28	17.11	20.37	24.77	34.34
(3)	17.84	23.69	30.70	36.47	41.97	47.49	53.21	59.44	67.02	80.50
(4)	55.98	67.39	79.20	88.27	96.71	105.1	113.6	122.8	133.6	151.9
(5)	117.0	134.3	151.0	163.5	174.9	186.1	197.5	209.6	223.8	247.0
(6)	201.4	224.8	246.4	262.2	276.6	290.7	304.9	320.0	337.6	365.8
(7)	309.3	338.8	365.4	384.6	401.9	418.9	436.0	454.1	475.1	508.2
(8)	440.7	476.3	508.0	530.5	550.9	570.7	590.6	611.7	636.1	674.2
(9)	595.7	637.5	674.2	700.0	723.4	746.0	768.9	793.0	820.7	863.8
(10)	774.2	822.2	863.9	893.2	919.5	945.0	970.7	997.8	1028	1077

TABLE 2.16 : T^* ($\alpha = 0.8$)

Q: MODE:	0.05	0.10	0.20	0.30	0.40	0.50	0.60	0.70	0.80	0.90
(1)	.2018	.4145	.8766	1.400	2.011	2.762	3.770	5.325	8.371	17.85
(2)	1.975	3.521	6.064	8.434	10.82	13.30	15.98	19.10	23.37	32.88
(3)	16.19	21.76	28.42	33.86	39.06	44.27	49.68	55.59	62.88	76.10
(4)	51.25	62.26	73.56	82.15	90.13	98.01	106.1	114.8	125.2	143.1
(5)	107.7	124.5	140.5	152.3	163.1	173.7	184.5	196.0	209.6	232.3
(6)	186.0	208.7	229.5	244.5	258.2	271.4	284.9	299.3	316.1	343.6
(7)	286.2	314.9	340.6	358.8	375.3	391.3	407.5	424.6	444.7	477.0
(8)	408.4	443.2	473.7	495.1	514.4	533.1	552.1	572.1	595.3	632.5
(9)	552.6	593.5	628.9	653.5	675.7	697.1	718.7	741.6	768.1	810.0
(10)	718.9	765.9	806.1	834.0	859.0	883.1	907.5	933.1	962.9	1009

TABLE 2.16 : T^* ($\alpha = 0.9$)

Q: MODE:	0.05	0.10	0.20	0.30	0.40	0.50	0.60	0.70	0.80	0.90
(1)	.1905	.3920	.8319	1.334	1.924	2.656	3.649	5.198	8.254	17.78
(2)	1.783	3.223	5.621	7.864	10.12	12.48	15.05	18.05	22.20	31.68
(3)	14.84	20.17	26.53	31.71	36.64	41.59	46.75	52.41	59.44	72.46
(4)	47.36	58.02	68.89	77.09	84.68	92.18	99.87	108.2	118.3	135.8
(5)	99.98	116.3	131.8	143.1	153.4	163.4	173.7	184.7	197.9	220.0
(6)	173.2	195.4	215.5	229.9	242.9	255.5	268.4	282.1	298.3	325.1
(7)	267.0	295.2	320.0	337.5	353.1	368.4	383.8	400.3	419.5	451.1
(8)	381.6	415.7	445.3	465.8	484.2	502.1	520.1	539.2	561.6	597.9
(9)	516.9	557.1	591.4	615.0	636.1	656.6	677.2	699.0	724.5	765.5
(10)	673.0	719.3	758.3	785.0	808.8	831.8	855.1	879.6	908.2	953.9

TABLE 2.16 : T^* ($\alpha = 1.0$)

Q: MODE:	0.05	0.10	0.20	0.30	0.40	0.50	0.60	0.70	0.80	0.90
(1)	.1810	.3730	.7942	1.277	1.850	2.567	3.548	5.092	8.159	17.72
(2)	1.626	2.975	5.249	7.383	9.538	11.79	14.26	17.15	21.22	30.67
(3)	13.71	18.83	24.94	29.89	34.60	39.33	44.27	49.72	56.54	69.39
(4)	44.09	54.45	64.94	72.81	80.07	87.24	94.62	102.6	112.4	129.6
(5)	93.48	109.4	124.5	135.3	145.1	154.8	164.6	175.2	187.9	209.7
(6)	162.4	184.1	203.7	217.5	229.9	242.0	254.4	267.5	283.2	309.6
(7)	250.8	278.5	302.6	319.4	334.4	349.0	363.8	379.6	398.3	429.2
(8)	358.9	392.5	421.3	441.1	458.7	475.8	493.1	511.4	533.1	568.6
(9)	486.8	526.3	559.7	582.5	602.7	622.3	642.0	663.0	687.6	727.8
(10)	634.3	679.8	717.9	743.6	766.4	788.5	810.7	834.3	861.9	906.7

TABLE 2.16 : T^* ($1/\alpha = 0.1$)

Q: MODE:	0.05	0.10	0.20	0.30	0.40	0.50	0.60	0.70	0.80	0.90
(1)	.0610	.1339	.3241	.5938	.9826	1.562	2.474	4.052	7.298	17.21
(2)	.1845	.4050	.9290	1.549	2.274	3.145	4.265	5.916	9.087	19.49
(3)	1.906	3.316	5.437	7.248	9.017	10.88	12.99	15.67	19.92	31.83
(4)	7.216	11.22	15.83	19.12	22.07	25.02	28.22	32.10	37.89	52.71
(5)	17.23	24.79	32.25	37.10	41.26	45.33	49.64	54.75	62.13	80.01
(6)	32.61	44.36	54.85	61.28	66.69	71.87	77.31	83.65	92.64	113.7
(7)	53.73	70.04	83.64	91.69	98.34	104.6	111.2	118.8	129.4	153.6
(8)	80.78	101.9	118.7	128.3	136.2	143.7	151.3	160.2	172.4	199.9
(9)	113.9	139.9	159.9	171.2	180.3	188.9	197.7	207.8	221.7	252.4
(10)	153.1	184.2	207.4	220.3	230.7	240.4	250.4	261.7	277.2	311.2

TABLE 2.16 : T^* ($1/\alpha = 0.2$)

Q: MODE:	0.05	0.10	0.20	0.30	0.40	0.50	0.60	0.70	0.80	0.90
(1)	.0840	.1794	.4117	.7169	1.131	1.723	2.632	4.192	7.403	17.27
(2)	.3629	.7705	1.647	2.591	3.620	4.782	6.176	8.062	11.32	21.23
(3)	3.609	5.889	8.962	11.46	13.84	16.29	18.97	22.19	26.90	38.62
(4)	13.10	18.89	24.97	29.25	33.09	36.89	40.92	45.60	52.13	67.06
(5)	30.15	40.27	49.62	55.74	61.07	66.24	71.65	77.82	86.20	104.5
(6)	55.41	70.31	83.05	91.04	97.86	104.4	111.2	118.9	129.1	150.8
(7)	89.16	109.1	125.3	135.1	143.5	151.4	159.6	168.8	180.9	206.1
(8)	131.5	156.7	176.3	188.1	197.9	207.2	216.8	227.5	241.5	270.2
(9)	182.6	213.0	236.2	249.8	261.1	271.9	282.8	295.0	310.9	343.1
(10)	242.5	278.2	304.8	320.4	333.2	345.3	357.7	371.4	389.2	424.9

TABLE 2.16 : T^* ($1/\alpha = 0.3$)

Q: MODE:	0.05	0.10	0.20	0.30	0.40	0.50	0.60	0.70	0.80	0.90
(1)	.1016	.2145	.4803	.8157	1.254	1.862	2.775	4.323	7.506	17.33
(2)	.5358	1.106	2.254	3.432	4.679	6.049	7.645	9.713	13.09	22.76
(3)	5.163	8.068	11.78	14.76	17.59	20.47	23.56	27.19	32.28	44.05
(4)	18.20	25.11	32.14	37.12	41.60	46.03	50.68	55.97	63.07	78.29
(5)	40.95	52.56	63.12	70.15	76.32	82.32	88.54	95.53	104.7	123.5
(6)	74.01	90.70	104.9	114.0	121.8	129.4	137.2	145.9	157.1	179.6
(7)	117.6	139.6	157.4	168.6	178.2	187.3	196.7	207.1	220.4	246.5
(8)	171.9	199.2	220.7	234.0	245.3	256.0	267.0	279.1	294.5	324.3
(9)	236.9	269.6	294.9	310.2	323.2	335.6	348.1	361.9	379.4	413.0
(10)	312.6	350.9	379.8	397.3	412.0	425.9	440.0	455.5	475.1	512.4

TABLE 2.16 : T^* ($1/\alpha = 0.4$)

Q: MODE:	0.05	0.10	0.20	0.30	0.40	0.50	0.60	0.70	0.80	0.90
(1)	.1165	.2441	.5387	.9006	1.362	1.986	2.906	4.448	7.606	17.38
(2)	.7037	1.418	2.790	4.157	5.581	7.122	8.883	11.11	14.60	24.14
(3)	6.601	9.993	14.21	17.57	20.76	23.99	27.44	31.41	36.82	48.72
(4)	22.76	30.48	38.24	43.78	48.79	53.74	58.91	64.71	72.30	87.84
(5)	50.42	63.08	74.56	82.32	89.19	95.88	102.8	110.5	120.3	139.6
(6)	90.14	108.0	123.3	133.3	142.1	150.5	159.2	168.7	180.8	203.9
(7)	142.1	165.4	184.6	196.8	207.4	217.6	228.0	239.4	253.7	280.7
(8)	206.5	235.3	258.3	272.8	285.2	297.2	309.4	322.6	339.2	370.0
(9)	283.2	317.6	344.4	361.2	375.6	389.3	403.2	418.3	437.2	471.9
(10)	372.4	412.3	443.1	462.1	478.4	493.8	509.5	526.5	547.6	586.2

TABLE 2.16 : T^* ($1/\alpha = 0.5$)

Q: MODE:	0.05	0.10	0.20	0.30	0.40	0.50	0.60	0.70	0.80	0.90
(1)	.1296	.2703	.5903	.9763	1.459	2.100	3.029	4.566	7.703	17.44
(2)	.8670	1.711	3.275	4.803	6.380	8.069	9.974	12.33	15.93	25.41
(3)	7.946	11.74	16.36	20.06	23.56	27.10	30.86	35.13	40.82	52.87
(4)	26.92	35.28	43.63	49.66	55.14	60.54	66.16	72.41	80.44	96.29
(5)	58.95	72.42	84.66	93.06	100.5	107.8	115.3	123.6	134.0	153.8
(6)	104.6	123.4	139.6	150.4	159.9	169.1	178.5	188.8	201.6	225.3
(7)	163.9	188.3	208.5	221.7	233.2	244.3	255.6	267.9	283.1	310.8
(8)	237.2	267.1	291.3	306.9	320.5	333.5	346.7	361.0	378.6	410.3
(9)	324.3	359.9	388.1	406.1	421.7	436.6	451.7	468.0	488.0	523.8
(10)	425.3	466.6	498.9	519.3	536.9	553.7	570.7	589.0	611.4	651.3

TABLE 2.16 : T^* ($1/\alpha = 0.6$)

Q: MODE:	0.05	0.10	0.20	0.30	0.40	0.50	0.60	0.70	0.80	0.90
(1)	.1414	.2939	.6372	1.045	1.548	2.205	3.143	4.680	7.798	17.50
(2)	1.026	1.988	3.722	5.393	7.105	8.926	10.96	13.44	17.15	26.59
(3)	9.213	13.34	18.33	22.32	26.10	29.92	33.95	38.49	44.44	56.65
(4)	30.78	39.65	48.53	54.98	60.88	66.69	72.71	79.37	87.79	103.9
(5)	66.78	80.91	93.82	102.8	110.8	118.7	126.7	135.5	146.4	166.6
(6)	117.7	137.3	154.4	165.9	176.0	185.9	196.0	206.9	220.4	244.7
(7)	183.8	209.0	230.2	244.2	256.5	268.4	280.5	293.6	309.6	338.1
(8)	265.1	296.0	321.3	337.8	352.3	366.3	380.4	395.6	414.2	446.8
(9)	361.6	398.2	427.6	446.7	463.4	479.4	495.5	512.9	534.0	570.8
(10)	473.3	515.7	549.3	571.0	589.8	607.8	626.0	645.5	669.2	710.1

TABLE 2.16 : T^* ($1/\alpha = 0.7$)

Q: MODE:	0.05	0.10	0.20	0.30	0.40	0.50	0.60	0.70	0.80	0.90
(1)	.1523	.3157	.6803	1.109	1.630	2.303	3.252	4.788	7.891	17.55
(2)	1.181	2.251	4.137	5.938	7.773	9.715	11.87	14.47	18.27	27.69
(3)	10.42	14.83	20.14	24.40	28.44	32.51	36.79	41.58	47.77	60.15
(4)	34.39	43.70	53.04	59.89	66.16	72.35	78.74	85.77	94.56	111.0
(5)	74.06	88.74	102.2	111.7	120.3	128.6	137.1	146.4	157.8	178.5
(6)	129.9	150.2	167.9	180.1	190.9	201.4	212.1	223.6	237.7	262.6
(7)	202.2	228.1	250.1	264.9	278.0	290.6	303.5	317.3	334.0	363.2
(8)	290.8	322.5	348.8	366.3	381.6	396.4	411.4	427.5	446.9	480.3
(9)	395.9	433.4	464.0	484.1	501.7	518.7	535.9	554.3	576.3	614.0
(10)	517.5	560.8	595.7	618.5	638.4	657.5	676.8	697.5	722.2	764.2

TABLE 2.16 : T^* ($1/\alpha = 0.8$)

Q: MODE:	0.05	0.10	0.20	0.30	0.40	0.50	0.60	0.70	0.80	0.90
(1)	.1625	.3360	.7205	1.168	1.708	2.396	3.355	4.893	7.982	17.61
(2)	1.333	2.502	4.528	6.448	8.397	10.45	12.71	15.42	19.31	28.74
(3)	11.56	16.24	21.83	26.34	30.62	34.92	39.43	44.46	50.87	63.41
(4)	37.79	47.49	57.24	64.45	71.08	77.61	84.36	91.73	100.9	117.6
(5)	80.89	96.05	110.1	120.1	129.1	137.9	146.9	156.6	168.5	189.5
(6)	141.4	162.2	180.6	193.3	204.7	215.7	227.0	239.1	253.8	279.2
(7)	219.3	245.9	268.7	284.2	297.9	311.3	324.8	339.3	356.8	386.5
(8)	314.9	347.3	374.4	392.7	408.9	424.5	440.3	457.2	477.4	511.6
(9)	428.0	466.3	497.8	518.9	537.4	555.3	573.4	592.7	615.7	654.3
(10)	558.8	602.9	638.9	662.7	683.7	703.8	724.2	745.9	771.6	814.6

TABLE 2.16 : T^* ($1/\alpha = 0.9$)

Q: MODE:	0.05	0.10	0.20	0.30	0.40	0.50	0.60	0.70	0.80	0.90
(1)	.1720	.3550	.7584	1.224	1.781	2.484	3.454	4.995	8.072	17.66
(2)	1.481	2.743	4.897	6.928	8.983	11.14	13.51	16.31	20.29	29.73
(3)	12.66	17.57	23.42	28.16	32.66	37.19	41.92	47.16	53.78	66.48
(4)	41.02	51.06	61.20	68.74	75.70	82.56	89.63	97.32	106.8	123.8
(5)	87.35	102.9	117.5	127.9	137.3	146.5	156.0	166.2	178.5	199.9
(6)	152.1	173.4	192.4	205.7	217.7	229.3	241.1	253.7	268.9	294.8
(7)	235.5	262.6	286.1	302.3	316.7	330.7	344.9	360.0	378.1	408.5
(8)	337.5	370.5	398.5	417.6	434.5	450.8	467.4	485.1	506.0	540.9
(9)	458.2	497.1	529.6	551.6	571.0	589.7	608.7	628.8	652.7	692.1
(10)	597.5	642.4	679.5	704.3	726.2	747.3	768.7	791.3	818.0	862.0

TABLE 2.17 : K^* ($\alpha = 0.1$)

Q: MODE:	0.05	0.10	0.20	0.30	0.40	0.50	0.60	0.70	0.80	0.90
(1)	1.651	3.897	10.83	22.50	41.11	69.50	112.8	183.1	317.6	703.5
(2)	7.655	18.03	52.73	114.8	205.0	324.5	486.4	727.5	1159	2340
(3)	15.16	37.60	133.9	351.7	644.8	987.9	1418	2028	3090	5941
(4)	21.92	55.70	232.4	770.7	1473	2231	3136	4385	6521	12207
(5)	28.37	72.84	338.4	1421	2817	4243	5892	8123	11899	21904
(6)	34.64	89.33	445.3	2353	4804	7210	9931	13563	19665	35790
(7)	40.80	105.4	550.5	3613	7557	11318	15499	21025	30262	54623
(8)	46.89	121.3	653.1	5254	11203	16752	22843	30831	44131	79161
(9)	52.94	136.9	752.9	7323	15866	23699	32209	43303	61715	110164
10)	58.96	152.5	850.2	9873	21674	32345	43845	58760	83455	148389

TABLE 2.17 : K^* ($\alpha = 0.2$)

Q: MODE:	0.05	0.10	0.20	0.30	0.40	0.50	0.60	0.70	0.80	0.90
(1)	1.461	3.554	10.47	22.45	40.90	67.72	107.1	169.3	285.8	614.1
(2)	4.735	11.91	42.72	110.0	204.6	323.7	481.0	711.7	1120	2226
(3)	8.370	21.72	97.95	335.7	643.2	987.3	1411	2007	3038	5792
(4)	11.57	30.25	156.0	737.5	1469	2230	3128	4360	6457	12022
(5)	14.68	38.38	212.6	1367	2812	4243	5883	8093	11823	21683
(6)	17.72	46.30	266.2	2274	4796	7210	9920	13528	19577	35533
(7)	20.74	54.10	317.0	3509	7547	11318	15487	20985	30161	54329
(8)	23.74	61.84	365.7	5121	11190	16752	22830	30787	44018	78832
(9)	26.73	69.53	412.7	7162	15852	23699	32195	43253	61589	109798
(10)	29.72	77.19	458.6	9683	21657	32344	43829	58706	83317	147987

TABLE 2.17 : K* (α = 0.3)

Q: MODE:	0.05	0.10	0.20	0.30	0.40	0.50	0.60	0.70	0.80	0.90
(1)	1.316	3.294	10.23	22.43	40.81	67.01	105.0	164.5	274.9	584.2
(2)	3.429	8.925	36.95	107.6	204.5	323.4	479.1	706.3	1107	2188
(3)	5.781	15.29	79.43	328.9	642.7	987.1	1409	2000	3021	5742
(4)	7.863	20.77	119.8	724.1	1468	2230	3126	4351	6436	11960
(5)	9.898	26.06	157.1	1346	2810	4243	5880	8083	11797	21610
(6)	11.91	31.25	191.4	2244	4793	7210	9917	13516	19547	35447
(7)	13.91	36.39	223.8	3470	7544	11317	15483	20972	30127	54232
(8)	15.90	41.50	254.8	5073	11186	16752	22826	30772	43980	78722
(9)	17.88	46.59	284.9	7105	15847	23699	32190	43237	61547	109676
(10)	19.86	51.67	314.4	9616	21651	32344	43824	58688	83270	147853

TABLE 2.17 : K* (α = 0.4)

Q: MODE:	0.05	0.10	0.20	0.30	0.40	0.50	0.60	0.70	0.80	0.90
(1)	1.201	3.088	10.06	22.41	40.75	66.64	103.9	162.0	269.5	569.2
(2)	2.688	7.144	33.04	106.3	204.4	323.3	478.1	703.5	1100	2169
(3)	4.416	11.80	67.64	325.0	642.4	987.0	1407	1997	3012	5717
(4)	5.954	15.81	97.97	716.8	1467	2230	3124	4347	6425	11930
(5)	7.467	19.73	125.1	1334	2809	4243	5878	8078	11784	21573
(6)	8.967	23.58	149.8	2229	4792	7210	9915	13510	19532	35404
(7)	10.46	27.41	173.2	3450	7542	11317	15481	20965	30110	54183
(8)	11.95	31.23	195.7	5049	11184	16752	22823	30764	43961	78667
(9)	13.43	35.04	217.6	7076	15844	23699	32188	43228	61526	109615
(10)	14.92	38.84	239.3	9582	21648	32344	43821	58679	83247	147786

TABLE 2.17 : K* (α = 0.5)

Q: MODE:	0.05	0.10	0.20	0.30	0.40	0.50	0.60	0.70	0.80	0.90
(1)	1.106	2.918	9.925	22.40	40.72	66.40	103.2	160.5	266.2	560.2
(2)	2.211	5.958	30.14	105.3	204.4	323.2	477.5	701.9	1096	2158
(3)	3.572	9.608	59.31	322.6	642.2	986.9	1407	1995	3007	5702
(4)	4.791	12.77	83.19	712.3	1467	2230	3124	4344	6419	11911
(5)	5.995	15.87	104.1	1327	2809	4243	5877	8075	11777	21551
(6)	7.190	18.94	123.2	2219	4791	7210	9914	13507	19523	35379
(7)	8.381	21.99	141.3	3438	7541	11317	15480	20961	30100	54153
(8)	9.570	25.03	158.9	5034	11183	16752	22822	30760	43949	78634
(9)	10.76	28.07	176.1	7058	15843	23699	32187	43223	61513	109578
(10)	11.94	31.11	193.2	9561	21647	32344	43820	58673	83233	147746

TABLE 2.17 : K* (α = 0.6)

Q: MODE:	0.05	0.10	0.20	0.30	0.40	0.50	0.60	0.70	0.80	0.90
(1)	1.027	2.775	9.820	22.39	40.69	66.24	102.8	159.5	264.0	554.2
(2)	1.877	5.111	27.88	104.7	204.3	323.1	477.1	700.8	1093	2150
(3)	2.999	8.104	53.03	320.9	642.1	986.8	1406	1993	3004	5692
(4)	4.008	10.71	72.44	709.1	1467	2230	3123	4343	6414	11899
(5)	5.007	13.27	89.26	1323	2808	4243	5877	8073	11772	21536
(6)	6.001	15.82	104.7	2213	4791	7210	9913	13504	19517	35361
(7)	6.992	18.36	119.4	3429	7540	11317	15479	20959	30093	54134
(8)	7.982	20.89	133.7	5023	11182	16752	22821	30757	43942	78612
(9)	8.970	23.42	147.9	7046	15842	23699	32186	43220	61505	109554
(10)	9.950	25.95	162.0	9547	21645	32344	43819	58670	83224	147719

TABLE 2.17 : K* (α = 0.7)

Q: MODE:	0.05	0.10	0.20	0.30	0.40	0.50	0.60	0.70	0.80	0.90
(1)	.9590	2.651	9.735	22.38	40.68	66.12	102.5	158.8	262.4	549.9
(2)	1.631	4.476	26.05	104.2	204.3	323.1	476.8	700.0	1091	2145
(3)	2.585	7.007	48.09	319.6	642.0	986.8	1406	1992	3001	5685
(4)	3.445	9.218	64.23	706.8	1467	2230	3123	4341	6411	11890
(5)	4.299	11.41	78.16	1319	2808	4243	5876	8071	11768	21525
(6)	5.150	13.59	90.98	2208	4790	7210	9913	13503	19513	35349
(7)	5.998	15.76	103.3	3423	7540	11317	15479	20957	30088	54120
(8)	6.846	17.92	115.5	5016	11181	16752	22821	30755	43936	78596
(9)	7.692	20.09	127.5	7037	15841	23699	32185	43218	61499	109537
(10)	8.538	22.25	139.5	9537	21645	32344	43818	58667	83218	147700

TABLE 2.17 : K* (α = 0.8)

Q: MODE:	0.05	0.10	0.20	0.30	0.40	0.50	0.60	0.70	0.80	0.90
(1)	.9003	2.542	9.664	22.38	40.66	66.03	102.2	158.3	261.2	546.6
(2)	1.442	3.981	24.52	103.8	204.3	323.0	476.6	699.4	1090	2141
(3)	2.271	6.171	44.08	318.6	641.9	986.8	1406	1992	2999	5680
(4)	3.020	8.093	57.74	705.1	1467	2230	3122	4340	6409	11883
(5)	3.767	10.00	69.53	1317	2808	4243	5876	8070	11765	21517
(6)	4.510	11.90	80.49	2204	4790	7210	9912	13501	19510	35340
(7)	5.252	13.80	91.11	3419	7539	11317	15478	20955	30085	54109
(8)	5.993	15.69	101.6	5010	11181	16752	22820	30753	43932	78584
(9)	6.733	17.59	112.0	7030	15841	23699	32184	43216	61494	109524
(10)	7.473	19.48	122.4	9529	21644	32344	43817	58665	83213	147685

TABLE 2.17 : K* (α = 0.9)

Q: MODE:	0.05	0.10	0.20	0.30	0.40	0.50	0.60	0.70	0.80	0.90
(1)	.8489	2.447	9.604	22.37	40.65	65.96	102.0	157.8	260.3	544.1
(2)	1.292	3.585	23.22	103.5	204.3	323.0	476.4	698.9	1089	2137
(3)	2.025	5.514	40.75	317.9	641.9	986.8	1405	1991	2998	5676
(4)	2.689	7.213	52.47	703.7	1466	2230	3122	4340	6407	11878
(5)	3.352	8.906	62.64	1315	2808	4243	5876	8069	11763	21511
(6)	4.011	10.59	72.17	2202	4790	7210	9912	13500	19507	35333
(7)	4.670	12.28	81.48	3415	7539	11317	15478	20954	30082	54101
(8)	5.329	13.96	90.71	5006	11181	16752	22820	30752	43929	78575
(9)	5.987	15.64	99.91	7025	15840	23699	32184	43215	61491	109513
(10)	6.644	17.32	109.1	9523	21644	32344	43817	58664	83209	147674

TABLE 2.17 : K* (α = 1.0)

Q: MODE:	0.05	0.10	0.20	0.30	0.40	0.50	0.60	0.70	0.80	0.90
(1)	.8034	2.361	9.553	22.37	40.64	65.91	101.9	157.5	259.5	542.1
(2)	1.171	3.261	22.10	103.2	204.3	323.0	476.3	698.5	1088	2135
(3)	1.827	4.983	37.92	317.3	641.8	986.7	1405	1991	2997	5672
(4)	2.423	6.505	48.11	702.6	1466	2230	3122	4339	6406	11874
(5)	3.019	8.025	56.99	1313	2807	4242	5875	8069	11761	21506
(6)	3.612	9.541	65.41	2199	4790	7210	9912	13500	19505	35327
(7)	4.205	11.05	73.69	3412	7539	11317	15478	20953	30080	54095
(8)	4.797	12.57	81.93	5003	11180	16752	22820	30751	43927	78568
(9)	5.389	14.08	90.16	7021	15840	23699	32184	43213	61488	109505
(10)	5.981	15.59	98.41	9519	21643	32344	43817	58662	83206	147665

TABLE 2.17 : K* (1/α = 0.1)

Q: MODE:	0.05	0.10	0.20	0.30	0.40	0.50	0.60	0.70	0.80	0.90
(1)	.1460	.8093	8.966	22.34	40.57	65.43	100.6	154.7	253.5	525.8
(2)	.1237	.3573	6.005	101.0	204.2	322.8	475.2	695.5	1081	2114
(3)	.1866	.5158	5.759	311.9	641.5	986.6	1404	1987	2987	5645
(4)	.2449	.6619	5.785	693.2	1466	2230	3120	4335	6394	11841
(5)	.3038	.8109	6.298	1299	2806	4242	5874	8063	11748	21467
(6)	.3628	.9607	6.957	2181	4788	7210	9910	13493	19489	35281
(7)	.4218	1.111	7.681	3389	7537	11317	15476	20946	30061	54042
(8)	.4808	1.261	8.441	4974	11178	16752	22817	30743	43906	78509
(9)	.5399	1.412	9.220	6988	15837	23698	32181	43205	61465	109439
(10)	.5990	1.563	10.01	9480	21640	32344	43814	58653	83181	147593

TABLE 2.17 : K* (1/α = 0.2)

Q: MODE:	0.05	0.10	0.20	0.30	0.40	0.50	0.60	0.70	0.80	0.90
(1)	.2655	1.182	9.053	22.34	40.58	65.49	100.8	155.0	254.2	527.6
(2)	.2459	.7069	9.310	101.3	204.2	322.8	475.3	695.8	1082	2117
(3)	.3723	1.028	10.74	312.5	641.5	986.6	1404	1987	2988	5648
(4)	.4893	1.321	11.29	694.2	1466	2230	3121	4335	6395	11844
(5)	.6072	1.620	12.44	1301	2806	4242	5874	8064	11749	21471
(6)	.7252	1.920	13.81	2183	4788	7210	9910	13494	19491	35286
(7)	.8433	2.221	15.29	3392	7537	11317	15476	20947	30064	54048
(8)	.9614	2.522	16.82	4977	11178	16752	22817	30744	43908	78515
(9)	1.080	2.823	18.39	6991	15838	23698	32181	43206	61468	109447
(10)	1.198	3.125	19.99	9484	21640	32344	43814	58654	83183	147601

TABLE 2.17 : K* (1/α = 0.3)

Q: MODE:	0.05	0.10	0.20	0.30	0.40	0.50	0.60	0.70	0.80	0.90
(1)	.3661	1.443	9.132	22.35	40.59	65.54	100.9	155.3	254.9	529.5
(2)	.3666	1.049	11.80	101.5	204.2	322.9	475.4	696.2	1082	2119
(3)	.5571	1.535	15.16	313.1	641.5	986.6	1404	1988	2989	5651
(4)	.7331	1.978	16.54	695.3	1466	2230	3121	4336	6397	11848
(5)	.9102	2.427	18.44	1302	2807	4242	5874	8064	11751	21475
(6)	1.087	2.878	20.57	2185	4789	7210	9910	13495	19493	35291
(7)	1.265	3.329	22.82	3394	7537	11317	15476	20948	30066	54054
(8)	1.442	3.781	25.15	4981	11179	16752	22818	30744	43911	78522
(9)	1.619	4.234	27.52	6995	15838	23698	32182	43207	61470	109454
(10)	1.796	4.686	29.92	9488	21641	32344	43814	58655	83186	147609

TABLE 2.17 : K* (1/α = 0.4)

Q: MODE:	0.05	0.10	0.20	0.30	0.40	0.50	0.60	0.70	0.80	0.90
(1)	.4528	1.645	9.206	22.35	40.59	65.59	101.0	155.6	255.5	531.3
(2)	.4857	1.385	13.85	101.8	204.2	322.9	475.6	696.5	1083	2121
(3)	.7411	2.039	19.17	313.7	641.6	986.7	1404	1988	2991	5654
(4)	.9763	2.632	21.57	696.4	1466	2230	3121	4336	6398	11852
(5)	1.213	3.232	24.31	1304	2807	4242	5874	8065	11752	21480
(6)	1.449	3.834	27.24	2187	4789	7210	9911	13495	19495	35296
(7)	1.685	4.436	30.29	3397	7538	11317	15476	20948	30068	54060
(8)	1.922	5.040	33.42	4984	11179	16752	22818	30745	43913	78528
(9)	2.158	5.643	36.60	6999	15838	23698	32182	43207	61473	109461
(10)	2.395	6.247	39.82	9493	21641	32344	43815	58656	83189	147617

TABLE 2.17 : K* (1/α = 0.5)

Q: MODE:	0.05	0.10	0.20	0.30	0.40	0.50	0.60	0.70	0.80	0.90
(1)	.5285	1.811	9.274	22.35	40.60	65.65	101.2	156.0	256.2	533.1
(2)	.6034	1.713	15.61	102.0	204.2	322.9	475.7	696.8	1084	2123
(3)	.9242	2.539	22.85	314.4	641.6	986.7	1404	1988	2992	5657
(4)	1.219	3.284	26.40	697.4	1466	2230	3121	4337	6399	11855
(5)	1.515	4.036	30.04	1305	2807	4242	5874	8066	11754	21484
(6)	1.810	4.789	33.81	2189	4789	7210	9911	13496	19497	35301
(7)	2.106	5.543	37.69	3399	7538	11317	15477	20949	30070	54065
(8)	2.402	6.297	41.64	4987	11179	16752	22818	30746	43915	78535
(9)	2.697	7.052	45.64	7003	15838	23698	32182	43208	61475	109469
(10)	2.993	7.806	49.68	9497	21641	32344	43815	58657	83192	147625

TABLE 2.17 : K* (1/α = 0.6)

Q: MODE:	0.05	0.10	0.20	0.30	0.40	0.50	0.60	0.70	0.80	0.90
(1)	.5956	1.951	9.337	22.36	40.61	65.70	101.3	156.3	256.9	534.9
(2)	.7197	2.035	17.16	102.3	204.2	322.9	475.8	697.2	1085	2126
(3)	1.106	3.035	26.25	314.9	641.7	986.7	1405	1989	2993	5660
(4)	1.461	3.933	31.05	698.5	1466	2230	3121	4337	6401	11859
(5)	1.816	4.837	35.65	1307	2807	4242	5875	8066	11755	21489
(6)	2.171	5.742	40.29	2191	4789	7210	9911	13497	19498	35307
(7)	2.526	6.647	45.02	3402	7538	11317	15477	20950	30072	54071
(8)	2.881	7.553	49.80	4990	11179	16752	22818	30747	43917	78542
(9)	3.236	8.459	54.63	7006	15839	23699	32183	43209	61478	109476
(10)	3.591	9.365	59.50	9501	21642	32344	43815	58658	83195	147633

TABLE 2.17 : K* (1/α = 0.7)

Q: MODE:	0.05	0.10	0.20	0.30	0.40	0.50	0.60	0.70	0.80	0.90
(1)	.6555	2.073	9.396	22.36	40.62	65.75	101.5	156.6	257.5	536.7
(2)	.8345	2.350	18.56	102.5	204.2	322.9	475.9	697.5	1086	2128
(3)	1.288	3.528	29.43	315.5	641.7	986.7	1405	1989	2994	5663
(4)	1.702	4.580	35.53	699.5	1466	2230	3121	4338	6402	11863
(5)	2.118	5.637	41.15	1308	2807	4242	5875	8067	11757	21493
(6)	2.532	6.694	46.70	2193	4789	7210	9911	13497	19500	35312
(7)	2.946	7.751	52.28	3405	7538	11317	15477	20951	30074	54077
(8)	3.361	8.808	57.91	4993	11180	16752	22819	30748	43920	78548
(9)	3.775	9.866	63.58	7010	15839	23699	32183	43210	61480	109483
(10)	4.189	10.92	69.28	9506	21642	32344	43816	58659	83197	147641

TABLE 2.17 : K* (1/α = 0.8)

Q: MODE:	0.05	0.10	0.20	0.30	0.40	0.50	0.60	0.70	0.80	0.90
(1)	.7096	2.180	9.452	22.36	40.63	65.80	101.6	156.9	258.2	538.5
(2)	.9480	2.660	19.84	102.8	204.2	323.0	476.1	697.8	1086	2130
(3)	1.468	4.017	32.42	316.1	641.7	986.7	1405	1990	2995	5666
(4)	1.943	5.224	39.86	700.5	1466	2230	3122	4338	6403	11867
(5)	2.418	6.435	46.53	1310	2807	4242	5875	8067	11758	21498
(6)	2.893	7.644	53.01	2195	4789	7210	9911	13498	19502	35317
(7)	3.366	8.854	59.48	3407	7538	11317	15477	20952	30076	54083
(8)	3.840	10.06	65.97	4996	11180	16752	22819	30749	43922	78555
(9)	4.313	11.27	72.48	7014	15839	23699	32183	43211	61483	109491
(10)	4.786	12.48	79.03	9510	21642	32344	43816	58660	83200	147649

TABLE 2.17 : K* (1/α = 0.9)

Q: MODE:	0.05	0.10	0.20	0.30	0.40	0.50	0.60	0.70	0.80	0.90
(1)	.7586	2.275	9.504	22.37	40.64	65.85	101.7	157.2	258.9	540.3
(2)	1.060	2.963	21.01	103.0	204.3	323.0	476.2	698.2	1087	2133
(3)	1.648	4.502	35.25	316.7	641.8	986.7	1405	1990	2996	5669
(4)	2.184	5.866	44.05	701.6	1466	2230	3122	4339	6404	11870
(5)	2.719	7.231	51.81	1311	2807	4242	5875	8068	11760	21502
(6)	3.253	8.593	59.25	2197	4790	7210	9912	13499	19504	35322
(7)	3.786	9.955	66.61	3410	7539	11317	15477	20952	30078	54089
(8)	4.319	11.32	73.97	5000	11180	16752	22819	30750	43924	78561
(9)	4.851	12.68	81.35	7017	15840	23699	32183	43212	61485	109498
(10)	5.384	14.04	88.73	9514	21643	32344	43816	58661	83203	147657

TABLE 2.18 : K* (α = 0.1)

Q: MODE:	0.05	0.10	0.20	0.30	0.40	0.50	0.60	0.70	0.80	0.90
(1)	.1813	.7401	3.212	8.268	17.55	33.54	60.05	105.3	194.6	454.6
(2)	5.955	13.80	38.58	81.02	144.1	230.6	351.2	533.9	865.2	1778
(3)	13.37	32.82	111.4	277.2	503.3	775.0	1121	1617	2484	4821
(4)	20.27	51.27	206.8	646.0	1222	1855	2619	3679	5498	10348
(5)	26.78	68.63	311.6	1234	2426	3658	5093	7042	10348	19116
(6)	33.08	85.25	418.7	2091	4240	6369	8787	12026	17475	31884
(7)	39.26	101.4	524.4	3265	6790	10174	13950	18952	27323	49409
(8)	45.37	117.3	627.7	4805	10201	15259	20827	28142	40333	72450
(9)	51.43	133.0	728.2	6763	14598	21810	29665	39917	56946	101766
(10)	57.46	148.6	826.1	9188	20108	30014	40710	54598	77606	138114

TABLE 2.18 : K* (α = 0.2)

Q: MODE:	0.05	0.10	0.20	0.30	0.40	0.50	0.60	0.70	0.80	0.90
(1)	.1667	.6871	3.083	8.208	17.51	32.60	56.19	94.87	168.9	380.4
(2)	3.912	9.715	32.56	78.20	143.9	229.8	345.9	519.0	828.7	1672
(3)	7.522	19.44	83.50	264.6	502.2	774.4	1114	1597	2436	4681
(4)	10.78	28.16	141.5	617.6	1219	1855	2611	3655	5437	10172
(5)	13.91	36.37	198.7	1185	2421	3658	5084	7013	10275	18905
(6)	16.96	44.33	253.1	2018	4233	6368	8777	11992	17390	31636
(7)	19.99	52.16	304.6	3166	6780	10174	13939	18914	27225	49125
(8)	22.99	59.91	353.7	4680	10189	15259	20814	28099	40222	72130
(9)	25.99	67.61	401.1	6609	14584	21810	29651	39869	56824	101409
(10)	28.97	75.27	447.2	9005	20092	30014	40694	54545	77471	137721

TABLE 2.18 : K* (α = 0.3)

Q: MODE:	0.05	0.10	0.20	0.30	0.40	0.50	0.60	0.70	0.80	0.90
(1)	.1544	.6427	2.982	8.168	17.49	32.19	54.72	91.14	160.1	355.4
(2)	2.916	7.535	28.96	76.79	143.9	229.4	344.1	513.9	816.4	1637
(3)	5.234	13.83	68.78	259.2	501.7	774.1	1112	1590	2419	4634
(4)	7.348	19.42	109.9	606.0	1218	1854	2609	3647	5417	10114
(5)	9.392	24.75	148.1	1166	2419	3657	5081	7003	10250	18834
(6)	11.41	29.96	183.1	1991	4231	6368	8774	11981	17361	31554
(7)	13.41	35.11	215.8	3130	6777	10173	13935	18901	27193	49030
(8)	15.40	40.22	247.1	4634	10185	15259	20810	28084	40186	72023
(9)	17.38	45.32	277.4	6554	14579	21810	29646	39853	56783	101290
(10)	19.37	50.40	307.1	8941	20086	30013	40689	54528	77426	137590

TABLE 2.18 : K* (α = 0.4)

Q: MODE:	0.05	0.10	0.20	0.30	0.40	0.50	0.60	0.70	0.80	0.90
(1)	.1439	.6046	2.898	8.139	17.47	31.96	53.93	89.22	155.7	343.0
(2)	2.325	6.165	26.46	75.94	143.8	229.2	343.1	511.3	810.1	1619
(3)	4.014	10.73	59.28	256.0	501.5	774.0	1111	1587	2411	4610
(4)	5.572	14.82	90.70	599.6	1218	1854	2608	3643	5407	10084
(5)	7.091	18.75	118.6	1156	2419	3657	5079	6998	10238	18799
(6)	8.592	22.62	143.8	1977	4229	6368	8772	11975	17347	31512
(7)	10.09	26.46	167.4	3111	6776	10173	13933	18894	27176	48983
(8)	11.58	30.28	190.1	4611	10183	15259	20808	28077	40167	71969
(9)	13.06	34.09	212.2	6526	14577	21810	29644	39845	56762	101230
(10)	14.55	37.89	233.9	8908	20083	30013	40687	54519	77403	137525

TABLE 2.18 : K* (α = 0.5)

Q: MODE:	0.05	0.10	0.20	0.30	0.40	0.50	0.60	0.70	0.80	0.90
(1)	.1347	.5715	2.827	8.117	17.46	31.82	53.45	88.05	153.0	335.4
(2)	1.933	5.222	24.58	75.37	143.8	229.1	342.3	509.7	806.4	1608
(3)	3.255	8.773	52.49	254.0	501.3	773.9	1110	1585	2406	4596
(4)	4.488	11.99	77.54	595.6	1217	1854	2607	3640	5401	10067
(5)	5.695	15.10	99.11	1150	2418	3657	5078	6995	10231	18778
(6)	6.892	18.17	118.5	1968	4229	6368	8771	11972	17339	31487
(7)	8.084	21.23	136.8	3100	6775	10173	13932	18890	27166	48955
(8)	9.273	24.27	154.5	4596	10182	15259	20807	28073	40156	71937
(9)	10.46	27.31	171.8	6509	14576	21810	29642	39840	56750	101195
(10)	11.65	30.35	188.9	8887	20082	30013	40685	54513	77389	137485

TABLE 2.18 : K* (α = 0.6)

Q: MODE:	0.05	0.10	0.20	0.30	0.40	0.50	0.60	0.70	0.80	0.90
(1)	.1267	.5424	2.766	8.100	17.45	31.72	53.11	87.25	151.2	330.4
(2)	1.654	4.531	23.10	74.96	143.8	229.1	342.1	508.7	803.9	1601
(3)	2.737	7.418	47.32	252.6	501.2	773.9	1110	1584	2403	4587
(4)	3.757	10.06	67.89	592.8	1217	1854	2606	3638	5396	10055
(5)	4.758	12.64	85.23	1145	2418	3657	5078	6993	10226	18764
(6)	5.753	15.19	100.9	1962	4228	6368	8770	11969	17333	31471
(7)	6.745	17.72	115.7	3092	6774	10173	13931	18888	27160	48936
(8)	7.734	20.26	130.2	4587	10181	15258	20806	28070	40149	71916
(9)	8.723	22.79	144.4	6497	14575	21810	29641	39837	56741	101171
(10)	9.711	25.31	158.5	8874	20081	30013	40684	54510	77380	137459

TABLE 2.18 : K* (α = 0.7)

Q: MODE:	0.05	0.10	0.20	0.30	0.40	0.50	0.60	0.70	0.80	0.90
(1)	.1196	.5164	2.714	8.086	17.45	31.64	52.87	86.68	149.9	326.8
(2)	1.446	4.003	21.88	74.65	143.8	229.0	341.9	507.9	802.1	1596
(3)	2.362	6.426	43.22	251.6	501.2	773.8	1109	1583	2401	4580
(4)	3.230	8.667	60.47	590.8	1217	1854	2606	3637	5394	10047
(5)	4.086	10.86	74.81	1142	2417	3657	5077	6992	10222	18753
(6)	4.937	13.04	87.84	1957	4228	6368	8770	11968	17329	31459
(7)	5.786	15.21	100.3	3086	6774	10173	13930	18886	27155	48922
(8)	6.634	17.38	112.4	4580	10181	15258	20805	28068	40143	71901
(9)	7.481	19.55	124.5	6489	14574	21810	29641	39835	56736	101154
(10)	8.327	21.71	136.5	8864	20080	30013	40684	54507	77374	137440

TABLE 2.18 : K* (α = 0.8)

Q: MODE:	0.05	0.10	0.20	0.30	0.40	0.50	0.60	0.70	0.80	0.90
(1)	.1132	.4932	2.667	8.075	17.45	31.58	52.69	86.25	148.9	324.2
(2)	1.284	3.585	20.87	74.41	143.8	229.0	341.6	507.3	800.7	1592
(3)	2.077	5.669	39.87	250.8	501.1	773.8	1109	1582	2399	4575
(4)	2.833	7.614	54.56	589.3	1217	1854	2606	3636	5391	10040
(5)	3.501	9.527	66.70	1140	2417	3657	5077	6991	10220	18746
(6)	4.324	11.43	77.79	1954	4228	6368	8770	11966	17326	31450
(7)	5.066	13.33	88.47	3082	6773	10173	13930	18885	27152	48912
(8)	5.807	15.22	98.99	4574	10180	15258	20805	28066	40139	71889
(9)	6.548	17.11	109.4	6482	14574	21810	29640	39833	56731	101141
(10)	7.288	19.01	119.8	8857	20079	30013	40683	54505	77369	137426

TABLE 2.18 : K* (α = 0.9)

Q: MODE:	0.05	0.10	0.20	0.30	0.40	0.50	0.60	0.70	0.80	0.90
(1)	.1076	.4722	2.625	8.065	17.44	31.54	52.55	85.92	148.2	322.1
(2)	1.155	3.247	20.00	74.22	143.8	228.9	341.5	506.9	799.7	1589
(3)	1.853	5.071	37.07	250.2	501.1	773.8	1109	1581	2397	4571
(4)	2.523	6.788	49.75	588.0	1217	1854	2605	3636	5390	10035
(5)	3.186	8.483	60.18	1138	2417	3657	5077	6990	10218	18740
(6)	3.847	10.17	69.81	1951	4227	6368	8769	11965	17323	31443
(7)	4.506	11.86	79.16	3078	6773	10173	13930	18883	27149	48904
(8)	5.164	13.54	88.40	4570	10180	15258	20804	28065	40136	71880
(9)	5.822	15.22	97.61	6478	14573	21810	29640	39831	56728	101131
(10)	6.480	16.90	106.8	8851	20079	30013	40683	54504	77365	137415

TABLE 2.18 : K* (α = 1.0)

Q: MODE:	0.05	0.10	0.20	0.30	0.40	0.50	0.60	0.70	0.80	0.90
(1)	.1024	.4531	2.588	8.057	17.44	31.50	52.43	85.65	147.6	320.4
(2)	1.049	2.968	19.24	74.06	143.8	228.9	341.3	506.5	798.8	1587
(3)	1.673	4.587	34.69	249.7	501.0	773.8	1109	1581	2396	4568
(4)	2.274	6.125	45.73	587.1	1217	1854	2605	3635	5388	10031
(5)	2.870	7.646	54.84	1136	2417	3657	5076	6989	10216	18735
(6)	3.464	9.162	63.33	1949	4227	6368	8769	11965	17321	31438
(7)	4.057	10.68	71.62	3076	6773	10173	13929	18883	27147	48898
(8)	4.649	12.19	79.87	4567	10179	15258	20804	28064	40134	71873
(9)	5.241	13.70	88.10	6474	14573	21810	29640	39830	56725	101123
(10)	5.833	15.22	96.35	8847	20078	30013	40682	54503	77362	137407

TABLE 2.18 : K* (1/α = 0.1)

Q: MODE:	0.05	0.10	0.20	0.30	0.40	0.50	0.60	0.70	0.80	0.90
(1)	.0196	.1061	1.935	7.971	17.41	31.18	51.46	83.43	142.6	306.8
(2)	.1138	.3403	7.982	72.61	143.7	228.7	340.3	503.6	792.0	1568
(3)	.1717	.4786	5.874	245.2	500.8	773.6	1107	1577	2387	4543
(4)	.2303	.6249	5.705	578.7	1216	1854	2604	3631	5377	10000
(5)	.2891	.7736	6.151	1124	2416	3657	5075	6984	10203	18697
(6)	.3480	.9232	6.784	1932	4226	6368	8767	11959	17306	31393
(7)	.4070	1.073	7.496	3053	6771	10173	13927	18876	27129	48846
(8)	.4661	1.224	8.248	4540	10177	15258	20802	28056	40114	71816
(9)	.5251	1.374	9.024	6441	14570	21810	29637	39822	56703	101059
(10)	.5842	1.525	9.814	8809	20075	30013	40679	54493	77338	137336

TABLE 2.18 : K* (1/α = 0.2)

Q: MODE:	0.05	0.10	0.20	0.30	0.40	0.50	0.60	0.70	0.80	0.90
(1)	.0355	.1820	2.078	7.983	17.42	31.22	51.57	83.68	143.2	308.3
(2)	.2255	.6694	10.40	72.78	143.7	228.7	340.4	504.0	792.8	1570
(3)	.3424	.9526	10.66	245.7	500.8	773.6	1108	1578	2388	4545
(4)	.4599	1.247	11.07	579.7	1216	1854	2604	3631	5378	10003
(5)	.5777	1.545	12.13	1125	2416	3657	5075	6985	10204	18701
(6)	.6957	1.845	13.46	1934	4226	6368	8767	11959	17308	31398
(7)	.8138	2.145	14.91	3056	6771	10173	13928	18876	27131	48852
(8)	.9319	2.446	16.44	4543	10178	15258	20802	28057	40116	71822
(9)	1.050	2.748	18.00	6445	14570	21810	29637	39823	56705	101066
(10)	1.168	3.050	19.59	8813	20076	30013	40680	54494	77341	137344

TABLE 2.18 : K* (1/α = 0.3)

Q: MODE:	0.05	0.10	0.20	0.30	0.40	0.50	0.60	0.70	0.80	0.90
(1)	.0488	.2405	2.186	7.994	17.42	31.26	51.68	83.93	143.7	309.8
(2)	.3351	.9880	12.16	72.95	143.7	228.8	340.5	504.3	793.5	1572
(3)	.5121	1.422	14.76	246.2	500.8	773.7	1108	1578	2389	4548
(4)	.6889	1.866	16.13	580.6	1216	1854	2604	3632	5380	10007
(5)	.8659	2.315	17.94	1126	2416	3657	5075	6985	10206	18706
(6)	1.043	2.765	20.02	1936	4226	6368	8768	11960	17309	31403
(7)	1.220	3.216	22.25	3058	6771	10173	13928	18877	27133	48858
(8)	1.397	3.668	24.56	4546	10178	15258	20802	28058	40118	71828
(9)	1.575	4.120	26.93	6449	14571	21810	29638	39824	56708	101073
(10)	1.752	4.573	29.32	8817	20076	30013	40680	54495	77343	137351

TABLE 2.18 : K* (1/α = 0.4)

Q: MODE:	0.05	0.10	0.20	0.30	0.40	0.50	0.60	0.70	0.80	0.90
(1)	.0599	.2874	2.273	8.005	17.42	31.29	51.79	84.17	144.3	311.3
(2)	.4427	1.297	13.60	73.12	143.7	228.8	340.6	504.6	794.3	1574
(3)	.6808	1.887	18.39	246.7	500.9	773.7	1108	1579	2390	4551
(4)	.9173	2.483	20.93	581.6	1216	1854	2604	3632	5381	10010
(5)	1.154	3.082	23.60	1128	2416	3657	5075	6986	10207	18710
(6)	1.390	3.683	26.49	1938	4226	6368	8768	11961	17311	31408
(7)	1.626	4.286	29.52	3061	6772	10173	13928	18878	27135	48863
(8)	1.863	4.889	32.63	4549	10178	15258	20803	28059	40121	71835
(9)	2.099	5.492	35.80	6452	14571	21810	29638	39825	56710	101081
(10)	2.336	6.096	39.01	8822	20076	30013	40680	54496	77346	137359

TABLE 2.18 : K* (1/α = 0.5)

Q: MODE:	0.05	0.10	0.20	0.30	0.40	0.50	0.60	0.70	0.80	0.90
(1)	.0695	.3263	2.346	8.015	17.43	31.33	51.90	84.42	144.8	312.8
(2)	.5484	1.596	14.82	73.29	143.7	228.8	340.7	504.9	795.1	1576
(3)	.8486	2.348	21.66	247.2	500.9	773.7	1108	1579	2391	4554
(4)	1.145	3.097	25.51	582.5	1216	1854	2604	3633	5382	10014
(5)	1.441	3.848	29.12	1129	2416	3657	5075	6986	10209	18714
(6)	1.737	4.600	32.86	1940	4226	6368	8768	11961	17313	31413
(7)	2.032	5.354	36.71	3063	6772	10173	13928	18879	27137	48869
(8)	2.328	6.108	40.64	4552	10178	15258	20803	28060	40123	71841
(9)	2.623	6.863	44.64	6456	14571	21810	29638	39826	56713	101088
(10)	2.919	7.618	48.67	8826	20077	30013	40681	54497	77349	137367

TABLE 2.18 : K* (1/α = 0.6)

Q: MODE:	0.05	0.10	0.20	0.30	0.40	0.50	0.60	0.70	0.80	0.90
(1)	.0778	.3592	2.407	8.024	17.43	31.36	52.01	84.67	145.4	314.4
(2)	.6522	1.886	15.89	73.45	143.7	228.8	340.9	505.3	795.8	1578
(3)	1.015	2.804	24.66	247.7	500.9	773.7	1108	1579	2392	4557
(4)	1.372	3.708	29.89	583.4	1216	1854	2604	3633	5383	10017
(5)	1.728	4.611	34.50	1131	2417	3657	5076	6987	10210	18718
(6)	2.083	5.516	39.13	1941	4227	6368	8768	11962	17314	31418
(7)	2.438	6.421	43.83	3066	6772	10173	13928	18879	27139	48875
(8)	2.793	7.327	48.60	4555	10179	15258	20803	28061	40125	71848
(9)	3.147	8.233	53.42	6459	14572	21810	29638	39826	56715	101095
(10)	3.502	9.139	58.28	8830	20077	30013	40681	54498	77351	137375

TABLE 2.18 : K* (1/α = 0.7)

Q: MODE:	0.05	0.10	0.20	0.30	0.40	0.50	0.60	0.70	0.80	0.90
(1)	.0851	.3875	2.461	8.033	17.43	31.40	52.11	84.91	145.9	315.9
(2)	.7542	2.168	16.85	73.60	143.7	228.9	341.0	505.6	796.6	1580
(3)	1.181	3.256	27.42	248.2	501.0	773.7	1108	1580	2393	4560
(4)	1.599	4.316	34.08	584.4	1216	1854	2605	3634	5385	10021
(5)	2.014	5.373	39.76	1132	2417	3657	5076	6988	10212	18723
(6)	2.429	6.430	45.31	1943	4227	6368	8768	11963	17316	31423
(7)	2.843	7.487	50.88	3068	6772	10173	13929	18880	27141	48881
(8)	3.257	8.544	56.50	4558	10179	15258	20803	28061	40127	71854
(9)	3.671	9.601	62.16	6463	14572	21810	29639	39827	56718	101102
(10)	4.085	10.66	67.85	8834	20077	30013	40681	54499	77354	137383

TABLE 2.18 : K* (1/α = 0.8)

Q: MODE:	0.05	0.10	0.20	0.30	0.40	0.50	0.60	0.70	0.80	0.90
(1)	.0915	.4121	2.508	8.042	17.43	31.43	52.22	85.16	146.5	317.4
(2)	.8543	2.442	17.72	73.76	143.8	228.9	341.1	505.9	797.3	1583
(3)	1.346	3.704	30.00	248.7	501.0	773.7	1108	1580	2394	4562
(4)	1.824	4.921	38.11	585.3	1216	1854	2605	3634	5386	10024
(5)	2.300	6.133	44.90	1134	2417	3657	5076	6988	10213	18727
(6)	2.774	7.342	51.40	1945	4227	6368	8769	11963	17318	31428
(7)	3.248	8.551	57.86	3071	6772	10173	13929	18881	27143	48886
(8)	3.721	9.760	64.34	4561	10179	15258	20804	28062	40129	71860
(9)	4.195	10.97	70.85	6466	14572	21810	29639	39828	56720	101109
(10)	4.668	12.18	77.39	8838	20078	30013	40682	54501	77357	137391

TABLE 2.18 : K* (1/α = 0.9)

Q: MODE:	0.05	0.10	0.20	0.30	0.40	0.50	0.60	0.70	0.80	0.90
(1)	.0973	.4338	2.550	8.050	17.44	31.47	52.32	85.40	147.0	318.9
(2)	.9527	2.709	18.51	73.91	143.8	228.9	341.2	506.2	798.1	1585
(3)	1.510	4.147	32.41	249.2	501.0	773.8	1109	1581	2395	4565
(4)	2.050	5.524	41.99	586.2	1217	1854	2605	3635	5387	10028
(5)	2.585	6.890	49.92	1135	2417	3657	5076	6989	10214	18731
(6)	3.119	8.253	57.40	1947	4227	6368	8769	11964	17320	31433
(7)	3.652	9.614	64.77	3073	6773	10173	13929	18882	27145	48892
(8)	4.185	10.98	72.13	4564	10179	15258	20804	28063	40132	71867
(9)	4.718	12.34	79.50	6470	14573	21810	29639	39829	56723	101116
(10)	5.251	13.70	86.89	8842	20078	30013	40682	54502	77360	137399

TABLE 2.19 : K* (α = 0.1)

Q: MODE:	0.05	0.10	0.20	0.30	0.40	0.50	0.60	0.70	0.80	0.90
(1)	.1280	.5156	2.122	5.020	9.620	16.73	28.01	47.41	86.79	205.6
(2)	.2465	1.007	4.386	11.38	24.42	47.04	84.24	146.5	266.2	607.0
(3)	8.005	18.83	54.80	118.0	208.9	328.7	491.2	733.2	1167	2354
(4)	15.12	37.50	133.6	351.1	644.2	987.4	1417	2028	3089	5939
(5)	21.92	55.71	232.5	770.8	1473	2231	3137	4385	6521	12207
(6)	28.37	72.84	338.4	1421	2817	4243	5892	8123	11899	21904
(7)	34.64	89.33	445.3	2353	4804	7210	9931	13563	19665	35790
(8)	40.80	105.4	550.5	3613	7557	11318	15499	21025	30262	54623
(9)	46.89	121.3	653.1	5254	11203	16752	22843	30831	44131	79161
(10)	52.94	136.9	752.9	7323	15866	23699	32209	43303	61715	110164

TABLE 2.19 : K* (α = 0.2)

Q: MODE:	0.05	0.10	0.20	0.30	0.40	0.50	0.60	0.70	0.80	0.90
(1)	.0774	.3118	1.282	3.026	5.781	10.01	16.69	28.09	51.09	120.2
(2)	.2046	.8463	3.876	10.79	24.29	46.82	81.78	138.0	242.9	535.7
(3)	4.997	12.60	45.11	113.7	208.6	327.9	485.5	717.0	1127	2239
(4)	8.347	21.66	97.57	335.1	642.7	986.8	1410	2007	3038	5790
(5)	11.58	30.25	156.0	737.6	1469	2230	3129	4360	6457	12022
(6)	14.68	38.38	212.6	1367	2812	4243	5883	8093	11823	21683
(7)	17.72	46.30	266.2	2274	4796	7210	9920	13528	19577	35533
(8)	20.74	54.10	317.0	3509	7547	11318	15487	20985	30161	54329
(9)	23.74	61.84	365.7	5121	11190	16752	22830	30787	44018	78832
(10)	26.73	69.53	412.7	7162	15852	23699	32195	43253	61589	109798

TABLE 2.19 : K* (α = 0.3)

Q: MODE:	0.05	0.10	0.20	0.30	0.40	0.50	0.60	0.70	0.80	0.90
(1)	.0610	.2456	1.009	2.377	4.529	7.819	12.98	21.73	39.30	91.80
(2)	.1737	.7265	3.486	10.37	24.22	46.72	80.83	134.9	234.9	511.8
(3)	3.635	9.504	39.52	111.6	208.5	327.5	483.5	711.5	1114	2201
(4)	5.765	15.24	79.05	328.2	642.1	986.5	1408	2000	3020	5741
(5)	7.864	20.77	119.8	724.2	1468	2230	3126	4351	6436	11960
(6)	9.898	26.06	157.1	1346	2810	4243	5880	8083	11797	21610
(7)	11.91	31.25	191.4	2244	47[illegible]	7210	9917	13516	19547	35447
(8)	13.91	36.39	223.8	3470	7544	11317	15483	20972	30127	54232
(9)	15.90	41.50	254.8	5073	11186	16752	22826	30772	43980	78722
(10)	17.88	46.59	284.9	7105	15847	23699	32190	43237	61547	109676

TABLE 2.19 : K* (α = 0.4)

Q: MODE:	0.05	0.10	0.20	0.30	0.40	0.50	0.60	0.70	0.80	0.90
(1)	.0530	.2134	.8754	2.059	3.916	6.742	11.15	18.59	33.45	77.69
(2)	.1504	.6349	3.177	10.06	24.17	46.66	80.31	133.3	230.9	499.7
(3)	2.857	7.643	35.73	110.4	208.4	327.4	482.5	708.7	1107	2181
(4)	4.404	11.76	67.27	324.3	641.8	986.4	1407	1996	3012	5716
(5)	5.954	15.82	98.00	716.9	1468	2230	3124	4347	6425	11930
(6)	7.467	19.73	125.1	1334	2809	4243	5878	8078	11784	21573
(7)	8.967	23.58	149.8	2229	4792	7210	9915	13510	19532	35404
(8)	10.46	27.41	173.2	3450	7542	11317	15481	20965	30110	54183
(9)	11.95	31.23	195.7	5049	11184	16752	22823	30764	43961	78667
(10)	13.43	35.04	217.6	7076	15844	23699	32188	43228	61526	109615

TABLE 2.19 : K* (α = 0.5)

Q: MODE:	0.05	0.10	0.20	0.30	0.40	0.50	0.60	0.70	0.80	0.90
(1)	.0483	.1944	.7970	1.872	3.555	6.106	10.07	16.73	29.97	69.26
(2)	.1323	.5629	2.922	9.804	24.14	46.62	80.00	132.3	228.4	492.5
(3)	2.353	6.395	32.92	109.6	208.4	327.3	481.9	707.0	1103	2170
(4)	3.562	9.576	58.95	321.8	641.6	986.4	1406	1994	3006	570[illegible]
(5)	4.791	12.77	83.22	712.3	1467	2230	3124	4344	6419	11911
(6)	5.995	15.87	104.1	1327	2809	4243	5877	8075	11777	21551
(7)	7.190	18.94	123.2	2219	4791	7210	9914	13507	19523	35379
(8)	8.381	21.99	141.3	3438	7541	11317	15480	20961	30100	54153
(9)	9.570	25.03	158.9	5034	11183	16752	22822	30760	43949	78634
(10)	10.76	28.07	176.1	7058	15843	23699	32187	43223	61513	109578

TABLE 2.19 : K* (α = 0.6)

Q: MODE:	0.05	0.10	0.20	0.30	0.40	0.50	0.60	0.70	0.80	0.90
(1)	.0452	.1820	.7456	1.750	3.318	5.688	9.357	15.50	27.67	63.66
(2)	.1180	.5051	2.709	9.598	24.11	46.60	79.78	131.7	226.8	407.7
(3)	2.001	5.499	30.73	109.0	208.4	327.2	481.5	705.8	1100	2162
(4)	2.991	8.076	52.68	320.1	641.5	986.3	1406	1993	3003	5691
(5)	4.008	10.71	72.47	709.2	1467	2230	3123	4343	6414	11899
(6)	5.007	13.27	89.26	1323	2808	4243	5877	8073	11772	21536
(7)	6.001	15.82	104.7	2213	4791	7210	9913	13504	19517	35361
(8)	6.992	18.36	119.4	3429	7540	11317	15479	20959	30093	54134
(9)	7.982	20.89	133.7	5023	11182	16752	22821	30757	43942	78612
(10)	8.970	23.42	147.9	7046	15842	23699	32186	43220	61505	109554

TABLE 2.19 : K* (α = 0.7)

Q: MODE:	0.05	0.10	0.20	0.30	0.40	0.50	0.60	0.70	0.80	0.90
(1)	.0431	.1732	.7095	1.664	3.151	5.393	8.854	14.63	26.04	59.68
(2)	.1064	.4579	2.528	9.426	24.09	46.58	79.62	131.2	225.6	484.2
(3)	1.740	4.824	28.95	108.6	208.4	327.1	481.2	705.0	1098	2157
(4)	2.577	6.982	47.74	318.8	641.4	986.3	1405	1992	3000	5684
(5)	3.445	9.220	64.26	706.9	1467	2230	3123	4341	6411	11890
(6)	4.299	11.41	78.15	1319	2808	4243	5876	8071	11768	21525
(7)	5.150	13.59	90.98	2208	4790	7210	9913	13503	19513	35349
(8)	5.998	15.76	103.3	3423	7540	11317	15479	20957	30088	54120
(9)	6.846	17.92	115.5	5016	11181	16752	22821	30755	43936	78596
(10)	7.692	20.09	127.5	7037	15841	23699	32185	43218	61499	109537

TABLE 2.19 : K* (α = 0.8)

Q: MODE:	0.05	0.10	0.20	0.30	0.40	0.50	0.60	0.70	0.80	0.90
(1)	.0415	.1668	.6827	1.600	3.027	5.174	8.480	13.98	24.82	56.70
(2)	.0968	.4185	2.371	9.279	24.08	46.56	79.50	130.8	224.7	481.6
(3)	1.540	4.297	27.46	108.3	208.3	327.1	481.0	704.4	1097	2152
(4)	2.264	6.150	43.74	317.9	641.3	986.2	1405	1991	2999	5678
(5)	3.021	8.094	57.77	705.2	1467	2230	3122	4341	6409	11883
(6)	3.767	10.00	69.53	1317	2808	4243	5876	8070	11765	21517
(7)	4.510	11.90	80.49	2204	4790	7210	9912	13501	19510	35340
(8)	5.252	13.80	91.11	3419	7539	11317	15478	20955	30085	54109
(9)	5.993	15.69	101.6	5010	11181	16752	22820	30753	43932	78584
(10)	6.733	17.59	112.0	7030	15841	23699	32184	43216	61494	109524

TABLE 2.19 : K* (α = 0.9)

Q: MODE:	0.05	0.10	0.20	0.30	0.40	0.50	0.60	0.70	0.80	0.90
(1)	.0402	.1618	.6622	1.551	2.932	5.006	8.192	13.48	23.87	54.39
(2)	.0888	.3853	2.233	9.152	24.07	46.55	79.41	130.6	224.0	479.6
(3)	1.381	3.875	26.20	108.0	208.3	327.1	480.8	703.9	1095	2149
(4)	2.019	5.495	40.42	317.1	641.3	986.2	1405	1990	2997	5674
(5)	2.690	7.214	52.50	703.8	1466	2230	3122	4340	6407	11878
(6)	3.352	8.906	62.64	1315	2808	4243	5876	8069	11763	21511
(7)	4.011	10.59	72.17	2202	4790	7210	9912	13500	19507	35333
(8)	4.670	12.28	81.48	3415	7539	11317	15478	20954	30082	54101
(9)	5.329	13.96	90.71	5006	11181	16752	22820	30752	43929	78575
(10)	5.987	15.64	99.91	7025	15840	23699	32184	43215	61491	109513

TABLE 2.19 : K* (α = 1.0)

Q: MODE:	0.05	0.10	0.20	0.30	0.40	0.50	0.60	0.70	0.80	0.90
(1)	.0392	.1579	.6459	1.512	2.856	4.872	7.963	13.09	23.12	52.54
(2)	.0820	.3568	2.112	9.040	24.06	46.54	79.33	130.3	223.5	478.0
(3)	1.251	3.528	25.10	107.8	208.3	327.1	480.6	703.5	1095	2147
(4)	1.822	4.966	37.60	316.5	641.2	986.2	1405	1990	2996	5671
(5)	2.424	6.506	48.13	702.7	1466	2230	3122	4339	6406	11874
(6)	3.019	8.025	56.99	1313	2807	4242	5875	8069	11761	21506
(7)	3.612	9.541	65.41	2199	4790	7210	9912	13500	19505	35327
(8)	4.205	11.05	73.69	3412	7539	11317	15478	20953	30080	54095
(9)	4.797	12.57	81.93	5003	11180	16752	22820	30751	43927	78568
(10)	5.389	14.08	90.16	7021	15840	23699	32184	43213	61488	109505

TABLE 2.19 : K* (1/α = 0.1)

Q: MODE:	0.05	0.10	0.20	0.30	0.40	0.50	0.60	0.70	0.80	0.90
(1)	.0317	.1275	.5203	1.212	2.272	3.835	6.181	9.971	17.20	37.83
(2)	.0102	.0462	.3766	7.517	23.97	46.46	78.70	128.5	218.9	464.9
(3)	.1328	.3902	9.098	105.8	208.3	326.9	479.5	700.4	1087	2126
(4)	.1860	.5139	5.669	311.1	640.9	986.1	1403	1986	2987	5644
(5)	.2450	.6620	5.788	693.3	1466	2230	3120	4335	6394	11841
(6)	.3038	.8109	6.298	1299	2806	4242	5874	8063	11748	21467
(7)	.3628	.9607	6.957	2181	4788	7210	9910	13493	19489	35281
(8)	.4218	1.111	7.681	3389	7537	11317	15476	20946	30061	54042
(9)	.4808	1.261	8.441	4974	11178	16752	22817	30743	43906	78509
(10)	.5399	1.412	9.220	6988	15837	23698	32181	43205	61465	109439

TABLE 2.19 : K* (1/α = 0.2)

Q: MODE:	0.05	0.10	0.20	0.30	0.40	0.50	0.60	0.70	0.80	0.90
(1)	.0325	.1307	.5336	1.244	2.334	3.945	6.371	10.30	17.84	39.43
(2)	.0199	.0895	.6877	7.785	23.98	46.47	78.77	128.7	219.5	466.4
(3)	.2637	.7712	12.49	106.1	208.3	326.9	479.6	700.8	1088	2128
(4)	.3712	1.024	10.59	311.7	640.9	986.1	1403	1987	2988	5647
(5)	.4894	1.321	11.29	694.3	1466	2230	3121	4335	6369	11844
(6)	.6072	1.620	12.44	1301	2806	4242	5874	8064	11749	21471
(7)	.7252	1.920	13.81	2183	4788	7210	9910	13494	19491	35286
(8)	.8433	2.221	15.29	3392	7537	11317	15476	20947	30064	54048
(9)	.9614	2.522	16.82	4977	11178	16752	22817	30744	43908	78515
(10)	1.080	2.823	18.39	6991	15838	23698	32181	43206	61468	109447

TABLE 2.19 : K* (1/α = 0.3)

Q: MODE:	0.05	0.10	0.20	0.30	0.40	0.50	0.60	0.70	0.80	0.90
(1)	.0333	.1340	.5470	1.276	2.396	4.057	6.564	10.64	18.48	41.05
(2)	.0292	.1303	.9517	8.013	23.99	46.48	78.84	128.9	220.0	467.8
(3)	.3927	1.143	14.99	106.3	208.3	326.9	479.8	701.1	1089	2130
(4)	.5554	1.530	14.97	312.3	641.0	986.1	1404	1987	2989	5650
(5)	.7332	1.978	16.55	695.4	1466	2230	3121	4336	6397	11848
(6)	.9102	2.427	18.44	1302	2807	4242	5874	8064	11751	21475
(7)	1.087	2.878	20.57	2185	4789	7210	9910	13495	19493	35291
(8)	1.265	3.329	22.82	3394	7537	11317	15476	20948	30066	54054
(9)	1.442	3.781	25.15	4981	11179	16752	22818	30744	43911	78522
(10)	1.619	4.234	27.52	6995	15838	23698	32182	43207	61470	109454

TABLE 2.19 : K* (1/α = 0.4)

Q: MODE:	0.05	0.10	0.20	0.30	0.40	0.50	0.60	0.70	0.80	0.90
(1)	.0341	.1373	.5607	1.309	2.460	4.170	6.758	10.98	19.13	42.68
(2)	.0379	.1686	1.180	8.211	24.00	46.49	78.91	129.1	220.5	469.3
(3)	.5204	1.507	17.02	106.5	208.3	326.9	479.9	701.5	1090	2133
(4)	.7389	2.032	18.95	312.9	641.0	986.1	1404	1987	2990	5653
(5)	.9764	2.633	21.58	696.5	1466	2230	3121	4336	6398	11852
(6)	1.213	3.232	24.31	1304	2807	4242	5874	8065	11752	21480
(7)	1.449	3.834	27.24	2187	4789	7210	9911	13495	19495	35296
(8)	1.685	4.436	30.29	3397	7538	11317	15476	20948	30068	54060
(9)	1.922	5.040	33.42	4984	11179	16752	22818	30745	43913	78528
(10)	2.158	5.643	36.60	6999	15838	23698	32182	43207	61473	109461

TABLE 2.19 : K* (1/α = 0.5)

Q: MODE:	0.05	0.10	0.20	0.30	0.40	0.50	0.60	0.70	0.80	0.90
(1)	.0350	.1406	.5745	1.342	2.524	4.284	6.955	11.33	19.79	44.31
(2)	.0462	.2047	1.380	8.387	24.01	46.50	78.98	129.3	221.0	470.7
(3)	.6463	1.862	18.75	106.7	208.3	326.9	480.0	701.8	1090	2135
(4)	.9214	2.530	22.60	313.5	641.0	986.1	1404	1988	2991	5656
(5)	1.219	3.284	26.41	697.5	1466	2230	3121	4337	6399	11856
(6)	1.515	4.036	30.04	1305	2807	4242	5874	8066	11754	21484
(7)	1.810	4.789	33.81	2189	4789	7210	9911	13496	19497	35301
(8)	2.106	5.543	37.69	3399	7538	11317	15477	20949	30070	54065
(9)	2.402	6.297	41.64	4987	11179	16752	22818	30746	43915	78535
(10)	2.697	7.052	45.64	7003	15838	23698	32182	43208	61475	109469

TABLE 2.19 : K* (1/α = 0.6)

Q: MODE:	0.05	0.10	0.20	0.30	0.40	0.50	0.60	0.70	0.80	0.90
(1)	.0358	.1440	.5885	1.375	2.589	4.400	7.153	11.67	20.45	45.94
(2)	.0541	.2387	1.558	8.543	24.02	46.50	79.05	129.5	221.5	472.2
(3)	.7704	2.210	20.28	107.0	208.3	327.0	480.1	702.2	1091	2137
(4)	1.103	3.025	25.99	314.1	641.1	986.2	1404	1988	2992	5659
(5)	1.461	3.934	31.06	698.6	1466	2230	3121	4337	6401	11859
(6)	1.816	4.837	35.65	1307	2807	4242	5875	8066	11755	21489
(7)	2.171	5.742	40.29	2191	4789	7210	9911	13497	19498	35307
(8)	2.526	6.647	45.02	34.02	7538	11317	15477	20950	30072	54071
(9)	2.881	7.553	49.80	4990	11179	16752	22818	30747	43917	78542
(10)	3.236	8.459	54.63	7006	15839	23699	32183	43209	61478	109476

TABLE 2.19 : K* (1/α = 0.7)

Q: MODE:	0.05	0.10	0.20	0.30	0.40	0.50	0.60	0.70	0.80	0.90
(1)	.0367	.1474	.6026	1.409	2.655	4.516	7.353	12.02	21.11	47.59
(2)	.0616	.2708	1.718	8.685	24.03	46.51	79.12	129.7	222.0	473.6
(3)	.8930	2.550	21.65	107.2	208.3	327.0	480.3	702.5	1092	2140
(4)	1.284	3.515	29.15	314.7	641.1	986.2	1404	1989	2993	5662
(5)	1.703	4.581	35.55	699.6	1466	2230	3121	4338	6402	11863
(6)	2.118	5.637	41.15	1308	2807	4242	5875	8067	11757	21493
(7)	2.532	6.694	46.70	2193	4789	7210	9911	13497	19500	35312
(8)	2.946	7.751	52.28	3405	7538	11317	15477	20951	30074	54077
(9)	3.361	8.808	57.91	4993	11180	16752	22819	30748	43920	78548
(10)	3.775	9.866	63.58	7010	15839	23699	32183	43210	61480	109483

TABLE 2.19 : K* (1/α = 0.8)

Q: MODE:	0.05	0.10	0.20	0.30	0.40	0.50	0.60	0.70	0.80	0.90
(1)	.0375	.1509	.6169	1.443	2.721	4.634	7.555	12.38	21.78	49.23
(2)	.0687	.3011	1.861	8.814	24.04	46.52	79.19	129.9	222.5	475.1
(3)	1.014	2.883	22.90	107.4	208.3	327.0	480.4	702.8	1093	2142
(4)	1.464	4.002	32.13	315.3	641.2	986.2	1404	1989	2994	5665
(5)	1.944	5.225	39.88	700.6	1466	2230	3122	4338	6403	11867
(6)	2.418	6.435	46.53	1310	2807	4242	5875	8067	11758	21498
(7)	2.893	7.644	53.01	2195	4789	7210	9911	13498	19502	35317
(8)	3.366	8.854	59.48	3407	7538	11317	15477	20952	30076	54038
(9)	3.840	10.06	65.97	4996	11180	16752	22819	30749	43922	78555
(10)	4.313	11.27	72.48	7014	15839	23699	32183	43211	61483	109491

TABLE 2.19 : K* ($1/\alpha = 0.9$)

Q: MODE:	0.05	0.10	0.20	0.30	0.40	0.50	0.60	0.70	0.80	0.90
(1)	.0384	.1543	.6313	1.477	2.788	4.752	7.759	12.73	22.45	50.89
(2)	.0755	.3298	1.992	8.931	24.05	46.53	79.26	130.1	223.0	476.6
(3)	1.133	3.209	24.04	107.6	208.3	327.0	480.5	703.2	1094	2144
(4)	1.643	4.486	34.94	315.9	641.2	986.2	1404	1990	2995	5668
(5)	2.184	5.867	44.07	701.6	1466	2230	3122	4339	6405	11870
(6)	2.719	7.231	51.01	1311	2807	4242	5875	8068	11760	21502
(7)	3.253	8.593	59.25	2197	4790	7210	9912	13499	19504	35322
(8)	3.786	9.955	66.61	3410	7539	11317	15477	20952	30078	54089
(9)	4.319	11.32	73.97	5000	11180	16752	22819	30750	43924	78561
(10)	4.851	12.68	81.35	7017	15840	23699	32183	43212	61485	109498

TABLE 2.20 : K_1^* ($\alpha = 0.1$)

Q: MODE:	0.05	0.10	0.20	0.30	0.40	0.50	0.60	0.70	0.80	0.90
(1)	.8350	3.356	13.70	31.93	59.94	101.4	163.8	265.2	459.6	1018
(2)	.9563	3.924	17.65	52.00	154.1	379.7	720.0	1232	2131	4503
(3)	32.04	64.94	150.4	304.1	689.0	1554	2781	4473	7287	14493
(4)	120.3	229.1	471.6	846.6	1683	3735	6775	10753	17094	32958
(5)	302.3	561.8	1097	1862	3460	7464	13614	21405	33468	63211
(6)	611.4	1121	2128	3495	6231	13140	24043	37566	58071	108112
(7)	1081	1967	3667	5894	10225	21178	38826	60402	92598	170584
(8)	1745	3160	5816	9209	15668	31995	58728	91074	138742	253534
(9)	2636	4757	8676	13588	22785	46006	84512	130743	198189	359896
(10)	3790	6819	12351	19181	31803	63627	116945	180569	272646	492552

TABLE 2.20 : K_1^* ($\alpha = 0.2$)

Q: MODE:	0.05	0.10	0.20	0.30	0.40	0.50	0.60	0.70	0.80	0.90
(1)	.4302	1.730	7.067	16.50	31.05	52.69	85.54	139.4	243.7	546.3
(2)	.9269	3.787	16.65	45.15	106.2	217.1	387.8	651.7	1125	2397
(3)	30.10	61.52	141.2	269.5	507.1	917.6	1511	2370	3842	7704
(4)	112.9	216.9	444.7	767.2	1325	2304	3726	5715	9013	17498
(5)	283.4	531.5	1036	1701	2801	4733	7548	11398	17647	33540
(6)	572.7	1060	2012	3206	5127	8489	13407	20032	30623	57343
(7)	1012	1860	3468	5422	8498	13867	21744	32243	48835	90455
(8)	1633	2987	5502	8487	13111	21161	32999	48656	73176	134415
(9)	2468	4497	8209	12538	19162	30666	47613	69895	104536	190775
(10)	3547	6445	11689	17716	26848	42674	66027	96583	143817	261064

TABLE 2.20 : K_1^* ($\alpha = 0.3$)

Q: MODE:	0.05	0.10	0.20	0.30	0.40	0.50	0.60	0.70	0.80	0.90
(1)	.2967	1.193	4.878	11.41	21.51	36.63	59.72	97.85	172.3	389.9
(2)	.8952	3.644	15.69	40.32	85.83	161.5	277.5	460.0	792.7	1701
(3)	28.29	58.20	132.6	243.5	420.6	693.9	1088	1675	2705	5459
(4)	106.0	205.1	419.2	702.7	1129	1779	2706	4049	6345	12388
(5)	265.8	502.5	978.5	1566	2417	3699	5510	8088	12424	23734
(6)	537.1	1002	1900	2962	4456	6686	9823	14230	21560	40564
(7)	949.1	1758	3277	5017	7422	10983	15973	22924	34384	60974
(8)	1531	2822	5199	7863	11489	16829	24290	34615	51526	95049
(9)	2313	4248	7760	11628	16833	24464	35103	49749	73612	134888
(10)	3324	6087	11050	16440	23629	34128	48740	68773	101277	184568

TABLE 2.20 : K_1^* ($\alpha = 0.4$)

Q: MODE:	0.05	0.10	0.20	0.30	0.40	0.50	0.60	0.70	0.80	0.90
(1)	.2309	.9289	3.801	8.903	16.83	28.73	47.01	77.38	137.0	312.3
(2)	.8615	3.494	14.80	36.66	73.95	132.9	222.7	365.3	628.7	1356
(3)	26.61	55.04	124.6	222.9	367.2	576.8	877.5	1332	2144	4349
(4)	99.61	193.9	395.3	649.1	1000	1496	2194	3225	5028	9862
(5)	249.8	474.8	923.6	1452	2155	3131	4483	6450	9847	18887
(6)	504.4	947.1	1794	2751	3989	5683	8010	11358	17089	32272
(7)	891.2	1660	3095	4667	6661	9363	13047	18308	27255	50888
(8)	1437	2665	4912	7320	10331	14377	19866	27658	40845	75598
(9)	2171	4011	7333	10832	15158	20933	28737	39765	58355	107274
(10)	3120	5748	10443	15322	21300	29241	39934	54900	80289	146773

TABLE 2.20 : K_1^* ($\alpha = 0.5$)

Q: MODE:	0.05	0.10	0.20	0.30	0.40	0.50	0.60	0.70	0.80	0.90
(1)	.1924	.7741	3.170	7.434	14.07	24.09	39.54	65.32	116.2	266.1
(2)	.8264	3.342	13.96	33.75	65.94	115.3	190.1	309.4	531.8	1152
(3)	25.07	52.05	117.3	206.1	330.0	503.6	751.2	1129	1813	3691
(4)	93.77	183.3	373.0	603.7	906.3	1316	1885	2737	4252	8366
(5)	235.0	448.8	872.1	1354	1960	2763	3861	5479	8326	16018
(6)	474.6	894.9	1693	2368	3638	5029	6909	9654	14451	27366
(7)	838.3	1569	2924	4361	6084	8298	11267	15568	23048	43146
(8)	1352	2518	4642	6844	9446	12758	17170	23527	34541	64091
(9)	2042	3789	6930	10132	13871	18594	24853	33836	49350	90940
(10)	2934	5430	9870	14337	19505	25992	34554	46799	67901	124418

TABLE 2.20 : K_1^* ($\alpha = 0.6$)

Q: MODE:	0.05	0.10	0.20	0.30	0.40	0.50	0.60	0.70	0.80	0.90
(1)	.1675	.6741	2.762	6.484	12.29	21.08	34.68	57.44	102.5	235.6
(2)	.7904	3.188	13.19	31.36	60.09	103.3	168.5	272.7	468.4	1018
(3)	23.65	49.24	110.7	192.1	302.1	453.0	667.1	996.0	1596	3259
(4)	88.43	173.4	352.4	564.7	833.8	1188	1679	2416	3744	7384
(5)	221.6	424.4	824.2	1268	1808	2502	3443	4840	7331	14135
(6)	447.3	846.2	1602	2408	3360	4560	6168	8532	12724	24145
(7)	790.1	1483	2765	4091	5625	7533	10065	13764	20295	38065
(8)	1274	2380	4390	6424	8740	11591	15345	20805	30416	56540
(9)	1924	3582	6554	9513	12841	16903	22222	29926	43458	80221
(10)	2765	5133	9335	13465	18063	23638	30905	41398	59795	109749

TABLE 2.20 : K_1^* ($\alpha = 0.7$)

Q: MODE:	0.05	0.10	0.20	0.30	0.40	0.50	0.60	0.70	0.80	0.90
(1)	.1504	.6052	2.481	5.829	11.06	18.99	31.30	51.94	92.89	214.0
(2)	.7542	3.037	12.48	29.35	55.57	94.57	153.1	247.0	424.0	923.7
(3)	22.36	46.62	104.6	180.1	280.2	415.7	607.2	902.4	1445	2955
(4)	83.55	164.2	333.3	530.7	775.7	1093	1531	2191	3388	6694
(5)	209.3	401.7	779.9	1193	1684	2305	3142	4390	6635	12811
(6)	422.5	801.0	1516	2267	3133	4205	5632	7741	11516	21882
(7)	746.3	1404	2617	3852	5248	6951	9195	12490	18368	34496
(8)	1203	2253	4155	6051	8158	10700	14024	18883	27529	51235
(9)	1817	3390	6204	8962	11989	15609	20313	27165	39333	72693
(10)	2611	4858	8836	12687	16870	21835	28257	37583	54121	99447

TABLE 2.20 : K_1^* ($\alpha = 0.8$)

Q: MODE:	0.05	0.10	0.20	0.30	0.40	0.50	0.60	0.70	0.80	0.90
(1)	.1381	.5558	2.279	5.356	10.17	17.48	28.83	47.91	85.80	198.0
(2)	.7185	2.890	11.82	27.62	51.95	87.88	141.7	228.1	391.4	853.7
(3)	21.17	44.19	99.00	169.7	262.3	386.8	562.4	833.5	1333	2730
(4)	79.10	155.6	315.7	500.7	727.6	1019	1419	2024	3127	6184
(5)	198.1	380.7	739.0	1126	1581	2150	2915	4057	6123	11835
(6)	399.9	759.1	1437	2141	2943	3925	5226	7156	10628	20213
(7)	706.3	1330	2480	3639	4931	6490	8534	11547	16953	31863
(8)	1139	2135	3938	5717	7667	9992	13019	17459	25408	47324
(9)	1720	3212	5879	8469	11270	14580	18861	25118	36304	67142
(10)	2471	4603	8375	11990	15860	20398	26241	34753	49952	91852

TABLE 2.20 : K_1^* ($\alpha = 0.9$)

Q: MODE:	0.05	0.10	0.20	0.30	0.40	0.50	0.60	0.70	0.80	0.90
(1)	.1290	.5190	2.129	5.005	9.508	16.35	26.98	44.85	80.38	185.6
(2)	.6837	2.748	11.21	26.12	48.96	82.58	132.9	213.6	366.6	800.0
(3)	20.08	41.95	93.90	160.6	247.4	363.8	527.6	780.8	1248	2558
(4)	75.02	147.7	299.6	474.2	686.9	959.5	1332	1897	2928	5794
(5)	187.9	361.3	701.2	1067	1494	2024	2736	3802	5734	11087
(6)	379.3	720.4	1364	2028	2780	3696	4908	6706	9953	18937
(7)	669.8	1262	2354	3448	4660	6113	8015	10822	15875	29850
(8)	1080	2026	3737	5417	7246	9413	12229	16364	23793	44334
(9)	1631	3049	5580	8025	10652	13736	17717	23544	33997	62899
(10)	2344	4368	7948	11363	14991	19219	24650	32575	46778	86047

TABLE 2.20 : K_1^* ($\alpha = 1.0$)

Q: MODE:	0.05	0.10	0.20	0.30	0.40	0.50	0.60	0.70	0.80	0.90
(1)	.1220	.4910	2.014	4.735	8.997	15.47	25.54	42.46	76.11	175.8
(2)	.6504	2.614	10.66	24.81	46.44	78.27	125.9	202.3	347.1	757.6
(3)	19.08	39.87	89.24	152.5	234.8	344.8	499.8	739.4	1182	2422
(4)	71.30	140.4	284.7	450.4	651.9	909.8	1262	1796	2772	5486
(5)	178.6	343.5	666.5	1013	1418	1920	2593	3601	5429	10499
(6)	360.4	684.7	1296	1927	2639	3505	4650	6351	9424	17932
(7)	636.6	1200	2237	3276	4423	5797	7595	10249	15031	28267
(8)	1026	1926	3552	5147	6879	8928	11588	15498	22528	41982
(9)	1550	2898	5304	7625	10112	13028	16790	22298	32189	59562
(10)	2227	4152	7555	10796	14232	18230	23361	30852	44291	81481

TABLE 2.21 : T^* ($\alpha = 0.1$)

Q: MODE:	0.05	0.10	0.20	0.30	0.40	0.50	0.60	0.70	0.80	0.90
(1)	.1870	.8440	5.987	28.15	71.95	137.8	235.2	388.6	671.7	1454
(2)	2.391	7.141	82.74	384.7	734.4	1157	1714	2532	3977	7872
(3)	3.413	9.485	94.31	1243	2512	3874	5545	7894	11946	22729
(4)	4.597	12.46	106.4	2925	6097	9283	13030	18164	26899	50020
(5)	5.775	15.44	119.1	5664	12101	18301	25387	34932	51027	93497
(6)	6.949	18.44	133.2	9717	21154	31858	43851	59807	86543	156982
(7)	8.134	21.43	148.1	15339	33884	50886	69652	94392	135655	244244
(8)	9.288	24.45	163.6	22782	50920	76315	104028	140295	200582	359075
(9)	10.48	27.46	179.4	32301	72890	109074	148204	199124	283527	505296
(10)	11.66	30.47	195.3	44151	100415	150093	203413	272481	386689	686617

TABLE 2.21 : T^* ($\alpha = 0.2$)

Q: MODE:	0.05	0.10	0.20	0.30	0.40	0.50	0.60	0.70	0.80	0.90
(1)	.1702	.7482	4.415	15.78	37.35	70.22	119.1	196.3	339.2	734.9
(2)	2.342	6.856	53.32	196.9	370.0	580.9	859.6	1269	1993	3947
(3)	3.394	9.381	78.47	632.5	1261	1941	2776	3951	5980	11379
(4)	4.586	12.40	98.08	1480	3056	4647	6520	9088	13458	25029
(5)	5.766	15.40	114.2	2857	6060	9157	12699	17473	25524	46771
(6)	6.947	18.40	129.9	4891	10589	15937	21933	29912	43284	78518
(7)	8.129	21.42	145.7	7710	16957	25453	34835	47206	67841	122153
(8)	9.304	24.43	161.7	11440	25478	38169	52024	70159	100307	179572
(9)	10.48	27.42	177.9	16208	36465	54550	74113	99575	141782	252687
(10)	11.65	30.45	194.1	22142	50230	75061	101719	136255	193364	343351

TABLE 2.21 : T^* ($\alpha = 0.3$)

Q: MODE:	0.05	0.10	0.20	0.30	0.40	0.50	0.60	0.70	0.80	0.90
(1)	.1563	.6743	3.609	11.40	25.71	47.64	80.38	132.2	228.3	495.2
(2)	2.296	6.599	41.41	134.1	248.5	388.8	574.7	848.2	1332	2639
(3)	3.374	9.281	68.83	428.5	844.3	1296	1853	2637	3991	7596
(4)	4.572	12.34	91.59	998.5	2042	3101	4350	6062	8977	16698
(5)	5.757	15.35	109.9	1921	4047	6109	8470	11653	17023	31196
(6)	6.939	18.37	126.9	3283	7068	10630	14627	19947	28864	52363
(7)	8.123	21.39	143.4	5167	11315	16975	23229	31477	45237	81456
(8)	9.275	24.41	159.9	7659	16997	25454	34689	46780	66882	119738
(9)	10.48	27.43	176.4	10844	24323	36375	49416	66391	94533	168484
(10)	11.66	30.45	192.8	14805	33502	50050	67821	90846	128923	228930

TABLE 2.21 : T^* ($\alpha = 0.4$)

Q: MODE:	0.05	0.10	0.20	0.30	0.40	0.50	0.60	0.70	0.80	0.90
(1)	.1445	.6152	3.096	9.099	19.84	36.32	60.99	100.1	172.9	375.4
(2)	2.251	6.366	34.62	102.6	187.6	292.8	432.3	637.7	1001	1985
(3)	3.354	9.183	62.09	326.2	635.7	974.4	1391	1980	2996	5705
(4)	4.559	12.28	86.32	757.1	1535	2329	3265	4549	6737	12533
(5)	5.747	15.31	106.1	1453	3040	4585	6356	8743	12772	23408
(6)	6.931	18.33	124.1	2478	5307	7977	10974	14964	21654	39286
(7)	8.116	21.36	141.3	3895	8493	12736	17426	23613	33935	61107
(8)	9.299	24.39	158.2	5769	12756	19096	26022	35090	50170	89821
(9)	10.48	27.41	175.0	8161	18252	27288	37068	49800	70909	126382
(10)	11.66	30.43	191.6	11136	25137	37545	50872	68141	96702	171719

TABLE 2.21 : T^* ($\alpha = 0.5$)

Q: MODE:	0.05	0.10	0.20	0.30	0.40	0.50	0.60	0.70	0.80	0.90
(1)	.1344	.9664	2.732	7.655	16.28	29.51	49.35	80.88	139.6	303.4
(2)	2.208	6.152	30.13	83.52	151.1	235.1	346.8	511.4	803.5	1592
(3)	3.335	9.087	57.00	264.7	510.5	781.0	1115	1586	2400	4570
(4)	4.545	12.21	81.92	612.1	1231	1865	2614	3642	5393	10034
(5)	5.738	15.26	102.7	1171	2436	3671	5087	6997	10222	18736
(6)	6.923	18.30	121.5	1995	4250	6385	8782	11975	17328	31439
(7)	8.110	21.33	139.2	3132	6801	10193	13944	18894	27153	48898
(8)	9.295	24.36	156.5	4634	10212	15281	20822	28077	40142	71871
(9)	10.46	27.39	173.6	6551	14610	21836	29659	39845	56734	101121
(10)	11.63	30.41	190.4	8935	20119	30042	40703	54518	77369	137392

TABLE 2.21 : T^* ($\alpha = 0.6$)

Q: MODE:	0.05	0.10	0.20	0.30	0.40	0.50	0.60	0.70	0.80	0.90
(1)	.1257	.5254	2.457	6.654	13.88	24.95	41.57	68.04	117.4	255.5
(2)	2.167	5.956	26.88	70.77	126.8	196.7	289.8	427.2	671.2	1330
(3)	3.316	8.994	52.97	223.6	427.1	652.1	930.4	1323	2002	3813
(4)	4.533	12.15	78.16	515.3	1028	1556	2180	3037	4497	8368
(5)	5.729	15.22	99.66	984.3	2033	3061	4241	5834	8522	15621
(6)	6.917	18.27	119.0	1672	3546	5323	7321	9982	14444	26208
(7)	8.105	21.30	137.3	2623	5672	8497	11623	15748	22633	40759
(8)	9.289	24.34	154.9	3877	8516	12738	17355	23401	33457	59904
(9)	10.46	27.37	172.2	5478	12182	18201	24720	33208	47284	84281
(10)	11.64	30.39	189.3	7467	16773	25040	33923	45437	64481	114508

TABLE 2.21 : T^* ($\alpha = 0.7$)

Q: MODE:	0.05	0.10	0.20	0.30	0.40	0.50	0.60	0.70	0.80	0.90
(1)	.1180	.4904	2.240	5.914	12.15	21.68	36.01	58.86	101.6	221.2
(2)	2.128	5.774	24.40	61.61	109.4	169.2	249.1	367.1	576.7	1144
(3)	3.297	8.903	49.67	194.1	367.4	560.0	798.6	1135	1718	3273
(4)	4.522	12.09	74.90	446.0	883.8	1335	1870	2604	3857	7178
(5)	5.719	15.18	96.89	850.1	1745	2626	3637	5002	7307	13396
(6)	6.909	18.23	116.8	1442	3043	4565	6277	8558	12384	22472
(7)	8.098	21.28	135.4	2259	4866	7286	9965	13501	19403	34945
(8)	9.285	24.31	153.3	3337	7304	10922	14878	20061	28682	51356
(9)	10.47	27.35	170.9	4711	10447	15604	21191	28468	40535	72252
(10)	11.65	30.38	188.1	6418	14383	21467	29081	38950	55275	98162

TABLE 2.21 : T^* (α = 0.8)

Q: MODE:	0.05	0.10	0.20	0.30	0.40	0.50	0.60	0.70	0.80	0.90
(1)	.1113	.4600	2.063	5.340	10.84	19.22	31.83	51.97	89.70	195.5
(2)	2.090	5.605	22.43	54.69	96.28	148.6	218.5	321.9	505.8	1003
(3)	3.278	8.815	46.90	171.9	322.7	490.9	699.7	994.7	1505	2867
(4)	4.509	12.03	72.02	393.9	775.1	1169	1637	2280	3377	6285
(5)	5.710	15.13	94.35	749.4	1530	2299	3184	4379	6396	11727
(6)	6.901	18.20	114.6	1269	2666	3997	5494	7490	10839	19670
(7)	8.092	21.25	133.6	1986	4261	6378	8722	11816	16981	30584
(8)	9.280	24.29	151.8	2931	6395	9559	13021	17556	25101	44945
(9)	10.46	27.33	169.6	4136	9146	13657	18545	24912	35472	63230
(10)	11.63	30.36	187.0	5632	12591	18787	25449	34084	48371	85902

TABLE 2.21 : T^* (α = 0.9)

Q: MODE:	0.05	0.10	0.20	0.30	0.40	0.50	0.60	0.70	0.80	0.90
(1)	.1053	.4334	1.916	4.881	9.804	17.29	28.57	46.61	80.45	175.5
(2)	2.054	5.448	20.82	49.28	86.09	132.6	194.8	286.8	450.7	894.9
(3)	3.259	8.729	44.53	154.6	287.9	437.2	622.7	885.1	1339	2552
(4)	4.496	11.98	69.45	353.4	690.6	1040	1456	2028	3004	5591
(5)	5.700	15.09	92.01	670.9	1362	2045	2831	3894	5688	10429
(6)	6.894	18.16	112.6	1135	2372	3554	4885	6660	9637	17490
(7)	8.086	21.22	131.9	1774	3791	5671	7755	10505	15098	27193
(8)	9.273	24.27	150.4	2616	5689	8500	11576	15608	22315	39959
(9)	10.46	27.30	168.4	3688	8134	12142	16487	22147	31535	56213
(10)	11.64	30.34	185.9	5020	11197	16703	22624	30300	43001	76367

TABLE 2.21 : T^* (α = 1.0)

Q: MODE:	0.05	0.10	0.20	0.30	0.40	0.50	0.60	0.70	0.80	0.90
(1)	.0999	.4098	1.790	4.504	8.969	15.75	25.96	42.31	73.04	159.5
(2)	2.019	5.301	19.46	44.93	77.93	119.7	175.7	258.8	406.6	807.7
(3)	3.241	8.644	42.46	140.7	260.0	394.2	561.2	797.5	1206	2300
(4)	4.484	11.92	67.14	320.8	622.9	937.9	1311	1826	2705	5036
(5)	5.691	15.05	89.85	608.0	1227	1842	2550	3506	5121	9391
(6)	6.886	18.13	110.7	1027	2137	3200	4398	5996	8676	15746
(7)	8.080	21.19	130.3	1604	3415	5106	6981	9457	13591	24480
(8)	9.268	24.24	149.0	2363	5123	7652	10421	14049	20087	35970
(9)	10.46	27.28	167.1	3330	7325	10931	14841	19935	28385	50599
(10)	11.64	30.32	184.8	4531	10082	15035	20364	27273	38704	68739

TABLE 2.21 : T^* (1/α = 0.1)

Q: MODE:	0.05	0.10	0.20	0.30	0.40	0.50	0.60	0.70	0.80	0.90
(1)	.0180	.0727	.2988	.7058	1.351	2.349	3.939	6.696	12.35	29.54
(2)	.8156	1.726	3.870	6.550	9.965	14.47	20.77	30.52	48.74	100.7
(3)	2.167	4.852	11.86	20.76	31.42	44.50	61.68	86.94	132.2	256.9
(4)	3.598	8.483	23.77	45.85	71.74	101.7	139.2	192.5	285.5	537.6
(5)	4.968	12.10	38.50	83.72	136.7	195.1	265.5	363.0	530.6	980.3
(6)	6.285	15.60	55.20	136.5	232.3	333.9	452.8	614.5	889.6	1623
(7)	7.567	19.01	73.22	206.3	364.9	527.4	713.5	963.2	1384	2503
(8)	8.824	22.33	92.04	295.3	540.5	784.9	1059	1425	2037	3659
(9)	10.06	25.59	111.4	405.9	765.4	1115	1504	2016	2870	5130
(10)	11.29	28.81	130.9	540.5	1045	1529	2059	2752	3906	6951

TABLE 2.21 : T^* (1/α = 0.2)

Q: MODE:	0.05	0.10	0.20	0.30	0.40	0.50	0.60	0.70	0.80	0.90
(1)	.0331	.1330	.5437	1.271	2.397	4.081	6.659	10.93	19.34	44.14
(2)	1.213	2.700	6.544	11.61	18.11	26.63	38.38	56.22	88.79	179.4
(3)	2.651	6.344	18.30	36.13	57.85	83.96	117.6	166.2	251.8	484.1
(4)	4.039	10.05	34.44	79.89	134.3	195.3	269.9	374.4	554.6	1037
(5)	5.345	13.55	53.02	146.9	259.4	378.7	519.7	712.5	1040	1914
(6)	6.603	16.90	72.84	241.8	445.5	650.0	891.6	1212	1754	3192
(7)	7.844	20.15	93.16	369.1	705.2	1036	1410	1907	2740	4945
(8)	9.065	23.35	113.5	533.6	1051	1548	2100	2827	4043	7250
(9)	10.28	26.50	133.7	740.2	1495	2206	2986	4007	5705	10182
(10)	11.48	29.63	153.6	993.8	2051	3030	4093	5477	7772	13816

TABLE 2.21 : T^* (1/α = 0.3)

Q: MODE:	0.05	0.10	0.20	0.30	0.40	0.50	0.60	0.70	0.80	0.90
(1)	.0458	.1843	.7548	1.768	3.334	5.666	9.201	14.98	26.16	58.64
(2)	1.451	3.358	8.739	16.20	25.87	38.47	55.70	81.66	128.6	238.0
(3)	2.867	7.109	23.10	50.24	83.56	123.0	173.2	245.2	371.3	711.2
(4)	4.213	10.74	41.81	111.7	195.9	288.3	400.3	556.0	823.5	1537
(5)	5.484	14.13	62.33	206.9	380.9	561.9	773.6	1061	1551	2849
(6)	6.720	17.39	83.38	342.8	657.5	971.7	1330	1810	2620	4761
(7)	7.940	20.57	104.3	526.7	1044	1545	2106	2850	4097	7387
(8)	9.149	23.71	124.8	765.9	1560	2311	3140	4230	6048	10840
(9)	10.35	26.82	144.9	1067	2224	3297	4468	5998	8540	15234
(10)	11.55	29.91	164.4	1439	3055	4531	6127	8201	11639	20682

TABLE 2.21 : T^* (1/α = 0.4)

Q: MODE:	0.05	0.10	0.20	0.30	0.40	0.50	0.60	0.70	0.80	0.90
(1)	.0567	.2287	.9420	2.221	4.212	7.180	11.67	18.95	32.91	73.09
(2)	1.612	3.842	10.66	20.57	33.46	50.18	72.92	107.0	168.4	336.6
(3)	2.990	7.579	27.01	63.77	109.0	161.8	228.7	324.2	490.7	938.2
(4)	4.306	11.13	47.49	142.5	257.1	381.2	530.6	737.6	1092	2037
(5)	5.556	14.44	69.10	265.4	502.2	744.9	1027	1410	2061	3784
(6)	6.777	17.64	90.64	442.1	869.2	1290	1768	2408	3485	6331
(7)	7.989	20.79	111.6	682.3	1383	2054	2803	3794	5453	9829
(8)	9.191	23.90	132.0	995.8	2069	3074	4180	5633	8054	14430
(9)	10.39	26.98	151.7	1392	2953	4388	5950	7989	11375	20286
(10)	11.58	30.06	170.9	1883	4059	6031	8161	10926	15505	27547

TABLE 2.21 : T^* (1/α = 0.5)

Q: MODE:	0.05	0.10	0.20	0.30	0.40	0.50	0.60	0.70	0.80	0.90
(1)	.0663	.2677	1.111	2.644	5.052	8.654	14.09	22.88	39.63	87.52
(2)	1.727	4.217	12.39	24.80	40.97	61.82	90.10	132.3	208.1	415.1
(3)	3.069	7.898	30.35	76.96	134.3	200.6	284.1	403.1	610.0	1165
(4)	4.363	11.38	52.10	172.8	318.2	474.0	660.9	919.2	1361	2537
(5)	5.601	14.64	74.35	323.2	623.2	927.9	1281	1760	2571	4718
(6)	6.815	17.80	96.04	540.5	1080	1608	2206	3006	4350	7900
(7)	8.020	20.92	116.9	836.9	1722	2563	3499	4738	6809	12271
(8)	9.217	24.01	136.9	1224	2578	3837	5220	7036	10059	18020
(9)	10.41	27.08	156.3	1716	3682	5478	7432	9980	14210	25339
(10)	11.60	30.15	175.1	2325	5063	7532	10194	13650	19372	34412

TABLE 2.21 : T^* (1/α = 0.6)

Q: MODE:	0.05	0.10	0.20	0.30	0.40	0.50	0.60	0.70	0.80	0.90
(1)	.0746	.3023	1.267	3.045	5.866	10.10	16.49	26.79	46.33	101.9
(2)	1.814	4.518	13.98	28.93	48.41	73.43	107.3	157.6	247.8	493.6
(3)	3.124	8.129	33.29	89.94	159.5	239.4	339.6	482.0	729.4	1392
(4)	4.403	11.55	55.98	202.7	379.2	566.9	791.1	1100	1630	3037
(5)	5.631	14.77	78.58	380.6	744.2	1110	1534	2109	3081	5653
(6)	6.837	17.91	100.2	638.4	1292	1927	2645	3604	5215	9469
(7)	8.039	21.01	120.9	990.9	2060	3071	4195	5682	8166	14713
(8)	9.234	24.09	140.6	1453	3087	4600	6260	8438	12065	21610
(9)	10.42	27.15	159.7	2040	4410	6569	8913	11971	17045	30391
(10)	11.61	30.20	178.2	2767	6067	9033	12228	16375	23238	41278

TABLE 2.21 : T^* (1/α = 0.7)

Q: MODE:	0.05	0.10	0.20	0.30	0.40	0.50	0.60	0.70	0.80	0.90
(1)	.0820	.3333	1.411	3.429	6.661	11.53	18.87	30.68	53.02	116.3
(2)	1.882	4.765	15.46	32.99	55.82	85.02	124.4	182.9	287.5	572.1
(3)	3.165	8.304	35.91	102.8	184.7	278.1	395.0	560.9	848.7	1619
(4)	4.431	11.68	59.32	232.5	440.2	659.6	921.3	1282	1898	3536
(5)	5.651	14.87	82.09	437.7	865.2	1293	1788	2458	3591	6587
(6)	6.855	17.99	103.6	735.9	1503	2245	3083	4202	6081	11039
(7)	8.054	21.07	124.0	1144	2399	3580	4892	6625	9522	17154
(8)	9.246	24.14	143.4	1680	3596	5363	7300	9841	14070	25200
(9)	10.44	27.20	162.2	2363	5139	7659	10395	13962	19880	35443
(10)	11.62	30.25	180.5	3208	7070	10533	14262	19099	27105	48143

TABLE 2.21 : T^* (1/α = 0.8)

Q: MODE:	0.05	0.10	0.20	0.30	0.40	0.50	0.60	0.70	0.80	0.90
(1)	.0886	.3612	1.545	3.798	7.441	12.94	21.24	34.56	59.70	130.7
(2)	1.936	4.972	16.86	37.01	63.21	96.60	141.5	208.2	327.2	650.7
(3)	3.196	8.442	38.28	115.5	209.8	316.8	450.4	639.7	968.1	1846
(4)	4.453	11.78	62.24	262.0	501.1	752.4	1051	1463	2167	4036
(5)	5.667	14.94	85.07	494.6	986.1	1476	2042	2807	4101	7522
(6)	6.869	18.05	106.4	833.2	1715	2564	3522	4800	6946	12608
(7)	8.064	21.12	126.5	1297	2738	4089	5588	7569	10878	19596
(8)	9.256	24.18	145.6	1908	4105	6126	8341	11244	16076	28790
(9)	10.44	27.23	164.2	2685	5868	8750	11877	15953	22715	40495
(10)	11.63	30.28	182.2	3649	8074	12034	16296	21824	30971	55008

TABLE 2.21 : T^* (1/α = 0.9)

Q: MODE:	0.05	0.10	0.20	0.30	0.40	0.50	0.60	0.70	0.80	0.90
(1)	.0946	.3866	1.671	4.156	8.210	14.35	23.60	38.44	66.37	145.1
(2)	1.981	5.149	18.19	40.98	70.58	108.2	158.6	233.5	366.9	729.2
(3)	3.221	8.553	40.46	128.1	234.9	355.5	505.8	718.6	1087	2073
(4)	4.470	11.86	64.82	291.5	562.0	845.2	1181	1645	2436	4536
(5)	5.680	15.00	87.62	551.4	1107	1659	2296	3156	4611	8456
(6)	6.879	18.09	108.7	930.4	1926	2882	3960	5398	7811	14177
(7)	8.072	21.16	128.6	1451	3076	4597	6284	8513	12234	22038
(8)	9.263	24.22	147.5	2136	4614	6889	9381	12647	18081	32380
(9)	10.45	27.26	165.8	3008	6596	9840	13359	17944	25550	45547
(10)	11.64	30.30	183.7	4090	9078	13535	18330	24549	34838	61874

TABLE 2.22 : T_1^* (α = 0.1)

Q: MODE:	0.05	0.10	0.20	0.30	0.40	0.50	0.60	0.70	0.80	0.90
(1)	.6289	1.403	3.583	7.150	13.21	23.58	41.31	72.90	137.8	334.4
(2)	1.098	2.409	5.963	11.57	20.92	36.96	64.64	114.1	215.7	521.7
(3)	1.578	3.444	8.433	16.18	28.97	50.83	88.56	156.0	294.2	709.9
(4)	2.058	4.480	10.91	20.81	37.07	64.78	112.6	198.2	373.2	899.1
(5)	2.538	5.516	13.39	25.45	45.18	78.77	136.7	240.4	452.3	1088
(6)	3.018	6.554	15.87	30.09	53.31	92.78	160.9	282.7	531.6	1278
(7)	3.499	7.591	18.35	34.74	61.44	106.8	185.1	325.0	610.8	1468
(8)	3.980	8.627	20.83	39.39	69.58	120.8	209.2	367.3	690.1	1657
(9)	4.459	9.664	23.31	44.04	77.71	134.9	233.4	409.7	769.6	1847
(10)	4.940	10.70	25.80	48.69	85.86	148.9	257.6	452.0	848.9	2037

TABLE 2.22 : T_1^* (α = 0.2)

Q: MODE:	0.05	0.10	0.20	0.30	0.40	0.50	0.60	0.70	0.80	0.90
(1)	.5688	1.247	3.046	5.699	9.704	15.90	25.85	43.17	78.56	185.7
(2)	.9952	2.154	5.132	9.407	15.76	25.51	41.15	68.27	123.5	290.2
(3)	1.431	3.082	7.275	13.22	21.97	35.34	56.71	93.67	168.8	395.1
(4)	1.867	4.012	9.424	17.05	28.21	45.22	72.35	119.2	214.4	500.6
(5)	2.303	4.941	11.57	20.88	34.47	55.12	88.01	144.8	260.0	606.1
(6)	2.739	5.871	13.73	24.71	40.73	65.03	103.7	170.4	305.6	712.0
(7)	3.174	6.801	15.88	28.55	46.99	74.95	119.4	196.1	351.3	817.9
(8)	3.611	7.730	18.03	32.39	53.26	84.86	135.1	221.7	397.1	923.6
(9)	4.046	8.661	20.18	36.23	59.53	94.79	150.8	247.3	442.9	1029
(10)	4.482	9.591	22.34	40.06	65.80	104.7	166.5	273.0	488.6	1135

TABLE 2.22 : T_1^* (α = 0.3)

Q: MODE:	0.05	0.10	0.20	0.30	0.40	0.50	0.60	0.70	0.80	0.90
(1)	.5203	1.129	2.690	4.884	8.042	12.74	20.09	32.70	58.30	135.7
(2)	.9119	1.956	4.561	8.134	13.19	20.64	32.22	51.96	91.92	212.3
(3)	1.312	2.801	6.475	11.45	18.44	28.67	44.50	71.42	125.8	289.2
(4)	1.711	3.647	8.393	14.79	23.72	36.74	56.85	90.98	159.8	366.5
(5)	2.111	4.493	10.31	18.12	29.00	44.82	69.22	110.6	193.8	443.8
(6)	2.511	5.339	12.23	21.46	34.29	52.91	81.61	130.2	228.0	521.3
(7)	2.910	6.186	14.15	24.80	39.58	61.01	94.01	149.8	262.1	598.9
(8)	3.311	7.031	16.08	28.14	44.87	69.11	106.4	169.5	296.3	676.3
(9)	3.710	7.878	18.00	31.49	50.17	77.22	118.8	189.1	330.5	754.1
(10)	4.110	8.725	19.92	34.83	55.46	85.32	131.2	208.8	364.6	831.6

TABLE 2.22 : T_1^* (α = 0.4)

Q: MODE:	0.05	0.10	0.20	0.30	0.40	0.50	0.60	0.70	0.80	0.90
(1)	.4802	1.036	2.430	4.339	7.019	10.92	16.96	27.20	47.93	110.4
(2)	.8425	1.797	4.135	7.260	11.57	17.78	27.30	43.35	75.69	172.9
(3)	1.212	2.575	5.875	10.24	16.20	24.74	37.75	59.65	103.6	235.6
(4)	1.582	3.352	7.618	13.22	20.85	31.73	48.26	76.02	131.7	298.7
(5)	1.951	4.130	9.363	16.21	25.51	38.73	58.79	92.43	159.8	361.7
(6)	2.321	4.909	11.11	19.20	30.17	45.74	69.34	108.9	188.0	425.0
(7)	2.690	5.688	12.86	22.20	34.83	52.76	79.89	125.3	216.1	488.2
(8)	3.060	6.465	14.60	25.19	39.50	59.77	90.44	141.7	244.3	551.3
(9)	3.429	7.244	16.35	28.18	44.16	66.79	101.0	158.2	272.6	614.7
(10)	3.799	8.023	18.10	31.18	48.83	73.81	111.5	174.6	300.7	677.9

TABLE 2.22 : T_1^* (α = 0.5)

Q: MODE:	0.05	0.10	0.20	0.30	0.40	0.50	0.60	0.70	0.80	0.90
(1)	.4464	.9584	2.228	3.938	6.306	9.716	14.94	23.76	41.56	95.16
(2)	.7837	1.666	3.800	6.609	10.43	15.86	24.10	37.94	65.70	149.1
(3)	1.128	2.387	5.401	9.324	14.61	22.09	33.36	52.23	89.98	203.2
(4)	1.472	3.108	7.006	12.05	18.81	28.34	42.67	66.59	114.4	257.5
(5)	1.815	3.829	8.612	14.78	23.02	34.60	51.99	80.98	138.8	311.9
(6)	2.159	4.552	10.22	17.51	27.23	40.87	61.33	95.39	163.3	366.5
(7)	2.503	5.274	11.83	20.24	31.45	47.15	70.67	109.8	187.8	421.0
(8)	2.848	5.995	13.43	22.97	35.66	53.42	80.01	124.2	212.3	475.5
(9)	3.191	6.717	15.04	25.70	39.88	59.70	89.36	138.6	236.8	530.2
(10)	3.535	7.439	16.65	28.43	44.10	65.97	98.71	153.1	261.3	584.7

TABLE 2.22 : T_1^* (α = 0.6)

Q: MODE:	0.05	0.10	0.20	0.30	0.40	0.50	0.60	0.70	0.80	0.90
(1)	.4173	.8935	2.065	3.627	5.772	8.840	13.51	21.38	37.22	84.88
(2)	.7331	1.554	3.527	6.098	9.563	14.46	21.84	34.17	58.88	133.0
(3)	1.055	2.227	5.015	8.607	13.41	20.14	30.24	47.06	80.66	181.3
(4)	1.377	2.901	6.506	11.12	17.27	25.85	38.68	60.02	102.6	229.8
(5)	1.698	3.574	7.998	13.64	21.13	31.56	47.14	72.99	124.5	278.4
(6)	2.020	4.249	9.491	16.17	25.00	37.29	55.61	85.99	146.4	327.1
(7)	2.342	4.923	10.98	18.69	28.87	43.02	64.09	98.99	168.4	375.8
(8)	2.664	5.596	12.48	21.21	32.74	48.74	72.56	112.0	190.4	424.4
(9)	2.985	6.270	13.97	23.74	36.61	54.47	81.04	125.0	212.4	473.2
(10)	3.307	6.945	15.47	26.26	40.49	60.20	89.52	138.0	234.3	521.9

TABLE 2.22 : T_1^* ($\alpha = 0.7$)

Q: MODE:	0.05	0.10	0.20	0.30	0.40	0.50	0.60	0.70	0.80	0.90
(1)	.3921	.8380	1.929	3.376	5.352	8.167	12.44	19.62	34.06	77.47
(2)	.6890	1.458	3.298	5.682	8.878	13.37	20.12	31.38	53.91	121.4
(3)	.9915	2.090	4.691	8.022	12.45	18.63	27.87	43.23	73.85	165.5
(4)	1.294	2.722	6.086	10.37	16.04	23.92	35.66	55.13	93.91	209.8
(5)	1.596	3.355	7.482	12.72	19.63	29.21	43.46	67.06	114.0	254.2
(6)	1.899	3.988	8.880	15.07	23.22	34.51	51.28	79.01	134.1	298.6
(7)	2.201	4.620	10.28	17.43	26.82	39.81	59.09	90.95	154.2	343.1
(8)	2.504	5.252	11.67	19.78	30.42	45.11	66.91	102.9	174.3	387.5
(9)	2.806	5.885	13.07	22.13	34.02	50.42	74.74	114.9	194.5	432.1
(10)	3.108	6.518	14.47	24.49	37.62	55.72	82.56	126.8	214.6	476.5

TABLE 2.22 : T_1^* ($\alpha = 0.8$)

Q: MODE:	0.05	0.10	0.20	0.30	0.40	0.50	0.60	0.70	0.80	0.90
(1)	.3699	.7898	1.814	3.168	5.011	7.631	11.60	18.26	31.65	71.87
(2)	.6501	1.375	3.103	5.335	8.318	12.50	18.77	29.22	50.10	112.7
(3)	.9356	1.971	4.414	7.533	11.67	17.42	26.01	40.25	68.64	153.6
(4)	1.221	2.567	5.727	9.739	15.03	22.36	33.28	51.34	87.29	194.7
(5)	1.506	3.163	7.042	11.95	18.40	27.32	40.56	62.45	105.9	235.8
(6)	1.792	3.760	8.357	14.16	21.77	32.27	47.85	73.58	124.6	277.1
(7)	2.077	4.356	9.672	16.37	25.14	37.23	55.15	84.71	143.4	318.4
(8)	2.363	4.952	10.99	18.58	28.51	42.19	62.45	95.84	162.1	359.6
(9)	2.648	5.549	12.30	20.79	31.88	47.16	69.75	107.0	180.8	400.9
(10)	2.933	6.146	13.62	23.00	35.26	52.12	77.05	118.1	199.5	442.2

TABLE 2.22 : T_1^* ($\alpha = 0.9$)

Q: MODE:	0.05	0.10	0.20	0.30	0.40	0.50	0.60	0.70	0.80	0.90
(1)	.3503	.7474	1.715	2.992	4.727	7.190	10.92	17.18	29.73	67.48
(2)	.6156	1.301	2.935	5.039	7.848	11.78	17.67	27.48	47.08	105.8
(3)	.8860	1.865	4.174	7.116	11.01	16.42	24.49	37.86	64.51	144.2
(4)	1.156	2.430	5.416	9.200	14.18	21.08	31.34	48.30	82.03	182.8
(5)	1.427	2.994	6.659	11.29	17.36	25.75	38.19	58.75	99.57	221.4
(6)	1.697	3.559	7.903	13.37	20.54	30.42	45.06	69.22	117.1	260.2
(7)	1.967	4.124	9.147	15.46	23.72	35.10	51.94	79.69	134.7	298.9
(8)	2.238	4.688	10.39	17.55	26.91	39.78	58.81	90.16	152.3	337.6
(9)	2.507	5.253	11.64	19.64	30.09	44.46	65.69	100.6	169.9	376.5
(10)	2.778	5.818	12.88	21.73	33.28	49.13	72.56	111.1	187.5	415.2

TABLE 2.22 : T_1^* ($\alpha = 1.0$)

Q: MODE:	0.05	0.10	0.20	0.30	0.40	0.50	0.60	0.70	0.80	0.90
(1)	.3327	.7098	1.628	2.839	4.485	6.820	10.36	16.28	28.18	63.94
(2)	.5848	1.236	2.786	4.783	7.447	11.17	16.76	26.06	44.62	100.2
(3)	.8416	1.772	3.963	6.755	10.45	15.50	23.22	35.90	61.14	136.6
(4)	1.098	2.308	5.143	8.733	13.46	20.00	29.72	45.79	77.75	173.2
(5)	1.355	2.844	6.323	10.71	16.47	24.43	36.22	55.70	94.38	209.8
(6)	1.612	3.380	7.504	12.70	19.49	28.86	42.74	65.63	111.0	246.6
(7)	1.868	3.917	8.685	14.68	22.51	33.30	49.25	75.55	127.7	283.3
(8)	2.126	4.452	9.867	16.66	25.53	37.74	55.77	85.48	144.4	319.9
(9)	2.382	4.989	11.05	18.64	28.55	42.18	62.29	95.41	161.1	356.7
(10)	2.639	5.526	12.23	20.63	31.58	46.61	68.81	105.3	177.7	393.4

TABLE 2.23 : K* ($\alpha = 0.1$)

Q: MODE:	0.05	0.10	0.20	0.30	0.40	0.50	0.60	0.70	0.80	0.90
(1)	2.057	5.301	16.34	34.90	63.11	104.7	167.4	269.0	463.7	1022
(2)	3.878	10.01	33.13	94.62	254.3	512.1	874.0	1416	2376	4944
(3)	31.52	63.82	147.8	298.1	672.4	1528	2753	4442	7248	14424
(4)	120.4	229.2	471.9	847.3	1685	3738	6778	10757	17098	32966
(5)	302.3	561.7	1097	1862	3459	7464	13614	21404	33467	63211
(6)	611.4	1121	2128	3495	6231	13140	24043	37567	58071	108112
(7)	1081	1967	3667	5894	10225	21178	38826	60402	92598	170584
(8)	1745	3160	5816	9209	15668	31995	58728	91074	138742	253534
(9)	2636	4757	8676	13588	22785	46006	84512	130743	198189	359896
(10)	3790	6819	12351	19181	31803	63627	116945	180569	272646	492552

TABLE 2.23 : K* ($\alpha = 0.2$)

Q: MODE:	0.05	0.10	0.20	0.30	0.40	0.50	0.60	0.70	0.80	0.90
(1)	1.364	3.322	9.454	19.32	34.14	56.00	89.06	143.1	247.6	550.5
(2)	3.636	9.399	29.87	72.53	154.2	281.3	465.4	746.8	1254	2636
(3)	29.64	60.51	138.8	265.0	498.8	905.9	1497	2354	3822	7666
(4)	112.9	217.0	445.0	767.7	1326	2305	3728	5716	9015	17503
(5)	283.4	531.5	1036	1701	2801	4733	7548	11398	17647	33540
(6)	572.7	1060	2012	3206	5126	8489	13407	20032	30623	57343
(7)	1012	1860	3468	5422	8498	13867	21744	32243	48835	90455
(8)	1633	2987	5502	8487	13111	21161	32999	48656	73176	134415
(9)	2468	4497	8209	12538	19162	30666	47613	69895	104536	190775
(10)	3547	6445	11689	17716	26848	42674	66027	96583	143817	261064

TABLE 2.23 : K* ($\alpha = 0.3$)

Q: MODE:	0.05	0.10	0.20	0.30	0.40	0.50	0.60	0.70	0.80	0.90
(1)	1.086	2.583	7.074	14.09	24.52	39.87	63.16	101.5	176.1	393.9
(2)	3.411	8.834	27.28	61.15	118.1	204.2	330.2	525.8	883.8	1872
(3)	27.88	57.29	130.5	239.9	415.0	686.5	1079	1664	2690	5432
(4)	106.0	205.2	419.5	703.1	1129	1780	2707	4050	6346	12391
(5)	265.8	502.5	978.5	1566	2417	3698	5510	8088	12424	23734
(6)	537.1	1002	1900	2961	4456	6686	9823	14231	21560	40564
(7)	949.1	1758	3277	5017	7422	10983	15973	22924	34384	63974
(8)	1531	2822	5199	7863	11489	16829	24290	34615	51526	95049
(9)	2313	4248	7760	11628	16833	24464	35103	49749	73612	134888
(10)	3324	6087	11050	16440	23629	34128	48740	68773	101277	184568

TABLE 2.23 : K* ($\alpha = 0.4$)

Q: MODE:	0.05	0.10	0.20	0.30	0.40	0.50	0.60	0.70	0.80	0.90
(1)	.9286	2.181	5.844	11.46	19.72	31.87	50.35	80.87	140.7	316.1
(2)	3.205	8.310	25.15	53.84	98.88	165.4	263.3	416.8	700.9	1494
(3)	26.24	54.21	122.8	220.0	362.9	571.3	870.7	1323	2132	4327
(4)	99.65	194.0	395.5	649.5	1000	1497	2195	3226	5030	9864
(5)	249.7	474.8	923.6	1452	2155	3131	4483	6450	9847	18887
(6)	504.4	947.1	1794	2751	3989	5683	8010	11358	17089	32272
(7)	891.2	1660	3095	4667	6661	9363	13047	18308	27255	50888
(8)	1437	2665	4912	7320	10331	14377	19866	27658	40845	75598
(9)	2171	4011	7333	10832	15158	20933	28737	39765	58355	107274
(10)	3120	5748	10443	15322	21300	29241	39934	54988	80289	146773

TABLE 2.23 : K* ($\alpha = 0.5$)

Q: MODE:	0.05	0.10	0.20	0.30	0.40	0.50	0.60	0.70	0.80	0.90
(1)	.8243	1.922	5.083	9.862	16.86	27.12	42.75	68.67	119.6	269.7
(2)	3.016	7.828	23.36	48.62	86.71	142.1	223.7	352.4	593.0	1269
(3)	24.73	51.29	115.7	203.6	326.4	499.1	745.6	1122	1803	3673
(4)	93.81	183.4	373.2	604.0	906.7	1316	1886	2738	4253	8368
(5)	235.0	448.7	872.1	1354	1960	2763	3861	5479	8326	16018
(6)	474.6	894.9	1695	2568	3638	5029	6909	9654	14451	27366
(7)	838.3	1569	2924	4361	6084	8298	11267	15568	23048	43146
(8)	1352	2518	4642	6844	9446	12758	17170	23527	34541	64091
(9)	2042	3789	6930	10132	13871	18594	24853	33836	49350	90940
(10)	2934	5430	9870	14337	19505	25992	34554	46799	67901	124418

TABLE 2.23 : K* (α = 0.6)

Q: MODE:	0.05	0.10	0.20	0.30	0.40	0.50	0.60	0.70	0.80	0.90
(1)	.7491	1.739	4.562	8.792	14.96	23.99	37.76	60.65	105.8	239.0
(2)	2.844	7.385	21.83	44.64	78.21	126.5	197.6	310.4	522.3	1122
(3)	23.34	48.54	109.2	189.8	299.0	449.2	662.4	989.9	1588	3243
(4)	88.46	173.5	352.5	565.0	834.2	1189	1679	2417	3744	7386
(5)	221.6	424.4	824.2	1268	1808	2502	3443	4840	7331	14135
(6)	447.3	846.2	1602	2408	3360	4560	6168	8532	12724	24145
(7)	790.1	1483	2765	4091	5625	7533	10065	13764	20295	38065
(8)	1274	2380	4390	6424	8740	11591	15345	20805	30416	56540
(9)	1924	3582	6554	9513	12841	16903	22222	29926	43458	80221
(10)	2765	5133	9335	13465	18063	23638	30905	41398	59795	109749

TABLE 2.23 : K* (α = 0.7)

Q: MODE:	0.05	0.10	0.20	0.30	0.40	0.50	0.60	0.70	0.80	0.90
(1)	.6916	1.601	4.180	8.022	13.60	21.78	34.25	55.00	96.01	217.2
(2)	2.687	6.980	20.50	41.48	71.88	115.3	179.2	280.9	472.9	1018
(3)	22.06	45.97	103.2	178.0	277.4	412.2	603.0	897.0	1437	2940
(4)	83.58	164.2	333.5	530.9	776.0	1094	1531	2191	3389	6695
(5)	209.3	401.7	779.9	1193	1684	2305	3142	4390	6635	12811
(6)	422.5	801.0	1516	2267	3133	4205	5632	7741	11516	21882
(7)	746.3	1404	2617	3852	5248	6951	9195	12490	18368	34496
(8)	1203	2253	4155	6051	8158	10700	14024	18883	27529	51235
(9)	1817	3390	6204	8962	11989	15609	20313	27165	39333	72693
(10)	2611	4858	8836	12687	16870	21835	28257	37583	54121	99447

TABLE 2.23 : K* (α = 0.8)

Q: MODE:	0.05	0.10	0.20	0.30	0.40	0.50	0.60	0.70	0.80	0.90
(1)	.6459	1.493	3.886	7.441	12.59	20.14	31.65	50.82	88.77	201.0
(2)	2.543	6.608	19.34	38.88	66.96	106.9	165.6	259.2	436.5	941.1
(3)	20.89	43.58	97.71	167.8	259.8	383.7	558.5	828.5	1326	2716
(4)	79.12	155.7	315.9	501.0	727.8	1019	1419	2025	3128	6185
(5)	198.1	380.7	738.9	1126	1581	2150	2914	4057	6123	11835
(6)	399.9	759.1	1437	2141	2943	3925	5226	7156	10628	20213
(7)	706.3	1330	2480	3639	4931	6490	8534	11547	16953	31863
(8)	1139	2135	3938	5717	7667	9992	13019	17459	25408	47324
(9)	1720	3212	5879	8469	11270	14580	18861	25118	36304	67142
(10)	2471	4603	8375	11990	15860	20398	26241	34753	49952	91852

TABLE 2.23 : K* (α = 0.9)

Q: MODE:	0.05	0.10	0.20	0.30	0.40	0.50	0.60	0.70	0.80	0.90
(1)	.6085	1.405	3.653	6.985	11.81	18.87	29.65	47.62	83.20	188.5
(2)	2.412	6.269	18.31	36.69	62.99	100.3	155.2	242.8	408.8	881.9
(3)	19.82	41.37	92.68	158.8	245.1	360.8	524.0	776.1	1242	2545
(4)	75.05	147.8	299.7	474.4	687.2	959.8	1332	1897	2929	5795
(5)	187.9	361.3	701.2	1067	1493	2024	2736	3802	5734	11087
(6)	379.3	720.4	1364	2028	2780	3696	4908	6706	9953	18937
(7)	669.8	1262	2354	3448	4660	6113	8015	10822	15875	29850
(8)	1080	2026	3737	5417	7246	9413	12229	16364	23793	44334
(9)	1631	3049	5580	8025	10652	13736	17717	23544	33997	62899
(10)	2344	4368	7948	11363	14991	19219	24650	32575	46778	86047

TABLE 2.23 : K* (α = 1.0)

Q: MODE:	0.05	0.10	0.20	0.30	0.40	0.50	0.60	0.70	0.80	0.90
(1)	.5771	1.333	3.462	6.619	11.19	17.88	28.08	45.10	78.79	178.5
(2)	2.292	5.957	17.39	34.82	59.73	95.07	147.0	229.9	387.0	835.2
(3)	18.83	39.32	88.08	150.8	232.5	342.1	496.4	735.0	1176	2410
(4)	71.32	140.4	284.9	450.6	652.2	910.1	1262	1797	2773	5488
(5)	178.6	343.4	666.5	1013	1418	1920	2593	3600	5429	10499
(6)	360.4	684.7	1296	1927	2639	3505	4650	6351	9424	17932
(7)	636.6	1200	2237	3276	4423	5797	7595	10249	15031	28267
(8)	1026	1926	3552	5147	6879	8928	11588	15498	22528	41982
(9)	1550	2898	5304	7625	10112	13028	16790	22298	32189	59562
(10)	2227	4152	7555	10796	14232	18230	23361	30852	44291	81481

TABLE 2.23 : K* (1/α = 0.1)

Q: MODE:	0.05	0.10	0.20	0.30	0.40	0.50	0.60	0.70	0.80	0.90
(1)	.2057	.5301	1.634	3.490	6.311	10.47	16.74	26.90	46.37	102.3
(2)	.3878	1.001	3.313	9.462	25.43	51.21	87.40	141.6	237.6	494.5
(3)	3.152	6.382	14.77	29.81	67.24	152.9	275.3	444.3	724.8	1442
(4)	12.04	22.92	47.19	84.73	168.6	373.8	677.9	1075	1709	3296
(5)	30.23	56.17	109.8	186.2	346.0	746.5	1361	2140	3346	6321
(6)	61.14	112.2	212.9	349.5	623.1	1314	2404	3756	5807	10811
(7)	108.1	196.8	366.8	589.5	1022	2117	3882	6040	9259	17058
(8)	174.5	316.0	581.6	921.0	1566	3199	5872	9107	13874	25353
(9)	263.7	475.8	867.7	1358	2278	4600	8451	13074	19818	35989
(10)	379.1	682.0	1235	1918	3180	6362	11694	18056	27264	49255

TABLE 2.23 : K* (1/α = 0.2)

Q: MODE:	0.05	0.10	0.20	0.30	0.40	0.50	0.60	0.70	0.80	0.90
(1)	.2729	.6645	1.891	3.864	6.829	11.20	17.81	28.62	49.52	110.1
(2)	.7271	1.880	5.975	14.51	30.83	56.27	93.07	149.4	250.9	527.2
(3)	5.929	12.10	27.77	53.01	99.77	181.2	299.5	470.9	764.4	1533
(4)	22.58	43.40	89.00	153.5	265.2	461.2	745.7	1143	1803	3500
(5)	56.67	106.3	207.3	340.2	560.3	946.6	1509	2279	3529	6708
(6)	114.5	212.2	402.4	641.4	1025	1697	2681	4006	6124	11468
(7)	202.5	372.2	693.7	1084	1699	2773	4348	6448	9767	18091
(8)	326.8	597.5	1100	1697	2622	4232	6599	9731	14635	26882
(9)	493.7	899.5	1641	2507	3832	6133	9522	13979	20907	38155
(10)	709.5	1289	2337	3543	5369	8534	13205	19316	28763	52212

TABLE 2.23 : K* (1/α = 0.3)

Q: MODE:	0.05	0.10	0.20	0.30	0.40	0.50	0.60	0.70	0.80	0.90
(1)	.3259	.7750	2.122	4.227	7.355	11.96	18.95	30.44	52.83	118.2
(2)	1.023	2.650	8.183	18.34	35.43	61.25	99.06	157.7	265.1	561.7
(3)	8.364	17.19	39.16	71.98	124.5	205.9	323.8	499.4	807.3	1629
(4)	31.80	61.57	125.8	210.9	339.0	534.1	812.3	1215	1904	3717
(5)	79.75	150.7	293.5	470.0	725.3	1109	1653	2426	3727	7120
(6)	161.1	300.8	570.2	888.6	1337	2006	2946	4269	6468	12169
(7)	284.7	527.4	983.1	1505	2226	3295	4792	6877	10315	19192
(8)	459.4	846.7	1559	2358	3446	5048	7287	10384	15457	28514
(9)	694.0	1274	2328	3488	5050	7339	10530	14924	22083	40466
(10)	997.3	1826	3315	4932	7088	10238	14622	20632	30383	55370

TABLE 2.23 : K* (1/α = 0.4)

Q: MODE:	0.05	0.10	0.20	0.30	0.40	0.50	0.60	0.70	0.80	0.90
(1)	.3714	.8724	2.338	4.582	7.889	12.75	20.14	32.35	56.27	126.4
(2)	1.282	3.324	10.06	21.54	39.55	66.18	105.3	166.7	280.4	597.7
(3)	10.50	21.68	49.13	87.99	145.2	228.5	348.3	529.5	853.2	1731
(4)	39.86	77.60	158.2	259.8	400.3	598.9	878.1	1290	2012	3945
(5)	99.90	189.9	369.4	580.9	862.2	1252	1793	2580	3938	7554
(6)	201.8	378.8	717.9	1100	1595	2273	3204	4543	6835	12909
(7)	356.5	664.3	1238	1866	2664	3745	5219	7323	10902	20355
(8)	575.1	1066	1965	2928	4132	5750	7946	11063	16338	30239
(9)	868.6	1604	2933	4332	6063	8373	11495	15906	23342	42909
(10)	1248	2299	4177	6129	8520	11696	15973	21995	32115	58709

TABLE 2.23 : K* (1/α = 0.5)

Q: MODE:	0.05	0.10	0.20	0.30	0.40	0.50	0.60	0.70	0.80	0.90
(1)	.4122	.9610	2.542	4.931	8.429	13.56	21.38	34.34	59.82	134.8
(2)	1.508	3.914	11.68	24.31	43.35	71.06	111.8	176.2	296.5	634.9
(3)	12.36	25.64	57.85	101.8	163.2	249.6	372.8	561.1	901.8	1836
(4)	46.90	91.70	186.6	302.0	453.3	658.3	943.3	1369	2126	4184
(5)	117.5	224.4	436.1	677.0	980.5	1381	1930	2739	4163	8009
(6)	237.3	447.5	847.7	1284	1819	2514	3454	4827	7225	13683
(7)	419.1	784.5	1462	2180	3042	4149	5633	7784	11524	21573
(8)	676.1	1259	2321	3422	4723	6379	8585	11763	17270	32045
(9)	1021	1894	3465	5066	6935	9297	12426	16918	24675	45470
(10)	1467	2715	4935	7169	9752	12996	17277	23399	33950	62209

TABLE 2.23 : K* (1/α = 0.6)

Q: MODE:	0.05	0.10	0.20	0.30	0.40	0.50	0.60	0.70	0.80	0.90
(1)	.4494	1.043	2.737	5.275	8.974	14.39	22.66	36.39	63.47	143.4
(2)	1.706	4.431	13.10	26.79	46.93	75.91	118.5	186.2	313.4	673.3
(3)	14.00	29.12	65.51	113.9	179.4	269.5	397.4	594.0	952.9	1945
(4)	53.08	104.1	211.5	339.0	500.5	713.6	1007	1450	2246	4431
(5)	132.9	254.6	494.5	761.1	1085	1501	2066	2904	4398	8481
(6)	268.4	507.7	961.7	1445	2016	2736	3700	5119	7634	14487
(7)	474.1	890.1	1659	2455	3375	4520	6039	8258	12177	22839
(8)	764.7	1428	2634	3854	5244	6954	9207	12483	18250	33924
(9)	1154	2149	3932	5708	7704	10141	13333	17956	26075	48133
(10)	1659	3080	5601	8079	10838	14183	18543	24839	35877	65849

TABLE 2.23 : K* (1/α = 0.7)

Q: MODE:	0.05	0.10	0.20	0.30	0.40	0.50	0.60	0.70	0.80	0.90
(1)	.4841	1.121	2.926	5.616	9.523	15.24	23.97	38.50	67.21	152.0
(2)	1.881	4.886	14.35	29.04	50.32	80.73	125.4	196.6	331.0	712.7
(3)	15.44	32.18	72.23	124.6	194.2	288.6	422.1	627.9	1006	2058
(4)	58.51	115.0	233.4	371.6	543.2	766.0	1072	1534	2372	4686
(5)	146.5	281.2	545.9	835.3	1179	1613	2199	3073	4644	8968
(6)	295.8	560.7	1061	1587	2193	2944	3942	5419	8061	15318
(7)	522.4	982.8	1832	2697	3674	4866	6436	8743	12858	24147
(8)	842.6	1577	2908	4235	5710	7490	9816	13218	19270	35865
(9)	1272	2373	4342	6273	8392	10926	14219	19016	27533	50885
(10)	1828	3400	6185	8880	11809	15285	19780	26308	37884	69613

TABLE 2.23 : K* (1/α = 0.8)

Q: MODE:	0.05	0.10	0.20	0.30	0.40	0.50	0.60	0.70	0.80	0.90
(1)	.5167	1.194	3.109	5.953	10.08	16.11	25.32	40.66	71.01	160.8
(2)	2.035	5.287	15.47	31.10	53.56	85.53	132.5	207.4	349.2	752.9
(3)	16.71	34.86	78.16	134.2	207.8	306.9	446.8	662.8	1061	2173
(4)	63.30	124.5	252.7	400.8	582.3	815.9	1135	1620	2502	4948
(5)	158.5	304.6	591.2	901.4	1265	1720	2331	3246	4899	9468
(6)	319.9	607.3	1149	1713	2354	3140	4181	5724	8503	16171
(7)	565.0	1064	1984	2911	3945	5192	6827	9237	13562	25491
(8)	911.3	1708	3150	4574	6134	7994	10415	13967	20326	37859
(9)	1376	2570	4703	6775	9016	11664	15089	20094	29043	53714
(10)	1977	3682	6700	9592	12688	16319	20992	27802	39962	73482

TABLE 2.23 : K* (1/α = 0.9)

Q: MODE:	0.05	0.10	0.20	0.30	0.40	0.50	0.60	0.70	0.80	0.90
(1)	.5476	1.265	3.287	6.287	10.63	16.99	26.69	42.86	74.88	169.6
(2)	2.171	5.642	16.48	33.02	56.69	90.31	139.7	210.5	367.9	793.7
(3)	17.83	37.23	83.42	142.9	220.6	324.7	471.6	698.5	1117	2291
(4)	67.34	133.0	269.7	426.9	618.4	863.8	1199	1707	2636	5215
(5)	169.1	325.2	631.1	960.5	1344	1822	2463	3422	5161	9979
(6)	341.4	648.4	1227	1825	2502	3326	4417	6035	8957	17043
(7)	602.9	1136	2118	3104	4194	5501	7214	9740	14288	26865
(8)	972.3	1823	3363	4876	6521	8472	11006	14727	21414	39901
(9)	1468	2744	5022	7223	9587	12362	15945	21189	30597	56609
(10)	2109	3931	7153	10226	13492	17297	22185	29318	42100	77442

TABLE 2.24 : T^* (α = 0.1)

Q: MODE:	0.05	0.10	0.20	0.30	0.40	0.50	0.60	0.70	0.80	0.90
(1)	1.800	6.616	25.28	57.63	106.7	178.2	283.8	451.0	762.3	1630
(2)	2.130	6.226	63.74	360.6	714.6	1139	1697	2513	3954	7832
(3)	3.430	9.540	95.93	1248	2515	3877	5548	7897	11949	22734
(4)	4.596	12.46	106.4	2925	6097	9283	13029	18164	26899	50019
(5)	5.776	15.44	119.1	5664	12101	18301	25387	34932	51027	93497
(6)	6.950	18.44	133.2	9717	21154	31858	43851	59807	86543	156982
(7)	8.134	21.43	148.1	15339	33884	50886	69652	94392	135655	244244
(8)	9.288	24.45	163.6	22782	50920	76315	104028	140295	200582	359075
(9)	10.48	27.46	179.4	32301	72890	109074	148204	199124	283527	505296
(10)	11.66	30.47	195.3	44131	100413	150093	203413	272481	386689	686617

TABLE 2.24 : T^* (α = 0.2)

Q: MODE:	0.05	0.10	0.20	0.30	0.40	0.50	0.60	0.70	0.80	0.90
(1)	.9346	3.465	13.01	29.33	54.01	89.97	143.1	227.3	384.3	823.0
(2)	2.092	6.016	44.33	185.3	360.2	572.2	851.1	1259	1982	3927
(3)	3.411	9.435	79.51	634.9	1263	1942	2777	3953	5981	11382
(4)	4.583	12.40	98.02	1480	3056	4647	6519	9088	13458	25029
(5)	5.766	15.40	114.2	2857	6060	9157	12699	17473	25524	46771
(6)	6.947	18.40	129.9	4891	10589	15937	21933	29912	43284	78518
(7)	8.129	21.42	145.7	7710	16957	25453	34835	47206	67841	122153
(8)	9.304	24.43	161.7	11440	25478	38169	52024	70159	100307	179572
(9)	10.48	27.42	177.9	16208	36465	54550	74113	99575	141782	252607
(10)	11.65	30.45	194.1	22142	50230	75061	101719	136255	193364	343351

TABLE 2.24 : T^* (α = 0.3)

Q: MODE:	0.05	0.10	0.20	0.30	0.40	0.50	0.60	0.70	0.80	0.90
(1)	.6336	2.375	8.871	19.86	36.43	60.54	96.20	152.8	258.4	553.8
(2)	2.055	5.824	35.62	126.6	242.0	383.0	569.0	842.1	1324	2626
(3)	3.391	9.333	69.60	430.0	845.5	1297	1854	2638	3992	7598
(4)	4.571	12.33	91.53	998.3	2042	3101	4350	6062	8977	16698
(5)	5.757	15.35	109.9	1921	4047	6109	8470	11653	17023	31196
(6)	6.939	18.37	126.9	3283	7068	10630	14627	19947	28864	52363
(7)	8.122	21.39	143.4	5167	11315	16975	23229	31477	45237	81456
(8)	9.275	24.41	159.9	7659	16997	25454	34689	46780	66882	119738
(9)	10.48	27.43	176.4	10844	24323	36375	49416	66391	94533	168484
(10)	11.66	30.45	192.8	14805	33502	50050	67821	90846	128923	228930

TABLE 2.24 : T^* (α = 0.4)

Q: MODE:	0.05	0.10	0.20	0.30	0.40	0.50	0.60	0.70	0.80	0.90
(1)	.4798	1.815	6.778	15.10	27.61	45.81	72.73	115.5	195.4	419.3
(2)	2.020	5.646	30.39	97.13	182.9	288.5	428.0	633.1	996.1	1975
(3)	3.371	9.234	62.70	327.4	636.6	975.1	1392	1981	2997	5706
(4)	4.558	12.27	86.28	757.0	1535	2328	3265	4549	6737	12533
(5)	5.747	15.31	106.1	1453	3040	4585	6356	8743	12772	23408
(6)	6.932	18.33	124.1	2478	5307	7977	10974	14964	21654	39286
(7)	8.116	21.36	141.3	3895	8493	12736	17426	23613	33935	61107
(8)	9.299	24.39	158.2	5769	12756	19096	26022	35090	50170	89821
(9)	10.48	27.41	175.0	8161	18252	27288	37068	49800	70909	126382
(10)	11.66	30.43	191.6	11136	25137	37545	50872	68141	96702	171719

TABLE 2.24 : T^* ($\alpha = 0.5$)

Q: MODE:	0.05	0.10	0.20	0.30	0.40	0.50	0.60	0.70	0.80	0.90
(1)	.3862	1.472	5.507	12.24	22.31	36.96	58.64	93.07	157.6	338.6
(2)	1.986	5.481	26.82	79.31	147.4	231.7	343.4	507.7	798.8	1584
(3)	3.351	9.137	57.51	265.6	511.3	781.6	1115	1586	2400	4571
(4)	4.545	12.21	81.88	612.0	1231	1865	2614	3642	5393	10034
(5)	5.739	15.27	102.7	1171	2436	3671	5087	6997	10222	18736
(6)	6.923	18.30	121.5	1995	4250	6385	8782	11975	17328	31439
(7)	8.110	21.33	139.2	3132	6801	10193	13944	18894	27153	48898
(8)	9.295	24.36	156.5	4634	10212	15281	20822	28077	40142	71871
(9)	10.46	27.39	173.6	6551	14610	21836	29659	39845	56734	101121
(10)	11.63	30.41	190.4	8935	20119	30042	40703	54518	77369	137392

TABLE 2.24 : T^* ($\alpha = 0.6$)

Q: MODE:	0.05	0.10	0.20	0.30	0.40	0.50	0.60	0.70	0.80	0.90
(1)	.3233	1.239	4.651	10.31	18.77	31.06	49.23	78.14	132.3	284.7
(2)	1.953	5.328	24.18	67.35	123.7	193.8	287.0	424.2	667.3	1324
(3)	3.332	9.043	53.41	224.3	427.7	652.6	930.9	1323	2002	3814
(4)	4.532	12.15	78.13	515.2	1028	1556	2179	3037	4497	8368
(5)	5.728	15.22	99.66	984.3	2033	3061	4241	5834	8522	15621
(6)	6.916	18.27	119.0	1672	3546	5323	7321	9982	14444	26208
(7)	8.105	21.30	137.3	2623	5672	8497	11623	15748	22633	40759
(8)	9.289	24.34	154.9	3877	8516	12738	17355	23401	33457	59904
(9)	10.46	27.37	172.2	5478	12182	18201	24720	33208	47284	84281
(10)	11.64	30.39	189.3	7467	16773	25040	33923	45437	64481	114508

TABLE 2.24 : T^* ($\alpha = 0.7$)

Q: MODE:	0.05	0.10	0.20	0.30	0.40	0.50	0.60	0.70	0.80	0.90
(1)	.2780	1.071	4.032	8.933	16.24	26.83	42.51	67.48	114.3	246.3
(2)	1.921	5.185	22.13	58.75	106.8	166.8	246.7	364.4	573.4	1138
(3)	3.313	8.951	50.05	194.7	367.9	560.4	798.9	1135	1718	3273
(4)	4.521	12.09	74.86	445.9	883.7	1335	1869	2604	3857	7178
(5)	5.719	15.18	96.89	850.1	1745	2626	3637	5002	7307	13396
(6)	6.910	18.23	116.8	1442	3043	4565	6277	8558	12384	22472
(7)	8.098	21.28	135.4	2259	4866	7286	9965	13501	19403	34945
(8)	9.285	24.31	153.3	3337	7304	10922	14878	20061	28682	51356
(9)	10.47	27.35	170.9	4711	10447	15604	21191	28468	40535	72252
(10)	11.65	30.38	188.1	6418	14383	21467	29081	38950	55275	98162

TABLE 2.24 : T^* ($\alpha = 0.8$)

Q: MODE:	0.05	0.10	0.20	0.30	0.40	0.50	0.60	0.70	0.80	0.90
(1)	.2438	.9432	3.564	7.892	14.33	23.66	37.47	59.47	100.8	217.4
(2)	1.890	5.051	20.47	52.24	94.03	146.5	216.4	319.7	502.9	999.0
(3)	3.294	8.862	47.24	172.5	323.1	491.3	700.0	995.0	1505	2868
(4)	4.508	12.03	71.99	393.9	775.1	1169	1637	2280	3377	6285
(5)	5.710	15.13	94.35	749.4	1530	2299	3184	4379	6396	11727
(6)	6.901	18.20	114.6	1269	2666	3997	5494	7490	10839	19670
(7)	8.092	21.25	133.6	1986	4261	6378	8722	11816	16981	30584
(8)	9.280	24.29	151.8	2931	6395	9559	13021	17556	25101	44945
(9)	10.46	27.33	169.6	4136	9146	13657	18545	24912	35472	63230
(10)	11.63	30.36	187.0	5632	12591	18787	25449	34084	48371	85902

TABLE 2.24 : T^* ($\alpha = 0.9$)

Q: MODE:	0.05	0.10	0.20	0.30	0.40	0.50	0.60	0.70	0.80	0.90
(1)	.2172	.8430	3.197	7.077	12.84	21.19	33.54	53.24	90.31	195.0
(2)	1.861	4.924	19.10	47.15	84.12	130.7	192.9	284.8	448.1	890.5
(3)	3.275	8.774	44.83	155.1	288.2	437.5	623.0	885.4	1339	2552
(4)	4.495	11.97	69.42	353.3	690.5	1040	1456	2028	3004	5591
(5)	5.700	15.09	92.01	670.9	1362	2045	2831	3894	5688	10429
(6)	6.894	18.16	112.6	1135	2372	3554	4885	6660	9637	17490
(7)	8.086	21.22	131.9	1774	3791	5671	7755	10505	15098	27193
(8)	9.273	24.27	150.4	2616	5689	8500	11576	15608	22315	39959
(9)	10.46	27.30	168.4	3688	8134	12142	16487	22147	31535	56213
(10)	11.64	30.34	185.9	5020	11197	16703	22624	30300	43001	76367

TABLE 2.24 : T^* ($\alpha = 1.0$)

Q: MODE:	0.05	0.10	0.20	0.30	0.40	0.50	0.60	0.70	0.80	0.90
(1)	.1958	.7622	2.900	6.422	11.64	19.20	30.40	48.25	81.90	177.1
(2)	1.832	4.806	17.94	43.04	76.17	118.1	174.1	257.0	404.3	803.7
(3)	3.256	8.689	42.74	141.1	260.3	394.5	561.4	797.7	1207	2300
(4)	4.483	11.92	67.11	320.8	622.9	937.9	1311	1826	2705	5036
(5)	5.692	15.05	89.85	608.1	1227	1842	2550	3506	5121	9391
(6)	6.886	18.13	110.7	1027	2137	3200	4398	5996	8676	15746
(7)	8.080	21.19	130.3	1604	3415	5106	6981	9457	13591	24480
(8)	9.268	24.24	149.0	2363	5123	7652	10421	14049	20087	35970
(9)	10.46	27.28	167.1	3330	7325	10931	14841	19935	28385	50599
(10)	11.64	30.32	184.8	4531	10082	15035	20364	27273	38704	68739

TABLE 2.24 : T^* ($1/\alpha = 0.1$)

Q: MODE:	0.05	0.10	0.20	0.30	0.40	0.50	0.60	0.70	0.80	0.90
(1)	.0199	.0799	.3282	.7736	1.476	2.557	4.265	7.202	13.18	31.27
(2)	.7836	1.670	3.786	6.452	9.858	14.35	20.64	30.36	48.52	100.3
(3)	2.174	4.865	11.88	20.79	31.45	44.52	61.70	86.96	132.2	256.9
(4)	3.597	8.482	23.77	45.84	71.74	101.7	139.2	192.5	285.5	537.6
(5)	4.968	12.10	38.50	83.72	136.7	195.1	265.5	363.0	530.6	980.3
(6)	6.285	15.60	55.20	136.5	232.3	333.9	452.8	614.5	889.6	1623
(7)	7.567	19.01	73.22	206.3	364.9	527.4	713.5	963.2	1384	2503
(8)	8.824	22.33	92.04	295.3	540.5	784.9	1059	1425	2037	3659
(9)	10.06	25.59	111.4	405.9	765.4	1115	1504	2016	2870	5130
(10)	11.29	28.81	130.9	540.5	1045	1529	2059	2752	3906	6951

TABLE 2.24 : T^* ($1/\alpha = 0.2$)

Q: MODE:	0.05	0.10	0.20	0.30	0.40	0.50	0.60	0.70	0.80	0.90
(1)	.0396	.1588	.6411	1.475	2.735	4.587	7.389	12.00	21.03	47.61
(2)	1.143	2.568	6.327	11.35	17.85	26.36	38.09	55.88	88.34	178.6
(3)	2.661	6.367	18.35	36.19	57.91	84.02	117.6	166.3	251.9	484.2
(4)	4.039	10.05	34.43	79.88	134.3	195.2	269.9	374.4	554.6	1037
(5)	5.345	13.55	53.02	146.9	259.4	378.7	519.7	712.5	1040	1914
(6)	6.605	16.90	72.84	241.8	445.5	653.0	891.6	1212	1754	3192
(7)	7.844	20.15	93.16	369.1	705.2	1036	1410	1907	2740	4945
(8)	9.065	23.35	113.5	533.6	1051	1548	2100	2827	4043	7250
(9)	10.28	26.50	133.7	740.2	1495	2206	2986	4007	5705	10182
(10)	11.48	29.63	153.6	993.8	2051	3030	4093	5477	7772	13816

TABLE 2.24 : T^* ($1/\alpha = 0.3$)

Q: MODE:	0.05	0.10	0.20	0.30	0.40	0.50	0.60	0.70	0.80	0.90
(1)	.0594	.2368	.9430	2.139	3.918	6.502	10.36	16.63	28.73	63.85
(2)	1.353	3.157	8.375	15.78	25.44	38.03	55.24	81.14	127.9	256.8
(3)	2.879	7.138	23.18	50.34	83.65	123.1	173.3	245.3	371.4	711.3
(4)	4.212	10.74	41.80	111.7	195.8	288.3	400.3	556.0	823.5	1537
(5)	5.484	14.13	62.33	206.9	380.9	561.9	773.6	1061	1551	2849
(6)	6.720	17.39	83.38	342.8	657.5	971.7	1330	1810	2620	4761
(7)	7.940	20.57	104.3	526.7	1044	1545	2106	2850	4097	7387
(8)	9.149	23.71	124.8	765.9	1560	2311	3140	4230	6048	10840
(9)	10.35	26.82	144.9	1067	2224	3297	4468	5998	8540	15234
(10)	11.55	29.91	164.4	1439	3055	4531	6127	8201	11639	20682

TABLE 2.24 : T^* ($1/\alpha = 0.4$)

Q: MODE:	0.05	0.10	0.20	0.30	0.40	0.50	0.60	0.70	0.80	0.90
(1)	.0790	.3139	1.237	2.780	5.064	8.366	13.28	21.20	36.37	80.05
(2)	1.491	3.581	10.14	19.96	32.85	49.57	72.29	106.3	167.5	335.0
(3)	3.003	7.612	27.12	63.91	109.1	161.9	228.8	324.3	490.8	938.4
(4)	4.305	11.13	47.48	142.5	257.1	381.2	530.6	737.6	1092	2037
(5)	5.556	14.44	69.10	265.4	502.2	744.9	1027	1410	2061	3784
(6)	6.779	17.64	90.64	442.1	869.2	1290	1768	2408	3485	6331
(7)	7.989	20.79	111.6	682.3	1383	2054	2803	3794	5453	9829
(8)	9.191	23.90	132.0	995.8	2069	3074	4180	5633	8054	14430
(9)	10.39	26.98	151.7	1392	2953	4388	5950	7989	11375	20286
(10)	11.58	30.06	170.9	1883	4059	6031	8161	10926	15505	27547

TABLE 2.24 : T^* ($1/\alpha = 0.5$)

Q: MODE:	0.05	0.10	0.20	0.30	0.40	0.50	0.60	0.70	0.80	0.90
(1)	.0986	.3903	1.524	3.407	6.186	10.20	16.16	25.74	43.99	96.24
(2)	1.589	3.903	11.71	23.99	40.17	61.04	89.30	131.5	207.0	413.1
(3)	3.083	7.934	30.49	77.15	134.4	200.8	284.3	403.2	610.2	1165
(4)	4.362	11.37	52.09	172.8	318.2	474.0	660.8	919.2	1361	2537
(5)	5.601	14.64	74.35	323.2	623.2	927.9	1281	1760	2571	4718
(6)	6.814	17.80	96.04	540.5	1080	1608	2206	3006	4350	7900
(7)	8.020	20.92	116.9	836.9	1722	2563	3499	4738	6809	12271
(8)	9.217	24.01	136.9	1224	2578	3837	5220	7036	10059	18020
(9)	10.41	27.08	156.3	1716	3682	5478	7432	9980	14210	25339
(10)	11.60	30.15	175.1	2325	5063	7532	10194	13650	19372	34412

TABLE 2.24 : T^* ($1/\alpha = 0.6$)

Q: MODE:	0.05	0.10	0.20	0.30	0.40	0.50	0.60	0.70	0.80	0.90
(1)	.1182	.4659	1.806	4.023	7.294	12.02	19.03	30.26	51.58	112.4
(2)	1.662	4.158	13.13	27.91	47.43	72.48	106.3	156.6	246.4	491.3
(3)	3.138	8.168	33.45	90.17	159.7	239.5	339.7	482.1	729.6	1392
(4)	4.402	11.55	55.97	202.7	379.2	566.8	791.1	1100	1630	3037
(5)	5.631	14.77	78.58	380.6	744.2	1110	1534	2109	3081	5653
(6)	6.837	17.91	100.2	638.4	1292	1927	2645	3604	5215	9469
(7)	8.039	21.01	120.9	990.9	2060	3071	4195	5682	8166	14713
(8)	9.234	24.09	140.6	1453	3087	4600	6260	8438	12065	21610
(9)	10.42	27.15	159.7	2040	4410	6569	8913	11971	17045	30391
(10)	11.61	30.20	178.2	2767	6067	9033	12228	16375	23238	41278

TABLE 2.24 : T^* ($1/\alpha = 0.7$)

Q: MODE:	0.05	0.10	0.20	0.30	0.40	0.50	0.60	0.70	0.80	0.90
(1)	.1377	.5408	2.084	4.631	8.391	13.83	21.88	34.77	59.17	128.6
(2)	1.719	4.365	14.45	31.77	54.65	83.90	123.2	181.7	285.9	569.4
(3)	3.179	8.345	36.10	103.0	184.9	278.3	395.2	561.0	849.0	1619
(4)	4.430	11.68	59.30	232.4	440.2	659.6	921.3	1282	1898	3536
(5)	5.651	14.87	82.09	437.7	865.2	1293	1788	2458	3591	6587
(6)	6.856	17.99	103.6	735.9	1503	2245	3083	4202	6081	11039
(7)	8.054	21.07	124.0	1144	2399	3580	4892	6625	9522	17154
(8)	9.246	24.14	143.4	1680	3596	5363	7300	9841	14070	25200
(9)	10.44	27.20	162.2	2363	5139	7659	10395	13962	19680	35443
(10)	11.62	30.25	180.5	3208	7070	10533	14262	19099	27105	48143

TABLE 2.24 : T^* ($1/\alpha = 0.8$)

Q: MODE:	0.05	0.10	0.20	0.30	0.40	0.50	0.60	0.70	0.80	0.90
(1)	.1571	.6152	2.359	5.232	9.480	15.62	24.72	39.27	66.75	144.7
(2)	1.764	4.536	15.69	35.57	61.84	95.30	140.2	206.8	325.4	647.5
(3)	3.211	8.484	38.50	115.8	210.1	317.0	450.6	639.9	968.3	1846
(4)	4.452	11.77	62.22	262.0	501.1	752.4	1051	1463	2167	4036
(5)	5.668	14.94	85.07	494.6	986.1	1476	2042	2807	4101	7522
(6)	6.868	18.05	106.4	833.2	1715	2564	3522	4800	6946	12608
(7)	8.064	21.12	126.5	1297	2738	4089	5588	7569	10878	19596
(8)	9.256	24.18	145.6	1908	4105	6126	8341	11244	16076	28790
(9)	10.44	27.23	164.2	2685	5868	8750	11877	15953	22715	40495
(10)	11.63	30.28	182.2	3649	8074	12034	16296	21824	30971	55008

TABLE 2.24 : T^* ($1/\alpha = 0.9$)

Q: MODE:	0.05	0.10	0.20	0.30	0.40	0.50	0.60	0.70	0.80	0.90
(1)	.1765	.6890	2.631	5.829	10.56	17.42	27.56	43.76	74.33	160.9
(2)	1.801	4.681	16.84	39.32	69.01	106.7	157.1	231.9	364.8	725.6
(3)	3.236	8.596	40.70	128.5	235.2	355.8	506.0	718.8	1087	2073
(4)	4.469	11.85	64.80	291.4	562.0	845.1	1181	1645	2436	4536
(5)	5.680	15.00	87.62	551.4	1107	1659	2296	3156	4611	8456
(6)	6.879	18.09	108.7	930.4	1926	2882	3960	5398	7811	14177
(7)	8.072	21.16	128.6	1451	3076	4597	6284	8513	12234	22038
(8)	9.263	24.22	147.5	2136	4614	6889	9381	12647	18081	32380
(9)	10.45	27.26	165.8	3008	6596	9840	13359	17944	25550	45547
(10)	11.64	30.30	183.7	4090	9078	13535	18330	24549	34838	61874

TABLE 2.26: T^* ($\alpha = 0.1$)

Q: MODE:	0.05	0.10	0.20	0.30	0.40	0.50	0.60	0.70	0.80	0.90
(1)	.6162	1.369	3.397	6.372	10.75	17.32	27.65	45.41	81.59	191.2
(2)	.6119	1.365	3.486	6.970	12.94	23.30	41.18	73.16	139.0	338.0
(3)	1.099	2.411	5.969	11.58	20.94	36.98	64.64	114.1	215.6	521.4
(4)	1.578	3.444	8.432	16.18	28.97	50.83	88.56	156.0	294.2	709.9
(5)	2.058	4.480	10.91	20.81	37.07	64.78	112.6	198.2	373.2	899.1
(6)	2.538	5.516	13.39	25.45	45.18	78.77	136.7	240.4	452.3	1088
(7)	3.018	6.554	15.87	30.09	53.31	92.78	160.9	282.7	531.6	1278
(8)	3.499	7.591	18.35	34.74	61.44	106.8	185.1	325.0	610.8	1468
(9)	3.980	8.627	20.83	39.39	69.58	120.8	209.2	367.3	690.1	1657
(10)	4.459	9.664	23.31	44.04	77.71	134.9	233.4	409.7	769.6	1847

TABLE 2.26: T^* ($\alpha = 0.2$)

Q: MODE:	0.05	0.10	0.20	0.30	0.40	0.50	0.60	0.70	0.80	0.90
(1)	.4257	.9244	2.197	3.966	6.475	10.16	15.87	25.61	45.39	105.2
(2)	.5546	1.219	2.989	5.621	9.624	15.85	25.91	43.44	79.27	187.7
(3)	.9962	2.156	5.135	9.412	15.76	25.51	41.14	68.24	123.4	290.0
(4)	1.431	3.082	7.275	13.22	21.97	35.34	56.71	93.67	168.8	395.1
(5)	1.867	4.012	9.424	17.05	28.22	45.22	72.35	119.2	214.4	500.6
(6)	2.303	4.941	11.57	20.88	34.47	55.12	88.01	144.8	260.0	606.1
(7)	2.739	5.871	13.73	24.71	40.73	65.03	103.7	170.4	305.6	712.0
(8)	3.174	6.801	15.88	28.55	46.99	74.95	119.4	196.1	351.3	817.9
(9)	3.611	7.730	18.03	32.39	53.26	84.86	135.1	221.7	397.1	923.6
(10)	4.046	8.661	20.18	36.23	59.53	94.79	150.8	247.3	442.9	1029

TABLE 2.25: T^* ($\alpha = 0.3$)

Q: MODE:	0.05	0.10	0.20	0.30	0.40	0.50	0.60	0.70	0.80	0.90
(1)	.3445	.7413	1.731	3.074	4.945	7.655	11.82	18.90	33.21	76.42
(2)	.5080	1.107	2.652	4.842	8.014	12.75	20.19	32.94	58.87	137.2
(3)	.9127	1.958	4.563	8.136	13.19	20.63	32.21	51.94	91.87	212.1
(4)	1.312	2.801	6.475	11.45	18.44	28.67	44.50	71.42	125.8	289.2
(5)	1.711	3.647	8.393	14.79	23.72	36.73	56.85	90.98	159.8	366.5
(6)	2.111	4.493	10.31	18.12	29.00	44.82	69.22	110.6	193.8	443.8
(7)	2.511	5.339	12.23	21.46	34.29	52.91	81.61	130.2	228.0	521.3
(8)	2.910	6.186	14.15	24.80	39.58	61.01	94.01	149.8	262.1	598.9
(9)	3.311	7.031	16.08	28.14	44.87	69.11	106.4	169.5	296.3	676.3
(10)	3.710	7.878	18.00	31.49	50.17	77.22	118.8	189.1	330.5	754.1

TABLE 2.25: T^* ($\alpha - 0.4$)

Q: MODE:	0.05	0.10	0.20	0.30	0.40	0.50	0.60	0.70	0.80	0.90
(1)	.2969	.6360	1.472	2.591	4.134	6.351	9.743	15.48	27.05	61.97
(2)	.4693	1.016	2.401	4.313	7.011	10.96	17.06	27.43	48.41	111.7
(3)	.8433	1.799	4.137	7.261	11.57	17.78	27.28	43.33	75.65	172.8
(4)	1.212	2.575	5.875	10.24	16.20	24.74	37.75	59.65	103.6	235.6
(5)	1.582	3.352	7.618	13.22	20.85	31.73	48.26	76.03	131.7	298.7
(6)	1.951	4.130	9.363	16.21	25.51	38.73	58.79	92.43	159.8	361.7
(7)	2.321	4.909	11.11	19.20	30.17	45.74	69.34	108.9	188.0	425.0
(8)	2.690	5.688	12.86	22.20	34.83	52.76	79.89	125.3	216.1	488.2
(9)	3.060	6.465	14.60	25.19	39.50	59.77	90.44	141.7	244.3	551.3
(10)	3.429	7.244	16.35	28.18	44.16	66.79	101.0	158.2	272.6	614.7

TABLE 2.25: T^* ($\alpha - 0.5$)

Q: MODE:	0.05	0.10	0.20	0.30	0.40	0.50	0.60	0.70	0.80	0.90
(1)	.2649	.5658	1.302	2.281	3.622	5.539	8.461	13.39	23.32	53.26
(2)	.4365	.9416	2.205	3.921	6.308	9.755	15.04	23.97	41.99	96.26
(3)	.7844	1.667	3.801	6.609	10.43	15.86	24.09	37.92	65.66	149.0
(4)	1.128	2.386	5.401	9.324	14.61	22.09	33.36	52.23	89.98	203.2
(5)	1.472	3.108	7.006	12.05	18.81	28.34	42.67	66.59	114.4	257.5
(6)	1.815	3.829	8.612	14.78	23.02	34.60	51.99	80.98	138.8	311.9
(7)	2.159	4.552	10.22	17.51	27.23	40.87	61.33	95.39	163.3	366.5
(8)	2.503	5.274	11.83	20.24	31.45	47.15	70.67	109.8	187.8	421.0
(9)	2.848	5.995	13.43	22.97	35.66	53.42	80.01	124.2	212.3	475.5
(10)	3.191	6.717	15.04	25.70	39.88	59.70	89.36	138.6	236.8	530.2

TABLE 2.25: T^* ($\alpha - 0.6$)

Q: MODE:	0.05	0.10	0.20	0.30	0.40	0.50	0.60	0.70	0.80	0.90
(1)	.2414	.5148	1.181	2.062	3.265	4.979	7.585	11.97	20.81	47.43
(2)	.4083	.8785	2.046	3.615	5.779	8.881	13.61	21.58	37.61	85.86
(3)	.7337	1.555	3.528	6.098	9.562	14.45	21.83	34.16	58.85	132.9
(4)	1.055	2.227	5.015	8.606	13.41	20.14	30.24	47.06	80.66	181.3
(5)	1.377	2.901	6.506	11.12	17.27	25.85	38.68	60.02	102.6	229.8
(6)	1.698	3.574	7.998	13.64	21.13	31.56	47.14	72.99	124.5	278.4
(7)	2.020	4.249	9.491	16.17	25.00	37.29	55.61	85.99	146.4	327.1
(8)	2.342	4.923	10.98	18.69	28.87	43.02	64.09	98.99	168.4	375.8
(9)	2.664	5.596	12.48	21.21	32.74	48.74	72.56	112.0	190.4	424.4
(10)	2.985	6.270	13.97	23.74	36.61	54.47	81.04	125.0	212.4	473.2

TABLE 2.25: T^* ($\alpha - 0.7$)

Q: MODE:	0.05	0.10	0.20	0.30	0.40	0.50	0.60	0.70	0.80	0.90
(1)	.2233	.4757	1.089	1.898	3.000	4.566	6.945	10.95	19.00	43.24
(2)	.3837	.8243	1.913	3.368	5.362	8.209	12.53	19.81	34.42	78.37
(3)	.6895	1.459	3.299	5.682	8.876	13.37	20.11	31.37	53.87	121.3
(4)	.9915	2.090	4.691	8.022	12.45	18.63	27.87	43.23	73.86	165.5
(5)	1.294	2.722	6.086	10.37	16.04	23.92	35.66	55.14	93.91	209.8
(6)	1.596	3.355	7.482	12.72	19.63	29.21	43.46	67.06	114.0	254.2
(7)	1.899	3.988	8.880	15.07	23.22	34.51	51.28	79.01	134.1	298.6
(8)	2.201	4.620	10.28	17.43	26.82	39.81	59.09	90.95	154.2	343.1
(9)	2.504	5.252	11.67	19.78	30.42	45.11	66.91	102.9	174.3	387.5
(10)	2.806	5.885	13.07	22.13	34.02	50.42	74.74	114.9	194.5	432.1

TABLE 2.25: T^* ($\alpha - 0.8$)

Q: MODE:	0.05	0.10	0.20	0.30	0.40	0.50	0.60	0.70	0.80	0.90
(1)	.2087	.4444	1.107	1.770	2.793	4.248	6.455	10.16	17.63	40.09
(2)	.3621	.7770	1.799	3.161	5.022	7.671	11.69	18.44	31.98	72.71
(3)	.6507	1.376	3.104	5.335	8.316	12.49	18.76	29.20	50.07	112.6
(4)	.9357	1.971	4.414	7.533	11.67	17.42	26.01	40.25	68.64	153.6
(5)	1.221	2.567	5.727	9.739	15.03	22.36	33.28	51.34	87.28	194.7
(6)	1.506	3.163	7.042	11.95	18.40	27.32	40.56	62.45	105.9	235.8
(7)	1.792	3.760	8.357	14.16	21.77	32.27	47.85	73.58	124.6	277.1
(8)	2.077	4.356	9.672	16.37	25.14	37.23	55.15	84.71	143.4	318.4
(9)	2.363	4.952	10.99	18.58	28.51	42.19	62.45	95.84	162.1	359.6
(10)	2.648	5.549	12.30	20.79	31.88	47.16	69.75	107.0	180.8	400.9

TABLE 2.25: T^* ($\alpha - 0.9$)

Q: MODE:	0.05	0.10	0.20	0.30	0.40	0.50	0.60	0.70	0.80	0.90
(1)	.1967	.4188	.9573	1.666	2.628	3.994	6.067	9.548	16.55	37.63
(2)	.3429	.7355	1.701	2.985	4.737	7.229	11.00	17.34	30.05	68.27
(3)	.6162	1.302	2.935	5.039	7.846	11.78	17.67	27.47	47.05	105.7
(4)	.8860	1.865	4.174	7.116	11.01	16.42	24.49	37.86	64.51	144.2
(5)	1.156	2.430	5.416	9.200	14.18	21.08	31.34	48.30	82.03	182.8
(6)	1.427	2.994	6.659	11.29	17.36	25.75	38.19	58.75	99.57	221.4
(7)	1.697	3.559	7.903	13.37	20.54	30.42	45.06	69.22	117.1	260.2
(8)	1.967	4.124	9.147	15.46	23.72	35.10	51.94	79.69	134.7	298.9
(9)	2.238	4.688	10.39	17.55	26.91	39.78	58.81	90.16	152.3	337.6
(10)	2.507	5.253	11.64	19.64	30.09	44.46	65.69	100.6	169.9	376.5

TABLE 2.25: T^* ($\alpha - 1.0$)

Q: MODE:	0.05	0.10	0.20	0.30	0.40	0.50	0.60	0.70	0.80	0.90
(1)	.1866	.3972	.9079	1.580	2.491	3.786	5.750	9.049	15.68	35.65
(2)	.3257	.6985	1.615	2.834	4.495	6.857	10.44	16.44	28.48	64.69
(3)	.5853	1.237	2.787	4.783	7.445	11.17	16.75	26.04	44.59	100.2
(4)	.8416	1.772	3.963	6.755	10.45	15.58	23.22	35.90	61.14	136.6
(5)	1.098	2.308	5.142	8.733	13.46	20.00	29.72	45.79	77.75	173.2
(6)	1.355	2.844	6.323	10.71	16.47	24.43	36.22	55.70	94.38	209.8
(7)	1.612	3.380	7.504	12.70	19.49	28.86	42.74	65.63	111.0	246.6
(8)	1.868	3.917	8.685	14.68	22.51	33.30	49.25	75.55	127.7	283.3
(9)	2.126	4.452	9.867	16.66	25.53	37.74	55.77	85.48	144.4	319.9
(10)	2.382	4.989	11.05	18.64	28.55	42.18	62.29	95.41	161.1	356.7

TABLE 2.25: T^* ($1/\alpha - 0.1$)

Q: MODE:	0.05	0.10	0.20	0.30	0.40	0.50	0.60	0.70	0.80	0.90
(1)	.0616	.1369	.3397	.6372	1.075	1.732	2.765	4.541	8.159	19.12
(2)	.0612	.1365	.3486	.6970	1.294	2.330	4.118	7.316	13.90	33.80
(3)	.1099	.2411	.5969	1.158	2.094	3.698	6.464	11.41	21.56	52.14
(4)	.1578	.3444	.8432	1.618	2.897	5.083	8.856	15.60	29.42	70.99
(5)	.2058	.4480	1.091	2.081	3.707	6.478	11.26	19.82	37.32	89.91
(6)	.2538	.5516	1.339	2.545	4.518	7.877	13.67	24.04	45.23	108.8
(7)	.3018	.6554	1.587	3.009	5.331	9.278	16.09	28.27	53.16	127.8
(8)	.3499	.7591	1.835	3.474	6.144	10.68	18.51	32.50	61.08	146.8
(9)	.3980	.8627	2.083	3.939	6.958	12.08	20.92	36.73	69.01	165.8
(10)	.4459	.9664	2.331	4.404	7.771	13.49	23.34	40.97	76.96	184.8

TABLE 2.25: T^* ($1/\alpha - 0.2$)

Q: MODE:	0.05	0.10	0.20	0.30	0.40	0.50	0.60	0.70	0.80	0.90
(1)	.0851	.1849	.4393	.7931	1.295	2.031	3.174	5.123	9.078	21.04
(2)	.1109	.2438	.5979	1.124	1.925	3.171	5.183	8.688	15.85	37.54
(3)	.1992	.4311	1.027	1.882	3.153	5.103	8.229	13.65	24.69	58.00
(4)	.2863	.6164	1.455	2.644	4.395	7.068	11.34	18.73	33.76	79.03
(5)	.3734	.8023	1.885	3.409	5.643	9.044	14.47	23.84	42.87	100.1
(6)	.4606	.9882	2.315	4.175	6.893	11.02	17.60	28.96	51.99	121.2
(7)	.5477	1.174	2.745	4.942	8.146	13.01	20.74	34.09	61.13	142.4
(8)	.6349	1.360	3.176	5.710	9.398	14.99	23.88	39.21	70.27	163.6
(9)	.7223	1.546	3.606	6.477	10.65	16.97	27.02	44.34	79.42	184.7
(10)	.8093	1.732	4.037	7.245	11.91	18.96	30.17	49.47	88.57	205.9

TABLE 2.25: T^* ($1/\alpha = 0.3$)

Q: MODE:	0.05	0.10	0.20	0.30	0.40	0.50	0.60	0.70	0.80	0.90
(1)	.1033	.2224	.5193	.9223	1.484	2.297	3.547	5.669	9.962	22.93
(2)	.1524	.3320	.7955	1.453	2.404	3.825	6.056	9.883	17.66	41.16
(3)	.2738	.5873	1.369	2.441	3.958	6.190	9.662	15.58	27.56	63.64
(4)	.3935	.8404	1.942	3.436	5.533	8.600	13.35	21.43	37.73	86.75
(5)	.5134	1.094	2.518	4.436	7.115	11.02	17.06	27.30	47.93	109.9
(6)	.6333	1.348	3.094	5.437	8.700	13.45	20.77	33.17	58.15	133.1
(7)	.7532	1.602	3.670	6.439	10.29	15.87	24.48	39.06	68.39	156.4
(8)	.8731	1.856	4.246	7.441	11.87	18.30	28.20	44.95	78.63	179.7
(9)	.9933	2.109	4.823	8.443	13.46	20.73	31.92	50.84	88.88	202.9
(10)	1.113	2.363	5.400	9.446	15.05	23.17	35.64	56.74	99.14	226.2

TABLE 2.25: T^* ($1/\alpha = 0.4$)

Q: MODE:	0.05	0.10	0.20	0.30	0.40	0.50	0.60	0.70	0.80	0.90
(1)	.1188	.2544	.5886	1.036	1.654	2.541	3.897	6.191	10.82	24.79
(2)	.1877	.4065	.9605	1.725	2.805	4.382	6.823	10.97	19.37	44.68
(3)	.3373	.7194	1.655	2.904	4.629	7.112	10.91	17.33	30.26	69.13
(4)	.4849	1.030	2.350	4.094	6.481	9.896	15.10	23.86	41.45	94.25
(5)	.6327	1.341	3.047	5.288	8.341	12.69	19.31	30.41	52.68	119.5
(6)	.7805	1.652	3.745	6.484	10.20	15.49	23.52	36.97	63.92	144.7
(7)	.9282	1.964	4.444	7.681	12.07	18.30	27.73	43.55	75.19	170.0
(8)	1.076	2.275	5.142	8.879	13.93	21.10	31.95	50.12	86.46	195.3
(9)	1.224	2.586	5.841	10.08	15.80	23.91	36.17	56.70	97.73	220.5
(10)	1.372	2.898	6.540	11.27	17.67	26.72	40.40	63.27	109.0	245.9

TABLE 2.25: T^* ($1/\alpha = 0.5$)

Q: MODE:	0.05	0.10	0.20	0.30	0.40	0.50	0.60	0.70	0.80	0.90
(1)	.1324	.2829	.6511	1.141	1.811	2.769	4.231	6.695	11.66	26.63
(2)	.2183	.4708	1.103	1.961	3.154	4.878	7.520	11.99	20.99	48.13
(3)	.3922	.8333	1.901	3.305	5.214	7.929	12.05	18.96	32.83	74.49
(4)	.5639	1.193	2.701	4.662	7.306	11.04	16.68	26.11	44.99	101.6
(5)	.7358	1.554	3.503	6.024	9.407	14.17	21.34	33.29	57.19	128.8
(6)	.9077	1.915	4.306	7.388	11.51	17.30	26.00	40.49	69.41	156.0
(7)	1.080	2.276	5.110	8.753	13.62	20.44	30.67	47.69	81.65	183.2
(8)	1.251	2.637	5.913	10.12	15.72	23.57	35.34	54.90	93.89	210.5
(9)	1.424	2.997	6.717	11.48	17.83	26.71	40.01	62.11	106.1	237.8
(10)	1.595	3.359	7.521	12.85	19.94	29.85	44.68	69.32	118.4	265.1

TABLE 2.25: T^* ($1/\alpha = 0.6$)

Q: MODE:	0.05	0.10	0.20	0.30	0.40	0.50	0.60	0.70	0.80	0.90
(1)	.1448	.3089	.7087	1.237	1.959	2.987	4.551	7.184	12.49	28.46
(2)	.2450	.5271	1.227	2.169	3.467	5.329	8.166	12.95	22.57	51.52
(3)	.4402	.9331	2.117	3.659	5.737	0.671	13.10	20.49	35.31	79.75
(4)	.6330	1.336	3.009	5.164	8.044	12.08	18.14	28.24	48.40	108.8
(5)	.0260	1.740	3.903	6.674	10.36	15.51	23.21	36.01	61.53	137.9
(6)	1.019	2.145	4.799	8.187	12.68	18.94	28.28	43.80	74.68	167.0
(7)	1.212	2.549	5.695	9.700	15.00	22.37	33.37	51.60	87.86	196.2
(8)	1.406	2.964	6.590	11.21	17.32	25.81	38.45	59.39	101.0	225.5
(9)	1.598	3.358	7.487	12.73	19.65	29.25	43.54	67.19	114.2	254.6
(10)	1.791	3.762	8.383	14.24	21.97	32.68	48.63	75.00	127.4	283.9

TABLE 2.25: T^* ($1/\alpha = 0.7$)

Q: MODE:	0.05	0.10	0.20	0.30	0.40	0.50	0.60	0.70	0.80	0.90
(1)	.1563	.3330	.7624	1.329	2.100	3.196	4.862	7.663	13.30	30.27
(2)	.2606	.5770	1.339	2.357	3.753	5.746	8.774	13.86	24.10	54.86
(3)	.4827	1.022	2.310	3.977	6.213	9.356	14.08	21.96	37.71	84.94
(4)	.6941	1.463	3.283	5.615	8.715	13.04	19.51	30.26	61.70	115.9
(5)	.9057	1.906	4.260	7.259	11.23	16.74	24.96	38.60	65.74	146.9
(6)	1.117	2.348	5.238	8.904	13.74	20.45	30.42	46.94	79.79	177.9
(7)	1.329	2.791	6.216	10.55	16.26	24.16	35.89	55.31	93.87	209.0
(8)	1.541	3.234	7.194	12.20	18.77	27.87	41.37	63.67	108.0	240.2
(9)	1.753	3.677	8.172	13.85	21.29	31.58	46.84	72.03	122.0	271.3
(10)	1.964	4.120	9.151	15.49	23.81	35.29	52.32	80.40	136.2	302.4

TABLE 2.25: T^* ($1/\alpha = 0.8$)

Q: MODE:	0.05	0.10	0.20	0.30	0.40	0.50	0.60	0.70	0.80	0.90
(1)	.1670	.3556	.8132	1.416	2.235	3.399	5.164	8.132	14.10	32.07
(2)	.2897	.6216	1.439	2.529	4.017	6.137	9.352	14.75	25.59	58.17
(3)	.5205	1.101	2.483	4.268	6.653	9.995	15.01	23.36	40.05	90.07
(4)	.7485	1.576	3.531	6.027	9.333	13.94	20.80	32.20	54.91	122.9
(5)	.9768	2.053	4.582	7.791	12.02	17.89	26.62	41.08	69.83	155.8
(6)	1.205	2.530	5.633	9.558	14.72	21.85	32.45	49.96	84.76	188.7
(7)	1.433	3.008	6.686	11.33	17.41	25.82	38.28	58.86	99.72	221.7
(8)	1.662	3.485	7.738	13.09	20.11	29.79	44.12	67.76	114.7	254.7
(9)	1.890	3.962	8.790	14.86	22.81	33.75	49.96	76.67	129.7	287.7
(10)	2.118	4.439	9.843	16.63	25.51	37.73	55.80	85.58	144.6	320.7

TABLE 2.25: T^* ($1/\alpha = 0.9$)

Q: MODE:	0.05	0.10	0.20	0.30	0.40	0.50	0.60	0.70	0.80	0.90
(1)	.1771	.3769	.8616	1.499	2.365	3.595	5.460	8.594	14.90	33.87
(2)	.3086	.6619	1.531	2.687	4.264	6.506	9.904	15.60	27.05	61.44
(3)	.5545	1.172	2.642	4.535	7.062	10.60	15.90	24.72	42.34	95.14
(4)	.7974	1.679	3.757	6.405	9.908	14.78	22.04	34.08	58.06	129.8
(5)	1.041	2.187	4.874	8.280	12.76	18.97	28.20	43.47	73.83	164.5
(6)	1.284	2.695	5.993	10.16	15.62	23.18	34.37	52.88	89.62	199.3
(7)	1.527	3.203	7.113	12.04	18.49	27.38	40.56	62.30	105.4	234.2
(8)	1.770	3.711	8.232	13.92	21.35	31.59	46.74	71.72	121.3	269.0
(9)	2.014	4.219	9.352	15.80	24.22	35.80	52.93	81.14	137.1	303.9
(10)	2.257	4.728	10.47	17.68	27.08	40.01	59.12	90.57	152.9	338.8

TABLE 2.26: I_1^* ($\alpha = 0.1$)

Q: MODE:	0.05	0.10	0.20	0.30	0.40	0.50	0.60	0.70	0.80	0.90
(1)	.0359	3.356	13.70	31.93	59.94	101.4	163.8	265.2	459.6	1018
(2)	.9563	3.924	17.65	52.00	154.1	379.7	720.0	1232	2131	4503
(3)	32.04	64.94	150.4	304.1	689.0	1554	2781	4473	7287	14493
(4)	120.3	229.1	471.6	846.6	1683	3735	6775	10753	17094	32958
(5)	302.3	561.8	1097	1862	3460	7464	13614	21405	33468	63211
(6)	611.4	1121	2128	3495	6231	13140	24043	37567	58071	108112
(7)	1081	1967	3667	5894	10225	21178	38826	60402	92598	170584
(8)	1745	3160	5816	9209	15668	31995	58728	91074	138742	253534
(9)	2636	4757	8676	13588	22785	46006	84512	130743	198189	359896
(10)	3790	6819	12351	19180	31803	63627	116945	180569	272646	492552

TABLE 2.26: I_1^* ($\alpha = 0.2$)

Q: MODE:	0.05	0.10	0.20	0.30	0.40	0.50	0.60	0.70	0.80	0.90
(1)	.4302	1.730	7.067	16.50	31.05	52.69	85.54	139.4	243.7	546.3
(2)	.9269	3.787	16.65	45.15	106.2	217.1	387.8	651.7	1125	2397
(3)	30.10	61.52	141.2	269.5	507.1	917.6	1511	2370	3842	7704
(4)	112.9	216.9	444.7	767.2	1325	2304	3726	5715	9013	17498
(5)	283.4	531.5	1036	1701	2801	4733	7548	11398	17647	33540
(6)	572.7	1060	2012	3206	5127	8489	13407	20032	30623	57343
(7)	1012	1860	3468	5422	8498	13867	21744	32243	48835	90455
(8)	1633	2987	5502	8487	13111	21161	32999	48656	73176	134414
(9)	2468	4497	8209	12538	19162	30666	47613	69895	104536	190775
(10)	3547	6445	11688	17716	26848	42674	66027	96583	143816	261064

TABLE 2.26 : I_1^* (α = 0.3)

MODE \ Q:	0.05	0.10	0.20	0.30	0.40	0.50	0.60	0.70	0.80	0.90
(1)	.2967	1.193	4.878	11.41	21.51	36.63	59.72	97.85	172.3	389.9
(2)	.8952	3.644	15.69	40.32	85.83	161.5	277.5	460.0	792.7	1701
(3)	28.29	58.20	132.6	243.5	420.6	693.9	1088	1675	2705	5459
(4)	106.0	205.1	419.2	702.7	1129	1779	2706	4049	6345	12388
(5)	265.8	502.5	978.5	1566	2417	3699	5510	8088	12424	23734
(6)	537.1	1002	1900	2961	4456	6686	9823	14230	21560	40564
(7)	949.1	1758	3277	5017	7422	10983	15973	22924	34384	63974
(8)	1531	2822	5199	7863	11489	16829	24290	34615	51526	95049
(9)	2313	4247	7760	11627	16833	24464	35103	49749	73612	134888
(10)	3324	6087	11050	16440	23629	34128	48740	68773	101277	184568

TABLE 2.26 : I_1^* (α = 0.4)

MODE \ Q:	0.05	0.10	0.20	0.30	0.40	0.50	0.60	0.70	0.80	0.90
(1)	.2309	.9289	3.801	8.903	16.83	28.73	47.01	77.38	137.0	312.3
(2)	.8615	3.494	14.80	36.66	73.95	132.9	222.7	365.3	628.7	1356
(3)	26.61	55.04	124.6	222.9	367.2	576.8	877.5	1332	2144	4349
(4)	99.61	193.9	395.3	649.1	1000	1496	2194	3225	5028	9862
(5)	249.8	474.8	923.6	1452	2155	3131	4483	6450	9847	18887
(6)	504.4	947.1	1794	2751	3989	5683	8010	11358	17089	32272
(7)	891.2	1660	3095	4667	6661	9363	13047	18308	27255	50888
(8)	1437	2665	4912	7320	10331	14377	19866	27658	40845	75598
(9)	2171	4011	7333	10832	15158	20934	28737	39765	58355	107274
(10)	3120	5748	10443	15322	21300	29241	39934	54988	80289	146773

TABLE 2.26 : I_1^* (α = 0.5)

MODE \ Q:	0.05	0.10	0.20	0.30	0.40	0.50	0.60	0.70	0.80	0.90
(1)	.1924	.7741	3.170	7.434	14.07	24.09	39.54	65.32	116.2	266.1
(2)	.8264	3.342	13.96	33.75	65.94	115.3	190.1	309.4	531.8	1152
(3)	25.07	52.05	117.3	206.1	330.0	503.6	751.2	1129	1813	3691
(4)	93.77	183.3	373.0	603.7	906.3	1316	1885	2737	4252	8366
(5)	235.0	448.8	872.1	1354	1960	2763	3861	5479	8326	16018
(6)	474.6	894.9	1695	2568	3638	5029	6909	9654	14451	27366
(7)	838.3	1569	2924	4361	6084	8298	11267	15568	23048	43146
(8)	1352	2518	4642	6844	9446	12758	17170	23527	34541	64091
(9)	2042	3789	6930	10132	13871	18594	24853	33836	49350	90940
(10)	2934	5430	9870	14337	19505	25992	34554	46799	67901	124417

TABLE 2.26 : I_1^* (α = 0.6)

MODE \ Q:	0.05	0.10	0.20	0.30	0.40	0.50	0.60	0.70	0.80	0.90
(1)	.1675	.6741	2.762	6.484	12.29	21.08	34.68	57.44	102.5	235.6
(2)	.7904	3.188	13.19	31.36	60.09	103.3	168.5	272.7	468.4	1018
(3)	23.65	49.24	110.7	192.1	302.1	453.0	667.1	996.0	1596	3259
(4)	88.43	173.4	352.4	564.7	833.8	1188	1679	2416	3744	7384
(5)	221.6	424.4	824.2	1268	1808	2502	3443	4840	7331	14135
(6)	447.3	846.2	1602	2408	3360	4560	6168	8532	12724	24145
(7)	790.1	1483	2765	4091	5625	7533	10065	13764	20295	38065
(8)	1274	2380	4390	6424	8740	11591	15345	20805	30416	56540
(9)	1924	3582	6554	9513	12841	16903	22222	29926	43458	80221
(10)	2765	5133	9335	13465	18063	23638	30905	41398	59795	109749

TABLE 2.26 : I_1^* (α = 0.7)

MODE \ Q:	0.05	0.10	0.20	0.30	0.40	0.50	0.60	0.70	0.80	0.90
(1)	.1504	.6052	2.481	5.829	11.06	18.99	31.30	51.94	92.89	214.0
(2)	.7542	3.037	12.48	29.35	55.57	94.57	153.1	247.0	424.0	923.7
(3)	22.36	46.62	104.6	180.1	280.2	415.7	607.2	902.4	1445	2955
(4)	83.55	164.2	333.3	530.7	775.7	1093	1531	2191	3388	6694
(5)	209.3	401.7	779.9	1193	1684	2305	3142	4390	6635	12811
(6)	422.5	801.0	1516	2267	3133	4205	5632	7741	11516	21882
(7)	746.3	1404	2617	3852	5248	6951	9195	12490	18368	34496
(8)	1203	2253	4155	6051	8158	10700	14024	18883	27529	51235
(9)	1817	3390	6204	8962	11989	15609	20313	27165	39333	72693
(10)	2611	4858	8836	12687	16870	21835	28257	37583	54121	99447

TABLE 2.26 : I_1^* (α = 0.8)

MODE \ Q:	0.05	0.10	0.20	0.30	0.40	0.50	0.60	0.70	0.80	0.90
(1)	.1381	.5558	2.279	5.356	10.17	17.48	28.83	47.91	85.80	198.0
(2)	.7185	2.890	11.82	27.62	51.95	87.88	141.7	228.1	391.4	853.7
(3)	21.17	44.19	99.00	169.7	262.3	386.8	562.4	833.5	1333	2730
(4)	79.10	155.6	315.7	500.7	727.6	1019	1419	2024	3127	6184
(5)	198.1	380.7	739.0	1126	1581	2150	2915	4057	6123	11834
(6)	399.9	759.1	1437	2141	2943	3925	5226	7156	10628	20213
(7)	706.3	1330	2480	3639	4931	6490	8534	11547	16953	31863
(8)	1139	2135	3938	5717	7667	9992	13019	17459	25408	47324
(9)	1720	3212	5879	8469	11270	14580	18861	25118	36304	67142
(10)	2471	4603	8375	11990	15860	20396	26241	34753	49952	91852

TABLE 2.26 : I_1^* (α = 0.9)

MODE \ Q:	0.05	0.10	0.20	0.30	0.40	0.50	0.60	0.70	0.80	0.90
(1)	.1290	.5190	2.129	5.005	9.508	16.35	26.98	44.85	80.38	185.6
(2)	.6837	2.748	11.21	26.12	48.96	82.58	132.9	213.6	366.6	800.0
(3)	20.08	41.95	93.90	160.6	247.4	363.8	527.6	780.8	1248	2558
(4)	75.02	147.7	299.6	474.2	686.9	959.5	1332	1897	2928	5794
(5)	187.9	361.3	701.2	1067	1494	2024	2736	3802	5734	11087
(6)	379.3	720.4	1364	2028	2780	3696	4908	6706	9953	18937
(7)	669.8	1262	2354	3448	4660	6113	8015	10822	15875	29850
(8)	1080	2026	3737	5417	7246	9413	12229	16364	23793	44334
(9)	1631	3049	5580	8025	10652	13736	17717	23544	33996	62899
(10)	2344	4368	7948	11363	14991	19219	24650	32575	46778	86047

TABLE 2.26 : I_1^* (α = 1.0)

MODE \ Q:	0.05	0.10	0.20	0.30	0.40	0.50	0.60	0.70	0.80	0.90
(1)	.1220	.4910	2.014	4.735	8.997	15.47	25.54	42.46	76.11	175.8
(2)	.6504	2.614	10.66	24.81	46.44	78.27	125.9	202.3	347.1	757.6
(3)	19.08	39.87	89.24	152.5	234.8	344.8	499.8	739.4	1182	2422
(4)	71.30	140.4	284.7	450.4	651.9	909.8	1262	1796	2772	5486
(5)	178.6	343.5	666.5	1013	1418	1920	2593	3601	5429	10499
(6)	360.4	684.7	1296	1927	2639	3505	4650	6351	9424	17932
(7)	636.6	1200	2237	3276	4423	5797	7595	10249	15031	28267
(8)	1026	1926	3552	5147	6879	8928	11588	15498	22528	41982
(9)	1550	2898	5304	7625	10112	13028	16790	22298	32189	59562
(10)	2227	4152	7555	10796	14232	18230	23361	30852	44291	81481

TABLE 2.27 : η (α = 0.1)

MODE \ Q:	0.05	0.10	0.20	0.30	0.40	0.50	0.60	0.70	0.80	0.90
(1)	.1870	.8440	5.987	28.15	71.95	137.8	235.2	388.6	671.7	1454
(2)	2.390	7.141	82.74	384.7	734.4	1157	1714	2532	3977	7872
(3)	3.413	9.485	94.31	1243	2512	3874	5545	7894	11946	22729
(4)	4.598	12.46	106.4	2925	6097	9283	13030	18164	26899	50020
(5)	5.777	15.44	119.1	5664	12101	18301	25387	34932	51027	93497
(6)	6.950	18.44	133.2	9717	21154	31858	43851	59807	86543	156982
(7)	8.107	21.43	148.1	15339	33884	50886	69652	94392	135655	244245
(8)	9.312	24.44	163.6	22782	50920	76315	104028	140295	200582	359075
(9)	10.49	27.44	179.4	32301	72890	109074	148204	199124	283527	505296
(10)	11.65	30.48	195.3	44151	100415	150093	203413	272481	386689	686617

TABLE 2.27 : $\eta(\alpha = 0.2)$

Q: MODE:	0.05	0.10	0.20	0.30	0.40	0.50	0.60	0.70	0.80	0.90
(1)	.1702	.7482	4.415	15.78	37.35	70.22	119.1	196.3	339.2	734.9
(2)	2.342	6.856	53.32	196.9	370.0	580.9	859.6	1269	1993	3947
(3)	3.394	9.381	78.47	632.5	1261	1941	2776	3951	5980	11379
(4)	4.585	12.40	98.08	1480	3056	4647	6520	9088	13458	25029
(5)	5.765	15.40	114.2	2857	6060	9157	12699	17473	25524	46771
(6)	6.946	18.40	129.9	4891	10589	15937	21933	29912	43284	78518
(7)	8.112	21.42	145.7	7710	16957	25453	34835	47206	67841	122153
(8)	9.288	24.43	161.7	11440	25478	38169	52024	70159	100307	179572
(9)	10.47	27.42	177.9	16208	36465	54550	74113	99575	141782	252687
(10)	11.66	30.45	194.1	22142	50230	75061	101719	136255	193364	343351

TABLE 2.27 : $\eta(\alpha = 0.3)$

Q: MODE:	0.05	0.10	0.20	0.30	0.40	0.50	0.60	0.70	0.80	0.90
(1)	.1563	.6743	3.609	11.40	25.71	47.64	80.38	132.2	228.3	495.2
(2)	2.296	6.599	41.41	134.1	248.5	388.8	574.7	848.2	1332	2639
(3)	3.374	9.281	68.83	428.5	844.3	1296	1853	2637	3991	7596
(4)	4.571	12.34	91.59	998.5	2042	3101	4350	6062	8977	16698
(5)	5.757	15.35	109.9	1921	4047	6109	8470	11653	17023	31196
(6)	6.939	18.37	126.9	3283	7068	10630	14627	19947	28864	52363
(7)	8.122	21.39	143.4	5167	11315	16975	23229	31477	45237	81456
(8)	9.304	24.41	159.9	7659	16997	25454	34689	46780	66882	119738
(9)	10.48	27.43	176.4	10844	24323	36375	49416	66391	94533	168484
(10)	11.65	30.44	192.8	14805	33502	50050	67821	90846	128923	228930

TABLE 2.27 : $\eta(\alpha = 0.4)$

Q: MODE:	0.05	0.10	0.20	0.30	0.40	0.50	0.60	0.70	0.80	0.90
(1)	.1445	.6152	3.096	9.099	19.84	36.32	60.99	100.1	172.9	375.4
(2)	2.251	6.366	34.62	102.6	187.7	292.8	432.3	637.7	1001	1985
(3)	3.354	9.183	62.09	326.2	635.7	974.4	1391	1980	2996	5705
(4)	4.559	12.28	86.32	757.1	1535	2329	3265	4549	6737	12533
(5)	5.748	15.31	106.1	1453	3040	4585	6356	8743	12772	23408
(6)	6.932	18.33	124.1	2478	5307	7977	10974	14964	21654	39286
(7)	8.117	21.36	141.3	3895	8493	12736	17426	23613	33935	61107
(8)	9.291	24.38	158.2	5769	12756	19096	26022	35090	50170	89821
(9)	10.47	27.41	175.0	8161	18252	27288	37068	49800	70909	126382
(10)	11.65	30.43	191.6	11136	25137	37545	50872	68141	96702	171719

TABLE 2.27 : $\eta(\alpha = 0.5)$

Q: MODE:	0.05	0.10	0.20	0.30	0.40	0.50	0.60	0.70	0.80	0.90
(1)	.1344	.5664	2.732	7.655	16.28	29.51	49.35	80.88	139.6	303.4
(2)	2.208	6.152	30.13	83.52	151.1	235.1	346.8	511.4	803.5	1592
(3)	3.335	9.087	57.00	264.7	510.5	781.0	1115	1586	2400	4570
(4)	4.546	12.21	81.92	612.1	1231	1865	2614	3642	5393	10034
(5)	5.738	15.27	102.7	1171	2436	3671	5087	6997	10222	18736
(6)	6.924	18.30	121.5	1995	4250	6385	8782	11975	17328	31439
(7)	8.110	21.33	139.2	3132	6801	10193	13944	18894	27153	48898
(8)	9.294	24.36	156.5	4634	10212	15281	20822	28077	40142	71871
(9)	10.45	27.39	173.6	6551	14610	21836	29659	39845	56734	101121
(10)	11.64	30.41	190.4	8935	20119	30042	40703	54518	77369	137392

TABLE 2.27 : $\eta(\alpha = 0.6)$

Q: MODE:	0.05	0.10	0.20	0.30	0.40	0.50	0.60	0.70	0.80	0.90
(1)	.1257	.5254	2.437	6.654	13.88	24.95	41.57	68.04	117.4	255.5
(2)	2.167	5.956	26.88	70.77	126.8	196.7	289.8	427.2	671.2	1330
(3)	3.316	8.994	52.97	223.6	427.1	652.1	930.4	1323	2002	3813
(4)	4.532	12.15	78.16	515.3	1028	1556	2180	3037	4497	8368
(5)	5.728	15.22	99.66	984.3	2033	3061	4241	5834	8522	15621
(6)	6.916	18.27	119.0	1672	3546	5323	7321	9982	14444	26208
(7)	8.104	21.30	137.3	2623	5672	8497	11623	15748	22633	40759
(8)	9.290	24.34	154.9	3877	8516	12738	17355	23401	33457	59904
(9)	10.46	27.37	172.2	5478	12182	18201	24720	33208	47284	84281
(10)	11.63	30.39	189.3	7467	16773	25040	33923	45437	64481	114508

TABLE 2.27 : $\eta(\alpha = 0.7)$

Q: MODE:	0.05	0.10	0.20	0.30	0.40	0.50	0.60	0.70	0.80	0.90
(1)	.1180	.4904	2.240	5.914	12.15	21.68	36.01	58.86	101.6	221.7
(2)	2.128	5.774	24.40	61.61	109.4	169.2	249.1	367.1	576.7	1144
(3)	3.297	8.903	49.67	194.1	367.4	560.0	798.6	1135	1718	3273
(4)	4.522	12.09	74.90	446.0	883.8	1335	1870	2604	3857	7178
(5)	5.719	15.18	96.89	850.1	1745	2626	3637	5002	7307	13396
(6)	6.909	18.23	116.8	1442	3043	4565	6277	8558	12384	22477
(7)	8.097	21.28	135.4	2259	4866	7286	9965	13501	19403	34945
(8)	9.284	24.31	153.3	3337	7304	10922	14878	20061	28682	51356
(9)	10.47	27.35	170.9	4711	10447	15604	21191	28468	40535	72252
(10)	11.64	30.38	188.1	6418	14383	21467	29081	38950	55275	98162

TABLE 2.27 : $\eta(\alpha = 0.8)$

Q: MODE:	0.05	0.10	0.20	0.30	0.40	0.50	0.60	0.70	0.80	0.90
(1)	.1113	.4600	2.063	5.340	10.84	19.22	31.83	51.97	89.70	195.5
(2)	2.090	5.605	22.43	54.69	96.28	148.6	218.5	321.9	505.8	1003
(3)	3.278	8.815	46.90	171.9	322.7	490.9	699.7	994.6	1505	2867
(4)	4.509	12.03	72.02	393.9	775.1	1169	1637	2280	3377	6285
(5)	5.710	15.13	94.35	749.4	1530	2299	3184	4379	6396	11727
(6)	6.902	18.20	114.6	1269	2666	3997	5494	7490	10839	19670
(7)	8.092	21.25	133.6	1986	4261	6378	8722	11816	16981	30584
(8)	9.279	24.29	151.8	2931	6395	9559	13021	17556	25101	44945
(9)	10.46	27.33	169.6	4136	9146	13657	18545	24912	35472	63230
(10)	11.65	30.36	187.0	5632	12591	18787	25449	34084	48371	85902

TABLE 2.27 : $\eta(\alpha = 0.9)$

Q: MODE:	0.05	0.10	0.20	0.30	0.40	0.50	0.60	0.70	0.80	0.90
(1)	.1053	.4334	1.916	4.881	9.804	17.29	28.57	46.61	80.45	175.5
(2)	2.054	5.448	20.82	49.28	86.09	132.6	194.8	286.8	450.7	894.9
(3)	3.259	8.729	44.53	154.6	287.9	437.2	622.7	885.1	1339	2552
(4)	4.496	11.98	69.45	353.4	690.6	1040	1456	2028	3004	5591
(5)	5.700	15.09	92.01	670.9	1362	2045	2831	3894	5688	10429
(6)	6.894	18.16	112.6	1135	2372	3554	4885	6660	9637	17490
(7)	8.086	21.22	131.9	1774	3791	5671	7755	10505	15098	27193
(8)	9.273	24.27	150.4	2616	5689	8500	11576	15608	22315	39959
(9)	10.46	27.31	168.4	3688	8134	12142	16487	22147	31535	56213
(10)	11.64	30.34	185.9	5020	11197	16703	22624	30300	43001	76367

TABLE 2.27 : $\eta(\alpha = 1.0)$

Q: MODE:	0.05	0.10	0.20	0.30	0.40	0.50	0.60	0.70	0.80	0.90
(1)	.0999	.4098	1.790	4.504	8.969	15.75	25.96	42.31	73.04	159.5
(2)	2.019	5.301	19.46	44.93	77.93	119.7	175.7	258.8	406.6	807.7
(3)	3.241	8.644	42.46	140.7	260.0	394.2	561.2	797.5	1206	2300
(4)	4.484	11.92	67.14	320.8	622.9	937.9	1311	1826	2705	5036
(5)	5.691	15.05	89.85	608.0	1227	1842	2550	3506	5121	9391
(6)	6.887	18.13	110.7	1027	2137	3201	4398	5996	8676	15746
(7)	8.080	21.19	130.3	1604	3415	5106	6981	9457	13591	24480
(8)	9.269	24.24	149.0	2363	5123	7652	10421	14049	20087	35970
(9)	10.46	27.28	167.1	3330	7325	10931	14841	19935	28385	50599
(10)	11.64	30.32	184.8	4531	10082	15035	20364	27273	38704	68739

TABLE 2.27 : η (1/α = 0.1)

Q: MODE:	0.05	0.10	0.20	0.30	0.40	0.50	0.60	0.70	0.80	0.90
(1)	.0180	.0727	.2988	.7058	1.351	2.349	3.939	6.696	12.35	29.54
(2)	.8156	1.726	3.870	6.550	9.965	14.47	20.77	30.52	48.74	100.7
(3)	2.167	4.852	11.86	20.76	31.42	44.50	61.68	86.94	132.2	256.9
(4)	3.598	8.483	23.77	45.85	71.74	101.7	139.2	192.5	285.5	537.6
(5)	4.968	12.10	38.50	83.72	136.7	195.1	265.5	363.0	530.6	980.3
(6)	6.285	15.60	55.20	136.5	232.3	333.9	452.8	614.5	889.6	1623
(7)	7.567	19.01	73.22	206.3	364.9	527.4	713.5	963.2	1384	2503
(8)	8.824	22.33	92.04	295.3	540.5	784.9	1059	1425	2037	3659
(9)	10.06	25.59	111.4	405.9	765.4	1115	1504	2016	2870	5130
(10	11.29	28.81	130.9	540.5	1045	1529	2059	2752	3906	6951

TABLE 2.27 : η (1/α = 0.2)

Q: MODE:	0.05	0.10	0.20	0.30	0.40	0.50	0.60	0.70	0.80	0.90
(1)	.0331	.1330	.5437	1.271	2.397	4.081	6.659	10.93	19.34	44.14
(2)	1.213	2.700	6.544	11.61	18.11	26.63	38.38	56.22	88.79	179.4
(3)	2.651	6.344	18.30	36.13	57.85	83.96	117.6	166.2	251.8	484.1
(4)	4.039	10.05	34.44	79.89	134.3	195.3	269.9	374.4	554.6	1037
(5)	5.345	13.55	53.02	146.9	259.4	378.7	519.7	712.5	1040	1914
(6)	6.605	16.90	72.84	241.8	445.5	653.0	891.6	1212	1754	3192
(7)	7.844	20.15	93.16	369.1	705.2	1036	1410	1907	2740	4945
(8)	9.065	23.35	113.5	533.6	1051	1548	2100	2827	4043	7250
(9)	10.28	26.50	133.7	740.2	1495	2206	2986	4007	5705	10182
(10)	11.48	29.63	153.6	993.8	2051	3030	4093	5477	7772	13816

TABLE 2.27 : η (1/α = 0.3)

Q: MODE:	0.05	0.10	0.20	0.30	0.40	0.50	0.60	0.70	0.80	0.90
(1)	.0458	.1843	.7548	1.768	3.334	5.666	9.201	14.98	26.16	58.64
(2)	1.451	3.358	8.739	16.20	25.87	38.47	55.70	81.66	128.6	258.0
(3)	2.867	7.109	23.10	50.24	83.56	123.0	173.2	245.2	371.3	711.2
(4)	4.213	10.74	41.81	111.7	195.9	288.3	400.3	556.0	823.5	1537
(5)	5.484	14.13	62.33	206.9	380.9	561.9	773.6	1061	1551	2849
(6)	6.720	17.39	83.38	342.8	657.5	971.7	1330	1810	2620	4761
(7)	7.939	20.57	104.3	526.7	1044	1545	2106	2850	4097	7387
(8)	9.149	23.71	124.8	765.9	1560	2311	3140	4230	6048	10840
(9)	10.35	26.82	144.9	1067	2224	3297	4468	5998	8540	15234
(10)	11.55	29.91	164.4	1439	3055	4531	6127	8201	11639	20682

TABLE 2.27 : η (1/α = 0.4)

Q: MODE:	0.05	0.10	0.20	0.30	0.40	0.50	0.60	0.70	0.80	0.90
(1)	.0567	.2287	.9420	2.221	4.212	7.180	11.67	18.95	32.91	73.09
(2)	1.612	3.842	10.66	20.57	33.46	50.18	72.92	107.0	168.4	336.6
(3)	2.990	7.579	27.01	63.77	109.0	161.8	228.7	324.2	490.7	938.2
(4)	4.306	11.13	47.49	142.5	257.1	381.2	530.6	737.6	1092	2037
(5)	5.556	14.44	69.10	265.4	502.2	744.9	1027	1410	2061	3784
(6)	6.779	17.64	90.64	442.1	869.2	1290	1768	2408	3485	6331
(7)	7.989	20.79	111.6	682.3	1383	2054	2803	3794	5453	9829
(8)	9.192	23.90	132.0	995.8	2069	3074	4180	5633	8054	14430
(9)	10.39	26.98	151.7	1392	2953	4388	5950	7989	11375	20286
(10)	11.58	30.06	170.9	1883	4059	6031	8161	10926	15505	27547

TABLE 2.27 : η (1/α = 0.5)

Q: MODE:	0.05	0.10	0.20	0.30	0.40	0.50	0.60	0.70	0.80	0.90
(1)	.0663	.2677	1.111	2.644	5.052	8.654	14.09	22.88	39.63	87.52
(2)	1.727	4.217	12.39	24.80	40.97	61.82	90.10	132.3	208.1	415.1
(3)	3.069	7.898	30.35	76.96	134.3	200.6	284.1	403.1	610.0	1165
(4)	4.363	11.38	52.10	172.8	318.2	474.0	660.9	919.2	1361	2537
(5)	5.601	14.64	74.35	323.2	623.2	927.9	1281	1760	2571	4718
(6)	6.814	17.80	96.04	540.5	1080	1608	2206	3006	4350	7900
(7)	8.018	20.92	116.9	836.9	1722	2563	3499	4738	6809	12271
(8)	9.216	24.01	136.9	1224	2578	3837	5220	7036	10059	18020
(9)	10.41	27.08	156.3	1716	3682	5478	7432	9980	14210	25339
(10)	11.60	30.14	175.1	2325	5063	7532	10194	13650	19372	34412

TABLE 2.27 : η (1/α = 0.6)

Q: MODE:	0.05	0.10	0.20	0.30	0.40	0.50	0.60	0.70	0.80	0.90
(1)	.0746	.3023	1.267	3.045	5.866	10.10	16.49	26.79	46.33	101.9
(2)	1.814	4.518	13.98	28.93	48.41	73.43	107.3	157.6	247.8	493.6
(3)	3.124	8.129	33.29	89.94	159.5	239.4	339.6	482.0	729.4	1392
(4)	4.403	11.55	55.98	202.7	379.2	566.9	791.1	1100	1630	3037
(5)	5.631	14.77	78.58	380.6	744.2	1110	1534	2109	3081	5653
(6)	6.839	17.91	100.2	638.4	1292	1927	2645	3604	5215	9469
(7)	8.040	21.01	120.9	990.9	2060	3071	4195	5682	8166	14713
(8)	9.235	24.09	140.6	1453	3087	4600	6260	8439	12065	21610
(9)	10.42	27.15	159.7	2040	4410	6569	8913	11971	17045	30391
(10)	11.61	30.20	178.2	2767	6067	9033	12228	16375	23238	41278

TABLE 2.27 : η (1/α = 0.7)

Q: MODE:	0.05	0.10	0.20	0.30	0.40	0.50	0.60	0.70	0.80	0.90
(1)	.0820	.3333	1.411	3.429	6.661	11.53	18.87	30.68	53.02	116.3
(2)	1.882	4.765	15.46	32.99	55.82	85.02	124.4	182.9	287.5	572.1
(3)	3.165	8.304	35.91	102.8	184.7	278.1	395.0	560.9	848.7	1619
(4)	4.431	11.68	59.32	232.5	440.2	659.6	921.3	1282	1898	3536
(5)	5.651	14.87	82.09	437.7	865.2	1293	1788	2458	3591	6587
(6)	6.856	17.99	103.6	735.9	1503	2245	3083	4202	6081	11039
(7)	8.054	21.08	124.0	1144	2399	3580	4892	6625	9522	17154
(8)	9.247	24.14	143.4	1680	3596	5363	7300	9841	14070	25200
(9)	10.44	27.20	162.2	2363	5139	7659	10395	13962	19880	35443
(10)	11.62	30.25	180.5	3208	7070	10533	14262	19099	27105	48143

TABLE 2.27 : η (1/α = 0.8)

Q: MODE:	0.05	0.10	0.20	0.30	0.40	0.50	0.60	0.70	0.80	0.90
(1)	.0886	.3612	1.545	3.798	7.441	12.94	21.24	34.56	59.70	130.7
(2)	1.936	4.972	16.86	37.01	63.21	96.60	141.5	208.2	327.2	650.7
(3)	3.196	8.442	38.28	115.5	209.8	316.8	450.4	639.7	968.1	1846
(4)	4.453	11.78	62.24	262.0	501.1	752.4	1051	1463	2167	4036
(5)	5.667	14.94	85.07	494.6	986.1	1476	2042	2807	4101	7522
(6)	6.868	18.05	106.4	833.2	1715	2564	3522	4800	6946	12608
(7)	8.065	21.12	126.5	1297	2738	4089	5588	7569	10878	19596
(8)	9.256	24.18	145.6	1908	4105	6126	8341	11244	16076	28790
(9)	10.44	27.23	164.2	2685	5868	8750	11877	15953	22715	40495
(10)	11.63	30.28	182.2	3649	8074	12034	16296	21824	30971	55008

TABLE 2.27 : η (1/α = 0.9)

Q: MODE:	0.05	0.10	0.20	0.30	0.40	0.50	0.60	0.70	0.80	0.90
(1)	.0946	.3866	1.671	4.156	8.210	14.35	23.60	38.44	66.37	145.1
(2)	1.981	5.149	18.19	40.98	70.58	108.2	158.6	233.5	366.9	729.2
(3)	3.221	8.553	40.46	128.1	234.9	355.5	505.8	718.6	1087	2073
(4)	4.470	11.86	64.82	291.5	562.0	845.2	1181	1645	2436	4536
(5)	5.680	15.00	87.62	551.4	1107	1659	2296	3156	4611	8456
(6)	6.878	18.09	108.7	930.4	1926	2882	3960	5398	7811	14177
(7)	8.073	21.16	128.6	1451	3076	4597	6284	8513	12234	22038
(8)	9.262	24.22	147.5	2136	4614	6889	9381	12647	18081	32380
(9)	10.45	27.26	165.8	3008	6596	9840	13359	17944	25550	45547
(10)	11.64	30.30	183.7	4090	9078	13535	18330	24549	34838	61874

TABLE 2.28 : η_1 ($\alpha = 0.1$)

Q: MODE:	0.05	0.10	0.20	0.30	0.40	0.50	0.60	0.70	0.80	0.90
(1)	.6289	1.403	3.583	7.150	13.21	23.58	41.31	72.90	137.8	334.4
(2)	1.098	2.409	5.963	11.57	20.92	36.96	64.64	114.1	215.7	521.7
(3)	1.578	3.444	8.433	16.18	28.97	50.83	88.56	156.0	294.2	709.9
(4)	2.058	4.480	10.91	20.81	37.07	64.78	112.6	198.2	373.2	899.1
(5)	2.538	5.516	13.39	25.45	45.18	78.77	136.7	240.4	452.3	1088
(6)	3.018	6.554	15.87	30.09	53.31	92.78	160.9	282.7	531.6	1278
(7)	3.499	7.591	18.35	34.74	61.44	106.8	185.1	325.0	610.8	1468
(8)	3.980	8.627	20.83	39.39	69.58	120.8	209.2	367.3	690.2	1657
(9)	4.459	9.664	23.31	44.04	77.71	134.9	233.4	409.7	769.6	1847
(10)	4.939	10.70	25.80	48.69	85.86	148.9	257.6	452.0	848.9	2037

TABLE 2.28 : η_1 ($\alpha = 0.2$)

Q: MODE:	0.05	0.10	0.20	0.30	0.40	0.50	0.60	0.70	0.80	0.90
(1)	.5687	1.247	3.046	5.699	9.704	15.90	25.85	43.17	78.56	185.7
(2)	.9952	2.154	5.132	9.407	15.76	25.51	41.15	68.27	123.5	290.2
(3)	1.431	3.082	7.275	13.22	21.97	35.34	56.71	93.67	168.8	395.1
(4)	1.867	4.011	9.424	17.05	28.21	45.22	72.35	119.2	214.4	500.6
(5)	2.303	4.941	11.57	20.88	34.47	55.12	88.01	144.8	260.0	606.1
(6)	2.739	5.871	13.73	24.71	40.73	65.03	103.7	170.4	305.6	712.0
(7)	3.174	6.801	15.88	28.55	46.99	74.95	119.4	196.1	351.3	817.9
(8)	3.611	7.730	18.03	32.39	53.26	84.86	135.1	221.7	397.1	923.6
(9)	4.046	8.661	20.18	36.23	59.53	94.79	150.8	247.3	442.9	1029
(10)	4.482	9.591	22.34	40.06	65.80	104.7	166.5	273.0	488.6	1135

TABLE 2.28 : η_1 ($\alpha = 0.3$)

Q: MODE:	0.05	0.10	0.20	0.30	0.40	0.50	0.60	0.70	0.80	0.90
(1)	.5203	1.129	2.690	4.884	8.042	12.74	20.09	32.70	58.30	135.7
(2)	.9119	1.956	4.561	8.134	13.19	20.64	32.22	51.96	91.92	212.3
(3)	1.312	2.801	6.475	11.45	18.44	28.67	44.50	71.42	125.8	289.2
(4)	1.711	3.647	8.393	14.79	23.72	36.74	56.85	90.98	159.8	366.5
(5)	2.111	4.493	10.31	18.12	29.00	44.82	69.22	110.6	193.8	443.8
(6)	2.511	5.339	12.23	21.46	34.29	52.91	81.61	130.2	228.0	521.3
(7)	2.910	6.186	14.13	24.80	39.58	61.01	94.01	149.8	262.1	598.9
(8)	3.311	7.031	16.08	28.14	44.87	69.11	106.4	169.5	296.3	676.3
(9)	3.710	7.878	18.00	31.49	50.17	77.22	118.8	189.1	330.5	754.1
(10)	4.110	8.725	19.92	34.83	55.46	85.32	131.2	208.8	364.6	831.6

TABLE 2.28 : η_1 ($\alpha = 0.4$)

Q: MODE:	0.05	0.10	0.20	0.30	0.40	0.50	0.60	0.70	0.80	0.90
(1)	.4002	1.036	2.430	4.339	7.019	10.92	16.96	27.20	47.93	110.4
(2)	.8425	1.797	4.135	7.260	11.57	17.78	27.30	43.35	75.69	172.9
(3)	1.212	2.575	5.875	10.24	16.20	24.74	37.75	59.65	103.6	235.6
(4)	1.582	3.352	7.618	13.22	20.85	31.73	48.26	76.02	131.7	298.7
(5)	1.951	4.130	9.363	16.21	25.51	38.73	58.79	92.43	159.8	361.7
(6)	2.321	4.909	11.11	19.20	30.17	45.74	69.34	108.9	188.0	425.0
(7)	2.690	5.688	12.86	22.20	34.83	52.76	79.89	125.3	216.1	488.2
(8)	3.060	6.465	14.60	25.19	39.50	59.77	90.44	141.7	244.3	551.3
(9)	3.429	7.244	16.35	28.18	44.16	66.79	101.0	158.2	272.6	614.7
(10)	3.799	8.023	18.10	31.18	48.83	73.81	111.5	174.6	300.7	677.9

TABLE 2.28 : η_1 ($\alpha = 0.5$)

Q: MODE:	0.05	0.10	0.20	0.30	0.40	0.50	0.60	0.70	0.80	0.90
(1)	.4464	.9584	2.228	3.938	6.306	9.716	14.94	23.76	41.56	95.16
(2)	.7837	1.666	3.800	6.609	10.43	15.86	24.10	37.94	65.70	149.1
(3)	1.128	2.387	5.401	9.324	14.61	22.09	33.36	52.23	89.98	203.2
(4)	1.472	3.108	7.006	12.05	18.81	28.34	42.67	66.59	114.4	257.5
(5)	1.815	3.829	8.612	14.78	23.02	34.60	51.99	80.98	138.8	311.9
(6)	2.159	4.552	10.22	17.51	27.23	40.87	61.33	95.39	163.3	366.5
(7)	2.503	5.274	11.83	20.24	31.45	47.15	70.67	109.8	187.8	421.0
(8)	2.848	5.995	13.43	22.97	35.66	53.42	80.01	124.2	212.3	475.5
(9)	3.191	6.717	15.04	25.70	39.88	59.70	89.36	138.6	236.8	530.2
(10)	3.535	7.439	16.65	28.43	44.10	65.97	98.71	153.1	261.3	584.7

TABLE 2.28 : η_1 ($\alpha = 0.6$)

Q: MODE:	0.05	0.10	0.20	0.30	0.40	0.50	0.60	0.70	0.80	0.90
(1)	.4173	.8935	2.065	3.627	5.772	8.840	13.51	21.38	37.22	84.88
(2)	.7331	1.554	3.527	6.098	9.563	14.46	21.84	34.17	58.88	133.0
(3)	1.055	2.227	5.015	8.607	13.41	20.14	30.24	47.06	80.66	181.3
(4)	1.377	2.901	6.506	11.12	17.27	25.85	38.68	60.02	102.6	229.8
(5)	1.698	3.574	7.998	13.64	21.13	31.56	47.14	72.99	124.5	278.4
(6)	2.020	4.249	9.491	16.17	25.00	37.29	55.61	85.99	146.4	327.1
(7)	2.342	4.923	10.98	18.69	28.87	43.02	64.09	98.99	168.4	375.8
(8)	2.664	5.596	12.48	21.21	32.74	48.74	72.56	112.0	190.4	424.4
(9)	2.985	6.270	13.97	23.74	36.61	54.47	81.04	125.0	212.4	473.2
(10)	3.307	6.945	15.47	26.26	40.49	60.20	89.52	138.0	234.3	521.9

TABLE 2.28 : η_1 ($\alpha = 0.7$)

Q: MODE:	0.05	0.10	0.20	0.30	0.40	0.50	0.60	0.70	0.80	0.90
(1)	.3921	.8380	1.929	3.376	5.352	8.167	12.44	19.62	34.06	77.47
(2)	.6890	1.458	3.298	5.682	8.878	13.37	20.12	31.38	53.91	121.4
(3)	.9915	2.090	4.691	8.022	12.45	18.63	27.87	43.23	73.85	165.5
(4)	1.294	2.722	6.086	10.37	16.04	23.92	35.66	55.13	93.91	209.8
(5)	1.596	3.355	7.482	12.72	19.63	29.21	43.46	67.06	114.0	254.2
(6)	1.899	3.988	8.880	15.07	23.22	34.51	51.28	79.01	134.1	298.6
(7)	2.201	4.620	10.28	17.43	26.82	39.81	59.09	90.95	154.2	343.1
(8)	2.504	5.252	11.67	19.78	30.42	45.11	66.91	102.9	174.3	387.5
(9)	2.806	5.885	13.07	22.13	34.02	50.42	74.74	114.9	194.5	432.1
(10)	3.108	6.518	14.47	24.49	37.62	55.72	82.56	126.8	214.6	476.5

TABLE 2.28 : η_1 ($\alpha = 0.8$)

Q: MODE:	0.05	0.10	0.20	0.30	0.40	0.50	0.60	0.70	0.80	0.90
(1)	.3699	.7898	1.814	3.168	5.011	7.631	11.60	18.26	31.65	71.87
(2)	.6501	1.375	3.103	5.335	8.318	12.50	18.77	29.22	50.10	112.7
(3)	.9356	1.971	4.414	7.533	11.67	17.42	26.01	40.25	68.64	153.6
(4)	1.221	2.567	5.727	9.739	15.03	22.36	33.28	51.34	87.29	194.7
(5)	1.506	3.163	7.042	11.95	18.40	27.32	40.56	62.45	105.9	235.8
(6)	1.792	3.760	8.357	14.16	21.77	32.27	47.85	73.58	124.6	277.1
(7)	2.077	4.356	9.672	16.37	25.14	37.23	55.15	84.71	143.4	318.4
(8)	2.363	4.952	10.99	18.58	28.51	42.19	62.45	95.84	162.1	359.6
(9)	2.648	5.549	12.30	20.79	31.88	47.16	69.75	107.0	180.8	400.9
(10)	2.933	6.146	13.62	23.00	35.26	52.12	77.05	118.1	199.5	442.2

TABLE 2.28 : η_1 ($\alpha = 0.9$)

Q: MODE:	0.05	0.10	0.20	0.30	0.40	0.50	0.60	0.70	0.80	0.90
(1)	.3503	.7474	1.715	2.992	4.727	7.190	10.92	17.18	29.73	67.48
(2)	.6156	1.301	2.935	5.039	7.848	11.78	17.67	27.48	47.08	105.8
(3)	.8860	1.865	4.174	7.116	11.01	16.42	24.49	37.86	64.51	144.2
(4)	1.156	2.430	5.416	9.200	14.18	21.08	31.34	48.30	82.03	182.8
(5)	1.427	2.994	6.659	11.29	17.36	25.75	38.19	58.75	99.57	221.4
(6)	1.697	3.559	7.903	13.37	20.54	30.42	45.06	69.22	117.1	260.2
(7)	1.967	4.124	9.147	15.46	23.72	35.10	51.94	79.69	134.7	298.9
(8)	2.238	4.688	10.39	17.55	26.91	39.78	58.81	90.16	152.3	337.6
(9)	2.507	5.253	11.64	19.64	30.09	44.46	65.69	100.6	169.9	376.5
(10)	2.778	5.818	12.88	21.73	33.28	49.13	72.56	111.1	187.5	415.2

TABLE 2.28 : η_1 ($\alpha = 1.0$)

Q: MODE:	0.05	0.10	0.20	0.30	0.40	0.50	0.60	0.70	0.80	0.90
(1)	.3327	.7098	1.628	2.839	4.485	6.820	10.36	16.28	28.18	63.94
(2)	.5848	1.236	2.786	4.783	7.447	11.17	16.76	26.06	44.62	100.2
(3)	.8416	1.772	3.963	6.755	10.45	15.58	23.22	35.90	61.14	136.6
(4)	1.098	2.308	5.142	8.733	13.46	20.00	29.72	45.79	77.75	173.2
(5)	1.355	2.844	6.323	10.71	16.47	24.43	36.22	55.70	94.38	209.8
(6)	1.612	3.380	7.504	12.70	19.49	28.86	42.74	65.63	111.0	246.6
(7)	1.868	3.917	8.685	14.68	22.51	33.30	49.25	75.55	127.7	283.3
(8)	2.126	4.452	9.867	16.66	25.53	37.74	55.77	85.48	144.4	319.9
(9)	2.382	4.989	11.05	18.64	28.55	42.18	62.29	95.41	161.1	356.7
(10)	2.639	5.526	12.23	20.63	31.58	46.61	68.81	105.3	177.7	393.4

CHAPTER THREE

FREE VIBRATION OF DOUBLE SPAN BEAMS WITH CLASSICAL AND NONCLASSICAL BOUNDARY CONDITIONS

Part 1. Beams with Classical Boundary Conditions

It is readily shown that there exist six unique families of uniform beams with classical boundary conditions. To designate these beams we utilize a triple order of subscripts of the type used in Chapter 2. For example, subscripts *SSS* denote a beam with simple (pinned) support at each end and intermediate simple support. Similarly, subscripts *CSS* denote a beam clamped at one end and with simple support at the other end and at an intermediate point. All intermediate supports must of course be simple supports.

For illustrative purposes the analytical procedure for obtaining the frequency and modal shape is presented for a typical family of beams. We consider therefore a beam with simple-simple-simple support, as shown in the inset of Fig. 3.1. The vibratory motion can still be expressed as a product of a displacement function and a time function as for single span beams, however, now the displacement for each span is handled separately and compatible common boundary conditions are enforced.

The spatial coordinates measured along the spans as well as the displacements are nondimensionalized with respect to L, the beam overall length. The length of the first span is designated μ, where in general $0<\mu<1$, and the remaining span has length $\gamma = 1-\mu$.

The governing differential equation is satisfied by expressing the first and second span deflections, respectively, as

$$r_1(\xi) = A_1 \sin \beta\xi + B_1 \cos \beta\xi + C_1 \sinh \beta\xi + D_1 \cosh \beta\xi \tag{3.1}$$

and

$$r_2(\xi) = A_2 \sin \beta\xi + B_2 \cos \beta\xi + C_2 \sinh \beta\xi + D_2 \cosh \beta\xi \tag{3.2}$$

where, to simplify the problem of enforcing the boundary conditions, ξ is measured in each span from the beam outer end. Upon enforcing the boundary conditions at each end of the beam, it is readily shown that $B_1 = D_1 = B_2 = D_2 = 0$. There are four boundary conditions remaining to be satisfied, which are

1. $r_1(\mu) = 0$
2. $r_2(\gamma) = 0$
3. $\left.\dfrac{dr_1(\xi)}{d\xi}\right|_{\xi=\mu} = \left.\dfrac{-dr_2(\xi)}{d\xi}\right|_{\xi=\gamma}$
4. $\left.\dfrac{d^2r_1(\xi)}{d\xi^2}\right|_{\xi=\mu} = \left.\dfrac{d^2r_2(\xi)}{d\xi^2}\right|_{\xi=\gamma}$

Boundary conditions 3 and 4 arise from the requirement to satisfy continuity of slope and bending moment, respectively, at the intermediate support. It is readily shown that satisfaction of boundary conditions 1 and 2 requires that

$$C_1 = -A_1\frac{\sin\beta\mu}{\sinh\beta\mu} \qquad \text{and} \qquad C_2 = -A_2\frac{\sin\beta\gamma}{\sinh\beta\gamma}$$

Utilizing these latter two equations and boundary conditions 3 and 4 we obtain

$$A_1\left(\cos\beta\mu\frac{-\sin\beta\mu\cosh\beta\mu}{\sinh\beta\mu}\right) + A_2\left(\cos\beta\gamma\frac{-\sin\beta\gamma\cosh\beta\gamma}{\sinh\beta\gamma}\right) = 0 \tag{3.3}$$

and

$$A_1(-\sin\beta\mu) + A_2(\sin\beta\gamma) = 0 \tag{3.4}$$

Finally, given any value of μ, $0 < \mu < 1$, we determine the eigenvalues β of the system by requiring that the determinant of the coefficient matrix for Eqs. 3.3 and 3.4 vanish. We recall that the quantity γ appearing in the equations is known once we specify μ, since $\gamma = (1-\mu)$. Because of the symmetry in this particular problem, we are never interested in values of $\mu > 0.5$. We may always choose our axis so that $\mu \leqslant 0.5$. This symmetry manifests itself in the determinant equation which is seen to be symmetrical in μ and γ. Where this symmetry does not exist, we are interested in eigenvalues for the full range of μ.

A plot of β versus μ is given in Fig. 3.1 for the first five modes of vibration. It is noted that as μ approaches zero the eigenvalues approach those for clamped–simply supported single span beams (Table 1.1), as would be expected. However, as μ approaches 0.5, the values of β approach 2π, 4π, and so on, for the first and third mode, and so on. These are antisymmetric modes, and the modal shapes correspond to extensions of the well-known first and second modes of single span simple-simple beams. The symmetric modes (with $\mu = .5$) correspond to extensions of the clamped–simply supported single span modal shapes. The limiting cases, as μ approaches 0, 0.5, or 1.0 (where appropriate), can be readily identified from the single span analysis for all of the six double span cases presented. Each of the six cases is given a designation number along with a brief description. This is followed by information giving the modal shapes. In determining modal shapes, only one of the coefficients is arbitrarily set equal to unity, as in the case of single span beams. Eigenvalues for the first five flexural vibration modes are to be found for any beam from corresponding figures with corresponding designation numbers (see Figs. 3.1 to 3.6). It is noted that these modes are numbered 1 to 5 in each figure except Fig. 3.6 where they are numbered 2 to 6. The reason for this is that in Fig. 3.6, which corresponds to a free-simple-free beam, there exists a nonflexural mode (rigid body rotation) which we designate the first mode, corresponding to $\beta = 0$.

CASE 3.1. Beam with simple-simple-simple support.

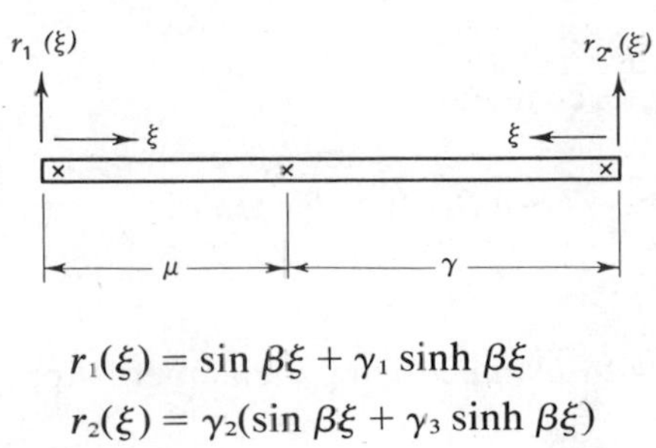

$$r_1(\xi) = \sin \beta\xi + \gamma_1 \sinh \beta\xi$$
$$r_2(\xi) = \gamma_2(\sin \beta\xi + \gamma_3 \sinh \beta\xi)$$

where

$$\gamma_1 = \frac{-\sin \beta\mu}{\sinh \beta\mu} \qquad \gamma_2 = \frac{\sin \beta\mu}{\sin \beta\gamma} \qquad \gamma_3 = \frac{-\sin \beta\gamma}{\sinh \beta\gamma}$$

CASE 3.2. Beam with simple-simple-free support.

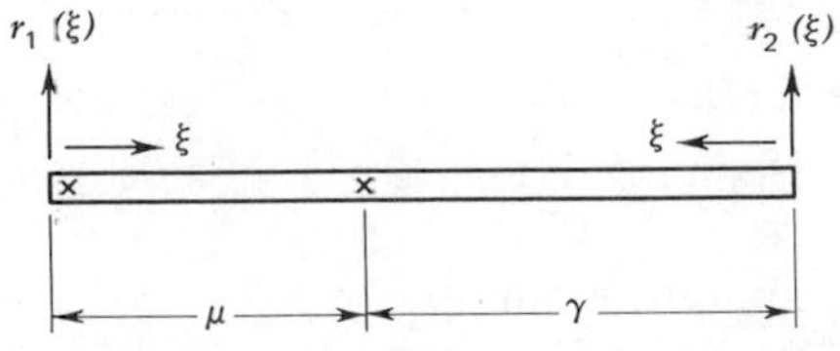

$$r_1(\xi) = \sin \beta\xi + \gamma_1 \sinh \beta\xi$$
$$r_2(\xi) = \gamma_2[\sin \beta\xi + \sinh \beta\xi + \gamma_3(\cos \beta\xi + \cosh \beta\xi)]$$

where

$$\gamma_1 = \frac{-\sin \beta\mu}{\sinh \beta\mu}$$

$$\gamma_2 = \frac{(\sin \beta\mu \cosh \beta\mu - \cos \beta\mu \sinh \beta\mu)(\cos \beta\gamma + \cosh \beta\gamma)}{2 \sinh \beta\mu (1 + \cos \beta\gamma \cosh \beta\gamma)}$$

$$\gamma_3 = -\frac{\sin \beta\gamma + \sinh \beta\gamma}{\cos \beta\gamma + \cosh \beta\gamma}$$

CASE 3.3. Beam with clamped-simple-simple support.

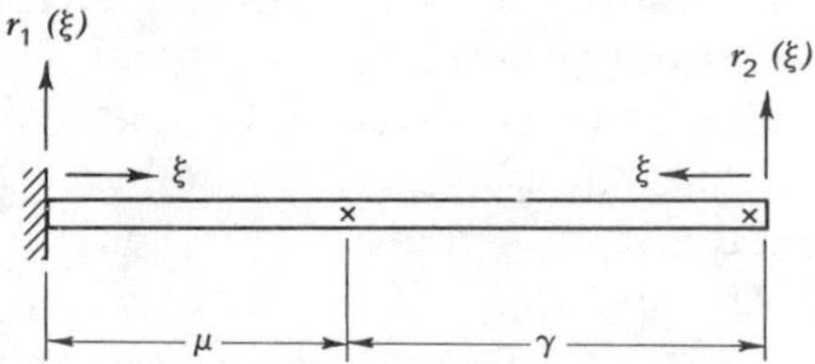

$$r_1(\xi) = \sin \beta\xi - \sinh \beta\xi + \gamma_1(\cos \beta\xi - \cosh \beta\xi)$$
$$r_2(\xi) = \gamma_2(\sin \beta\xi + \gamma_3 \sinh \beta\xi)$$

where

$$\gamma_1 = \frac{\sinh \beta\mu - \sin \beta\mu}{\cos \beta\mu - \cosh \beta\mu}$$

$$\gamma_2 = \frac{-2(1 - \cos \beta\mu \cosh \beta\mu)}{(\cos \beta\mu - \cosh \beta\mu)(\cos \beta\gamma \sinh \beta\gamma - \sin \beta\gamma \cosh \beta\gamma)}$$

$$\gamma_3 = \frac{-\sin \beta\gamma}{\sinh \beta\gamma}$$

CASE 3.4. Beam with clamped-simple-clamped support.

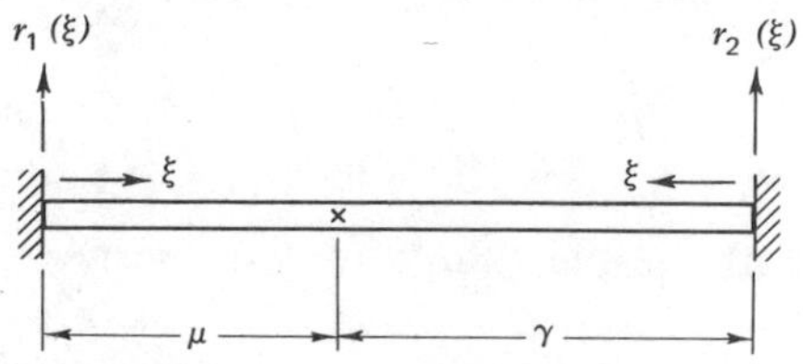

$$r_1(\xi) = \sin \beta\xi - \sinh \beta\xi + \gamma_1(\cos \beta\xi - \cosh \beta\xi)$$
$$r_2(\xi) = \gamma_2[\sin \beta\xi - \sinh \beta\xi + \gamma_3(\cos \beta\xi - \cosh \beta\xi)]$$

where

$$\gamma_1 = \frac{\sinh \beta\mu - \sin \beta\mu}{\cos \beta\mu - \cosh \beta\mu}$$

$$\gamma_2 = \frac{(1 - \cos \beta\mu \cosh \beta\mu)(\cosh \beta\gamma - \cos \beta\gamma)}{(\cos \beta\mu - \cosh \beta\mu)(1 - \cos \beta\gamma \cosh \beta\gamma)}$$

$$\gamma_3 = \frac{\sinh \beta\gamma - \sin \beta\gamma}{\cos \beta\gamma - \cosh \beta\gamma}$$

CASE 3.5. Beam with clamped-simple-free support.

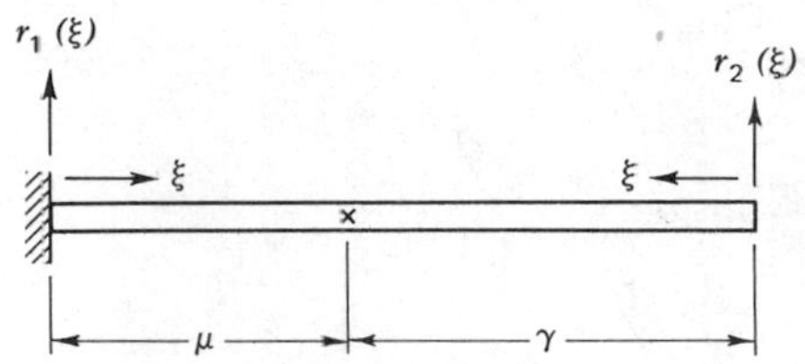

$$r_1(\xi) = \sin \beta\xi - \sinh \beta\xi + \gamma_1(\cos \beta\xi - \cosh \beta\xi)$$
$$r_2(\xi) = \gamma_2[\sin \beta\xi + \sinh \beta\xi + \gamma_3(\cos \beta\xi + \cosh \beta\xi)]$$

where

$$\gamma_1 = \frac{\sinh \beta\mu - \sin \beta\mu}{\cos \beta\mu - \cosh \beta\mu}$$
$$\gamma_2 = \frac{(1 - \cos \beta\mu \cosh \beta\mu)(\cos \beta\gamma + \cosh \beta\gamma)}{(\cosh \beta\mu - \cos \beta\mu)(1 + \cos \beta\gamma \cosh \beta\gamma)}$$
$$\gamma_3 = -\frac{\sin \beta\gamma + \sinh \beta\gamma}{\cos \beta\gamma + \cosh \beta\gamma}$$

CASE 3.6. Beam with free-simple-free support.

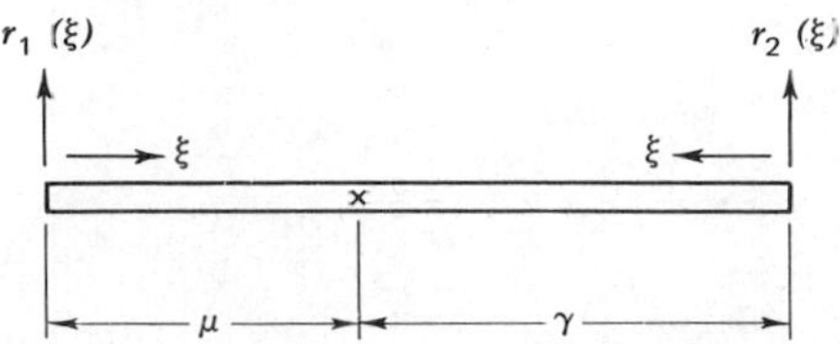

$$r_1(\xi) = \sin \beta\xi + \sinh \beta\xi + \gamma_1(\cos \beta\xi + \cosh \beta\xi)$$
$$r_2(\xi) = \gamma_2[\sin \beta\xi + \sinh \beta\xi + \gamma_3(\cos \beta\xi + \cosh \beta\xi)]$$

where

$$\gamma_1 = \frac{-(\sin \beta\mu + \sinh \beta\mu)}{\cos \beta\mu + \cosh \beta\mu}$$
$$\gamma_2 = -\frac{(1 + \cos \beta\mu \cosh \beta\mu)(\cos \beta\gamma + \cosh \beta\gamma)}{(\cos \beta\mu + \cosh \beta\mu)(1 + \cos \beta\gamma \cosh \beta\gamma)}$$
$$\gamma_3 = \frac{-(\sin \beta\gamma + \sinh \beta\gamma)}{\cos \beta\gamma + \cosh \beta\gamma}$$

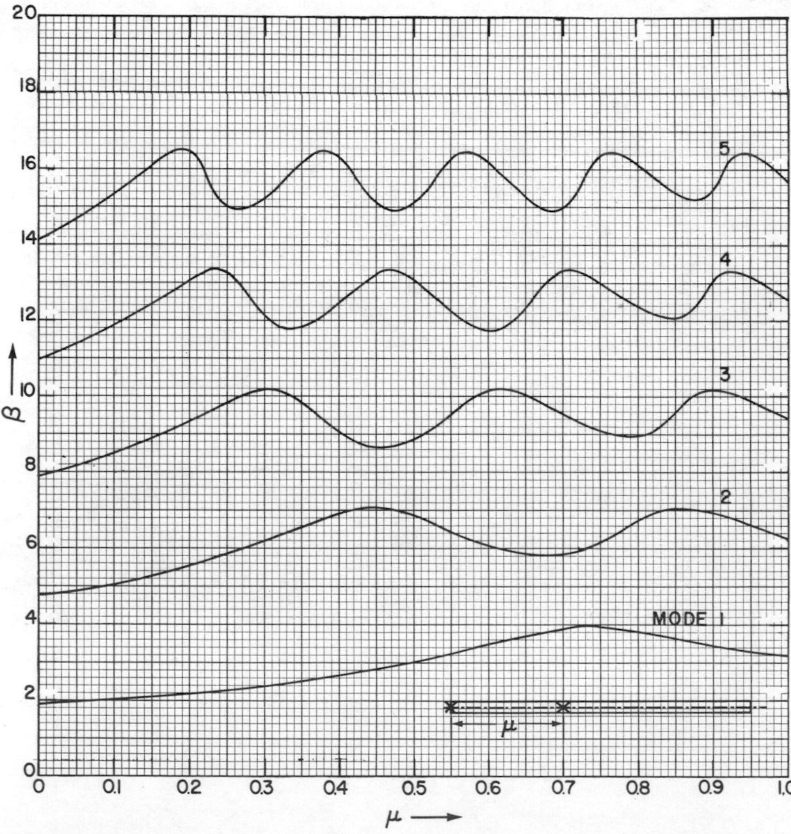

Fig. 3.1 Eigenvalues versus intermediate support spacing for beams with simple-simple-simple support.

Fig. 3.2 Eigenvalues versus intermediate support spacing for beams with simple-simple-free suport.

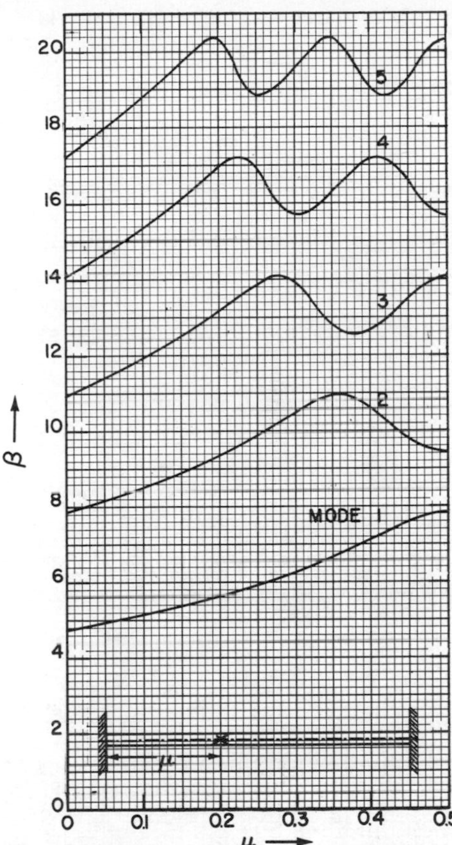

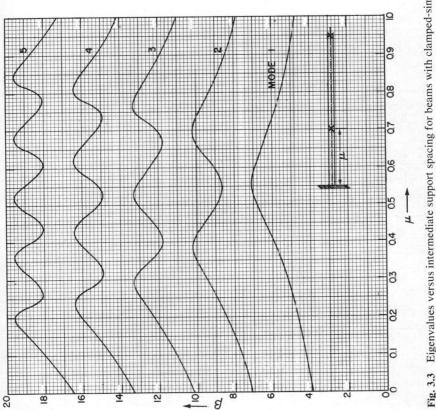

Fig. 3.3 Eigenvalues versus intermediate support spacing for beams with clamped-simple-simple support.

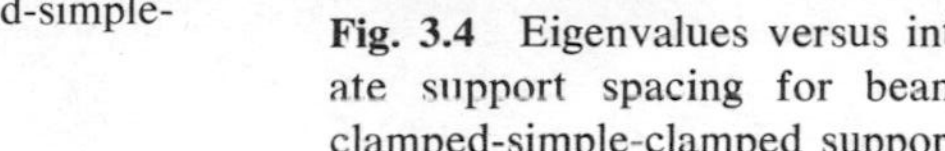

Fig. 3.4 Eigenvalues versus intermediate support spacing for beams with clamped-simple-clamped support.

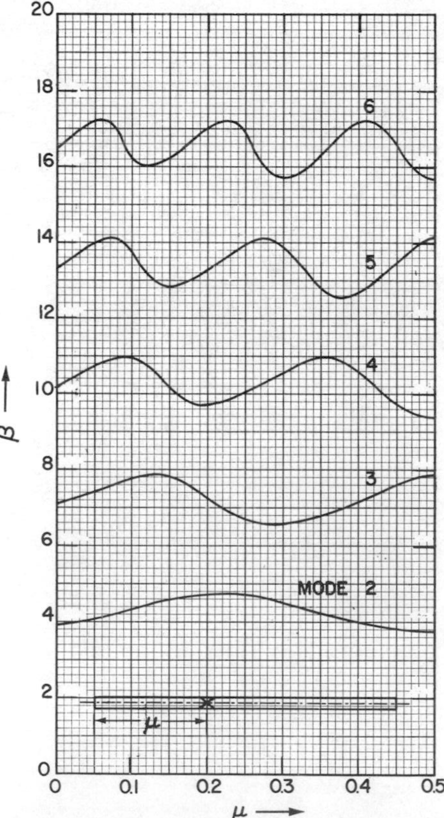

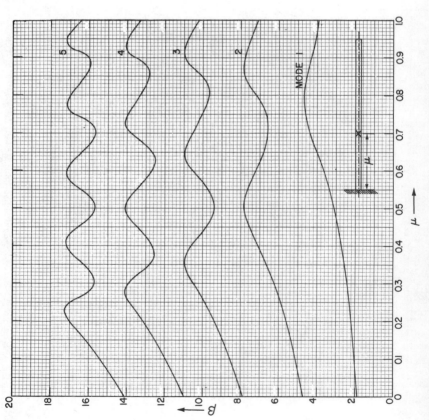

Fig. 3.5 Eigenvalues versus intermediate support spacing for beams with clamped-simple-free support.

Fig. 3.6 Eigenvalues versus intermediate support spacing for beams with free-simple-free support.

Part 2. Beams with Single Nonclassical Boundary Conditions Including Arbitrarily Located Springs and Masses

Data presented in this part of the chapter permit the rapid evaluation of frequencies and modal shapes for a large family of beams of practical industrial interest. The general procedure followed is virtually identical to that of the previous chapter, however, it will be seen that in general the upper and/or lower limits are double span beams with classical boundary conditions. These limits are immediately available of course from Part 1 of this chapter.

To illustrate the technique utilized for acquiring the tabulated data, the problem of analyzing the free vibration of a cantilever (clamped-free) beam with mass M located distance μ from the clamped end is worked out in detail (see inset figure for Case 3.7). As in the case for double span beams with classical boundary conditions, two functions $r_1(\xi)$ and $r_2(\xi)$ are used to describe the displacement mode in the first and second spans, respectively. Utilizing the boundary conditions at the extreme ends of the beam it is evident that we may write

$$r_1(\xi) = \frac{dr_1(\xi)}{d\xi} = \frac{d^2r_2(\xi)}{d\xi^2} = \frac{d^3r_2(\xi)}{d\xi^3} = 0\Big|_{\xi=0} \tag{3.5}$$

Equation 3.5 permits elimination of two constants from each of the displacement functions, and we obtain

$$r_1(\xi) = A_1(\sin\beta\xi - \sinh\beta\xi) + B_1(\cos\beta\xi - \cosh\beta\xi)$$

and

$$r_2(\xi) = A_2(\sin\beta\xi + \sinh\beta\xi) + B_2(\cos\beta\xi + \cosh\beta\xi)$$

There remain four interspan boundary conditions which we write:

1. From continuity of displacement.

$$r_1(\mu) = r_2(\gamma)$$

2. From continuity of slope.

$$\frac{dr_1(\xi)}{d\xi}\Big|_{\xi=\mu} = -\frac{dr_2(\xi)}{d\xi}\Big|_{\xi=\gamma}$$

3. From continuity of bending moment.

$$\frac{d^2r_1(\xi)}{d\xi^2}\Big|_{\xi=\mu} = \frac{d^2r_2(\xi)}{d\xi^2}\Big|_{\xi=\gamma}$$

4. From condition of dynamic equilibrium between motion of the concentrated mass and adjacent shear forces.

$$\frac{d^3r_1(\xi)}{d\xi^3}\Big|_{\xi=\mu} + \frac{d^3r_2(\xi)}{d\xi^3}\Big|_{\xi=\gamma} = \frac{-\beta^4}{\eta}\, r_1(\mu)$$

The resulting four equations may be written:

$$A_1(\sin\beta\mu - \sinh\beta\mu) + B_1(\cos\beta\mu - \cosh\beta\mu) - A_2(\sin\beta\gamma + \sinh\beta\gamma) - B_2(\cos\beta\gamma + \cosh\beta\gamma) = 0 \quad (3.6)$$

$$A_1(\cos\beta\mu - \cosh\beta\mu) - B_1(\sin\beta\mu + \sinh\beta\mu) + A_2(\cos\beta\gamma + \cosh\beta\gamma) + B_2(\sinh\beta\gamma - \sin\beta\gamma) = 0 \quad (3.7)$$

$$A_1(\sin\beta\mu + \sinh\beta\mu) + B_1(\cos\beta\mu + \cosh\beta\mu) + A_2(\sinh\beta\gamma - \sin\beta\gamma) + B_2(\cosh\beta\gamma - \cos\beta\gamma) = 0 \quad (3.8)$$

$$A_1\left[\frac{\beta}{\eta}(\sin\beta\mu - \sinh\beta\mu) - (\cos\beta\mu + \cosh\beta\mu)\right] + B_1\left[\frac{\beta}{\eta}(\cos\beta\mu - \cosh\beta\mu) + (\sin\beta\mu - \sinh\beta\mu)\right] + A_2(\cosh\beta\gamma - \cos\beta\gamma) + B_2(\sin\beta\gamma + \sinh\beta\gamma) = 0 \quad (3.9)$$

Given values for μ and β, we wish to determine that value of η that will make the determinant of the coefficient matrix for A_1, B_1, and so on, equal to zero. We concentrate on the elements of the fourth row (Eq. 3.9). By designating the minor determinant associated with the first element of this row $D1$, the second $D2$, and so on, it is readily shown after some rearrangement that

$$\eta = \frac{\beta[(\sinh\beta\mu - \sin\beta\mu)D1 + (\cos\beta\mu - \cosh\beta\mu)D2]}{\theta + \phi} \quad (3.10)$$

where

$$\theta = (\sinh\beta\mu - \sin\beta\mu)D2 - (\cos\beta\mu + \cosh\beta\mu)D1$$

and

$$\phi = (\cosh\beta\gamma - \cos\beta\gamma)D3 - (\sin\beta\gamma + \sinh\beta\gamma)D4$$

For the particular beam under study, it is evident that as η approaches zero ($M \to \infty$) the frequency also approaches zero for the first mode. For all other modes the point on the beam where the mass is located approaches a nodal point when M approaches infinity. Therefore the lower limits coincide with those of clamped-simple-free beams. In each mode the upper limits are evidently those of a clamped-free beam. Having selected a mode of interest and a value for the Q factor, the corresponding value for β, hence η, is obtained as in Chapter 1.

In determining the modal shapes it is permissible to arbitrarily set one of the nonzero constants A_1, B_1, and so on, equal to one. We chose therefore to set the constant A_1 equal to 1. It is seen that Eqs. (3.6) through (3.8) may then be combined to form three simultaneous nonhomogeneous linear algebraic equations for the other three constants (see Case 3.7). These equations are readily solved utilizing Cramer's rule, and we obtain expressions for the functions $r_1(\xi)$ and $r_2(\xi)$. It is appreciated that there may arise cases for which the only permissible value of the constant A_1 is zero. This situation manifests itself in the obtaining of infinite or indeterminate values for the other constants. In such a case, since both A_1 and B_1 cannot both equal zero, it is evident that B_1 should have been set equal to 1. It will be seen that solutions for A_2 and B_2 can easily be obtained in such cases by setting the right-hand side of the nonhomogeneous equations provided equal to zero, setting B_1 equal to one, and solving for A_2 and B_2 from the first two equations provided. In some beam systems of Part 2, Section 1, and all of Section 2 (which involve rotary masses and torsion springs), three of the constants appearing in the modal shapes can be eliminated immediately by means of classical boundary conditions. For these cases explicit expressions are provided for the modal shapes.

In presentation of the data, each case is given a reference number and a brief description. Corresponding sets of tables with the same reference number are also provided. In general, there are nine tables in each set, each table with a distinct value of μ, where $\mu = .1, .2$, and so on, up to .9. Linear interpolation may be used to solve problems when μ takes on values intermediate to those given. For symmetric cases the range of μ is restricted to an upper limit of 0.5. Expressions for the modal shapes are also provided, along with the linear equations needed to solve for the constants appearing in these expressions. This is followed by expressions for determination of free vibration frequencies. In certain cases it is found that the upper and lower frequency limits do not differ by more than 1 percent. This can happen, for example, when concentrated masses or lateral coil springs are located near nodal points. In such cases either the upper or lower limit may be utilized as the beam frequency, and no data are given in the corresponding row of the table. The tables contain data for the first four modes.

1. *Concentrated Masses or Lateral Coil Springs*

CASE 3.7. Beam clamped at one end and free at the other with intermediate concentrated mass.

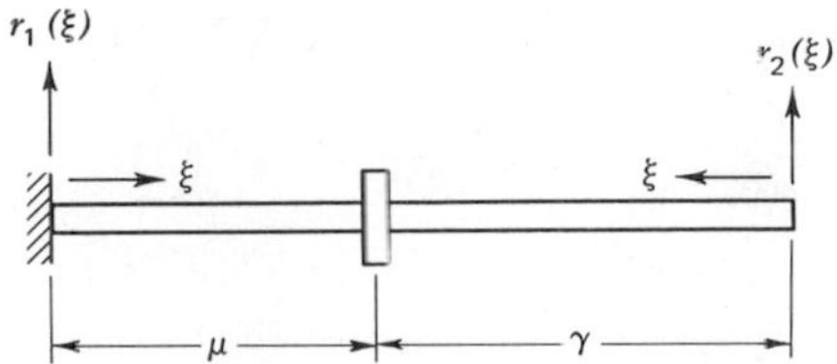

$$r_1(\xi) = \sin\beta\xi - \sinh\beta\xi + B_1(\cos\beta\xi - \cosh\beta\xi)$$
$$r_2(\xi) = A_2(\sin\beta\xi + \sinh\beta\xi) + B_2(\cos\beta\xi + \cosh\beta\xi)$$
$$B_1(\cos\beta\mu - \cosh\beta\mu) - A_2(\sin\beta\gamma + \sinh\beta\gamma) - B_2(\cos\beta\gamma + \cosh\beta\gamma) = \sinh\beta\mu - \sin\beta\mu$$
$$-B_1(\sin\beta\mu + \sinh\beta\mu) + A_2(\cos\beta\gamma + \cosh\beta\gamma) + B_2(\sinh\beta\gamma - \sin\beta\gamma) = \cosh\beta\mu - \cos\beta\mu$$
$$B_1(\cos\beta\mu + \cosh\beta\mu) + A_2(\sinh\beta\gamma - \sin\beta\gamma) + B_2(\cosh\beta\gamma - \cos\beta\gamma) = -\sin\beta\mu - \sinh\beta\mu$$
$$f_n = Qf_{CFn} \qquad n = 1$$
$$= f_{CSF(n-1)} + Q(f_{CFn} - f_{CSF(n-1)}) \qquad n > 1$$

ILLUSTRATIVE PROBLEM 3.1

The reference aluminum beam (Illustrative Problem 1.1) is clamped at one end, free at the other, and has a concentrated mass of 4.50 lb located at a distance of 12 in. from the clamped end (see inset, Case 3.7).

(a) Determine the first mode frequency and modal shape. The first mode limits are evidently zero and clamped-free.

$$f_{CF} = 17.8 \text{ Hz} \qquad \text{(Illustrative Problem 1.1)}$$
$$\eta = \frac{1.463}{4.50} = .325 \qquad \mu = \frac{12}{30} = 0.4$$
$$Q = .70 + \frac{(.325 - .229)}{(.410 - .229)} \times .1 = .70 + .053 = .753 \qquad \text{(Table 3.7)}$$
$$\therefore f = Qf_{CF} = .753 \times 17.8 = 13.4 \text{ Hz}$$
$$\beta = \beta_{CF}\sqrt{\frac{f}{f_{CF}}} = 1.875\sqrt{\frac{13.4}{17.8}} = 1.63$$

Solve modal shape coefficient equations of Case 3.7 to obtain

$$B_1 = -.821, \qquad A_2 = -.385, \qquad B_2 = .490$$
$$\therefore r_1(\xi) = \sin 1.63\xi - \sinh 1.63\xi - .821(\cos 1.63\xi - \cosh 1.63\xi)$$
$$r_2(\xi) = -.385(\sin 1.63\xi + \sinh 1.63\xi) + .490(\cos 1.63\xi + \cosh 1.63\xi)$$

(b) Determine the second mode frequency. The upper and lower limits are f_{CF2} and f_{CSF1}, respectively (Case 3.7).

$$f_{CF2} = f_{CF1} \times \frac{\beta^2_{CF2}}{\beta^2_{CF1}} = 17.8 \times \frac{4.694^2}{1.875^2} = 111.6 \text{ Hz}$$
$$\beta_{CSF1} = 2.75$$
$$f_{CSF1} = 17.8 \times \frac{2.75^2}{1.875^2} = 38.3 \text{ Hz}$$

$$Q = .10 + \frac{.325 - .155}{.345 - .155} \times .1 = .10 + .089 = .189$$

$$\therefore f = f_{CSF1} + Q(f_{CF2} - f_{CSF1}) = 38.3 + .189(111.6 - 38.3) = 52.2 \text{ Hz}$$

(c) Determine the eigenvalue β for this problem.

$$\beta = \beta_{CSF1}\sqrt{\frac{f}{f_{CSF1}}} = 2.75\sqrt{\frac{52.2}{38.3}} = 3.21$$

CASE 3.8. Beam clamped at one end and simply supported at the other end, and with intermediate concentrated mass.

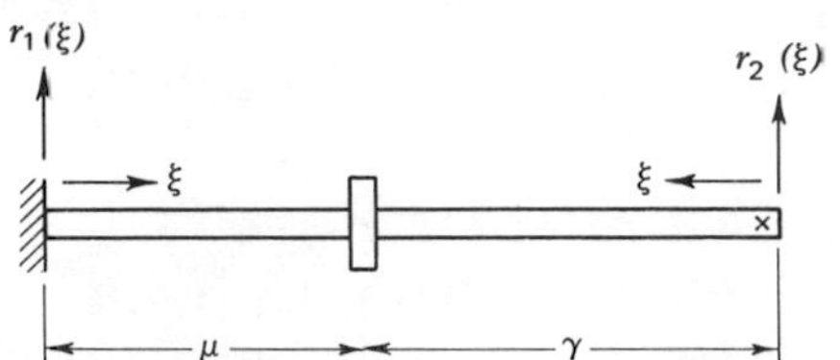

$$r_1(\xi) = \sin\beta\xi - \sinh\beta\xi + B_1(\cos\beta\xi - \cosh\beta\xi)$$
$$r_2(\xi) = A_2\sin\beta\xi + C_2\sinh\beta\xi$$

$$B_1(\cos\beta\mu - \cosh\beta\mu) - A_2\sin\beta\gamma - C_2\sinh\beta\gamma = \sinh\beta\mu - \sin\beta\mu$$
$$-B_1(\sin\beta\mu + \sinh\beta\mu) + A_2\cos\beta\gamma + C_2\cosh\beta\gamma = \cosh\beta\mu - \cos\beta\mu$$
$$-B_1(\cos\beta\mu + \cosh\beta\mu) + A_2\sin\beta\gamma - C_2\sinh\beta\gamma = \sin\beta\mu + \sinh\beta\mu$$

$$f_n = Qf_{CSn} \qquad n = 1$$
$$= f_{CSS(n-1)} + Q(f_{CSn} - f_{CSS(n-1)}) \qquad n > 1$$

CASE 3.9. Beam clamped at each end, and with intermediate concentrated mass.

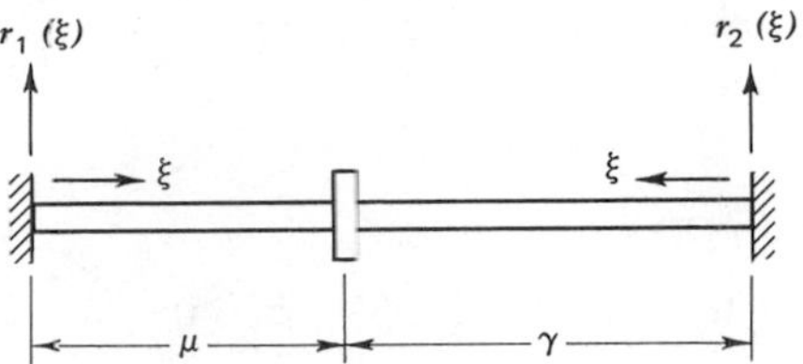

$$r_1(\xi) = \sin\beta\xi - \sinh\beta\xi + B_1(\cos\beta\xi - \cosh\beta\xi)$$
$$r_2(\xi) = A_2(\sin\beta\xi - \sinh\beta\xi) + B_2(\cos\beta\xi - \cosh\beta\xi)$$

$$B_1(\cos\beta\mu - \cosh\beta\mu) + A_2(\sinh\beta\gamma - \sin\beta\gamma) + B_2(\cosh\beta\gamma - \cos\beta\gamma) = \sinh\beta\mu - \sin\beta\mu$$

$$-B_1(\sin\beta\mu + \sinh\beta\mu) + A_2(\cos\beta\gamma - \cosh\beta\gamma) - B_2(\sin\beta\gamma + \sinh\beta\gamma) = \cosh\beta\mu - \cos\beta\mu$$

$$- B_1(\cos \beta\mu + \cosh \beta\mu) + A_2(\sin \beta\gamma + \sinh \beta\gamma) + B_2(\cos \beta\gamma + \cosh \beta\gamma) = \sin \beta\mu + \sinh \beta\mu$$

$$f_n = Qf_{CCn} \qquad n = 1$$
$$= f_{CSC(n-1)} + Q(f_{CCn} - f_{CSC(n-1)}) \qquad n > 1$$

CASE 3.10. Beam with simple support at each end, and with intermediate concentrated mass.

$$r_1(\xi) = \sin \beta\xi + C_1 \sinh \beta\xi$$
$$r_2(\xi) = A_2 \sin \beta\xi + C_2 \sinh \beta\xi$$
$$C_1 \sinh \beta\mu - A_2 \sin \beta\gamma - C_2 \sinh \beta\gamma = -\sin \beta\mu$$
$$C_1 \cosh \beta\mu + A_2 \cos \beta\gamma + C_2 \cosh \beta\gamma = -\cos \beta\mu$$
$$C_1 \sinh \beta\mu + A_2 \sin \beta\gamma - C_2 \sinh \beta\gamma = \sin \beta\mu$$
$$f_n = Qf_{SSn} \qquad n = 1$$
$$= f_{SSS(n-1)} + Q(f_{SSn} - f_{SSS(n-1)}) \qquad n > 1$$

CASE 3.11. Beam with simple support at one end and free at the other, and with intermediate concentrated mass.

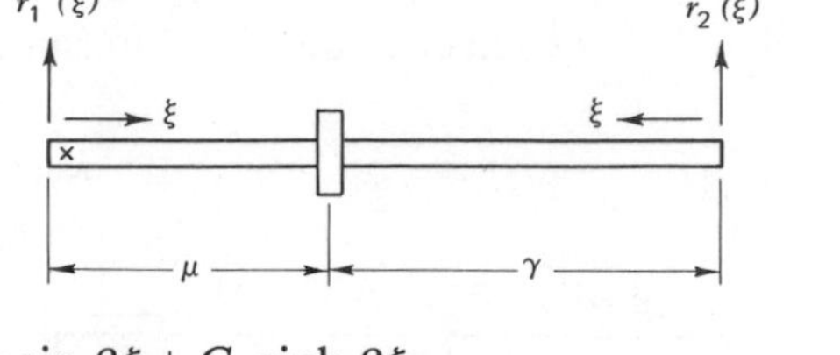

$$r_1(\xi) = \sin \beta\xi + C_1 \sinh \beta\xi$$
$$r_2(\xi) = A_2(\sin \beta\xi + \sinh \beta\xi) + B_2(\cos \beta\xi + \cosh \beta\xi)$$
$$C_1 \sinh \beta\mu - A_2(\sin \beta\gamma + \sinh \beta\gamma) - B_2(\cos \beta\gamma + \cosh \beta\gamma) = -\sin \beta\mu$$
$$C_1 \cosh \beta\mu + A_2(\cos \beta\gamma + \cosh \beta\gamma) + B_2(\sinh \beta\gamma - \sin \beta\gamma) = -\cos \beta\mu$$
$$C_1 \sinh \beta\mu + A_2(\sin \beta\gamma - \sinh \beta\gamma) + B_2(\cos \beta\gamma - \cosh \beta\gamma) = \sin \beta\mu$$
$$f_n = f_{SSFn} + Q(f_{SF(n+1)} - f_{SSFn})$$

CASE 3.12. Beam clamped at one end, with concentrated mass at the other, and with intermediate simple support.

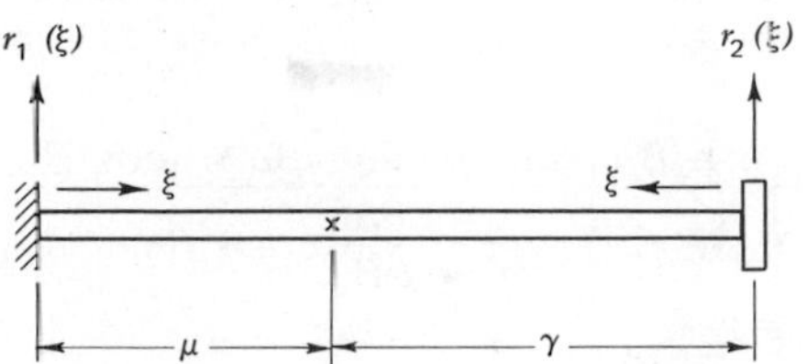

$$r_1(\xi) = \sin \beta\xi - \sinh \beta\xi + \frac{\sinh \beta\mu - \sin \beta\mu}{\cos \beta\mu - \cosh \beta\mu}(\cos \beta\xi - \cosh \beta\xi)$$

$$r_2(\xi) = \gamma_1\left[\sin \beta\xi + \theta(\cos \beta\xi + \cosh \beta\xi) + \frac{\eta + 2\theta\beta}{\eta - 2\phi\beta}(\sinh \beta\xi + \phi \cos \beta\xi + \phi \cosh \beta\xi)\right]$$

where

$$\gamma_1 = \frac{\cosh \beta\mu - \cos \beta\mu + \dfrac{\sinh \beta\mu - \sin \beta\mu}{\cos \beta\mu - \cosh \beta\mu}(\sin \beta\mu + \sinh \beta\mu)}{\cos \beta\gamma + \theta(\sinh \beta\gamma - \sin \beta\gamma) + \dfrac{\eta + 2\theta\beta}{\eta - 2\phi\beta}(\cosh \beta\gamma + \phi \sinh \beta\gamma - \phi \sin \beta\gamma)}$$

$$\theta = \frac{-\sin \beta\gamma}{\cos \beta\gamma + \cosh \beta\gamma} \qquad \phi = \frac{-\sinh \beta\gamma}{\cos \beta\gamma + \cosh \beta\gamma}$$

$$f_n = Qf_{CSFn} \qquad n = 1$$

$$= f_{CSSn} + Q(f_{CSFn} - f_{CSSn}) \qquad n > 1$$

CASE 3.13. Beam simply supported at one end, with concentrated mass at the other, and with intermediate simple support.

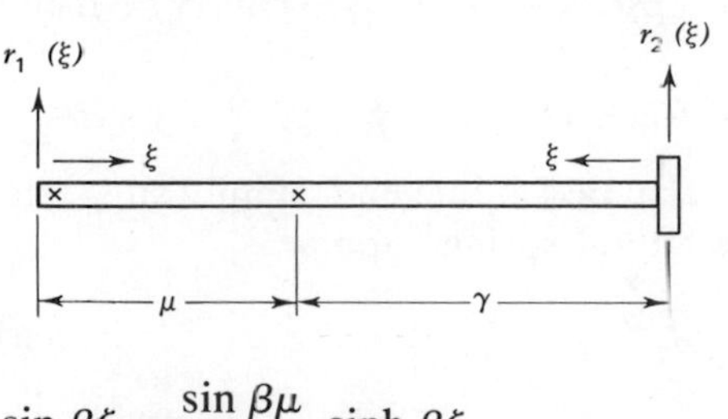

$$r_1(\xi) = \sin \beta\xi - \frac{\sin \beta\mu}{\sinh \beta\mu} \sinh \beta\xi$$

$$r_2(\xi) = \gamma_1\left[\sin \beta\xi + \theta(\cos \beta\xi + \cosh \beta\xi) + \frac{\eta + 2\theta\beta}{\eta - 2\phi\beta} \times (\sinh \beta\xi + \phi \cos \beta\xi + \phi \cosh \beta\xi)\right]$$

where

$$\gamma_1 = \frac{\cosh \beta\mu (\sin \beta\mu / \sinh \beta\mu) - \cos \beta\mu}{\cos \beta\gamma + \theta(\sinh \beta\gamma - \sin \beta\gamma) + \dfrac{\eta + 2\theta\beta}{\eta - 2\phi\beta}(\cosh \beta\gamma + \phi \sinh \beta\gamma - \phi \sin \beta\gamma)}$$

$$\theta = \frac{-\sin \beta\gamma}{\cos \beta\gamma + \cosh \beta\gamma} \qquad \phi = \frac{-\sinh \beta\gamma}{\cos \beta\gamma + \cosh \beta\gamma}$$

$$f_n = Q f_{SSFn} \qquad n = 1$$

$$= f_{SSSn} + Q(f_{SSFn} - f_{SSSn}) \qquad n > 1$$

CASE 3.14. Beam clamped at one end, free at the other, and with intermediate lateral spring support.

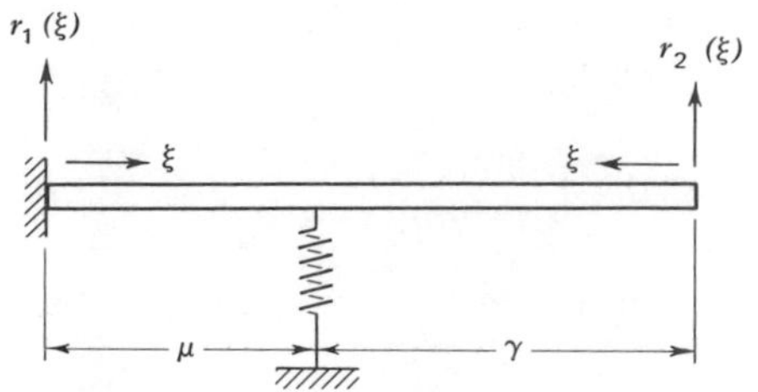

$$r_1(\xi) = \sin \beta\xi - \sinh \beta\xi + B_1(\cos \beta\xi - \cosh \beta\xi)$$

$$r_2(\xi) = A_2(\sin \beta\xi + \sinh \beta\xi) + B_2(\cos \beta\xi + \cosh \beta\xi)$$

$$B_1(\cos \beta\mu - \cosh \beta\mu) - A_2(\sin \beta\gamma + \sinh \beta\gamma) - B_2(\cos \beta\gamma + \cosh \beta\gamma) = \sinh \beta\mu - \sin \beta\mu$$

$$-B_1(\sin \beta\mu + \sinh \beta\mu) + A_2(\cos \beta\gamma + \cosh \beta\gamma) + B_2(\sinh \beta\gamma - \sin \beta\gamma) = \cosh \beta\mu - \cos \beta\mu$$

$$B_1(\cos \beta\mu + \cosh \beta\mu) + A_2(\sinh \beta\gamma - \sin \beta\gamma) + B_2(\cosh \beta\gamma - \cos \beta\gamma) = -\sin \beta\mu - \sinh \beta\mu$$

$$f_n = f_{CFn} + Q(f_{CSFn} - f_{CFn})$$

CASE 3.15. Beam clamped at one end, simply supported at the other, and with intermediate lateral spring support.

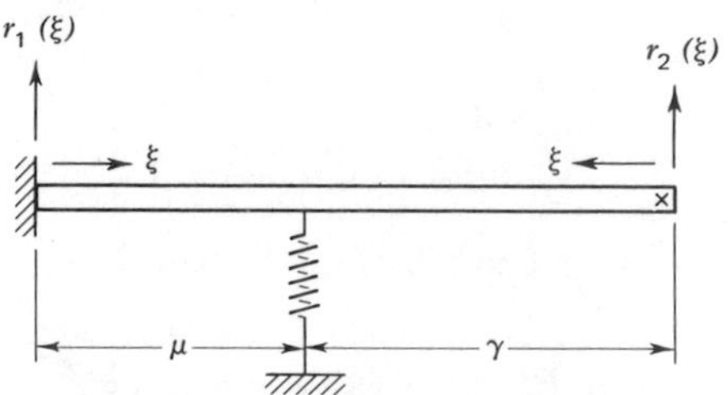

$$r_1(\xi) = \sin \beta\xi - \sinh \beta\xi + B_1(\cos \beta\mu - \cosh \beta\mu)$$
$$r_2(\xi) = A_2 \sin \beta\xi + C_2 \sinh \beta\xi$$
$$B_1(\cos \beta\mu - \cosh \beta\mu) - A_2 \sin \beta\gamma - C_2 \sinh \beta\gamma = \sinh \beta\mu - \sin \beta\mu$$
$$-B_1(\sin \beta\mu + \sinh \beta\mu) + A_2 \cos \beta\gamma + C_2 \cosh \beta\gamma = \cosh \beta\mu - \cos \beta\mu$$
$$-B_1(\cos \beta\mu + \cosh \beta\mu) + A_2 \sin \beta\gamma - C_2 \sinh \beta\gamma = \sin \beta\mu + \sinh \beta\mu$$
$$f_n = f_{CSn} + Q(f_{CSSn} - f_{CSn})$$

CASE 3.16. Beam clamped at each end and with intermediate lateral spring support.

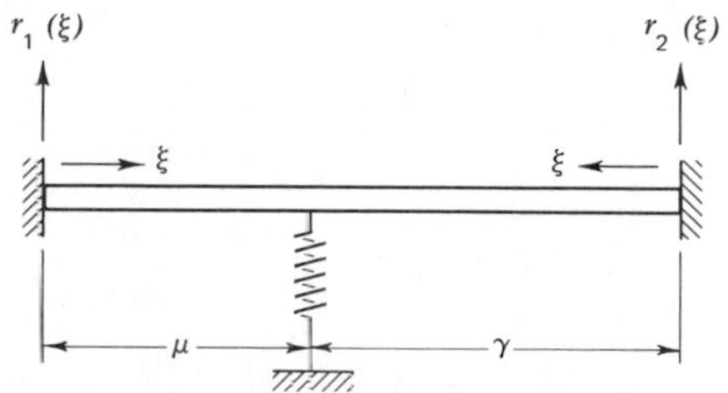

$$r_1(\xi) = \sin \beta\xi - \sinh \beta\xi + B_1(\cos \beta\xi - \cosh \beta\xi)$$
$$r_2(\xi) = A_2(\sin \beta\xi - \sinh \beta\xi) + B_2(\cos \beta\xi - \cosh \beta\xi)$$
$$B_1(\cos \beta\mu - \cosh \beta\mu) + A_2(\sinh \beta\gamma - \sin \beta\gamma) + B_2(\cosh \beta\gamma - \cos \beta\gamma) = \sinh \beta\mu - \sin \beta\mu$$
$$-B_1(\sin \beta\mu + \sinh \beta\mu) + A_2(\cos \beta\gamma - \cosh \beta\gamma) - B_2(\sin \beta\gamma + \sinh \beta\gamma) = \cosh \beta\mu - \cos \beta\mu$$
$$-B_1(\cos \beta\mu + \cosh \beta\mu) + A_2(\sin \beta\gamma + \sinh \beta\gamma) + B_2(\cos \beta\gamma + \cosh \beta\gamma) = \sin \beta\mu + \sinh \beta\mu$$
$$f_n = f_{CCn} + Q(f_{CSCn} - f_{CCn})$$

CASE 3.17. Beam with simple support at one end, free at the other, and with intermediate lateral spring support.

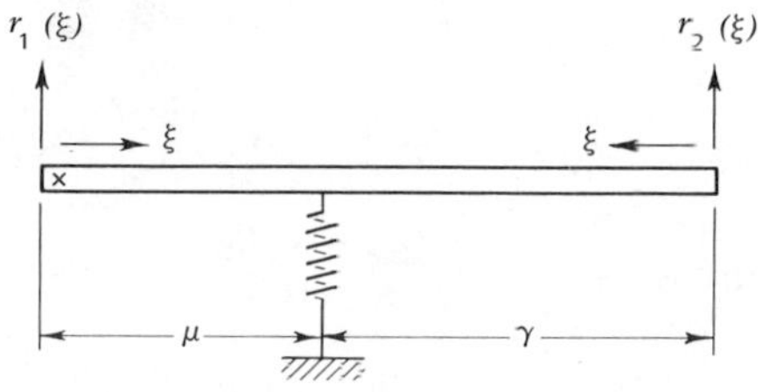

$$r_1(\xi) = \sin \beta\xi + C_1 \sinh \beta\xi$$
$$r_2(\xi) = A_2(\sin \beta\xi + \sinh \beta\xi) + B_2(\cos \beta\xi + \cosh \beta\xi)$$

$$C_1 \sinh \beta\mu - A_2(\sin \beta\gamma + \sinh \beta\gamma) - B_2(\cos \beta\gamma + \cosh \beta\gamma) = -\sin \beta\mu$$
$$C_1 \cosh \beta\mu + A_2(\cos \beta\gamma + \cosh \beta\gamma) - B_2(\sin \beta\gamma - \sinh \beta\gamma) = -\cos \beta\mu$$
$$C_1 \sinh \beta\mu + A_2(\sin \beta\gamma - \sinh \beta\gamma) + B_2(\cos \beta\gamma - \cosh \beta\gamma) = \sin \beta\mu$$
$$f_n = f_{SFn} + Q(f_{SSFn} - f_{SFn})$$

CASE 3.18. Beam with simple support at each end, and with intermediate lateral spring support.

$$r_1(\xi) = \sin \beta\xi + C_1 \sinh \beta\xi$$
$$r_2(\xi) = A_2 \sin \beta\xi + C_2 \sinh \beta\xi$$
$$C_1 \sinh \beta\mu - A_2 \sin \beta\gamma - C_2 \sinh \beta\gamma = -\sin \beta\mu$$
$$C_1 \cosh \beta\mu + A_2 \cos \beta\gamma + C_2 \cosh \beta\gamma = -\cos \beta\mu$$
$$C_1 \sinh \beta\mu + A_2 \sin \beta\gamma - C_2 \sinh \beta\gamma = \sin \beta\mu$$
$$f_n = f_{SSn} + Q(f_{SSSn} - f_{SSn})$$

CASE 3.19. Beam clamped at one end, with lateral spring support at the other, and with intermediate simple support.

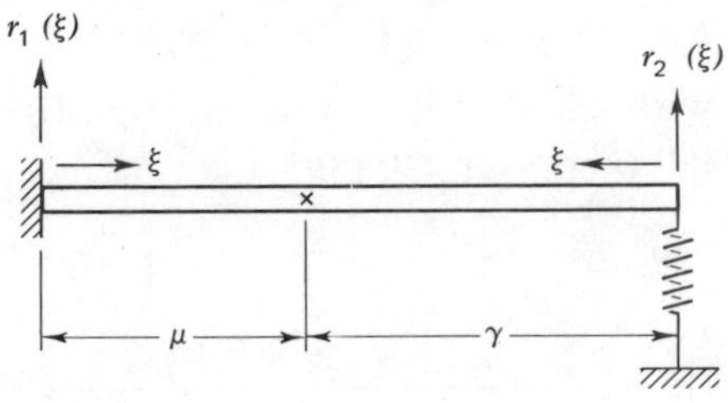

$$r_1(\xi) = \sin \beta\xi - \sinh \beta\xi + \frac{\sinh \beta\mu - \sin \beta\mu}{\cos \beta\mu - \cosh \beta\mu}(\cos \beta\xi - \cosh \beta\xi)$$

$$r_2(\xi) = \gamma_1 \Big[\sin \beta\xi + \theta(\cos \beta\xi + \cosh \beta\xi) + \frac{\beta^3 - 2\theta K^*}{\beta^3 + 2\phi K^*}(\sinh \beta\xi + \phi \cos \beta\xi + \phi \cosh \beta\xi)\Big]$$

where

$$\gamma_1 = \frac{\cosh\beta\mu - \cos\beta\mu + \dfrac{\sinh\beta\mu - \sin\beta\mu}{\cos\beta\mu - \cosh\beta\mu}(\sin\beta\mu + \sinh\beta\mu)}{\cos\beta\gamma + \theta(\sinh\beta\gamma - \sin\beta\gamma) + \dfrac{\beta^3 - 2\theta K^*}{\beta^3 + 2\phi K^*}(\cosh\beta\gamma + \phi\sinh\beta\gamma - \phi\sinh\beta\gamma)}$$

$$\theta = \frac{-\sin\beta\gamma}{\cos\beta\gamma + \cosh\beta\gamma} \qquad \phi = \frac{-\sinh\beta\gamma}{\cos\beta\gamma + \cosh\beta\gamma}$$

$$f_n = f_{CSFn} + Q(f_{CSSn} - f_{CSFn})$$

CASE 3.20. Beam with simple support at one end, with lateral spring support at the other, and with intermediate simple support.

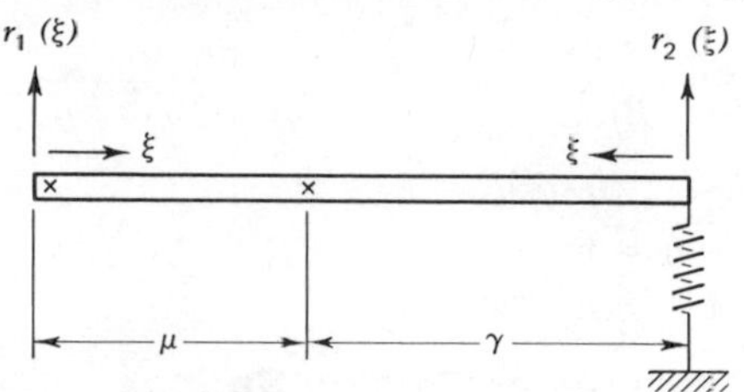

$$r_1(\xi) = \sin\beta\xi - \frac{\sin\beta\mu}{\sinh\beta\mu}\sinh\beta\xi$$

$$r_2(\xi) = \gamma_1\left[\sin\beta\xi + \theta(\cos\beta\xi + \cosh\beta\xi) + \frac{\beta^3 - 2\theta K^*}{\beta^3 + 2\phi K^*}(\sinh\beta\xi + \phi\cos\beta\xi + \phi\cosh\beta\xi)\right]$$

where

$$\gamma_1 = \frac{\cosh\beta\mu\,(\sin\beta\mu/\sinh\beta\mu) - \cos\beta\mu}{\theta^* + \dfrac{\beta^3 - 2\theta K^*}{\beta^3 + 2\phi K^*}(\cosh\beta\gamma + \phi\sinh\beta\gamma - \phi\sin\beta\gamma)}$$

$$\theta^* = \cos\beta\gamma + \theta(\sinh\beta\gamma - \sin\beta\gamma)$$

$$\theta = \frac{-\sin\beta\gamma}{\cos\beta\gamma + \cosh\beta\gamma} \qquad \phi = \frac{-\sinh\beta\gamma}{\cos\beta\gamma + \cosh\beta\gamma}$$

$$f_n = f_{SSFn} + Q(f_{SSSn} - f_{SSFn})$$

CASE 3.21. Beam with lateral spring support at one end, free at the other end, and with intermediate simple support.

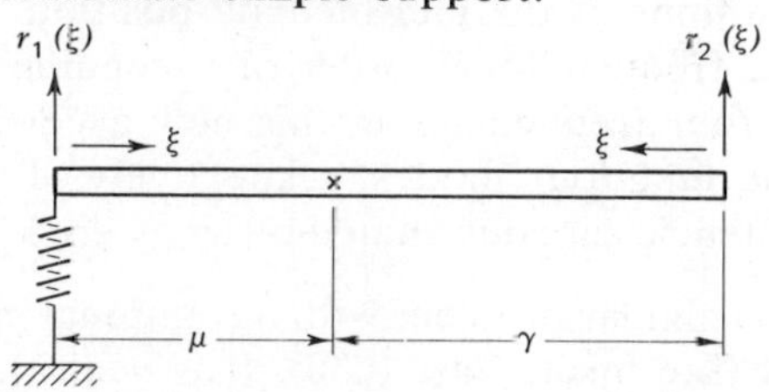

$$r_1(\xi) = \sin \beta\xi + \theta(\cos \beta\xi + \cosh \beta\xi) + \frac{\beta^3 - 2\theta K^*}{\beta^3 + 2\phi K^*}(\sinh \beta\xi + \phi \cos \beta\xi + \phi \cosh \beta\xi)$$

$$r_2(\xi) = \gamma_1 \left[\sin \beta\xi + \sinh \beta\xi - \frac{\sin \beta\gamma + \sinh \beta\gamma}{\cos \beta\gamma + \cosh \beta\gamma}(\cos \beta\xi + \cosh \beta\xi)\right]$$

where

$$\gamma_1 = \frac{\theta^* + \frac{\beta^3 - 2\theta K^*}{\beta^3 + 2\phi K^*}(\phi \sin \beta\mu - \phi \sinh \beta\mu - \cosh \beta\mu)}{\cos \beta\gamma + \cosh \beta\gamma - \frac{\sin \beta\gamma + \sinh \beta\gamma}{\cos \beta\gamma + \cosh \beta\gamma}(\sinh \beta\gamma - \sin \beta\gamma)}$$

where $\theta^* = \theta(\sin \beta\mu - \sinh \beta\mu) - \cos \beta\mu$

$$\theta = \frac{-\sin \beta\mu}{\cos \beta\mu + \cosh \beta\mu} \qquad \phi = \frac{-\sinh \beta\mu}{\cos \beta\mu + \cosh \beta\mu}$$

$$f_n = f_{FSFn} + Q(f_{SSFn} - f_{FSFn})$$

Note: $f_{FSFn} = 0, \quad n = 1.$

2. *Rotational Masses or Torsion Springs*

The analyst will recognize that there appears initially to be some uncertainty in the selection of frequency limits for these beams. We look first at the case of a beam clamped at one end, with intermediate torsion spring support, and free at the other end (see inset, Case 3.22). The lower limit frequencies for this beam are evidently those for clamped-simple-free beams. However, as the spring constant approaches infinity and the intermediate point acquires what is essentially clamped support, we can associate two distinct frequencies with this limit. With the left span we associate the frequency of a clamped-clamped beam of dimensionless length μ. With the right span we associate the frequency of a clamped-free beam of dimensionless length $\gamma = 1 - \mu$. An intensive study of the variation of the beam eigenvalues with T^*, the dimensionless torsion spring constant, reveals the correct procedure for analysing these beams. The steps to be taken are:

1. Establish numerical values for the first four frequencies of each span with clamped conditions at the intermediate position.
2. Arrange these frequencies in order of ascending magnitude.
3. Use the first four frequencies in this new ascending order for the upper limits of the first four modes, respectively. To avoid confusion these ordered frequencies are designated by an asterisk: f_1^*, f_2^*, and so on.

We look next at a similar problem with a rotational mass located at the intermediate point (see inset, Case 3.30). It is seen that now the upper limits are well defined, but the lower limits appear to present difficulty. For the first mode the lower limit is evidently zero. For all the higher modes the intermediate point takes on a clamped condition. Here it is necessary to establish the first three numerical frequencies for each of the individual spans, with a clamped intermediate condition. These frequencies are then located in ascending order and, denoting them f_1^*, f_2^*, and f_3^*, they are utilized for the lower limits for the second, third, and fourth modes.

CASE 3.22. Beam clamped at one end, free at the other, with intermediate torsion spring support.

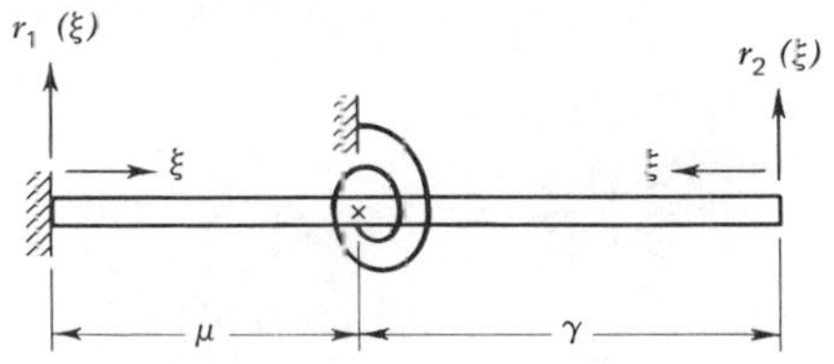

$$r_1(\xi) = \sin \beta\xi - \sinh \beta\xi + \frac{\sinh \beta\mu - \sin \beta\mu}{\cos \beta\mu - \cosh \beta\mu}(\cos \beta\xi - \cosh \beta\xi)$$

$$r_2(\xi) = \gamma_1 \left[\sin \beta\xi + \sinh \beta\xi - \frac{\sin \beta\gamma + \sinh \beta\gamma}{\cos \beta\gamma + \cosh \beta\gamma}(\cos \beta\xi + \cosh \beta\xi)\right]$$

$$\gamma_1 = \frac{(\cos \beta\gamma + \cosh \beta\gamma)(\cos \beta\mu \cosh \beta\mu - 1)}{(\cos \beta\mu - \cosh \beta\mu)(\cos \beta\gamma \cosh \beta\gamma - 1)}$$

$$f_n = f_{CSFn} + Q(f_n^* - f_{CSFn})$$

with f^* based on f_{cc} for beam of length μL
and f_{cF} for beam of length γL

ILLUSTRATIVE PROBLEM 3.2.

The reference aluminum beam (Illustrative Problem 1.1) is clamped at one end, free at the other, and pinned with intermediate torsion spring support at a distance of 18 in. from the clamped end. The torsion spring stiffness is 100 in.-lb per degree of rotation. Determine the first mode frequency of vibration of the beam.

For the 18-in. span the limits with $T^* = \infty$ are

$$f_{CCn} = \frac{\sqrt{EIg/\rho A}}{2\pi L^2}\beta_{CCn}^2 \qquad \text{where } L = 18 \text{ in.}$$
$$= 14.1\beta_{CCn}^2$$

$$\therefore f_{CC1} = 315 \qquad f_{CC2} = 870 \qquad f_{CC3} = 1705 \qquad f_{CC4} = 2818$$

For the 12-in. span the limits are $31.7\beta^2_{CFn}$

$$\therefore f_{CF1} = 111 \qquad f_{CF2} = 698 \qquad f_{CF3} = 1956 \qquad f_{CF4} = 3833$$

The ordered upper limits are therefore

$$f_1^* = 111 \qquad f_2^* = 315 \qquad f_3^* = 698 \qquad f_4^* = 870$$

The torsion spring stiffness $= (100 \times 180)/\pi = 5730$ in.-lb/unit slope

$$T^* = \frac{TL}{EI} = \frac{5730 \times 30}{10 \times 10^6 \times 1.04 \times 10^{-2}} = 16.5$$

$$\mu = \frac{18}{30} = 0.6 \qquad \beta_{CSF1} = 3.67 \qquad \text{(Fig. 3.5)}$$

$$\therefore f_{CSF1} = 5.07 \times 3.67^2 = 68.3$$

$$Q = 0.50 + \frac{16.5 - 13.1}{19.7 - 13.1} \times 0.1 = 0.55 \qquad \text{(Table 3.22)}$$

$$\therefore f_1 = f_{CSF1} + Q(f_1^* - f_{CSF1}) = 68.3 + 0.55(111 - 68.3) = 91.8 \text{ Hz}$$

CASE 3.23. Beam clamped at one end, with simple support at the other, and with intermediate torsion spring support.

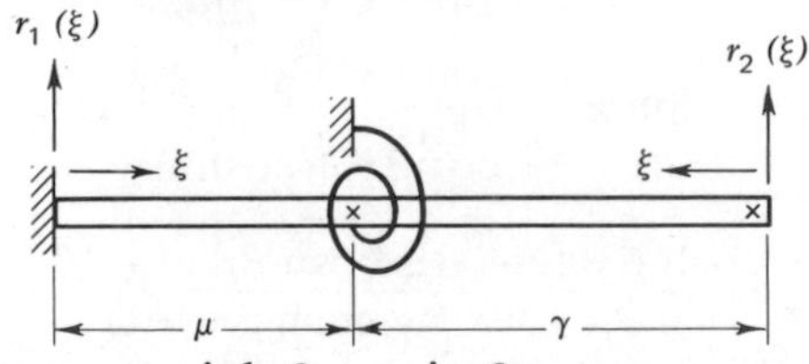

$$r_1(\xi) = \sin\beta\xi - \sinh\beta\xi + \frac{\sinh\beta\mu - \sin\beta\mu}{\cos\beta\mu - \cosh\beta\mu}(\cos\beta\xi - \cosh\beta\xi)$$

$$r_2(\xi) = \gamma_1\left(\sin\beta\xi - \frac{\sin\beta\gamma}{\sinh\beta\gamma}\sinh\beta\xi\right)$$

$$\gamma_1 = \frac{\cosh\beta\mu - \cos\beta\mu + \dfrac{\sinh\beta\mu - \sin\beta\mu}{\cos\beta\mu - \cosh\beta\mu}(\sin\beta\mu + \sinh\beta\mu)}{\cos\beta\gamma - \cosh\beta\gamma(\sin\beta\gamma/\sinh\beta\gamma)}$$

$$f_n = f_{CSSn} + Q(f_n^* - f_{CSSn})$$

with f^* based on f_{cc} for beam of length μL, and f_{cs} for beam of length γL.

CASE 3.24. Beam clamped at each end, and with intermediate torsion spring support.

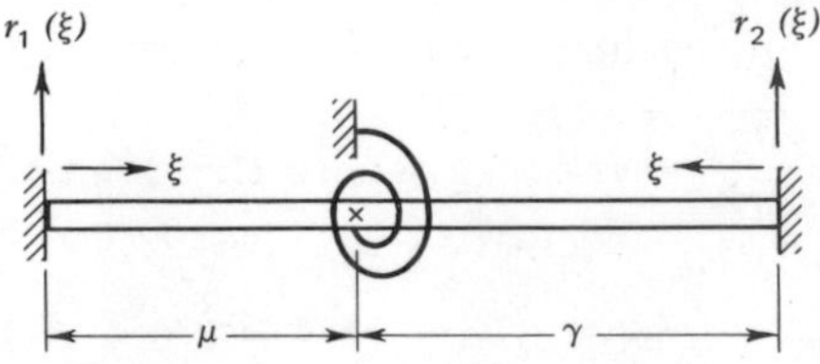

$$r_1(\xi) = \sin \beta\xi - \sinh \beta\xi + \frac{\sinh \beta\mu - \sin \beta\mu}{\cos \beta\mu - \cosh \beta\mu}(\cos \beta\xi - \cosh \beta\xi)$$

$$r_2(\xi) = \gamma_1\left[\sin \beta\xi - \sinh \beta\xi + \frac{\sinh \beta\gamma - \sin \beta\gamma}{\cos \beta\gamma - \cosh \beta\gamma}(\cos \beta\xi - \cosh \beta\xi)\right]$$

$$\gamma_1 = \frac{\cosh \beta\mu - \cos \beta\mu + \dfrac{\sinh \beta\mu - \sin \beta\mu}{\cos \beta\mu - \cosh \beta\mu}(\sin \beta\mu + \sinh \beta\mu)}{\cos \beta\gamma - \cosh \beta\gamma + \dfrac{\sinh \beta\gamma - \sin \beta\gamma}{\cosh \beta\gamma - \cos \beta\gamma}(\sin \beta\gamma + \sinh \beta\gamma)}$$

$$f_n = f_{CSCn} + Q(f_n^* - f_{CSCn})$$

with f^* based on f_{cc} for beam of length μL, and f_{cc} for beam of length γL.

CASE 3.25. Beam with simple support at one end, free at the other, and with intermediate torsion spring support.

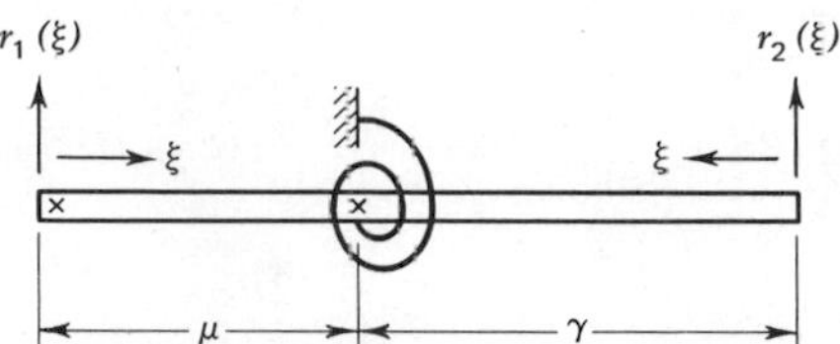

$$r_1(\xi) = \sin \beta\xi - \frac{\sin \beta\mu}{\sinh \beta\mu}\sinh \beta\xi$$

$$r_2(\xi) = \gamma_1\left[\sin \beta\xi + \sinh \beta\xi - \frac{\sin \beta\gamma + \sinh \beta\gamma}{\cos \beta\gamma + \cosh \beta\gamma}(\cos \beta\xi + \cosh \beta\xi)\right]$$

$$\gamma_1 = \frac{\cosh \beta\mu(\sin \beta\mu / \sinh \beta\mu) - \cos \beta\mu}{\cos \beta\gamma + \cosh \beta\gamma - \dfrac{\sin \beta\gamma + \sinh \beta\gamma}{\cos \beta\gamma + \cosh \beta\gamma}(\sinh \beta\gamma - \sin \beta\gamma)}$$

$$f_n = f_{SSFn} + Q(f_n^* - f_{SSFn})$$

with f^* based on f_{cs} for beam of length μL, and f_{cF} for beam of length γL.

CASE 3.26. Beam with simple support at each end, and with intermediate torsion spring support.

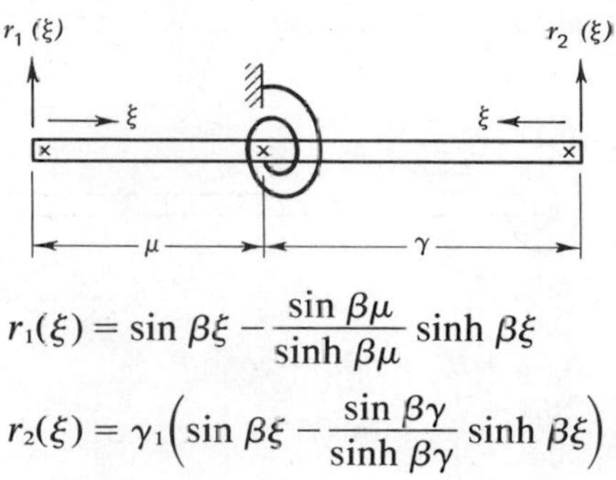

$$r_1(\xi) = \sin \beta\xi - \frac{\sin \beta\mu}{\sinh \beta\mu}\sinh \beta\xi$$

$$r_2(\xi) = \gamma_1\left(\sin \beta\xi - \frac{\sin \beta\gamma}{\sinh \beta\gamma}\sinh \beta\xi\right)$$

where

$$\gamma_1 = \frac{\cosh \beta\mu(\sin \beta\mu/\sinh \beta\mu) - \cos \beta\mu}{\cos \beta\gamma - \cosh \beta\gamma(\sin \beta\gamma/\sinh \beta\gamma)}$$

$$f_n = f_{SSSn} + Q(f^* - f_{SSSn})$$

with f^* based on f_{cs} for beam of length μL, and f_{cs} for beam of length γL.

CASE 3.27. Beam clamped at one end, with torsional spring support at the other and with intermediate simple support.

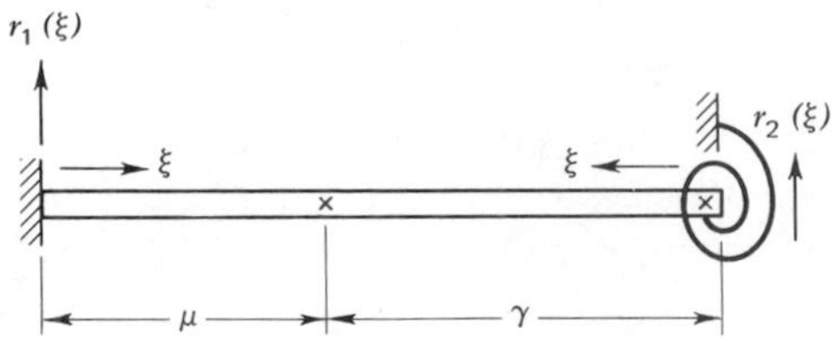

$$r_1(\xi) = \sin \beta\xi - \sinh \beta\xi + \frac{\sinh \beta\mu - \sin \beta\mu}{\cos \beta\mu - \cosh \beta\mu}(\cos \beta\xi - \cosh \beta\xi)$$

$$r_2(\xi) = \gamma_1\left[\sin \beta\xi + \theta(\cos \beta\xi - \cosh \beta\xi) - \frac{2\theta\beta + T^*}{2\phi\beta + T^*}\right.$$

$$\left.\times(\sinh \beta\xi + \phi \cos \beta\xi - \phi \cosh \beta\xi)\right]$$

where

$$\gamma_1 = \frac{2(\cos \beta\mu \cosh \beta\mu - 1)}{(\cos \beta\mu - \cosh \beta\mu)\{\cos \beta\gamma - \theta(\sin \beta\gamma + \sinh \beta\gamma) - \theta^*\}}$$

where $\theta^* = \dfrac{(2\theta\beta + T^*)(\cosh \beta\gamma - \phi \sin \beta\gamma - \phi \sinh \beta\gamma)}{(2\phi\beta + T^*)}$

$$\theta = \frac{\sin \beta\gamma}{\cosh \beta\gamma - \cos \beta\gamma}, \qquad \phi = \frac{\sinh \beta\gamma}{\cosh \beta\gamma - \cos \beta\gamma}$$

$$f_n = f_{CSSn} + Q(f_{CSCn} - f_{CSSn})$$

CASE 3.28. Beam with torsion spring support at one end, with simple support at the other, and with intermediate simple support.

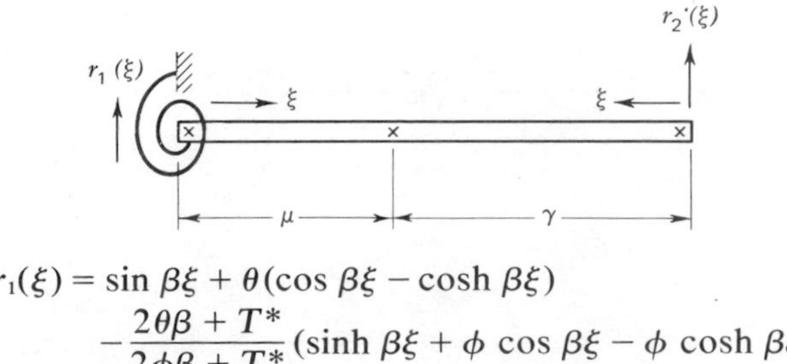

$$r_1(\xi) = \sin \beta\xi + \theta(\cos \beta\xi - \cosh \beta\xi)$$
$$- \frac{2\theta\beta + T^*}{2\phi\beta + T^*}(\sinh \beta\xi + \phi \cos \beta\xi - \phi \cosh \beta\xi)$$

$$r_2(\xi) = \gamma_1\left[\sin \beta\xi - \frac{\sin \beta\gamma}{\sinh \beta\gamma} \sinh \beta\xi\right]$$

where

$$\gamma_1 = \frac{\theta(\sin \beta\mu + \sinh \beta\mu) - \cos \beta\mu + \theta^*}{\cos \beta\gamma - (\sin \beta\gamma/\sinh \beta\gamma)\cosh \beta\gamma}$$

where $$\theta^* = \frac{(2\theta\beta + T^*)(\cosh \beta\mu - \phi \sin \beta\mu - \phi \sinh \beta\mu)}{(2\phi\beta + T^*)}$$

$$\theta = \frac{\sin \beta\mu}{\cosh \beta\mu - \cos \beta\mu}, \qquad \phi = \frac{\sinh \beta\mu}{\cosh \beta\mu - \cos \beta\mu}$$

$$f_n = f_{SSSn} + Q(f_{CSSn} - f_{SSSn})$$

CASE 3.29. Beam with torsion spring support at one end, free at the other, and with intermediate simple support.

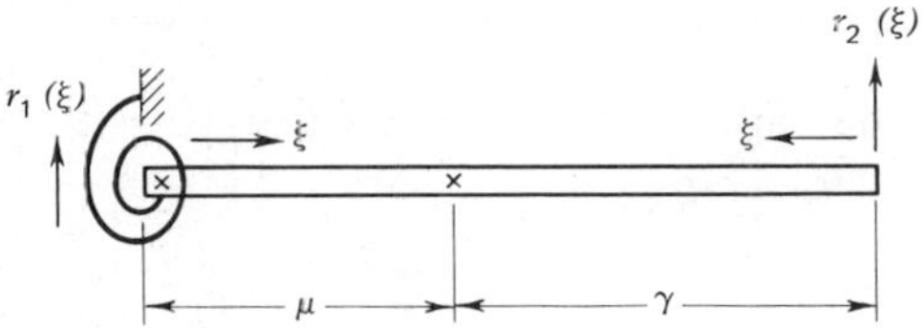

$$r_1(\xi) = \sin \beta\xi + \theta(\cos \beta\xi - \cosh \beta\xi) - \frac{2\theta\beta + T^*}{2\phi\beta + T^*}(\sinh \beta\xi + \phi \cos \beta\xi - \phi \cosh \beta\xi)$$

$$r_2(\xi) = \gamma_1\left[\sin \beta\xi + \sinh \beta\xi - \frac{\sin \beta\gamma - \sinh \beta\gamma}{\cos \beta\gamma + \cosh \beta\gamma}(\cos \beta\xi + \cosh \beta\xi)\right]$$

where

$$\gamma_1 = \frac{(\cos \beta\gamma + \cosh \beta\gamma)\{\theta(\sin \beta\mu + \sinh \beta\mu) - \cos \beta\mu + \theta^*\}}{2(1 + \cos \beta\gamma \cosh \beta\gamma)}$$

where $$\theta^* = \frac{(2\theta\beta + T^*)(\cosh \beta\mu - \phi \sin \beta\mu - \phi \sinh \beta\mu)}{(2\phi\beta + T^*)}$$

$$\theta = \frac{\sin \beta\mu}{\cosh \beta\mu - \cos \beta\mu} \qquad \phi = \frac{\sinh \beta\mu}{\cosh \beta\mu - \cos \beta\mu}$$

$$f_n = f_{SSFn} + Q(f_{CSFn} - f_{SSFn})$$

CASE 3.30. Beam clamped at one end, free at the other, and with intermediate rotational mass.

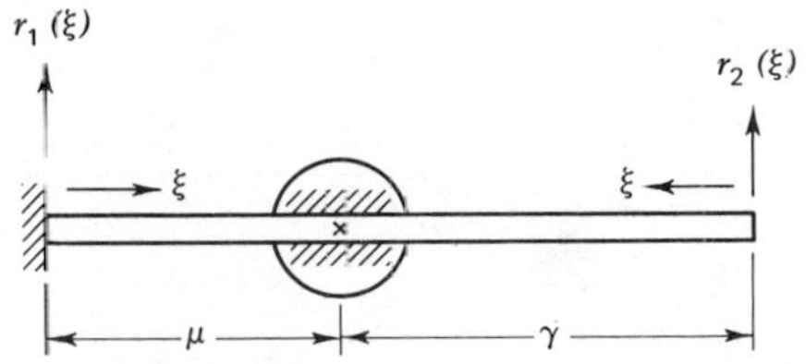

$r_1(\xi)$, $r_2(\xi)$; Use the same expressions as those for Case 3.22

$$f_n = Qf_{CSFn} \qquad n = 1$$
$$f_n = f^*_{n-1} + Q(f_{CSFn} - f^*_{n-1}) \qquad n > 1$$

with f^* based on f_{cc} for beam of length μL, and f_{cF} for beam of length γL.

CASE 3.31. Beam clamped at one end, with simple support at the other, and with intermediate rotational mass.

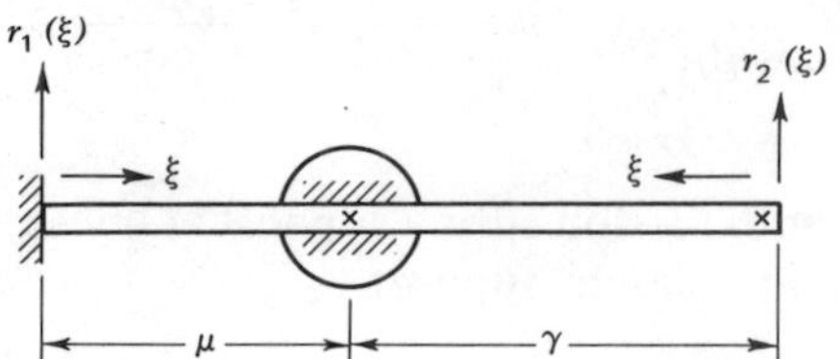

$r_1(\xi)$, $r_2(\xi)$; use the same expressions as those for Case 3.23.

$$f_n = Qf_{CSSn} \qquad n = 1$$
$$= f^*_{n-1} + Q(f_{CSSn} - f^*_{n-1}) \qquad n > 1$$

with f^* based on f_{cc} for beam of length μL, and f_{cs} for beam of length γL.

CASE 3.32. Beam clamped at each end, and with intermediate rotational mass.

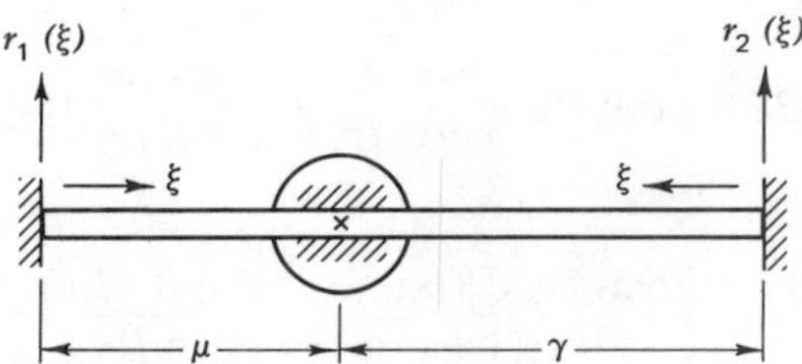

$r_1(\xi)$, $r_2(\xi)$; use the same expressions as those for Case 3.24.

$$f_n = Qf_{CSFn} \qquad n = 1$$
$$= f^*_{n-1} + Q(f_{CSFn} - f^*_{n-1}) \qquad n > 1$$

With f^* based on f_{cc} for beam of length μL, and f_{cc} for beam of length γL.

CASE 3.33. Beam with simple support at one end, free at the other, and with intermediate rotational mass.

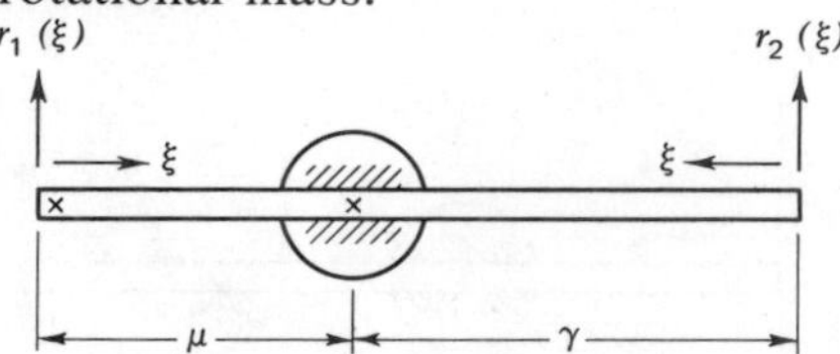

$r_1(\xi)$, $r_2(\xi)$; use the same expressions as those for Case 3.25.

$$f_n = Qf_{SSFn} \qquad n = 1$$
$$= f^*_{n-1} + Q(f_{SSFn} - f^*_{n-1}) \qquad n > 1$$

with f^* based on f_{cs} for beam of length μL, and f_{cF} for beam of length γL.

CASE 3.34. Beam with simple support at each end, and with intermediate rotational mass.

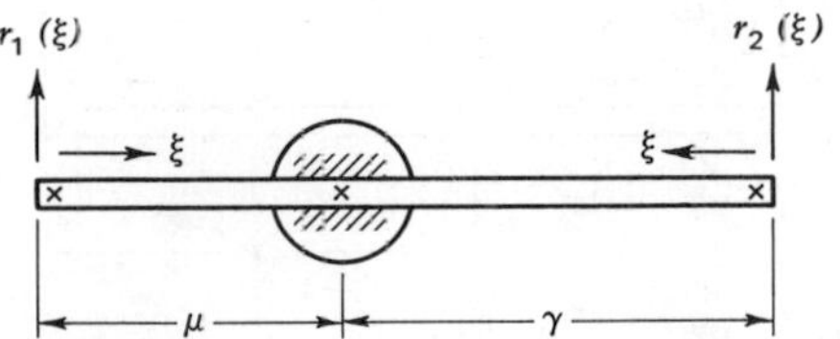

$r_1(\xi)$, $r_2(\xi)$; use the same expressions as those for Case 3.26.

$$f_n = Qf_{sss_n} \qquad n = 1$$
$$= f^*_{n-1} + Q(f_{sss_n} - f^*_{n-1}) \qquad n > 1$$

with f^* based on f_{cs} for beam of length μL, and f_{cs} for beam of length γL.

CASE 3.35. Beam clamped at one end, with rotational mass at the other, and with intermediate simple support.

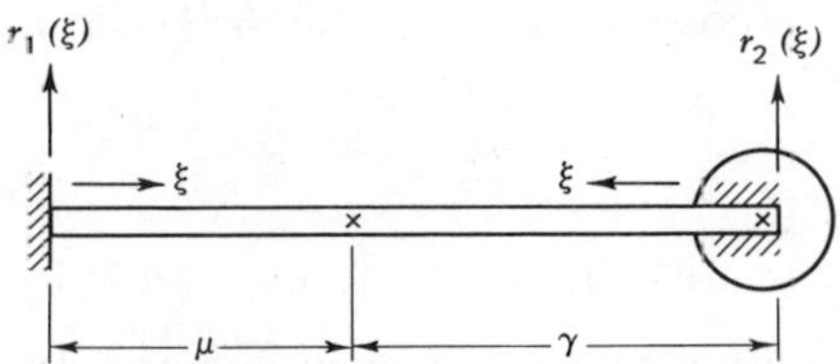

$$r_1(\xi) = \sin \beta\xi - \sinh \beta\xi + \frac{\sinh \beta\mu - \sin \beta\mu}{\cos \beta\mu - \cosh \beta\mu}(\cos \beta\xi - \cosh \beta\xi)$$

$$r_2(\xi) = \gamma_1\Big[\sin \beta\xi + \theta(\cos \beta\xi - \cosh \beta\xi) - \frac{2\theta I^* - \beta^3}{2\phi I^* - \beta^3} \times (\sinh \beta\xi + \phi \cos \beta\xi - \phi \cosh \beta\xi)\Big]$$

where

$$\gamma_1 = \frac{\cosh \beta\mu - \cos \beta\mu + \dfrac{\sinh \beta\mu - \sin \beta\mu}{\cos \beta\mu - \cosh \beta\mu}(\sin \beta\mu + \sinh \beta\mu)}{\cos \beta\gamma - \theta(\sin \beta\gamma + \sinh \beta\gamma) - \dfrac{2\theta I^* - \beta^3}{2\phi I^* - \beta^3}(\cosh \beta\gamma - \phi \sin \beta\gamma - \phi \sinh \beta\gamma)}$$

$$\theta = \frac{\sin \beta\gamma}{\cosh \beta\gamma - \cos \beta\gamma} \qquad \phi = \frac{\sinh \beta\gamma}{\cosh \beta\gamma - \cos \beta\gamma}$$

$$f_n = Qf_{CSSn} \qquad n = 1$$
$$= f_{CSC(n-1)} + Q(f_{CSSn} - f_{CSC(n-1)}) \qquad n > 1$$

CASE 3.36. Beam with rotational mass at one end, with simple support at the other, and with intermediate simple support.

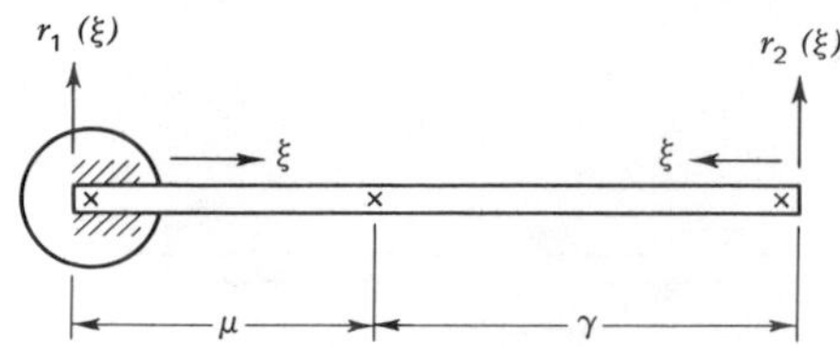

$$r_1(\xi) = \sin \beta\xi + \theta(\cos \beta\xi - \cosh \beta\xi) - \frac{2\theta I^* - \beta^3}{2\phi I^* - \beta^3}(\sinh \beta\xi + \phi \cos \beta\xi - \phi \cosh \beta\xi)$$

$$r_2(\xi) = \gamma_1\left(\sin \beta\xi - \frac{\sin \beta\gamma}{\sinh \beta\gamma} \sinh \beta\xi\right)$$

where

$$\gamma_1 = \frac{\theta(\sin \beta\mu + \sinh \beta\mu) - \cos \beta\mu + \dfrac{2\theta I^* - \beta^3}{2\phi I^* - \beta^3}(\cosh \beta\mu - \phi \sin \beta\mu - \phi \sinh \beta\mu)}{\cos \beta\gamma - (\sin \beta\gamma / \sinh \beta\gamma) \cosh \beta\gamma}$$

$$\theta = \frac{\sin \beta\mu}{\cosh \beta\mu - \cos \beta\mu} \qquad \phi = \frac{\sinh \beta\mu}{\cosh \beta\mu - \cos \beta\mu}$$

$$f_n = Qf_{SSSn} \qquad n = 1$$
$$= f_{CSS(n-1)} + Q(f_{SSSn} - f_{CSS(n-1)}) \qquad n > 1$$

CASE 3.37. Beam with rotational mass at one end, free at the other, and with intermediate simple support.

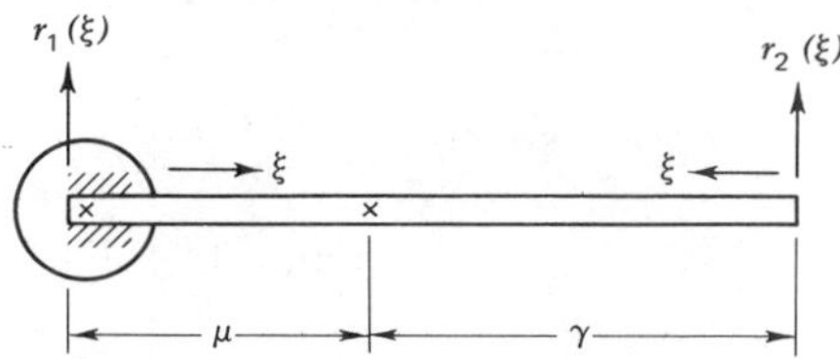

$$r_1(\xi) = \sin \beta\xi + \theta(\cos \beta\xi - \cosh \beta\xi) - \frac{2\theta I^* - \beta^3}{2\phi I^* - \beta^3}(\sinh \beta\xi + \phi \cos \beta\xi - \phi \cosh \beta\xi)$$

$$r_2(\xi) = \gamma_1\left[\sin \beta\xi + \sinh \beta\xi - \frac{\sin \beta\gamma + \sinh \beta\gamma}{\cos \beta\gamma + \cosh \beta\gamma}(\cos \beta\xi + \cosh \beta\xi)\right]$$

where

$$\gamma_1 = \frac{\theta(\sin\beta\mu + \sinh\beta\mu) - \cos\beta\mu + \dfrac{2\theta I^* - \beta^3}{2\phi I^* - \beta^3}(\cosh\beta\mu - \phi\sin\beta\mu - \phi\sinh\beta\mu)}{\cos\beta\gamma + \cosh\beta\gamma - \dfrac{\sin\beta\gamma + \sinh\beta\gamma}{\cos\beta\gamma + \cosh\beta\gamma}(\sinh\beta\gamma - \sin\beta\gamma)}$$

$$\theta = \frac{\sin\beta\mu}{\cosh\beta\mu - \cos\beta\mu} \qquad \phi = \frac{\sinh\beta\mu}{\cosh\beta\mu - \cos\beta\mu}$$

$$\begin{aligned} f_n &= Qf_{SSFn} \qquad n = 1 \\ &= f_{CSF(n-1)} + Q(f_{SSFn} - f_{CSF(n-1)}) \qquad n > 1 \end{aligned}$$

TABLE 3.7 : η ($\mu = 0.1$)

Q: MODE:	0.05	0.10	0.20	0.30	0.40	0.50	0.60	0.70	0.80	0.90
(1)	.0000	.0000	.0002	.0004	.0007	.0011	.0017	.0026	.0039	.0072
(2)	.0040	.0069	.0136	.0221	.0331	.0473	.0664	.0937	.1401	.2582
(3)	.0502	.0876	.1539	.2232	.3040	.4047	.5390	.7375	1.091	2.047
(4)	.1714	.3170	.5800	.8442	1.142	1.507	2.000	2.738	4.107	7.939

TABLE 3.7 : η ($\mu = 0.2$)

Q: MODE:	0.05	0.10	0.20	0.30	0.40	0.50	0.60	0.70	0.80	0.90
(1)	.0001	.0003	.0013	.0031	.0058	.0096	.0152	.0239	.0398	.0832
(2)	.0208	.0398	.0829	.1383	.2117	.3117	.4554	.6817	1.108	2.317
(3)	.1521	.3000	.6068	.9591	1.396	1.974	2.805	4.137	6.715	14.26
(4)	.3366	.6891	1.466	2.392	3.559	5.123	7.392	11.09	18.35	39.89

TABLE 3.7 : η ($\mu = 0.3$)

Q: MODE:	0.05	0.10	0.20	0.30	0.40	0.50	0.60	0.70	0.80	0.90
(1)	.0003	.0011	.0046	.0107	.0201	.0341	.0552	.0898	.1563	.3479
(2)	.0470	.0952	.2065	.3505	.5441	.8152	1.219	1.882	3.184	7.029
(3)	.2069	.4278	.9280	1.543	2.337	3.426	5.034	7.685	12.95	28.65
(4)	.2603	.5444	1.212	2.055	3.176	4.720	7.025	10.84	18.46	41.24

TABLE 3.7 : η ($\mu = 0.4$)

Q: MODE:	0.05	0.10	0.20	0.30	0.40	0.50	0.60	0.70	0.80	0.90
(1)	.0007	.0027	.0109	.0256	.0487	.0836	.1379	.2290	.4098	.9444
(2)	.0747	.1554	.3454	.5919	.9248	1.396	2.106	3.293	5.668	12.78
(3)	.1777	.3726	.8289	1.405	2.162	3.210	4.766	7.340	12.46	27.76
(4)	.1689	.3543	.7896	1.338	2.057	3.051	4.524	6.951	11.77	26.23

TABLE 3.7 : η ($\mu = 0.5$)

Q: MODE:	0.05	0.10	0.20	0.30	0.40	0.50	0.60	0.70	0.80	0.90
(1)	.0013	.0052	.0214	.0505	.0965	.1673	.2790	.4698	.8548	2.010
(2)	.0946	.1991	.4468	.7670	1.197	1.803	2.718	4.249	7.318	16.54
(3)	UPPER and LOWER LIMITS AGREE WITHIN 1%									
(4)	.2105	.4429	.9910	1.689	2.612	3.895	5.806	8.975	15.29	34.16

TABLE 3.7 : η ($\mu = 0.6$)

Q: MODE:	0.05	0.10	0.20	0.30	0.40	0.50	0.60	0.70	0.80	0.90
(1)	.0022	.0090	.0370	.0877	.1684	.2935	.4928	.8368	1.538	3.659
(2)	.0982	.2069	.4637	.7924	1.229	1.838	2.749	4.263	7.285	16.33
(3)	.1646	.3462	.7725	1.314	2.028	3.019	4.494	6.938	11.80	26.36
(4)	.1632	.3431	.7633	1.295	1.990	2.947	4.368	6.713	11.37	25.24

TABLE 3.7 : η ($\mu = 0.7$)

Q: MODE:	0.05	0.10	0.20	0.30	0.40	0.50	0.60	0.70	0.80	0.90
(1)	.0035	.0143	.0589	.1396	.2687	.4698	.7919	1.350	2.494	5.967
(2)	.0842	.1774	.3972	.6776	1.049	1.566	2.336	3.616	6.165	13.80
(3)	.1686	.3543	.7907	1.345	2.078	3.095	4.612	7.126	12.13	27.11
(4)	.2329	.4913	1.099	1.871	2.896	4.322	6.428	9.952	16.95	37.83

TABLE 3.7 : η ($\mu = 0.8$)

Q: MODE:	0.05	0.10	0.20	0.30	0.40	0.50	0.60	0.70	0.80	0.90
(1)	.0053	.0213	.0879	.2086	.4017	.7029	1.186	2.025	3.746	8.979
(2)	UPPER and LOWER LIMITS AGREE WITHIN 1%									
(3)	.1194	.2509	.5598	.9511	1.466	2.178	3.233	4.978	8.438	18.78
(4)	.2385	.4995	1.103	1.856	2.835	4.181	6.162	9.424	15.89	35.08

TABLE 3.7 : η ($\mu = 0.9$)

Q: MODE:	0.05	0.10	0.20	0.30	0.40	0.50	0.60	0.70	0.80	0.90
(1)	.0075	.0303	.1251	.2968	.5713	.9990	1.684	2.873	5.308	12.71
(2)	.2037	.4300	.9669	1.656	2.575	3.859	5.785	8.991	15.40	34.64
(3)	.2264	.4603	1.066	1.836	2.839	4.274	6.381	9.905	17.01	38.56
(4)	UPPER and LOWER LIMITS AGREE WITHIN 1%									

TABLE 3.8 : η ($\mu = 0.1$)

Q: MODE:	0.05	0.10	0.20	0.30	0.40	0.50	0.60	0.70	0.80	0.90
(1)	.0002	.0006	.0025	.0058	.0106	.0172	.0264	.0395	.0612	.1138
(2)	.0317	.0545	.0959	.1409	.1945	.2617	.3514	.4829	.7140	1.332
(3)	.1348	.2467	.4455	.6447	.8694	1.146	1.516	2.074	3.096	5.953
(4)	.2960	.5668	1.080	1.610	2.212	2.957	3.970	5.528	8.450	16.81

TABLE 3.8 : η ($\mu = 0.2$)

Q: MODE:	0.05	0.10	0.20	0.30	0.40	0.50	0.60	0.70	0.80	0.90
(1)	.0010	.0039	.0158	.0366	.0683	.1144	.1824	.2896	.4880	1.038
(2)	.1089	.2121	.4246	.6701	.9765	1.384	1.969	2.903	4.700	9.927
(3)	.2940	.5976	1.257	2.033	3.003	4.295	6.164	9.194	15.13	32.69
(4)	.4244	.8834	1.934	3.233	4.909	7.198	10.57	16.10	27.05	59.69

TABLE 3.8 : η ($\mu = 0.3$)

Q: MODE:	0.05	0.10	0.20	0.30	0.40	0.50	0.60	0.70	0.80	0.90
(1)	.0024	.0098	.0400	.0939	.1776	.3038	.4977	.8194	1.449	3.286
(2)	.1677	.3441	.7391	1.222	1.846	2.701	3.964	6.046	10.17	22.47
(3)	.2677	.5612	1.246	2.109	3.241	4.804	7.125	10.96	18.60	41.41
(4)	UPPER and LOWER LIMITS AGREE WITHIN 1%									

TABLE 3.8 : η ($\mu = 0.4$)

Q: MODE:	0.05	0.10	0.20	0.30	0.40	0.50	0.60	0.70	0.80	0.90
(1)	.0041	.0166	.0682	.1610	.3077	.5331	.8885	1.495	2.715	6.371
(2)	.1719	.3598	.7965	1.346	2.067	3.064	4.546	7.001	11.89	26.51
(3)	UPPER and LOWER LIMITS AGREE WITHIN 1%									
(4)	.2590	.5435	1.209	2.050	3.154	4.681	6.945	10.68	18.11	40.28

TABLE 3.8 : η ($\mu = 0.5$)

Q: MODE:	0.05	0.10	0.20	0.30	0.40	0.50	0.60	0.70	0.80	0.90
(1)	.0054	.0219	.0902	.2136	.4104	.7159	1.203	2.045	3.764	8.966
(2)	.1170	.2461	.5501	.9365	1.447	2.155	3.210	4.958	8.439	18.85
(3)	.2098	.4400	.9775	1.656	2.547	3.780	5.610	8.637	14.65	32.64
(4)	.2145	.4507	1.007	1.709	2.638	3.927	5.847	9.019	15.29	34.17

TABLE 3.8 : η ($\mu = 0.6$)

Q: MODE:	0.05	0.10	0.20	0.30	0.40	0.50	0.60	0.70	0.80	0.90
(1)	.0058	.0235	.0969	.2297	.4416	.7710	1.297	2.208	4.070	9.714
(2)	.1173	.2471	.5539	.9461	1.467	2.192	3.276	5.077	8.676	19.43
(3)	.1822	.3823	.8499	1.439	2.211	3.274	4.847	7.439	12.57	27.87
(4)	.3078	.6456	1.429	2.413	3.696	5.467	8.090	12.42	21.00	46.57

TABLE 3.8 : η ($\mu = 0.7$)

Q: MODE:	0.05	0.10	0.20	0.30	0.40	0.50	0.60	0.70	0.80	0.90
(1)	.0051	.0204	.0838	.1983	.3800	.6606	1.106	1.869	3.417	8.078
(2)	.1782	.3741	.8333	1.415	2.182	3.246	4.831	7.460	12.70	28.38
(3)	UPPER and LOWER LIMITS AGREE WITHIN 1%									
(4)	.2080	.4355	.9653	1.630	2.492	3.673	5.408	8.237	13.91	30.33

TABLE 3.8 : η ($\mu = 0.8$)

Q: MODE:	0.05	0.10	0.20	0.30	0.40	0.50	0.60	0.70	0.80	0.90
(1)	.0033	.0131	.0537	.1265	.2408	.4149	.6866	1.145	2.057	4.761
(2)	.1675	.3477	.7592	1.268	1.927	2.830	4.162	6.352	10.69	23.60
(3)	.2888	.6027	1.335	2.259	3.466	5.130	7.606	11.69	19.82	44.13
(4)	.4459	.7253	1.571	2.417	3.722	5.699	8.530	13.33	22.94	48.38

TABLE 3.8 : η ($\mu = 0.9$)

Q: MODE:	0.05	0.10	0.20	0.30	0.40	0.50	0.60	0.70	0.80	0.90
(1)	.0011	.0045	.0184	.0431	.0813	.1383	.2252	.3677	.6435	1.441
(2)	.0801	.1621	.3400	.5494	.8108	1.157	1.652	2.443	3.968	8.416
(3)	.2155	.4378	.9184	1.479	2.171	3.084	4.386	6.470	10.52	22.44
(4)	.3962	.7869	1.649	2.686	3.979	5.734	8.246	12.27	20.05	52.80

TABLE 3.9 : η (μ = 0.1)

Q: MODE:	0.05	0.10	0.20	0.30	0.40	0.50	0.60	0.70	0.80	0.90
(1)	.0003	.0012	.0049	.0113	.0207	.0337	.0516	.0776	.1205	.2253
(2)	.0511	.0893	.1566	.2266	.3081	.4094	.5444	.7440	1.100	2.064
(3)	.1704	.3164	.5794	.8436	1.141	1.506	2.000	2.738	4.106	7.937
(4)	.3589	.6686	1.284	1.933	2.656	3.540	4.963	6.477	10.06	20.07

TABLE 3.9 : η (μ = 0.2)

Q: MODE:	0.05	0.10	0.20	0.30	0.40	0.50	0.60	0.70	0.80	0.90
(1)	.0017	.0069	.0280	.0652	.1218	.2046	.3275	.5227	.8864	1.899
(2)	.1553	.3062	.6189	.9768	1.419	2.005	2.845	4.192	6.800	14.43
(3)	.3347	.6866	1.463	2.388	3.552	5.116	7.384	11.07	18.32	39.85
(4)	.4834	.9813	2.085	3.453	5.220	7.761	11.37	17.34	29.25	63.23

TABLE 3.9 : η (μ = 0.3)

Q: MODE:	0.05	0.10	0.20	0.30	0.40	0.50	0.60	0.70	0.80	0.90
(1)	.0039	.0156	.0639	.1501	.2849	.4890	.8050	1.333	2.374	5.429
(2)	.2119	.4379	.9491	1.576	2.386	3.494	5.129	7.823	13.17	29.12
(3)	.2536	.5413	1.205	2.052	3.168	4.714	7.016	10.81	18.39	40.99
(4)	.1894	.4108	.8401	1.464	2.187	3.231	4.940	7.605	12.34	26.55

TABLE 3.9 : η (μ = 0.4)

Q: MODE:	0.05	0.10	0.20	0.30	0.40	0.50	0.60	0.70	0.80	0.90
(1)	.0058	.0233	.0958	.2266	.4341	.7544	1.262	2.132	3.895	9.197
(2)	.1821	.3819	.8491	1.439	2.214	3.286	4.879	7.512	12.75	28.41
(3)	.1680	.3528	.7857	1.331	2.049	3.039	4.509	6.944	11.77	26.06
(4)	.3111	.6519	1.443	2.432	3.722	5.498	8.136	12.48	21.10	46.88

TABLE 3.9 : η (μ = 0.5)

Q: MODE:	0.05	0.10	0.20	0.30	0.40	0.50	0.60	0.70	0.80	0.90
(1)	.0065	.0263	.1085	.2571	.4941	.8621	1.450	2.465	4.539	10.82
(2)	UPPER and LOWER LIMITS AGREE WITHIN 1%									
(3)	.2132	.4488	1.004	1.709	2.642	3.938	5.868	9.067	15.44	34.49
(4)	UPPER and LOWER LIMITS AGREE WITHIN 1%									

TABLE 3.10 : η (μ = 0.1)

Q: MODE:	0.05	0.10	0.20	0.30	0.40	0.50	0.60	0.70	0.80	0.90
(1)	.0007	.0026	.0108	.0254	.0480	.0819	.1337	.2195	.3869	.8745
(2)	.0541	.1099	.2328	.3803	.5669	.8161	1.174	1.746	2.849	6.060
(3)	.1779	.3605	.7539	1.211	1.776	2.519	3.578	5.274	8.557	18.18
(4)	.3349	.6856	1.457	2.368	3.506	5.015	7.184	10.68	17.50	37.63

TABLE 3.10 : η (μ = 0.2)

Q: MODE:	0.05	0.10	0.20	0.30	0.40	0.50	0.60	0.70	0.80	0.90
(1)	.0021	.0084	.0344	.0810	.1542	.2659	.4401	.7344	1.321	3.064
(2)	.1259	.2614	.5719	.9580	1.461	2.152	3.172	4.851	8.173	18.06
(3)	.2665	.5569	1.229	2.070	3.166	4.675	6.909	10.60	17.92	39.78
(4)	.3024	.6365	1.424	2.427	3.753	5.599	8.353	12.92	22.03	49.31

TABLE 3.10 : η (μ = 0.3)

Q: MODE:	0.05	0.10	0.20	0.30	0.40	0.50	0.60	0.70	0.80	0.90
(1)	.0036	.0145	.0595	.1406	.2692	.4675	.7815	1.320	2.409	5.685
(2)	.1533	.3217	.7162	1.216	1.877	2.795	4.165	6.441	10.98	24.57
(3)	.1662	.3502	.7856	1.343	2.083	3.114	4.637	7.222	12.34	27.69
(4)	.1789	.3747	.8297	1.399	2.140	3.153	4.644	7.086	11.91	26.23

TABLE 3.10 : η (μ = 0.4)

Q: MODE:	0.05	0.10	0.20	0.30	0.40	0.50	0.60	0.70	0.80	0.90
(1)	.0047	.0189	.0778	.1844	.3544	.6185	1.040	1.769	3.256	7.764
(2)	.1260	.2651	.5929	1.010	1.562	2.328	3.471	5.367	9.144	20.45
(3)	.1541	.3233	.7182	1.215	1.866	2.761	4.085	6.266	10.59	23.47
(4)	.2903	.6067	1.339	2.256	3.452	5.100	7.540	11.57	19.57	43.48

TABLE 3.10 : η (μ = 0.5)

Q: MODE:	0.05	0.10	0.20	0.30	0.40	0.50	0.60	0.70	0.80	0.90
(1)	.0051	.0205	.0845	.2004	.3857	.6740	1.136	1.936	3.574	8.550
(2)	UPPER and LOWER LIMITS AGREE WITHIN 1%									
(3)	.1781	.3756	.8428	1.441	2.234	3.340	4.993	7.737	13.21	29.58
(4)	UPPER and LOWER LIMITS AGREE WITHIN 1%									

TABLE 3.11 : η (μ = 0.1)

Q: MODE:	0.05	0.10	0.20	0.30	0.40	0.50	0.60	0.70	0.80	0.90
(1)	.0096	.0209	.0492	.0871	.1382	.2092	.3136	.4824	.8094	1.763
(2)	.0789	.1596	.3348	.5414	.7998	1.143	1.633	2.416	3.926	8.330
(3)	.2168	.4393	.9198	1.480	2.173	3.085	4.388	6.475	10.52	22.43
(4)	.3539	.7415	1.621	2.641	3.936	5.656	8.168	12.15	20.04	52.76

TABLE 3.11 : η (μ = 0.2)

Q: MODE:	0.05	0.10	0.20	0.30	0.40	0.50	0.60	0.70	0.80	0.90
(1)	.0316	.0681	.1591	.2807	.4463	.6798	1.029	1.608	2.752	6.148
(2)	.1640	.3406	.7442	1.244	1.893	2.783	4.095	6.255	10.53	23.27
(3)	.2889	.6036	1.340	2.265	3.472	5.141	7.617	11.71	19.85	44.10
(4)	.1923	.4488	1.221	2.138	3.453	5.173	7.735	11.97	20.31	44.45

TABLE 3.11 : η (μ = 0.3)

Q: MODE:	0.05	0.10	0.20	0.30	0.40	0.50	0.60	0.70	0.80	0.90
(1)	.0564	.1209	.2797	.4909	.7789	1.188	1.806	2.841	4.910	11.11
(2)	.1741	.3656	.8147	1.384	2.136	3.179	4.735	7.315	12.46	27.85
(3)	UPPER and LOWER LIMITS AGREE WITHIN 1 %									
(4)	.2079	.4359	.9658	1.629	2.491	3.673	5.409	8.238	13.91	30.90

TABLE 3.11 : η (μ = 0.4)

Q: MODE:	0.05	0.10	0.20	0.30	0.40	0.50	0.60	0.70	0.80	0.90
(1)	.0759	.1617	.3703	.6449	1.017	1.546	2.345	3.685	6.375	14.46
(2)	.1140	.2402	.5390	.9200	1.426	2.130	3.184	4.932	8.418	18.87
(3)	.1823	.3826	.8502	1.439	2.211	3.274	4.847	7.437	12.57	27.86
(4)	.3086	.6477	1.434	2.419	3.707	5.482	8.105	12.43	21.03	46.71

TABLE 3.11 : η (μ = 0.5)

Q: MODE:	0.05	0.10	0.20	0.30	0.40	0.50	0.60	0.70	0.80	0.90
(1)	.0838	.1775	.4020	.6932	1.084	1.634	2.461	3.843	6.609	14.91
(2)	.1083	.2279	.5100	.8692	1.344	2.005	2.989	4.623	7.877	17.61
(3)	.2045	.4290	.9543	1.618	2.491	3.700	5.495	8.466	14.37	32.03
(4)	.2141	.4508	1.007	1.715	2.644	3.942	5.855	9.046	15.38	34.38

TABLE 3.11 : η (μ = 0.6)

Q: MODE:	0.05	0.10	0.20	0.30	0.40	0.50	0.60	0.70	0.80	0.90
(1)	.0773	.1631	.3663	.6265	.9723	1.455	2.176	3.375	5.767	12.93
(2)	.1517	.3193	.7140	1.217	1.883	2.810	4.194	6.493	11.08	24.81
(3)	.1288	.2719	.6097	1.042	1.618	2.419	3.624	5.619	9.604	21.54
(4)	.2574	.5400	1.201	2.035	3.132	4.646	6.893	10.60	17.97	39.97

TABLE 3.11 : η (μ = 0.7)

Q: MODE:	0.05	0.10	0.20	0.30	0.40	0.50	0.60	0.70	0.80	0.90
(1)	.0694	.1470	.3313	.5681	.8832	1.324	1.985	3.086	5.287	11.88
(2)	.1355	.2858	.6416	1.097	1.701	2.542	3.797	5.877	10.02	22.41
(3)	.2393	.5027	1.120	1.903	2.935	4.364	6.492	10.02	17.04	38.03
(4)	UPPER and LOWER LIMITS AGREE WITHIN 1 %									

TABLE 3.11 : η (μ = 0.8)

Q: MODE:	0.05	0.10	0.20	0.30	0.40	0.50	0.60	0.70	0.80	0.90
(1)	.1054	.2225	.5004	.8578	1.334	2.000	2.998	4.663	7.989	17.96
(2)	.1063	.2238	.5009	.8537	1.320	1.968	2.935	4.537	7.726	17.27
(3)	.1967	.4120	.9127	1.541	2.361	3.487	5.150	7.884	13.29	29.39
(4)	.3563	.7452	1.647	2.776	4.249	6.277	9.277	14.22	24.05	53.37

TABLE 3.11 : η (μ = 0.9)

Q: MODE:	0.05	0.10	0.20	0.30	0.40	0.50	0.60	0.70	0.80	0.90
(1)	.1801	.3812	.8594	1.475	2.298	3.450	5.180	8.065	13.84	31.16
(2)	.2239	.4738	1.067	1.829	2.844	4.263	6.392	9.939	17.03	38.28
(3)	UPPER and LOWER LIMITS AGREE WITHIN 1 %									
(4)	.1737	.3660	.8203	1.401	2.170	3.242	4.844	7.503	12.80	28.69

TABLE 3.12 : η (μ = 0.1)

Q: MODE:	0.05	0.10	0.20	0.30	0.40	0.50	0.60	0.70	0.80	0.90
(1)	.0112	.0451	.1857	.4402	.8462	1.477	2.484	4.227	7.787	18.58
(2)	.3983	.8389	1.879	3.206	4.965	7.413	11.07	17.13	29.22	65.40
(3)	.6661	1.398	3.112	5.280	8.126	12.06	17.90	27.56	46.73	104.2
(4)	.9478	1.988	4.423	7.433	11.38	16.76	25.07	38.92	64.42	143.6

TABLE 3.12 : η (μ = 0.2)

Q: MODE:	0.05	0.10	0.20	0.30	0.40	0.50	0.60	0.70	0.80	0.90
(1)	.0122	.0494	.2034	.4821	.9267	1.618	2.721	4.630	8.530	20.35
(2)	.4379	.9224	2.066	3.526	5.461	8.155	12.18	18.85	32.16	71.98
(3)	.7351	1.543	3.437	5.830	8.976	13.33	19.78	30.44	51.61	115.1
(4)	1.048	2.204	4.868	8.232	12.72	18.73	27.69	42.47	72.80	166.4

TABLE 3.12 : η (μ = 0.3)

Q: MODE:	0.05	0.10	0.20	0.30	0.40	0.50	0.60	0.70	0.80	0.90
(1)	.0136	.0548	.2260	.5356	1.030	1.797	3.024	5.146	9.482	22.63
(2)	.4887	1.029	2.306	3.937	6.098	9.109	13.60	21.06	35.94	80.46
(3)	.8158	1.713	3.811	6.462	9.949	14.76	21.91	33.72	57.19	127.3
(4)	.9198	1.927	4.265	7.175	10.95	16.09	23.78	36.28	60.43	134.9

TABLE 3.12 : η (μ = 0.4)

Q: MODE:	0.05	0.10	0.20	0.30	0.40	0.50	0.60	0.70	0.80	0.90
(1)	.0154	.0620	.2557	.6060	1.165	2.034	3.423	5.826	10.74	25.63
(2)	.5500	1.159	2.596	4.431	6.862	10.25	15.30	23.68	40.39	90.39
(3)	.6258	1.313	2.911	4.914	7.528	11.12	16.40	25.08	42.21	93.25
(4)	1.209	2.520	5.529	9.261	14.10	20.74	30.53	46.65	78.56	174.1

TABLE 3.12 : η (μ = 0.5)

Q: MODE:	0.05	0.10	0.20	0.30	0.40	0.50	0.60	0.70	0.80	0.90
(1)	.0178	.0717	.2957	.7009	1.348	2.353	3.960	6.741	12.43	29.67
(2)	.5271	1.106	2.455	4.149	6.355	9.373	13.80	21.03	35.24	77.28
(3)	1.182	2.466	5.420	9.091	13.86	20.38	30.01	45.87	77.34	171.3
(4)	.9411	1.969	4.342	7.293	11.10	16.27	23.79	36.04	60.02	130.8

TABLE 3.12 : η (μ = 0.6)

Q: MODE:	0.05	0.10	0.20	0.30	0.40	0.50	0.60	0.70	0.80	0.90
(1)	.0209	.0842	.3472	.8230	1.582	2.762	4.646	7.907	14.57	34.77
(2)	.4272	.8980	2.007	3.415	5.274	7.853	11.69	18.05	30.69	68.47
(3)	.8616	1.805	4.001	6.765	10.39	15.38	22.79	35.00	59.22	131.6
(4)	1.298	2.740	6.117	10.39	16.12	24.03	35.52	54.77	92.96	208.3

TABLE 3.12 : η (μ = 0.7)

Q: MODE:	0.05	0.10	0.20	0.30	0.40	0.50	0.60	0.70	0.80	0.90
(1)	.0227	.0914	.3763	.8900	1.705	2.964	4.959	8.381	15.31	36.13
(2)	.7221	1.515	3.376	5.734	8.837	13.14	19.55	30.16	51.31	114.5
(3)	.4947	1.046	2.327	3.950	6.075	9.006	13.35	20.49	34.65	76.65
(4)	1.156	2.407	5.268	8.805	13.37	19.56	28.70	43.59	72.58	158.5

TABLE 3.12 : η (μ = 0.8)

Q: MODE:	0.05	0.10	0.20	0.30	0.40	0.50	0.60	0.70	0.80	0.90
(1)	.0128	.0516	.2108	.4917	.9231	1.560	2.516	4.050	6.941	15.05
(2)	1.035	2.074	4.294	6.899	10.19	14.64	21.16	31.88	53.11	116.4
(3)	1.166	2.462	5.479	9.286	14.33	21.24	31.62	48.68	82.74	184.3
(4)	1.238	2.201	4.604	8.025	12.11	17.84	26.50	41.09	70.99	196.5

TABLE 3.12 : η (μ = 0.9)

Q: MODE:	0.05	0.10	0.20	0.30	0.40	0.50	0.60	0.70	0.80	0.90
(1)	.0023	.0093	.0377	.0872	.1619	.2697	.4267	.6709	1.116	2.331
(2)	.2628	.5019	.9730	1.489	2.099	2.863	3.889	5.417	8.158	15.63
(3)	1.246	2.335	4.357	6.447	8.844	11.84	15.88	22.04	33.51	65.69
(4)	3.043	5.814	11.41	17.67	24.95	34.59	48.25	72.09	113.1	203.0

TABLE 3.13 : η (μ = 0.1)

Q: MODE:	0.05	0.10	0.20	0.30	0.40	0.50	0.60	0.70	0.80	0.90
(1)	.0111	.0447	.1843	.4369	.8397	1.466	2.465	4.195	7.728	18.44
(2)	.3956	.8333	1.866	3.185	4.932	7.364	10.99	17.02	29.03	65.00
(3)	.6621	1.391	3.095	5.251	8.081	11.99	17.80	27.39	46.50	103.7
(4)	.9453	1.980	4.366	7.375	11.40	16.94	25.12	38.52	66.69	151.6

TABLE 3.13 : η (μ = 0.2)

Q: MODE:	0.05	0.10	0.20	0.30	0.40	0.50	0.60	0.70	0.80	0.90
(1)	.0121	.0487	.2006	.4755	.9139	1.595	2.684	4.567	8.414	20.08
(2)	.4330	.9120	2.043	3.487	5.402	8.067	12.05	18.65	31.83	71.26
(3)	.7280	1.528	3.402	5.774	8.889	13.20	19.59	30.16	51.14	113.7
(4)	1.025	2.153	4.768	8.105	12.32	18.16	26.90	41.32	68.70	152.7

TABLE 3.13 : η (μ = 0.3)

Q: MODE:	0.05	0.10	0.20	0.30	0.40	0.50	0.60	0.70	0.80	0.90
(1)	.0133	.0538	.2217	.5255	1.010	1.764	2.968	5.051	9.307	22.21
(2)	.4801	1.011	2.266	3.869	5.993	8.952	13.37	20.70	35.32	79.10
(3)	.7558	1.586	3.522	5.959	9.147	13.54	20.03	30.70	51.88	114.6
(4)	1.054	2.214	4.918	8.326	12.79	18.96	28.07	43.21	73.02	161.8

TABLE 3.13 : η (μ = 0.4)

Q: MODE:	0.05	0.10	0.20	0.30	0.40	0.50	0.60	0.70	0.80	0.90
(1)	.0150	.0605	.2496	.5916	1.137	1.986	3.342	5.690	10.49	25.04
(2)	.5194	1.093	2.446	4.169	6.446	9.607	14.31	22.10	37.60	83.89
(3)	.7521	1.581	3.528	5.999	9.255	13.77	20.49	31.62	53.75	120.0
(4)	1.000	2.096	4.657	7.880	12.11	17.94	26.58	40.84	69.15	153.6

TABLE 3.13 : η (μ = 0.5)

Q: MODE:	0.05	0.10	0.20	0.30	0.40	0.50	0.60	0.70	0.80	0.90
(1)	.0172	.0692	.2854	.6767	1.301	2.272	3.823	6.508	12.00	28.64
(2)	.3718	.7788	1.724	2.906	4.442	6.539	9.617	14.65	24.57	54.00
(3)	.9333	1.937	4.228	7.058	10.73	15.76	23.19	35.45	59.80	132.5
(4)	.6996	1.463	3.232	5.434	8.285	12.17	17.86	27.16	45.44	99.58

TABLE 3.13 : η (μ = 0.6)

Q: MODE:	0.05	0.10	0.20	0.30	0.40	0.50	0.60	0.70	0.80	0.90
(1)	.0193	.0778	.3206	.7596	1.459	2.545	4.277	7.270	13.37	31.85
(2)	.4777	1.006	2.251	3.838	5.939	8.861	13.22	20.46	34.89	78.08
(3)	.7204	1.510	3.346	5.647	8.639	12.73	18.72	28.49	47.65	104.3
(4)	1.623	3.389	7.463	12.54	19.13	28.16	41.50	63.46	107.0	237.1

TABLE 3.13 : η (μ = 0.7)

Q: MODE:	0.05	0.10	0.20	0.30	0.40	0.50	0.60	0.70	0.80	0.90
(1)	.0171	.0690	.2832	.6674	1.272	2.195	3.636	6.070	10.91	25.24
(2)	.7478	1.558	3.421	5.736	8.746	12.88	19.00	29.09	49.16	109.1
(3)	.5510	1.160	2.595	4.423	6.840	10.20	15.21	23.51	40.06	89.60
(4)	.8904	1.856	4.071	6.792	10.26	14.91	21.59	32.27	52.83	112.6

TABLE 3.13 : η (μ = 0.8)

Q: MODE:	0.05	0.10	0.20	0.30	0.40	0.50	0.60	0.70	0.80	0.90
(1)	.0071	.0287	.1169	.2722	.5095	.8580	1.377	2.207	3.765	8.133
(2)	.7353	1.457	2.988	4.805	7.128	10.30	14.97	22.63	37.75	82.63
(3)	1.260	2.636	5.824	9.816	15.02	22.19	32.79	50.29	85.05	188.9
(4)	.9351	1.970	4.419	7.550	11.70	17.50	26.17	40.57	69.33	155.4

TABLE 3.13 ; n (μ = 0.9)

Q: MODE:	0.05	0.10	0.20	0.30	0.40	0.50	0.60	0.70	0.80	0.90
(1)	.0012	.0049	.0201	.0466	.0869	.1457	.2325	.3699	.6261	1.341
(2)	.1531	.2968	.5917	.9275	1.334	1.850	2.548	3.591	5.461	10.56
(3)	.9044	1.693	3.157	4.673	6.414	8.574	11.48	15.88	23.89	46.20
(4)	2.525	4.876	9.448	14.33	20.06	27.40	37.69	53.98	85.27	176.5

TABLE 3.14 : K* (μ = 0.3)

Q: MODE:	0.05	0.10	0.20	0.30	0.40	0.50	0.60	0.70	0.80	0.90
(1)	12.35	26.14	59.12	101.8	158.9	239.2	359.6	560.5	962.2	2166
(2)	35.56	74.35	163.9	274.8	416.9	608.4	885.1	1330	2194	4731
(3)	131.5	262.7	532.4	823.9	1152	1538	2022	2689	3799	6623
(4)	940.2	1614	2677	3633	4619	5718	7027	0724	11282	16932

TABLE 3.14 : K* (μ = 0.6)

Q: MODE:	0.05	0.10	0.20	0.30	0.40	0.50	0.60	0.70	0.80	0.90
(1)	4.517	9.804	23.03	40.63	64.23	96.76	144.1	220.2	366.3	789.9
(2)	46.65	92.62	188.6	296.7	424.9	584.3	795.1	1103	1647	3112
(3)	300.0	547.1	981.8	1407	1871	2415	3102	4064	5680	9817
(4)	1011	1831	3216	4520	5944	7685	10054	13746	20763	41166

TABLE 3.14 : K* (μ = 0.9)

Q: MODE:	0.05	0.10	0.20	0.30	0.40	0.50	0.60	0.70	0.80	0.90
(1)	2.026	4.479	10.74	18.99	29.59	43.21	61.20	86.68	129.1	235.5
(2)	56.46	95.82	164.0	233.0	308.4	393.0	490.0	605.4	756.4	1027
(3)	654.6	868.3	1160	1427	1706	2006	2333	2693	3096	3577
(4)	4036	4440	5182	5998	6956	8129	9661	11850	15608	25541

TABLE 3.15 : K* (μ = 0.3)

Q: MODE:	0.05	0.10	0.20	0.30	0.40	0.50	0.60	0.70	0.80	0.90
(1)	26.47	55.72	124.5	211.5	325.3	481.4	710.8	1085	1023	4009
(2)	90.11	183.0	381.7	606.0	869.1	1192	1618	2241	3348	6361
(3)	534.3	971.7	1724	2428	3148	3920	4776	5749	6894	8343
(4)	9444	10518	12278	14261	16748	20078	24895	32680	47869	92650

TABLE 3.15 : K* (μ = 0.6)

Q: MODE:	0.05	0.10	0.20	0.30	0.40	0.50	0.60	0.70	0.80	0.90
(1)	23.40	49.49	110.3	183.7	271.8	377.9	507.9	674.3	912.2	1388
(2)	553.3	844.2	1260	1653	2102	2672	3466	4710	7079	13933
(3)	392.9	762.3	1474	2204	3014	3983	5246	7118	10509	19968
(4)	1072	1985	3574	5055	6567	8219	10148	12598	16225	24114

TABLE 3.14 : K* (μ = 0.1)

Q: MODE:	0.05	0.10	0.20	0.30	0.40	0.50	0.60	0.70	0.80	0.90
(1)	197.8	417.6	939.9	1612	2508	3763	5647	8786	15066	33900
(2)	262.2	553.1	1241	2124	3297	4937	7389	11467	19617	43994
(3)	345.8	728.5	1624	2759	4251	6318	9390	14460	24576	54967
(4)	475.0	992.9	2184	3602	5517	8041	11764	17804	29779	63563

TABLE 3.14 : K* (μ = 0.4)

Q: MODE:	0.05	0.10	0.20	0.30	0.40	0.50	0.60	0.70	0.80	0.90
(1)	7.099	15.09	34.35	59.44	93.15	140.5	211.6	329.7	565.5	1270
(2)	34.74	72.04	156.1	256.4	379.3	535.5	746.5	1061	1628	3184
(3)	255.8	476.5	878.9	1281	1721	2236	2880	3769	5235	8907
(4)	1072	1912	3298	4592	5998	7718	10055	13712	20650	40819

TABLE 3.14 : K* (μ = 0.7)

Q: MODE:	0.05	0.10	0.20	0.30	0.40	0.50	0.60	0.70	0.80	0.90
(1)	4.394	9.740	23.48	41.77	65.52	96.37	137.5	196.2	294.0	537.2
(2)	97.90	172.7	303.2	438.0	597.1	801.6	1088	1539	2399	4890
(3)	171.4	333.3	648.1	972.7	1328	1738	2245	2935	4065	6912
(4)	1075	1819	2961	3977	5025	6202	7624	9506	12434	19202

TABLE 3.15 : K* (μ = 0.1)

Q: MODE:	0.05	0.10	0.20	0.30	0.40	0.50	0.60	0.70	0.80	0.90
(1)	243.8	514.5	1156	1980	3078	4613	6912	10741	18392	41330
(2)	321.6	676.8	1513	2576	3981	5931	8836	13646	23225	51863
(3)	435.9	912.4	2017	3399	5195	7655	11276	17218	28961	63902
(4)	610.5	1266	2754	4564	6861	9943	14399	21608	35692	77275

TABLE 3.15 : K* (μ = 0.4)

Q: MODE:	0.05	0.10	0.20	0.30	0.40	0.50	0.60	0.70	0.80	0.90
(1)	21.89	46.09	102.6	173.0	262.9	382.3	551.4	817.3	1321	2770
(2)	146.4	284.6	556.7	841.6	1153	1505	1911	2396	3010	3946
(3)	3078	3949	5057	6119	7381	9037	11414	15246	22722	44779
(4)	736.0	1424	2727	4009	5350	6831	8581	10862	14415	22880

TABLE 3.15 : K* (μ = 0.7)

Q: MODE:	0.05	0.10	0.20	0.30	0.40	0.50	0.60	0.70	0.80	0.90
(1)	20.11	42.35	94.19	158.1	238.3	342.6	486.5	706.0	1109	2236
(2)	171.3	322.4	603.5	885.9	1187	1518	1891	2316	2818	3455
(3)	3863	4375	5219	6164	7339	8899	11130	14694	21584	41703
(4)	831.5	1611	3161	4781	6543	8753	11593	15982	24437	48149

TABLE 3.14 : K* (μ = 0.2)

Q: MODE:	0.05	0.10	0.20	0.30	0.40	0.50	0.60	0.70	0.80	0.90
(1)	31.66	66.91	150.8	259.0	403.4	605.9	909.9	1416	2431	5473
(2)	59.35	124.8	278.4	473.0	728.9	1082	1606	2470	4184	9294
(3)	118.0	244.1	527.4	868.0	1293	1857	2659	3934	6397	13604
(4)	286.4	570.8	1156	1786	2512	3397	4547	6249	9297	17639

TABLE 3.14 : K* (μ = 0.5)

Q: MODE:	0.05	0.10	0.20	0.30	0.40	0.50	0.60	0.70	0.80	0.90
(1)	5.182	11.09	25.56	44.59	70.22	106.1	159.6	248.0	422.7	941.7
(2)	43.55	89.15	189.3	304.6	438.4	594.2	776.5	991.7	1248	1564
(3)	2058	2244	2644	3134	3765	4621	5868	7894	11862	23588
(4)	457.3	885.7	1701	2511	3354	4261	5264	6402	7726	9315

TABLE 3.14 : K* (μ = 0.8)

Q: MODE:	0.05	0.10	0.20	0.30	0.40	0.50	0.60	0.70	0.80	0.90
(1)	3.482	7.782	18.87	33.36	51.36	73.07	98.72	128.0	164.1	208.7
(2)	262.9	313.7	417.8	541.6	695.0	892.1	1160	1562	2284	4256
(3)	299.6	563.0	1043	1528	2074	2752	3680	5117	7036	15664
(4)	397.7	775.4	1506	2252	3082	4055	5321	7160	10467	19850

TABLE 3.15 : K* (μ = 0.2)

Q: MODE:	0.05	0.10	0.20	0.30	0.40	0.50	0.60	0.70	0.80	0.90
(1)	50.15	105.7	237.1	404.8	627.0	936.0	1396	2159	3679	8222
(2)	97.70	203.2	443.8	737.9	1112	1613	2337	3502	5770	12441
(3)	224.6	453.6	936.7	1476	2110	2899	3964	5578	8559	16992
(4)	668.3	1271	2375	3440	4544	5768	7230	9179	12339	20291

TABLE 3.15 : K* (μ = 0.5)

Q: MODE:	0.05	0.10	0.20	0.30	0.40	0.50	0.60	0.70	0.80	0.90
(1)	23.95	50.66	113.0	189.0	281.7	396.0	541.9	740.7	1054	1790
(2)	360.9	617.6	1028	1420	1862	2417	3184	4386	6678	13328
(3)	393.4	758.1	1449	2140	2877	3705	4693	5979	7949	12471
(4)	2463	4124	6568	8709	10996	13785	17574	23485	34724	67411

TABLE 3.15 : K* (μ = 0.8)

Q: MODE:	0.05	0.10	0.20	0.30	0.40	0.50	0.60	0.70	0.80	0.90
(1)	23.76	50.04	111.7	189.3	290.0	426.5	624.5	943.8	1563	3380
(2)	95.60	192.1	394.7	620.2	884.8	1212	1649	2301	3485	6785
(3)	400.5	746.4	1359	1939	2538	3185	3943	4900	6377	9762
(4)	2569	3968	5821	7348	8853	10472	12280	14416	17180	21781

TABLE 3.15 : K* (μ − 0.9)

Q: MODE:	0.05	0.10	0.20	0.30	0.40	0.50	0.60	0.70	0.80	0.90
(1)	55.04	116.1	260.2	444.0	686.4	1021	1518	2337	3958	8783
(2)	134.8	279.8	606.7	1004	1506	2179	3151	4715	7771	16757
(3)	274.2	555.3	1174	1883	2738	3853	5412	7849	12590	26357
(4)	409.4	1045	2145	3056	4321	6238	7799	11397	15525	27743

TABLE 3.16 : K* (μ − 0.1)

Q: MODE:	0.05	0.10	0.20	0.30	0.40	0.50	0.60	0.70	0.80	0.90
(1)	259.0	546.9	1228	2101	3263	4886	7313	11353	19427	43622
(2)	345.6	728.7	1623	2759	4252	6320	9398	14489	24646	54688
(3)	474.9	995.7	2186	3717	5647	8224	12235	18702	31671	71812
(4)	653.7	1662	2424	4261	5987	9369	13888	19346	27780	58422

TABLE 3.16 : K* (μ − 0.2)

Q: MODE:	0.05	0.10	0.20	0.30	0.40	0.50	0.60	0.70	0.80	0.90
(1)	57.72	121.5	271.5	461.8	712.4	1058	1572	2419	4100	9110
(2)	118.3	244.6	528.2	869.1	1295	1858	2660	3936	6401	13606
(3)	287.1	573.4	1157	1799	2527	3414	4577	6330	9460	18237
(4)	901.5	1689	3011	4213	5673	6828	8236	10037	12865	18551

TABLE 3.16 : K* (μ − 0.3)

Q: MODE:	0.05	0.10	0.20	0.30	0.40	0.50	0.60	0.70	0.80	0.90
(1)	33.79	70.87	157.0	264.2	402.1	587.9	856.3	1288	2125	4580
(2)	132.5	264.2	534.7	826.7	1155	1542	2026	2695	3805	6633
(3)	938.9	1613	2676	3632	4618	5717	7027	8725	11284	16937
(4)	4000	6553	10130	13345	16735	20962	26822	35831	53761	106450

TABLE 3.16 : K* (μ − 0.4)

Q: MODE:	0.05	0.10	0.20	0.30	0.40	0.50	0.60	0.70	0.80	0.90
(1)	31.75	66.48	146.1	242.2	360.4	510.7	713.1	1013	1550	3013
(2)	262.2	485.6	889.3	1291	1731	2245	2888	3775	5238	8897
(3)	1078	1923	3315	4609	6016	7741	10084	13742	20714	40976
(4)	944.2	1805	3391	4905	6438	8058	9827	11866	14375	18175

TABLE 3.16 : K* (μ − 0.5)

Q: MODE:	0.05	0.10	0.20	0.30	0.40	0.50	0.60	0.70	0.80	0.90
(1)	36.63	76.87	168.9	277.6	405.1	554.1	728.6	933.9	1177	1471
(2)	1983	2146	2525	2996	3604	4427	5625	7568	11368	22580
(3)	461.8	893.2	1712	2524	3367	4275	5282	6417	7745	9339
(4)	12023	12783	14536	16709	19513	23318	28865	37819	55518	107521

TABLE 3.17 : K* (μ − 0.1)

Q: MODE:	0.05	0.10	0.20	0.30	0.40	0.50	0.60	0.70	0.80	0.90
(1)	1.370	5.520	22.74	53.84	103.3	180.0	302.2	512.7	941.5	2238
(2)	59.35	124.6	277.4	470.4	723.7	1072	1589	2439	4122	9134
(3)	134.5	278.7	605.1	1001	1503	2174	3143	4707	7746	16706
(4)	267.6	554.4	1164	1871	2735	3825	5366	7740	12360	25667

TABLE 3.17 : K* (μ − 0.2)

Q: MODE:	0.05	0.10	0.20	0.30	0.40	0.50	0.60	0.70	0.80	0.90
(1)	.4717	1.900	7.819	18.49	35.40	61.51	102.9	173.8	317.6	750.6
(2)	26.23	54.78	120.7	202.5	307.8	450.1	656.3	989.0	1635	3533
(3)	94.90	190.9	392.8	618.0	881.8	1208	1644	2296	3480	6765
(4)	399.1	743.6	1354	1932	2525	3171	3916	4866	6299	9496

TABLE 3.17 : K* (μ − 0.3)

Q: MODE:	0.05	0.10	0.20	0.30	0.40	0.50	0.60	0.70	0.80	0.90
(1)	.3026	1.218	5.007	11.81	22.54	38.98	64.82	108.8	197.2	461.5
(2)	23.03	47.69	103.5	170.9	254.7	363.4	513.6	743.6	1168	2362
(3)	168.5	318.3	598.6	881.0	1182	1514	1886	2312	2812	3447
(4)	3870	4377	5217	6162	7337	8896	11125	14691	21581	41709

TABLE 3.17 : K* (μ − 0.4)

Q: MODE:	0.05	0.10	0.20	0.30	0.40	0.50	0.60	0.70	0.80	0.90
(1)	.2607	1.049	4.301	10.10	19.18	32.92	54.23	89.96	160.7	369.9
(2)	28.55	58.37	124.4	201.7	293.4	403.7	539.4	715.5	973.5	1513
(3)	516.4	809.7	1237	1638	2094	2671	3472	4726	7113	14021
(4)	390.6	758.8	1468	2197	3004	3971	5235	7103	10490	19919

TABLE 3.17 : K* (μ − 0.5)

Q: MODE:	0.05	0.10	0.20	0.30	0.40	0.50	0.60	0.70	0.80	0.90
(1)	.2747	1.105	4.510	10.52	19.78	33.50	54.25	88.03	153.0	339.8
(2)	34.06	67.71	139.4	221.9	319.9	439.1	589.1	790.7	1102	1813
(3)	418.3	700.6	1138	1551	2015	2600	3413	4691	7140	14270
(4)	387.3	748.2	1435	2125	2861	3689	4676	5963	7936	12468

TABLE 3.17 : K* (μ − 0.6)

Q: MODE:	0.05	0.10	0.20	0.30	0.40	0.50	0.60	0.70	0.80	0.90
(1)	.3348	1.344	5.450	12.57	23.20	38.33	59.94	92.70	150.5	303.0
(2)	46.06	85.54	161.2	244.3	344.3	473.2	652.2	929.7	1449	2931
(3)	171.2	326.7	622.9	926.1	1255	1626	2057	2579	3260	4378
(4)	2815	3813	5064	6211	7544	9276	11750	15735	23511	46451

TABLE 3.17 : K* (μ − 0.7)

Q: MODE:	0.05	0.10	0.20	0.30	0.40	0.50	0.60	0.70	0.80	0.90
(1)	.3877	1.553	6.240	14.16	25.47	40.46	59.62	83.84	115.4	163.8
(2)	153.4	201.1	281.0	371.2	484.8	636.5	854.4	1202	1873	3827
(3)	126.5	248.1	490.3	747.4	1037	1386	1837	2488	3631	6720
(4)	619.4	1108	1927	2680	3445	4266	5181	6234	7504	9224

TABLE 3.17 : K* (μ − 0.8)

Q: MODE:	0.05	0.10	0.20	0.30	0.40	0.50	0.60	0.70	0.80	0.90
(1)	.2784	1.115	4.477	10.15	18.25	29.02	42.89	60.88	85.98	132.8
(2)	94.32	136.6	211.2	294.4	394.1	517.5	676.9	899.3	1263	2155
(3)	356.4	620.0	1049	1460	1921	2499	3297	4545	6923	13821
(4)	337.0	659.4	1292	1951	2689	3580	4753	6504	9702	18690

TABLE 3.17 : K* (μ − 0.9)

Q: MODE:	0.05	0.10	0.20	0.30	0.40	0.50	0.60	0.70	0.80	0.90
(1)	.1515	.6078	2.461	5.658	10.40	17.09	26.53	40.66	65.28	129.4
(2)	27.56	49.61	91.26	135.9	186.4	244.8	313.9	399.9	520.5	764.3
(3)	366.8	523.6	743.8	944.1	1150	1371	1609	1868	2150	2459
(4)	3002	3244	3767	4361	5048	5864	6875	8231	10362	15382

TABLE 3.18 : K* (μ − 0.1)

Q: MODE:	0.05	0.10	0.20	0.30	0.40	0.50	0.60	0.70	0.80	0.90
(1)	43.56	92.10	207.7	356.2	553.6	828.6	1238	1917	3265	7290
(2)	111.1	231.1	505.0	841.2	1271	1851	2694	4060	6732	14621
(3)	230.3	471.4	999.7	1618	2378	3369	4768	6983	11234	23620
(4)	450.5	903.9	1846	2888	4108	5634	7713	10906	16889	34029

TABLE 3.18 : K* (μ − 0.2)

Q: MODE:	0.05	0.10	0.20	0.30	0.40	0.50	0.60	0.70	0.80	0.90
(1)	17.29	36.59	82.51	141.3	218.7	325.4	482.4	739.0	1242	2732
(2)	68.41	139.3	292.6	469.3	682.1	952.7	1322	1888	2939	5917
(3)	272.3	519.8	982.0	1439	1926	2476	3140	4031	5472	9074
(4)	1519	2505	3935	5129	6281	7470	8741	10134	11692	13466

TABLE 3.18 : K* (μ − 0.3)

Q: MODE:	0.05	0.10	0.20	0.30	0.40	0.50	0.60	0.70	0.80	0.90
(1)	13.16	27.93	63.09	107.7	165.5	243.2	354.3	530.2	865.5	1832
(2)	97.68	190.9	376.4	572.4	788.4	1033	1321	1678	2178	3173
(3)	1326	1955	2807	3579	4438	5496	6923	9087	13073	24272
(4)	996.2	1908	3616	5337	7251	9591	12760	17656	26932	53807

TABLE 3.18 : K* (μ = 0.4)

Q: MODE:	0.05	0.10	0.20	0.30	0.40	0.50	0.60	0.70	0.80	0.90
(1)	13.75	29.36	66.70	113.5	172.2	247.3	347.5	492.7	743.1	1397
(2)	182.5	329.2	588.0	848.7	1143	1501	1975	2674	3921	7321
(3)	463.9	872.9	1619	2362	3190	4204	5576	7688	11667	23129
(4)	726.5	1382	2583	3734	4905	6147	7504	9030	10793	12895

TABLE 3.18 : K* (μ = 0.5)

Q: MODE:	0.05	0.10	0.20	0.30	0.40	0.50	0.60	0.70	0.80	0.90
(1)	15.78	33.96	77.81	132.4	198.8	278.7	374.1	488.1	625.0	791.0
(2)	1085	1184	1413	1699	2070	2575	3311	4508	6851	13775
(3)	295.8	576.4	1120	1672	2256	2893	3605	4418	5368	6511
(4)	8447	9004	10288	11882	13937	16725	20790	27403	40371	78741

TABLE 3.19 : K* (μ = 0.1)

Q: MODE:	0.05	0.10	0.20	0.30	0.40	0.50	0.60	0.70	0.80	0.90
(1)	1.460	3.192	7.572	13.42	21.20	31.77	46.86	70.49	114.7	239.8
(2)	18.60	36.08	71.30	110.3	156.8	215.3	294.5	413.8	630.9	1233
(3)	78.10	147.4	276.6	408.0	555.3	732.4	963.9	1307	1926	3654
(4)	201.8	380.7	702.7	1017	1355	1827	2357	3033	4434	8419

TABLE 3.19 : K* (μ = 0.2)

Q: MODE:	0.05	0.10	0.20	0.30	0.40	0.50	0.60	0.70	0.80	0.90
(1)	1.925	4.212	10.01	17.75	28.07	42.11	62.17	93.59	152.5	319.0
(2)	24.83	48.20	95.36	147.7	210.1	288.8	395.3	555.9	848.2	1659
(3)	105.3	198.8	373.3	550.7	748.6	988.9	1303	1764	2605	4930
(4)	275.3	528.3	946.0	1369	1825	2360	3130	4181	5904	11012

TABLE 3.19 : K* (μ = 0.3)

Q: MODE:	0.05	0.10	0.20	0.30	0.40	0.50	0.60	0.70	0.80	0.90
(1)	2.649	5.807	13.83	24.58	38.94	58.51	86.49	130.4	212.6	445.6
(2)	34.55	67.13	132.9	206.1	293.2	403.0	551.7	775.4	1181	2309
(3)	131.2	249.3	468.5	687.0	925.5	1204	1562	2069	2951	5289
(4)	258.9	492.9	919.9	1336	1779	2299	2964	3969	5789	11077

TABLE 3.19 : K* (μ = 0.4)

Q: MODE:	0.05	0.10	0.20	0.30	0.40	0.50	0.60	0.70	0.80	0.90
(1)	3.832	8.417	20.11	35.84	56.89	85.60	126.7	191.1	311.9	653.9
(2)	41.82	81.38	159.9	244.0	340.0	455.2	602.6	809.6	1155	2030
(3)	167.1	312.8	571.6	820.8	1090	1413	1846	2513	3770	7410
(4)	270.5	502.0	920.3	1334	1788	2335	3045	4092	5960	11134

TABLE 3.19 : K* (μ = 0.5)

Q: MODE:	0.05	0.10	0.20	0.30	0.40	0.50	0.60	0.70	0.80	0.90
(1)	5.691	12.50	29.83	53.07	84.02	126.0	185.5	277.8	448.6	925.1
(2)	26.09	51.96	104.4	159.9	222.3	296.9	394.3	538.5	800.4	1534
(3)	152.7	268.3	471.4	679.8	919.3	1215	1612	2205	3277	6235
(4)	151.4	299.1	591.6	894.9	1226	1616	2121	2865	4214	8007

TABLE 3.19 : K* (μ = 0.6)

Q: MODE:	0.05	0.10	0.20	0.30	0.40	0.50	0.60	0.70	0.80	0.90
(1)	6.154	13.16	29.95	50.94	77.02	109.7	151.9	209.9	302.0	516.7
(2)	101.3	175.4	295.9	411.8	543.3	709.3	940.9	1306	2010	4063
(3)	62.58	123.2	241.3	359.9	484.0	620.6	780.9	988.8	1314	2108
(4)	571.1	879.5	1338	1776	2269	2877	3690	4908	7122	13242

TABLE 3.19 : K* (μ = 0.7)

Q: MODE:	0.05	0.10	0.20	0.30	0.40	0.50	0.60	0.70	0.80	0.90
(1)	4.625	9.597	20.72	33.74	49.26	68.36	93.17	128.7	189.8	351.0
(2)	65.02	112.2	196.1	284.7	387.3	512.2	672.3	894.8	1256	2127
(3)	290.7	525.1	917.2	1291	1704	2217	2929	4053	6224	12586
(4)	95.39	193.6	384.7	586.3	801.8	1048	1363	1797	2547	4652

TABLE 3.19 : K* (μ = 0.8)

Q: MODE:	0.05	0.10	0.20	0.30	0.40	0.50	0.60	0.70	0.80	0.90
(1)	6.825	14.31	31.73	53.53	81.85	120.5	177.2	269.0	431.9	992.0
(2)	17.92	36.19	74.50	116.4	163.6	219.2	288.6	384.3	543.8	953.8
(3)	164.6	275.2	449.3	610.3	779.0	966.8	1187	1468	1879	2793
(4)	949.0	1420	2056	2625	3247	4051	4926	6481	9233	16220

TABLE 3.19 : K* (μ = 0.9)

Q: MODE:	0.05	0.10	0.20	0.30	0.40	0.50	0.60	0.70	0.80	0.90
(1)	30.89	65.16	146.4	250.6	389.2	582.5	871.6	1352	2311	5181
(2)	46.01	96.91	215.8	365.8	562.2	833.2	1235	1899	3214	7132
(3)	54.55	112.9	246.6	401.0	610.9	889.0	1300	1971	3236	7111
(4)	95.76	163.4	307.0	463.1	657.7	901.4	1220	1666	2425	4360

TABLE 3.20 : K* (μ = 0.1)

Q: MODE:	0.05	0.10	0.20	0.30	0.40	0.50	0.60	0.70	0.80	0.90
(1)	1.428	3.123	7.410	13.13	20.75	31.11	45.90	69.06	112.4	235.1
(2)	18.25	35.40	69.98	108.3	154.0	211.5	289.4	406.6	620.2	1212
(3)	76.82	145.0	272.6	401.9	546.3	720.2	950.2	1288	1896	3589
(4)	206.6	377.1	694.3	1002	1334	1754	2276	3064	4352	8197

TABLE 3.20 : K* (μ = 0.2)

Q: MODE:	0.05	0.10	0.20	0.30	0.40	0.50	0.60	0.70	0.80	0.90
(1)	1.850	4.051	9.635	17.11	27.08	40.65	60.04	90.45	147.4	308.7
(2)	24.07	46.73	92.49	143.4	204.0	280.5	384.2	540.4	824.7	1613
(3)	101.7	192.7	361.8	534.0	725.6	957.5	1261	1710	2513	4756
(4)	255.0	472.9	892.5	1261	1673	2152	2768	3777	5259	9614

TABLE 3.20 : K* (μ = 0.3)

Q: MODE:	0.05	0.10	0.20	0.30	0.40	0.50	0.60	0.70	0.80	0.90
(1)	2.505	5.499	13.13	23.37	37.08	55.77	82.52	124.5	203.2	426.2
(2)	32.26	62.71	124.1	192.1	272.7	373.8	509.8	713.1	1080	2089
(3)	79.76	154.2	294.2	431.4	575.2	736.5	934.1	1207	1670	2889
(4)	335.3	575.2	952.1	1293	1658	2097	2674	3561	5183	9814

TABLE 3.20 : K* (μ = 0.4)

Q: MODE:	0.05	0.10	0.20	0.30	0.40	0.50	0.60	0.70	0.80	0.90
(1)	3.525	7.754	18.56	33.12	52.61	79.17	117.1	176.6	287.7	601.5
(2)	24.75	48.54	95.13	142.9	194.5	253.4	324.9	421.1	576.9	964.4
(3)	177.7	294.8	475.5	644.8	833.1	1065	1381	1869	2781	5397
(4)	220.4	417.2	779.1	1137	1526	1979	2554	3366	4758	8406

TABLE 3.20 : K* (μ = 0.5)

Q: MODE:	0.05	0.10	0.20	0.30	0.40	0.50	0.60	0.70	0.80	0.90
(1)	4.515	9.840	23.17	40.67	63.48	93.59	134.9	196.4	304.2	587.1
(2)	37.75	73.32	142.2	214.0	295.8	397.6	537.2	755.1	1170	2381
(3)	91.34	173.7	331.7	498.0	687.0	914.8	1207	1623	2324	4108
(4)	274.6	526.4	995.8	1461	1974	2595	3431	4720	7169	14288

TABLE 3.20 : K* (μ = 0.6)

Q: MODE:	0.05	0.10	0.20	0.30	0.40	0.50	0.60	0.70	0.80	0.90
(1)	3.658	7.697	17.02	28.24	41.78	58.40	79.54	108.5	155.2	268.3
(2)	66.82	111.2	185.6	263.1	355.5	475.1	642.6	905.3	1403	2836
(3)	44.94	89.67	180.4	276.0	382.0	506.8	666.7	898.3	1311	2451
(4)	293.4	500.1	845.2	1184	1559	2004	2571	3363	4678	7945

TABLE 3.20 : K* (μ = 0.7)

Q: MODE:	0.05	0.10	0.20	0.30	0.40	0.50	0.60	0.70	0.80	0.90
(1)	3.115	6.483	14.13	23.32	34.73	49.53	70.08	101.9	161.5	331.5
(2)	29.36	55.52	107.1	163.5	228.4	305.8	401.8	529.1	723.4	1157
(3)	292.5	481.0	765.7	1031	1330	1708	2236	3069	4670	9338
(4)	91.43	184.6	380.0	595.6	845.7	1155	1571	2207	3395	6798

TABLE 3.20 : K* (μ = 0.8)

Q: / MODE:	0.05	0.10	0.20	0.30	0.40	0.50	0.60	0.70	0.80	0.90
(1)	6.186	13.02	29.12	49.57	76.52	113.8	169.1	260.5	441.7	982.4
(2)	12.24	25.03	52.68	84.03	120.8	165.9	225.2	312.1	467.0	891.5
(3)	89.94	160.6	282.2	398.9	522.0	658.9	819.0	1020	1316	1959
(4)	679.5	1016	1478	1889	2329	2846	3500	4410	5917	9671

TABLE 3.20 : K* (μ = 0.9)

Q: / MODE:	0.05	0.10	0.20	0.30	0.40	0.50	0.60	0.70	0.80	0.90
(1)	26.85	56.77	127.9	219.5	341.6	512.5	768.5	1194	2046	4598
(2)	43.34	91.11	203.1	344.9	531.4	789.4	1172	1806	3065	6826
(3)	51.63	107.8	237.2	397.5	604.4	885.7	1297	1969	3292	7219
(4)	63.65	130.3	275.3	442.3	643.4	899.6	1252	1798	2828	5791

TABLE 3.21 : K* (μ = 0.1)

Q: / MODE:	0.05	0.10	0.20	0.30	0.40	0.50	0.60	0.70	0.80	0.90
(1)	1.000	4.031	16.62	39.39	75.75	132.3	222.6	379.1	699.0	1669
(2)	32.56	68.61	153.7	262.4	406.6	607.2	907.0	1404	2397	5367
(3)	46.34	97.04	215.7	365.5	561.6	832.9	1233	1894	3208	7149
(4)	48.20	108.1	237.6	400.3	595.8	867.9	1268	1915	3153	6772

TABLE 3.21 : K* (μ = 0.2)

Q: / MODE:	0.05	0.10	0.20	0.30	0.40	0.50	0.60	0.70	0.80	0.90
(1)	.2453	.9886	4.074	9.656	18.56	32.39	54.48	92.68	170.7	407.2
(2)	7.166	15.00	33.18	55.86	85.28	125.4	184.3	280.4	469.8	1031
(3)	17.79	35.96	74.13	115.9	163.2	218.7	288.1	383.8	543.2	952.1
(4)	164.5	274.9	448.6	609.1	777.1	963.5	1183	1461	1867	2747

TABLE 3.21 : K* (μ = 0.3)

Q: / MODE:	0.05	0.10	0.20	0.30	0.40	0.50	0.60	0.70	0.80	0.90
(1)	.1120	.4511	1.857	4.390	8.411	14.62	24.46	41.37	75.65	179.0
(2)	4.924	10.13	21.57	34.73	50.26	69.28	93.99	129.5	191.1	355.1
(3)	63.49	109.7	192.7	280.7	382.9	507.5	667.3	889.6	1250	2119
(4)	290.9	524.3	918.5	1292	1705	2218	2931	4055	6230	12616

TABLE 3.21 : K* (μ = 0.4)

Q: / MODE:	0.05	0.10	0.20	0.30	0.40	0.50	0.60	0.70	0.80	0.90
(1)	.0730	.2938	1.205	2.831	5.376	9.235	15.23	25.28	45.24	104.3
(2)	7.874	16.15	34.60	56.49	82.90	115.4	156.9	213.6	303.4	514.0
(3)	103.2	177.1	295.9	410.1	540.0	704.3	933.8	1296	1993	4029
(4)	62.71	123.3	241.4	360.0	484.4	621.1	781.5	989.6	1316	2110

TABLE 3.21 : K* (μ = 0.5)

Q: / MODE:	0.05	0.10	0.20	0.30	0.40	0.50	0.60	0.70	0.80	0.90
(1)	.0687	.2760	1.125	2.616	4.892	8.233	13.21	21.19	36.25	78.83
(2)	10.76	21.15	43.72	71.05	105.8	151.7	216.2	316.2	501.7	1023
(3)	26.14	52.09	104.7	160.7	223.7	299.5	398.9	546.9	817.3	1578
(4)	148.4	262.2	463.3	670.5	909.1	1204	1599	2189	3256	6197

TABLE 3.21 : K* (μ = 0.6)

Q: / MODE:	0.05	0.10	0.20	0.30	0.40	0.50	0.60	0.70	0.80	0.90
(1)	.0938	.3763	1.527	3.526	6.520	10.80	16.94	26.31	43.00	87.37
(2)	12.41	23.26	44.32	67.63	95.93	132.7	184.2	264.7	417.1	854.4
(3)	42.99	83.28	162.5	246.5	341.6	455.2	599.7	801.6	1137	1979
(4)	175.5	326.3	590.9	842.5	1114	1441	1879	2555	3832	7532

TABLE 3.21 : K* (μ = 0.7)

Q: / MODE:	0.05	0.10	0.20	0.30	0.40	0.50	0.60	0.70	0.80	0.90
(1)	.1436	.5766	2.344	5.428	10.08	16.80	26.57	41.74	69.27	143.8
(2)	10.76	20.90	40.80	61.98	86.48	117.1	158.7	222.7	342.4	684.6
(3)	43.13	81.38	154.8	233.6	325.8	441.2	596.9	831.6	1260	2453
(4)	131.8	250.6	470.2	688.2	924.9	1202	1554	2054	2917	5183

TABLE 3.21 : K* (μ = 0.8)

Q: / MODE:	0.05	0.10	0.20	0.30	0.40	0.50	0.60	0.70	0.80	0.90
(1)	.1449	.5824	2.373	5.519	10.32	17.36	27.84	44.59	76.13	165.0
(2)	11.57	22.74	45.30	69.85	98.22	132.9	178.2	244.1	359.4	668.4
(3)	54.35	100.4	181.8	261.4	348.9	454.7	595.3	807.7	1199	2307
(4)	121.0	225.6	415.2	604.4	812.6	1063	1390	1876	2758	5193

TABLE 3.21 : K* (μ = 0.9)

Q: / MODE:	0.05	0.10	0.20	0.30	0.40	0.50	0.60	0.70	0.80	0.90
(1)	.1107	.4447	1.813	4.220	7.901	13.31	21.40	34.39	58.99	128.7
(2)	11.94	23.51	47.62	75.21	108.7	151.4	209.7	297.9	458.8	905.5
(3)	57.78	109.8	207.6	307.7	419.9	555.3	733.2	996.2	1469	2771
(4)	144.8	275.0	510.1	734.7	973.4	1252	1609	2122	3031	5539

TABLE 3.22 : T* (μ = 0.1)

Q: / MODE:	0.05	0.10	0.20	0.30	0.40	0.50	0.60	0.70	0.80	0.90
(1)	2.272	4.796	10.79	18.50	28.78	43.17	64.76	100.7	172.7	388.6
(2)	2.480	5.235	11.78	20.19	31.40	47.09	70.62	109.8	188.3	423.5
(3)	2.681	5.658	12.72	21.80	33.89	50.80	76.15	118.4	202.8	456.1
(4)	2.894	6.106	13.72	23.50	36.50	54.69	81.93	127.3	217.9	489.8

TABLE 3.22 : T* (μ = 0.2)

Q: / MODE:	0.05	0.10	0.20	0.30	0.40	0.50	0.60	0.70	0.80	0.90
(1)	1.238	2.613	5.880	10.08	15.69	23.53	35.31	54.93	94.19	212.0
(2)	1.479	3.121	7.018	12.02	18.69	28.02	42.01	65.31	111.9	251.6
(3)	1.703	3.592	8.064	13.79	21.41	32.05	47.96	74.44	127.3	285.8
(4)	1.897	3.996	8.947	15.26	23.63	35.27	52.64	81.47	139.0	311.1

TABLE 3.22 : T* (μ = 0.3)

Q: / MODE:	0.05	0.10	0.20	0.30	0.40	0.50	0.60	0.70	0.80	0.90
(1)	.9093	1.920	4.323	7.415	11.54	17.32	25.99	40.45	69.38	156.2
(2)	1.184	2.498	5.612	9.607	14.92	22.35	33.47	51.97	88.95	199.8
(3)	1.374	2.888	6.444	10.95	16.89	25.11	37.31	57.50	97.61	217.4
(4)	2.566	5.364	11.84	19.93	30.41	44.67	65.43	98.72	161.5	328.2

TABLE 3.22 : T* (μ = 0.4)

Q: / MODE:	0.05	0.10	0.20	0.30	0.40	0.50	0.60	0.70	0.80	0.90
(1)	.7626	1.611	3.629	6.229	9.700	14.57	21.87	34.05	58.43	131.6
(2)	1.065	2.243	5.021	8.561	13.24	19.75	29.44	45.50	77.50	173.2
(3)	1.721	3.600	7.947	13.36	20.35	29.85	43.70	66.20	110.2	239.7
(4)	3.944	8.226	18.08	30.33	46.19	67.90	99.91	152.6	257.0	568.4

TABLE 3.22 : T* (μ = 0.5)

Q: / MODE:	0.05	0.10	0.20	0.30	0.40	0.50	0.60	0.70	0.80	0.90
(1)	.6958	1.471	3.316	5.696	8.876	13.34	20.03	31.20	53.55	120.6
(2)	1.392	2.930	6.553	11.16	17.21	25.54	37.77	57.55	95.23	197.6
(3)	UPPER and LOWER LIMITS AGREE WITHIN 1 %									
(4)	2.451	5.146	11.45	19.42	29.88	44.34	65.78	101.2	171.3	379.3

TABLE 3.22 : T* (μ = 0.6)

Q: / MODE:	0.05	0.10	0.20	0.30	0.40	0.50	0.60	0.70	0.80	0.90
(1)	.6848	1.448	3.267	5.612	8.743	13.13	19.69	30.62	52.45	117.8
(2)	1.197	2.515	5.604	9.515	14.66	21.78	32.37	49.87	84.68	188.7
(3)	1.755	3.669	8.096	13.60	20.73	30.42	44.57	67.59	112.7	245.8
(4)	3.750	7.828	17.23	28.94	44.14	64.97	95.71	146.3	246.7	546.2

TABLE 3.22 : T* (μ = 0.7)

Q: / MODE:	0.05	0.10	0.20	0.30	0.40	0.50	0.60	0.70	0.80	0.90
(1)	.8054	1.708	3.869	6.659	10.37	15.50	23.05	35.32	59.03	127.2
(2)	4.827	10.04	21.97	36.67	55.59	81.34	119.1	181.1	303.6	668.7
(3)	1.481	3.112	6.933	11.77	18.12	26.91	39.96	61.51	104.3	232.2
(4)	2.652	5.548	12.26	20.67	31.62	46.63	68.70	104.7	174.7	374.5

TABLE 3.22 : T* (μ = 0.8)

Q: MODE:	0.05	0.10	0.20	0.30	0.40	0.50	0.60	0.70	0.80	0.90
(1)	.4482	.9402	2.085	3.515	5.365	7.873	11.52	17.43	28.94	62.84
(2)	1.947	3.986	8.528	14.04	21.12	30.73	44.78	67.59	111.9	240.4
(3)	9.529	19.93	44.00	74.06	113.2	166.8	246.0	376.3	634.7	1405
(4)	2.266	4.769	10.66	18.16	28.07	41.85	62.38	96.43	164.3	367.3

TABLE 3.22 : T* (μ = 0.9)

Q: MODE:	0.05	0.10	0.20	0.30	0.40	0.50	0.60	0.70	0.80	0.90
(1)	.3840	.8100	1.819	3.111	4.825	7.212	10.78	16.69	28.46	63.68
(2)	.6673	1.396	3.085	5.195	7.937	11.69	17.21	26.27	44.17	97.42
(3)	.9459	1.957	4.229	6.958	10.38	14.91	21.39	31.75	51.81	110.6
(4)	1.940	3.894	7.942	12.35	17.37	23.39	31.14	42.23	61.45	76.50

TABLE 3.23 : T* (μ = 0.1)

Q: MODE:	0.05	0.10	0.20	0.30	0.40	0.50	0.60	0.70	0.80	0.90
(1)	2.426	5.122	11.52	19.75	30.72	46.08	69.12	107.5	184.3	414.6
(2)	2.629	5.549	12.48	21.38	33.25	49.84	74.73	116.2	199.1	447.7
(3)	2.841	5.994	13.47	23.07	35.85	53.72	80.49	125.1	214.2	481.4
(4)	3.055	6.443	14.47	24.76	38.45	57.58	86.21	133.9	229.1	514.5

TABLE 3.23 : T* (μ = 0.2)

Q: MODE:	0.05	0.10	0.20	0.30	0.40	0.50	0.60	0.70	0.80	0.90
(1)	1.415	2.988	6.721	11.52	17.91	26.86	40.28	62.65	107.4	241.5
(2)	1.647	3.475	7.805	13.36	20.74	31.06	46.51	72.23	123.6	277.6
(3)	1.856	3.910	8.762	14.96	23.18	34.62	51.72	80.12	136.8	306.4
(4)	1.977	4.155	9.265	15.74	24.27	36.07	53.61	82.61	140.3	312.6

TABLE 3.23 : T* (μ = 0.3)

Q: MODE:	0.05	0.10	0.20	0.30	0.40	0.50	0.60	0.70	0.80	0.90
(1)	1.113	2.350	5.284	9.053	14.07	21.09	31.62	49.15	84.19	189.3
(2)	1.338	2.818	6.305	10.75	16.62	24.79	36.96	57.14	97.34	217.6
(3)	1.680	3.491	7.594	12.54	18.72	26.78	38.02	55.43	87.85	179.1
(4)	7.046	14.19	29.35	46.79	68.26	96.72	137.8	204.7	336.3	727.7

TABLE 3.23 : T* (μ = 0.4)

Q: MODE:	0.05	0.10	0.20	0.30	0.40	0.50	0.60	0.70	0.80	0.90
(1)	.9936	2.095	4.705	8.047	12.49	18.69	27.96	43.37	74.13	166.3
(2)	1.788	3.758	8.375	14.21	21.87	32.40	47.88	72.99	121.2	253.3
(3)	UPPER and LOWER LIMITS AGREE WITHIN 1 %									
(4)	2.176	4.557	10.09	17.00	26.00	38.30	56.40	86.05	144.6	318.5

TABLE 3.23 : T* (μ = 0.5)

Q: MODE:	0.05	0.10	0.20	0.30	0.40	0.50	0.60	0.70	0.80	0.90
(1)	.9873	2.074	4.614	7.813	11.99	17.70	26.10	39.80	66.73	146.4
(2)	2.318	4.832	10.61	17.78	27.07	39.79	58.56	89.47	150.8	333.7
(3)	1.888	3.948	8.710	14.63	22.26	32.63	47.74	72.29	120.3	262.0
(4)	3.973	8.271	18.12	30.28	45.98	67.42	98.97	150.8	253.5	559.6

TABLE 3.23 : T* (μ = 0.6)

Q: MODE:	0.05	0.10	0.20	0.30	0.40	0.50	0.60	0.70	0.80	0.90
(1)	.9206	1.932	4.289	7.249	11.10	16.36	24.08	36.68	61.42	134.6
(2)	2.187	4.546	9.936	16.58	25.17	36.89	54.15	82.52	138.7	306.4
(3)	2.100	4.408	9.809	16.63	25.59	37.98	56.36	86.74	147.1	327.5
(4)	2.608	5.446	11.97	20.03	30.34	44.18	64.06	95.74	136.2	329.5

TABLE 3.23 : T* (μ = 0.7)

Q: MODE:	0.05	0.10	0.20	0.30	0.40	0.50	0.60	0.70	0.80	0.90
(1)	.9860	2.079	4.667	7.980	12.38	18.52	27.70	42.97	73.43	164.7
(2)	1.221	2.548	5.592	9.333	14.10	20.50	29.69	44.45	73.00	156.5
(3)	3.525	7.233	15.45	25.31	37.79	54.61	79.16	119.3	198.5	434.4
(4)	3.105	6.522	14.53	24.66	38.00	56.46	83.91	129.3	219.7	490.0

TABLE 3.23 : T* (μ = 0.8)

Q: MODE:	0.05	0.10	0.20	0.30	0.40	0.50	0.60	0.70	0.80	0.90
(1)	1.204	2.541	5.713	9.787	15.21	22.80	34.17	53.12	90.98	204.5
(2)	1.430	3.014	6.757	11.54	17.89	26.73	39.95	61.91	105.7	237.0
(3)	1.567	3.292	7.332	12.44	19.15	28.42	42.17	64.87	110.0	244.6
(4)	1.857	3.840	8.276	13.54	20.03	28.41	40.04	58.02	72.10	165.1

TABLE 3.23 : T* (μ = 0.9)

Q: MODE:	0.05	0.10	0.20	0.30	0.40	0.50	0.60	0.70	0.80	0.90
(1)	1.946	4.109	9.243	15.84	24.64	36.95	55.41	86.17	147.7	332.2
(2)	2.161	4.559	10.25	17.55	27.28	40.88	61.26	95.20	163.0	366.5
(3)	2.374	5.008	11.25	19.24	29.88	44.73	66.97	104.0	177.9	399.5
(4)	2.581	5.441	41.11	49.75	61.23	77.25	101.2	141.1	220.6	459.0

TABLE 3.24 : T* (μ = 0.1)

Q: MODE:	0.05	0.10	0.20	0.30	0.40	0.50	0.60	0.70	0.80	0.90
(1)	2.471	5.217	11.74	20.12	31.29	46.93	70.38	109.5	187.6	422.1
(2)	2.682	5.660	12.73	21.80	33.90	50.81	76.17	118.4	202.9	456.2
(3)	2.894	6.106	13.72	23.50	36.50	54.69	81.93	127.3	217.9	489.7
(4)	42.56	46.01	54.17	64.63	78.53	97.96	127.0	175.4	272.1	562.0

TABLE 3.24 : T* (μ = 0.2)

Q: MODE:	0.05	0.10	0.20	0.30	0.40	0.50	0.60	0.70	0.80	0.90
(1)	1.468	3.097	6.966	11.94	18.56	27.82	41.71	64.85	111.1	249.9
(2)	1.704	3.594	8.069	13.80	21.42	32.06	47.99	74.48	127.4	286.0
(3)	1.897	3.996	8.947	15.26	23.63	35.26	52.63	81.47	139.0	311.0
(4)	1.984	4.164	9.264	15.70	24.14	46.52	63.59	91.86	148.2	316.6

TABLE 3.24 : T* (μ = 0.3)

Q: MODE:	0.05	0.10	0.20	0.30	0.40	0.50	0.60	0.70	0.80	0.90
(1)	1.170	2.467	5.545	9.493	14.75	22.09	33.09	51.39	87.97	197.6
(2)	1.375	2.891	6.451	10.96	16.91	25.13	37.35	57.55	97.71	217.7
(3)	2.565	5.364	11.84	19.93	30.41	44.67	65.43	98.72	161.6	328.2
(4)	UPPER and LOWER LIMITS AGREE WITHIN 1 %									

TABLE 3.24 : T* (μ = 0.4)

Q: MODE:	0.05	0.10	0.20	0.30	0.40	0.50	0.60	0.70	0.80	0.90
(1)	1.045	2.202	4.930	8.407	13.01	19.40	28.93	44.72	76.16	170.2
(2)	1.728	3.614	7.975	13.40	20.41	29.93	43.82	66.37	110.4	240.2
(3)	3.950	8.240	18.11	30.37	46.26	68.01	100.1	152.8	257.3	569.2
(4)	2.234	4.658	10.21	17.01	25.65	37.18	53.69	80.02	130.6	277.6

TABLE 3.24 : T* (μ = 0.5)

Q: MODE:	0.05	0.10	0.20	0.30	0.40	0.50	0.60	0.70	0.80	0.90
(1)	1.385	2.923	6.571	11.25	17.48	26.16	39.13	60.73	103.8	232.6
(2)	UPPER and LOWER LIMITS AGREE WITHIN 1 %									
(3)	2.457	5.158	11.48	19.47	29.96	44.47	66.01	101.6	172.3	383.5
(4)	UPPER and LOWER LIMITS AGREE WITHIN 1 %									

TABLE 3.25 : T* (μ = 0.1)

Q: MODE:	0.05	0.10	0.20	0.30	0.40	0.50	0.60	0.70	0.80	0.90
(1)	1.745	3.684	8.289	14.21	22.11	33.16	49.75	77.39	132.7	298.5
(2)	1.956	4.128	9.285	15.91	24.75	37.11	55.65	86.55	148.3	333.7
(3)	2.160	4.558	10.25	17.55	27.27	40.87	61.24	95.17	163.0	366.4
(4)	2.374	5.008	11.25	19.24	29.88	44.73	66.97	104.0	177.9	399.5

TABLE 3.25 : T* (μ = 0.2)

Q: MODE:	0.05	0.10	0.20	0.30	0.40	0.50	0.60	0.70	0.80	0.90
(1)	.9729	2.054	4.624	7.929	12.34	18.51	27.78	43.23	74.12	166.8
(2)	1.216	2.566	5.768	9.880	15.36	23.01	34.48	53.59	91.78	206.3
(3)	1.429	3.012	6.752	11.53	17.88	26.71	39.92	61.87	105.7	236.8
(4)	1.567	3.292	7.333	12.44	19.15	28.42	42.17	64.87	110.0	244.6

TABLE 3.25 : T* (μ = 0.3)

Q: MODE:	0.05	0.10	0.20	0.30	0.40	0.50	0.60	0.70	0.80	0.90
(1)	.7310	1.544	3.477	5.966	9.289	13.95	20.93	32.59	55.91	125.9
(2)	1.002	2.113	4.740	8.104	12.57	18.80	28.11	43.59	74.49	167.0
(3)	1.220	2.545	5.585	9.322	14.09	20.47	29.66	44.40	72.90	156.3
(4)	3.529	7.241	15.47	25.33	37.82	54.65	79.22	119.4	198.7	434.6

TABLE 3.25 : T* (μ = 0.4)

Q: MODE:	0.05	0.10	0.20	0.30	0.40	0.50	0.60	0.70	0.80	0.90
(1)	.6262	1.323	2.983	5.123	7.983	11.99	18.01	28.06	48.16	108.5
(2)	.9404	1.973	4.378	7.379	11.33	16.70	24.59	37.48	62.83	137.9
(3)	2.140	4.452	9.744	16.29	24.74	36.31	53.34	81.35	136.9	302.4
(4)	2.098	4.404	9.800	16.62	25.57	37.94	56.31	86.66	147.0	327.3

TABLE 3.25 : T* (μ = 0.5)

Q: MODE:	0.05	0.10	0.20	0.30	0.40	0.50	0.60	0.70	0.80	0.90
(1)	.5841	1.235	2.788	4.791	7.469	11.22	16.86	26.26	45.05	101.4
(2)	1.051	2.204	4.886	8.245	12.61	18.57	27.30	41.53	69.45	152.0
(3)	2.399	4.998	10.97	18.37	27.95	41.06	60.40	92.23	155.4	343.7
(4)	1.883	3.937	8.688	14.59	22.22	32.56	47.65	72.15	120.1	261.5

TABLE 3.25 : T* (μ = 0.6)

Q: MODE:	0.05	0.10	0.20	0.30	0.40	0.50	0.60	0.70	0.80	0.90
(1)	.6006	1.271	2.868	4.923	7.656	11.46	17.13	26.51	45.11	100.5
(2)	1.411	2.956	6.553	11.08	17.00	25.18	37.32	57.38	97.28	216.5
(3)	1.839	3.863	8.606	14.61	22.48	33.34	49.36	75.51	126.3	269.8
(4)	UPPER and LOWER LIMITS AGREE WITHIN 1 %									

TABLE 3.25 : T* (μ = 0.7)

Q: MODE:	0.05	0.10	0.20	0.30	0.40	0.50	0.60	0.70	0.80	0.90
(1)	.5950	1.255	2.807	4.753	7.245	10.54	15.13	22.06	34.31	65.99
(2)	5.814	11.43	22.72	35.08	49.85	69.09	96.65	141.2	228.9	489.4
(3)	1.522	3.201	7.144	12.15	18.75	27.90	41.51	64.06	108.9	243.1
(4)	1.748	3.630	7.892	13.02	19.40	27.71	39.24	56.97	89.71	181.1

TABLE 3.25 : T* (μ = 0.8)

Q: MODE:	0.05	0.10	0.20	0.30	0.40	0.50	0.60	0.70	0.80	0.90
(1)	.3725	.7847	1.756	2.989	4.611	6.849	10.16	15.60	26.34	58.30
(2)	1.288	2.659	5.734	9.416	13.98	19.84	27.77	39.29	58.35	101.8
(3)	19.56	36.27	66.01	95.60	129.4	172.4	233.5	332.2	525.8	1101
(4)	2.424	5.100	11.40	19.41	29.99	44.70	66.62	103.0	175.4	392.2

TABLE 3.25 : T* (μ = 0.9)

Q: MODE:	0.05	0.10	0.20	0.30	0.40	0.50	0.60	0.70	0.80	0.90
(1)	.3250	.6867	1.547	2.654	4.129	6.190	9.276	14.41	24.65	55.31
(2)	.6009	1.259	2.793	4.720	7.236	10.70	15.81	24.22	40.87	90.48
(3)	.8648	1.796	3.912	6.489	9.762	14.15	20.49	30.74	50.73	109.7
(4)	1.481	3.004	6.242	9.873	14.13	19.39	26.40	36.88	55.91	108.9

TABLE 3.26 : T* (μ = 0.1)

Q: MODE:	0.05	0.10	0.20	0.30	0.40	0.50	0.60	0.70	0.80	0.90
(1)	1.901	4.012	9.027	15.47	24.07	36.09	54.14	84.20	144.3	324.7
(2)	2.107	4.447	9.998	17.13	26.62	39.90	59.81	92.97	159.2	358.0
(3)	2.321	4.896	11.00	18.82	29.23	43.78	65.56	101.8	174.2	391.4
(4)	2.531	5.336	11.97	20.47	31.76	47.50	71.06	110.2	188.5	422.9

TABLE 3.26 : T* (μ = 0.2)

Q: MODE:	0.05	0.10	0.20	0.30	0.40	0.50	0.60	0.70	0.80	0.90
(1)	1.152	2.432	5.470	9.373	14.58	21.85	32.77	50.95	87.30	196.3
(2)	1.379	2.907	6.523	11.15	17.30	25.88	38.71	60.04	102.6	230.2
(3)	1.546	3.250	7.257	12.34	19.05	28.35	42.18	65.07	110.6	246.8
(4)	1.651	3.435	7.502	12.47	18.77	27.19	39.30	58.76	96.51	207.3

TABLE 3.26 : T* (μ = 0.3)

Q: MODE:	0.05	0.10	0.20	0.30	0.40	0.50	0.60	0.70	0.80	0.90
(1)	.9342	1.971	4.430	7.586	11.79	17.65	26.45	41.08	70.32	158.0
(2)	1.135	2.379	5.277	8.914	13.65	20.15	29.71	45.40	76.36	168.4
(3)	2.418	5.023	10.97	18.27	27.67	40.41	59.04	89.42	149.1	325.9
(4)	4.178	8.753	19.40	32.78	50.27	74.38	110.1	169.1	286.2	636.0

TABLE 3.26 : T* (μ = 0.4)

Q: MODE:	0.05	0.10	0.20	0.30	0.40	0.50	0.60	0.70	0.80	0.90
(1)	.8542	1.799	4.026	6.862	10.61	15.81	23.55	36.37	61.86	138.0
(2)	1.548	3.240	7.165	12.07	18.46	27.21	40.10	61.26	103.0	227.3
(3)	2.560	5.361	11.88	20.07	30.78	45.55	67.42	103.5	175.3	389.7
(4)	2.226	4.639	10.15	16.85	25.29	36.39	51.91	75.86	119.9	240.9

TABLE 3.26 : T* (μ = 0.5)

Q: MODE:	0.05	0.10	0.20	0.30	0.40	0.50	0.60	0.70	0.80	0.90
(1)	1.170	2.472	5.573	9.566	14.89	22.35	33.52	52.11	89.24	200.5
(2)	UPPER and LOWER LIMITS AGREE WITHIN 1 %									
(3)	2.197	4.615	10.29	17.47	26.92	40.00	59.44	91.59	155.5	346.5
(4)	UPPER and LOWER LIMITS AGREE WITHIN 1 %									

TABLE 3.27 : T* (μ = 0.1)

Q: MODE:	0.05	0.10	0.20	0.30	0.40	0.50	0.60	0.70	0.80	0.90
(1)	.3756	.7928	1.783	3.052	4.741	7.098	10.62	16.48	28.15	63.11
(2)	.6674	1.401	3.119	5.289	8.141	12.08	17.93	27.60	46.82	104.2
(3)	.9471	1.985	4.405	7.448	11.43	16.91	25.03	38.42	64.98	144.2
(4)	1.228	2.573	5.698	9.619	14.74	21.77	32.17	49.29	83.23	184.4

TABLE 3.27 : T* (μ = 0.2)

Q: MODE:	0.05	0.10	0.20	0.30	0.40	0.50	0.60	0.70	0.80	0.90
(1)	.4135	.8728	1.963	3.361	5.221	7.818	11.70	18.15	31.02	69.54
(2)	.7368	1.547	3.443	5.840	8.990	13.34	19.81	30.49	51.72	115.1
(3)	1.046	2.194	4.867	8.230	12.63	18.69	27.66	42.45	71.80	159.4
(4)	1.348	2.823	6.253	10.55	16.17	23.88	35.27	54.03	91.20	202.0

TABLE 3.27 : T* (μ = 0.3)

Q: MODE:	0.05	0.10	0.20	0.30	0.40	0.50	0.60	0.70	0.80	0.90
(1)	.4625	.9763	2.196	3.761	5.842	8.748	13.09	20.31	34.72	77.84
(2)	.8173	1.716	3.818	6.475	9.964	14.78	21.94	33.76	57.23	127.3
(3)	.9171	1.919	4.242	7.140	10.90	16.04	23.57	35.91	60.24	132.5
(4)	1.566	3.264	7.165	12.00	18.26	26.82	39.42	60.14	101.2	223.6

TABLE 3.27 : T* (μ = 0.4)

Q: MODE:	0.05	0.10	0.20	0.30	0.40	0.50	0.60	0.70	0.80	0.90
(1)	.5220	1.102	2.477	4.241	6.585	9.857	14.75	22.87	39.06	87.54
(2)	.6278	1.316	2.917	4.926	7.545	11.14	16.43	25.13	42.33	93.50
(3)	1.213	2.526	5.542	9.283	14.14	20.78	30.60	46.75	78.79	174.4
(4)	1.285	2.690	5.953	10.03	15.34	22.60	33.29	50.82	85.42	188.3

TABLE 3.27 : T* (μ = 0.5)

Q: MODE:	0.05	0.10	0.20	0.30	0.40	0.50	0.60	0.70	0.80	0.90
(1)	.4936	1.037	2.307	3.907	5.993	8.851	13.05	19.90	33.36	73.19
(2)	1.159	2.416	5.305	8.889	13.53	19.90	29.28	44.73	75.38	166.8
(3)	.9439	1.974	4.355	7.314	11.13	16.32	23.87	36.14	60.15	131.0
(4)	1.987	4.135	9.058	15.14	22.99	33.71	49.49	75.41	126.7	279.8

TABLE 3.27 : T* (μ = 0.6)

Q: MODE:	0.05	0.10	0.20	0.30	0.40	0.50	0.60	0.70	0.80	0.90
(1)	.3871	.8154	1.827	3.118	4.828	7.210	10.77	16.67	28.44	63.67
(2)	.8085	1.697	3.777	6.407	9.863	14.64	21.73	33.42	56.62	125.8
(3)	1.314	2.763	6.166	10.49	16.18	24.09	35.86	55.36	94.20	210.4
(4)	.9589	1.999	4.378	7.293	11.00	15.96	23.10	34.53	56.64	121.3

TABLE 3.27 : T* (μ = 0.7)

Q: MODE:	0.05	0.10	0.20	0.30	0.40	0.50	0.60	0.70	0.80	0.90
(1)	.5733	1.210	2.722	4.664	7.252	10.87	16.31	25.36	43.45	97.73
(2)	.4545	.9552	2.130	3.618	5.576	8.283	12.30	18.94	32.14	71.55
(3)	1.076	2.242	4.923	8.248	12.55	18.40	26.97	40.92	68.28	149.0
(4)	2.323	4.854	10.71	18.02	27.54	40.59	59.89	91.67	154.7	343.0

TABLE 3.27 : T* (μ = 0.8)

Q: MODE:	0.05	0.10	0.20	0.30	0.40	0.50	0.60	0.70	0.80	0.90
(1)	.8598	1.815	4.084	7.000	10.89	16.33	24.50	38.10	65.32	147.0
(2)	.8540	1.803	4.055	6.949	10.81	16.20	24.30	37.79	64.76	145.7
(3)	.7549	1.592	3.577	6.123	9.510	14.24	21.33	33.13	56.70	127.4
(4)	.5891	1.232	2.718	7.382	8.004	10.21	14.96	22.72	37.98	83.23

TABLE 3.27 : T* (μ = 0.9)

Q: MODE:	0.05	0.10	0.20	0.30	0.40	0.50	0.60	0.70	0.80	0.90
(1)	1.657	3.499	7.872	13.49	20.99	31.49	47.23	73.47	125.9	283.4
(2)	1.692	3.572	8.036	13.78	21.43	32.14	48.21	74.99	128.5	289.2
(3)	UPPER and LOWER LIMITS AGREE WITHIN 1 %									
(4)	UPPER and LOWER LIMITS AGREE WITHIN 1 %									

TABLE 3.28 : T* (μ = 0.1)

Q: MODE:	0.05	0.10	0.20	0.30	0.40	0.50	0.60	0.70	0.80	0.90
(1)	1.649	3.481	7.832	13.43	20.88	31.33	46.99	73.09	125.3	281.9
(2)	1.684	3.555	8.000	13.71	21.33	32.00	47.99	74.65	128.0	287.9
(3)	1.710	3.610	8.123	13.92	21.66	32.49	48.72	75.79	129.9	292.3
(4)	UPPER and LOWER LIMITS AGREE WITHIN 1 %									

TABLE 3.28 : T* (μ = 0.2)

Q: MODE:	0.05	0.10	0.20	0.30	0.40	0.50	0.60	0.70	0.80	0.90
(1)	.8550	1.805	4.061	6.962	10.83	16.24	24.36	37.90	64.97	146.2
(2)	.8616	1.819	4.091	7.013	10.91	16.36	24.53	38.15	65.39	147.1
(3)	.7922	1.672	3.757	6.435	10.00	14.99	22.46	34.90	59.77	134.4
(4)	.6055	1.272	2.834	4.809	7.404	10.99	16.31	25.08	42.52	94.56

TABLE 3.28 : T* (μ = 0.3)

Q: MODE:	0.05	0.10	0.20	0.30	0.40	0.50	0.60	0.70	0.80	0.90
(1)	.5804	1.225	2.756	4.724	7.347	11.02	16.53	25.70	44.05	99.10
(2)	.4930	1.039	2.330	3.979	6.168	9.219	13.78	21.36	36.47	81.76
(3)	.8209	1.714	3.766	6.293	9.520	13.84	20.05	29.95	49.00	104.3
(4)	2.571	5.289	11.34	18.63	27.86	40.32	58.51	88.27	147.1	322.2

TABLE 3.28 : T* (μ = 0.4)

Q: MODE:	0.05	0.10	0.20	0.30	0.40	0.50	0.60	0.70	0.80	0.90
(1)	.4187	.8834	1.985	3.399	5.280	7.909	11.85	18.41	31.51	70.80
(2)	.6717	1.410	3.133	5.299	8.120	11.98	17.63	26.84	44.90	98.21
(3)	1.611	3.359	7.384	12.38	18.86	27.73	40.82	62.35	105.0	232.5
(4)	.7801	1.633	3.609	6.075	9.273	13.64	20.04	30.52	51.18	112.5

TABLE 3.28 : T* (μ = 0.5)

Q: MODE:	0.05	0.10	0.20	0.30	0.40	0.50	0.60	0.70	0.80	0.90
(1)	.3440	.7215	1.602	2.707	4.148	6.123	9.031	13.80	23.21	51.19
(2)	.8816	1.832	4.011	6.714	10.23	15.05	22.20	33.99	57.42	127.4
(3)	.7019	1.468	3.242	5.451	8.310	12.20	17.90	27.21	45.52	99.82
(4)	1.586	3.285	7.143	11.88	17.97	26.29	38.52	58.64	98.50	217.4

TABLE 3.28 : T* (μ = 0.6)

Q: MODE:	0.05	0.10	0.20	0.30	0.40	0.50	0.60	0.70	0.80	0.90
(1)	.4927	1.039	2.332	3.985	6.175	9.223	13.76	21.29	36.26	81.00
(2)	.7528	1.582	3.529	5.998	9.250	13.76	20.46	31.57	53.67	119.8
(3)	1.003	2.102	4.667	7.897	12.13	17.97	26.63	40.92	69.29	154.0
(4)	.9822	2.053	4.530	7.607	11.58	16.99	24.90	37.80	63.15	138.3

TABLE 3.28 : T* (μ = 0.7)

Q: MODE:	0.05	0.10	0.20	0.30	0.40	0.50	0.60	0.70	0.80	0.90
(1)	.4552	.9610	2.161	3.701	5.750	8.610	12.89	19.99	34.17	76.60
(2)	.7577	1.589	3.530	5.972	9.168	13.57	20.07	30.77	51.97	115.1
(3)	1.056	2.215	4.921	8.332	12.80	18.98	28.13	43.25	73.28	163.0
(4)	1.248	2.613	5.785	9.765	14.96	22.11	32.68	50.09	84.62	187.6

TABLE 3.28 : T* (μ = 0.8)

Q: MODE:	0.05	0.10	0.20	0.30	0.40	0.50	0.60	0.70	0.80	0.90
(1)	.4092	.8638	1.943	3.327	5.169	7.739	11.58	17.97	30.71	68.86
(2)	.7296	1.532	3.410	5.783	8.902	13.21	19.61	30.19	51.21	114.0
(3)	1.030	2.159	4.790	8.099	12.43	18.39	27.20	41.74	70.57	156.6
(4)	1.219	2.550	5.639	9.499	14.52	21.39	31.49	48.08	80.84	178.3

TABLE 3.28 : T* (μ = 0.9)

Q: MODE:	0.05	0.10	0.20	0.30	0.40	0.50	0.60	0.70	0.80	0.90
(1)	.3732	.7877	1.771	3.033	4.711	7.053	10.55	16.37	27.98	62.72
(2)	.6637	1.393	3.101	5.260	8.096	12.02	17.84	27.45	46.57	103.6
(3)	.9425	1.976	4.384	7.412	11.37	16.83	24.91	38.24	64.67	143.5
(4)	1.223	2.561	5.673	9.577	14.67	21.68	32.03	49.08	82.87	183.6

TABLE 3.29 : T* (μ = 0.1)

Q: MODE:	0.05	0.10	0.20	0.30	0.40	0.50	0.60	0.70	0.80	0.90
(1)	1.617	3.414	7.681	13.17	20.48	30.73	46.09	71.70	122.9	276.5
(2)	1.659	3.502	7.880	13.51	21.01	31.52	47.28	73.55	126.1	283.7
(3)	1.692	3.571	8.035	13.77	21.43	32.14	48.21	74.98	128.5	289.2
(4)	UPPER and LOWER LIMITS AGREE WITHIN 1 %									

TABLE 3.29 : T* (μ = 0.2)

Q: MODE:	0.05	0.10	0.20	0.30	0.40	0.50	0.60	0.70	0.80	0.90
(1)	.8279	1.748	3.932	6.742	10.49	15.73	23.60	36.71	62.93	141.6
(2)	.8616	1.819	4.092	7.015	10.91	16.37	24.55	38.19	65.46	147.3
(3)	.8538	1.802	4.054	6.947	10.80	16.20	24.30	37.78	64.75	145.6
(4)	.7549	1.592	3.577	6.123	9.510	14.24	21.33	33.13	56.70	127.4

TABLE 3.29 : T* (μ = 0.3)

Q: MODE:	0.05	0.10	0.20	0.30	0.40	0.50	0.60	0.70	0.80	0.90
(1)	.5645	1.192	2.681	4.597	7.151	10.73	16.09	25.03	42.91	96.55
(2)	.5756	1.215	2.733	4.683	7.282	10.92	16.37	25.46	43.63	98.13
(3)	.4544	.9550	2.130	3.617	5.574	8.281	12.30	18.94	32.13	71.52
(4)	1.076	2.242	4.924	8.250	12.55	18.41	26.97	40.93	68.29	149.1

TABLE 3.29 : T* (μ = 0.4)

Q: MODE:	0.05	0.10	0.20	0.30	0.40	0.50	0.60	0.70	0.80	0.90
(1)	.4319	.9119	2.052	3.517	5.471	8.207	12.31	19.15	32.83	73.87
(2)	.3889	.8193	1.836	3.135	4.856	7.254	10.83	16.78	28.64	64.15
(3)	.8042	1.689	3.760	6.381	9.827	14.59	21.66	33.33	56.47	125.5
(4)	1.313	2.760	6.161	10.48	16.17	24.07	35.83	55.32	94.13	210.2

TABLE 3.29 : T* (μ = 0.5)

Q: MODE:	0.05	0.10	0.20	0.30	0.40	0.50	0.60	0.70	0.80	0.90
(1)	.3507	.7402	1.665	2.854	4.439	6.657	9.983	15.53	26.61	59.85
(2)	.4904	1.029	2.288	3.868	5.926	8.739	12.86	19.58	32.77	71.75
(3)	1.175	2.445	5.359	8.966	13.63	20.02	29.42	44.91	75.61	167.2
(4)	.9441	1.974	4.356	7.316	11.14	16.32	23.88	36.16	60.19	131.1

TABLE 3.29 : T* (μ = 0.6)

Q: MODE:	0.05	0.10	0.20	0.30	0.40	0.50	0.60	0.70	0.80	0.90
(1)	.2955	.6234	1.401	2.398	3.724	5.578	8.354	12.97	22.20	49.88
(2)	.5467	1.152	2.581	4.407	6.827	10.20	15.22	23.56	40.19	89.92
(3)	.6236	1.307	2.900	4.900	7.509	11.09	16.38	25.07	42.26	93.45
(4)	1.196	2.492	5.470	9.168	13.97	20.55	30.26	46.26	77.98	172.6

TABLE 3.29 : T* (μ = 0.7)

Q: MODE:	0.05	0.10	0.20	0.30	0.40	0.50	0.60	0.70	0.80	0.90
(1)	.2870	.6044	1.352	2.305	3.563	5.309	7.908	12.21	20.77	46.33
(2)	.5733	1.202	2.669	4.520	6.950	10.31	15.31	23.57	40.01	89.15
(3)	.8264	1.734	3.857	6.538	10.06	14.92	22.13	34.03	57.68	128.3
(4)	.9034	1.891	4.179	7.035	10.74	15.80	23.24	35.41	59.40	130.7

TABLE 3.29 : T* (μ = 0.8)

Q: MODE:	0.05	0.10	0.20	0.30	0.40	0.50	0.60	0.70	0.80	0.90
(1)	.3409	.7198	1.619	2.774	4.310	6.453	9.658	14.98	25.60	57.36
(2)	.5424	1.139	2.533	4.292	6.600	9.787	14.51	22.31	37.80	84.04
(3)	.8167	1.711	3.790	6.403	9.820	14.53	21.50	33.00	55.83	124.0
(4)	1.060	2.222	4.928	8.330	12.78	18.90	27.97	42.91	72.55	161.0

TABLE 3.29 : T* (μ = 0.9)

Q: MODE:	0.05	0.10	0.20	0.30	0.40	0.50	0.60	0.70	0.80	0.90
(1)	.3233	.6833	1.540	2.644	4.115	6.175	9.260	14.39	24.65	55.36
(2)	.6003	1.261	2.810	4.770	7.349	10.92	16.22	24.98	42.40	94.44
(3)	.8560	1.794	3.981	6.731	10.33	15.28	22.61	34.68	58.63	130.1
(4)	1.069	2.239	4.957	8.364	12.81	18.91	27.91	42.73	72.07	159.5

TABLE 3.30 : I* (μ = 0.1)

Q: MODE:	0.05	0.10	0.20	0.30	0.40	0.50	0.60	0.70	0.80	0.90
(1)	.0011	.0042	.0170	.0385	.0691	.1096	.1613	.2280	.3202	.4911
(2)	.5629	.8852	1.652	2.638	3.868	5.375	7.222	9.551	12.79	19.07
(3)	11.86	16.85	24.98	33.42	42.90	53.86	66.91	83.37	107.1	157.2
(4)	62.89	91.66	130.4	164.4	199.6	238.8	285.0	344.1	433.5	638.1

TABLE 3.30 : I* (μ = 0.2)

Q: MODE:	0.05	0.10	0.20	0.30	0.40	0.50	0.60	0.70	0.80	0.90
(1)	.0030	.0121	.0486	.1107	.2005	.3220	.4840	.7066	1.051	1.821
(2)	1.344	2.254	4.345	7.018	10.40	14.64	20.05	27.33	38.59	64.24
(3)	23.29	36.78	58.96	81.61	107.3	137.8	175.7	227.1	309.1	506.9
(4)	109.7	179.1	282.7	376.0	474.4	587.7	728.4	922.5	1244	2066

TABLE 3.30 : I* (μ = 0.3)

Q: MODE:	0.05	0.10	0.20	0.30	0.40	0.50	0.60	0.70	0.80	0.90
(1)	.0067	.0270	.1092	.2500	.4565	.7429	1.138	1.710	2.663	5.022
(2)	2.542	4.507	8.969	14.67	21.95	31.30	43.61	60.97	89.59	160.4
(3)	38.50	65.27	111.6	159.7	215.7	284.5	374.8	505.0	729.3	1316
(4)	132.5	241.9	432.8	619.8	828.8	1087	1441	1990	3030	6039

TABLE 3.30 : I* (μ = 0.4)

Q: MODE:	0.05	0.10	0.20	0.30	0.40	0.50	0.60	0.70	0.80	0.90
(1)	.0142	.0571	.2314	.5327	.9814	1.617	2.521	3.885	6.290	12.63
(2)	4.533	8.429	17.31	28.71	43.51	62.99	89.59	128.9	197.5	378.0
(3)	42.00	77.70	142.7	209.3	285.7	380.9	510.6	709.4	1079	2132
(4)	31.06	61.92	124.1	189.1	260.4	343.5	449.0	601.0	871.0	1615

TABLE 3.30 : I* (μ = 0.5)

Q: MODE:	0.05	0.10	0.20	0.30	0.40	0.50	0.60	0.70	0.80	0.90
(1)	.0305	.1224	.4976	1.152	2.141	3.572	5.664	8.939	14.96	31.57
(2)	7.286	14.05	29.30	48.74	74.30	108.9	158.4	236.0	381.6	794.8
(3)	UPPER and LOWER LIMITS AGREE WITHIN 1 %									
(4)	64.22	119.1	229.7	352.4	498.3	682.0	930.6	1305	1987	3882

TABLE 3.30 : I* (μ = 0.6)

Q: MODE:	0.05	0.10	0.20	0.30	0.40	0.50	0.60	0.70	0.80	0.90
(1)	.0691	.2778	1.134	2.642	4.958	8.380	13.53	21.86	37.76	83.19
(2)	6.460	12.80	26.00	40.89	58.62	80.79	110.3	153.9	231.3	440.9
(3)	40.93	75.55	138.4	202.7	276.4	368.1	492.8	683.7	1038	2047
(4)	32.80	65.18	130.0	197.2	270.5	355.5	463.2	617.8	891.8	1646

TABLE 3.30 : I* (μ = 0.7)

Q: MODE:	0.05	0.10	0.20	0.30	0.40	0.50	0.60	0.70	0.80	0.90
(1)	.1604	.6455	2.645	6.211	11.78	20.20	33.23	55.02	98.08	225.0
(2)	1.426	2.897	6.007	9.423	13.30	17.92	23.82	32.31	47.24	87.90
(3)	28.50	47.41	85.17	128.2	179.6	243.4	326.9	446.1	648.8	1171
(4)	129.3	235.1	418.4	597.7	798.2	1046	1385	1912	2910	5795

TABLE 3.30 : I* (μ = 0.8)

Q: MODE:	0.05	0.10	0.20	0.30	0.40	0.50	0.60	0.70	0.80	0.90
(1)	.2408	.9686	3.962	9.273	17.50	29.80	48.54	79.32	138.8	310.6
(2)	12.34	25.18	53.23	86.29	127.5	182.3	261.2	388.6	637.7	1373
(3)	3.058	6.178	12.65	19.56	27.13	35.75	46.24	60.44	83.96	144.7
(4)	64.75	100.6	168.2	240.3	321.3	416.0	532.6	689.1	938.5	1543

TABLE 3.30 : I* (μ = 0.9)

Q: MODE:	0.05	0.10	0.20	0.30	0.40	0.50	0.60	0.70	0.80	0.90
(1)	.2011	.8078	3.290	7.644	14.27	23.95	38.29	61.06	103.6	222.3
(2)	19.38	37.91	75.72	118.1	169.2	234.6	324.5	462.6	718.4	1439
(3)	73.17	142.4	279.3	427.2	601.1	822.2	1128	1606	2511	5116
(4)	132.7	267.3	551.9	876.1	1270	1786	2521	3702	6002	12791

TABLE 3.31 : I* (μ = 0.1)

Q: MODE:	0.05	0.10	0.20	0.30	0.40	0.50	0.60	0.70	0.80	0.90
(1)	.0189	.0758	.3043	.6887	1.236	1.958	2.883	4.070	5.701	8.680
(2)	6.587	9.398	14.38	19.82	26.08	33.42	42.20	53.26	69.01	101.5
(3)	44.79	64.73	92.37	117.4	143.8	173.4	208.3	252.8	319.4	469.3
(4)	145.3	217.1	309.7	385.4	460.6	542.7	639.2	764.4	958.8	1422

TABLE 3.31 : I* (μ = 0.2)

Q: MODE:	0.05	0.10	0.20	0.30	0.40	0.50	0.60	0.70	0.80	0.90
(1)	.0503	.2013	.8104	1.844	3.335	5.348	8.015	11.64	17.15	29.10
(2)	13.50	21.28	34.85	49.43	66.36	86.72	112.2	146.5	200.7	329.6
(3)	79.95	129.3	203.7	272.0	345.0	429.3	534.1	677.7	914.0	1509
(4)	239.8	402.3	646.3	860.5	1082	1337	1656	2107	2880	4915

TABLE 3.31 : I* (μ = 0.3)

Q: MODE:	0.05	0.10	0.20	0.30	0.40	0.50	0.60	0.70	0.80	0.90
(1)	.1050	.4207	1.698	3.881	7.072	11.46	17.46	25.97	39.77	72.61
(2)	22.97	38.79	67.23	98.04	134.5	179.6	238.5	322.5	464.9	830.4
(3)	113.6	201.4	350.6	497.2	663.1	869.4	1150	1583	2391	4693
(4)	55.32	113.0	237.9	381.5	554.7	776.7	1085	1571	2501	5210

TABLE 3.31 : I* (μ = 0.4)

Q: MODE:	0.05	0.10	0.20	0.30	0.40	0.50	0.60	0.70	0.80	0.90
(1)	.2053	.8235	3.334	7.664	14.08	23.10	35.76	54.51	86.56	168.0
(2)	32.51	58.59	107.0	159.5	222.9	304.7	418.5	595.1	926.3	1869
(3)	UPPER and LOWER LIMITS AGREE WITHIN 1 %									
(4)	102.1	191.3	364.0	545.0	749.2	994.4	1310	1763	2544	4597

TABLE 3.31 : I* (μ = 0.5)

Q: MODE:	0.05	0.10	0.20	0.30	0.40	0.50	0.60	0.70	0.80	0.90
(1)	.3793	1.523	6.197	14.36	26.72	44.62	70.81	111.7	186.6	391.0
(2)	13.77	27.43	55.13	84.58	117.7	157.5	209.9	287.7	429.9	830.3
(3)	74.99	133.0	235.4	340.5	461.0	609.9	809.1	1106	1645	3131
(4)	75.45	149.2	295.4	446.3	611.6	806.1	1057	1427	2098	3987

TABLE 3.31 : I* (μ = 0.6)

Q: MODE:	0.05	0.10	0.20	0.30	0.40	0.50	0.60	0.70	0.80	0.90
(1)	.4053	1.628	6.619	15.33	28.49	47.50	75.20	118.2	196.1	406.3
(2)	18.22	36.17	72.42	111.2	155.4	209.6	282.4	393.4	600.5	1193
(3)	51.99	97.42	180.2	262.0	350.8	454.5	586.2	773.6	1098	1964
(4)	203.5	360.4	621.9	871.2	1145	1479	1927	2608	3873	7467

TABLE 3.31 : I* (μ = 0.7)

Q: MODE:	0.05	0.10	0.20	0.30	0.40	0.50	0.60	0.70	0.80	0.90
(1)	.2361	.9470	3.830	8.789	16.11	26.32	40.50	61.16	95.72	181.0
(2)	38.03	67.37	119.9	175.5	241.4	325.2	440.2	615.8	939.3	1843
(3)	46.53	93.37	190.5	297.0	421.3	576.9	790.2	1121	1751	3578
(4)	76.08	146.1	276.5	403.6	536.2	684.2	863.3	1106	1512	2562

TABLE 3.32 : I* (μ = 0.1)

Q: MODE:	0.05	0.10	0.20	0.30	0.40	0.50	0.60	0.70	0.80	0.90
(1)	.0390	.1559	.6259	1.417	2.543	4.031	5.935	8.379	11.74	17.84
(2)	12.09	17.19	25.41	33.88	43.37	54.32	67.35	83.80	107.5	157.8
(3)	62.87	91.61	130.4	164.4	199.6	238.7	284.9	344.1	433.4	638.0
(4)	183.7	276.6	395.8	491.5	585.7	687.9	808.0	964.5	1209	1799

TABLE 3.32 : I* (μ = 0.4)

Q: MODE:	0.05	0.10	0.20	0.30	0.40	0.50	0.60	0.70	0.80	0.90
(1)	.3923	1.574	6.379	14.69	27.04	44.49	69.14	105.8	168.9	329.5
(2)	42.18	78.31	144.3	211.9	289.1	385.4	516.4	717.2	1091	2156
(3)	30.99	61.78	123.9	188.7	259.9	342.8	448.1	599.7	869.2	1612
(4)	227.5	376.2	621.8	866.3	1143	1482	1934	2606	3820	7166

TABLE 3.33 : I* (μ = 0.2)

Q: MODE:	0.05	0.10	0.20	0.30	0.40	0.50	0.60	0.70	0.80	0.90
(1)	.0038	.0151	.0609	.1390	.2528	.4087	.6202	.9189	1.400	2.540
(2)	1.543	2.664	5.217	8.478	12.62	17.88	24.69	34.09	49.13	85.01
(3)	24.94	40.87	67.63	95.09	126.4	164.1	212.0	278.6	388.3	663.2
(4)	112.6	191.3	315.5	431.1	555.9	703.3	892.4	1163	1632	2878

TABLE 3.33 : I* (μ = 0.5)

Q: MODE:	0.05	0.10	0.20	0.30	0.40	0.50	0.60	0.70	0.80	0.90
(1)	.0343	.1380	.5624	1.308	2.446	4.117	6.607	10.60	18.12	39.41
(2)	5.380	10.58	21.86	35.53	52.89	75.86	108.1	158.1	250.9	511.6
(3)	13.07	26.04	52.37	80.36	111.9	149.7	199.5	273.5	408.7	789.5
(4)	74.22	131.1	231.6	335.3	454.5	602.2	799.7	1094	1628	3097

TABLE 3.33 : I* (μ = 0.8)

Q: MODE:	0.05	0.10	0.20	0.30	0.40	0.50	0.60	0.70	0.80	0.90
(1)	.1427	.5740	2.347	5.487	10.34	17.59	28.60	46.64	81.48	182.0
(2)	10.32	20.90	43.94	71.30	105.8	152.3	219.7	329.3	544.0	1178
(3)	3.525	7.357	16.17	27.06	41.08	60.10	87.86	133.1	222.2	486.3
(4)	27.09	52.61	103.7	157.9	217.9	287.1	371.1	482.5	658.2	1081

TABLE 3.31 : I* (μ = 0.8)

Q: MODE:	0.05	0.10	0.20	0.30	0.40	0.50	0.60	0.70	0.80	0.90
(1)	.1210	.4851	1.955	4.459	8.095	13.05	19.71	28.97	43.45	76.27
(2)	25.30	41.54	68.70	96.38	127.9	165.7	213.7	280.4	390.6	667.2
(3)	112.6	191.2	315.4	430.9	555.7	703.0	892.1	1163	1632	2877
(4)	267.5	484.0	852.6	1205	1593	2068	2714	3710	5588	10984

TABLE 3.32 : I* (μ = 0.2)

Q: MODE:	0.05	0.10	0.20	0.30	0.40	0.50	0.60	0.70	0.80	0.90
(1)	.1012	.4055	1.632	3.714	6.719	10.78	16.15	23.45	34.48	58.25
(2)	23.66	37.42	59.91	82.73	108.5	139.0	177.1	228.5	310.7	509.7
(3)	109.7	179.0	282.5	375.8	474.2	587.5	728.2	922.2	1244	2066
(4)	297.9	504.9	819.0	1094	1381	1710	2127	2721	3752	6501

TABLE 3.32 : I* (μ = 0.5)

Q: MODE:	0.05	0.10	0.20	0.30	0.40	0.50	0.60	0.70	0.80	0.90
(1)	.5951	2.391	9.733	22.59	42.12	70.56	112.5	178.9	302.3	645.8
(2)	UPPER and LOWER LIMITS AGREE WITHIN 1 %									
(3)	38.63	114.0	225.5	348.7	494.8	670.5	927.0	1301	1902	3073
(4)	UPPER and LOWER LIMITS AGREE WITHIN 1 %									

TABLE 3.33 : I* (μ = 0.3)

Q: MODE:	0.05	0.10	0.20	0.30	0.40	0.50	0.60	0.70	0.80	0.90
(1)	.0082	.0327	.1325	.3044	.5589	.9168	1.420	2.169	3.465	6.814
(2)	2.772	5.071	10.29	16.98	25.59	36.79	51.83	73.60	110.7	205.8
(3)	37.60	66.42	118.1	173.1	238.3	321.5	435.8	610.3	931.5	1828
(4)	46.48	93.28	190.3	296.8	421.1	576.7	789.9	1121	1750	3578

TABLE 3.33 : I* (μ = 0.6)

Q: MODE:	0.05	0.10	0.20	0.30	0.40	0.50	0.60	0.70	0.80	0.90
(1)	.0724	.2911	1.191	2.790	5.270	8.991	14.69	24.11	42.51	96.16
(2)	3.282	6.627	13.68	21.51	30.60	41.73	56.39	77.98	116.7	223.1
(3)	30.93	54.95	99.86	149.5	209.9	288.1	397.0	565.9	882.0	1781
(4)	UPPER and LOWER LIMITS AGREE WITHIN 1 %									

TABLE 3.33 : I* (μ = 0.9)

Q: MODE:	0.05	0.10	0.20	0.30	0.40	0.50	0.60	0.70	0.80	0.90
(1)	.1105	.4442	1.812	4.219	7.904	13.33	21.46	34.55	59.38	129.9
(2)	11.90	23.52	47.97	76.29	111.1	156.1	218.5	314.4	491.8	990.3
(3)	56.28	109.1	213.4	326.0	458.4	626.1	857.4	1215	1888	3813
(4)	123.6	246.2	499.6	782.6	1122	1563	2188	3185	5118	10799

TABLE 3.31 : I* (μ = 0.9)

Q: MODE:	0.05	0.10	0.20	0.30	0.40	0.50	0.60	0.70	0.80	0.90
(1)	.0489	.1959	.7869	1.784	3.210	5.106	7.560	10.77	15.32	24.14
(2)	13.84	20.41	30.96	41.72	53.80	67.85	84.83	106.8	139.7	213.1
(3)	68.93	104.6	154.1	197.4	242.5	292.9	353.4	432.7	556.7	852.6
(4)	196.3	307.6	457.3	579.5	700.4	833.0	991.4	1202	1543	2390

TABLE 3.32 : I* (μ = 0.3)

Q: MODE:	0.05	0.10	0.20	0.30	0.40	0.50	0.60	0.70	0.80	0.90
(1)	.2071	.8303	3.351	7.663	13.97	22.65	34.50	51.30	78.43	142.5
(2)	39.00	66.29	113.3	161.9	218.2	287.4	378.0	508.7	734.1	1324
(3)	132.5	241.9	432.7	619.5	828.5	1087	1440	1989	3029	6036
(4)	UPPER and LOWER LIMITS AGREE WITHIN 1 %									

TABLE 3.33 : I* (μ = 0.1)

Q: MODE:	0.05	0.10	0.20	0.30	0.40	0.50	0.60	0.70	0.80	0.90
(1)	.0014	.0055	.0220	.0499	.0897	.1427	.2114	.3017	.4314	.6892
(2)	.6895	1.106	2.083	3.337	4.908	6.848	9.256	12.36	16.84	26.09
(3)	13.39	20.03	30.43	41.16	53.22	67.27	84.26	106.2	139.0	212.1
(4)	68.95	104.6	154.1	197.5	242.6	293.0	353.5	432.8	556.8	852.8

TABLE 3.33 : I* (μ = 0.4)

Q: MODE:	0.05	0.10	0.20	0.30	0.40	0.50	0.60	0.70	0.80	0.90
(1)	.0167	.0669	.2719	.6284	1.165	1.935	3.053	4.785	7.932	16.51
(2)	4.616	8.835	18.49	30.97	47.39	69.50	100.6	148.5	236.0	477.2
(3)	18.57	36.75	73.24	112.0	156.2	210.2	282.8	393.2	599.1	1187
(4)	52.06	97.59	180.5	262.5	351.4	455.4	587.3	775.0	1100	1968

TABLE 3.33 : I* (μ = 0.7)

Q: MODE:	0.05	0.10	0.20	0.30	0.40	0.50	0.60	0.70	0.80	0.90
(1)	.1332	.5362	2.200	5.176	9.846	16.95	28.02	46.69	83.88	194.2
(2)	1.472	3.054	6.628	10.95	16.39	23.62	33.99	50.64	82.98	178.1
(3)	13.80	26.55	52.83	82.41	117.2	159.8	214.6	292.0	421.9	754.1
(4)	111.0	195.0	336.6	476.5	635.6	834.3	1105	1523	2305	4534

TABLE 3.34 : I* (μ = 0.1)

Q: MODE:	0.05	0.10	0.20	0.30	0.40	0.50	0.60	0.70	0.80	0.90
(1)	.0239	.0958	.3850	.8727	1.570	2.497	3.696	5.265	7.495	11.84
(2)	7.665	11.31	17.68	24.58	32.54	41.95	53.37	68.09	89.87	137.4
(3)	49.53	74.53	109.9	141.8	175.5	213.6	259.3	318.9	411.3	628.6
(4)	156.1	242.8	359.6	456.2	552.7	659.0	785.9	954.6	1224	1891

TABLE 3.34 : I* ($\mu = 0.2$)

Q: MODE:	0.05	0.10	0.20	0.30	0.40	0.50	0.60	0.70	0.80	0.90
(1)	.0607	.2432	.9802	2.235	4.058	6.543	9.884	14.53	21.84	38.55
(2)	14.66	23.96	40.38	58.04	78.68	103.8	135.7	180.0	252.2	430.5
(3)	82.88	139.3	228.5	312.4	403.6	511.2	648.5	843.2	1175	2045
(4)	230.3	407.8	700.5	975.4	1275	1637	2121	2852	4198	7977

TABLE 3.34 : I* ($\mu = 0.3$)

Q: MODE:	0.05	0.10	0.20	0.30	0.40	0.50	0.60	0.70	0.80	0.90
(1)	.1215	.4872	1.970	4.516	8.268	13.49	20.74	31.30	48.99	92.85
(2)	23.40	40.99	73.47	109.1	152.1	206.8	280.9	391.4	588.7	1122
(3)	62.65	121.0	232.6	347.6	477.5	637.8	856.2	1194	1835	3690
(4)	41.50	82.17	162.5	244.2	331.0	428.8	548.0	712.1	990.8	1730

TABLE 3.34 : I* ($\mu = 0.4$)

Q: MODE:	0.05	0.10	0.20	0.30	0.40	0.50	0.60	0.70	0.80	0.90
(1)	.2239	.8988	3.649	8.428	15.60	25.85	40.58	63.07	103.0	208.9
(2)	20.37	38.81	73.98	110.9	153.4	206.6	278.9	389.6	596.0	1183
(3)	29.65	58.07	113.3	169.2	229.6	299.3	387.2	512.7	733.3	1334
(4)	170.9	290.0	487.3	683.1	906.0	1182	1555	2121	3164	6094

TABLE 3.34 : I* ($\mu = 0.5$)

Q: MODE:	0.05	0.10	0.20	0.30	0.40	0.50	0.60	0.70	0.80	0.90
(1)	.3252	1.307	5.329	12.40	23.22	39.14	62.93	101.2	173.5	378.8
(2)	UPPER and LOWER LIMITS AGREE WITHIN 1 %									
(3)	35.65	70.19	142.4	225.2	326.0	454.9	631.1	898.3	1386	2743
(4)	UPPER and LOWER LIMITS AGREE WITHIN 1 %									

TABLE 3.35 : I* ($\mu = 0.1$)

Q: MODE:	0.05	0.10	0.20	0.30	0.40	0.50	0.60	0.70	0.80	0.90
(1)	.1896	.7617	3.101	7.198	13.42	22.49	35.87	57.03	96.38	206.0
(2)	18.74	36.43	72.09	111.5	158.3	217.2	296.8	416.8	635.1	1241
(3)	77.96	147.5	276.9	408.4	555.2	732.3	965.3	1310	1932	3656
(4)	204.9	383.7	704.9	1017	1354	1751	2265	3019	4374	8125

TABLE 3.35 : I* ($\mu = 0.2$)

Q: MODE:	0.05	0.10	0.20	0.30	0.40	0.50	0.60	0.70	0.80	0.90
(1)	.2515	1.010	4.114	9.549	17.81	29.84	47.62	75.74	128.1	273.9
(2)	25.03	48.68	96.40	149.3	212.1	291.2	398.3	559.7	853.7	1670
(3)	105.0	198.7	373.4	550.9	749.3	988.8	1303	1770	2611	4941
(4)	275.6	516.2	948.8	1369	1822	2356	3045	4055	5868	10878

TABLE 3.35 : I* ($\mu = 0.3$)

Q: MODE:	0.05	0.10	0.20	0.30	0.40	0.50	0.60	0.70	0.80	0.90
(1)	.3501	1.406	5.726	13.29	24.80	41.56	66.34	105.6	178.7	382.5
(2)	34.83	67.79	134.4	208.2	295.9	406.4	555.7	780.6	1189	2324
(3)	131.1	249.1	468.2	687.1	925.6	1205	1562	2071	2952	5291
(4)	255.7	489.2	919.0	1336	1779	2298	2972	3978	5836	11123

TABLE 3.35 : I* ($\mu = 0.4$)

Q: MODE:	0.05	0.10	0.20	0.30	0.40	0.50	0.60	0.70	0.80	0.90
(1)	.5134	2.063	8.398	19.50	36.36	60.95	97.27	154.8	261.8	560.3
(2)	42.15	82.14	161.6	246.5	343.1	459.2	607.5	816.0	1165	2050
(3)	165.7	310.6	568.7	817.1	1085	1408	1840	2504	3759	7390
(4)	270.1	502.2	920.4	1334	1790	2335	3047	4095	5976	11162

TABLE 3.35 : I* ($\mu = 0.5$)

Q: MODE:	0.05	0.10	0.20	0.30	0.40	0.50	0.60	0.70	0.80	0.90
(1)	.7586	3.047	12.39	28.72	53.44	89.23	141.6	223.5	373.2	782.0
(2)	27.54	54.87	110.3	169.1	235.4	315.1	419.7	575.4	859.8	1660
(3)	150.0	265.9	470.9	681.0	922.0	1219	1618	2213	3290	6262
(4)	150.9	298.4	590.7	892.5	1223	1612	2114	2853	4197	7974

TABLE 3.35 : I* ($\mu = 0.6$)

Q: MODE:	0.05	0.10	0.20	0.30	0.40	0.50	0.60	0.70	0.80	0.90
(1)	.6751	2.707	10.93	24.99	45.53	73.67	111.5	163.4	241.1	399.0
(2)	130.7	215.0	343.2	463.9	601.2	775.7	1020	1408	2156	4343
(3)	63.10	124.3	244.0	364.6	491.4	631.4	796.6	1013	1356	2203
(4)	555.3	863.6	1325	1765	2260	2869	3682	4901	7114	13235

TABLE 3.35 : I* ($\mu = 0.7$)

Q: MODE:	0.05	0.10	0.20	0.30	0.40	0.50	0.60	0.70	0.80	0.90
(1)	.3371	1.350	5.417	12.26	22.00	34.82	51.12	71.69	98.77	143.1
(2)	124.9	169.0	244.9	328.8	429.1	552.8	711.9	931.8	1283	2104
(3)	348.3	614.5	1046	1447	1888	2437	3200	4407	6742	13605
(4)	98.35	196.3	394.5	601.9	828.8	1091	1421	1890	2709	4935

TABLE 3.35 : I* ($\mu = 0.8$)

Q: MODE:	0.05	0.10	0.20	0.30	0.40	0.50	0.60	0.70	0.80	0.90
(1)	.1512	.6052	2.425	5.473	9.776	15.39	22.40	31.06	42.03	58.67
(2)	58.58	76.18	107.1	140.3	177.8	220.5	269.4	327.1	400.5	523.4
(3)	388.3	495.9	646.7	791.4	948.2	1125	1332	1587	1940	2633
(4)	1290	1791	2429	3000	3631	4398	5414	6906	9541	16571

TABLE 3.35 : I* ($\mu = 0.9$)

Q: MODE:	0.05	0.10	0.20	0.30	0.40	0.50	0.60	0.70	0.80	0.90
(1)	.0543	.2174	.8706	1.962	3.498	5.488	7.955	10.95	14.61	19.69
(2)	22.50	28.15	38.56	49.91	62.62	76.87	92.91	111.2	133.3	166.3
(3)	146.5	177.3	222.0	265.0	310.3	359.4	413.3	474.4	548.8	667.8
(4)	517.2	623.4	754.5	868.9	984.7	1107	1240	1391	1616	1900

TABLE 3.36 : I* ($\mu = 0.1$)

Q: MODE:	0.05	0.10	0.20	0.30	0.40	0.50	0.60	0.70	0.80	0.90
(1)	.0259	.1037	.4153	.9361	1.669	2.618	3.796	5.225	6.979	9.422
(2)	11.48	14.72	21.21	28.59	37.04	46.66	57.59	70.16	85.28	107.5
(3)	99.01	120.3	152.8	185.0	219.4	256.9	298.2	345.1	402.0	491.5
(4)	391.2	471.6	574.1	665.6	759.1	858.5	966.8	1089	1241	1495

TABLE 3.36 : I* ($\mu = 0.2$)

Q: MODE:	0.05	0.10	0.20	0.30	0.40	0.50	0.60	0.70	0.80	0.90
(1)	.0720	.2883	1.155	2.607	4.656	7.328	10.67	14.80	20.07	28.16
(2)	29.95	39.80	58.66	79.93	104.4	132.6	165.2	203.6	252.1	331.2
(3)	261.2	332.7	437.5	540.5	652.6	778.7	924.2	1100	1335	1774
(4)	1019	1363	1805	2206	2646	3168	3830	4750	6261	9971

TABLE 3.36 : I* ($\mu = 0.3$)

Q: MODE:	0.05	0.10	0.20	0.30	0.40	0.50	0.60	0.70	0.80	0.90
(1)	.1604	.6421	2.575	5.823	10.43	16.48	24.13	33.74	46.36	67.11
(2)	64.36	88.01	132.8	184.2	245.8	320.4	413.1	533.8	710.0	1071
(3)	379.4	590.4	889.6	1164	1477	1874	2432	3316	5018	9989
(4)	97.72	197.5	407.6	640.5	912.2	1250	1708	2410	3729	7522

TABLE 3.36 : I* ($\mu = 0.4$)

Q: MODE:	0.05	0.10	0.20	0.30	0.40	0.50	0.60	0.70	0.80	0.90
(1)	.3356	1.344	5.410	12.30	22.21	35.49	52.77	75.43	107.2	165.6
(2)	100.4	146.6	219.3	296.3	390.1	513.0	686.4	959.6	1478	2974
(3)	46.78	93.46	188.4	289.2	401.4	534.6	706.5	957.8	1409	2665
(4)	285.2	492.0	840.2	1182	1559	2007	2575	3370	4691	7977

TABLE 3.36 : I* ($\mu = 0.5$)

Q: MODE:	0.05	0.10	0.20	0.30	0.40	0.50	0.60	0.70	0.80	0.90
(1)	.5689	2.283	9.262	21.36	39.43	65.06	101.4	155.6	248.3	479.7
(2)	42.98	82.73	158.4	236.1	324.1	433.4	583.4	817.8	1265	2571
(3)	91.60	174.7	334.4	502.3	692.8	922.0	1216	1634	2341	4142
(4)	273.1	523.6	990.8	1455	1966	2584	3417	4704	7146	14243

TABLE 3.36 : I* (μ = 0.6)

Q: MODE:	0.05	0.10	0.20	0.30	0.40	0.50	0.60	0.70	0.80	0.90
(1)	.4759	1.912	7.782	18.06	33.66	56.37	89.86	142.7	240.8	513.1
(2)	25.08	49.21	96.55	145.1	197.7	257.7	330.8	429.7	590.8	994.8
(3)	174.1	290.7	471.7	641.5	830.2	1062	1379	1866	2779	5392
(4)	220.4	417.0	779.0	1137	1526	1980	2555	3369	4765	8421

TABLE 3.36 : I* (μ = 0.7)

Q: MODE:	0.05	0.10	0.20	0.30	0.40	0.50	0.60	0.70	0.80	0.90
(1)	.3342	1.343	5.467	12.69	23.68	39.69	63.37	100.9	170.7	365.7
(2)	32.53	63.32	125.4	194.1	275.2	376.9	513.6	718.0	1087	2103
(3)	79.79	154.2	294.1	431.1	574.6	736.0	933.4	1206	1668	2885
(4)	335.7	576.7	953.9	1296	1662	2100	2683	3565	5202	9856

TABLE 3.36 : I* (μ = 0.8)

Q: MODE:	0.05	0.10	0.20	0.30	0.40	0.50	0.60	0.70	0.80	0.90
(1)	.2430	.9761	3.974	9.225	17.21	28.84	46.02	73.22	123.9	265.0
(2)	24.25	47.18	93.49	144.9	205.9	282.9	387.1	544.1	830.0	1624
(3)	101.7	192.5	361.8	533.8	725.8	957.5	1261	1712	2523	4766
(4)	253.7	476.5	877.0	1263	1676	2156	2768	3653	5218	9486

TABLE 3.36 : I* (μ = 0.9)

Q: MODE:	0.05	0.10	0.20	0.30	0.40	0.50	0.60	0.70	0.80	0.90
(1)	.1857	.7461	3.037	7.050	13.15	22.03	35.14	55.87	94.44	201.8
(2)	18.39	35.75	70.75	109.5	155.4	213.3	291.6	409.5	624.2	1220
(3)	76.71	145.1	272.5	402.0	546.6	721.1	950.6	1290	1904	3602
(4)	202.0	378.3	695.1	1003	1335	1727	2234	2979	4316	8018

TABLE 3.37 : I* (μ = 0.1)

Q: MODE:	0.05	0.10	0.20	0.30	0.40	0.50	0.60	0.70	0.80	0.90
(1)	.0014	.0055	.0219	.0494	.0880	.1382	.2005	.2764	.3703	.5045
(2)	.8245	1.237	2.265	3.593	5.232	7.194	9.504	12.22	15.51	20.21
(3)	21.99	27.55	37.91	49.26	61.99	76.29	92.40	110.8	133.0	166.1
(4)	146.5	177.4	222.1	265.1	310.4	359.5	413.4	474.4	548.9	667.8

TABLE 3.37 : I* (μ = 0.2)

Q: MODE:	0.05	0.10	0.20	0.30	0.40	0.50	0.60	0.70	0.80	0.90
(1)	.0038	.0151	.0604	.1364	.2439	.3846	.5617	.7833	1.073	1.550
(2)	2.174	3.352	6.241	9.971	14.59	20.17	26.81	34.78	44.84	60.78
(3)	57.46	74.72	105.4	138.6	176.1	218.8	267.9	325.9	399.5	522.4
(4)	388.3	496.1	647.0	791.7	948.5	1125	1332	1587	1940	2633

TABLE 3.37 : I* (μ = 0.3)

Q: MODE:	0.05	0.10	0.20	0.30	0.40	0.50	0.60	0.70	0.80	0.90
(1)	.0002	.0026	.1310	.2963	.5312	.8409	1.236	1.742	2.434	3.690
(2)	4.563	7.233	13.74	22.19	32.76	45.69	61.35	80.61	105.9	149.2
(3)	123.2	166.2	241.1	324.6	424.4	547.9	706.9	926.5	1277	2095
(4)	347.6	613.5	1045	1446	1888	2437	3199	4406	6741	13604

TABLE 3.37 : I* (μ = 0.4)

Q: MODE:	0.05	0.10	0.20	0.30	0.40	0.50	0.60	0.70	0.80	0.90
(1)	.0167	.0667	.2679	.6071	1.092	1.736	2.568	3.657	5.206	8.235
(2)	8.976	14.56	28.19	46.25	69.57	99.36	137.7	189.2	266.2	427.2
(3)	134.3	218.0	344.0	462.9	598.4	771.1	1013	1398	2140	4311
(4)	63.09	124.3	244.0	364.7	491.5	631.6	796.9	1013	1357	2203

TABLE 3.37 : I* (μ = 0.5)

Q: MODE:	0.05	0.10	0.20	0.30	0.40	0.50	0.60	0.70	0.80	0.90
(1)	.0343	.1374	.5526	1.256	2.267	3.624	5.408	7.800	11.35	18.77
(2)	14.46	23.04	42.03	66.61	99.13	143.1	205.6	302.4	481.4	980.8
(3)	27.68	55.17	111.0	170.5	237.7	318.8	425.9	586.1	880.5	1712
(4)	145.9	260.0	463.0	672.0	912.0	1208	1605	2197	3268	6222

TABLE 3.37 : I* (μ = 0.6)

Q: MODE:	0.05	0.10	0.20	0.30	0.40	0.50	0.60	0.70	0.80	0.90
(1)	.0723	.2898	1.169	2.670	4.858	7.857	11.91	17.59	26.57	47.15
(2)	17.08	28.28	48.36	70.74	98.49	135.1	186.8	268.1	421.9	863.2
(3)	43.31	84.04	164.1	249.0	344.9	459.3	604.7	808.2	1146	1999
(4)	173.9	323.7	587.2	838.6	1109	1435	1873	2546	3820	7510

TABLE 3.37 : I* (μ = 0.7)

Q: MODE:	0.05	0.10	0.20	0.30	0.40	0.50	0.60	0.70	0.80	0.90
(1)	.1331	.5342	2.168	5.003	9.254	15.32	24.01	37.22	60.56	122.0
(2)	12.30	23.37	44.35	66.29	91.59	123.2	166.4	233.0	357.8	714.6
(3)	43.20	81.79	156.0	235.6	328.4	444.5	601.0	836.9	1267	2468
(4)	131.9	250.5	469.9	688.2	925.3	1202	1554	2055	2917	5198

TABLE 3.37 : I* (μ = 0.8)

Q: MODE:	0.05	0.10	0.20	0.30	0.40	0.50	0.60	0.70	0.80	0.90
(1)	.1427	.5733	2.336	5.430	10.15	17.05	27.31	43.67	74.38	160.7
(2)	11.76	23.04	45.80	70.54	99.18	134.2	180.1	246.9	364.1	678.7
(3)	53.83	99.64	181.0	260.7	348.3	454.2	594.9	807.2	1199	2308
(4)	121.0	225.8	415.7	604.6	813.1	1063	1392	1881	2764	5221

TABLE 3.37 : I* (μ = 0.9)

Q: MODE:	0.05	0.10	0.20	0.30	0.40	0.50	0.60	0.70	0.80	0.90
(1)	.1105	.4441	1.811	4.214	7.889	13.29	21.37	34.33	58.87	128.4
(2)	11.95	23.52	47.64	75.23	108.7	151.5	209.8	298.0	459.0	905.8
(3)	57.78	109.8	207.6	307.7	419.9	555.5	733.6	996.7	1469	2773
(4)	146.7	276.1	509.2	734.3	974.3	1253	1608	2123	3037	5538

CHAPTER FOUR

FREE VIBRATION OF TRIPLE SPAN BEAMS WITH CLASSICAL BOUNDARY CONDITIONS

The problems encountered in providing frequency and modal shape data for triple span beams are much more difficult to handle than those encountered with the double span beams of the preceding chapter. These difficulties are related to the fact that the eigenvalues for triple span beams are a function of two variables, as will be seen. In mathematical terms we say that they generate a surface, and the information necessary for vibration analysis cannot be presented in the form of simple two-dimensional graphs as was done for double span beams.

Before entering into discussion of the method adopted for storage of the necessary information, it is conducive to a better understanding of the problem to look at a typical beam.

We consider the beam shown in the inset of Case 4.1. The extreme ends are simply supported. The first and second spans are of dimensionless length μ_1 and μ_2, respectively, where the span lengths have been nondimensionalized with respect to the beam overall length. The length of the remaining span is γ, where of course $\gamma = 1 - (\mu_1 + \mu_2)$.

Following the practice established in Chapter 3, the deflection of each span is represented by a separate function $r(\xi)$, the origin for the outer spans being selected at the extreme outer ends, and for the intermediate span, at its left end (inset, Case 4.1).

The deflection expression for the first span is written as

$$r_1(\xi) = A_1 \sin \beta\xi + B_1 \cos \beta\xi + C_1 \sinh \beta\xi + D_1 \cosh \beta\xi \tag{4.1}$$

with similar expressions for the other spans.

The boundary conditions

$$r_1(0) = \left.\frac{d^2 r_1(\xi)}{d\xi^2}\right|_{\xi=0} = 0$$

require that $B_1 = D_1 = 0$

$$\therefore\ r_1(\xi) = A_1 \sin \beta\xi + C_1 \sinh \beta\xi \tag{4.2}$$

But $r_1(\mu_1) = 0$

$$\therefore A_1 \sin \beta\mu_1 + C_1 \sinh \beta\mu_1 = 0 \tag{4.3}$$

hence

$$C_1 = -A_1 \frac{\sin \beta\mu_1}{\sinh \beta\mu_1} \tag{4.4}$$

$$\therefore r_1(\xi) = A_1\left(\sin \beta\xi - \frac{\sin \beta\mu_1}{\sinh \beta\mu_1} \sinh \beta\xi\right) \tag{4.5}$$

By virtue of an identical argument for the third span, we obtain

$$r_3(\xi) = A_3\left(\sin \beta\xi - \frac{\sin \beta\gamma}{\sinh \beta\gamma} \sinh \beta\xi\right) \tag{4.6}$$

Finally, looking at the intermediate span we have, $r_2(0) = 0$, which requires that $B_2 = -D_2$, hence

$$r_2(\xi) = A_2 \sin \beta\xi + B_2(\cos \beta\xi - \cosh \beta\xi) + C_2 \sinh \beta\xi \tag{4.7}$$

Also, $r_2(\mu_2) = 0$, hence

$$A_2 \sin \beta\mu_2 + B_2(\cos \beta\mu_2 - \cosh \beta\mu_2) + C_2 \sinh \beta\mu_2 = 0 \tag{4.8}$$

Equation 4.8 can be rearranged to obtain

$$C_2 = -(A_2\theta + B_2\phi) \tag{4.9}$$

where

$$\theta = \frac{\sin \beta\mu_2}{\sinh \beta\mu_2} \quad \text{and} \quad \phi = \frac{\cosh \beta\mu_2 - \cos \beta\mu_2}{\sinh \beta\mu_2}$$

$$\therefore r_2(\xi) = A_2(\sin \beta\xi - \theta \sinh \beta\xi) + B_2(\cos \beta\xi - \cosh \beta\xi + \phi \sinh \beta\xi) \tag{4.10}$$

We are left therefore with four unknowns. We may write four equations involving continuity of slope and bending moment across the two interior supports. We write these equations in the above order for the left and right interior supports, respectively, as:

$$A_1\left(\cos \beta\mu_1 - \frac{\sin \beta\mu_1}{\sinh \beta\mu_1}\right) + A_2(\theta - 1) + B_2(-\phi) + A_3(0) = 0 \tag{4.11}$$

$$A_1\left(-\sin \beta\mu_1 - \frac{\sin \beta\mu_1}{\sinh \beta\mu_1}\right) + A_2(0) + B_2(2) + A_3(0) = 0 \tag{4.12}$$

$$A_1(0) + A_2(-\cos \beta\mu_2 + \theta \cosh \beta\mu_2) + B_2(\sin \beta\mu_2 + \sinh \beta\mu_2 - \phi \cosh \beta\mu_2) + A_3((\sin \beta\gamma/\sinh \beta\gamma) \cosh \beta\gamma - \cos \beta\gamma) = 0 \tag{4.13}$$

$$A_1(0) + A_2(\sin \beta\mu_2 + \theta \sinh \beta\mu_2) + B_2(\cos \beta\mu_2 + \cosh \beta\mu_2 - \phi \sinh \beta\mu_2) + A_3(-2 \sin \beta\gamma) = 0 \tag{4.14}$$

The problem confronting the vibration analyst then, is to find those values of β that permit the determinant of the coefficient matrix in the above four equations to vanish. It is seen that the values of β that permit this to occur depend on the values of μ_1 and μ_2.

Data for frequency analysis of the first four modes of all six possible beams of the above type are presented in Tables 4.1 through 4.6. The beams have been divided into 40 equal increments, giving 39 possible stations for the intermediate supports. It has been found that following this procedure the error encountered in using linear interpolation to obtain frequencies for a beam whose supports may lie between any of these stations is ≤ 2 percent for the first mode and ≤ 1 percent for the higher modes.

It is important to note that μ_2 is always less than $1-\mu_1$, or $\mu_1+\mu_2$ is always less than one. It is for this reason that tables provided for each mode are in triangular form. It is seen therefore that information for analyzing two modes may be contained in one rectangular array (see, for example, Table 4.4). Use of the tables is demonstrated later by means of an illustrative problem.

It is further appreciated that certain families of beams possess symmetry because they have identical outer end supports. In problems of this type we are never interested in situations in which $\mu_1 > 0.5$, since we can always choose our axes in such a way that $\mu_1 \leq 0.5$. It will be seen in the corresponding tables that redundant data associated with $\mu_1 > 0.5$ are not provided.

Finally, it is necessary to discuss briefly the matter of modal shapes. Since we can arbitrarily set one of the nonzero unknown constants equal to one, we choose to assign this value to the coefficient A_1 associated with the first span. This permits obtaining a solution for the other unknowns, hence provides exact expressions for the modal shapes associated with each eigenvalue β. This information is given below for all six cases. The case number and the associated eigenvalue table numbers are identical.

Because of the nature of the equations, two expressions are available for A_3 (Eq. 4.13 or 4.14). Since under certain conditions one of the expressions may take on an indeterminate form, both expressions are given and either may be used.

ILLUSTRATIVE PROBLEM 4.1

A heat exchanger is fitted with straight internal tubes. The tubes receive simple support at two intermediate positions where they pass through baffle plates. Each tube is essentially clamped at one end, where it is rolled into a tube sheet, and is given simple support at the other end (see Fig. 4.1).

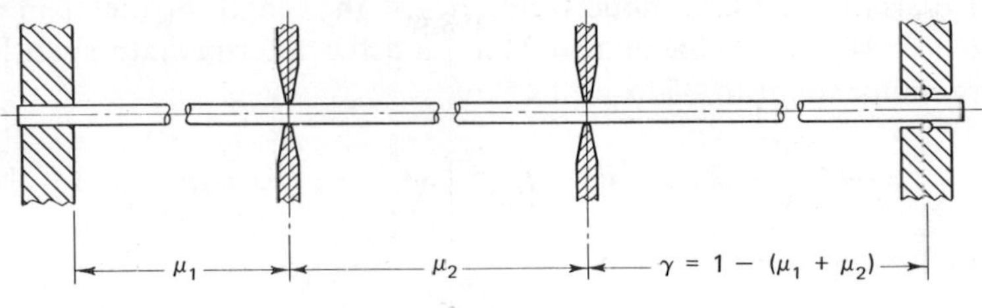

Fig. 4.1

Pertinent data are: tube material: steel; tube o.d. = 1/2 in.; tube wall thickness = 0.020 in.; length between end supports = 100 in.; length of span at clamped end = 25 in.; length of intermediate span = 35 in.; Young's modulus of tube material = 30.0×10^6 lb/in.^{2}; density of tube material = 0.283 lb/in.^{3}

(a) Determine the first mode lateral vibration frequency and modal shape of the tubes in a dry condition (no liquid present in tubes or shell).

$$\mu_1 = \frac{25}{100} = .25 \qquad \mu_2 = \frac{35}{100} = .35 \qquad \gamma = 1 - (\mu_1 + \mu_2) = .40$$

$$\beta = 8.526 \qquad \text{(Table 4.4, mode 1)}$$

$$f = \frac{\beta^2}{2\pi L^2}\sqrt{\frac{EI}{\rho A}}$$

where

$$\rho = \frac{0.283}{386} = .733 \times 10^{-3} \text{ lb sec}^2/\text{in.}^4$$

$$\therefore f = \frac{8.526^2}{2\pi \times 100^2}\sqrt{\frac{30.0 \times 10^6 \times 8.70 \times 10^{-4}}{.733 \times 10^{-3} \times 30 \times 10^{-3}}} = 40.1 \text{ Hz}$$

$$\beta\mu_1 = 2.132 \qquad \beta\mu_2 = 2.984 \qquad \beta\gamma = 3.410$$

Substituting in expressions of case 4.4, we obtain.

$$\theta = 0.016 \qquad \phi = 1.105 \qquad A_2 = -2.746$$

$$B_2 = 1.213 \qquad A_3 = 6.133$$

hence

$$r_1(\xi) = \sin\beta\xi - \sinh\beta\xi - 0.688(\cos\beta\xi - \cosh\beta\xi)$$

$$r_2(\xi) = -2.746(\sin\beta\xi - 0.016\sinh\beta\xi) + 1.213(\cos\beta\xi - \cosh\beta\xi + 1.105\sinh\beta\xi)$$

$$r_3(\xi) = 6.133(\sin\beta\xi + 0.018\sinh\beta\xi)$$

(b) Determine the first mode frequency if the length of the span at the clamped end is changed to 24 in. and the intermediate span length remains equal to 35 in.

$$\mu_1 = \frac{24}{100} = .24 \qquad \mu_2 = \frac{35}{100} = .35 \qquad \gamma = 1 - (\mu_1 + \mu_2) = .41$$

For $\mu_1 = .25$ and $\mu_2 = .35$,

$$\beta = 8.526 \qquad \text{[part (a) above]}$$

For $\mu_1 = .225$ and $\mu_2 = .35$,

$$\beta = 8.110 \qquad \text{(Table 4.4, mode 1)}$$

Interpolating linearly for β, we obtain

$$\beta = 8.110 + \frac{.240 - .225}{.250 - .225}(8.526 - 8.110) = 8.360$$

Utilizing the results of part (a) and recalling that f varies inversely with β^2,

$$f = \frac{8.360^2}{8.526^2} \times 40.1 = 38.6 \text{ Hz}$$

(c) Determine the second mode frequency for the heat exchanger tubes of part (a).

$$\beta = 11.42 \qquad \text{(table 4.4, mode 2)}$$

$$f = \frac{11.42^2}{8.526^2} \times 40.1 = 71.9 \text{ Hz}$$

CASE 4.1. Beam with simple support at each end.

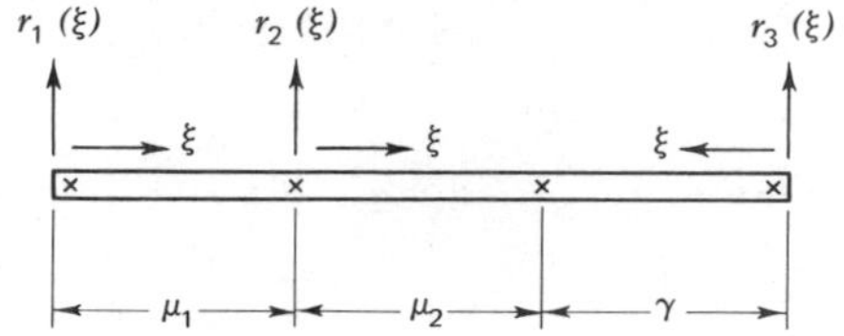

$$r_1(\xi) = \sin \beta\xi - \frac{\sin \beta\mu_1}{\sinh \beta\mu_1} \sinh \beta\xi$$

$$r_2(\xi) = A_2(\sin \beta\xi - \theta \sinh \beta\xi) + B_2(\cos \beta\xi - \cosh \beta\xi + \phi \sinh \beta\xi)$$

$$r_3(\xi) = A_3\left(\sin \beta\xi - \frac{\sin \beta\gamma}{\sinh \beta\gamma} \sinh \beta\xi\right)$$

$$A_2 = \frac{\sin \beta\mu_1 \cosh \beta\mu_1 - \cos \beta\mu_1 \sinh \beta\mu_1 + \phi \sin \beta\mu_1 \sinh \beta\mu_1}{(\theta - 1) \sinh \beta\mu_1}$$

$$B_2 = \sin \beta\mu_1$$

$$A_3 = \frac{A_2(\cos \beta\mu_2 - \theta \cosh \beta\mu_2) + B_2(-\sin \beta\mu_2 - \sinh \beta\mu_2 + \phi \cosh \beta\mu_2)}{(\sin \beta\gamma \cosh \beta\gamma / \sinh \beta\gamma) - \cos \beta\gamma}$$

or

$$A_3 = A_2 \frac{\sin \beta\mu_2}{\sin \beta\gamma} + B_2 \frac{\cos \beta\mu_2}{\sin \beta\gamma}$$

where

$$\theta = \frac{\sin \beta\mu_2}{\sinh \beta\mu_2}, \qquad \text{and } \phi = \frac{\cosh \beta\mu_2 - \cos \beta\mu_2}{\sinh \beta\mu_2}$$

CASE 4.2. Beam with outer ends clamped.

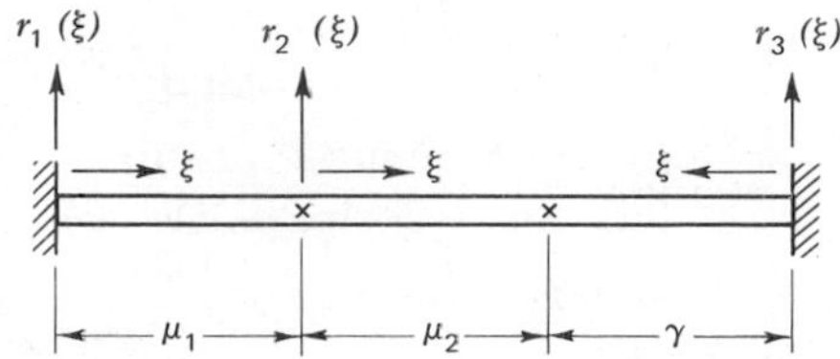

$$r_1(\xi) = \sin \beta\xi - \sinh \beta\xi + \frac{\sin \beta\mu_1 - \sinh \beta\mu_1}{\cosh \beta\mu_1 - \cos \beta\mu_1} (\cos \beta\xi - \cosh \beta\xi)$$

$$r_2(\xi) = A_2(\sin \beta\xi - \theta \sinh \beta\xi) + B_2(\cos \beta\xi - \cosh \beta\xi + \phi \sinh \beta\xi)$$

$$r_3(\xi) = A_3\left[\sin \beta\xi - \sinh \beta\xi + \frac{\sin \beta\gamma - \sinh \beta\gamma}{\cosh \beta\gamma - \cos \beta\gamma} (\cos \beta\xi - \cosh \beta\xi)\right]$$

$$A_2 = \frac{2(1 - \cosh \beta\mu_1 \cos \beta\mu_1) - \phi(\sinh \beta\mu_1 \cos \beta\mu_1 - \sin \beta\mu_1 \cosh \beta\mu_1)}{(\theta - 1)(\cosh \beta\mu_1 - \cos \beta\mu_1)}$$

$$B_2 = \frac{\sin \beta\mu_1 \cosh \beta\mu_1 - \sinh \beta\mu_1 \cos \beta\mu_1}{\cosh \beta\mu_1 - \cos \beta\mu_1}$$

$$A_3 = \frac{A_2 (\sinh \beta\mu_2 \cos \beta\mu_2 - \sin \beta\mu_2 \cosh \beta\mu_2)}{2 \sinh \beta\mu_2[(1 - \cosh \beta\gamma \cos \beta\gamma)/(\cosh \beta\gamma - \cos \beta\gamma)]}$$
$$\quad + \frac{B_2(1 - \cos \beta\mu_2 \cosh \beta\mu_2 - \sin \beta\mu_2 \sinh \beta\mu_2)}{2 \sinh \beta\mu_2[(1 - \cosh \beta\gamma \cos \beta\gamma)/(\cosh \beta\gamma - \cos \beta\gamma)]}$$

or

$$A_3 = \frac{A_2 \sin \beta\mu_2 + B_2 \cos \beta\mu_2}{(\sin \beta\gamma \cosh \beta\gamma - \sinh \beta\gamma \cos \beta\gamma)/(\cosh \beta\gamma - \cos \beta\gamma)}$$

where

$$\theta = \frac{\sin \beta\mu_2}{\sinh \beta\mu_2} \qquad \text{and} \qquad \phi = \frac{\cosh \beta\mu_2 - \cos \beta\mu_2}{\sinh \beta\mu_2}$$

CASE 4.3. Beam with outer ends free.

$$r_1(\xi) = \sin \beta\xi + \sinh \beta\xi - \frac{\sin \beta\mu_1 + \sinh \beta\mu_1}{\cos \beta\mu_1 + \cosh \beta\mu_1}(\cos \beta\xi + \cosh \beta\xi)$$

$$r_2(\xi) = A_2(\sin \beta\xi - \theta \sinh \beta\xi) + B_2(\cos \beta\xi - \cosh \beta\xi + \phi \sinh \beta\xi)$$

$$r_3(\xi) = A_3\left[\sin \beta\xi + \sinh \beta\xi - \frac{\sin \beta\gamma + \sinh \beta\gamma}{\cos \beta\gamma + \cosh \beta\gamma}(\cos \beta\xi + \cosh \beta\xi)\right]$$

$$A_2 = \frac{2(1 + \cos \beta\mu_1 \cosh \beta\mu_1) + \phi(\sinh \beta\mu_1 \cos \beta\mu_1 - \sin \beta\mu_1 \cosh \beta\mu_1)}{(\cos \beta\mu_1 + \cosh \beta\mu_1)(1 - \theta)}$$

$$B_2 = \frac{\sin \beta\mu_1 \cosh \beta\mu_1 - \sinh \beta\mu_1 \cos \beta\mu_1}{\cos \beta\mu_1 + \cosh \beta\mu_1}$$

$$A_3 = \frac{A_2(\sin \beta\mu_2 \cosh \beta\mu_2 - \sinh \beta\mu_2 \cos \beta\mu_2)}{2 \sinh \beta\mu_2[(1 + \cos \beta\gamma \cosh \beta\gamma)/(\cos \beta\gamma + \cosh \beta\gamma)]}$$
$$+ \frac{B_2(\sin \beta\mu_2 \sinh \beta\mu_2 + \cos \beta\mu_2 \cosh \beta\mu_2 - 1)}{2 \sinh \beta\mu_2[(1 + \cos \beta\gamma \cosh \beta\gamma)/(\cos \beta\gamma + \cosh \beta\gamma)]}$$

or

$$A_3 = \frac{A_2(\sin \beta\mu_2 \sinh \beta\mu_2) + B_2(\cos \beta\mu_2 \sinh \beta\mu_2)}{\sinh \beta\mu_2[(\sin \beta\gamma \cosh \beta\gamma - \sinh \beta\gamma \cos \beta\gamma)/(\cos \beta\gamma + \cosh \beta\gamma)]}$$

where

$$\theta = \frac{\sin \beta\mu_2}{\sinh \beta\mu_2}, \qquad \text{and } \phi = \frac{\cosh \beta\mu_2 - \cos \beta\mu_2}{\sinh \beta\mu_2}$$

CASE 4.4. Beam with one outer end clamped and the other simply supported.

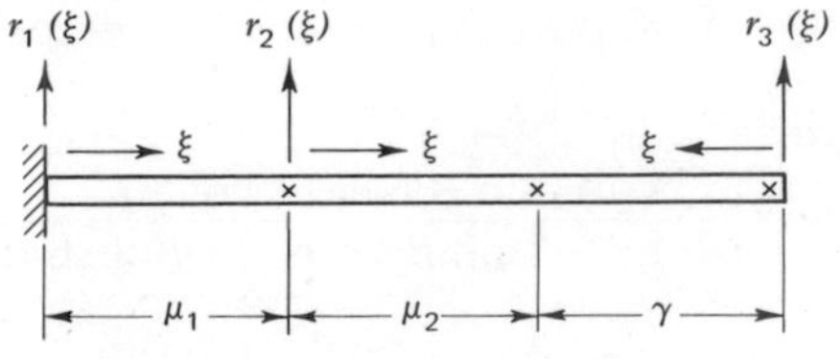

$r_1(\xi)$, $r_2(\xi)$, A_2, B_2, as for Case 4.2.

$r_3(\xi)$, A_3; as for Case 4.1.

CASE 4.5. Beam with one outer end clamped and the other free.

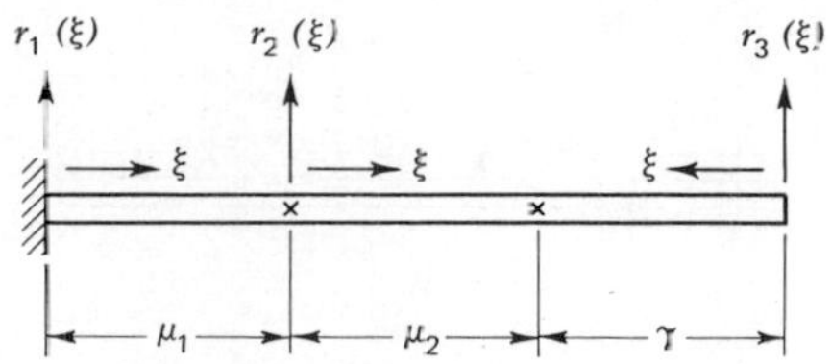

$r_1(\xi)$, $r_2(\xi)$, A_2, B_2; as for Case 4.2.
$r_3(\xi)$, A_3; as for Case 4.3.

CASE 4.6. Beam with one end simply supported and the other free.

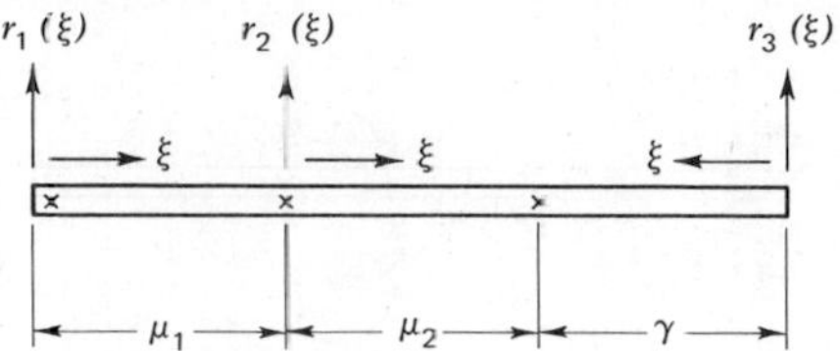

$r_1(\xi)$, $r_2(\xi)$, A_2, B_2; as for Case 4.1.
$r_3(\xi)$, A_3; as for Case 4.3.

TABLE 4.1 (MODES 1 AND 2)

MODE 1 $\mu_2 \longrightarrow$

μ_1 ↓	0.025	0.050	0.075	0.100	0.125	0.150	0.175	0.200	0.225	0.250	0.275	0.300	0.325	0.350	0.375	0.400	0.425	0.450	0.475	0.500	0.525	0.550	0.575	0.600	0.625	0.650	0.675	0.700	0.725	0.750	0.775	0.800	0.825	0.850	0.875	0.900	0.925	0.950
0.025	4.103	4.186	4.274	4.369	4.469	4.576	4.690	4.812	4.942	5.081	5.230	5.390	5.561	5.745	5.941	6.151	6.373	6.602	6.830	7.034	7.168	7.182	7.082	6.918	6.730	6.538	6.351	6.172	6.002	5.843	5.694	5.554	5.425	5.303	5.190	5.085	4.987	4.896
0.050	4.211	4.297	4.390	4.490	4.596	4.709	4.830	4.960	5.099	5.248	5.407	5.579	5.763	5.961	6.173	6.398	6.633	6.873	7.097	7.268	7.326	7.254	7.098	6.906	6.707	6.510	6.322	6.144	5.977	5.820	5.674	5.538	5.412	5.294	5.184	5.082	4.987	
0.075	4.327	4.417	4.514	4.619	4.731	4.851	4.980	5.119	5.267	5.427	5.599	5.783	5.982	6.195	6.423	6.664	6.913	7.156	7.361	7.468	7.434	7.292	7.099	6.892	6.686	6.487	6.299	6.123	5.958	5.805	5.662	5.529	5.405	5.291	5.184	5.085		
0.100	4.449	4.543	4.646	4.757	4.876	5.004	5.141	5.289	5.449	5.620	5.805	6.005	6.219	6.449	6.695	6.951	7.210	7.445	7.603	7.619	7.500	7.310	7.095	6.878	6.668	6.469	6.282	6.108	5.946	5.795	5.655	5.525	5.405	5.294	5.190			
0.125	4.578	4.678	4.786	4.904	5.030	5.167	5.314	5.473	5.644	5.829	6.029	6.244	6.476	6.725	6.989	7.260	7.521	7.726	7.802	7.720	7.538	7.316	7.088	6.866	6.655	6.456	6.271	6.099	5.940	5.792	5.655	5.529	5.412	5.303				
0.150	4.715	4.821	4.936	5.061	5.196	5.342	5.499	5.670	5.855	6.054	6.271	6.504	6.756	7.025	7.307	7.588	7.835	7.976	7.947	7.782	7.558	7.318	7.082	6.857	6.645	6.448	6.266	6.096	5.940	5.795	5.662	5.538	5.425					
0.175	4.861	4.973	5.095	5.228	5.373	5.529	5.699	5.883	6.082	6.299	6.534	6.788	7.061	7.351	7.649	7.931	8.134	8.174	8.042	7.819	7.568	7.318	7.077	6.851	6.641	6.446	6.266	6.099	5.946	5.805	5.674	5.554						
0.200	5.016	5.135	5.266	5.408	5.563	5.731	5.914	6.113	6.329	6.565	6.820	7.097	7.394	7.705	8.013	8.273	8.391	8.310	8.099	7.840	7.574	7.317	7.075	6.849	6.641	6.448	6.271	6.108	5.958	5.820	5.694							
0.225	5.182	5.309	5.448	5.601	5.767	5.949	6.146	6.362	6.598	6.854	7.134	7.435	7.757	8.086	8.390	8.587	8.580	8.394	8.131	7.850	7.576	7.316	7.075	6.851	6.645	6.456	6.282	6.123	5.977	5.843								
0.250	5.358	5.494	5.644	5.808	5.987	6.184	6.398	6.633	6.890	7.171	7.476	7.805	8.151	8.489	8.756	8.836	8.696	8.437	8.145	7.853	7.576	7.317	7.077	6.857	6.655	6.469	6.299	6.144	6.002									
0.275	5.548	5.693	5.854	6.031	6.225	6.438	6.672	6.929	7.210	7.518	7.852	8.209	8.573	8.895	9.064	8.989	8.749	8.451	8.145	7.850	7.574	7.318	7.082	6.866	6.668	6.487	6.322	6.172										
0.300	5.751	5.908	6.081	6.272	6.483	6.715	6.971	7.252	7.561	7.898	8.263	8.644	9.004	9.245	9.244	9.039	8.749	8.437	8.131	7.840	7.568	7.318	7.088	6.878	6.686	6.510	6.351											
0.325	5.969	6.139	6.326	6.534	6.763	7.017	7.297	7.606	7.945	8.313	8.703	9.083	9.357	9.400	9.244	8.989	8.696	8.394	8.099	7.819	7.558	7.316	7.095	6.892	6.707	6.538												
0.350	6.205	6.388	6.592	6.818	7.069	7.346	7.653	7.992	8.360	8.750	9.117	9.343	9.357	9.245	9.064	8.836	8.580	8.310	8.042	7.782	7.538	7.310	7.099	6.906	6.730													
0.375	6.460	6.659	6.881	7.127	7.401	7.705	8.039	8.401	8.767	9.041	9.117	9.083	9.004	8.895	8.756	8.587	8.391	8.174	7.947	7.720	7.500	7.292	7.098	6.918														
0.400	6.737	6.954	7.195	7.464	7.762	8.087	8.423	8.688	8.767	8.750	8.703	8.644	8.573	8.489	8.390	8.273	8.134	7.976	7.802	7.619	7.434	7.254	7.082															
0.425	7.039	7.275	7.537	7.825	8.126	8.364	8.423	8.401	8.360	8.313	8.263	8.209	8.151	8.086	8.013	7.931	7.835	7.726	7.603	7.468	7.326	7.182																
0.450	7.368	7.623	7.892	8.101	8.126	8.087	8.039	7.992	7.945	7.898	7.852	7.805	7.757	7.705	7.649	7.588	7.521	7.445	7.361	7.268	7.168																	
0.475	7.726	7.922	7.892	7.825	7.762	7.705	7.653	7.606	7.561	7.518	7.476	7.435	7.394	7.351	7.307	7.260	7.210	7.156	7.097	7.034																		
0.500	7.726	7.623	7.537	7.464	7.401	7.346	7.297	7.252	7.210	7.171	7.134	7.097	7.061	7.025	6.989	6.952	6.913	6.873	6.830																			

MODE 2 $\longleftarrow \mu_2$

0.950	0.925	0.900	0.875	0.850	0.825	0.800	0.775	0.750	0.725	0.700	0.675	0.650	0.625	0.600	0.575	0.550	0.525	0.500	0.475	0.450	0.425	0.400	0.375	0.350	0.325	0.300	0.275	0.250	0.225	0.200	0.175	0.150	0.125	0.100	0.075	0.050	0.025	μ_1 ↑
																			9.181	9.477	9.819	10.21	10.64	11.12	11.61	12.07	12.42	12.57	12.31	11.66	10.96	10.33	9.769	9.279	8.850	8.474	8.144	0.500
																		8.930	9.180	9.475	9.817	10.21	10.66	11.15	11.66	12.13	12.39	12.17	11.55	10.87	10.25	9.693	9.204	8.775	8.406	8.133	8.144	0.475
																	8.811	9.001	9.233	9.514	9.846	10.23	10.67	11.15	11.63	11.99	11.94	11.42	10.79	10.18	9.630	9.148	8.739	8.453	8.406	8.474	8.583	0.450
																8.877	9.002	9.161	9.362	9.614	9.920	10.28	10.70	11.14	11.53	11.64	11.28	10.70	10.12	9.584	9.131	8.818	8.739	8.775	8.850	8.951	9.078	0.425
															9.120	9.202	9.303	9.430	9.593	9.801	10.06	10.38	10.75	11.11	11.31	11.11	10.62	10.07	9.581	9.237	9.131	9.148	9.204	9.279	9.373	9.489	9.634	0.400
														9.486	9.551	9.626	9.712	9.815	9.943	10.10	10.31	10.56	10.84	11.05	10.97	10.57	10.09	9.716	9.581	9.584	9.630	9.693	9.769	9.859	9.966	10.10	10.26	0.375
													9.906	9.980	10.06	10.13	10.22	10.31	10.42	10.54	10.69	10.85	10.99	10.95	10.63	10.26	10.09	10.07	10.12	10.18	10.25	10.32	10.42	10.52	10.63	10.78	10.96	0.350
												10.27	10.39	10.50	10.61	10.71	10.81	10.90	11.00	11.10	11.20	11.27	11.18	10.88	10.63	10.57	10.62	10.70	10.79	10.87	10.96	11.05	11.14	11.23	11.25	11.04	10.75	0.325
											10.39	10.59	10.79	10.98	11.16	11.31	11.45	11.57	11.68	11.76	11.77	11.57	11.18	10.95	10.97	11.11	11.28	11.42	11.55	11.66	11.74	11.78	11.63	11.31	10.96	10.64	10.35	0.300
										10.27	10.50	10.76	11.03	11.31	11.58	11.83	12.06	12.23	12.33	12.23	11.77	11.27	10.99	11.05	11.31	11.64	11.94	12.17	12.31	12.30	12.03	11.65	11.26	10.90	10.56	10.26	9.990	0.275
									10.04	10.28	10.55	10.83	11.15	11.47	11.81	12.14	12.42	12.57	12.33	11.76	11.20	10.85	10.84	11.11	11.53	11.99	12.39	12.57	12.39	12.02	11.61	11.21	10.84	10.50	10.18	9.902	9.649	0.250
								9.788	10.02	10.28	10.56	10.87	11.20	11.55	11.90	12.25	12.42	12.23	11.68	11.10	10.69	10.56	10.75	11.14	11.63	12.13	12.42	12.30	11.96	11.56	11.16	10.79	10.44	10.12	9.831	9.568	9.331	0.225
							9.536	9.757	10.00	10.27	10.56	10.88	11.22	11.57	11.90	12.14	12.06	11.57	11.00	10.54	10.31	10.38	10.70	11.15	11.66	12.07	12.13	11.87	11.50	11.11	10.74	10.39	10.07	9.773	9.501	9.255	9.033	0.200
						9.293	9.502	9.734	9.989	10.27	10.56	10.88	11.22	11.55	11.81	11.83	11.45	10.90	10.42	10.10	10.06	10.28	10.67	11.15	11.61	11.87	11.74	11.42	11.06	10.69	10.35	10.02	9.723	9.446	9.193	8.962	8.753	0.175
					9.062	9.261	9.480	9.720	9.982	10.26	10.56	10.88	11.20	11.47	11.58	11.31	10.81	10.31	9.943	9.801	9.920	10.23	10.66	11.12	11.50	11.56	11.33	11.00	10.64	10.30	9.981	9.679	9.399	9.141	8.904	8.687	8.491	0.150
				8.847	9.035	9.243	9.470	9.716	9.982	10.27	10.56	10.87	11.15	11.31	11.16	10.71	10.22	9.815	9.593	9.614	9.846	10.21	10.64	11.07	11.30	11.21	10.93	10.59	10.26	9.941	9.640	9.359	9.097	8.856	8.633	8.429	8.243	0.125
			8.646	8.825	9.022	9.237	9.470	9.720	9.989	10.27	10.56	10.83	11.03	10.98	10.61	10.13	9.712	9.430	9.362	9.514	9.817	10.21	10.63	10.97	11.04	10.84	10.54	10.22	9.904	9.605	9.323	9.060	8.815	8.588	8.378	8.186	8.010	0.100
		8.461	8.631	8.818	9.022	9.243	9.480	9.734	10.00	10.28	10.55	10.76	10.79	10.50	10.06	9.626	9.303	9.161	9.233	9.475	9.819	10.22	10.59	10.80	10.73	10.47	10.17	9.867	9.571	9.291	9.027	8.780	8.551	8.337	8.139	7.957	7.790	0.075
	8.290	8.453	8.631	8.825	9.035	9.261	9.502	9.757	10.02	10.28	10.50	10.59	10.39	9.980	9.551	9.202	9.002	9.001	9.180	9.477	9.842	10.22	10.52	10.58	10.40	10.12	9.830	9.540	9.261	8.998	8.751	8.519	8.303	8.102	7.915	7.742	7.583	0.050
8.133	8.290	8.461	8.646	8.847	9.062	9.293	9.536	9.788	10.04	10.27	10.39	10.27	9.906	9.486	9.120	8.877	8.811	8.930	9.181	9.511	9.878	10.22	10.39	10.30	10.07	9.792	9.509	9.234	8.972	8.724	8.492	8.275	8.072	7.882	7.704	7.540	7.387	0.025

TABLE 4.1 (MODES 3 AND 4)

MODE 3 μ_2 →, μ_1 ↓

μ_1	0.025	0.050	0.075	0.100	0.125	0.150	0.175	0.200	0.225	0.250	0.275	0.300	0.325	0.350	0.375	0.400	0.425	0.450	0.475
0.025	10.67	10.90	11.14	11.41	11.69	11.99	12.31	12.65	13.00	13.33	13.57	13.46	12.94	12.39	12.05	12.06	12.34	12.74	13.18
0.050	10.95	11.19	11.45	11.73	12.03	12.35	12.69	13.05	13.41	13.73	13.85	13.50	12.90	12.41	12.24	12.43	12.81	13.26	13.69
0.075	11.26	11.50	11.77	12.07	12.39	12.73	13.09	13.47	13.83	14.12	14.05	13.50	12.89	12.53	12.57	12.90	13.34	13.82	14.14
0.100	11.57	11.83	12.12	12.43	12.77	13.14	13.52	13.91	14.28	14.48	14.17	13.51	12.96	12.78	13.01	13.43	13.93	14.37	14.44
0.125	11.91	12.18	12.49	12.82	13.19	13.57	13.98	14.39	14.73	14.77	14.22	13.54	13.12	13.16	13.53	14.03	14.54	14.82	14.59
0.150	12.27	12.56	12.88	13.24	13.62	14.04	14.46	14.88	15.17	14.94	14.24	13.61	13.40	13.65	14.12	14.67	15.12	15.08	14.64
0.175	12.65	12.96	13.30	13.68	14.09	14.53	14.98	15.38	15.49	14.96	14.23	13.76	13.81	14.22	14.77	15.32	15.50	15.15	14.63
0.200	13.05	13.38	13.75	14.15	14.59	15.04	15.47	15.71	15.41	14.78	14.20	14.02	14.32	14.85	15.42	15.71	15.47	15.02	14.54
0.225	13.48	13.83	14.21	14.63	15.05	15.30	15.24	15.01	14.69	14.33	14.13	14.35	14.82	15.24	15.29	15.12	14.87	14.57	14.27
0.250	13.94	14.28	14.51	14.44	14.30	14.15	14.01	13.86	13.74	13.77	14.08	14.43	14.50	14.38	14.23	14.07	13.92	13.79	13.79
0.275	13.89	13.62	13.42	13.26	13.12	13.01	12.95	13.03	13.37	13.81	14.12	14.03	13.75	13.49	13.29	13.14	13.08	13.24	13.79
0.300	12.78	12.55	12.37	12.25	12.21	12.39	12.81	13.30	13.82	14.23	14.19	13.74	13.26	12.90	12.68	12.67	13.08	13.79	14.27
0.325	11.82	11.64	11.62	11.92	12.34	12.82	13.35	13.90	14.42	14.63	14.23	13.60	13.05	12.70	12.68	13.14	13.92	14.57	14.54
0.350	11.21	11.55	11.94	12.38	12.87	13.41	13.99	14.57	15.02	14.89	14.24	13.56	13.05	12.90	13.29	14.07	14.87	15.02	14.63
0.375	11.64	12.03	12.46	12.94	13.48	14.07	14.69	15.27	15.46	14.97	14.24	13.60	13.26	13.49	14.23	15.12	15.47	15.15	14.64
0.400	12.14	12.56	13.03	13.56	14.15	14.78	15.40	15.71	15.46	14.89	14.23	13.74	13.75	14.38	15.29	15.71	15.50	15.08	14.59
0.425	12.68	13.14	13.65	14.23	14.85	15.35	15.40	15.27	15.02	14.63	14.19	14.03	14.50	15.24	15.42	15.32	15.12	14.82	14.44
0.450	13.28	13.77	14.32	14.80	14.85	14.78	14.69	14.57	14.42	14.23	14.12	14.43	14.82	14.85	14.77	14.67	14.54	14.37	14.14
0.475	13.92	14.34	14.32	14.23	14.15	14.07	13.99	13.90	13.82	13.81	14.08	14.35	14.32	14.22	14.12	14.03	13.93	13.82	13.69
0.500	13.92	13.77	13.65	13.56	13.48	13.41	13.35	13.30	13.37	13.77	14.13	14.02	13.81	13.65	13.53	13.43	13.34	13.26	13.18

μ_1	0.500	0.525	0.550	0.575	0.600	0.625	0.650	0.675	0.700	0.725	0.750	0.775	0.800	0.825	0.850	0.875	0.900	0.925	0.950
0.025	13.53	13.54	13.21	12.80	12.42	12.11	11.99	12.24	12.82	13.43	13.58	13.36	13.06	12.74	12.43	12.14	11.87	11.62	11.39
0.050	13.85	13.60	13.19	12.77	12.42	12.23	12.38	12.93	13.60	13.86	13.67	13.36	13.03	12.70	12.40	12.11	11.85	11.62	
0.075	14.02	13.62	13.17	12.78	12.51	12.56	13.05	13.76	14.15	14.01	13.69	13.35	13.01	12.69	12.39	12.11	11.87		
0.100	14.09	13.62	13.19	12.85	12.78	13.18	13.91	14.45	14.38	14.06	13.70	13.34	13.00	12.69	12.40	12.14			
0.125	14.12	13.64	13.24	13.06	13.34	14.05	14.72	14.77	14.46	14.08	13.70	13.34	13.01	12.70	12.43				
0.150	14.13	13.68	13.40	13.54	14.20	14.97	15.16	14.86	14.47	14.08	13.70	13.35	13.03	12.74					
0.175	14.13	13.77	13.77	14.34	15.18	15.51	15.26	14.86	14.46	14.06	13.69	13.36	13.06						
0.200	14.14	14.01	14.47	15.31	15.71	15.51	15.16	14.77	14.38	14.01	13.67	13.36							
0.225	14.14	14.49	15.15	15.31	15.18	14.97	14.72	14.45	14.15	13.86	13.58								
0.250	14.14	14.49	14.47	14.34	14.20	14.05	13.91	13.76	13.60	13.43									
0.275	14.14	14.01	13.77	13.54	13.34	13.18	13.05	12.93	12.82										
0.300	14.14	13.77	13.40	13.06	12.78	12.56	12.38	12.24											
0.325	14.13	13.68	13.24	12.85	12.51	12.23	11.99												
0.350	14.13	13.64	13.19	12.78	12.42	12.11													
0.375	14.12	13.62	13.17	12.77	12.42														
0.400	14.09	13.62	13.19	12.80															
0.425	14.02	13.60	13.21																
0.450	13.85	13.54																	
0.475	13.53																		
0.500																			

MODE 4 ← μ_2, μ_1 ↑

μ_1	0.950	0.925	0.900	0.875	0.850	0.825	0.800	0.775	0.750	0.725	0.700	0.675	0.650	0.625	0.600	0.575	0.550	0.525	0.500
0.500																			
0.475																			15.23
0.450																		15.30	15.51
0.425																	15.80	15.91	16.06
0.400																16.43	16.56	16.68	16.81
0.375															16.79	17.09	17.34	17.53	17.69
0.350														16.55	16.99	17.46	17.91	18.29	18.55
0.325													16.10	16.55	17.05	17.59	18.15	18.64	18.76
0.300												15.65	16.07	16.54	17.07	17.62	18.16	18.39	17.54
0.275											15.29	15.66	16.09	16.57	17.08	17.60	17.92	17.32	16.26
0.250										15.23	15.50	15.83	16.22	16.66	17.12	17.47	17.11	16.11	15.71
0.225									15.84	16.00	16.18	16.41	16.68	16.99	17.26	17.03	16.26	16.11	16.26
0.200								16.67	16.91	17.13	17.36	17.37	17.73	17.87	17.34	17.03	17.11	17.32	17.34
0.175							16.73	17.09	17.50	17.91	18.32	18.67	18.74	17.87	17.26	17.47	17.92	18.39	18.74
0.150						16.40	16.78	17.20	17.65	18.12	18.55	18.67	17.75	16.99	17.12	17.60	18.16	18.64	18.55
0.125					16.02	16.38	16.78	17.22	17.68	18.12	18.32	17.57	16.68	16.66	17.08	17.62	18.15	18.29	17.69
0.100				15.64	15.98	16.37	16.78	17.22	17.65	17.91	17.36	16.41	16.22	16.57	17.07	17.59	17.91	17.53	16.81
0.075			15.28	15.61	15.97	16.37	16.78	17.20	17.50	17.15	16.18	15.83	16.09	16.54	17.05	17.46	17.34	16.68	16.06
0.050		14.95	15.26	15.61	15.98	16.38	16.78	17.09	16.91	16.00	15.50	15.66	16.07	16.55	16.99	17.09	16.56	15.91	15.51
0.025	15.63	16.21	16.72	16.70	16.27	15.78	15.37	15.21	15.64	16.44	16.79	16.53	16.13	15.72	15.32	14.94	14.58	14.26	13.96

μ_1	0.475	0.450	0.425	0.400	0.375	0.350	0.325	0.300	0.275	0.250	0.225	0.200	0.175	0.150	0.125	0.100	0.075	0.050	0.025
0.500	15.63	16.17	16.81	17.50	18.19	18.71	18.79	17.73	16.67	16.64	17.39	18.28	18.81	18.62	17.76	16.84	16.02	15.30	14.67
0.475	15.65	16.19	16.82	17.53	18.21	18.47	17.50	16.34	16.00	16.50	17.40	18.30	18.38	17.62	16.73	15.91	15.19	14.65	14.67
0.450	15.84	16.29	16.87	17.50	17.91	17.26	16.14	15.73	15.96	16.55	17.38	17.91	17.44	16.63	15.85	15.26	15.19	15.30	15.46
0.425	16.27	16.58	17.00	17.39	17.05	16.18	16.01	16.13	16.38	16.81	17.33	17.24	16.58	15.97	15.85	15.91	16.02	16.16	16.36
0.400	16.95	17.13	17.31	17.12	16.66	16.65	16.76	16.89	17.07	17.28	17.24	16.78	16.58	16.63	16.73	16.84	16.97	17.12	17.31
0.375	17.82	17.90	17.63	17.20	17.31	17.51	17.67	17.81	17.90	17.68	17.24	17.24	17.44	17.62	17.76	17.87	17.84	17.36	16.81
0.350	18.62	17.87	17.19	17.42	17.92	18.34	18.60	18.55	17.90	17.28	17.33	17.91	18.38	18.62	18.48	17.89	17.25	16.67	16.15
0.325	17.76	16.85	16.93	17.51	18.20	18.74	18.60	17.81	17.07	16.81	17.39	18.30	18.81	18.45	17.81	17.17	16.56	16.02	15.54
0.300	16.51	16.37	16.85	17.54	18.20	18.34	17.67	16.89	16.38	16.55	17.40	18.28	18.27	17.71	17.08	16.48	15.92	15.42	14.97
0.275	15.92	16.26	16.95	17.51	17.93	17.51	16.76	16.13	15.96	16.50	17.39	17.93	17.58	17.00	16.41	15.85	15.34	14.87	14.46
0.250	15.92	16.37	16.93	17.42	17.31	16.65	16.01	15.73	16.00	16.64	17.35	17.40	16.92	16.35	15.81	15.32	14.96	15.00	15.28
0.225	16.51	16.85	17.19	17.20	16.66	16.18	16.14	16.34	16.67	17.09	17.26	16.90	16.39	16.00	15.93	16.07	16.27	16.54	16.90
0.200	17.74	17.87	17.63	17.12	17.03	17.26	17.30	17.73	17.87	17.67	17.23	16.99	17.13	17.38	17.61	17.81	17.83	17.48	17.06
0.175	18.62	17.90	17.31	17.39	17.91	18.47	18.79	18.50	17.91	17.40	17.31	17.90	18.57	18.77	18.40	17.90	17.41	16.95	16.54
0.150	17.82	17.13	17.00	17.50	18.21	18.71	18.44	17.85	17.29	16.98	17.37	18.32	18.71	18.35	17.84	17.34	16.87	16.43	16.05
0.125	16.95	16.58	16.87	17.53	18.19	18.25	17.75	17.17	16.71	16.66	17.39	18.25	18.21	17.77	17.28	16.80	16.36	15.95	15.58
0.100	16.27	16.29	16.82	17.50	17.91	17.62	17.07	16.55	16.22	16.48	17.36	17.91	17.66	17.21	16.74	16.29	15.87	15.48	15.14
0.075	15.84	16.19	16.81	17.39	17.43	16.97	16.44	16.00	15.87	16.41	17.28	17.47	17.12	16.68	16.24	15.81	15.42	15.05	14.72
0.050	15.65	16.17	16.79	17.13	16.85	16.35	15.87	15.54	15.67	16.41	17.09	17.00	16.61	16.18	15.76	15.36	14.99	14.64	14.33
0.025	15.63	16.21	16.72	16.70	16.27	15.78	15.37	15.21	15.64	16.44	16.79	16.53	16.13	15.72	15.32	14.94	14.58	14.26	13.96

TABLE 4.2 (MODES 1 AND 2)

MODE 1 $\mu_2 \longrightarrow$ (μ_1 ↓)

μ_1 \ μ_2	0.025	0.050	0.075	0.100	0.125	0.150	0.175	0.200	0.225	0.250	0.275	0.300	0.325	0.350	0.375	0.400	0.425	0.450	0.475	0.500	0.525	0.550	0.575	0.600	0.625	0.650	0.675	0.700	0.725	0.750	0.775	0.800	0.825	0.850	0.875	0.900	0.925	0.950
0.025	4.943	5.044	5.152	5.267	5.390	5.520	5.660	5.809	5.969	6.139	6.322	6.517	6.725	6.946	7.179	7.419	7.657	7.867	7.994	7.975	7.822	7.602	7.362	7.122	6.892	6.673	6.468	6.276	6.097	5.929	5.772	5.625	5.487	5.358	5.237	5.124	5.017	4.916
0.050	5.074	5.179	5.292	5.414	5.544	5.683	5.831	5.990	6.160	6.343	6.539	6.748	6.971	7.208	7.455	7.706	7.942	8.119	8.165	8.051	7.839	7.592	7.341	7.097	6.866	6.649	6.446	6.256	6.079	5.913	5.759	5.614	5.479	5.352	5.234	5.122	5.017	
0.075	5.213	5.323	5.442	5.570	5.708	5.856	6.014	6.184	6.366	6.562	6.772	6.997	7.236	7.489	7.751	8.008	8.228	8.342	8.285	8.092	7.842	7.579	7.321	7.076	6.845	6.629	6.428	6.240	6.065	5.902	5.750	5.608	5.475	5.350	5.234	5.124		
0.100	5.360	5.476	5.601	5.737	5.883	6.041	6.210	6.392	6.587	6.798	7.024	7.265	7.523	7.793	8.066	8.320	8.499	8.516	8.359	8.112	7.838	7.565	7.304	7.058	6.828	6.614	6.415	6.229	6.057	5.896	5.746	5.606	5.475	5.352	5.237			
0.125	5.516	5.638	5.771	5.915	6.071	6.238	6.419	6.615	6.825	7.052	7.295	7.556	7.832	8.118	8.399	8.632	8.733	8.636	8.403	8.120	7.831	7.552	7.289	7.044	6.815	6.603	6.406	6.223	6.052	5.894	5.746	5.608	5.479	5.358				
0.150	5.681	5.810	5.951	6.105	6.271	6.450	6.645	6.854	7.081	7.326	7.589	7.870	8.166	8.467	8.744	8.926	8.913	8.712	8.427	8.122	7.824	7.542	7.279	7.034	6.807	6.596	6.401	6.220	6.052	5.896	5.750	5.614	5.487					
0.175	5.857	5.994	6.144	6.307	6.485	6.678	6.887	7.113	7.358	7.623	7.908	8.211	8.527	8.836	9.088	9.175	9.036	8.758	8.440	8.121	7.818	7.534	7.271	7.028	6.802	6.594	6.401	6.223	6.057	5.902	5.759	5.625						
0.200	6.044	6.189	6.349	6.528	6.715	6.923	7.148	7.393	7.658	7.946	8.254	8.581	8.914	9.219	9.408	9.362	9.113	8.786	8.446	8.120	7.814	7.530	7.268	7.026	6.802	6.596	6.406	6.229	6.065	5.913	5.772							
0.225	6.243	6.398	6.570	6.757	6.962	7.186	7.430	7.696	7.985	8.297	8.631	8.982	9.326	9.600	9.673	9.485	9.160	8.801	8.449	8.119	7.811	7.528	7.268	7.028	6.807	6.603	6.415	6.240	6.079	5.929								
0.250	6.456	6.622	6.806	7.008	7.229	7.471	7.736	8.025	8.340	8.680	9.042	9.414	9.753	9.947	9.861	9.559	9.186	8.809	8.451	8.118	7.811	7.530	7.271	7.034	6.815	6.614	6.428	6.256	6.097									
0.275	6.684	6.862	7.060	7.278	7.517	7.781	8.069	8.384	8.728	9.099	9.489	9.873	10.17	10.22	9.974	9.597	9.197	8.811	8.451	8.119	7.814	7.534	7.279	7.044	6.828	6.629	6.446	6.276										
0.300	6.928	7.120	7.334	7.570	7.830	8.117	8.432	8.777	9.153	9.556	9.970	10.34	10.53	10.38	10.02	9.609	9.197	8.809	8.449	8.120	7.818	7.542	7.289	7.058	6.845	6.649	6.468											
0.325	7.192	7.399	7.630	7.886	8.170	8.483	8.828	9.207	9.617	10.05	10.47	10.76	10.74	10.44	10.02	9.597	9.186	8.801	8.446	8.121	7.824	7.552	7.304	7.076	6.866	6.673												
0.350	7.476	7.700	7.951	8.230	8.540	8.883	9.262	9.676	10.12	10.56	10.91	10.96	10.74	10.38	9.974	9.559	9.160	8.786	8.440	8.122	7.831	7.565	7.321	7.097	6.892													
0.375	7.783	8.027	8.300	8.605	8.944	9.320	9.733	10.18	10.61	10.89	10.91	10.76	10.53	10.22	9.861	9.485	9.113	8.758	8.427	8.120	7.838	7.579	7.341	7.122														
0.400	8.117	8.382	8.680	9.013	9.383	9.788	10.21	10.53	10.61	10.56	10.47	10.34	10.17	9.947	9.673	9.362	9.036	8.712	8.403	8.112	7.842	7.592	7.362															
0.425	8.481	8.770	9.094	9.453	9.833	10.14	10.21	10.18	10.12	10.05	9.970	9.873	9.753	9.600	9.408	9.175	8.913	8.636	8.359	8.092	7.839	7.602																
0.450	8.878	9.191	9.527	9.797	9.833	9.788	9.733	9.676	9.617	9.556	9.489	9.414	9.326	9.219	9.088	8.926	8.733	8.516	8.285	8.051	7.822																	
0.475	9.309	9.555	9.527	9.453	9.383	9.320	9.262	9.207	9.153	9.099	9.042	8.982	8.914	8.836	8.744	8.632	8.499	8.342	8.165	7.975																		
0.500	9.309	9.191	9.094	9.013	8.944	8.883	8.828	8.777	8.728	8.680	8.631	8.581	8.527	8.467	8.399	8.320	8.228	8.119	7.994																			

MODE 2 $\longleftarrow \mu_2$ (μ_1 ↑)

0.950	0.925	0.900	0.875	0.850	0.825	0.800	0.775	0.750	0.725	0.700	0.675	0.650	0.625	0.600	0.575	0.550	0.525	0.500	0.475	0.450	0.425	0.400	0.375	0.350	0.325	0.300	0.275	0.250	0.225	0.200	0.175	0.150	0.125	0.100	0.075	0.050	0.025	μ_2 / μ_1
																			9.661	9.916	10.23	10.61	11.06	11.58	12.17	12.80	13.41	13.89	14.13	13.92	13.23	12.48	11.80	11.20	10.68	10.22	9.812	0.500
																		9.688	9.849	10.06	10.34	10.69	11.11	11.62	12.21	12.85	13.47	13.89	13.75	13.11	12.39	11.72	11.12	10.59	10.14	9.799	9.812	0.475
																	9.917	10.02	10.14	10.31	10.53	10.83	11.22	11.69	12.25	12.85	13.37	13.46	12.97	12.30	11.64	11.05	10.55	10.19	10.14	10.22	10.34	0.450
																10.28	10.35	10.43	10.54	10.67	10.84	11.07	11.39	11.81	12.30	12.81	13.08	12.80	12.20	11.58	11.03	10.64	10.55	10.59	10.68	10.79	10.94	0.425
															10.69	10.77	10.85	10.93	11.02	11.13	11.27	11.45	11.69	12.01	12.39	12.70	12.61	12.13	11.58	11.15	11.03	11.05	11.12	11.20	11.31	11.44	11.61	0.400
														11.07	11.19	11.30	11.41	11.50	11.60	11.70	11.81	11.95	12.12	12.34	12.54	12.53	12.16	11.74	11.58	11.58	11.64	11.72	11.80	11.90	12.02	12.17	12.36	0.375
													11.21	11.43	11.64	11.83	11.99	12.12	12.24	12.36	12.47	12.58	12.70	12.80	12.73	12.41	12.16	12.13	12.20	12.30	12.39	12.48	12.58	12.68	12.79	12.73	12.41	0.350
												11.07	11.36	11.66	11.96	12.25	12.52	12.74	12.93	13.08	13.21	13.31	13.34	13.14	12.73	12.53	12.61	12.80	12.97	13.11	13.23	13.32	13.33	13.08	12.68	12.29	11.94	0.325
											10.80	11.10	11.42	11.76	12.13	12.50	12.88	13.23	13.54	13.78	13.92	13.84	13.34	12.80	12.54	12.70	13.08	13.46	13.75	13.92	13.87	13.51	13.05	12.61	12.20	11.84	11.51	0.300
										10.50	10.79	11.10	11.44	11.81	12.21	12.63	13.06	13.50	13.89	14.13	13.92	13.31	12.70	12.34	12.39	12.81	13.37	13.89	14.13	13.91	13.47	13.00	12.55	12.13	11.75	11.41	11.10	0.275
									10.20	10.48	10.78	11.10	11.45	11.83	12.24	12.68	13.13	13.57	13.89	13.78	13.21	12.58	12.12	12.01	12.30	12.85	13.47	13.89	13.79	13.39	12.94	12.49	12.07	11.68	11.33	11.01	10.72	0.250
								9.918	10.18	10.46	10.76	11.09	11.45	11.84	12.26	12.70	13.13	13.50	13.54	13.08	12.47	11.95	11.69	11.81	12.25	12.85	13.41	13.57	13.29	12.87	12.43	12.01	11.62	11.26	10.93	10.64	10.37	0.225
							9.645	9.889	10.15	10.44	10.75	11.09	11.45	11.85	12.26	12.68	13.06	13.23	12.93	12.36	11.81	11.45	11.39	11.69	12.21	12.80	13.21	13.13	12.78	12.37	11.96	11.57	11.21	10.87	10.57	10.29	10.04	0.200
						9.389	9.619	9.868	10.14	10.43	10.75	11.09	11.45	11.84	12.24	12.63	12.88	12.74	12.24	11.70	11.27	11.07	11.22	11.62	12.17	12.69	12.89	12.67	12.30	11.90	11.52	11.16	10.82	10.51	10.22	9.961	9.726	0.175
					9.149	9.366	9.601	9.856	10.13	10.43	10.75	11.09	11.45	11.83	12.21	12.50	12.52	12.12	11.60	11.13	10.84	10.83	11.11	11.58	12.11	12.51	12.51	12.21	11.85	11.47	11.11	10.77	10.46	10.17	9.900	9.656	9.435	0.150
				8.924	9.130	9.352	9.593	9.852	10.13	10.43	10.75	11.09	11.45	11.81	12.13	12.25	11.99	11.50	11.02	10.67	10.53	10.69	11.06	11.55	12.03	12.25	12.10	11.78	11.42	11.07	10.73	10.42	10.12	9.850	9.599	9.369	9.160	0.125
			8.714	8.909	9.120	9.348	9.593	9.856	10.14	10.44	10.76	11.10	11.44	11.76	11.96	11.83	11.41	10.93	10.54	10.31	10.34	10.61	11.03	11.51	11.89	11.93	11.69	11.37	11.02	10.69	10.38	10.08	9.807	9.552	9.316	9.098	8.901	0.100
		8.518	8.704	8.905	9.120	9.352	9.601	9.868	10.15	10.46	10.78	11.10	11.42	11.66	11.64	11.30	10.85	10.43	10.14	10.06	10.23	10.58	11.02	11.46	11.69	11.58	11.30	10.98	10.65	10.34	10.05	9.771	9.512	9.272	9.050	8.844	8.657	0.075
	8.335	8.513	8.704	8.909	9.130	9.366	9.619	9.889	10.18	10.48	10.79	11.10	11.36	11.43	11.19	10.77	10.35	10.02	9.849	9.916	10.19	10.58	11.02	11.37	11.42	11.22	10.93	10.61	10.31	10.01	9.738	9.478	9.236	9.010	8.799	8.605	8.426	0.050
8.164	8.335	8.518	8.714	8.924	9.149	9.389	9.645	9.918	10.20	10.50	10.80	11.07	11.21	11.07	10.69	10.28	9.917	9.688	9.661	9.849	10.18	10.59	10.99	11.21	11.12	10.87	10.57	10.27	9.984	9.708	9.448	9.204	8.976	8.763	8.564	8.379	8.208	0.025

TABLE 4.2 (MODES 3 AND 4)

MODE 3 $\mu_2 \rightarrow$ (rows: $\mu_1 \downarrow$)

μ_1	0.025	0.050	0.075	0.100	0.125	0.150	0.175	0.200	0.225	0.250	0.275	0.300	0.325	0.350	0.375	0.400	0.425	0.450	0.475	0.500	0.525	0.550	0.575	0.600	0.625	0.650	0.675	0.700	0.725	0.750	0.775	0.800	0.825	0.850	0.875	0.900	0.925	0.950
0.025	11.49	11.74	12.00	12.29	12.60	12.93	13.27	13.64	14.00	14.31	14.39	13.95	13.31	12.89	12.87	13.16	13.58	14.03	14.38	14.32	13.93	13.48	13.08	12.84	12.95	13.45	14.08	14.41	14.25	13.92	13.55	13.19	12.85	12.52	12.22	11.94	11.68	11.43
0.050	11.80	12.06	12.34	12.64	12.97	13.31	13.68	14.06	14.43	14.70	14.56	13.94	13.34	13.10	13.27	13.66	14.13	14.56	14.71	14.39	13.92	13.48	13.16	13.14	13.57	14.22	14.69	14.61	14.28	13.90	13.52	13.16	12.82	12.50	12.21	11.93	11.68	
0.075	12.12	12.39	12.69	13.01	13.36	13.73	14.11	14.51	14.88	15.05	14.65	13.95	13.46	13.43	13.75	14.21	14.70	15.04	14.88	14.42	13.93	13.54	13.39	13.71	14.36	14.96	14.99	14.67	14.28	13.88	13.50	13.14	12.81	12.49	12.21	11.94		
0.100	12.47	12.75	13.06	13.40	13.77	14.16	14.58	14.99	15.33	15.32	14.70	14.01	13.70	13.88	14.30	14.82	15.28	15.36	14.96	14.44	13.98	13.71	13.88	14.49	15.19	15.39	15.10	14.69	14.27	13.87	13.49	13.14	12.81	12.50	12.22			
0.125	12.83	13.13	13.46	13.82	14.22	14.63	15.07	15.49	15.78	15.50	14.74	14.15	14.07	14.41	14.92	15.46	15.78	15.54	15.01	14.48	14.10	14.10	14.63	15.39	15.78	15.55	15.14	14.70	14.27	13.87	13.49	13.14	12.82	12.52				
0.150	13.2[illegible]	13.53	13.88	14.27	14.69	15.14	15.59	16.01	16.17	15.61	14.81	14.40	14.56	15.02	15.59	16.09	16.11	15.62	15.04	14.57	14.40	14.79	15.57	16.16	16.03	15.61	15.15	14.70	14.27	13.87	13.50	13.16	12.85					
0.175	13.62	13.96	14.34	14.75	15.20	15.67	16.15	16.34	16.45	15.66	14.94	14.79	15.14	15.70	16.29	16.60	16.25	15.66	15.11	14.78	14.99	15.73	16.48	16.52	16.11	15.63	15.15	14.70	14.27	13.88	13.52	13.19						
0.200	14.06	14.42	14.82	15.26	15.74	16.24	16.72	17.00	16.55	15.69	15.17	15.30	15.81	16.43	16.95	16.85	16.29	15.68	15.23	15.24	15.89	16.75	16.97	16.60	16.11	15.61	15.14	14.69	14.28	13.90	13.55							
0.225	14.52	14.90	15.33	15.80	16.31	16.82	17.23	17.11	16.41	15.70	15.53	15.92	16.53	17.12	17.22	16.78	16.21	15.70	15.53	16.04	16.91	17.27	16.97	16.52	16.03	15.55	15.10	14.67	14.28	13.92								
0.250	15.01	15.42	15.87	16.36	16.80	16.92	16.73	16.40	15.96	15.71	15.97	16.31	16.91	16.86	16.62	16.29	15.93	15.73	16.10	16.79	16.91	16.75	16.48	16.16	15.78	15.39	14.99	14.61	14.25									
0.275	15.54	15.93	16.10	15.98	15.83	15.68	15.52	15.37	15.38	15.71	16.06	16.07	15.92	15.75	15.59	15.44	15.40	15.75	16.10	16.04	15.89	15.73	15.57	15.39	15.19	14.96	14.69	14.41										
0.300	15.38	15.12	14.92	14.76	14.63	14.54	14.58	14.94	15.43	15.72	15.53	15.20	14.94	14.76	14.67	14.82	15.40	15.73	15.53	15.24	14.99	14.79	14.63	14.49	14.36	14.22	14.08											
0.325	14.24	14.01	13.95	13.75	13.87	14.32	14.88	15.47	15.90	15.74	15.15	14.63	14.30	14.24	14.67	15.44	15.93	15.70	15.23	14.78	14.40	14.10	13.88	13.71	13.57	13.45												
0.350	13.25	13.11	13.34	13.80	14.34	14.93	15.56	16.15	16.33	15.74	14.99	14.44	14.30	14.76	15.59	16.29	16.21	15.68	15.11	14.57	14.10	13.71	13.39	13.14	12.95													
0.375	12.94	13.38	13.86	14.41	15.01	15.66	16.33	16.81	16.56	15.74	14.99	14.63	14.94	15.75	16.62	16.78	16.29	15.66	15.04	14.48	13.98	13.54	13.16	12.84														
0.400	13.49	13.96	14.49	15.09	15.75	16.45	17.08	17.17	16.56	15.74	15.15	15.20	15.92	16.86	17.22	16.85	16.25	15.62	15.01	14.44	13.93	13.48	13.08															
0.425	14.09	14.61	15.19	15.84	16.54	17.08	17.08	16.81	16.33	15.74	15.53	16.07	16.91	17.12	16.95	16.60	16.11	15.54	14.96	14.42	13.92	13.48																
0.450	14.75	15.31	15.93	16.48	16.54	16.45	16.33	16.15	15.90	15.72	16.06	16.51	16.53	16.43	16.29	16.09	15.78	15.36	14.88	14.39	13.93																	
0.475	15.47	15.95	15.93	15.84	15.75	15.66	15.56	15.47	15.43	15.71	15.97	15.92	15.81	15.70	15.59	15.46	15.28	15.04	14.71	14.31																		
0.500	15.47	15.31	15.19	15.09	13.01	14.93	14.88	14.94	15.38	15.71	15.53	15.30	15.14	15.02	14.92	14.82	14.70	14.56	14.38																			

$\leftarrow \mu_2$ MODE 4 (rows: $\mu_1 \uparrow$)

0.950	0.925	0.900	0.875	0.850	0.825	0.800	0.775	0.750	0.725	0.700	0.675	0.650	0.625	0.600	0.575	0.550	0.525	0.500	0.475	0.450	0.425	0.400	0.375	0.350	0.325	0.300	0.275	0.250	0.225	0.200	0.175	0.150	0.125	0.100	0.075	0.050	0.025	μ_1
																			16.06	16.56	17.18	17.92	18.75	19.58	20.23	20.37	19.31	18.25	18.31	19.23	20.13	20.41	19.73	18.74	17.82	17.00	16.30	0.500
																		16.13	16.37	16.74	17.28	17.98	18.80	19.62	20.01	19.08	17.90	17.57	18.19	19.26	20.02	19.54	18.61	17.70	16.89	16.28	16.30	0.475
																	16.65	16.77	16.93	17.17	17.55	18.12	18.82	19.37	18.83	17.71	17.32	17.60	18.27	19.15	19.25	18.49	17.63	16.97	16.89	17.00	17.18	0.450
																17.29	17.43	17.56	17.69	17.85	18.09	18.44	18.04	18.63	17.85	17.75	17.89	18.14	18.57	18.89	18.41	17.77	17.63	17.70	17.82	17.97	18.17	0.425
															17.62	17.96	18.23	18.42	18.58	18.73	18.88	19.03	18.87	18.46	18.50	18.64	18.79	18.95	19.04	18.67	18.41	18.49	18.61	18.74	18.87	19.01	18.86	0.400
														17.31	17.83	18.35	18.85	19.24	19.52	19.71	19.80	19.39	18.87	19.11	19.43	19.65	19.79	19.65	19.04	18.89	19.25	19.54	19.73	19.80	19.38	18.72	18.11	0.375
													16.82	17.33	17.89	18.51	19.14	19.75	20.25	20.37	19.40	18.59	18.86	19.54	20.17	20.40	19.79	18.95	18.57	19.15	20.02	20.41	20.02	19.31	18.60	17.96	17.40	0.350
												16.35	16.82	17.35	17.93	18.57	19.24	19.88	20.14	19.21	18.20	18.20	18.85	19.66	20.17	19.65	18.79	18.14	18.27	19.26	20.13	19.89	19.21	18.51	17.86	17.26	16.74	0.325
											16.05	16.44	16.89	17.41	17.98	18.61	19.23	19.64	18.98	17.90	17.62	18.11	18.85	19.54	19.43	18.64	17.89	17.60	18.19	19.23	19.61	19.09	18.42	17.77	17.17	16.62	16.14	0.300
										16.28	16.50	16.78	17.15	17.59	18.11	18.66	19.11	18.75	17.72	17.32	17.62	18.20	18.86	19.11	18.50	17.75	17.32	17.57	18.31	19.07	18.93	18.34	17.71	17.12	16.63	16.53	16.78	0.275
									17.10	17.27	17.44	17.64	17.86	18.15	18.49	18.82	18.64	17.85	17.72	17.90	18.20	18.59	18.87	18.46	17.85	17.71	17.90	18.25	18.72	18.81	18.33	17.79	17.52	17.61	17.79	18.04	18.39	0.250
								17.62	17.97	18.30	18.61	18.87	19.10	19.31	19.45	19.14	18.64	18.75	18.98	19.21	19.40	19.39	18.87	18.63	18.83	19.08	19.31	19.45	19.18	18.71	18.62	18.86	19.11	19.33	19.45	19.13	18.67	0.225
							17.40	17.83	18.30	18.79	19.30	19.80	20.22	20.34	19.45	18.82	19.11	19.64	20.14	20.37	19.80	19.03	18.04	19.37	20.01	20.37	20.02	19.38	18.87	19.00	19.80	20.35	20.16	19.63	19.07	18.54	18.08	0.200
						16.99	17.41	17.88	18.39	18.93	19.49	20.02	20.22	19.31	18.49	18.66	19.23	19.88	20.25	19.71	18.88	18.44	18.82	19.62	20.23	19.95	19.30	18.70	18.47	19.18	20.16	20.08	19.56	18.99	18.46	17.96	17.52	0.175
					16.55	16.96	17.41	17.89	18.41	18.96	19.49	19.80	19.10	18.15	18.11	18.61	19.24	19.75	19.52	18.73	18.09	18.12	18.80	19.58	19.73	19.18	18.56	18.08	18.19	19.19	19.80	19.45	18.92	18.38	17.87	17.41	16.99	0.150
				16.13	16.52	16.94	17.40	17.90	18.41	18.93	19.30	18.87	17.86	17.59	17.98	18.57	19.14	19.24	18.58	17.85	17.55	17.98	18.75	19.30	19.03	18.44	17.88	17.57	18.03	19.06	19.26	18.83	18.31	17.80	17.33	16.89	16.50	0.125
			15.74	16.10	16.50	16.94	17.40	17.89	18.39	18.79	18.61	17.64	17.15	17.41	17.93	18.51	18.85	18.42	17.69	17.17	17.28	17.92	18.64	18.80	18.32	17.75	17.28	17.19	17.94	18.79	18.69	18.23	17.74	17.26	16.81	16.40	16.03	0.100
		15.37	15.71	16.09	16.50	16.94	17.41	17.88	18.30	18.30	17.44	16.78	16.89	17.35	17.89	18.35	18.23	17.56	16.93	16.74	17.18	17.88	18.39	18.18	17.65	17.12	16.77	16.98	17.88	18.41	18.13	17.67	17.20	16.75	16.33	15.94	15.59	0.075
	15.02	15.35	15.71	16.10	16.52	16.96	17.41	17.83	17.97	17.27	16.50	16.44	16.82	17.33	17.83	17.96	17.43	16.77	16.37	16.56	17.15	17.80	17.96	17.54	17.02	16.56	16.38	16.89	17.78	17.95	17.59	17.15	16.70	16.28	15.88	15.51	15.17	0.050
14.70	15.02	15.37	15.74	16.13	16.55	16.99	17.40	17.62	17.10	16.28	16.05	16.35	16.82	17.31	17.62	17.29	16.65	16.13	16.06	16.52	17.16	17.61	17.41	16.93	16.44	16.09	16.16	16.90	17.59	17.48	17.08	16.65	16.23	15.83	15.45	15.10	14.78	0.025

TABLE 4.3 (MODES 1 AND 2)

MODE 1 $\mu_2 \longrightarrow$ (μ_1 ↓)

μ_1	0.025	0.050	0.075	0.100	0.125	0.150	0.175	0.200	0.225	0.250	0.275	0.300	0.325	0.350	0.375	0.400	0.425	0.450	0.475	0.500	0.525	0.550	0.575	0.600	0.625	0.650	0.675	0.700	0.725	0.750	0.775	0.800	0.825	0.850	0.875	0.900	0.925	0.950
0.025	1.957	1.992	2.029	2.068	2.109	2.152	2.198	2.246	2.297	2.351	2.409	2.470	2.535	2.605	2.679	2.759	2.845	2.938	3.037	3.143	3.257	3.378	3.504	3.634	3.760	3.875	3.965	4.017	4.025	3.995	3.936	3.858	3.770	3.676	3.581	3.487	3.395	3.307
0.050	2.009	2.046	2.086	2.127	2.170	2.216	2.264	2.316	2.370	2.428	2.490	2.555	2.626	2.701	2.782	2.868	2.962	3.062	3.170	3.286	3.410	3.541	3.676	3.812	3.939	4.043	4.110	4.132	4.109	4.053	3.974	3.883	3.785	3.685	3.586	3.489	3.395	
0.075	2.065	2.104	2.145	2.189	2.235	2.284	2.335	2.390	2.448	2.510	2.577	2.647	2.723	2.805	2.892	2.986	3.088	3.198	3.316	3.442	3.577	3.718	3.862	3.999	4.119	4.203	4.238	4.225	4.172	4.094	3.999	3.897	3.792	3.688	3.586	3.487		
0.100	2.123	2.165	2.209	2.255	2.304	2.356	2.411	2.470	2.532	2.599	2.670	2.746	2.828	2.917	3.012	3.115	3.226	3.346	3.475	3.613	3.759	3.910	4.058	4.191	4.291	4.343	4.341	4.293	4.215	4.118	4.012	3.902	3.792	3.685	3.581			
0.125	2.185	2.229	2.276	2.325	2.377	2.433	2.492	2.554	2.621	2.693	2.770	2.853	2.942	3.038	3.142	3.254	3.376	3.507	3.649	3.799	3.956	4.113	4.259	4.375	4.442	4.452	4.411	4.334	4.236	4.126	4.012	3.897	3.785	3.676				
0.150	2.250	2.297	2.347	2.399	2.455	2.515	2.578	2.645	2.717	2.795	2.878	2.968	3.065	3.170	3.283	3.406	3.540	3.684	3.838	4.000	4.165	4.321	4.451	4.534	4.556	4.521	4.447	4.348	4.236	4.118	3.999	3.883	3.770					
0.175	2.320	2.370	2.423	2.479	2.539	2.602	2.670	2.743	2.821	2.905	2.995	3.093	3.198	3.313	3.437	3.572	3.719	3.876	4.042	4.212	4.377	4.518	4.612	4.644	4.616	4.545	4.447	4.334	4.215	4.094	3.974	3.858						
0.200	2.394	2.447	2.504	2.564	2.628	2.696	2.769	2.848	2.932	3.023	3.121	3.228	3.343	3.469	3.605	3.753	3.912	4.081	4.256	4.425	4.571	4.670	4.706	4.683	4.616	4.521	4.411	4.293	4.172	4.053	3.936							
0.225	2.473	2.530	2.590	2.655	2.724	2.797	2.876	2.961	3.053	3.152	3.259	3.375	3.501	3.639	3.788	3.948	4.118	4.293	4.462	4.605	4.698	4.730	4.706	4.644	4.556	4.452	4.341	4.225	4.109	3.995								
0.250	2.557	2.618	2.683	2.752	2.826	2.906	2.991	3.084	3.183	3.291	3.408	3.535	3.673	3.822	3.982	4.151	4.322	4.482	4.609	4.682	4.698	4.670	4.612	4.534	4.442	4.343	4.238	4.132	4.025									
0.275	2.648	2.713	2.783	2.857	2.937	3.023	3.116	3.216	3.325	3.442	3.569	3.707	3.856	4.014	4.177	4.337	4.475	4.568	4.609	4.605	4.571	4.518	4.451	4.375	4.291	4.203	4.110	4.017										
0.300	2.745	2.815	2.890	2.971	3.057	3.150	3.251	3.360	3.478	3.605	3.742	3.888	4.041	4.193	4.329	4.426	4.475	4.482	4.462	4.425	4.377	4.321	4.259	4.191	4.119	4.043	3.965											
0.325	2.849	2.925	3.006	3.093	3.187	3.288	3.397	3.515	3.642	3.777	3.918	4.060	4.188	4.282	4.329	4.337	4.322	4.293	4.256	4.212	4.165	4.113	4.058	3.999	3.939	3.875												
0.350	2.961	3.043	3.131	3.226	3.327	3.437	3.554	3.680	3.811	3.943	4.062	4.148	4.188	4.193	4.177	4.151	4.118	4.081	4.042	4.000	3.956	3.910	3.862	3.812	3.760													
0.375	3.083	3.172	3.267	3.370	3.479	3.596	3.719	3.842	3.954	4.031	4.062	4.060	4.041	4.014	3.982	3.948	3.912	3.876	3.838	3.799	3.759	3.718	3.676	3.634														
0.400	3.215	3.312	3.415	3.525	3.641	3.758	3.863	3.933	3.954	3.943	3.918	3.888	3.856	3.822	3.788	3.753	3.719	3.684	3.649	3.613	3.577	3.541	3.504															
0.425	3.359	3.463	3.574	3.687	3.791	3.855	3.863	3.842	3.811	3.777	3.742	3.707	3.673	3.639	3.605	3.572	3.540	3.507	3.475	3.442	3.410	3.378																
0.450	3.516	3.628	3.735	3.798	3.791	3.758	3.719	3.680	3.642	3.605	3.569	3.535	3.501	3.469	3.437	3.406	3.376	3.346	3.316	3.286	3.257																	
0.475	3.686	3.763	3.735	3.687	3.641	3.596	3.554	3.515	3.478	3.442	3.408	3.375	3.343	3.313	3.283	3.254	3.226	3.198	3.170	3.143																		
0.500	3.686	3.628	3.574	3.525	3.479	3.437	3.397	3.360	3.325	3.291	3.259	3.228	3.198	3.170	3.142	3.115	3.088	3.062	3.037																			

MODE 2 $\longleftarrow \mu_2$ (μ_1 ↑)

0.950	0.925	0.900	0.875	0.850	0.825	0.800	0.775	0.750	0.725	0.700	0.675	0.650	0.625	0.600	0.575	0.550	0.525	0.500	0.475	0.450	0.425	0.400	0.375	0.350	0.325	0.300	0.275	0.250	0.225	0.200	0.175	0.150	0.125	0.100	0.075	0.050	0.025	μ_1
																			7.094	7.353	7.581	7.754	7.847	7.806	7.528	7.059	6.566	6.117	5.723	5.381	5.083	4.822	4.591	4.385	4.201	4.036	3.886	0.500
																		6.897	7.172	7.433	7.650	7.773	7.723	7.424	6.962	6.482	6.042	5.656	5.319	5.025	4.767	4.539	4.337	4.157	3.999	3.881	3.886	0.475
																	6.683	6.952	7.217	7.447	7.592	7.563	7.292	6.860	6.400	5.973	5.594	5.263	4.974	4.721	4.499	4.305	4.141	4.025	3.999	4.036	4.095	0.450
																6.476	6.730	6.984	7.212	7.369	7.371	7.147	6.756	6.320	5.908	5.539	5.216	4.934	4.688	4.478	4.304	4.184	4.141	4.157	4.201	4.261	4.331	0.425
															6.290	6.525	6.761	6.980	7.140	7.171	6.998	6.652	6.245	5.849	5.492	5.178	4.907	4.676	4.489	4.360	4.304	4.305	4.337	4.385	4.445	4.516	4.596	0.400
														6.134	6.347	6.564	6.769	6.927	6.978	6.852	6.552	6.176	5.799	5.456	5.156	4.902	4.698	4.558	4.489	4.478	4.499	4.539	4.591	4.652	4.723	4.804	4.896	0.375
													6.018	6.208	6.404	6.591	6.742	6.806	6.717	6.462	6.117	5.763	5.438	5.159	4.935	4.779	4.698	4.676	4.688	4.721	4.767	4.822	4.885	4.958	5.040	5.133	5.238	0.350
												5.956	6.122	6.294	6.460	6.599	6.668	6.607	6.391	6.078	5.748	5.449	5.204	5.031	4.935	4.902	4.907	4.934	4.974	5.025	5.083	5.150	5.224	5.308	5.402	5.509	5.631	0.325
											5.965	6.105	6.252	6.395	6.517	6.582	6.539	6.354	6.073	5.773	5.511	5.317	5.204	5.159	5.156	5.178	5.216	5.263	5.319	5.381	5.451	5.528	5.615	5.711	5.820	5.944	6.086	0.300
										6.062	6.179	6.301	6.419	6.521	6.576	6.539	6.376	6.123	5.859	5.646	5.511	5.449	5.438	5.456	5.492	5.539	5.594	5.656	5.723	5.797	5.878	5.967	6.065	6.175	6.299	6.438	6.575	0.275
									6.255	6.355	6.457	6.554	6.636	6.677	6.637	6.483	6.251	6.024	5.859	5.773	5.748	5.763	5.799	5.849	5.908	5.973	6.042	6.117	6.196	6.280	6.371	6.467	6.567	6.656	6.663	6.548	6.402	0.250
								6.532	6.627	6.720	6.805	6.873	6.899	6.846	6.683	6.455	6.251	6.123	6.073	6.078	6.117	6.176	6.245	6.320	6.400	6.482	6.566	6.651	6.737	6.819	6.884	6.892	6.801	6.653	6.493	6.338	6.192	0.225
							6.851	6.958	7.059	7.147	7.210	7.222	7.138	6.934	6.683	6.483	6.376	6.354	6.391	6.462	6.552	6.652	6.756	6.860	6.962	7.059	7.147	7.211	7.220	7.135	6.978	6.797	6.617	6.444	6.283	6.133	5.995	0.200
						7.121	7.261	7.390	7.498	7.567	7.560	7.415	7.138	6.846	6.637	6.539	6.539	6.607	6.717	6.852	6.998	7.147	7.292	7.424	7.528	7.576	7.526	7.377	7.180	6.973	6.771	6.579	6.401	6.235	6.082	5.940	5.810	0.175
					7.244	7.418	7.582	7.720	7.802	7.771	7.560	7.222	6.899	6.677	6.576	6.582	6.668	6.806	6.978	7.171	7.371	7.563	7.723	7.806	7.763	7.606	7.394	7.170	6.950	6.740	6.544	6.361	6.192	6.036	5.891	5.758	5.635	0.150
				7.218	7.409	7.592	7.748	7.841	7.802	7.567	7.210	6.873	6.636	6.521	6.517	6.599	6.742	6.927	7.140	7.369	7.592	7.773	7.847	7.773	7.594	7.373	7.144	6.920	6.708	6.510	6.325	6.153	5.995	5.848	5.712	5.587	5.471	0.125
			7.106	7.299	7.485	7.647	7.748	7.720	7.498	7.147	6.805	6.554	6.419	6.395	6.460	6.591	6.769	6.980	7.212	7.447	7.650	7.754	7.709	7.547	7.335	7.109	6.888	6.676	6.477	6.291	6.118	5.957	5.808	5.671	5.543	5.425	5.316	0.100
		6.953	7.142	7.325	7.485	7.592	7.582	7.390	7.059	6.720	6.457	6.301	6.252	6.294	6.404	6.564	6.761	6.984	7.217	7.433	7.581	7.594	7.474	7.283	7.069	6.853	6.643	6.445	6.259	6.085	5.923	5.773	5.633	5.504	5.384	5.273	5.170	0.075
	6.785	6.966	7.142	7.299	7.409	7.418	7.261	6.958	6.627	6.355	6.179	6.105	6.122	6.208	6.347	6.525	6.730	6.952	7.172	7.353	7.431	7.373	7.218	7.021	6.814	6.609	6.413	6.228	6.054	5.891	5.740	5.599	5.468	5.346	5.233	5.128	5.031	0.050
6.612	6.785	6.953	7.106	7.218	7.244	7.121	6.851	6.532	6.255	6.062	5.965	5.956	6.018	6.134	6.290	6.476	6.683	6.897	7.094	7.225	7.238	7.133	6.964	6.770	6.573	6.380	6.197	6.024	5.861	5.709	5.567	5.435	5.312	5.198	5.091	4.992	4.900	0.025

TABLE 4.3 (MODES 3 AND 4)

MODE 3 $\mu_2 \longrightarrow$ (μ_1 ↓)

μ_1	0.025	0.050	0.075	0.100	0.125	0.150	0.175	0.200	0.225	0.250	0.275	0.300	0.325	0.350	0.375	0.400	0.425	0.450	0.475	0.500	0.525	0.550	0.575	0.600	0.625	0.650	0.675	0.700	0.725	0.750	0.775	0.800	0.825	0.850	0.875	0.900	0.925	0.950
0.025	8.201	8.360	8.534	8.723	8.927	9.147	9.381	9.628	9.884	10.14	10.36	10.47	10.31	9.928	9.501	9.136	8.898	8.846	8.981	9.247	9.592	9.971	10.31	10.47	10.36	10.13	9.865	9.609	9.387	9.224	9.174	9.333	9.768	10.28	10.47	10.37	10.16	9.915
0.050	8.421	8.589	8.773	8.973	9.190	9.423	9.672	9.934	10.20	10.46	10.67	10.70	10.42	9.982	9.556	9.227	9.065	9.112	9.334	9.663	10.05	10.44	10.69	10.67	10.46	10.18	9.911	9.667	9.479	9.396	9.514	9.926	10.47	10.72	10.63	10.42	10.16	
0.075	8.653	8.831	9.025	9.238	9.468	9.715	9.978	10.25	10.53	10.78	10.92	10.79	10.40	9.948	9.560	9.307	9.261	9.424	9.729	10.12	10.53	10.85	10.91	10.72	10.45	10.17	9.917	9.710	9.599	9.679	10.06	10.63	10.92	10.85	10.63	10.37		
0.100	8.898	9.086	9.292	9.516	9.759	10.02	10.29	10.57	10.83	10.98	10.89	10.56	10.15	9.766	9.483	9.377	9.490	9.771	10.15	10.56	10.90	10.98	10.82	10.57	10.31	10.06	9.851	9.730	9.785	10.14	10.69	10.98	10.92	10.72	10.47			
0.125	9.157	9.355	9.571	9.804	10.05	10.31	10.54	10.69	10.64	10.43	10.17	9.878	9.607	9.399	9.330	9.454	9.733	10.10	10.46	10.68	10.66	10.51	10.32	10.12	9.921	9.755	9.659	9.719	10.04	10.51	10.69	10.63	10.47	10.28				
0.150	9.430	9.635	9.848	10.05	10.14	10.05	9.884	9.700	9.516	9.336	9.168	9.031	8.965	9.034	9.250	9.557	9.881	10.10	10.13	10.02	9.873	9.718	9.567	9.432	9.325	9.283	9.381	9.698	10.04	10.14	10.06	9.926	9.768					
0.175	9.711	9.769	9.548	9.316	9.111	8.929	8.767	8.624	8.502	8.417	8.405	8.516	8.748	9.052	9.381	9.666	9.790	9.717	9.551	9.367	9.194	9.040	8.916	8.835	8.839	9.009	9.381	9.719	9.785	9.679	9.514	9.333						
0.200	9.021	8.734	8.499	8.302	8.136	7.999	7.895	7.845	7.899	8.075	8.335	8.644	8.983	9.328	9.620	9.738	9.631	9.412	9.174	8.953	8.765	8.622	8.547	8.592	8.839	9.283	9.659	9.730	9.599	9.396	9.174							
0.225	8.055	7.827	7.642	7.495	7.398	7.393	7.520	7.737	8.002	8.303	8.633	8.988	9.352	9.678	9.853	9.772	9.526	9.241	8.972	8.741	8.569	8.484	8.547	8.835	9.325	9.755	9.851	9.710	9.479	9.224								
0.250	7.276	7.101	7.001	7.063	7.242	7.467	7.722	8.005	8.316	8.655	9.020	9.401	9.767	10.02	10.02	9.786	9.470	9.158	8.886	8.678	8.569	8.622	8.916	9.432	9.921	10.06	9.917	9.667	9.387									
0.275	6.686	6.836	7.031	7.247	7.484	7.745	8.032	8.345	8.688	9.060	9.456	9.855	10.19	10.30	10.12	9.792	9.446	9.134	8.886	8.741	8.765	9.040	9.567	10.12	10.31	10.17	9.911	9.609										
0.300	6.879	7.073	7.288	7.524	7.784	8.070	8.384	8.729	9.105	9.511	9.934	10.33	10.56	10.48	10.16	9.794	9.446	9.158	8.972	8.953	9.194	9.718	10.32	10.57	10.45	10.18	9.865											
0.325	7.138	7.346	7.578	7.834	8.117	8.430	8.775	9.153	9.566	10.00	10.44	10.77	10.79	10.53	10.16	9.792	9.470	9.241	9.174	9.367	9.873	10.51	10.82	10.72	10.46	10.13												
0.350	7.420	7.644	7.894	8.173	8.483	8.826	9.204	9.619	10.07	10.52	10.89	10.98	10.79	10.48	10.12	9.786	9.526	9.412	9.551	10.02	10.66	10.98	10.91	10.67	10.36													
0.375	7.725	7.968	8.240	8.544	8.883	9.259	9.673	10.12	10.56	10.86	10.89	10.77	10.56	10.30	10.02	9.772	9.631	9.717	10.13	10.68	10.90	10.85	10.69	10.47														
0.400	8.056	8.320	8.617	8.950	9.319	9.725	10.15	10.48	10.56	10.52	10.44	10.33	10.19	10.02	9.853	9.738	9.790	10.10	10.46	10.36	10.33	10.44	10.31															
0.425	8.417	8.704	9.028	9.386	9.767	10.08	10.15	10.12	10.07	10.00	9.934	9.855	9.767	9.678	9.620	9.666	9.881	10.10	10.15	10.12	10.05	9.971																
0.450	8.811	9.122	9.458	9.730	9.767	9.725	9.673	9.619	9.566	9.511	9.456	9.401	9.352	9.328	9.381	9.557	9.733	9.771	9.729	9.663	9.592																	
0.475	9.239	9.485	9.458	9.386	9.319	9.259	9.204	9.153	9.105	9.060	9.020	8.988	8.983	9.052	9.250	9.454	9.490	9.424	9.334	9.247																		
0.500	9.239	9.122	9.028	8.950	8.883	8.826	8.775	8.729	8.688	8.655	8.633	8.644	8.748	9.034	9.330	9.377	9.261	9.112	8.981																			

$\longleftarrow \mu_2$ MODE 4 (μ_1 ↑)

0.950	0.925	0.900	0.875	0.850	0.825	0.800	0.775	0.750	0.725	0.700	0.675	0.650	0.625	0.600	0.575	0.550	0.525	0.500	0.475	0.450	0.425	0.400	0.375	0.350	0.325	0.300	0.275	0.250	0.225	0.200	0.175	0.150	0.125	0.100	0.075	0.050	0.025	μ_1
																			13.59	13.98	14.14	13.69	12.71	12.32	12.50	12.96	13.49	13.93	14.14	13.87	13.15	12.40	11.72	11.12	10.60	10.14	9.737	0.500
																		13.20	13.68	13.93	13.44	12.43	11.93	12.01	12.40	12.95	13.53	13.91	13.71	13.04	12.31	11.64	11.04	10.51	10.06	9.725	9.737	0.475
																	12.75	13.23	13.50	13.12	12.16	11.60	11.59	11.89	12.36	12.92	13.40	13.44	12.91	12.22	11.57	10.98	10.47	10.11	10.06	10.14	10.26	0.450
																11.33	11.78	13.07	12.81	11.95	11.40	11.34	11.55	11.90	12.37	12.84	13.08	12.75	12.14	11.51	10.95	10.56	10.47	10.51	10.60	10.71	10.86	0.425
															12.09	12.44	12.70	12.54	11.83	11.41	11.38	11.50	11.72	12.03	12.41	12.70	12.57	12.07	11.50	11.08	10.95	10.98	11.04	11.12	11.23	11.36	11.52	0.400
														12.12	12.35	12.53	12.43	11.94	11.72	11.72	11.80	11.93	12.10	12.31	12.52	12.48	12.08	11.67	11.50	11.51	11.57	11.64	11.72	11.82	11.94	12.08	12.27	0.375
													12.51	12.64	12.74	12.66	12.35	12.22	12.25	12.33	12.42	12.53	12.65	12.74	12.66	12.33	12.08	12.07	12.14	12.22	12.31	12.40	12.50	12.60	12.71	12.71	12.41	0.350
												13.10	13.21	13.28	13.18	12.83	12.71	12.79	12.92	13.04	13.15	13.25	13.28	13.07	12.66	12.40	12.57	12.75	12.91	13.04	13.15	13.24	13.28	13.07	12.68	12.30	11.95	0.325
											13.60	13.80	13.89	13.66	13.04	12.87	13.04	13.30	13.55	13.76	13.88	13.79	13.28	12.74	12.52	12.70	13.08	13.44	13.71	13.87	13.84	13.51	13.06	12.61	12.21	11.84	11.51	0.300
										13.65	13.98	14.13	13.71	12.91	12.68	12.86	13.19	13.58	13.93	14.13	13.88	13.25	12.65	12.31	12.41	12.84	13.40	13.91	14.14	13.92	13.47	13.00	12.55	12.14	11.75	11.41	11.10	0.275
									13.39	13.76	13.95	13.53	12.69	12.39	12.53	12.85	13.25	13.65	13.93	13.76	13.15	12.53	12.10	12.03	12.37	12.92	13.53	13.93	13.81	13.40	12.94	12.50	12.07	11.69	11.33	11.01	10.72	0.250
								13.07	13.43	13.63	13.29	12.47	12.10	12.19	12.48	12.85	13.25	13.58	13.55	13.04	12.42	11.93	11.72	11.90	12.36	12.95	13.49	13.61	13.31	12.88	12.44	12.02	11.63	11.27	10.94	10.64	10.37	0.225
							12.76	13.10	13.31	13.05	12.27	11.85	11.89	12.14	12.40	12.85	13.19	13.30	12.92	12.33	11.80	11.50	11.55	11.89	12.40	12.96	13.30	13.18	12.81	12.39	11.98	11.59	11.22	10.89	10.58	10.30	10.05	0.200
						12.50	12.82	13.03	12.83	12.11	11.67	11.68	11.89	12.19	12.53	12.86	13.04	12.79	12.25	11.72	11.38	11.34	11.59	12.01	12.50	12.94	13.01	12.73	12.34	11.94	11.55	11.19	10.86	10.56	10.29	10.10	10.28	0.175
					12.39	12.68	12.87	12.71	12.06	11.67	11.67	11.85	12.10	12.39	12.68	12.87	12.71	12.22	11.72	11.41	11.40	11.60	11.93	12.32	12.71	12.87	12.68	12.32	11.93	11.56	11.22	10.93	10.77	10.84	11.09	11.44	11.89	0.150
				12.65	12.88	13.03	12.88	12.32	12.06	12.11	12.27	12.47	12.69	12.91	13.04	12.83	12.35	11.94	11.83	11.95	12.16	12.43	12.71	12.97	13.01	12.76	12.40	12.04	11.75	11.64	11.74	11.96	12.24	12.56	12.89	13.02	12.81	0.125
			13.38	13.61	13.74	13.48	12.88	12.71	12.83	13.05	13.29	13.53	13.71	13.66	13.18	12.66	12.43	12.54	12.81	13.12	13.44	13.69	13.70	13.41	13.01	12.64	12.41	12.46	12.78	13.18	13.55	13.74	13.60	13.31	13.01	12.72	12.46	0.100
		13.69	13.99	14.14	13.74	13.03	12.87	13.03	13.31	13.63	13.95	14.13	13.89	13.28	12.74	12.53	12.70	13.07	13.50	13.93	14.14	13.93	13.51	13.08	12.72	12.54	12.73	13.28	13.89	14.14	13.94	13.62	13.29	12.96	12.65	12.37	12.12	0.075
	13.52	13.83	13.99	13.61	12.88	12.68	12.82	13.10	13.43	13.76	13.98	13.80	13.21	12.64	12.35	12.44	12.78	13.23	13.68	13.98	13.85	13.46	13.03	12.65	12.39	12.43	12.89	13.56	13.98	13.87	13.57	13.24	12.91	12.59	12.30	12.03	11.79	0.050
13.22	13.52	13.69	13.38	12.65	12.39	12.50	12.76	13.07	13.39	13.65	13.60	13.10	12.51	12.12	12.09	12.35	12.75	13.20	13.59	13.66	13.36	12.95	12.55	12.23	12.08	[illegible]	12.85	13.48	13.68	13.48	13.17	12.85	12.54	12.24	11.96	11.71	11.48	0.025

TABLE 4.4 (MODES 1 AND 2)

MODE 1 μ_2 →

μ_1 ↓	0.025	0.050	0.075	0.100	0.125	0.150	0.175	0.200	0.225	0.250	0.275	0.300	0.325	0.350	0.375	0.400	0.425	0.450	0.475	0.500	0.525	0.550	0.575	0.600	0.625	0.650	0.675	0.700	0.725	0.750	0.775	0.800	0.825	0.850	0.875	0.900	0.925	0.950	μ_1 ↑
0.025	4.103	4.187	4.276	4.370	4.471	4.578	4.692	4.813	4.943	5.083	5.232	5.391	5.563	5.747	5.943	6.153	6.375	6.606	6.835	7.041	7.179	7.197	7.100	6.937	6.748	6.556	6.367	6.187	6.017	5.857	5.707	5.567	5.437	5.315	5.202	5.096	4.998	4.906	
0.050	4.212	4.299	4.392	4.492	4.598	4.712	4.833	4.963	5.102	5.251	5.410	5.582	5.767	5.965	6.177	6.402	6.639	6.880	7.108	7.285	7.352	7.286	7.132	6.941	6.740	6.542	6.352	6.173	6.004	5.846	5.699	5.562	5.434	5.315	5.205	5.102	5.007		
0.075	4.327	4.418	4.516	4.621	4.734	4.855	4.984	5.122	5.271	5.431	5.603	5.788	5.987	6.201	6.429	6.671	6.922	7.169	7.381	7.500	7.478	7.340	7.148	6.940	6.731	6.531	6.341	6.162	5.996	5.840	5.696	5.561	5.436	5.320	5.213	5.113		8.207	0.950
0.100	4.449	4.545	4.648	4.759	4.879	5.007	5.145	5.294	5.453	5.625	5.810	6.010	6.225	6.456	6.703	6.962	7.224	7.466	7.637	7.669	7.560	7.371	7.156	6.936	6.724	6.522	6.333	6.156	5.992	5.839	5.697	5.566	5.444	5.331	5.226		8.378	8.425	0.925
0.125	4.579	4.679	4.788	4.907	5.034	5.171	5.318	5.477	5.649	5.834	6.035	6.251	6.484	6.734	6.999	7.274	7.541	7.759	7.855	7.788	7.610	7.389	7.159	6.933	6.719	6.517	6.329	6.155	5.993	5.843	5.704	5.575	5.456	5.347		8.562	8.602	8.656	0.900
0.150	4.716	4.822	4.938	5.064	5.199	5.346	5.504	5.675	5.860	6.061	6.278	6.512	6.765	7.036	7.321	7.608	7.866	8.028	8.020	7.865	7.642	7.400	7.160	6.932	6.717	6.516	6.330	6.158	5.999	5.852	5.716	5.591	5.475		8.761	8.796	8.842	8.900	0.875
0.175	4.861	4.974	5.098	5.231	5.377	5.534	5.704	5.889	6.089	6.306	6.542	6.797	7.072	7.365	7.668	7.959	8.183	8.249	8.132	7.914	7.662	7.407	7.163	6.933	6.718	6.519	6.336	6.166	6.010	5.866	5.734	5.612		8.974	9.006	9.046	9.096	9.159	0.850
0.200	5.017	5.137	5.268	5.411	5.567	5.736	5.919	6.119	6.336	6.573	6.829	7.108	7.407	7.722	8.039	8.316	8.464	8.409	8.205	7.946	7.676	7.414	7.167	6.937	6.724	6.528	6.347	6.180	6.028	5.887	5.758		9.202	9.232	9.263	9.312	9.366	9.434	0.825
0.225	5.182	5.310	5.450	5.604	5.771	5.954	6.152	6.369	6.606	6.863	7.144	7.448	7.773	8.110	8.430	8.657	8.685	8.515	8.252	7.967	7.687	7.421	7.174	6.945	6.734	6.541	6.363	6.200	6.051	5.914		9.446	9.475	9.508	9.548	9.595	9.653	9.726	0.800
0.250	5.359	5.496	5.646	5.811	5.992	6.189	6.405	6.641	6.899	7.182	7.489	7.822	8.174	8.526	8.823	8.948	8.836	8.582	8.283	7.983	7.697	7.430	7.184	6.957	6.750	6.560	6.386	6.227	6.082		9.706	9.734	9.766	9.803	9.846	9.897	9.959	10.04	0.775
0.275	5.548	5.695	5.857	6.035	6.230	6.444	6.680	6.938	7.221	7.531	7.869	8.232	8.609	8.960	9.185	9.161	8.931	8.624	8.304	7.996	7.708	7.442	7.198	6.975	6.771	6.585	6.415	6.260		9.981	10.01	10.04	10.08	10.12	10.16	10.22	10.28	10.37	0.750
0.300	5.751	5.909	6.084	6.276	6.489	6.722	6.979	7.262	7.574	7.916	8.287	8.682	9.074	9.387	9.472	9.293	8.986	8.649	8.318	8.008	7.720	7.457	7.216	6.997	6.798	6.616	6.451		10.27	10.30	10.34	10.37	10.41	10.45	10.50	10.56	10.63	10.72	0.725
0.325	5.970	6.140	6.329	6.538	6.770	7.025	7.307	7.619	7.963	8.340	8.748	9.169	9.549	9.748	9.649	9.359	9.011	8.661	8.327	8.018	7.734	7.475	7.240	7.026	6.831	6.655		10.57	10.61	10.65	10.69	10.72	10.77	10.81	10.87	10.93	11.00	11.10	0.700
0.350	6.205	6.390	6.595	6.823	7.076	7.357	7.668	8.013	8.393	8.809	9.250	9.677	9.970	9.965	9.717	9.374	9.012	8.661	8.332	8.028	7.750	7.498	7.268	7.060	6.872		10.86	10.92	10.97	11.02	11.06	11.10	11.15	11.20	11.25	11.32	11.40	11.50	0.675
0.375	6.461	6.661	6.885	7.134	7.412	7.722	8.066	8.448	8.868	9.321	9.773	10.11	10.17	9.986	9.685	9.340	8.988	8.648	8.330	8.036	7.768	7.523	7.302	7.102		11.11	11.21	11.29	11.35	11.41	11.46	11.51	11.56	11.61	11.67	11.73	11.81	11.77	0.650
0.400	6.738	6.956	7.201	7.475	7.781	8.123	8.504	8.925	9.380	9.828	10.12	10.15	10.03	9.824	9.556	9.251	8.932	8.618	8.318	8.040	7.784	7.551	7.340		11.18	11.40	11.56	11.67	11.76	11.83	11.89	11.94	12.00	12.05	12.10	12.08	11.58	10.98	0.625
0.425	7.040	7.279	7.547	7.849	8.187	8.565	8.981	9.422	9.793	9.916	9.888	9.806	9.688	9.532	9.333	9.096	8.832	8.559	8.290	8.035	7.798	7.580		10.96	11.31	11.64	11.89	12.07	12.19	12.28	12.35	12.41	12.45	12.46	12.11	11.42	10.80	10.27	0.600
0.450	7.370	7.632	7.928	8.260	8.630	9.032	9.405	9.562	9.558	9.514	9.453	9.379	9.290	9.179	9.041	8.873	8.675	8.459	8.234	8.012	7.800		10.56	10.95	11.38	11.81	12.18	12.45	12.63	12.75	12.82	12.86	12.70	12.02	11.29	10.65	10.11	9.637	0.575
0.475	7.733	8.021	8.345	8.701	9.051	9.216	9.212	9.174	9.127	9.076	9.021	8.960	8.890	8.810	8.713	8.598	8.462	8.305	8.134	7.957		10.15	10.51	10.93	11.41	11.91	12.40	12.79	13.05	13.19	13.19	12.71	11.91	11.18	10.54	9.980	9.500	9.084	0.550
0.500	8.132	8.445	8.767	8.925	8.903	8.856	8.807	8.759	8.712	8.665	8.616	8.565	8.510	8.448	8.379	8.299	8.206	8.100	7.981		9.820	10.13	10.50	10.93	11.42	11.96	12.52	13.00	13.31	13.24	12.60	11.81	11.08	10.44	9.882	9.401	9.003	8.894	0.525
0.525	8.568	8.718	8.654	8.586	8.524	8.469	8.419	8.372	8.328	8.285	8.242	8.199	8.153	8.105	8.053	7.995	7.931	7.859		9.640	9.872	10.16	10.52	10.94	11.44	11.98	12.54	12.99	13.03	12.46	11.71	11.00	10.36	9.813	9.370	9.178	9.221	9.322	0.500
0.550	8.476	8.374	8.291	8.220	8.160	8.107	8.059	8.016	7.975	7.936	7.898	7.860	7.822	7.783	7.742	7.699	7.653		9.675	9.821	10.01	10.27	10.59	10.99	11.46	11.98	12.47	12.67	12.29	11.61	10.92	10.30	9.800	9.537	9.529	9.591	9.683	9.803	0.475
0.575	8.116	8.023	7.945	7.879	7.823	7.773	7.728	7.688	7.650	7.614	7.581	7.548	7.515	7.482	7.449	7.416		9.909	9.998	10.11	10.27	10.47	10.74	11.09	11.51	11.95	12.25	12.08	11.50	10.86	10.30	9.961	9.906	9.943	10.01	10.10	10.20	10.34	0.450
0.600	7.782	7.697	7.626	7.564	7.511	7.465	7.423	7.385	7.351	7.318	7.288	7.259	7.231	7.203	7.176		10.27	10.34	10.42	10.51	10.63	10.79	11.00	11.27	11.60	11.88	11.87	11.42	10.86	10.45	10.34	10.36	10.41	10.48	10.56	10.66	10.78	10.93	0.425
0.625	7.475	7.397	7.330	7.273	7.224	7.180	7.142	7.107	7.075	7.045	7.018	6.992	6.967	6.944		10.68	10.76	10.83	10.91	11.00	11.10	11.22	11.38	11.56	11.75	11.77	11.44	11.03	10.84	10.83	10.88	10.94	11.01	11.09	11.18	11.29	11.43	11.60	0.400
0.650	7.191	7.119	7.057	7.004	6.958	6.917	6.881	6.848	6.819	6.792	6.767	6.744	6.723		11.05	11.17	11.28	11.38	11.48	11.57	11.67	11.77	11.88	11.97	11.95	11.67	11.38	11.32	11.36	11.44	11.52	11.61	11.69	11.78	11.87	11.96	11.93	11.63	0.375
0.675	6.928	6.861	6.803	6.754	6.711	6.673	6.639	6.609	6.582	6.558	6.535	6.515		11.19	11.40	11.61	11.79	11.95	12.09	12.21	12.31	12.40	12.46	12.36	11.99	11.72	11.73	11.86	12.01	12.14	12.25	12.34	12.42	12.45	12.26	11.89	11.51	11.17	0.350
0.700	6.683	6.621	6.567	6.521	6.481	6.446	6.414	6.387	6.362	6.340	6.320		11.04	11.31	11.60	11.90	12.19	12.45	12.68	12.86	12.98	13.00	12.68	12.12	11.79	11.83	12.11	12.42	12.68	12.88	12.99	12.98	12.67	12.24	11.81	11.42	11.07	10.75	0.325
0.725	6.455	6.397	6.347	6.304	6.267	6.234	6.205	6.180	6.157	6.137		10.78	11.05	11.36	11.69	12.04	12.41	12.77	13.09	13.31	13.26	12.72	12.07	11.64	11.60	11.91	12.37	12.85	13.22	13.34	13.09	12.65	12.19	11.75	11.35	10.98	10.65	10.36	0.300
0.750	6.242	6.188	6.141	6.101	6.067	6.037	6.010	5.987	5.967		10.48	10.74	11.04	11.37	11.72	12.10	12.50	12.88	13.19	13.16	12.63	11.96	11.45	11.28	11.49	11.94	12.49	13.01	13.24	13.01	12.59	12.13	11.69	11.28	10.91	10.57	10.26	9.991	0.275
0.775	6.043	5.992	5.949	5.912	5.880	5.852	5.828	5.807		10.18	10.43	10.72	11.03	11.37	11.73	12.12	12.51	12.84	12.92	12.50	11.85	11.29	11.00	11.06	11.42	11.94	12.50	12.91	12.86	12.50	12.07	11.64	11.23	10.85	10.51	10.19	9.905	9.650	0.250
0.800	5.856	5.809	5.768	5.734	5.705	5.680	5.658		9.895	10.13	10.40	10.69	11.01	11.36	11.72	12.10	12.45	12.62	12.33	11.73	11.16	10.77	10.69	10.94	11.38	11.92	12.42	12.61	12.38	11.99	11.58	11.18	10.80	10.45	10.13	9.836	9.570	9.331	0.225
0.825	5.680	5.636	5.599	5.568	5.541	5.519		9.624	9.849	10.10	10.37	10.67	11.00	11.34	11.70	12.04	12.27	12.14	11.62	11.04	10.59	10.40	10.51	10.86	11.34	11.86	12.24	12.22	11.90	11.52	11.12	10.75	10.40	10.08	9.778	9.505	9.257	9.033	0.200
0.850	5.515	5.474	5.440	5.411	5.388		9.368	9.581	9.817	10.07	10.36	10.66	10.99	11.32	11.66	11.92	11.91	11.49	10.93	10.45	10.16	10.16	10.41	10.82	11.31	11.77	11.96	11.79	11.44	11.07	10.70	10.35	10.03	9.728	9.451	9.196	6.964	8.754	0.175
0.875	5.359	5.321	5.290	5.265		9.129	9.330	9.552	9.795	10.06	10.35	10.65	10.97	11.29	11.56	11.65	11.35	10.83	10.34	9.935	9.871	10.02	10.35	10.79	11.26	11.61	11.62	11.36	11.01	10.66	10.31	9.988	9.685	9.405	9.145	8.907	8.689	8.491	0.150
0.900	5.212	5.178	5.150		8.905	9.096	9.305	9.535	9.784	10.05	10.34	10.64	10.95	11.22	11.37	11.20	10.74	10.24	9.8[illegible]6	9.644	9.690	9.943	10.32	10.76	11.18	11.37	11.24	10.94	10.61	10.27	9.949	9.647	9.364	9.102	8.860	8.636	8.431	8.244	0.125
0.925	5.073	5.043		8.696	8.877	9.075	9.292	9.527	9.781	10.05	10.34	10.63	10.90	11.09	11.02	10.63	10.15	9.735	9.466	9.419	9.590	9.907	10.31	10.73	11.05	11.08	10.86	10.55	10.23	9.911	9.611	9.329	9.065	8.820	8.592	8.382	8.188	8.011	0.100
0.950	4.942		8.501	8.673	8.861	9.067	9.290	9.529	9.785	10.06	10.33	10.60	10.81	10.83	10.53	10.07	9.643	9.328	9.201	9.290	9.543	9.896	10.30	10.67	10.86	10.75	10.49	10.18	9.874	9.577	9.296	9.032	8.785	8.555	8.341	8.142	7.959	7.791	0.075
		8.318	8.483	8.662	8.858	9.069	9.297	9.540	9.796	10.06	10.32	10.54	10.62	10.41	9.993	9.563	9.218	9.026	9.038	9.227	9.531	9.900	10.28	10.57	10.60	10.41	10.13	9.836	9.544	9.265	9.002	8.754	8.523	8.307	8.105	7.918	7.744	7.584	0.050
	8.149	8.306	8.477	8.664	8.865	9.082	9.313	9.557	9.810	10.06	10.29	10.41	10.28	9.913	9.492	9.128	8.888	8.829	8.954	9.209	9.543	9.911	10.25	10.41	10.31	10.08	9.796	9.512	9.236	8.974	8.727	8.495	8.277	8.074	7.884	7.706	7.541	7.388	0.025
	0.950	0.925	0.900	0.875	0.850	0.825	0.800	0.775	0.750	0.725	0.700	0.675	0.650	0.625	0.600	0.575	0.550	0.525	0.500	0.475	0.450	0.425	0.400	0.375	0.350	0.325	0.300	0.275	0.250	0.225	0.200	0.175	0.150	0.125	0.100	0.075	0.050	0.025	

← μ_2 MODE 2

TABLE 4.4 (MODES 3 AND 4)

MODE 3 μ_2 ⟶

μ_1 ↓ (Mode 3)	0.025	0.050	0.075	0.100	0.125	0.150	0.175	0.200	0.225	0.250	0.275	0.300	0.325	0.350	0.375	0.400	0.425	0.450	0.475	0.500	0.525	0.550	0.575	0.600	0.625	0.650	0.675	0.700	0.725	0.750	0.775	0.800	0.825	0.850	0.875	0.900	0.925	0.950	μ_1 ↑ (Mode 4)
0.025	10.67	10.90	11.14	11.41	11.69	12.00	12.32	12.66	13.00	13.34	13.59	13.50	12.99	12.43	12.08	12.07	12.34	12.75	13.19	13.55	13.57	13.25	12.84	12.45	12.14	12.01	12.25	12.83	13.45	13.60	13.39	13.08	12.76	12.45	12.16	11.89	11.64	11.41	
0.050	10.96	11.19	11.45	11.73	12.03	12.35	12.70	13.05	13.42	13.74	13.90	13.58	12.98	12.47	12.28	12.45	12.82	13.28	13.71	13.90	13.66	13.25	12.83	12.47	12.27	12.40	12.95	13.62	13.90	13.72	13.41	13.07	12.75	12.44	12.15	11.89	11.65		
0.075	11.26	11.50	11.78	12.07	12.39	12.74	13.10	13.48	13.85	14.15	14.14	13.62	13.00	12.60	12.60	12.91	13.36	13.84	14.18	14.10	13.70	13.25	12.85	12.57	12.60	13.07	13.78	14.20	14.07	13.76	13.41	13.07	12.74	12.44	12.17	11.92		14.78	0.950
0.100	11.57	11.84	12.12	12.44	12.78	13.14	13.53	13.93	14.30	14.54	14.29	13.64	13.06	12.84	13.03	13.45	13.95	14.41	14.52	14.19	13.72	13.27	12.93	12.84	13.21	13.94	14.50	14.45	14.14	13.77	13.41	13.07	12.75	12.46	12.20		15.10	15.17	0.925
0.125	11.91	12.19	12.49	12.83	13.19	13.58	13.99	14.40	14.77	14.88	14.38	13.69	13.22	13.21	13.55	14.05	14.57	14.90	14.70	14.23	13.75	13.34	13.14	13.38	14.09	14.78	14.85	14.54	14.16	13.78	13.42	13.08	12.78	12.50		15.45	15.51	15.59	0.900
0.150	12.27	12.56	12.89	13.25	13.63	14.05	14.48	14.91	15.24	15.12	14.44	13.77	13.49	13.69	14.15	14.70	15.18	15.22	14.79	14.27	13.80	13.50	13.59	14.24	15.04	15.26	14.98	14.58	14.18	13.80	13.44	13.12	12.82		15.82	15.87	15.94	16.03	0.875
0.175	12.65	12.96	13.31	13.69	14.10	14.54	15.00	15.44	15.67	15.27	14.50	13.93	13.89	14.26	14.81	15.39	15.67	15.37	14.83	14.31	13.92	13.86	14.40	15.26	15.67	15.44	15.03	14.60	14.20	13.82	13.47	13.16		16.22	16.27	16.33	16.40	16.50	0.850
0.200	13.05	13.38	13.75	14.16	14.61	15.08	15.56	15.97	15.99	15.32	14.56	14.20	14.41	14.92	15.53	16.02	15.93	15.42	14.86	14.39	14.19	14.58	15.45	16.06	15.89	15.48	15.04	14.61	14.21	13.85	13.52		16.65	16.70	16.73	16.81	16.89	16.99	0.825
0.225	13.48	13.84	14.23	14.67	15.14	15.64	16.12	16.41	16.05	15.29	14.67	14.61	15.03	15.65	16.24	16.38	15.95	15.39	14.88	14.56	14.78	15.61	16.36	16.29	15.90	15.46	15.02	14.61	14.23	13.89		17.08	17.14	17.20	17.26	17.32	17.40	17.51	0.800
0.250	13.94	14.32	14.74	15.21	15.70	16.17	16.42	16.20	15.71	15.13	14.86	15.14	15.71	16.27	16.39	16.12	15.71	15.25	14.90	14.97	15.71	16.38	16.34	16.06	15.70	15.31	14.93	14.56	14.22		17.46	17.58	17.66	17.73	17.80	17.87	17.95	18.06	0.775
0.275	14.43	14.83	15.26	15.65	15.73	15.61	15.45	15.24	14.99	14.85	15.08	15.53	15.74	15.67	15.51	15.32	15.10	14.90	14.95	15.49	15.74	15.67	15.52	15.35	15.15	14.91	14.66	14.39		17.54	17.92	18.11	18.22	18.30	18.37	18.43	18.33	16.92	0.750
0.300	14.94	15.05	14.89	14.73	14.59	14.47	14.33	14.27	14.36	14.72	15.05	15.02	14.84	14.67	14.52	14.40	14.32	14.49	14.97	15.06	14.91	14.75	14.60	14.47	14.34	14.21	14.07		16.83	17.68	18.34	18.65	18.80	18.89	18.93	18.14	16.56	15.31	0.725
0.325	14.23	14.00	13.82	13.68	13.57	13.51	13.63	14.05	14.58	14.95	14.81	14.42	14.09	13.86	13.72	13.73	14.17	14.83	14.93	14.64	14.32	14.05	13.84	13.68	13.55	13.44		16.11	16.78	17.73	18.67	19.18	19.39	19.35	17.90	16.32	15.50	15.56	0.700
0.350	13.24	13.04	12.90	12.83	13.01	13.48	14.04	14.64	15.15	15.20	14.67	14.08	13.64	13.40	13.52	14.18	15.01	13.23	14.90	14.45	14.01	13.64	13.35	13.11	12.94		16.07	16.33	16.86	17.75	18.82	19.53	19.29	17.67	16.22	15.94	16.01	16.12	0.675
0.375	12.38	12.37	12.52	12.98	13.51	14.10	14.73	15.35	15.71	15.33	14.60	13.96	13.56	13.64	14.31	15.23	15.71	15.41	14.89	14.37	13.89	13.47	13.11	12.82		16.44	16.55	16.73	17.09	17.81	18.72	18.82	17.48	16.49	16.45	16.51	16.61	16.73	0.650
0.400	12.15	12.57	13.05	13.58	14.17	14.82	15.50	16.08	16.06	15.37	14.60	14.02	13.91	14.48	15.44	16.15	15.97	15.45	14.88	14.34	13.85	13.42	13.05		16.92	17.01	17.11	17.24	17.48	17.92	18.23	17.42	16.99	17.01	17.07	17.15	17.25	17.39	0.625
0.425	12.69	13.14	13.67	14.26	14.91	15.62	16.27	16.47	16.05	15.34	14.64	14.31	14.68	15.61	16.42	16.38	15.95	15.41	14.86	14.33	13.86	13.44		17.40	17.53	17.63	17.73	17.85	18.00	18.11	17.73	17.54	17.60	17.67	17.75	17.84	17.95	18.08	0.600
0.450	13.28	13.78	14.36	15.00	15.70	16.22	16.23	16.05	15.71	15.20	14.77	14.91	15.71	16.25	16.21	16.01	15.70	15.28	14.80	14.33	13.89		17.57	17.93	18.15	18.30	18.41	18.51	18.54	18.18	18.01	18.16	18.30	18.40	18.48	18.55	18.21	17.36	0.575
0.475	13.93	14.48	15.10	15.64	15.70	15.62	15.52	15.38	15.19	14.97	15.00	15.51	15.71	15.65	15.55	15.42	15.24	14.99	14.66	14.29		17.10	17.72	18.31	18.74	18.99	19.13	19.11	18.31	17.97	18.44	18.84	19.04	19.15	18.98	18.06	17.15	16.36	0.550
0.500	14.65	15.14	15.15	15.07	14.98	14.90	14.82	14.73	14.65	14.72	15.08	15.18	15.09	14.99	14.90	14.80	14.68	14.54	14.36		16.47	17.07	17.77	18.51	19.17	19.57	19.40	18.08	17.36	17.82	18.74	19.42	19.60	18.93	17.92	16.99	16.18	15.48	0.525
0.525	14.74	14.59	14.48	14.39	14.31	14.24	14.18	14.15	14.31	14.80	14.89	14.69	14.51	14.39	14.29	14.20	14.11	14.02		16.03	16.49	17.08	17.79	18.56	19.24	19.14	17.82	16.86	16.97	17.78	18.82	19.37	18.76	17.79	16.87	16.06	15.47	15.50	0.500
0.550	14.09	13.95	13.85	13.76	13.69	13.64	13.66	14.07	14.88	15.12	14.72	14.30	14.02	13.85	13.73	13.65	13.57		16.12	16.33	16.68	17.18	17.82	18.49	18.67	17.60	16.62	16.58	17.02	17.81	18.64	18.51	17.67	16.79	16.12	16.02	16.13	16.29	0.475
0.575	13.49	13.36	13.26	13.19	13.15	13.29	14.09	15.13	15.71	15.31	14.62	14.03	13.64	13.40	13.25	13.15		16.65	16.76	16.90	17.12	17.45	17.91	18.19	17.47	16.85	16.87	17.05	17.39	17.92	18.17	17.38	16.88	16.69	16.75	16.85	16.99	17.18	0.450
0.600	12.93	12.82	12.74	12.73	13.21	14.20	15.36	16.24	16.08	15.37	14.57	13.89	13.39	13.06	12.86		17.28	17.42	17.54	17.67	17.81	18.00	18.12	17.71	17.46	17.54	17.66	17.81	18.01	18.12	17.74	17.43	17.47	17.57	17.67	17.80	17.94	18.12	0.425
0.625	12.42	12.34	12.49	13.29	14.34	15.55	16.43	16.39	16.01	15.33	14.55	13.83	13.27	12.86		17.59	17.92	18.19	18.39	18.55	18.67	18.70	18.22	18.03	18.23	18.43	18.58	18.70	18.64	18.17	18.03	18.24	18.43	18.57	18.69	18.67	18.14	17.54	0.400
0.650	11.99	12.57	13.44	14.50	15.69	16.15	16.10	15.96	15.70	15.21	14.53	13.84	13.26		17.27	17.76	18.28	18.77	19.16	19.41	19.33	18.29	17.95	18.39	18.93	19.31	19.45	18.99	18.20	17.97	18.52	19.11	19.41	19.34	18.72	18.03	17.38	16.82	0.375
0.675	12.78	13.64	14.68	15.64	15.70	15.65	15.57	15.47	15.29	14.98	14.48	13.89		16.78	17.26	17.81	18.40	19.00	19.49	19.32	18.06	17.46	17.84	18.54	19.26	19.55	18.91	18.03	17.50	17.83	18.79	19.53	19.31	18.64	17.93	17.27	16.68	16.16	0.350
0.700	13.91	14.90	15.29	15.24	15.18	15.12	15.06	14.97	14.85	14.67	14.36		16.31	16.75	17.26	17.83	18.42	18.97	18.98	17.83	17.01	17.18	17.81	18.56	19.10	18.73	17.88	17.17	17.06	17.79	18.80	19.07	18.52	17.84	17.18	16.58	16.03	15.55	0.325
0.725	14.97	14.89	14.81	14.74	14.68	14.62	14.57	14.50	14.42	14.30		16.02	16.38	16.82	17.31	17.85	18.38	18.56	17.63	16.69	16.69	17.17	17.83	18.46	18.47	17.73	16.98	16.61	16.97	17.02	18.36	18.36	17.75	17.11	16.51	15.96	15.50	15.42	0.300
0.750	14.52	14.41	14.33	14.26	14.21	14.16	14.11	14.05	13.99		16.26	16.47	16.73	17.08	17.49	17.93	18.17	17.48	16.66	16.65	16.94	17.39	17.92	18.16	17.63	16.92	16.61	16.80	17.28	17.94	18.14	17.67	17.08	16.52	16.21	16.27	16.47	16.77	0.275
0.775	14.05	13.96	13.88	13.82	13.77	13.72	13.68	13.63		17.09	17.25	17.42	17.60	17.81	18.04	18.18	17.69	17.32	17.40	17.58	17.80	18.05	18.17	17.75	17.34	17.37	17.55	17.79	18.07	18.16	17.76	17.33	17.25	17.38	17.56	17.76	18.00	18.17	0.250
0.800	13.62	13.53	13.46	13.40	13.35	13.31	13.27		17.59	17.93	18.26	18.56	18.82	19.01	19.01	18.27	18.05	18.32	18.63	18.89	19.06	18.84	18.21	18.06	18.38	18.71	18.97	19.05	18.67	18.16	18.06	18.41	18.77	19.01	19.01	18.59	18.09	17.63	0.225
0.825	13.21	13.12	13.06	13.00	12.96	12.92		17.36	17.78	18.23	18.71	19.19	19.57	19.39	18.26	17.94	18.33	18.88	19.41	19.61	18.98	18.18	17.94	18.42	19.12	19.62	19.37	18.73	18.14	17.95	18.60	19.45	19.57	19.10	18.54	18.01	17.51	17.07	0.200
0.850	12.83	12.75	12.68	12.64	12.60		16.95	17.36	17.81	18.30	18.80	19.25	19.17	18.04	17.52	17.83	18.39	18.99	19.35	18.85	18.02	17.53	17.81	18.52	19.22	19.22	18.63	18.01	17.57	17.79	18.78	19.35	19.01	18.47	17.93	17.43	16.96	16.54	0.175
0.875	12.46	12.39	12.33	12.29		16.52	16.91	17.34	17.81	18.30	18.75	18.82	17.82	17.11	17.29	17.79	18.38	18.85	18.65	17.86	17.21	17.17	17.74	18.49	18.89	18.50	17.89	17.34	17.11	17.70	18.69	18.82	18.38	17.86	17.35	16.88	16.44	16.05	0.150
0.900	12.12	12.05	12.00		16.10	16.47	16.88	17.32	17.79	18.23	18.41	17.61	16.75	16.77	17.20	17.75	18.28	18.37	17.72	16.99	16.67	17.01	17.70	18.33	18.31	17.77	17.20	16.76	16.80	17.63	18.40	18.25	17.78	17.29	16.81	16.36	15.95	15.58	0.125
0.925	11.80	11.74		15.71	16.05	16.44	16.86	17.30	17.73	17.99	17.40	16.45	16.30	16.66	17.17	17.69	17.99	17.56	16.83	16.32	16.39	16.95	17.63	17.99	17.64	17.09	16.58	16.28	16.63	17.54	17.98	17.68	17.22	16.75	16.30	15.88	15.49	15.14	0.100
0.950	11.49		15.34	15.67	16.03	16.43	16.85	17.26	17.56	17.18	16.22	15.89	16.16	16.62	17.13	17.53	17.37	16.70	16.09	15.90	16.28	16.91	17.48	17.46	16.99	16.46	16.03	15.94	16.55	17.40	17.51	17.14	16.69	16.24	15.82	15.42	15.05	14.72	0.075
		14.99	15.31	15.66	16.03	16.43	16.83	17.14	16.94	16.02	15.54	15.72	16.13	16.61	17.05	17.12	16.58	15.93	15.55	15.72	16.25	16.87	17.18	16.87	16.36	15.89	15.57	15.75	16.53	17.15	17.02	16.62	16.19	15.77	15.37	14.99	14.64	14.33	0.050
	14.68	14.98	15.31	15.67	16.05	16.43	16.76	16.69	15.86	15.25	15.32	15.68	16.13	16.59	16.82	16.45	15.81	15.31	15.26	15.68	16.26	16.76	16.71	16.27	15.79	15.38	15.25	15.70	16.51	16.81	16.54	16.13	15.72	15.32	14.94	14.59	14.26	13.96	0.025
	0.950	0.925	0.900	0.875	0.850	0.825	0.800	0.775	0.750	0.725	0.700	0.675	0.650	0.625	0.600	0.575	0.550	0.525	0.500	0.475	0.450	0.425	0.400	0.375	0.350	0.325	0.300	0.275	0.250	0.225	0.200	0.175	0.150	0.125	0.100	0.075	0.050	0.025	

⟵ μ_2 MODE 4

TABLE 4.5 (MODES 1 AND 2)

MODE 1 μ_2 →, μ_1 ↓ (left labels and top headers); MODE 2 ← μ_2, μ_1 ↑ (right labels and bottom headers)

μ_1 (Mode 1)	0.025	0.050	0.075	0.100	0.125	0.150	0.175	0.200	0.225	0.250	0.275	0.300	0.325	0.350	0.375	0.400	0.425	0.450	0.475	0.500	0.525	0.550	0.575	0.600	0.625	0.650	0.675	0.700	0.725	0.750	0.775	0.800	0.825	0.850	0.875	0.900	0.925	0.950	μ_1 (Mode 2)
0.025	1.959	1.998	2.039	2.083	2.128	2.176	2.227	2.280	2.337	2.397	2.460	2.528	2.600	2.677	2.760	2.848	2.944	3.047	3.158	3.278	3.409	3.551	3.705	3.871	4.049	4.234	4.418	4.585	4.712	4.777	4.778	4.728	4.647	4.547	4.439	4.327	4.215	4.107	
0.050	2.011	2.052	2.095	2.140	2.188	2.239	2.293	2.349	2.409	2.473	2.541	2.614	2.691	2.774	2.863	2.959	3.063	3.175	3.296	3.428	3.572	3.728	3.897	4.078	4.269	4.462	4.642	4.785	4.866	4.877	4.833	4.751	4.649	4.537	4.420	4.304	4.191		
0.075	2.066	2.108	2.154	2.201	2.252	2.306	2.363	2.423	2.487	2.555	2.628	2.706	2.789	2.879	2.975	3.080	3.192	3.315	3.448	3.593	3.751	3.923	4.108	4.305	4.506	4.699	4.859	4.958	4.982	4.943	4.862	4.758	4.641	4.520	4.398	4.280		8.199	0.950
0.100	2.124	2.169	2.216	2.267	2.320	2.377	2.438	2.502	2.570	2.643	2.722	2.805	2.895	2.992	3.097	3.211	3.334	3.469	3.615	3.775	3.949	4.138	4.341	4.551	4.756	4.934	5.052	5.091	5.060	4.980	4.873	4.752	4.625	4.499	4.375		8.358	8.419	0.925
0.125	2.186	2.233	2.283	2.336	2.393	2.454	2.518	2.587	2.660	2.738	2.823	2.913	3.011	3.116	3.230	3.355	3.490	3.638	3.800	3.977	4.169	4.377	4.595	4.813	5.008	5.148	5.205	5.183	5.106	4.996	4.870	4.738	4.605	4.476		8.532	8.587	8.651	0.900
0.150	2.251	2.301	2.354	2.410	2.471	2.535	2.604	2.678	2.756	2.841	2.932	3.030	3.136	3.251	3.376	3.513	3.662	3.826	4.005	4.201	4.414	4.640	4.870	5.083	5.245	5.323	5.313	5.239	5.127	4.997	4.858	4.719	4.583		8.721	8.771	8.829	8.896	0.875
0.175	2.321	2.373	2.429	2.490	2.554	2.623	2.696	2.775	2.860	2.952	3.050	3.157	3.272	3.399	3.536	3.687	3.852	4.034	4.233	4.451	4.685	4.926	5.158	5.344	5.446	5.451	5.382	5.268	5.132	4.987	4.840	4.697		8.925	8.972	9.024	9.084	9.155	0.850
0.200	2.395	2.450	2.510	2.574	2.643	2.717	2.796	2.881	2.973	3.072	3.179	3.295	3.422	3.561	3.713	3.880	4.064	4.267	4.489	4.730	4.983	5.232	5.443	5.572	5.596	5.533	5.418	5.277	5.124	4.969	4.818		9.145	9.188	9.236	9.290	9.353	9.428	0.825
0.225	2.474	2.533	2.597	2.665	2.739	2.818	2.904	2.995	3.095	3.203	3.320	3.447	3.587	3.740	3.909	4.095	4.301	4.528	4.775	5.039	5.305	5.542	5.701	5.747	5.695	5.579	5.432	5.271	5.108	4.948		9.379	9.422	9.466	9.514	9.569	9.633	9.709	0.800
0.250	2.558	2.621	2.690	2.763	2.842	2.928	3.020	3.120	3.228	3.345	3.474	3.615	3.769	3.939	4.128	4.336	4.567	4.821	5.095	5.378	5.640	5.832	5.906	5.866	5.752	5.599	5.430	5.257	5.088		9.627	9.670	9.713	9.757	9.802	9.846	9.768	9.021	0.775
0.275	2.649	2.716	2.789	2.868	2.954	3.046	3.146	3.255	3.373	3.502	3.644	3.799	3.971	4.162	4.373	4.608	4.868	5.151	5.450	5.738	5.964	6.071	6.049	5.937	5.778	5.600	5.416	5.237		9.883	9.932	9.976	10.02	10.05	10.04	9.548	8.734	8.055	0.750
0.300	2.745	2.818	2.897	2.982	3.074	3.175	3.284	3.403	3.532	3.675	3.832	4.005	4.197	4.411	4.650	4.915	5.208	5.521	5.834	6.097	6.242	6.242	6.135	5.971	5.783	5.588	5.396		10.14	10.20	10.25	10.29	10.31	10.14	9.317	8.499	7.828	7.276	0.725
0.325	2.850	2.928	3.013	3.105	3.206	3.315	3.434	3.565	3.708	3.866	4.041	4.235	4.451	4.694	4.964	5.265	5.592	5.928	6.228	6.416	6.445	6.347	6.178	5.979	5.772	5.568		10.36	10.46	10.53	10.57	10.54	10.05	9.112	8.303	7.643	7.103	6.718	0.700
0.350	2.962	3.047	3.139	3.239	3.349	3.468	3.599	3.743	3.902	4.079	4.274	4.493	4.739	5.015	5.323	5.663	6.021	6.357	6.593	6.658	6.572	6.400	6.191	5.970	5.752		10.47	10.67	10.78	10.83	10.69	9.884	8.930	8.138	7.498	7.016	6.885	6.931	0.675
0.375	3.084	3.176	3.276	3.385	3.505	3.637	3.782	3.942	4.119	4.317	4.538	4.786	5.067	5.382	5.733	6.112	6.483	6.769	6.878	6.811	6.637	6.417	6.182	5.949		10.31	10.70	10.92	10.98	10.64	9.701	8.769	8.002	7.408	7.102	7.082	7.127	7.193	0.650
0.400	3.216	3.316	3.425	3.546	3.677	3.823	3.984	4.162	4.361	4.585	4.836	5.121	5.443	5.805	6.201	6.606	6.942	7.102	7.060	6.888	6.657	6.408	6.159		9.932	10.42	10.79	10.89	10.43	9.517	8.626	7.902	7.421	7.291	7.300	7.342	7.402	7.477	0.625
0.425	3.360	3.469	3.589	3.722	3.868	4.029	4.209	4.410	4.635	4.889	5.178	5.506	5.877	6.290	6.724	7.109	7.323	7.312	7.147	6.909	6.646	6.382		9.508	9.987	10.40	10.56	10.17	9.339	8.508	7.863	7.563	7.519	7.538	7.580	7.634	7.702	7.784	0.600
0.450	3.518	3.638	3.770	3.916	4.079	4.259	4.461	4.689	4.946	5.238	5.571	5.951	6.378	6.838	7.266	7.531	7.554	7.402	7.164	6.892	6.615		9.150	9.566	9.955	10.15	9.883	9.174	8.429	7.932	7.786	7.776	7.800	7.840	7.892	7.953	8.028	8.118	0.575
0.475	3.691	3.823	3.970	4.133	4.314	4.518	4.747	5.006	5.302	5.640	6.027	6.467	6.947	7.407	7.704	7.752	7.622	7.399	7.130	6.848		8.924	9.246	9.573	9.777	9.616	9.042	8.429	8.120	8.054	8.059	8.087	8.127	8.176	8.233	8.301	8.383	8.481	0.550
0.500	3.882	4.029	4.192	4.374	4.579	4.810	5.071	5.370	5.712	6.106	6.555	7.049	7.520	7.804	7.847	7.748	7.566	7.329	7.061		8.886	9.099	9.334	9.504	9.418	8.989	8.555	8.384	8.356	8.371	8.402	8.442	8.489	8.544	8.607	8.681	8.770	8.878	0.525
0.525	4.094	4.257	4.440	4.646	4.878	5.142	5.443	5.790	6.188	6.644	7.141	7.580	7.780	7.793	7.725	7.601	7.428	7.211		9.031	9.160	9.304	9.415	9.365	9.072	8.794	8.695	8.686	8.709	8.745	8.787	8.835	8.887	8.946	9.014	9.094	9.190	9.308	0.500
0.550	4.330	4.514	4.720	4.954	5.220	5.524	5.873	6.276	6.733	7.212	7.547	7.638	7.627	7.577	7.499	7.390	7.245		9.302	9.388	9.477	9.541	9.502	9.296	9.098	9.032	9.039	9.073	9.116	9.164	9.213	9.266	9.323	9.384	9.452	9.522	9.515	9.240	0.475
0.575	4.596	4.803	5.038	5.305	5.612	5.964	6.368	6.819	7.234	7.412	7.434	7.412	7.372	7.318	7.247	7.156		9.646	9.720	9.786	9.827	9.786	9.603	9.421	9.370	9.397	9.450	9.509	9.568	9.625	9.681	9.736	9.788	9.827	9.760	9.462	9.123	8.811	0.450
0.600	4.896	5.132	5.401	5.709	6.063	6.466	6.891	7.166	7.222	7.213	7.186	7.151	7.108	7.057	6.995		10.02	10.10	10.17	10.20	10.15	9.916	9.690	9.644	9.709	9.805	9.900	9.985	10.06	10.12	10.18	10.20	10.10	9.772	9.387	9.027	8.704	8.417	0.425
0.625	5.238	5.509	5.819	6.174	6.569	6.915	7.020	7.018	6.995	6.964	6.931	6.894	6.853	6.806		10.34	10.48	10.57	10.61	10.50	10.12	9.789	9.734	9.855	10.04	10.22	10.37	10.48	10.57	10.61	10.51	10.15	9.725	9.318	8.948	8.616	8.319	8.056	0.400
0.650	5.631	5.944	6.299	6.667	6.841	6.843	6.817	6.786	6.753	6.720	6.686	6.650	6.611		10.47	10.71	10.87	10.92	10.69	10.12	9.698	9.605	9.746	10.00	10.30	10.57	10.78	10.91	10.88	10.56	10.12	9.670	9.256	8.881	8.542	8.238	7.967	7.725	0.375
0.675	6.087	6.441	6.698	6.699	6.664	6.626	6.590	6.555	6.521	6.488	6.454	6.420		10.34	10.65	10.90	10.97	10.65	10.00	9.522	9.371	9.480	9.741	10.08	10.45	10.78	10.97	10.88	10.51	10.06	9.615	9.200	8.821	8.479	8.170	7.892	7.642	7.419	0.350
0.700	6.593	6.595	6.541	6.492	6.447	6.407	6.370	6.335	6.302	6.269	6.237		10.09	10.42	10.69	10.79	10.49	9.847	9.328	9.117	9.175	9.402	9.728	10.11	10.48	10.76	10.75	10.43	9.996	9.559	9.147	8.768	8.423	8.111	7.828	7.573	7.343	7.137	0.325
0.725	6.451	6.386	6.330	6.281	6.237	6.198	6.162	6.128	6.095	6.064		9.805	10.13	10.40	10.53	10.28	9.680	9.140	8.875	8.876	9.057	9.349	9.705	10.09	10.41	10.52	10.31	9.920	9.500	9.095	8.719	8.374	8.059	7.773	7.514	7.279	7.067	6.876	0.300
0.750	6.240	6.180	6.127	6.081	6.039	6.001	5.966	5.933	5.902		9.518	9.828	10.10	10.24	10.06	9.512	8.965	8.654	8.690	8.733	8.984	9.307	9.668	10.02	10.23	10.15	9.829	9.437	9.043	8.672	8.328	8.013	7.725	7.463	7.224	7.007	6.810	6.633	0.275
0.775	6.041	5.985	5.936	5.891	5.851	5.815	5.781	5.750		9.242	9.536	9.798	9.951	9.828	9.345	8.802	8.454	8.350	8.436	8.646	8.933	9.267	9.612	9.887	9.940	9.718	9.366	8.990	8.626	8.285	7.971	7.682	7.418	7.176	6.956	6.755	6.572	6.407	0.250
0.800	5.854	5.802	5.755	5.713	5.675	5.640	5.608		8.979	9.258	9.510	9.671	9.598	9.179	8.651	8.276	8.126	8.168	8.337	8.589	8.893	9.221	9.522	9.680	9.576	9.284	8.933	8.579	8.244	7.931	7.643	7.378	7.135	6.912	6.707	6.520	6.350	6.195	0.225
0.825	5.678	5.629	5.585	5.545	5.509	5.475		8.732	8.996	9.238	9.403	9.370	9.016	8.510	8.115	7.926	7.926	8.058	8.276	8.549	8.855	9.158	9.383	9.397	9.185	8.869	8.531	8.203	7.894	7.607	7.341	7.097	6.872	6.666	6.476	6.302	6.142	5.998	0.200
0.850	5.513	5.466	5.424	5.386	5.352		8.500	8.751	8.982	9.148	9.148	8.854	8.377	7.971	7.748	7.709	7.805	7.991	8.236	8.517	8.811	9.066	9.179	9.064	8.797	8.480	8.162	7.858	7.573	7.308	7.063	6.838	6.629	6.437	6.260	6.098	5.948	5.812	0.175
0.875	5.357	5.313	5.273	5.237		8.283	8.521	8.742	8.909	8.934	8.694	8.251	7.840	7.589	7.514	7.578	7.734	7.951	8.208	8.486	8.751	8.928	8.913	8.711	8.424	8.119	7.822	7.540	7.277	7.032	6.806	6.597	6.403	6.224	6.059	5.907	5.766	5.638	0.150
0.900	5.209	5.168	5.130		8.080	8.306	8.519	8.684	8.728	8.536	8.130	7.720	7.447	7.340	7.372	7.500	7.692	7.926	8.186	8.448	8.661	8.733	8.608	8.361	8.074	7.785	7.508	7.247	7.004	6.778	6.568	6.373	6.193	6.026	5.871	5.727	5.595	5.474	0.125
0.925	5.070	5.030		7.890	8.106	8.310	8.473	8.532	8.382	8.013	7.609	7.319	7.184	7.187	7.289	7.457	7.670	7.911	8.163	8.392	8.526	8.484	8.288	8.026	7.748	7.477	7.219	6.977	6.751	6.541	6.346	6.165	5.996	5.840	5.694	5.559	5.434	5.319	0.100
0.950	4.937		7.713	7.919	8.115	8.275	8.346	8.232	7.900	7.507	7.203	7.044	7.020	7.098	7.243	7.435	7.659	7.898	8.130	8.304	8.337	8.203	7.972	7.709	7.445	7.191	6.951	6.726	6.517	6.321	6.140	5.970	5.813	5.666	5.529	5.401	5.283	5.173	0.075
		7.548	7.745	7.934	8.091	8.170	8.086	7.791	7.411	7.099	6.918	6.870	6.924	7.049	7.222	7.428	7.654	7.882	8.077	8.170	8.102	7.911	7.667	7.413	7.164	6.926	6.703	6.494	6.299	6.117	5.947	5.789	5.641	5.502	5.373	5.253	5.140	5.035	0.050
	7.393	7.583	7.765	7.918	8.004	7.946	7.685	7.321	7.004	6.805	6.734	6.766	6.872	7.027	7.217	7.429	7.650	7.854	7.989	7.985	7.841	7.622	7.379	7.136	6.902	6.681	6.473	6.278	6.096	5.926	5.767	5.619	5.480	5.349	5.227	5.113	5.006	4.906	0.025
μ_2 (Mode 2)	0.950	0.925	0.900	0.875	0.850	0.825	0.800	0.775	0.750	0.725	0.700	0.675	0.650	0.625	0.600	0.575	0.550	0.525	0.500	0.475	0.450	0.425	0.400	0.375	0.350	0.325	0.300	0.275	0.250	0.225	0.200	0.175	0.150	0.125	0.100	0.075	0.050	0.025	

← μ_2 MODE 2

TABLE 4.5 (MODES 3 AND 4)

MODE 3 $\mu_2 \longrightarrow$

μ_1 ↓ (Mode 3)	0.025	0.050	0.075	0.100	0.125	0.150	0.175	0.200	0.225	0.250	0.275	0.300	0.325	0.350	0.375	0.400	0.425	0.430	0.475	0.300	0.525	0.550	0.575	0.600	0.625	0.650	0.675	0.700	0.725	0.750	0.775	0.800	0.825	0.850	0.875	0.900	0.925	0.950	μ_1 ↑ (Mode 4)
0.025	8.210	8.381	8.566	8.765	8.978	9.206	9.450	9.710	9.986	10.28	10.57	10.87	11.12	11.21	10.99	10.59	10.18	9.834	9.632	9.642	9.862	10.22	10.64	11.03	11.21	11.09	10.84	10.56	10.30	10.08	9.937	9.957	10.28	10.85	11.20	11.15	10.94	10.68	
0.050	8.427	8.606	8.801	9.011	9.237	9.480	9.740	10.02	10.31	10.62	10.93	11.22	11.43	11.36	11.01	10.57	10.17	9.893	9.808	9.961	10.29	10.71	11.15	11.42	11.38	11.14	10.85	10.57	10.33	10.16	10.14	10.43	11.02	11.42	11.39	11.18	10.91		
0.075	8.658	8.846	9.051	9.274	9.514	9.772	10.05	10.34	10.66	10.98	11.30	11.58	11.69	11.46	11.02	10.57	10.21	10.03	10.09	10.37	10.79	11.25	11.62	11.67	11.45	11.16	10.86	10.59	10.40	10.35	10.60	11.19	11.65	11.65	11.43	11.15		14.77	0.950
0.100	8.903	9.100	9.317	9.554	9.809	10.08	10.38	10.69	11.03	11.37	11.70	11.93	11.89	11.51	11.03	10.60	10.31	10.26	10.48	10.87	11.35	11.79	11.96	11.79	11.49	11.18	10.89	10.67	10.58	10.78	11.37	11.90	11.93	11.71	11.41		15.07	15.16	0.925
0.125	9.162	9.370	9.601	9.852	10.12	10.42	10.73	11.07	11.42	11.78	12.10	12.25	12.03	11.54	11.05	10.67	10.49	10.61	10.96	11.44	11.94	12.24	12.15	11.86	11.52	11.21	10.96	10.83	10.99	11.56	12.16	12.23	12.00	11.70		15.40	15.48	15.58	0.900
0.150	9.436	9.658	9.902	10.17	10.46	10.78	11.11	11.47	11.85	12.22	12.51	12.51	12.11	11.57	11.10	10.80	10.79	11.07	11.53	12.07	12.49	12.32	12.25	11.90	11.56	11.28	11.11	11.21	11.76	12.43	12.55	12.32	12.00		15.77	15.83	15.91	16.01	0.875
0.175	9.728	9.963	10.22	10.51	10.82	11.16	11.52	11.91	12.30	12.67	12.89	12.69	12.16	11.61	11.19	11.03	11.21	11.63	12.18	12.71	12.89	12.66	12.30	11.94	11.62	11.42	11.46	11.97	12.70	12.88	12.66	12.32		16.15	16.22	16.27	16.19	14.16	0.850
0.200	10.04	10.29	10.57	10.87	11.21	11.57	11.96	12.37	12.78	13.14	13.20	12.79	12.20	11.68	11.36	11.39	11.75	12.29	12.88	13.23	13.09	12.73	12.35	12.00	11.75	11.73	12.19	12.98	13.23	13.01	12.66		16.54	16.62	16.65	15.80	13.63	13.22	0.825
0.225	10.37	10.64	10.94	11.26	11.62	12.01	12.43	12.87	13.29	13.57	13.41	12.84	12.24	11.78	11.64	11.89	12.40	13.02	13.52	13.52	13.18	12.77	12.40	12.11	12.03	12.42	13.25	13.59	13.38	13.01		16.86	17.00	17.00	15.31	13.60	13.55	13.62	0.800
0.250	10.72	11.01	11.33	11.68	12.07	12.49	12.94	13.40	13.80	13.89	13.47	12.84	12.29	11.97	12.06	12.51	13.14	13.74	13.90	13.61	13.20	12.80	12.48	12.34	12.66	13.51	13.92	13.73	13.36		16.74	17.23	17.05	14.91	13.90	13.90	13.96	14.06	0.775
0.275	11.10	11.41	11.75	12.13	12.55	13.00	13.47	13.91	14.13	13.89	13.37	12.80	12.37	12.28	12.63	13.24	13.88	14.13	13.92	13.56	13.17	12.83	12.65	12.89	13.70	14.13	13.97	13.63		15.87	16.74	16.51	14.70	14.27	14.29	14.34	14.42	14.52	0.750
0.300	11.51	11.84	12.21	12.61	13.06	13.51	13.87	13.92	13.75	13.46	13.07	12.69	12.50	12.73	13.28	13.81	13.93	13.80	13.58	13.31	13.04	12.86	13.03	13.68	13.93	13.84	13.63		15.31	15.87	15.85	14.85	14.69	14.71	14.76	14.82	14.90	15.01	0.725
0.325	11.95	12.30	12.68	13.08	13.33	13.32	13.23	13.11	12.97	12.79	12.60	12.50	12.68	13.10	13.34	13.32	13.22	13.10	12.97	12.83	12.74	12.87	13.24	13.35	13.28	13.16		15.42	15.62	15.63	15.21	15.12	15.15	15.20	15.26	15.33	15.42	15.54	0.700
0.350	12.41	12.73	12.79	12.60	12.50	12.48	12.39	12.29	12.20	12.12	12.13	12.37	12.71	12.80	12.72	12.60	12.49	12.39	12.31	12.28	12.40	12.72	12.80	12.71	12.58		15.80	15.89	15.89	15.64	15.57	15.62	15.68	15.74	15.81	15.87	15.92	15.27	0.675
0.375	12.36	12.17	12.02	11.90	11.80	11.72	11.64	11.58	11.56	11.71	12.11	12.51	12.56	12.36	12.16	11.99	11.87	11.79	11.78	11.99	12.47	12.57	12.39	12.18		16.26	16.34	16.33	16.04	15.95	16.06	16.16	16.24	16.31	16.35	16.01	15.01	14.13	0.650
0.400	11.61	11.44	11.31	11.20	11.12	11.05	11.02	11.12	11.52	12.07	12.58	12.72	12.44	12.07	11.77	11.55	11.43	11.45	11.86	12.55	12.72	12.46	12.12		16.71	16.82	16.79	16.18	15.97	16.29	16.57	16.73	16.82	16.78	15.88	14.81	13.91	13.15	0.625
0.425	10.94	10.79	10.68	10.59	10.54	10.60	10.97	11.51	12.14	12.75	13.08	12.86	12.38	11.92	11.57	11.37	11.42	11.96	12.81	13.07	12.79	12.36		16.89	17.20	17.09	15.88	15.43	15.87	16.54	17.03	17.23	16.85	15.73	14.66	13.75	13.04	12.94	0.600
0.450	10.34	10.21	10.13	10.16	10.48	10.98	11.57	12.22	12.91	13.44	13.41	12.92	12.37	11.90	11.61	11.61	12.17	13.13	13.51	13.23	12.76		16.46	17.04	16.00	15.46	14.04	15.12	15.81	16.62	17.12	16.67	15.39	14.34	13.67	13.32	13.37	13.49	0.373
0.475	9.811	9.767	10.07	10.52	11.04	11.64	12.31	13.04	13.71	13.91	13.53	12.96	12.41	12.01	11.93	12.43	13.44	13.93	13.68	13.20		15.82	16.43	16.37	15.10	14.49	14.62	15.09	15.78	16.45	16.35	15.44	14.46	13.84	13.80	13.86	13.96	14.09	0.550
0.500	9.739	10.14	10.60	11.12	11.72	12.40	13.15	13.87	14.14	13.93	13.49	12.96	12.51	12.32	12.72	13.69	14.14	13.90	13.59		15.36	15.04	15.00	14.94	14.58	14.63	14.86	13.23	13.73	13.93	13.31	14.33	14.31	14.34	14.41	14.49	14.60	14.75	0.525
0.525	10.26	10.71	11.23	11.82	12.50	13.24	13.84	13.91	13.81	13.61	13.30	12.94	12.71	12.97	13.70	13.92	13.85	13.66		15.38	15.62	15.66	15.16	15.02	15.06	15.17	15.33	15.55	15.71	15.34	14.91	14.88	14.93	15.01	15.09	15.19	15.31	15.47	0.500
0.550	10.86	11.36	11.94	12.60	13.28	13.50	13.47	13.40	13.31	13.18	13.01	12.87	13.01	13.41	13.50	13.46	13.36		15.83	15.94	15.95	15.65	15.56	15.62	15.72	15.82	15.93	15.97	15.67	15.41	15.46	15.56	15.66	15.75	15.84	15.93	15.95	15.48	0.475
0.575	11.52	12.00	12.71	13.07	13.05	13.00	12.94	12.88	12.81	12.73	12.67	12.76	13.01	13.07	13.03	12.95		16.42	16.53	16.52	16.11	15.99	16.15	16.31	16.44	16.54	16.50	16.00	15.69	15.89	16.15	16.33	16.45	16.54	16.48	15.94	15.31	14.76	0.450
0.600	12.27	12.71	12.68	12.61	12.55	12.49	12.44	12.39	12.34	12.31	12.40	12.64	12.71	12.64	12.55		16.88	17.10	17.02	16.14	15.92	16.25	16.65	16.97	17.13	16.86	15.98	15.47	15.72	16.32	16.81	17.08	17.08	16.54	15.84	15.19	14.61	14.10	0.425
0.625	12.41	12.29	12.20	12.13	12.07	12.02	11.97	11.94	11.93	12.04	12.41	12.54	12.39	12.23		16.74	17.17	17.02	15.84	15.48	15.81	16.34	16.89	17.23	16.80	15.83	15.12	15.11	15.73	16.56	17.17	17.08	16.45	15.75	15.09	14.50	13.96	13.49	0.400
0.650	11.94	11.84	11.75	11.68	11.63	11.58	11.55	11.55	11.75	12.46	12.73	12.42	12.08		16.28	16.76	16.66	15.49	15.01	15.26	15.76	16.34	16.80	16.57	15.67	14.86	14.58	14.97	15.74	16.56	16.82	16.33	15.66	15.01	14.41	13.86	13.38	12.95	0.375
0.675	11.51	11.41	11.33	11.27	11.22	11.19	11.21	11.64	12.78	13.28	12.89	12.28		15.80	16.25	16.24	15.18	14.63	14.79	15.22	15.74	16.24	16.26	15.51	14.68	14.25	14.42	14.98	15.74	16.33	16.15	15.57	14.94	14.34	13.80	13.34	13.11	13.25	0.350
0.700	11.10	11.01	10.94	10.89	10.86	10.93	11.74	13.18	13.89	13.56	12.85		15.41	15.82	15.86	14.97	14.47	14.57	14.87	15.29	15.73	15.92	15.37	14.60	14.20	14.29	14.62	15.14	15.74	15.90	15.47	14.88	14.32	13.88	13.76	13.85	14.01	14.24	0.325
0.725	10.72	10.64	10.58	10.55	10.77	11.96	13.55	14.14	13.98	13.48		15.35	15.64	15.68	15.06	14.80	14.87	15.04	15.27	15.55	15.73	15.35	14.78	14.66	14.76	14.94	15.19	15.52	15.72	15.44	14.94	14.58	14.54	14.63	14.76	14.92	15.12	15.38	0.300
0.750	10.37	10.30	10.28	10.84	12.24	13.74	13.94	13.87	13.68		15.91	16.06	16.07	15.66	15.53	15.62	15.76	15.91	16.05	16.09	15.71	15.37	15.43	15.59	15.75	15.92	16.07	16.06	15.71	15.38	15.37	15.52	15.68	15.83	15.98	16.10	15.93	15.54	0.275
0.775	10.04	10.10	11.09	12.56	13.60	13.62	13.57	13.48		16.70	16.90	16.85	16.17	16.03	16.25	16.52	16.76	16.92	16.75	16.03	15.69	15.92	16.29	16.62	16.86	16.91	16.52	15.97	15.71	15.96	16.40	16.73	16.92	16.80	16.36	15.87	15.42	15.01	0.250
0.800	10.28	11.44	12.89	13.31	13.29	13.24	13.17		16.87	17.22	17.08	16.09	15.87	16.16	16.50	17.01	17.27	16.85	15.96	15.40	15.60	16.20	16.78	17.22	17.12	16.53	15.92	15.53	15.70	16.41	17.11	17.23	16.82	16.31	15.80	15.33	14.90	14.52	0.225
0.825	11.89	13.02	13.01	12.96	12.91	12.85		16.55	16.94	16.83	15.81	15.50	15.76	16.19	16.64	16.99	16.69	15.81	15.18	15.21	15.67	16.28	16.85	16.95	16.43	15.81	15.30	15.17	15.69	16.55	17.00	16.72	16.24	15.74	15.26	14.82	14.42	14.06	0.200
0.850	12.81	12.72	12.65	12.59	12.54		16.14	16.52	16.47	15.50	15.11	15.32	15.71	16.16	16.54	16.45	15.66	14.92	14.75	15.09	15.65	16.25	16.60	16.29	15.70	15.14	14.79	14.94	15.66	16.45	16.54	16.15	15.67	15.20	14.75	14.34	13.96	13.62	0.175
0.875	12.46	12.37	12.30	12.24		15.72	16.09	16.10	15.20	14.73	14.88	15.24	15.66	16.06	16.14	15.50	14.72	14.35	14.56	15.04	15.61	16.10	16.09	15.59	15.02	14.56	14.40	14.81	15.61	16.17	16.01	15.59	15.14	14.69	14.27	13.88	13.53	13.21	0.150
0.900	12.12	12.03	11.96		15.32	15.68	15.73	14.92	14.37	14.47	14.80	15.19	15.58	15.78	15.33	14.56	14.05	14.08	14.47	15.00	15.53	15.78	15.46	14.92	14.41	14.07	14.15	14.74	15.50	15.78	15.49	15.07	14.63	14.22	13.82	13.46	13.13	12.83	0.125
0.925	11.79	11.71		14.94	15.29	15.37	14.65	14.05	14.09	14.38	14.75	15.13	15.40	15.14	14.42	13.81	13.68	13.97	14.44	14.96	15.36	15.28	14.82	14.31	13.88	13.70	14.01	14.70	15.32	15.33	14.99	14.58	14.16	13.77	13.40	13.06	12.75	12.47	0.100
0.950	11.48		14.59	14.93	15.03	14.40	13.76	13.74	14.00	14.34	14.71	15.01	14.92	14.29	13.64	13.35	13.52	13.93	14.42	14.88	15.04	14.70	14.21	13.75	13.44	13.46	13.95	14.65	15.05	14.88	14.51	14.11	13.73	13.36	13.01	12.69	12.39	12.12	0.075
		14.26	14.59	14.71	14.17	13.49	13.43	13.65	13.97	14.32	14.63	14.68	14.16	13.50	13.09	13.14	13.47	13.92	14.39	14.71	14.56	14.13	13.66	13.27	13.10	13.34	13.94	14.56	14.70	14.43	14.06	13.60	13.31	12.97	12.64	12.34	12.06	11.80	0.050
	13.96	14.27	14.41	13.95	13.25	13.14	13.33	13.62	13.96	14.27	14.41	14.03	13.38	12.90	12.81	13.07	13.47	13.93	14.32	14.38	14.03	13.58	13.16	12.88	12.89	13.31	13.95	14.39	14.31	14.00	13.64	13.27	12.93	12.60	12.29	12.00	11.74	11.49	0.025
	0.950	0.925	0.900	0.875	0.850	0.825	0.800	0.775	0.750	0.725	0.700	0.675	0.650	0.625	0.600	0.575	0.550	0.525	0.500	0.475	0.450	0.425	0.400	0.375	0.350	0.325	0.300	0.275	0.250	0.225	0.200	0.175	0.150	0.125	0.100	0.075	0.050	0.025	

$\longleftarrow \mu_2$ MODE 4

TABLE 4.6 (MODES 1 AND 2)

MODE 1 $\mu_2 \longrightarrow$

$\mu_1 \downarrow$ (Mode 1)	0.025	0.050	0.075	0.100	0.125	0.150	0.175	0.200	0.225	0.250	0.275	0.300	0.325	0.350	0.375	0.400	0.425	0.450	0.475	0.500	0.525	0.550	0.575	0.600	0.625	0.650	0.675	0.700	0.725	0.750	0.775	0.800	0.825	0.850	0.875	0.900	0.925	0.950	$\mu_1 \uparrow$ (Mode 2)
0.025	1.959	1.998	2.039	2.082	2.128	2.175	2.226	2.279	2.336	2.396	2.459	2.527	2.599	2.676	2.758	2.847	2.942	3.045	3.156	3.277	3.407	3.549	3.703	3.869	4.046	4.230	4.413	4.579	4.704	4.768	4.768	4.718	4.637	4.537	4.429	4.318	4.207	4.098	
0.050	2.011	2.051	2.094	2.139	2.187	2.237	2.291	2.348	2.408	2.471	2.539	2.612	2.689	2.772	2.861	2.956	3.060	3.172	3.293	3.425	3.568	3.724	3.892	4.073	4.262	4.454	4.632	4.771	4.849	4.859	4.813	4.731	4.630	4.518	4.402	4.287	4.175		
0.075	2.066	2.108	2.152	2.200	2.250	2.304	2.361	2.421	2.485	2.553	2.625	2.703	2.786	2.875	2.972	3.076	3.188	3.310	3.443	3.588	3.745	3.916	4.100	4.296	4.495	4.685	4.840	4.934	4.955	4.914	4.833	4.729	4.613	4.493	4.373	4.256		7.380	0.950
0.100	2.124	2.168	2.215	2.265	2.318	2.375	2.435	2.499	2.567	2.640	2.718	2.802	2.891	2.988	3.093	3.206	3.329	3.463	3.609	3.768	3.941	4.129	4.330	4.537	4.739	4.911	5.022	5.057	5.023	4.943	4.836	4.716	4.591	4.466	4.343		7.522	7.578	0.925
0.125	2.186	2.232	2.282	2.335	2.391	2.451	2.515	2.583	2.656	2.734	2.818	2.908	3.006	3.111	3.225	3.348	3.483	3.631	3.792	3.967	4.159	4.364	4.580	4.793	4.982	5.114	5.164	5.139	5.060	4.951	4.826	4.696	4.565	4.437		7.676	7.728	7.786	0.900
0.150	2.251	2.300	2.353	2.409	2.469	2.533	2.601	2.674	2.752	2.836	2.927	3.024	3.130	3.245	3.369	3.505	3.654	3.816	3.995	4.189	4.400	4.623	4.848	5.055	5.208	5.277	5.262	5.186	5.075	4.945	4.809	4.671	4.537		7.844	7.891	7.945	8.007	0.875
0.175	2.321	2.373	2.428	2.488	2.552	2.620	2.693	2.772	2.856	2.947	3.045	3.151	3.266	3.391	3.528	3.678	3.842	4.023	4.221	4.436	4.667	4.904	5.128	5.304	5.395	5.393	5.321	5.208	5.073	4.930	4.786	4.645		8.026	8.069	8.119	8.175	8.240	0.850
0.200	2.395	2.450	2.509	2.573	2.641	2.714	2.792	2.877	2.968	3.066	3.173	3.289	3.415	3.553	3.704	3.870	4.053	4.254	4.474	4.711	4.959	5.201	5.401	5.518	5.532	5.466	5.351	5.211	5.060	4.908	4.760		8.221	8.262	8.307	8.358	8.418	8.487	0.825
0.225	2.474	2.532	2.596	2.664	2.737	2.815	2.900	2.991	3.090	3.197	3.313	3.440	3.579	3.731	3.899	4.083	4.287	4.512	4.756	5.015	5.273	5.499	5.644	5.679	5.621	5.504	5.359	5.200	5.040	4.884		8.430	8.470	8.512	8.559	8.611	8.673	8.747	0.800
0.250	2.558	2.621	2.688	2.761	2.840	2.924	3.016	3.115	3.222	3.339	3.467	3.606	3.760	3.929	4.116	4.323	4.551	4.802	5.071	5.346	5.596	5.772	5.833	5.785	5.669	5.518	5.351	5.182	5.016		8.652	8.692	8.732	8.775	8.822	8.875	8.934	8.947	0.775
0.275	2.648	2.715	2.788	2.866	2.951	3.043	3.142	3.250	3.367	3.495	3.636	3.790	3.961	4.150	4.359	4.592	4.848	5.127	5.418	5.693	5.902	5.992	5.960	5.845	5.688	5.513	5.333	5.158		8.886	8.928	8.968	9.009	9.049	9.089	9.104	8.706	8.052	0.750
0.300	2.745	2.817	2.895	2.980	3.072	3.171	3.279	3.397	3.526	3.667	3.823	3.995	4.186	4.398	4.634	4.896	5.183	5.489	5.789	6.032	6.157	6.143	6.032	5.870	5.686	5.496	5.309		9.127	9.176	9.219	9.258	9.291	9.308	9.148	8.473	7.818	7.272	0.725
0.325	2.849	2.927	3.012	3.103	3.203	3.311	3.429	3.559	3.701	3.857	4.031	4.223	4.438	4.678	4.945	5.241	5.560	5.882	6.161	6.323	6.334	6.230	6.063	5.870	5.668	5.471		9.364	9.429	9.480	9.519	9.543	9.504	9.028	8.273	7.623	7.083	6.630	0.700
0.350	2.962	3.046	3.138	3.237	3.346	3.464	3.594	3.737	3.894	4.069	4.263	4.480	4.723	4.995	5.298	5.630	5.974	6.286	6.489	6.529	6.434	6.265	6.063	5.851	5.642		9.569	9.671	9.740	9.785	9.792	9.6[illegible]	8.868	8.100	7.458	6.924	6.477	6.097	0.675
0.375	3.084	3.175	3.274	3.383	3.502	3.632	3.776	3.934	4.110	4.306	4.525	4.771	5.047	5.357	5.700	6.063	6.406	6.648	6.720	6.639	6.471	6.263	6.041	5.820		9.664	9.857	9.973	10.03	9.997	9.549	8.703	7.949	7.317	6.793	6.358	6.023	5.976	0.650
0.400	3.216	3.315	3.424	3.543	3.674	3.818	3.977	4.154	4.351	4.572	4.821	5.102	5.418	5.770	6.148	6.518	6.794	6.894	6.830	6.669	6.460	6.232	6.003		9.535	9.878	10.10	10.20	10.06	9.407	8.558	7.814	7.197	6.691	6.2[illegible]	6.133	6.150	6.207	0.625
0.425	3.360	3.469	3.588	3.719	3.864	4.024	4.202	4.400	4.623	4.874	5.158	5.479	5.840	6.231	6.618	6.912	7.024	6.976	6.831	6.634	6.411	6.181		9.213	9.645	9.989	10.15	9.938	9.234	8.415	7.695	7.100	6.632	6.360	6.313	6.342	6.394	6.462	0.600
0.450	3.518	3.637	3.768	3.913	4.074	4.253	4.453	4.678	4.931	5.218	5.543	5.909	6.308	6.697	6.974	7.067	7.028	6.913	6.749	6.551	6.335		8.840	9.272	9.657	9.871	9.691	9.049	8.281	7.594	7.035	6.654	6.524	6.520	6.552	6.601	6.663	6.738	0.575
0.475	3.691	3.822	3.968	4.129	4.309	4.510	4.736	4.992	5.281	5.610	5.979	6.376	6.739	6.945	6.990	6.954	6.874	6.757	6.607	6.430		8.497	8.890	9.261	9.501	9.401	8.861	8.158	7.516	7.027	6.786	6.737	6.750	6.784	6.831	6.888	6.958	7.040	0.550
0.500	3.882	4.028	4.190	4.370	4.573	4.800	5.057	5.349	5.679	6.047	6.426	6.711	6.816	6.822	6.789	6.732	6.655	6.555	6.431		8.227	8.563	8.895	9.135	9.107	8.680	8.055	7.482	7.116	6.997	6.984	7.004	7.039	7.084	7.139	7.203	7.280	7.370	0.525
0.525	4.094	4.256	4.438	4.641	4.870	5.129	5.422	5.751	6.109	6.434	6.595	6.627	6.614	6.582	6.539	6.485	6.418	6.335		8.073	8.335	8.609	8.827	8.846	8.523	7.992	7.530	7.309	7.255	7.259	7.285	7.321	7.365	7.417	7.478	7.5[illegible]	7.633	7.733	0.500
0.550	4.330	4.512	4.717	4.947	5.207	5.500	5.825	6.153	6.367	6.422	6.419	6.397	6.367	6.330	6.288	6.239	6.182		8.074	8.253	8.453	8.626	8.662	8.429	8.012	7.692	7.572	7.551	7.565	7.594	7.632	7.676	7.727	7.784	7.851	7.9[illegible]	8.021	[illegible].132	0.475
0.575	4.595	4.801	5.033	5.294	5.586	5.897	6.148	6.230	6.232	6.212	6.186	6.155	6.122	6.087	6.047	6.004		8.229	8.345	8.474	8.589	8.618	8.452	8.153	7.948	7.880	7.877	7.901	7.935	7.976	8.021	8.071	8.127	8.189	8.261	8.344	8.444	8.[illegible]	[illegible]
0.600	4.895	5.128	5.391	5.679	5.949	6.062	6.064	6.042	6.014	5.985	5.954	5.923	5.891	5.857	5.821		8.499	8.582	8.667	8.739	8.751	8.626	8.404	8.258	8.217	8.230	8.264	8.306	8.353	8.401	8.453	8.508	8.566	8.630	8.699	8.756	8.664	8.412	0.4[illegible]
0.625	5.236	5.500	5.776	5.928	5.924	5.895	5.862	5.829	5.797	5.766	5.735	5.705	5.674	5.642		8.843	8.916	8.983	9.033	9.032	8.915	8.711	8.582	8.560	8.593	8.645	8.703	8.761	8.818	8.874	8.929	8.983	9.029	9.034	8.875	8.593	8.312	8.054	0.400
0.650	5.626	5.835	5.817	5.774	5.732	5.693	5.656	5.622	5.590	5.560	5.530	5.501	5.471		9.217	9.302	9.372	9.417	9.402	9.251	8.997	8.851	8.848	8.919	9.011	9.102	9.185	9.260	9.327	9.383	9.420	9.385	9.166	8.844	8.525	8.230	7.963	7.724	0.375
0.675	5.744	5.681	5.626	5.578	5.535	5.496	5.461	5.427	5.396	5.366	5.338	5.310		9.538	9.673	9.773	9.827	9.784	9.521	9.148	8.958	8.972	9.100	9.266	9.430	9.574	9.691	9.780	9.828	9.773	9.514	9.155	8.797	8.464	8.161	7.886	7.640	7.419	0.350
0.700	5.545	5.487	5.435	5.389	5.348	5.311	5.276	5.244	5.214	5.186	5.158		9.664	9.887	10.06	10.14	10.03	9.605	9.127	8.886	8.890	9.047	9.282	9.542	9.789	9.993	10.12	10.11	9.876	9.504	9.116	8.749	8.410	8.102	7.822	7.569	7.341	7.137	0.325
0.725	5.356	5.303	5.254	5.211	5.172	5.136	5.103	5.073	5.044	5.016		9.567	9.846	10.08	10.20	10.05	9.540	9.016	8.729	8.695	8.838	9.085	9.388	9.708	10.00	10.18	10.14	9.845	9.461	9.071	8.702	8.361	8.050	7.766	7.509	7.276	7.065	6.875	0.300
0.750	5.180	5.130	5.085	5.044	5.007	4.973	4.941	4.912	4.884		9.362	9.652	9.906	10.05	9.911	9.416	8.878	8.550	8.471	8.579	8.803	9.099	9.433	9.763	10.01	10.02	9.768	9.404	9.022	8.656	8.316	8.003	7.718	7.457	7.220	7.004	6.809	6.633	0.275
0.775	5.015	4.968	4.925	4.887	4.851	4.819	4.789	4.760		9.125	9.409	9.662	9.817	9.720	9.270	8.735	8.375	8.250	8.317	8.508	8.778	9.096	9.429	9.716	9.823	9.659	9.334	8.970	8.611	8.274	7.961	7.674	7.411	7.171	6.952	6.752	6.571	6.406	0.250
0.800	4.860	4.815	4.775	4.739	4.705	4.674	4.645		8.884	9.157	9.404	9.566	9.510	9.117	8.596	8.211	8.045	8.070	8.225	8.464	8.757	9.076	9.380	9.569	9.513	9.251	8.912	8.565	8.232	7.922	7.635	7.371	7.129	6.907	6.704	6.518	6.349	6.195	0.225
0.825	4.714	4.672	4.634	4.599	4.567	4.538		8.650	8.910	9.149	9.315	9.294	8.960	8.462	8.062	7.859	7.844	7.963	8.170	8.435	8.734	9.036	9.275	9.328	9.148	8.847	8.516	8.191	7.884	7.599	7.334	7.091	6.867	6.661	6.472	6.299	6.141	5.997	0.200
0.850	4.576	4.537	4.501	4.468	4.437		8.428	8.676	8.904	9.072	9.080	8.802	8.335	7.925	7.691	7.639	7.724	7.901	8.138	8.413	8.704	8.965	9.103	9.020	8.772	8.464	8.150	7.848	7.564	7.301	7.057	6.832	6.624	6.433	6.257	6.095	5.947	5.812	0.175
0.875	4.447	4.409	4.375	4.344		8.219	8.455	8.674	8.841	8.873	8.646	8.213	7.800	7.541	7.455	7.508	7.656	7.866	8.118	8.392	8.658	8.851	8.864	8.683	8.407	8.107	7.812	7.532	7.270	7.026	6.800	6.592	6.399	6.220	6.056	5.904	5.765	5.637	0.150
0.900	4.324	4.289	4.256		8.024	8.249	8.459	8.624	8.673	8.493	8.096	7.686	7.407	7.291	7.314	7.435	7.620	7.849	8.105	8.365	8.586	8.678	8.576	8.343	8.062	7.776	7.500	7.240	6.997	6.772	6.562	6.368	6.189	6.022	5.868	5.725	5.594	5.473	0.125
0.925	4.208	4.175		7.843	8.057	8.259	8.422	8.485	8.344	7.983	7.581	7.286	7.144	7.140	7.235	7.398	7.605	7.843	8.093	8.324	8.471	8.450	8.268	8.013	7.739	7.469	7.212	6.971	6.746	6.536	6.341	6.160	5.992	5.836	5.691	5.557	5.433	5.319	0.100
0.950	4.099		7.675	7.880	8.074	8.234	8.307	8.200	7.876	7.485	7.179	7.014	6.984	7.056	7.197	7.385	7.605	7.842	8.075	8.255	8.303	8.183	7.959	7.700	7.438	7.185	6.946	6.722	6.512	6.317	6.136	5.967	5.809	5.662	5.526	5.399	5.281	5.173	0.075
		7.521	7.717	7.905	8.061	8.142	8.062	7.773	7.395	7.082	6.898	6.845	6.895	7.017	7.187	7.391	7.615	7.842	8.039	8.141	8.084	7.900	7.660	7.407	7.159	6.922	6.699	6.491	6.296	6.114	5.944	5.786	5.638	5.500	5.371	5.251	5.138	5.035	0.050
	7.379	7.568	7.749	7.902	7.989	7.933	7.675	7.312	6.995	6.795	6.721	6.752	6.855	7.009	7.197	7.408	7.629	7.833	7.971	7.973	7.834	7.617	7.376	7.134	6.900	6.679	6.471	6.276	6.094	5.924	5.765	5.617	5.478	5.348	5.226	5.112	5.005	4.905	0.025
μ_2 (Mode 2)	0.950	0.925	0.900	0.875	0.850	0.825	0.800	0.775	0.750	0.725	0.700	0.675	0.650	0.625	0.600	0.575	0.550	0.525	0.500	0.475	0.450	0.425	0.400	0.375	0.350	0.325	0.300	0.275	0.250	0.225	0.200	0.175	0.150	0.125	0.100	0.075	0.050	0.025	

$\longleftarrow \mu_2$ MODE 2

TABLE 4.6 (MODES 3 AND 4)

MODE 3 $\mu_2 \longrightarrow$

$\mu_1 \downarrow$ (Mode 3)	0.025	0.050	0.075	0.100	0.125	0.150	0.175	0.200	0.225	0.250	0.275	0.300	0.325	0.350	0.375	0.400	0.425	0.450	0.475	0.500	0.525	0.550	0.575	0.600	0.625	0.650	0.675	0.700	0.725	0.750	0.775	0.800	0.825	0.850	0.875	0.900	0.925	0.950	$\mu_1 \uparrow$ (Mode 4)
0.025	8.20[a]	8.379	8.564	8.763	8.976	9.204	9.448	9.707	9.983	10.27	10.57	10.86	11.11	11.18	10.95	10.55	10.14	9.805	9.610	9.629	9.854	10.21	10.63	11.02	11.19	11.07	10.82	10.54	10.28	10.06	9.917	9.941	10.27	10.84	11.18	11.13	10.92	10.66	
0.050	8.426	8.604	8.798	9.008	9.234	9.476	9.736	10.01	10.30	10.61	10.92	11.21	11.40	11.31	10.95	10.51	10.12	9.848	9.779	9.944	10.28	10.70	11.13	11.39	11.33	11.09	10.80	10.53	10.29	10.12	10.11	10.41	10.99	11.38	11.35	11.14	10.87		
0.075	8.657	8.844	9.048	9.270	9.510	9.768	10.04	10.34	10.65	10.97	11.29	11.56	11.64	11.38	10.93	10.49	10.14	9.978	10.06	10.36	10.78	11.24	11.59	11.61	11.39	11.10	10.80	10.54	10.35	10.31	10.56	11.16	11.61	11.60	11.38	11.10		13.95	0.950
0.100	8.902	9.098	9.314	9.549	9.804	10.08	10.37	10.69	11.02	11.36	11.68	11.89	11.81	11.40	10.92	10.51	10.24	10.22	10.45	10.85	11.33	11.76	11.90	11.72	11.42	11.11	10.82	10.61	10.53	10.74	11.33	11.85	11.87	11.65	11.36		14.23	14.32	0.925
0.125	9.161	9.368	9.597	9.847	10.12	10.41	10.73	11.06	11.41	11.76	12.07	12.18	11.91	11.42	10.93	10.57	10.43	10.57	10.94	11.42	11.91	12.18	12.07	11.77	11.44	11.13	10.89	10.77	10.93	11.51	12.10	12.16	11.93	11.63		14.54	14.62	14.71	0.900
0.150	9.436	9.655	9.899	10.17	10.46	10.77	11.11	11.46	11.83	12.19	12.45	12.39	11.96	11.43	10.98	10.71	10.73	11.04	11.51	12.04	12.43	12.43	12.14	11.80	11.47	11.19	11.04	11.15	11.71	12.36	12.46	12.24	11.92		14.89	14.95	15.03	15.12	0.875
0.175	9.727	9.961	10.22	10.50	10.82	11.15	11.51	11.89	12.28	12.63	12.79	12.51	11.98	11.45	11.07	10.95	11.16	11.60	12.15	12.65	12.78	12.53	12.18	11.82	11.52	11.33	11.38	11.90	12.62	12.78	12.56	12.22		15.25	15.31	15.37	15.41	14.15	0.850
0.200	10.04	10.29	10.56	10.87	11.20	11.56	11.95	12.35	12.75	13.05	13.00	12.54	11.97	11.49	11.23	11.32	11.70	12.25	12.81	13.09	12.91	12.56	12.19	11.86	11.63	11.63	12.10	12.87	13.08	12.87	12.53		15.62	15.69	15.74	15.53	13.55	12.30	0.825
0.225	10.37	10.63	10.93	11.26	11.61	12.00	12.41	12.83	13.19	13.31	12.99	12.46	11.94	11.57	11.51	11.81	12.34	12.93	13.30	13.21	12.89	12.52	12.18	11.93	11.88	12.29	13.06	13.32	13.12	12.79		15.96	16.07	16.11	15.20	13.15	12.59	12.65	0.800
0.250	10.72	11.01	11.32	11.67	12.05	12.45	12.86	13.19	13.24	13.03	12.67	12.24	11.86	11.71	11.91	12.39	12.95	13.25	13.18	12.96	12.68	12.38	12.14	12.07	12.41	13.08	13.26	13.12	12.87		16.08	16.38	16.38	14.78	13.06	12.92	12.97	13.05	0.775
0.275	11.10	11.40	11.74	12.10	12.46	12.71	12.71	12.60	12.46	12.29	12.07	11.85	11.74	11.90	12.31	12.66	12.72	12.64	12.52	12.37	12.21	12.06	12.02	12.27	12.66	12.72	12.64	12.50		15.49	16.26	16.19	14.41	13.31	13.27	13.32	13.39	13.48	0.750
0.300	11.50	11.81	12.08	12.11	12.02	11.91	11.81	11.71	11.61	11.50	11.41	11.42	11.63	11.97	12.12	12.08	11.97	11.87	11.76	11.67	11.59	11.60	11.81	12.09	12.11	12.02	11.90		14.71	15.45	15.53	14.19	13.66	13.66	13.70	13.76	13.84	13.94	0.725
0.325	11.77	11.58	11.42	11.29	11.18	11.08	10.99	10.92	10.86	10.85	10.99	11.34	11.70	11.79	11.66	11.48	11.32	11.20	11.10	11.05	11.10	11.40	11.76	11.77	11.61	11.43		14.50	14.85	14.97	14.29	14.06	14.07	14.11	14.17	14.24	14.32	14.43	0.700
0.350	10.98	10.80	10.65	10.54	10.44	10.36	10.30	10.28	10.41	10.78	11.26	11.70	11.84	11.64	11.33	11.06	10.85	10.71	10.65	10.78	11.28	11.79	11.80	11.56	11.27		14.75	14.88	14.93	14.63	14.49	14.51	14.56	14.61	14.67	14.74	14.83	14.93	0.675
0.375	10.27	10.11	9.980	9.881	9.809	9.785	9.916	10.28	10.77	11.31	11.82	12.11	11.94	11.55	11.16	10.83	10.61	10.53	10.74	11.38	12.01	12.07	11.79	11.42		15.15	15.23	15.25	15.04	14.91	14.95	15.02	15.08	15.15	15.21	15.26	14.96	14.12	0.650
0.400	9.637	9.499	9.398	9.355	9.488	9.842	10.29	10.81	11.38	11.96	12.40	12.40	12.00	11.52	11.10	10.79	10.67	10.88	11.56	12.32	12.44	12.15	11.73		15.60	15.68	15.69	15.42	15.21	15.32	15.46	15.57	15.64	15.70	15.62	14.78	13.89	13.14	0.625
0.425	9.083	8.992	9.123	9.463	9.874	10.34	10.87	11.45	12.08	12.65	12.86	12.55	12.03	11.54	11.15	10.95	11.11	11.78	12.64	12.85	12.56	12.12		16.00	16.15	16.15	15.53	15.08	15.35	15.74	15.99	16.13	16.17	15.65	14.63	13.72	12.94	12.29	0.600
0.450	8.830	9.153	9.521	9.937	10.41	10.94	11.53	12.18	12.82	13.21	13.05	12.58	12.06	11.60	11.32	11.40	12.02	12.93	13.21	12.95	12.52		16.04	16.42	16.39	15.20	14.57	14.80	13.42	16.03	16.43	16.39	13.33	14.50	13.59	12.81	12.20	12.15	0.575
0.475	9.252	9.611	10.02	10.49	11.02	11.61	12.27	12.94	13.33	13.26	12.96	12.53	12.07	11.74	11.72	12.25	13.11	13.34	13.17	12.83		15.56	16.13	16.13	14.91	14.07	14.16	14.69	15.42	16.10	16.17	15.37	14.38	13.48	12.76	12.51	12.57	12.69	0.550
0.500	9.729	10.13	10.58	11.11	11.70	12.35	12.96	13.09	13.02	12.88	12.67	12.37	12.09	12.00	12.41	12.99	13.09	13.00	12.83		14.97	15.54	15.63	14.38	13.77	13.76	14.10	14.67	13.35	15.71	15.17	14.27	13.44	12.99	12.98	13.04	13.14	13.28	0.525
0.525	10.26	10.70	11.21	11.79	12.39	12.67	12.65	12.59	12.51	12.41	12.28	12.13	12.08	12.32	12.62	12.67	12.61	12.52		14.55	15.01	15.15	14.36	13.80	13.80	13.97	14.27	14.73	15.15	14.95	14.20	13.58	13.47	13.51	13.58	13.67	13.78	13.93	0.500
0.550	10.85	11.34	11.89	12.24	12.24	12.19	12.13	12.07	12.01	11.94	11.87	11.86	12.02	12.22	12.25	12.20	12.12		14.56	14.81	14.91	14.45	14.19	14.21	14.30	14.44	14.64	14.88	14.83	14.31	14.03	14.04	14.10	14.17	14.26	14.36	14.48	14.65	0.475
0.575	11.51	11.91	11.88	11.82	11.75	11.70	11.64	11.59	11.54	11.50	11.31	11.66	11.88	11.91	11.83	11.74		14.98	15.10	15.14	14.87	14.71	14.75	14.84	14.94	15.04	15.14	15.06	14.69	14.56	14.63	14.73	14.82	14.91	15.00	15.10	15.14	14.75	0.450
0.600	11.63	11.51	11.42	11.35	11.29	11.23	11.19	11.15	11.12	11.14	11.34	11.70	11.73	11.58	11.43		15.57	15.68	15.69	15.37	15.16	15.26	15.42	15.55	15.65	15.71	15.50	15.00	14.93	15.15	15.35	15.50	15.62	15.70	15.64	15.15	14.60	14.09	0.425
0.625	11.17	11.07	10.98	10.91	10.86	10.81	10.77	10.75	10.80	11.16	11.81	11.89	11.58	11.27		16.05	16.26	16.25	15.54	15.16	15.38	15.74	16.05	16.25	16.26	15.67	14.95	14.78	15.18	15.71	16.07	16.27	16.22	15.70	15.07	14.48	13.95	13.49	0.400
0.650	10.75	10.65	10.57	10.51	10.46	10.42	10.41	10.51	11.19	12.15	12.36	11.95	11.42		15.99	16.39	16.36	15.33	14.81	15.02	15.47	15.99	16.40	16.36	15.59	14.76	14.37	14.64	15.33	16.05	16.47	16.23	15.63	14.99	14.39	13.85	13.36	12.94	0.375
0.675	10.36	10.27	10.19	10.14	10.10	10.10	10.34	11.37	12.58	12.95	12.55	11.89		15.59	16.04	16.05	15.03	14.40	14.53	14.96	15.49	15.99	16.11	15.44	14.59	14.02	14.04	14.59	15.37	16.05	16.05	15.52	14.91	14.31	13.76	13.27	12.82	12.43	0.350
0.700	9.991	9.908	9.845	9.805	9.833	10.36	11.61	12.95	13.33	13.07	12.49		15.14	15.58	15.65	14.74	14.03	14.08	14.45	14.93	15.43	15.71	15.26	14.44	13.79	13.60	13.94	14.59	15.33	15.71	15.38	14.82	14.24	13.70	13.20	12.74	12.34	11.99	0.325
0.725	9.651	9.578	9.535	9.652	10.52	11.87	13.11	13.26	13.15	12.87		14.74	15.15	15.25	14.48	13.77	13.76	14.05	14.46	14.92	15.27	15.05	14.32	13.65	13.36	13.54	14.01	14.64	15.20	15.19	14.73	14.18	13.65	13.16	12.73	12.49	12.57	12.78	0.300
0.750	9.335	9.292	9.694	10.76	12.14	12.96	12.98	12.91	12.79		14.50	14.86	14.96	14.34	13.76	13.76	13.95	14.24	14.60	14.92	14.87	14.26	13.67	13.50	13.63	13.91	14.31	14.77	14.97	14.65	14.13	13.66	13.29	13.21	13.29	13.44	13.64	13.91	0.275
0.775	9.111	9.928	11.05	12.38	12.68	12.65	12.60	12.53		14.72	14.93	15.00	14.58	14.28	14.32	14.44	14.60	14.80	14.98	14.93	14.44	14.14	14.17	14.30	14.47	14.68	14.91	15.00	14.72	14.31	14.07	14.09	14.20	14.34	14.50	14.68	14.90	14.97	0.250
0.800	10.27	11.41	12.36	12.38	12.33	12.28	12.23		15.51	15.66	15.68	15.26	15.02	15.12	15.30	15.47	15.62	15.71	15.45	14.92	14.81	15.00	15.23	15.44	15.61	15.71	15.51	15.08	14.80	14.88	15.13	15.36	15.55	15.69	15.64	15.29	14.89	14.52	0.225
0.825	11.87	12.15	12.08	12.03	11.98	11.92		16.09	16.37	16.35	15.55	15.21	15.41	15.74	16.09	16.37	16.35	15.67	15.00	14.90	15.24	15.71	16.15	16.42	16.25	15.71	15.18	14.89	15.12	15.71	16.24	16.43	16.15	15.70	15.24	14.81	14.41	14.05	0.200
0.850	11.90	11.81	11.74	11.69	11.63		15.95	16.32	16.29	15.38	14.97	15.15	15.52	15.94	16.31	16.31	15.59	14.84	14.60	14.89	15.41	15.97	16.38	16.21	15.66	15.09	14.69	14.72	15.31	16.08	16.39	16.10	15.65	15.18	14.74	14.33	13.96	13.62	0.175
0.875	11.57	11.48	11.42	11.36		15.61	15.98	15.99	15.13	14.63	14.77	15.12	15.53	15.93	16.04	15.46	14.67	14.27	14.43	14.89	15.45	15.95	16.01	15.56	14.99	14.52	14.30	14.62	15.37	16.01	15.96	15.57	15.12	14.68	14.26	13.88	13.53	13.21	0.150
0.900	11.25	11.17	11.11		15.24	15.60	15.65	14.86	14.30	14.39	14.70	15.09	15.49	15.70	15.29	14.53	13.99	13.99	14.36	14.88	15.41	15.70	15.42	14.90	14.39	14.02	14.03	14.57	15.33	15.70	15.47	15.06	14.62	14.21	13.82	13.46	13.12	12.83	0.125
0.925	10.95	10.87		14.88	15.22	15.31	14.61	13.99	14.02	14.31	14.67	15.05	15.32	15.11	14.39	13.78	13.61	13.88	14.34	14.86	15.28	15.24	14.80	14.29	13.85	13.64	13.89	14.55	15.21	15.29	14.97	14.57	14.16	13.77	13.40	13.06	12.75	12.46	0.100
0.950	10.66		14.54	14.87	14.98	14.36	13.71	13.69	13.94	14.28	14.64	14.94	14.89	14.27	13.61	13.30	13.45	13.84	14.33	14.80	14.99	14.68	14.20	13.74	13.40	13.39	13.84	14.53	14.98	14.85	14.50	14.11	13.72	13.35	13.00	12.68	12.39	12.12	0.075
		14.22	14.58	14.67	14.14	13.65	13.38	13.60	13.92	14.27	14.58	14.64	14.15	13.48	13.06	13.09	13.41	13.86	14.33	14.66	14.54	14.11	13.65	13.26	13.06	13.27	13.83	14.48	14.67	14.42	14.05	13.68	13.31	12.96	12.63	12.33	12.05	11.80	0.050
	13.94	14.25	14.39	13.93	13.23	13.11	13.30	13.60	13.93	14.25	14.39	14.02	13.37	12.88	12.79	13.04	13.44	13.89	14.29	14.36	14.03	13.58	13.16	12.86	12.86	13.26	13.89	14.36	14.30	13.99	13.63	13.27	12.92	12.60	12.29	12.00	11.74	11.49	0.025
	0.950	0.925	0.900	0.875	0.850	0.825	0.800	0.775	0.750	0.725	0.700	0.675	0.650	0.625	0.600	0.575	0.550	0.525	0.500	0.475	0.450	0.425	0.400	0.375	0.350	0.325	0.300	0.275	0.250	0.225	0.200	0.175	0.150	0.125	0.100	0.075	0.050	0.025	

$\longleftarrow \mu_2$ MODE 4

CHAPTER FIVE

FREE VIBRATION ANALYSIS OF FOUR SPAN BEAMS WITH CLASSICAL BOUNDARY CONDITIONS

The problem of establishing frequencies and modal shapes for beams of this type is almost identical to that encountered with the triple span beams in Chapter 4. No attempt should be made to utilize the material of this chapter therefore without first studying the preceding one.

It is appreciated that the eigenvalues for four span beams are a function of the three dimensionless span lengths μ_1, μ_2, and μ_3 (see inset, Case 5.1). Utilizing for illustrative purposes a beam with simple support at the extreme outer ends, as in Chapter 4, it readily follows that the expressions for the first and second spans (counting from the left end) are unchanged from those in Chapter 4. Following identical procedures it is shown that we may write

$$r_3(\xi) = A_3(\sin \beta\xi - \theta_3 \sinh \beta\xi) + B_3(\cos \beta\xi - \cosh \beta\xi + \phi_3 \sinh \beta\xi) \tag{5.1}$$

Also, for the fourth span,

$$r_4(\xi) = A_4\left(\sin \beta\xi - \frac{\sin \beta\gamma}{\sinh \beta\gamma} \sinh \beta\xi\right) \tag{5.2}$$

where

$$\theta_3 = \frac{\sin \mu_3}{\sinh \mu_3} \qquad \phi_3 = \frac{\cosh \mu_3 - \cos \mu_3}{\sinh \mu_3}$$

We are now left with six unknowns, A_1, A_2, B_2, A_3, B_3, and A_4, and we may write six equations involving continuity of slope and bending moment across the three interior supports.

The equations pertaining to the first intermediate support are of course Eqs. 4.11 and 4.12. The equations pertaining to the second and third intermediate supports are:

$$A_2(\cos\beta\mu_2 - \theta_2\cosh\beta\mu_2) + B_2(-\sin\beta\mu_2 - \sinh\beta\mu_2 + \phi_2\cosh\beta\mu_2) + A_3(\theta_3 - 1) + B_3(-\phi_3) = 0 \quad (5.3)$$

$$A_2(-\sin\beta\mu_2 - \theta_2\sinh\beta\mu_2) + B_2(-\cos\beta\mu_2 - \cosh\beta\mu_2 + \phi_2\sinh\beta\mu_2) + A_3(0) + B_3(2) = 0 \quad (5.4)$$

where

$$\theta_2 = \frac{\sin\beta\mu_2}{\sinh\beta\mu_2} \qquad \text{and} \qquad \phi_2 = \frac{\cosh\beta\mu_2 - \cos\beta\mu_2}{\sinh\beta\mu_2}$$

$$A_3(\cos\beta\mu_3 - \theta_3\cosh\beta\mu_3) + B_3(-\sin\beta\mu_3 - \sinh\beta\mu_3 + \phi_3\cosh\beta\mu_3) + A_4\left(\cos\beta\gamma \frac{-\sin\beta\gamma}{\sinh\beta\gamma}\cosh\beta\gamma\right) = 0 \quad (5.5)$$

$$A_3(-\sin\beta\mu_3 - \theta_3\sinh\beta\mu_3) + B_3(-\cos\beta\mu_3 - \cosh\beta\mu_3 + \phi_3\sinh\beta\mu_3) + A_4(2\sin\beta\gamma) = 0 \quad (5.6)$$

The Eqs. 4.11, 4.12, and 5.3 through 5.6 form a set of six simultaneous homogeneous algebraic equations from which values of the eigenvalues β may be obtained. Similar sets of equations are obtained for the other five possible sets of boundary conditions.

The eigenvalues to be stored are seen to be functions of the parameters μ_1, μ_2, and μ_3. The next matter to be decided upon is the number of stations to be selected along the dimensionless beam for the purpose of data storage. The more stations utilized, the smaller will be the station-to-station increments in the parameters μ, and the greater will be the accuracy if interpolation is used to establish frequencies of beams whose internal supports do not lie exactly on these stations.

Experience has indicated that the number of stations required to insure 1 percent accuracy on interpolation, even for the first mode, is very large. The associated storage space is considered excessive. For this reason the beam has been divided into 20 segments with 19 internal stations. Data are provided in the tables for any beam so long as its internal supports lie on any of these stations. Accurate analysis of beams that do not fall into this category, that is, beams with supports located between these stations, can be guaranteed only by utilizing the appropriate equations, such as those provided above for the beam with simple end supports, and obtaining an exact solution for the eigenvalues. It is not necessary to develop these equations, however, since expressions for the elements in the determinant matrix for multispan beams (three or more spans) with arbitrary support spacing are provided in Chapter 6. It is expected that most practical problems can be solved using the data provided here.

Data for frequency analysis of the first four modes of all six possible four span beams are presented in Tables 5.1 through 5.6. Each table has a value of μ_1 associated with it. Associated with the table and its value of μ_1 is a triangular array of data so that eigenvalues may be obtained for any possible combination of the values μ_2 and μ_3 for the first four modes. By combining first and second mode data and third and fourth mode data, four triangular arrays are combined to form two square arrays.

Modal shape expressions are given for each of the six cases. The case number and the associated eigenvalue table numbers are identical.

Values for the constants appearing in the modal shape expressions have been obtained in a manner following that discussed for the three span beams of Chapter 4. In each case the coefficient A_1 has been set equal to one. The equations have permitted solution for the other five constants.

CASE 5.1. Beam with simple support at each end.

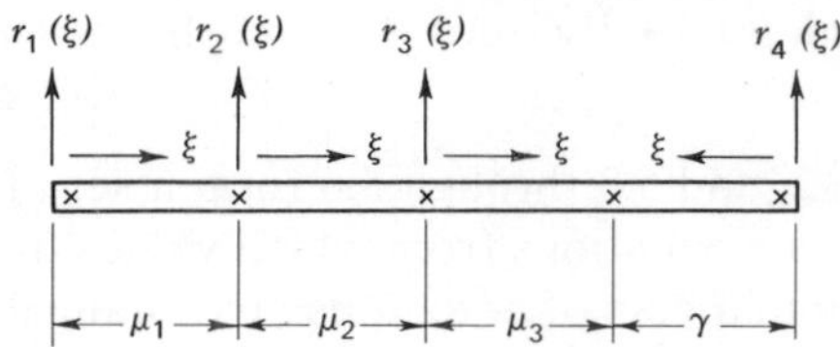

$$r_1(\xi) = \sin \beta\xi \frac{-\sin \beta\mu_1}{\sinh \beta\mu_1} \sinh \beta\xi$$

$$r_2(\xi) = A_2(\sin \beta\xi - \theta_2 \sinh \beta\xi) + B_2(\cos \beta\xi - \cosh \beta\xi + \phi_2 \sinh \beta\xi)$$

$$r_3(\xi) = A_3(\sin \beta\xi - \theta_3 \sinh \beta\xi) + B_3(\cos \beta\xi - \cosh \beta\xi + \phi_3 \sinh \beta\xi)$$

$$r_4(\xi) = A_4\left(\sin \beta\xi \frac{-\sin \beta\gamma}{\sinh \beta\gamma} \sinh \beta\xi\right)$$

where

$$A_2 = \frac{\sin \beta\mu_1 \cosh \beta\mu_1 - \cos \beta\mu_1 \sinh \beta\mu_1 + \phi_2 \sin \beta\mu_1 \sinh \beta\mu_1}{(\theta_2 - 1)\sinh \beta\mu_1}$$

$$B_2 = \sin \beta\mu_1$$

$$B_3 = \frac{A_2}{2}(\sin \beta\mu_2 + \theta_2 \sinh \beta\mu_2) + \frac{B_2}{2}(\cos \beta\mu_2 + \cosh \beta\mu_2 - \phi_2 \sinh \beta\mu_2)$$

$$A_3 = \frac{A_2}{1-\theta_3}(\cos \beta\mu_2 - \theta_2 \cosh \beta\mu_2) + \frac{B_2}{1-\theta_3}(-\sin \beta\mu_2 - \sinh \beta\mu_2 + \phi_2 \cosh \beta\mu_2) - \frac{B_3\phi_3}{1-\theta_3}$$

$$A_4 = \frac{A_3(\cos \beta\mu_3 - \theta_3 \cosh \beta\mu_3) + B_3(-\sin \beta\mu_3 - \sinh \beta\mu_3 + \phi_3 \cosh \beta\mu_3)}{(\sin \beta\gamma \cosh \beta\gamma / \sinh \beta\gamma) - \cos \beta\gamma}$$

or

$$A_4 = A_3 \frac{\sin \beta\mu_3}{\sin \beta\gamma} + B_3 \frac{\cos \beta\mu_3}{\sin \beta\gamma}$$

where

$$\theta_2 = \frac{\sin \beta\mu_2}{\sinh \beta\mu_2} \qquad \phi_2 = \frac{\cosh \beta\mu_2 - \cos \beta\mu_2}{\sinh \beta\mu_2}$$

$$\theta_3 = \frac{\sin \beta\mu_3}{\sinh \beta\mu_3} \qquad \text{and } \phi_3 = \frac{\cosh \beta\mu_3 - \cos \beta\mu_3}{\sinh \beta\mu_3}$$

CASE 5.2. Beam with outer ends clamped.

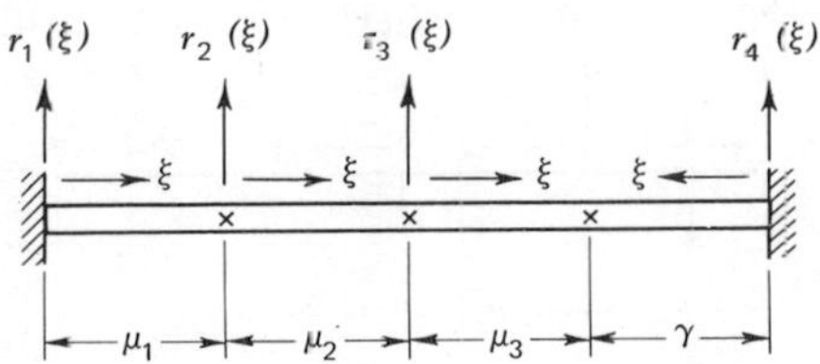

$$r_1(\xi) = \sin \beta\xi - \sinh \beta\xi + \frac{\sin \beta\mu_1 - \sinh \beta\mu_1}{\cosh \beta\mu_1 - \cos \beta\mu_1}(\cos \beta\xi - \cosh \beta\xi)$$

$$r_2(\xi) = A_2(\sin \beta\xi - \theta_2 \sinh \beta\xi) + B_2(\cos \beta\xi - \cosh \beta\xi + \phi_2 \sinh \beta\xi)$$

$$r_3(\xi) = A_3(\sin \beta\xi - \theta_3 \sinh \beta\xi) + B_3(\cos \beta\xi - \cosh \beta\xi + \phi_3 \sinh \beta\xi)$$

$$r_4(\xi) = A_4\left[\sin \beta\xi - \sinh \beta\xi + \frac{\sin \beta\gamma - \sinh \beta\gamma}{\cosh \beta\gamma - \cos \beta\gamma}(\cos \beta\xi - \cosh \beta\xi)\right]$$

$$A_2 = \frac{2(1 - \cosh \beta\mu_1 \cos \beta\mu_1) - \phi_2(\sinh \beta\mu_1 \cos \beta\mu_1 - \sin \beta\mu_1 \cosh \beta\mu_1)}{(\theta_2 - 1)(\cosh \beta\mu_1 - \cos \beta\mu_1)}$$

$$B_2 = \frac{\sin \beta\mu_1 \cosh \beta\mu_1 - \sinh \beta\mu_1 \cos \beta\mu_1}{\cosh \beta\mu_1 - \cos \beta\mu_1}$$

$$B_3 = \frac{A_2}{2}(\sin \beta\mu_2 + \theta_2 \sinh \beta\mu_2) + \frac{B_2}{2}(\cos \beta\mu_2 + \cosh \beta\mu_2 - \phi_2 \sinh \beta\mu_2)$$

$$A_3 = \frac{A_2}{1 - \theta_3}(\cos \beta\mu_2 - \theta_2 \cosh \beta\mu_2) + \frac{B_2}{1 - \theta_3}(-\sin \beta\mu_2 - \sinh \beta\mu_2 + \phi_2 \cosh \beta\mu_2) - \frac{B_3\phi_3}{1 - \theta_3}$$

$$A_4 = \frac{A_3(\sinh \beta\mu_3 \cos \beta\mu_3 - \sin \beta\mu_3 \cosh \beta\mu_3)}{2 \sinh \beta\mu_3(1 - \cosh \beta\gamma \cos \beta\gamma)/(\cosh \beta\gamma - \cos \beta\gamma)} + \frac{B_3(1 - \cos \beta\mu_3 \cosh \beta\mu_3 - \sin \beta\mu_3 \sinh \beta\mu_3)}{2 \sinh \beta\mu_3(1 - \cosh \beta\gamma \cos \beta\gamma)/(\cosh \beta\gamma - \cos \beta\gamma)}$$

or

$$A_4 = \frac{A_3 \sin \beta\mu_3 + B_3 \cos \beta\mu_3}{(\sin \beta\gamma \cosh \beta\gamma - \sinh \beta\gamma \cos \beta\gamma)/(\cosh \beta\gamma - \cos \beta\gamma)}$$

where

$$\theta_2 = \frac{\sin \beta\mu_2}{\sinh \beta\mu_2} \qquad \text{and} \qquad \phi_2 = \frac{\cosh \beta\mu_2 - \cos \beta\mu_2}{\sinh \beta\mu_2}$$

$$\theta_3 = \frac{\sin \beta\mu_3}{\sinh \beta\mu_3} \qquad \text{and} \qquad \phi_3 = \frac{\cosh \beta\mu_3 - \cos \beta\mu_3}{\sinh \beta\mu_3}$$

CASE 5.3. Beam with outer ends free.

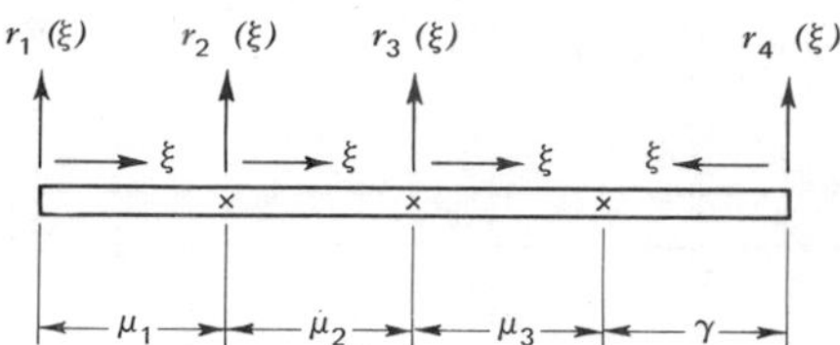

$$r_1(\xi) = \sin \beta\xi + \sinh \beta\xi - \frac{\sin \beta\mu_1 + \sinh \beta\mu_1}{\cos \beta\mu_1 + \cosh \beta\mu_1}(\cos \beta\xi + \cosh \beta\xi)$$

$$r_2(\xi) = A_2(\sin \beta\xi - \theta_2 \sinh \beta\xi) + B_2(\cos \beta\xi - \cosh \beta\xi + \phi_2 \sinh \beta\xi)$$

$$r_3(\xi) = A_3(\sin \beta\xi - \theta_3 \sinh \beta\xi) + B_3(\cos \beta\xi - \cosh \beta\xi + \phi_3 \sinh \beta\xi)$$

$$r_4(\xi) = A_4\left[\sin \beta\xi + \sinh \beta\xi - \frac{\sin \beta\gamma + \sinh \beta\gamma}{\cos \beta\gamma + \cosh \beta\gamma}(\cos \beta\xi + \cosh \beta\xi)\right]$$

$$A_2 = \frac{2(1 + \cos \beta\mu_1 \cosh \beta\mu_1) + \phi_2(\sinh \beta\mu_1 \cos \beta\mu_1 - \sin \beta\mu_1 \cosh \beta\mu_1)}{(\cos \beta\mu_1 + \cosh \beta\mu_1)(1 - \theta_2)}$$

$$B_2 = \frac{\sin \beta\mu_1 \cosh \beta\mu_1 - \sinh \beta\mu_1 \cos \beta\mu_1}{\cos \beta\mu_1 + \cosh \beta\mu_1}$$

$$B_3 = \frac{A_2}{2}(\sin \beta\mu_2 + \theta_2 \sinh \beta\mu_2) + \frac{B_2}{2}(\cos \beta\mu_2 + \cosh \beta\mu_2 - \phi_2 \sinh \beta\mu_2)$$

$$A_3 = \frac{A_2}{1 - \theta_3}(\cos \beta\mu_2 - \theta_2 \cosh \beta\mu_2) + \frac{B_2}{1 - \theta_3}(\sin \beta\mu_2 - \sinh \beta\mu_2 + \phi_2 \cosh \beta\mu_2) - \frac{B_3\phi_3}{1 - \theta_3}$$

$$A_4 = \frac{A_3(\sin \beta\mu_3 \cosh \beta\mu_3 - \sinh \beta\mu_3 \cos \beta\mu_3)}{2 \sinh \beta\mu_3[(1 + \cos \beta\gamma \cosh \beta\gamma)/(\cos \beta\gamma + \cosh \beta\gamma)]} + \frac{B_3(\sin \beta\mu_3 \sinh \beta\mu_3 + \cos \beta\mu_3 \cosh \beta\mu_3 - 1)}{2 \sinh \beta\mu_3[(1 + \cos \beta\gamma \cosh \beta\gamma)/(\cos \beta\gamma + \cosh \beta\gamma)]}$$

or

$$A_4 = \frac{A_3\{\sin \beta\mu_3 \sinh \beta\mu_3\} + B_3\{\cos \beta\mu_3 \sinh \beta\mu_3\}}{\sinh \beta\mu_3\{(\sin \beta\gamma \cosh \beta\gamma - \sinh \beta\gamma \cos \beta\gamma)/(\cos \beta\gamma + \cosh \beta\gamma)\}}$$

where

$$\theta_2 = \frac{\sin \beta\mu_2}{\sinh \beta\mu_2}, \qquad \phi_2 = \frac{\cosh \beta\mu_2 - \cos \beta\mu_2}{\sinh \beta\mu_2}$$

$$\theta_3 = \frac{\sin \beta\mu_3}{\sinh \beta\mu_3}, \qquad \text{and } \phi_3 = \frac{\cosh \beta\mu_3 - \cos \beta\mu_3}{\sinh \beta\mu_3}$$

CASE 5.4. Beam with one outer end clamped and the other simply supported.

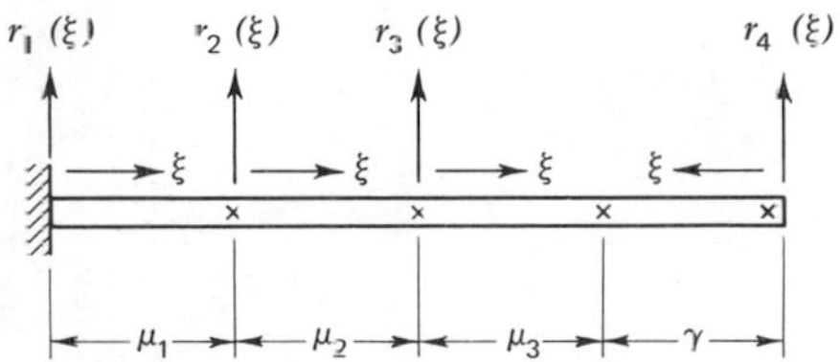

$r_1(\xi)$, $r_2(\xi)$, $r_3(\xi)$, A_2, B_2, A_3, B_3; as for Case 5.2.
$r_4(\xi)$, A_4; as for Case 5.1.

CASE 5.5. Beam with one outer end free and the other clamped.

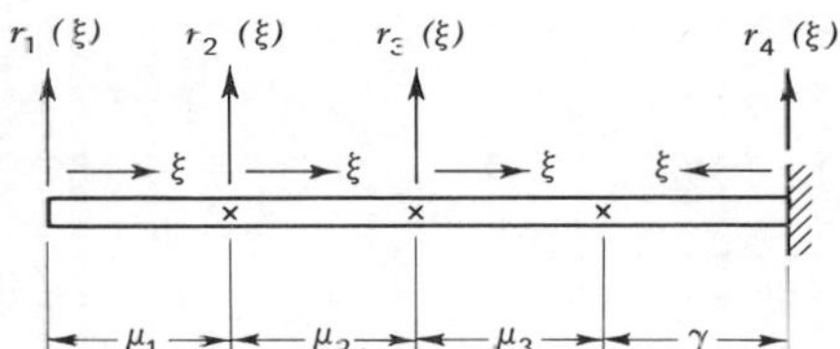

$r_1(\xi)$, $r_2(\xi)$, $r_3(\xi)$, A_2, B_2, A_3, B_3; as for Case 5.3.
$r_4(\xi)$, A_4; as for Case 5.2.

CASE 5.6. Beam with one end simply supported and the other free.

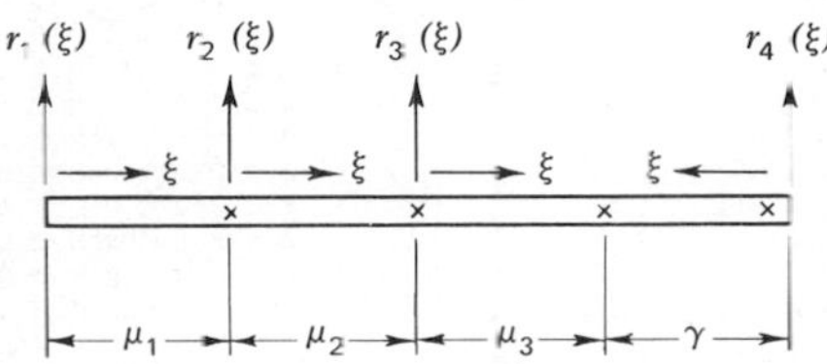

$r_1(\xi)$, $r_2(\xi)$, $r_3(\xi)$, A_2, B_2, A_3, B_3; as for Case 5.3.
$r_4(\xi)$, A_4; as for Case 5.1.

TABLE 5.1 (μ_1 = .05)

MODE 1

μ_3 →

μ_2 ↓	.050	.100	.150	.200	.250	.300	.350	.400	.450	.500	.550	.600	.650	.700	.750	.800	.850	μ_2 ↑
.05	4.547	4.764	5.013	5.300	5.632	6.018	6.466	6.976	7.495	7.739	7.460	7.020	6.598	6.224	5.900	5.621	5.382	
.10	4.823	5.067	5.350	5.680	6.067	6.519	7.045	7.625	8.073	7.939	7.473	6.999	6.577	6.213	5.902	5.637		
.15	5.138	5.413	5.739	6.123	6.578	7.115	7.733	8.343	8.470	8.014	7.476	6.993	6.578	6.227	5.930		8.952	.85
.20	5.496	5.813	6.192	6.645	7.187	7.829	8.542	8.997	8.650	8.046	7.486	7.007	6.604	6.268		9.393	9.489	.80
.25	5.910	6.278	6.725	7.267	7.922	8.693	9.424	9.372	8.722	8.069	7.510	7.044	6.658		9.907	9.985	10.10	.75
.30	6.391	6.825	7.360	8.018	8.820	9.715	10.10	9.496	8.749	8.095	7.553	7.107		10.50	10.57	10.66	10.79	.70
.35	6.957	7.477	8.129	8.940	9.904	10.57	10.19	9.464	8.745	8.126	7.617		11.15	11.24	11.33	11.44	11.59	.65
.40	7.634	8.266	9.065	9.972	10.21	10.04	9.726	9.242	8.678	8.148		11.76	11.97	12.11	12.22	12.33	12.38	.60
.45	8.453	9.190	9.472	9.384	9.277	9.151	8.990	8.762	8.453		11.56	12.39	12.87	13.09	13.19	12.16	10.80	.55
.50	8.925	8.761	8.638	8.538	8.448	8.358	8.261	8.148		10.66	11.57	12.71	13.67	13.58	11.94	10.55	9.515	.50
.55	8.168	8.021	7.912	7.825	7.750	7.682	7.617		10.19	10.74	11.61	12.74	13.14	11.74	10.40	9.774	9.905	.45
.60	7.518	7.392	7.297	7.222	7.161	7.107		10.66	10.86	11.21	11.82	12.43	11.59	10.68	10.69	10.84	11.06	.40
.65	6.963	6.854	6.773	6.709	6.658		11.56	11.78	11.98	12.20	12.40	11.92	11.71	11.86	12.04	12.23	12.40	.35
.70	6.485	6.391	6.321	6.268		11.76	12.39	12.98	13.38	13.42	12.38	12.42	13.06	13.45	13.25	12.33	11.52	.30
.75	6.069	5.988	5.930		11.15	11.83	12.63	13.46	13.49	12.13	11.76	12.70	13.70	13.17	12.21	11.35	10.65	.25
.80	5.704	5.637		10.50	11.10	11.82	12.62	13.01	11.87	11.05	11.54	12.66	12.91	12.08	11.23	10.51	9.907	.20
.85	5.382		9.907	10.44	11.07	11.78	12.34	11.64	10.62	10.59	11.46	12.33	11.92	11.13	10.40	9.781	9.259	.15
		9.393	9.863	10.42	11.05	11.64	11.39	10.36	9.937	10.46	11.37	11.66	11.02	10.32	9.691	9.150	8.691	.10
	8.952	9.374	9.870	10.43	11.00	11.08	10.18	9.523	9.705	10.45	11.16	10.89	10.24	9.620	9.073	8.599	8.192	.05
	.850	.800	.750	.700	.650	.600	.550	.500	.450	.400	.350	.300	.250	.200	.150	.100	.050	

MODE 2 ← μ_3

TABLE 5.1 (μ_1 = .10)

MODE 1

μ_3 →

μ_2 ↓	.050	.100	.150	.200	.250	.300	.350	.400	.450	.500	.550	.600	.650	.700	.750	.800	μ_2 ↑
.05	4.827	5.072	5.358	5.689	6.076	6.530	7.060	7.649	8.136	8.049	7.584	7.099	6.667	6.294	5.975	5.704	
.10	5.139	5.417	5.745	6.130	6.586	7.124	7.748	8.377	8.553	8.108	7.563	7.071	6.649	6.290	5.988		
.15	5.498	5.816	6.196	6.651	7.194	7.839	8.562	9.055	8.735	8.126	7.558	7.071	6.663	6.321		9.374	.80
.20	5.911	6.281	6.729	7.272	7.929	8.706	9.463	9.458	8.805	8.142	7.573	7.100	6.709		9.863	9.968	.75
.25	6.391	6.827	7.364	8.024	8.831	9.744	10.21	9.607	8.837	8.167	7.613	7.161		10.44	10.52	10.65	.70
.30	6.958	7.480	8.134	8.951	9.940	10.80	10.46	9.630	8.855	8.208	7.682		11.10	11.18	11.28	11.43	.65
.35	7.635	8.271	9.081	10.09	10.93	10.80	10.25	9.540	8.849	8.261		11.83	11.94	12.05	12.17	12.33	.60
.40	8.456	9.237	10.09	10.18	10.06	9.885	9.624	9.237	8.762		12.39	12.74	12.92	13.06	13.17	12.51	.55
.45	9.422	9.444	9.315	9.202	9.096	8.983	8.849	8.678		11.78	12.84	13.68	14.03	13.89	12.19	10.81	.50
.50	8.759	8.593	8.471	8.374	8.288	8.208	8.126		10.86	11.76	12.97	14.20	13.72	11.96	10.57	9.747	.45
.55	8.012	7.870	7.765	7.682	7.613	7.553		10.74	11.16	11.89	12.96	13.25	11.78	10.64	10.61	10.81	.40
.60	7.381	7.260	7.170	7.100	7.044		11.57	11.76	12.01	12.42	12.77	11.89	11.63	11.76	11.95	12.22	.35
.65	6.843	6.739	6.663	6.604		12.39	12.84	13.17	13.42	13.44	12.70	12.88	13.20	13.43	13.39	12.54	.30
.70	6.379	6.290	6.227		11.97	12.74	13.61	14.38	13.89	12.57	12.91	14.06	14.37	13.36	12.35	11.52	.25
.75	5.975	5.902		11.24	11.94	12.78	13.67	13.59	12.18	11.89	12.94	13.91	13.20	12.22	11.36	10.65	.20
.80	5.621		10.57	11.18	11.92	12.73	13.08	11.90	11.12	11.67	12.82	12.95	12.09	11.24	10.51	9.909	.15
		9.985	10.52	11.16	11.88	12.43	11.67	10.67	10.71	11.61	12.43	11.95	11.14	10.41	9.787	9.261	.10
	9.489	9.968	10.54	11.18	11.77	11.44	10.41	10.04	10.63	11.56	11.72	11.04	10.33	9.702	9.159	8.696	.05
	.800	.750	.700	.650	.600	.550	.500	.450	.400	.350	.300	.250	.200	.150	.100	.050	

MODE 2 ← μ_3

TABLE 5.1 (μ_1 = .05)

MODE 3

μ_3 →

μ_2 ↓	.050	.100	.150	.200	.250	.300	.350	.400	.450	.500	.550	.600	.650	.700	.750	.800	.850	μ_2 ↑
.05	11.84	12.45	13.15	13.94	14.62	13.85	12.94	13.48	14.46	14.33	13.41	12.92	13.98	14.56	13.88	13.17	12.55	
.10	12.56	13.25	14.05	14.93	15.26	13.92	13.72	14.73	15.32	14.38	13.58	14.27	15.35	14.68	13.88	13.19		
.15	13.39	14.17	15.09	16.02	15.54	14.30	14.95	16.10	15.57	14.50	14.62	16.16	15.60	14.71	13.93		16.16	.85
.20	14.32	15.22	16.27	16.99	15.59	15.24	16.49	16.78	15.60	15.12	16.80	16.56	15.60	14.74		17.00	17.14	.80
.25	15.39	16.30	16.27	15.82	15.47	16.31	16.27	15.83	15.46	16.39	16.22	15.80	15.25		17.96	18.09	18.25	.75
.30	14.50	14.17	14.02	14.75	15.54	14.79	14.29	14.46	15.54	14.93	14.39	14.07		18.68	19.14	19.31	18.49	.70
.35	12.71	13.60	14.84	16.17	15.64	14.26	14.55	16.30	15.64	14.53	13.65		17.16	19.18	20.35	17.94	16.31	.65
.40	13.79	15.02	16.51	16.99	15.62	15.05	16.90	16.73	15.60	14.50		17.44	17.81	19.07	17.72	17.42	17.60	.60
.45	15.26	16.17	16.03	15.76	15.47	16.17	16.04	15.77	15.26		18.52	18.81	19.03	18.61	18.70	18.90	19.07	.55
.50	14.93	14.71	14.57	14.60	15.51	15.00	14.69	14.50		18.00	19.50	20.28	18.53	19.13	20.25	19.13	17.16	.50
.55	13.63	13.48	14.26	16.43	15.63	14.26	13.65		16.97	18.06	19.51	17.92	17.41	19.18	18.82	16.94	16.55	.45
.60	12.70	14.52	16.89	16.76	15.58	14.07		18.00	18.27	18.69	18.16	18.15	18.51	18.62	18.00	18.20	18.48	.40
.65	14.93	16.06	15.94	15.74	15.25		18.52	19.57	20.17	18.77	19.35	20.14	19.26	19.01	20.07	19.62	18.16	.35
.70	15.14	14.99	14.87	14.74		17.44	18.61	19.79	18.15	18.07	19.60	18.98	17.65	19.22	19.39	17.95	16.69	.30
.75	14.17	14.03	13.93		17.16	17.81	18.69	17.82	17.42	18.31	18.62	17.32	17.78	18.85	17.83	16.91	17.21	.25
.80	13.31	13.19		18.68	19.45	19.98	18.83	19.27	19.93	19.24	19.03	19.85	19.57	18.78	19.70	19.82	18.74	.20
.85	12.55		17.96	18.97	19.95	18.38	18.60	19.83	19.06	18.16	19.61	19.48	18.23	19.17	19.70	18.56	17.52	.15
		17.00	17.92	18.86	17.87	17.40	18.52	18.70	17.27	17.91	18.99	17.91	17.23	18.89	18.40	17.36	16.44	.10
	16.16	16.98	17.86	17.46	16.42	17.31	18.09	16.87	16.54	17.81	17.68	16.62	16.87	18.07	17.24	16.31	15.49	.05
	.850	.800	.750	.700	.650	.600	.550	.500	.450	.400	.350	.300	.250	.200	.150	.100	.050	

MODE 4 ← μ_3

TABLE 5.1 (μ_1 = .10)

MODE 3

μ_3 →

μ_2 ↓	.050	.100	.150	.200	.250	.300	.350	.400	.450	.500	.550	.600	.650	.700	.750	.800	μ_2 ↑
.05	12.57	13.26	14.07	14.96	15.45	14.15	13.79	14.76	15.47	14.56	13.72	14.32	15.48	14.82	14.01	13.31	
.10	13.39	14.18	15.10	16.07	15.77	14.42	14.98	16.17	15.73	14.63	14.67	16.25	15.72	14.82	14.03		
.15	14.33	15.23	16.30	17.20	15.89	15.32	16.54	17.05	15.81	15.22	16.93	16.77	15.76	14.87		16.98	.80
.20	15.41	16.45	17.61	17.49	16.07	16.78	17.91	17.00	16.02	17.35	17.64	16.68	15.74		17.92	18.06	.75
.25	16.47	16.20	15.88	15.65	16.33	16.31	15.94	15.69	16.44	16.25	15.91	15.58		18.97	19.13	19.30	.70
.30	14.12	13.94	14.90	16.26	16.02	14.79	14.74	16.46	15.91	14.92	14.26		19.45	20.28	20.49	18.53	.65
.35	13.80	15.04	16.57	17.58	15.96	15.18	17.12	17.15	15.83	14.69		17.81	19.65	21.30	18.03	17.38	.60
.40	15.28	16.78	17.47	17.02	16.12	17.28	17.38	16.78	15.77		18.61	18.89	19.64	18.61	18.67	18.88	.55
.45	16.20	15.97	15.78	15.64	16.27	16.07	15.83	15.60		19.57	20.17	20.41	19.63	20.11	20.37	19.51	.50
.50	14.69	14.49	14.58	16.59	15.99	14.92	14.53		18.27	19.98	21.27	18.69	19.67	21.24	19.17	17.19	.45
.55	13.44	14.55	17.25	17.61	15.91	14.39		18.06	18.67	19.84	18.30	18.38	19.65	18.92	17.92	18.16	.40
.60	14.95	17.06	16.96	16.68	15.80		19.50	19.98	20.26	19.54	19.93	20.25	19.62	19.83	20.17	19.89	.35
.65	16.06	15.89	15.76	15.60		18.81	20.17	21.23	19.00	19.98	21.22	19.41	19.66	21.19	19.68	18.17	.30
.70	14.97	14.82	14.71		17.81	18.89	20.03	18.30	18.45	19.91	19.05	18.02	19.67	19.45	17.99	16.88	.25
.75	14.01	13.88		19.18	19.65	20.06	19.23	19.56	20.02	19.40	19.45	19.96	19.63	19.27	19.75	20.03	.20
.80	13.17		19.14	20.28	21.19	19.24	20.15	21.18	19.50	19.94	21.17	19.79	19.63	21.15	19.96	18.75	.15
		18.09	19.13	20.12	18.46	18.77	20.03	19.11	18.33	19.86	19.52	18.31	19.54	19.75	18.58	17.53	.10
	17.14	18.06	19.02	17.94	17.57	18.71	18.76	17.36	18.15	19.09	17.95	17.42	19.11	18.43	17.38	16.45	.05
	.800	.750	.700	.650	.600	.550	.500	.450	.400	.350	.300	.250	.200	.150	.100	.050	

MODE 4 ← μ_3

TABLE 5.1 ($\mu_1 = .15$)

MODE 1 ($\mu_3 \rightarrow$)

$\mu_2 \downarrow$	.050	.100	.150	.200	.250	.300	.350	.400	.450	.500	.550	.600	.650	.700	.750	$\mu_2 \uparrow$
.05	5.143	5.424	5.753	6.140	6.597	7.138	7.768	8.422	8.669	8.242	7.686	7.181	6.746	6.379	6.069	
.10	5.500	5.821	6.203	6.658	7.204	7.852	8.586	9.128	8.846	8.232	7.653	7.156	6.739	6.391		
.15	5.912	6.284	6.734	7.279	7.938	8.721	9.504	9.555	8.904	8.230	7.651	7.170	6.773		9.870	.75
.20	6.392	6.830	7.369	8.031	8.842	9.770	10.30	9.717	8.931	8.247	7.682	7.222		10.42	10.54	.70
.25	6.959	7.483	8.139	8.960	9.963	10.91	10.65	9.766	8.956	8.288	7.750		11.07	11.16	11.30	.65
.30	7.636	8.275	9.091	10.12	11.27	11.43	10.64	9.751	8.983	8.358		11.82	11.92	12.03	12.19	.60
.35	8.459	9.251	10.26	11.10	11.03	10.74	10.26	9.624	8.990		12.63	12.78	12.91	13.04	13.22	.55
.40	9.473	10.21	10.16	10.04	9.908	9.751	9.540	9.242		12.98	13.61	13.90	14.06	14.08	12.53	.50
.45	9.495	9.306	9.168	9.057	8.956	8.855	8.745		11.98	13.17	14.44	15.06	14.09	12.20	10.82	.45
.50	8.619	8.458	8.340	8.247	8.167	8.095		11.21	12.01	13.22	14.54	13.79	11.99	10.70	10.62	.40
.55	7.888	7.752	7.651	7.573	7.510		11.61	11.89	12.41	13.29	13.38	11.95	11.58	11.73	11.98	.35
.60	7.272	7.156	7.071	7.007		12.71	12.97	13.22	13.49	13.49	12.84	12.97	13.18	13.41	13.60	.30
.65	6.746	6.649	6.578		12.87	13.68	14.44	14.91	14.20	13.34	14.13	14.83	14.66	13.54	12.55	.25
.70	6.294	6.213		12.11	12.92	13.90	14.82	14.00	12.69	13.20	14.60	14.52	13.38	12.36	11.53	.20
.75	5.900		11.33	12.05	12.91	13.81	13.64	12.23	12.01	13.13	14.03	13.23	12.23	11.37	10.66	.15
		10.66	11.28	12.03	12.85	13.16	11.94	11.21	11.83	12.99	13.00	12.11	11.25	10.52	9.912	.10
	10.10	10.65	11.30	12.03	12.56	11.71	10.74	10.88	11.83	12.56	11.98	11.16	10.43	9.797	9.267	.05
	.750	.700	.650	.600	.550	.500	.450	.400	.350	.300	.250	.200	.150	.100	.050	

MODE 2 ($\leftarrow \mu_3$)

TABLE 5.1 ($\mu_1 = .20$)

MODE 1 ($\mu_3 \rightarrow$)

$\mu_2 \downarrow$	.050	.100	.150	.200	.250	.300	.350	.400	.450	.500	.550	.600	.650	.700	$\mu_2 \uparrow$
.05	5.504	5.828	6.213	6.670	7.217	7.871	8.618	9.222	9.001	8.379	7.783	7.272	6.843	6.485	
.10	5.914	6.289	6.742	7.288	7.950	8.740	9.553	9.675	9.030	8.343	7.752	7.260	6.854		
.15	6.394	6.834	7.375	8.040	8.854	9.797	10.39	9.837	9.038	8.340	7.765	7.297		10.43	.70
.20	6.960	7.487	8.146	8.970	9.982	10.98	10.80	9.891	9.057	8.374	7.825		11.05	11.18	.65
.25	7.637	8.279	9.099	10.14	11.36	11.79	10.88	9.908	9.096	8.448		11.78	11.88	12.03	.60
.30	8.461	9.260	10.30	11.57	12.05	11.57	10.74	9.885	9.151		12.62	12.73	12.85	13.02	.55
.35	9.481	10.47	11.15	11.06	10.88	10.64	10.25	9.726		13.46	13.67	13.81	13.95	14.12	.50
.40	10.36	10.17	10.02	9.891	9.766	9.630	9.464		13.38	14.38	14.82	14.99	14.42	12.54	.45
.45	9.356	9.171	9.038	8.931	8.837	8.749		12.20	13.42	14.91	15.71	14.14	12.22	10.86	.40
.50	8.498	8.343	8.230	8.142	8.069		11.82	12.42	13.49	14.76	13.84	12.07	11.53	11.74	.35
.55	7.783	7.653	7.558	7.486		12.74	12.96	13.29	13.84	13.64	12.77	12.06	13.07	13.35	.30
.60	7.181	7.071	6.993		13.67	14.20	14.54	14.76	14.33	14.02	14.41	14.67	14.74	13.76	.25
.65	6.667	6.577		13.09	14.03	15.06	15.71	14.36	13.65	14.84	15.71	14.78	13.55	12.55	.20
.70	6.224		12.22	13.06	14.06	14.99	14.05	12.78	13.41	14.85	14.56	13.40	12.37	11.53	.15
		11.44	12.17	13.04	13.95	13.70	12.28	12.16	13.34	14.13	13.25	12.24	11.37	10.66	.10
	10.79	11.43	12.19	13.02	13.26	11.99	11.34	12.07	13.22	13.05	12.13	11.27	10.53	9.919	.05
	.700	.650	.600	.550	.500	.450	.400	.350	.300	.250	.200	.150	.100	.050	

MODE 2 ($\leftarrow \mu_3$)

TABLE 5.1 ($\mu_1 = .15$)

MODE 3 ($\mu_3 \rightarrow$)

$\mu_2 \downarrow$	.050	.100	.150	.200	.250	.300	.350	.400	.450	.500	.550	.600	.650	.700	.750	$\mu_2 \uparrow$
.05	13.40	14.19	15.12	16.12	16.06	14.61	15.02	16.25	15.94	14.81	14.75	16.38	15.89	14.97	14.17	
.10	14.33	15.24	16.31	17.31	16.15	15.39	16.58	17.23	15.98	15.32	17.01	16.92	15.89	14.99		
.15	15.41	16.47	17.70	18.17	16.38	16.86	18.30	17.42	16.24	17.50	18.08	16.96	15.94		17.86	.75
.20	16.66	17.85	18.24	17.46	17.13	18.30	17.96	17.12	17.72	18.22	17.61	16.76		18.86	19.02	.70
.25	16.17	15.78	15.59	16.47	16.78	16.05	15.74	16.80	16.58	15.99	15.63		19.95	20.12	20.29	.65
.30	13.96	15.07	16.61	17.85	16.31	15.39	17.22	17.41	16.07	15.00		19.98	21.19	21.40	18.56	.60
.35	15.29	16.83	18.55	18.09	16.50	17.68	18.55	17.38	16.04		18.69	20.06	21.68	18.51	18.56	.55
.40	17.09	17.44	17.22	16.91	17.33	17.41	17.15	16.73		19.79	20.03	20.41	19.77	19.92	20.13	.50
.45	15.96	15.73	15.61	16.81	16.58	15.91	15.64		20.17	21.23	21.47	20.27	21.25	21.47	19.54	.45
.50	14.49	14.69	17.35	18.22	16.25	14.93		18.69	20.26	21.57	18.93	20.06	21.46	19.21	17.82	.40
.55	14.97	17.75	18.08	17.64	16.22		19.51	19.84	20.40	19.53	19.77	20.37	19.53	19.59	19.87	.35
.60	17.09	16.92	16.77	16.56		20.28	21.27	21.57	20.36	21.30	21.54	20.33	21.32	21.52	19.94	.30
.65	15.89	15.72	15.60		19.03	20.41	21.47	19.17	20.27	21.41	19.49	20.06	21.33	19.70	18.19	.25
.70	14.82	14.68		19.07	19.64	20.41	19.12	19.49	20.38	19.34	19.34	20.32	19.59	18.95	19.34	.20
.75	13.88		20.35	21.30	21.68	20.41	21.34	21.63	20.38	21.39	21.54	20.38	21.43	21.48	20.17	.15
		19.31	20.49	21.40	19.39	20.40	21.36	19.59	20.25	21.31	19.84	20.02	21.25	19.97	18.76	.10
	18.25	19.30	20.29	18.56	18.98	20.23	19.16	18.55	20.12	19.56	18.41	19.91	19.78	18.59	17.53	.05
	.750	.700	.650	.600	.550	.500	.450	.400	.350	.300	.250	.200	.150	.100	.050	

MODE 4 ($\leftarrow \mu_3$)

TABLE 5.1 ($\mu_1 = .20$)

MODE 3 ($\mu_3 \rightarrow$)

$\mu_2 \downarrow$	.050	.100	.150	.200	.250	.300	.350	.400	.450	.500	.550	.600	.650	.700	$\mu_2 \uparrow$
.05	14.34	15.26	16.34	17.39	16.47	15.49	16.63	17.41	16.20	15.45	17.09	17.09	16.06	15.14	
.10	15.42	16.48	17.67	17.72	16.55	16.89	17.81	17.43	16.37	17.49	17.75	17.06	16.06		
.15	16.67	17.38	17.32	17.19	17.11	17.37	17.30	17.10	17.33	17.35	17.25	16.89		17.46	.70
.20	16.73	16.58	16.46	16.64	16.90	16.63	16.48	16.88	16.81	16.59	16.43		17.87	17.94	.65
.25	13.28	15.25	16.62	17.65	16.49	15.72	17.20	17.33	16.27	15.51		18.38	18.46	18.56	.60
.30	15.30	16.84	17.72	17.54	16.60	17.54	17.68	17.28	16.17		18.83	19.24	19.39	18.51	.55
.35	17.09	17.22	17.14	17.01	17.20	17.22	17.12	16.90		17.82	19.23	20.41	18.00	17.67	.50
.40	16.54	16.40	16.30	16.88	16.80	16.46	16.30		18.15	18.30	19.12	18.17	18.16	18.25	.45
.45	15.41	15.30	17.33	17.72	16.44	15.54		18.77	19.00	19.17	18.82	18.92	19.07	19.19	.40
.50	15.00	17.49	17.50	17.35	16.39		18.16	19.54	20.36	18.58	19.20	20.33	19.14	17.38	.35
.55	17.09	17.01	16.93	16.88		17.92	18.30	19.53	18.11	18.06	19.22	18.83	17.87	17.94	.30
.60	16.38	16.25	16.16		18.53	18.69	18.93	18.58	18.64	18.84	10.00	18.36	18.69	18.86	.25
.65	15.48	15.35		18.61	19.63	20.27	18.82	19.41	20.24	19.28	19.09	20.18	19.62	18.17	.20
.70	14.56		17.72	18.61	19.77	18.17	18.14	19.58	18.97	17.83	19.18	19.38	17.96	17.49	.15
		17.94	18.03	18.51	18.00	17.97	18.20	18.53	17.95	18.04	18.68	17.97	17.94	17.99	.10
	18.49	18.53	18.56	18.51	18.52	18.55	18.56	18.51	18.53	18.57	18.50	18.52	18.54	18.57	.05
	.700	.650	.600	.550	.500	.450	.400	.350	.300	.250	.200	.150	.100	.050	

MODE 4 ($\leftarrow \mu_3$)

TABLE 5.1 ($\mu_1 = .25$)

$\mu_3 \rightarrow$ MODE 1

$\mu_2 \downarrow$	.050	.100	.150	.200	.250	.300	.350	.400	.450	.500	.550	.600	.650	
.05	5.919	6.298	6.753	7.302	7.968	8.765	9.614	9.838	9.204	8.498	7.888	7.381	6.963	
.10	6.397	6.841	7.384	8.051	8.870	9.828	10.50	9.983	9.171	8.458	7.870	7.392		
.15	6.962	7.492	8.153	8.981	10.00	11.04	10.94	10.02	9.168	8.471	7.912		11.00	.65
.20	7.639	8.284	9.107	10.16	11.41	11.97	11.06	10.04	9.202	8.538		11.64	11.77	.60
.25	8.463	9.266	10.32	11.65	12.57	12.05	11.03	10.06	9.277		12.34	12.43	12.56	.55
.30	9.485	10.51	11.75	11.97	11.79	11.43	10.80	10.04		13.01	13.08	13.16	13.26	.50
.35	10.76	11.08	10.94	10.80	10.65	10.46	10.19		13.49	13.59	13.64	13.70	13.77	.45
.40	10.18	9.983	9.837	9.717	9.607	9.496		13.42	13.89	14.00	14.05	14.06	12.55	.40
.45	9.204	9.030	8.904	8.805	8.722		12.40	13.44	14.20	14.36	14.09	12.25	11.40	.35
.50	8.379	8.232	8.126	8.046		12.43	12.77	13.49	14.33	13.81	12.39	12.38	12.56	.30
.55	7.686	7.563	7.476		13.14	13.25	13.38	13.64	13.57	13.17	13.23	13.31	13.42	.25
.60	7.099	6.999		13.58	13.72	13.79	13.84	13.81	13.67	13.74	13.79	13.84	13.75	.20
.65	6.598		13.18	13.89	14.09	14.14	14.09	13.67	14.04	14.12	14.15	13.56	12.56	.15
		12.33	13.17	14.08	14.42	14.06	12.87	13.56	14.37	14.41	13.41	12.38	11.53	.10
	11.59	12.33	13.22	14.12	13.77	12.36	12.38	13.61	14.22	13.27	12.25	11.39	10.67	.05
	.650	.600	.550	.500	.450	.400	.350	.300	.250	.200	.150	.100	.050	$\uparrow \mu_2$

MODE 2 $\leftarrow \mu_3$

TABLE 5.1 ($\mu_1 = .30$)

$\mu_3 \rightarrow$ MODE 1

$\mu_2 \downarrow$	.050	.100	.150	.200	.250	.300	.350	.400	.450	.500	.550	.600	
.05	6.403	6.851	7.398	8.068	8.892	9.868	10.63	10.18	9.356	8.619	8.012	7.518	
.10	6.965	7.499	8.164	8.995	10.02	11.10	11.08	10.17	9.306	8.593	8.021		
.15	7.641	8.290	9.117	10.17	11.42	11.75	11.15	10.16	9.315	8.638		11.08	.60
.20	8.465	9.272	10.33	11.57	11.65	11.57	11.10	10.18	9.384		11.39	11.44	.55
.25	9.488	10.52	11.42	11.41	11.36	11.27	10.93	10.21		11.64	11.67	11.71	.50
.30	10.78	11.10	11.04	10.98	10.91	10.80	10.57		11.87	11.90	11.94	11.99	.45
.35	10.63	10.50	10.39	10.30	10.21	10.10		12.13	12.18	12.23	12.28	12.36	.40
.40	9.838	9.675	9.555	9.458	9.372		12.38	12.57	12.69	12.78	12.87	12.51	.35
.45	9.001	8.846	8.735	8.650		11.92	12.70	13.34	13.65	13.67	12.19	11.22	.30
.50	8.242	8.108	8.014		11.59	11.89	12.84	14.02	13.67	11.97	11.51	11.55	.25
.55	7.584	7.473		11.74	11.78	11.95	12.77	13.17	11.85	11.74	11.76	11.79	.20
.60	7.020		11.94	11.96	11.99	12.07	12.39	11.97	11.95	11.96	11.98	12.01	.15
		12.16	12.19	12.20	12.22	12.25	12.19	12.18	12.20	12.21	12.23	12.25	.10
	12.38	12.51	12.53	12.54	12.55	12.51	12.52	12.53	12.54	12.55	12.38	11.54	.05
	.600	.550	.500	.450	.400	.350	.300	.250	.200	.150	.100	.050	$\uparrow \mu_2$

MODE 2 $\leftarrow \mu_3$

TABLE 5.1 ($\mu_1 = .25$)

$\mu_3 \rightarrow$ MODE 3

$\mu_2 \downarrow$	.050	.100	.150	.200	.250	.300	.350	.400	.450	.500	.550	.600	.650	
.05	14.99	14.97	14.96	14.95	14.95	14.99	14.96	14.95	14.95	15.00	14.97	14.95	14.93	
.10	14.57	14.54	14.52	14.69	15.26	14.63	14.54	14.55	15.30	14.69	14.55	14.52		
.15	14.25	14.21	14.84	16.16	15.60	14.44	14.57	16.30	15.61	14.58	14.26		16.42	.65
.20	13.94	15.03	16.54	17.30	15.69	15.08	17.01	16.91	15.64	14.60		17.40	17.57	.60
.25	15.27	16.56	16.52	16.16	15.71	16.60	16.50	16.12	15.47		18.60	18.77	18.98	.55
.30	15.40	15.24	15.13	15.08	15.72	15.39	15.18	15.05		19.27	20.15	20.40	18.52	.50
.35	14.61	14.53	14.57	16.48	15.74	14.74	14.55		17.42	19.56	21.34	17.97	16.67	.45
.40	14.18	14.55	17.10	17.12	15.69	14.46		18.07	18.45	19.49	18.14	18.09	18.34	.40
.45	14.95	16.37	16.24	16.02	15.46		19.35	19.98	20.27	19.41	19.89	20.22	19.50	.35
.50	15.45	15.32	15.22	15.12		18.15	19.93	21.30	18.64	19.58	21.27	19.16	17.17	.30
.55	14.75	14.67	14.62		17.41	18.38	19.77	18.06	17.85	19.54	18.88	17.15	17.23	.25
.60	14.32	14.27		19.13	19.67	20.06	19.20	19.58	20.03	19.38	19.41	19.88	19.85	.20
.65	13.98		18.70	20.11	21.25	18.92	19.89	21.23	19.38	19.52	21.20	19.67	18.17	.15
		17.42	18.67	19.92	18.16	18.09	19.76	19.00	17.64	19.41	19.41	17.95	16.69	.10
	16.31	17.38	18.56	17.67	16.67	18.00	18.56	16.99	17.02	18.76	17.77	16.50	15.43	.05
	.650	.600	.550	.500	.450	.400	.350	.300	.250	.200	.150	.100	.050	$\uparrow \mu_2$

MODE 4 $\leftarrow \mu_3$

TABLE 5.1 ($\mu_1 = .30$)

$\mu_3 \rightarrow$ MODE 3

$\mu_2 \downarrow$	.050	.100	.150	.200	.250	.300	.350	.400	.450	.500	.550	.600	
.05	12.58	12.57	13.44	14.65	15.39	14.01	13.10	14.18	15.41	14.49	13.44	12.70	
.10	12.57	13.58	14.85	16.22	15.78	14.17	14.53	16.40	15.73	14.49	13.48		
.15	13.79	15.03	16.57	17.70	15.91	15.13	17.14	17.22	15.78	14.57		17.31	.60
.20	15.28	16.80	18.26	17.54	16.16	17.54	18.09	17.02	15.76		18.52	18.71	.55
.25	16.52	16.19	15.91	15.69	16.49	16.31	15.96	15.62		19.83	20.03	20.23	.50
.30	14.41	14.17	14.44	16.63	16.05	14.79	14.26		19.93	21.18	21.36	18.55	.45
.35	13.10	14.54	17.30	17.96	15.94	14.29		18.31	20.02	21.63	18.20	18.00	.40
.40	14.95	17.43	17.42	17.00	15.83		19.60	19.91	20.38	19.58	19.76	20.02	.35
.45	16.20	15.98	15.81	15.60		20.14	21.22	21.41	20.24	21.23	21.41	19.53	.30
.50	14.81	14.63	14.50		18.51	20.25	21.54	18.84	20.03	21.45	19.20	17.24	.25
.55	13.72	13.58		19.18	19.65	20.37	19.22	19.54	20.34	19.30	19.23	19.63	.20
.60	12.92		20.25	21.24	21.46	20.33	21.27	21.45	20.30	21.28	21.45	19.94	.15
		18.90	20.37	21.47	19.07	20.22	21.41	19.45	19.96	21.33	19.69	18.17	.10
	17.60	18.88	20.13	18.25	18.34	20.02	19.05	17.83	19.80	19.46	17.97	16.70	.05
	.600	.550	.500	.450	.400	.350	.300	.250	.200	.150	.100	.050	$\uparrow \mu_2$

MODE 4 $\leftarrow \mu_3$

TABLE 5.1 ($\mu_1 = .35$)

MODE 1 ($\mu_3 \rightarrow$ across top, $\mu_2 \downarrow$ at left); MODE 2 ($\leftarrow \mu_3$ across bottom, $\mu_2 \uparrow$ at right)

μ_2	.050	.100	.150	.200	.250	.300	.350	.400	.450	.500	.550	μ_2
.05	6.972	7.511	8.180	9.015	10.05	10.78	10.76	10.36	9.495	8.759	8.168	
.10	7.645	8.298	9.128	10.18	10.52	10.51	10.47	10.21	9.444	8.761		
.15	8.467	9.278	10.26	10.33	10.32	10.30	10.26	10.09	9.472		10.18	.55
.20	9.490	10.18	10.17	10.16	10.14	10.12	10.09	9.972		10.36	10.41	.50
.25	10.05	10.02	10.00	9.982	9.963	9.940	9.904		10.62	10.67	10.74	.45
.30	9.868	9.828	9.797	9.770	9.744	9.715		11.05	11.12	11.21	11.34	.40
.35	9.614	9.553	9.504	9.463	9.424		11.76	11.89	12.01	12.16	12.38	.35
.40	9.222	9.128	9.055	8.997		12.42	12.91	13.20	13.41	13.56	12.52	.30
.45	8.669	8.553	8.470		11.71	12.88	14.13	14.84	14.04	12.18	10.80	.25
.50	8.049	7.939		10.68	11.63	12.97	14.41	13.74	11.95	10.55	10.22	.20
.55	7.460		10.40	10.64	11.58	12.86	13.23	11.74	10.43	10.37	10.39	.15
		10.55	10.57	10.70	11.53	12.38	11.51	10.55	10.55	10.56	10.58	.10
	10.80	10.81	10.82	10.86	11.40	11.22	10.80	10.81	10.81	10.82	10.83	.05
	.550	.500	.450	.400	.350	.300	.250	.200	.150	.100	.050	

TABLE 5.1 ($\mu_1 = .40$)

MODE 1 ($\mu_3 \rightarrow$ across top, $\mu_2 \downarrow$ at left); MODE 2 ($\leftarrow \mu_3$ across bottom, $\mu_2 \uparrow$ at right)

μ_2	.050	.100	.150	.200	.250	.300	.350	.400	.450	.500	μ_2
.05	7.653	8.313	9.146	9.490	9.488	9.485	9.481	9.473	9.422	8.925	
.10	8.471	9.237	9.278	9.272	9.266	9.260	9.251	9.237	9.190		
.15	9.146	9.128	9.117	9.107	9.099	9.091	9.081	9.065		9.523	.50
.20	9.015	8.995	8.981	8.970	8.960	8.951	8.940		9.937	10.04	.45
.25	8.892	8.870	8.854	8.842	8.831	8.820		10.59	10.71	10.88	.40
.30	8.765	8.740	8.721	8.706	8.693		11.54	11.67	11.83	12.07	.35
.35	8.618	8.586	8.562	8.542		12.70	12.94	13.13	13.34	13.61	.30
.40	8.422	8.377	8.343		13.06	14.06	14.60	14.85	14.37	12.53	.25
.45	8.136	8.073		11.86	13.20	14.83	15.71	14.12	12.20	10.81	.20
.50	7.739		10.69	11.76	13.18	14.67	13.79	11.96	10.55	9.502	.15
		9.774	10.61	11.73	13.07	13.31	11.76	10.37	9.348	9.312	.10
	9.515	9.747	10.62	11.74	12.56	11.55	10.22	9.502	9.504	9.514	.05
	.500	.450	.400	.350	.300	.250	.200	.150	.100	.050	

TABLE 5.1 ($\mu_1 = .35$)

MODE 3 ($\mu_3 \rightarrow$ across top, $\mu_2 \downarrow$ at left); MODE 4 ($\leftarrow \mu_3$ across bottom, $\mu_2 \uparrow$ at right)

μ_2	.050	.100	.150	.200	.250	.300	.350	.400	.450	.500	.550	μ_2
.05	12.57	13.60	14.87	16.29	16.11	14.41	14.61	16.54	15.96	14.69	13.63	
.10	13.79	15.04	16.60	17.89	16.19	15.24	17.22	17.44	15.97	14.71		
.15	15.28	16.84	18.55	18.26	16.52	17.72	18.55	17.47	16.03		18.09	.55
.20	17.12	17.89	17.70	17.30	17.65	17.85	17.58	16.99		18.70	18.76	.50
.25	16.11	15.78	15.60	16.90	16.78	16.02	15.64		19.06	19.11	19.16	.45
.30	14.01	14.63	17.37	18.30	16.31	14.79		19.24	19.50	19.59	18.56	.40
.35	14.96	17.81	18.30	17.91	16.27		18.62	19.40	20.38	18.53	18.56	.35
.40	17.41	17.23	17.05	16.78		18.98	19.05	19.34	18.97	19.00	19.05	.30
.45	15.94	15.73	15.57		19.26	19.41	19.49	19.28	19.38	19.45	19.43	.25
.50	14.56	14.38		18.62	19.62	20.33	18.80	19.38	20.30	19.16	18.17	.20
.55	13.41		18.82	18.92	19.53	18.83	18.88	19.30	18.96	18.82	18.87	.15
		19.13	19.17	19.21	19.14	19.16	19.20	19.16	19.15	19.18	19.21	.10
	19.07	19.51	19.54	19.19	19.50	19.53	19.43	19.47	19.53	19.53	18.19	.05
	.550	.500	.450	.400	.350	.300	.250	.200	.150	.100	.050	

TABLE 5.1 ($\mu_1 = .40$)

MODE 3 ($\mu_3 \rightarrow$ across top, $\mu_2 \downarrow$ at left); MODE 4 ($\leftarrow \mu_3$ across bottom, $\mu_2 \uparrow$ at right)

μ_2	.050	.100	.150	.200	.250	.300	.350	.400	.450	.500	μ_2
.05	13.81	15.06	16.63	17.12	16.52	15.40	17.09	17.09	16.20	14.93	
.10	15.29	16.78	16.84	16.80	16.56	16.84	16.83	16.78	16.17		
.15	16.63	16.60	16.57	16.54	16.62	16.61	16.57	16.51		16.87	.50
.20	16.29	16.22	16.16	16.64	16.47	16.26	16.17		17.27	17.36	.45
.25	15.39	15.26	17.11	17.13	16.33	15.54		18.16	18.33	18.55	.40
.30	14.99	16.89	16.86	16.78	16.31		19.03	19.94	20.25	18.51	.35
.35	16.63	16.58	16.54	16.49		17.32	19.45	21.39	17.95	16.99	.30
.40	16.25	16.17	16.10		17.65	18.02	19.34	17.83	17.64	17.83	.25
.45	15.47	15.32		19.01	19.66	20.06	19.09	19.52	19.96	19.47	.20
.50	14.33		18.00	19.83	21.32	18.56	19.41	21.28	19.15	17.16	.15
		16.94	17.92	19.59	17.87	17.15	19.23	18.82	16.94	16.89	.10
	17.16	17.19	17.82	17.38	17.17	17.24	18.17	17.16	17.16	17.18	.05
	.500	.450	.400	.350	.300	.250	.200	.150	.100	.050	

TABLE 5.1 ($\mu_1 = .45$)

MODE 1: $\mu_3 \rightarrow$ (top), $\mu_2 \downarrow$ (left); MODE 2: $\leftarrow \mu_3$ (bottom), $\mu_2 \uparrow$ (right)

μ_2 ↓ (Mode 1)	.050	.100	.150	.200	.250	.300	.350	.400	.450	μ_2 ↑ (Mode 2)
.05	8.453	8.471	8.467	8.465	8.463	8.461	8.459	8.456	8.453	
.10	8.313	8.298	8.290	8.284	8.279	8.275	8.271	8.266		
.15	8.180	8.164	8.153	8.146	8.139	8.134	8.129		9.705	.45
.20	8.068	8.051	8.040	8.031	8.024	8.018		10.46	10.63	.40
.25	7.968	7.950	7.938	7.929	7.922		11.46	11.61	11.83	.35
.30	7.871	7.852	7.839	7.829		12.66	12.82	12.99	13.22	.30
.35	7.768	7.748	7.733		13.70	13.91	14.03	14.13	14.22	.25
.40	7.649	7.625		13.45	14.37	14.52	14.56	14.41	12.54	.20
.45	7.495		12.04	13.43	14.66	14.78	14.15	12.21	10.81	.15
		10.84	11.95	13.41	14.74	13.84	11.98	10.56	9.504	.10
	9.905	10.81	11.98	13.35	13.42	11.79	10.39	9.312	8.512	.05
	.450	.400	.350	.300	.250	.200	.150	.100	.050	

TABLE 5.1 ($\mu_1 = .50$)

MODE 1: $\mu_3 \rightarrow$ (top), $\mu_2 \downarrow$ (left); MODE 2: $\leftarrow \mu_3$ (bottom), $\mu_2 \uparrow$ (right)

μ_2 ↓ (Mode 1)	.050	.100	.150	.200	.250	.300	.350	.400	μ_2 ↑ (Mode 2)
.05	7.653	7.645	7.641	7.639	7.637	7.636	7.635	7.634	
.10	7.511	7.499	7.492	7.487	7.483	7.480	7.477		
.15	7.398	7.384	7.375	7.369	7.364	7.360		10.45	.40
.20	7.302	7.288	7.279	7.272	7.267		11.37	11.56	.35
.25	7.217	7.204	7.194	7.187		12.33	12.43	12.56	.30
.30	7.138	7.124	7.115		12.91	12.95	13.00	13.05	.25
.35	7.060	7.045		13.17	13.20	13.23	13.25	13.27	.20
.40	6.976		13.25	13.36	13.38	13.40	13.41	12.55	.15
		12.23	13.39	13.54	13.55	13.56	12.23	10.82	.10
	11.06	12.22	13.60	13.76	13.75	12.01	10.58	9.514	.05
	.400	.350	.300	.250	.200	.150	.100	.050	

TABLE 5.1 ($\mu_1 = .45$)

MODE 3: $\mu_3 \rightarrow$ (top), $\mu_2 \downarrow$ (left); MODE 4: $\leftarrow \mu_3$ (bottom), $\mu_2 \uparrow$ (right)

μ_2 ↓ (Mode 3)	.050	.100	.150	.200	.250	.300	.350	.400	.450	μ_2 ↑ (Mode 4)
.05	15.26	15.29	15.28	15.28	15.27	15.30	15.29	15.28	15.26	
.10	15.06	15.04	15.03	15.03	15.25	15.07	15.04	15.02		
.15	14.87	14.85	14.84	16.46	15.59	14.90	14.84		16.54	.45
.20	14.65	14.69	17.19	17.46	15.65	14.75		17.91	18.15	.40
.25	14.95	16.55	16.38	16.07	15.47		19.61	19.86	20.12	.35
.30	15.49	15.39	15.32	15.24		19.85	21.17	21.31	18.53	.30
.35	15.02	14.98	14.95		17.78	19.96	21.54	18.04	17.02	.25
.40	14.76	14.73		19.22	19.67	20.32	19.18	19.41	19.80	.20
.45	14.46		20.07	21.19	21.33	20.18	21.20	21.33	19.53	.15
		18.20	20.17	21.52	18.69	19.88	21.45	19.18	17.16	.10
	16.55	18.16	19.87	17.94	17.23	19.63	18.87	16.89	15.35	.05
	.450	.400	.350	.300	.250	.200	.150	.100	.050	

TABLE 5.1 ($\mu_1 = .50$)

MODE 3: $\mu_3 \rightarrow$ (top), $\mu_2 \downarrow$ (left); MODE 4: $\leftarrow \mu_3$ (bottom), $\mu_2 \uparrow$ (right)

μ_2 ↓ (Mode 3)	.050	.100	.150	.200	.250	.300	.350	.400	μ_2 ↑ (Mode 4)
.05	13.81	13.79	13.79	13.94	15.28	13.96	13.80	13.79	
.10	13.60	13.58	14.21	16.58	15.78	13.94	13.60		
.15	13.44	14.52	17.32	18.24	15.88	14.02		17.81	.40
.20	14.95	17.72	18.17	17.49	15.82		18.99	19.09	.35
.25	16.47	16.15	15.89	15.59		19.48	19.52	19.56	.30
.30	14.61	14.42	14.30		19.57	19.79	19.84	18.57	.25
.35	13.79	13.72		18.85	19.63	20.38	18.68	18.76	.20
.40	13.48		19.39	19.45	19.59	19.38	19.41	19.46	.15
		19.62	19.68	19.70	19.62	19.67	19.69	19.53	.10
	18.48	19.89	19.94	18.86	19.85	19.94	19.21	17.18	.05
	.400	.350	.300	.250	.200	.150	.100	.050	

TABLE 5.2 ($\mu_1 = .05$)

MODE 1 ($\mu_3 \rightarrow$, $\mu_2 \downarrow$); MODE 2 ($\leftarrow \mu_3$, $\mu_2 \uparrow$)

$\mu_2 \downarrow$	.050	.100	.150	.200	.250	.300	.350	.400	.450	.500	.550	.600	.650	.700	.750	.800	.850	$\mu_2 \uparrow$
.05	5.479	5.743	6.047	6.399	6.806	7.276	7.807	8.351	8.598	8.219	7.665	7.147	6.693	6.300	5.960	5.664	5.406	
.10	5.812	6.108	6.455	6.861	7.333	7.880	8.484	8.973	8.802	8.212	7.623	7.106	6.661	6.279	5.949	5.663		
.15	6.190	6.527	6.927	7.398	7.953	8.591	9.244	9.434	8.870	8.197	7.598	7.087	6.651	6.279	5.959		8.987	.85
.20	6.623	7.010	7.475	8.031	8.688	9.429	10.00	9.647	8.889	8.188	7.591	7.088	6.663	6.301		9.432	9.528	.80
.25	7.121	7.572	8.121	8.783	9.568	10.38	10.50	9.715	8.892	8.188	7.600	7.109	6.695		9.950	10.03	10.14	.75
.30	7.701	8.233	8.889	9.689	10.61	11.22	10.62	9.707	8.886	8.197	7.626	7.149		10.55	10.62	10.71	10.84	.70
.35	8.384	9.019	9.812	10.75	11.37	11.08	10.42	9.615	8.858	8.209	7.664		11.21	11.30	11.39	11.50	11.65	.65
.40	9.197	9.949	10.57	10.51	10.37	10.16	9.843	9.352	8.767	8.203		11.83	12.04	12.17	12.29	12.41	12.58	.60
.45	9.905	9.734	9.592	9.472	9.357	9.230	9.068	8.838	3.513		11.65	12.50	12.97	13.19	13.33	13.45	12.99	.55
.50	9.002	8.830	8.704	8.601	8.510	8.420	8.321	8.203		10.73	11.70	12.89	13.89	14.32	14.21	12.71	11.44	.50
.55	8.219	8.073	7.964	7.877	7.802	7.733	7.664		10.26	10.85	11.77	13.03	14.35	14.05	12.50	11.22	10.26	.45
.60	7.561	7.436	7.342	7.267	7.204	7.149		10.73	10.95	11.32	12.01	13.08	13.57	12.33	11.17	10.95	11.15	.40
.65	7.000	6.893	6.812	6.748	6.695		11.65	11.88	12.09	12.33	12.70	13.10	12.38	12.03	12.16	12.36	12.64	.35
.70	6.518	6.425	6.355	6.301		11.83	12.49	13.11	13.55	13.84	13.87	13.10	13.28	13.66	13.90	13.66	12.80	.30
.75	6.098	6.019	5.959		11.21	11.92	12.76	13.70	14.57	14.19	12.94	13.09	14.30	14.54	13.57	12.62	11.84	.25
.80	5.730	5.663		10.55	11.18	11.93	12.81	13.74	13.86	12.62	12.11	13.05	14.06	13.42	12.50	11.69	11.01	.20
.85	5.406		9.950	10.51	11.17	11.93	12.78	13.31	12.38	11.49	11.81	12.94	13.17	12.38	11.58	10.88	10.29	.15
		9.432	9.928	10.51	11.18	11.93	12.59	12.15	11.16	10.92	11.72	12.61	12.24	11.48	10.78	10.17	9.658	.10
	8.987	9.433	9.949	10.54	11.21	11.88	11.88	10.96	10.38	10.76	11.68	11.99	11.38	10.70	10.09	9.559	9.103	.05
	.850	.800	.750	.700	.650	.600	.550	.500	.450	.400	.350	.300	.250	.200	.150	.100	.050	

TABLE 5.2 ($\mu_1 = .10$)

MODE 1 ($\mu_3 \rightarrow$, $\mu_2 \downarrow$); MODE 2 ($\leftarrow \mu_3$, $\mu_2 \uparrow$)

$\mu_2 \downarrow$	.050	.100	.150	.200	.250	.300	.350	.400	.450	.500	.550	.600	.650	.700	.750	.800	$\mu_2 \uparrow$
.05	5.816	6.115	6.464	6.870	7.344	7.893	8.506	9.036	8.929	8.340	7.738	7.207	6.751	6.360	6.022	5.730	
.10	6.193	6.532	6.933	7.406	7.962	8.605	9.275	9.525	8.980	8.298	7.688	7.166	6.723	6.344	6.019		
.15	6.624	7.014	7.480	8.037	8.697	9.446	10.05	9.748	8.985	8.272	7.665	7.154	6.723	6.355		9.433	.80
.20	7.122	7.575	8.125	8.790	9.579	10.41	10.60	9.821	8.982	8.264	7.667	7.167	6.748		9.928	10.03	.75
.25	7.702	8.236	8.895	9.700	10.64	11.36	10.80	9.832	8.980	8.273	7.689	7.204		10.51	10.60	10.72	.70
.30	8.385	9.024	9.825	10.80	11.79	11.68	10.75	9.798	8.976	8.295	7.733		11.18	11.26	11.37	11.51	.65
.35	9.200	9.974	10.92	11.51	11.37	11.02	10.43	9.676	8.950	8.321		11.92	12.04	12.14	12.26	12.43	.60
.40	10.18	10.61	10.48	10.34	10.20	10.02	9.751	9.348	8.838		12.49	12.85	13.03	13.17	13.31	13.49	.55
.45	9.760	9.572	9.429	9.312	9.203	9.088	8.930	8.767		11.88	12.99	13.83	14.19	14.37	14.39	13.02	.50
.50	8.850	8.685	8.563	8.464	8.378	8.295	8.209		10.95	11.90	13.18	14.60	15.35	14.39	12.73	11.45	.45
.55	8.088	7.948	7.843	7.760	7.689	7.626		10.85	11.28	12.06	13.28	14.66	14.11	12.53	11.28	10.95	.40
.60	7.448	7.328	7.238	7.167	7.109		11.70	11.90	12.16	12.63	13.46	13.72	12.46	11.97	12.11	12.38	.35
.65	6.902	6.799	6.723	6.663		12.50	12.99	13.35	13.63	13.90	13.90	13.28	13.42	13.65	13.09	13.00	.30
.70	6.431	6.344	6.279		12.04	12.85	13.77	14.69	15.30	14.49	13.60	14.43	15.27	14.81	13.73	12.81	.25
.75	6.022	5.949		11.30	12.04	12.91	13.93	14.90	14.27	13.02	13.33	14.72	14.64	13.59	12.63	11.84	.20
.80	5.664		10.62	11.26	12.03	12.92	13.87	13.92	12.65	12.21	13.23	14.16	13.44	12.51	11.69	11.01	.15
		10.03	10.60	11.26	12.04	12.89	13.39	12.41	11.55	11.95	13.10	13.22	12.39	11.58	10.88	10.29	.10
	9.528	10.03	10.62	11.31	12.07	12.72	12.20	11.21	11.06	11.92	12.74	12.26	11.50	10.79	10.18	9.664	.05
	.800	.750	.700	.650	.600	.550	.500	.450	.400	.350	.300	.250	.200	.150	.100	.050	

TABLE 5.2 ($\mu_1 = .05$)

MODE 3 ($\mu_3 \rightarrow$, $\mu_2 \downarrow$); MODE 4 ($\leftarrow \mu_3$, $\mu_2 \uparrow$)

$\mu_2 \downarrow$	.050	.100	.150	.200	.250	.300	.350	.400	.450	.500	.550	.600	.650	.700	.750	.800	.850	$\mu_2 \uparrow$
.05	12.75	13.41	14.18	15.01	15.48	14.21	13.93	14.85	15.49	14.60	13.82	14.53	15.50	14.82	13.99	13.24	12.60	
.10	13.53	14.28	15.15	16.05	15.80	14.50	15.05	16.14	15.76	14.69	14.83	16.23	15.72	14.80	13.96	13.24		
.15	14.42	15.27	16.26	17.13	15.94	15.36	16.47	17.05	15.85	15.31	16.82	16.77	15.74	14.80	13.99		16.22	.85
.20	15.43	16.40	17.50	17.69	16.14	16.70	17.92	17.12	16.09	17.21	17.76	16.73	15.70	14.81		17.06	17.20	.80
.25	16.56	16.82	16.49	16.12	16.54	16.83	16.48	16.13	16.78	16.74	16.39	15.94	15.35		18.02	18.15	18.31	.75
.30	14.64	14.37	15.01	16.24	16.16	15.07	14.98	16.45	16.00	15.09	14.51	14.18		18.78	19.22	19.40	19.57	.70
.35	13.97	15.11	16.49	17.54	16.04	15.31	16.97	17.17	15.83	14.64	13.73		17.29	19.41	20.55	20.73	18.06	.65
.40	15.32	16.69	17.80	17.32	16.23	17.27	17.70	16.93	15.72	14.58		17.52	17.93	19.60	20.69	17.80	17.68	.60
.45	16.55	16.34	16.15	15.97	16.50	16.40	16.16	15.87	15.35		18.61	18.91	19.18	19.64	18.87	19.00	19.22	.55
.50	15.02	14.81	14.87	16.53	16.08	15.13	14.78	14.58		18.11	19.66	20.50	20.66	19.66	20.47	20.69	19.06	.50
.55	13.73	14.79	17.06	17.71	15.91	14.39	13.73		17.06	18.21	19.95	20.74	18.42	19.67	20.75	18.77	17.06	.45
.60	15.13	17.22	17.25	16.91	15.76	14.18		18.11	18.39	18.89	19.71	18.44	18.70	19.65	18.67	18.34	18.61	.40
.65	16.33	16.16	16.01	15.82	15.35		18.61	19.71	20.40	20.60	19.71	20.37	20.61	19.64	20.30	20.63	19.55	.35
.70	15.21	15.05	14.93	14.81		17.52	18.73	20.16	20.76	18.78	19.99	20.78	19.05	19.67	20.79	19.33	17.97	.30
.75	14.22	14.09	13.99		17.29	17.94	18.95	19.86	18.15	18.58	19.82	18.73	18.16	19.67	19.15	17.84	17.41	.25
.80	13.36	13.24		18.78	19.61	20.25	20.52	19.60	20.21	20.54	19.56	20.14	20.56	19.64	20.00	20.55	19.83	.20
.85	12.60		18.02	19.08	20.27	20.77	19.08	20.16	20.78	19.18	19.95	20.81	19.47	19.62	20.82	19.66	18.55	.15
		17.06	18.01	19.09	19.91	18.22	18.76	19.90	18.76	18.27	19.77	19.21	18.15	19.48	19.48	18.39	17.41	.10
	16.22	17.06	18.03	18.92	17.69	17.55	18.67	18.47	17.26	18.13	18.87	17.78	17.38	18.96	18.25	17.27	16.41	.05
	.850	.800	.750	.700	.650	.600	.550	.500	.450	.400	.350	.300	.250	.200	.150	.100	.050	

TABLE 5.2 ($\mu_1 = .10$)

MODE 3 ($\mu_3 \rightarrow$, $\mu_2 \downarrow$); MODE 4 ($\leftarrow \mu_3$, $\mu_2 \uparrow$)

$\mu_2 \downarrow$	.050	.100	.150	.200	.250	.300	.350	.400	.450	.500	.550	.600	.650	.700	.750	.800	$\mu_2 \uparrow$
.05	13.54	14.29	15.16	16.09	16.07	14.67	15.08	16.20	15.95	14.85	14.89	16.34	15.87	14.94	14.09	13.36	
.10	14.42	15.28	16.27	17.21	16.18	15.42	16.50	17.20	16.01	15.39	16.88	16.91	15.87	14.91	14.09		
.15	15.43	16.42	17.55	18.15	16.40	16.76	18.11	17.45	16.27	17.30	18.07	16.95	15.86	14.93		17.06	.80
.20	16.59	17.72	18.77	17.91	17.12	18.42	18.49	17.34	17.63	18.74	17.89	16.81	15.82		18.01	18.16	.75
.25	16.88	16.46	16.17	16.57	17.15	16.60	16.26	16.92	16.96	16.48	16.12	15.76		19.08	19.23	19.41	.70
.30	14.37	15.15	16.52	17.75	16.43	15.54	17.06	17.42	16.12	15.06	14.39		19.61	20.41	20.64	20.82	.65
.35	15.33	16.72	18.35	18.29	16.55	17.46	18.73	17.41	15.97	14.78		17.94	19.92	21.85	21.53	18.12	.60
.40	16.97	17.92	17.70	17.30	17.52	17.87	17.57	16.94	15.87		18.73	19.03	20.16	20.87	18.85	19.01	.55
.45	16.36	16.12	15.98	16.80	16.80	16.24	15.97	15.72		19.71	20.33	20.63	20.75	20.32	20.56	20.79	.50
.50	14.82	14.95	17.13	18.22	16.28	15.06	14.64		18.39	20.17	21.89	21.54	20.26	22.05	21.28	19.08	.45
.55	15.15	17.50	18.46	17.89	16.12	14.51		18.21	18.84	20.31	20.91	18.89	20.18	20.87	18.84	18.34	.40
.60	17.40	17.25	17.09	16.81	15.94		19.66	20.17	20.53	20.72	20.19	20.52	20.72	20.11	20.41	20.73	.35
.65	16.17	16.00	15.86	15.70		18.91	20.33	21.84	21.51	20.40	22.04	21.31	20.29	22.22	21.19	19.57	.30
.70	15.06	14.91	14.80		17.93	19.03	20.44	20.91	19.04	20.35	20.90	19.17	20.16	20.89	19.36	18.01	.25
.75	14.09	13.96		19.41	19.92	20.43	20.70	19.93	20.40	20.72	19.88	20.35	20.73	19.81	20.12	20.67	.20
.80	13.24		19.22	20.41	21.73	21.47	20.48	21.90	21.29	20.38	22.08	21.21	20.31	22.20	21.13	19.86	.15
		18.15	19.23	20.46	20.88	19.23	20.39	20.88	19.27	20.23	20.89	19.52	20.00	20.88	19.68	18.56	.10
	17.20	18.16	19.25	20.05	18.32	18.95	20.05	18.81	18.48	19.98	19.25	18.25	19.80	19.51	18.41	17.42	.05
	.800	.750	.700	.650	.600	.550	.500	.450	.400	.350	.300	.250	.200	.150	.100	.050	

TABLE 5.2 ($\mu_1 = .15$) MODE 1

$\mu_2 \downarrow$ \ $\mu_3 \rightarrow$	.050	.100	.150	.200	.250	.300	.350	.400	.450	.500	.550	.600	.650	.700	.750	$\mu_2 \uparrow$
.05	6.197	6.539	6.943	7.417	7.975	8.624	9.315	9.645	9.134	8.438	7.812	7.276	6.820	6.431	6.098	
.10	6.627	7.019	7.487	8.046	8.709	9.467	10.12	9.874	9.107	8.381	7.761	7.240	6.799	6.425		
.15	7.124	7.579	8.131	8.798	9.592	10.45	10.71	9.939	9.085	8.354	7.746	7.238	6.812		9.949	.75
.20	7.704	8.240	8.901	9.709	10.66	11.45	10.94	9.951	9.077	8.354	7.760	7.267		10.51	10.62	.70
.25	8.387	9.029	9.834	10.83	11.91	12.02	10.98	9.942	9.080	8.378	7.802		11.17	11.26	11.40	.65
.30	9.203	9.984	10.97	12.12	12.46	11.82	10.85	9.898	9.088	8.420		11.93	12.03	12.14	12.30	.60
.35	10.19	11.12	11.58	11.44	11.24	10.93	10.43	9.751	9.068		12.76	12.91	13.04	13.17	13.36	.55
.40	10.72	10.50	10.34	10.20	10.06	9.898	9.676	9.352		13.11	13.77	14.06	14.23	14.39	14.57	.50
.45	9.628	9.437	9.297	9.183	9.080	8.976	8.858		12.09	13.35	14.67	15.37	15.61	14.70	13.03	.45
.50	8.727	8.567	8.449	8.354	8.273	8.197		11.32	12.16	13.44	15.09	16.06	14.44	12.75	11.48	.40
.55	7.982	7.846	7.746	7.667	7.600		11.77	12.06	12.60	13.63	14.92	14.17	12.60	11.97	12.18	.35
.60	7.354	7.240	7.154	7.088		12.89	13.18	13.44	13.76	14.25	14.06	13.32	13.44	13.67	14.00	.30
.65	6.820	6.723	6.651		12.97	13.83	14.67	15.29	15.60	14.77	14.56	15.22	15.59	15.02	13.94	.25
.70	6.360	6.279		12.17	13.03	14.06	15.23	16.00	14.61	13.85	15.10	15.96	14.88	13.75	12.81	.20
.75	5.960		11.39	12.14	13.04	14.09	15.07	14.32	13.09	13.52	14.96	14.68	13.60	12.64	11.84	.15
		10.71	11.37	12.14	13.05	14.01	13.97	12.69	12.34	13.43	14.26	13.46	12.52	11.70	11.02	.10
	10.14	10.72	11.40	12.19	13.06	13.49	12.45	11.65	12.16	13.32	13.27	12.41	11.60	10.89	10.30	.05
	.750	.700	.650	.600	.550	.500	.450	.400	.350	.300	.250	.200	.150	.100	.050	

MODE 2 $\leftarrow \mu_3$

TABLE 5.2 ($\mu_1 = .20$) MODE 1

$\mu_2 \downarrow$ \ $\mu_3 \rightarrow$	.050	.100	.150	.200	.250	.300	.350	.400	.450	.500	.550	.600	.650	.700	$\mu_2 \uparrow$
.05	6.632	7.028	7.499	8.059	8.725	9.494	10.20	10.05	9.276	8.529	7.891	7.354	6.902	6.518	
.10	7.127	7.585	8.140	8.809	9.608	10.48	10.83	10.08	9.215	8.468	7.846	7.328	6.893		
.15	7.706	8.244	8.908	9.720	10.68	11.53	11.09	10.08	9.186	8.449	7.843	7.342		10.54	.70
.20	8.388	9.033	9.842	10.84	11.96	12.23	11.15	10.07	9.183	8.464	7.877		11.18	11.31	.65
.25	9.205	9.991	10.99	12.22	13.13	12.33	11.11	10.06	9.203	8.510		11.93	12.04	12.19	.60
.30	10.20	11.17	12.36	12.77	12.47	11.85	10.93	10.02	9.230		12.81	12.92	13.05	13.23	.55
.35	11.39	11.64	11.47	11.30	11.11	10.85	10.43	9.843		13.70	13.93	14.09	14.24	14.45	.50
.40	10.59	10.37	10.21	10.07	9.942	9.798	9.615		13.55	14.69	15.23	15.47	15.64	15.09	.45
.45	9.507	9.321	9.186	9.077	8.980	8.886		12.33	13.63	15.29	16.58	16.72	14.74	13.04	.40
.50	8.623	8.468	8.354	8.264	8.188		12.01	12.63	13.76	15.44	16.32	14.48	12.79	12.03	.35
.55	7.891	7.761	7.665	7.591		13.03	13.28	13.63	14.30	15.27	14.30	13.32	13.45	13.76	.30
.60	7.276	7.166	7.087		13.89	14.60	15.09	15.44	15.69	14.99	15.03	15.36	15.63	15.27	.25
.65	6.751	6.661		13.19	14.19	15.37	16.58	16.74	15.07	15.40	16.85	16.41	15.07	13.95	.20
.70	6.300		12.29	13.17	14.23	15.47	16.20	14.67	14.03	15.44	16.09	14.90	13.76	12.81	.15
		11.50	12.26	13.17	14.24	15.22	14.36	13.17	13.73	15.16	14.71	13.62	12.65	11.85	.10
	10.84	11.51	12.30	13.23	14.19	14.04	12.75	12.53	13.70	14.36	13.48	12.53	11.71	11.02	.05
	.700	.650	.600	.550	.500	.450	.400	.350	.300	.250	.200	.150	.100	.050	

MODE 2 $\leftarrow \mu_3$

TABLE 5.2 ($\mu_1 = .15$) MODE 3

$\mu_2 \downarrow$ \ $\mu_3 \rightarrow$	.050	.100	.150	.200	.250	.300	.350	.400	.450	.500	.550	.600	.650	.700	.750	$\mu_2 \uparrow$
.05	14.43	15.29	16.29	17.29	16.50	15.52	16.54	17.38	16.22	15.51	16.95	17.08	16.03	15.06	14.22	
.10	15.44	16.43	17.58	18.36	16.63	16.80	18.19	17.67	16.42	17.35	18.24	17.11	16.00	15.05		
.15	16.60	17.75	19.04	18.81	17.33	18.58	19.28	17.76	17.78	19.44	18.35	17.09	16.01		18.03	.75
.20	17.94	19.10	18.90	18.07	18.66	19.10	18.46	18.13	19.18	18.70	17.89	16.91		19.09	19.25	.70
.25	16.55	16.18	16.66	18.00	17.27	16.50	17.25	17.87	16.89	16.28	15.91		20.27	20.46	20.65	.65
.30	15.35	16.74	18.43	18.81	16.86	17.54	19.15	17.71	16.24	15.13		20.25	21.73	22.03	22.06	.60
.35	16.99	18.69	19.53	18.56	18.05	19.58	18.96	17.57	16.16		18.95	20.43	23.09	21.66	18.87	.55
.40	18.02	17.75	17.52	17.60	18.04	17.71	17.41	16.93		20.16	20.44	21.08	21.14	20.35	20.57	.50
.45	16.17	15.98	17.21	18.60	16.89	16.12	15.83		20.40	21.84	22.32	22.00	21.97	22.34	21.69	.45
.50	15.20	17.55	19.75	18.70	16.48	15.09		18.89	20.53	22.70	21.71	20.71	23.14	21.34	19.11	.40
.55	17.99	18.55	18.35	17.89	16.39		19.95	20.31	21.06	21.18	20.33	21.13	21.06	20.13	20.43	.35
.60	17.30	17.11	16.95	16.73		20.50	21.89	22.70	22.08	22.13	22.80	21.79	22.30	22.83	21.48	.30
.65	16.03	15.87	15.74		19.18	20.63	22.32	21.65	20.74	22.66	21.40	20.71	22.79	21.23	19.59	.25
.70	14.94	14.80		19.60	20.16	21.08	21.19	20.17	21.12	21.10	20.12	21.19	21.04	19.80	20.16	.20
.75	13.99		20.55	21.85	23.09	22.07	22.13	23.31	21.76	22.41	23.16	21.64	22.83	22.80	21.36	.15
		19.40	20.64	22.03	21.58	20.73	22.27	21.36	20.68	22.46	21.27	20.65	22.44	21.16	19.87	.10
	18.31	19.41	20.65	20.99	19.40	20.62	20.95	19.38	20.53	20.95	19.58	20.41	20.93	19.70	18.57	.05
	.750	.700	.650	.600	.550	.500	.450	.400	.350	.300	.250	.200	.150	.100	.050	

MODE 4 $\leftarrow \mu_3$

TABLE 5.2 ($\mu_1 = .20$) MODE 3

$\mu_2 \downarrow$ \ $\mu_3 \rightarrow$	.050	.100	.150	.200	.250	.300	.350	.400	.450	.500	.550	.600	.650	.700	$\mu_2 \uparrow$
.05	15.45	16.44	17.60	18.54	16.93	16.86	18.27	17.93	16.61	17.41	18.43	17.30	16.17	15.21	
.10	16.61	17.76	19.09	19.18	17.47	18.62	19.53	17.99	17.85	19.63	18.55	17.25	16.16		
.15	17.96	19.31	20.34	19.13	18.94	20.34	19.54	18.57	20.32	19.75	18.46	17.25		18.92	.70
.20	18.94	18.59	18.27	18.60	18.88	18.48	18.29	18.98	18.60	18.22	17.71		19.91	20.05	.65
.25	16.16	16.80	18.48	19.15	17.38	17.66	19.36	18.04	16.80	16.08		20.77	20.88	20.99	.60
.30	17.00	18.74	20.36	19.22	18.23	20.34	19.58	17.87	16.40		20.52	21.47	21.58	21.64	.55
.35	19.01	19.29	19.05	18.74	19.36	19.15	18.73	17.70		19.86	20.70	22.07	21.63	19.88	.50
.40	17.67	17.43	17.57	18.98	17.87	17.42	17.17		20.76	20.91	21.19	21.19	20.88	20.99	.45
.45	15.98	17.59	20.32	19.18	16.96	16.00		20.60	21.51	21.65	21.61	21.54	21.64	21.61	.40
.50	18.04	19.63	19.44	18.74	16.74		19.71	20.72	22.08	21.66	20.88	22.17	21.33	19.64	.35
.55	18.43	18.24	18.07	17.76		20.74	20.91	21.18	21.22	20.93	21.22	21.12	20.88	21.00	.30
.60	17.08	16.91	16.77		20.66	21.54	21.71	21.66	21.58	21.73	21.58	21.59	21.72	21.47	.25
.65	15.87	15.72		19.64	20.75	22.00	21.61	20.85	22.14	21.39	20.86	22.25	21.23	19.64	.20
.70	14.82		20.69	20.87	21.14	21.19	20.88	21.16	21.13	20.88	21.19	21.08	20.82	20.96	.15
		20.73	21.53	21.66	21.63	21.56	21.67	21.57	21.58	21.67	21.52	21.59	21.67	21.36	.10
	19.57	20.82	22.06	21.64	20.94	22.18	21.41	20.93	22.24	21.31	20.96	22.26	21.18	19.88	.05
	.700	.650	.600	.550	.500	.450	.400	.350	.300	.250	.200	.150	.100	.050	

MODE 4 $\leftarrow \mu_3$

TABLE 5.2 ($\mu_1 = .25$) MODE 1

$\mu_3 \rightarrow$

$\mu_2 \downarrow$	.050	.100	.150	.200	.250	.300	.350	.400	.450	.500	.550	.600	.650	$\mu_2 \uparrow$
.05	7.133	7.595	8.153	8.825	9.630	10.53	10.99	10.29	9.394	8.623	7.982	7.448	7.000	
.10	7.709	8.252	8.918	9.734	10.71	11.61	11.26	10.24	9.321	8.567	7.948	7.436		
.15	8.391	9.039	9.851	10.86	12.01	12.39	11.31	10.21	9.297	8.563	7.964		11.21	.65
.20	9.207	9.997	11.01	12.26	13.37	12.62	11.30	10.20	9.312	8.601		11.93	12.07	.60
.25	10.20	11.18	12.45	13.78	13.57	12.47	11.24	10.20	9.357		12.78	12.89	13.06	.55
.30	11.43	12.58	12.81	12.62	12.33	11.82	11.02	10.16		13.74	13.87	14.01	14.19	.50
.35	11.75	11.50	11.31	11.15	10.98	10.75	10.42		14.57	14.90	15.07	15.22	15.40	.45
.40	10.45	10.24	10.08	9.951	9.832	9.707		13.84	15.30	16.00	16.20	16.29	15.12	.40
.45	9.394	9.215	9.085	9.982	8.892		12.70	13.90	15.60	16.74	16.79	14.76	13.05	.35
.50	8.529	8.381	8.272	8.188		13.08	13.46	14.25	15.69	16.38	14.53	13.26	13.43	.30
.55	7.812	7.688	7.598		14.35	14.66	14.92	15.27	15.71	14.84	14.80	15.00	15.27	.25
.60	7.207	7.106		14.32	15.35	16.06	16.32	16.38	15.72	16.11	16.32	16.38	15.31	.20
.65	6.693		13.33	14.37	15.61	16.72	16.79	15.18	15.74	16.06	16.45	15.09	13.96	.15
		12.41	13.31	14.39	15.64	16.29	14.72	14.20	15.69	16.15	14.92	13.77	12.82	.10
	11.65	12.43	13.36	14.45	15.40	14.41	13.29	14.01	15.38	14.75	13.63	12.66	11.86	.05
	.650	.600	.550	.500	.450	.400	.350	.300	.250	.200	.150	.100	.050	

MODE 2 $\leftarrow \mu_3$

TABLE 5.2 ($\mu_1 = .30$) MODE 1

$\mu_3 \rightarrow$

$\mu_2 \downarrow$	.050	.100	.150	.200	.250	.300	.350	.400	.450	.500	.550	.600	$\mu_2 \uparrow$
.05	7.716	8.264	8.934	9.753	10.74	11.71	11.48	10.45	9.507	8.727	8.088	7.561	
.10	8.395	9.047	9.863	10.87	12.05	12.55	11.50	10.37	9.437	8.685	8.073		
.15	9.210	10.00	11.02	12.28	13.46	12.81	11.47	10.34	9.429	8.704		11.88	.60
.20	10.20	11.19	12.48	13.84	13.78	12.77	11.44	10.34	9.472		12.59	12.72	.55
.25	11.44	12.68	13.46	13.37	13.13	12.46	11.37	10.37		13.31	13.39	13.49	.50
.30	12.70	12.55	12.39	12.23	12.02	11.68	11.08		13.86	13.92	13.97	14.04	.45
.35	11.48	11.26	11.09	10.94	10.80	10.62		14.19	14.27	14.32	14.36	14.41	.40
.40	10.29	10.08	9.939	9.821	9.715		13.87	14.49	14.61	14.67	14.72	14.76	.35
.45	9.276	9.107	8.985	8.889		13.10	13.90	14.77	15.07	15.18	14.71	13.11	.30
.50	8.438	8.298	8.197		13.57	13.72	14.06	14.99	15.72	14.46	13.66	13.74	.25
.55	7.738	7.623		14.05	14.11	14.17	14.30	14.84	14.30	14.13	14.16	14.22	.20
.60	7.147		14.21	14.39	14.44	14.48	14.53	14.46	14.43	14.45	14.48	14.52	.15
		13.45	14.39	14.70	14.74	14.76	14.71	14.68	14.73	14.75	14.76	13.96	.10
	12.58	13.49	14.57	15.09	15.12	14.76	14.41	15.08	15.11	14.93	13.78	12.83	.05
	.600	.550	.500	.450	.400	.350	.300	.250	.200	.150	.100	.050	

MODE 2 $\leftarrow \mu_3$

TABLE 5.2 ($\mu_1 = .25$) MODE 3

$\mu_3 \rightarrow$

$\mu_2 \downarrow$	.050	.100	.150	.200	.250	.300	.350	.400	.450	.500	.550	.600	.650	$\mu_2 \uparrow$
.05	16.62	17.77	18.04	18.00	17.59	18.04	18.03	17.90	17.85	18.04	17.99	17.40	16.33	
.10	17.61	17.57	17.54	17.51	17.61	17.57	17.53	17.55	17.59	17.55	17.50	17.22		
.15	17.20	17.15	17.12	17.78	17.29	17.16	17.28	17.57	17.21	17.13	17.06		17.69	.65
.20	16.62	16.80	18.36	18.35	17.09	17.52	18.74	17.60	16.80	16.53		18.22	18.32	.60
.25	17.00	18.25	18.16	17.90	17.85	18.23	18.05	17.52	16.50		19.08	19.23	19.40	.55
.30	17.62	17.55	17.49	17.52	17.66	17.54	17.46	17.27		19.60	20.48	20.73	20.94	.50
.35	17.12	17.05	17.28	18.29	17.25	17.06	16.97		18.15	19.93	22.13	21.56	18.17	.45
.40	16.48	17.55	18.57	18.13	16.92	16.45		18.78	19.04	20.17	20.88	18.87	19.02	.40
.45	17.85	17.85	17.78	17.63	16.78		19.71	20.40	20.74	20.85	20.38	20.67	20.93	.35
.50	17.41	17.35	17.30	17.21		18.44	20.19	22.13	21.58	20.28	22.40	21.29	19.08	.30
.55	16.95	16.88	16.82		18.42	18.89	20.33	20.93	18.92	20.21	20.88	18.85	18.54	.25
.60	16.34	16.23		19.66	20.26	20.71	20.87	20.28	20.71	20.86	20.19	20.58	20.98	.20
.65	15.50		18.87	20.32	21.97	21.54	20.38	22.24	21.32	20.27	22.47	21.19	19.57	.15
		17.80	18.85	20.35	20.88	18.87	20.22	20.87	19.09	19.96	20.87	19.35	17.98	.10
	18.06	18.12	18.87	19.88	18.17	18.45	19.86	18.70	18.18	19.69	19.14	18.08	18.10	.05
	.650	.600	.550	.500	.450	.400	.350	.300	.250	.200	.150	.100	.050	

MODE 4 $\leftarrow \mu_3$

TABLE 5.2 ($\mu_1 = .30$) MODE 3

$\mu_3 \rightarrow$

$\mu_2 \downarrow$	.050	.100	.150	.200	.250	.300	.350	.400	.450	.500	.550	.600	$\mu_2 \uparrow$
.05	15.17	15.15	15.15	16.25	16.16	15.20	15.17	16.48	15.98	15.20	15.15	15.13	
.10	14.81	15.12	16.51	17.76	16.28	15.40	17.05	17.43	15.98	14.95	14.79		
.15	15.33	16.72	18.40	18.63	16.60	17.49	19.05	17.52	15.98	14.87		17.55	.60
.20	16.98	18.64	19.10	18.11	17.90	19.22	18.56	17.30	15.97		18.76	18.95	.55
.25	17.13	16.81	16.60	17.09	17.38	16.86	16.55	16.23		20.16	20.39	20.62	.50
.30	15.48	15.40	17.16	18.48	16.50	15.54	15.31		20.21	21.90	22.27	22.18	.45
.35	15.17	17.53	19.54	18.46	16.26	14.98		18.58	20.40	23.31	21.67	18.45	.40
.40	17.90	17.99	17.76	17.34	16.13		19.99	20.35	21.12	21.16	20.22	20.51	.35
.45	16.61	16.42	16.27	16.09		20.37	22.04	22.66	22.14	22.24	22.70	21.70	.30
.50	15.51	15.39	15.31		18.70	20.52	22.80	21.73	20.71	23.26	21.34	19.10	.25
.55	14.89	14.83		19.67	20.18	21.13	21.22	20.21	21.22	21.08	19.90	20.31	.20
.60	14.53		20.47	22.05	23.14	22.17	22.40	23.26	21.85	22.71	23.20	21.48	.15
		19.00	20.56	22.34	21.64	20.67	22.70	21.39	20.61	22.82	21.22	19.58	.10
	17.68	19.01	20.57	20.99	19.02	20.51	20.96	19.17	20.37	20.93	19.36	17.99	.05
	.600	.550	.500	.450	.400	.350	.300	.250	.200	.150	.100	.050	

MODE 4 $\leftarrow \mu_3$

TABLE 5.2 ($\mu_1 = .35$)

MODE 1 (upper left, $\mu_3 \rightarrow$, $\mu_2 \downarrow$); MODE 2 (lower right, $\leftarrow \mu_3$, $\mu_2 \uparrow$)

$\mu_2 \downarrow$ \ $\mu_3 \rightarrow$	.050	.100	.150	.200	.250	.300	.350	.400	.450	.500	.550	$\mu_2 \uparrow$
.05	8.402	9.061	9.881	10.90	12.10	12.70	11.75	10.59	9.628	8.850	8.219	
.10	9.215	10.01	11.03	12.29	12.68	12.58	11.64	10.50	9.572	8.830		
.15	10.21	11.20	12.41	12.48	12.45	12.36	11.58	10.48	9.592		11.88	.55
.20	11.44	12.29	12.28	12.26	12.22	12.12	11.51	10.51		12.15	12.20	.50
.25	12.10	12.05	12.01	11.96	11.91	11.79	11.37		12.38	12.41	12.45	.45
.30	11.71	11.61	11.53	11.45	11.36	11.22		12.62	12.65	12.69	12.75	.40
.35	10.99	10.83	10.71	10.60	10.50		12.94	13.02	13.09	13.17	13.29	.35
.40	10.05	9.874	9.748	9.647		13.10	13.60	13.85	14.03	14.20	14.41	.30
.45	9.134	8.980	8.870		12.38	13.28	14.56	15.40	15.74	14.68	13.03	.25
.50	8.340	8.212		12.33	12.46	13.32	15.03	16.11	14.43	12.74	12.35	.20
.55	7.665		12.50	12.53	12.60	13.32	14.80	14.13	12.60	12.53	12.55	.15
		12.71	12.73	12.75	12.79	13.26	13.66	12.74	12.74	12.75	12.78	.10
	12.99	13.02	13.03	13.04	13.05	13.11	13.03	13.03	13.03	13.04	13.06	.05
$\leftarrow \mu_3$	.550	.500	.450	.400	.350	.300	.250	.200	.150	.100	.050	

TABLE 5.2 ($\mu_1 = .40$)

MODE 1 (upper left, $\mu_3 \rightarrow$, $\mu_2 \downarrow$); MODE 2 (lower right, $\leftarrow \mu_3$, $\mu_2 \uparrow$)

$\mu_2 \downarrow$ \ $\mu_3 \rightarrow$	.050	.100	.150	.200	.250	.300	.350	.400	.450	.500	$\mu_2 \uparrow$
.05	9.224	10.03	11.05	11.44	11.44	11.43	11.39	10.72	9.768	9.002	
.10	10.21	11.16	11.20	11.19	11.18	11.17	11.12	10.61	9.734		
.15	11.05	11.03	11.02	11.01	10.99	10.97	10.92	10.57		10.96	.50
.20	10.90	10.87	10.86	10.84	10.83	10.80	10.75		11.16	11.21	.45
.25	10.74	10.71	10.68	10.66	10.64	10.61		11.49	11.55	11.65	.40
.30	10.53	10.48	10.45	10.41	10.38		12.11	12.21	12.34	12.53	.35
.35	10.20	10.12	10.05	10.00		13.09	13.33	13.52	13.73	14.01	.30
.40	9.645	9.525	9.434		13.28	14.43	15.10	15.44	15.69	15.08	.25
.45	8.929	8.802		12.03	13.42	15.22	16.85	16.86	14.73	13.03	.20
.50	8.219		11.17	11.97	13.44	15.36	16.32	14.45	12.74	11.45	.15
		11.22	11.28	11.97	13.45	15.00	14.16	12.53	11.28	11.24	.10
	11.44	11.45	11.48	12.03	13.43	13.74	12.35	11.45	11.46	11.47	.05
$\leftarrow \mu_3$	.500	.450	.400	.350	.300	.250	.200	.150	.100	.050	

TABLE 5.2 ($\mu_1 = .35$)

MODE 3 (upper left, $\mu_3 \rightarrow$, $\mu_2 \downarrow$); MODE 4 (lower right, $\leftarrow \mu_3$, $\mu_2 \uparrow$)

$\mu_2 \downarrow$ \ $\mu_3 \rightarrow$	.050	.100	.150	.200	.250	.300	.350	.400	.450	.500	.550	$\mu_2 \uparrow$
.05	13.98	15.13	16.54	17.91	16.59	15.48	17.12	17.67	16.17	14.82	13.73	
.10	15.33	16.74	18.45	18.99	16.81	17.55	19.29	17.75	16.12	14.81		
.15	16.99	18.73	20.40	19.10	18.16	20.36	19.53	17.70	16.15		18.67	.55
.20	18.95	18.99	18.63	18.35	19.15	18.81	18.29	17.32		19.90	20.05	.50
.25	16.59	16.28	17.29	18.88	17.27	16.43	16.04		20.78	20.88	20.95	.45
.30	15.20	17.57	20.34	19.10	16.60	15.07		20.54	21.29	21.36	21.41	.40
.35	18.03	19.53	19.28	18.49	16.48		19.82	20.72	21.76	21.57	19.86	.35
.40	17.93	17.67	17.45	17.12		20.78	20.90	21.10	21.13	20.87	20.96	.30
.45	16.22	16.01	15.85		20.61	21.31	21.40	21.39	21.32	21.39	21.43	.25
.50	14.85	14.69		19.65	20.72	21.79	21.58	20.86	21.85	21.32	19.53	.20
.55	13.82		20.75	20.87	21.06	21.12	20.88	21.08	21.06	20.84	20.93	.15
		20.69	21.28	21.34	21.33	21.29	21.34	21.32	21.29	21.33	21.35	.10
	19.22	20.79	21.69	21.61	20.93	21.70	21.43	20.96	21.70	21.24	19.59	.05
$\leftarrow \mu_3$	.550	.500	.450	.400	.350	.300	.250	.200	.150	.100	.050	

TABLE 5.2 ($\mu_1 = .40$)

MODE 3 (upper left, $\mu_3 \rightarrow$, $\mu_2 \downarrow$); MODE 4 (lower right, $\leftarrow \mu_3$, $\mu_2 \uparrow$)

$\mu_2 \downarrow$ \ $\mu_3 \rightarrow$	.050	.100	.150	.200	.250	.300	.350	.400	.450	.500	$\mu_2 \uparrow$
.05	15.35	16.76	18.48	18.95	17.13	17.62	19.01	18.02	16.36	15.02	
.10	17.00	18.67	18.73	18.64	18.25	18.74	18.69	17.92	16.34		
.15	18.48	18.45	18.40	18.36	18.48	18.43	18.35	17.80		18.47	.50
.20	17.91	17.76	17.78	18.60	18.00	17.75	17.54		18.76	18.81	.45
.25	16.16	17.61	18.94	18.66	17.15	16.16		19.18	19.27	19.38	.40
.30	18.04	18.62	18.58	18.42	16.83		19.56	20.38	20.68	20.93	.35
.35	18.27	18.19	18.11	17.92		18.73	19.88	22.41	21.58	18.70	.30
.40	17.38	17.20	17.05		19.05	19.17	20.12	20.88	19.09	19.17	.25
.45	15.95	15.76		19.64	20.29	20.71	20.86	20.27	20.61	20.96	.20
.50	14.60		18.67	20.11	22.30	21.59	20.19	22.71	21.29	19.08	.15
		18.77	18.84	20.13	20.88	18.85	19.90	20.84	18.84	18.79	.10
	19.06	19.08	19.11	19.64	19.08	19.10	19.53	19.08	19.08	19.10	.05
$\leftarrow \mu_3$	.500	.450	.400	.350	.300	.250	.200	.150	.100	.050	

TABLE 5.2 ($\mu_1 = .45$) MODE 1 / MODE 2

$\mu_2 \downarrow$ \ $\mu_3 \rightarrow$ (MODE 1)	.050	.100	.150	.200	.250	.300	.350	.400	.450	$\mu_2 \uparrow$ (MODE 2)
.05	10.19	10.21	10.21	10.20	10.20	10.20	10.19	10.18	9.905	
.10	10.03	10.01	10.00	9.997	9.991	9.984	9.974	9.949		
.15	9.881	9.863	9.851	9.842	9.834	9.825	9.812		10.38	.45
.20	9.753	9.734	9.720	9.709	9.700	9.689		10.92	11.06	.40
.25	9.630	9.608	9.592	9.579	9.568		11.81	11.95	12.16	.35
.30	9.494	9.467	9.446	9.429		13.05	13.23	13.43	13.70	.30
.35	9.315	9.275	9.244		14.30	14.72	14.96	15.16	15.38	.25
.40	9.036	8.973		13.66	15.27	15.96	16.09	16.15	15.11	.20
.45	8.598		12.16	13.65	15.59	16.41	16.45	14.75	13.03	.15
		10.95	12.11	13.67	15.63	16.38	14.48	12.75	11.46	.10
	10.26	10.95	12.18	13.76	15.27	14.22	12.55	11.24	10.26	.05
MODE 2 $\leftarrow \mu_3$	.450	.400	.350	.300	.250	.200	.150	.100	.050	

TABLE 5.2 ($\mu_1 = .50$) MODE 1 / MODE 2

$\mu_2 \downarrow$ \ $\mu_3 \rightarrow$ (MODE 1)	.050	.100	.150	.200	.250	.300	.350	.400	$\mu_2 \uparrow$ (MODE 2)
.05	9.224	9.215	9.210	9.207	9.205	9.203	9.200	9.197	
.10	9.061	9.047	9.039	9.033	9.029	9.024	9.019		
.15	8.934	8.918	8.908	8.901	8.895	8.889		10.76	.40
.20	8.825	8.809	8.798	8.790	8.783		11.72	11.92	.35
.25	8.725	8.709	8.697	8.688		12.94	13.10	13.32	.30
.30	8.624	8.605	8.591		14.06	14.16	14.26	14.36	.25
.35	8.506	8.484		14.54	14.64	14.68	14.71	14.75	.20
.40	8.351		13.90	14.81	14.88	14.90	14.92	14.93	.15
		12.36	13.89	15.02	15.07	15.09	14.76	13.04	.10
	11.15	12.38	14.00	15.27	15.31	14.52	12.78	11.47	.05
MODE 2 $\leftarrow \mu_3$	.400	.350	.300	.250	.200	.150	.100	.050	

TABLE 5.2 ($\mu_1 = .45$) MODE 3 / MODE 4

$\mu_2 \downarrow$ \ $\mu_3 \rightarrow$ (MODE 3)	.050	.100	.150	.200	.250	.300	.350	.400	.450	$\mu_2 \uparrow$ (MODE 4)
.05	16.97	17.00	16.99	16.98	17.00	17.00	16.99	16.97	16.55	
.10	16.76	16.74	16.72	16.80	16.80	16.74	16.72	16.69		
.15	16.54	16.51	17.12	18.27	16.66	16.52	16.49		17.26	.45
.20	16.25	17.51	19.13	18.07	16.57	16.24		18.27	18.48	.40
.25	17.59	17.47	17.33	17.12	16.54		19.95	20.23	20.53	.35
.30	16.86	16.80	16.76	16.70		20.14	22.08	22.46	22.24	.30
.35	16.54	16.50	16.47		18.16	20.35	23.16	21.67	18.18	.25
.40	16.20	16.14		19.67	20.16	21.19	21.19	19.96	20.37	.20
.45	15.49		20.30	22.22	22.79	22.25	22.47	22.82	21.70	.15
		18.34	20.41	22.83	21.72	20.58	23.20	21.33	19.08	.10
	17.06	18.34	20.43	21.00	18.54	20.31	20.93	18.79	17.06	.05
MODE 4 $\leftarrow \mu_3$	.450	.400	.350	.300	.250	.200	.150	.100	.050	

TABLE 5.2 ($\mu_1 = .50$) MODE 3 / MODE 4

$\mu_2 \downarrow$ \ $\mu_3 \rightarrow$ (MODE 3)	.050	.100	.150	.200	.250	.300	.350	.400	$\mu_2 \uparrow$ (MODE 4)
.05	15.35	15.33	15.33	16.62	16.16	15.35	15.33	15.32	
.10	15.13	15.12	17.15	18.59	16.18	15.15	15.11		
.15	15.15	17.54	20.34	18.90	16.17	15.01		18.13	.40
.20	18.00	19.18	18.81	17.91	16.12		19.77	19.98	.35
.25	16.93	16.63	16.40	16.14		20.81	20.89	20.95	.30
.30	15.52	15.42	15.36		20.56	21.21	21.27	21.31	.25
.35	15.08	15.05		19.67	20.73	21.64	21.52	19.69	.20
.40	14.85		20.79	20.89	21.04	21.08	20.87	20.93	.15
		20.63	21.19	21.23	21.23	21.19	21.22	21.24	.10
	18.61	20.73	21.48	21.47	20.98	21.48	21.35	19.10	.05
MODE 4 $\leftarrow \mu_3$	.400	.350	.300	.250	.200	.150	.100	.050	

TABLE 5.3 ($\mu_1 = .05$)

MODE 1 (μ_3 →, μ_2 ↓)

μ_2	.050	.100	.150	.200	.250	.300	.350	.400	.450	.500	.550	.600	.650	.700	.750	.800	.850	μ_2
.05	2.170	2.269	2.380	2.505	2.647	2.810	2.998	3.217	3.475	3.783	4.148	4.565	4.958	5.128	5.021	4.791	4.535	
.10	2.301	2.411	2.536	2.679	2.842	3.032	3.253	3.515	3.829	4.205	4.646	5.093	5.340	5.258	5.015	4.736		
.15	2.450	2.574	2.717	2.881	3.072	3.295	3.561	3.880	4.267	4.730	5.232	5.572	5.533	5.277	4.969		8.162	.85
.20	2.621	2.763	2.927	3.119	3.344	3.613	3.937	4.333	4.817	5.371	5.820	5.849	5.582	5.241		8.545	8.661	.80
.25	2.818	2.981	3.173	3.400	3.672	4.001	4.407	4.909	5.510	6.075	6.209	5.937	5.556		8.997	9.093	9.217	.75
.30	3.046	3.238	3.466	3.740	4.074	4.487	5.006	5.649	6.328	6.607	6.345	5.920		9.514	9.601	9.667	8.726	.70
.35	3.316	3.544	3.820	4.158	4.578	5.111	5.788	6.569	7.018	6.795	6.329		10.07	10.17	9.922	8.289	7.087	.65
.40	3.637	3.915	4.256	4.682	5.227	5.931	6.784	7.356	7.202	6.743		10.42	10.67	9.657	7.969	6.826	6.400	.60
.45	4.028	4.372	4.804	5.359	6.080	6.940	7.350	7.251	6.966		9.863	10.44	9.294	7.739	6.871	6.833	6.947	.55
.50	4.513	4.950	5.512	6.233	6.880	6.945	6.870	6.743		9.060	9.653	8.940	7.669	7.322	7.363	7.466	7.616	.50
.55	5.130	5.697	6.352	6.523	6.476	6.409	6.329		8.693	9.085	8.730	8.019	7.949	8.013	8.112	8.244	8.426	.45
.60	5.930	6.202	6.130	6.055	5.987	5.920		9.060	9.218	9.025	8.691	8.710	8.805	8.914	9.036	9.175	9.092	.40
.65	5.891	5.781	5.694	5.621	5.556		9.863	9.981	9.700	9.325	9.481	9.702	9.875	9.981	9.662	8.938	8.317	.35
.70	5.484	5.387	5.308	5.241		10.42	10.69	9.904	9.260	9.569	10.17	10.65	10.44	9.599	8.815	8.167	7.641	.30
.75	5.126	5.041	4.969		10.07	10.46	9.648	8.835	8.999	9.627	10.33	10.28	9.491	8.714	8.056	7.512	7.066	.25
.80	4.812	4.736		9.514	9.929	9.332	8.439	8.413	8.904	9.580	9.921	9.361	8.623	7.969	7.417	6.955	6.572	.20
.85	4.535		8.997	9.403	9.016	8.115	7.926	8.276	8.855	9.383	9.185	8.531	7.894	7.341	6.872	6.476	6.142	.15
		8.545	8.934	8.711	7.854	7.536	7.763	8.242	8.785	8.931	8.431	7.825	7.279	6.808	6.405	6.060	5.767	.10
	8.162	8.531	8.425	7.641	7.230	7.350	7.743	8.243	8.586	8.310	7.759	7.226	6.757	6.352	6.001	5.698	5.436	.05
	.850	.800	.750	.700	.650	.600	.550	.500	.450	.400	.350	.300	.250	.200	.150	.100	.050	

MODE 2 (← μ, μ_2 ↑)

TABLE 5.3 ($\mu_1 = .10$)

MODE 1 (μ_3 →, μ_2 ↓)

μ_2	.050	.100	.150	.200	.250	.300	.350	.400	.450	.500	.550	.600	.650	.700	.750	.800	μ_2
.05	2.303	2.415	2.542	2.686	2.851	3.041	3.264	3.528	3.844	4.224	4.673	5.138	5.414	5.345	5.099	4.812	
.10	2.452	2.577	2.721	2.887	3.079	3.304	3.571	3.892	4.282	4.751	5.267	5.634	5.612	5.354	5.041		
.15	2.622	2.765	2.930	3.123	3.350	3.621	3.947	4.346	4.834	5.399	5.872	5.921	5.654	5.308		8.531	.80
.20	2.818	2.983	3.176	3.405	3.678	4.009	4.417	4.923	5.533	6.119	6.276	6.007	5.621		8.934	9.051	.75
.25	3.047	3.240	3.469	3.745	4.080	4.496	5.018	5.667	6.367	6.675	6.419	5.987		9.403	9.499	9.622	.70
.30	3.316	3.546	3.823	4.162	4.585	5.121	5.805	6.606	7.103	6.889	6.409		9.929	10.01	10.04	8.733	.65
.35	3.638	3.916	4.259	4.687	5.236	5.948	6.830	7.510	7.376	6.870		10.46	10.57	10.05	8.300	7.095	.60
.40	4.028	4.373	4.808	5.366	6.099	7.029	7.720	7.647	7.251		10.69	10.97	9.699	7.992	6.937	6.859	.55
.45	4.513	4.952	5.519	6.264	7.146	7.447	7.376	7.202		9.981	10.56	9.334	7.813	7.298	7.348	7.479	.50
.50	5.131	5.705	6.442	6.963	6.959	6.889	6.795		9.218	9.765	9.023	7.981	7.873	7.941	8.055	8.216	.45
.55	5.940	6.551	6.563	6.491	6.419	6.345		9.085	9.365	9.009	8.571	8.577	8.657	8.762	8.897	9.083	.40
.60	6.296	6.179	6.085	6.007	5.937		9.653	9.765	9.544	9.301	9.376	9.492	9.606	9.716	9.726	9.121	.35
.65	5.846	5.740	5.654	5.582		10.44	10.56	10.10	9.730	10.02	10.33	10.52	10.48	9.724	8.948	8.319	.30
.70	5.448	5.354	5.277		10.67	10.97	10.02	9.396	9.775	10.48	10.97	10.52	9.618	8.823	8.171	7.643	.25
.75	5.099	5.015		10.17	10.57	9.705	8.911	9.109	9.766	10.47	10.33	9.507	8.723	8.062	7.516	7.067	.20
.80	4.791		9.601	10.01	9.384	8.502	8.508	9.020	9.704	9.990	9.382	8.634	7.977	7.423	6.959	6.573	.15
		9.093	9.499	9.077	8.179	8.022	8.398	8.992	9.495	9.223	8.548	7.905	7.351	6.880	6.481	6.145	.10
	8.661	9.051	8.791	7.923	7.640	7.898	8.400	8.942	9.007	8.458	7.842	7.293	6.820	6.415	6.068	5.771	.05
	.800	.750	.700	.650	.600	.550	.500	.450	.400	.350	.300	.250	.200	.150	.100	.050	

MODE 2 (← μ_3, μ_2 ↑)

TABLE 5.3 ($\mu_1 = .05$)

MODE 3 (μ_3 →, μ_2 ↓)

μ_2	.050	.100	.150	.200	.250	.300	.350	.400	.450	.500	.550	.600	.650	.700	.750	.800	.850	μ_2
.05	9.104	9.559	10.09	10.70	11.38	11.98	11.64	10.71	10.32	10.89	11.83	11.88	11.26	10.74	10.83	11.96	11.78	
.10	9.658	10.17	10.78	11.48	12.22	12.55	11.63	10.83	11.08	12.08	12.55	11.93	11.31	11.23	12.45	12.35		
.15	10.29	10.87	11.57	12.37	13.14	12.79	11.68	11.39	12.29	13.23	12.73	12.00	11.73	12.98	13.01		15.37	.85
.20	11.01	11.68	12.49	13.39	13.83	12.78	11.94	12.50	13.71	13.56	12.76	12.31	13.48	13.68		16.13	16.13	.80
.25	11.84	12.60	13.42	13.48	13.05	12.44	12.63	13.53	13.41	13.01	12.66	13.40	13.47		16.94	15.30	13.34	.75
.30	12.35	12.05	11.83	11.64	11.55	12.08	12.39	12.08	11.83	11.75	12.35	12.27		16.71	14.63	14.10	14.24	.70
.35	10.73	10.48	10.37	10.90	12.02	12.61	11.85	11.20	11.16	12.47	12.35		15.51	15.08	15.00	15.11	15.28	.65
.40	9.487	9.973	10.96	12.21	13.40	12.85	11.80	11.53	13.09	13.17		16.23	15.94	15.94	16.13	16.25	15.01	.60
.45	10.13	11.12	12.39	13.80	13.76	12.85	12.25	13.60	13.83		17.14	15.86	15.82	16.96	16.83	14.66	12.98	.55
.50	11.35	12.58	13.22	13.10	12.89	12.65	13.15	13.17		16.40	15.07	14.52	15.74	16.34	14.44	13.52	13.68	.50
.55	12.43	12.27	12.15	12.04	11.98	12.37	12.35		15.52	15.01	14.87	15.16	15.62	14.74	14.70	14.87	15.10	.45
.60	11.47	11.31	11.21	11.20	12.36	12.27		16.40	16.00	16.02	16.31	16.39	15.59	16.00	16.30	16.39	15.31	.40
.65	10.62	10.49	10.65	13.14	13.47		17.14	15.82	15.77	16.85	16.78	15.10	15.68	17.12	16.45	15.09	13.96	.35
.70	9.895	10.80	13.61	13.68		16.23	15.15	14.73	15.71	16.25	14.66	14.32	15.69	16.14	14.93	13.79	12.87	.30
.75	11.43	13.10	13.01		15.51	14.87	14.56	15.06	15.62	14.59	14.39	14.91	15.62	14.86	14.12	14.27	14.65	.25
.80	12.49	12.35		16.71	16.05	16.09	16.56	16.62	15.59	16.10	16.64	16.44	15.60	16.15	16.69	16.34	15.42	.20
.85	11.78		16.94	15.81	15.77	16.65	16.69	15.18	15.67	16.85	16.44	15.31	15.69	17.00	16.24	15.26	14.42	.15
		16.13	15.22	14.92	15.70	16.17	14.73	14.60	15.66	16.11	15.03	14.44	15.68	16.03	15.14	14.27	13.53	.10
	15.37	14.71	14.17	14.84	15.48	14.45	13.76	14.55	15.46	14.84	13.92	14.16	15.44	15.01	14.17	13.41	12.75	.05
	.850	.800	.750	.700	.650	.600	.550	.500	.450	.400	.350	.300	.250	.200	.150	.100	.050	

MODE 4 (← μ_3, μ_2 ↑)

TABLE 5.3 ($\mu_1 = .10$)

MODE 3 (μ_3 →, μ_2 ↓)

μ_2	.050	.100	.150	.200	.250	.300	.350	.400	.450	.500	.550	.600	.650	.700	.750	.800	μ_2
.05	9.665	10.18	10.79	11.49	12.26	12.71	11.87	11.00	11.14	12.13	12.70	12.10	11.46	11.34	12.56	12.49	
.10	10.29	10.88	11.58	12.39	13.18	12.95	11.84	11.46	12.32	13.30	12.85	12.11	11.82	13.05	13.10		
.15	11.01	11.69	12.50	13.38	13.69	12.88	12.03	12.53	13.66	13.51	12.83	12.39	13.46	13.61		14.71	.80
.20	11.84	12.61	13.18	13.11	12.91	12.54	12.66	13.18	13.10	12.91	12.69	13.12	13.14		15.22	15.31	.75
.25	12.43	12.26	12.12	11.99	11.89	12.18	12.44	12.25	12.09	12.02	12.40	12.36		15.81	15.19	13.22	.70
.30	11.28	11.08	10.95	11.06	12.04	12.61	11.95	11.44	11.36	12.48	12.37		16.05	14.45	13.76	13.85	.65
.35	10.14	10.11	10.98	12.22	13.37	12.84	11.86	11.59	13.07	13.15		14.87	14.47	14.35	14.42	14.52	.60
.40	10.14	11.12	12.40	13.64	13.53	12.85	12.29	13.48	13.60		15.15	15.02	14.96	15.05	15.13	14.96	.55
.45	11.35	12.59	13.12	13.03	12.88	12.69	13.07	13.09		15.82	15.44	15.27	15.70	15.84	14.64	12.98	.50
.50	12.54	12.43	12.33	12.24	12.18	12.48	12.47		16.00	14.92	14.26	15.37	16.07	14.39	13.32	13.42	.45
.55	11.77	11.63	11.54	11.51	12.40	12.35		15.01	14.47	14.19	14.46	15.13	14.30	14.03	14.12	14.25	.40
.60	11.00	10.88	10.92	13.12	13.40		15.07	14.92	14.84	14.96	15.10	14.78	14.77	14.89	15.00	15.10	.35
.65	10.30	10.84	13.46	13.48		15.86	15.44	15.30	15.72	15.88	14.94	15.13	15.70	15.88	15.07	13.96	.30
.70	11.44	13.05	12.98		15.94	15.02	14.54	15.42	16.02	14.59	14.10	15.29	15.98	14.90	13.77	12.85	.25
.75	12.56	12.45		15.08	14.47	13.89	14.44	15.19	14.32	13.66	14.09	15.10	14.72	13.69	13.42	13.55	.20
.80	11.96		14.63	14.45	14.33	14.44	14.65	14.39	14.28	14.38	14.57	14.59	14.24	14.30	14.38	14.50	.15
		15.30	15.19	15.13	15.24	15.31	14.99	15.07	15.23	15.30	15.10	15.02	15.21	15.29	15.20	14.42	.10
	16.13	15.31	15.05	15.79	16.16	14.79	14.77	15.78	16.11	15.07	14.60	15.83	16.04	15.15	14.29	13.54	.05
	.800	.750	.700	.650	.600	.550	.500	.450	.400	.350	.300	.250	.200	.150	.100	.050	

MODE 4 (← μ_3, μ_2 ↑)

TABLE 5.3 (μ_1 = .15)

MODE 1

μ_2 ↓ / μ_3 →	.050	.100	.150	.200	.250	.300	.350	.400	.450	.500	.550	.600	.650	.700	.750	
.05	2.454	2.581	2.727	2.895	3.088	3.315	3.584	3.908	4.301	4.777	5.309	5.709	5.707	5.448	5.126	
.10	2.623	2.768	2.935	3.130	3.359	3.631	3.939	4.361	4.855	5.432	5.932	6.006	5.740	5.387		
.15	2.819	2.985	3.180	3.410	3.685	4.019	4.429	4.939	5.558	6.167	6.351	6.085	5.694		8.425	.75
.20	3.048	3.242	3.473	3.750	4.087	4.505	5.031	5.687	6.404	6.740	6.491	6.055		8.711	8.791	.70
.25	3.317	3.547	3.826	4.167	4.592	5.131	5.821	6.635	7.160	6.959	6.476		9.016	9.077	9.155	.65
.30	3.638	3.918	4.262	4.693	5.244	5.961	6.856	7.562	7.447	6.945		9.332	9.384	9.429	8.729	.60
.35	4.029	4.375	4.811	5.373	6.112	7.059	7.771	7.720	7.350		9.648	9.705	9.668	8.297	7.100	.55
.40	4.514	4.955	5.525	6.278	7.209	7.562	7.510	7.356		9.904	10.02	9.600	7.991	7.058	7.067	.50
.45	5.132	5.710	6.466	7.129	7.160	7.103	7.018		9.700	10.10	9.274	7.818	7.433	7.481	7.584	.45
.50	5.943	6.655	6.797	6.740	6.675	6.607		9.025	9.544	8.932	7.961	7.881	7.928	8.005	8.110	.40
.55	6.544	6.438	6.351	6.276	6.209		8.730	9.009	8.747	8.352	8.342	8.384	8.442	8.514	8.611	.35
.60	6.111	6.006	5.921	5.849		8.940	9.023	8.932	8.770	8.782	8.824	8.870	8.920	8.979	9.026	.30
.65	5.707	5.612	5.533		9.294	9.334	9.274	9.135	9.173	9.230	9.276	9.314	9.335	8.940	8.320	.25
.70	5.345	5.258		9.657	9.699	9.600	9.270	9.428	9.586	9.661	9.698	9.555	8.824	8.173	7.644	.20
.75	5.021		9.922	10.05	9.668	8.953	9.151	9.674	9.987	10.04	9.505	8.729	8.068	7.520	7.069	.15
		9.667	10.04	9.429	8.573	8.612	9.135	9.785	10.01	9.400	8.646	7.986	7.430	6.965	6.576	.10
	9.217	9.622	9.155	8.261	8.148	8.555	9.167	9.630	9.267	8.569	7.920	7.363	6.891	6.490	6.149	.05
← μ_3	.750	.700	.650	.600	.550	.500	.450	.400	.350	.300	.250	.200	.150	.100	.050	↑ μ_2

MODE 2

TABLE 5.3 (μ_1 = .15)

MODE 3

μ_2 ↓ / μ_3 →	.050	.100	.150	.200	.250	.300	.350	.400	.450	.500	.550	.600	.650	.700	.750	
.05	10.30	10.89	11.43	11.43	11.42	11.40	11.36	11.33	11.44	11.43	11.42	11.39	11.38	11.44	11.43	
.10	10.84	10.80	10.77	10.75	10.73	10.71	10.75	10.84	10.79	10.76	10.73	10.74	10.84	10.80		
.15	10.36	10.30	10.26	10.23	10.22	10.52	10.92	10.54	10.34	10.28	10.33	10.92	10.65		14.17	.75
.20	9.923	9.859	9.820	9.858	10.64	11.50	11.17	10.44	10.10	10.35	11.51	11.20		14.92	15.05	.70
.25	9.510	9.455	9.809	10.78	11.90	12.20	11.29	10.57	10.75	12.18	11.98		15.77	15.13	12.04	.65
.30	9.188	9.932	10.93	12.14	12.87	12.26	11.41	11.28	12.69	12.65		16.09	14.33	12.65	12.76	.60
.35	10.13	11.09	12.19	12.29	12.12	11.83	11.67	12.29	12.25		14.56	13.89	13.45	13.54	13.70	.55
.40	11.32	11.57	11.48	11.39	11.30	11.28	11.59	11.53		14.73	14.54	14.39	14.52	14.66	14.76	.50
.45	10.98	10.86	10.78	10.72	10.75	11.36	11.16		15.77	15.30	15.04	15.59	15.81	14.61	12.94	.45
.50	10.41	10.32	10.26	10.35	12.02	11.75		16.02	14.84	13.99	15.27	16.09	14.36	12.67	11.81	.40
.55	9.946	9.882	10.33	12.69	12.66		14.87	14.19	13.25	13.90	15.01	14.12	12.74	12.69	12.89	.35
.60	9.557	10.74	12.39	12.31		14.52	14.26	13.99	14.21	14.55	14.20	13.81	13.97	14.18	14.46	.30
.65	11.38	11.82	11.73		15.82	15.27	15.04	15.60	15.87	14.80	14.78	15.54	15.87	15.04	13.95	.25
.70	11.34	11.23		15.94	14.96	14.39	15.35	16.03	14.54	13.87	15.17	15.99	14.89	13.75	12.81	.20
.75	10.83		15.00	14.35	13.45	14.21	15.11	14.25	13.05	13.56	14.99	14.68	13.61	12.65	11.85	.15
		14.10	13.76	12.65	13.20	14.07	13.93	12.64	12.37	13.47	14.27	13.47	12.52	11.71	11.04	.10
	13.34	13.22	12.04	12.38	13.13	13.49	12.39	11.72	12.23	13.35	13.27	12.42	11.62	11.48	11.51	.05
← μ_3	.750	.700	.650	.600	.550	.500	.450	.400	.350	.300	.250	.200	.150	.100	.050	↑ μ_2

MODE 4

TABLE 5.3 (μ_1 = .20)

MODE 1

μ_2 ↓ / μ_3 →	.050	.100	.150	.200	.250	.300	.350	.400	.450	.500	.550	.600	.650	.700	
.05	2.626	2.773	2.942	3.139	3.370	3.644	3.975	4.380	4.880	5.471	6.005	6.111	5.846	5.484	
.10	2.821	2.989	3.186	3.418	3.695	4.031	4.444	4.959	5.587	6.222	6.438	6.179	5.781		
.15	3.048	3.244	3.477	3.756	4.096	4.517	5.046	5.709	6.441	6.797	6.563	6.130		7.641	.70
.20	3.317	3.550	3.830	4.173	4.601	5.143	5.836	6.652	7.129	6.963	6.523		7.854	7.923	.65
.25	3.639	3.920	4.266	4.699	5.252	5.971	6.839	7.209	7.146	6.880		8.115	8.179	8.261	.60
.30	4.029	4.377	4.815	5.379	6.116	6.934	7.059	7.029	6.940		8.439	8.502	8.573	8.588	.55
.35	4.515	4.956	5.528	6.275	6.839	6.856	6.830	6.784		8.835	8.911	8.953	8.253	7.084	.50
.40	5.132	5.711	6.435	6.652	6.635	6.606	6.569		9.260	9.396	9.270	7.933	6.818	6.580	.45
.45	5.943	6.457	6.441	6.404	6.367	6.328		9.325	9.730	9.135	7.673	6.813	6.785	6.830	.40
.50	6.290	6.222	6.167	6.119	6.075		8.691	9.301	8.770	7.466	6.990	6.992	7.023	7.069	.35
.55	6.005	5.932	5.872	5.820		8.019	8.571	8.352	7.366	7.195	7.202	7.225	7.257	7.302	.30
.60	5.709	5.634	5.572		7.669	7.981	7.961	7.466	7.413	7.422	7.439	7.462	7.493	7.536	.25
.65	5.414	5.340		7.739	7.813	7.818	7.673	7.650	7.659	7.674	7.691	7.712	7.740	7.782	.20
.70	5.128		7.969	7.992	7.991	7.933	7.921	7.931	7.944	7.958	7.972	7.988	7.996	7.642	.15
		8.289	8.300	8.297	8.253	8.246	8.263	8.277	8.288	8.297	8.301	8.062	7.524	7.072	.10
	8.726	8.733	8.729	8.588	8.610	8.699	8.721	8.730	8.734	8.640	8.000	7.443	6.974	6.581	.05
← μ_3	.700	.650	.600	.550	.500	.450	.400	.350	.300	.250	.200	.150	.100	.050	↑ μ_2

MODE 2

TABLE 5.3 (μ_1 = .20)

MODE 3

μ_2 ↓ / μ_3 →	.050	.100	.150	.200	.250	.300	.350	.400	.450	.500	.550	.600	.650	.700	
.05	8.792	8.770	8.759	8.753	8.773	9.447	10.18	10.11	9.423	8.930	8.853	9.557	10.30	9.895	
.10	8.388	8.356	8.348	8.763	9.555	10.44	10.87	10.20	9.482	9.185	9.882	10.88	10.49		
.15	8.069	8.210	8.853	9.661	10.63	11.52	11.18	10.28	9.755	10.26	11.54	11.21		14.84	.70
.20	8.330	8.972	9.778	10.78	11.92	12.29	11.31	10.51	10.72	12.24	12.04		15.70	15.79	.65
.25	9.138	9.922	10.92	12.16	13.17	12.47	11.45	11.30	12.88	12.89		16.65	15.24	12.38	.60
.30	10.12	11.10	12.32	12.95	12.68	12.14	11.83	12.85	12.85		16.56	14.44	13.20	13.34	.55
.35	11.33	11.89	11.75	11.60	11.45	11.41	11.86	11.80		15.06	14.44	14.21	14.34	14.54	.50
.40	10.94	10.74	10.61	10.51	10.57	11.44	11.20		15.71	15.42	15.35	15.55	15.71	14.98	.45
.45	9.984	9.836	9.755	10.10	12.09	11.83		16.85	15.72	15.60	16.65	16.70	14.64	12.94	.40
.50	9.260	9.105	10.28	12.91	13.01		16.31	14.96	14.21	15.57	16.28	14.39	12.72	12.25	.35
.55	8.853	10.73	12.83	12.76		15.16	14.46	13.90	14.45	15.32	14.25	13.47	13.62	13.92	.30
.60	11.39	12.11	12.00		15.74	15.37	15.27	15.57	15.78	15.02	15.14	15.48	15.74	15.28	.25
.65	11.46	11.31		16.96	15.70	15.59	16.65	16.70	15.00	15.41	16.89	16.42	15.08	13.95	.20
.70	10.74		16.13	15.05	14.52	15.55	16.18	14.59	14.00	15.45	16.09	14.91	13.76	12.82	.15
		15.11	14.42	13.54	14.34	15.24	14.29	13.11	13.73	15.16	14.71	13.62	12.65	11.85	.10
	14.24	13.85	12.76	13.34	14.23	13.99	12.68	12.51	13.70	14.36	13.49	12.54	11.71	11.03	.05
← μ_3	.700	.650	.600	.550	.500	.450	.400	.350	.300	.250	.200	.150	.100	.050	↑ μ_2

MODE 4

TABLE 5.3 ($\mu_1 = .25$)

$\mu_3 \rightarrow$ MODE 1

$\mu_2 \downarrow$	.050	.100	.150	.200	.250	.300	.350	.400	.450	.500	.550	.600	.650	
.05	2.823	2.995	3.194	3.429	3.708	4.046	4.463	4.983	5.624	6.290	6.544	6.296	5.891	
.10	3.050	3.248	3.484	3.765	4.108	4.531	5.065	5.735	6.457	6.655	6.551	6.202		
.15	3.318	3.553	3.836	4.181	4.611	5.156	5.849	6.435	6.466	6.442	6.352		7.230	.65
.20	3.640	3.922	4.270	4.705	5.260	5.960	6.275	6.278	6.264	6.233		7.536	7.640	.60
.25	4.030	4.379	4.819	5.381	6.024	6.116	6.112	6.099	6.080		7.926	8.022	8.148	.55
.30	4.515	4.958	5.524	5.960	5.971	5.961	5.948	5.931		8.413	8.508	8.612	8.610	.50
.35	5.132	5.691	5.849	5.836	5.821	5.805	5.788		8.999	9.109	9.151	8.246	7.077	.45
.40	5.762	5.735	5.709	5.687	5.667	5.649		9.569	9.775	9.428	7.921	6.771	5.965	.40
.45	5.624	5.587	5.558	5.533	5.510		9.481	10.02	9.173	7.650	6.540	5.926	5.927	.35
.50	5.471	5.432	5.399	5.371		8.710	9.376	8.782	7.413	6.366	6.042	6.051	6.078	.30
.55	5.309	5.267	5.232		7.949	8.577	8.342	7.195	6.283	6.182	6.190	6.209	6.238	.25
.60	5.138	5.093		7.322	7.873	7.881	6.990	6.366	6.343	6.350	6.363	6.383	6.413	.20
.65	4.958		6.871	7.298	7.433	6.813	6.540	6.534	6.539	6.548	6.561	6.579	6.608	.15
		6.826	6.937	7.058	6.818	6.771	6.770	6.774	6.779	6.786	6.796	6.810	6.836	.10
	7.087	7.095	7.100	7.084	7.077	7.078	7.079	7.082	7.084	7.087	7.091	7.097	7.063	.05
	.650	.600	.550	.500	.450	.400	.350	.300	.250	.200	.150	.100	.050	$\uparrow \mu_2$

MODE 2 $\leftarrow \mu_3$

TABLE 5.3 ($\mu_1 = .30$)

$\mu_3 \rightarrow$ MODE 1

$\mu_2 \downarrow$	.050	.100	.150	.200	.250	.300	.350	.400	.450	.500	.550	.600	
.05	3.053	3.255	3.494	3.778	4.123	4.551	5.089	5.762	5.943	5.943	5.940	5.930	
.10	3.320	3.557	3.843	4.192	4.625	5.172	5.691	5.711	5.710	5.705	5.697		
.15	3.641	3.926	4.277	4.714	5.260	5.524	5.528	5.525	5.519	5.512		7.350	.60
.20	4.031	4.382	4.822	5.317	5.381	5.379	5.373	5.366	5.359		7.763	7.898	.55
.25	4.516	4.955	5.260	5.260	5.252	5.244	5.236	5.227		8.276	8.398	8.555	.50
.30	5.119	5.172	5.156	5.143	5.131	5.121	5.111		8.904	9.020	9.135	8.699	.45
.35	5.089	5.065	5.046	5.031	5.018	5.006		9.627	9.766	9.674	8.263	7.078	.40
.40	4.983	4.959	4.939	4.923	4.909		10.17	10.48	9.586	7.931	6.770	5.954	.35
.45	4.880	4.855	4.834	4.817		9.702	10.33	9.230	7.659	6.534	5.735	5.220	.30
.50	4.777	4.751	4.730		8.805	9.492	8.824	7.422	6.343	5.573	5.311	5.325	.25
.55	4.673	4.646		8.013	8.657	8.384	7.202	6.182	5.492	5.425	5.435	5.456	.20
.60	4.565		7.363	7.941	7.928	6.992	6.042	5.573	5.563	5.570	5.583	5.604	.15
		6.833	7.348	7.481	6.785	5.926	5.735	5.734	5.738	5.745	5.755	5.774	.10
	6.400	6.859	7.067	6.580	5.965	5.954	5.954	5.956	5.958	5.961	5.967	5.978	.05
	.600	.550	.500	.450	.400	.350	.300	.250	.200	.150	.100	.050	$\uparrow \mu_2$

MODE 2 $\leftarrow \mu_3$

TABLE 5.3 ($\mu_1 = .25$)

$\mu_3 \rightarrow$ MODE 3

$\mu_2 \downarrow$	.050	.100	.150	.200	.250	.300	.350	.400	.450	.500	.550	.600	.650	
.05	7.132	7.542	8.096	8.766	9.571	10.48	11.01	10.37	9.604	9.260	9.946	11.00	10.62	
.10	7.652	8.193	8.857	9.670	10.65	11.58	11.32	10.41	9.836	10.32	11.63	11.31		
.15	8.328	8.974	9.784	10.79	11.95	12.42	11.44	10.61	10.78	12.33	12.15		15.48	.65
.20	9.139	9.926	10.93	12.19	13.36	12.72	11.60	11.39	13.03	13.10		16.17	16.16	.60
.25	10.12	11.11	12.38	13.76	13.65	12.68	12.12	13.53	13.76		16.69	15.31	13.13	.55
.30	11.35	12.53	12.89	12.72	12.47	12.26	12.84	12.85		16.62	14.65	14.07	14.23	.50
.35	11.85	11.62	11.44	11.31	11.29	11.95	11.85		15.62	15.19	15.11	15.24	15.41	.45
.40	10.60	10.41	10.28	10.44	12.25	12.08		16.25	16.02	16.03	16.18	16.26	15.01	.40
.45	9.604	9.482	10.34	13.10	13.41		16.78	15.88	15.87	16.70	16.72	14.66	12.96	.35
.50	8.930	10.76	13.51	13.56		16.39	15.10	14.55	15.78	16.34	14.44	13.31	13.50	.30
.55	11.42	12.85	12.73		15.62	15.13	15.01	15.32	15.71	14.84	14.84	15.03	15.29	.25
.60	12.10	11.93		16.34	16.07	16.09	16.28	16.34	15.69	16.08	16.28	16.34	15.31	.20
.65	11.26		16.83	15.84	15.81	16.70	16.72	15.11	15.73	16.79	16.45	15.09	13.96	.15
		16.25	15.13	14.66	15.71	16.26	14.63	14.17	15.69	16.15	14.92	13.77	12.82	.10
	15.28	14.52	13.70	14.54	15.41	14.34	13.23	14.00	15.38	14.75	13.64	12.66	11.86	.05
	.650	.600	.550	.500	.450	.400	.350	.300	.250	.200	.150	.100	.050	$\uparrow \mu_2$

MODE 4 $\leftarrow \mu_3$

TABLE 5.3 ($\mu_1 = .30$)

$\mu_3 \rightarrow$ MODE 3

$\mu_2 \downarrow$	.050	.100	.150	.200	.250	.300	.350	.400	.450	.500	.550	.600	
.05	7.658	8.203	8.871	9.688	10.67	11.67	11.54	10.60	9.984	10.41	11.77	11.47	
.10	8.332	8.982	9.795	10.81	11.99	12.57	11.62	10.74	10.86	12.43	12.27		
.15	9.141	9.933	10.94	12.21	13.44	12.89	11.75	11.48	13.12	13.22		14.45	.60
.20	10.13	11.11	12.40	13.79	13.76	12.95	12.29	13.64	13.80		14.73	14.79	.55
.25	11.35	12.60	13.44	13.36	13.17	12.87	13.37	13.40		15.18	14.99	13.49	.50
.30	12.69	12.57	12.42	12.29	12.20	12.61	12.61		15.59	14.39	13.93	13.99	.45
.35	11.54	11.32	11.18	11.17	12.44	12.39		14.59	14.32	14.25	14.29	14.34	.40
.40	10.37	10.20	10.54	13.18	13.53		14.66	14.59	14.54	14.59	14.63	14.68	.35
.45	9.423	10.79	13.66	13.71		15.10	14.94	14.80	15.00	15.11	14.61	13.04	.30
.50	11.43	13.30	13.23		15.59	14.78	14.20	15.02	15.69	14.37	13.62	13.70	.25
.55	12.70	12.55		14.74	14.30	14.12	14.25	14.84	14.22	14.05	14.09	14.14	.20
.60	11.88		14.44	14.39	14.36	14.39	14.44	14.37	14.34	14.36	14.39	14.42	.15
		14.66	14.64	14.61	14.64	14.66	14.61	14.59	14.63	14.65	14.66	13.96	.10
	15.01	14.96	14.76	14.98	15.01	14.68	14.38	14.97	15.00	14.91	13.78	12.83	.05
	.600	.550	.500	.450	.400	.350	.300	.250	.200	.150	.100	.050	$\uparrow \mu_2$

MODE 4 $\leftarrow \mu_3$

TABLE 5.3 ($\mu_1 = .35$)

MODE 1 ($\mu_1 \rightarrow$)

$\mu_2 \downarrow$	.050	.100	.150	.200	.250	.300	.350	.400	.450	.500	.550	$\mu_2 \uparrow$
.05	3.324	3.566	3.855	4.208	4.644	5.119	5.132	5.132	5.132	5.131	5.130	
.10	3.643	3.932	4.285	4.721	4.955	4.958	4.956	4.955	4.952	4.950		
.15	4.033	4.385	4.779	4.822	4.819	4.815	4.811	4.808	4.804		7.743	.55
.20	4.515	4.721	4.714	4.705	4.699	4.693	4.687	4.682		8.242	8.400	.50
.25	4.644	4.625	4.611	4.601	4.592	4.585	4.578		8.855	8.992	9.167	.45
.30	4.551	4.531	4.517	4.505	4.496	4.487		9.580	9.704	9.785	8.721	.40
.35	4.463	4.444	4.429	4.417	4.407		10.33	10.47	9.987	8.277	7.079	.35
.40	4.380	4.361	4.346	4.333		10.65	10.97	9.661	[illegible]	6.774	5.954	.30
.45	4.301	4.282	4.267		9.875	10.52	9.276	7.674	6.539	5.734	5.140	.25
.50	4.224	4.205		8.914	9.606	8.870	7.439	6.350	5.563	4.981	4.755	.20
.55	4.148		8.112	8.762	8.442	7.225	6.190	5.425	4.896	4.854	4.867	.15
		7.466	8.055	8.005	7.023	6.051	5.311	4.981	4.978	4.985	4.999	.10
	6.947	7.479	7.584	6.830	5.927	5.220	5.140	5.140	5.142	5.146	5.154	.05
	.550	.500	.450	.400	.350	.300	.250	.200	.150	.100	.050	

MODE 2 ($\leftarrow \mu_3$)

TABLE 5.3 ($\mu_1 = .40$)

MODE 1 ($\mu_3 \rightarrow$)

$\mu_2 \downarrow$	.050	.100	.150	.200	.250	.300	.350	.400	.450	.500	$\mu_2 \uparrow$
.05	3.648	3.941	4.299	4.515	4.516	4.515	4.515	4.514	4.513	4.513	
.10	4.035	4.361	4.385	4.382	4.379	4.377	4.375	4.373	4.372		
.15	4.299	4.285	4.277	4.270	4.266	4.262	4.259	4.256		8.243	.50
.20	4.208	4.192	4.181	4.173	4.167	4.162	4.158		8.785	8.942	.45
.25	4.123	4.108	4.096	4.087	4.080	4.074		9.383	9.495	9.630	.40
.30	4.046	4.031	4.019	4.009	4.001		9.921	9.990	10.01	8.730	.35
.35	3.975	3.959	3.947	3.937		10.28	10.33	10.04	8.288	7.082	.30
.40	3.908	3.892	3.880		10.44	10.52	9.698	7.958	6.779	5.956	.25
.45	3.844	3.829		9.981	10.48	9.314	7.691	6.548	5.738	5.140	.20
.50	3.783		9.036	9.716	8.920	7.462	6.363	5.570	4.978	4.524	.15
		8.244	8.897	8.514	7.257	6.209	5.435	4.854	4.432	4.409	.10
	7.616	8.216	8.110	7.069	6.078	5.325	4.755	4.524	4.525	4.531	.05
	.500	.450	.400	.350	.300	.250	.200	.150	.100	.050	

MODE 2 ($\leftarrow \mu_3$)

TABLE 5.3 ($\mu_1 = .35$)

MODE 3 ($\mu_3 \rightarrow$)

$\mu_2 \downarrow$	.050	.100	.150	.200	.250	.300	.350	.400	.450	.500	.550	$\mu_2 \uparrow$
.05	8.339	8.995	9.812	10.83	12.04	12.69	11.85	10.94	10.98	12.54	12.43	
.10	9.146	9.942	10.96	12.22	12.60	12.53	11.89	11.57	12.59	12.58		
.15	10.13	11.12	12.33	12.40	12.38	12.32	12.19	12.40	12.39		13.76	.55
.20	11.36	12.22	12.21	12.19	12.16	12.14	12.22	12.21		14.60	14.77	.50
.25	12.04	11.99	11.95	11.92	11.90	12.04	12.02		15.67	15.07	12.39	.45
.30	11.67	11.58	11.52	11.50	12.18	12.08		16.10	14.28	12.64	12.68	.40
.35	11.01	10.87	10.92	12.66	12.63		14.39	13.66	13.05	13.11	13.23	.35
.40	10.11	10.84	12.53	12.50		14.32	14.10	13.87	14.00	14.17	14.38	.30
.45	11.44	12.32	12.29		15.68	15.13	14.78	15.41	15.73	14.59	12.93	.25
.50	12.13	12.08		16.00	14.77	13.81	15.14	16.08	14.34	12.65	12.28	.20
.55	11.83		14.70	14.03	12.74	13.47	14.84	14.05	12.52	12.45	12.47	.15
		13.52	13.32	12.67	12.72	13.31	13.62	12.65	12.65	12.66	12.68	.10
	12.98	12.98	12.94	12.94	12.96	13.04	12.93	12.93	12.94	12.95	12.96	.05
	.550	.500	.450	.400	.350	.300	.250	.200	.150	.100	.050	

MODE 4 ($\leftarrow \mu_3$)

TABLE 5.3 ($\mu_1 = .40$)

MODE 3 ($\mu_3 \rightarrow$)

$\mu_2 \downarrow$	.050	.100	.150	.200	.250	.300	.350	.400	.450	.500	$\mu_2 \uparrow$
.05	9.154	9.958	10.98	11.36	11.35	11.35	11.33	11.32	11.35	11.35	
.10	10.14	11.08	11.12	11.11	11.11	11.10	11.09	11.12	11.12		
.15	10.98	10.96	10.94	10.93	10.92	10.93	10.98	10.96		14.55	.50
.20	10.83	10.81	10.79	10.78	10.78	11.06	10.90		15.66	15.78	.45
.25	10.67	10.65	10.63	10.64	11.89	11.55		16.85	15.23	11.72	.40
.30	10.48	10.44	10.52	12.54	12.44		16.64	14.38	12.37	12.51	.35
.35	10.18	10.75	12.03	11.94		14.91	14.09	13.56	13.73	14.00	.30
.40	11.33	11.46	11.39		15.69	15.29	15.17	15.45	15.69	14.97	.25
.45	11.14	11.08		17.12	15.70	15.54	16.89	16.79	14.63	12.93	.20
.50	10.89		16.30	14.89	13.97	15.48	16.28	14.36	12.65	11.37	.15
		14.87	14.12	12.69	13.62	15.03	14.09	12.45	11.20	11.16	.10
	13.68	13.42	11.81	12.25	13.50	13.70	12.28	11.37	11.37	11.38	.05
	.500	.450	.400	.350	.300	.250	.200	.150	.100	.050	

MODE 4 ($\leftarrow \mu_3$)

TABLE 5.3 ($\mu_1 = .45$) MODE 1

$\mu_2 \downarrow$ \ $\mu_3 \rightarrow$	.050	.100	.150	.200	.250	.300	.350	.400	.450	$\mu_2 \uparrow$
.05	4.027	4.035	4.033	4.031	4.030	4.029	4.029	4.028	4.028	
.10	3.941	3.932	3.926	3.922	3.920	3.918	3.916	3.915		
.15	3.855	3.843	3.836	3.830	3.826	3.823	3.820		8.586	.45
.20	3.778	3.765	3.756	3.750	3.745	3.740		8.931	9.007	.40
.25	3.708	3.695	3.685	3.678	3.672		9.185	9.223	9.267	.35
.30	3.644	3.631	3.621	3.613		9.361	9.382	9.400	8.734	.30
.35	3.584	3.571	3.561		9.491	9.507	9.505	8.297	7.084	.25
.40	3.528	3.515		9.599	9.618	9.555	7.972	6.786	5.958	.20
.45	3.475		9.662	9.724	9.335	7.712	6.561	5.745	5.142	.15
		9.175	9.726	8.979	7.493	6.383	5.583	4.985	4.525	.10
	8.426	9.083	8.611	7.302	6.238	5.456	4.867	4.409	4.056	.05
	.450	.400	.350	.300	.250	.200	.150	.100	.050	

MODE 2 $\leftarrow \mu_3$

TABLE 5.3 ($\mu_1 = .50$) MODE 1

$\mu_2 \downarrow$ \ $\mu_3 \rightarrow$	.050	.100	.150	.200	.250	.300	.350	.400	$\mu_2 \uparrow$
.05	3.648	3.643	3.641	3.640	3.639	3.638	3.638	3.637	
.10	3.566	3.557	3.553	3.550	3.547	3.546	3.544		
.15	3.494	3.484	3.477	3.473	3.469	3.466		8.310	.40
.20	3.429	3.418	3.410	3.405	3.400		8.431	8.458	.35
.25	3.370	3.359	3.350	3.344		8.531	8.548	8.569	.30
.30	3.315	3.304	3.295		8.623	8.634	8.646	8.640	.25
.35	3.264	3.253		8.714	8.723	8.729	8.301	7.087	.20
.40	3.217		8.815	8.823	8.824	7.988	6.796	5.961	.15
		8.938	8.948	8.940	7.740	6.579	5.755	5.146	.10
	9.092	9.121	9.026	7.536	6.413	5.604	4.999	4.531	.05
	.400	.350	.300	.250	.200	.150	.100	.050	

MODE 2 $\leftarrow \mu_3$

TABLE 5.3 ($\mu_1 = .45$) MODE 3

$\mu_2 \downarrow$ \ $\mu_3 \rightarrow$	.050	.100	.150	.200	.250	.300	.350	.400	.450	$\mu_2 \uparrow$
.05	10.11	10.14	10.13	10.13	10.12	10.12	10.13	10.14	10.13	
.10	9.958	9.942	9.933	9.926	9.922	9.932	10.11	9.973		
.15	9.812	9.795	9.784	9.778	9.809	10.95	10.37		15.46	.45
.20	9.688	9.670	9.661	9.858	11.99	11.64		16.11	16.11	.40
.25	9.571	9.555	10.22	12.91	13.05		16.44	15.30	12.23	.35
.30	9.447	10.71	12.88	12.78		16.44	14.57	13.47	13.70	.30
.35	11.36	11.84	11.68		15.62	15.10	14.99	15.16	15.38	.25
.40	11.00	10.83		16.14	15.98	15.99	16.09	16.15	15.00	.20
.45	10.32		16.45	15.88	15.87	16.42	16.45	14.65	12.94	.15
		16.39	15.00	14.18	15.74	16.34	14.39	12.66	11.37	.10
	15.10	14.25	12.89	13.92	15.29	14.14	12.47	11.16	10.18	.05
	.450	.400	.350	.300	.250	.200	.150	.100	.050	

MODE 4 $\leftarrow \mu_3$

TABLE 5.3 ($\mu_1 = .50$) MODE 3

$\mu_2 \downarrow$ \ $\mu_3 \rightarrow$	.050	.100	.150	.200	.250	.300	.350	.400	$\mu_2 \uparrow$
.05	9.154	9.146	9.141	9.139	9.138	9.188	10.14	9.487	
.10	8.995	8.982	8.974	8.972	9.455	11.08	10.48		
.15	8.871	8.857	8.853	9.820	12.12	11.83		14.84	.40
.20	8.766	8.763	10.23	13.11	13.48		15.03	15.07	.35
.25	8.773	10.73	13.69	13.83		15.31	15.10	13.35	.30
.30	11.40	12.95	12.79		15.60	14.59	14.27	14.36	.25
.35	11.87	11.63		14.86	14.72	14.68	14.71	14.75	.20
.40	10.71		14.93	14.90	14.89	14.91	14.92	14.91	.15
.45		15.09	15.07	15.04	15.08	15.09	14.66	12.95	.10
	15.31	15.10	14.46	15.28	15.31	14.42	12.68	11.38	.05
	.400	.350	.300	.250	.200	.150	.100	.050	

MODE 4 $\leftarrow \mu_3$

TABLE 5.4 ($\mu_1 = .05$)

MODE 1 ($\mu_3 \rightarrow$ across top, $\mu_2 \downarrow$ at left); MODE 2 ($\leftarrow \mu_3$ across bottom, $\mu_2 \uparrow$ at right)

μ_2	.050	.100	.150	.200	.250	.300	.350	.400	.450	.500	.550	.600	.650	.700	.750	.800	.850	μ_2
.05	4.547	4.764	5.013	5.300	5.633	6.018	6.466	6.976	7.496	7.742	7.463	7.023	6.600	6.226	5.902	5.623	5.384	
.10	4.824	5.067	5.350	5.681	6.067	6.520	7.046	7.626	8.075	7.943	7.477	7.002	6.580	6.216	5.904	5.639		
.15	5.138	5.414	5.739	6.124	6.578	7.115	7.734	8.344	8.474	8.018	7.481	6.997	6.582	6.230	5.933		8.985	.85
.20	5.496	5.813	6.192	6.645	7.187	7.830	8.543	9.001	8.655	8.051	7.490	7.011	6.608	6.271		9.429	9.526	.80
.25	5.910	6.278	6.725	7.267	7.922	8.694	9.427	9.378	8.728	8.074	7.514	7.048	6.661		9.946	10.03	10.14	.75
.30	6.391	6.825	7.361	8.019	8.821	9.717	10.11	9.505	8.756	8.101	7.557	7.111		10.54	10.61	10.71	10.84	.70
.35	6.957	7.478	8.129	8.941	9.908	10.60	10.22	9.479	8.754	8.133	7.622		11.20	11.29	11.38	11.49	11.64	.65
.40	7.634	8.267	9.066	9.989	10.27	10.10	9.772	9.270	8.694	8.159		11.81	12.03	12.16	12.27	12.39	12.40	.60
.45	8.453	9.198	9.529	9.445	9.336	9.208	9.041	8.803	8.482		11.58	12.42	12.93	13.15	13.24	12.17	10.80	.55
.50	8.978	8.815	8.691	8.590	8.498	8.408	8.308	8.192		10.67	11.59	12.73	13.71	13.60	11.94	10.55	9.518	.50
.55	8.215	8.067	7.957	7.869	7.794	7.725	7.659		10.22	10.77	11.63	12.75	13.15	11.74	10.41	9.831	9.970	.45
.60	7.558	7.431	7.336	7.261	7.198	7.145		10.72	10.92	11.26	11.85	12.45	11.60	10.74	10.76	10.91	11.14	.40
.65	6.999	6.889	6.806	6.742	6.691		11.63	11.85	12.06	12.28	12.47	11.97	11.79	11.95	12.13	12.32	12.44	.35
.70	6.516	6.421	6.351	6.297		11.78	12.43	13.04	13.46	13.48	12.40	12.46	13.14	13.54	13.27	12.34	11.52	.30
.75	6.096	6.015	5.956		11.16	11.84	12.64	13.48	13.50	12.14	11.77	12.72	13.72	13.17	12.21	11.35	10.65	.25
.80	5.729	5.661		10.50	11.11	11.83	12.63	13.01	11.08	11.06	11.33	12.67	12.92	12.08	11.24	10.51	9.907	.20
.85	5.404		9.911	10.44	11.07	11.79	12.35	11.64	10.63	10.60	11.46	12.33	11.92	11.13	10.41	9.782	9.259	.15
		9.396	9.867	10.42	11.06	11.65	11.39	10.37	9.941	10.47	11.38	11.66	11.03	10.32	9.692	9.150	8.691	.10
	8.954	9.376	9.873	10.44	11.01	11.08	10.18	9.525	9.709	10.46	11.16	10.89	10.24	9.621	9.073	8.599	8.192	.05
	.850	.800	.750	.700	.650	.600	.550	.500	.450	.400	.350	.300	.250	.200	.150	.100	.050	

TABLE 5.4 ($\mu_1 = .10$)

MODE 1 ($\mu_3 \rightarrow$ across top, $\mu_2 \downarrow$ at left); MODE 2 ($\leftarrow \mu_3$ across bottom, $\mu_2 \uparrow$ at right)

μ_2	.050	.100	.150	.200	.250	.300	.350	.400	.450	.500	.550	.600	.650	.700	.750	.800	μ_2
.05	4.827	5.073	5.358	5.689	6.076	6.530	7.060	7.650	8.138	8.052	7.587	7.102	6.669	6.295	5.977	5.706	
.10	5.139	5.417	5.745	6.130	6.586	7.125	7.749	8.379	8.558	8.113	7.568	7.076	6.653	6.294	5.992		
.15	5.498	5.816	6.197	6.651	7.194	7.840	8.563	9.060	8.742	8.133	7.564	7.077	6.667	6.325		9.430	.80
.20	5.911	6.281	6.729	7.272	7.930	8.708	9.466	9.466	8.813	8.149	7.579	7.106	6.714		9.924	10.03	.75
.25	6.391	6.828	7.365	8.025	8.832	9.747	10.22	9.618	8.847	8.175	7.620	7.167		10.50	10.59	10.72	.70
.30	6.958	7.480	8.135	8.952	9.943	10.81	10.49	9.647	8.867	8.217	7.690		11.17	11.26	11.36	11.51	.65
.35	7.635	8.271	9.082	10.09	11.00	10.88	10.31	9.569	8.867	8.274		11.91	12.03	12.13	12.25	12.42	.60
.40	8.457	9.239	10.13	10.29	10.16	9.982	9.704	9.291	8.795		12.46	12.82	13.02	13.15	13.26	12.52	.55
.45	9.438	9.537	9.410	9.296	9.188	9.071	8.931	8.749		11.80	12.88	13.76	14.13	13.93	12.19	10.81	.50
.50	8.845	8.676	8.553	8.454	8.368	8.286	8.202		10.89	11.78	13.00	14.24	13.73	11.96	10.58	9.831	.45
.55	8.086	7.943	7.836	7.753	7.683	7.621		10.83	11.22	11.93	12.99	13.26	11.79	10.71	10.72	10.93	.40
.60	7.446	7.323	7.232	7.162	7.105		11.69	11.88	12.12	12.51	12.83	11.96	11.76	11.90	12.09	12.36	.35
.65	6.900	6.795	6.718	6.659		12.46	12.94	13.31	13.56	13.55	12.78	13.02	13.36	13.59	13.44	12.54	.30
.70	6.430	6.340	6.276		11.99	12.76	13.65	14.44	13.91	12.59	12.95	14.14	14.40	13.36	12.35	11.52	.25
.75	6.021	5.946		11.25	11.95	12.79	13.68	13.59	12.19	11.90	12.95	13.93	13.21	12.22	11.36	10.66	.20
.80	5.663		10.58	11.19	11.92	12.73	13.09	11.90	11.13	11.68	12.83	12.96	12.09	11.24	10.51	9.909	.15
		9.990	10.53	11.17	11.89	12.44	11.67	10.67	10.71	11.62	12.43	11.95	11.14	10.41	9.787	9.262	.10
	9.492	9.971	10.54	11.18	11.77	11.44	10.41	10.05	10.64	11.56	11.72	11.04	10.33	9.702	9.159	8.697	.05
	.800	.750	.700	.650	.600	.550	.500	.450	.400	.350	.300	.250	.200	.150	.100	.050	

TABLE 5.4 ($\mu_1 = .05$)

MODE 3 ($\mu_3 \rightarrow$ across top, $\mu_2 \downarrow$ at left); MODE 4 ($\leftarrow \mu_3$ across bottom, $\mu_2 \uparrow$ at right)

μ_2	.050	.100	.150	.200	.250	.300	.350	.400	.450	.500	.550	.600	.650	.700	.750	.800	.850	μ_2
.05	11.84	12.45	13.16	13.94	14.62	13.86	12.94	13.48	14.46	14.34	13.41	12.92	13.98	14.57	13.89	13.17	12.56	
.10	12.56	13.25	14.06	14.93	15.27	13.92	13.73	14.73	15.32	14.39	13.59	14.27	15.36	14.68	13.89	13.20		
.15	13.39	14.17	15.09	16.03	15.55	14.31	14.95	16.10	15.58	14.51	14.62	16.17	15.61	14.71	13.94		16.21	.85
.20	14.32	15.22	16.27	17.01	15.61	15.25	16.49	16.80	15.62	15.13	16.81	16.58	15.61	14.75		17.06	17.19	.80
.25	15.39	16.34	16.41	15.94	15.52	16.38	16.39	15.93	15.52	16.49	16.33	15.89	15.30		18.01	18.15	18.31	.75
.30	14.63	14.29	14.12	14.77	15.57	14.87	14.38	14.51	15.57	14.99	14.47	14.16		18.73	19.20	19.38	18.49	.70
.35	12.77	13.81	14.85	16.17	15.66	14.28	14.55	16.31	15.65	14.54	13.68		17.19	19.21	20.41	17.94	16.37	.65
.40	13.79	15.02	16.52	17.04	15.65	15.06	16.93	16.77	15.61	14.51		17.51	17.87	19.10	17.75	17.49	17.67	.60
.45	13.26	16.25	16.11	15.84	15.52	16.24	16.12	15.84	15.31		18.59	18.88	19.11	18.67	18.78	18.98	19.14	.55
.50	15.01	14.79	14.64	14.66	15.55	15.06	14.76	14.57		18.01	19.54	20.36	18.54	19.17	20.33	19.14	17.16	.50
.55	13.70	13.54	14.27	16.44	15.65	14.30	13.71		17.02	18.08	19.53	17.93	17.45	19.20	18.82	16.96	16.64	.45
.60	12.73	14.52	16.93	16.82	15.60	14.08		18.09	18.36	18.76	18.24	18.25	18.60	18.67	18.09	18.30	18.59	.40
.65	14.94	16.12	16.00	15.80	15.29		18.55	19.63	20.26	18.79	19.41	20.23	19.28	19.06	20.18	19.62	18.16	.35
.70	15.20	15.04	14.92	14.80		17.45	18.62	19.81	18.16	18.08	19.62	18.98	17.66	19.25	19.39	17.93	16.69	.30
.75	14.22	14.08	13.98		17.26	17.87	18.73	17.85	17.51	18.36	18.64	17.39	17.87	18.88	17.84	17.03	17.37	.25
.80	13.36	13.24		18.73	19.54	20.11	18.88	19.37	20.07	19.27	19.12	20.00	19.61	18.85	19.87	19.85	18.74	.20
.85	12.60		17.96	18.98	19.96	18.38	18.61	19.85	19.07	18.17	19.62	19.48	18.24	19.20	19.71	18.57	17.52	.15
		17.01	17.92	18.87	17.87	17.41	18.52	18.70	17.27	17.92	18.99	17.91	17.23	18.90	18.40	17.36	16.44	.10
	16.17	16.98	17.86	17.46	16.42	17.32	18.09	16.87	16.55	17.82	17.68	16.62	16.87	18.07	17.24	16.31	15.50	.05
	.850	.800	.750	.700	.650	.600	.550	.500	.450	.400	.350	.300	.250	.200	.150	.100	.050	

TABLE 5.4 ($\mu_1 = .10$)

MODE 3 ($\mu_3 \rightarrow$ across top, $\mu_2 \downarrow$ at left); MODE 4 ($\leftarrow \mu_3$ across bottom, $\mu_2 \uparrow$ at right)

μ_2	.050	.100	.150	.200	.250	.300	.350	.400	.450	.500	.550	.600	.650	.700	.750	.800	μ_2
.05	12.57	13.26	14.07	14.96	15.45	14.15	13.79	14.76	15.47	14.57	13.72	14.32	15.48	14.82	14.01	13.31	
.10	13.39	14.18	15.10	16.07	15.78	14.43	14.98	16.17	15.74	14.64	14.68	16.26	15.73	14.82	14.03		
.15	14.33	15.23	16.30	17.21	15.90	15.32	16.55	17.07	15.82	15.23	16.94	16.78	15.77	14.88		17.06	.80
.20	15.41	16.45	17.63	17.59	16.11	16.80	17.98	17.05	16.05	17.38	17.71	16.72	15.77		18.00	18.15	.75
.25	16.56	16.42	16.09	15.81	16.43	16.49	16.13	15.85	16.58	16.42	16.09	15.73		19.07	19.22	19.40	.70
.30	14.30	14.07	14.91	16.27	16.07	16.88	14.78	16.48	16.05	14.90	14.36		13.31	20.39	20.80	18.53	.65
.35	13.80	15.04	16.57	17.61	15.99	15.20	17.13	17.18	15.85	14.71		17.88	19.69	21.36	18.04	17.49	.60
.40	15.28	16.79	17.61	17.13	16.16	17.35	17.50	16.86	15.80		18.72	19.00	19.71	18.71	18.79	19.00	.55
.45	16.33	16.10	15.91	15.76	16.36	16.19	15.95	15.71		19.65	20.30	20.54	19.71	20.25	20.51	19.51	.50
.50	14.80	14.60	14.65	16.60	16.04	15.00	14.62		18.30	20.01	21.32	18.71	19.71	21.28	19.17	17.19	.45
.55	13.54	14.56	17.26	17.67	15.94	14.43		18.19	18.76	19.88	18.38	18.50	19.71	18.94	18.07	18.32	.40
.60	14.95	17.15	17.06	16.77	15.85		19.61	20.14	20.43	19.65	20.10	20.42	19.71	20.01	20.36	19.91	.35
.65	16.15	15.99	15.85	15.69		18.83	20.20	21.28	19.01	20.01	21.25	19.42	19.71	21.22	19.68	18.17	.30
.70	15.05	14.90	14.79		17.88	18.93	20.06	18.33	18.50	19.94	19.06	18.09	19.72	19.45	17.99	17.03	.25
.75	14.08	13.95		19.37	19.88	20.28	19.41	19.80	20.26	19.53	19.71	20.22	19.72	19.54	20.06	20.09	.20
.80	13.24		19.16	20.31	21.23	19.26	20.18	21.21	19.51	19.97	21.19	19.80	19.67	21.16	19.96	18.75	.15
		18.09	19.14	20.13	18.47	18.78	20.04	19.11	18.34	19.88	19.52	18.31	19.56	19.75	18.58	17.53	.10
	17.14	18.07	19.02	17.94	17.57	18.71	18.76	17.36	18.16	19.10	17.95	17.43	19.11	18.43	17.38	16.45	.05
	.800	.750	.700	.650	.600	.550	.500	.450	.400	.350	.300	.250	.200	.150	.100	.050	

TABLE 5.4 ($\mu_1 = .15$)

MODE 1 (top $\mu_3 \rightarrow$, left $\mu_2 \downarrow$); MODE 2 (bottom $\leftarrow \mu_3$, right $\mu_2 \uparrow$)

μ_2	.050	.100	.150	.200	.250	.300	.350	.400	.450	.500	.550	.600	.650	.700	.750	μ_2
.05	5.143	5.424	5.754	6.140	6.598	7.139	7.769	8.423	8.671	8.245	7.689	7.183	6.748	6.381	6.071	
.10	5.500	5.821	6.203	6.659	7.204	7.853	8.587	9.132	8.852	8.238	7.658	7.161	6.743	6.395		
.15	5.912	6.284	6.734	7.279	7.939	8.723	9.508	9.563	8.912	8.237	7.658	7.176	6.778		9.946	.75
.20	6.393	6.831	7.369	8.032	8.843	9.773	10.31	9.728	8.940	8.255	7.690	7.229		10.50	10.62	.70
.25	6.959	7.483	8.140	8.961	9.966	10.92	10.67	9.782	8.967	8.298	7.758		11.16	11.26	11.40	.65
.30	7.636	8.276	9.092	10.12	11.29	11.49	10.68	9.774	8.999	8.370		11.92	12.02	12.14	12.30	.60
.35	8.459	9.253	10.27	11.21	11.17	10.87	10.34	9.668	9.016		12.74	12.90	13.03	13.16	13.34	.55
.40	9.475	10.29	10.30	10.17	10.04	9.873	9.645	9.321		13.04	13.72	14.04	14.21	14.18	12.53	.50
.45	9.616	9.424	9.285	9.172	9.068	8.964	8.848		12.00	13.21	14.53	15.20	14.10	12.21	10.82	.45
.50	8.724	8.561	8.441	8.346	8.265	8.192		11.27	12.06	13.25	14.58	13.79	11.99	10.76	10.77	.40
.55	7.979	7.841	7.739	7.660	7.596		11.76	12.02	12.51	13.35	13.41	12.01	11.74	11.90	12.16	.35
.60	7.352	7.235	7.149	7.084		12.86	13.15	13.41	13.68	13.61	13.01	13.17	13.40	13.63	13.68	.30
.65	6.818	6.719	6.648		12.91	13.74	14.57	15.11	14.24	13.39	14.27	15.05	14.70	13.54	12.55	.25
.70	6.358	6.276		12.12	12.94	13.92	14.85	14.01	12.70	13.23	14.64	14.52	13.39	12.36	11.53	.20
.75	5.958		11.34	12.06	12.92	13.82	13.65	12.23	12.02	13.15	14.04	13.23	12.23	11.37	10.66	.15
		10.67	11.29	12.04	12.86	13.16	11.94	11.21	11.84	13.00	13.00	12.11	11.25	10.52	9.912	.10
	10.10	10.65	11.30	12.04	12.56	11.72	10.74	10.88	11.83	12.56	11.98	11.16	10.43	9.797	9.268	.05
	.750	.700	.650	.600	.550	.500	.450	.400	.350	.300	.250	.200	.150	.100	.050	

TABLE 5.4 ($\mu_1 = .20$)

MODE 1 (top $\mu_3 \rightarrow$, left $\mu_2 \downarrow$); MODE 2 (bottom $\leftarrow \mu_3$, right $\mu_2 \uparrow$)

μ_2	.050	.100	.150	.200	.250	.300	.350	.400	.450	.500	.550	.600	.650	.700	μ_2
.05	5.504	5.829	6.213	6.671	7.218	7.871	8.619	9.224	9.004	8.382	7.786	7.274	6.845	6.487	
.10	5.914	6.290	6.742	7.289	7.951	8.741	9.555	9.682	9.036	8.349	7.757	7.264	6.858		
.15	6.394	6.835	7.376	8.041	8.855	9.799	10.40	9.847	9.047	8.348	7.772	7.303		10.54	.70
.20	6.960	7.487	8.146	8.971	9.984	10.99	10.82	9.905	9.068	8.383	7.833		11.17	11.30	.65
.25	7.637	8.280	9.100	10.14	11.37	11.83	10.92	9.928	9.111	8.459		11.93	12.03	12.19	.60
.30	8.461	9.261	10.30	11.59	12.23	11.70	10.81	9.918	9.173		12.80	12.91	13.04	13.22	.55
.35	9.482	10.48	11.33	11.26	11.08	10.80	10.37	9.793		13.68	13.92	14.08	14.23	14.39	.50
.40	10.51	10.34	10.19	10.06	9.926	9.780	9.598		13.46	14.61	15.19	15.41	14.45	12.54	.45
.45	9.501	9.313	9.176	9.067	8.970	8.878		12.25	13.48	15.06	16.01	14.15	12.23	10.89	.40
.50	8.620	8.462	8.347	8.257	8.183		11.98	12.54	13.58	14.85	13.86	12.12	11.76	11.98	.35
.55	7.889	7.756	7.660	7.587		13.02	13.25	13.57	14.07	13.76	13.06	13.20	13.42	13.74	.30
.60	7.273	7.162	7.083		13.84	14.54	15.04	15.32	14.54	14.35	14.93	15.28	14.92	13.77	.25
.65	6.749	6.658		13.12	14.08	15.20	15.96	14.39	13.70	15.01	15.94	14.79	13.56	12.55	.20
.70	6.298		12.23	13.07	14.09	15.03	14.06	12.79	13.44	14.90	14.57	13.40	12.37	11.53	.15
		11.45	12.18	13.05	13.96	13.70	12.28	12.17	13.36	14.14	13.25	12.24	11.37	10.66	.10
	10.80	11.43	12.19	13.03	13.26	11.99	11.35	12.07	13.22	13.05	12.13	11.27	10.53	9.919	.05
	.700	.650	.600	.550	.500	.450	.400	.350	.300	.250	.200	.150	.100	.050	

TABLE 5.4 ($\mu_1 = .15$)

MODE 3 (top $\mu_3 \rightarrow$, left $\mu_2 \downarrow$); MODE 4 (bottom $\leftarrow \mu_3$, right $\mu_2 \uparrow$)

μ_2	.050	.100	.150	.200	.250	.300	.350	.400	.450	.500	.550	.600	.650	.700	.750	μ_2
.05	13.40	14.19	15.12	16.12	16.07	14.61	15.02	16.25	15.95	14.82	14.75	16.38	15.89	14.97	14.17	
.10	14.33	15.24	16.32	17.31	16.17	15.39	16.59	17.25	16.00	15.33	17.01	16.93	15.90	14.99		
.15	15.41	16.47	17.71	18.23	16.40	16.87	18.32	17.45	16.26	17.51	18.12	16.99	15.95		18.02	.75
.20	16.67	17.88	18.67	17.74	17.20	18.57	18.29	17.27	17.82	18.58	17.81	16.85		19.08	19.25	.70
.25	16.53	16.12	15.87	16.54	16.98	16.32	15.99	16.90	16.78	16.24	15.89		20.26	20.45	20.64	.65
.30	14.12	15.08	16.61	17.89	16.37	15.44	17.23	17.45	16.12	15.07		20.11	21.68	21.94	18.57	.60
.35	15.29	16.84	18.63	18.24	16.54	17.71	18.73	17.44	16.07		18.91	20.21	21.85	18.72	18.81	.55
.40	17.11	17.70	17.48	17.12	17.51	17.66	17.38	16.89		20.14	20.41	20.76	20.12	20.32	20.55	.50
.45	16.16	15.92	15.79	16.86	16.71	16.08	15.81		20.30	21.76	22.10	20.43	21.86	21.97	19.54	.45
.50	14.65	14.75	17.36	18.31	16.31	15.03		18.82	20.36	21.69	19.02	20.20	21.54	19.23	18.10	.40
.55	14.97	17.79	18.30	17.81	16.27		19.93	20.27	20.74	19.94	20.24	20.73	19.90	20.08	20.41	.35
.60	17.27	17.10	16.94	16.71		20.42	21.79	22.28	20.51	21.94	21.99	20.48	22.11	21.74	19.95	.30
.65	16.03	15.86	15.74		19.10	20.49	21.56	19.22	20.37	21.48	19.52	20.20	21.38	19.71	18.20	.25
.70	14.93	14.79		19.58	20.10	20.74	19.60	20.02	20.73	19.67	19.93	20.73	19.76	19.61	20.13	.20
.75	13.98		20.48	21.73	22.31	20.56	21.90	21.98	20.53	22.12	21.74	20.52	22.37	21.56	20.17	.15
		19.33	20.53	21.44	19.41	20.44	21.40	19.60	20.30	21.34	19.85	20.09	21.27	19.98	18.76	.10
	18.25	19.31	20.30	18.57	18.99	20.24	19.16	18.56	20.13	19.56	18.42	19.93	19.79	18.59	17.54	.05
	.750	.700	.650	.600	.550	.500	.450	.400	.350	.300	.250	.200	.150	.100	.050	

TABLE 5.4 ($\mu_1 = .20$)

MODE 3 (top $\mu_3 \rightarrow$, left $\mu_2 \downarrow$); MODE 4 (bottom $\leftarrow \mu_3$, right $\mu_2 \uparrow$)

μ_2	.050	.100	.150	.200	.250	.300	.350	.400	.450	.500	.550	.600	.650	.700	μ_2
.05	14.34	15.26	16.34	17.41	16.51	15.50	16.63	17.45	16.22	15.46	17.10	17.12	16.08	15.15	
.10	15.42	16.48	17.73	18.48	16.64	16.92	18.43	17.68	16.42	17.58	18.31	17.15	16.09		
.15	16.67	17.93	19.31	18.74	17.44	18.90	19.24	17.75	18.04	19.48	18.35	17.18		18.92	.70
.20	18.12	18.53	18.18	17.75	18.53	18.41	17.98	18.06	18.53	18.18	17.67		19.91	20.04	.65
.25	16.12	15.82	16.71	18.12	17.05	16.23	17.38	17.80	16.72	16.05		20.76	20.88	20.98	.60
.30	15.31	16.86	18.76	18.80	16.84	17.81	19.23	17.73	16.32		20.39	21.46	21.56	18.73	.55
.35	17.14	18.95	18.99	18.33	18.25	19.09	18.67	17.62		19.84	20.55	21.69	19.73	19.87	.50
.40	17.65	17.40	17.20	17.49	17.76	17.40	17.15		20.75	20.90	21.04	20.75	20.87	20.98	.45
.45	15.93	15.76	17.44	18.69	16.82	15.97		20.50	21.49	21.60	20.65	21.50	21.60	19.56	.40
.50	15.03	17.89	19.38	18.61	16.62		19.68	20.59	21.60	19.72	20.54	21.50	19.53	19.58	.35
.55	18.25	18.22	18.05	17.72		20.73	20.90	21.05	20.74	20.90	21.05	20.72	20.86	20.99	.30
.60	17.08	16.90	16.76		20.59	21.51	21.65	20.68	21.54	21.64	20.67	21.56	21.61	19.96	.25
.65	15.87	15.72		19.59	20.64	21.50	19.63	20.58	21.45	19.73	20.51	21.37	19.76	19.05	.20
.70	14.81		20.67	20.85	21.01	20.69	20.85	21.01	20.69	20.85	21.01	20.66	20.81	20.95	.15
		20.67	21.50	21.62	20.76	21.52	21.61	20.73	21.55	21.58	20.73	21.56	21.52	20.18	.10
	19.51	20.71	21.57	19.57	20.66	21.49	19.69	20.58	21.41	19.91	20.48	21.33	19.99	18.77	.05
	.700	.650	.600	.550	.500	.450	.400	.350	.300	.250	.200	.150	.100	.050	

TABLE 5.4 ($\mu_1 = .25$)

MODE 1 (upper values: $\mu_3 \rightarrow$ across top, $\mu_2 \downarrow$ at left); MODE 2 (lower values: $\leftarrow \mu_3$ across bottom, $\mu_2 \uparrow$ at right)

μ_2 ↓	.050	.100	.150	.200	.250	.300	.350	.400	.450	.500	.550	.600	.650	μ_2 ↑
.05	5.919	6.299	6.754	7.302	7.968	8.766	9.615	9.841	9.207	8.501	7.890	7.383	6.965	
.10	6.397	6.841	7.385	8.052	8.871	9.830	10.50	9.993	9.179	8.465	7.875	7.396		
.15	6.962	7.492	8.154	8.982	10.00	11.05	10.96	10.04	9.180	8.480	7.919		11.21	.65
.20	7.639	8.284	9.109	10.16	11.42	12.02	11.09	10.06	9.217	8.550		11.92	12.07	.60
.25	8.463	9.267	10.32	11.66	12.84	12.23	11.09	10.09	9.297		12.77	12.89	13.06	.55
.30	9.486	10.51	11.84	12.50	12.26	11.73	10.92	10.10		13.73	13.86	14.00	14.18	.50
.35	10.77	11.42	11.28	11.13	10.95	10.72	10.39		14.53	14.88	15.06	15.20	14.88	.45
.40	10.44	10.22	10.07	9.939	9.820	9.697		13.74	15.18	15.96	16.12	14.49	12.55	.40
.45	9.390	9.208	9.077	8.973	8.886		12.64	13.76	15.36	16.22	14.19	12.30	11.80	.35
.50	8.527	8.375	8.266	8.183		13.06	13.41	14.11	15.10	13.98	13.00	13.14	13.42	.30
.55	7.809	7.683	7.594		14.32	14.63	14.89	15.12	14.58	14.53	14.76	14.97	15.11	.25
.60	7.205	7.102		14.25	15.26	16.00	16.19	14.88	15.17	16.03	16.18	15.00	13.78	.20
.65	6.691		13.26	14.26	15.44	16.15	14.45	13.89	15.35	16.08	14.82	13.57	12.56	.15
		12.35	13.21	14.24	15.18	14.10	12.89	13.65	15.10	14.60	13.41	12.38	11.53	.10
	11.60	12.34	13.23	14.15	13.78	12.36	12.39	13.64	14.24	13.28	12.26	11.39	10.67	.05
	.650	.600	.550	.500	.450	.400	.350	.300	.250	.200	.150	.100	.050	

TABLE 5.4 ($\mu_1 = .30$)

MODE 1 (upper values: $\mu_3 \rightarrow$ across top, $\mu_2 \downarrow$ at left); MODE 2 (lower values: $\leftarrow \mu_3$ across bottom, $\mu_2 \uparrow$ at right)

μ_2 ↓	.050	.100	.150	.200	.250	.300	.350	.400	.450	.500	.550	.600	μ_2 ↑
.05	6.403	6.852	7.398	8.068	8.892	9.869	10.64	10.19	9.361	8.623	8.014	7.520	
.10	6.966	7.499	8.164	8.996	10.03	11.12	11.12	10.19	9.318	8.601	8.027		
.15	7.642	8.290	9.118	10.17	11.45	12.17	11.25	10.20	9.333	8.651		11.87	.60
.20	8.465	9.273	10.33	11.70	13.02	12.50	11.28	10.23	9.411		12.59	12.72	.55
.25	9.489	10.52	11.90	13.19	13.05	12.32	11.25	10.30		13.30	13.38	13.49	.50
.30	10.79	12.07	12.34	12.20	11.99	11.63	11.03		13.86	13.91	13.97	14.04	.45
.35	11.46	11.24	11.07	10.93	10.78	10.61		14.18	14.27	14.31	14.35	14.40	.40
.40	10.28	10.08	9.929	9.811	9.707		13.79	14.47	14.60	14.66	14.44	12.57	.35
.45	9.273	9.100	8.977	8.884		13.06	13.80	14.70	15.01	14.17	12.70	12.80	.30
.50	8.435	8.292	8.192		13.56	13.70	13.99	14.65	13.96	13.56	13.63	13.74	.25
.55	7.735	7.619		14.04	14.10	14.16	14.25	14.14	14.07	14.12	14.16	14.21	.20
.60	7.144		14.18	14.39	14.44	14.46	14.40	14.36	14.42	14.45	14.47	13.78	.15
		13.39	14.31	14.68	14.73	14.45	14.02	14.65	14.72	14.70	13.58	12.56	.10
	12.52	13.40	14.43	15.04	14.16	13.05	13.94	15.02	14.64	13.43	12.39	11.54	.05
	.600	.550	.500	.450	.400	.350	.300	.250	.200	.150	.100	.050	

TABLE 5.4 ($\mu_1 = .25$)

MODE 3 (upper values: $\mu_3 \rightarrow$ across top, $\mu_2 \downarrow$ at left); MODE 4 (lower values: $\leftarrow \mu_3$ across bottom, $\mu_2 \uparrow$ at right)

μ_2 ↓	.050	.100	.150	.200	.250	.300	.350	.400	.450	.500	.550	.600	.650	μ_2 ↑
.05	15.43	16.50	17.75	17.97	16.92	16.97	18.00	17.77	16.61	17.62	17.97	17.31	16.26	
.10	16.68	17.56	17.53	17.47	17.34	17.56	17.52	17.40	17.52	17.54	17.49	17.18		
.15	17.20	17.14	17.09	17.08	17.23	17.15	17.09	17.19	17.20	17.12	17.06		17.69	.65
.20	16.61	16.49	16.73	17.76	16.90	16.59	17.26	17.43	16.76	16.52		18.22	18.32	.60
.25	15.46	16.86	18.07	17.79	16.83	17.65	18.00	17.44	16.44		19.07	19.22	19.39	.55
.30	17.13	17.54	17.47	17.37	17.49	17.53	17.45	17.25		19.50	20.44	20.69	18.53	.50
.35	17.11	17.04	16.98	17.15	17.19	17.05	16.97		18.12	19.69	21.46	18.12	17.97	.45
.40	16.47	16.36	17.38	17.95	16.84	16.44		18.77	19.01	19.71	18.76	18.83	19.01	.40
.45	15.48	17.68	17.76	17.60	16.69		19.65	20.36	20.64	19.71	20.30	20.61	19.51	.35
.50	17.40	17.35	17.29	17.20		18.36	20.02	21.39	18.72	19.71	21.34	19.18	17.65	.30
.55	16.95	16.88	16.82		18.41	18.82	19.89	18.50	18.62	19.72	18.95	18.36	18.53	.25
.60	16.33	16.23		19.62	20.22	20.57	19.66	20.17	20.57	19.72	20.07	20.51	19.91	.20
.65	15.50		18.79	20.19	21.32	18.98	19.98	21.29	19.40	19.65	21.24	19.68	18.17	.15
		17.77	18.73	19.95	18.22	18.24	19.80	19.01	17.89	19.48	19.42	17.97	17.62	.10
	18.06	18.11	18.63	18.09	18.08	18.27	18.59	18.07	18.12	18.81	18.07	18.07	18.09	.05
	.650	.600	.550	.500	.450	.400	.350	.300	.250	.200	.150	.100	.050	

TABLE 5.4 ($\mu_1 = .30$)

MODE 3 (upper values: $\mu_3 \rightarrow$ across top, $\mu_2 \downarrow$ at left); MODE 4 (lower values: $\leftarrow \mu_3$ across bottom, $\mu_2 \uparrow$ at right)

μ_2 ↓	.050	.100	.150	.200	.250	.300	.350	.400	.450	.500	.550	.600	μ_2 ↑
.05	15.17	15.15	15.14	15.14	15.48	15.17	15.14	15.14	15.49	15.18	15.14	15.13	
.10	14.81	14.78	14.90	16.23	15.86	14.87	14.84	16.42	15.79	14.91	14.79		
.15	14.53	15.04	16.58	17.74	16.01	15.21	17.16	17.27	15.86	14.82		17.55	.60
.20	15.28	16.82	18.46	17.81	16.31	17.61	18.37	17.18	15.89		18.75	18.94	.55
.25	16.93	16.78	16.54	16.29	16.86	16.81	16.53	16.20		20.15	20.38	20.60	.50
.30	15.47	15.33	15.28	16.69	16.27	15.50	15.30		20.06	21.85	22.14	18.55	.45
.35	14.80	14.81	17.32	18.09	16.08	14.95		18.53	20.15	21.89	18.34	18.31	.40
.40	14.97	17.61	17.72	17.28	16.03		19.97	20.31	20.75	19.95	20.18	20.49	.35
.45	16.59	16.41	16.26	16.08		20.26	21.94	22.34	20.40	22.10	22.06	19.54	.30
.50	15.50	15.39	15.31		18.62	20.34	21.72	18.90	20.16	21.55	19.21	17.38	.25
.55	14.89	14.83		19.65	20.13	20.75	19.68	20.07	20.74	19.64	19.81	20.28	.20
.60	14.53		20.38	21.91	22.47	20.47	22.15	22.05	20.44	22.43	21.75	19.95	.15
		18.92	20.41	21.55	19.09	20.26	21.47	19.46	20.02	21.37	19.69	18.18	.10
	17.61	18.89	20.15	18.26	18.36	20.04	19.05	17.84	19.82	19.46	17.97	16.70	.05
	.600	.550	.500	.450	.400	.350	.300	.250	.200	.150	.100	.050	

TABLE 5.4 ($\mu_1 = .35$) MODE 1

$\mu_2 \downarrow$ \ $\mu_3 \rightarrow$	.050	.100	.150	.200	.250	.300	.350	.400	.450	.500	.550	$\mu_2 \uparrow$
.05	6.972	7.512	8.181	9.016	10.05	11.20	11.34	10.41	9.506	8.765	8.172	
.10	7.645	8.299	9.131	10.19	11.49	12.25	11.42	10.36	9.475	8.776		
.15	8.468	9.281	10.34	11.71	12.42	12.30	11.42	10.37	9.530		11.88	.55
.20	9.492	10.53	11.91	12.25	12.21	12.09	11.41	10.44		12.15	12.19	.50
.25	10.80	11.98	12.00	11.96	11.90	11.77	11.31		12.38	12.41	12.45	.45
.30	11.69	11.60	11.52	11.44	11.35	11.21		12.62	12.65	12.69	12.75	.40
.35	10.99	10.82	10.70	10.59	10.49		12.93	13.01	13.08	13.16	13.28	.35
.40	10.04	9.867	9.740	9.641		13.05	13.56	13.83	14.00	14.04	12.54	.30
.45	9.131	8.974	8.865		12.35	13.16	14.40	15.18	14.09	12.26	12.11	.25
.50	8.337	8.207		12.33	12.44	13.14	14.49	13.77	12.32	12.32	12.35	.20
.55	7.662		12.50	12.53	12.58	13.04	13.28	12.51	12.51	12.53	12.55	.15
		12.71	12.73	12.74	12.77	12.83	12.72	12.72	12.74	12.75	12.77	.10
	12.99	13.02	13.03	13.03	13.04	13.01	13.02	13.03	13.03	13.04	12.57	.05
$\leftarrow \mu_3$	.550	.500	.450	.400	.350	.300	.250	.200	.150	.100	.050	

MODE 2

TABLE 5.4 ($\mu_1 = .35$) MODE 3

$\mu_2 \downarrow$ \ $\mu_3 \rightarrow$	.050	.100	.150	.200	.250	.300	.350	.400	.450	.500	.550	$\mu_2 \uparrow$
.05	13.06	13.60	14.87	16.30	16.13	14.43	14.62	16.55	15.97	14.70	13.67	
.10	13.80	15.05	16.60	17.93	16.22	15.25	17.23	17.47	15.99	14.74		
.15	15.29	16.84	18.73	18.64	16.57	17.75	19.15	17.56	16.06		18.66	.55
.20	17.13	18.74	18.55	17.80	18.03	18.73	18.22	17.25		19.89	20.05	.50
.25	16.57	16.22	16.00	17.04	17.08	16.38	16.02		20.77	20.87	20.95	.45
.30	14.49	14.71	17.40	18.58	16.41	15.00		20.41	21.28	21.35	18.64	.40
.35	14.97	17.86	19.20	18.35	16.35		19.80	20.58	21.52	19.66	19.83	.35
.40	17.83	17.65	17.44	17.09		20.77	20.89	20.99	20.77	20.86	20.95	.30
.45	16.22	16.00	15.84		20.52	21.29	21.37	20.66	21.30	21.37	19.56	.25
.50	14.85	14.68		19.61	20.61	21.47	19.66	20.56	21.42	19.45	19.44	.20
.55	13.82		20.74	20.86	20.96	20.75	20.86	20.96	20.74	20.83	20.92	.15
		20.61	21.27	21.32	20.72	21.28	21.32	20.72	21.28	21.32	19.96	.10
	19.14	20.64	21.57	19.27	20.57	21.51	19.55	20.45	21.42	19.72	18.19	.05
$\leftarrow \mu_3$	.550	.500	.450	.400	.350	.300	.250	.200	.150	.100	.050	

MODE 4

TABLE 5.4 ($\mu_1 = .40$) MODE 1

$\mu_2 \downarrow$ \ $\mu_3 \rightarrow$	.050	.100	.150	.200	.250	.300	.350	.400	.450	.500	$\mu_2 \uparrow$
.05	7.653	8.314	9.150	10.22	11.40	11.42	11.37	10.58	9.669	8.946	
.10	8.473	9.291	10.36	11.18	11.18	11.16	11.10	10.50	9.671		
.15	9.495	10.54	11.01	11.00	10.99	10.97	10.91	10.51		10.96	.50
.20	10.78	10.87	10.85	10.84	10.82	10.80	10.75		11.16	11.21	.45
.25	10.73	10.70	10.68	10.66	10.64	10.61		11.49	11.55	11.65	.40
.30	10.53	10.48	10.44	10.41	10.38		12.10	12.21	12.33	12.52	.35
.35	10.20	10.12	10.05	9.996		13.07	13.31	13.51	13.71	13.97	.30
.40	9.643	9.520	9.430		13.20	14.34	15.04	15.36	14.42	12.53	.25
.45	8.926	8.797		11.95	13.26	14.97	16.05	14.13	12.20	10.92	.20
.50	8.216		11.15	11.84	13.22	14.72	13.80	11.97	11.05	11.06	.15
		11.22	11.27	11.81	13.09	13.32	11.77	11.21	11.22	11.24	.10
	11.44	11.45	11.47	11.81	12.57	11.58	11.45	11.45	11.46	11.47	.05
$\leftarrow \mu_3$	.500	.450	.400	.350	.300	.250	.200	.150	.100	.050	

MODE 2

TABLE 5.4 ($\mu_1 = .40$) MODE 3

$\mu_2 \downarrow$ \ $\mu_3 \rightarrow$	.050	.100	.150	.200	.250	.300	.350	.400	.450	.500	$\mu_2 \uparrow$
.05	13.81	15.06	16.64	18.12	16.59	15.41	17.32	17.73	16.23	14.94	
.10	15.29	16.86	18.64	18.56	16.82	17.82	18.66	17.77	16.25		
.15	17.14	18.43	18.39	18.23	18.22	18.42	18.34	17.72		18.47	.50
.20	17.90	17.75	17.59	17.73	17.94	17.73	17.52		18.76	18.81	.45
.25	16.12	15.86	17.48	18.48	17.02	16.12		19.17	19.26	19.37	.40
.30	15.01	17.90	18.57	18.38	16.70		19.47	20.35	20.63	18.58	.35
.35	18.21	18.19	18.10	17.90		18.72	19.65	21.51	18.67	18.68	.30
.40	17.38	17.19	17.04		19.04	19.16	19.66	19.03	19.07	19.17	.25
.45	15.95	15.75		19.58	20.25	20.58	19.63	20.18	20.54	19.51	.20
.50	14.59		18.65	19.94	21.41	18.79	19.60	21.34	19.17	18.50	.15
		18.76	18.83	19.66	18.77	18.80	19.38	18.90	18.76	18.79	.10
	19.06	19.08	19.10	19.06	19.07	19.09	19.07	19.06	19.08	19.09	.05
$\leftarrow \mu_3$	.500	.450	.400	.350	.300	.250	.200	.150	.100	.050	

MODE 4

TABLE 5.4 ($\mu_1 = .45$) MODE 1

$\mu_2 \downarrow$ \ $\mu_3 \rightarrow$	.050	.100	.150	.200	.250	.300	.350	.400	.450	$\mu_2 \uparrow$
.05	8.482	9.308	10.20	10.20	10.20	10.20	10.19	10.17	9.848	
.10	9.499	10.01	10.00	9.996	9.990	9.983	9.972	9.946		
.15	9.880	9.862	9.850	9.841	9.833	9.823	9.811		10.37	.45
.20	9.752	9.733	9.719	9.708	9.699	9.688		10.91	11.05	.40
.25	9.629	9.607	9.591	9.578	9.567		11.80	11.94	12.16	.35
.30	9.493	9.465	9.445	9.428		13.04	13.21	13.41	13.69	.30
.35	9.314	9.272	9.243		14.25	14.70	14.94	15.13	14.86	.25
.40	9.035	8.971		13.55	15.13	15.93	16.05	14.48	12.54	.20
.45	8.596		12.07	13.49	15.32	16.19	14.17	12.21	10.81	.15
		10.88	11.97	13.44	14.94	13.85	11.98	10.56	10.04	.10
	10.25	10.84	12.00	13.37	13.42	11.79	10.40	10.22	10.22	.05
$\leftarrow \mu_3$	.450	.400	.350	.300	.250	.200	.150	.100	.050	

MODE 2

TABLE 5.4 ($\mu_1 = .45$) MODE 3

$\mu_2 \downarrow$ \ $\mu_3 \rightarrow$	.050	.100	.150	.200	.250	.300	.350	.400	.450	$\mu_2 \uparrow$
.05	15.31	16.86	16.99	16.97	16.87	16.99	16.99	16.96	16.47	
.10	16.75	16.73	16.72	16.70	16.74	16.74	16.72	16.68		
.15	16.54	16.51	16.49	16.71	16.61	16.52	16.49		17.25	.45
.20	16.24	16.18	17.28	17.78	16.50	16.23		18.27	18.48	.40
.25	15.48	17.34	17.31	17.10	16.48		19.93	20.22	20.51	.35
.30	16.85	16.80	16.76	16.70		19.97	22.02	22.32	18.54	.30
.35	16.54	16.50	16.47		18.09	20.08	21.92	18.11	17.71	.25
.40	16.20	16.14		19.65	20.11	20.72	19.60	19.90	20.34	.20
.45	15.49		20.19	22.10	22.50	20.32	22.32	22.10	19.53	.15
		18.23	20.21	21.71	18.71	19.93	21.54	19.18	17.17	.10
	17.05	18.18	19.88	17.95	17.34	19.65	18.87	17.02	17.02	.05
$\leftarrow \mu_3$	.450	.400	.350	.300	.250	.200	.150	.100	.050	

MODE 4

TABLE 5.4 ($\mu_1 = .50$) MODE 1

$\mu_2 \downarrow$ \ $\mu_3 \rightarrow$	.050	.100	.150	.200	.250	.300	.350	.400	$\mu_2 \uparrow$
.05	9.221	9.214	9.210	9.207	9.205	9.202	9.200	9.197	
.10	9.061	9.047	9.038	9.033	9.028	9.024	9.019		
.15	8.933	8.918	8.908	8.900	8.894	8.889		10.75	.40
.20	8.825	8.809	8.798	8.790	8.783		11.71	11.91	.35
.25	8.725	8.708	8.696	8.687		12.93	13.09	13.31	.30
.30	8.623	8.604	8.591		14.05	14.16	14.25	14.36	.25
.35	8.506	8.483		14.52	14.63	14.67	14.70	14.72	.20
.40	8.350		13.79	14.80	14.88	14.90	14.50	12.55	.15
		12.26	13.72	14.99	15.07	14.19	12.23	10.82	.10
	11.07	12.23	13.76	15.12	13.90	12.01	10.58	9.518	.05
$\leftarrow \mu_3$	.400	.350	.300	.250	.200	.150	.100	.050	

MODE 2

TABLE 5.4 ($\mu_1 = .50$) MODE 3

$\mu_2 \downarrow$ \ $\mu_3 \rightarrow$	.050	.100	.150	.200	.250	.300	.350	.400	$\mu_2 \uparrow$
.05	15.34	15.33	15.33	15.32	15.47	15.34	15.33	15.32	
.10	15.13	15.11	15.10	16.60	15.86	15.14	15.11		
.15	14.95	14.95	17.34	18.36	15.97	15.00		18.12	.40
.20	14.98	17.78	18.71	17.79	16.00		19.76	19.97	.35
.25	16.89	16.61	16.38	16.12		20.80	20.88	20.95	.30
.30	15.51	15.42	15.35		20.42	21.21	21.26	18.59	.25
.35	15.08	15.04		19.63	20.60	21.43	19.40	19.65	.20
.40	14.85		20.79	20.88	20.96	20.79	20.86	20.93	.15
		20.53	21.18	21.21	20.70	21.18	21.21	19.55	.10
	18.50	20.54	21.45	18.88	20.42	21.43	19.21	17.18	.05
$\leftarrow \mu_3$	.400	.350	.300	.250	.200	.150	.100	.050	

MODE 4

TABLE 5.4 ($\mu_1 = .55$) MODE 1

$\mu_2 \downarrow$ \ $\mu_3 \rightarrow$	.050	.100	.150	.200	.250	.300	.350	$\mu_2 \uparrow$
.05	8.402	8.394	8.391	8.388	8.386	8.385	8.384	
.10	8.263	8.251	8.244	8.239	8.236	8.233		
.15	8.153	8.139	8.131	8.125	8.120		11.68	.35
.20	8.059	8.046	8.037	8.030		12.61	12.74	.30
.25	7.975	7.961	7.952		13.17	13.21	13.26	.25
.30	7.893	7.879		13.42	13.44	13.46	13.48	.20
.35	7.807		13.57	13.59	13.60	13.61	13.63	.15
		13.64	13.73	13.75	13.76	13.76	12.56	.10
	12.54	13.84	13.94	13.95	13.94	12.26	10.83	.05
$\leftarrow \mu_3$	.350	.300	.250	.200	.150	.100	.050	

MODE 2

TABLE 5.4 ($\mu_1 = .55$) MODE 3

$\mu_2 \downarrow$ \ $\mu_3 \rightarrow$	.050	.100	.150	.200	.250	.300	.350	$\mu_2 \uparrow$
.05	13.97	13.96	14.26	16.74	16.19	14.22	13.97	
.10	13.79	14.55	17.40	18.72	16.23	14.27		
.15	14.96	17.88	19.03	18.66	16.34		18.87	.35
.20	18.31	18.34	18.13	17.64		19.21	19.25	.30
.25	16.49	16.16	15.92		19.47	19.52	19.58	.25
.30	14.66	14.50		19.58	20.28	20.59	18.74	.20
.35	13.93		19.14	19.63	21.54	19.11	19.14	.15
		19.33	19.36	19.49	19.33	19.34	19.36	.10
	19.55	19.57	19.58	19.55	19.57	19.58	19.53	.05
$\leftarrow \mu_3$	.350	.300	.250	.200	.150	.100	.050	

MODE 4

TABLE 5.4 ($\mu_1 = .60$)

$\mu_3 \rightarrow$ MODE 1

$\mu_2 \downarrow$	.050	.100	.150	.200	.250	.300	
.05	7.715	7.709	7.705	7.703	7.702	7.701	
.10	7.595	7.585	7.579	7.575	7.572		
.15	7.499	7.487	7.480	7.475		11.99	.30
.20	7.417	7.405	7.398		12.23	12.26	.25
.25	7.344	7.333		12.38	12.39	12.41	.20
.30	7.275		12.50	12.51	12.52	12.53	.15 $\uparrow$
		12.62	12.63	12.64	12.65	12.66	.10 μ_2
	12.80	12.80	12.81	12.81	12.82	12.57	.05
	.300	.250	.200	.150	.100	.050	

MODE 2 $\leftarrow \mu_3$

TABLE 5.4 ($\mu_1 = .60$)

$\mu_3 \rightarrow$ MODE 3

$\mu_2 \downarrow$	.050	.100	.150	.200	.250	.300	
.05	12.84	14.58	17.47	17.94	16.65	14.52	
.10	14.97	17.73	17.75	17.71	16.67		
.15	17.60	17.57	17.55	17.50		17.78	.30
.20	17.29	17.21	17.12		18.14	18.25	.25
.25	16.07	15.79		19.59	19.97	20.38	.20
.30	14.20		19.82	22.14	22.36	18.53	.15 $\uparrow$
		17.83	19.80	21.92	17.98	17.78	.10 μ_2
	17.97	18.00	19.37	17.98	17.97	17.98	.05
	.300	.250	.200	.150	.100	.050	

MODE 4 $\leftarrow \mu_3$

TABLE 5.4 ($\mu_1 = .65$)

$\mu_3 \rightarrow$ MODE 1

$\mu_2 \downarrow$	.050	.100	.150	.200	.250	
.05	7.133	7.127	7.124	7.122	7.121	
.10	7.028	7.019	7.013	7.010		
.15	6.942	6.932	6.926		11.38	.25
.20	6.870	6.860		11.48	11.49	.20
.25	6.806		11.58	11.58	11.60	.15 $\uparrow$
		11.69	11.69	11.70	11.71	.10 μ_2
	11.84	11.84	11.84	11.85	11.86	.05
	.250	.200	.150	.100	.050	

MODE 2 $\leftarrow \mu_3$

TABLE 5.4 ($\mu_1 = .65$)

$\mu_3 \rightarrow$ MODE 3

$\mu_2 \downarrow$	.050	.100	.150	.200	.250	
.05	15.00	16.61	16.60	16.59	16.54	
.10	16.44	16.43	16.42	16.40		
.15	16.29	16.27	16.26		17.37	.25
.20	16.09	16.05		19.46	19.79	.20
.25	15.47		20.81	20.88	20.93	.15 $\uparrow$
		20.37	21.13	21.15	18.55	.10 μ_2
	17.26	20.35	21.34	17.99	16.62	.05
	.250	.200	.150	.100	.050	

MODE 4 $\leftarrow \mu_3$

TABLE 5.4 ($\mu_1 = .70$)

$\mu_3 \rightarrow$ MODE 1

$\mu_2 \downarrow$	.050	.100	.150	.200	
.05	6.632	6.627	6.624	6.623	
.10	6.539	6.531	6.527		
.15	6.464	6.455		10.70	.20
.20	6.399		10.78	10.79	.15 $\uparrow$
		10.88	10.88	10.89	.10 μ_2
	11.01	11.01	11.02	11.02	.05
	.200	.150	.100	.050	

MODE 2 $\leftarrow \mu_3$

TABLE 5.4 ($\mu_1 = .70$)

$\mu_3 \rightarrow$ MODE 3

$\mu_2 \downarrow$	.050	.100	.150	.200	
.05	15.45	15.44	15.43	15.43	
.10	15.29	15.28	15.27		
.15	15.16	15.15		18.95	.20
.20	15.01		19.48	19.51	.15 $\uparrow$
		19.66	19.68	19.69	.10 μ_2
	19.83	19.86	19.87	18.58	.05
	.200	.150	.100	.050	

MODE 4 $\leftarrow \mu_3$

TABLE 5.4 ($\mu_1 = .75$)

$\mu_3 \rightarrow$ MODE 1

$\mu_2 \downarrow$	.050	.100	.150	
.05	6.197	6.192	6.190	
.10	6.115	6.108		
.15	6.047		10.09	.15 $\uparrow$
		10.17	10.18	.10 μ_2
	10.29	10.29	10.30	.05
	.150	.100	.050	

MODE 2 $\leftarrow \mu_3$

TABLE 5.4 ($\mu_1 = .75$)

$\mu_3 \rightarrow$ MODE 3

$\mu_2 \downarrow$	.050	.100	.150	
.05	14.43	14.42	14.42	
.10	14.29	14.28		
.15	14.18		18.25	.15 $\uparrow$
		18.39	18.40	.10 μ_2
	18.55	18.56	18.57	.05
	.150	.100	.050	

MODE 4 $\leftarrow \mu_3$

TABLE 5.4 ($\mu_1 = .80$)

$\mu_3 \rightarrow$ MODE 1

$\mu_2 \downarrow$	.050	.100	
.05	5.815	5.812	
.10	5.742		
		9.559	.10 $\uparrow$
	9.658	9.664	.05 μ_2
	.100	.050	

MODE 2 $\leftarrow \mu_3$

TABLE 5.4 ($\mu_1 = .80$)

$\mu_3 \rightarrow$ MODE 3

$\mu_2 \downarrow$	.050	.100	
.05	13.54	13.53	
.10	13.41		
		17.27	.10 $\uparrow \mu_2$
	17.41	17.42	.05
	.100	.050	

MODE 4 $\leftarrow \mu_3$

TABLE 5.4 ($\mu_1 = .85$)

$\mu_3 \rightarrow$ MODE 1

$\mu_2 \downarrow$	.050	
.05	5.479	
	9.103	.05 $\uparrow \mu_2$
	.050	

MODE 2 $\leftarrow \mu_3$

TABLE 5.4 ($\mu_1 = .85$)

$\mu_3 \rightarrow$ MODE 3

$\mu_2 \downarrow$	.050	
.05	12.75	
	16.41	.05 $\uparrow \mu_2$
	.050	

MODE 4 $\leftarrow \mu_3$

TABLE 5.5 ($\mu_1 = .05$)

MODE 1: $\mu_2 \downarrow$ (left), $\mu_3 \rightarrow$ (top). MODE 2: $\mu_2 \uparrow$ (right), $\leftarrow \mu_3$ (bottom).

μ_2	.050	.100	.150	.200	.250	.300	.350	.400	.450	.500	.550	.600	.650	.700	.750	.800	.850	μ_2
.05	5.478	5.741	6.046	6.397	6.804	7.273	7.804	8.344	8.579	8.193	7.641	7.126	6.674	6.284	5.945	5.651	5.393	
.10	5.811	6.106	6.452	6.857	7.329	7.874	8.473	8.945	8.747	8.157	7.573	7.062	6.622	6.243	5.917	5.634		
.15	6.189	6.524	6.923	7.393	7.946	8.581	9.219	9.362	8.786	8.120	7.530	7.026	6.596	6.229	5.913		8.180	.850
.20	6.622	7.007	7.470	8.023	8.677	9.408	9.928	9.526	8.779	8.092	7.506	7.013	6.595	6.239		8.584	8.675	.800
.25	7.120	7.568	8.114	8.772	9.547	10.31	10.29	9.531	8.746	8.068	7.498	7.019	6.614		9.055	9.131	9.237	.750
.30	7.699	8.227	8.876	9.659	10.48	10.63	10.11	9.389	8.669	8.035	7.494	7.037		9.597	9.667	9.755	9.878	.700
.35	8.380	9.002	9.716	9.983	9.867	9.679	9.402	8.993	8.481	7.956	7.476		10.21	10.29	10.37	10.47	10.61	.650
.40	9.152	9.177	9.034	8.908	8.790	8.666	8.520	8.329	8.070	7.744		10.85	10.98	11.09	11.19	11.31	11.47	.600
.45	8.426	8.242	8.106	7.997	7.902	7.811	7.717	7.612	7.485		11.16	11.63	11.87	12.02	12.14	12.27	12.44	.550
.50	7.615	7.461	7.345	7.252	7.174	7.102	7.034	6.965		10.48	11.40	12.31	12.85	13.09	13.22	12.66	11.44	.500
.55	6.944	6.814	6.715	6.635	6.569	6.511	6.458		9.721	10.48	11.48	12.69	13.72	13.84	12.47	11.20	10.21	.450
.60	6.381	6.270	6.185	6.117	6.061	6.012		9.509	9.936	10.61	11.56	12.77	13.42	12.28	11.03	10.04	9.496	.400
.65	5.903	5.808	5.734	5.676	5.628		10.14	10.32	10.58	11.03	11.77	12.59	12.08	10.96	10.38	10.48	10.73	.350
.70	5.492	5.409	5.347	5.297		11.04	11.32	11.56	11.79	12.05	12.38	12.12	11.57	11.64	11.83	12.05	12.35	.300
.75	5.134	5.064	5.011		11.01	11.65	12.32	12.94	13.39	13.54	12.70	12.46	13.05	13.48	13.41	12.60	11.83	.250
.80	4.821	4.762		10.45	11.07	11.80	12.64	13.52	13.75	12.57	11.98	12.78	13.83	13.39	12.49	11.68	11.01	.200
.85	4.545		9.889	10.44	11.09	11.85	12.68	13.23	12.36	11.45	11.69	12.80	13.14	12.37	11.57	10.87	10.29	.150
		9.391	9.884	10.46	11.12	11.87	12.54	12.14	11.14	10.86	11.63	12.55	12.22	11.47	10.78	10.17	9.657	.100
	8.969	9.412	9.927	10.52	11.18	11.85	11.87	10.96	10.36	10.72	11.64	11.98	11.38	10.70	10.09	9.557	9.102	.050
	.850	.800	.750	.700	.650	.600	.550	.500	.450	.400	.350	.300	.250	.200	.150	.100	.050	

TABLE 5.5 ($\mu_1 = .05$)

MODE 3: $\mu_2 \downarrow$ (left), $\mu_3 \rightarrow$ (top). MODE 4: $\mu_2 \uparrow$ (right), $\leftarrow \mu_3$ (bottom).

μ_2	.050	.100	.150	.200	.250	.300	.350	.400	.450	.500	.550	.600	.650	.700	.750	.800	.850	μ_2
.05	12.75	13.41	14.17	15.01	15.44	14.16	13.91	14.84	15.46	14.56	13.79	14.52	15.47	14.78	13.96	13.21	12.57	
.10	13.53	14.27	15.14	16.03	15.68	14.44	15.03	16.10	15.67	14.61	14.81	16.18	15.65	14.73	13.90	13.19		
.15	14.42	15.26	16.24	17.00	15.69	15.30	16.43	16.85	15.68	15.24	16.75	16.61	15.61	14.69	13.90		15.40	.85
.20	15.42	16.34	16.69	16.15	15.60	16.44	16.64	16.11	15.62	16.65	16.55	16.00	15.29	14.55		16.19	16.32	.80
.25	14.65	14.27	14.12	14.86	15.62	14.91	14.39	14.65	15.61	15.01	14.46	14.10	13.82		17.08	17.20	17.35	.75
.30	12.87	13.78	14.93	16.14	15.69	14.34	14.74	16.28	15.64	14.50	13.56	12.88		17.91	18.20	18.35	18.52	.70
.35	13.96	15.09	16.45	17.12	15.69	15.18	16.84	16.81	15.58	14.40	13.42		16.87	18.78	19.44	19.62	18.04	.65
.40	15.31	16.39	16.30	16.00	15.63	16.40	16.29	15.95	15.27	14.34		16.47	17.15	19.18	20.35	17.60	16.52	.60
.45	15.10	14.87	14.70	14.77	15.62	15.13	14.81	14.60	14.36		17.48	17.69	18.04	18.99	17.75	17.71	17.91	.55
.50	13.68	13.53	14.52	16.38	15.65	14.26	13.65	13.39		17.78	18.78	19.16	19.36	18.79	19.07	19.29	19.01	.50
.55	13.07	14.76	16.89	16.94	15.56	13.93	12.92		16.52	17.91	19.58	20.37	18.25	19.22	20.41	18.74	17.00	.45
.60	15.11	16.25	16.12	15.90	15.23	13.87		16.66	17.10	18.10	19.36	17.70	17.55	19.15	18.48	16.92	16.88	.40
.65	15.28	15.11	14.98	14.83	14.55		17.91	18.38	18.69	19.01	18.42	18.60	18.92	18.65	18.44	18.70	18.98	.35
.70	14.23	14.08	13.97	13.87		17.31	18.50	19.75	20.36	18.59	19.54	20.40	18.95	19.14	20.41	19.31	17.96	.30
.75	13.32	13.19	13.09		16.38	17.38	18.59	19.63	17.89	18.09	19.54	18.64	17.58	19.22	19.09	17.77	16.62	.25
.80	12.51	12.40		17.07	17.47	18.04	18.77	17.76	17.75	18.51	18.43	17.55	18.12	18.78	17.74	17.37	17.81	.20
.85	11.80		17.83	18.79	19.79	20.32	18.82	19.63	20.34	19.03	19.36	20.35	19.38	19.00	20.32	19.63	18.54	.15
		17.00	17.93	19.00	19.83	18.17	18.65	19.80	18.74	18.17	19.64	19.19	18.10	19.28	19.46	18.38	17.41	.10
	16.19	17.03	17.99	18.89	17.68	17.51	18.62	18.45	17.23	18.07	18.85	17.77	17.32	18.91	18.25	17.27	16.41	.05
	.850	.800	.750	.700	.650	.600	.550	.500	.450	.400	.350	.300	.250	.200	.150	.100	.050	

TABLE 5.5 ($\mu_1 = .10$)

MODE 1: $\mu_2 \downarrow$ (left), $\mu_3 \rightarrow$ (top). MODE 2: $\mu_2 \uparrow$ (right), $\leftarrow \mu_3$ (bottom).

μ_2	.050	.100	.150	.200	.250	.300	.350	.400	.450	.500	.550	.600	.650	.700	.750	.800	μ_2
.05	5.815	6.114	6.462	6.869	7.342	7.891	8.503	9.026	8.909	8.320	7.720	7.192	6.738	6.348	6.012	5.720	
.10	6.192	6.530	6.930	7.402	7.957	8.599	9.260	9.479	8.926	8.250	7.646	7.130	6.691	6.315	5.992		
.15	6.623	7.011	7.476	8.032	8.689	9.430	10.00	9.653	8.900	8.201	7.604	7.100	6.675	6.312		8.549	.80
.20	7.121	7.571	8.120	8.781	9.563	10.36	10.44	9.672	8.866	8.170	7.586	7.098	6.685		8.974	9.067	.75
.25	7.700	8.231	8.885	9.679	10.57	10.96	10.41	9.595	8.816	8.148	7.587	7.117		9.462	9.538	9.645	.70
.30	8.382	9.013	9.788	10.50	10.51	10.31	9.935	9.362	8.715	8.114	7.594		10.01	10.08	10.17	10.29	.65
.35	9.189	9.756	9.716	9.601	9.480	9.339	9.148	8.864	8.469	8.016		10.61	10.69	10.77	10.87	11.00	.60
.40	9.083	8.896	8.758	8.645	8.543	8.441	8.326	8.183	7.994		11.20	11.34	11.43	11.52	11.63	11.77	.55
.45	8.216	8.051	7.929	7.830	7.745	7.666	7.586	7.500		11.35	11.86	12.10	12.22	12.32	12.43	12.55	.50
.50	7.477	7.335	7.228	7.143	7.071	7.007	6.947		10.57	11.50	12.40	12.85	13.03	13.12	12.67	11.44	.45
.55	6.850	6.728	6.636	6.562	6.501	6.448		9.892	10.60	11.56	12.70	13.51	13.66	12.48	11.20	10.21	.40
.60	6.315	6.210	6.131	6.067	6.015		9.960	10.26	10.79	11.65	12.77	13.38	12.29	11.05	10.16	10.14	.35
.65	5.855	5.765	5.697	5.643		10.70	10.85	11.04	11.34	11.89	12.60	12.10	11.10	10.95	11.08	11.28	.30
.70	5.457	5.380	5.322		11.39	11.69	11.90	12.06	12.23	12.43	12.22	11.90	11.99	12.12	12.26	12.43	.25
.75	5.108	5.044		11.14	11.78	12.42	12.87	13.09	13.18	12.71	12.55	12.92	13.11	13.18	12.60	11.84	.20
.80	4.803		10.54	11.17	11.90	12.72	13.48	13.68	12.60	12.07	12.88	13.69	13.38	12.49	11.68	11.01	.15
		9.988	10.55	11.20	11.97	12.80	13.31	12.39	11.51	11.85	12.96	13.18	12.38	11.58	10.88	10.29	.10
	9.512	10.01	10.60	11.28	12.04	12.69	12.19	11.20	11.03	11.88	12.72	12.26	11.49	10.79	10.18	9.663	.05
	.800	.750	.700	.650	.600	.550	.500	.450	.400	.350	.300	.250	.200	.150	.100	.050	

TABLE 5.5 ($\mu_1 = .10$)

MODE 3: $\mu_2 \downarrow$ (left), $\mu_3 \rightarrow$ (top). MODE 4: $\mu_2 \uparrow$ (right), $\leftarrow \mu_3$ (bottom).

μ_2	.050	.100	.150	.200	.250	.300	.350	.400	.450	.500	.550	.600	.650	.700	.750	.800	μ_2
.05	13.54	14.29	15.15	16.04	15.83	14.60	15.06	16.11	15.78	14.78	14.87	16.17	15.75	14.88	14.05	13.33	
.10	14.42	15.20	15.29	15.21	15.02	15.10	15.30	15.23	15.07	15.02	15.31	15.24	15.08	14.63	13.95		
.15	14.50	14.38	14.30	14.24	14.59	14.57	14.38	14.29	14.42	14.64	14.43	14.30	14.20	14.04		14.73	.80
.20	13.55	13.42	13.69	14.72	15.11	14.10	13.67	14.39	15.17	14.38	13.74	13.45	13.30		15.27	15.36	.75
.25	12.85	13.77	14.90	15.98	15.30	14.13	14.67	16.03	15.36	14.30	13.42	12.81		15.90	15.98	16.09	.70
.30	13.95	15.07	15.88	15.70	15.14	15.01	15.88	15.70	15.16	14.26	13.37		16.58	16.73	16.84	16.98	.65
.35	15.10	15.00	14.89	14.77	14.81	15.09	14.95	14.81	14.60	14.12		16.37	17.46	17.75	17.90	17.87	.60
.40	14.25	14.12	14.03	14.35	15.12	14.43	14.14	14.00	13.87		15.43	16.44	18.29	18.94	17.53	15.37	.55
.45	13.42	13.33	14.48	16.09	15.28	13.98	13.43	13.24		15.97	16.15	16.74	18.39	17.18	16.07	16.19	.50
.50	13.06	14.73	15.84	15.69	15.10	13.80	12.91		16.72	16.99	17.19	17.48	17.24	17.04	17.18	17.37	.45
.55	15.04	15.13	15.04	14.93	14.65	13.77		16.16	17.39	18.24	18.56	17.87	17.97	18.46	18.49	16.98	.40
.60	14.52	14.41	14.33	14.24	14.09		15.42	16.23	17.59	18.92	17.51	16.79	18.54	18.39	16.73	15.40	.35
.65	13.85	13.75	13.67	13.60		16.05	16.25	16.68	17.63	17.15	16.39	16.96	17.86	16.61	16.20	16.38	.30
.70	13.21	13.11	13.04		16.77	17.35	17.69	17.96	17.62	17.55	17.83	18.00	17.40	17.63	17.89	17.88	.25
.75	12.58	12.49		16.03	17.03	18.17	19.12	17.73	17.63	18.92	18.52	17.27	18.41	18.96	17.74	16.60	.20
.80	11.98		15.20	16.01	17.06	18.15	17.33	16.41	17.57	18.15	16.79	16.59	18.25	17.56	16.43	15.44	.15
		15.39	15.56	16.10	17.01	16.93	15.72	16.25	17.29	16.54	15.72	16.47	17.25	16.29	15.46	15.49	.10
	16.48	16.56	16.61	16.69	16.72	16.58	16.63	16.73	16.62	16.59	16.64	16.78	16.58	16.59	16.63	16.69	.05
	.800	.750	.700	.650	.600	.550	.500	.450	.400	.350	.300	.250	.200	.150	.100	.050	

TABLE 5.5 ($\mu_1 = .15$) MODE 1

$\mu_2 \downarrow$ / $\mu_3 \rightarrow$	.050	.100	.150	.200	.250	.300	.350	.400	.450	.500	.550	.600	.650	.700	.750	$\mu_2 \uparrow$
.05	6.196	6.538	6.941	7.415	7.973	8.621	9.307	9.620	9.106	8.417	7.795	7.262	6.808	6.421	6.089	
.10	6.626	7.017	7.484	8.042	8.702	9.449	10.03	9.741	9.024	8.323	7.717	7.203	6.769	6.398		
.15	7.123	7.575	8.125	8.787	9.558	10.05	9.976	9.610	8.932	8.257	7.674	7.180	6.762		8.439	.75
.20	7.702	8.233	8.885	9.593	9.698	9.658	9.572	9.343	8.823	8.208	7.657	7.185		8.738	8.801	.70
.25	8.382	9.001	9.339	9.313	9.273	9.223	9.146	8.993	8.654	8.153	7.654		9.053	9.101	9.168	.65
.30	9.042	8.980	8.919	8.867	8.817	8.763	8.692	8.581	8.379	8.042		9.380	9.421	9.472	9.543	.60
.35	8.611	8.513	8.439	8.377	8.321	8.264	8.198	8.114	7.991		9.722	9.763	9.809	9.866	9.944	.55
.40	8.109	8.002	7.919	7.852	7.792	7.735	7.677	7.611		10.08	10.14	10.19	10.25	10.31	10.40	.50
.45	7.582	7.472	7.387	7.318	7.259	7.206	7.154		10.40	10.54	10.63	10.71	10.78	10.86	10.98	.45
.50	7.061	6.954	6.872	6.806	6.751	6.702		10.22	10.78	11.10	11.27	11.38	11.47	11.58	11.40	.40
.55	6.569	6.470	6.395	6.334	6.285		9.555	10.23	11.04	11.73	12.10	12.28	12.22	11.17	10.20	.35
.60	6.120	6.031	5.963	5.910		9.216	9.581	10.24	11.15	12.19	12.86	12.20	11.00	10.00	9.240	.30
.65	5.716	5.638	5.579		9.396	9.493	9.716	10.26	11.17	12.16	11.95	10.85	9.873	9.459	9.510	.25
.70	5.355	5.288		9.741	9.785	9.845	9.969	10.34	11.12	11.52	10.70	9.882	9.821	9.859	9.923	.20
.75	5.034		10.13	10.18	10.21	10.25	10.32	10.51	10.91	10.56	10.22	10.23	10.26	10.30	10.36	.15
		10.43	10.62	10.69	10.72	10.75	10.79	10.84	10.76	10.71	10.73	10.75	10.77	10.80	10.84	.10
	10.11	10.66	11.21	11.38	11.41	11.43	11.44	11.35	11.36	11.40	11.42	11.43	11.43	10.89	10.30	.05
	.750	.700	.650	.600	.550	.500	.450	.400	.350	.300	.250	.200	.150	.100	.050	$\leftarrow \mu_3$

MODE 2

TABLE 5.5 ($\mu_1 = .20$) MODE 1

$\mu_2 \downarrow$ / $\mu_3 \rightarrow$	.050	.100	.150	.200	.250	.300	.350	.400	.450	.500	.550	.600	.650	.700	$\mu_2 \uparrow$
.05	6.631	7.026	7.497	8.055	8.678	8.734	8.729	8.720	8.689	8.435	7.860	7.335	6.888	6.507	
.10	7.125	7.580	8.115	8.302	8.297	8.288	8.276	8.259	8.224	8.108	7.732	7.267	6.851		
.15	7.699	8.002	7.988	7.972	7.957	7.943	7.927	7.907	7.874	7.799	7.587	7.219		7.651	.70
.20	7.782	7.740	7.711	7.690	7.672	7.655	7.637	7.615	7.585	7.531	7.411		7.877	7.932	.65
.25	7.536	7.492	7.461	7.437	7.416	7.397	7.378	7.356	7.328	7.286		8.153	8.202	8.272	.60
.30	7.302	7.256	7.223	7.196	7.173	7.152	7.132	7.109	7.083		8.495	8.545	8.610	8.700	.55
.35	7.069	7.021	6.984	6.955	6.930	6.907	6.885	6.862		8.928	8.985	9.052	9.137	9.255	.50
.40	6.828	6.777	6.737	6.704	6.676	6.652	6.628		9.466	9.548	9.629	9.717	9.828	9.982	.45
.45	6.573	6.517	6.474	6.438	6.408	6.381		10.03	10.22	10.36	10.48	10.60	10.74	10.94	.40
.50	6.300	6.241	6.194	6.156	6.134		10.02	10.67	11.13	11.41	11.39	11.75	11.89	11.41	.35
.55	6.012	5.950	5.902	5.864		9.324	10.11	11.04	11.99	12.65	12.95	12.39	11.17	10.20	.30
.60	5.716	5.655	5.608		8.619	9.278	10.12	11.17	12.38	13.18	12.23	11.00	9.994	9.206	.25
.65	5.422	5.365		8.108	8.582	9.254	10.12	11.18	12.25	11.97	10.84	9.846	9.038	8.392	.20
.70	5.140		8.050	8.189	8.587	9.254	10.12	11.11	11.54	10.69	9.725	8.915	8.265	8.069	.15
		8.316	8.339	8.396	8.646	9.280	10.12	10.85	10.49	9.614	8.821	8.353	8.356	8.388	.10
	8.739	8.745	8.752	8.767	8.832	9.326	10.07	10.21	9.500	8.791	8.754	8.759	8.770	8.792	.05
	.700	.650	.600	.550	.500	.450	.400	.350	.300	.250	.200	.150	.100	.050	$\leftarrow \mu_3$

MODE 2

TABLE 5.5 ($\mu_1 = .15$) MODE 3

$\mu_2 \downarrow$ / $\mu_3 \rightarrow$	.050	.100	.150	.200	.250	.300	.350	.400	.450	.500	.550	.600	.650	.700	.750	$\mu_2 \uparrow$
.05	11.51	11.48	11.62	12.42	13.27	13.35	12.24	11.75	12.46	13.51	13.09	12.25	11.62	11.48	11.45	
.10	11.04	11.71	12.52	13.46	14.27	13.47	12.40	12.71	13.98	14.03	13.09	12.20	11.48	11.01		
.15	11.85	12.64	13.61	14.68	14.99	13.58	13.11	14.33	15.10	14.12	13.08	12.20	11.48		14.20	.75
.20	12.81	13.75	14.89	15.99	15.17	13.90	14.63	16.04	15.28	14.11	13.09	12.24		14.98	15.11	.70
.25	13.95	15.04	15.87	15.54	14.79	14.87	15.87	15.57	14.88	13.97	13.08		15.89	16.01	16.17	.65
.30	14.46	14.18	13.97	13.81	14.25	14.54	14.18	13.92	13.67	13.34		16.85	17.07	17.23	17.41	.60
.35	12.89	12.69	12.78	14.19	14.98	13.80	13.00	12.66	12.47		16.44	17.97	18.44	18.64	18.00	.55
.40	11.82	12.76	14.45	16.11	15.17	13.55	12.41	11.85		15.23	16.53	18.75	19.81	17.55	15.23	.50
.45	13.03	14.71	15.81	15.57	14.83	13.45	12.24		16.01	16.28	16.98	18.63	17.23	16.17	16.36	.45
.50	14.79	14.66	14.52	14.35	14.04	13.30		17.04	17.55	17.84	18.16	17.71	17.67	17.89	18.15	.40
.55	13.70	13.54	13.41	13.30	13.14		16.19	17.60	19.02	19.66	18.07	18.63	19.62	18.70	16.99	.35
.60	12.75	12.62	12.52	12.43		15.24	16.28	17.73	19.17	17.55	16.91	18.84	18.43	16.74	15.36	.30
.65	11.98	11.87	11.79		16.17	16.31	17.04	17.94	17.29	16.70	17.40	18.03	16.72	16.47	16.84	.25
.70	11.36	11.27		16.97	17.97	18.84	19.32	18.19	18.59	19.25	18.71	18.20	19.08	19.15	17.95	.20
.75	10.85		16.03	17.11	18.36	19.46	17.78	17.78	19.30	18.58	17.34	18.84	19.05	17.75	16.60	.15
		15.06	16.01	17.12	18.25	17.35	16.43	17.68	18.20	16.81	16.64	18.37	17.58	16.43	15.44	.10
	14.23	15.07	16.04	17.09	16.96	15.51	16.23	17.39	16.54	15.52	16.51	17.29	16.29	15.29	14.43	.05
	.750	.700	.650	.600	.550	.500	.450	.400	.350	.300	.250	.200	.150	.100	.050	$\leftarrow \mu_3$

MODE 4

TABLE 5.5 ($\mu_1 = .20$) MODE 3

$\mu_2 \downarrow$ / $\mu_3 \rightarrow$	.050	.100	.150	.200	.250	.300	.350	.400	.450	.500	.550	.600	.650	.700	$\mu_2 \uparrow$
.05	11.02	11.71	12.53	13.49	14.36	13.70	12.53	12.75	14.04	14.19	13.23	12.31	11.52	10.85	
.10	11.85	12.65	13.62	14.71	15.16	13.74	13.17	14.36	15.22	14.25	13.18	12.27	11.51		
.15	12.81	13.76	14.91	16.09	15.46	14.04	14.67	16.20	15.48	14.24	13.18	12.30		14.87	.70
.20	13.95	15.08	16.41	16.89	15.43	15.08	16.76	16.61	15.39	14.20	13.20		15.77	15.92	.65
.25	15.28	15.74	15.48	15.15	15.06	15.78	15.55	15.20	14.68	13.95		16.81	16.94	17.12	.60
.30	13.92	13.62	13.48	14.32	15.30	14.38	13.76	13.44	13.20		17.78	18.10	18.27	18.46	.55
.35	12.25	12.80	14.48	16.32	15.46	13.80	12.72	12.18		16.76	18.76	19.45	19.63	18.04	.50
.40	13.04	14.74	16.75	16.62	15.32	13.65	12.38		16.05	16.98	19.18	20.30	17.59	16.04	.45
.45	15.09	15.71	15.54	15.30	14.75	13.58		17.21	17.47	17.90	18.99	17.60	17.47	17.71	.40
.50	14.54	14.33	14.18	14.03	13.79		17.71	18.75	19.16	19.37	18.76	19.06	19.30	19.01	.35
.55	13.34	13.16	13.04	12.93		16.41	17.88	19.58	20.32	18.24	19.23	20.34	18.74	17.00	.30
.60	12.33	12.19	12.08		16.15	16.84	18.05	19.37	17.66	17.41	19.15	18.48	16.81	16.26	.25
.65	11.48	11.36		17.75	18.26	18.64	19.00	18.33	18.52	18.91	18.62	18.32	18.64	18.97	.20
.70	10.76		17.25	18.47	19.75	20.30	18.57	19.54	20.32	18.95	19.14	20.32	19.31	17.96	.15
		16.16	17.26	18.55	19.63	17.85	17.99	19.53	18.62	17.48	19.19	19.09	17.76	16.61	.10
	15.21	16.17	17.30	18.43	17.41	16.62	17.93	18.27	16.86	16.93	18.55	17.60	16.44	15.45	.05
	.700	.650	.600	.550	.500	.450	.400	.350	.300	.250	.200	.150	.100	.050	$\leftarrow \mu_3$

MODE 4

TABLE 5.5 ($\mu_1 = .25$) MODE 1

$\mu_2 \downarrow$ \ $\mu_3 \rightarrow$	.050	.100	.150	.200	.250	.300	.350	.400	.450	.500	.550	.600	.650	
.05	7.095	7.098	7.091	7.087	7.084	7.081	7.079	7.077	7.074	7.070	7.063	7.044	6.917	
.10	6.836	6.810	6.796	6.786	6.779	6.773	6.767	6.762	6.756	6.749	6.738	6.718		
.15	6.608	6.579	6.560	6.547	6.537	6.529	6.521	6.513	6.505	6.496	6.484		7.244	.65
.20	6.412	6.382	6.362	6.346	6.334	6.324	6.315	6.306	6.297	6.287		7.571	7.653	.60
.25	6.238	6.208	6.186	6.170	6.157	6.145	6.135	6.126	6.116		7.983	8.058	8.163	.55
.30	6.078	6.047	6.025	6.008	5.994	5.982	5.971	5.961		8.498	8.572	8.668	8.801	.50
.35	5.925	5.895	5.872	5.854	5.839	5.827	5.815		9.136	9.219	9.314	9.433	9.599	.45
.40	5.777	5.746	5.722	5.704	5.688	5.674		9.891	10.01	10.13	10.25	10.40	10.60	.40
.45	5.628	5.596	5.571	5.551	5.535		10.54	10.88	11.11	11.28	11.43	11.61	11.86	.35
.50	5.474	5.441	5.415	5.395		10.20	11.07	11.89	12.41	12.71	12.89	12.60	11.43	.30
.55	5.313	5.279	5.253		9.382	10.22	11.27	12.50	13.61	13.80	12.46	11.18	10.20	.25
.60	5.143	5.109		8.627	9.332	10.21	11.32	12.63	13.38	12.26	11.01	9.998	9.208	.20
.65	4.966		7.996	8.585	9.314	10.22	11.32	12.39	12.01	10.86	9.852	9.040	8.391	.15
		7.481	7.975	8.585	9.334	10.24	11.26	11.61	10.71	9.735	8.920	8.253	7.710	.10
	7.160	7.481	7.997	8.632	9.400	10.29	10.99	10.53	9.630	8.826	8.154	7.598	7.154	.05
	.650	.600	.550	.500	.450	.400	.350	.300	.250	.200	.150	.100	.050	$\leftarrow \mu_3$ \ $\mu_2 \uparrow$

MODE 2

TABLE 5.5 ($\mu_1 = .30$) MODE 1

$\mu_2 \downarrow$ \ $\mu_3 \rightarrow$	.050	.100	.150	.200	.250	.300	.350	.400	.450	.500	.550	.600	
.05	5.978	5.966	5.961	5.958	5.955	5.954	5.952	5.951	5.950	5.949	5.947	5.946	
.10	5.774	5.755	5.744	5.737	5.732	5.728	5.724	5.721	5.718	5.716	5.713		
.15	5.604	5.582	5.568	5.558	5.551	5.545	5.540	5.536	5.531	5.528		7.369	.60
.20	5.456	5.433	5.418	5.407	5.398	5.391	5.385	5.379	5.374		7.808	7.914	.55
.25	5.324	5.302	5.286	5.273	5.264	5.256	5.249	5.243		8.348	8.443	8.576	.50
.30	5.204	5.182	5.165	5.153	5.143	5.134	5.127		9.010	9.102	9.219	9.382	.45
.35	5.092	5.070	5.054	5.041	5.031	5.022		9.809	9.913	10.03	10.17	10.36	.40
.40	4.985	4.964	4.948	4.935	4.924		10.69	10.87	11.01	11.15	11.31	11.54	.35
.45	4.882	4.860	4.844	4.831		11.11	11.72	12.06	12.26	12.42	12.57	12.72	.30
.50	4.779	4.758	4.742		10.38	11.38	12.47	13.13	13.35	13.44	12.68	11.44	.25
.55	4.676	4.655		9.481	10.35	11.45	12.78	13.74	13.81	12.48	11.19	10.20	.20
.60	4.570		8.712	9.436	10.34	11.48	12.81	13.46	12.28	11.02	10.00	9.210	.15
		8.081	8.691	9.442	10.37	11.50	12.55	12.05	10.87	9.863	9.048	8.395	.10
	7.565	8.092	8.730	9.509	10.45	11.49	11.71	10.74	9.753	8.934	8.264	7.716	.05
	.600	.550	.500	.450	.400	.350	.300	.250	.200	.150	.100	.050	$\leftarrow \mu_3$ \ $\mu_2 \uparrow$

MODE 2

TABLE 5.5 ($\mu_1 = .25$) MODE 3

$\mu_2 \downarrow$ \ $\mu_3 \rightarrow$	.050	.100	.150	.200	.250	.300	.350	.400	.450	.500	.550	.600	.650	
.05	11.86	12.66	13.63	14.75	15.38	14.01	13.29	14.41	15.40	14.45	13.36	12.43	11.65	
.10	12.82	13.77	14.92	16.15	15.69	14.20	14.72	16.29	15.64	14.39	13.31	12.41		
.15	13.96	15.09	16.44	16.80	15.74	15.18	16.75	16.68	15.61	14.37	13.33		15.51	.65
.20	15.31	16.34	16.28	16.09	15.72	16.34	16.28	16.04	15.35	14.33		16.21	16.31	.60
.25	15.29	15.03	14.84	14.87	15.71	15.29	14.95	14.70	14.39		16.76	16.82	16.89	.55
.30	13.50	13.33	14.53	16.37	15.69	14.27	13.51	13.15		17.13	17.21	17.26	17.32	.50
.35	13.06	14.76	16.74	16.69	15.61	13.91	12.73		16.75	17.54	17.69	17.77	17.79	.45
.40	15.12	16.26	16.17	15.99	15.31	13.85		16.41	16.87	18.06	18.50	17.54	16.44	.40
.45	15.41	15.24	15.09	14.92	14.59		16.90	16.98	17.17	18.26	17.24	16.97	17.03	.35
.50	14.22	14.04	13.91	13.78		17.19	17.36	17.44	17.56	17.43	17.39	17.45	17.52	.30
.55	13.11	12.95	12.83		16.48	17.46	17.96	18.14	17.80	17.81	18.06	18.17	16.99	.25
.60	12.13	11.99		16.48	16.74	17.57	18.71	17.51	17.03	18.31	18.35	16.78	16.56	.20
.65	11.29		16.97	17.04	17.12	17.52	17.22	17.07	17.20	17.76	17.03	17.05	17.10	.15
		17.17	17.39	17.44	17.48	17.44	17.42	17.45	17.49	17.40	17.42	17.45	17.49	.10
	16.33	17.39	17.87	17.91	17.80	17.80	17.90	17.91	17.57	17.87	17.91	17.76	16.62	.05
	.650	.600	.550	.500	.450	.400	.350	.300	.250	.200	.150	.100	.050	$\leftarrow \mu_3$ \ $\mu_2 \uparrow$

MODE 4

TABLE 5.5 ($\mu_1 = .30$) MODE 3

$\mu_2 \downarrow$ \ $\mu_3 \rightarrow$	.050	.100	.150	.200	.250	.300	.350	.400	.450	.500	.550	.600	
.05	12.83	13.78	14.91	15.00	14.97	14.41	14.75	15.01	14.98	14.55	13.49	12.58	
.10	13.96	14.66	14.65	14.63	14.59	14.62	14.66	14.64	14.60	14.35	13.45		
.15	14.42	14.39	14.36	14.34	14.38	14.44	14.39	14.35	14.31	14.16		14.46	.60
.20	14.14	14.09	14.06	14.26	14.81	14.23	14.10	14.04	13.98		14.76	14.82	.55
.25	13.70	13.62	14.45	15.69	14.95	14.02	13.68	13.54		15.25	15.34	15.46	.50
.30	13.11	14.70	15.10	14.99	14.70	13.87	13.10		16.06	16.24	16.39	16.59	.45
.35	14.68	14.63	14.58	14.53	14.41	13.84		16.12	17.33	17.75	17.98	17.90	.40
.40	14.34	14.28	14.25	14.20	14.12		14.93	16.25	18.45	19.53	17.53	15.16	.35
.45	13.99	13.92	13.87	13.82		15.24	15.48	16.48	18.48	17.16	15.33	15.41	.30
.50	13.48	13.37	13.30		16.19	16.52	16.84	17.36	17.07	16.56	16.78	17.11	.25
.55	12.73	12.61		15.96	17.29	18.55	19.21	17.89	18.10	19.09	18.64	16.98	.20
.60	11.91		14.83	15.97	17.52	19.05	17.49	16.60	18.63	18.40	16.72	15.33	.15
		14.69	14.88	15.97	17.42	17.05	15.39	16.27	17.76	16.51	15.12	14.71	.10
	15.02	15.04	15.10	15.97	16.48	15.07	15.09	16.16	16.25	15.06	15.04	15.06	.05
	.600	.550	.500	.450	.400	.350	.300	.250	.200	.150	.100	.050	$\leftarrow \mu_3$ \ $\mu_2 \uparrow$

MODE 4

TABLE 5.5 (μ_1 = .35) MODE 1

μ_2 ↓ \ μ_3 →	.050	.100	.150	.200	.250	.300	.350	.400	.450	.500	.550	μ_2 ↑
.05	5.154	5.146	5.142	5.139	5.137	5.136	5.135	5.134	5.134	5.133	5.132	
.10	4.999	4.984	4.976	4.970	4.966	4.963	4.961	4.959	4.957	4.956		
.15	4.867	4.850	4.839	4.832	4.826	4.821	4.818	4.815	4.812		7.766	.55
.20	4.751	4.733	4.721	4.712	4.706	4.700	4.696	4.692		8.295	8.419	.50
.25	4.647	4.629	4.617	4.607	4.600	4.594	4.589		8.937	9.045	9.195	.45
.30	4.552	4.535	4.522	4.513	4.505	4.499		9.698	9.797	9.920	10.09	.40
.35	4.464	4.448	4.435	4.425	4.418		10.53	10.63	10.73	10.85	11.00	.35
.40	4.382	4.365	4.353	4.344		11.22	11.35	11.44	11.51	11.58	11.67	.30
.45	4.303	4.287	4.276		11.36	11.75	11.86	11.91	11.95	11.99	12.04	.25
.50	4.226	4.212		10.52	11.51	12.07	12.16	12.19	12.21	12.22	11.44	.20
.55	4.151		9.595	10.48	11.57	12.29	12.38	12.40	12.36	11.20	10.21	.15
		8.833	9.574	10.50	11.64	12.51	12.60	12.29	11.03	10.01	9.215	.10
	8.220	8.852	9.629	10.59	11.75	12.68	12.10	10.90	9.882	9.062	8.402	.05
MODE 2 ← μ_3	.550	.500	.450	.400	.350	.300	.250	.200	.150	.100	.050	

TABLE 5.5 (μ_1 = .35) MODE 3

μ_2 ↓ \ μ_3 →	.050	.100	.150	.200	.250	.300	.350	.400	.450	.500	.550	μ_2 ↑
.05	12.96	12.95	12.94	12.93	12.93	13.03	12.96	12.94	12.93	12.92	12.90	
.10	12.68	12.66	12.65	12.66	13.65	13.24	12.70	12.66	12.64	12.62		
.15	12.47	12.45	12.57	14.13	14.80	13.30	12.53	12.45	12.42		13.79	.55
.20	12.28	12.74	14.43	16.11	15.03	13.30	12.40	12.26		14.67	14.84	.50
.25	13.03	14.68	15.73	15.38	14.54	13.25	12.34		15.84	16.00	16.22	.45
.30	14.39	14.17	13.99	13.81	13.55	13.06		17.12	17.45	17.67	17.92	.40
.35	13.23	13.10	13.02	12.95	12.87		16.48	18.49	19.28	19.52	18.03	.35
.40	12.68	12.62	12.58	12.55		15.06	16.59	19.10	20.34	17.57	15.20	.30
.45	12.38	12.34	12.31		16.03	16.42	17.27	18.88	17.29	16.27	16.58	.25
.50	12.14	12.10		17.32	18.28	18.81	19.14	18.35	18.63	18.99	18.95	.20
.55	11.84		16.15	17.70	19.53	20.37	18.16	19.10	20.40	18.73	16.99	.15
		14.81	16.12	17.75	19.29	17.55	16.81	18.99	18.45	16.74	15.33	.10
	13.73	14.82	16.17	17.67	17.12	15.48	16.59	17.91	16.54	15.13	13.98	.05
MODE 4 ← μ_3	.550	.500	.450	.400	.350	.300	.250	.200	.150	.100	.050	

TABLE 5.5 (μ_1 = .40) MODE 1

μ_2 ↓ \ μ_3 →	.050	.100	.150	.200	.250	.300	.350	.400	.450	.500	μ_2 ↑
.05	4.531	4.524	4.521	4.519	4.517	4.516	4.516	4.515	4.515	4.514	
.10	4.408	4.397	4.390	4.386	4,382	4.380	4.378	4.377	4.375		
.15	4.302	4.289	4.280	4.274	4.270	4.266	4.264	4.261		8.268	.50
.20	4.209	4.195	4.185	4.178	4.173	4.169	4.165		8.838	8.962	.45
.25	4.124	4.110	4.100	4.093	4.087	4.082		9.452	9.539	9.654	.40
.30	4.047	4.033	4.023	4.016	4.010		9.988	10.04	10.10	10.17	.35
.35	3.976	3.963	3.953	3.945		10.34	10.37	10.40	10.44	10.48	.30
.40	3.909	3.896	3.887		10.56	10.58	10.60	10.62	10.64	10.67	.25
.45	3.846	3.834		10.69	10.74	10.76	10.77	10.79	10.80	10.83	.20
.50	3.785		10.55	10.86	10.90	10.92	10.93	10.94	10.96	10.98	.15
		9.734	10.60	11.05	11.09	11.10	11.11	11.12	11.10	10.21	.10
	9.002	9.769	10.71	11.32	11.34	11.35	11.36	11.05	10.03	9.224	.05
MODE 2 ← μ_3	.500	.450	.400	.350	.300	.250	.200	.150	.100	.050	

TABLE 5.5 (μ_1 = .40) MODE 3

μ_2 ↓ \ μ_3 →	.050	.100	.150	.200	.250	.300	.350	.400	.450	.500	μ_2 ↑
.05	11.38	11.37	11.37	12.35	13.74	13.43	12.02	11.39	11.37	11.36	
.10	11.16	11.25	12.53	14.16	15.00	13.45	11.96	11.21	11.14		
.15	11.45	12.74	14.45	16.32	15.36	13.44	11.96	11.11		14.59	.50
.20	13.03	14.73	16.86	16.85	15.22	13.42	12.02		15.75	15.95	.45
.25	15.08	15.69	15.44	15.09	14.42	13.28		17.05	17.20	17.38	.40
.30	14.00	13.72	13.51	13.32	13.07		17.92	18.11	18.20	18.27	.35
.35	12.50	12.31	12.18	12.07		16.83	18.42	18.58	18.63	18.04	.30
.40	11.60	11.50	11.44		16.15	17.15	18.66	18.95	17.61	16.16	.25
.45	11.15	11.10		17.54	17.75	18.00	18.60	17.78	17.76	17.91	.20
.50	10.90		17.80	18.35	18.43	18.48	18.36	18.40	18.45	18.49	.15
		16.34	17.92	18.69	18.74	18.25	18.64	18.73	18.68	17.00	.10
	15.02	16.36	18.02	19.01	17.62	17.13	18.95	18.48	16.76	15.35	.05
MODE 4 ← μ_3	.500	.450	.400	.350	.300	.250	.200	.150	.100	.050	

TABLE 5.5 (μ_1 = .45) MODE 1

μ_2 ↓ \ μ_3 →	.050	.100	.150	.200	.250	.300	.350	.400	.450	μ_2 ↑
.05	4.042	4.036	4.034	4.032	4.031	4.030	4.030	4.029	4.029	
.10	3.943	3.933	3.920	3.924	3.922	3.920	3.918	3.917		
.15	3.856	3.845	3.838	3.833	3.830	3.827	3.825		8.605	.45
.20	3.779	3.767	3.760	3.754	3.750	3.746		8.958	9.016	.40
.25	3.709	3.697	3.689	3.683	3.679		9.209	9.237	9.274	.35
.30	3.645	3.633	3.625	3.619		9.380	9.396	9.416	9.441	.30
.35	3.585	3.574	3.567		9.311	9.522	9.534	9.549	9.570	.25
.40	3.529	3.520		9.627	9.637	9.646	9.656	9.669	9.688	.20
.45	3.477		9.746	9.757	9.766	9.774	9.783	9.794	9.812	.15
		9.881	9.903	9.912	9.919	9.925	9.932	9.942	9.958	.10
	9.891	10.10	10.12	10.12	10.12	10.13	10.13	10.14	10.14	.05
MODE 2 ← μ_3	.450	.400	.350	.300	.250	.200	.150	.100	.050	

TABLE 5.5 (μ_1 = .45) MODE 3

μ_2 ↓ \ μ_3 →	.050	.100	.150	.200	.250	.300	.350	.400	.450	μ_2 ↑
.05	10.24	11.24	12.55	14.22	15.27	13.76	12.17	10.95	10.20	
.10	11.46	12.75	14.48	16.38	15.63	13.67	12.11	10.95		
.15	13.03	14.75	16.45	16.41	15.59	13.65	12.16		15.49	.45
.20	15.11	16.15	16.09	15.96	15.27	13.66		16.14	16.21	.40
.25	15.38	15.15	14.96	14.72	14.29		16.47	16.51	16.55	.35
.30	13.69	13.42	13.22	13.04		16.71	16.76	16.81	16.86	.30
.35	12.15	11.94	11.79		16.54	17.13	17.33	17.48	17.59	.25
.40	11.03	10.89		16.24	16.57	18.07	19.13	17.51	16.25	.20
.45	10.33		16.49	16.53	16.67	18.27	17.12	16.52	16.54	.15
		16.69	16.72	16.74	16.81	16.80	16.72	16.74	16.76	.10
	16.55	16.97	16.99	17.00	17.00	16.98	16.99	17.00	16.97	.05
MODE 4 ← μ_3	.450	.400	.350	.300	.250	.200	.150	.100	.050	

TABLE 5.5 (μ_1 = .50) MODE 1

μ_2 ↓ \ μ_3 →	.050	.100	.150	.200	.250	.300	.350	.400	μ_2 ↑
.05	3.648	3.644	3.642	3.640	3.639	3.639	3.638	3.638	
.10	3.566	3.559	3.554	3.551	3.549	3.548	3.546		
.15	3.494	3.485	3.479	3.475	3.472	3.470		8.316	.40
.20	3.430	3.420	3.413	3.409	3.405		8.440	8.462	.35
.25	3.370	3.361	3.354	3.349		8.541	8.554	8.572	.30
.30	3.316	3.307	3.300		8.633	8.642	8.653	8.669	.25
.35	3.265	3.257		8.725	8.731	8.739	8.750	8.765	.20
.40	3.218		8.827	8.833	8.839	8.846	8.855	8.870	.15
		8.954	8.959	8.963	8.968	8.973	8.981	8.995	.10
	9.129	9.132	9.134	9.136	9.138	9.141	9.145	9.154	.05
MODE 2 ← μ_3	.400	.350	.300	.250	.200	.150	.100	.050	

TABLE 5.5 (μ_1 = .50) MODE 3

μ_2 ↓ \ μ_3 →	.050	.100	.150	.200	.250	.300	.350	.400	μ_2 ↑
.05	11.47	12.78	14.52	15.31	15.28	14.00	12.38	11.15	
.10	13.04	14.76	15.09	15.08	15.02	13.89	12.36		
.15	14.93	14.92	14.91	14.88	14.82	13.90		14.85	.40
.20	14.75	14.71	14.68	14.64	14.54		15.05	15.08	.35
.25	14.36	14.26	14.16	14.06		15.36	15.42	15.52	.30
.30	13.32	13.10	12.93		16.15	16.40	16.63	16.93	.25
.35	11.91	11.71		16.12	17.91	18.81	19.18	18.00	.20
.40	10.74		15.01	16.17	18.90	20.34	17.54	15.15	.15
		15.11	15.15	16.18	18.59	17.15	15.13	15.13	.10
	15.32	15.33	15.35	16.16	16.62	15.33	15.34	15.35	.05
MODE 4 ← μ_3	.400	.350	.300	.250	.200	.150	.100	.050	

TABLE 5.5 (μ_1 = .55) MODE 1

μ_2 ↓ \ μ_3 →	.050	.100	.150	.200	.250	.300	.350	μ_2 ↑
.05	3.324	3.321	3.319	3.318	3.317	3.317	3.316	
.10	3.256	3.249	3.246	3.243	3.242	3.240		
.15	3.195	3.187	3.182	3.179	3.177		7.762	.35
.20	3.140	3.132	3.126	3.122		7.831	7.844	.30
.25	3.089	3.081	3.076		7.900	7.909	7.922	.25
.30	3.042	3.035		7.976	7.982	7.991	8.004	.20
.35	2.999		8.063	8.068	8.074	8.082	8.095	.15
		8.173	8.176	8.179	8.184	8.191	8.203	.10
	8.321	8.322	8.324	8.325	8.328	8.331	8.339	.05
MODE 2 ← μ_3	.350	.300	.250	.200	.150	.100	.050	

TABLE 5.5 (μ_1 = .55) MODE 3

μ_2 ↓ \ μ_3 →	.050	.100	.150	.200	.250	.300	.350	μ_2 ↑
.05	13.06	13.96	13.96	13.95	13.94	13.88	12.64	
.10	13.78	13.77	13.76	13.75	13.73	13.66		
.15	13.64	13.62	13.60	13.59	13.57		13.93	.35
.20	13.49	13.46	13.44	13.42		14.50	14.67	.30
.25	13.27	13.22	13.18		15.94	16.18	16.50	.25
.30	12.74	12.61		17.69	18.15	18.36	18.54	.20
.35	11.68		16.49	18.77	19.04	19.09	18.04	.15
		14.37	16.46	19.10	19.31	17.57	15.15	.10
	13.98	14.37	16.55	18.94	17.20	14.81	13.98	.05
MODE 4 ← μ_3	.350	.300	.250	.200	.150	.100	.050	

TABLE 5.5 ($\mu_1 = .60$) MODE 1

$\mu_2 \downarrow$ \ $\mu_3 \rightarrow$	.050	.100	.150	.200	.250	.300	$\mu_2 \uparrow$
.05	3.054	3.050	3.049	3.048	3.047	3.047	
.10	2.995	2.990	2.987	2.985	2.983		
.15	2.943	2.937	2.932	2.930		7.228	.30
.20	2.895	2.889	2.884		7.284	7.294	.25
.25	2.852	2.845		7.347	7.354	7.365	.20
.30	2.811		7.422	7.427	7.434	7.445	.15
		7.516	7.519	7.523	7.529	7.539	.10
	7.643	7.644	7.646	7.648	7.651	7.657	.05
MODE 2 $\leftarrow \mu_3$	.300	.250	.200	.150	.100	.050	

TABLE 5.5 ($\mu_1 = .60$) MODE 3

$\mu_2 \downarrow$ \ $\mu_3 \rightarrow$	.050	.100	.150	.200	.250	.300	$\mu_2 \uparrow$
.05	12.83	12.82	12.82	12.81	12.81	12.80	
.10	12.66	12.65	12.64	12.64	12.63		
.15	12.54	12.52	12.51	12.50		14.21	.30
.20	12.41	12.40	12.38		15.80	16.07	.25
.25	12.27	12.24		17.13	17.21	17.29	.20
.30	11.99		17.50	17.55	17.58	17.60	.15
		16.82	17.72	17.75	17.76	17.77	.10
	14.64	16.88	17.94	17.96	17.61	15.17	.05
MODE 4 $\leftarrow \mu_3$	.300	.250	.200	.150	.100	.050	

TABLE 5.5 ($\mu_1 = .65$) MODE 1

$\mu_2 \downarrow$ \ $\mu_3 \rightarrow$	.050	.100	.150	.200	.250	$\mu_2 \uparrow$
.05	2.824	2.821	2.820	2.819	2.818	
.10	2.773	2.769	2.766	2.764		
.15	2.728	2.723	2.719		6.759	.25
.20	2.687	2.681		6.812	6.821	.20
.25	2.649		6.877	6.883	6.892	.15
		6.958	6.962	6.967	6.976	.10
	7.068	7.069	7.070	7.073	7.079	.05
MODE 2 $\leftarrow \mu_3$	.250	.200	.150	.100	.050	

TABLE 5.5 ($\mu_1 = .65$) MODE 3

$\mu_2 \downarrow$ \ $\mu_3 \rightarrow$	.050	.100	.150	.200	.250	$\mu_2 \uparrow$
.05	11.86	11.85	11.85	11.84	11.84	
.10	11.71	11.70	11.69	11.69		
.15	11.60	11.59	11.58		15.48	.25
.20	11.50	11.48		16.05	16.09	.20
.25	11.39		16.26	16.27	16.29	.15
		16.40	16.42	16.43	16.44	.10
	16.56	16.59	16.60	16.61	16.62	.05
MODE 4 $\leftarrow \mu_3$	.250	.200	.150	.100	.050	

TABLE 5.5 ($\mu_1 = .70$) MODE 1

$\mu_2 \downarrow$ \ $\mu_3 \rightarrow$	.050	.100	.150	.200	$\mu_2 \uparrow$
.05	2.626	2.624	2.622	2.622	
.10	2.582	2.578	2.576		
.15	2.543	2.538		6.353	.20
.20	2.506		6.408	6.417	.15
		6.479	6.483	6.491	.10
	6.573	6.574	6.577	6.582	.05
MODE 2 $\leftarrow \mu_3$	.200	.150	.100	.050	

TABLE 5.5 ($\mu_1 = .70$) MODE 3

$\mu_2 \downarrow$ \ $\mu_3 \rightarrow$	.050	.100	.150	.200	$\mu_2 \uparrow$
.05	11.03	11.02	11.01	11.01	
.10	10.89	10.88	10.88		
.15	10.79	10.78		15.01	.20
.20	10.71		15.15	15.16	.15
		15.27	15.28	15.29	.10
	15.43	15.43	15.44	15.45	.05
MODE 4 $\leftarrow \mu_3$	.200	.150	.100	.050	

TABLE 5.5 ($\mu_1 = .75$) MODE 1

$\mu_2 \downarrow$ \ $\mu_3 \rightarrow$	.050	.100	.150	$\mu_2 \uparrow$
.05	2.454	2.452	2.451	
.10	2.415	2.412		
.15	2.381		6.003	.15
		6.063	6.069	.10
	6.144	6.146	6.150	.05
MODE 2 $\leftarrow \mu_3$	.150	.100	.050	

TABLE 5.5 ($\mu_1 = .75$) MODE 3

$\mu_2 \downarrow$ \ $\mu_3 \rightarrow$	.050	.100	.150	$\mu_2 \uparrow$
.05	10.30	10.29	10.29	
.10	10.18	10.18		
.15	10.09		14.18	.15
		14.28	14.29	.10
	14.42	14.42	14.43	.05
MODE 4 $\leftarrow \mu_3$	.150	.100	.050	

TABLE 5.5 ($\mu_1 = .80$) MODE 1

$\mu_2 \downarrow$ \ $\mu_3 \rightarrow$	.050	.100	$\mu_2 \uparrow$
.05	2.303	2.302	
.10	2.269		
		5.700	.10
	5.768	5.772	.05
MODE 2 $\leftarrow \mu_3$	.100	.050	

TABLE 5.5 ($\mu_1 = .80$) MODE 3

$\mu_2 \downarrow$ \ $\mu_3 \rightarrow$	.050	.100	$\mu_2 \uparrow$
.05	9.666	9.660	
.10	9.561		
		13.41	.10
	13.53	13.54	.05
MODE 4 $\leftarrow \mu_3$	.100	.050	

TABLE 5.5 ($\mu_1 = .85$) MODE 1

$\mu_2 \downarrow$ \ $\mu_3 \rightarrow$	.050	$\mu_2 \uparrow$
.05	2.170	
	5.437	05
MODE 2 $\leftarrow \mu_3$	.050	

TABLE 5.5 ($\mu_1 = .85$) MODE 3

$\mu_2 \downarrow$ \ $\mu_3 \rightarrow$	.050	$\mu_2 \uparrow$
.05	9.105	
	12.75	.05
MODE 4 $\leftarrow \mu_3$	.050	

TABLE 5.6 ($\mu_1 = .05$)

MODE 1 ($\mu_3 \rightarrow$ across top, $\mu_2 \downarrow$ at left); MODE 2 ($\leftarrow \mu_3$ across bottom, $\mu_2 \uparrow$ at right)

μ_2	.050	.100	.150	.200	.250	.300	.350	.400	.450	.500	.550	.600	.650	.700	.750	.800	.850	μ_2
.05	4.546	4.763	5.012	5.299	5.631	6.016	6.464	6.973	7.489	7.725	7.441	7.002	6.582	6.210	5.887	5.610	5.372	
.10	4.823	5.065	5.348	5.677	6.063	6.515	7.040	7.615	8.047	7.895	7.430	6.959	6.541	6.181	5.872	5.610		
.15	5.137	5.411	5.736	6.119	6.573	7.108	7.722	8.316	8.409	7.946	7.414	6.937	6.528	6.180	5.887		8.178	.85
.20	5.495	5.810	6.188	6.639	7.179	7.818	8.519	8.928	8.555	7.958	7.407	6.937	6.541	6.210		8.581	8.673	.80
.25	5.908	6.275	6.720	7.259	7.910	8.672	9.354	9.228	8.589	7.957	7.413	6.959	6.581		9.051	9.126	9.235	.75
.30	6.389	6.821	7.353	8.006	8.794	9.621	9.788	9.222	8.551	7.944	7.428	7.001		9.592	9.661	9.750	9.876	.70
.35	6.955	7.471	8.113	8.893	9.587	9.585	9.324	8.910	8.400	7.890	7.438		10.20	10.28	10.36	10.47	10.61	.65
.40	7.629	8.242	8.835	8.861	8.762	8.641	8.492	8.295	8.031	7.715		10.84	10.97	11.08	11.18	11.29	11.45	.60
.45	8.331	8.217	8.089	7.983	7.888	7.797	7.703	7.597	7.473		11.11	11.59	11.85	12.00	12.11	12.01	10.79	.55
.50	7.610	7.453	7.336	7.243	7.164	7.093	7.026	6.959		10.41	11.29	12.21	12.78	12.95	11.90	10.53	9.494	.50
.55	6.942	6.809	6.708	6.628	6.562	6.505	6.454		9.669	10.39	11.34	12.44	12.96	11.70	10.34	9.300	8.563	.45
.60	6.380	6.266	6.180	6.111	6.055	6.009		9.483	9.875	10.50	11.37	12.18	11.47	10.22	9.327	9.260	9.448	.40
.65	5.902	5.804	5.730	5.671	5.625		10.12	10.30	10.54	10.93	11.47	11.28	10.34	10.16	10.28	10.46	10.72	.35
.70	5.490	5.406	5.343	5.294		11.01	11.29	11.53	11.74	11.93	11.66	11.22	11.39	11.59	11.79	11.95	11.50	.30
.75	5.133	5.061	5.008		10.97	11.58	12.23	12.83	13.04	12.01	11.48	12.14	12.88	12.98	12.17	11.34	10.65	.25
.80	4.820	4.760		10.41	11.00	11.70	12.47	12.89	11.84	10.97	11.38	12.45	12.84	12.06	11.23	10.50	9.905	.20
.85	4.544		9.849	10.37	11.00	11.70	12.27	11.62	10.59	10.51	11.34	12.24	11.90	11.12	10.40	9.778	9.257	.15
		9.356	9.823	10.38	11.00	11.59	11.37	10.35	9.897	10.39	11.30	11.64	11.02	10.32	9.688	9.147	8.690	.10
	8.936	9.356	9.851	10.41	10.98	11.07	10.18	9.510	9.680	10.42	11.14	10.88	10.24	9.618	9.071	8.597	8.191	.05
	.850	.800	.750	.700	.650	.600	.550	.500	.450	.400	.350	.300	.250	.200	.150	.100	.050	

MODE 2 $\leftarrow \mu_3$

TABLE 5.6 ($\mu_1 = .10$)

MODE 1 ($\mu_3 \rightarrow$ across top, $\mu_2 \downarrow$ at left); MODE 2 ($\leftarrow \mu_3$ across bottom, $\mu_2 \uparrow$ at right)

μ_2	.050	.100	.150	.200	.250	.300	.350	.400	.450	.500	.550	.600	.650	.700	.750	.800	μ_2
.05	4.826	5.072	5.357	5.688	6.075	6.529	7.058	7.646	8.128	8.035	7.570	7.086	6.655	6.284	5.966	5.696	
.10	5.139	5.416	5.742	6.127	6.583	7.120	7.742	8.363	8.518	8.068	7.528	7.040	6.621	6.266	5.966		
.15	5.497	5.814	6.193	6.646	7.189	7.832	8.546	9.007	8.666	8.064	7.503	7.023	6.620	6.282		8.547	.80
.20	5.910	6.278	6.725	7.266	7.921	8.690	9.412	9.347	8.703	8.057	7.501	7.037	6.652		8.970	9.064	.75
.25	6.390	6.824	7.358	8.015	8.813	9.687	9.994	9.408	8.691	8.053	7.519	7.080		9.458	9.534	9.643	.70
.30	6.956	7.475	8.123	8.926	9.826	10.13	9.816	9.252	8.620	8.042	7.553		10.01	10.08	10.16	10.29	.65
.35	7.632	8.258	9.023	9.498	9.439	9.305	9.108	8.813	8.413	7.979		10.61	10.68	10.76	10.86	10.99	.60
.40	8.441	8.826	8.732	8.627	8.528	8.425	8.307	8.162	7.976		11.19	11.33	11.43	11.52	11.62	11.75	.55
.45	8.208	8.041	7.918	7.820	7.735	7.655	7.576	7.493		11.30	11.83	12.08	12.21	12.30	12.08	10.80	.50
.50	7.474	7.330	7.221	7.135	7.063	7.000	6.942		10.51	11.39	12.31	12.80	12.93	11.91	10.54	9.499	.45
.55	6.848	6.724	6.630	6.556	6.495	6.444		9.846	10.50	11.42	12.48	12.95	11.71	10.36	9.351	9.117	.40
.60	6.314	6.206	6.125	6.062	6.012		9.942	10.22	10.70	11.47	12.21	11.50	10.28	9.838	9.935	10.13	.35
.65	5.854	5.762	5.693	5.640		10.69	10.83	11.01	11.27	11.64	11.39	10.77	10.79	10.91	11.06	11.27	.30
.70	5.455	5.376	5.319		11.36	11.67	11.88	12.03	12.15	11.87	11.61	11.80	11.96	12.09	12.13	11.51	.25
.75	5.107	5.042		11.09	11.71	12.35	12.80	12.94	12.06	11.63	12.30	12.83	12.92	12.19	11.35	10.65	.20
.80	4.802		10.50	11.09	11.80	12.55	12.94	11.87	11.05	11.52	12.57	12.87	12.07	11.23	10.51	9.907	.15
		9.948	10.48	11.11	11.82	12.37	11.65	10.65	10.65	11.52	12.36	11.93	11.14	10.41	9.784	9.260	.10
	9.476	9.953	10.52	11.16	11.75	11.43	10.40	10.03	10.60	11.53	11.71	11.04	10.33	9.700	9.157	8.696	.05
	.800	.750	.700	.650	.600	.550	.500	.450	.400	.350	.300	.250	.200	.150	.100	.050	

MODE 2 $\leftarrow \mu_3$

TABLE 5.6 ($\mu_1 = .05$)

MODE 3 ($\mu_3 \rightarrow$ across top, $\mu_2 \downarrow$ at left); MODE 4 ($\leftarrow \mu_3$ across bottom, $\mu_2 \uparrow$ at right)

μ_2	.050	.100	.150	.200	.250	.300	.350	.400	.450	.500	.550	.600	.650	.700	.750	.800	.850	μ_2
.05	11.84	12.45	13.15	13.94	14.60	13.80	12.92	13.47	14.45	14.30	13.38	12.90	13.97	14.54	13.86	13.15	12.53	
.10	12.56	13.25	14.05	14.92	15.18	13.83	13.70	14.71	15.25	14.31	13.53	14.25	15.30	14.62	13.83	13.14		
.15	13.38	14.16	15.08	15.97	15.32	14.20	14.92	16.02	15.42	14.39	14.58	16.06	15.48	14.61	13.85		15.39	.85
.20	14.32	15.20	16.12	15.99	15.00	15.10	16.18	15.95	15.14	14.92	16.24	15.91	15.22	14.50		16.18	16.31	.80
.25	14.62	14.24	13.97	13.86	14.58	14.72	14.26	13.97	14.24	14.81	14.40	14.07	13.80		17.08	17.20	17.35	.75
.30	12.42	12.49	13.45	14.62	15.14	13.93	13.20	14.15	15.21	14.36	13.49	12.84		17.88	18.18	18.33	18.31	.70
.35	12.56	13.57	14.82	16.06	15.32	13.96	14.46	16.12	15.41	14.30	13.36		16.75	18.65	19.37	17.89	15.37	.65
.40	13.78	15.00	16.11	15.90	15.10	14.87	16.13	15.89	15.19	14.27		16.45	17.04	18.71	17.47	16.35	16.50	.60
.45	15.01	14.84	14.67	14.49	14.55	15.00	14.78	14.58	14.34		17.46	17.67	17.95	17.68	17.53	17.69	17.89	.55
.50	13.67	13.47	13.36	14.00	15.07	14.14	13.61	13.38		17.70	18.72	19.11	18.30	18.41	19.02	18.97	17.15	.50
.55	12.49	12.49	14.19	16.21	15.32	13.80	12.87		16.45	17.77	19.23	17.82	16.95	18.81	18.76	16.87	15.39	.45
.60	12.56	14.49	16.04	15.85	15.12	13.78		16.64	17.05	17.89	17.45	16.77	17.32	18.16	16.83	16.62	16.87	.40
.65	14.87	15.09	14.96	14.81	14.52		17.88	18.35	18.63	18.19	18.18	18.54	18.61	17.97	18.38	18.65	18.13	.35
.70	14.23	14.08	13.96	13.86		17.25	18.39	19.48	18.06	17.82	19.25	18.91	17.49	18.78	19.31	17.93	16.68	.30
.75	13.31	13.18	13.09		16.32	17.28	18.36	17.62	16.64	17.80	18.46	16.97	16.91	18.54	17.75	16.50	15.45	.25
.80	12.51	12.40		17.05	17.43	17.92	17.59	17.20	17.63	18.08	17.19	17.34	17.92	17.71	17.02	17.30	17.78	.20
.85	11.80		17.78	18.71	19.57	18.26	18.33	19.41	18.98	17.94	19.12	19.37	18.14	18.60	19.56	18.54	17.51	.15
		16.94	17.85	18.79	17.84	17.33	18.42	18.67	17.23	17.80	18.93	17.90	17.15	18.76	18.39	17.36	16.44	.10
	16.14	16.95	17.82	17.44	16.39	17.28	18.06	16.86	16.50	17.77	17.67	16.60	16.81	18.05	17.23	16.31	15.49	.05
	.850	.800	.750	.700	.650	.600	.550	.500	.450	.400	.350	.300	.250	.200	.150	.100	.050	

MODE 4 $\leftarrow \mu_3$

TABLE 5.6 ($\mu_1 = .10$)

MODE 3 ($\mu_3 \rightarrow$ across top, $\mu_2 \downarrow$ at left); MODE 4 ($\leftarrow \mu_3$ across bottom, $\mu_2 \uparrow$ at right)

μ_2	.050	.100	.150	.200	.250	.300	.350	.400	.450	.500	.550	.600	.650	.700	.750	.800	μ_2
.05	12.57	13.26	14.06	14.95	15.35	14.08	13.77	14.75	15.38	14.50	13.68	14.30	15.41	14.76	13.97	13.28	
.10	13.39	14.17	15.03	15.17	14.90	14.23	14.87	15.19	14.99	14.39	14.56	15.20	15.05	14.57	13.91		
.15	14.29	14.37	14.28	14.19	14.10	14.38	14.35	14.25	14.14	14.25	14.39	14.29	14.19	14.03		14.73	.80
.20	13.55	13.40	13.30	13.44	14.24	13.91	13.51	13.37	13.86	14.17	13.69	13.43	13.29		15.26	15.36	.75
.25	12.48	12.50	13.43	14.57	14.87	13.70	13.10	14.09	14.99	14.17	13.35	12.78		15.89	15.97	16.09	.70
.30	12.56	13.57	14.79	15.53	14.90	13.85	14.40	15.53	15.06	14.17	13.31		16.57	16.72	16.83	16.96	.65
.35	13.78	14.84	14.87	14.73	14.47	14.56	14.90	14.79	14.57	14.08		16.27	17.41	17.73	17.63	15.00	.60
.40	14.24	14.10	14.01	13.92	14.08	14.31	14.11	13.99	13.86		15.40	16.30	18.02	17.33	15.22	15.28	.55
.45	13.41	13.28	13.21	13.91	14.80	13.87	13.41	13.23		15.96	16.13	16.59	16.79	15.96	16.04	16.18	.50
.50	12.59	12.56	14.17	15.54	14.95	13.69	12.87		16.69	16.98	17.16	17.14	16.81	17.01	17.16	17.07	.45
.55	12.56	14.46	15.02	14.91	14.60	13.69		16.09	17.28	18.15	17.62	16.51	17.79	18.25	16.83	15.29	.40
.60	14.48	14.40	14.32	14.23	14.08		15.39	16.14	17.35	17.20	15.59	16.35	17.78	16.61	15.21	15.18	.35
.65	13.85	13.74	13.67	13.59		16.04	16.23	16.59	16.76	16.10	16.28	16.73	16.49	16.04	16.17	16.37	.30
.70	13.21	13.11	13.03		16.73	17.31	17.65	17.49	16.99	17.47	17.73	17.06	17.16	17.58	17.69	16.66	.25
.75	12.58	12.49		15.97	16.93	17.97	17.47	16.26	17.34	18.21	16.86	16.39	18.00	17.68	16.47	15.42	.20
.80	11.98		15.16	15.93	16.89	16.99	15.41	16.02	17.18	16.58	15.45	16.17	17.28	16.32	15.26	14.55	.15
		15.38	15.53	15.99	16.38	15.42	15.53	16.05	16.26	15.43	15.52	16.15	16.13	15.38	15.39	15.49	.10
	16.47	16.55	16.60	16.62	16.51	16.57	16.61	16.59	16.52	16.58	16.62	16.54	16.53	16.58	16.62	16.43	.05
	.800	.750	.700	.650	.600	.550	.500	.450	.400	.350	.300	.250	.200	.150	.100	.050	

MODE 4 $\leftarrow \mu_3$

TABLE 5.6 ($\mu_1 = .15$) MODE 1

$\mu_2 \downarrow$ \ $\mu_3 \rightarrow$	.050	.100	.150	.200	.250	.300	.350	.400	.450	.500	.550	.600	.650	.700	.750	
.05	5.143	5.423	5.752	6.139	6.596	7.137	7.766	8.417	8.653	8.225	7.673	7.169	6.737	6.370	6.062	
.10	5.499	5.819	6.201	6.656	7.200	7.847	8.574	9.082	8.784	8.184	7.615	7.125	6.713	6.368		
.15	5.911	6.282	6.730	7.274	7.930	8.703	9.414	9.360	8.778	8.145	7.588	7.119	6.729		8.437	.75
.20	6.391	6.827	7.363	8.021	8.814	9.506	9.507	9.250	8.714	8.117	7.589	7.148		8.736	8.800	.70
.25	6.957	7.478	8.126	8.903	9.234	9.202	9.123	8.952	8.584	8.086	7.614		9.051	9.098	9.166	.65
.30	7.633	8.257	8.836	8.848	8.805	8.751	8.678	8.559	8.343	8.009		9.378	9.418	9.469	9.541	.60
.35	8.424	8.494	8.427	8.368	8.312	8.255	8.188	8.101	7.978		9.719	9.760	9.805	9.863	9.942	.55
.40	8.105	7.996	7.913	7.845	7.785	7.728	7.669	7.605		10.08	10.13	10.19	10.24	10.31	10.40	.50
.45	7.580	7.467	7.381	7.312	7.253	7.200	7.150		10.38	10.53	10.62	10.70	10.77	10.84	10.75	.45
.50	7.059	6.950	6.867	6.800	6.746	6.699		10.17	10.73	11.07	11.25	11.35	11.38	10.51	9.491	.40
.55	6.568	6.467	6.390	6.330	6.281		9.509	10.15	10.93	11.63	11.98	11.58	10.32	9.286	8.652	.35
.60	6.119	6.028	5.959	5.907		9.196	9.523	10.13	10.98	11.79	11.37	10.16	9.189	9.013	9.081	.30
.65	5.714	5.634	5.576		9.391	9.478	9.670	10.14	10.88	11.01	10.03	9.389	9.389	9.436	9.506	.25
.70	5.353	5.286		9.738	9.779	9.834	9.938	10.20	10.51	9.962	9.757	9.775	9.809	9.854	9.920	.20
.75	5.032		10.12	10.17	10.21	10.25	10.30	10.36	10.21	10.17	10.19	10.22	10.25	10.29	10.35	.15
		10.41	10.61	10.68	10.72	10.75	10.76	10.64	10.64	10.69	10.72	10.74	10.76	10.50	9.908	.10
	10.07	10.60	11.15	11.37	11.41	11.39	10.71	10.79	11.31	11.39	11.41	11.14	10.42	9.795	9.266	.05
	.750	.700	.650	.600	.550	.500	.450	.400	.350	.300	.250	.200	.150	.100	.050	$\mu_2 \uparrow$

MODE 2 $\leftarrow \mu_3$

TABLE 5.6 ($\mu_1 = .20$) MODE 1

$\mu_2 \downarrow$ \ $\mu_3 \rightarrow$	.050	.100	.150	.200	.250	.300	.350	.400	.450	.500	.550	.600	.650	.700	
.05	5.503	5.828	6.212	6.669	7.216	7.866	8.581	8.705	8.659	8.315	7.757	7.256	6.831	6.476	
.10	5.913	6.287	6.738	7.282	7.931	8.271	8.269	8.252	8.211	8.062	7.656	7.207	6.817		
.15	6.393	6.830	7.364	7.908	7.948	7.938	7.923	7.901	7.865	7.778	7.544	7.185		7.650	.70
.20	6.957	7.466	7.695	7.683	7.667	7.651	7.633	7.611	7.578	7.520	7.395		7.875	7.930	.65
.25	7.516	7.486	7.456	7.433	7.413	7.394	7.374	7.352	7.323	7.281		8.150	8.199	8.270	.60
.30	7.300	7.254	7.220	7.193	7.170	7.149	7.128	7.106	7.081		8.492	8.541	8.606	8.699	.55
.35	7.068	7.019	6.982	6.952	6.927	6.904	6.882	6.860		8.924	8.980	9.046	9.132	9.253	.50
.40	6.828	6.775	6.734	6.701	6.673	6.649	6.627		9.459	9.540	9.620	9.709	9.820	9.977	.45
.45	6.573	6.515	6.471	6.435	6.405	6.379		10.01	10.20	10.34	10.46	10.58	10.71	10.71	.40
.50	6.300	6.238	6.191	6.153	6.122		9.970	10.60	11.08	11.37	11.55	11.54	10.50	9.485	.35
.55	6.011	5.948	5.899	5.861		9.268	10.01	10.91	11.83	12.37	11.62	10.31	9.266	8.464	.30
.60	5.715	5.652	5.606		8.572	9.189	9.989	10.97	11.87	11.38	10.15	9.107	8.288	7.659	.25
.65	5.421	5.363		8.079	8.510	9.144	9.955	10.85	11.01	9.991	8.979	8.165	7.765	7.789	.20
.70	5.138		8.045	8.160	8.506	9.123	9.892	10.42	9.809	8.867	8.107	8.007	8.026	8.064	.15
		8.315	8.335	8.382	8.572	9.114	9.720	9.572	8.762	8.322	8.320	8.332	8.352	8.387	.10
	8.739	8.744	8.751	8.763	8.804	9.082	9.253	8.772	8.743	8.746	8.751	8.757	8.769	8.791	.05
	.700	.650	.600	.550	.500	.450	.400	.350	.300	.250	.200	.150	.100	.050	$\mu_2 \uparrow$

MODE 2 $\leftarrow \mu_3$

TABLE 5.6 ($\mu_1 = .15$) MODE 3

$\mu_2 \downarrow$ \ $\mu_3 \rightarrow$	.050	.100	.150	.200	.250	.300	.350	.400	.450	.500	.550	.600	.650	.700	.750	
.05	11.51	11.47	11.46	11.46	11.99	12.59	11.96	11.52	11.47	11.76	12.59	12.10	11.59	11.48	11.45	
.10	10.87	10.83	11.27	12.12	13.01	13.05	11.95	11.35	11.96	13.18	12.89	12.10	11.42	10.99		
.15	10.67	11.37	12.23	13.23	14.06	13.21	12.10	12.25	13.66	13.86	12.96	12.12	11.44		14.20	.75
.20	11.53	12.37	13.39	14.54	14.71	13.31	12.74	14.02	14.89	13.97	12.99	12.19		14.97	15.11	.70
.25	12.55	13.55	14.74	15.31	14.50	13.52	14.29	15.34	14.77	13.88	13.02		15.88	16.01	16.17	.65
.30	13.75	14.12	13.94	13.72	13.50	13.93	14.10	13.89	13.64	13.31		16.83	17.06	17.21	17.37	.60
.35	12.88	12.67	12.52	12.50	13.50	13.55	12.94	12.64	12.46		16.32	17.89	18.40	17.80	14.98	.55
.40	11.77	11.64	12.06	13.81	14.64	13.37	12.32	11.82		15.18	16.36	18.34	17.37	14.89	14.90	.50
.45	11.04	12.22	14.12	15.38	14.67	13.32	12.16		16.00	16.25	16.81	16.90	15.98	16.12	16.35	.45
.50	12.54	14.31	14.49	14.32	13.99	13.23		17.00	17.52	17.80	17.61	17.26	17.62	17.84	17.12	.40
.55	13.68	13.52	13.40	13.29	13.13		16.11	17.47	18.79	17.72	16.58	18.30	18.65	16.84	15.29	.35
.60	12.75	12.61	12.51	12.43		15.19	16.17	17.48	17.24	15.51	16.43	17.91	16.62	15.10	14.53	.30
.65	11.98	11.87	11.79		16.15	16.47	16.94	17.00	16.25	16.56	17.13	16.62	16.17	16.43	16.81	.25
.70	11.35	11.27		16.91	17.89	18.70	17.86	17.34	18.42	18.63	17.24	17.87	18.80	17.89	16.67	.20
.75	10.85		15.97	17.00	18.14	17.52	16.28	17.47	18.33	16.87	16.43	18.25	17.71	16.47	15.41	.15
		15.01	15.92	16.94	17.02	15.34	16.01	17.26	16.59	15.40	16.19	17.32	16.32	15.24	14.33	.10
	14.18	14.98	15.90	16.38	14.76	14.83	15.96	16.25	15.03	14.63	16.09	16.12	15.12	14.19	13.40	.05
	.750	.700	.650	.600	.550	.500	.450	.400	.350	.300	.250	.200	.150	.100	.050	$\mu_2 \uparrow$

MODE 4 $\leftarrow \mu_3$

TABLE 5.6 ($\mu_1 = .20$) MODE 3

$\mu_2 \downarrow$ \ $\mu_3 \rightarrow$	.050	.100	.150	.200	.250	.300	.350	.400	.450	.500	.550	.600	.650	.700	
.05	9.920	10.53	11.27	12.13	13.05	13.23	12.08	11.35	11.99	13.27	13.03	12.20	11.44	10.81	
.10	10.66	11.37	12.24	13.25	14.14	13.37	12.19	12.29	13.71	13.97	13.06	12.18	11.46		
.15	11.53	12.37	13.40	14.57	14.91	13.45	12.80	14.06	15.04	14.10	13.08	12.24		14.87	.70
.20	12.55	13.56	14.79	15.95	15.04	13.72	14.39	15.98	15.22	14.10	13.14		15.76	15.92	.65
.25	13.77	14.94	15.39	15.05	14.44	14.58	15.42	15.14	14.62	13.90		16.80	16.94	17.11	.60
.30	13.90	13.60	13.38	13.24	13.82	14.16	13.71	13.41	13.18		17.75	18.08	18.25	18.26	.55
.35	12.22	12.01	12.19	13.87	14.86	13.63	12.65	12.15		16.63	18.62	19.38	17.89	15.04	.50
.40	11.04	12.24	14.16	16.03	15.09	13.51	12.30		16.02	16.85	18.70	17.45	15.82	16.00	.45
.45	12.54	14.45	15.48	15.26	14.66	13.50		17.20	17.44	17.80	17.52	17.25	17.45	17.69	.40
.50	14.46	14.32	14.17	14.01	13.77		17.63	18.69	19.10	18.26	18.35	19.01	18.96	17.14	.35
.55	13.33	13.16	13.03	12.92		16.33	17.74	19.23	17.81	16.85	18.80	18.76	16.86	15.31	.30
.60	12.33	12.18	12.08		16.12	16.76	17.82	17.39	16.29	17.09	18.13	16.71	15.92	16.23	.25
.65	11.48	11.36		17.71	18.23	18.57	18.09	18.03	18.45	18.56	17.80	18.24	18.58	18.12	.20
.70	10.76		17.18	18.36	19.49	18.04	17.76	19.25	18.90	17.44	18.75	19.31	17.93	16.68	.15
		16.10	17.15	18.31	17.58	16.42	17.69	18.43	16.92	16.64	18.49	17.73	16.48	15.42	.10
	15.15	16.08	17.12	17.10	15.46	16.22	17.46	16.63	15.50	16.51	17.41	16.34	15.26	14.34	.05
	.700	.650	.600	.550	.500	.450	.400	.350	.300	.250	.200	.150	.100	.050	$\mu_2 \uparrow$

MODE 4 $\leftarrow \mu_3$

TABLE 5.6 ($\mu_1 = .25$)

MODE 1 (top/left: $\mu_2 \downarrow$, $\mu_3 \rightarrow$); MODE 2 (bottom/right: $\mu_2 \uparrow$, $\leftarrow \mu_3$)

μ_2 ↓ \ μ_3 →	.050	.100	.150	.200	.250	.300	.350	.400	.450	.500	.550	.600	.650	
.05	5.919	6.297	6.749	7.080	7.082	7.080	7.078	7.076	7.073	7.069	7.061	7.038	6.893	
.10	6.394	6.773	6.791	6.784	6.777	6.771	6.766	6.761	6.755	6.747	6.736	6.715		
.15	6.605	6.576	6.558	6.545	6.536	6.527	6.520	6.512	6.504	6.495	6.483		7.242	.65
.20	6.411	6.381	6.360	6.345	6.333	6.323	6.314	6.305	6.296	6.286		7.568	7.650	.60
.25	6.237	6.207	6.185	6.168	6.155	6.144	6.134	6.124	6.115		7.979	8.054	8.161	.55
.30	6.077	6.046	6.024	6.006	5.992	5.980	5.970	5.960		8.493	8.566	8.663	8.798	.50
.35	5.925	5.894	5.871	5.853	5.838	5.825	5.814		9.130	9.211	9.306	9.427	9.595	.45
.40	5.777	5.745	5.721	5.702	5.686	5.673		9.881	10.00	10.12	10.24	10.39	10.59	.40
.45	5.627	5.595	5.570	5.550	5.534		10.50	10.85	11.08	11.26	11.41	11.52	10.70	.35
.50	5.473	5.440	5.414	5.393		10.14	10.97	11.80	12.34	12.58	11.85	10.51	9.487	.30
.55	5.312	5.277	5.251		9.322	10.11	11.11	12.25	12.86	11.67	10.32	9.268	8.464	.25
.60	5.143	5.108		8.576	9.238	10.08	11.11	12.03	11.42	10.16	9.110	8.286	7.640	.20
.65	4.965		7.952	8.504	9.198	10.05	10.97	11.06	10.00	8.984	8.156	7.495	6.966	.15
		7.444	7.904	8.484	9.193	10.00	10.51	9.832	8.873	8.055	7.390	6.880	6.840	.10
	7.149	7.422	7.907	8.510	9.214	9.845	9.617	8.767	7.970	7.310	7.097	7.102	7.117	.05
← μ_3	.650	.600	.550	.500	.450	.400	.350	.300	.250	.200	.150	.100	.050	μ_2 ↑

TABLE 5.6 ($\mu_1 = .30$)

MODE 1 (top/left: $\mu_2 \downarrow$, $\mu_3 \rightarrow$); MODE 2 (bottom/right: $\mu_2 \uparrow$, $\leftarrow \mu_3$)

μ_2 ↓ \ μ_3 →	.050	.100	.150	.200	.250	.300	.350	.400	.450	.500	.550	.600	
.05	5.977	5.966	5.960	5.957	5.955	5.953	5.952	5.951	5.950	5.948	5.947	5.946	
.10	5.773	5.754	5.743	5.736	5.731	5.727	5.724	5.721	5.718	5.715	5.712		
.15	5.603	5.581	5.567	5.557	5.550	5.544	5.539	5.535	5.531	5.527		7.367	.60
.20	5.456	5.433	5.417	5.406	5.397	5.390	5.384	5.379	5.374		7.804	7.912	.55
.25	5.324	5.301	5.285	5.272	5.263	5.255	5.248	5.243		8.344	8.438	8.574	.50
.30	5.204	5.181	5.165	5.152	5.142	5.134	5.127		9.005	9.095	9.213	9.379	.45
.35	5.092	5.069	5.053	5.040	5.030	5.021		9.802	9.903	10.02	10.16	10.36	.40
.40	4.985	4.963	4.946	4.934	4.924		10.68	10.85	11.00	11.14	11.30	11.51	.35
.45	4.001	4.060	4.843	4.831		11.06	11.67	12.03	12.24	12.37	12.07	10.79	.30
.50	4.779	4.757	4.741		10.31	11.26	12.34	13.05	13.19	11.90	10.52	9.489	.25
.55	4.675	4.654		9.420	10.24	11.29	12.51	13.02	11.70	10.33	9.274	8.465	.20
.60	4.569		8.660	9.340	10.20	11.26	12.17	11.45	10.17	9.119	8.291	7.642	.15
		8.035	8.608	9.323	10.20	11.12	11.12	10.03	8.996	8.165	7.500	6.966	.10
	7.525	8.018	8.625	9.363	10.19	10.64	9.870	8.893	8.069	7.399	6.853	6.404	.05
← μ_3	.600	.550	.500	.450	.400	.350	.300	.250	.200	.150	.100	.050	μ_2 ↑

TABLE 5.6 ($\mu_1 = .25$)

MODE 3 (top/left: $\mu_2 \downarrow$, $\mu_3 \rightarrow$); MODE 4 (bottom/right: $\mu_2 \uparrow$, $\leftarrow \mu_3$)

μ_2 ↓ \ μ_3 →	.050	.100	.150	.200	.250	.300	.350	.400	.450	.500	.550	.600	.650	
.05	10.67	11.39	12.26	13.28	14.24	13.64	12.39	12.36	13.78	14.15	13.23	12.34	11.60	
.10	11.53	12.38	13.41	14.60	15.10	13.66	12.90	14.10	15.18	14.24	13.21	12.36		
.15	12.56	13.57	14.82	16.07	15.35	13.90	14.45	16.14	15.44	14.27	13.27		15.50	.65
.20	13.78	15.00	16.16	16.01	15.18	14.89	16.17	15.99	15.26	14.26		16.21	16.31	.60
.25	15.12	15.00	14.80	14.57	14.61	15.14	14.92	14.67	14.36		16.76	16.82	16.89	.55
.30	13.49	13.22	13.08	13.99	15.11	14.14	13.46	13.13		17.12	17.21	17.26	17.31	.50
.35	11.91	12.31	14.19	16.21	15.36	13.77	12.67		16.66	17.52	17.67	17.63	15.49	.45
.40	12.55	14.49	16.11	15.95	15.19	13.75		16.40	16.80	17.89	17.36	16.33	16.43	.40
.45	14.89	15.21	15.08	14.90	14.56		16.89	16.97	17.12	17.09	16.91	16.96	17.03	.35
.50	14.22	14.03	13.90	13.77		17.17	17.35	17.43	17.40	17.28	17.38	17.44	17.13	.30
.55	13.10	12.94	12.82		16.42	17.39	17.91	17.62	16.80	17.71	17.99	16.85	15.47	.25
.60	12.13	11.99		16.47	16.70	17.39	17.23	16.53	16.84	17.72	16.70	16.44	16.55	.20
.65	11.28		16.97	17.03	17.10	17.10	17.00	17.05	17.14	17.00	17.00	17.04	17.10	.15
		17.14	17.38	17.43	17.42	17.31	17.41	17.44	17.27	17.36	17.42	17.44	16.68	.10
	16.26	17.30	17.86	17.61	16.61	17.70	17.88	16.97	16.91	17.85	17.73	16.50	15.43	.05
← μ_3	.650	.600	.550	.500	.450	.400	.350	.300	.250	.200	.150	.100	.050	μ_2 ↑

TABLE 5.6 ($\mu_1 = .30$)

MODE 3 (top/left: $\mu_2 \downarrow$, $\mu_3 \rightarrow$); MODE 4 (bottom/right: $\mu_2 \uparrow$, $\leftarrow \mu_3$)

μ_2 ↓ \ μ_3 →	.050	.100	.150	.200	.250	.300	.350	.400	.450	.500	.550	.600	
.05	11.54	12.39	13.43	14.63	14.93	13.93	13.05	14.16	14.95	14.42	13.40	12.52	
.10	12.56	13.58	14.62	14.62	14.56	14.01	14.42	14.63	14.59	14.29	13.39		
.15	13.78	14.38	14.36	14.33	14.28	14.32	14.38	14.35	14.30	14.13		14.46	.60
.20	14.13	14.08	14.04	14.01	14.07	14.17	14.09	14.03	13.98		14.76	14.82	.55
.25	13.69	13.59	13.53	13.94	14.61	13.95	13.66	13.53		15.25	15.33	15.45	.50
.30	12.81	12.72	14.16	14.93	14.63	13.78	13.06		16.04	16.23	16.38	16.57	.45
.35	12.57	14.42	14.57	14.52	14.39	13.77		16.02	17.26	17.71	17.60	14.97	.40
.40	14.32	14.28	14.24	14.20	14.12		14.89	16.07	18.08	17.32	14.75	14.71	.35
.45	13.98	13.92	13.87	13.81		15.23	15.45	16.25	16.69	15.21	15.26	15.40	.30
.50	13.47	13.37	13.29		16.17	16.49	16.78	16.84	16.26	16.50	16.75	16.92	.25
.55	12.73	12.60		15.88	17.17	18.36	17.60	16.30	17.80	18.45	16.82	15.28	.20
.60	11.90		14.78	15.85	17.26	17.16	15.20	16.00	17.74	16.58	15.04	14.43	.15
		14.69	14.84	15.78	16.42	14.76	14.78	15.85	16.23	14.88	14.68	14.70	.10
	15.02	15.03	15.08	15.47	15.03	15.03	15.07	15.46	15.03	15.02	15.04	15.06	.05
← μ_3	.600	.550	.500	.450	.400	.350	.300	.250	.200	.150	.100	.050	μ_2 ↑

TABLE 5.5 ($\mu_1 = .35$) MODE 1 / MODE 2

μ_2 ↓ \ μ_3 →	.050	.100	.150	.200	.250	.300	.350	.400	.450	.500	.550	↑ μ_2
.05	5.154	5.146	5.142	5.139	5.137	5.136	5.135	5.134	5.134	5.133	5.132	
.10	4.999	4.984	4.976	4.970	4.966	4.963	4.961	4.959	4.957	4.956		
.15	4.867	4.850	4.839	4.832	4.826	4.821	4.818	4.815	4.812		7.766	.55
.20	4.751	4.733	4.721	4.712	4.706	4.700	4.696	4.692		8.295	8.419	.50
.25	4.647	4.629	4.617	4.607	4.600	4.594	4.589		8.937	9.045	9.195	.45
.30	4.552	4.535	4.522	4.513	4.505	4.499		9.698	9.797	9.920	10.09	.40
.35	4.464	4.448	4.435	4.425	4.418		10.53	10.63	10.73	10.85	11.00	.35
.40	4.382	4.365	4.353	4.344		11.22	11.35	11.44	11.51	11.58	11.67	.30
.45	4.303	4.287	4.276		11.36	11.75	11.86	11.91	11.95	11.99	12.04	.25
.50	4.226	4.212		10.52	11.51	12.07	12.16	12.19	12.21	12.22	11.44	.20
.55	4.151		9.595	10.48	11.57	12.29	12.38	12.40	12.36	11.20	10.21	.15
		8.833	9.574	10.50	11.64	12.51	12.60	12.29	11.03	10.01	9.215	.10
	8.220	8.852	9.629	10.59	11.75	12.68	12.10	10.90	9.882	9.062	8.402	.05
← μ_3 (MODE 2)	.550	.500	.450	.400	.350	.300	.250	.200	.150	.100	.050	

TABLE 5.5 ($\mu_1 = .35$) MODE 3 / MODE 4

μ_2 ↓ \ μ_3 →	.050	.100	.150	.200	.250	.300	.350	.400	.450	.500	.550	↑ μ_2
.05	12.96	12.95	12.94	12.93	12.93	13.03	12.96	12.94	12.93	12.92	12.90	
.10	12.68	12.66	12.65	12.66	13.65	13.24	12.70	12.66	12.64	12.62		
.15	12.47	12.45	12.57	14.13	14.80	13.30	12.53	12.45	12.42		13.79	.55
.20	12.28	12.74	14.43	16.11	15.03	13.30	12.40	12.26		14.67	14.84	.50
.25	13.03	14.68	15.73	15.38	14.54	13.25	12.34		15.84	16.00	16.22	.45
.30	14.39	14.17	13.99	13.81	13.55	13.06		17.12	17.45	17.67	17.92	.40
.35	13.23	13.10	13.02	12.95	12.87		16.48	18.49	19.28	19.52	18.03	.35
.40	12.68	12.62	12.58	12.55		15.06	16.59	19.10	20.34	17.57	15.20	.30
.45	12.38	12.34	12.31		16.03	16.42	17.27	18.88	17.29	16.27	16.58	.25
.50	12.14	12.10		17.32	18.28	18.81	19.14	18.35	18.63	18.99	18.95	.20
.55	11.84		16.15	17.70	19.53	20.37	18.16	19.10	20.40	18.73	16.99	.15
		14.81	16.12	17.75	19.29	17.55	16.81	18.99	18.45	16.74	15.33	.10
	13.73	14.82	16.17	17.67	17.12	15.48	16.59	17.91	16.54	15.13	13.98	.05
← μ_3 (MODE 4)	.550	.500	.450	.400	.350	.300	.250	.200	.150	.100	.050	

TABLE 5.5 ($\mu_1 = .40$) MODE 1 / MODE 2

μ_2 ↓ \ μ_3 →	.050	.100	.150	.200	.250	.300	.350	.400	.450	.500	↑ μ_2
.05	4.531	4.524	4.521	4.519	4.517	4.516	4.516	4.515	4.515	4.514	
.10	4.408	4.397	4.390	4.386	4.382	4.380	4.378	4.377	4.375		
.15	4.302	4.289	4.280	4.274	4.270	4.266	4.264	4.261		8.268	.50
.20	4.209	4.195	4.185	4.178	4.173	4.169	4.165		8.838	8.962	.45
.25	4.124	4.110	4.100	4.093	4.087	4.082		9.452	9.539	9.654	.40
.30	4.047	4.033	4.023	4.016	4.010		9.988	10.04	10.10	10.17	.35
.35	3.976	3.963	3.953	3.945		10.34	10.37	10.40	10.44	10.48	.30
.40	3.909	3.896	3.887		10.56	10.58	10.60	10.62	10.64	10.67	.25
.45	3.846	3.834		10.69	10.74	10.76	10.77	10.79	10.80	10.83	.20
.50	3.785		10.55	10.86	10.90	10.92	10.93	10.94	10.96	10.98	.15
		9.734	10.60	11.05	11.09	11.10	11.11	11.12	11.10	10.21	.10
	9.002	9.769	10.71	11.32	11.34	11.35	11.36	11.05	10.03	9.224	.05
← μ_3 (MODE 2)	.500	.450	.400	.350	.300	.250	.200	.150	.100	.050	

TABLE 5.5 ($\mu_1 = .40$) MODE 3 / MODE 4

μ_2 ↓ \ μ_3 →	.050	.100	.150	.200	.250	.300	.350	.400	.450	.500	↑ μ_2
.05	11.38	11.37	11.37	12.35	13.74	13.43	12.02	11.39	11.37	11.36	
.10	11.16	11.25	12.53	14.16	15.00	13.45	11.96	11.21	11.14		
.15	11.45	12.74	14.45	16.32	15.36	13.44	11.96	11.11		14.59	.50
.20	13.03	14.73	16.86	16.85	15.22	13.42	12.02		15.75	15.95	.45
.25	15.08	15.69	15.44	15.09	14.42	13.28		17.05	17.20	17.38	.40
.30	14.00	13.72	13.51	13.32	13.07		17.92	18.11	18.20	18.27	.35
.35	12.50	12.31	12.18	12.07		16.83	18.42	18.58	18.63	18.04	.30
.40	11.60	11.50	11.44		16.15	17.15	18.66	18.95	17.61	16.16	.25
.45	11.15	11.10		17.54	17.75	18.00	18.60	17.78	17.76	17.91	.20
.50	10.90		17.80	18.35	18.43	18.48	18.36	18.40	18.45	18.49	.15
		16.34	17.92	18.69	18.74	18.25	18.64	18.73	18.68	17.00	.10
	15.02	16.36	18.02	19.01	17.62	17.13	18.95	18.48	16.76	15.35	.05
← μ_3 (MODE 4)	.500	.450	.400	.350	.300	.250	.200	.150	.100	.050	

TABLE 5.5 ($\mu_1 = .45$) MODE 1 / MODE 2

μ_2 ↓ \ μ_3 →	.050	.100	.150	.200	.250	.300	.350	.400	.450	↑ μ_2
.05	4.042	4.036	4.034	4.032	4.031	4.030	4.030	4.029	4.029	
.10	3.943	3.933	3.928	3.924	3.922	3.920	3.918	3.917		
.15	3.856	3.845	3.838	3.833	3.830	3.827	3.825		8.605	.45
.20	3.779	3.767	3.760	3.754	3.750	3.746		8.958	9.016	.40
.25	3.709	3.697	3.689	3.683	3.679		9.209	9.237	9.274	.35
.30	3.645	3.633	3.625	3.619		9.380	9.396	9.416	9.441	.30
.35	3.585	3.574	3.567		9.511	9.522	9.534	9.549	9.570	.25
.40	3.529	3.520		9.627	9.637	9.646	9.656	9.669	9.688	.20
.45	3.477		9.746	9.757	9.766	9.774	9.783	9.794	9.812	.15
		9.881	9.903	9.912	9.919	9.925	9.932	9.942	9.958	.10
	9.891	10.10	10.12	10.12	10.12	10.13	10.13	10.14	10.14	.05
← μ_3 (MODE 2)	.450	.400	.350	.300	.250	.200	.150	.100	.050	

TABLE 5.5 ($\mu_1 = .45$) MODE 3 / MODE 4

μ_2 ↓ \ μ_3 →	.050	.100	.150	.200	.250	.300	.350	.400	.450	↑ μ_2
.05	10.24	11.24	12.55	14.22	15.27	13.76	12.17	10.95	10.20	
.10	11.46	12.75	14.48	16.38	15.63	13.67	12.11	10.95		
.15	13.03	14.75	16.45	16.41	15.59	13.65	12.16		15.49	.45
.20	15.11	16.15	16.09	15.96	15.27	13.66		16.14	16.21	.40
.25	15.38	15.15	14.96	14.72	14.29		16.47	16.51	16.55	.35
.30	13.69	13.42	13.22	13.04		16.71	16.76	16.81	16.86	.30
.35	12.15	11.94	11.79		16.54	17.13	17.33	17.48	17.59	.25
.40	11.03	10.89		16.24	16.57	18.07	19.13	17.51	16.25	.20
.45	10.33		16.49	16.53	16.67	18.27	17.12	16.52	16.54	.15
		16.69	16.72	16.74	16.81	16.80	16.72	16.74	16.76	.10
	16.55	16.97	16.99	17.00	17.00	16.98	16.99	17.00	16.97	.05
← μ_3 (MODE 4)	.450	.400	.350	.300	.250	.200	.150	.100	.050	

TABLE 5.5 ($\mu_1 = .50$) MODE 1 / MODE 2

μ_2 ↓ \ μ_3 →	.050	.100	.150	.200	.250	.300	.350	.400	↑ μ_2
.05	3.648	3.644	3.642	3.640	3.639	3.639	3.638	3.638	
.10	3.566	3.559	3.554	3.551	3.549	3.548	3.546		
.15	3.494	3.485	3.479	3.475	3.472	3.470		8.316	.40
.20	3.430	3.420	3.413	3.409	3.405		8.440	8.462	.35
.25	3.370	3.361	3.354	3.349		8.541	8.554	8.572	.30
.30	3.316	3.307	3.300		8.633	8.642	8.653	8.669	.25
.35	3.265	3.257		8.725	8.731	8.739	8.750	8.765	.20
.40	3.218		8.827	8.833	8.839	8.846	8.855	8.870	.15
		8.954	8.959	8.963	8.968	8.973	8.981	8.995	.10
	9.129	9.132	9.134	9.136	9.138	9.141	9.145	9.154	.05
← μ_3 (MODE 2)	.400	.350	.300	.250	.200	.150	.100	.050	

TABLE 5.5 ($\mu_1 = .50$) MODE 3 / MODE 4

μ_2 ↓ \ μ_3 →	.050	.100	.150	.200	.250	.300	.350	.400	↑ μ_2
.05	11.47	12.78	14.52	15.31	15.28	14.00	12.38	11.15	
.10	13.04	14.76	15.09	15.08	15.02	13.89	12.36		
.15	14.93	14.92	14.91	14.88	14.82	13.90		14.85	.40
.20	14.75	14.71	14.68	14.64	14.54		15.05	15.08	.35
.25	14.36	14.26	14.16	14.06		15.36	15.42	15.52	.30
.30	13.32	13.10	12.93		16.15	16.40	16.63	16.93	.25
.35	11.91	11.71		16.12	17.91	18.81	19.18	18.00	.20
.40	10.74		15.01	16.17	18.90	20.34	17.54	15.15	.15
		15.11	15.15	16.18	18.59	17.15	15.13	15.13	.10
	15.32	15.33	15.35	16.16	16.62	15.33	15.34	15.35	.05
← μ_3 (MODE 4)	.400	.350	.300	.250	.200	.150	.100	.050	

TABLE 5.5 ($\mu_1 = .55$) MODE 1 / MODE 2

μ_2 ↓ \ μ_3 →	.050	.100	.150	.200	.250	.300	.350	↑ μ_2
.05	3.324	3.321	3.319	3.318	3.317	3.317	3.316	
.10	3.256	3.249	3.246	3.243	3.242	3.240		
.15	3.195	3.187	3.182	3.179	3.177		7.762	.35
.20	3.140	3.132	3.126	3.122		7.831	7.844	.30
.25	3.089	3.081	3.076		7.900	7.909	7.922	.25
.30	3.042	3.035		7.976	7.982	7.991	8.004	.20
.35	2.999		8.063	8.068	8.074	8.082	8.095	.15
		8.173	8.176	8.179	8.184	8.191	8.203	.10
	8.321	8.322	8.324	8.325	8.328	8.331	8.339	.05
← μ_3 (MODE 2)	.350	.300	.250	.200	.150	.100	.050	

TABLE 5.5 ($\mu_1 = .55$) MODE 3 / MODE 4

μ_2 ↓ \ μ_3 →	.050	.100	.150	.200	.250	.300	.350	↑ μ_2
.05	13.06	13.96	13.96	13.95	13.94	13.88	12.64	
.10	13.78	13.77	13.76	13.75	13.73	13.66		
.15	13.64	13.62	13.60	13.59	13.57		13.93	.35
.20	13.49	13.46	13.44	13.42		14.50	14.67	.30
.25	13.27	13.22	13.18		15.94	16.18	16.50	.25
.30	12.74	12.61		17.69	18.15	18.36	18.54	.20
.35	11.68		16.49	18.77	19.04	19.09	18.04	.15
		14.37	16.46	19.10	19.31	17.57	15.15	.10
	13.98	14.37	16.55	18.94	17.20	14.81	13.98	.05
← μ_3 (MODE 4)	.350	.300	.250	.200	.150	.100	.050	

TABLE 5.6 ($\mu_1 = .60$)

$\mu_2 \downarrow$ / $\mu_3 \rightarrow$ MODE 1	.050	.100	.150	.200	.250	.300	$\uparrow \mu_2$
.05	3.054	3.050	3.049	3.048	3.047	3.047	
.10	2.995	2.990	2.986	2.984	2.983		
.15	2.943	2.936	2.932	2.930		7.228	.30
.20	2.895	2.888	2.884		7.283	7.294	.25
.25	2.852	2.845		7.346	7.353	7.365	.20
.30	2.811		7.421	7.426	7.433	7.444	.15
		7.516	7.519	7.523	7.529	7.539	.10
	7.643	7.644	7.645	7.647	7.651	7.657	.05
MODE 2 $\leftarrow \mu_3$	.300	.250	.200	.150	.100	.050	

TABLE 5.6 ($\mu_1 = .60$)

$\mu_2 \downarrow$ / $\mu_3 \rightarrow$ MODE 3	.050	.100	.150	.200	.250	.300	$\uparrow \mu_2$
.05	12.57	12.82	12.82	12.81	12.81	12.80	
.10	12.66	12.65	12.64	12.63	12.63		
.15	12.54	12.52	12.51	12.50		14.20	.30
.20	12.41	12.40	12.38		15.79	16.07	.25
.25	12.26	12.24		17.12	17.21	17.29	.20
.30	11.99		17.50	17.55	17.57	17.60	.15
		16.67	17.71	17.75	17.73	14.97	.10
	14.52	16.65	17.94	17.47	14.58	12.84	.05
MODE 4 $\leftarrow \mu_3$	.300	.250	.200	.150	.100	.050	

TABLE 5.6 ($\mu_1 = .65$)

$\mu_2 \downarrow$ / $\mu_3 \rightarrow$ MODE 1	.050	.100	.150	.200	.250	$\uparrow \mu_2$
.05	2.823	2.821	2.820	2.819	2.818	
.10	2.773	2.768	2.766	2.764		
.15	2.728	2.722	2.719		6.759	.25
.20	2.687	2.681		6.812	6.821	.20
.25	2.648		6.876	6.882	6.892	.15
		6.958	6.961	6.967	6.975	.10
	7.068	7.069	7.070	7.073	7.079	.05
MODE 2 $\leftarrow \mu_3$	.250	.200	.150	.100	.050	

TABLE 5.6 ($\mu_1 = .65$)

$\mu_2 \downarrow$ / $\mu_3 \rightarrow$ MODE 3	.050	.100	.150	.200	.250	$\uparrow \mu_2$
.05	11.86	11.85	11.85	11.84	11.84	
.10	11.71	11.70	11.69	11.69		
.15	11.60	11.59	11.58		15.47	.25
.20	11.50	11.48		16.05	16.09	.20
.25	11.38		16.26	16.27	16.29	.15
		16.40	16.41	16.43	16.44	.10
	16.54	16.59	16.60	16.61	15.00	.05
MODE 4 $\leftarrow \mu_3$	.250	.200	.150	.100	.050	

TABLE 5.6 ($\mu_1 = .70$)

$\mu_2 \downarrow$ / $\mu_3 \rightarrow$ MODE 1	.050	.100	.150	.200	$\uparrow \mu_2$
.05	2.626	2.623	2.622	2.622	
.10	2.582	2.578	2.576		
.15	2.542	2.538		6.353	.20
.20	2.506		6.408	6.416	.15
		6.479	6.483	6.491	.10
	6.573	6.574	6.577	6.582	.05
MODE 2 $\leftarrow \mu_3$	.200	.150	.100	.050	

TABLE 5.6 ($\mu_1 = .70$)

$\mu_2 \downarrow$ / $\mu_3 \rightarrow$ MODE 3	.050	.100	.150	.200	$\uparrow \mu_2$
.05	11.02	11.02	11.01	11.01	
.10	10.89	10.88	10.88		
.15	10.79	10.78		15.01	.20
.20	10.70		15.15	15.16	.15
		15.27	15.28	15.29	.10
	15.43	15.43	15.44	15.45	.05
MODE 4 $\leftarrow \mu_3$	.200	.150	.100	.050	

TABLE 5.6 ($\mu_1 = .75$)

$\mu_2 \downarrow$ / $\mu_3 \rightarrow$ MODE 1	.050	.100	.150	$\uparrow \mu_2$
.05	2.454	2.452	2.451	
.10	2.415	2.412		
.15	2.381		6.003	.15
		6.063	6.069	.10
	6.144	6.146	6.150	.05
MODE 2 $\leftarrow \mu_3$	.150	.100	.050	

TABLE 5.6 ($\mu_1 = .75$)

$\mu_2 \downarrow$ / $\mu_3 \rightarrow$ MODE 3	.050	.100	.150	$\uparrow \mu_2$
.05	10.30	10.29	10.29	
.10	10.18	10.17		
.15	10.09		14.18	.15
		14.28	14.29	.10
	14.42	14.42	14.43	.05
MODE 4 $\leftarrow \mu_3$	.150	.100	.050	

TABLE 5.6 ($\mu_1 = .80$)

$\mu_2 \downarrow$ / $\mu_3 \rightarrow$ MODE 1	.050	.100	$\uparrow \mu_2$
.05	2.303	2.301	
.10	2.269		
		5.699	.10
	5.768	5.771	.05
MODE 2 $\leftarrow \mu_3$	.100	.050	

TABLE 5.6 ($\mu_1 = .80$)

$\mu_2 \downarrow$ / $\mu_3 \rightarrow$ MODE 3	.050	.100	$\uparrow \mu_2$
.05	9.666	9.660	
.10	9.561		
		13.41	.10
	13.53	13.54	.05
MODE 4 $\leftarrow \mu_3$	.100	.050	

TABLE 5.6 ($\mu_1 = .85$)

$\mu_2 \downarrow$ / $\mu_3 \rightarrow$ MODE 1	.050	$\uparrow \mu_2$
.05	2.170	
	5.437	.05
MODE 2 $\leftarrow \mu_3$	.050	

TABLE 5.6 ($\mu_1 = .85$)

$\mu_2 \downarrow$ / $\mu_3 \rightarrow$ MODE 3	.050	$\uparrow \mu_2$
.05	9.105	
	12.75	.05
MODE 4 $\leftarrow \mu_3$	.050	

CHAPTER SIX

FREE VIBRATION OF MULTISPAN BEAMS WITH CLASSICAL BOUNDARY CONDITIONS

Part 1. Beams with Arbitrary Intermediate Support Spacing

This is the most general of the multispan beam vibration problems. While all interior supports must be simple supports there are, as seen in the previous chapters, six unique combinations of outer boundary conditions that may be imposed.

The modal shape is expressed as a function of ξ for each span, where ξ equals the distance along the span divided by the beam overall length. Subscripts of the displacement functions indicate the span number, where the spans are numbered from the left end (Fig. 6.1). It is noted (Fig. 6.1) that all displacement functions, except the one for the extreme right outer span, have their origins at the left end of the span. The beam is considered to have N spans, where $N \geqslant 3$. Each span has its length expressed by the dimensionless subscripted quantity μ, where μ equals span length divided by the beam overall length. The quantity μ_n then, equals the dimensionless length of span n.

For illustrative purposes we develop the equations for a beam with simple-simple support, that is, a beam with simple support at each outer end. The boundary condition

$$r_1(\xi)\bigg|_{\xi=0} = \frac{d^2 r_1(\xi)}{d\xi^2}\bigg|_{\xi=0} = 0 \tag{6.1}$$

requires that

$$B_1 = D_1 = 0 \tag{6.2}$$

A further requirement is that displacement equal zero at $\xi = \mu_1$, that is,

$$r_1(\xi)|_{\xi=\mu_1} = 0 \tag{6.3}$$

Enforcing the boundary condition expressed by Eq. 6.3 we obtain

$$A_1 \sin \beta\mu_1 + C_1 \sinh \beta\mu_1 = 0 \tag{6.4}$$

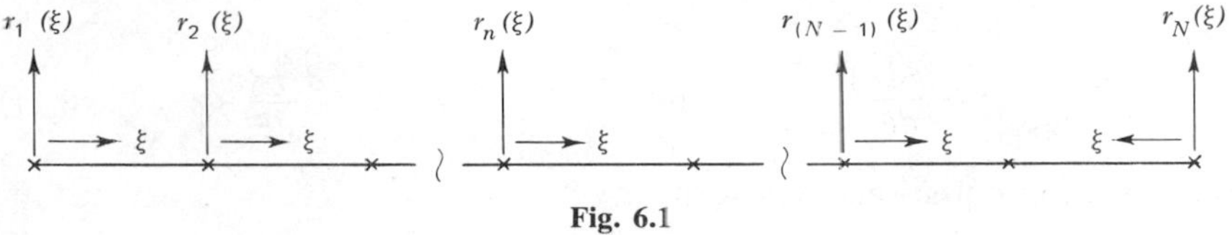

Fig. 6.1

from which we obtain

$$C_1 = -A_1 \frac{\sin \beta\mu_1}{\sinh \beta\mu_1} \tag{6.5}$$

hence

$$r_1(\xi) = A_1\left(\sin \beta\xi - \frac{\sin \beta\mu_1}{\sinh \beta\mu_1} \sinh \beta\xi\right) \tag{6.6}$$

The function $r_N(\xi)$ is evidently subjected to identical boundary conditions. We may therefore write

$$r_N(\xi) = A_N\left(\sin \beta\xi - \frac{\sin \beta\mu_N}{\sinh \beta\mu_N} \sinh \beta\xi\right) \tag{6.7}$$

For interior spans $(1 < n < N)$ we have the boundary condition

$$r_n(\xi)\big|_{\xi=0} = 0$$

To satisfy this boundary condition we must set $B_n = -D_n$ and therefore obtain

$$r_n(\xi) = A_n \sin \beta\xi + B_n(\cos \beta\xi - \cosh \beta\xi) + C_n \sinh \beta\xi \tag{6.8}$$

The constant C_n may be eliminated by utilizing the boundary condition

$$r_n(\xi)\big|_{\xi=\mu_n} = 0 \tag{6.9}$$

after which we obtain

$$r_n(\xi) = A_n(\sin \beta\xi - \theta_n \sinh \beta\xi) + B_n(\cos \beta\xi - \cosh \beta\xi + \phi_n \sinh \beta\xi) \tag{6.10}$$

where

$$\theta_n = \frac{\sin \beta\mu_n}{\sinh \beta\mu_n} \qquad \text{and} \qquad \phi_n = \frac{\cosh \beta\mu_n - \cos \beta\mu_n}{\sinh \beta\mu_n}$$

There remain, then, $2(N-1)$ unknown constants A_1, A_2, B_2, A_3, B_3, and so on. There remain also $2(N-1)$ equations arising from the requirement to satisfy continuity of slope and bending moment at the interior supports. Solution of the vibration problem involves finding those values of β for which the determinant of the coefficient matrix vanishes.

The first two equations of the set are obtained by enforcing the condition

$$\left.\frac{dr_1(\xi)}{d\xi}\right|_{\xi=\mu_1} = \left.\frac{dr_2(\xi)}{d\xi}\right|_{\xi=0} \quad \text{and} \quad \left.\frac{d^2r_1(\xi)}{d\xi^2}\right|_{\xi=\mu_1} = \left.\frac{d^2r_2(\xi)}{d\xi^2}\right|_{\xi=0} \tag{6.11}$$

The resulting equations are, respectively,

$$A_1\left(\cos \beta\mu_1 - \frac{\sin \beta\mu_1}{\sinh \beta\mu_1}\cosh \beta\mu_1\right) + A_2(\theta_2 - 1) + B_2(-\phi_2) = 0 \tag{6.12}$$

and

$$A_1(-2 \sin \beta\mu_1) + A_2(0) + B_2(2) = 0 \tag{6.13}$$

Continuity of slope and bending moment across internal supports require, respectively, for $2 \leqslant n \leqslant N-2$

$$\begin{aligned} A_n(\cos \beta\mu_n - \theta_n \cosh \beta\mu_n) + B_n(-\sin \beta\mu_n - \sinh \beta\mu_n + \phi_n \cosh \beta\mu_n) \\ + A_{n+1}(\theta_{n+1} - 1) + B_{n+1}(-\phi_{n+1}) = 0 \end{aligned} \tag{6.14}$$

and

$$\begin{aligned} A_n(-\sin \beta\mu_n - \theta_n \sinh \beta\mu_n) + B_n(-\cos \beta\mu_n - \cosh \beta\mu_n + \phi_n \sinh \beta\mu_n) \\ + A_{n+1}\{0\} + B_{n+1}\{2\} = 0 \end{aligned} \tag{6.15}$$

The final two equations relate to satisfying the requirement of continuity of slope and bending moment, respectively, across the internal support adjacent to the extreme right-hand span. They are

$$\begin{aligned} A_{N-1}(\cos \beta\mu_{N-1} + \theta_{N-1} \cosh \beta\mu_{N-1}) \\ + B_{N-1}(-\sin \beta\mu_{N-1} - \sinh \beta\mu_{N-1} + \phi_{N-1} \cosh \beta\mu_{N-1}) \\ + A_N\left(\frac{-\sin \beta\mu_N \cosh \beta\mu_N + \cos \beta\mu_N \sinh \beta\mu_N}{\sinh \beta\mu_N}\right) = 0 \end{aligned} \tag{6.16}$$

and

$$\begin{aligned} A_{N-1}(-\sin \beta\mu_{N-1} - \theta_{N-1} \sinh \beta\mu_{N-1}) \\ + B_{N-1}(-\cos \beta\mu_{N-1} - \cosh \beta\mu_{N-1} + \phi_{N-1} \sinh \beta\mu_{N-1}) \\ + A_N(2 \sin \beta\mu_N) = 0 \end{aligned} \tag{6.17}$$

For illustrative purposes the coefficient matrix for a four-span beam is represented below in matrix form, by the double subscripted quantities. The single subscripted quantities above each matrix column indicate the constants by which the elements in the column are multiplied.

It is extremely important to note that only the expressions for the two elements of the first column and the two elements of the last column are dependent on the boundary conditions at the beam outer ends. This is true regardless of the number of spans in the beam. For any beam whatever, then, the elements of the matrix, other than the first and last columns, are

TABLE 6.1. COEFFICIENT MATRIX FOR A FOUR SPAN BEAM

A_1	A_2	B_2	A_3	B_3	A_4
α_{11}	α_{12}	α_{13}			
α_{21}	α_{22}	α_{23}			
	α_{32}	α_{33}	α_{34}	α_{35}	
	α_{42}	α_{43}	α_{44}	α_{45}	
			α_{54}	α_{55}	α_{56}
			α_{64}	α_{65}	α_{66}

readily obtained from Eqs. 6.12 through 6.17. For completeness the elements of the first and last column are presented below for the three possible end conditions that may prevail.

1. Left end simply supported.

$$\alpha_{11} = \cos \beta\mu_1 \frac{-\sin \beta\mu_1}{\sinh \beta\mu_1} \cosh \beta\mu_1$$

$$\alpha_{21} = -2 \sin \beta\mu_1$$

2. Left end clamped.

$$\alpha_{11} = \frac{-2(1 - \cos \beta\mu_1 \cosh \beta\mu_1)}{\cosh \beta\mu_1 - \cos \beta\mu_1}$$

$$\alpha_{21} = \frac{2(\sinh \beta\mu_1 \cos \beta\mu_1 - \sin \beta\mu_1 \cosh \beta\mu_1)}{\cosh \beta\mu_1 - \cos \beta\mu_1}$$

3. Left end free.

$$\alpha_{11} = \frac{2(1 + \cos \beta\mu_1 \cosh \beta\mu_1)}{\cos \beta\mu_1 + \cosh \beta\mu_1}$$

$$\alpha_{21} = \frac{2(\sinh \beta\mu_1 \cos \beta\mu_1 - \sin \beta\mu_1 \cosh \beta\mu_1)}{\cos \beta\mu_1 + \cosh \beta\mu_1}$$

4. Right end simply supported.

$$\alpha_{2N-3,2(N-1)} = \cos \beta\mu_N \frac{-(\sin \beta\mu_N \cosh \beta\mu_N)}{\sinh \beta\mu_N}$$

$$\alpha_{2(N-1),2(N-1)} = 2 \sin \beta\mu_N$$

5. Right end clamped.

$$\alpha_{2N-3,2(N-1)} = \frac{-2(1 - \cos \beta\mu_N \cosh \beta\mu_N)}{\cosh \beta\mu_N - \cos \beta\mu_N}$$

$$\alpha_{2(N-1),2(N-1)} = \frac{-2(\sinh \beta\mu_N \cos \beta\mu_N - \sin \beta\mu_N \cosh \beta\mu_N)}{\cosh \beta\mu_N - \cos \beta\mu_N}$$

6. Right end free.

$$\alpha_{2N-3,2(N-1)} = \frac{2(1+\cos\beta\mu_N \cosh\beta\mu_N)}{\cos\beta\mu_N + \cosh\beta\mu_N}$$

$$\alpha_{2(N-1),2(N-1)} = \frac{-2(\sinh\beta\mu_N \cos\beta\mu_N - \sin\beta\mu_N \cosh\beta\mu_N)}{\cos\beta\mu_N + \cosh\beta\mu_N}$$

The information presented above is sufficient for writing the coefficient matrix, hence obtaining the frequencies of lateral vibration of any multispan beam regardless of the beam outer end supports. Because no restriction has been imposed on the spacing of the intermediate supports, the problem is quite general and it is not practical to store all the direct information necessary for establishing frequencies of beams in this class. It appears that the most significant contribution that can be made here is to provide information for establishing the elements of the matrices and to require the analyst to seek out the eigenvalues, that is, those values of β that cause the determinant of the matrix to vanish. This can be done by using standard existing programs on most digital computers. Recursion formulas for generating matrix elements are presented later.

Having looked at the means for establishing frequencies, we next turn our attention to the modal shapes. It will be seen that they are readily obtained for any frequency once the associated eigenvalue is determined.

Consider, for example, the four span beam for which the matrix elements were represented in Table 6.1 by doubly subscripted quantities α_{11}, α_{12}, and so on. Let us suppose that an eigenvalue has been evaluated and the elements are based on this eigenvalue. The elements then represent the coefficients for a set of homogeneous simultaneous algebraic equations. In order to effect a solution for the modal shapes we begin by arbitrarily assigning to A_1 the value of unity. A_1 can never be zero, and we are at liberty to set it equal to any nonzero value we wish for the purpose of obtaining modal shapes. The first two equations then automatically become nonhomogeneous and may be written as

$$\begin{aligned} \alpha_{12}A_2 + \alpha_{13}B_2 &= -\alpha_{11} \\ \alpha_{22}A_2 + \alpha_{23}B_2 &= -\alpha_{21} \end{aligned} \tag{6.18}$$

By utilizing Cramer's rule it is obvious that the solutions for A_2 and B_2 become, respectively

$$A_2 = \frac{\alpha_{21}\alpha_{13} - \alpha_{11}\alpha_{23}}{\alpha_{12}\alpha_{23} - \alpha_{13}\alpha_{22}}$$

and (6.19)

$$B_2 = \frac{\alpha_{11}\alpha_{22} - \alpha_{21}\alpha_{12}}{\alpha_{12}\alpha_{23} - \alpha_{13}\alpha_{22}}$$

The second pair of equations can now be made nonhomogeneous by introducing the values for A_2 and B_2. The newly acquired values for A_3 and B_3 can then be utilized to solve the next pair of equations, and so on. In the general case we can write solutions for the intermediate pair of equations (not including the first or last pair in the set) as follows for each value of n

$$3 \leq n \leq (N-1)$$

Introduce

$$m = 3 + 2(n-3)$$

and

$$D = \alpha_{m,m+1}\alpha_{m+1,m+2} - \alpha_{m+1,m+1}\alpha_{m,m+2}$$

$$A_n = \frac{A_{n-1}}{D}(\alpha_{m,m+2}\alpha_{m+1,m-1} - \alpha_{m+1,m+2}\alpha_{m,m-1}) + \frac{B_{n-1}}{D}(\alpha_{m,m+2}\alpha_{m+1,m} - \alpha_{m+1,m+2}\alpha_{m,m}) \quad (6.20)$$

$$B_n = \frac{A_{n-1}}{D}(\alpha_{m+1,m+1}\alpha_{m,m-1} - \alpha_{m,m+1}\alpha_{m+1,m-1}) + \frac{B_{n-1}}{D}(\alpha_{m+1,m+1}\alpha_{m,m} - \alpha_{m,m+1}\alpha_{m+1,m})$$

There are two final equations available for establishing the value of the constant A_N. They may be used as follows,

$$A_N = \frac{-(A_{N-1}\alpha_{2N-3,2(N-2)} + B_{N-1}\alpha_{2N-3,2N-3})}{\alpha_{2N-3,2(N-1)}}$$

or (6.21)

$$A_N = \frac{-(A_{N-1}\alpha_{2(N-1),2(N-2)} + B_{N-1}\alpha_{2(N-1),2N-3})}{\alpha_{2(N-1),2(N-1)}}$$

While either of Eqs. 6.21 may be utilized to evaluate A_N it may be found that for some problems the right-hand side of one of the equations takes on an indeterminate form. In such a case the alternate equation may be utilized.

Having established values for the constants A_2, B_2, A_3, B_3, etc., based on a value of unity for A_1, it is then possible to write expressions for the modal shape. For all internal spans, i.e., for $2 \leq n \leq (N-1)$, the displacement is expressed according to Eq. 6.10. For the outer two spans the displacement expressions depend on the adjacent beam outer end condition. For convenience they are summarized as follows.

1. Left end simply supported.

$$r_1(\xi) = \sin\beta\xi \frac{-\sin\beta\mu_1}{\sinh\beta\mu_1}\sinh\beta\xi$$

2. Left end clamped.

$$r_1(\xi) = \sin \beta\xi - \sinh \beta\xi + \frac{\sin \beta\mu_1 - \sinh \beta\mu_1}{\cosh \beta\mu_1 - \cos \beta\mu_1}(\cos \beta\xi - \cosh \beta\xi)$$

3. Left end free.

$$r_1(\xi) = \sin \beta\xi + \sinh \beta\xi \frac{-(\sin \beta\mu_1 + \sinh \beta\mu_1)}{\cos \beta\mu_1 + \cosh \beta\mu_1}(\cos \beta\xi + \cosh \beta\xi)$$

4. Right end simply supported.

$$r_N(\xi) = A_N\left(\sin \beta\xi - \frac{\sin \beta\mu_N}{\sinh \beta\mu_N} \sinh \beta\xi\right)$$

5. Right end clamped.

$$r_N(\xi) = A_N\left[\sin \beta\xi - \sinh \beta\xi + \frac{\sin \beta\mu_N - \sinh \beta\mu_N}{\cosh \beta\mu_N - \cos \beta\mu_N}(\cos \beta\xi - \cosh \beta\xi)\right]$$

6. Right end free.

$$r_N(\xi) = A_N\left[\sin \beta\xi + \sinh \beta\xi - \frac{\sin \beta\mu_N + \sinh \beta\mu_N}{\cos \beta\mu_N + \cosh \beta\mu_N}(\cos \beta\xi + \cosh \beta\xi)\right]$$

This completes the analysis provided here for multispan beams with arbitrary interior support spacing.

To construct the matrix, then, the elements of the first and last columns may be taken from the appropriate expressions above. The remaining elements in the first and last two rows are, respectively (Eqs. 6.12, 6.13, 6.16, and 6.17),

$$\alpha_{12} = \theta_2 - 1 \qquad \alpha_{13} = -\phi_2$$
$$\alpha_{22} = 0 \qquad \alpha_{23} = 2$$

and

$$\alpha_{2N-3,\,2(N-2)} = \cos \beta\mu_{N-1} + \theta_{N-1} \cosh \beta\mu_{N-1}$$
$$\alpha_{2N-3,\,2N-3} = -\sin \beta\mu_{N-1} - \sinh \beta\mu_{N-1} + \phi_{N-1} \cosh \beta\mu_{N-1}$$
$$\alpha_{2(N-1),\,2(N-2)} = -\sin \beta\mu_{N-1} - \theta_{N-1} \sinh \beta\mu_{N-1}$$
$$\alpha_{2(N-1),\,2N-3} = -\cos \beta\mu_{N-1} - \cosh \beta\mu_{N-1} + \phi_{N-1} \sinh \beta\mu_{N-1}$$

The rows of the matrix, other than the first two and last two rows, may be generated as follows:

for $n = 2, 3, 4, \ldots$ etc. $\ldots$ to $N - 2$

set $m = 2n - 1$

the mth row has elements

$$\alpha_{m,m-1} = \cos \beta\mu_n - \theta_n \cosh \beta\mu_n$$

$$\alpha_{m,m} = -\sin \beta\mu_n - \sinh \beta\mu_n + \phi_n \cosh \beta\mu_n$$

$$\alpha_{m,m+1} = \theta_{n+1} - 1; \qquad \alpha_{m,m+2} = -\phi_{n+1}$$

the $m+1$ row has elements

$$\alpha_{m+1,m-1} = -\sin \beta\mu_n - \theta_n \sinh \beta\mu_n$$

$$\alpha_{m+1,m} = -\cos \beta\mu_n - \cosh \beta\mu_n + \phi_n \sinh \beta\mu_n$$

$$\alpha_{m+1,m+1} = 0; \qquad \alpha_{m+1,m+2} = 2$$

Frequencies are obtained using Eq. 1.18, i.e.,

$$f = \frac{\beta^2}{2\pi L^2}\sqrt{\frac{EI}{\rho A}}$$

Part 2. Beams with Uniform Support Spacing

This family of problems constitutes a subset of the more general set discussed in Part 1. It will be seen that the equations from which the eigenvalues are obtained are greatly simplified when support spacing is uniform. It will also be seen that eigenvalues and modal shape expression information for lateral vibration analysis of any beam of this type are easily stored.

The nondimensional span length μ is now constant, and we have

$$\mu = \frac{1}{N} \tag{6.22}$$

where, as in Part 1, N equals number of spans. Introducing the parameter β^*, where $\beta^* = \beta\mu$, it is seen that for beams with simple support at each end the modal shape expressions become (Eqs. 6.6, 6.7, and 6.10)

$$r_1(\xi) = A_1\left(\sin \beta\xi - \frac{\sin \beta^*}{\sinh \beta^*} \sinh \beta\xi\right) \tag{6.23}$$

$$r_N(\xi) = A_N\left(\sin \beta\xi - \frac{\sin \beta^*}{\sinh \beta^*} \sinh \beta\xi\right) \tag{6.24}$$

and for $1 < n < N$,

$$r_n(\xi) = A_n(\sin \beta\xi - \theta \sinh \beta\xi) + B_n(\cos \beta\xi - \cosh \beta\xi + \phi \sinh \beta\xi) \tag{6.25}$$

where

$$\theta = \frac{\sin \beta^*}{\sinh \beta^*} \qquad \text{and} \qquad \phi = \frac{\cosh \beta^* - \cos \beta^*}{\sinh \beta^*}$$

The homogeneous algebraic equations relating the unknown constants are obtained in a manner identical to that of the general case in Part 1 (Eqs. 6.12 through 6.17) by satisfying the continuity of slope and bending moment across the internal supports. They are, for the first interior support,

$$A_1\left(\cos\beta^* - \frac{\sin\beta^*}{\sinh\beta^*}\cosh\beta^*\right) + A_2(\theta - 1) + B_2(-\phi) = 0 \qquad (6.26)$$

$$A_1(-2\sin\beta^*) + A_2(0) + B_2(2) = 0 \qquad (6.27)$$

and for $2 \leqslant n \leqslant (N-2)$

$$A_n(\cos\beta^* - \theta\cosh\beta^*) + B_n(-\sin\beta^* - \sinh\beta^* + \phi\cosh\beta^*) + A_{n+1}(\theta - 1) + B_{n+1}(-\phi) = 0 \qquad (6.28)$$

$$A_n(-\sin\beta^* - \theta\sinh\beta^*) + B_n(-\cos\beta^* - \cosh\beta^* + \phi\sinh\beta^*) + A_{n+1}(0) + B_{n+1}(2) = 0 \qquad (6.29)$$

and for the internal support adjacent to the extreme right-hand span

$$A_{N-1}(\cos\beta^* - \theta\cosh\beta^*) + B_{N-1}(-\sin\beta^* - \sinh\beta^* + \phi\cosh\beta^*) + A_N\left(\frac{\cos\beta^*\sinh\beta^* - \sin\beta^*\cosh\beta^*}{\sinh\beta^*}\right) = 0 \qquad (6.30)$$

$$A_{N-1}(-\sin\beta^* - \theta\sinh\beta^*) + B_{N-1}(-\cos\beta^* - \cosh\beta^* + \phi\sinh\beta^*) + A_n(2\sin\beta^*) = 0 \qquad (6.31)$$

The coefficient matrix in the above equations has the same form as that pertaining to the equations of Part 1. The matrix for a four-span beam is depicted in Table 6.1. Again, only the expressions for the elements of the

TABLE 6.2 VALUES OF β FOR PINNED-PINNED BEAMS

	Mode					
Number of spans, N	1	2	3	4	5	6
3	9.425	10.67	12.89	18.85	20.12	22.29
4	12.57	13.57	15.71	17.85	25.13	26.18
5	15.71	16.55	18.50	20.76	22.75	31.42
6	18.85	19.56	21.34	23.56	25.79	27.61
7	21.99	22.61	24.22	26.35	28.62	30.76
8	25.13	25.68	27.15	29.16	31.41	33.66
9	28.27	28.76	30.10	32.01	34.20	36.48
10	31.42	31.86	33.09	34.88	37.00	39.27
11	34.56	34.96	36.10	37.79	39.83	42.05
12	37.70	38.07	39.13	40.72	42.68	44.85
13	40.84	41.18	42.17	43.67	45.55	47.66
14	43.98	44.30	45.22	46.65	48.44	50.50
15	47.12	47.42	48.29	49.64	51.36	53.35

first and last columns will depend on the boundary conditions imposed at the ends of the beam. Expressions for the elements of the internal columns will now depend on N, the number of spans, only.

In view of the two latter observations it is seen that the eigenvalues will be a function of the beam outer boundary conditions and the number of spans. Due to this great simplicity the eigenvalues for this family of problems can be easily stored. They are presented for all possible

TABLE 6.3 VALUES OF β FOR CLAMPED-CLAMPED BEAMS

	Mode					
Number of spans, N	1	2	3	4	5	6
3	10.67	12.89	14.19	20.12	22.29	23.56
4	13.57	15.71	17.85	18.92	26.18	28.27
5	16.55	18.50	20.76	22.75	23.65	32.30
6	19.56	21.34	23.56	25.79	27.61	28.38
7	22.61	24.22	26.35	28.62	30.76	32.44
8	25.68	27.15	29.16	31.41	33.66	35.71
9	28.76	30.10	32.01	34.20	36.48	38.68
10	31.86	33.09	34.88	37.00	39.27	41.53
11	34.96	36.10	37.79	39.83	42.05	44.33
12	38.07	39.13	40.72	42.68	44.85	47.12
13	41.18	42.17	43.67	45.55	47.66	49.91
14	44.30	45.22	46.65	48.44	50.50	52.70
15	47.42	48.29	49.64	51.36	53.35	55.51

TABLE 6.4 VALUES OF β FOR CLAMPED-PINNED BEAMS

	Mode					
Number of spans, N	1	2	3	4	5	6
3	9.782	11.78	13.80	19.23	21.21	23.18
4	12.84	14.58	16.83	18.62	25.43	27.18
5	15.93	17.44	19.63	21.83	23.41	31.66
6	19.04	20.36	22.43	24.69	26.78	28.18
7	22.15	23.32	25.25	27.49	29.73	31.69
8	25.27	26.32	28.11	30.27	32.55	34.73
9	28.40	29.34	31.00	33.08	35.34	37.60
10	31.53	32.39	33.93	35.92	38.13	40.41
11	34.66	35.45	36.89	38.78	40.93	43.19
12	37.79	38.52	39.87	41.66	43.74	45.98
13	40.93	41.60	42.86	44.57	46.58	48.77
14	44.06	44.69	45.88	47.51	49.44	51.58
15	47.20	47.79	48.91	50.46	52.33	54.41

combinations of beam outer end boundary conditions, and with the number of spans varying from 3 through 15. They are discussed at greater length later, and illustrative problems are solved, but before doing so it is desirable to look at the modal shapes associated with the vibration frequencies.

TABLE 6.5 VALUES OF β FOR CLAMPED-FREE BEAMS

	Mode					
Number of spans, N	1	2	3	4	5	6
3	4.624	10.71	12.85	14.16	20.12	22.29
4	6.157	13.61	15.71	17.80	18.89	26.18
5	7.695	16.58	18.53	20.74	22.69	23.62
6	9.234	19.59	21.38	23.56	25.75	27.55
7	10.77	22.63	24.26	26.37	28.60	30.72
8	12.31	25.70	27.19	29.19	31.41	33.63
9	13.85	28.78	30.14	32.04	34.22	36.46
10	15.39	31.87	33.12	34.92	37.03	39.27
11	16.93	34.97	36.13	37.83	39.86	42.06
12	18.47	38.08	39.15	40.76	42.71	44.87
13	20.01	41.19	42.19	43.71	45.59	47.69
14	21.55	44.31	45.25	46.68	48.48	50.53
15	23.08	47.43	48.31	49.67	51.40	53.38

TABLE 6.6 VALUES OF β FOR PINNED-FREE BEAMS

	Mode					
Number of spans, N	1	2	3	4	5	6
3	4.609	9.810	11.78	13.74	19.23	21.21
4	6.155	12.86	14.61	16.80	18.56	25.43
5	7.695	15.94	17.48	19.63	21.79	23.34
6	9.234	19.04	20.40	22.45	24.67	26.73
7	10.77	22.16	23.36	25.28	27.49	29.69
8	12.31	25.28	26.35	28.15	30.29	32.53
9	13.85	28.40	29.37	31.04	33.11	35.34
10	15.39	31.53	32.41	33.97	35.95	38.14
11	16.93	34.66	35.47	36.93	38.81	40.95
12	18.47	37.80	38.54	39.90	41.70	43.77
13	20.01	40.93	41.62	42.90	44.61	46.62
14	21.55	44.06	44.71	45.91	47.54	49.48
15	23.08	47.20	47.80	48.94	50.50	52.36

TABLE 6.7 VALUES OF β FOR FREE-FREE BEAMS

Number of spans, N	Mode 1	2	3	4	5	6
3	4.235	4.943	10.74	12.82	14.12	20.12
4	6.024	6.283	13.65	15.71	17.75	18.85
5	7.650	7.740	16.62	18.55	20.72	22.64
6	9.219	9.249	19.62	21.41	23.56	25.71
7	10.7676	10.7775	22.66	24.30	26.39	28.59
8	12.3099	12.3130	25.72	27.23	29.22	31.41
9	13.8499	13.8509	28.80	30.18	32.08	34.23
10	15.3892	15.3895	31.89	33.16	34.96	37.05
11	16.9282	16.9283	34.98	36.16	37.87	39.89
12	18.4672	18.4672	38.09	39.18	40.80	42.75
13	20.0061	20.0061	41.20	42.22	43.75	45.63
14	21.5451	21.5451	44.32	45.27	46.72	48.52
15	23.0840	23.0840	47.44	48.33	49.70	51.44

Modal shapes are established in a manner identical to that described in Part 1. The unknown constant for the left-hand span, A_1, is set equal to unity, thereby making the first two equations of the set nonhomogeneous (Table 6.1). After solving for A_2 and B_2, these values are utilized to solve for A_3 and B_3, and so on. Expressions for the left-hand span and also for the constants A_2 and B_2 are summarized below for the benefit of the analyst. They cover all possible boundary conditions that can prevail at the left end and are independent of the boundary conditions prevailing at the right.

1. Left end pinned.

$$r_1(\xi) = \sin \beta\xi - \frac{\sin \beta^*}{\sinh \beta^*} \sinh \beta\xi$$

$$A_2 = \frac{2 \sin \beta^* \cosh \beta^* - (\sin \beta^* + \sinh \beta^*) \cos \beta^*}{\sin \beta^* - \sinh \beta^*}$$

$$B_2 = \sin \beta^*$$

2. Left end clamped.

$$r_1(\xi) = \sin \beta\xi - \sinh \beta\xi + \frac{\sin \beta^* - \sinh \beta^*}{\cosh \beta^* - \cos \beta^*} (\cos \beta\xi - \cosh \beta\xi)$$

$$B_2 = \frac{\sin \beta^* \cosh \beta^* - \sinh \beta^* \cos \beta^*}{\cosh \beta^* - \cos \beta^*}$$

$$A_2 = \frac{B_2(\cosh \beta^* - \cos \beta^*)}{\sin \beta^* - \sinh \beta^*} + \frac{2 \sinh \beta^*(1 - \cos \beta^* \cosh \beta^*)}{(\sin \beta^* - \sinh \beta^*)(\cosh \beta^* - \cos \beta^*)}$$

3. Left end free.

$$r_1(\xi) = \sin \beta\xi + \sinh \beta\xi - \frac{\sin \beta^* + \sinh \beta^*}{\cos \beta^* + \cosh \beta^*}(\cos \beta\xi + \cosh \beta\xi)$$

$$B_2 = \frac{\sin \beta^* \cosh \beta^* - \sinh \beta^* \cos \beta^*}{\cos \beta^* + \cosh \beta^*}$$

$$A_2 = \frac{B_2(\cosh^2 \beta^* - \cos^2 \beta^*) - 2 \sinh \beta^*(1 + \cos \beta^* \cosh \beta^*)}{(\sin \beta^* - \sinh \beta^*)(\cos \beta^* + \cosh \beta^*)}$$

For the nth span where, $1 < n < N$

$$r_n(\xi) = A_n(\sin \beta\xi - \theta \sinh \beta\xi) + B_n(\cos \beta\xi - \cosh \beta\xi + \phi \sinh \beta\xi)$$

where

$$\theta = \frac{\sin \beta^*}{\sinh \beta^*} \qquad \text{and } \phi = \frac{\cosh \beta^* - \cos \beta^*}{\sinh \beta^*}$$

as shown earlier, and, for $2 < n < N$

$$A_n = A_{n-1}\frac{\theta \cosh \beta^* - \cos \beta^* + \phi \sin \beta^*}{\theta - 1} + B_{n-1}\frac{\sin \beta^* + \sinh \beta^* + \phi(\cos \beta^* - \cosh \beta^*)}{\theta - 1} \tag{6.32}$$

and

$$B_n = A_{n-1} \sin \beta^* + B_{n-1} \cos \beta^* \tag{6.33}$$

Finally, at the right end of the beam two pieces of information are required. They are the Nth span modal shape expression, which depends on the boundary condition prevailing at the outer right-hand end of the beam only, and the value for A_N. Two expressions are available for this latter quantity and they are both given since one may give rise to an indeterminate form.

1. Right end pinned.

$$r_N(\xi) = A_N\left(\sin \beta\xi - \frac{\sin \beta^*}{\sinh \beta^*} \sinh \beta\xi\right)$$

$$A_N = A_{N-1}(-1) + B_{N-1}\frac{-1 + \sin \beta^* \sinh \beta^* + \cos \beta^* \cosh \beta^*}{\cos \beta^* \sinh \beta^* - \sin \beta^* \cosh \beta^*}$$

or

$$A_N = A_{N-1} + B_{N-1}\frac{\cos \beta^*}{\sin \beta^*}$$

2. Right end clamped.

$$r_N(\xi) = A_N\left[\sin \beta\xi - \sinh \beta\xi + \frac{\sin \beta^* - \sinh \beta^*}{\cosh \beta^* - \cos \beta^*}(\cos \beta\xi - \cosh \beta\xi)\right]$$

$$A_N = \frac{\cosh \beta^* - \cos \beta^*}{2 \sinh \beta^*(\cos \beta^* \cosh \beta^* - 1)} \times [A_{N-1}(\sin \beta^* \cosh \beta^* - \cos \beta^* \sinh \beta^*) + B_{N-1}(\sin \beta^* \sinh \beta^* + \cos \beta^* \cosh \beta^* - 1)]$$

or

$$A_N = (\cos \beta^* - \cosh \beta^*)\frac{A_{N-1} \sin \beta^* + B_{N-1} \cos \beta^*}{\sinh \beta^* \cos \beta^* - \sin \beta^* \cosh \beta^*}$$

3. Right end free.

$$r_N(\xi) = A_N\left[\sin \beta\xi + \sinh \beta\xi - \frac{\sin \beta^* + \sinh \beta^*}{\cos \beta^* + \cosh \beta^*}(\cos \beta\xi + \cosh \beta\xi)\right]$$

$$A_N = \frac{\cos \beta^* - \cosh \beta^*}{2 \sinh \beta^*(1 + \cos \beta^* \cosh \beta^*)} \times [A_{N-1}(\sin \beta^* \cosh \beta^* - \cos \beta^* \sinh \beta^*) + B_{N-1}(\sin \beta^* \sinh \beta^* + \cos \beta^* \cosh \beta^* - 1)]$$

or

$$A_N = (\cos \beta^* + \cosh \beta^*)\frac{A_{N-1} \sin \beta^* + B_{N-1} \cos \beta^*}{\sin \beta^* \cosh \beta^* - \sinh \beta^* \cos \beta^*}$$

The procedure then for determining frequencies and modal shapes of any beam of the family under discussion is:

1. For the mode of interest determine the eigenvalue β from the appropriate table.
2. Write the modal shape expression for the first (left-hand) span.
3. Using appropriate expressions provided, evaluate A_2 and B_2.
4. Utilize the recursion formulas (Eqs. 6.32 and 6.33) to evaluate A_n and B_n for $2 < n < N$.
5. Write the modal shape expressions for all the internal spans, that is, all spans except the extreme outer spans.
6. Evaluate A_N from the appropriate expression.
7. Write the modal shape expression for the Nth span using the appropriate expression.

The tables and modal shape expressions discussed above permit the immediate evaluation of frequencies and modal shapes for any beam of the family under discussion. In order to obtain a better understanding of this vibration problem, however, it is wise to observe how eigenvalues for beams with a large number of spans can sometimes be inferred from eigenvalues of beams with a smaller number of spans. For beams with

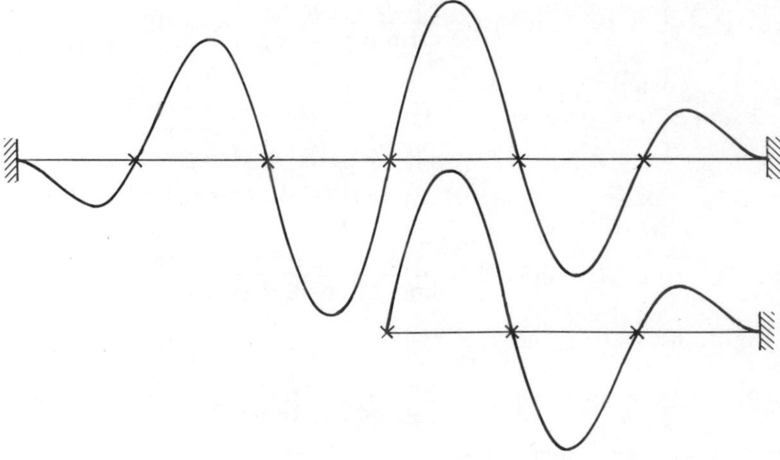

Fig. 6.2

identical boundary conditions at each end and an even number of spans, the modal shapes for odd-numbered modes are antisymmetric about the middle of the beam. For the same beams the modal shapes for even-numbered modes are symmetric about the middle of the beam.

Let us consider then, for illustrative purposes, the first mode vibration of a six span clamped-clamped beam as shown in Fig. 6.2. We recognize that the first mode motion of the right half of this beam should be no different than that of a simply supported–clamped beam of three spans as shown also in Fig. 6.2. This condition exists because no displacement or bending moment is possible at the left end of the three span beam or the center support of the six span beam. The three span beam has half the length of the six span beam and, since their frequencies are identical, the

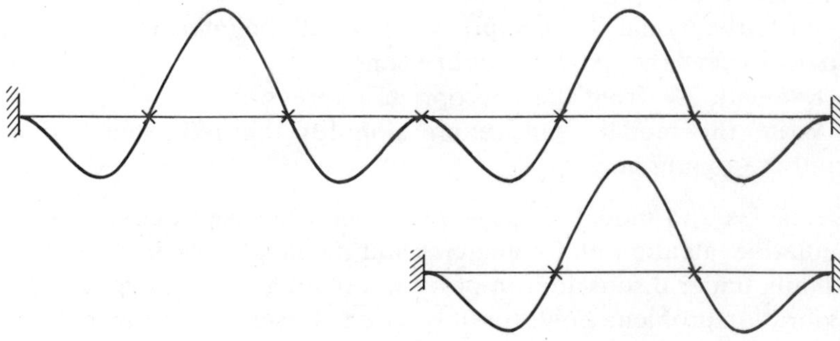

Fig. 6.3

eigenvalue of the shorter beam must be half that of the longer one (Eq. 1.18). This is in keeping with the observation in the tables that the eigenvalues are, respectively, 9.782 and 19.56.

The modal shape for second mode vibration of the six span clamped-clamped beam is shown in Fig. 6.3. Following arguments similar to those presented above, it is seen that the second mode eigenvalue for this beam should be equal to twice the first mode eigenvalue for a three span clamped-clamped beam. That this is in fact the case is readily established by consulting the tables.

There are numerous other examples in the data presented in which eigenvalues of one beam can be inferred from another. These cases are too numerous to be discussed here, but they can be utilized, and have been utilized, to check on the computer search for eigenvalues. It is noted that the first mode eigenvalues for beams with one free end, and the first two eigenvalues for beams with two free ends, are substantially reduced from those of other beams. This is particularly obvious if we study the last row of numbers in each table, that is, the eigenvalues for 15-span beams. For beams with one free end, the first mode motion primarily involves the span with the free end. Focusing our attention on this span for the moment, and recalling that the first mode eigenvalue for a clamped-free beam is 1.875, we recognize that the first mode eigenvalue for a beam of N spans cannot exceed $1.875N$. This condition follows because, for first mode motion, the bending moment at the inner end of the span in question is somewhat less than that provided by a clamped condition. For higher mode vibration of beams with one free end, the limit discussed above is no longer applicable and the eigenvalues are not so greatly influenced by the existence of the free end, particularly as the number of spans in the beam increases. This is easily verified by comparing the final row of numbers of Tables 6.5 and 6.6 with those of Tables 6.2 through 6.4.

Let us suppose we have a multispan beam with clamped-pinned or pinned-pinned end conditions and we decide to add one more span to the pinned end, the additional span having a free outer end. Then, while the additional span has the effect of bringing about some change in existing frequencies, its major effect is essentially the introduction of a new and much lower fundamental frequency. A similar change could be effected by removing the simple support from one end of a multispan clamped-pinned or pinned-pinned beam. The first and second mode eigenvalues for free-free multispan beams are very close together (Table 6.7). This is in keeping with the observations made above. For free-free beams of an even number of spans the first modal shape is antisymmetric about the center of the beam. The second modal shape is symmetric about the center. For beams of an odd number of spans the reverse situation prevails, that is, the first modal shape is symmetric about the center of the beam and the second mode antisymmetric, and so on. We observe that, regardless of the number of spans, the first and second mode frequencies of free-free beams are determined primarily by the end spans. The major difference in the two modes is that there is a switch from a symmetric to an antisymmetric mode in going from first to second mode, or from antisymmetric to symmetric, depending on the number of spans. In either case there is not much difference in the associated bending energy for a multispan beam, hence the eigenvalues of the two modes are not very different. There is no difficulty in obtaining the first and second mode frequencies with error less than 1 percent using Table 6.7. The problem of establishing exact modal shapes for first and second mode vibrations of free-free multispan beams is more difficult because of the high dependence of the modal shape on the eigenvalue. For beams with an even number of spans the problem can be reduced to analyzing a pinned-free or clamped-free beam of half the length, as discussed earlier. For beams with an odd number of spans the latter procedure cannot be utilized to advantage, since it involves analyzing beams with one half-span. It is recognized, however, that since the modes are either symmetric or antisymmetric about the center of the beam, depending on whether the mode is odd-numbered or even-numbered, one need only determine the shape of one half of the beam. It is evidently advantageous to work with the left half of the beam (Fig. 6.1), since the recursion formulas need be used less often and less arithmetical error is introduced.

ILLUSTRATIVE PROBLEM 6.1

A heat exchanger tube of the type described in Illustrative Problem 4.1 is 200 in. long. It is essentially clamped at each end (rolled in) and receives simple support at nine equispaced internal baffle plates.

(a) Determine the first mode lateral vibration frequency. There are 10 spans in this problem and consulting Table 6.3, the applicable table, it is seen that $\beta = 31.86$. By referring to Illustrative Problem 4.1, it is seen that the frequency is

$$f = 40.1 \times \frac{100^2}{200^2} \times \frac{31.86^2}{8.526^2} = 140 \text{ Hz}$$

(b) Suppose the number of internal baffles is reduced to seven. Determine the new frequency. There are now eight spans in the heat exchanger and, consulting Table 6.3, it is seen that

$$\beta = 25.68$$

Referring to part (a) above it is seen that frequency

$$f = 140 \times \frac{25.68^2}{31.86^2} = 91 \text{ Hz}$$

It is observed that a strong relationship exists between tube fundamental lateral vibration frequency and the number of internal baffles even for a heat exchanger with nine such baffles.

CHAPTER SEVEN

TORSIONAL VIBRATION OF CIRCULAR SHAFTS

Underlying Theory

It is known from the theory of elasticity that the torque at any position along a circular shaft may be written as

$$T = GI_p \frac{d\phi}{dx} \tag{7.1}$$

where G = shear modulus of elasticity of the shaft material = $E/[2(1+\nu)]$
E = Young's modulus of the material
ν = Poisson ratio of the material
I_p = polar moment of inertia of shaft cross-sectional area = $\pi(D_o^4 - D_i^4)/32$
ϕ = angular departure of the shaft from its neutral configuration and is a function of x, the distance measured along the shaft (see Fig. 7.1 for sign convention)
D_o = shaft outer diameter
D_i = inner diameter of hollow shaft

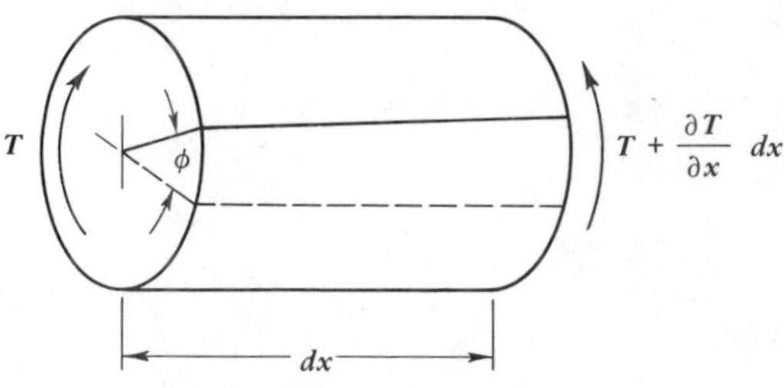

Fig. 7.1

If we consider the dynamic equilibrium of a small differential element of the shaft of length dx, it follows that the net torque acting on the element at any time is

$$\frac{\partial T}{\partial x}\,dx = GI_p\frac{\partial^2\phi}{\partial x^2}\,dx \tag{7.2}$$

and this torque is opposed by the inertial torque

$$-I_p\rho\frac{\partial^2\phi}{\partial t^2}\,dx \tag{7.3}$$

Since the sum of these two torques must be equal to zero, we obtain

$$\frac{\partial^2\phi}{\partial t^2}=\frac{G}{\rho}\frac{\partial^2\phi}{\partial x^2} \tag{7.4}$$

which is the differential equation governing the torsional vibration of circular shafts.

This equation, like the one for lateral vibration of beams, is readily solved by the method of separation of variables. We therefore express the angular displacement of the shaft as

$$\phi(x,t)=p(t)\theta(x) \tag{7.5}$$

Substituting into Eq. 7.4 we obtain

$$\theta(x)\frac{d^2p(t)}{dt^2}=\frac{G}{\rho}p(t)\frac{d^2\theta(x)}{dx^2} \tag{7.6}$$

Equation 7.6 may then be divided by $\theta(x)p(t)$ to obtain

$$\frac{1}{p(t)}\frac{d^2p(t)}{dt^2}=\frac{G}{\rho}\frac{1}{\theta(x)}\frac{d^2\theta(x)}{dx^2} \tag{7.7}$$

It is observed that, as for Eq. 1.4, the left-hand side of Eq. 7.7 is a function of time only, while the right-hand side is a function of distance x. Since the two quantities are equal, they must in turn be equal to a constant which we denote as $-\kappa^2$. The left-hand side can then be written as

$$\frac{d^2p(t)}{dt^2}+\kappa^2p(t)=0 \tag{7.8}$$

for which the solution is

$$p(t)=\cos(\omega t-\alpha) \tag{7.9}$$

where α is a phase angle depending on starting conditions and ω equals κ which is the circular frequency of vibration. The right-hand side of Eq. 7.7 becomes

$$\frac{d^2\theta(x)}{dx^2}+\omega^2\frac{\rho}{G}\theta(x)=0 \tag{7.10}$$

for which the solution is

$$\theta(x) = A \sin \omega \sqrt{\frac{\rho}{G}}\, x + B \cos \omega \sqrt{\frac{\rho}{G}}\, x \tag{7.11}$$

where A and B are constants to be determined.

Equation 7.11 represents the modal shape expression for torsional vibration of shafts. It is required to satisfy prescribed boundary conditions for any problems of interest. There are two types of boundary conditions that are commonly utilized, and they are discussed as follows.

Classical Boundary Conditions

1. Free boundaries. These are boundaries where no torque acts on the shaft. This condition is represented mathematically as

$$T = GI_p \frac{d\theta}{dx} = 0$$

that is,

$$\frac{d\theta}{dx} = 0 \tag{7.12}$$

2. Clamped boundaries. These are boundaries where no oscillatory motion is permitted. It is important to note that boundaries where the shaft rotates with prescribed nonoscillatory motion fulfill these conditions. Examples are shafts driven at constant speed. This condition is represented mathematically as

$$\theta(x) = 0 \tag{7.13}$$

Nonclassical Boundary Conditions

1. Boundary with torsion spring restoring moment. This is a type of boundary condition that is physically achievable with the aid of a torsion spring. With the spring at the left end of the shaft, as shown in Fig. 7.2, the mathematical formulation of the boundary condition becomes

$$T = GI_p \frac{d\theta(x)}{dx} = K\theta(x)|_{x=0} \tag{7.14}$$

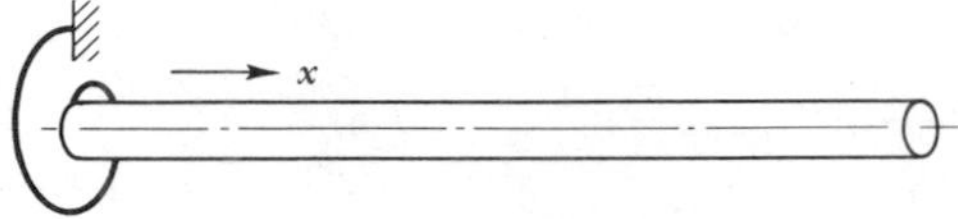

Fig. 7.2

where K is the torsional stiffness of the spring. With the spring at the right-hand end of the shaft the formulation becomes

$$T = GI_p \frac{d\theta(x)}{dx} = -K\theta(x)|_{x=L} \tag{7.15}$$

2. Boundary with attached rotational mass. This is the most commonly encountered boundary condition in torsional vibration of shafts.

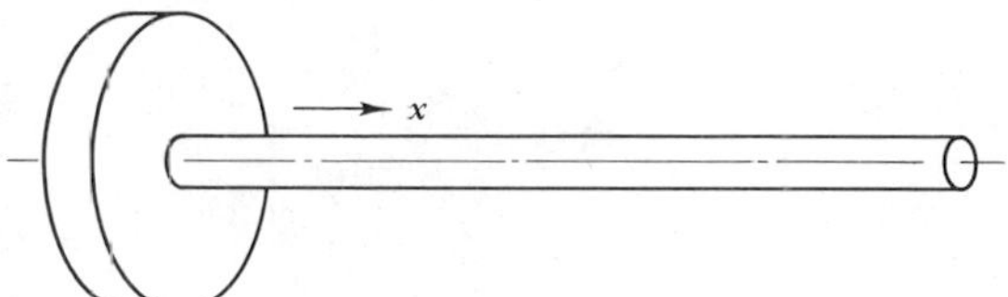

Fig. 7.3

Dynamic equilibrium for a rotational mass located at the left end of the shaft, as shown in Fig. 7.3, requires that

$$GI_p \frac{d\theta(x)}{dx} = -I_0\omega^2\theta(x)|_{x=0} \tag{7.16}$$

and for a shaft with rotational mass at the right end,

$$GI_p \frac{d\theta(x)}{dx} = I_0\omega^2\theta(x)|_{x=L} \tag{7.17}$$

In the usual approach to the torsional vibration analysis of shafts the problem consists of utilizing Eq. 7.11, enforcing the prescribed boundary conditions, obtaining two simultaneous, homogeneous algebraic equations involving the constants A and B, and looking for those values of the parameter $\omega\sqrt{\rho/G}$ that permit a nontrivial solution for the constants.

In the work undertaken here it is advantageous to express the distance measured along the shaft in dimensionless form. This is achieved by introducing the dimensionless distance ξ where $\xi = x/L$. Equation 7.11 then becomes

$$\theta(\xi) = A \sin \beta\xi + B \cos \beta\xi \tag{7.18}$$

where

$$\beta = \omega L \sqrt{\frac{\rho}{G}} \tag{7.19}$$

and

$$0 \leq \xi \leq 1$$

In dimensionless form then, the classical boundary conditions are written, for free boundaries,

$$\frac{d\theta(\xi)}{d\xi} = 0 \tag{7.20}$$

and for clamped boundaries,

$$\theta(\xi) = 0 \tag{7.21}$$

The nonclassical boundary conditions are formulated as follows.

1. Boundary with torsional spring attached (Eq. 7.14) may be rewritten as

$$\frac{GI_p}{L}\frac{d\theta}{d(x/L)} = K\theta\bigg|_{x=0}$$

or

$$\frac{d\theta(\xi)}{d\xi} = K^*\theta(\xi)\bigg|_{\xi=0} \tag{7.22}$$

where

$$K^* = \frac{KL}{GI_p}$$

Similarily, Eq. 7.15 becomes

$$\frac{d\theta(\xi)}{d\xi} = -K^*\theta(\xi)\bigg|_{\xi=1} \tag{7.23}$$

2. Boundary with rotational mass attached. Equation 7.16 may be rewritten as

$$\frac{GI_p}{L}\frac{d\theta(x)}{d(x/L)} = -I_0\omega^2\theta\bigg|_{x=0}$$

or

$$\frac{d\theta(\xi)}{d\xi} = -\frac{I_0\omega^2 L}{GI_p}\,\theta(\xi)\bigg|_{\xi=0} = -\frac{\beta^2\theta(\xi)}{I_0^*}\bigg|_{\xi=0} \tag{7.24}$$

where

$$I_0^* = \frac{I_p\rho L}{I_0}$$

Similarily, Eq. 7.17 becomes

$$\frac{d\theta(\xi)}{d\xi} = \frac{\beta^2\theta(\xi)}{I_0^*}\bigg|_{\xi=1} \tag{7.25}$$

In all problems to be discussed here only the differential equation and boundary conditions expressed with dimensionless distance ξ are utilized.

Part 1. Shafts with Classical Boundary Conditions

There are only three types of shafts with classical boundary conditions to be considered. For illustrative purposes the free torsional vibration of a clamped-clamped shaft, that is, one with vibratory motion forbidden at each end, is discussed in detail.

The boundary condition at the origin may be expressed as

$$\theta(\xi) = A \sin \beta\xi + B \cos \xi \Big|_{\xi=0} = 0 \tag{7.26}$$

This condition requires that we set $B = 0$. From the second boundary condition we obtain

$$\theta(\xi) = A \sin \beta\xi \Big|_{\xi=1} = 0 \tag{7.27}$$

This condition is satisfied only if

$$\beta = n\pi$$

when $n = 1, 2, 3$, and so on, and increasing values of n correspond to higher modes. With the appropriate value of β obtained, the frequency of oscillation may be obtained from Eq. 7.19 which may be rearranged as

$$\omega = 2\pi f = \frac{\beta}{L}\sqrt{\frac{G}{\rho}} \quad \text{or} \quad f = \frac{\beta}{2\pi L}\sqrt{\frac{G}{\rho}} \tag{7.28}$$

The eigenvalues β for the three cases of shaft torsional vibration with classical boundary conditions are equally spaced and are summarized in Table 7.1.

TABLE 7.1. EIGENVALUES FOR SHAFTS WITH CLASSICAL BOUNDARY CONDITIONS.

Mode	Clamped-clamped	Clamped-free	Free-free
1	π	$\pi/2$	π
2	2π	$3\pi/2$	2π
3	3π	$5\pi/2$	3π
4	4π	$7\pi/2$	4π
and so on			

The modal shapes are, for clamped-clamped shaft,

$$\theta(\xi) = \sin \beta\xi \tag{7.29}$$

for clamped-free shaft,

$$\theta(\xi) = \sin \beta\xi \tag{7.30}$$

and for free-free shaft,

$$\theta(\xi) = \cos \beta\xi \tag{7.31}$$

In each case the nonzero constant in the modal shape expression has been arbitrarily set equal to unity.

Part 2. Shafts with Single Nonclassical Boundary Conditions

For illustrative purposes and in order to obtain a better understanding of the theory underlying the tables presented, we choose to look at a specific problem in detail. Consider the determination of the torsional vibration frequencies and modal shapes of a shaft which is clamped at one end ($\xi = 0$) and has a torsion spring attached at the other.

The appropriate boundary conditions to be enforced are

$$(1)\ \theta(\xi) = 0\Big|_{\xi=0} \qquad \text{and} \qquad (2)\ \frac{d\theta(\xi)}{d\xi} = -K^*\theta(\xi)\Big|_{\xi=1}$$

The general expression for the angular displacement is

$$\theta(\xi) = A \sin \beta\xi + B \cos \beta\xi$$

and it is evident that the first boundary condition above requires that $B = 0$.

From the second boundary condition we obtain

$$\beta \cos \beta = -K^* \sin \beta$$

or

$$K^* = \frac{-\beta}{\tan \beta} \tag{7.32}$$

It is found that as K^* approaches zero this equation approaches that governing the vibration of a clamped-free shaft. As K^* approaches infinity, the equation approaches that of a clamped-clamped shaft. Solutions for both of these classical cases are readily obtained from Table 7.1.

The frequencies for any shaft of the type under consideration lie somewhere between those associated with the classical limits. Again, as in the case of lateral beam vibration, we introduce a factor Q which is a measure of how far the actual frequencies lie between the well-defined limits. For the fundamental frequency of the shaft being considered, we may write

$$f = f_{CF} + Q(f_{CC} - f_{CF}) \tag{7.33}$$

where subscripts CF and CC denote clamped-free and clamped-clamped. The parameter Q is dependent on the dimensionless spring constant K^* only, and if this relationship is established for the modes of interest it is obvious that the frequency of any shaft can be quickly determined. This relationship is obtained by means of relationships between Q and β, and β and K^*. It is seen from the definition of β (Eq. 7.19) that it is proportional to f, hence

$$\frac{\beta}{\beta_{CF}} = \frac{f}{f_{CF}}$$

and utilizing Eq. 7.33 we obtain

$$\beta = \beta_{CF}\frac{f}{f_{CF}} = \beta_{CF}\left[1 + Q\left(\frac{f_{CC}}{f_{CF}} - 1\right)\right]$$

or

$$\beta = \beta_{CF}\left[1 + Q\left(\frac{\beta_{CC}}{\beta_{CF}} - 1\right)\right] \tag{7.34}$$

With Q selected we obtain a corresponding value of β, from which in turn we obtain a corresponding value of K^* from Eq. 7.32. With a plot of Q versus K^*, similar to that for lateral beam vibration as shown in Fig. 1.3, we could readily establish the fundamental frequency of torsional vibration for any shaft whatever of the type under consideration. It is found, however, as in the case of lateral vibration, that it is advantageous to present the data in tabulated form as in Table 7.2. Equation 7.33 is valid for any mode and any applicable type of boundary conditions provided the proper upper and lower limits, and Q versus K^* tables, are employed.

The tables have been presented with a view toward permitting analysis of any shaft whatever with an accuracy of about 1 percent. For values of Q up to 0.8, linear interpolation may be used to analyze shafts with parameters lying part way between the tabulated values. For values of Q greater than 0.8, intermediate values may be obtained according to the formula

$$Q = 1 - .2\frac{K^*_{Q=.8}}{K^*} \tag{7.35}$$

This formula is identical to that utilized in lateral beam vibration analysis. The term $K^*_{Q=.8}$ indicates the value of K^* associated with $Q = .8$. For shafts with rotational masses, K^* is of course replaced by I^*_0. Use of the tables is illustrated with examples.

In the presentation of data for the vibration analysis each case is given a reference number and a brief description. In addition, the eigenvalue equation and modal shape expressions are given along with the appropriate formula for obtaining frequencies for any of the first six modes once the value of the Q factor is obtained from the tables. Each case has an associated table provided, the table number and case number being identical. Subscripts n denote mode number.

CASE 7.2. Shaft clamped at one end, and with in-plane torsional spring attached at the other.

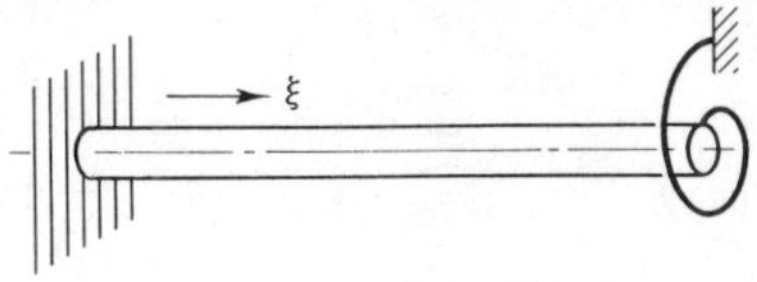

$$K^* = \frac{-\beta}{\tan \beta}$$

$$\theta(\xi) = \sin \beta\xi$$

$$f_n = f_{CFn} + Q(f_{CCn} - f_{CFn})$$

CASE 7.3. Shaft free at one end, and with in-plane torsional spring attached at the other.

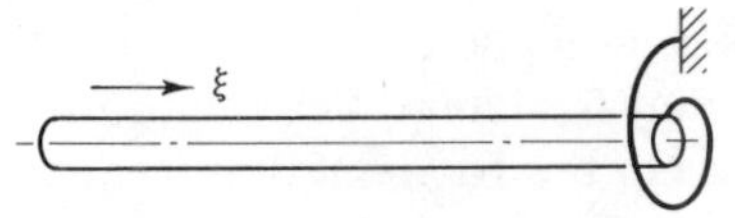

$$K^* = \beta \tan \beta$$

$$\theta(\xi) = \cos \beta\xi$$

$$f_n = Qf_{CF_n} \qquad n = 1$$

$$= f_{FF_{n-1}} + Q(f_{CF_n} - f_{FF_{n-1}}) \qquad n > 1$$

CASE 7.4. Shaft clamped at one end, and with rotational mass attached at the other.

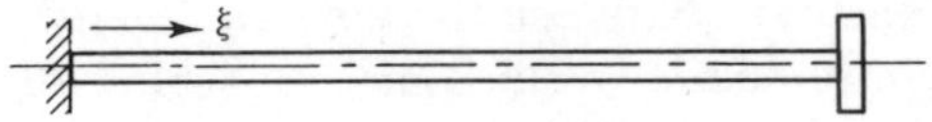

$$I_0^* = \beta \tan \beta$$

$$\theta(\xi) = \sin \beta\xi$$

$$f_n = Qf_{CFn} \qquad n = 1$$

$$= f_{CC_{n-1}} + Q(f_{CF_n} - f_{CC_{n-1}}) \qquad n > 1$$

CASE 7.5. Shaft free at one end, and with rotational mass attached at the other.

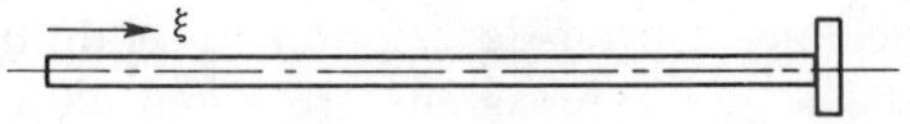

$$I_0^* = \frac{-\beta}{\tan \beta}$$

$$\theta(\xi) = \cos \beta\xi$$

$$f_n = f_{CF_n} + Q(f_{FF_n} - f_{CF_n})$$

ILLUSTRATIVE PROBLEM 7.1

A rock-drilling device consists of a steel circular tube attached to a cutting tool. The tube is 20 ft in length, with an i.d. and o.d. of $\frac{1}{2}$ and $1\frac{1}{2}$ in., respectively. The moment of inertia of the cutting tool is 375 lb in.^{2}.

The density of steel may be taken as .283 lb/in.^{3} and its shear modulus of elasticity is 12.0×10^6 lb/in.^{2}.

(a) Determine the fundamental frequency of torsional vibration assuming that oscillatory motion is forbidden at the driven end.

The polar moment of inertia of the steel tube

$$= \pi \frac{(D_o^4 - D_i^4)}{32} = .491 \text{ in.}^4$$

where D_o and D_i are outside and inside diameters, respectively.

The dimensionless parameter I_0^*

$$= \frac{I_p \rho L}{I_0} = \frac{.491 \times .283 \times 20 \times 12}{375} = .089$$

The value of the associated Q factor is obtained from Table 7.4 using linear interpolation

$$Q = .15 + \frac{.089 - .0566}{.1021 - .0566} \times .05 = .185$$

The upper limit

$$f_{CF} = \frac{\beta_{CF}}{2\pi L} \sqrt{\frac{G}{\rho}}$$

where $\beta_{CF} = \pi/2$ (Table 7.1).

$$\therefore f_{CF} = \frac{1}{4L} \sqrt{\frac{G}{\rho}} = \frac{1}{4 \times 240} \sqrt{\frac{12.0 \times 10^6 \times 386}{.283}} = 133 \text{ Hz}$$

where the factor 386 has been introduced to convert density to compatible units of lb sec^2/in.^{4}.

Then

$$f = Qf_{CF} = .185 \times 133 = 24.6 \text{ Hz}$$

(b) Determine the modal shape of the oscillatory motion. Since the frequency is proportional to the eigenvalue β we may write

$$\frac{f}{f_{CF}} = \frac{\beta}{\beta_{CF}}$$

$$\therefore \beta = \frac{f}{f_{CF}} \beta_{CF} = \frac{24.6}{133} \times \frac{\pi}{2} = .290$$

$$\therefore \theta(\xi) = \sin .290\xi$$

or, in terms of inches x measured from the driven end, replacing ξ by x/L we obtain

$$\theta(x) = \sin\frac{.290x}{240} = \sin 1.21 \times 10^{-3}x$$

(c) Determine the first mode frequency if oscillatory motion is permitted at the driven end. This problem corresponds to Case 7.5, and referring to Table 7.5 it is found that

$$Q = \frac{.089}{.1298} \times .05 = .342 \times 10^{-1}$$

$$f_{FF} = f_{CF}\frac{\beta_{FF}}{\beta_{CF}} = f_{CF} \times 2 = 266 \text{ Hz}$$

$$f = f_{CF} + Q(f_{FF} - f_{CF}) = 133 + .342 \times 10^{-1}(266 - 133) = 138 \text{ Hz}$$

(d) Determine the Q factor for part (a) if the moment of inertia of the cutting tool is increased to 2250 lb in.^{2}.

$$I_0^* = \frac{.089 \times 375}{2250} = .015$$

Referring to Table 7.4 and interpolating, obtain

$$Q = .05 + \frac{.015 - .0062}{.0249 - .0062} \times .05 = .074$$

Part 3. Shafts with Two Nonclassical Boundary Conditions

This family of problems constitutes numerous different cases, some of which are fairly commonly encountered in practical engineering work. For illustrative purposes two of the cases are worked out in detail.

We consider first a circular shaft with rotational masses of moments of inertia I_{01} and I_{02} attached to the left and right ends, respectively (see inset, Case 7.6). The boundary conditions are expressed mathematically as

$$\frac{d\theta(\xi)}{d\xi} = -\frac{\beta^2}{I_{01}^*}\,\theta(\xi)\bigg|_{\xi=0} \quad \text{and} \quad \frac{d\theta(\xi)}{d\xi} = \frac{\beta^2}{I_{02}^*}\,\theta(\xi)\bigg|_{\xi=1} \tag{7.36}$$

Differentiating and substituting the appropriate values of the variable ξ we obtain,

$$A(1) + B\frac{\beta}{I_{01}^*} = 0$$

and (7.37)

$$A\left(\cos\beta - \frac{\beta}{I_{02}^*}\sin\beta\right) - B\left(\sin\beta + \frac{\beta}{I_{02}^*}\cos\beta\right) = 0$$

Existence of a nontrivial solution for the constants A and B of Eqs. 7.37 requires the determinant of their coefficient matrix to vanish, that is,

$$\sin\beta + \frac{\beta}{I_{02}^*}\cos\beta + \frac{\beta}{I_{01}^*}\cos\beta - \frac{\beta^2}{I_{01}^* I_{02}^*}\sin\beta = 0 \tag{7.38}$$

Introducing $\alpha = I_{02}^*/I_{01}^*$, that is, $I_{02}^* = \alpha I_{01}^*$, substituting in Eq. 7.38, and rearranging we obtain

$$I_{01}^{*2} + \frac{\beta(1+\alpha)}{\alpha\tan\beta} I_{01}^* - \frac{\beta^2}{\alpha} = 0 \tag{7.39}$$

This is a quadratic equation in I_{01}^* and, once values of β and α are selected, two associated values of I_{01}^* are obtained. The negative value is rejected.

For any fixed value of α, the lower limit frequency for first mode vibration is zero and is approached as the moments of inertia of the rotational masses approach infinity. This corresponds to a mode of vibration where the masses rotate in opposite directions. For all higher modes in the limit as the rotational masses approach infinite moment of inertia, the shaft ends become nodal points. Therefore, the lower limits for the higher modes are clamped-clamped conditions. For all modes the upper limits are represented by free-free end conditions.

We now look for the appropriate expression for the modal shape. We are at liberty to divide the general modal shape expression (Eq. 7.18) by the constant A. Having done this, and recognizing from Eqs. 7.37 that we may write $B/A = -I_{01}^*/\beta$, we obtain for the modal shape of the shaft under study

$$\theta(\xi) = \sin\beta\xi - \frac{I_{01}^*}{\beta}\cos\beta\xi \tag{7.40}$$

We next look at a problem involving a shaft with a torsion spring–connected rotational mass and work out a detailed solution. The torsion spring–connected rotational mass consists of a rotational mass mounted on a shaft and permitted to undergo constrained rotation relative to the shaft. The relative motion of the mass is opposed by a spring system which opposes relative motion with a restoring torque proportional to the departure from a neutral configuration. This type of attachment is often utilized in conjunction with an energy dissipating material to damp out torsional shaft vibration. Such a damper without the energy-dissipating device is illustrated in Fig. 7.4.

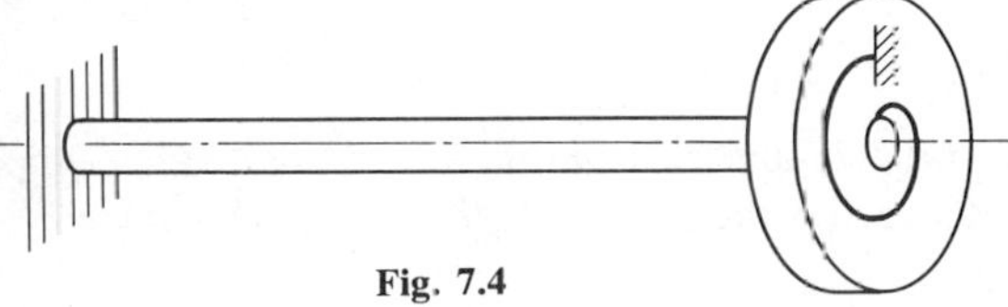

Fig. 7.4

No damping effects on the frequency and modal shapes of vibration are considered here, however, experience has shown that for light damping the frequencies and modal shapes are only slightly changed from those of free vibration, that is, with no damping action.

We consider the free torsional vibration of a shaft clamped at one end and with a torsion spring–connected rotational mass at the other (see inset, Case 7.11). By referring to the section on underlying theory, it is seen that, choosing the time scale properly, we may write

$$\phi(\xi,t) = (A \sin \beta\xi + B \cos \beta\xi) \cos \omega t \tag{7.41}$$

The absolute angular motion of the rotational mass is sinusoidal in nature, of the same frequency as the end of the shaft, and may be expressed as

$$\eta(t) = C \cos(\omega t + \alpha) \tag{7.42}$$

Equating the torque at the shaft outer end to the restoring moment provided by the spring, we obtain

$$\left.\frac{\partial \phi(\xi,t)}{\partial \xi}\right|_{\xi=1} = K^*[\eta(t) - \phi(\xi,t)]\Big|_{\xi=1} \tag{7.43}$$

The condition at the clamped end requires that $B = 0$, and substituting for ϕ and η in Eq. 7.43, and rearranging we obtain

$$A\{\beta \cos \beta + K^* \sin \beta\} \cos \omega t = CK^* \cos(\omega t + \alpha) \tag{7.44}$$

It is evident that for the equality stated in Eq. 7.44 to be satisfied the phase angle α must equal zero. The motion of the rotating mass and the end of the shaft are therefore in phase. Equation 7.44 may therefore be rewritten as

$$A\left(\frac{\beta \cos \beta}{K^*} + \sin \beta\right) - C = 0 \tag{7.45}$$

The requirement of dynamic equilibrium between the torque at the shaft outer end and the angular acceleration of the rotational mass permits us to write

$$-\left(\frac{GI_p}{L}\right)\left(\frac{\partial \phi(\xi,t)}{\partial \xi}\right)\Big|_{\xi=1} = I_0 \frac{\partial^2 \eta(t)}{\partial t^2} \tag{7.46}$$

Substituting for ϕ and η and rearranging we obtain

$$A\{\cos \beta\} - C\left\{\frac{\beta}{I_0^*}\right\} = 0 \tag{7.47}$$

Equations 7.45 and 7.47 are the homogeneous algebraic equations required for solving the vibration problem. The eigenvalue equation is obtained by setting the determinant of the coefficient matrix of these equations equal to zero. It may be written as

$$I_0^{*2} - I_0^*(\beta \tan \beta) - \frac{\beta^2}{\alpha} = 0 \tag{7.48}$$

where

$$\alpha = \frac{K^*}{I_0^*}$$

For α fixed, Eq. 7.48 is quadratic in I_0^*. The two values of I_0^* associated with any value of β are easily obtained, and the negative one is rejected.

Data for the torsional vibrational analysis of shafts with two nonclassical boundary conditions are presented in Tables 7.6 through 7.12. Each case is described and given a case number identical to the corresponding table number, as in Part 1. The major difference is that it is necessary to provide a separate table for each value of α, the ratio of the dimensionless parameters involved. While in practice this ratio can take on an infinity of values, experience has indicated that tables corresponding to values of α ranging from .1 to 1, in intervals of .1, and of $1/\alpha$ from .1 to 1 with similar increments, permit establishment of frequencies with error not greater than about 1 percent when linear interpolation is used. This interpolation is demonstrated later with illustrative examples.

It is recognized that in certain cases, such as the first one discussed in detail above, the symmetry of the problem eliminates the need for providing tables with the ratio α greater than one.

CASE 7.6. Shaft with rotational mass attached at each end.

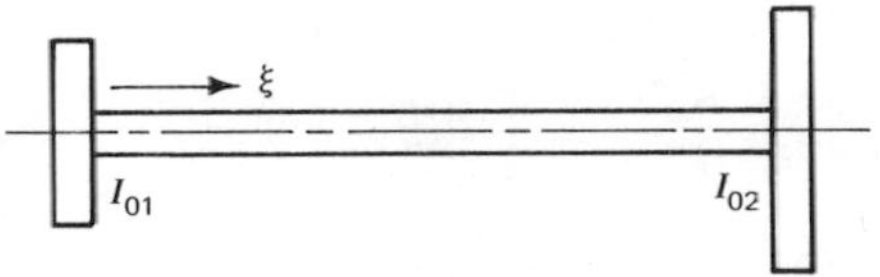

$$I_{01}^{*2} + I_{01}^* \frac{\beta(1+\alpha)}{\alpha \tan \beta} - \frac{\beta^2}{\alpha} = 0$$

where

$$\alpha = \frac{I_{02}^*}{I_{01}^*}$$

$$\theta(\xi) = \sin \beta\xi - \frac{I_{01}^*}{\beta} \cos \beta\xi$$

$$f_n = Qf_{FF} \qquad n = 1$$

$$= f_{CC_{n-1}} + Q(f_{FFn} - f_{CC_{n-1}}) \qquad n > 1$$

CASE 7.7. Shaft with in-plane torsion spring attached at each end.

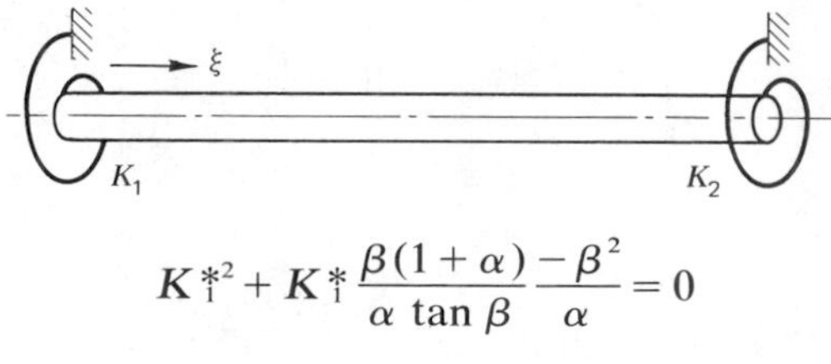

$$K_1^{*2} + K_1^* \frac{\beta(1+\alpha)}{\alpha \tan \beta} - \frac{\beta^2}{\alpha} = 0$$

where

$$\alpha = \frac{K_2^*}{K_1^*}$$

$$\theta(\xi) = \sin \beta\xi + \frac{\beta}{K_1^*} \cos \beta\xi$$

$$\begin{aligned} f_n &= Qf_{CCn} \qquad n = 1 \\ &= f_{FF_{n-1}} + Q(f_{CC} - f_{FF_{n-1}}) \qquad n > 1 \end{aligned}$$

CASE 7.8. Shaft with torsion spring attached at one end, and with rotational mass at the other.

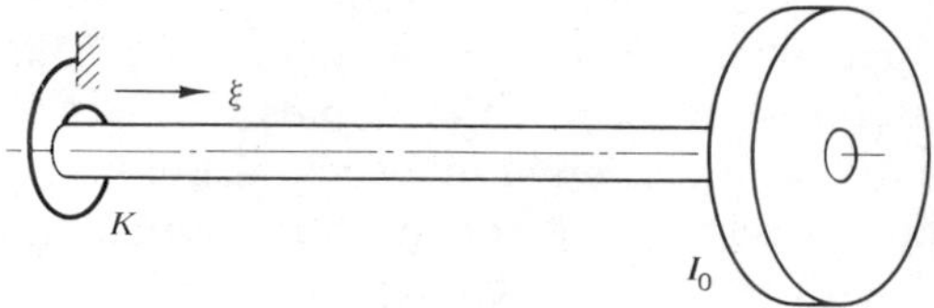

$$I_0^{*2} - I_0^* \beta \tan \beta \frac{1+\alpha}{\alpha} - \frac{\beta^2}{\alpha} = 0$$

where

$$\alpha = \frac{K^*}{I_0^*}$$

$$\theta(\xi) = \sin \beta\xi + \frac{\beta}{K^*} \cos \beta\xi$$

$$\begin{aligned} f_n &= Qf_{CFn} \qquad n = 1 \\ &= f_{CF_{n-1}} + Q(f_{CF_n} - f_{CF_{n-1}}) \qquad n > 1 \end{aligned}$$

CASE 7.9. Shaft clamped at one end, and with in-plane torsion spring and rotational mass at the other.

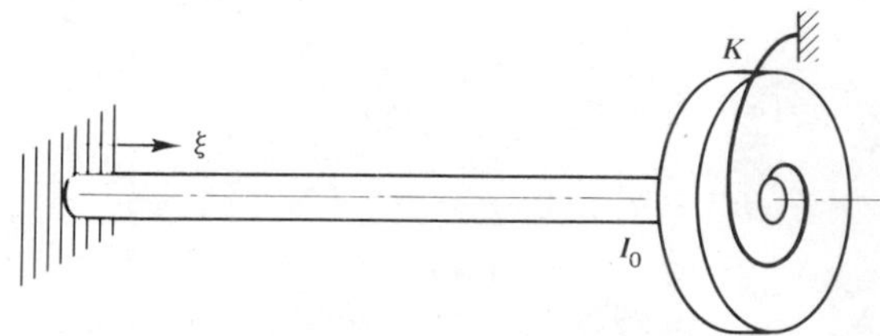

$$I_0^{*2} + I_0^* \frac{\beta}{\alpha \tan \beta} - \frac{\beta^2}{\alpha} = 0$$

where

$$\alpha = \frac{K^*}{I_0^*}$$

$$\theta(\xi) = \sin \beta\xi$$

$$f_n = Qf_{CCn} \qquad n = 1$$

$$= f_{CC_{n-1}} + Q(f_{CC_n} - f_{CC_{n-1}}) \qquad n > 1$$

CASE 7.10. Shaft free at one end, and with in-plane torsion spring and rotational mass at the other.

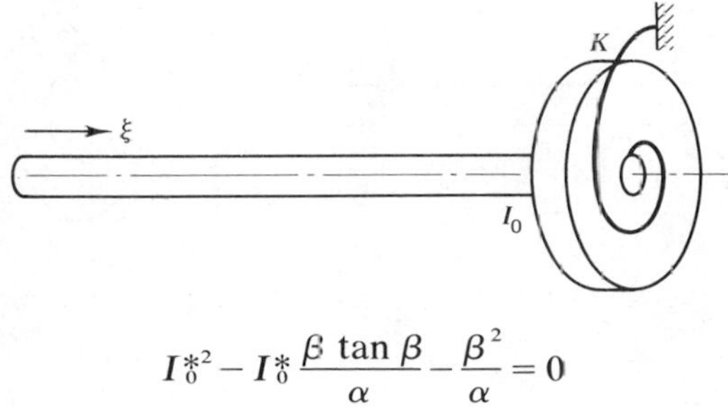

$$I_0^{*2} - I_0^* \frac{\beta \tan \beta}{\alpha} - \frac{\beta^2}{\alpha} = 0$$

where

$$\alpha = \frac{K^*}{I_0^*}$$

$$\theta(\xi) = \cos \beta\xi$$

$$f_n = Qf_{CF_n} \qquad n = 1$$

$$= f_{CF_{n-1}} + Q(f_{CF_n} - f_{CF_{n-1}}) \qquad n > 1$$

CASE 7.11. Shaft clamped at one end, and with torsion spring–connected rotational mass at the other.

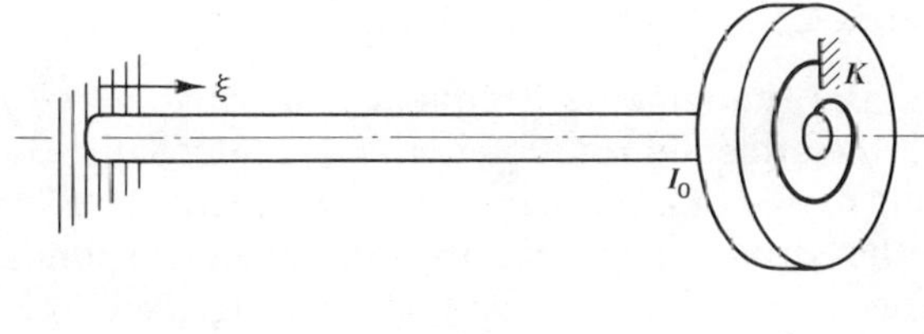

$$I_0^{*2} - I_0^* \beta \tan \beta - \frac{\beta}{\alpha} = 0$$

where

$$\alpha = \frac{K^*}{I_0^*}$$

$$\theta(\xi) = \sin \beta\xi$$

$$C = \text{rotational mass vibration amplitude} = I_0^* \frac{\cos \beta}{\beta}$$

$$f_n = Qf_{CF_n} \qquad n = 1$$
$$= f_{CF_{n-1}} + Q(f_{CF_n} - f_{CF_{n-1}}) \qquad n > 1$$

CASE 7.12. Shaft free at one end, and with torsion spring–connected rotational mass at the other.

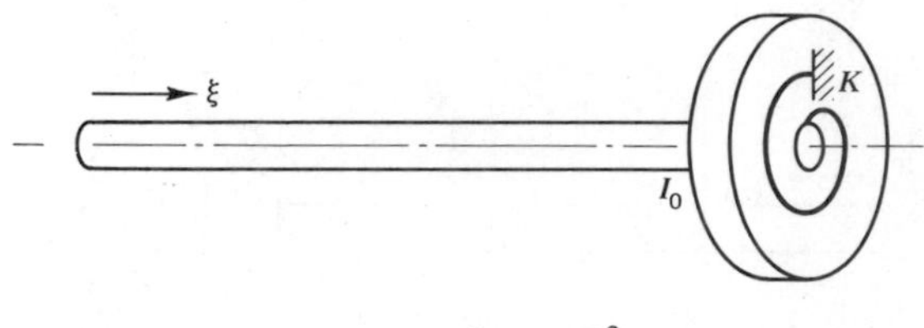

$$I_0^{*2} + I_0^* \frac{\beta}{\tan \beta} \frac{-\beta^2}{\alpha} = 0$$

where

$$\alpha = \frac{K^*}{I_0^*}$$

$$\theta(\xi) = \cos \beta\xi$$

$$C = \text{rotational mass vibration amplitude} = -I_0^* \frac{\sin \beta}{\beta}$$

$$f = Qf_{FF_n} \qquad n = 1$$
$$= f_{FF_{n-1}} + Q(f_{FF_n} - f_{FF_{n-1}}) \qquad n > 1$$

ILLUSTRATIVE PROBLEM 7.2

The hollow circular steel shaft of Illustrative Problem 7.1 has a cutting tool of moment of inertia of 375 lb in.^{2} at one end and an attached driving gear of moment of inertia of 312.5 lb in.^{2} at the other end.

(a) Determine the fundamental frequency of torsional vibration of the shaft with the attached rotational masses. (Note that this problem is comparable to that of an electric motor connected by a shaft to a pump rotor or other driven device). This problem corresponds to Case 7.6. We assure α is less than one by setting $I_{01} = 312.5$ lb in.^{2} and $I_{02} = 375$ lb in.^{2}.

Then

$$I_{02}^* = .089 \qquad \text{(Illustrative Problem 7.1)}$$

$$I_{01}^* = \frac{375}{312.5} \times .089 = .107$$

and

$$\alpha = \frac{.089}{.107} = .832$$

Referring to Table 7.6, and interpolating, we obtain for $\alpha = 0.8$,

$$Q = 0.10 + \frac{.107 - .0553}{.1258 - .0553} \times .05 = .137$$

for $\alpha = 0.9$,

$$Q = 0.10 + \frac{.107 - .0524}{.1191 - .0524} \times .05 = .141$$

Interpolating, for $\alpha = .832$,

$$Q = .137 + \frac{.832 - .800}{0.9 - 0.8}(.141 - .137) = .138$$

$$f = Qf_{FF}$$

where

$$f_{FF} = 266 \text{ Hz} \qquad \text{[Illustrative Problem 7.1(c)]}$$

$$\therefore f = .138 \times 266 = 36.7 \text{ Hz}$$

(b) Determine the modal shape for part (a).

$$\beta = \frac{f}{f_{FF}} \beta_{FF} = \frac{36.7}{266} \times \pi = .433$$

$$\therefore \theta(\xi) = \sin \beta\xi - \frac{I_{01}^*}{\beta} \cos \beta\xi$$

or

$$\theta(\xi) = \sin .433\xi - .247 \cos .433\xi$$

ILLUSTRATIVE PROBLEM 7.3

The circular steel tube of Illustrative Problem 7.1 is clamped at one end and has a cutting tool of moment of inertia 375 lb in.^{2} torsion spring–connected to the other end. The torsion spring has a stiffness of 19.1 in. lb/degree.

(a) Determine the first mode vibrational frequency in torsion. This problem corresponds to Case 7.11. The value of $I_0^* = .089$ and $f_{CF} = 133$ Hz (Illustrative Problem 7.1). We require the value of K in in.-lb/radian $= 19.10 \times (180/\pi) = 1.094 \times 10^3$.

Then

$$K^* = \frac{KL}{GI_p} = \frac{1.094 \times 10^3 \times 240}{12.0 \times 10^6 \times .491} = 44.5 \times 10^{-3}$$

and

$$\alpha = \frac{K^*}{I_0^*} = .500$$

Referring to Table 7.11,

$$Q = \frac{.089}{.1142} \times .05 = .039$$

$$\therefore f = .039 \times 133 = 5.19 \text{ Hz}$$

It is interesting to note that the presence of the torsion spring has reduced the frequency from 24.6 to 5.19 Hz (Illustrative Problem 7.1(a)).

(b) Determine the modal shape for part (a)

$$\beta = \frac{5.19}{f_{CF}} \times \frac{\pi}{2} = \frac{5.19 \times \pi}{133 \times 2} = .061$$

$$\therefore \theta(\xi) = \sin .061\xi$$

$$C = \text{rotational mass vibration amplitude}$$

$$= I_0^* \frac{\cos \beta}{\beta} = \frac{.089 \cos (.061)}{.061} = 1.43$$

It is seen that the ratio of amplitude of rotational mass vibration to shaft vibration at the same end ($\xi = 1$) is equal to 1.43/sin .061 = 23.5.

Part 4. Shafts with Intermediately Located Torsion Springs or Rotational Masses

Typical of the problems falling into this class is the problem of a shaft that is clamped at one end, free at the other, and with a rotational mass located dimensionless distance μ from the clamped end (see inset, Case 7.13). The value of μ is equal to the actual distance between the clamped end and the mass, divided by the shaft overall length. The displacements of the two sections of the shaft may be represented as (inset, Case 7.13)

$$\theta_1(\xi) = A_1 \sin \beta\xi + B_1 \cos \beta\xi \tag{7.49}$$

and

$$\theta_2(\xi) = A_2 \sin \beta\xi + B_2 \cos \beta\xi \tag{7.50}$$

where subscripts 1 and 2 refer to the left- and right-hand segments of the shaft, respectively. Denoting the dimensionless length of the second segment as γ, where of course $\gamma = 1 - \mu$, and enforcing the shaft outer end boundary conditions

$$\theta(\xi) = 0|_{\xi=0} \qquad \text{and} \qquad \frac{d\theta_2(\xi)}{d\xi} = 0|_{\xi=\gamma}$$

we obtain

$$B_1 = 0 \qquad \text{and} \qquad A_2 \cos\beta\gamma - B_2 \sin\beta\gamma = 0$$

or

$$A_2 = B_2 \tan\beta\gamma$$

$$\therefore\ \theta_1(\xi) = A_1 \sin\beta\xi \qquad \text{and} \qquad \theta_2(\xi) = B_2(\tan\beta\gamma \sin\beta\xi + \cos\beta\xi)$$

Two other boundary conditions arise as a result of continuity of angular displacement and dynamic equilibrium of the rotating mass where the segments join. They may be written, respectively, as

$$\theta_1(\xi)|_{\xi=\mu} = \theta_2(\xi)|_{\xi=0}$$

that is,

$$A_1(\sin\beta) - B_2 = 0 \tag{7.51}$$

and

$$\left.\frac{d\theta_1(\xi)}{d\xi}\right|_{\xi=\mu} - \left.\frac{d\theta_2(\xi)}{d\xi}\right|_{\xi=0} = \left.\frac{\beta^2}{I_0^*}\theta_1(\xi)\right|_{\xi=\mu}$$

Substituting for this latter boundary condition and rearranging, we obtain

$$A_1\left(\cos\beta\mu - \frac{\beta}{I_0^*}\sin\beta\mu\right) - B_2(\tan\beta\gamma) = 0 \tag{7.52}$$

Setting the determinant of the coefficient matrix of Eqs. 7.51 and 7.52 equal to zero gives rise to the eigenvalue equation

$$I_0^* = \frac{\beta \sin\beta\mu}{\cos\beta\mu - \tan\beta\gamma \sin\beta\mu} \tag{7.53}$$

To obtain expressions for the modal shapes we set A_1 equal to one, and using Eq. 7.51 to solve for B_2 we obtain

$$\theta_1(\xi) = \sin\beta\xi \tag{7.54}$$

and

$$\theta_2(\xi) = \sin\beta(\tan\beta\gamma \sin\beta\xi + \cos\beta\xi) \tag{7.55}$$

As in analogous cases for lateral vibration of beams, there appears initially to be some uncertainty in the selection of frequency limits for this shaft. The upper limit frequencies are evidently those for a clamped-free shaft. The first mode lower limit frequency is evidently zero, since as I_0 approaches zero (I^* approaching ∞) the rotational mass completely dominates the character of the oscillation. However, for all higher modes the intermediate point on the shaft approaches a nodal point as I_0 approaches infinity. We associate two distinct frequencies with this limit. With the left segment we associate the frequency of a clamped-clamped shaft of dimensionless length μ. With the right segment we associate the frequency of a clamped-free shaft of dimensionless length $\gamma = 1 - \mu$.

An intensive study of the variation of shaft eigenvalues with I_0^*, the dimensionless rotational mass constant, reveals the correct procedure for analyzing these shafts. The steps to be taken are as follows.

1. Establish numerical values for the first five frequencies of each segment with a clamped condition at the intermediate position.
2. Arrange these frequencies in order of ascending magnitude.
3. Use the first five frequencies in this new ascending order for the lower limits of modes 2 through 6. To avoid confusion these ordered frequencies will be designated by an asterisk: f_1^*, f_2^*, and so on.

We look next at a similar problem with a torsion spring located at the intermediate point and opposing departure of the shaft from the equilibrium configuration. It is seen that now the lower limits are well defined, but the upper limits appear to present difficulty since the intermediate position approaches a clamped condition as K^* approaches infinity. Here it is necessary to establish the first six numerical frequencies for each of the individual segments, with a clamped intermediate condition. These frequencies are then located in an ascending order and, denoting them as f_1^*, f_2^*, and so on, they are utilized for the upper limits of the first, second, and third modes, and so on. This procedure is illustrated later with an illustrative example.

We also wish to look at the problem of a shaft of dimensionless length one and of two segments of length μ and $\gamma = 1 - \mu$, as shown in the inset for Case 7.19 where the shaft segments are coupled by a torsion spring. The spring represents a torsion spring coupling where the mass of the associated flanges may be neglected. We recognize that the upper limit frequencies for the various modes of this shaft correspond to those of a clamped-free shaft, that is, the spring stiffness is infinite. The lower limit frequencies are not quite so obvious. It is necessary to establish ordered limits as in the case of concentrated masses. As the spring constant approaches zero, we have a set of easily established clamped-free frequencies to associate with the segment of length μ. For the other segment of length γ we associate a set of free-free frequencies. Having established both sets of frequencies, we then put them in an algebraically increasing order which we denote as f_n^*. The frequencies f_n^* then form a set of lower limits for this problem and are utilized according to the formulas provided. This procedure is also demonstrated later with an illustrative example.

It is evident that solutions to these problems are dependent on two parameters, namely, μ and K^* or I_0^*. Each case is given a case number which is identical to the corresponding table number. Tables are provided for values of μ varying from .1 to .9, with the spacing of intermediate values of μ selected so as to keep error associated with linear interpolation approximately below 1 percent. It is noted that, where possible, advantage is taken of symmetry and tables for $\mu > .5$ are not given.

CASE 7.13. Shaft clamped at one end, free at the other, and with intermediate rotational mass.

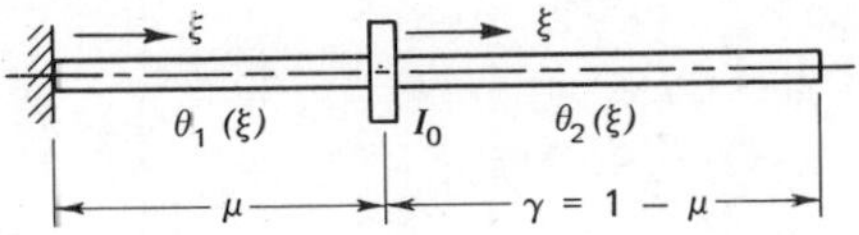

$$I_0^* = \frac{\beta \sin \beta\mu}{\cos \beta\mu - \tan \beta\gamma \sin \beta\mu}$$

$$\theta_1(\xi) = \sin \beta\xi$$

$$\theta_2(\xi) = \sin \beta\mu (\tan \beta\gamma \sin \beta\xi + \cos \beta\xi)$$

$$f_n = Q f_{CF_n} \qquad n = 1$$

$$= f_{n-1}^* + Q(f_{CF_n} - f_{n-1}^*) \qquad n > 1$$

with f_n^* based on f_{CC} for shaft segment of length μ, and f_{CF} for shaft segment of length $\gamma = 1 - \mu$.

CASE 7.14. Shaft free at both ends, and with intermediate rotational mass.

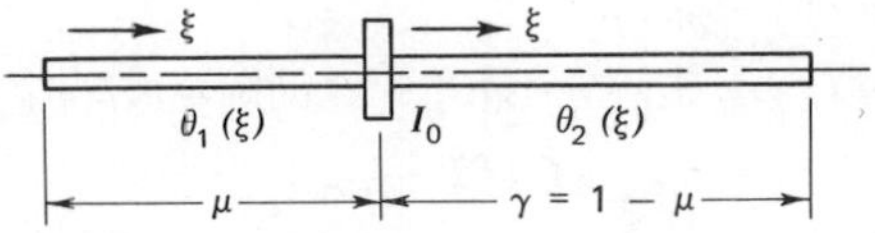

$$I_0^* = \frac{-\beta \cos \beta\mu}{\sin \beta\mu + \cos \beta\mu \tan \beta\gamma}$$

$$\theta_1(\xi) = \cos \beta\xi$$

$$\theta_2(\xi) = \cos \beta\mu (\tan \beta\gamma \sin \beta\xi + \cos \beta\xi)$$

$$f_n = f_n^* + Q(f_{FF_n} - f_n^*)$$

with f_n^* based on f_{CF} for shaft segment of length μ, and f_{CF} for shaft segment of length $\gamma = 1 - \mu$.

CASE 7.15. Shaft clamped at both ends, and with intermediate rotational mass.

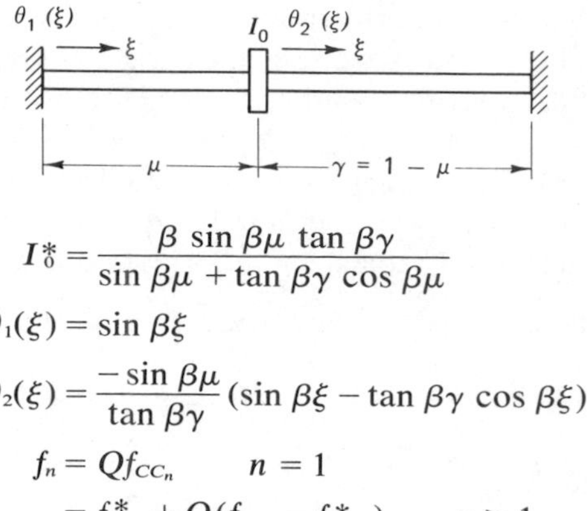

$$I_0^* = \frac{\beta \sin \beta\mu \tan \beta\gamma}{\sin \beta\mu + \tan \beta\gamma \cos \beta\mu}$$

$$\theta_1(\xi) = \sin \beta\xi$$

$$\theta_2(\xi) = \frac{-\sin \beta\mu}{\tan \beta\gamma}(\sin \beta\xi - \tan \beta\gamma \cos \beta\xi)$$

$$f_n = Qf_{CC_n} \qquad n = 1$$

$$= f_{n-1}^* + Q(f_{CC_n} - f_{n-1}^*) \qquad n > 1$$

with f_n^* based on f_{CC} for shaft segment of length μ, and f_{CC} for shaft segment of length $\gamma = 1 - \mu$.

CASE 7.16. Shaft clamped at one end, free at the other, and with in-plane intermediate torsion spring.

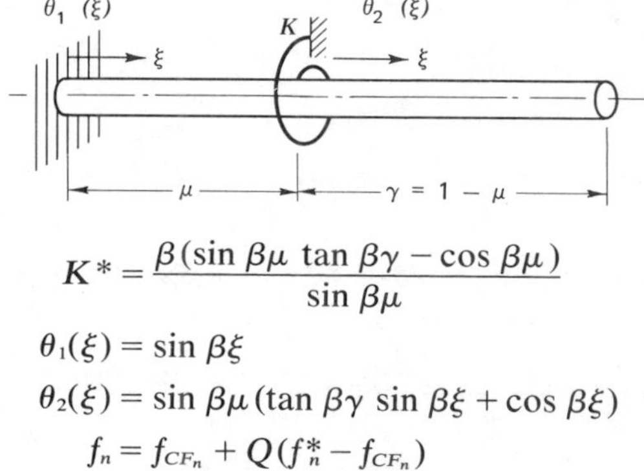

$$K^* = \frac{\beta(\sin \beta\mu \tan \beta\gamma - \cos \beta\mu)}{\sin \beta\mu}$$

$$\theta_1(\xi) = \sin \beta\xi$$

$$\theta_2(\xi) = \sin \beta\mu (\tan \beta\gamma \sin \beta\xi + \cos \beta\xi)$$

$$f_n = f_{CF_n} + Q(f_n^* - f_{CF_n})$$

with f_n^* based on f_{CC} for shaft segment of length μ, and f_{CF} for shaft segment of length $\gamma = 1 - \mu$.

CASE 7.17. Shaft free at each end, and with intermediate in-plane torsion spring.

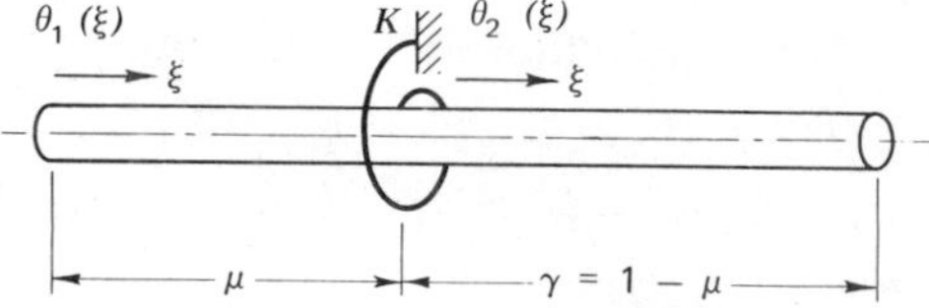

$$K^* = \frac{\beta(\sin\beta\mu + \cos\beta\mu \tan\beta\gamma)}{\cos\beta\mu}$$

$$\theta_1(\xi) = \cos\beta\xi$$

$$\theta_2(\xi) = \cos\beta\mu(\tan\beta\gamma \sin\beta\xi + \cos\beta\xi)$$

$$f_n = Qf_n^* \qquad n = 1$$

$$= f_{FF_{n-1}} + Q(f_n^* - f_{FF_{n-1}}) \qquad n > 1$$

with f^* based on f_{CF} for shaft segment of length μ, and f_{CF} for shaft segment of length $\gamma = 1 - \mu$.

CASE 7.18. Shaft clamped at each end, and with intermediate in-plane torsion spring.

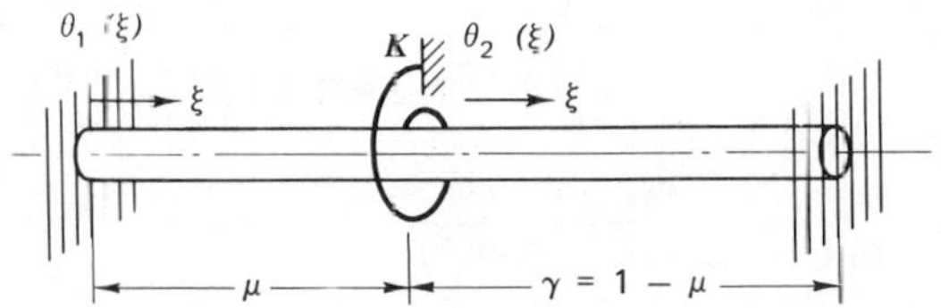

$$K^* = -\beta\frac{\sin\beta\mu + \tan\beta\gamma \cos\beta\mu}{\sin\beta\mu \tan\beta\gamma}$$

$$\theta_1(\xi) = \sin\beta\xi$$

$$\theta_2(\xi) = \frac{-\sin\beta\mu}{\tan\beta\gamma}(\sin\beta\xi - \tan\beta\gamma \cos\beta\xi)$$

$$f_n = f_{CC_n} + Q(f_n^* - f_{CC_n})$$

with f_n^* based on f_{CC} for shaft segment of length μ, and f_{CC} for shaft segment of length $\gamma = 1 - \mu$.

CASE 7.19. Double segment shaft clamped at one end, free at the other, and with segments joined by an in-plane torsion spring.

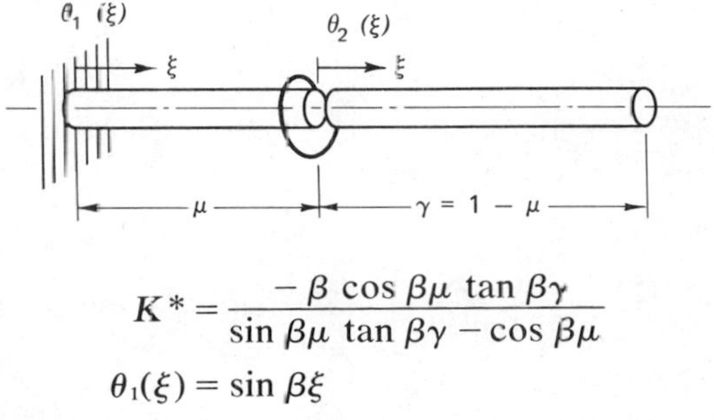

$$K^* = \frac{-\beta\cos\beta\mu \tan\beta\gamma}{\sin\beta\mu \tan\beta\gamma - \cos\beta\mu}$$

$$\theta_1(\xi) = \sin\beta\xi$$

$$\theta_2(\xi) = \left(\frac{\beta}{K^*}\cos\beta\mu + \sin\beta\mu\right)(\tan\beta\gamma\sin\beta\xi + \cos\beta\xi)$$

$$f_n = Qf_{CF_n} \qquad n = 1$$

$$f_n = f^*_{n-1} + Q(f_{CF_n} - f^*_{n-1}) \qquad n > 1$$

with f^* based on f_{CF} for shaft of length μ, and f_{FF} for shaft of length $\gamma = 1 - \mu$.

CASE 7.20. Double segment shaft free at both ends, and with segments joined by an in-plane torsion spring.

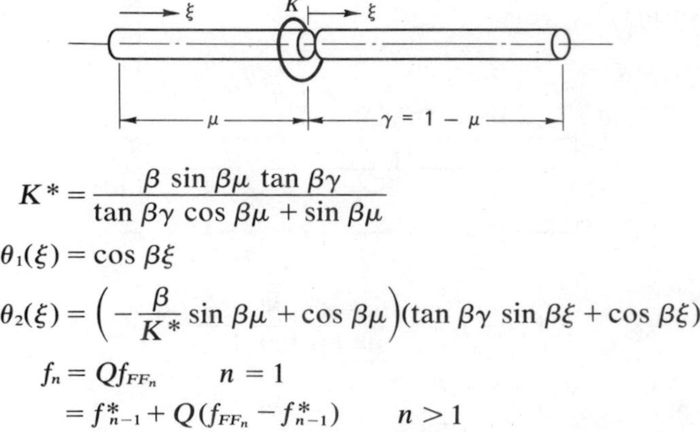

$$K^* = \frac{\beta\sin\beta\mu\tan\beta\gamma}{\tan\beta\gamma\cos\beta\mu + \sin\beta\mu}$$

$$\theta_1(\xi) = \cos\beta\xi$$

$$\theta_2(\xi) = \left(-\frac{\beta}{K^*}\sin\beta\mu + \cos\beta\mu\right)(\tan\beta\gamma\sin\beta\xi + \cos\beta\xi)$$

$$f_n = Qf_{FF_n} \qquad n = 1$$

$$= f^*_{n-1} + Q(f_{FF_n} - f^*_{n-1}) \qquad n > 1$$

with f^*_n based on f_{FF} for shaft segment of length μ, and f_{FF} for shaft segment of length $\gamma = 1 - \mu$.

CASE 7.21. Double segment shaft clamped at both ends, and with segments joined by an in-plane torsion spring.

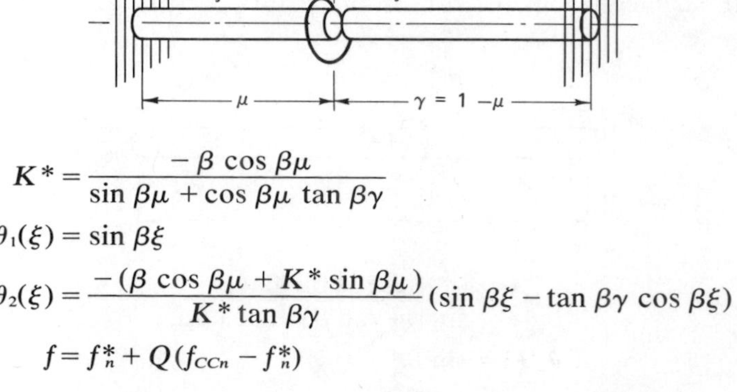

$$K^* = \frac{-\beta\cos\beta\mu}{\sin\beta\mu + \cos\beta\mu\tan\beta\gamma}$$

$$\theta_1(\xi) = \sin\beta\xi$$

$$\theta_2(\xi) = \frac{-(\beta\cos\beta\mu + K^*\sin\beta\mu)}{K^*\tan\beta\gamma}(\sin\beta\xi - \tan\beta\gamma\cos\beta\xi)$$

$$f = f^*_n + Q(f_{CCn} - f^*_n)$$

with f^* based on f_{CF} for shaft segment of length μ, and f_{CF} for shaft segment of length $\gamma = 1 - \mu$.

ILLUSTRATIVE PROBLEM 7.4

The circular steel bar of Illustrative Problem 7.1 is clamped at one end and has a rotational mass of moment of inertia of 375 lb in.^{2} attached 8 ft., 0 in. from the clamped end. It is free at the other end. Determine the first and second mode frequencies of the shaft.

This problem corresponds to that of Case 7.13. The value of $\mu = 8/20 = 0.4$. The dimensionless parameter $I_0^* = .089$ (Illustrative Problem 7.1). Referring to Table 7.13 ($\mu = 0.4$),

$$Q = .25 + \frac{.089 - .0647}{.0952 - .0647} \times .05 = .290$$

For first mode frequency,

$$f_{CF_1} = 133 \text{ Hz} \qquad \text{(Illustrative Problem 7.1)}$$

$$\therefore f_1 = Qf_{CF_1} = .290 \times 133 = 38.6 \text{ Hz.}$$

To obtain the ordered frequencies we proceed as follows. The frequencies are proportional to the eigenvalue and inversely proportional to shaft length. The first mode clamped-clamped frequency of the full length shaft equals the first mode free-free frequency which is 266 Hz [Illustrative Problem 7.1(c)], therefore the clamped-clamped frequencies of the shaft segment of length μ are

$$f_1 = \frac{266}{.4} = 665 \text{ Hz}$$

$$f_2 = \frac{665 \times 2\pi}{\pi} = 1330 \text{ Hz} \qquad f_3 = 665 \times \frac{3\pi}{\pi} = 1995 \text{ Hz}$$

and so on. In summary, the limit frequencies associated with the span of length μ are 665, 1330, 1995, 2660, and 3325 Hz.

The limit frequencies associated with the segment of length $\gamma = 1 - \mu$ are obtained in an identical manner, considering the segment to be in a clamped-free condition, and are

$$f_1 = \frac{133}{.6} = 222 \text{ Hz} \qquad \text{[Illustrative Problem 7.1(a)]}$$

$$f_2 = 222 \times \frac{3\pi/2}{\pi/2} = 666 \text{ Hz} \qquad f_3 = 222 \times \frac{5\pi/2}{\pi/2} = 1110 \text{ Hz}$$

and so on.

In summary, these limits are 222, 666, 1110, 1554, and 1998 Hz. Putting these limit frequencies in ascending order, we obtain

$$f_1^* = 222 \qquad f_2^* = 665 \qquad f_3^* = 666 \qquad f_4^* = 1110 \qquad f_5^* = 1330 \text{ Hz}$$

Referring to Table 7.13 ($\mu = .4$) for second mode vibration, we obtain

$$Q = \frac{.089 \times .05}{.1658 - 0.0} = .027$$

and since

$$f_{CF_2} = f_{CF_1} \frac{3\pi/2}{\pi/2} = 399 \text{ Hz} \qquad \text{[Illustrative Problem 7.1(a)]}$$

$$\therefore f_2 = 222 + .027(399 - 222) = 227 \text{ Hz}$$

ILLUSTRATIVE PROBLEM 7.5

The circular steel bar of Illustrative Problem 7.1 is clamped at one end, free at the other, and is composed of two segments joined at distance 8 ft, 0 in. from the clamped end with a torsion spring coupling of stiffness 19.1 in. lb/degree. Determine the first and second mode frequencies of the shaft. This problem corresponds to Case 7.19. The value of $K^* = 44.5 \times 10^{-3}$ [Illustrative Problem 7.3(a)]. The value of $\mu = 8/20 = 0.4$. $f_{CF_1} = 133$ Hz (Illustrative Problem 7.1). From Table 7.19 ($\mu = 0.4$),

$$Q = .15 + \frac{44.5 \times 10^{-3} - .0340}{.0614 - .0340} \times .05 = .169$$

for first mode vibration. Then

$$f = Qf_{CF_1} = .169 \times 133 = 22.5 \text{ Hz}$$

For second mode vibration,

$$Q = \frac{44.5 \times 10^{-3}}{.0633} \times .05 = .035$$

The limit frequencies associated with the clamped end segment are those for a clamped-free shaft of length μ. They are obtained in the same manner as those of Illustrative Problem 7.4 and are 333, 998, 1663, 2328, and 2993 Hz. The limit frequencies associated with the other segment are those for a free-free shaft of length γ and are 443, 886, 1329, 1772, and 2215 Hz. The ordered frequencies are therefore

$f_1^* = 333 \qquad f_2^* = 443 \qquad f_3^* = 886 \qquad f_4^* = 998 \qquad f_5^* = 1329$ and $f_{CF_2} =$ 399 Hz (Illustrative Problem 7.4).

Then

$$f_2 = 333 + .035(399 - 333) = 335 \text{ Hz}.$$

Part 5. Shafts with Intermediately Located Rotational Masses or Torsion Spring Couplings, and Rotational Masses Attached at One End

Typical of this family of problems is that of a shaft fixed at one end, with an intermediately located torsion spring coupling, and with a rotational mass attached at the other end (see inset, Case 7.22). Utilizing the general expressions

$$\theta_1(\xi) = A_1 \sin \beta\xi + B_1 \cos \beta\xi \tag{7.56}$$

and

$$\theta_2(\xi) = A_2 \sin \beta\xi + B_2 \cos \beta\xi \tag{7.57}$$

and enforcing the clamped boundary condition

$$\theta_1(\xi)\big|_{\xi=0} = 0$$

we obtain $B_1 = 0$, hence

$$\theta_1(\xi) = A_1 \sin \beta\xi \tag{7.58}$$

To satisfy dynamic equilibrium at the right end, we write

$$\frac{d\theta_2(\xi)}{d\xi} = \frac{\beta^2}{I_0^*}\,\theta_2(\xi)\bigg|_{\xi=\gamma}$$

or

$$A_2 \cos \beta\gamma - B_2 \sin \beta\gamma = \frac{\beta}{I_0^*}(A_2 \sin \beta\gamma + B_2 \cos \beta\gamma) \tag{7.59}$$

Solving for B_2, and substituting in Eq. 7.57, we obtain

$$\theta_2(\xi) = A_2\left(\sin \beta\xi + \frac{I_0^* \cos \beta\gamma - \beta \sin \beta\gamma}{I_0^* \sin \beta\gamma + \beta \cos \beta\gamma} \cos \beta\xi\right) \tag{7.60}$$

There are two other boundary conditions that may be utilized. From continuity of torque across the torsion spring, we have

$$\frac{d\theta_1(\xi)}{d\xi}\bigg|_{\xi=\mu} = \frac{d\theta_2(\xi)}{d\xi}\bigg|_{\xi=0}$$

or, substituting for $\theta_1(\xi)$ and $\theta_2(\xi)$,

$$A_1 \cos \beta\mu - A_2 = 0 \tag{7.61}$$

Also, the torque exerted by the torsion spring is proportional to $\theta_2(\xi) - \theta_1(\xi)$ at the junction of the shaft segments, and we obtain

$$\frac{d\theta_1(\xi)}{d\xi}\bigg|_{\xi=\mu} = -K^*\left[\theta_1(\xi)\bigg|_{\xi=\mu} - \theta_2(\xi)\bigg|_{\xi=0}\right]$$

or, substituting for $\theta_1(\xi)$ and $\theta_2(\xi)$,

$$A_1(\beta \cos \beta\mu + K^* \sin \beta\mu) - A_2 \frac{K^*(I_0^* \cos \beta\gamma - \beta \sin \beta\gamma)}{I_0^* \sin \beta\gamma + \beta \cos \beta\gamma} = 0 \qquad (7.62)$$

Introducing $\alpha = K^*/I_0^*$ and setting the determinant of the coefficient matrix of Eqs. 7.61 and 7.62 equal to zero, we obtain the eigenvalue equation

$$I_0^{*2} + I_0^*\beta \frac{(1+\alpha)\cos \beta\mu \sin \beta\gamma + \alpha \cos \beta\gamma \sin \beta\mu}{\alpha(\sin \beta\mu \sin \beta\gamma - \cos \beta\mu \cos \beta\gamma)} + \frac{\beta^2 \cos \beta\mu \cos \beta\gamma}{\alpha(\sin \beta\mu \sin \beta\gamma - \cos \beta\mu \cos \beta\gamma)} = 0 \qquad (7.63)$$

There are three parameters involved in the quadratic Eq. 7.63, namely, I_0^*, α, and μ. For fixed α, the upper limits of frequency for any mode of shaft torsional vibration are evidently f_{CF} (I_0^* and K^* approaching infinity). For the first mode the lower limit is zero (as I_0^* and K^* approach zero). For the higher modes the lower limits are denoted by f^* and are those frequencies associated with each segment as I_0^* and K^* approach zero, the frequencies being arranged in ascending order as discussed in Part 3. As K^* approaches zero, the frequencies associated with the left segment are the clamped-free frequencies of a shaft of length μ. As K^* and I_0^* approach zero, the frequencies associated with the right segment are the clamped-free frequencies of a shaft of length γ (the right end approaches a nodal point as I_0^* approaches zero for all the higher modes).

Each case in the family of problems is given a number and the corresponding table has the same number. Data are presented for values of μ varying between .2 and .8 in intervals of .1.

CASE 7.22. Shaft clamped at one end, with rotational mass at the other, and with intermediate torsion spring coupling.

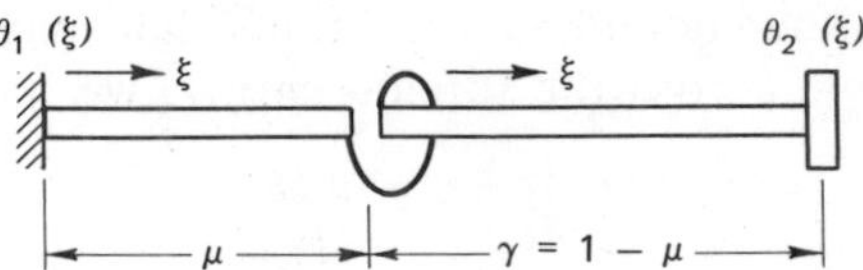

$$I_0^{*2} + I_0^*\beta \frac{(1+\alpha)\cos \beta\mu \sin \beta\mu + \alpha \cos \beta\gamma \sin \beta\mu}{\alpha(\sin \beta\mu \sin \beta\gamma - \cos \beta\mu \cos \beta\gamma)} + \frac{\beta^2 \cos \beta\mu \cos \beta\gamma}{\alpha(\sin \beta\mu \sin \beta\gamma - \cos \beta\mu \cos \beta\gamma)} = 0$$

where

$$\alpha = \frac{K^*}{I_0^*}$$

$$\theta_1(\xi) = \sin \beta\xi$$

$$\theta_2(\xi) = \cos \beta\mu\left(\sin \beta\xi + \frac{I_0^* \cos \beta\gamma - \beta \sin \beta\gamma}{I_0^* \sin \beta\gamma + \beta \cos \beta\gamma} \cos \beta\xi\right)$$

$$f_n = Q f_{CF_n} \qquad n = 1$$

$$= f_{n-1}^* + Q(f_{CF_n} - f_{n-1}^*) \qquad n > 1$$

with f_n^* based on f_{CF} for shaft segment of length μ, and f_{CF} for shaft segment of length $\gamma = 1 - \mu$.

CASE 7.23. Shaft free at one end, with rotational mass at the other, and with intermediate torsion spring coupling.

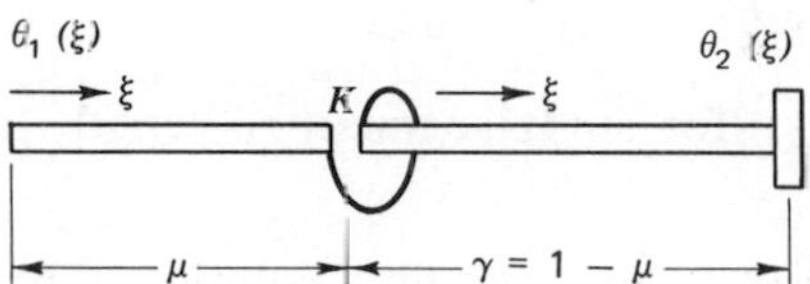

$$I_0^{*2} + I_0^*\left[\frac{\alpha\beta \cos \beta\mu \cos \beta\gamma - \beta(1+\alpha) \sin \beta\mu \sin \beta\gamma}{\alpha(\cos \beta\mu \sin \beta\gamma + \sin \beta\mu \cos \beta\gamma)}\right]$$

$$- \frac{\beta^2 \sin \beta\mu \cos \beta\gamma}{\alpha\,(\cos \beta\mu \sin \beta\gamma + \sin \beta\mu \cos \beta\gamma)} = 0$$

where

$$\alpha = \frac{K^*}{I_0^*}$$

$$\theta_1(\xi) = \cos \beta\xi$$

$$\theta_2(\xi) = -\sin \beta\mu\left(\sin \beta\xi + \frac{I_0^* \cos \beta\gamma - \beta \sin \beta\gamma}{I_0^* \sin \beta\gamma + \beta \cos \beta\gamma} \cos \beta\xi\right)$$

$$f_n = Q f_{FF_n} \qquad n = 1$$

$$= f_{n-1}^* + Q(f_{FF_n} - f_{n-1}^*) \qquad n > 1$$

with f_n^* based on f_{FF} for shaft segment of length μ, and f_{CF} for shaft segment of length $\gamma = 1 - \mu$.

CASE 7.24. Shaft clamped at one end, with rotational mass at the other, and with intermediate rotational mass.

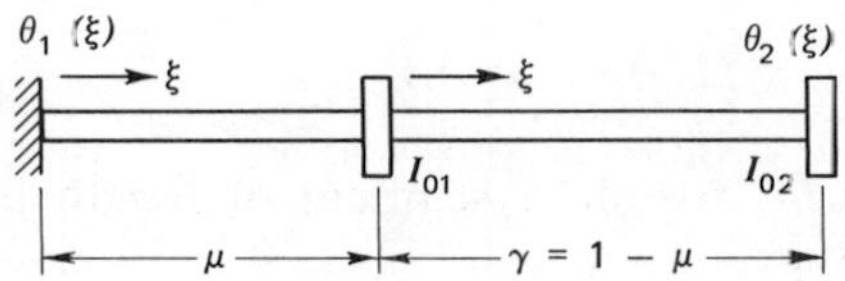

$$I_{02}^{*2} + I_{02}^{*}\beta \frac{(1+\alpha)\sin\beta\mu\cos\beta\gamma + \alpha\cos\beta\mu\sin\beta\gamma}{\alpha(\sin\beta\mu\sin\beta\gamma - \cos\beta\mu\cos\beta\gamma)} - \frac{\beta^2\sin\beta\mu\sin\beta\gamma}{\alpha(\sin\beta\mu\sin\beta\gamma - \cos\beta\mu\cos\beta\gamma)} = 0$$

where

$$\alpha = \frac{I_{01}^{*}}{I_{02}^{*}}$$

$$\theta_1(\xi) = \sin\beta\xi$$

$$\theta_2(\xi) = \left(\cos\beta\mu \frac{-\beta}{I_{01}^{*}}\sin\beta\mu\right)\left(\sin\beta\xi + \frac{I_{02}^{*}\cos\beta\gamma - \beta\sin\beta\gamma}{I_{02}^{*}\sin\beta\gamma + \beta\cos\beta\gamma}\cos\beta\xi\right)$$

$$f_n = Qf_{CF_n} \qquad n = 1$$

$$= f_{n-1}^{*} + Q(f_{CF_n} - f_{n-1}^{*}) \qquad n > 1$$

with f_n^{*} based on f_{CC} for shaft segment of length μ, and f_{CC} for shaft segment of length γ.

CASE 7.25. Shaft free at one end, with rotational mass at the other, and with intermediate rotational mass.

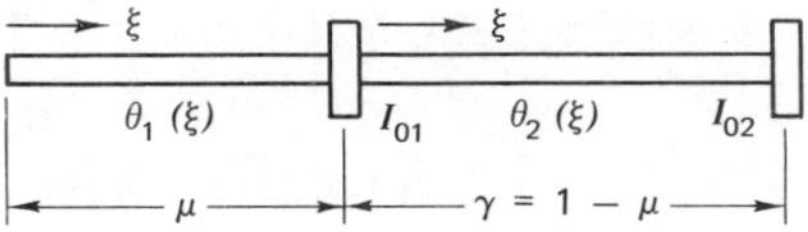

$$I_{02}^{*2} + I_{02}^{*}\beta \frac{(1+\alpha)\cos\beta\mu\cos\beta\gamma - \alpha\sin\beta\mu\sin\beta\gamma}{\alpha(\cos\beta\mu\sin\beta\gamma + \sin\beta\mu\cos\beta\gamma)} - \frac{\beta^2\cos\beta\mu\sin\beta\gamma}{\alpha(\cos\beta\mu\sin\beta\gamma + \sin\beta\mu\cos\beta\gamma)} = 0$$

where

$$\alpha = \frac{I_{01}^{*}}{I_{02}^{*}}$$

$$\theta_1(\xi) = \cos\beta\xi$$

$$\theta_2(\xi) = -\left(\frac{\beta}{I_{01}^{*}}\cos\beta\mu + \sin\beta\mu\right)\left(\sin\beta\xi + \frac{I_{02}^{*}\cos\beta\gamma - \beta\sin\beta\gamma}{I_{02}^{*}\sin\beta\gamma + \beta\cos\beta\gamma}\cos\beta\xi\right)$$

$$f_n = Qf_{FF_n} \qquad n = 1$$

$$= f_{n-1}^{*} + Q(f_{FF_n} - f_{n-1}^{*}) \qquad n > 1$$

with f^{*} based on f_{CF} for shaft segment of length μ, and f_{CC} for shaft segment of length γ.

ILLUSTRATIVE PROBLEM 7.6

The circular steel bar of Illustrative Problem 7.1 is free at one end, has a rotational mass of moment of inertia of 375 lb in.^{2} attached at the other, and is composed of two segments joined by a torsion spring coupling at a distance 4 ft, 0 in. from the rotational mass. The stiffness of the spring coupling is 251 in. lb/degree.

(a) Determine the first mode frequency.
This problem corresponds to Case 7.23. It is representative of a centrifugal pump impeller, or other rotational device, driven through a segmented, torsion spring–coupled shaft, where the driver attachment has negligible effect on the vibration.

$$K^* = \frac{251}{19.10} \times 44.5 \times 10^{-3} = .585 \qquad \text{[Illustrative Problem 7.3(a)]}$$

$$\mu = 16/20 = 0.8$$

$$I_0^* = .089 \qquad \text{[Illustrative Problem 7.1(a)]}$$

$$\alpha = \frac{K^*}{I_0^*} = 6.57$$

$$\therefore \frac{1}{\alpha} = .152$$

$$f_{FF_1} = 266 \text{ Hz} \qquad \text{[Illustrative Problem 7.1(c)]}$$

For $1/\alpha = 0.1$ ($\mu = .8$),

$$Q = .25 + \frac{.089 - .0606}{.0964 - .0606} \times .05 = .290 \qquad \text{(Table 7.23)}$$

For $1/\alpha = 0.2$ ($\mu = .8$),

$$Q = .20 + \frac{.089 - .0697}{.1151 - .0697} \times .05 = .221$$

Interpolating for $1/\alpha = .152$,

$$Q = .290 + .52(.221 - .290) = .254$$

$$\therefore f = Qf_{FF_1} = .254 \times 266 = 67.6 \text{ Hz}$$

(b) What has been the effect of the torsion spring coupling on the first mode frequency?
The first mode natural frequency for the same shaft and rotational mass, without the torsion-spring coupling, was found to be 138 Hz [Illustrative Problem 7.1(c)]. Therefore the coupling has had the effect of reducing the frequency from 138 to 67.6 Hz, that is, reducing it by approximately 50 percent.

Part 6. Shafts with Intermediately Located Torsion-Spring Coupling and Rotational Mass at Each End

This family of problems is one that is very likely to be encountered in practical engineering. A typical application is one in which an electric motor is used to drive a centrifugal pump. It may be desired to change the torsional frequencies of the system, and this can be accomplished by introducing a torsionspring coupling at an intermediate position along the shaft. The system is illustrated in the inset of Case 7.26. In analyzing this system the assumption is made here that the effect of the mass of the coupling is negligible compared to that of the shaft and the rotary masses at the ends of the shaft. Referring to the inset of Case 7.26 and utilizing the general expressions of Eqs. 7.56 and 7.57, we proceed to enforce the boundary conditions. For dynamic equilibrium at the left end of the shaft,

$$\left.\frac{d\theta_1(\xi)}{d\xi}\right|_{\xi=0} = -\frac{\beta^2}{I_{01}^*}\,\theta_1(\xi)\bigg|_{\xi=0}$$

from which we obtain

$$B_1 = -\frac{A_1 I_{01}^*}{\beta}$$

$$\therefore\ \theta_1(\xi) = A_1 \sin\beta\xi - \frac{I_{01}^*}{\beta}\cos\beta\xi \tag{7.64}$$

For dynamic equilibrium at the right end of the shaft

$$\left.\frac{d\theta_2(\xi)}{d\xi}\right|_{\xi=\gamma} = \frac{\beta^2}{I_{02}}\,\theta_2(\xi)\bigg|_{\xi=\gamma}$$

from which we obtain

$$B_2 = A_2\,\frac{I_{02}^*\cos\beta\gamma - \beta\sin\beta\gamma}{I_{02}^*\sin\beta\gamma + \beta\cos\beta\gamma}$$

$$\therefore\ \theta_2(\xi) = A_2\left(\sin\beta\xi + \frac{I_{02}^*\cos\beta\gamma - \beta\sin\beta\gamma}{I_{02}^*\sin\beta\gamma + \beta\cos\beta\gamma}\cos\beta\xi\right) \tag{7.65}$$

For continuity of torque across the torsion spring

$$\left.\frac{d\theta_1(\xi)}{d\xi}\right|_{\xi=\mu} = \left.\frac{d\theta_2(\xi)}{d\xi}\right|_{\xi=0}$$

from which we obtain

$$A_1\left(\cos\beta\mu + \frac{I_{01}^*}{\beta}\sin\beta\mu\right) + A_2(-1) = 0 \tag{7.66}$$

and for continuity of torque at the inner end of the left shaft segment

$$\left.\frac{d\theta_1(\xi)}{d\xi}\right|_{\xi=\mu} = -K^*[\theta_1(\mu) - \theta_2(0)]$$

from which we obtain

$$A_1\left[\left(\beta - \frac{K^* I_{01}^*}{\beta}\right)\cos\beta\mu + (I_{01}^* + K^*)\sin\beta\mu\right] - A_2 K^*\left(\frac{I_{02}^*\cos\beta\gamma - \beta\sin\beta\gamma}{I_{02}^*\sin\beta\gamma + \beta\cos\beta\gamma}\right) = 0 \quad (7.67)$$

Solution of the problem is obtained by requiring the determinant of the coefficient matrix of the unknowns A_1 and A_2, of Eqs. 7.66 and 7.67, to vanish.

Introducing $\alpha_1 = I_{02}^*/I_{01}^*$, $\alpha_2 = K^*/I_{01}^*$ and expanding the determinant we obtain

$$\overset{3}{I_{01}^*} + \overset{2}{I_{01}^*}\beta\frac{\alpha_2(1+\alpha_1)\cos\beta\gamma\cos\beta\mu - (\alpha_1\alpha_2 + \alpha_2 + \alpha_1)\sin\beta\gamma\sin\beta\mu}{\alpha_1\alpha_2(\cos\beta\mu\sin\beta\gamma + \cos\beta\gamma\sin\beta\mu)}$$

$$- I_{01}^*\beta\frac{\beta(\alpha_1+\alpha_2)\sin\beta\gamma\cos\beta\mu + \beta(\alpha_2+1)\sin\beta\mu\cos\beta\gamma}{\alpha_1\alpha_2(\cos\beta\mu\sin\beta\gamma + \sin\beta\mu\cos\beta\gamma)}$$

$$-\frac{\beta^3\cos\beta\gamma\cos\beta\mu}{\alpha_1\alpha_2(\cos\beta\mu\sin\beta\gamma + \sin\beta\mu\cos\beta\gamma)} = 0 \quad (7.68)$$

Equation 7.68 is seen to be a cubic equation in I_{01}^*, the quantity to be determined. It is found, as would be expected, that there is only one positive value of I_{01}^* which satisfies the equation.

It will be observed that there are four variables involved in this problem, namely I_{01}^*, I_{02}^*, K^* and μ. The table storage space required would become excessive if a format similar to that employed for previous problems was followed. To help minimize storage space requirements the following steps were taken.

1. Data are provided for the first mode of vibration only.
2. Values of μ are restricted to those between 0.1 and 0.9 in intervals of 0.2.

It is recognized that in general, the first mode of vibration is the one of greatest interest to designers. In special cases where an interest in higher modes exists, Eq. 7.68 may be used to solve for the appropriate eigenvalues and frequencies.

Since we restrict ourselves here to first mode vibration there can be only one nodal point along the shaft and it is reasonable to expect that greater spacing in the tables of the values of μ would be permissible. A study of the results indicates this to be so. To further minimize space requirements only 10 values of the Q factor are given in each table. Linear interpolation may be used for $Q < .8$, and for values of $Q > .8$ the asymptotic approximation discussed at the beginning of the chapter may be used. On examining a typical table of the set 7.26, it is seen that the first column of each table gives values for α_1. Beside each table heading are found the pertinent values of α_2 and μ. The value of I^*_{01} associated with any value of Q is found in the column below that value of Q, and across from the pertinent value of α_1. An illustrative example is presented. A study of the results has indicated that error associated with linear interpolation in general is not greater than 1 percent, except for interpolation between adjacent values of Q for $Q > .2$. In the latter situation error up to about 3 percent may be encountered. This small error could be reduced considerably, if desired, by plotting adjacent points of the Q curve and "fairing in" a curve that would be an improvement over the linear interpolation.

We next look at the lower and upper limits for this shaft system as I^*_{01} ranges from zero to infinity for fixed α_1 and α_2. As I^*_{01} approaches zero, I_{01} and I_{02} approach infinity, and K, the torsion spring constant, approaches zero. The lower limit frequency is then evidently equal to zero with the rotational masses rotating in opposite directions. As I^*_{01} approaches infinity, I_{01} and I_{02} approach zero and K approaches infinity. The upper limit is then evidently equal to f_{FF}, the first torsional frequency of the full length shaft with free ends. For any such shaft system the frequency can therefore be written as

$$f = Qf_{cc} \tag{7.69}$$

and the eigenvalue is given by

$$\beta = Q \times \pi \tag{7.70}$$

Following established practices, this case is given a case number and a small sketch of the beam along with modal shape information is given below. The associated tables have the same number. It will be appreciated that the axis can always be selected so that $I^*_{01} > I^*_{02}$ and therefore tables for values of $\alpha_1 > 1.0$ are not required.

CASE 7.26. Shaft with rotational mass at each end and intermediately located torsionspring coupling.

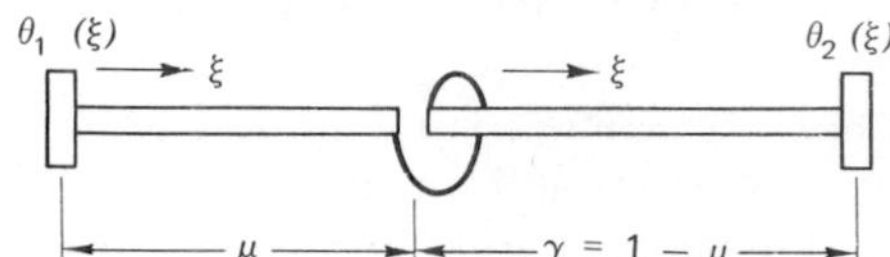

Determinant equation: Eq. 7.68

$$\theta_1(\xi) = \sin \beta\xi - \frac{I_{01}^*}{\beta} \cos \beta\xi$$

$$\theta_2(\xi) = \gamma_1 \left(\sin \beta\xi + \frac{I_{02}^* \cos \beta\gamma - \beta \sin \beta\gamma}{I_{02}^* \sin \beta\gamma + \beta \cos \beta\gamma} \cos \beta\xi \right)$$

where

$$\gamma_1 = \cos \beta\mu + \frac{I_{01}^*}{\beta} \sin \beta\mu$$

$$f = \text{first mode frequency} = Qf_{FF}$$

$$\alpha_1 = \frac{I_{02}^*}{I_{01}^*}$$

$$\alpha_2 = \frac{K^*}{I_{01}^*}$$

ILLUSTRATIVE PROBLEM 7.7

Determine the first mode vibration frequency of the hollow circular steel shaft of Illustrative Problem 7.2 if a torsion spring coupling of stiffness 459.3 in. lb/degree is located at the mid-shaft position.

$$I_{01}^* = .107 \qquad \alpha_1 = \frac{I_{02}^*}{I_{01}^*} = .832 \text{ (from Ex. 7.2a)}$$

$$K^* = \frac{459.3 \times 44.5 \times 10^{-3}}{19.1} = 1.070 \text{ (from Ex. 7.3a)}$$

$$\therefore \frac{1}{\alpha_2} = \frac{K^*}{I_{01}^*} = \frac{1.070}{.107} = 10$$

and

$$\frac{1}{\alpha_2} = 0.1$$

$$f_{FF} = 266 \text{ Hz (from Ex. 7.1c)}$$

Referring to Table 7.26,

$$\left(\frac{1}{\alpha_2} = 0.1, \ \mu = .5\right)$$

$$\text{For } \alpha_1 = .800; \ Q = .05 + \frac{.107 - .0449}{.1084 - .0449} \times .05 = .099$$

$$\text{For } \alpha_1 = .900; \ Q = .10 + \frac{.107 - .1046}{.1872 - .1046} \times .05 = .102$$

Interpolating for $\alpha_1 = .832$, $Q = .099 + .32(.102 - .099) = .100$

$$\therefore f = Qf_{FF} = .100 \times 266 = 26.6 \text{ Hz}$$

It will be noted that without the torsion spring coupling the first mode frequency was 36.7 Hz (from Ex. 7.2a). The effect of the introduction of the torsion spring coupling has therefore been to reduce this frequency by 27.5 percent.

TABLE 7.2 : K^*

Q: MODE:	0.05	0.10	0.15	0.20	0.25	0.30	0.35	0.40	0.45	0.50	0.55	0.60	0.70	0.80	0.90
(1)	.1298	.2737	.4337	.6125	.8133	1.040	1.299	1.598	1.945	2.356	2.851	3.459	5.241	8.702	18.84
(2)	.3770	.7712	1.188	1.633	2.115	2.641	3.225	3.880	4.628	5.498	6.529	7.783	11.41	18.37	38.68
(3)	.6243	1.269	1.942	2.654	3.416	4.242	5.150	6.163	7.312	8.639	10.21	12.11	17.57	28.04	58.51
(4)	.8715	1.766	2.696	3.675	4.717	5.843	7.075	8.445	9.995	11.78	13.89	16.43	23.74	37.71	78.35
(5)	1.119	2.264	3.451	4.695	6.018	7.443	9.000	10.73	12.68	14.92	17.56	20.76	29.90	47.38	98.18
(6)	1.366	2.761	4.204	5.716	7.319	9.044	10.92	13.01	15.36	18.06	21.24	25.08	36.07	57.04	118.0

TABLE 7.3 : K^*

Q: MODE:	0.05	0.10	0.15	0.20	0.25	0.30	0.35	0.40	0.45	0.50	0.55	0.60	0.70	0.80	0.90
(1)	.0062	.0249	.0566	.1021	.1627	.2401	.3369	.4565	.6037	.7854	1,012	1.297	2.158	3.868	8.926
(2)	.2534	.5225	.8108	1.123	1.464	1.841	2.262	2.739	3.287	3.927	4.690	5.621	8.324	13.54	28.76
(3)	.5007	1.020	1.565	2.144	2.765	3.442	4.187	5.021	5.970	7.069	8.368	9.945	14.49	23.20	48.60
(4)	.7479	1.518	2.319	3.164	4.066	5.042	6.112	7.304	8.653	10.21	12.05	14.27	20.65	32.87	68.43
(5)	.9951	2.015	3.073	4.185	5.368	6.643	8.038	9.586	11.34	13.35	15.72	18.59	26.82	42.54	88.26
(6)	1.242	2.513	3.828	5.206	6.669	8.243	9.962	11.87	14.02	16.49	19.40	22.92	32.99	52.21	108.1

TABLE 7.4 : I_0^*

Q: MODE:	0.05	0.10	0.15	0.20	0.25	0.30	0.35	0.40	0.45	0.50	0.55	0.60	0.70	0.80	0.90
(1)	.0062	.0249	.0566	.1021	.1627	.2401	.3369	.4565	.6037	.7854	1.012	1.297	2.158	3.868	8.926
(2)	.2534	.5225	.8108	1.123	1.464	1.841	2.262	2.739	3.287	3.927	4.690	5.621	8.324	13.54	28.76
(3)	.5007	1.020	1.565	2.144	2.765	3.442	4.187	5.021	5.970	7.069	8.368	9.945	14.49	23.20	48.60
(4)	.7479	1.518	2.319	3.164	4.066	5.042	6.112	7.304	8.653	10.21	12.05	14.27	20.65	32.87	68.43
(5)	.9951	2.015	3.073	4.185	5.368	6.643	8.038	9.586	11.34	13.35	15.72	18.59	26.82	42.54	88.26
(6)	1.242	2.513	3.828	5.206	6.669	8.243	9.962	11.87	14.02	16.49	19.40	22.92	32.99	52.21	108.1

TABLE 7.5 : I_0^*

Q: MODE:	0.05	0.10	0.15	0.20	0.25	0.30	0.35	0.40	0.45	0.50	0.55	0.60	0.70	0.80	0.90
(1)	.1298	.2737	.4337	.6125	.8133	1.040	1.299	1.598	1.945	2.356	2.851	3.459	5.241	8.702	18.84
(2)	.3770	.7712	1.188	1.633	2.115	2.641	3.225	3.880	4.628	5.498	6.529	7.783	11.41	18.37	38.68
(3)	.6243	1.269	1.942	2.654	3.416	4.242	5.150	6.163	7.312	8.639	10.21	12.11	17.57	28.04	58.51
(4)	.8715	1.766	2.696	3.675	4.717	5.843	7.075	8.445	9.995	11.78	13.89	16.43	23.74	37.71	78.35
(5)	1.119	2.264	3.451	4.695	6.018	7.443	9.000	10.73	12.68	14.92	17.56	20.76	29.90	47.38	98.18
(6)	1.366	2.761	4.204	5.716	7.319	9.044	10.92	13.01	15.36	18.06	21.24	25.08	36.07	57.04	118.0

TABLE 7.6 : I_0^* ($\alpha = .10$)

Q: MODE:	0.05	0.10	0.15	0.20	0.25	0.30	0.35	0.40	0.45	0.50	0.55	0.60	0.70	0.80	0.90
(1)	.0226	.0920	.2138	.3983	.6631	1.037	1.565	2.319	3.406	4.967	7.173	10.22	19.99	39.64	96.55
(2)	.4740	1.012	1.639	2.390	3.316	4.492	6.035	8.116	10.97	14.90	20.21	27.24	48.56	89.20	203.8
(3)	.9254	1.932	3.064	4.382	5.968	7.947	10.51	13.91	18.54	24.84	33.26	44.27	77.12	138.8	311.1
(4)	1.377	2.852	4.490	6.373	8.620	11.40	14.98	19.71	26.11	34.77	46.30	61.29	105.7	188.3	418.4
(5)	1.828	3.772	5.915	8.365	11.27	14.86	19.45	25.51	33.68	44.71	59.34	78.32	134.2	237.9	525.6
(6)	2.280	4.692	7.340	10.36	13.92	18.31	23.92	31.30	41.24	54.64	72.38	95.34	162.8	287.4	632.9

TABLE 7.6 : I_0^* ($\alpha = .20$)

Q: MODE:	0.05	0.10	0.15	0.20	0.25	0.30	0.35	0.40	0.45	0.50	0.55	0.60	0.70	0.80	0.90
(1)	.0207	.0839	.1934	.3560	.5825	.8888	1.298	1.840	2.560	3.512	4.771	6.435	11.66	22.18	52.97
(2)	.4339	.9224	1.482	2.136	2.912	3.851	5.005	6.441	8.249	10.54	13.45	17.16	28.32	49.90	111.8
(3)	.8471	1.761	2.771	3.916	5.242	6.814	8.712	11.04	13.94	17.56	22.12	27.89	44.97	77.63	170.7
(4)	1.260	2.599	4.060	5.696	7.572	9.776	12.42	15.64	19.63	24.59	30.79	38.61	61.63	105.4	229.5
(5)	1.673	3.438	5.349	7.476	9.902	12.74	16.13	20.24	25.32	31.61	39.47	49.34	78.29	133.1	288.4
(6)	2.087	4.276	6.638	9.256	12.23	15.70	19.83	24.84	31.00	38.64	48.14	60.06	94.95	160.8	347.2

TABLE 7.6 : I_0^* ($\alpha = .30$)

Q: MODE:	0.05	0.10	0.15	0.20	0.25	0.30	0.35	0.40	0.45	0.50	0.55	0.60	0.70	0.80	0.90
(1)	.0191	.0771	.1769	.3233	.5236	.7884	1.132	1.574	2.141	2.868	3.803	5.015	8.763	16.28	38.40
(2)	.4001	.8481	1.356	1.940	2.618	3.416	4.367	5.510	6.899	8.604	10.72	13.37	21.28	36.64	01.07
(3)	.7812	1.619	2.535	3.556	4.712	6.044	7.601	9.446	11.66	14.34	17.63	21.73	33.80	56.99	123.7
(4)	1.162	2.390	3.715	5.172	6.807	8.672	10.84	13.38	16.42	20.07	24.55	30.09	46.32	77.34	166.4
(5)	1.543	3.161	4.894	6.789	8.901	11.30	14.07	17.32	21.17	25.81	31.46	38.45	58.84	97.70	209.1
(6)	1.924	3.932	6.073	8.405	11.00	13.93	17.30	21.25	25.93	31.55	38.37	46.81	71.36	118.0	251.7

TABLE 7.6 : I_0^* ($\alpha = .40$)

Q: MODE.	0.05	0.10	0.15	0.20	0.25	0.30	0.35	0.40	0.45	0.50	0.55	0.60	0.70	0.80	0.90
(1)	.0177	.0714	.1633	.2969	.4779	.7139	1.016	1.397	1.878	2.484	3.253	4.239	7.238	13.29	31.10
(2)	.3713	.7855	1.252	1.782	2.390	3.094	3.917	4.889	6.050	7.451	9.166	11.30	17.63	29.91	65.65
(3)	.7249	1.499	2.340	3.266	4.301	5.473	6.818	8.382	10.22	12.42	15.08	18.37	27.99	46.53	100.2
(4)	1.078	2.214	3.429	4.751	6.213	7.853	9.720	11.87	14.39	17.39	20.99	25.43	38.36	63.15	134.8
(5)	1.432	2.928	4.517	6.236	8.124	10.23	12.62	15.37	18.57	22.35	26.91	32.50	48.73	79.77	169.3
(6)	1.786	3.642	5.605	7.720	10.04	12.61	15.52	18.86	22.74	27.32	32.82	39.56	59.10	96.39	203.9

TABLE 7.6 : I_0^* ($\alpha = .50$)

Q: MODE:	0.05	0.10	0.15	0.20	0.25	0.30	0.35	0.40	0.45	0.50	0.55	0.60	0.70	0.80	0.90
(1)	.0165	.0665	.1518	.2752	.4410	.6556	.9272	1.267	1.691	2.221	2.888	3.738	6.323	11.48	26.70
(2)	.3464	.7318	1.164	1.651	2.205	2.841	3.576	4.435	5.450	6.664	8.140	9.969	15.36	25.83	56.38
(3)	.6763	1.397	2.175	3.027	3.969	5.026	6.225	7.604	9.209	11.11	13.39	16.20	24.39	40.17	86.05
(4)	1.006	2.062	3.187	4.402	5.734	7.211	8.874	10.77	12.97	15.55	18.64	22.43	33.42	54.52	115.7
(5)	1.336	2.728	4.199	5.778	7.498	9.397	11.52	13.94	16.73	19.99	23.89	28.66	42.45	68.87	145.4
(6)	1.666	3.393	5.211	7.154	9.262	11.58	14.17	17.11	20.49	24.44	29.15	34.89	51.49	83.22	175.1

TABLE 7.6 : I_0^* ($\alpha = .60$)

Q. MODE:	0.05	0.10	0.15	0.20	0.25	0.30	0.35	0.40	0.45	0.50	0.55	0.60	0.70	0.80	0.90
(1)	.0155	.0623	.1419	.2567	.4104	.6082	.8571	1.167	1.551	2.028	2.625	3.383	5.680	10.25	23.77
(2)	.3246	.6852	1.088	1.540	2.052	2.635	3.306	4.084	4.997	6.084	7.398	9.022	13.79	23.07	50.17
(3)	.6338	1.308	2.034	2.824	3.694	4.663	5.755	7.001	8.443	10.14	12.17	14.66	21.91	35.88	76.58
(4)	.9430	1.931	2.980	4.108	5.336	6.690	8.203	9.918	11.89	14.20	16.94	20.30	30.02	48.69	103.0
(5)	1.252	2.554	3.926	5.391	6.977	8.717	10.65	12.83	15.34	18.25	21.72	25.94	38.14	61.51	129.4
(6)	1.561	3.177	4.872	6.675	8.619	10.74	13.10	15.75	18.78	22.31	26.49	31.58	46.25	74.32	155.8

TABLE 7.6 : I_0^* (α = .70)

Q: MODE:	0.05	0.10	0.15	0.20	0.25	0.30	0.35	0.40	0.45	0.50	0.55	0.60	0.70	0.80	0.90
(1)	.0145	.0586	.1333	.2409	.3845	.5686	.7996	1.086	1.440	1.877	2.424	3.116	5.207	9.365	21.66
(2)	.3055	.6444	1.022	1.445	1.922	2.464	3.084	3.801	4.639	5.632	6.831	8.310	12.65	21.07	45.73
(3)	.5964	1.230	1.911	2.650	3.460	4.359	5.368	6.515	7.838	9.387	11.24	13.50	20.08	32.78	69.79
(4)	.8873	1.816	2.800	3.854	4.998	6.255	7.653	9.230	11.04	13.14	15.65	18.70	27.52	44.48	93.86
(5)	1.178	2.402	3.689	5.059	6.536	8.150	9.937	11.94	14.24	16.90	20.05	23.89	34.96	56.19	117.9
(6)	1.469	2.988	4.577	6.263	8.074	10.05	12.22	14.66	17.43	20.65	24.46	29.08	42.40	67.89	142.0

TABLE 7.6 : I_0^* (α = .80)

Q: MODE:	0.05	0.10	0.15	0.20	0.25	0.30	0.35	0.40	0.45	0.50	0.55	0.60	0.70	0.80	0.90
(1)	.0137	.0553	.1258	.2271	.3621	.5349	.7512	1.019	1.349	1.756	2.264	2.906	4.843	8.692	20.08
(2)	.2885	.6083	.9644	1.363	1.811	2.318	2.898	3.566	4.346	5.269	6.381	7.750	11.76	19.56	42.38
(3)	.5632	1.161	1.803	2.498	3.259	4.101	5.044	6.113	7.343	8.781	10.50	12.59	18.68	30.42	64.69
(4)	.8379	1.714	2.642	3.634	4.708	5.884	7.190	8.660	10.34	12.29	14.61	17.44	25.60	41.29	87.00
(5)	1.113	2.267	3.480	4.769	6.156	7.667	9.336	11.21	13.34	15.81	18.73	22.28	32.52	52.15	109.3
(6)	1.387	2.820	4.319	5.905	7.604	9.450	11.48	13.75	16.33	19.32	22.85	27.12	39.44	63.01	131.6

TABLE 7.6 : I_0^* (α = .90)

Q: MODE:	0.05	0.10	0.15	0.20	0.25	0.30	0.35	0.40	0.45	0.50	0.55	0.60	0.70	0.80	0.90
(1)	.0130	.0524	.1191	.2150	.3426	.5058	.7098	.9620	1.272	1.656	2.133	2.736	4.553	8.163	18.84
(2)	.2733	.5762	.9131	1.290	1.713	2.192	2.738	3.367	4.100	4.967	6.011	7.296	11.06	18.37	39.78
(3)	.5335	1.100	1.707	2.364	3.083	3.878	4.766	5.772	6.928	8.279	9.889	11.86	17.56	28.57	60.71
(4)	.7938	1.624	2.501	3.439	4.454	5.564	6.794	8.177	9.756	11.59	13.77	16.42	24.07	38.77	81.65
(5)	1.054	2.148	3.295	4.514	5.824	7.249	8.822	10.58	12.58	14.90	17.65	20.98	30.57	48.98	102.6
(6)	1.314	2.671	4.089	5.589	7.194	8.935	10.85	12.99	15.41	18.21	21.52	25.54	37.07	59.18	123.5

TABLE 7.6 : I_0^* (α = 1.0)

Q: MODE:	0.05	0.10	0.15	0.20	0.25	0.30	0.35	0.40	0.45	0.50	0.55	0.60	0.70	0.80	0.90
(1)	.0124	.0498	.1131	.2042	.3253	.4802	.6738	.9130	1.207	1.571	2.023	2.594	4.316	7.735	17.85
(2)	.2596	.5473	.8674	1.225	1.627	2.081	2.599	3.195	3.891	4.712	5.701	6.918	10.48	17.40	37.69
(3)	.5069	1.045	1.622	2.246	2.928	3.682	4.524	5.478	6.574	7.854	9.380	11.24	16.65	27.07	57.52
(4)	.7541	1.542	2.376	3.266	4.229	5.282	6.449	7.760	9.257	11.00	13.06	15.57	22.81	36.74	77.36
(5)	1.001	2.040	3.130	4.287	5.530	6.883	8.374	10.04	11.94	14.14	16.74	19.89	28.98	46.41	97.19
(6)	1.249	2.538	3.884	5.308	6.832	8.484	10.30	12.33	14.62	17.28	20.41	24.21	35.14	56.08	117.0

TABLE 7.7 : K_1^* (α = .10)

Q: MODE:	0.05	0.10	0.15	0.20	0.25	0.30	0.35	0.40	0.45	0.50	0.55	0.60	0.70	0.80	0.90
(1)	.0226	.0920	.2138	.3983	.6631	1.037	1.565	2.319	3.406	4.967	7.173	10.22	19.99	39.64	96.55
(2)	.4740	1.012	1.639	2.390	3.316	4.492	6.035	8.116	10.97	14.90	20.21	27.24	48.56	89.20	203.8
(3)	.9254	1.932	3.064	4.382	5.968	7.947	10.51	13.91	18.54	24.84	33.26	44.27	77.12	138.8	311.1
(4)	1.377	2.852	4.490	6.373	8.620	11.40	14.98	19.71	26.11	34.77	46.30	61.29	105.7	188.3	418.4
(5)	1.828	3.772	5.915	8.365	11.27	14.86	19.45	25.51	33.68	44.71	59.34	78.32	134.2	237.9	525.6
(6)	2.280	4.692	7.340	10.36	13.92	18.31	23.92	31.30	41.24	54.64	72.38	95.34	162.8	287.4	632.9

TABLE 7.7 : K_1^* (α = .20)

Q: MODE:	0.05	0.10	0.15	0.20	0.25	0.30	0.35	0.40	0.45	0.50	0.55	0.60	0.70	0.80	0.90
(1)	.0207	.0839	.1934	.3560	.5825	.8888	1.298	1.840	2.560	3.512	4.771	6.435	11.66	22.18	52.97
(2)	.4339	.9224	1.482	2.136	2.912	3.851	5.005	6.441	8.249	10.54	13.45	17.16	28.32	49.90	111.8
(3)	.8471	1.761	2.771	3.916	5.242	6.814	8.712	11.04	13.94	17.56	22.12	27.89	44.97	77.63	170.7
(4)	1.260	2.599	4.060	5.696	7.572	9.776	12.42	15.64	19.63	24.59	30.79	38.61	61.63	105.4	229.5
(5)	1.673	3.438	5.349	7.476	9.902	12.74	16.13	20.24	25.32	31.61	39.47	49.34	78.29	133.1	288.4
(6)	2.087	4.276	6.638	9.256	12.23	15.70	19.83	24.84	31.00	38.64	48.14	60.06	94.95	160.8	347.2

TABLE 7.7 : K_1^* (α = .30)

Q: MODE:	0.05	0.10	0.15	0.20	0.25	0.30	0.35	0.40	0.45	0.50	0.55	0.60	0.70	0.80	0.90
(1)	.0191	.0771	.1769	.3233	.5236	.7884	1.132	1.574	2.141	2.868	3.803	5.015	8.763	16.28	38.40
(2)	.4001	.8481	1.356	1.940	2.618	3.416	4.367	5.510	6.899	8.604	10.72	13.37	21.28	36.64	81.07
(3)	.7812	1.619	2.535	3.556	4.712	6.044	7.601	9.446	11.66	14.34	17.63	21.73	33.80	56.99	123.7
(4)	1.162	2.390	3.715	5.172	6.807	8.672	10.84	13.38	16.42	20.07	24.55	30.09	46.32	77.34	166.4
(5)	1.543	3.161	4.894	6.789	8.901	11.30	14.07	17.32	21.17	25.81	31.46	38.45	58.84	97.70	209.1
(6)	1.924	3.932	6.073	8.405	11.00	13.93	17.30	21.25	25.93	31.55	38.37	46.81	71.36	118.0	251.7

TABLE 7.7 : K_1^* (α = .40)

Q: MODE:	0.05	0.10	0.15	0.20	0.25	0.30	0.35	0.40	0.45	0.50	0.55	0.60	0.70	0.80	0.90
(1)	.0177	.0714	.1633	.2969	.4779	.7139	1.016	1.397	1.878	2.484	3.253	4.239	7.258	13.29	31.10
(2)	.3713	.7855	1.252	1.782	2.390	3.094	3.917	4.889	6.050	7.451	9.166	11.30	17.63	29.91	65.65
(3)	.7249	1.499	2.340	3.266	4.301	5.473	6.818	8.382	10.22	12.42	15.08	18.37	27.99	46.53	100.2
(4)	1.078	2.214	3.429	4.751	6.213	7.853	9.720	11.87	14.39	17.39	20.99	25.43	38.36	63.15	134.8
(5)	1.432	2.928	4.517	6.236	8.124	10.23	12.62	15.37	18.57	22.35	26.91	32.50	48.73	79.77	169.3
(6)	1.786	3.642	5.605	7.720	10.04	12.61	15.52	18.86	22.74	27.32	32.82	39.56	59.10	96.39	203.9

TABLE 7.7 : K_1^* (α = .50)

Q: MODE:	0.05	0.10	0.15	0.20	0.25	0.30	0.35	0.40	0.45	0.50	0.55	0.60	0.70	0.80	0.90
(1)	.0165	.0665	.1518	.2752	.4410	.6556	.9272	1.267	1.691	2.221	2.888	3.738	6.323	11.48	26.70
(2)	.3464	.7318	1.164	1.651	2.205	2.841	3.576	4.435	5.450	6.664	8.140	9.969	15.36	25.83	56.38
(3)	.6763	1.397	2.175	3.027	3.969	5.026	6.225	7.604	9.209	11.11	13.39	16.20	24.39	40.17	86.05
(4)	1.006	2.062	3.187	4.402	5.734	7.211	8.874	10.77	12.97	15.55	18.64	22.43	33.42	54.52	115.7
(5)	1.336	2.728	4.199	5.778	7.498	9.397	11.52	13.94	16.73	19.99	23.89	28.66	42.45	68.87	145.4
(6)	1.666	3.393	5.211	7.154	9.262	11.58	14.17	17.11	20.49	24.44	29.15	34.89	51.49	83.22	175.1

TABLE 7.7 : K_1^* (α = .60)

Q: MODE:	0.05	0.10	0.15	0.20	0.25	0.30	0.35	0.40	0.45	0.50	0.55	0.60	0.70	0.80	0.90
(1)	.0155	.0623	.1419	.2567	.4104	.6082	.8571	1.167	1.551	2.028	2.625	3.383	5.680	10.25	23.77
(2)	.3246	.6852	1.088	1.540	2.052	2.635	3.306	4.084	4.997	6.084	7.398	9.022	13.79	23.07	50.17
(3)	.6338	1.308	2.034	2.824	3.694	4.663	5.755	7.001	8.443	10.14	12.17	14.66	21.91	35.88	76.58
(4)	.9430	1.931	2.980	4.108	5.336	6.690	8.203	9.918	11.89	14.20	16.94	20.30	30.02	48.69	103.0
(5)	1.252	2.554	3.926	5.391	6.977	8.717	10.65	12.83	15.34	18.25	21.72	25.94	38.14	61.51	129.4
(6)	1.561	3.177	4.872	6.675	8.619	10.74	13.10	15.75	18.78	22.31	26.49	31.58	46.25	74.32	155.8

TABLE 7.7 : K_1^* (α = .70)

Q:	0.05	0.10	0.15	0.20	0.25	0.30	0.35	0.40	0.45	0.50	0.55	0.60	0.70	0.80	0.90
MODE:															
(1)	.0145	.0586	.1333	.2409	.3845	.5686	.7996	1.086	1.440	1.877	2.424	3.116	5.207	9.365	21.66
(2)	.3055	.6444	1.022	1.445	1.922	2.464	3.084	3.801	4.639	5.632	6.831	8.310	12.65	21.07	45.73
(3)	.5964	1.230	1.911	2.650	3.460	4.359	5.368	6.515	7.838	9.387	11.24	13.50	20.08	32.78	69.79
(4)	.8873	1.816	2.800	3.854	4.998	6.255	7.653	9.230	11.04	13.14	15.65	18.70	27.52	44.48	93.86
(5)	1.178	2.402	3.689	5.059	6.536	8.150	9.937	11.94	14.24	16.90	20.05	23.89	34.96	56.19	117.9
(6)	1.469	2.988	4.577	6.263	8.074	10.05	12.22	14.66	17.43	20.65	24.46	29.08	42.40	67.89	142.0

TABLE 7.7 : K_1^* (α = .80)

Q:	0.05	0.10	0.15	0.20	0.25	0.30	0.35	0.40	0.45	0.50	0.55	0.60	0.70	0.80	0.90
MODE:															
(1)	.0137	.0553	.1258	.2271	.3621	.5349	.7512	1.019	1.349	1.756	2.264	2.906	4.843	8.692	20.08
(2)	.2885	.6083	.9644	1.363	1.811	2.318	2.898	3.566	4.346	5.269	6.381	7.750	11.76	19.56	42.38
(3)	.5632	1.161	1.803	2.498	3.259	4.101	5.044	6.113	7.343	8.781	10.50	12.59	18.68	30.42	64.69
(4)	.8379	1.714	2.642	3.634	4.708	5.884	7.190	8.660	10.34	12.29	14.61	17.44	25.60	41.29	87.00
(5)	1.113	2.267	3.480	4.769	6.156	7.667	9.336	11.21	13.34	15.81	18.73	22.28	32.52	52.15	109.3
(6)	1.387	2.820	4.319	5.905	7.604	9.450	11.48	13.75	16.33	19.32	22.85	27.12	39.44	63.01	131.6

TABLE 7.7 : K_1^* (α = .90)

Q:	0.05	0.10	0.15	0.20	0.25	0.30	0.35	0.40	0.45	0.50	0.55	0.60	0.70	0.80	0.90
MODE:															
(1)	.0130	.0524	.1191	.2150	.3426	.5058	.7098	.9620	1.272	1.656	2.133	2.736	4.553	8.163	18.84
(2)	.2733	.5762	.9131	1.290	1.713	2.192	2.738	3.367	4.100	4.967	6.011	7.296	11.06	18.37	39.78
(3)	.5335	1.100	1.707	2.364	3.083	3.878	4.766	5.772	6.928	8.279	9.889	11.86	17.56	28.57	60.71
(4)	.7938	1.624	2.501	3.439	4.454	5.564	6.794	8.177	9.756	11.59	13.77	16.42	24.07	38.77	81.65
(5)	1.054	2.148	3.295	4.514	5.824	7.249	8.822	10.58	12.58	14.90	17.65	20.98	30.57	48.98	102.6
(6)	1.314	2.671	4.089	5.589	7.194	8.935	10.85	12.99	15.41	18.21	21.52	25.54	37.07	59.18	123.5

TABLE 7.7 : K_1^* (α = 1.0)

Q:	0.05	0.10	0.15	0.20	0.25	0.30	0.35	0.40	0.45	0.50	0.55	0.60	0.70	0.80	0.90
MODE:															
(1)	.0124	.0498	.1131	.2042	.3253	.4802	.6738	.9130	1.207	1.571	2.023	2.594	4.316	7.735	17.85
(2)	.2596	.5473	.8674	1.225	1.627	2.081	2.599	3.195	3.891	4.712	5.701	6.918	10.48	17.40	37.69
(3)	.5069	1.045	1.622	2.246	2.928	3.682	4.524	5.478	6.574	7.854	9.380	11.24	16.65	27.07	57.52
(4)	.7541	1.542	2.376	3.266	4.229	5.282	6.449	7.760	9.257	11.00	13.06	15.57	22.81	36.74	77.36
(5)	1.001	2.040	3.130	4.287	5.530	6.883	8.374	10.04	11.94	14.14	16.74	19.89	28.98	46.41	97.19
(6)	1.249	2.538	3.884	5.308	6.832	8.484	10.30	12.33	14.62	17.28	20.41	24.21	35.14	56.08	117.0

TABLE 7.8 : I_0^* (α = .10)

Q:	0.05	0.10	0.15	0.20	0.25	0.30	0.35	0.40	0.45	0.50	0.55	0.60	0.70	0.80	0.90
MODE:															
(1)	.2847	.6521	1.119	1.703	2.425	3.312	4.394	5.713	7.323	9.302	11.76	14.87	24.24	42.91	98.39
(2)	.2483	.5520	.9264	1.394	1.989	2.764	3.800	5.217	7.190	9.935	13.69	18.73	34.28	64.42	150.2
(3)	.6997	1.472	2.352	3.386	4.642	6.220	8.270	11.01	14.76	19.87	26.73	35.75	62.84	114.0	257.5
(4)	1.151	2.392	3.777	5.377	7.294	9.675	12.74	16.81	22.33	29.80	39.78	52.78	91.40	163.5	364.7
(5)	1.602	3.312	5.202	7.369	9.946	13.13	17.21	22.61	29.89	39.74	52.82	69.80	120.0	213.1	472.0
(6)	2.054	4.232	6.627	9.360	12.60	16.59	21.68	28.40	37.46	49.67	65.86	86.83	148.5	262.6	579.3

TABLE 7.8 : I_0^* (α = .20)

Q:	0.05	0.10	0.15	0.20	0.25	0.30	0.35	0.40	0.45	0.50	0.55	0.60	0.70	0.80	0.90
MODE:															
(1)	.1951	.4337	.7232	1.073	1.493	1.997	2.602	3.331	4.215	5.295	6.632	8.317	13.40	23.54	53.74
(2)	.2273	.5031	.8379	1.246	1.747	2.370	3.151	4.141	5.404	7.025	9.108	11.80	19.99	36.04	82.39
(3)	.6405	1.342	2.127	3.026	4.077	5.332	6.858	8.742	11.09	14.05	17.78	22.52	36.65	63.77	141.2
(4)	1.054	2.180	3.416	4.806	6.407	8.295	10.57	13.34	16.78	21.07	26.46	33.25	53.30	91.49	200.1
(5)	1.467	3.019	4.705	6.586	8.737	11.26	14.27	17.94	22.47	28.10	35.13	43.97	69.96	119.2	258.9
(6)	1.880	3.857	5.994	8.366	11.07	14.22	17.98	22.54	28.16	35.12	43.81	54.70	86.62	146.9	317.8

TABLE 7.8 : I_0^* (α = .30)

Q:	0.05	0.10	0.15	0.20	0.25	0.30	0.35	0.40	0.45	0.50	0.55	0.60	0.70	0.80	0.90
MODE:															
(1)	.1574	.3457	.5699	.8359	1.151	1.526	1.971	2.504	3.146	3.927	4.892	6.106	9.764	17.07	38.85
(2)	.2096	.4626	.7665	1.131	1.571	2.102	2.749	3.542	4.520	5.736	7.260	9.195	15.02	26.46	59.74
(3)	.5906	1.234	1.946	2.748	3.665	4.730	5.984	7.478	9.278	11.47	14.17	17.55	27.54	46.81	102.4
(4)	.9717	2.005	3.125	4.364	5.760	7.358	9.218	11.41	14.04	17.21	21.09	25.91	40.06	67.17	145.1
(5)	1.353	2.776	4.304	5.980	7.854	9.986	12.45	15.35	18.79	22.94	28.00	34.27	52.58	87.52	187.7
(6)	1.734	3.547	5.484	7.597	9.948	12.61	15.69	19.28	23.55	28.68	34.92	42.63	65.10	107.9	230.4

TABLE 7.8 : I_0^* (α = .40)

Q:	0.05	0.10	0.15	0.20	0.25	0.30	0.35	0.40	0.45	0.50	0.55	0.60	0.70	0.80	0.90
MODE:															
(1)	.1335	.2957	.4845	.7065	.9677	1.276	1.640	2.074	2.594	3.227	4.006	4.986	7.934	13.82	31.40
(2)	.1945	.4284	.7075	1.039	1.434	1.904	2.466	3.143	3.964	4.967	6.209	7.772	12.44	21.60	48.38
(3)	.5481	1.142	1.796	2.524	3.345	4.283	5.368	6.636	8.136	9.935	12.12	14.84	22.81	38.22	82.93
(4)	.9017	1.857	2.884	4.009	5.257	6.663	8.269	10.13	12.31	14.90	18.04	21.90	33.18	54.84	117.5
(5)	1.255	2.571	3.973	5.493	7.169	9.043	11.17	13.62	16.48	19.87	23.95	28.97	43.55	71.46	152.0
(6)	1.609	3.285	5.061	6.978	9.080	11.42	14.07	17.11	20.65	24.84	29.86	36.03	53.91	88.08	186.6

TABLE 7.8 : I_0^* (α = .50)

Q:	0.05	0.10	0.15	0.20	0.25	0.30	0.35	0.40	0.45	0.50	0.55	0.60	0.70	0.80	0.90
MODE:															
(1)	.1207	.2626	.4287	.6230	.8506	1.118	1.433	1.807	2.254	2.797	3.465	4.304	6.828	11.87	26.93
(2)	.1814	.3992	.6577	.9630	1.323	1.748	2.252	2.851	3.571	4.443	5.514	6.853	10.84	18.65	41.54
(3)	.5113	1.064	1.669	2.339	3.087	3.933	4.901	6.020	7.330	8.886	10.77	13.08	19.87	33.00	71.21
(4)	.8412	1.730	2.681	3.715	4.851	6.119	7.550	9.188	11.09	13.33	16.02	19.31	28.90	47.35	100.9
(5)	1.171	2.395	3.693	5.090	6.616	8.304	10.20	12.36	14.85	17.77	21.27	25.54	37.94	61.70	130.6
(6)	1.501	3.060	4.705	6.466	8.380	10.49	12.85	15.52	18.61	22.21	26.52	31.77	46.97	76.04	160.2

TABLE 7.8 : I_0^* (α = .60)

Q:	0.05	0.10	0.15	0.20	0.25	0.30	0.35	0.40	0.45	0.50	0.55	0.60	0.70	0.80	0.90
MODE:															
(1)	.1100	.2387	.3888	.5639	.7683	1.008	1.289	1.623	2.022	2.505	3.099	3.844	6.086	10.56	23.94
(2)	.1700	.3738	.6149	.8986	1.231	1.622	2.081	2.625	3.274	4.056	5.012	6.203	9.737	16.66	36.97
(3)	.4792	.9967	1.561	2.182	2.873	3.649	4.530	5.542	6.720	8.112	9.785	11.84	17.85	29.47	63.37
(4)	.7884	1.620	2.507	3.466	4.515	5.676	6.979	8.459	10.17	12.17	14.56	17.48	25.96	42.29	89.78
(5)	1.098	2.243	3.453	4.750	6.157	7.704	9.428	11.38	13.61	16.22	19.33	23.12	34.08	55.10	116.2
(6)	1.407	2.865	4.399	6.033	7.798	9.731	11.88	14.29	17.06	20.28	24.10	28.76	42.19	67.91	142.6

TABLE 7.8 : I_0^* (α = .70)

Q: MODE:	0.05	0.10	0.15	0.20	0.25	0.30	0.35	0.40	0.45	0.50	0.55	0.60	0.70	0.80	0.90
(1)	.1017	.2204	.3586	.5194	.7067	.9258	1.183	1.488	1.852	2.292	2.833	3.512	5.552	9.627	21.81
(2)	.1600	.3515	.5777	.8431	1.153	1.516	1.942	2.443	3.039	3.755	4.628	5.713	8.926	15.22	33.69
(3)	.4509	.9373	1.467	2.048	2.691	3.412	4.226	5.158	6.238	7.510	9.035	10.91	16.36	26.92	57.76
(4)	.7419	1.523	2.355	3.252	4.229	5.307	6.511	7.873	9.437	11.26	13.44	16.10	23.80	38.63	81.83
(5)	1.033	2.109	3.244	4.457	5.767	7.203	8.795	10.59	12.64	15.02	17.85	21.29	31.24	50.33	105.9
(6)	1.324	2.695	4.133	5.661	7.305	9.098	11.08	13.30	15.83	18.77	22.26	26.49	38.68	62.04	130.0

TABLE 7.8 : I_0^* (α = .80)

Q: MODE:	0.05	0.10	0.15	0.20	0.25	0.30	0.35	0.40	0.45	0.50	0.55	0.60	0.70	0.80	0.90
(1)	.0950	.2058	.3346	.4844	.6587	.8622	1.101	1.384	1.721	2.129	2.631	3.259	5.149	8.923	20.21
(2)	.1511	.3318	.5451	.7949	1.086	1.426	1.824	2.292	2.847	3.512	4.322	5.328	8.302	14.12	31.23
(3)	.4258	.8848	1.384	1.930	2.535	3.210	3.971	4.839	5.844	7.025	8.439	10.17	15.22	24.99	53.54
(4)	.7006	1.438	2.222	3.066	3.983	4.993	6.117	7.386	8.841	10.54	12.56	15.02	22.14	35.85	75.85
(5)	.9753	1.991	3.061	4.201	5.432	6.776	8.263	9.933	11.84	14.05	16.67	19.86	29.06	46.72	98.15
(6)	1.250	2.544	3.899	5.337	6.880	8.559	10.41	12.48	14.83	17.56	20.79	24.70	35.98	57.58	120.5

TABLE 7.8 : I_0^* (α = .90)

Q: MODE:	0.05	0.10	0.15	0.20	0.25	0.30	0.35	0.40	0.45	0.50	0.55	0.60	0.70	0.80	0.90
(1)	.0896	.1939	.3152	.4560	.6198	.8111	1.036	1.301	1.618	2.001	2.471	3.061	4.834	8.374	18.96
(2)	.1431	.3143	.5161	.7523	1.028	1.349	1.724	2.164	2.686	3.312	4.072	5.016	7.805	13.26	29.31
(3)	.4034	.8381	1.310	1.827	2.398	3.035	3.752	4.569	5.514	6.623	7.950	9.576	14.31	23.47	50.25
(4)	.6637	1.362	2.104	2.902	3.768	4.721	5.780	6.974	8.342	9.935	11.83	14.14	20.81	33.67	71.18
(5)	.9239	1.886	2.898	3.977	5.139	6.407	7.808	9.379	11.17	13.25	15.71	18.70	27.32	43.87	92.12
(6)	1.184	2.409	3.692	5.051	6.509	8.092	9.836	11.78	14.00	16.56	19.58	23.26	33.82	54.08	113.0

TABLE 7.8 : I_0^* (α = 1.0)

Q: MODE:	0.05	0.10	0.15	0.20	0.25	0.30	0.35	0.40	0.45	0.50	0.55	0.60	0.70	0.80	0.90
(1)	.0850	.1839	.2989	.4324	.5877	.7690	.9817	1.233	1.533	1.896	2.342	2.901	4.580	7.934	17.96
(2)	.1360	.2985	.4902	.7145	.9760	1.281	1.636	2.054	2.549	3.142	3.862	4.756	7.399	12.57	27.77
(3)	.3832	.7961	1.244	1.735	2.277	2.881	3.562	4.337	5.232	6.283	7.541	9.080	13.56	22.24	47.60
(4)	.6305	1.294	1.999	2.756	3.579	4.482	5.487	6.619	7.915	9.425	11.22	13.40	19.73	31.91	67.44
(5)	.8777	1.791	2.753	3.777	4.880	6.083	7.412	8.902	10.60	12.57	14.90	17.73	25.90	41.58	87.27
(6)	1.125	2.289	3.507	4.798	6.181	7.683	9.337	11.18	13.28	15.71	18.58	22.05	32.06	51.24	107.1

TABLE 7.8 : I_0^* ($1/\alpha$ = .10)

Q: MODE:	0.05	0.10	0.15	0.20	0.25	0.30	0.35	0.40	0.45	0.50	0.55	0.60	0.70	0.80	0.90
(1)	.0285	.0652	.1119	.1703	.2425	.3312	.4394	.5713	.7323	.9302	1.176	1.487	2.424	4.291	9.839
(2)	.0248	.0552	.0926	.1394	.1989	.2764	.3800	.5217	.7190	.9935	1.369	1.873	3.428	6.442	15.02
(3)	.0700	.1472	.2352	.3386	.4642	.6220	.8270	1.101	1.476	1.987	2.673	3.575	6.284	11.40	25.75
(4)	.1151	.2392	.3777	.5377	.7294	.9675	1.274	1.681	2.233	2.980	3.978	5.278	9.140	16.35	36.47
(5)	.1602	.3312	.5202	.7369	.9946	1.313	1.721	2.261	2.989	3.974	5.282	6.980	12.00	21.31	47.20
(6)	.2054	.4232	.6627	.9360	1.260	1.659	2.168	2.840	3.746	4.967	6.586	8.683	14.85	26.26	57.93

TABLE 7.8 : I_0^* ($1/\alpha$ = .20)

Q: MODE:	0.05	0.10	0.15	0.20	0.25	0.30	0.35	0.40	0.45	0.50	0.55	0.60	0.70	0.80	0.90
(1)	.0390	.0867	.1446	.2145	.2985	.3993	.5204	.6663	.8430	1.059	1.326	1.663	2.680	4.708	10.75
(2)	.0455	.1006	.1676	.2492	.3495	.4740	.6302	.8282	1.081	1.405	1.822	2.360	3.998	7.208	16.48
(3)	.1281	.2683	.4254	.6052	.8155	1.066	1.372	1.748	2.219	2.810	3.556	4.505	7.329	12.75	28.25
(4)	.2107	.4360	.6832	.9612	1.281	1.659	2.113	2.669	3.356	4.215	5.291	6.650	10.66	18.30	40.02
(5)	.2934	.6037	.9410	1.317	1.747	2.251	2.855	3.589	4.494	5.620	7.026	8.795	13.99	23.84	51.79
(6)	.3760	.7714	1.199	1.673	2.213	2.844	3.596	4.509	5.632	7.025	8.761	10.94	17.32	29.39	63.56

TABLE 7.8 : I_0^* ($1/\alpha$ = .30)

Q: MODE:	0.05	0.10	0.15	0.20	0.25	0.30	0.35	0.40	0.45	0.50	0.55	0.60	0.70	0.80	0.90
(1)	.0472	.1037	.1710	.2508	.3454	.4577	.5913	.7511	.9437	1.178	1.468	1.832	2.929	5.120	11.65
(2)	.0629	.1388	.2300	.3394	.4712	.6307	.8248	1.063	1.356	1.721	2.178	2.758	4.507	7.938	17.92
(3)	.1772	.3701	.5837	.8243	1.100	1.419	1.795	2.243	2.783	3.441	4.252	5.266	8.262	14.04	30.72
(4)	.2915	.6014	.9375	1.309	1.728	2.207	2.766	3.424	4.211	5.162	6.327	7.774	12.02	20.15	43.52
(5)	.4058	.8327	1.291	1.794	2.356	2.996	3.736	4.605	5.638	6.883	8.401	10.28	15.77	26.26	56.32
(6)	.5201	1.064	1.645	2.279	2.984	3.784	4.706	5.785	7.066	8.604	10.47	12.79	19.53	32.36	69.12

TABLE 7.8 : I_0^* ($1/\alpha$ = .40)

Q: MODE:	0.05	0.10	0.15	0.20	0.25	0.30	0.35	0.40	0.45	0.50	0.55	0.60	0.70	0.80	0.90
(1)	.0542	.1183	.1938	.2826	.3871	.5102	.6560	.8295	1.038	1.291	1.602	1.994	3.174	5.529	12.56
(2)	.0778	.1714	.2830	.4157	.5735	.7615	.9865	1.257	1.585	1.987	2.484	3.109	4.977	8.642	19.35
(3)	.2192	.4570	.7184	1.010	1.338	1.713	2.147	2.654	3.254	3.974	4.849	5.935	9.124	15.29	33.17
(4)	.3607	.7426	1.154	1.603	2.103	2.665	3.308	4.051	4.923	5.961	7.215	8.761	13.27	21.94	46.99
(5)	.5021	1.028	1.589	2.197	2.867	3.617	4.468	5.448	6.592	7.948	9.580	11.59	17.42	28.58	60.82
(6)	.6436	1.314	2.025	2.791	3.632	4.569	5.629	6.845	8.261	9.934	11.95	14.41	21.57	35.23	74.63

TABLE 7.8 : I^* ($1/\alpha$ = .50)

Q: MODE:	0.05	0.10	0.15	0.20	0.25	0.30	0.35	0.40	0.45	0.50	0.55	0.60	0.70	0.80	0.90
(1)	.0604	.1313	.2144	.3115	.4253	.5588	.7163	.9033	1.127	1.399	1.733	2.152	3.414	5.934	13.46
(2)	.0907	.1996	.3288	.4815	.6616	.8741	1.126	1.426	1.785	2.221	2.757	3.427	5.420	9.326	20.77
(3)	.2557	.5322	.8347	1.169	1.544	1.967	2.450	3.010	3.665	4.443	5.383	6.542	9.936	16.50	35.61
(4)	.4206	.8648	1.341	1.857	2.426	3.059	3.775	4.594	5.544	6.664	8.009	9.657	14.45	23.67	50.44
(5)	.5855	1.197	1.847	2.545	3.308	4.152	5.100	6.178	7.424	8.886	10.63	12.77	18.97	30.85	65.28
(6)	.7505	1.530	2.352	3.233	4.190	5.245	6.424	7.762	9.303	11.11	13.26	15.89	23.48	38.02	80.11

TABLE 7.8 : I_0^* ($1/\alpha$ = .60)

Q: MODE:	0.05	0.10	0.15	0.20	0.25	0.30	0.35	0.40	0.45	0.50	0.55	0.60	0.70	0.80	0.90
(1)	.0660	.1432	.2333	.3383	.4610	.6046	.7735	.9737	1.213	1.503	1.859	2.307	3.651	6.338	14.36
(2)	.1020	.2243	.3689	.5391	.7388	.9731	1.249	1.575	1.964	2.433	3.007	3.722	5.842	9.995	22.18
(3)	.2875	.5980	.9366	1.309	1.724	2.189	2.718	3.325	4.032	4.867	5.871	7.105	10.71	17.68	38.02
(4)	.4730	.9718	1.504	2.080	2.709	3.406	4.187	5.076	6.100	7.300	8.735	10.49	15.58	25.37	53.87
(5)	.6585	1.346	2.072	2.850	3.694	4.622	5.657	6.826	8.168	9.734	11.60	13.87	20.45	33.06	69.71
(6)	.8440	1.719	2.639	3.620	4.679	5.839	7.126	8.576	10.24	12.17	14.46	17.26	25.31	40.75	85.55

TABLE 7.8 : I_0^* ($1/\alpha$ = .70)

Q: MODE:	0.05	0.10	0.15	0.20	0.25	0.30	0.35	0.40	0.45	0.50	0.55	0.60	0.70	0.80	0.90
(1)	.0712	.1543	.2510	.3636	.4947	.6481	.8282	1.041	1.296	1.604	1.983	2.458	3.886	6.739	15.27
(2)	.1120	.2460	.4044	.5902	.8074	1.061	1.359	1.710	2.127	2.628	3.239	3.999	6.240	10.65	23.59
(3)	.3157	.6561	1.027	1.433	1.884	2.388	2.958	3.611	4.367	5.257	6.325	7.635	11.46	18.85	40.43
(4)	.5193	1.066	1.649	2.276	2.961	3.715	4.557	5.511	6.606	7.885	9.410	11.27	16.66	27.04	57.28
(5)	.7230	1.476	2.271	3.120	4.037	5.042	6.157	7.411	8.845	10.51	12.49	14.91	21.87	35.23	74.13
(6)	.9266	1.886	2.893	3.963	5.114	6.369	7.756	9.311	11.08	13.14	15.58	18.54	27.08	43.43	90.97

TABLE 7.8 : I_0^* ($1/\alpha$ = .80)

Q: MODE:	0.05	0.10	0.15	0.20	0.25	0.30	0.35	0.40	0.45	0.50	0.55	0.60	0.70	0.80	0.90
(1)	.0760	.1647	.2677	.3875	.5269	.6898	.8809	1.107	1.377	1.703	2.104	2.607	4.119	7.138	16.17
(2)	.1209	.2655	.4361	.6359	.8691	1.141	1.460	1.834	2.278	2.810	3.458	4.262	6.642	11.30	24.98
(3)	.3407	.7079	1.107	1.544	2.028	2.568	3.177	3.871	4.675	5.620	6.751	8.137	12.18	19.99	42.83
(4)	.5605	1.150	1.778	2.453	3.187	3.994	4.894	5.909	7.073	8.430	10.04	12.01	17.71	28.68	60.68
(5)	.7803	1.593	2.449	3.361	4.345	5.421	6.611	7.947	9.470	11.24	13.34	15.89	23.25	37.37	78.52
(6)	1.000	2.035	3.120	4.270	5.504	6.847	8.328	9.984	11.87	14.05	16.63	19.76	28.78	46.06	96.37

TABLE 7.8 : I_0^* ($1/\alpha$ = .90)

Q: MODE:	0.05	0.10	0.15	0.20	0.25	0.30	0.35	0.40	0.45	0.50	0.55	0.60	0.70	0.80	0.90
(1)	.0806	.1745	.2836	.4104	.5579	.7300	.9320	1.171	1.456	1.801	2.224	2.755	4.350	7.537	17.06
(2)	.1288	.2829	.4645	.6771	.9250	1.214	1.551	1.948	2.418	2.980	3.665	4.514	7.025	11.94	26.38
(3)	.3631	.7543	1.179	1.644	2.158	2.731	3.377	4.112	4.963	5.961	7.155	8.618	12.88	21.12	45.22
(4)	.5973	1.226	1.894	2.612	3.392	4.249	5.202	6.277	7.508	8.941	10.65	12.72	18.73	30.30	64.06
(5)	.8315	1.697	2.608	3.579	4.625	5.766	7.027	8.441	10.05	11.92	14.14	16.83	24.59	39.49	82.90
(6)	1.066	2.169	3.323	4.546	5.858	7.283	8.852	10.61	12.60	14.90	17.63	20.93	30.44	48.67	101.7

TABLE 7.9 : I_0^* (α = .10)

Q: MODE:	0.05	0.10	0.15	0.20	0.25	0.30	0.35	0.40	0.45	0.50	0.55	0.60	0.70	0.80	0.90
(1)	.0248	.1010	.2342	.4347	.7195	1.115	1.664	2.426	3.489	4.967	7.001	9.764	18.58	36.33	87.93
(2)	.5211	1.111	1.795	2.608	3.597	4.834	6.418	8.491	11.24	14.90	19.73	26.04	45.12	81.74	185.6
(3)	1.017	2.121	3.357	4.781	6.475	8.552	11.17	14.56	19.00	24.84	32.46	42.31	71.67	127.2	283.3
(4)	1.514	3.132	4.918	6.954	9.353	12.27	15.93	20.62	26.75	34.77	45.19	58.58	98.21	172.6	381.0
(5)	2.010	4.142	6.479	9.128	12.23	15.99	20.68	26.69	34.50	44.71	57.92	74.85	124.8	218.0	478.7
(6)	2.506	5.152	8.040	11.30	15.11	19.71	25.43	32.75	42.26	54.64	70.65	91.13	151.3	263.4	576.4

TABLE 7.9 : I_0^* (α = .20)

Q: MODE:	0.05	0.10	0.15	0.20	0.25	0.30	0.35	0.40	0.45	0.50	0.55	0.60	0.70	0.80	0.90
(1)	.0248	.1000	.2288	.4164	.6708	1.003	1.429	1.969	2.651	3.512	4.608	6.016	10.33	18.96	44.41
(2)	.5199	1.100	1.754	2.498	3.354	4.347	5.512	6.891	8.541	10.54	12.99	16.04	25.09	42.66	93.75
(3)	1.015	2.100	3.279	4.580	6.037	7.691	9.595	11.81	14.43	17.56	21.36	26.07	39.84	66.37	143.1
(4)	1.510	3.100	4.805	6.662	8.720	11.04	13.68	16.73	20.32	24.59	29.74	36.09	54.60	90.07	192.4
(5)	2.005	4.100	6.330	8.744	11.40	14.38	17.76	21.66	26.21	31.61	38.12	46.12	69.36	113.8	241.8
(6)	2.500	5.100	7.855	10.83	14.09	17.72	21.84	26.58	32.10	38.64	46.50	56.14	84.11	137.5	291.1

TABLE 7.9 : I_0^* (α = .30)

Q: MODE:	0.05	0.10	0.15	0.20	0.25	0.30	0.35	0.40	0.45	0.50	0.55	0.60	0.70	0.80	0.90
(1)	.0247	.0990	.2239	.4008	.6326	.9235	1.280	1.713	2.235	2.868	3.644	4.610	7.481	13.13	29.90
(2)	.5186	1.089	1.716	2.405	3.163	4.002	4.938	5.994	7.201	8.604	10.27	12.29	18.17	29.55	63.12
(3)	1.012	2.080	3.209	4.409	5.693	7.080	8.596	10.28	12.17	14.34	16.89	19.98	28.85	45.97	96.34
(4)	1.506	3.070	4.701	6.412	8.223	10.16	12.25	14.56	17.13	20.07	23.52	27.66	39.54	62.39	129.6
(5)	2.000	4.060	6.193	8.416	10.75	13.24	15.91	18.84	22.10	25.81	30.14	35.35	50.23	78.80	162.8
(6)	2.494	5.050	7.685	10.42	13.28	16.32	19.57	23.12	27.06	31.55	36.77	43.03	60.91	95.22	196.0

TABLE 7.9 : I_0^* (α = .40)

Q: MODE:	0.05	0.10	0.15	0.20	0.25	0.30	0.35	0.40	0.45	0.50	0.55	0.60	0.70	0.80	0.90
(1)	.0246	.0981	.2193	.3872	.6013	.8626	1.174	1.541	1.973	2.484	3.095	3.843	6.007	10.20	22.64
(2)	.5173	1.079	1.681	2.323	3.006	3.738	4.528	5.394	6.357	7.451	8.723	10.25	14.59	22.94	47.79
(3)	1.010	2.060	3.143	4.259	5.411	6.613	7.882	9.246	10.74	12.42	14.35	16.65	23.17	35.69	72.94
(4)	1.503	3.041	4.605	6.195	7.817	9.488	11.24	13.10	15.13	17.39	19.98	23.06	31.75	48.43	98.10
(5)	1.995	4.022	6.067	8.130	10.22	12.36	14.59	16.95	19.51	22.35	25.61	29.46	40.33	61.18	123.2
(6)	2.488	5.003	7.529	10.07	12.63	15.24	17.95	20.80	23.89	27.32	31.24	35.87	48.91	73.92	148.4

TABLE 7.9 : I_0^* (α = .50)

Q: MODE:	0.05	0.10	0.15	0.20	0.25	0.30	0.35	0.40	0.45	0.50	0.55	0.60	0.70	0.80	0.90
(1)	.0246	.0972	.2151	.3751	.5750	.8137	1.093	1.415	1.788	2.221	2.733	3.348	5.094	8.419	18.28
(2)	.5161	1.069	1.649	2.231	2.875	3.526	4.214	4.953	5.761	6.664	7.701	8.927	12.37	18.94	38.59
(3)	1.008	2.041	3.083	4.126	5.175	6.236	7.336	8.491	9.734	11.11	12.67	14.51	19.65	29.47	58.90
(4)	1.499	3.013	4.517	6.002	7.474	8.951	10.46	12.03	13.71	15.55	17.64	20.09	26.93	39.99	79.21
(5)	1.990	3.985	5.951	7.878	9.774	11.66	13.58	15.57	17.68	19.99	22.61	25.67	34.20	50.51	99.52
(6)	2.482	4.957	7.385	9.753	12.07	14.38	16.70	19.10	21.65	24.44	27.57	31.24	41.48	61.04	119.8

TABLE 7.9 : I_0^* (α = .60)

Q: MODE:	0.05	0.10	0.15	0.20	0.25	0.30	0.35	0.40	0.45	0.50	0.55	0.60	0.70	0.80	0.90
(1)	.0245	.0963	.2112	.3644	.5523	.7733	1.027	1.317	1.648	2.028	2.470	2.997	4.467	7.223	15.37
(2)	.5148	1.060	1.619	2.186	2.762	3.351	3.963	4.611	5.310	6.084	6.962	7.991	10.85	16.25	32.45
(3)	1.005	2.023	3.027	4.008	4.971	5.928	6.899	7.904	8.973	10.14	11.45	12.99	17.23	25.28	49.53
(4)	1.495	2.986	4.435	5.830	7.180	8.506	9.834	11.20	12.63	14.20	15.95	17.98	23.61	34.31	66.60
(5)	1.986	3.949	5.843	7.652	9.390	11.08	12.77	14.49	16.30	18.25	20.44	22.98	29.99	43.34	83.68
(6)	2.476	4.912	7.250	9.474	11.60	13.66	15.71	17.78	19.96	22.31	24.93	27.97	36.38	52.36	100.8

TABLE 7.9 : I_0^* (α = .70)

Q: MODE:	0.05	0.10	0.15	0.20	0.25	0.30	0.35	0.40	0.45	0.50	0.55	0.60	0.70	0.80	0.90
(1)	.0245	.0955	.2075	.3547	.5326	.7390	.9736	1.238	1.537	1.877	2.270	2.732	4.007	6.360	13.29
(2)	.5136	1.050	1.591	2.128	2.663	3.202	3.755	4.334	4.954	5.632	6.397	7.287	9.731	14.31	28.06
(3)	1.003	2.005	2.974	3.901	4.793	5.665	6.537	7.430	8.370	9.387	10.52	11.84	15.45	22.26	42.82
(4)	1.492	2.960	4.358	5.675	6.924	8.129	9.319	10.53	11.79	13.14	14.65	16.39	21.18	30.21	57.59
(5)	1.981	3.915	5.741	7.448	9.054	10.59	12.10	13.62	15.20	16.90	18.78	20.95	26.90	38.16	72.36
(6)	2.470	4.869	7.124	9.221	11.18	13.05	14.88	16.72	18.62	20.65	22.91	25.50	32.63	46.11	87.12

TABLE 7.9 : I_0^* (α = .80)

MODE: \ Q:	0.05	0.10	0.15	0.20	0.25	0.30	0.35	0.40	0.45	0.50	0.55	0.60	0.70	0.80	0.90
(1)	.0244	.0947	.2041	.3459	.5151	.7093	.9281	1.173	1.447	1.756	2.110	2.525	3.652	5.707	11.73
(2)	.5124	1.041	1.565	2.075	2.576	3.074	3.580	4.105	4.662	5.269	5.948	6.733	8.870	12.84	24.76
(3)	1.000	1.988	2.925	3.804	4.636	5.438	6.231	7.037	7.877	8.781	9.785	10.94	14.09	19.98	37.79
(4)	1.488	2.935	4.286	5.534	6.697	7.803	8.883	9.968	11.09	12.29	13.62	15.15	19.31	27.11	50.83
(5)	1.976	3.881	5.646	7.263	8.757	10.17	11.53	12.90	14.31	15.81	17.46	19.36	24.52	34.24	63.86
(6)	2.464	4.828	7.007	8.992	10.82	12.53	14.19	15.83	17.52	19.32	21.30	23.56	29.74	41.38	76.89

TABLE 7.9 : I_0^* (α = .90)

MODE: \ Q:	0.05	0.10	0.15	0.20	0.25	0.30	0.35	0.40	0.45	0.50	0.55	0.60	0.70	0.80	0.90
(1)	.0243	.0939	.2009	.3378	.4995	.6834	.8888	1.117	1.371	1.656	1.980	2.356	3.370	5.195	10.51
(2)	.5112	1.033	1.540	2.027	2.497	2.961	3.428	3.910	4.418	4.967	5.579	6.283	8.184	11.69	22.20
(3)	.9980	1.971	2.879	3.715	4.495	5.239	5.968	6.702	7.464	8.279	9.179	10.21	13.00	18.18	33.88
(4)	1.485	2.910	4.218	5.404	6.493	7.517	8.508	9.495	10.51	11.59	12.78	14.14	17.81	24.67	45.56
(5)	1.972	3.849	5.557	7.093	8.491	9.795	11.05	12.29	13.56	14.90	16.38	18.06	22.63	31.17	57.24
(6)	2.458	4.787	6.896	8.782	10.49	12.07	13.59	15.08	16.60	18.21	19.98	21.99	27.44	37.66	68.92

TABLE 7.9 : I_0^* (α = 1.0)

MODE: \ Q:	0.05	0.10	0.15	0.20	0.25	0.30	0.35	0.40	0.45	0.50	0.55	0.60	0.70	0.80	0.90
(1)	.0243	.0931	.1978	.3303	.4854	.6604	.8546	1.069	1.306	1.571	1.870	2.216	3.139	4.781	9.540
(2)	.5100	1.024	1.516	1.982	2.427	2.862	3.296	3.741	4.209	4.712	5.270	5.909	7.622	10.76	20.14
(3)	.9956	1.955	2.835	3.634	4.369	5.063	5.738	6.414	7.111	7.854	8.671	9.602	12.11	16.73	30.74
(4)	1.481	2.886	4.154	5.285	6.310	7.264	8.179	9.086	10.01	11.00	12.07	13.30	16.59	22.71	41.34
(5)	1.967	3.817	5.473	6.937	8.252	9.465	10.62	11.76	12.92	14.14	15.47	16.99	21.07	28.68	51.94
(6)	2.453	4.748	6.791	8.588	10.19	11.67	13.06	14.43	15.82	17.28	18.87	20.68	25.56	34.66	62.54

TABLE 7.9 : I_0^* (1/α = .10)

MODE: \ Q:	0.05	0.10	0.15	0.20	0.25	0.30	0.35	0.40	0.45	0.50	0.55	0.60	0.70	0.80	0.90
(1)	.0206	.0621	.1098	.1601	.2122	.2658	.3208	.3775	.4360	.4967	.5603	.6275	.7799	.9863	1.429
(2)	.4326	.6835	.8417	.9606	1.061	1.152	1.237	1.321	1.405	1.490	1.579	1.673	1.894	2.219	3.018
(3)	.8446	1.305	1.574	1.761	1.910	2.037	2.154	2.265	2.374	2.484	2.598	2.719	3.008	3.452	4.606
(4)	1.257	1.926	2.306	2.562	2.758	2.923	3.071	3.209	3.343	3.477	3.616	3.765	4.122	4.685	6.194
(5)	1.669	2.548	3.037	3.362	3.607	3.809	3.987	4.152	4.312	4.471	4.635	4.811	5.236	5.918	7.783
(6)	2.081	3.169	3.769	4.163	4.456	4.695	4.904	5.096	5.280	5.464	5.653	5.857	6.350	7.151	9.371

TABLE 7.9 : I_0^* (1/α = .15)

MODE: \ Q:	0.05	0.10	0.15	0.20	0.25	0.30	0.35	0.40	0.45	0.50	0.55	0.60	0.70	0.80	0.90
(1)	.0217	.0691	.1259	.1870	.2509	.3173	.3859	.4570	.5310	.6084	.6900	.7774	.9799	1.267	1.927
(2)	.4559	.7604	.9651	1.122	1.255	1.375	1.488	1.600	1.711	1.825	1.945	2.073	2.380	2.850	4.069
(3)	.8901	1.452	1.804	2.057	2.258	2.432	2.591	2.742	2.891	3.042	3.199	3.369	3.780	4.434	6.211
(4)	1.324	2.143	2.644	2.992	3.262	3.490	3.694	3.885	4.071	4.259	4.454	4.665	5.180	6.017	8.352
(5)	1.759	2.834	3.483	3.927	4.266	4.547	4.796	5.027	5.251	5.475	5.709	5.960	6.580	7.601	10.49
(6)	2.193	3.525	4.322	4.861	5.269	5.605	5.899	6.170	6.431	6.692	6.963	7.256	7.979	9.184	12.64

TABLE 7.9 : I_0^* (1/α = .20)

MODE: \ Q:	0.05	0.10	0.15	0.20	0.25	0.30	0.35	0.40	0.45	0.50	0.55	0.60	0.70	0.80	0.90
(1)	.0224	.0739	.1377	.2075	.2814	.3585	.4389	.5226	.6102	.7025	.8006	.9064	1.156	1.522	2.405
(2)	.4695	.8125	1.055	1.245	1.407	1.554	1.693	1.829	1.966	2.107	2.256	2.417	2.808	3.424	5.078
(3)	.9167	1.551	1.973	2.283	2.532	2.749	2.947	3.136	3.322	3.512	3.712	3.928	4.459	5.327	7.750
(4)	1.364	2.290	2.891	3.320	3.658	3.944	4.201	4.442	4.678	4.917	5.167	5.439	6.111	7.229	10.42
(5)	1.811	3.028	3.809	4.358	4.783	5.139	5.455	5.749	6.035	6.322	6.623	6.949	7.763	9.131	13.09
(6)	2.258	3.767	4.726	5.395	5.909	6.334	6.709	7.056	7.391	7.727	8.079	8.460	9.414	11.03	15.77

TABLE 7.9 : I_0^* (1/α = .25)

MODE: \ Q:	0.05	0.10	0.15	0.20	0.25	0.30	0.35	0.40	0.45	0.50	0.55	0.60	0.70	0.80	0.90
(1)	.0228	.0773	.1468	.2241	.3066	.3934	.4842	.5793	.6794	.7854	.8988	1.022	1.317	1.761	2.871
(2)	.4785	.8507	1.126	1.345	1.533	1.705	1.868	2.028	2.189	2.356	2.533	2.726	3.199	3.963	6.062
(3)	.9342	1.624	2.105	2.465	2.759	3.016	3.251	3.476	3.699	3.927	4.167	4.429	5.081	6.165	9.253
(4)	1.390	2.397	3.084	3.586	3.986	4.327	4.634	4.924	5.209	5.498	5.801	6.133	6.963	8.366	12.44
(5)	1.846	3.171	4.063	4.707	5.212	5.638	6.018	6.373	6.719	7.069	7.436	7.836	8.844	10.57	15.63
(6)	2.301	3.944	5.042	5.828	6.439	6.949	7.401	7.821	8.229	8.639	9.070	9.540	10.73	12.77	18.82

TABLE 7.9 : I_0^* (1/α = .30)

MODE: \ Q:	0.05	0.10	0.15	0.20	0.25	0.30	0.35	0.40	0.45	0.50	0.55	0.60	0.70	0.80	0.90
(1)	.0231	.0800	.1543	.2381	.3282	.4236	.5240	.6298	.7415	.8604	.9883	1.128	1.468	1.990	3.331
(2)	.4848	.8801	1.183	1.428	1.641	1.836	2.021	2.204	2.389	2.581	2.785	3.009	3.565	4.477	7.031
(3)	.9466	1.680	2.212	2.619	2.954	3.248	3.519	3.779	4.037	4.302	4.582	4.890	5.661	6.965	10.73
(4)	1.408	2.480	3.240	3.809	4.267	4.660	5.016	5.353	5.685	6.023	6.379	6.770	7.758	9.452	14.43
(5)	1.870	3.280	4.269	4.999	5.580	6.072	6.513	6.927	7.332	7.743	8.176	8.651	9.855	11.94	18.13
(6)	2.332	4.080	5.297	6.190	6.892	7.484	8.010	8.502	8.980	9.464	9.973	10.53	11.95	14.43	21.83

TABLE 7.9 : I_0^* (1/α = .40)

MODE: \ Q:	0.05	0.10	0.15	0.20	0.25	0.30	0.35	0.40	0.45	0.50	0.55	0.60	0.70	0.80	0.90
(1)	.0235	.0839	.1658	.2604	.3639	.4747	.5923	.7173	.8504	.9935	1.149	1.321	1.747	2.425	4.236
(2)	.4932	.9227	1.271	1.563	1.819	2.057	2.285	2.511	2.740	2.980	3.238	3.522	4.242	5.457	8.942
(3)	.9630	1.762	2.376	2.865	3.275	3.639	3.977	4.304	4.630	4.967	5.327	5.724	6.737	8.489	13.65
(4)	1.433	2.600	3.482	4.167	4.731	5.221	5.670	6.097	6.520	6.954	7.416	7.925	9.232	11.52	18.35
(5)	1.903	3.439	4.587	5.469	6.186	6.803	7.362	7.890	8.410	8.941	9.505	10.13	11.73	14.55	23.06
(6)	2.372	4.278	5.692	6.771	7.642	8.385	9.054	9.683	10.30	10.93	11.59	12.33	14.22	17.58	27.77

TABLE 7.9 : I_0^* (1/α = .50)

MODE: \ Q:	0.05	0.10	0.15	0.20	0.25	0.30	0.35	0.40	0.45	0.50	0.55	0.60	0.70	0.80	0.90
(1)	.0237	.0866	.1744	.2779	.3927	.5169	.6500	.7923	.9452	1.111	1.292	1.495	2.005	2.841	5.130
(2)	.4986	.9523	1.337	1.667	1.963	2.240	2.507	2.773	3.046	3.332	3.641	3.986	4.869	6.393	10.83
(3)	.9734	1.818	2.499	3.057	3.534	3.963	4.364	4.754	5.146	5.554	5.991	6.477	7.733	9.944	16.53
(4)	1.448	2.684	3.662	4.446	5.105	5.686	6.221	6.735	7.247	7.775	8.340	8.968	10.60	13.50	22.23
(5)	1.923	3.549	4.824	5.836	6.676	7.409	8.078	8.716	9.347	9.996	10.69	11.46	13.46	17.05	27.93
(6)	2.398	4.415	5.986	7.225	8.247	9.131	9.935	10.70	11.45	12.22	13.04	13.95	16.33	20.60	33.63

TABLE 7.9 : I_0^* ($1/\alpha = .60$)

Q:	0.05	0.10	0.15	0.20	0.25	0.30	0.35	0.40	0.45	0.50	0.55	0.60	0.70	0.80	0.90
MODE:															
(1)	.0239	.0886	.1810	.2921	.4168	.5530	.7001	.8586	1.030	1.217	1.423	1.655	2.249	3.244	6.018
(2)	.5023	.9741	1.388	1.752	2.084	2.396	2.700	3.005	3.319	3.650	4.010	4.414	5.462	7.299	12.70
(3)	.9806	1.860	2.595	3.213	3.751	4.239	4.700	5.151	5.607	6.084	6.598	7.173	8.674	11.35	19.39
(4)	1.459	2.745	3.802	4.673	5.418	6.083	6.701	7.298	7.896	8.517	9.185	9.932	11.89	15.41	26.08
(5)	1.937	3.631	5.009	6.134	7.085	7.926	8.701	9.444	10.18	10.95	11.77	12.69	15.10	19.46	32.77
(6)	2.416	4.516	6.216	7.594	8.752	9.769	10.70	11.59	12.47	13.38	14.36	15.45	18.31	23.52	39.45

TABLE 7.9 : I_0^* ($1/\alpha = .70$)

Q:	0.05	0.10	0.15	0.20	0.25	0.30	0.35	0.40	0.45	0.50	0.55	0.60	0.70	0.80	0.90
MODE:															
(1)	.0240	.0901	.1864	.3039	.4374	.5845	.7445	.9181	1.107	1.314	1.545	1.806	2.482	3.637	6.902
(2)	.5050	.9909	1.429	1.824	2.187	2.533	2.872	3.213	3.567	3.943	4.353	4.816	6.028	8.184	14.57
(3)	.9859	1.892	2.672	3.343	3.937	4.481	4.999	5.509	6.027	6.571	7.161	7.826	9.574	12.73	22.24
(4)	1.467	2.793	3.915	4.863	5.686	6.429	7.126	7.804	8.487	9.200	9.970	10.84	13.12	17.28	29.91
(5)	1.948	3.693	5.158	6.382	7.436	8.378	9.253	10.10	10.95	11.83	12.78	13.85	16.67	21.82	37.58
(6)	2.429	4.594	6.400	7.902	9.185	10.33	11.38	12.39	13.41	14.46	15.59	16.86	20.21	26.37	45.25

TABLE 7.9 : I_0^* ($1/\alpha = .80$)

Q:	0.05	0.10	0.15	0.20	0.25	0.30	0.35	0.40	0.45	0.50	0.55	0.60	0.70	0.00	0.90
MODE:															
(1)	.0241	.0913	.1909	.3140	.4554	.6125	.7846	.9725	1.178	1.405	1.659	1.949	2.707	4.023	7.783
(2)	.5070	1.004	1.463	1.884	2.277	2.654	3.026	3.404	3.796	4.215	4.675	5.196	6.575	9.053	16.43
(3)	.9899	1.917	2.736	3.454	4.098	4.696	5.268	5.835	6.414	7.025	7.691	8.444	10.44	14.08	25.08
(4)	1.473	2.830	4.008	5.024	5.920	6.737	7.510	8.266	9.032	9.835	10.71	11.69	14.31	19.11	33.73
(5)	1.956	3.743	5.281	6.594	7.741	8.779	9.751	10.70	11.65	12.64	13.72	14.94	18.18	24.14	42.37
(6)	2.439	4.656	6.553	8.164	9.563	10.82	11.99	13.13	14.27	15.45	16.74	18.19	22.04	29.17	51.02

TABLE 7.9 : I_0^* ($1/\alpha = .90$)

Q:	0.05	0.10	0.15	0.20	0.25	0.30	0.35	0.40	0.45	0.50	0.55	0.60	0.70	0.80	0.90
MODE:															
(1)	.0242	.0923	.1946	.3227	.4712	.6376	.8211	1.022	1.244	1.490	1.767	2.085	2.926	4.404	8.662
(2)	.5087	1.015	1.492	1.936	2.356	2.763	3.167	3.579	4.009	4.471	4.980	5.560	7.105	9.909	18.29
(3)	.9931	1.938	2.789	3.550	4.241	4.888	5.513	6.135	6.774	7.451	8.192	9.035	11.28	15.41	27.91
(4)	1.477	2.861	4.087	5.163	6.126	7.013	7.859	8.691	9.539	10.43	11.41	12.51	15.46	20.92	37.54
(5)	1.962	3.784	5.384	6.777	8.011	9.139	10.20	11.25	12.30	13.41	14.62	15.98	19.64	26.42	47.16
(6)	2.446	4.707	6.681	8.390	9.896	11.26	12.55	13.80	15.07	16.39	17.83	19.46	23.82	31.93	56.78

TABLE 7.10 : I_0^* ($\alpha = .10$)

Q:	0.05	0.10	0.15	0.20	0.25	0.30	0.35	0.40	0.45	0.50	0.55	0.60	0.70	0.80	0.90
MODE:															
(1)	.2812	.6365	1.080	1.627	2.298	3.114	4.105	5.309	6.775	8.573	10.81	13.62	22.13	39.08	89.48
(2)	.2730	.6061	1.015	1.521	2.158	2.975	4.041	5.459	7.366	9.935	13.37	17.90	31.85	59.04	136.8
(3)	.7693	1.616	2.576	3.695	5.036	6.693	8.795	11.52	15.12	19.87	26.09	34.17	58.39	104.5	234.5
(4)	1.266	2.627	4.137	5.868	7.914	10.41	13.55	17.59	22.87	29.80	38.82	50.45	84.94	149.9	332.2
(5)	1.762	3.637	5.698	8.041	10.79	14.13	18.30	23.65	30.63	39.74	51.55	66.72	111.5	195.3	429.9
(6)	2.258	4.647	7.259	10.21	13.67	17.85	23.06	29.72	38.38	49.67	64.28	82.99	138.0	240.7	527.5

TABLE 7.10 : I_0^* ($\alpha = .20$)

Q:	0.05	0.10	0.15	0.20	0.25	0.30	0.35	0.40	0.45	0.50	0.55	0.60	0.70	0.80	0.90
MODE:															
(1)	.1918	.4189	.6869	1.003	1.374	1.813	2.332	2.951	3.695	4.598	5.711	7.111	11.32	19.74	44.85
(2)	.2723	.6000	.9914	1.457	2.012	2.675	3.470	4.430	5.596	7.025	8.797	11.03	17.71	30.81	69.08
(3)	.7674	1.600	2.517	3.539	4.696	6.019	7.553	9.352	11.49	14.05	17.17	21.05	32.46	54.51	118.4
(4)	1.263	2.600	4.042	5.621	7.379	9.364	11.64	14.27	17.38	21.07	25.55	31.08	47.22	78.22	167.8
(5)	1.758	3.600	5.567	7.703	10.06	12.71	15.72	19.20	23.27	28.10	33.93	41.11	61.98	101.9	217.1
(6)	2.253	4.600	7.092	9.785	12.75	16.05	19.80	24.12	29.16	35.12	42.31	51.13	76.73	125.6	266.4

TABLE 7.10 : I_0^* ($\alpha = .30$)

Q:	0.05	0.10	0.15	0.20	0.25	0.30	0.35	0.40	0.45	0.50	0.55	0.60	0.70	0.80	0.90
MODE:															
(1)	.1541	.3312	.5347	.7684	1.038	1.349	1.712	2.137	2.643	3.251	3.995	4.925	7.716	13.29	29.97
(2)	.2716	.5942	.9700	1.403	1.898	2.463	3.109	3.853	4.718	5.736	6.956	8.452	12.82	21.34	46.51
(3)	.7655	1.585	2.462	3.407	4.428	5.541	6.767	8.135	9.684	11.47	13.58	16.14	23.51	37.76	79.73
(4)	1.259	2.575	3.955	5.411	6.958	8.620	10.43	12.42	14.65	17.21	20.21	23.82	34.20	54.18	112.9
(5)	1.753	3.565	5.447	7.414	9.488	11.70	14.08	16.70	19.62	22.94	26.83	31.50	44.88	70.59	146.2
(6)	2.247	4.555	6.939	9.418	12.02	14.78	17.74	20.98	24.58	28.68	33.45	39.19	55.57	87.01	179.4

TABLE 7.10 : I_0^* ($\alpha = .40$)

Q:	0.05	0.10	0.15	0.20	0.25	0.30	0.35	0.40	0.45	0.50	0.55	0.60	0.70	0.80	0.90
MODE:															
(1)	.1321	.2814	.4499	.6405	.8567	1.103	1.387	1.716	2.103	2.565	3.126	3.824	5.907	10.06	22.54
(2)	.2710	.5886	.9503	1.355	1.804	2.300	2.851	3.467	4.165	4.967	5.909	7.045	10.30	16.57	35.21
(3)	.7637	1.570	2.412	3.291	4.209	5.175	6.205	7.320	8.549	9.935	11.54	13.45	18.88	29.32	60.37
(4)	1.256	2.550	3.874	5.227	6.614	8.051	9.560	11.17	12.93	14.90	17.17	19.85	27.46	42.06	85.52
(5)	1.749	3.531	5.336	7.163	9.019	10.93	12.91	15.03	17.32	19.87	22.79	26.26	36.04	54.81	110.7
(6)	2.242	4.512	6.798	9.098	11.42	13.80	16.27	18.88	21.70	24.84	28.42	32.66	44.62	67.55	135.8

TABLE 7.10 : I_0^* ($\alpha = .50$)

Q:	0.05	0.10	0.15	0.20	0.25	0.30	0.35	0.40	0.45	0.50	0.55	0.60	0.70	0.80	0.90
MODE:															
(1)	.1174	.2484	.3946	.5579	.7414	.9485	1.184	1.455	1.772	2.146	2.598	3.157	4.818	8.124	18.07
(2)	.2703	.5831	.9321	1.313	1.725	2.170	2.653	3.184	3.774	4.443	5.217	6.137	8.733	13.68	28.43
(3)	.7618	1.555	2.366	3.189	4.025	4.882	5.775	6.722	7.747	8.886	10.18	11.72	16.01	24.20	48.74
(4)	1.253	2.527	3.800	5.064	6.324	7.595	8.897	10.26	11.72	13.33	15.15	17.30	23.29	34.73	69.05
(5)	1.745	3.499	5.234	6.940	8.624	10.31	12.02	13.80	15.69	17.77	20.12	22.88	30.56	45.25	89.36
(6)	2.236	4.471	6.668	8.816	10.92	13.02	15.14	17.34	19.67	22.21	25.09	28.45	37.84	55.77	109.7

TABLE 7.10 : I_0^* ($\alpha = .60$)

Q:	0.05	0.10	0.15	0.20	0.25	0.30	0.35	0.40	0.45	0.50	0.55	0.60	0.70	0.80	0.90
MODE:															
(1)	.1067	.2246	.3550	.4995	.6603	.8405	1.044	1.276	1.545	1.861	2.241	2.709	4.089	6.831	15.10
(2)	.2697	.5779	.9151	1.275	1.657	2.062	2.495	2.964	3.479	4.056	4.716	5.494	7.658	11.74	23.91
(3)	.7600	1.541	2.323	3.097	3.866	4.640	5.431	6.257	7.141	8.112	9.208	10.49	14.04	20.77	40.99
(4)	1.250	2.504	3.731	4.919	6.076	7.217	8.366	9.551	10.80	12.17	13.70	15.48	20.42	29.79	58.06
(5)	1.741	3.467	5.139	6.741	8.285	9.795	11.30	12.84	14.47	16.22	18.19	20.48	26.80	38.82	75.14
(6)	2.231	4.431	6.546	8.563	10.49	12.37	14.24	16.14	18.13	20.28	22.68	25.47	33.18	47.85	92.22

TABLE 7.10 : I_0^* (α = .70)

Q: MODE:	0.05	0.10	0.15	0.20	0.25	0.30	0.35	0.40	0.45	0.50	0.55	0.60	0.70	0.80	0.90
(1)	.0984	.2064	.3249	.4554	.5997	.7603	.9404	1.145	1.380	1.655	1.983	2.385	3.567	5.907	12.97
(2)	.2690	.5729	.8992	1.241	1.598	1.971	2.364	2.786	3.245	3.755	4.333	5.010	6.869	10.34	20.67
(3)	.7582	1.528	2.283	3.015	3.728	4.434	5.146	5.882	6.662	7.510	8.461	9.564	12.59	18.29	35.44
(4)	1.247	2.482	3.666	4.788	5.858	6.897	7.928	8.978	10.08	11.26	12.59	14.12	18.32	26.24	50.21
(5)	1.736	3.437	5.050	6.562	7.989	9.360	10.71	12.07	13.49	15.02	16.71	18.67	24.04	34.19	64.98
(6)	2.226	4.392	6.433	8.335	10.12	11.82	13.49	15.17	16.91	18.77	20.84	23.23	29.76	42.14	79.74

TABLE 7.10 : I_0^* (α = .80)

Q: MODE:	0.05	0.10	0.15	0.20	0.25	0.30	0.35	0.40	0.45	0.50	0.55	0.60	0.70	0.80	0.90
(1)	.0918	.1919	.3011	.4208	.5523	.6979	.8603	1.044	1.253	1.497	1.787	2.140	3.174	5.213	11.38
(2)	.2684	.5680	.8844	1.210	1.545	1.892	2.254	2.639	3.054	3.512	4.029	4.629	6.261	9.275	18.25
(3)	.7564	1.515	2.245	2.940	3.606	4.256	4.906	5.571	6.270	7.025	7.866	8.836	11.48	16.41	31.28
(4)	1.244	2.461	3.605	4.669	5.666	6.621	7.557	8.502	9.485	10.54	11.70	13.04	16.70	23.54	44.31
(5)	1.732	3.408	4.966	6.398	7.727	8.985	10.21	11.43	12.70	14.05	15.54	17.25	21.91	30.68	57.34
(6)	2.220	4.354	6.326	8.127	9.787	11.35	12.86	14.37	15.91	17.56	19.38	21.46	27.13	37.81	70.37

TABLE 7.10 : I_0^* (α = .90)

Q: MODE:	0.05	0.10	0.15	0.20	0.25	0.30	0.35	0.40	0.45	0.50	0.55	0.60	0.70	0.80	0.90
(1)	.0863	.1800	.2818	.3927	.5141	.6477	.7962	.9628	1.152	1.372	1.632	1.948	2.866	4.673	10.14
(2)	.2678	.5632	.8704	1.182	1.498	1.822	2.159	2.513	2.894	3.312	3.779	4.320	5.777	8.441	16.35
(3)	.7546	1.502	2.209	2.871	3.496	4.100	4.698	5.306	5.941	6.623	7.379	8.246	10.59	14.93	28.04
(4)	1.241	2.441	3.548	4.560	5.494	6.378	7.238	8.099	8.987	9.935	10.98	12.17	15.40	21.43	39.72
(5)	1.728	3.379	4.887	6.249	7.492	8.656	9.777	10.89	12.03	13.25	14.58	16.10	20.22	27.92	51.40
(6)	2.215	4.318	6.226	7.938	9.490	10.93	12.32	13.68	15.08	16.56	18.18	20.03	25.03	34.41	63.08

TABLE 7.10 : I_0^* (α = 1.0)

Q: MODE:	0.05	0.10	0.15	0.20	0.25	0.30	0.35	0.40	0.45	0.50	0.55	0.60	0.70	0.80	0.90
(1)	.0817	.1700	.2656	.3693	.4824	.6063	.7435	.8967	1.070	1.271	1.507	1.793	2.620	4.240	9.144
(2)	.2671	.5587	.8571	1.156	1.456	1.761	2.075	2.405	2.758	3.142	3.570	4.062	5.380	7.768	14.84
(3)	.7528	1.490	2.176	2.808	3.398	3.962	4.517	5.078	5.660	6.283	6.970	7.756	9.864	13.74	25.44
(4)	1.238	2.421	3.495	4.459	5.339	6.163	6.958	7.750	8.563	9.425	10.37	11.45	14.35	19.72	36.04
(5)	1.724	3.352	4.813	6.111	7.281	8.365	9.400	10.42	11.47	12.57	13.77	15.14	18.83	25.70	46.64
(6)	2.210	4.283	6.132	7.763	9.223	10.57	11.84	13.09	14.37	15.71	17.17	18.83	23.31	31.67	57.24

TABLE 7.10 : I_0^* (1/α = .10)

Q: MODE:	0.05	0.10	0.15	0.20	0.25	0.30	0.35	0.40	0.45	0.50	0.55	0.60	0.70	0.80	0.90
(1)	.0251	.0509	.0774	.1046	.1326	.1615	.1915	.2228	.2557	.2907	.3284	.3699	.4720	.6353	1.078
(2)	.2266	.3728	.4757	.5604	.6365	.7087	.7791	.8494	.9204	.9935	1.070	1.150	1.337	1.603	2.224
(3)	.6386	.9942	1.208	1.361	1.485	1.595	1.696	1.793	1.889	1.987	2.088	2.196	2.451	2.836	3.812
(4)	1.051	1.616	1.940	2.161	2.334	2.480	2.612	2.737	2.858	2.980	3.107	3.242	3.565	4.069	5.400
(5)	1.463	2.237	2.671	2.962	3.183	3.366	3.529	3.681	3.827	3.974	4.126	4.288	4.679	5.302	6.988
(6)	1.875	2.858	3.403	3.762	4.031	4.252	4.446	4.624	4.796	4.967	5.144	5.334	5.793	6.534	8.577

TABLE 7.10 : I_0^* (1/α = .15)

Q: MODE:	0.05	0.10	0.15	0.20	0.25	0.30	0.35	0.40	0.45	0.50	0.55	0.60	0.70	0.80	0.90
(1)	.0309	.0627	.0956	.1296	.1648	.2014	.2397	.2800	.3228	.3687	.4190	.4751	.6174	.8566	1.534
(2)	.2388	.4148	.5455	.6544	.7528	.8460	.9372	1.028	1.121	1.217	1.317	1.425	1.680	2.059	2.998
(3)	.6730	1.106	1.385	1.589	1.757	1.904	2.040	2.171	2.301	2.433	2.572	2.721	3.080	3.642	5.140
(4)	1.107	1.797	2.224	2.524	2.760	2.961	3.142	3.313	3.481	3.650	3.827	4.017	4.480	5.226	7.281
(5)	1.541	2.489	3.063	3.459	3.764	4.019	4.245	4.456	4.661	4.867	5.081	5.312	5.880	6.809	9.423
(6)	1.976	3.180	3.902	4.394	4.768	5.076	5.348	5.599	5.841	6.084	6.336	6.608	7.279	8.393	11.56

TABLE 7.10 : I_0^* (1/α = .20)

Q: MODE:	0.05	0.10	0.15	0.20	0.25	0.30	0.35	0.40	0.45	0.50	0.55	0.60	0.70	0.80	0.90
(1)	.0357	.0728	.1112	.1511	.1926	.2361	.2819	.3303	.3822	.4385	.5005	.5707	.7528	1.069	1.986
(2)	.2459	.4432	.5965	.7263	.8441	.9561	1.066	1.176	1.288	1.405	1.528	1.662	1.982	2.473	3.741
(3)	.6931	1.182	1.514	1.764	1.970	2.151	2.320	2.483	2.644	2.810	2.984	3.173	3.634	4.376	6.414
(4)	1.140	1.920	2.432	2.801	3.095	3.346	3.574	3.789	4.000	4.215	4.440	4.683	5.285	6.278	9.086
(5)	1.587	2.659	3.350	3.839	4.221	4.541	4.828	5.096	5.357	5.620	5.895	6.194	6.937	8.180	11.76
(6)	2.035	3.398	4.267	4.877	5.346	5.737	6.082	6.402	6.713	7.025	7.351	7.705	8.588	10.08	14.43

TABLE 7.10 : I_0^* (1/α = .25)

Q: MODE:	0.05	0.10	0.15	0.20	0.25	0.30	0.35	0.40	0.45	0.50	0.55	0.60	0.70	0.80	0.90
(1)	.0401	.0817	.1251	.1704	.2177	.2675	.3202	.3764	.4369	.5030	.5765	.6605	.8821	1.276	2.437
(2)	.2506	.4640	.6363	.7845	.9198	1.049	1.176	1.304	1.434	1.571	1.716	1.874	2.258	2.862	4.467
(3)	.7063	1.237	1.615	1.905	2.146	2.360	2.559	2.752	2.944	3.142	3.350	3.577	4.140	5.064	7.657
(4)	1.162	2.011	2.594	3.026	3.373	3.671	3.943	4.200	4.454	4.712	4.984	5.281	6.022	7.266	10.85
(5)	1.618	2.784	3.573	4.147	4.599	4.982	5.326	5.649	5.964	6.283	6.619	6.985	7.904	9.467	14.04
(6)	2.073	3.557	4.552	5.267	5.826	6.294	6.709	7.097	7.474	7.854	8.253	8.688	9.785	11.67	17.23

TABLE 7.10 : I_0^* (1/α = .30)

Q: MODE:	0.05	0.10	0.15	0.20	0.25	0.30	0.35	0.40	0.45	0.50	0.55	0.60	0.70	0.80	0.90
(1)	.0440	.0898	.1378	.1881	.2409	.2966	.3559	.4194	.4882	.5638	.6487	.7463	1.007	1.480	2.886
(2)	.2540	.4800	.6686	.8332	.9846	1.130	1.273	1.417	1.565	1.721	1.887	2.069	2.516	3.234	5.181
(3)	.7157	1.280	1.697	2.024	2.297	2.542	2.770	2.991	3.213	3.441	3.684	3.949	4.613	5.721	8.882
(4)	1.177	2.080	2.726	3.214	3.610	3.954	4.267	4.566	4.861	5.162	5.481	5.830	6.710	8.209	12.58
(5)	1.639	2.880	3.755	4.404	4.923	5.366	5.765	6.140	6.508	6.883	7.278	7.711	8.807	10.70	16.28
(6)	2.101	3.680	4.783	5.594	6.236	6.778	7.262	7.715	8.156	8.604	9.075	9.591	10.90	13.18	19.98

TABLE 7.10 : I_0^* (1/α = .40)

Q: MODE:	0.05	0.10	0.15	0.20	0.25	0.30	0.35	0.40	0.45	0.50	0.55	0.60	0.70	0.80	0.90
(1)	.0509	.1044	.1608	.2202	.2830	.3499	.4216	.4990	.5838	.6781	.7850	.9095	1.250	1.883	3.782
(2)	.2584	.5033	.7185	.9115	1.092	1.266	1.439	1.614	1.795	1.987	2.193	2.422	2.994	3.941	6.589
(3)	.7281	1.342	1.824	2.214	2.547	2.848	3.131	3.407	3.685	3.974	4.282	4.623	5.489	6.973	11.30
(4)	1.198	2.181	2.929	3.516	4.003	4.430	4.823	5.200	5.575	5.961	6.371	6.825	7.985	10.00	16.00
(5)	1.668	3.020	4.034	4.818	5.458	6.012	6.516	6.994	7.465	7.948	8.460	9.026	10.48	13.04	20.71
(6)	2.137	3.859	5.140	6.120	6.914	7.594	8.208	8.787	9.355	9.935	10.55	11.23	12.97	16.07	25.41

TABLE 7.10 : I_0^* ($1/\alpha$ = .50)

Q:	0.05	0.10	0.15	0.20	0.25	0.30	0.35	0.40	0.45	0.50	0.55	0.60	0.70	0.80	0.90
MODE:															
(1)	.0571	.1175	.1813	.2491	.3213	.3986	.4820	.5728	.6730	.7854	.9141	1.065	1.486	2.280	4.677
(2)	.2612	.5194	.7556	.9726	1.170	1.378	1.578	1.783	1.995	2.221	2.467	2.740	3.437	4.617	7.980
(3)	.7360	1.385	1.918	2.362	2.749	3.101	3.435	3.764	4.096	4.443	4.816	5.232	6.301	8.168	13.68
(4)	1.211	2.251	3.080	3.752	4.320	4.824	5.292	5.744	6.197	6.664	7.165	7.723	9.165	11.72	19.38
(5)	1.686	3.117	4.243	5.141	5.890	6.547	7.150	7.725	8.297	8.886	9.515	10.21	12.03	15.27	25.08
(6)	2.160	3.982	5.405	6.531	7.461	8.270	9.007	9.706	10.40	11.11	11.86	12.71	14.89	18.82	30.78

TABLE 7.10 : I_0^* ($1/\alpha$ = .60)

Q:	0.05	0.10	0.15	0.20	0.25	0.30	0.35	0.40	0.45	0.50	0.55	0.60	0.70	0.80	0.90
MODE:															
(1)	.0627	.1294	.2003	.2759	.3569	.4441	.5388	.6425	.7578	.8880	1.038	1.216	1.717	2.675	5.571
(2)	.2631	.5313	.7845	1.022	1.250	1.475	1.700	1.932	2.174	2.433	2.717	3.035	3.855	5.271	9.362
(3)	.7414	1.417	1.991	2.483	2.917	3.318	3.700	4.078	4.463	4.867	5.304	5.794	7.068	9.326	16.05
(4)	1.220	2.302	3.198	3.943	4.585	5.161	5.701	6.225	6.752	7.300	7.891	8.553	10.28	13.38	22.74
(5)	1.698	3.188	4.405	5.404	6.252	7.004	7.701	8.371	9.041	9.734	10.48	11.31	13.49	17.44	29.42
(6)	2.176	4.074	5.612	6.864	7.919	8.847	9.701	10.52	11.33	12.17	13.07	14.07	16.71	21.49	36.11

TABLE 7.10 : I_0^* ($1/\alpha$ = .70)

Q:	0.05	0.10	0.15	0.20	0.25	0.30	0.35	0.40	0.45	0.50	0.55	0.60	0.70	0.80	0.90
MODE:															
(1)	.0679	.1404	.2179	.3010	.3904	.4872	.5928	.7092	.8393	.9872	1.159	1.364	1.946	3.068	6.464
(2)	.2645	.5405	.8078	1.064	1.312	1.559	1.808	2.066	2.337	2.628	2.949	3.311	4.255	5.910	10.74
(3)	.7454	1.441	2.051	2.583	3.062	3.507	3.935	4.361	4.797	5.257	5.757	6.321	7.801	10.46	18.41
(4)	1.226	2.342	3.293	4.103	4.811	5.455	6.063	6.656	7.257	7.885	8.565	9.331	11.35	15.00	26.07
(5)	1.707	3.243	4.536	5.622	6.561	7.403	8.190	8.952	9.717	10.51	11.37	12.34	14.89	19.55	33.74
(6)	2.188	4.144	5.779	7.142	8.311	9.352	10.32	11.25	12.18	13.14	14.18	15.35	18.44	24.10	41.41

TABLE 7.10 : I_0^* ($1/\alpha$ = .80)

Q:	0.05	0.10	0.15	0.20	0.25	0.30	0.35	0.40	0.45	0.50	0.55	0.60	0.70	0.80	0.90
MODE:															
(1)	.0728	.1508	.2346	.3248	.4223	.5283	.6446	.7735	.9183	1.084	1.277	1.509	2.172	3.459	7.358
(2)	.2656	.5478	.8271	1.099	1.366	1.633	1.905	2.188	2.487	2.810	3.167	3.573	4.641	6.538	12.11
(3)	.7485	1.461	2.100	2.669	3.188	3.675	4.147	4.619	5.105	5.620	6.183	6.820	8.509	11.57	20.76
(4)	1.231	2.374	3.372	4.239	5.009	5.716	6.389	7.050	7.723	8.430	9.199	10.07	12.38	16.60	29.40
(5)	1.714	3.287	4.644	5.809	6.831	7.758	8.630	9.481	10.34	11.24	12.21	13.32	16.24	21.6[illegible]	38.05
(6)	2.197	4.200	5.917	7.379	8.652	9.799	10.87	11.91	12.96	14.05	15.23	16.56	20.11	26.65	46.70

TABLE 7.10 : I_0^* ($1/\alpha$ = .90)

Q:	0.05	0.10	0.15	0.20	0.25	0.30	0.35	0.40	0.45	0.50	0.55	0.60	0.70	0.80	0.90
MODE:															
(1)	.0773	.1606	.2504	.3475	.4529	.5680	.6948	.8359	.9952	1.178	1.393	1.652	2.396	3.850	8.251
(2)	.2664	.5537	.8433	1.129	1.414	1.700	1.994	2.301	2.627	2.980	3.373	3.822	5.015	7.157	13.47
(3)	.7509	1.477	2.141	2.743	3.299	3.825	4.340	4.857	5.391	5.961	6.586	7.297	9.195	12.66	23.10
(4)	1.235	2.399	3.438	4.356	5.184	5.951	6.686	7.413	8.156	8.941	9.799	10.77	13.37	18.17	32.72
(5)	1.720	3.322	4.735	5.970	7.069	8.076	9.032	9.969	10.92	11.92	13.01	14.25	17.55	23.67	42.35
(6)	2.204	4.245	6.033	7.583	8.953	10.20	11.38	12.53	13.69	14.90	16.22	17.72	21.73	29.18	51.97

TABLE 7.11 : I_0^* (α = .10)

Q:	0.05	0.10	0.15	0.20	0.25	0.30	0.35	0.40	0.45	0.50	0.55	0.60	0.70	0.80	0.90
MODE:															
(1)	.2515	.5093	.7739	1.046	1.326	1.615	1.915	2.228	2.557	2.907	3.284	3.699	4.720	6.353	10.78
(2)	2.266	3.728	4.757	5.604	6.365	7.087	7.791	8.494	9.204	9.935	10.70	11.50	13.37	16.03	22.24
(3)	6.386	9.942	12.08	13.61	14.85	15.95	16.96	17.93	18.89	19.87	20.88	21.96	24.51	28.36	38.12
(4)	10.51	16.16	19.40	21.61	23.34	24.80	26.12	27.37	28.58	29.80	31.07	32.42	35.65	40.69	54.00
(5)	14.63	22.37	26.71	29.62	31.83	33.66	35.29	36.81	38.27	39.74	41.26	42.88	46.79	53.02	69.88
(6)	18.75	28.58	34.03	37.62	40.31	42.52	44.46	46.24	47.96	49.67	51.44	53.34	57.93	65.34	85.77

TABLE 7.11 : I_0^* (α = .15)

Q:	0.05	0.10	0.15	0.20	0.25	0.30	0.35	0.40	0.45	0.50	0.55	0.60	0.70	0.80	0.90
MODE:															
(1)	.2059	.4182	.6373	.8638	1.099	1.343	1.598	1.867	2.152	2.458	2.793	3.167	4.116	5.711	10.23
(2)	1.592	2.765	3.637	4.363	5.019	5.640	6.248	6.855	7.473	8.112	8.782	9.502	11.20	13.72	19.99
(3)	4.487	7.374	9.231	10.60	11.71	12.69	13.60	14.47	15.34	16.22	17.15	18.14	20.53	24.28	34.27
(4)	7.381	11.98	14.83	16.83	18.40	19.74	20.95	22.09	23.21	24.33	25.51	26.78	29.86	34.84	48.54
(5)	10.28	16.59	20.42	23.06	25.09	26.79	28.30	29.71	31.07	32.45	33.87	35.42	39.20	45.39	62.82
(6)	13.17	21.20	26.02	29.29	31.78	33.84	35.65	37.32	38.94	40.56	42.24	44.05	48.53	55.95	77.10

TABLE 7.11 : I_0^* (α = .20)

Q:	0.05	0.10	0.15	0.20	0.25	0.30	0.35	0.40	0.45	0.50	0.55	0.60	0.70	0.80	0.90
MODE:															
(1)	.1787	.3639	.5559	.7554	.9632	1.181	1.409	1.652	1.911	2.192	2.503	2.854	3.764	5.345	9.932
(2)	1.230	2.216	2.983	3.632	4.221	4.781	5.329	5.880	6.441	7.025	7.642	8.309	9.910	12.37	18.71
(3)	3.466	5.909	7.571	8.820	9.848	10.76	11.60	12.41	13.22	14.05	14.92	15.86	18.17	21.88	32.07
(4)	5.701	9.602	12.16	14.01	15.48	16.73	17.87	18.95	20.00	21.07	22.20	23.42	26.43	31.39	45.43
(5)	7.937	13.30	16.75	19.20	21.10	22.71	24.14	25.48	26.78	28.10	29.48	30.97	34.68	40.90	58.79
(6)	10.17	16.99	21.34	24.38	26.73	28.68	30.41	32.01	33.56	35.12	36.75	38.52	42.94	50.41	72.15

TABLE 7.11 : I_0^* (α = .30)

Q:	0.05	0.10	0.15	0.20	0.25	0.30	0.35	0.40	0.45	0.50	0.55	0.60	0.70	0.80	0.90
MODE:															
(1)	.1465	.2995	.4594	.6269	.8029	.9887	1.186	1.398	1.627	1.879	2.162	2.488	3.358	4.934	9.618
(2)	.8465	1.600	2.229	2.777	3.282	3.766	4.242	4.723	5.218	5.736	6.289	6.896	8.387	10.78	17.27
(3)	2.386	4.267	5.658	6.745	7.658	8.472	9.233	9.971	10.71	11.47	12.28	13.16	15.38	19.07	29.61
(4)	3.925	6.934	9.086	10.71	12.03	13.18	14.22	15.22	16.20	17.21	18.27	19.43	22.37	27.36	41.94
(5)	5.464	9.601	12.52	14.68	16.41	17.89	19.22	20.47	21.69	22.94	24.26	25.70	29.36	35.65	54.28
(6)	7.003	12.27	15.94	18.65	20.79	22.59	24.21	25.72	27.19	28.68	30.25	31.97	36.34	43.95	66.61

TABLE 7.11 : I_0^* (α = .40)

Q:	0.05	0.10	0.15	0.20	0.25	0.30	0.35	0.40	0.45	0.50	0.55	0.60	0.70	0.80	0.90
MODE:															
(1)	.1273	.2611	.4019	.5504	.7075	.8748	1.054	1.248	1.460	1.695	1.962	2.274	3.125	4.706	9.454
(2)	.6459	1.258	1.796	2.279	2.729	3.164	3.596	4.035	4.488	4.967	5.483	6.054	7.486	9.853	16.47
(3)	1.820	3.355	4.559	5.534	6.368	7.120	7.827	8.518	9.213	9.935	10.71	11.56	13.72	17.43	28.24
(4)	2.995	5.452	7.323	8.790	10.01	11.08	12.06	13.00	13.94	14.90	15.93	17.06	19.96	25.01	40.00
(5)	4.169	7.549	10.09	12.04	13.65	15.03	16.29	17.48	18.66	19.87	21.15	22.57	26.20	32.59	51.77
(6)	5.343	9.646	12.85	15.30	17.28	18.99	20.52	21.97	23.39	24.84	26.37	28.07	32.44	40.17	63.53

TABLE 7.11 : I_0^* (α = .50)

Q: MODE:	0.05	0.10	0.15	0.20	0.25	0.30	0.35	0.40	0.45	0.50	0.55	0.60	0.70	0.80	0.90
(1)	.1142	.2349	.3627	.4982	.6426	.7972	.9640	1.146	1.346	1.571	1.828	2.131	2.972	4.560	9.353
(2)	.5223	1.039	1.511	1.945	2.356	2.757	3.157	3.566	3.991	4.443	4.934	5.481	6.874	9.234	15.96
(3)	1.472	2.770	3.836	4.724	5.498	6.203	6.871	7.527	8.192	8.886	9.632	10.46	12.60	16.34	27.36
(4)	2.422	4.502	6.161	7.503	8.639	9.648	10.58	11.49	12.39	13.33	14.33	15.45	18.33	23.44	38.76
(5)	3.371	6.233	8.486	10.28	11.78	13.09	14.30	15.45	16.59	17.77	19.03	20.43	24.06	30.54	50.16
(6)	4.321	7.965	10.81	13.06	14.92	16.54	18.01	19.41	20.80	22.21	23.73	25.41	29.79	37.65	61.56

TABLE 7.11 : I_0^* (α = .60)

Q: MODE:	0.05	0.10	0.15	0.20	0.25	0.30	0.35	0.40	0.45	0.50	0.55	0.60	0.70	0.80	0.90
(1)	.1045	.2156	.3338	.4598	.5948	.7402	.8979	1.071	1.263	1.480	1.730	2.027	2.862	4.458	9.285
(2)	.4385	.8856	1.308	1.704	2.084	2.458	2.834	3.220	3.624	4.056	4.528	5.058	6.425	8.785	15.60
(3)	1.236	2.362	3.319	4.138	4.862	5.530	6.167	6.797	7.438	8.112	8.840	9.656	11.78	15.54	26.75
(4)	2.033	3.837	5.331	6.572	7.641	8.602	9.501	10.37	11.25	12.17	13.15	14.25	17.13	22.30	37.89
(5)	2.830	5.313	7.342	9.006	10.42	11.67	12.83	13.95	15.07	16.22	17.46	18.85	22.49	29.06	49.04
(6)	3.627	6.789	9.354	11.44	13.20	14.75	16.17	17.53	18.88	20.28	21.78	23.45	27.84	35.82	60.18

TABLE 7.11 : I_0^* (α = .70)

Q: MODE:	0.05	0.10	0.15	0.20	0.25	0.30	0.35	0.40	0.45	0.50	0.55	0.60	0.70	0.80	0.90
(1)	.0970	.2006	.3111	.4300	.5577	.6959	.8468	1.013	1.199	1.410	1.656	1.948	2.779	4.382	9.235
(2)	.3779	.7722	1.154	1.520	1.875	2.227	2.583	2.951	3.339	3.755	4.213	4.730	6.079	8.443	15.34
(3)	1.065	2.059	2.930	3.690	4.374	5.010	5.622	6.230	6.853	7.510	8.224	9.030	11.14	14.94	26.29
(4)	1.752	3.346	4.705	5.861	6.873	7.793	8.661	9.509	10.37	11.26	12.24	13.33	16.21	21.43	37.25
(5)	2.439	4.633	6.480	8.032	9.373	10.58	11.70	12.79	13.88	15.02	16.25	17.63	21.28	27.93	48.20
(6)	3.126	5.920	8.256	10.20	11.87	13.36	14.74	16.07	17.40	18.77	20.26	21.93	26.34	34.42	59.16

TABLE 7.11 : I_0^* (α = .80)

Q: MODE:	0.05	0.10	0.15	0.20	0.25	0.30	0.35	0.40	0.45	0.50	0.55	0.60	0.70	0.80	0.90
(1)	.0910	.1885	.2932	.4060	.5279	.6604	.8058	.9669	1.148	1.355	1.596	1.886	2.715	4.324	9.197
(2)	.3320	.6847	1.034	1.374	1.708	2.042	2.382	2.735	3.109	3.512	3.958	4.466	5.801	8.172	15.13
(3)	.9356	1.826	2.624	3.336	3.984	4.593	5.184	5.774	6.381	7.025	7.728	8.525	10.64	14.46	25.94
(4)	1.539	2.967	4.215	5.299	6.261	7.145	7.986	8.813	9.654	10.54	11.50	12.58	15.47	20.75	36.75
(5)	2.143	4.108	5.806	7.261	8.538	9.697	10.79	11.85	12.93	14.05	15.27	16.64	20.30	27.03	47.56
(6)	2.746	5.250	7.396	9.224	10.81	12.25	13.59	14.89	16.20	17.56	19.04	20.70	25.14	33.32	58.37

TABLE 7.11 : I_0^* (α = .90)

Q: MODE:	0.05	0.10	0.15	0.20	0.25	0.30	0.35	0.40	0.45	0.50	0.55	0.60	0.70	0.80	0.90
(1)	.0859	.1785	.2783	.3861	.5032	.6311	.7720	.9288	1.106	1.309	1.547	1.835	2.663	4.278	9.168
(2)	.2960	.6153	.9370	1.255	1.571	1.889	2.216	2.556	2.918	3.312	3.748	4.247	5.573	7.952	14.97
(3)	.8343	1.641	2.379	3.048	3.665	4.251	4.822	5.396	5.991	6.623	7.318	8.108	10.22	14.07	25.67
(4)	1.373	2.666	3.820	4.841	5.760	6.612	7.429	8.237	9.063	9.935	10.89	11.97	14.86	20.19	36.36
(5)	1.911	3.692	5.262	6.633	7.854	8.973	10.04	11.08	12.13	13.25	14.46	15.83	19.50	26.30	47.05
(6)	2.449	4.717	6.703	8.426	9.948	11.33	12.64	13.92	15.21	16.56	18.03	19.69	24.15	32.42	57.75

TABLE 7.11 : I_0^* (α = 1.0)

Q: MODE:	0.05	0.10	0.15	0.20	0.25	0.30	0.35	0.40	0.45	0.50	0.55	0.60	0.70	0.80	0.90
(1)	.0817	.1700	.2656	.3693	.4824	.6063	.7435	.8967	1.070	1.271	1.507	1.793	2.620	4.240	9.144
(2)	.2671	.5587	.8571	1.156	1.456	1.761	2.075	2.405	2.758	3.142	3.570	4.062	5.380	7.768	14.84
(3)	.7528	1.490	2.176	2.808	3.398	3.962	4.517	5.078	5.660	6.283	6.970	7.756	9.864	13.74	25.44
(4)	1.238	2.421	3.495	4.459	5.339	6.163	6.958	7.750	8.563	9.425	10.37	11.45	14.35	19.72	36.04
(5)	1.724	3.352	4.813	6.111	7.281	8.365	9.400	10.42	11.47	12.57	13.77	15.14	18.83	25.70	46.64
(6)	2.210	4.283	6.132	7.763	9.223	10.57	11.84	13.09	14.37	15.71	17.17	18.83	23.31	31.67	57.24

TABLE 7.11 : I_0^* ($1/\alpha$ = .10)

Q: MODE:	0.05	0.10	0.15	0.20	0.25	0.30	0.35	0.40	0.45	0.50	0.55	0.60	0.70	0.80	0.90
(1)	.0281	.0636	.1080	.1627	.2298	.3114	.4105	.5309	.6775	.8573	1.081	1.362	2.213	3.908	8.948
(2)	.0273	.0606	.1015	.1521	.2158	.2975	.4041	.5459	.7366	.9935	1.337	1.790	3.185	5.904	13.68
(3)	.0769	.1616	.2576	.3695	.5036	.6693	.8795	1.152	1.512	1.987	2.609	3.417	5.839	10.45	23.45
(4)	.1266	.2627	.4137	.5868	.7914	1.041	1.355	1.759	2.287	2.980	3.882	5.045	8.494	14.99	33.22
(5)	.1762	.3637	.5698	.8041	1.079	1.413	1.830	2.365	3.063	3.974	5.155	6.672	11.15	19.53	42.99
(6)	.2258	.4647	.7259	1.021	1.367	1.785	2.306	2.972	3.838	4.967	6.428	8.299	13.80	24.07	52.75

TABLE 7.11 : I_0^* ($1/\alpha$ = .20)

Q: MODE:	0.05	0.10	0.15	0.20	0.25	0.30	0.35	0.40	0.45	0.50	0.55	0.60	0.70	0.80	0.90
(1)	.0384	.0838	.1374	.2005	.2749	.3626	.4665	.5903	.7389	.9196	1.142	1.422	2.265	3.948	8.970
(2)	.0545	.1200	.1983	.2915	.4025	.5351	.6941	.8860	1.119	1.405	1.759	2.206	3.542	6.163	13.82
(3)	.1535	.3200	.5033	.7079	.9391	1.204	1.511	1.870	2.297	2.810	3.435	4.211	6.493	10.90	23.68
(4)	.2525	.5200	.8084	1.124	1.476	1.873	2.327	2.855	3.475	4.215	5.111	6.216	9.444	15.64	33.55
(5)	.3515	.7200	1.113	1.541	2.012	2.542	3.144	3.839	4.653	5.620	6.786	8.221	12.40	20.38	43.42
(6)	.4505	.9201	1.418	1.957	2.549	3.210	3.960	4.824	5.831	7.025	8.462	10.23	15.35	25.12	53.29

TABLE 7.11 : I_0^* ($1/\alpha$ = .30)

Q: MODE:	0.05	0.10	0.15	0.20	0.25	0.30	0.35	0.40	0.45	0.50	0.55	0.60	0.70	0.80	0.90
(1)	.0462	.0994	.1604	.2305	.3113	.4047	.5135	.6412	.7928	.9752	1.198	1.478	2.315	3.986	8.992
(2)	.0815	.1783	.2910	.4208	.5693	.7388	.9328	1.156	1.415	1.721	2.087	2.536	3.847	6.403	13.95
(3)	.2297	.4754	.7387	1.022	1.328	1.662	2.030	2.440	2.905	3.441	4.074	4.841	7.053	11.33	23.92
(4)	.3778	.7725	1.186	1.623	2.087	2.586	3.128	3.725	4.395	5.162	6.062	7.146	10.26	16.25	33.88
(5)	.5260	1.070	1.634	2.224	2.847	3.509	4.225	5.009	5.885	6.883	8.049	9.451	13.47	21.18	43.85
(6)	.6741	1.367	2.082	2.825	3.606	4.433	5.322	6.294	7.375	8.604	10.04	11.76	16.67	26.10	53.81

TABLE 7.11 : I_0^* ($1/\alpha$ = .40)

Q: MODE:	0.05	0.10	0.15	0.20	0.25	0.30	0.35	0.40	0.45	0.50	0.55	0.60	0.70	0.80	0.90
(1)	.0529	.1126	.1800	.2562	.3427	.4414	.5548	.6865	.8413	1.026	1.250	1.530	2.363	4.024	9.014
(2)	.1084	.2354	.3801	.5420	.7215	.9201	1.140	1.387	1.666	1.987	2.364	2.818	4.119	6.628	14.09
(3)	.3055	.6278	.9649	1.316	1.684	2.070	2.482	2.928	3.420	3.974	4.615	5.380	7.552	11.73	24.15
(4)	.5025	1.020	1.550	2.091	2.646	3.220	3.824	4.469	5.173	5.961	6.866	7.942	10.98	16.82	34.21
(5)	.6996	1.413	2.135	2.865	3.608	4.370	5.166	6.010	6.927	7.948	9.117	10.50	14.42	21.92	44.27
(6)	.8967	1.805	2.719	3.639	4.570	5.520	6.507	7.551	8.680	9.935	11.37	13.07	17.85	27.02	54.33

TABLE 7.11 : I_0^* (1/α = .50)

Q: MODE:	0.05	0.10	0.15	0.20	0.25	0.30	0.35	0.40	0.45	0.50	0.55	0.60	0.70	0.80	0.90
(1)	.0587	.1242	.1973	.2790	.3707	.4742	.5921	.7277	.8858	1.073	1.299	1.579	2.409	4.062	9.036
(2)	.1352	.2916	.4660	.6565	.8624	1.085	1.327	1.592	1.887	2.221	2.608	3.069	4.366	6.840	14.22
(3)	.3809	.7775	1.183	1.594	2.012	2.441	2.888	3.361	3.874	4.443	5.092	5.858	8.005	12.10	24.37
(4)	.6266	1.263	1.900	2.532	3.162	3.797	4.448	5.130	5.860	6.664	7.577	8.648	11.64	17.36	34.53
(5)	.8724	1.749	2.617	3.470	4.312	5.154	6.009	6.899	7.847	8.886	10.06	11.44	15.28	22.63	44.68
(6)	1.118	2.235	3.334	4.408	5.462	6.510	7.570	8.668	9.833	11.11	12.54	14.23	18.92	27.89	54.83

TABLE 7.11 : I_0^* (1/α = .60)

Q: MODE:	0.05	0.10	0.15	0.20	0.25	0.30	0.35	0.40	0.45	0.50	0.55	0.60	0.70	0.80	0.90
(1)	.0640	.1347	.2130	.2997	.3962	.5043	.6264	.7658	.9271	1.117	1.345	1.625	2.454	4.099	9.050
(2)	.1618	.3467	.5491	.7652	.9942	1.237	1.497	1.778	2.087	2.433	2.830	3.296	4.595	7.042	14.35
(3)	.4560	.9247	1.394	1.858	2.320	2.784	3.258	3.754	4.285	4.867	5.525	6.293	8.424	12.46	24.59
(4)	.7502	1.503	2.238	2.951	3.645	4.330	5.020	5.730	6.482	7.300	8.220	9.290	12.25	17.88	34.84
(5)	1.044	2.080	3.083	4.045	4.971	5.877	6.781	7.706	8.680	9.734	10.91	12.29	16.08	23.29	45.08
(6)	1.339	2.658	3.928	5.138	6.297	7.423	8.542	9.682	10.88	12.17	13.61	15.28	19.91	28.71	55.33

TABLE 7.11 : I_0^* (1/α = .70)

Q: MODE:	0.05	0.10	0.15	0.20	0.25	0.30	0.35	0.40	0.45	0.50	0.55	0.60	0.70	0.80	0.90
(1)	.0689	.1444	.2274	.3188	.4198	.5322	.6583	.8014	.9658	1.158	1.388	1.670	2.497	4.135	9.080
(2)	.1883	.4010	.6295	.8690	1.118	1.379	1.655	1.950	2.272	2.628	3.033	3.507	4.808	7.235	14.47
(3)	.5307	1.069	1.598	2.110	2.610	3.104	3.602	4.118	4.663	5.257	5.922	6.695	8.815	12.80	24.81
(4)	.8731	1.738	2.566	3.352	4.101	4.828	5.550	6.285	7.055	7.885	8.811	9.883	12.82	18.37	35.15
(5)	1.215	2.406	3.535	4.593	5.592	6.552	7.497	8.452	9.446	10.51	11.70	13.07	16.83	23.93	45.48
(6)	1.558	3.074	4.503	5.834	7.083	8.276	9.444	10.62	11.84	13.14	14.59	16.26	20.83	29.50	55.82

TABLE 7.11 : I_0^* (1/α = .80)

Q: MODE:	0.05	0.10	0.15	0.20	0.25	0.30	0.35	0.40	0.45	0.50	0.55	0.60	0.70	0.80	0.90
(1)	.0734	.1535	.2409	.3366	.4419	.5583	.6882	.8348	1.002	1.197	1.429	1.712	2.539	4.170	9.101
(2)	.2147	.4544	.7075	.9684	1.236	1.513	1.803	2.111	2.444	2.810	3.223	3.703	5.009	7.420	14.60
(3)	.6051	1.212	1.796	2.352	2.885	3.405	3.924	4.456	5.016	5.620	6.293	7.069	9.183	13.13	25.02
(4)	.9955	1.969	2.884	3.735	4.533	5.296	6.046	6.802	7.588	8.430	9.363	10.44	13.36	18.83	35.45
(5)	1.386	2.726	3.973	5.119	6.181	7.188	8.167	9.148	10.16	11.24	12.43	13.80	17.53	24.54	45.87
(6)	1.776	3.484	5.061	6.502	7.830	9.080	10.29	11.49	12.73	14.05	15.50	17.17	21.70	30.25	56.30

TABLE 7.11 : I_0^* (1/α = .90)

Q: MODE:	0.05	0.10	0.15	0.20	0.25	0.30	0.35	0.40	0.45	0.50	0.55	0.60	0.70	0.80	0.90
(1)	.0777	.1620	.2536	.3534	.4627	.5829	.7165	.8665	1.037	1.235	1.469	1.753	2.580	4.205	9.123
(2)	.2410	.5069	.7833	1.064	1.349	1.640	1.943	2.262	2.605	2.980	3.402	3.888	5.199	7.597	14.72
(3)	.6791	1.352	1.988	2.584	3.147	3.690	4.228	4.775	5.347	5.961	6.641	7.422	9.532	13.44	25.23
(4)	1.117	2.197	3.194	4.104	4.945	5.740	6.514	7.289	8.089	8.941	9.881	10.96	13.86	19.29	35.75
(5)	1.555	3.041	4.399	5.624	6.743	7.791	8.800	9.802	10.83	11.92	13.12	14.49	18.20	25.13	46.26
(6)	1.994	3.886	5.604	7.144	8.541	9.841	11.09	12.32	13.57	14.90	16.36	18.02	22.53	30.97	56.77

TABLE 7.12 : I_0^* (α = .10)

Q: MODE:	0.05	0.10	0.15	0.20	0.25	0.30	0.35	0.40	0.45	0.50	0.55	0.60	0.70	0.80	0.90
(1)	.2060	.6214	1.098	1.601	2.122	2.658	3.208	3.775	4.360	4.967	5.603	6.275	7.799	9.863	14.29
(2)	4.326	6.835	8.417	9.606	10.61	11.52	12.37	13.21	14.05	14.90	15.79	16.73	18.94	22.19	30.18
(3)	8.446	13.05	15.74	17.61	19.10	20.37	21.54	22.65	23.74	24.84	25.98	27.19	30.08	34.52	46.06
(4)	12.57	19.26	23.06	25.62	27.58	29.23	30.71	32.09	33.43	34.77	36.16	37.65	41.22	46.85	61.94
(5)	16.69	25.48	30.37	33.62	36.07	38.09	39.87	41.52	43.12	44.71	46.35	48.11	52.36	59.18	77.83
(6)	20.81	31.69	37.69	41.63	44.56	46.95	49.04	50.96	52.80	54.64	56.53	58.57	63.50	71.51	93.71

TABLE 7.12 : I_0^* (α = .15)

Q: MODE:	0.05	0.10	0.15	0.20	0.25	0.30	0.35	0.40	0.45	0.50	0.55	0.60	0.70	0.80	0.90
(1)	.1447	.4609	.8392	1.247	1.673	2.115	2.573	3.047	3.540	4.056	4.600	5.183	6.533	8.445	12.85
(2)	3.039	5.069	6.434	7.479	8.364	9.165	9.923	10.66	11.41	12.17	12.96	13.82	15.87	19.00	27.13
(3)	5.934	9.678	12.03	13.71	15.06	16.22	17.27	18.28	19.27	20.28	21.33	22.46	25.20	29.56	41.40
(4)	8.829	14.29	17.62	19.94	21.75	23.27	24.62	25.90	27.14	28.39	29.69	31.10	34.53	40.12	55.68
(5)	11.72	18.89	23.22	26.18	28.44	30.32	31.98	33.52	35.01	36.50	38.06	39.73	43.86	50.67	69.96
(6)	14.62	23.50	28.81	32.41	35.13	37.37	39.33	41.13	42.87	44.61	46.42	48.37	53.20	61.23	84.23

TABLE 7.12 : I_0^*(α = .20)

Q: MODE:	0.05	0.10	0.15	0.20	0.25	0.30	0.35	0.40	0.45	0.50	0.55	0.60	0.70	0.80	0.90
(1)	.1118	.3693	.6883	1.038	1.407	1.793	2.194	2.613	3.051	3.512	4.003	4.532	5.781	7.610	12.03
(2)	2.348	4.062	5.277	6.226	7.034	7.768	8.464	9.146	9.832	10.54	11.28	12.09	14.04	17.12	25.39
(3)	4.583	7.756	9.866	11.41	12.66	13.74	14.73	15.68	16.61	17.56	18.56	19.64	22.30	26.63	38.75
(4)	6.819	11.45	14.45	16.60	18.29	19.72	21.00	22.21	23.39	24.59	25.84	27.19	30.56	36.15	52.11
(5)	9.055	15.14	19.04	21.79	23.92	25.70	27.27	28.74	30.17	31.61	33.11	34.75	38.81	45.66	65.47
(6)	11.29	18.83	23.63	26.98	29.54	31.67	33.54	35.28	36.95	38.64	40.39	42.30	47.07	55.17	78.83

TABLE 7.12 : I_0^*(α = .30)

Q: MODE:	0.05	0.10	0.15	0.20	0.25	0.30	0.35	0.40	0.45	0.50	0.55	0.60	0.70	0.80	0.90
(1)	.0770	.2667	.5143	.7935	1.094	1.412	1.747	2.099	2.472	2.868	3.294	3.761	4.893	6.633	11.10
(2)	1.616	2.934	3.943	4.761	5.470	6.119	6.738	7.347	7.964	8.604	9.284	10.03	11.88	14.93	23.44
(3)	3.155	5.601	7.372	8.729	9.846	10.83	11.73	12.60	13.46	14.34	15.27	16.30	18.87	23.22	35.77
(4)	4.694	8.267	10.80	12.70	14.22	15.53	16.72	17.84	18.95	20.08	21.26	22.57	25.86	31.51	48.11
(5)	6.233	10.93	14.23	16.66	18.60	20.24	21.71	23.09	24.44	25.81	27.25	28.84	32.85	39.80	60.44
(6)	7.773	13.60	17.66	20.63	22.97	24.95	26.70	28.34	29.93	31.55	33.24	35.11	39.84	48.09	72.78

TABLE 7.12 : I_0^*(α = .40)

Q: MODE:	0.05	0.10	0.15	0.20	0.25	0.30	0.35	0.40	0.45	0.50	0.55	0.60	0.70	0.80	0.90
(1)	.0587	.2097	.4145	.6511	.9097	1.187	1.481	1.793	2.126	2.484	2.872	3.302	4.367	6.064	10.59
(2)	1.233	2.307	3.178	3.906	4.549	5.142	5.712	6.276	6.851	7.451	8.095	8.806	10.60	13.64	22.36
(3)	2.408	4.404	5.941	7.162	8.188	9.098	9.943	10.76	11.58	12.42	13.32	14.31	16.84	21.22	34.12
(4)	3.582	6.501	8.704	10.42	11.83	13.05	14.17	15.24	16.30	17.39	18.54	19.81	23.08	28.80	45.89
(5)	4.756	8.598	11.47	13.67	15.47	17.01	18.40	19.73	21.03	22.35	23.76	25.32	29.32	36.38	57.65
(6)	5.931	10.69	14.23	16.93	19.10	20.96	22.64	24.21	25.75	27.32	28.98	30.82	35.56	43.96	69.42

TABLE 7.12 : I_0^* (α = .50)

MODE \ Q:	0.05	0.10	0.15	0.20	0.25	0.30	0.35	0.40	0.45	0.50	0.55	0.60	0.70	0.80	0.90
(1)	.0475	.1731	.3487	.5558	.7854	1.304	1.300	1.585	1.890	2.221	2.584	2.989	4.010	5.682	10.26
(2)	.9972	1.905	2.674	3.335	3.927	4.480	5.014	5.546	6.092	6.664	7.283	7.972	9.738	12.79	21.66
(3)	1.947	3.636	4.998	6.114	7.069	7.925	8.728	9.508	10.29	11.11	11.98	12.95	15.47	19.89	33.06
(4)	2.897	5.368	7.323	8.893	10.21	11.37	12.44	13.47	14.49	15.55	16.68	17.94	21.19	26.99	44.46
(5)	3.846	7.099	9.648	11.67	13.35	14.82	16.16	17.43	18.69	19.99	21.38	22.92	26.92	34.09	55.86
(6)	4.796	8.830	11.97	14.45	16.49	18.26	19.87	21.39	22.90	24.44	26.08	27.90	32.65	41.20	67.26

TABLE 7.12 : I_0^* (α = .60)

MODE \ Q:	0.05	0.10	0.15	0.20	0.25	0.30	0.35	0.40	0.45	0.50	0.55	0.60	0.70	0.80	0.90
(1)	.0399	.1476	.3017	.4868	.6946	.9216	1.167	1.431	1.717	2.028	2.372	2.759	3.748	5.406	10.03
(2)	.8371	1.624	2.313	2.921	3.473	3.994	4.500	5.008	5.531	6.084	6.684	7.357	9.103	12.16	21.17
(3)	1.634	3.100	4.325	5.355	6.252	7.066	7.834	8.586	9.346	10.14	11.00	11.96	14.46	18.92	32.32
(4)	2.432	4.575	6.336	7.789	9.030	10.14	11.17	12.16	13.16	14.20	15.31	16.55	19.81	25.68	43.46
(5)	3.229	6.051	8.348	10.22	11.81	13.21	14.50	15.74	16.97	18.25	19.62	21.15	25.17	32.44	54.61
(6)	4.026	7.527	10.36	12.66	14.59	16.28	17.83	19.32	20.79	22.31	23.93	25.75	30.52	39.20	65.75

TABLE 7.12 : I_0^* (α = .70)

MODE \ Q:	0.05	0.10	0.15	0.20	0.25	0.30	0.35	0.40	0.45	0.50	0.55	0.60	0.70	0.80	0.90
(1)	.0344	.1287	.2663	.4342	.6249	.8350	1.064	1.312	1.581	1.877	2.207	2.580	3.546	5.196	9.860
(2)	.7214	1.416	2.042	2.605	3.124	3.618	4.103	4.591	5.096	5.632	6.219	6.880	8.612	11.69	20.82
(3)	1.408	2.703	3.817	4.776	5.624	6.402	7.141	7.870	8.610	9.387	10.23	11.18	13.68	18.19	31.77
(4)	2.095	3.989	5.593	6.947	8.123	9.185	10.18	11.15	12.12	13.14	14.24	15.48	18.74	24.68	42.73
(5)	2.782	5.276	7.368	9.118	10.62	11.97	13.22	14.43	15.64	16.90	18.25	19.78	23.81	31.18	53.68
(6)	3.469	6.563	9.143	11.29	13.12	14.75	16.26	17.71	19.15	20.65	22.27	24.08	28.87	37.67	64.64

TABLE 7.12 : I_0^* (α = .80)

MODE \ Q:	0.05	0.10	0.15	0.20	0.25	0.30	0.35	0.40	0.45	0.50	0.55	0.60	0.70	0.80	0.90
(1)	.0302	.1141	.2386	.3925	.5692	.7656	.9807	1.216	1.473	1.756	2.073	2.436	3.384	5.029	9.729
(2)	.6338	1.255	1.829	2.355	2.846	3.317	3.783	4.254	4.745	5.269	5.843	6.495	8.218	11.32	20.54
(3)	1.237	2.397	3.420	4.317	5.123	5.869	6.585	7.293	8.017	8.781	9.613	10.56	13.05	17.60	31.35
(4)	1.841	3.538	5.010	6.280	7.400	8.421	9.387	10.33	11.29	12.29	13.38	14.61	17.89	23.89	42.16
(5)	2.445	4.679	6.601	8.242	9.677	10.97	12.19	13.37	14.56	15.81	17.15	18.67	22.72	30.17	52.97
(6)	3.048	5.820	8.191	10.20	11.95	13.52	14.99	16.41	17.83	19.32	20.92	22.73	27.56	36.46	63.78

TABLE 7.12 : I_0^* (α = .90)

MODE \ Q:	0.05	0.10	0.15	0.20	0.25	0.30	0.35	0.40	0.45	0.50	0.55	0.60	0.70	0.80	0.90
(1)	.0269	.1025	.2162	.3586	.5236	.7084	.9123	1.136	1.382	1.656	1.963	2.317	3.251	4.893	9.625
(2)	.5652	1.128	1.658	2.151	2.618	3.070	3.519	3.976	4.454	4.967	5.533	6.178	7.895	11.01	20.32
(3)	1.103	2.153	3.099	3.944	4.712	5.431	6.125	6.817	7.527	8.279	9.103	10.04	12.54	17.13	31.01
(4)	1.642	3.179	4.541	5.737	6.807	7.793	8.732	9.657	10.60	11.59	12.67	13.90	17.18	23.24	41.71
(5)	2.180	4.204	5.982	7.530	8.901	10.15	11.34	12.50	13.67	14.90	16.24	17.76	21.83	29.36	52.40
(6)	2.718	5.230	7.424	9.322	11.00	12.52	13.94	15.34	16.74	18.21	19.81	21.62	26.47	35.48	63.09

TABLE 7.12 : I_0^* (α = 1.0)

MODE \ Q:	0.05	0.10	0.15	0.20	0.25	0.30	0.35	0.40	0.45	0.50	0.55	0.60	0.70	0.80	0.90
(1)	.0243	.0931	.1978	.3303	.4854	.6604	.8546	1.069	1.306	1.571	1.870	2.216	3.139	4.781	9.540
(2)	.5100	1.024	1.516	1.982	2.427	2.862	3.296	3.741	4.209	4.712	5.270	5.909	7.622	10.76	20.14
(3)	.9956	1.955	2.835	3.634	4.369	5.063	5.738	6.414	7.111	7.854	8.671	9.602	12.11	16.73	30.74
(4)	1.481	2.886	4.154	5.285	6.310	7.264	8.179	9.086	10.01	11.00	12.07	13.30	16.59	22.71	41.34
(5)	1.967	3.817	5.473	6.937	8.252	9.465	10.62	11.76	12.92	14.14	15.47	16.99	21.07	28.68	51.94
(6)	2.453	4.748	6.791	8.588	10.19	11.67	13.06	14.43	15.82	17.28	18.87	20.68	25.56	34.66	62.54

TABLE 7.12 : I_0^* ($1/\alpha$ = .10)

MODE \ Q:	0.05	0.10	0.15	0.20	0.25	0.30	0.35	0.40	0.45	0.50	0.55	0.60	0.70	0.80	0.90
(1)	.0025	.0101	.0234	.0435	.0719	.1115	.1664	.2426	.3489	.4967	.7001	.9764	1.858	3.633	8.793
(2)	.0251	.1111	.1795	.2608	.3597	.4834	.6418	.8491	1.124	1.490	1.973	2.604	4.512	8.174	18.56
(3)	.1017	.2121	.3357	.4781	.6475	.8552	1.117	1.456	1.900	2.484	3.246	4.231	7.167	12.72	28.33
(4)	.1514	.3132	.4918	.6954	.9353	1.227	1.593	2.062	2.675	3.477	4.519	5.858	9.821	17.26	38.10
(5)	.2010	.4142	.6479	.9128	1.223	1.599	2.068	2.669	3.450	4.471	5.792	7.485	12.48	21.80	47.87
(6)	.2506	.5152	.8040	1.130	1.511	1.971	2.543	3.275	4.226	5.464	7.065	9.113	15.13	26.34	57.64

TABLE 7.12 : I_0^* ($1/\alpha$ = .20)

MODE \ Q:	0.05	0.10	0.15	0.20	0.25	0.30	0.35	0.40	0.45	0.50	0.55	0.60	0.70	0.80	0.90
(1)	.0050	.0200	.0458	.0833	.1342	.2006	.2858	.3938	.5301	.7025	.9216	1.203	2.066	3.792	8.882
(2)	.1040	.2200	.3508	.4997	.6708	.8695	1.102	1.378	1.708	2.107	2.597	3.208	5.017	8.533	18.75
(3)	.2030	.4200	.6559	.9161	1.207	1.538	1.919	2.363	2.886	3.512	4.273	5.213	7.969	13.27	28.62
(4)	.3020	.6200	.9609	1.332	1.744	2.207	2.736	3.347	4.064	4.917	5.948	7.219	10.92	18.01	38.49
(5)	.4010	.8201	1.266	1.749	2.281	2.876	3.552	4.331	5.242	6.322	7.624	9.224	13.87	22.75	48.36
(6)	.5000	1.020	1.571	2.165	2.817	3.545	4.369	5.316	6.420	7.727	9.299	11.23	16.82	27.49	58.22

TABLE 7.12 : I_0^* ($1/\alpha$ = .30)

MODE \ Q:	0.05	0.10	0.15	0.20	0.25	0.30	0.35	0.40	0.45	0.50	0.55	0.60	0.70	0.80	0.90
(1)	.0074	.0297	.0672	.1202	.1898	.2771	.3841	.5138	.6704	.8604	1.093	1.383	2.244	3.940	8.969
(2)	.1556	.3268	.5149	.7214	.9488	1.201	1.481	1.798	2.160	2.581	3.080	3.688	5.450	8.865	18.94
(3)	.3037	.6239	.9626	1.323	1.708	2.124	2.579	3.083	3.650	4.302	5.068	5.993	8.656	13.79	28.90
(4)	.4519	.9210	1.410	1.924	2.467	3.048	3.676	4.367	5.140	6.022	7.055	8.299	11.86	18.72	38.87
(5)	.6001	1.218	1.858	2.525	3.226	3.971	4.774	5.651	6.630	7.743	9.043	10.60	15.07	23.64	48.83
(6)	.7482	1.515	2.306	3.126	3.985	4.895	5.871	6.936	8.119	9.464	11.03	12.91	18.27	28.57	58.80

TABLE 7.12 : I_0^* ($1/\alpha$ = .40)

MODE \ Q:	0.05	0.10	0.15	0.20	0.25	0.30	0.35	0.40	0.45	0.50	0.55	0.60	0.70	0.80	0.90
(1)	.0099	.0392	.0877	.1549	.2405	.3450	.4696	.6164	.7891	.9935	1.238	1.537	2.403	4.079	9.055
(2)	.2069	.4316	.6725	.9292	1.203	1.495	1.811	2.157	2.543	2.980	3.489	4.099	5.835	9.177	19.12
(3)	.4040	.8240	1.257	1.704	2.165	2.645	3.153	3.698	4.296	4.967	5.741	6.661	9.268	14.28	29.18
(4)	.6011	1.216	1.842	2.478	3.127	3.795	4.495	5.240	6.050	6.954	7.992	9.222	12.70	19.37	39.24
(5)	.7981	1.609	2.427	3.252	4.089	4.945	5.836	6.781	7.804	8.941	10.24	11.78	16.13	24.47	49.30
(6)	.9952	2.001	3.012	4.026	5.051	6.095	7.178	8.321	9.557	10.93	12.49	14.35	19.57	29.57	59.36

TABLE 7.12 : I_0^* ($1/\alpha$ = .50)

Q: MODE:	0.05	0.10	0.15	0.20	0.25	0.30	0.35	0.40	0.45	0.50	0.55	0.60	0.70	0.80	0.90
(1)	.0123	.0486	.1075	.1876	.2875	.4069	.5463	.7076	.8939	1.111	1.366	1.674	2.547	4.209	9.139
(2)	.2580	.5346	.8245	1.125	1.437	1.763	2.107	2.477	2.880	3.332	3.850	4.464	6.186	9.471	19.29
(3)	.5038	1.021	1.542	2.063	2.587	3.119	3.668	4.245	4.867	5.554	6.334	7.253	9.824	14.73	29.45
(4)	.7495	1.506	2.258	3.001	3.737	4.475	5.229	6.014	6.854	7.775	8.819	10.04	13.46	19.99	39.60
(5)	.9952	1.992	2.975	3.939	4.887	5.832	6.790	7.783	8.840	9.996	11.30	12.83	17.10	25.26	49.76
(6)	1.241	2.478	3.692	4.877	6.037	7.188	8.351	9.552	10.83	12.22	13.79	15.62	20.74	30.52	59.91

TABLE 7.12 : I_0^* ($1/\alpha$ = .60)

Q: MODE:	0.05	0.10	0.15	0.20	0.25	0.30	0.35	0.40	0.45	0.50	0.55	0.60	0.70	0.80	0.90
(1)	.0147	.0578	.1267	.2186	.3314	.4640	.6165	.7904	.9888	1.217	1.482	1.798	2.680	4.334	9.222
(2)	.3089	.6357	.9714	1.312	1.657	2.010	2.378	2.766	3.186	3.650	4.177	4.795	6.509	9.751	19.47
(3)	.6031	1.214	1.816	2.405	2.983	3.557	4.139	4.742	5.384	6.084	6.872	7.792	10.34	15.17	29.72
(4)	.8972	1.792	2.661	3.498	4.308	5.104	5.900	6.718	7.581	8.517	9.567	10.79	14.17	20.59	39.96
(5)	1.191	2.369	3.506	4.591	5.634	6.650	7.662	8.695	9.778	10.95	12.26	13.79	18.00	26.00	50.21
(6)	1.486	2.947	4.350	5.684	6.959	8.197	9.423	10.67	11.98	13.38	14.96	16.78	21.83	31.42	60.45

TABLE 7.12 : I_0^* ($1/\alpha$ = .70)

Q: MODE:	0.05	0.10	0.15	0.20	0.25	0.30	0.35	0.40	0.45	0.50	0.55	0.60	0.70	0.80	0.90
(1)	.0171	.0668	.1453	.2483	.3728	.5173	.6815	.8669	1.076	1.314	1.589	1.913	2.805	4.452	9.303
(2)	.3595	.7352	1.114	1.490	1.864	2.242	2.629	3.034	3.468	3.943	4.478	5.101	6.811	10.02	19.64
(3)	.7019	1.404	2.082	2.731	3.355	3.966	4.576	5.201	5.859	6.571	7.367	8.289	10.82	15.58	29.98
(4)	1.044	2.072	3.050	3.972	4.847	5.690	6.523	7.368	8.250	9.200	10.26	11.48	14.82	21.15	40.31
(5)	1.387	2.740	4.019	5.214	6.338	7.414	8.470	9.535	10.64	11.83	13.14	14.66	18.83	26.71	50.65
(6)	1.729	3.408	4.987	6.455	7.829	9.138	10.42	11.70	13.03	14.46	16.03	17.85	22.84	32.28	60.99

TABLE 7.12 : I_0^* ($1/\alpha$ = .80)

Q: MODE:	0.05	0.10	0.15	0.20	0.25	0.30	0.35	0.40	0.45	0.50	0.55	0.60	0.70	0.80	0.90
(1)	.0195	.0757	.1633	.2767	.4121	.5675	.7425	.9382	1.157	1.405	1.688	2.020	2.922	4.566	9.383
(2)	.4099	.8330	1.252	1.660	2.060	2.459	2.864	3.284	3.730	4.215	4.758	5.386	7.096	10.27	19.81
(3)	.8003	1.590	2.340	3.043	3.709	4.351	4.985	5.629	6.302	7.025	7.828	8.752	11.27	15.98	30.24
(4)	1.191	2.348	3.429	4.427	5.357	6.242	7.106	7.975	8.874	9.835	10.90	12.12	15.44	21.69	40.66
(5)	1.581	3.105	4.517	5.810	7.006	8.134	9.228	10.32	11.45	12.64	13.97	15.48	19.62	27.40	51.09
(6)	1.971	3.862	5.605	7.194	8.654	10.03	11.35	12.67	14.02	15.45	17.04	18.85	23.79	33.10	61.51

TABLE 7.12 : I_0^* ($1/\alpha$ = .90)

Q: MODE:	0.05	0.10	0.15	0.20	0.25	0.30	0.35	0.40	0.45	0.50	0.55	0.60	0.70	0.80	0.90
(1)	.0219	.0845	.1808	.3040	.4495	.6150	.8000	1.005	1.234	1.490	1.782	2.120	3.033	4.675	9.462
(2)	.4600	.9293	1.386	1.824	2.248	2.665	3.086	3.519	3.976	4.471	5.021	5.655	7.365	10.52	19.98
(3)	.8982	1.774	2.591	3.344	4.046	4.715	5.371	6.032	6.718	7.451	8.261	9.189	11.70	16.36	30.49
(4)	1.336	2.619	3.796	4.864	5.844	6.766	7.657	8.545	9.460	10.43	11.50	12.72	16.03	22.21	41.00
(5)	1.774	3.464	5.001	6.384	7.642	8.816	9.942	11.06	12.20	13.41	14.74	16.26	20.36	28.05	51.52
(6)	2.213	4.309	6.206	7.904	9.440	10.87	12.23	13.57	14.94	16.39	17.98	19.79	24.70	33.89	62.03

TABLE 7.13 : I_0^* (μ = .10)

Q: MODE:	0.05	0.10	0.15	0.20	0.25	0.30	0.35	0.40	0.45	0.50	0.55	0.60	0.70	0.80	0.90
(1)	.0006	.0025	.0056	.0100	.0157	.0227	.0312	.0412	.0528	.0663	.0818	.0998	.1459	.2170	.3744
(2)	.1495	.2407	.3196	.3972	.4777	.5637	.6569	.7592	.8727	1.000	1.145	1.312	1.753	2.490	4.344
(3)	.5298	.9484	1.310	1.644	1.968	2.293	2.630	2.988	3.377	3.811	4.306	4.885	6.472	9.330	17.22
(4)	.8355	1.590	2.299	2.985	3.670	4.371	5.107	5.896	6.762	7.736	8.859	10.19	13.92	20.87	40.65
(5)	1.035	2.044	3.048	4.067	5.120	6.231	7.423	8.727	10.18	11.84	13.78	16.11	22.76	35.38	71.94
(6)	1.118	2.263	3.450	4.698	6.027	7.464	9.041	10.80	12.79	15.10	17.83	21.15	30.76	49.28	103.6

TABLE 7.13 : I_0^* (μ = .15)

Q: MODE:	0.05	0.10	0.15	0.20	0.25	0.30	0.35	0.40	0.45	0.50	0.55	0.60	0.70	0.80	0.90
(1)	.0009	.0037	.0084	.0150	.0236	.0344	.0474	.0629	.0811	.1025	.1276	.1572	.2363	.3671	.6863
(2)	.1743	.3031	.4196	.5355	.6566	.7865	.9288	1.087	1.265	1.468	1.704	1.984	2.755	4.131	7.853
(3)	.5129	.9826	1.431	1.873	2.323	2.792	3.293	3.839	4.447	5.139	5.948	6.917	9.666	14.85	29.72
(4)	.6883	1.382	2.094	2.835	3.619	4.462	5.384	6.410	7.572	8.914	10.50	12.43	18.01	28.76	60.27
(5)	.6778	1.396	2.165	2.995	3.902	4.905	6.026	7.299	8.767	10.49	12.55	15.09	22.52	37.08	80.14
(6)	.4754	.9924	1.558	2.182	2.877	3.657	4.543	5.561	6.749	8.156	9.857	11.96	18.20	30.50	67.11

TABLE 7.13 : I_0^* (μ = .20)

Q: MODE:	0.05	0.10	0.15	0.20	0.25	0.30	0.35	0.40	0.45	0.50	0.55	0.60	0.70	0.80	0.90
(1)	.0012	.0050	.0112	.0201	.0317	.0463	.0640	.0853	.1106	.1406	.1762	.2190	.3368	.5419	1.073
(2)	.1860	.3414	.4884	.6372	.7941	.9639	1.151	1.361	1.600	1.877	2.204	2.597	3.712	5.788	11.63
(3)	.4572	.9108	1.371	1.847	2.348	2.887	3.475	4.130	4.873	5.732	6.750	7.988	11.57	18.49	38.74
(4)	.4753	.9801	1.521	2.107	2.748	3.457	4.251	5.154	6.196	7.419	8.887	10.69	15.99	26.37	57.11
(5)	.2248	.4696	.7378	1.034	1.364	1.734	2.155	2.638	3.202	3.869	4.675	5.672	8.621	14.43	31.70
(6)	.2482	.5005	.7587	1.025	1.300	1.589	1.893	2.218	2.568	2.952	3.380	3.869	5.148	7.315	13.00

TABLE 7.13 : I_0^* (μ = .25)

Q: MODE:	0.05	0.10	0.15	0.20	0.25	0.30	0.35	0.40	0.45	0.50	0.55	0.60	0.70	0.80	0.90
(1)	.0015	.0062	.0140	.0252	.0399	.0583	.0809	.1083	.1410	.1802	.2273	.2846	.4463	.7386	1.528
(2)	.1892	.3612	.5301	.7042	.8898	1.092	1.317	1.571	1.862	2.203	2.608	3.102	4.529	7.255	15.13
(3)	.3756	.7658	1.176	1.614	2.087	2.606	3.181	3.831	4.576	5.447	6.488	7.763	11.49	18.77	40.27
(4)	.2135	.4447	.6967	.9735	1.280	1.623	2.011	2.455	2.970	3.578	4.311	5.214	7.873	13.10	28.57
(5)	.2486	.5022	.7630	1.033	1.316	1.614	1.931	2.273	2.647	3.063	3.534	4.080	5.546	8.125	15.11
(6)	1.258	2.287	3.208	4.085	4.957	5.858	6.813	7.854	9.013	10.34	11.88	13.74	19.02	29.02	57.83

TABLE 7.13 : I_0^* (μ = .30)

Q: MODE:	0.05	0.10	0.15	0.20	0.25	0.30	0.35	0.40	0.45	0.50	0.55	0.60	0.70	0.80	0.90
(1)	.0019	.0074	.0169	.0303	.0481	.0705	.0981	.1317	.1722	.2211	.2805	.3535	.5634	.9542	2.042
(2)	.1863	.3658	.5476	.7383	.9433	1.168	1.420	1.705	2.035	2.421	2.885	3.454	5.116	8.341	17.79
(3)	.2721	.5618	.8730	1.211	1.580	1.990	2.449	2.970	3.572	4.278	5.125	6.165	9.214	15.18	32.80
(4)	.0836	.1704	.2608	.3555	.4554	.5616	.6755	.7990	.9346	1.086	1.259	1.460	2.006	2.982	5.672
(5)	.8482	1.452	1.991	2.522	3.070	3.652	4.285	4.987	5.778	6.687	7.755	9.041	12.70	19.60	39.37
(6)	.8699	1.761	2.685	3.659	4.698	5.823	7.061	8.445	10.02	11.85	14.02	16.67	24.37	39.33	83.37

TABLE 7.13 : I_c^* (μ = .40)

Q: MODE:	0.05	0.10	0.15	0.20	0.25	0.30	0.35	0.40	0.45	0.50	0.55	0.60	0.70	0.80	0.90
(1)	.0025	.0099	.0226	.0407	.0647	.0952	.1331	.1796	.2364	.3058	.3914	.4984	.8148	1.428	3.202
(2)	.1658	.3371	.5178	7.118	.9233	1.157	1.420	1.719	2.066	2.474	2.964	3.568	5.343	8.817	19.08
(3)	UPPER AND LOWER LIMITS AGREE WITHIN 1 %														
(4)	.3026	.6208	.9587	1.321	1.714	2.145	2.622	3.159	3.771	4.482	5.323	6.343	9.274	14.86	31.03
(5)	.4210	.8646	1.336	1.842	2.390	2.990	3.655	4.404	5.259	6.253	7.433	8.871	13.04	21.09	44.68
(6)	.4968	1.004	1.528	2.074	2.651	3.267	3.935	4.669	5.489	6.422	7.508	8.804	12.45	19.28	38.78

TABLE 7.13 : I_0^* (μ = .50)

Q: MODE:	0.05	0.10	0.15	0.20	0.25	0.30	0.35	0.40	0.45	0.50	0.55	0.60	0.70	0.80	0.90
(1)	.0031	.0124	.0283	.0510	.0813	.1201	.1685	.2282	.3019	.3927	.5058	.6486	1.079	1.934	4.463
(2)	.1267	.2612	.4054	.5614	.7320	.9204	1.131	1.369	1.643	1.963	2.345	2.811	4.162	6.768	14.38
(3)	.2503	.5100	.7825	1.072	1.383	1.721	2.094	2.511	2.985	3.534	4.184	4.973	7.245	11.60	24.30
(4)	.3740	.7588	1.160	1.582	2.033	2.521	3.056	3.652	4.327	5.105	6.023	7.135	10.33	16.44	34.22
(5)	.4976	1.008	1.537	2.093	2.684	3.321	4.019	4.793	5.668	6.676	7.862	9.297	13.41	21.27	44.13
(6)	.6212	1.256	1.914	2.603	3.335	4.122	4.981	5.934	7.010	8.247	9.701	11.46	16.49	26.10	54.05

TABLE 7.13 : I_0^* (μ = .60)

Q: MODE:	0.05	0.10	0.15	0.20	0.25	0.30	0.35	0.40	0.45	0.50	0.55	0.60	0.70	0.80	0.90
(1)	.0037	.0149	.0340	.0614	.0980	.1449	.2083	.2769	.3673	.4796	.6201	.7988	1.343	2.439	5.724
(2)	.0633	.1304	.2017	.2781	.3605	.4502	.5487	.6581	.7814	.9226	1.088	1.285	1.840	2.876	5.824
(3)	.3767	.7149	1.042	1.375	1.725	2.103	2.520	2.988	3.524	4.149	4.896	5.809	8.471	13.64	28.81
(4)	.1696	.3508	.5455	.7564	.9869	1.241	1.524	1.845	2.211	2.639	3.147	3.767	5.564	9.032	19.18
(5)	.5391	1.061	1.582	2.115	2.674	3.269	3.915	4.628	5.429	6.348	7.426	8.722	12.41	19.39	39.48
(6)	.7455	1.509	2.300	3.132	4.018	4.976	6.027	7.200	8.530	10.07	11.89	14.11	20.53	32.93	69.31

TABLE 7.13 : I_0^* (μ = .70)

Q: MODE:	0.05	0.10	0.15	0.20	0.25	0.30	0.35	0.40	0.45	0.50	0.55	0.60	0.70	0.80	0.90
(1)	.0043	.0174	.0397	.0717	.1146	.1696	.2388	.3248	.4315	.5643	.7311	.9437	1.595	2.913	6.884
(2)	.0366	.0763	.1195	.1669	.2195	.2782	.3444	.4201	.5078	.6111	.7352	.8879	1.336	2.213	4.803
(3)	.1998	.3962	.5973	.8085	1.034	1.279	1.548	1.847	2.184	2.571	3.023	3.563	5.079	7.872	15.70
(4)	.6230	1.234	1.849	2.480	3.142	3.851	4.623	5.481	6.453	7.578	8.911	10.53	15.24	24.37	51.19
(5)	.3279	.6800	1.061	1.476	1.932	2.439	3.008	3.656	4.404	5.282	6.334	7.626	11.41	18.80	40.61
(6)	.3725	.7521	1.142	1.547	1.971	2.421	2.902	3.424	3.998	4.643	5.380	6.245	8.612	12.88	24.73

TABLE 7.13 : I_0^* (μ = .80)

Q: MODE:	0.05	0.10	0.15	0.20	0.25	0.30	0.35	0.40	0.45	0.50	0.55	0.60	0.70	0.80	0.90
(1)	.0049	.0199	.0454	.0820	.1309	.1938	.2729	.3712	.4931	.6448	.8353	1.078	1.821	3.326	7.852
(2)	.1288	.2695	.4244	.5961	.7881	1.005	1.252	1.536	1.869	2.265	2.745	3.340	5.105	8.598	19.00
(3)	UPPER AND LOWER LIMITS AGREE WITHIN 1 %														
(4)	.2016	.4128	.6353	.8714	1.123	1.395	1.689	2.011	2.369	2.770	3.228	3.762	5.197	7.688	14.27
(5)	1.032	1.953	2.816	3.654	4.497	5.367	6.291	7.293	8.408	9.676	11.16	12.93	17.99	27.57	55.27
(6)	.9941	2.012	3.069	4.181	5.369	6.655	8.069	9.651	11.45	13.54	16.02	19.05	27.84	44.89	95.09

TABLE 7.13 : I_c^* (μ = .90)

Q: MODE:	0.05	0.10	0.15	0.20	0.25	0.30	0.35	0.40	0.45	0.50	0.55	0.60	0.70	0.80	0.90
(1)	.0056	.0224	.0510	.0921	.1470	.2174	.3057	.4153	.5509	.7191	.9297	1.197	2.012	3.650	8.551
(2)	.1995	.4162	.6534	.9150	1.206	1.534	1.906	2.334	2.834	3.426	4.143	5.030	7.655	12.83	28.22
(3)	.2855	.5964	.9374	1.314	1.735	2.208	2.746	3.367	4.091	4.952	5.994	7.286	11.12	18.70	41.28
(4)	.2580	.5410	.8536	1.201	1.591	2.032	2.537	3.120	3.804	4.619	5.609	6.839	10.50	17.77	39.46
(5)	.1148	.2413	.3814	.5376	.7133	.9126	1.141	1.406	1.716	2.087	2.538	3.099	4.772	8.096	18.03
(6)	.1241	.2498	.3777	.5082	.6419	.7796	.9221	1.071	1.226	1.392	1.569	1.763	2.231	2.930	4.535

TABLE 7.14 : I_0^* (μ = .10)

Q: MODE:	0.05	0.10	0.15	0.20	0.25	0.30	0.35	0.40	0.45	0.50	0.55	0.60	0.70	0.80	0.90
(1)	.1155	.2440	.3875	.5487	.7310	.9386	1.177	1.454	1.780	2.168	2.640	3.226	4.968	8.411	18.64
(2)	.2565	.5347	.8390	1.174	1.548	1.967	2.443	2.990	3.629	4.386	5.301	6.435	9.792	16.42	36.14
(3)	.2861	.5990	.9433	1.325	1.753	2.235	2.785	3.420	4.164	5.048	6.121	7.452	11.41	19.25	42.64
(4)	.2009	.4218	.6664	.9389	1.245	1.592	1.989	2.450	2.990	3.634	4.418	5.392	8.295	14.06	31.29
(5)	UPPER AND LOWER LIMITS AGREE WITHIN 1%														
(6)	.2244	.4544	.6908	.9351	1.189	1.453	1.730	2.023	2.335	2.670	3.035	3.439	4.431	5.940	9.406

TABLE 7.14 : I_0^* (μ = .20)

Q: MODE:	0.05	0.10	0.15	0.20	0.25	0.30	0.35	0.40	0.45	0.50	0.55	0.60	0.70	0.80	0.90
(1)	.0973	.2053	.3257	.4609	.6135	.7874	.9873	1.220	1.493	1.820	2.218	2.713	4.189	7.122	15.87
(2)	.0965	.2020	.3180	.4465	.5899	.7516	.9356	1.148	1.395	1.689	2.045	2.486	3.791	6.370	14.04
(3)	.1247	.2528	.3850	.5223	.6661	.8176	.9786	1.152	1.339	1.546	1.778	2.044	2.742	3.929	7.037
(4)	.8813	1.539	2.103	2.629	3.147	3.679	4.242	4.852	5.528	6.295	7.184	8.243	11.21	16.71	32.23
(5)	1.057	2.086	3.111	4.155	5.238	6.386	7.624	8.989	10.52	12.28	14.36	16.86	24.08	37.94	78.47
(6)	.8532	1.755	2.716	3.753	4.882	6.126	7.515	9.088	10.90	13.02	15.56	18.67	27.79	45.62	98.33

TABLE 7.14 : I_0^* (μ = .30)

Q: MODE:	0.05	0.10	0.15	0.20	0.25	0.30	0.35	0.40	0.45	0.50	0.55	0.60	0.70	0.80	0.90
(1)	.0738	.1550	.2449	.3449	.4571	.5841	.7290	.8965	1.092	1.325	1.607	1.957	2.993	5.039	11.12
(2)	.0839	.1715	.2635	.3606	.4639	.5746	.6944	.8254	.9706	1.134	1.322	1.542	2.145	3.230	6.231
(3)	.5512	1.007	1.427	1.838	2.260	2.708	3.193	3.732	4.343	5.048	5.882	6.894	9.805	15.37	31.51
(4)	.5386	1.102	1.696	2.332	3.019	3.772	4.607	5.548	6.626	7.885	9.385	11.22	16.58	27.02	57.80
(5)	UPPER AND LOWER LIMITS AGREE WITHIN 1%														
(6)	.5242	1.064	1.627	2.217	2.845	3.519	4.252	5.060	5.962	6.989	8.182	9.600	13.56	20.84	41.29

TABLE 7.14 : I_0^* (μ = .40)

Q: MODE:	0.05	0.10	0.15	0.20	0.25	0.30	0.35	0.40	0.45	0.50	0.55	0.60	0.70	0.80	0.90
(1)	.0427	.0890	.1394	.1947	.2559	.3240	.4008	.4883	.5893	.7080	.8500	1.024	1.532	2.518	5.416
(2)	.1832	.3674	.5579	.7594	.9762	1.213	1.476	1.771	2.109	2.501	2.966	3.530	5.153	8.245	17.17
(3)	.3746	.7618	1.167	1.597	2.058	2.561	3.116	3.738	4.449	5.275	6.258	7.456	10.94	17.69	37.52
(4)	.1888	.3860	.5934	.8129	1.047	1.299	1.574	1.877	2.215	2.600	3.047	3.579	5.069	7.835	15.70
(5)	.9280	1.736	2.497	3.250	4.026	4.846	5.736	6.723	7.838	9.129	10.66	12.51	17.88	28.25	58.62
(6)	.2965	.6125	.9515	1.318	1.718	2.158	2.648	3.201	3.833	4.569	5.444	6.510	9.595	15.54	32.95

TABLE 7.14 : I_0^* (μ = .50)

Q: MODE:	0.05	0.10	0.15	0.20	0.25	0.30	0.35	0.40	0.45	0.50	0.55	0.60	0.70	0.80	0.90
(1)	UPPER AND LOWER LIMITS AGREE WITHIN 1%														
(2)	.1298	.2737	.4337	.6125	.8133	1.040	1.299	1.598	1.945	2.356	2.851	3.459	5.241	8.702	18.84
(3)	UPPER AND LOWER LIMITS AGREE WITHIN 1%														
(4)	.3771	.7712	1.188	1.633	2.115	2.641	3.225	3.880	4.628	5.498	6.529	7.783	11.41	18.37	38.68
(5)	UPPER AND LOWER LIMITS AGREE WITHIN 1%														
(6)	.6243	1.269	1.942	2.654	3.416	4.242	5.150	6.163	7.312	8.639	10.21	12.11	17.57	28.04	58.51

TABLE 7.15 : I_0^* (μ = .10)

Q: MODE:	0.05	0.10	0.15	0.20	0.25	0.30	0.35	0.40	0.45	0.50	0.55	0.60	0.70	0.80	0.90
(1)	.0022	.0089	.0201	.0360	.0566	.0822	.1132	.1498	.1920	.2427	.3007	.3685	.5442	.8193	1.435
(2)	.3442	.5872	.7895	.9752	1.156	1.340	1.533	1.739	1.965	2.216	2.501	2.834	3.729	5.295	9.482
(3)	.6950	1.288	1.824	2.331	2.827	3.329	3.850	4.405	5.011	5.688	6.465	7.380	9.916	14.57	27.67
(4)	.9492	1.844	2.711	3.573	4.448	5.357	6.321	7.366	8.522	9.831	11.35	13.16	18.29	27.94	55.68
(5)	1.091	2.184	3.297	4.447	5.656	6.946	8.348	9.896	11.64	13.64	16.00	18.85	27.02	42.68	88.34
(6)	1.115	2.277	3.501	4.806	6.213	7.751	9.453	11.37	13.56	16.10	19.14	22.84	33.63	54.57	116.2

TABLE 7.15 : I_0^* (μ = .15)

Q: MODE:	0.05	0.10	0.15	0.20	0.25	0.30	0.35	0.40	0.45	0.50	0.55	0.60	0.70	0.80	0.90
(1)	.0031	.0126	.0286	.0512	.0808	.1179	.1630	.2170	.2810	.3566	.4460	.5527	.8416	1.329	2.534
(2)	.3607	.6639	.9402	1.207	1.475	1.754	2.052	2.375	2.736	3.145	3.621	4.188	5.777	8.713	16.98
(3)	.6232	1.227	1.827	2.436	3.068	3.736	4.458	5.251	6.142	7.163	8.362	9.809	13.95	21.87	44.87
(4)	.7068	1.440	2.211	3.030	3.912	4.875	5.941	7.139	8.510	10.11	12.01	14.33	21.10	34.27	73.05
(5)	.6007	1.247	1.948	2.715	3.561	4.506	5.572	6.790	8.205	9.874	11.88	14.36	21.68	36.07	78.79
(6)	.3029	.6341	.9986	1.403	1.854	2.364	2.944	3.614	4.397	5.327	6.454	7.851	12.00	20.20	44.63

TABLE 7.15 : I_0^* (μ = .20)

Q: MODE:	0.05	0.10	0.15	0.20	0.25	0.30	0.35	0.40	0.45	0.50	0.55	0.60	0.70	0.80	0.90
(1)	.0040	.0159	.0360	.0645	.1021	.1494	.2074	.2774	.3613	.4616	.5821	.7282	1.137	1.864	3.778
(2)	.3507	.6766	.9932	1.312	1.642	1.992	2.371	2.791	3.265	3.812	4.455	5.234	7.468	11.73	24.04
(3)	.4996	1.016	1.557	2.132	2.750	3.424	4.171	5.011	5.972	7.093	8.429	10.06	14.83	24.11	51.48
(4)	.3837	.7979	1.248	1.743	2.289	2.900	3.591	4.382	5.300	6.384	7.691	9.304	14.06	23.42	51.20
(5)	UPPER AND LOWER LIMITS AGREE WITHIN 1%														
(6)	.3993	.8096	1.235	1.678	2.145	2.640	3.171	3.745	4.376	5.078	5.874	6.796	9.257	13.51	24.76

TABLE 7.15 : I_0^* (μ = .25)

Q: MODE:	0.05	0.10	0.15	0.20	0.25	0.30	0.35	0.40	0.45	0.50	0.55	0.60	0.70	0.80	0.90
(1)	.0046	.0186	.0422	.0759	.1204	.1766	.2460	.3303	.4322	.5554	.7049	.8887	1.416	2.390	5.067
(2)	.3241	.6446	.9700	1.308	1.666	2.053	2.478	2.955	3.498	4.129	4.881	5.799	8.466	13.64	28.81
(3)	.3391	.7014	1.092	1.516	1.982	2.499	3.080	3.741	4.506	5.405	6.484	7.812	11.71	19.36	42.00
(4)	UPPER AND LOWER LIMITS AGREE WITHIN 1%														
(5)	.3753	.7634	1.168	1.594	2.047	2.532	3.057	3.634	4.274	4.998	5.832	6.813	9.507	14.34	27.58
(6)	1.272	2.486	3.677	4.875	6.109	7.408	8.806	10.34	12.06	14.04	16.37	19.18	27.28	42.86	88.42

TABLE 7.15 : I_0^* (μ = .30)

Q: MODE:	0.05	0.10	0.15	0.20	0.25	0.30	0.35	0.40	0.45	0.50	0.55	0.60	0.70	0.80	0.90
(1)	.0052	.0209	.0474	.0853	.1355	.1992	.2782	.3748	.4923	.6354	.8108	1.029	1.665	2.875	6.299
(2)	.2844	.5771	.8831	1.208	1.558	1.942	2.368	2.848	3.400	4.045	4.816	5.762	8.530	13.93	29.89
(3)	.1462	.3040	.4755	.6633	.8707	1.102	1.362	1.660	2.003	2.408	2.894	3.490	5.241	8.663	18.77
(4)	.3274	.6557	.9902	1.336	1.699	2.084	2.499	2.952	3.454	4.020	4.674	5.445	7.578	11.46	22.28
(5)	1.009	1.954	2.869	3.786	4.727	5.719	6.786	7.962	9.284	10.81	12.60	14.77	21.05	33.14	68.57
(6)	.5803	1.198	1.862	2.580	3.366	4.234	5.205	6.306	7.572	9.056	10.83	13.00	19.36	31.76	68.32

TABLE 7.15 : I_0^* (μ = .40)

Q: MODE:	0.05	0.10	0.15	0.20	0.25	0.30	0.35	0.40	0.45	0.50	0.55	0.60	0.70	0.80	0.90
(1)	.0059	.0239	.0543	.0979	.1558	.2298	.3220	.4357	.5753	.7470	.9598	1.228	2.029	3.603	8.214
(2)	.1694	.3500	.5440	.7541	.9840	1.238	1.522	1.843	2.213	2.645	3.161	3.792	5.630	9.196	19.67
(3)	.2479	.5070	.7751	1.057	1.358	1.681	2.034	2.424	2.863	3.364	3.950	4.651	6.635	10.35	20.99
(4)	.6442	1.274	1.905	2.553	3.232	3.957	4.749	5.627	6.623	7.775	9.140	10.80	15.62	24.95	52.36
(5)	UPPER AND LOWER LIMITS AGREE WITHIN 1%														
(6)	.5993	1.218	1.863	2.544	3.272	4.060	4.923	5.883	6.967	8.216	9.685	11.46	16.52	26.12	53.84

TABLE 7.15 : I_0^* (μ = .50)

Q: MODE:	0.05	0.10	0.15	0.20	0.25	0.30	0.35	0.40	0.45	0.50	0.55	0.60	0.70	0.80	0.90
(1)	.0062	.0249	.0566	.1021	.1627	.2401	.3369	.4565	.6037	.7854	1.012	1.297	2.158	3.868	8.926
(2)	UPPER AND LOWER LIMITS AGREE WITHIN 1%														
(3)	.2534	.5225	.8108	1.123	1.464	1.841	2.262	2.739	3.287	3.927	4.690	5.621	8.324	13.54	28.76
(4)	UPPER AND LOWER LIMITS AGREE WITHIN 1%														
(5)	.5007	1.020	1.565	2.144	2.765	3.442	4.187	5.022	5.970	7.069	8.368	9.945	14.49	23.21	48.60
(6)	UPPER AND LOWER LIMITS AGREE WITHIN 1%														

TABLE 7.16 : K^* (μ = .10)

Q: MODE:	0.05	0.10	0.15	0.20	0.25	0.30	0.35	0.40	0.45	0.50	0.55	0.60	0.70	0.80	0.90
(1)	.5894	1.244	1.975	2.796	3.727	4.790	6.016	7.446	9.134	11.16	13.63	16.72	25.99	44.52	100.1
(2)	.6279	1.321	2.091	2.952	3.923	5.026	6.292	7.762	9.491	11.56	14.07	17.21	26.56	45.18	100.8
(3)	.7146	1.494	2.350	3.297	4.353	5.542	6.894	8.450	10.27	12.42	15.02	18.24	27.77	46.56	102.4
(4)	.8733	1.808	2.817	3.913	5.117	6.451	7.946	9.644	11.60	13.89	16.63	19.99	29.80	48.87	105.0
(5)	1.158	2.363	3.630	4.975	6.417	7.983	9.704	11.62	13.79	16.30	19.25	22.81	33.03	52.50	109.0
(6)	1.693	3.386	5.103	6.866	8.702	10.64	12.71	14.97	17.47	20.29	23.55	27.42	38.23	58.30	115.4

TABLE 7.16 : K^* (μ = .20)

Q: MODE:	0.05	0.10	0.15	0.20	0.25	0.30	0.35	0.40	0.45	0.50	0.55	0.60	0.70	0.80	0.90
(1)	.3396	.7159	1.136	1.606	2.139	2.746	3.445	4.258	5.218	6.367	7.769	9.519	14.75	25.20	56.48
(2)	.4411	.9200	1.443	2.019	2.657	3.371	4.179	5.102	6.173	7.436	8.952	10.82	16.29	26.98	58.52
(3)	.7820	1.591	2.437	3.329	4.278	5.301	6.415	7.645	9.023	10.59	12.42	14.60	20.71	32.07	64.29
(4)	2.216	4.298	6.311	8.303	10.31	12.38	14.54	16.82	19.29	21.99	25.01	28.45	37.37	52.04	88.31
(5)	10.30	18.82	26.39	33.51	40.52	47.70	55.28	63.54	72.77	83.37	95.88	111.1	155.2	240.7	492.9
(6)	3.691	7.698	12.08	16.90	22.27	28.29	35.12	42.97	52.12	62.96	76.06	92.28	140.3	235.2	517.4

TABLE 7.16 : K^* (μ = .30)

Q: MODE:	0.05	0.10	0.15	0.20	0.25	0.30	0.35	0.40	0.45	0.50	0.55	0.60	0.70	0.80	0.90
(1)	.2694	.5675	.8990	1.270	1.689	2.165	2.711	3.346	4.092	4.984	6.068	7.418	11.44	19.44	43.31
(2)	.5021	1.036	1.608	2.224	2.893	3.625	4.434	5.339	6.363	7.540	8.920	10.58	15.24	23.93	48.55
(3)	1.997	3.925	5.842	7.801	9.846	12.02	14.39	17.00	19.95	23.33	27.30	32.07	45.61	70.87	141.7
(4)	5.198	10.76	16.76	23.28	30.47	38.46	47.45	57.71	69.59	83.61	100.5	121.3	182.6	303.4	662.1
(5)	1.431	2.935	4.530	6.233	8.071	10.07	12.28	14.75	17.54	20.77	24.57	29.17	42.33	67.41	140.1
(6)	3.120	6.178	9.237	12.35	15.57	18.95	22.56	26.46	30.74	35.53	40.96	47.27	63.98	91.78	157.8

TABLE 7.16 : K^* (μ = .40)

Q: MODE:	0.05	0.10	0.15	0.20	0.25	0.30	0.35	0.40	0.45	0.50	0.55	0.60	0.70	0.80	0.90
(1)	.2500	.5262	.8330	1.176	1.561	1.998	2.499	3.077	3.754	4.560	5.536	6.746	10.33	17.39	38.35
(2)	.8395	1.730	2.687	3.725	4.865	6.130	7.550	9.165	11.03	13.21	15.83	19.03	28.35	46.39	99.21
(3)	UPPER AND LOWER LIMITS AGREE WITHIN 1%														
(4)	1.296	2.645	4.060	5.558	7.158	8.885	10.77	12.85	15.18	17.84	20.94	24.62	34.95	54.06	108.0
(5)	3.174	6.310	9.471	12.72	16.11	19.71	23.62	27.92	32.76	38.32	44.86	52.76	75.45	118.9	245.7
(6)	2.668	5.459	8.405	11.55	14.94	18.65	22.76	27.37	32.63	38.76	46.04	54.93	80.75	130.8	278.0

TABLE 7.16 : K^* (μ = .50)

Q: MODE:	0.05	0.10	0.15	0.20	0.25	0.30	0.35	0.40	0.45	0.50	0.55	0.60	0.70	0.80	0.90
(1)	.2596	.5473	.8674	1.225	1.627	2.081	2.599	3.195	3.891	4.712	5.701	6.918	10.48	17.40	37.69
(2)	.7541	1.542	2.376	3.266	4.229	5.282	6.449	7.760	9.257	11.00	13.06	15.57	22.81	36.74	77.36
(3)	1.249	2.538	3.884	5.308	6.832	8.484	10.30	12.33	14.62	17.28	20.41	24.21	35.14	56.08	117.0
(4)	1.743	3.533	5.393	7.349	9.434	11.69	14.15	16.89	19.99	23.56	27.77	32.86	47.48	75.42	156.7
(5)	2.238	4.528	6.901	9.391	12.04	14.89	18.00	21.46	25.36	29.84	35.13	41.51	59.81	94.75	196.4
(6)	2.732	5.523	8.409	11.43	14.64	18.09	21.85	26.02	30.72	36.13	42.48	50.16	72.14	114.1	236.0

TABLE 7.16 : K^* (μ = .60)

Q: MODE:	0.05	0.10	0.15	0.20	0.25	0.30	0.35	0.40	0.45	0.50	0.55	0.60	0.70	0.80	0.90
(1)	.2997	.6361	1.014	1.439	1.920	2.465	3.087	3.802	4.633	5.611	6.780	8.205	12.30	20.02	41.82
(2)	1.323	2.718	4.205	5.808	7.556	9.486	11.64	14.09	16.92	20.23	24.21	29.11	43.50	71.74	155.5
(3)	1.047	2.138	3.286	4.503	5.804	7.209	8.741	10.43	12.32	14.45	16.92	19.82	27.73	41.67	78.52
(4)	4.446	8.810	13.19	17.66	22.34	27.32	32.73	38.72	45.49	53.32	62.60	73.90	106.8	170.7	359.5
(5)	1.728	3.531	5.428	7.444	9.609	11.96	14.54	17.42	20.68	24.44	28.87	34.21	49.54	78.75	163.5
(6)	3.087	6.167	9.291	12.51	15.88	19.46	23.33	27.57	32.30	37.68	43.93	51.39	72.30	111.0	219.8

TABLE 7.16 : K^* (μ = .70)

Q: MODE:	0.05	0.10	0.15	0.20	0.25	0.30	0.35	0.40	0.45	0.50	0.55	0.60	0.70	0.80	0.90
(1)	.3074	.6549	1.047	1.489	1.989	2.554	3.195	3.927	4.769	5.744	6.888	8.250	11.96	18.27	33.28
(2)	4.781	9.193	13.42	17.59	21.85	26.31	31.09	36.36	42.30	49.16	57.28	67.20	96.13	152.7	320.2
(3)	.9088	1.880	2.926	4.061	5.305	6.682	8.224	9.973	11.99	14.34	17.16	20.61	30.69	50.29	108.0
(4)	1.398	2.822	4.288	5.811	7.412	9.112	10.94	12.92	15.11	17.57	20.37	23.64	32.50	48.10	90.00
(5)	5.130	9.897	14.48	19.03	23.67	28.51	33.69	39.37	45.73	53.01	61.58	71.95	101.8	159.4	328.0
(6)	2.906	6.008	9.344	12.97	16.94	21.34	26.29	31.91	38.40	46.02	55.17	66.42	99.45	164.2	355.8

TABLE 7.16 : K^* (μ = .80)

Q: MODE:	0.05	0.10	0.15	0.20	0.25	0.30	0.35	0.40	0.45	0.50	0.55	0.60	0.70	0.80	0.90
(1)	.2155	.4544	.7189	1.012	1.338	1.701	2.107	2.564	3.084	3.682	4.378	5.206	7.492	11.59	22.71
(2)	1.980	3.762	5.475	7.202	9.004	10.94	13.06	15.45	18.18	21.38	25.20	29.88	43.59	70.34	149.2
(3)	UPPER AND LOWER LIMITS AGREE WITHIN 1%														
(4)	1.297	2.701	4.230	5.909	7.768	9.848	12.20	14.89	18.02	21.72	26.16	31.65	47.83	79.63	173.9
(5)	1.255	2.573	3.970	5.461	7.068	8.820	10.75	12.91	15.35	18.18	21.50	25.52	37.04	58.99	122.6
(6)	2.220	4.405	6.587	8.798	11.07	13.43	15.92	18.59	21.50	24.72	28.37	32.61	44.04	64.25	119.3

TABLE 7.16 : K^* (μ = .90)

Q: MODE:	0.05	0.10	0.15	0.20	0.25	0.30	0.35	0.40	0.45	0.50	0.55	0.60	0.70	0.80	0.90
(1)	.1624	.3417	.5400	.7599	1.005	1.279	1.589	1.940	2.345	2.816	3.375	4.051	5.984	9.627	20.03
(2)	.6613	1.308	1.954	2.610	3.286	3.995	4.749	5.562	6.455	7.453	8.592	9.925	13.56	20.07	38.03
(3)	1.902	3.579	5.119	6.582	8.010	9.435	10.89	12.40	14.00	15.74	17.65	19.82	25.44	34.84	59.19
(4)	6.574	11.48	15.55	19.17	22.58	25.91	29.27	32.73	36.38	40.29	44.57	49.34	61.12	78.73	115.9
(5)	33.22	53.97	69.86	83.69	96.82	110.1	124.1	139.5	156.8	176.9	200.8	230.1	316.2	485.4	987.7
(6)	6.442	13.54	21.39	30.14	39.98	51.13	63.91	78.71	96.10	116.9	142.1	173.5	267.1	453.3	1009

TABLE 7.17 : K^* (μ = .10)

Q: MODE:	0.05	0.10	0.15	0.20	0.25	0.30	0.35	0.40	0.45	0.50	0.55	0.60	0.70	0.80	0.90
(1)	.0076	.0307	.0697	.1256	.1998	.2942	.4117	.5560	.7326	.9490	1.216	1.551	2.548	4.493	10.17
(2)	.3669	.7436	1.136	1.548	1.987	2.461	2.976	3.545	4.183	4.908	5.751	6.753	9.548	14.68	29.11
(3)	1.121	2.166	3.168	4.153	5.140	6.150	7.199	8.310	9.507	10.82	12.30	14.00	18.53	26.40	47.57
(4)	3.373	6.144	8.569	10.79	12.91	14.99	17.08	19.22	21.47	23.87	26.48	29.41	36.75	48.41	76.65
(5)	15.39	25.38	33.33	40.47	47.43	54.59	62.25	70.72	80.32	91.50	104.8	121.2	169.2	263.4	542.8
(6)	UPPER AND LOWER LIMITS AGREE WITHIN 1%														

TABLE 7.17 : K^* (μ = .15)

Q: MODE:	0.05	0.10	0.15	0.20	0.25	0.30	0.35	0.40	0.45	0.50	0.55	0.60	0.70	0.80	0.90
(1)	.0086	.0344	.0781	.1406	.2234	.3287	.4593	.6194	.8145	1.053	1.346	1.712	2.793	4.883	10.92
(2)	.4748	.9544	1.447	1.960	2.500	3.077	3.698	4.377	5.129	5.972	6.938	8.067	11.12	16.51	31.02
(3)	2.429	4.513	6.418	8.244	10.06	11.91	13.84	15.89	18.11	20.55	23.28	26.38	34.28	46.56	73.18
(4)	16.66	29.02	39.30	48.60	57.56	66.64	76.22	86.66	98.39	111.9	128.0	147.6	205.1	317.9	652.6
(5)	2.537	5.311	8.363	11.75	15.53	19.80	24.67	30.28	36.86	44.67	54.15	65.91	100.9	170.1	376.6
(6)	1.505	3.131	4.900	6.839	8.983	11.38	14.08	17.17	20.75	24.97	30.04	36.29	54.67	90.68	197.3

TABLE 7.17 : K^* (μ = .20)

Q: MODE:	0.05	0.10	0.15	0.20	0.25	0.30	0.35	0.40	0.45	0.50	0.55	0.60	0.70	0.80	0.90
(1)	.0097	.0388	.0881	.1585	.2517	.3699	.5162	.6950	.9124	1.177	1.501	1.904	3.085	5.345	11.81
(2)	.6526	1.303	1.966	2.655	3.380	4.154	4.989	5.900	6.905	8.030	9.306	10.78	14.65	21.06	36.82
(3)	4.611	8.481	11.97	15.30	18.62	22.05	25.72	29.74	34.27	39.49	45.68	53.22	75.18	117.9	243.9
(4)	2.016	4.208	6.608	9.256	12.20	15.52	19.28	23.61	28.67	34.67	41.92	50.90	77.54	130.2	287.0
(5)	1.167	2.416	3.762	5.223	6.825	8.598	10.58	12.83	15.42	18.45	22.07	26.50	39.40	64.46	138.1
(6)	1.554	3.144	4.785	6.498	8.305	10.23	12.31	14.60	17.13	20.01	23.34	27.30	38.34	58.79	116.8

TABLE 7.17 : K^* ($\mu = .25$)

Q: MODE:	0.05	0.10	0.15	0.20	0.25	0.30	0.35	0.40	0.45	0.50	0.55	0.60	0.70	0.80	0.90
(1)	.0110	.0441	.1001	.1801	.2858	.4197	.5851	.7867	1.031	1.328	1.690	2.138	3.440	5.903	12.86
(2)	.9680	1.928	2.917	3.964	5.096	6.343	7.740	9.329	11.17	13.33	15.93	19.12	28.50	46.80	100.7
(3)	UPPER AND LOWER LIMITS AGREE WITHIN 1%														
(4)	1.020	2.114	3.297	4.586	6.002	7.575	9.340	11.35	13.66	16.38	19.64	23.64	35.33	58.13	125.4
(5)	1.329	2.693	4.106	5.584	7.145	8.813	10.61	12.59	14.78	17.26	20.13	23.52	32.93	50.17	98.60
(6)	4.656	8.941	13.06	17.18	21.40	25.86	30.67	35.98	41.97	48.87	57.03	66.93	95.55	150.8	312.7

TABLE 7.17 : K^* ($\mu = .30$)

Q: MODE:	0.05	0.10	0.15	0.20	0.25	0.30	0.35	0.40	0.45	0.50	0.55	0.60	0.70	0.80	0.90
(1)	.0126	.0507	.1149	.2066	.3277	.4808	.6698	.8999	1.178	1.515	1.924	2.429	3.883	6.598	14.15
(2)	.9353	1.858	2.797	3.774	4.813	5.939	7.180	8.571	10.16	12.00	14.19	16.85	24.54	39.26	81.88
(3)	1.525	3.162	4.933	6.867	8.999	11.38	14.06	17.12	20.67	24.86	29.91	36.15	54.55	90.77	198.4
(4)	.9565	1.961	3.025	4.160	5.382	6.710	8.170	9.796	11.63	13.74	16.22	19.20	27.65	43.57	89.29
(5)	2.936	5.782	8.623	11.53	14.58	17.84	21.39	25.33	29.79	34.95	41.05	48.45	69.86	111.1	231.7
(6)	UPPER AND LOWER LIMITS AGREE WITHIN 1%														

TABLE 7.17 : K^* ($\mu = .40$)

Q: MODE:	0.05	0.10	0.15	0.20	0.25	0.30	0.35	0.40	0.45	0.50	0.55	0.60	0.70	0.80	0.90
(1)	.0172	.0690	.1565	.2814	.4465	.6554	.9132	1.227	1.606	2.065	2.621	3.303	5.247	8.772	18.12
(2)	1.282	2.562	3.867	5.225	6.663	8.215	9.920	11.83	14.01	16.55	19.57	23.28	34.13	55.36	118.2
(3)	.7733	1.591	2.462	3.399	4.415	5.529	6.765	8.154	9.738	11.58	13.75	16.39	24.00	38.57	80.94
(4)	1.694	3.402	5.153	6.975	8.900	10.96	13.21	15.68	18.46	21.64	25.36	29.80	42.30	65.50	130.8
(5)	3.514	7.198	11.10	15.28	19.81	24.79	30.32	36.58	43.77	52.18	62.24	74.59	110.8	181.6	391.1
(6)	2.077	4.208	6.415	8.724	11.16	13.76	16.57	19.63	23.03	26.84	31.21	36.34	50.22	74.56	138.9

TABLE 7.17 : K^* ($\mu = .50$)

Q: MODE:	0.05	0.10	0.15	0.20	0.25	0.30	0.35	0.40	0.45	0.50	0.55	0.60	0.70	0.80	0.90
(1)	.0247	.0995	.2263	.4083	.6506	.9604	1.348	1.826	2.415	3.142	4.046	5.189	8.632	15.47	35.70
(2)	UPPER AND LOWER LIMITS AGREE WITHIN 1%														
(3)	1.014	2.090	3.243	4.491	5.856	7.363	9.048	10.96	13.15	15.71	18.76	22.48	33.29	54.14	115.0
(4)	UPPER AND LOWER LIMITS AGREE WITHIN 1%														
(5)	2.003	4.080	6.260	8.574	11.06	13.77	16.75	20.09	23.88	28.27	33.47	39.78	57.96	92.82	194.4
(6)	UPPER AND LOWER LIMITS AGREE WITHIN 1%														

TABLE 7.18 : K^* ($\mu = .10$)

Q: MODE:	0.05	0.10	0.15	0.20	0.25	0.30	0.35	0.40	0.45	0.50	0.55	0.60	0.70	0.80	0.90
(1)	.6035	1.272	2.017	2.854	3.799	4.877	6.118	7.562	9.266	11.31	13.80	16.90	26.20	44.76	100.4
(2)	.6642	1.394	2.200	3.097	4.104	5.244	6.547	8.053	9.819	11.92	14.48	17.65	27.08	45.77	101.5
(3)	.7826	1.629	2.551	3.563	4.684	5.937	7.352	8.971	10.85	13.06	15.73	19.01	28.66	47.58	103.5
(4)	.9945	2.046	3.166	4.371	5.680	7.117	8.712	10.51	12.56	14.95	17.79	21.24	31.23	50.48	106.8
(5)	1.381	2.793	4.252	5.778	7.392	9.121	11.00	13.07	15.38	18.03	21.12	24.82	35.30	55.04	111.8
(6)	2.142	4.229	6.297	8.378	10.50	12.71	15.04	17.54	20.27	23.31	26.79	30.87	42.10	62.57	120.1

TABLE 7.18 : K^* ($\mu = .20$)

Q: MODE:	0.05	0.10	0.15	0.20	0.25	0.30	0.35	0.40	0.45	0.50	0.55	0.60	0.70	0.80	0.90
(1)	.3739	.7853	1.240	1.748	2.317	2.961	3.699	4.551	5.550	6.739	8.182	9.973	15.29	25.83	57.20
(2)	.5617	1.159	1.800	2.493	3.248	4.078	5.002	6.042	7.230	8.609	10.24	12.23	17.95	28.89	60.68
(3)	1.218	2.430	3.653	4.905	6.202	7.565	9.014	10.58	12.29	14.20	16.37	18.91	25.76	37.92	71.03
(4)	5.173	9.641	13.77	17.79	21.87	26.13	30.72	35.78	41.50	48.10	55.93	65.45	93.13	146.8	304.6
(5)	UPPER AND LOWER LIMITS AGREE WITHIN 1%														
(6)	2.220	4.616	7.218	10.07	13.22	16.73	20.70	25.24	30.49	36.69	44.14	53.33	80.37	133.4	290.6

TABLE 7.18 : K^* ($\mu = .30$)

Q: MODE:	0.05	0.10	0.15	0.20	0.25	0.30	0.35	0.40	0.45	0.50	0.55	0.60	0.70	0.80	0.90
(1)	.3370	.7046	1.108	1.553	2.047	2.601	3.228	3.945	4.776	5.754	6.928	8.370	12.59	20.79	44.87
(2)	.9413	1.904	2.900	3.944	5.049	6.232	7.512	8.913	10.46	12.21	14.20	16.53	22.74	33.37	60.52
(3)	4.908	9.449	13.80	18.11	22.51	27.10	32.03	37.44	43.52	50.53	58.80	68.87	98.12	155.0	322.8
(4)	1.681	3.474	5.402	7.493	9.782	12.32	15.15	18.38	22.09	26.44	31.66	38.05	56.78	93.34	201.3
(5)	1.778	3.598	5.478	7.438	9.503	11.70	14.07	16.65	19.50	22.72	26.41	30.76	42.72	64.33	124.2
(6)	5.461	10.57	15.51	20.42	25.44	30.68	36.29	42.40	49.22	57.01	66.10	77.04	108.2	167.3	337.7

TABLE 7.18 : K^* ($\mu = .40$)

Q: MODE:	0.05	0.10	0.15	0.20	0.25	0.30	0.35	0.40	0.45	0.50	0.55	0.60	0.70	0.80	0.90
(1)	.3789	.7909	1.241	1.734	2.278	2.882	3.557	4.321	5.194	6.206	7.399	8.839	12.92	20.55	42.15
(2)	1.420	2.843	4.296	5.805	7.400	9.115	10.99	13.07	15.43	18.15	21.37	25.28	36.56	58.24	121.6
(3)	1.459	2.993	4.619	6.363	8.253	10.33	12.63	15.22	18.19	21.66	25.79	30.84	45.54	74.10	158.2
(4)	2.194	4.434	6.758	9.204	11.82	14.64	17.75	21.22	25.17	29.73	35.13	41.68	60.56	96.82	202.5
(5)	UPPER AND LOWER LIMITS AGREE WITHIN 1%														
(6)	2.212	4.498	6.879	9.383	12.04	14.89	17.98	21.38	25.17	29.48	34.47	40.41	56.98	87.57	173.9

TABLE 7.18 : K^* ($\mu = .50$)

Q: MODE:	0.05	0.10	0.15	0.20	0.25	0.30	0.35	0.40	0.45	0.50	0.55	0.60	0.70	0.80	0.90
(1)	.5192	1.095	1.735	2.450	3.253	4.162	5.198	6.391	7.781	9.425	11.40	13.84	20.96	34.81	75.37
(2)	UPPER AND LOWER LIMITS AGREE WITHIN 1%														
(3)	1.508	3.085	4.752	6.533	8.458	10.56	12.90	15.52	18.51	21.99	26.12	31.13	45.63	73.48	154.7
(4)	UPPER AND LOWER LIMITS AGREE WITHIN 1%														
(5)	2.497	5.075	7.768	10.62	13.66	16.97	20.60	24.65	29.25	34.56	40.83	48.43	70.29	112.2	234.0
(6)	UPPER AND LOWER LIMITS AGREE WITHIN 1%														

TABLE 7.19 : K^* ($\mu = .10$)

Q: MODE:	0.05	0.10	0.15	0.20	0.25	0.30	0.35	0.40	0.45	0.50	0.55	0.60	0.70	0.80	0.90
(1)	.0056	.0224	.0510	.0921	.1470	.2174	.3057	.4153	.5509	.7191	.9297	1.197	2.012	3.650	8.551
(2)	.1995	.4162	.6534	.9150	1.206	1.534	1.906	2.334	2.834	3.426	4.143	5.030	7.655	12.83	28.22
(3)	.2855	.5964	.9374	1.314	1.735	2.208	2.746	3.367	4.091	4.952	5.994	7.286	11.12	18.70	41.28
(4)	.2580	.5410	.8536	1.201	1.591	2.032	2.537	3.120	3.804	4.620	5.609	6.839	10.50	17.77	39.46
(5)	.1148	.2413	.3814	.5377	.7134	.9127	1.141	1.406	1.716	2.087	2.538	3.099	4.772	8.096	18.03
(6)	.1241	.2498	.3776	.5081	.6419	.7796	.9221	1.071	1.226	1.392	1.569	1.763	2.230	2.930	4.534

TABLE 7.19 : K^* (μ = .20)

Q: MODE:	0.05	0.10	0.15	0.20	0.25	0.30	0.35	0.40	0.45	0.50	0.55	0.60	0.70	0.80	0.90
(1)	.0049	.0199	.0454	.0820	.1309	.1938	.2729	.3712	.4931	.6448	.8353	1.078	1.821	3.326	7.852
(2)	.1288	.2695	.4244	.5961	.7881	1.005	1.252	1.536	1.869	2.265	2.745	3.340	5.105	8.598	19.00
(3)	UPPER AND LOWER LIMITS AGREE WITHIN 1%														
(4)	.2016	.4128	.6353	.8713	1.123	1.395	1.689	2.011	2.369	2.770	3.228	3.762	5.197	7.688	14.27
(5)	1.032	1.953	2.816	3.654	4.497	5.367	6.291	7.293	8.408	9.676	11.16	12.93	17.99	27.57	55.26
(6)	.9941	2.012	3.069	4.181	5.369	6.655	8.069	9.651	11.45	13.54	16.02	19.05	27.84	44.89	95.09

TABLE 7.19 : K^* (μ = .30)

Q: MODE:	0.05	0.10	0.15	0.20	0.25	0.30	0.35	0.40	0.45	0.50	0.55	0.60	0.70	0.80	0.90
(1)	.0043	.0174	.0397	.0717	.1146	.1696	.2388	.3248	.4315	.5643	.7311	.9437	1.595	2.913	6.884
(2)	.0366	.0763	.1195	.1669	.2195	.2782	.3444	.4201	.5078	.6111	.7353	.8880	1.336	2.213	4.803
(3)	.1998	.3962	.5973	.8085	1.034	1.279	1.548	1.847	2.184	2.571	3.023	3.563	5.079	7.872	15.70
(4)	.6230	1.234	1.849	2.480	3.142	3.851	4.623	5.481	6.453	7.578	8.911	10.53	15.24	24.37	51.19
(5)	.3280	.6800	1.061	1.476	1.932	2.439	3.008	3.656	4.404	5.282	6.334	7.626	11.41	18.80	40.61
(6)	.3725	.7521	1.142	1.547	1.971	2.421	2.902	3.424	3.998	4.643	5.380	6.245	8.612	12.88	24.73

TABLE 7.19 : K^* (μ = .40)

Q: MODE:	0.05	0.10	0.15	0.20	0.25	0.30	0.35	0.40	0.45	0.50	0.55	0.60	0.70	0.80	0.90
(1)	.0037	.0149	.0340	.0614	.0980	.1449	.2038	.2769	.3673	.4796	.6201	.7988	1.343	2.439	5.724
(2)	.0633	.1304	.2017	.2781	.3605	.4502	.5486	.6581	.7814	.9226	1.088	1.285	1.840	2.876	5.824
(3)	.3767	.7149	1.042	1.375	1.725	2.103	2.520	2.988	3.524	4.149	4.896	5.809	8.471	13.64	28.81
(4)	.1696	.3508	.5455	.7564	.9869	1.241	1.524	1.845	2.211	2.639	3.147	3.767	5.564	9.032	19.18
(5)	.5391	1.061	1.582	2.115	2.674	3.269	3.915	4.628	5.429	6.348	7.426	8.722	12.41	19.39	39.48
(6)	.7455	1.509	2.300	3.132	4.018	4.976	6.027	7.200	8.530	10.07	11.89	14.11	20.53	32.93	69.31

TABLE 7.19 : K^* (μ = .50)

Q: MODE:	0.05	0.10	0.15	0.20	0.25	0.30	0.35	0.40	0.45	0.50	0.55	0.60	0.70	0.80	0.90
(1)	.0031	.0124	.0283	.0510	.0813	.1201	.1685	.2282	.3019	.3927	.5058	.6486	1.079	1.934	4.463
(2)	.1267	.2612	.4054	.5614	.7320	.9204	1.131	1.369	1.643	1.963	2.345	2.811	4.162	6.768	14.38
(3)	.2503	.5100	.7825	1.072	1.383	1.721	2.094	2.511	2.985	3.534	4.184	4.973	7.245	11.60	24.30
(4)	.3740	.7588	1.160	1.582	2.033	2.521	3.056	3.652	4.327	5.105	6.023	7.135	10.33	16.44	34.21
(5)	.4976	1.008	1.537	2.093	2.684	3.321	4.019	4.793	5.668	6.676	7.862	9.297	13.41	21.27	44.13
(6)	.6212	1.256	1.914	2.603	3.334	4.122	4.981	5.934	7.010	8.247	9.701	11.46	16.49	26.10	54.05

TABLE 7.19 : K^* (μ = .60)

Q: MODE:	0.05	0.10	0.15	0.20	0.25	0.30	0.35	0.40	0.45	0.50	0.55	0.60	0.70	0.80	0.90
(1)	.0025	.0099	.0226	.0407	.0647	.0952	.1331	.1796	.2364	.3058	.3914	.4984	.8148	1.428	3.202
(2)	.1658	.3371	.5178	.7118	.9233	1.157	1.420	1.719	2.066	2.474	2.964	3.568	5.343	8.817	19.08
(3)	UPPER AND LOWER LIMITS AGREE WITHIN 1%														
(4)	.3026	.6208	.9587	1.321	1.714	2.145	2.622	3.159	3.771	4.482	5.323	6.343	9.274	14.86	31.03
(5)	.4210	.8646	1.336	1.842	2.390	2.990	3.655	4.404	5.259	6.252	7.433	8.871	13.04	21.09	44.68
(6)	.4968	1.004	1.528	2.074	2.651	3.267	3.935	4.669	5.489	6.422	7.508	8.805	12.46	19.28	38.78

TABLE 7.19 : K^* (μ = .70)

Q: MODE:	0.05	0.10	0.15	0.20	0.25	0.30	0.35	0.40	0.45	0.50	0.55	0.60	0.70	0.80	0.90
(1)	.0019	.0074	.0169	.0303	.0481	.0705	.0981	.1317	.1722	.2211	.2805	.3535	.5634	.9542	2.042
(2)	.1863	.3658	.5476	.7383	.9433	1.168	1.420	1.705	2.035	2.421	2.885	3.454	5.116	8.341	17.79
(3)	.2721	.5618	.8730	1.211	1.580	1.990	2.449	2.970	3.572	4.278	5.125	6.165	9.214	15.18	32.80
(4)	.0836	.1704	.2608	.3555	.4554	.5616	.6755	.7990	.9347	1.086	1.259	1.460	2.006	2.982	5.672
(5)	.8482	1.452	1.991	2.522	3.070	3.652	4.286	4.987	5.778	6.687	7.755	9.041	12.70	19.60	39.37
(6)	.8699	1.761	2.685	3.659	4.698	5.823	7.061	8.445	10.02	11.85	14.02	16.67	24.37	39.33	83.37

TABLE 7.19 : K^* (μ = .80)

Q: MODE:	0.05	0.10	0.15	0.20	0.25	0.30	0.35	0.40	0.45	0.50	0.55	0.60	0.70	0.80	0.90
(1)	.0012	.0050	.0112	.0201	.0317	.0463	.0640	.0853	.1106	.1406	.1762	.2190	.3368	.5419	1.073
(2)	.1860	.3414	.4884	.6372	.7941	.9639	1.151	1.361	1.600	1.877	2.204	2.597	3.712	5.788	11.64
(3)	.4572	.9108	1.371	1.847	2.348	2.887	3.475	4.130	4.873	5.732	6.750	7.988	11.57	18.49	38.74
(4)	.4753	.9801	1.521	2.107	2.748	3.457	4.251	5.154	6.196	7.419	8.887	10.69	15.99	26.37	57.11
(5)	.2248	.4696	.7378	1.034	1.364	1.734	2.155	2.638	3.201	3.869	4.675	5.672	8.621	14.43	31.70
(6)	.2482	.5006	.7587	1.025	1.300	1.589	1.893	2.218	2.568	2.952	3.380	3.869	5.148	7.315	13.00

TABLE 7.19 : K^* (μ = .90)

Q: MODE:	0.05	0.10	0.15	0.20	0.25	0.30	0.35	0.40	0.45	0.50	0.55	0.60	0.70	0.80	0.90
(1)	.0006	.0025	.0056	.0100	.0157	.0227	0.312	.0412	.0528	.0663	.0818	.0998	.1459	.2170	.3744
(2)	.1495	.2407	.3196	.3972	.4777	.5637	.6569	.7592	.8727	1.000	1.145	1.312	1.753	2.490	4.344
(3)	.5298	.9484	1.310	1.644	1.968	2.293	2.630	2.988	3.377	3.811	4.306	4.885	6.472	9.330	17.22
(4)	.8355	1.590	2.299	2.985	3.670	4.371	5.107	5.896	6.762	7.736	8.859	10.19	13.92	20.87	40.66
(5)	1.035	2.044	3.048	4.067	5.120	6.231	7.423	8.727	10.18	11.84	13.78	16.11	22.76	35.38	71.94
(6)	1.118	2.263	3.450	4.698	6.027	7.464	9.040	10.80	12.79	15.10	17.83	21.15	30.76	49.28	103.6

TABLE 7.20 : K^* (μ = .10)

Q: MODE:	0.05	0.10	0.15	0.20	0.25	0.30	0.35	0.40	0.45	0.50	0.55	0.60	0.70	0.80	0.90
(1)	.0022	.0089	.0201	.0360	.0566	.0822	.1132	.1498	.1928	.2427	.3007	.3685	.5442	.8193	1.435
(2)	.3442	.5872	.7895	.9752	1.156	1.340	1.533	1.739	1.965	2.216	2.501	2.834	3.729	5.295	9.482
(3)	.6950	1.288	1.824	2.331	2.827	3.329	3.850	4.405	5.011	5.688	6.465	7.380	9.916	14.57	27.67
(4)	.9492	1.844	2.711	3.573	4.448	5.357	6.321	7.366	8.522	9.831	11.35	13.16	18.29	27.94	55.68
(5)	1.091	2.184	3.297	4.447	5.656	6.946	8.348	9.896	11.64	13.64	16.00	18.85	27.02	42.68	88.34
(6)	1.115	2.277	3.501	4.806	6.213	7.751	9.453	11.37	13.56	16.10	19.14	22.84	33.63	54.57	116.2

TABLE 7.20 : K^* (μ = .20)

Q: MODE:	0.05	0.10	0.15	0.20	0.25	0.30	0.35	0.40	0.45	0.50	0.55	0.60	0.70	0.80	0.90
(1)	.0040	.0159	.0360	.0645	.1021	.1494	.2074	.2774	.3613	.4616	.5821	.7282	1.137	1.864	3.778
(2)	.3507	.6766	.9932	1.312	1.642	1.992	2.371	2.791	3.265	3.812	4.455	5.234	7.468	11.73	24.04
(3)	.4996	1.016	1.557	2.132	2.750	3.424	4.171	5.011	5.972	7.093	8.429	10.06	14.83	24.11	51.48
(4)	.3837	.7979	1.248	1.743	2.289	2.900	3.591	4.381	5.300	6.384	7.691	9.304	14.06	23.42	51.20
(5)	UPPER AND LOWER LIMITS AGREE WITHIN 1%														
(6)	.3992	.8096	1.235	1.678	2.145	2.640	3.171	3.745	4.376	5.078	5.874	6.796	9.257	13.51	24.76

TABLE 7.20 : K^* (μ = .30)

Q: MODE:	0.05	0.10	0.15	0.20	0.25	0.30	0.35	0.40	0.45	0.50	0.55	0.60	0.70	0.80	0.90
(1)	.0052	.0209	.0474	.0853	.1355	.1992	.2782	.3748	.4923	.6354	.8108	1.029	1.665	2.875	6.299
(2)	.2844	.5771	.8831	1.208	1.558	1.942	2.368	2.848	3.400	4.045	4.816	5.762	8.530	13.93	29.89
(3)	.1462	.3040	.4755	.6633	.8707	1.102	1.362	1.660	2.003	2.408	2.893	3.490	5.241	8.663	18.77
(4)	.3274	.6557	.9902	1.336	1.699	2.084	2.499	2.952	3.454	4.020	4.674	5.445	7.578	11.46	22.28
(5)	1.009	1.954	2.869	3.786	4.727	5.719	6.786	7.962	9.284	10.81	12.60	14.77	21.05	33.14	68.57
(6)	.5805	1.199	1.862	2.580	3.366	4.234	5.205	6.306	7.573	9.056	10.83	13.01	19.36	31.76	68.33

TABLE 7.20 : K^* (μ = .40)

Q: MODE:	0.05	0.10	0.15	0.20	0.25	0.30	0.35	0.40	0.45	0.50	0.55	0.60	0.70	0.80	0.90
(1)	.0059	.0239	.0543	.0979	.1558	.2298	.3220	.4357	.5753	.7470	.9598	1.228	2.029	3.603	8.214
(2)	.1694	.3500	.5440	.7541	.9840	1.238	1.522	1.843	2.213	2.645	3.161	3.792	5.630	9.196	19.67
(3)	.2497	.5070	.7751	1.057	1.358	1.681	2.034	2.424	2.863	3.364	3.950	4.651	6.635	10.35	20.99
(4)	.6442	1.274	1.905	2.553	3.232	3.957	4.749	5.627	6.623	7.775	9.140	10.80	15.62	24.95	52.36
(5)	UPPER AND LOWER LIMITS AGREE WITHIN 1%														
(6)	.5993	1.218	1.863	2.544	3.272	4.060	4.923	5.883	6.967	8.216	9.685	11.46	16.52	26.12	53.84

TABLE 7.20 : K^* (μ = .50)

Q: MODE:	0.05	0.10	0.15	0.20	0.25	0.30	0.35	0.40	0.45	0.50	0.55	0.60	0.70	0.80	0.90
(1)	.0062	.0249	.0566	.1021	.1627	.2401	.3369	.4565	.6037	.7854	1.012	1.297	2.158	3.868	8.926
(2)	UPPER AND LOWER LIMITS AGREE WITHIN 1%														
(3)	.2534	.5225	.8108	1.123	1.464	1.841	2.262	2.739	3.287	3.927	4.690	5.621	8.324	13.54	28.76
(4)	UPPER AND LOWER LIMITS AGREE WITHIN 1%														
(5)	.5007	1.020	1.565	2.144	2.765	3.442	4.187	5.021	5.970	7.069	8.368	9.945	14.49	23.21	48.60
(6)	UPPER AND LOWER LIMITS AGREE WITHIN 1%														

TABLE 7.21 : K^* (μ = .10)

Q: MODE:	0.05	0.10	0.15	0.20	0.25	0.30	0.35	0.40	0.45	0.50	0.55	0.60	0.70	0.80	0.90
(1)	.1155	.2440	.3875	.5487	.7310	.9306	1.177	1.454	1.780	2.168	2.640	3.226	4.968	8.411	18.64
(2)	.2565	.5347	.8390	1.174	1.548	1.967	2.443	2.990	3.629	4.386	5.301	6.435	9.792	16.42	36.14
(3)	.2861	.5990	.9433	1.325	1.753	2.235	2.785	3.420	4.164	5.048	6.121	7.452	11.41	19.25	42.64
(4)	.2009	.4218	.6664	.9389	1.245	1.592	1.989	2.450	2.990	3.634	4.418	5.392	8.295	14.06	31.29
(5)	UPPER AND LOWER LIMITS AGREE WITHIN 1%														
(6)	.2244	.4544	.6908	.9351	1.189	1.453	1.730	2.023	2.335	2.670	3.035	3.439	4.431	5.940	9.406

TABLE 7.21 : K^* (μ = .20)

Q: MODE:	0.05	0.10	0.15	0.20	0.25	0.30	0.35	0.40	0.45	0.50	0.55	0.60	0.70	0.80	0.90
(1)	.0973	.2053	.3257	.4609	.6135	.7874	.9873	1.220	1.493	1.820	2.218	2.713	4.189	7.122	15.87
(2)	.0965	.2020	.3180	.4465	.5899	.7516	.9356	1.148	1.395	1.689	2.045	2.486	3.791	6.370	14.04
(3)	.1247	.2528	.3850	.5223	.6661	.8176	.9786	1.152	1.339	1.546	1.778	2.044	2.742	3.929	7.037
(4)	.8813	1.539	2.103	2.629	3.147	3.679	4.242	4.852	5.528	6.295	7.184	8.243	11.21	16.71	32.23
(5)	1.057	2.086	3.111	4.155	5.238	6.386	7.624	8.989	10.52	12.28	14.36	16.86	24.08	37.94	78.47
(6)	.8532	1.755	2.716	3.753	4.882	6.126	7.515	9.088	10.90	13.02	15.56	18.67	27.79	45.62	98.33

TABLE 7.21 : K^* (μ = .30)

Q: MODE:	0.05	0.10	0.15	0.20	0.25	0.30	0.35	0.40	0.45	0.50	0.55	0.60	0.70	0.80	0.90
(1)	.0738	.1550	.2449	.3449	.4571	.5841	.7290	.8965	1.092	1.325	1.607	1.957	2.993	5.039	11.12
(2)	.0839	.1715	.2635	.3606	.4639	.5746	.6944	.8254	.9706	1.134	1.322	1.542	2.145	3.230	6.231
(3)	.5512	1.007	1.427	1.838	2.260	2.708	3.193	3.732	4.343	5.048	5.882	6.894	9.805	15.37	31.51
(4)	.5386	1.102	1.696	2.332	3.019	3.772	4.607	5.548	6.626	7.885	9.385	11.22	16.58	27.02	57.80
(5)	UPPER AND LOWER LIMITS AGREE WITHIN 1%														
(6)	.5242	1.064	1.627	2.217	2.845	3.519	4.252	5.060	5.962	6.989	8.182	9.600	13.56	20.84	41.29

TABLE 7.21 : K^* (μ = .40)

Q: MODE:	0.05	0.10	0.15	0.20	0.25	0.30	0.35	0.40	0.45	0.50	0.55	0.60	0.70	0.80	0.90
(1)	.0427	.0890	.1394	.1947	.2559	.3240	.4008	.4883	.5893	.7080	.8500	1.024	1.532	2.518	5.416
(2)	.1832	.3674	.5579	.7594	.9762	1.213	1.476	1.771	2.109	2.501	2.966	3.530	5.153	8.245	17.17
(3)	.3746	.7618	1.167	1.597	2.058	2.561	3.116	3.738	4.449	5.275	6.258	7.456	10.94	17.69	37.52
(4)	.1888	.3860	.5934	.8129	1.047	1.299	1.574	1.877	2.215	2.600	3.047	3.579	5.069	7.835	15.70
(5)	.9280	1.736	2.497	3.250	4.026	4.846	5.736	6.723	7.838	9.129	10.66	12.51	17.88	28.25	58.62
(6)	.2965	.6125	.9515	1.318	1.718	2.158	2.648	3.201	3.833	4.569	5.444	6.510	9.595	15.54	32.95

TABLE 7.21 : K^* (μ = .50)

Q: MODE:	0.05	0.10	0.15	0.20	0.25	0.30	0.35	0.40	0.45	0.50	0.55	0.60	0.70	0.80	0.90
(1)	UPPER AND LOWER LIMITS AGREE WITHIN 1%														
(2)	.1298	.2737	.4337	.6125	.8133	1.040	1.299	1.598	1.945	2.356	2.851	3.459	5.241	8.702	18.84
(3)	UPPER AND LOWER LIMITS AGREE WITHIN 1%														
(4)	.3771	.7712	1.188	1.633	2.115	2.641	3.225	3.880	4.628	5.498	6.529	7.783	11.41	18.37	38.68
(5)	UPPER AND LOWER LIMITS AGREE WITHIN 1%														
(6)	.6243	1.269	1.942	2.654	3.416	4.242	5.150	6.163	7.312	8.639	10.21	12.11	17.57	28.04	58.51

TABLE 7.22 : I_0^* (α = .10 , μ = .20)

Q: MODE:	0.05	0.10	0.15	0.20	0.25	0.30	0.35	0.40	0.45	0.50	0.55	0.60	0.70	0.80	0.90
(1)	.2779	.6222	1.046	1.564	2.194	2.958	3.886	5.015	6.394	8.095	10.22	12.92	21.15	37.82	88.04
(2)	.2107	.4534	.7336	1.059	1.439	1.887	2.421	3.069	3.866	4.866	6.151	7.845	13.37	25.45	63.36
(3)	.4289	.8777	1.351	1.855	2.397	2.985	3.630	4.347	5.154	6.079	7.160	8.454	12.11	19.00	39.07

TABLE 7.22 : I_0^* (α = .10 , μ = .30)

Q: MODE:	0.05	0.10	0.15	0.20	0.25	0.30	0.35	0.40	0.45	0.50	0.55	0.60	0.70	0.80	0.90
(1)	.2745	.6075	1.010	1.494	2.077	2.777	3.620	4.640	5.880	7.405	9.304	11.71	19.05	33.89	78.58
(2)	.1864	.3947	.6279	.8896	1.184	1.518	1.899	2.336	2.843	3.438	4.149	5.012	7.475	12.04	24.65
(3)	2.129	4.340	6.610	8.966	11.46	14.13	17.06	20.32	24.00	28.25	33.23	39.24	56.33	88.52	180.8

TABLE 7.22 : I_0^* (α = .10 , μ = .40)

Q: MODE:	0.05	0.10	0.15	0.20	0.25	0.30	0.35	0.40	0.45	0.50	0.55	0.60	0.70	0.80	0.90
(1)	.2711	.5929	.9741	1.426	1.960	2.595	3.352	4.258	5.352	6.687	8.339	10.42	16.73	29.37	67.22
(2)	.1558	.3250	.5089	.7090	.9276	1.167	1.431	1.726	2.057	2.441	2.905	3.516	6.181	13.88	36.76
(3)	2.644	4.211	5.575	7.084	8.860	10.98	13.54	16.60	20.28	24.73	30.16	36.88	56.65	94.99	206.9

TABLE 7.22 . I_0^* (α = .10 μ = .50)

Q: MODE:	0.05	0.10	0.15	0.20	0.25	0.30	0.35	0.40	0.45	0.50	0.55	0.60	0.70	0.80	0.90
(1)	.2678	.5785	.9390	1.358	1.846	2.416	3.086	3.877	4.822	5.961	7.357	9.101	14.30	24.57	54.90
(2)	1.407	2.901	4.505	6.245	8.152	10.27	12.64	15.34	18.46	22.12	26.51	31.89	47.67	78.46	169.3
(3)	.3597	.7775	1.264	1.834	2.506	3.305	4.269	5.446	6.905	8.743	11.10	14.16	23.72	42.65	96.03

TABLE 7.22 : I_0^* (α = .10 , μ = .60)

Q: MODE:	0.05	0.10	0.15	0.20	0.25	0.30	0.35	0.40	0.45	0.50	0.55	0.60	0.70	0.80	0.90
(1)	.2645	.5642	.9045	1.292	1.734	2.241	2.826	3.506	4.304	5.251	6.395	7.802	11.91	19.80	42.62
(2)	1.776	3.674	5.679	7.826	10.16	12.74	15.63	18.92	22.72	27.21	32.60	39.25	58.82	97.24	211.0
(3)	.2947	.6202	.9800	1.379	1.822	2.315	2.868	3.490	4.194	4.998	5.925	7.005	9.837	14.33	24.55

TABLE 7.22 : I_0^* (α = .10 , μ = .70)

Q: MODE:	0.05	0.10	0.15	0.20	0.25	0.30	0.35	0.40	0.45	0.50	0.55	0.60	0.70	0.80	0.90
(1)	.2612	.5502	.8706	1.227	1.625	2.072	2.576	3.150	3.810	4.577	5.483	6.576	9.664	15.38	31.38
(2)	2.050	4.138	6.250	8.441	10.77	13.31	16.12	19.30	22.95	27.23	32.35	38.62	56.93	92.45	196.5
(3)	.1917	.3934	.6055	.8289	1.064	1.313	1.576	1.855	2.155	2.488	2.912	5.105	23.26	56.38	149.2

TABLE 7.22 : I_0^* (α = .10 , μ = .80)

Q: MODE:	0.05	0.10	0.15	0.20	0.25	0.30	0.35	0.40	0.45	0.50	0.55	0.60	0.70	0.80	0.90
(1)	.2579	.5363	.8376	1.164	1.521	1.910	2.340	2.816	3.349	3.954	4.650	5.467	7.682	11.58	22.03
(2)	2.218	4.287	6.224	8.130	10.09	12.16	14.40	16.89	19.69	22.91	26.68	31.22	44.05	67.90	135.2
(3)	4.930	9.953	15.05	20.32	25.87	31.81	38.30	45.52	53.69	63.15	74.35	87.96	127.4	203.4	426.2

TABLE 7.22 : I_0^* (α = .15 , μ = .20)

Q: MODE:	0.05	0.10	0.15	0.20	0.25	0.30	0.35	0.40	0.45	0.50	0.55	0.60	0.70	0.80	0.90
(1)	.2234	.4928	.8164	1.204	1.668	2.224	2.891	3.696	4.675	5.875	7.369	9.264	15.03	26.70	61.84
(2)	.2016	.4334	.7003	1.009	1.366	1.785	2.278	2.868	3.581	4.459	5.560	6.973	11.39	20.59	48.80
(3)	.4109	.8419	1.297	1.783	2.306	2.874	3.498	4.193	4.977	5.878	6.935	8.206	11.83	18.73	38.89

TABLE 7.22 : I_0^* (α = .15 , μ = .30)

Q: MODE:	0.05	0.10	0.15	0.20	0.25	0.30	0.35	0.40	0.45	0.50	0.55	0.60	0.70	0.80	0.90
(1)	.2212	.4833	.7933	1.160	1.594	2.109	2.722	3.458	4.347	5.435	6.783	8.489	13.67	24.13	55.59
(2)	.1784	.3778	.6007	.8505	1.131	1.448	1.808	2.219	2.694	3.250	3.908	4.705	6.963	11.13	22.76
(3)	1.427	2.935	4.507	6.152	7.893	9.767	11.82	14.09	16.67	19.64	23.15	27.38	39.47	62.44	128.8

TABLE 7.22 : I_0^* (α = .15 , μ = .40)

Q: MODE:	0.05	0.10	0.15	0.20	0.25	0.30	0.35	0.40	0.45	0.50	0.55	0.60	0.70	0.80	0.90
(1)	.2190	.4738	.7702	1.116	1.519	1.993	2.551	3.214	4.009	4.973	6.162	7.656	12.16	21.16	48.07
(2)	.1492	.3114	.4880	.6807	.8916	1.124	1.381	1.670	2.000	2.389	2.870	3.508	5.932	12.06	30.87
(3)	1.877	3.254	4.429	5.641	6.993	8.548	10.36	12.50	15.04	18.10	21.82	26.44	40.05	66.60	144.4

TABLE 7.22 : I_0^* (α = .15 , μ = .50)

Q: MODE:	0.05	0.10	0.15	0.20	0.25	0.30	0.35	0.40	0.45	0.50	0.55	0.60	0.70	0.80	0.90
(1)	.2168	.4643	.7474	1.072	1.445	1.877	2.380	2.969	3.667	4.505	5.527	6.798	10.57	18.00	39.91
(2)	.9902	2.043	3.173	4.400	5.747	7.242	8.922	10.84	13.05	15.66	18.79	22.64	33.94	56.09	121.6
(3)	.3405	.7349	1.192	1.723	2.342	3.068	3.928	4.956	6.196	7.713	9.593	11.96	19.04	32.66	70.98

TABLE 7.22 : I_0^* (α = .15 , μ = .60)

Q: MODE:	0.05	0.10	0.15	0.20	0.25	0.30	0.35	0.40	0.45	0.50	0.55	0.60	0.70	0.80	0.90
(1)	.2146	.4549	.7247	1.029	1.372	1.763	2.211	2.728	3.331	4.043	4.899	5.949	9.000	14.85	31.77
(2)	1.202	2.517	3.916	5.416	7.047	8.847	10.86	13.16	15.82	18.94	22.71	27.35	41.02	67.87	147.5
(3)	.2817	.5922	.9349	1.313	1.732	2.197	2.715	3.295	3.947	4.687	5.534	6.516	9.073	13.17	22.99

TABLE 7.22 : I_0^* (α = .15 , μ = .70)

Q: MODE:	0.05	0.10	0.15	0.20	0.25	0.30	0.35	0.40	0.45	0.50	0.55	0.60	0.70	0.80	0.90
(1)	.2124	.4456	.7024	.9860	1.301	1.652	2.047	2.494	3.006	3.599	4.299	5.141	7.518	11.92	24.29
(2)	1.389	2.846	4.335	5.883	7.529	9.316	11.29	13.52	16.08	19.07	22.65	27.03	39.81	64.59	137.2
(3)	.1834	.3764	.5796	.7936	1.019	1.258	1.512	1.783	2.079	2.417	2.882	4.483	16.94	40.64	107.5

TABLE 7.22 : I_0^* (α = .15 , μ = .80)

Q: MODE:	0.05	0.10	0.15	0.20	0.25	0.30	0.35	0.40	0.45	0.50	0.55	0.60	0.70	0.80	0.90
(1)	.2102	.4364	.6803	.9443	1.231	1.544	1.889	2.271	2.698	3.183	3.741	4.398	6.188	9.368	18.02
(2)	1.512	2.988	4.394	5.781	7.202	8.700	10.32	12.10	14.11	16.41	19.10	22.33	31.45	48.40	96.21
(3)	3.359	6.852	10.41	14.09	17.97	22.12	26.65	31.69	37.40	43.99	51.80	61.29	88.76	141.8	297.0

TABLE 7.22 : I_0^* (α = .20 , μ = .20)

Q: MODE:	0.05	0.10	0.15	0.20	0.25	0.30	0.35	0.40	0.45	0.50	0.55	0.60	0.70	0.80	0.90
(1)	.1918	.4196	.6896	1.009	1.388	1.837	2.373	3.016	3.793	4.744	5.924	7.418	11.96	21.12	48.72
(2)	.1932	.4153	.6705	.9642	1.304	1.698	2.160	2.707	3.361	4.157	5.142	6.389	10.19	17.93	41.32
(3)	.3946	.8098	1.250	1.721	2.229	2.782	3.391	4.071	4.841	5.729	6.774	8.036	11.65	18.57	38.79

TABLE 7.22 : I_0^* (α = .20 , μ = .30)

Q: MODE:	0.05	0.10	0.15	0.20	0.25	0.30	0.35	0.40	0.45	0.50	0.55	0.60	0.70	0.80	0.90
(1)	.1902	.4126	.6726	.9768	1.334	1.754	2.251	2.843	3.556	4.424	5.498	6.853	10.96	19.22	44.07
(2)	.1710	.3624	.5763	.8158	1.085	1.388	1.731	2.123	2.575	3.102	3.726	4.479	6.615	10.57	21.70
(3)	1.073	2.222	3.435	4.716	6.079	7.550	9.161	10.95	12.98	15.33	18.10	21.45	31.08	49.46	102.9

TABLE 7.22 : I_0^* (α = .20 , μ = .40)

Q:	0.05	0.10	0.15	0.20	0.25	0.30	0.35	0.40	0.45	0.50	0.55	0.60	0.70	0.80	0.90
MODE:															
(1)	.1885	.4056	.6557	.9445	1.279	1.669	2.126	2.665	3.309	4.088	5.043	6.242	9.841	17.02	38.47
(2)	.1431	.2992	.4694	.6557	.8604	1.087	1.339	1.626	1.957	2.352	2.847	3.504	5.835	11.30	28.13
(3)	1.444	2.617	3.649	4.687	5.812	7.077	8.529	10.22	12.21	14.59	17.49	21.07	31.65	52.34	113.1

TABLE 7.22 : I_0^* (α = .20 , μ = .50)

Q:	0.05	0.10	0.15	0.20	0.25	0.30	0.35	0.40	0.45	0.50	0.55	0.60	0.70	0.80	0.90
MODE:															
(1)	.1869	.3986	.6387	.9122	1.225	1.584	2.000	2.486	3.058	3.743	4.576	5.610	8.670	14.68	32.38
(2)	.7848	1.619	2.516	3.490	4.559	5.748	7.086	8.613	10.38	12.46	14.97	18.05	27.14	44.97	97.80
(3)	.3221	.6944	1.124	1.620	2.193	2.860	3.640	4.558	5.650	6.962	8.561	10.54	16.34	27.34	58.21

TABLE 7.22 : I_0^* (α = .20 , μ = .60)

Q:	0.05	0.10	0.15	0.20	0.25	0.30	0.35	0.40	0.45	0.50	0.55	0.60	0.70	0.80	0.90
MODE:															
(1)	.1853	.3916	.6219	.8800	1.171	1.500	1.875	2.307	2.809	3.401	4.111	4.980	7.501	12.33	26.29
(2)	.9117	1.929	3.022	4.197	5.477	6.889	8.472	10.27	12.35	14.81	17.76	21.40	32.12	53.20	115.7
(3)	.2697	.5666	.8935	1.254	1.652	2.092	2.581	3.126	3.738	4.429	5.219	6.131	8.511	12.38	22.06

TABLE 7.22 : I_0^* (α = .20 , μ = .70)

Q:	0.05	0.10	0.15	0.20	0.25	0.30	0.35	0.40	0.45	0.50	0.55	0.60	0.70	0.80	0.90
MODE:															
(1)	.1836	.3846	.6051	.8482	1.117	1.417	1.752	2.132	2.566	3.069	3.662	4.375	6.391	10.13	20.68
(2)	1.053	2.184	3.354	4.576	5.876	7.286	8.845	10.60	12.61	14.96	17.77	21.21	31.22	50.64	107.5
(3)	.1758	.3609	.5559	.7616	.9790	1.210	1.455	1.721	2.014	2.360	2.862	4.276	13.92	32.91	86.78

TABLE 7.22 : I_0^* (α = .20 , μ = .80)

Q:	0.05	0.10	0.15	0.20	0.25	0.30	0.35	0.40	0.45	0.50	0.55	0.60	0.70	0.80	0.90
MODE:															
(1)	.1820	.3776	.5885	.8167	1.065	1.335	1.633	1.963	2.334	2.754	3.239	3.812	5.381	8.195	15.93
(2)	1.149	2.311	3.438	4.557	5.703	6.909	8.210	9.642	11.25	13.09	15.24	17.81	25.08	38.57	76.65
(3)	2.561	5.280	8.067	10.95	13.99	17.25	20.80	24.75	29.22	34.39	40.51	47.94	69.44	110.9	232.4

TABLE 7.22 : I_0^* (α = .25 , μ = .20)

Q:	0.05	0.10	0.15	0.20	0.25	0.30	0.35	0.40	0.45	0.50	0.55	0.60	0.70	0.80	0.90
MODE:															
(1)	.1707	.3714	.6071	.8840	1.210	1.594	2.050	2.595	3.252	4.053	5.045	6.297	10.10	17.77	40.85
(2)	.1856	.3988	.6435	.9246	1.248	1.623	2.059	2.573	3.184	3.921	4.825	5.959	9.375	16.22	36.71
(3)	.3796	.7808	1.208	1.666	2.162	2.703	3.302	3.972	4.734	5.614	6.654	7.913	11.53	18.47	38.74

TABLE 7.22 : I_0^* (α = .25 , μ = .30)

Q:	0.05	0.10	0.15	0.20	0.25	0.30	0.35	0.40	0.45	0.50	0.55	0.60	0.70	0.80	0.90
MODE:															
(1)	.1694	.3658	.5938	.8587	1.168	1.529	1.955	2.461	3.067	3.803	4.712	5.855	9.314	16.27	37.15
(2)	.1643	.3483	.5542	.7047	1.043	1.335	1.665	2.041	2.475	2.981	3.580	4.304	6.358	10.19	21.02
(3)	.8600	1.789	2.781	3.838	4.972	6.200	7.548	9.052	10.76	12.73	15.07	17.90	26.06	41.72	87.38

TABLE 7.22 : I_0^* (α = .25 , μ = .40)

Q:	0.05	0.10	0.15	0.20	0.25	0.30	0.35	0.40	0.45	0.50	0.55	0.60	0.70	0.80	0.90
MODE:															
(1)	.1681	.3602	.5804	.8332	1.125	1.463	1.857	2.321	2.874	3.539	4.355	5.375	8.434	14.52	32.69
(2)	.1376	.2880	.4526	.6334	.8329	1.055	1.304	1.589	1.922	2.325	2.831	3.501	5.784	10.90	26.57
(3)	1.171	2.179	3.093	4.009	4.987	6.074	7.307	8.731	10.40	12.39	14.79	17.77	26.55	43.74	94.34

TABLE 7.22 : I_0^* (α = .25 , μ = .50)

Q:	0.05	0.10	0.15	0.20	0.25	0.30	0.35	0.40	0.45	0.50	0.55	0.60	0.70	0.80	0.90
MODE:															
(1)	.1667	.3547	.5670	.8076	1.082	1.396	1.758	2.180	2.676	3.268	3.986	4.876	7.506	12.66	27.84
(2)	.6636	1.369	2.128	2.952	3.858	4.866	6.001	7.297	8.800	10.57	12.71	15.34	23.10	38.35	83.61
(3)	.3046	.6562	1.061	1.525	2.060	2.677	3.392	4.227	5.210	6.380	7.792	9.527	14.54	23.97	50.40

TABLE 7.22 : I_0^* (α = .25 , μ = .60)

Q:	0.05	0.10	0.15	0.20	0.25	0.30	0.35	0.40	0.45	0.50	0.55	0.60	0.70	0.80	0.90
MODE:															
(1)	.1654	.3491	.5535	.7820	1.039	1.329	1.659	2.038	2.478	2.996	3.616	4.375	6.576	10.79	22.98
(2)	.7353	1.572	2.478	3.458	4.526	5.706	7.028	8.532	10.27	12.32	14.79	17.83	26.79	44.41	96.67
(3)	.2587	.5430	.8556	1.200	1.579	1.998	2.462	2.978	3.557	4.210	4.955	5.817	8.073	11.80	21.43

TABLE 7.22 : I_0^* (α = .25 , μ = .70)

Q:	0.05	0.10	0.15	0.20	0.25	0.30	0.35	0.40	0.45	0.50	0.55	0.60	0.70	0.80	0.90
MODE:															
(1)	.1641	.3435	.5401	.7565	.9958	1.262	1.560	1.898	2.283	2.730	3.256	3.890	5.685	9.026	18.48
(2)	.8483	1.779	2.753	3.776	4.866	6.048	7.355	8.825	10.51	12.48	14.82	17.69	26.05	42.25	89.71
(3)	.1688	.3467	.5343	.7325	.9423	1.166	1.405	1.666	1.959	2.314	2.848	4.179	12.20	28.37	74.47

TABLE 7.22 : I_0^* (α = .25 , μ = .80)

Q:	0.05	0.10	0.15	0.20	0.25	0.30	0.35	0.40	0.45	0.50	0.55	0.60	0.70	0.80	0.90
MODE:															
(1)	.1628	.3379	.5268	.7312	.9535	1.196	1.464	1.761	2.095	2.474	2.914	3.434	4.867	7.458	14.65
(2)	.9282	1.892	2.843	3.794	4.771	5.799	6.907	8.125	9.490	11.05	12.87	15.06	21.21	32.63	64.87
(3)	2.077	4.324	6.644	9.052	11.59	14.31	17.28	20.57	24.30	28.62	33.72	39.92	57.84	92.43	193.7

TABLE 7.22 : I_0^* (α = .30 , μ = .20)

Q:	0.05	0.10	0.15	0.20	0.25	0.30	0.35	0.40	0.45	0.50	0.55	0.60	0.70	0.80	0.90
MODE:															
(1)	.1552	.3366	.5483	.7956	1.085	1.426	1.828	2.307	2.883	3.583	4.449	5.542	8.853	15.52	35.59
(2)	.1785	.3837	.6190	.8888	1.199	1.557	1.972	2.459	3.036	3.728	4.572	5.626	8.771	15.01	33.57
(3)	.3659	.7544	1.170	1.617	2.103	2.635	3.226	3.889	4.645	5.522	6.559	7.818	11.44	18.40	38.70

TABLE 7.22 : I_0^* (α = .30 , μ = .30)

Q:	0.05	0.10	0.15	0.20	0.25	0.30	0.35	0.40	0.45	0.50	0.55	0.60	0.70	0.80	0.90
MODE:															
(1)	.1542	.3320	.5373	.7749	1.051	1.372	1.750	2.197	2.732	3.380	4.178	5.181	8.208	14.29	32.53
(2)	.1581	.3354	.5340	.7565	1.006	1.288	1.607	1.971	2.390	2.880	3.460	4.162	6.160	9.901	20.54
(3)	.7176	1.498	2.339	3.244	4.222	5.286	6.460	7.772	9.265	10.99	13.04	15.54	22.73	36.60	77.13

TABLE 7.22 : I_0^* (α = .30 , μ = .40)

Q:	0.05	0.10	0.15	0.20	0.25	0.30	0.35	0.40	0.45	0.50	0.55	0.60	0.70	0.80	0.90
MODE:															
(1)	.1531	.3274	.5263	.7539	1.016	1.318	1.670	2.083	2.573	3.163	3.885	4.786	7.483	12.83	28.83
(2)	.1325	.2778	.4374	.6133	.8084	1.027	1.274	1.558	1.894	2.303	2.819	3.499	5.753	10.65	25.58
(3)	.9838	1.862	2.680	3.502	4.376	5.338	6.424	7.670	9.124	10.85	12.93	15.51	23.11	37.98	81.80

TABLE 7.22 : I_0^* (α = .30 , μ = .50)

Q:	0.05	0.10	0.15	0.20	0.25	0.30	0.35	0.40	0.45	0.50	0.55	0.60	0.70	0.80	0.90
MODE:															
(1)	.1520	.3228	.5152	.7327	.9799	1.262	1.588	1.966	2.410	2.939	3.580	4.374	6.716	11.30	24.80
(2)	.5843	1.206	1.874	2.600	3.399	4.288	5.290	6.435	7.763	9.331	11.22	13.55	20.43	33.97	74.18
(3)	.2883	.6205	1.002	1.439	1.939	2.514	3.177	3.946	4.847	5.912	7.189	8.750	13.23	21.62	45.09

TABLE 7.22 : I_0^* (α = .30 , μ = .60)

Q: MODE:	0.05	0.10	0.15	0.20	0.25	0.30	0.35	0.40	0.45	0.50	0.55	0.60	0.70	0.80	0.90
(1)	.1509	.3181	.5040	.7115	.9442	1.207	1.505	1.848	2.245	2.713	3.273	3.958	5.942	9.745	20.75
(2)	.6166	1.330	2.111	2.959	3.886	4.911	6.060	7.367	8.879	10.66	12.80	15.45	23.23	38.55	83.98
(3)	.2485	.5212	.8206	1.150	1.512	1.912	2.355	2.847	3.398	4.020	4.730	5.552	7.719	11.35	20.97

TABLE 7.22 : I_0^* (α = .30 , μ = .70)

Q: MODE:	0.05	0.10	0.15	0.20	0.25	0.30	0.35	0.40	0.45	0.50	0.55	0.60	0.70	0.80	0.90
(1)	.1498	.3135	.4928	.6902	.9084	1.151	1.423	1.731	2.083	2.490	2.972	3.552	5.197	8.269	16.99
(2)	.7108	1.503	2.344	3.231	4.180	5.210	6.348	7.627	9.092	10.80	12.84	15.33	22.59	36.65	77.81
(3)	.1624	.3337	.5145	.7058	.9089	1.126	1.360	1.617	1.910	2.275	2.838	4.124	11.12	25.41	66.33

TABLE 7.22 : I_0^* (α = .30 , μ = .80)

Q: MODE:	0.05	0.10	0.15	0.20	0.25	0.30	0.35	0.40	0.45	0.50	0.55	0.60	0.70	0.80	0.90
(1)	.1487	.3088	.4816	.6689	.8728	1.096	1.342	1.616	1.924	2.275	2.683	3.166	4.506	6.946	13.77
(2)	.7788	1.605	2.433	3.269	4.129	5.036	6.013	7.087	8.290	9.664	11.27	13.19	18.60	28.64	56.98
(3)	1.749	3.679	5.684	7.772	9.975	12.34	14.91	17.77	21.01	24.75	29.18	34.56	50.10	80.08	167.9

TABLE 7.22 : I_0^* (α = .40 , μ = .20)

Q: MODE:	0.05	0.10	0.15	0.20	0.25	0.30	0.35	0.40	0.45	0.50	0.55	0.60	0.70	0.80	0.90
(1)	.1339	.2890	.4687	.6773	.9200	1.204	1.538	1.933	2.407	2.982	3.691	4.583	7.280	12.70	29.01
(2)	.1659	.3569	.5758	.8266	1.114	1.445	1.828	2.274	2.800	3.428	4.189	5.133	7.923	13.40	29.54
(3)	.3416	.7080	1.103	1.533	2.003	2.522	3.102	3.757	4.508	5.381	6.419	7.681	11.32	18.31	38.65

TABLE 7.22 : I_0^* (α = .40 , μ = .30)

Q: MODE:	0.05	0.10	0.15	0.20	0.25	0.30	0.35	0.40	0.45	0.50	0.55	0.60	0.70	0.80	0.90
(1)	.1331	.2856	.4607	.6621	.8948	1.165	1.481	1.854	2.298	2.834	3.493	4.320	6.809	11.79	26.73
(2)	.1471	.3126	.4985	.7073	.9423	1.208	1.509	1.854	2.252	2.718	3.272	3.944	5.869	9.503	19.91
(3)	.5391	1.131	1.778	2.487	3.264	4.121	5.074	6.149	7.378	8.809	10.51	12.59	18.61	30.25	64.40

TABLE 7.22 : I_0^* (α = .40 , μ = .40)

Q: MODE:	0.05	0.10	0.15	0.20	0.25	0.30	0.35	0.40	0.45	0.50	0.55	0.60	0.70	0.80	0.90
(1)	.1322	.2822	.4525	.6466	.8689	1.125	1.422	1.770	2.181	2.675	3.278	4.030	6.275	10.73	23.98
(2)	.1234	.2597	.4106	.5783	.7663	.9792	1.223	1.509	1.851	2.271	2.803	3.497	5.716	10.36	24.38
(3)	.7450	1.439	2.111	2.795	3.523	4.321	5.215	6.237	7.423	8.824	10.51	12.60	18.73	30.73	66.09

TABLE 7.22 : I_0^* (α = .40 , μ = .50)

Q: MODE:	0.05	0.10	0.15	0.20	0.25	0.30	0.35	0.40	0.45	0.50	0.55	0.60	0.70	0.80	0.90
(1)	.1314	.2787	.4442	.6309	.8425	1.084	1.361	1.683	2.060	2.509	3.052	3.724	5.704	9.575	20.96
(2)	.4881	1.007	1.565	2.173	2.841	3.585	4.424	5.384	6.498	7.814	9.401	11.36	17.15	28.56	62.47
(3)	.2588	.5567	.8977	1.287	1.731	2.239	2.821	3.493	4.274	5.193	6.290	7.622	11.43	18.51	38.30

TABLE 7.22 : I_0^* (α = .40 , μ = .60)

Q: MODE:	0.05	0.10	0.15	0.20	0.25	0.30	0.35	0.40	0.45	0.50	0.55	0.60	0.70	0.80	0.90
(1)	.1306	.2752	.4358	.6149	.8157	1.042	1.299	1.594	1.937	2.340	2.822	3.412	5.124	8.408	17.93
(2)	.4667	1.022	1.643	2.325	3.075	3.906	4.839	5.902	7.131	8.579	10.32	12.47	18.80	31.24	68.14
(3)	.2303	.4823	.7585	1.062	1.395	1.763	2.170	2.623	3.131	3.705	4.362	5.128	7.171	10.69	20.35

TABLE 7.22 : I_0^* (α = .40 , μ = .70)

Q: MODE:	0.05	0.10	0.15	0.20	0.25	0.30	0.35	0.40	0.45	0.50	0.55	0.60	0.70	0.80	0.90
(1)	.1298	.2717	.4274	.5988	.7886	1.000	1.237	1.506	1.813	2.171	2.594	3.104	4.559	7.291	15.09
(2)	.5372	1.152	1.819	2.532	3.300	4.137	5.062	6.103	7.294	8.683	10.34	12.36	18.24	29.63	62.93
(3)	.1509	.3104	.4791	.6584	.8498	1.056	1.281	1.533	1.829	2.213	2.823	4.064	9.857	21.81	56.28

TABLE 7.22 : I_0^* (α = .40 , μ = .80)

Q: MODE:	0.05	0.10	0.15	0.20	0.25	0.30	0.35	0.40	0.45	0.50	0.55	0.60	0.70	0.80	0.90
(1)	.1290	.2682	.4189	.5827	.7615	.9579	1.175	1.418	1.692	2.006	2.372	2.808	4.028	6.278	12.64
(2)	.5895	1.235	1.902	2.585	3.294	4.045	4.854	5.744	6.741	7.879	9.208	10.80	15.27	23.59	47.05
(3)	1.334	2.856	4.464	6.149	7.932	9.845	11.93	14.25	16.87	19.91	23.49	27.84	40.41	64.65	135.6

TABLE 7.22 : I_0^* (α = .50 , μ = .20)

Q: MODE:	0.05	0.10	0.15	0.20	0.25	0.30	0.35	0.40	0.45	0.50	0.55	0.60	0.70	0.80	0.90
(1)	.1195	.2573	.4164	.6003	.8136	1.062	1.354	1.698	2.110	2.609	3.223	3.995	6.324	11.00	25.05
(2)	.1551	.3338	.5390	.7741	1.044	1.354	1.712	2.129	2.619	3.202	3.908	4.781	7.347	12.35	27.04
(3)	.3206	.6682	1.047	1.462	1.921	2.432	3.005	3.656	4.405	5.280	6.320	7.587	11.25	18.26	38.62

TABLE 7.22 : I_0^* (α = .50 , μ = .30)

Q: MODE:	0.05	0.10	0.15	0.20	0.25	0.30	0.35	0.40	0.45	0.50	0.55	0.60	0.70	0.80	0.90
(1)	.1188	.2546	.4100	.5883	.7938	1.032	1.309	1.636	2.025	2.494	3.069	3.790	5.955	10.28	23.24
(2)	.1376	.2930	.4681	.6655	.8885	1.141	1.429	1.760	2.143	2.592	3.129	3.783	5.664	9.237	19.51
(3)	.4318	.9085	1.436	2.022	2.674	3.402	4.223	5.156	6.230	7.487	8.987	10.82	16.16	26.50	56.85

TABLE 7.22 : I_0^* (α = .50 , μ = .40)

Q: MODE:	0.05	0.10	0.15	0.20	0.25	0.30	0.35	0.40	0.45	0.50	0.55	0.60	0.70	0.80	0.90
(1)	.1182	.2519	.4035	.5760	.7733	1.000	1.263	1.570	1.933	2.369	2.900	3.561	5.533	9.438	21.05
(2)	.1156	.2442	.3876	.5487	.7312	.9402	1.183	1.471	1.818	2.248	2.792	3.495	5.695	10.20	23.70
(3)	.5993	1.172	1.739	2.326	2.954	3.644	4.417	5.299	6.321	7.529	8.982	10.77	16.04	26.33	56.65

TABLE 7.22 : I_0^* (α = .50 , μ = .50)

Q: MODE:	0.05	0.10	0.15	0.20	0.25	0.30	0.35	0.40	0.45	0.50	0.55	0.60	0.70	0.80	0.90
(1)	.1175	.2491	.3969	.5635	.7522	.9674	1.215	1.501	1.837	2.237	2.720	3.318	5.078	8.520	18.64
(2)	.4328	.8930	1.388	1.927	2.519	3.179	3.924	4.775	5.764	6.933	8.341	10.08	15.23	25.37	55.51
(3)	.2335	.5022	.8096	1.160	1.560	2.016	2.538	3.140	3.840	4.661	5.639	6.827	10.21	16.51	34.10

TABLE 7.22 : I_0^* (α = .50 , μ = .60)

Q: MODE:	0.05	0.10	0.15	0.20	0.25	0.30	0.35	0.40	0.45	0.50	0.55	0.60	0.70	0.80	0.90
(1)	.1169	.2463	.3902	.5507	.7307	.9339	1.165	1.430	1.738	2.101	2.535	3.067	4.613	7.585	16.21
(2)	.3758	.8328	1.355	1.935	2.578	3.293	4.097	5.014	6.075	7.324	8.827	10.68	16.14	26.86	58.65
(3)	.2145	.4487	.7049	.9861	1.295	1.637	2.015	2.438	2.912	3.451	4.070	4.797	6.761	10.23	19.95

TABLE 7.22 : I_0^* (α = .50 , μ = .70)

Q: MODE:	0.05	0.10	0.15	0.20	0.25	0.30	0.35	0.40	0.45	0.50	0.55	0.60	0.70	0.80	0.90
(1)	.1162	.2435	.3834	.5377	.7089	.8999	1.115	1.358	1.638	1.964	2.350	2.818	4.156	6.682	13.93
(2)	.4320	.9362	1.494	2.099	2.755	3.473	4.269	5.165	6.191	7.388	8.813	10.55	15.61	25.39	53.99
(3)	.1410	.2902	.4487	.6177	.7992	.9962	1.214	1.462	1.762	2.165	2.814	4.032	9.168	19.74	50.35

TABLE 7.22 : I_0^* (α = .50 , μ = .80)

Q: MODE:	0.05	0.10	0.15	0.20	0.25	0.30	0.35	0.40	0.45	0.50	0.55	0.60	0.70	0.80	0.90
(1)	.1155	.2407	.3765	.5247	.6869	.8657	1.064	1.287	1.539	1.829	2.169	2.576	3.722	5.857	11.95
(2)	.4744	1.006	1.568	2.154	2.767	3.420	4.125	4.902	5.773	6.768	7.929	9.319	13.23	20.51	41.05
(3)	1.081	2.351	3.715	5.156	6.685	8.329	10.12	12.12	14.37	16.98	20.06	23.80	34.59	55.38	116.2

TABLE 7.22 : I_0^* (α = .60 , μ = .20)

Q: MODE:	0.05	0.10	0.15	0.20	0.25	0.30	0.35	0.40	0.45	0.50	0.55	0.60	0.70	0.80	0.90
(1)	.1089	.2343	.3787	.5453	.7382	.9626	1.225	1.536	1.906	2.354	2.904	3.596	5.680	9.859	22.40
(2)	.1456	.3137	.5071	.7290	.9836	1.277	1.615	2.009	2.472	3.023	3.690	4.512	6.925	11.61	25.33
(3)	.3023	.6336	.9985	1.402	1.052	2.357	2.926	3.576	4.325	5.202	6.246	7.519	11.19	18.22	38.60

TABLE 7.22 : I_0^* (α = .60 , μ = .30)

Q: MODE:	0.05	0.10	0.15	0.20	0.25	0.30	0.35	0.40	0.45	0.50	0.55	0.60	0.70	0.80	0.90
(1)	.1084	.2321	.3734	.5355	.7219	.9377	1.189	1.485	1.836	2.259	2.778	3.428	5.377	9.268	20.91
(2)	.1293	.2758	.4417	.6295	.8424	1.085	1.362	1.681	2.053	2.491	3.016	3.657	5.511	9.046	19.23
(3)	.3601	.7594	1.205	1.706	2.271	2.912	3.642	4.480	5.453	6.598	7.969	9.649	14.55	24.04	51.87

TABLE 7.22 : I_0^* (α = .60 , μ = .40)

Q: MODE:	0.05	0.10	0.15	0.20	0.25	0.30	0.35	0.40	0.45	0.50	0.55	0.60	0.70	0.80	0.90
(1)	.1078	.2298	.3680	.5253	.7050	.9115	1.151	1.430	1.760	2.156	2.638	3.239	5.028	8.568	19.09
(2)	.1088	.2306	.3677	.5232	.7012	.9075	1.150	1.440	1.793	2.231	2.784	3.494	5.682	10.10	23.25
(3)	.5012	.9873	1.478	1.991	2.545	3.157	3.844	4.629	5.540	6.616	7.912	9.509	14.20	23.36	50.33

TABLE 7.22 : I_0^* (α = .60 , μ = .50)

Q: MODE:	0.05	0.10	0.15	0.20	0.25	0.30	0.35	0.40	0.45	0.50	0.55	0.60	0.70	0.80	0.90
(1)	.1073	.2275	.3625	.5149	.6875	.8843	1.110	1.373	1.680	2.046	2.489	3.037	4.650	7.805	17.08
(2)	.3974	.8200	1.275	1.769	2.313	2.919	3.602	4.384	5.291	6.363	7.655	9.252	13.97	23.27	50.90
(3)	.2119	.4558	.7350	1.054	1.417	1.832	2.308	2.857	3.495	4.245	5.140	6.227	9.327	15.10	31.22

TABLE 7.22 : I_0^* (α = .60 , μ = .60)

Q: MODE:	0.05	0.10	0.15	0.20	0.25	0.30	0.35	0.40	0.45	0.50	0.55	0.60	0.70	0.80	0.90
(1)	.1068	.2252	.3569	.5042	.6695	.8563	1.069	1.313	1.598	1.933	2.334	2.827	4.261	7.022	15.05
(2)	.3146	.7045	1.158	1.670	2.240	2.878	3.596	4.416	5.366	6.484	7.829	9.487	14.37	23.95	52.34
(3)	.2008	.4193	.6583	.9205	1.209	1.529	1.884	2.281	2.729	3.240	3.831	4.529	6.439	9.879	19.67

TABLE 7.22 : I_0^* (α = .60 , μ = .70)

Q: MODE:	0.05	0.10	0.15	0.20	0.25	0.30	0.35	0.40	0.45	0.50	0.55	0.60	0.70	0.80	0.90
(1)	.1062	.2220	.3513	.4933	.6511	.8277	1.027	1.253	1.513	1.817	2.178	2.617	3.875	6.264	13.14
(2)	.3613	.7894	1.271	1.801	2.380	3.018	3.726	4.525	5.441	6.509	7.781	9.334	13.85	22.56	48.02
(3)	.1323	.2726	.4220	.5821	.7551	.9445	1.156	1.402	1.707	2.127	2.807	4.013	8.741	18.41	46.45

TABLE 7.22 : I_0^* (α = .60 , μ = .80)

Q: MODE:	0.05	0.10	0.15	0.20	0.25	0.30	0.35	0.40	0.45	0.50	0.55	0.60	0.70	0.80	0.90
(1)	.1056	.2204	.3455	.4823	.6325	.7987	.9838	1.192	1.429	1.702	2.024	2.410	3.507	5.565	11.47
(2)	.3970	.8500	1.338	1.854	2.400	2.984	3.618	4.318	5.103	6.001	7.049	8.305	11.84	18.42	37.02
(3)	.9094	2.006	3.205	4.480	5.839	7.304	8.904	10.68	12.69	15.01	17.76	21.09	30.70	49.19	103.2

TABLE 7.22 : I_0^* (α = .70 , μ = .20)

Q: MODE:	0.05	0.10	0.15	0.20	0.25	0.30	0.35	0.40	0.45	0.50	0.55	0.60	0.70	0.80	0.90
(1)	.1008	.2167	.3500	.5037	.6815	.8881	1.130	1.415	1.755	2.166	2.671	3.305	5.214	9.038	20.51
(2)	.1372	.2961	.4791	.6896	.9316	1.210	1.533	1.909	2.351	2.877	3.513	4.298	6.598	11.06	24.08
(3)	.2861	.6031	.9559	1.350	1.793	2.293	2.860	3.510	4.261	5.141	6.189	7.466	11.15	18.19	38.58

TABLE 7.22 : I_0^* (α = .70 , μ = .30)

Q: MODE:	0.05	0.10	0.15	0.20	0.25	0.30	0.35	0.40	0.45	0.50	0.55	0.60	0.70	0.80	0.90
(1)	.1003	.2148	.3455	.4953	.6677	.8670	1.099	1.372	1.696	2.087	2.565	3.163	4.957	8.536	19.23
(2)	.1219	.2607	.4185	.5979	.8022	1.036	1.304	1.615	1.978	2.407	2.923	3.556	5.391	8.902	19.03
(3)	.3088	.6524	1.039	1.477	1.978	2.553	3.217	3.988	4.890	5.957	7.240	8.814	13.41	22.30	48.34

TABLE 7.22 : I_0^* (α = .70 , μ = .40)

Q: MODE:	0.05	0.10	0.15	0.20	0.25	0.30	0.35	0.40	0.45	0.50	0.55	0.60	0.70	0.80	0.90
(1)	.0999	.2128	.3409	.4867	.6532	.8447	1.067	1.326	1.632	1.999	2.446	3.003	4.660	7.940	17.68
(2)	.1027	.2186	.3502	.5009	.6752	.8794	1.122	1.415	1.773	2.219	2.778	3.493	5.673	10.03	22.94
(3)	.4306	.8529	1.284	1.741	2.237	2.788	3.410	4.123	4.952	5.933	7.115	8.574	12.86	21.22	45.81

TABLE 7.22 : I_0^* (α = .70 , μ = .50)

Q: MODE:	0.05	0.10	0.15	0.20	0.25	0.30	0.35	0.40	0.45	0.50	0.55	0.60	0.70	0.80	0.90
(1)	.0994	.2109	.3362	.4777	.6382	.8214	1.032	1.277	1.563	1.905	2.318	2.830	4.337	7.286	15.96
(2)	.3730	.7698	1.196	1.660	2.171	2.739	3.380	4.113	4.963	5.967	7.178	8.673	13.09	21.79	47.63
(3)	.1935	.4163	.6716	.9634	1.297	1.678	2.117	2.623	3.214	3.909	4.740	5.752	8.645	14.04	29.12

TABLE 7.22 : I_0^* (α = .70 , μ = .60)

Q: MODE:	0.05	0.10	0.15	0.20	0.25	0.30	0.35	0.40	0.45	0.50	0.55	0.60	0.70	0.80	0.90
(1)	.0989	.2089	.3314	.4685	.6228	.7973	.9963	1.225	1.492	1.807	2.185	2.649	4.002	6.613	14.22
(2)	.2707	.6113	1.015	1.476	1.994	2.576	3.234	3.986	4.856	5.882	7.115	8.634	13.10	21.87	47.84
(3)	.1886	.3934	.6173	.8632	1.134	1.435	1.770	2.146	2.572	3.061	3.630	4.305	6.177	9.604	19.46

TABLE 7.22 : I_0^* (α = .70 , μ = .70)

Q: MODE:	0.05	0.10	0.15	0.20	0.25	0.30	0.35	0.40	0.45	0.50	0.55	0.60	0.70	0.80	0.90
(1)	.0985	.2068	.3265	.4591	.6069	.7725	.9596	1.173	1.419	1.707	2.049	2.466	3.668	5.957	12.57
(2)	.3105	.6830	1.109	1.583	2.105	2.684	3.329	4.058	4.894	5.870	7.033	8.452	12.57	20.53	43.74
(3)	.1246	.2570	.3985	.5507	.7162	.8990	1.106	1.350	1.659	2.096	2.801	3.999	8.455	17.49	43.70

TABLE 7.22 : I_0^* (α = .70 , μ = .80)

Q: MODE:	0.05	0.10	0.15	0.20	0.25	0.30	0.35	0.40	0.45	0.50	0.55	0.60	0.70	0.80	0.90
(1)	.0980	.2048	.3215	.4496	.5907	.7473	.9222	1.120	1.345	1.606	1.914	2.286	3.346	5.350	11.13
(2)	.3413	.7363	1.169	1.633	2.128	2.661	3.242	3.885	4.608	5.435	6.403	7.562	10.83	16.91	34.12
(3)	.7855	1.755	2.832	3.988	5.226	6.561	8.022	9.645	11.48	13.60	16.11	19.15	27.91	44.77	94.01

TABLE 7.22 : I_0^* (α = .80 , μ = .20)

Q: MODE:	0.05	0.10	0.15	0.20	0.25	0.30	0.35	0.40	0.45	0.50	0.55	0.60	0.70	0.80	0.90
(1)	.0943	.2026	.3272	.4709	.6369	.8299	1.056	1.322	1.639	2.022	2.493	3.084	4.861	8.418	19.08
(2)	.1297	.2804	.4544	.6549	.8860	1.153	1.462	1.823	2.248	2.754	3.367	4.123	6.337	10.63	23.13
(3)	.2717	.5760	.9182	1.304	1.742	2.238	2.804	3.454	4.208	5.091	6.143	7.425	11.12	18.17	38.57

TABLE 7.22 : I_0^* (α = .80 , μ = .30)

Q: MODE:	0.05	0.10	0.15	0.20	0.25	0.30	0.35	0.40	0.45	0.50	0.55	0.60	0.70	0.80	0.90
(1)	.0939	.2010	.3233	.4636	.6250	.8116	1.029	1.285	1.588	1.954	2.401	2.961	4.639	7.983	17.98
(2)	.1154	.2473	.3979	.5699	.7668	.9929	1.254	1.557	1.913	2.336	2.845	3.472	5.294	8.790	18.88
(3)	.2703	.5719	.9128	1.303	1.754	2.279	2.892	3.613	4.462	5.473	6.692	8.190	12.56	21.01	45.72

TABLE 7.22 : I_0^* (α = .80 , μ = .40)

Q: MODE:	0.05	0.10	0.15	0.20	0.25	0.30	0.35	0.40	0.45	0.50	0.55	0.60	0.70	0.80	0.90
(1)	.0935	.1993	.3193	.4560	.6124	.7922	1.001	1.244	1.532	1.877	2.298	2.821	4.380	7.463	16.62
(2)	.0973	.2079	.3346	.4810	.6522	.8549	1.098	1.394	1.757	2.208	2.774	3.493	5.666	9.976	22.71
(3)	.3775	.7506	1.136	1.547	1.997	2.499	3.069	3.724	4.490	5.398	6.495	7.849	11.83	19.60	42.41

TABLE 7.22 : I_0^* (α = .80 , μ = .50)

Q: MODE:	0.05	0.10	0.15	0.20	0.25	0.30	0.35	0.40	0.45	0.50	0.55	0.60	0.70	0.80	0.90
(1)	.0930	.1975	.3152	.4482	.5993	.7718	.9704	1.201	1.472	1.795	2.186	2.670	4.097	6.891	15.12
(2)	.3554	.7334	1.140	1.582	2.068	2.609	3.219	3.915	4.724	5.678	6.828	8.248	12.44	20.69	45.19
(3)	.1777	.3825	.6174	.8864	1.194	1.548	1.955	2.427	2.978	3.630	4.411	5.364	8.099	13.21	27.51

TABLE 7.22 : I_0^* (α = .80 , μ = .60)

Q: MODE:	0.05	0.10	0.15	0.20	0.25	0.30	0.35	0.40	0.45	0.50	0.55	0.60	0.70	0.80	0.90
(1)	.0926	.1958	.3110	.4401	.5856	.7506	.9390	1.156	1.409	1.709	2.068	2.511	3.802	6.300	13.58
(2)	.2375	.5400	.9053	1.327	1.806	2.347	2.959	3.659	4.471	5.428	6.578	7.994	12.16	20.32	44.47
(3)	.1778	.3705	.5811	.8125	1.068	1.352	1.670	2.028	2.436	2.906	3.457	4.115	5.958	9.381	19.30

TABLE 7.22 : I_0^* (α = .80 , μ = .70)

Q: MODE:	0.05	0.10	0.15	0.20	0.25	0.30	0.35	0.40	0.45	0.50	0.55	0.60	0.70	0.80	0.90
(1)	.0922	.1940	.3067	.4318	.5716	.7287	.9065	1.110	1.345	1.620	1.948	2.349	3.507	5.722	12.14
(2)	.2723	.6022	.9843	1.415	1.894	2.427	3.024	3.700	4.476	5.383	6.463	7.783	11.61	19.00	40.53
(3)	.1177	.2432	.3776	.5229	.6817	.8586	1.061	1.304	1.618	2.070	2.797	3.990	8.250	16.82	41.67

TABLE 7.22 : I_0^* (α = .80 , μ = .80)

Q: MODE:	0.05	0.10	0.15	0.20	0.25	0.30	0.35	0.40	0.45	0.50	0.55	0.60	0.70	0.80	0.90
(1)	.0918	.1922	.3023	.4234	.5573	.7063	.8733	1.062	1.279	1.530	1.828	2.189	3.222	5.184	10.87
(2)	.2994	.6496	1.039	1.462	1.917	2.411	2.951	3.550	4.225	4.998	5.904	6.991	10.06	15.77	31.92
(3)	.6917	1.563	2.547	3.613	4.758	5.997	7.353	8.861	10.57	12.54	14.86	17.69	25.82	41.45	87.09

TABLE 7.22 : I_0^* (α = .90 , μ = .20)

Q: MODE:	0.05	0.10	0.15	0.20	0.25	0.30	0.35	0.40	0.45	0.50	0.55	0.60	0.70	0.80	0.90
(1)	.0889	.1911	.3086	.4441	.6008	.7829	.9958	1.247	1.546	1.908	2.352	2.909	4.583	7.934	17.97
(2)	.1230	.2663	.4323	.6240	.8456	1.102	1.400	1.749	2.159	2.649	3.243	3.975	6.123	10.28	22.38
(3)	.2588	.5516	.8844	1.263	1.696	2.190	2.756	3.407	4.163	5.049	6.106	7.392	11.09	18.15	38.56

TABLE 7.22 : I_0^* (α = .90 , μ = .30)

Q: MODE:	0.05	0.10	0.15	0.20	0.25	0.30	0.35	0.40	0.45	0.50	0.55	0.60	0.70	0.80	0.90
(1)	.0885	.1896	.3052	.4377	.5903	.7668	.9725	1.214	1.502	1.847	2.271	2.801	4.388	7.550	16.99
(2)	.1095	.2352	.3794	.5449	.7352	.9548	1.210	1.507	1.857	2.275	2.779	3.401	5.215	8.699	18.76
(3)	.2404	.5090	.8142	1.166	1.577	2.061	2.635	3.315	4.125	5.093	6.265	7.706	11.91	20.02	43.70

TABLE 7.22 : I_0^* (α = .90 , μ = .40)

Q: MODE:	0.05	0.10	0.15	0.20	0.25	0.30	0.35	0.40	0.45	0.50	0.55	0.60	0.70	0.80	0.90
(1)	.0882	.1881	.3016	.4310	.5791	.7496	.9473	1.179	1.452	1.780	2.179	2.677	4.159	7.090	15.79
(2)	.0925	.1984	.3206	.4633	.6317	.8332	1.077	1.376	1.743	2.200	2.770	3.492	5.661	9.936	22.53
(3)	.3360	.6701	1.018	1.391	1.803	2.265	2.792	3.402	4.116	4.966	5.996	7.269	11.02	18.33	39.76

TABLE 7.22 : I_0^* (α = .90 , μ = .50)

Q: MODE:	0.05	0.10	0.15	0.20	0.25	0.30	0.35	0.40	0.45	0.50	0.55	0.60	0.70	0.80	0.90
(1)	.0878	.1866	.2980	.4240	.5674	.7315	.9205	1.140	1.399	1.707	2.080	2.542	3.907	6.581	14.45
(2)	.3422	.7060	1.097	1.522	1.990	2.510	3.096	3.766	4.542	5.458	6.562	7.924	11.94	19.85	43.30
(3)	.1641	.3534	.5708	.8201	1.106	1.436	1.816	2.259	2.778	3.394	4.134	5.040	7.649	12.54	26.24

TABLE 7.22 : I_0^* (α = .90 , μ = .60)

Q: MODE:	0.05	0.10	0.15	0.20	0.25	0.30	0.35	0.40	0.45	0.50	0.55	0.60	0.70	0.80	0.90
(1)	.0874	.1850	.2942	.4168	.5553	.7125	.8923	1.100	1.342	1.629	1.975	2.400	3.644	6.053	13.09
(2)	.2116	.4845	.8185	1.210	1.658	2.165	2.742	3.403	4.170	5.074	6.160	7.496	11.42	19.12	41.85
(3)	.1682	.3501	.5488	.7675	1.010	1.279	1.582	1.925	2.317	2.771	3.306	3.949	5.772	9.196	19.17

TABLE 7.22 : I_0^* (α = .90 , μ = .70)

Q: MODE:	0.05	0.10	0.15	0.20	0.25	0.30	0.35	0.40	0.45	0.50	0.55	0.60	0.70	0.80	0.90
(1)	.0871	.1834	.2903	.4094	.5427	.6928	.8631	1.058	1.284	1.550	1.867	2.255	3.379	5.536	11.80
(2)	.2424	.5386	.8858	1.281	1.725	2.222	2.781	3.416	4.145	4.998	6.015	7.256	10.86	17.80	38.03
(3)	.1116	.2308	.3589	.4978	.6507	.8223	1.021	1.263	1.582	2.048	2.794	3.982	8.097	16.31	40.10

TABLE 7.22 : I_0^* (α = .90 , μ = .80)

Q: MODE:	0.05	0.10	0.15	0.20	0.25	0.30	0.35	0.40	0.45	0.50	0.55	0.60	0.70	0.80	0.90
(1)	.0867	.1818	.2864	.4018	.5298	.6726	.8332	1.015	1.225	1.469	1.759	2.110	3.122	5.053	10.66
(2)	.2666	.5814	.9359	1.325	1.748	2.209	2.717	3.281	3.918	4.649	5.507	6.536	9.443	14.86	30.21
(3)	.6180	1.411	2.321	3.315	4.388	5.552	6.827	8.246	9.851	11.70	13.89	16.54	24.19	38.86	81.70

TABLE 7.22 : I_0^* (α = 1.0 , μ = .20)

Q: MODE:	0.05	0.10	0.15	0.20	0.25	0.30	0.35	0.40	0.45	0.50	0.55	0.60	0.70	0.80	0.90
(1)	.0843	.1814	.2931	.4219	.5708	.7440	.9466	1.186	1.470	1.814	2.237	2.767	4.359	7.545	17.08
(2)	.1170	.2537	.4124	.5963	.8094	1.057	1.345	1.683	2.082	2.558	3.136	3.849	5.942	9.992	21.76
(3)	.2471	.5296	.8539	1.227	1.655	2.147	2.714	3.367	4.125	5.014	6.074	7.364	11.07	18.14	38.56

TABLE 7.22 : I_0^* (α = 1.0 , μ = .30)

Q: MODE:	0.05	0.10	0.15	0.20	0.25	0.30	0.35	0.40	0.45	0.50	0.55	0.60	0.70	0.80	0.90
(1)	.0840	.1801	.2900	.4161	.5614	.7296	.9258	1.156	1.430	1.761	2.165	2.670	4.185	7.201	16.20
(2)	.1042	.2244	.3628	.5224	.7068	.9207	1.170	1.462	1.808	2.221	2.722	3.340	5.148	8.624	18.66
(3)	.2164	.4586	.7349	1.056	1.434	1.884	2.425	3.073	3.851	4.787	5.922	7.320	11.39	19.23	42.09

TABLE 7.22 : I_0^* (α = 1.0 , μ = .40)

Q: MODE:	0.05	0.10	0.15	0.20	0.25	0.30	0.35	0.40	0.45	0.50	0.55	0.60	0.70	0.80	0.90
(1)	.0837	.1787	.2868	.4101	.5514	.7142	.9031	1.124	1.386	1.700	2.083	2.560	3.980	6.788	15.12
(2)	.0881	.1897	.3080	.4472	.6133	.8139	1.059	1.360	1.732	2.193	2.768	3.492	5.657	9.905	22.39
(3)	.3027	.6052	.9217	1.264	1.644	2.072	2.564	3.134	3.806	4.609	5.584	6.792	10.36	17.30	37.63

TABLE 7.22 : I_0^* ($\alpha = 1.0$, $\mu = .50$)

Q: MODE:	0.05	0.10	0.15	0.20	0.25	0.30	0.35	0.40	0.45	0.50	0.55	0.60	0.70	0.80	0.90
(1)	.0834	.1773	.2835	.4038	.5409	.6978	.8790	1.090	1.338	1.634	1.993	2.438	3.753	6.330	13.92
(2)	.3318	.6846	1.064	1.476	1.929	2.433	3.001	3.649	4.400	5.286	6.353	7.669	11.55	19.18	41.80
(3)	.1523	.3281	.5303	.7626	1.030	1.338	1.696	2.114	2.605	3.190	3.896	4.763	7.270	11.99	25.20

TABLE 7.22 : I_0^* ($\alpha = 1.0$, $\mu = .60$)

Q: MODE:	0.05	0.10	0.15	0.20	0.25	0.30	0.35	0.40	0.45	0.50	0.55	0.60	0.70	0.80	0.90
(1)	.0830	.1759	.2801	.3973	.5299	.6807	.8535	1.053	1.287	1.564	1.898	2.309	3.515	5.853	12.69
(2)	.1908	.4394	.7479	1.114	1.537	2.018	2.567	3.197	3.928	4.789	5.824	7.098	10.84	18.15	39.76
(3)	.1595	.3318	.5199	.7272	.9571	1.214	1.503	1.832	2.210	2.651	3.172	3.804	5.610	9.041	19.06

TABLE 7.22 : I_0^* ($\alpha = 1.0$, $\mu = .70$)

Q: MODE:	0.05	0.10	0.15	0.20	0.25	0.30	0.35	0.40	0.45	0.50	0.55	0.60	0.70	0.80	0.90
(1)	.0827	.1745	.2766	.3906	.5185	.6628	.8270	1.015	1.234	1.491	1.800	2.177	3.274	5.384	11.52
(2)	.2185	.4873	.8058	1.172	1.588	2.055	2.583	3.184	3.876	4.685	5.651	6.830	10.25	16.84	36.03
(3)	.1061	.2196	.3420	.4753	.6227	.7896	.9851	1.227	1.550	2.029	2.792	3.977	7.978	15.91	38.86

TABLE 7.22 : I_0^* ($\alpha = 1.0$, $\mu = .80$)

Q: MODE:	0.05	0.10	0.15	0.20	0.25	0.30	0.35	0.40	0.45	0.50	0.55	0.60	0.70	0.80	0.90
(1)	.0824	.1730	.2730	.3836	.5067	.6444	.7996	.9763	1.180	1.418	1.701	2.045	3.040	4.946	10.50
(2)	.2403	.5262	.8518	1.213	1.609	2.044	2.524	3.059	3.666	4.363	5.181	6.165	8.945	14.13	28.83
(3)	.5587	1.288	2.137	3.073	4.088	5.192	6.402	7.749	9.274	11.03	13.11	15.63	22.88	36.79	77.39

TABLE 7.22 : I_0^* ($1/\alpha = .10$, $\mu = .20$)

Q: MODE:	0.05	0.10	0.15	0.20	0.25	0.30	0.35	0.40	0.45	0.50	0.55	0.60	0.70	0.80	0.90
(1)	.0284	.0650	.1114	.1695	.2413	.3295	.4371	.5683	.7284	.9252	1.170	1.478	2.409	4.262	9.764
(2)	.0220	.0496	.0852	.1321	.1958	.2849	.4122	.5953	.8535	1.203	1.659	2.240	3.953	7.198	16.40
(3)	.0512	.1316	.2723	.5286	.9401	1.484	2.128	2.860	3.692	4.646	5.763	7.103	10.89	18.03	38.50

TABLE 7.22 : I_0^* ($1/\alpha = .10$, $\mu = .30$)

Q: MODE:	0.05	0.10	0.15	0.20	0.25	0.30	0.35	0.40	0.45	0.50	0.55	0.60	0.70	0.80	0.90
(1)	.0284	.0649	.1111	.1689	.2405	.3282	.4352	.5657	.7249	.9204	1.163	1.470	2.393	4.230	9.680
(2)	.0198	.0454	.0793	.1265	.1949	.2986	.4580	.6941	1.019	1.438	1.958	2.600	4.450	7.929	17.82
(3)	.0217	.0463	.0756	.1123	.1629	.2420	.3910	.7132	1.306	2.116	3.093	4.256	7.461	13.29	29.63

TABLE 7.22 : I_0^* ($1/\alpha = .10$, $\mu = .40$)

Q: MODE:	0.05	0.10	0.15	0.20	0.25	0.30	0.35	0.40	0.45	0.50	0.55	0.60	0.70	0.80	0.90
(1)	.0283	.0647	.1108	.1683	.2394	.3266	.4329	.5623	.7203	.9142	1.155	1.458	2.373	4.189	9.573
(2)	.0173	.0413	.0783	.1424	.2643	.4787	.7839	1.161	1.605	2.125	2.742	3.489	5.625	9.668	21.30
(3)	.0305	.0622	.0971	.1372	.1851	.2443	.3207	.4242	.5726	.7999	1.167	1.764	3.960	8.284	20.05

TABLE 7.22 : I_0^* ($1/\alpha = .10$, $\mu = .50$)

Q: MODE:	0.05	0.10	0.15	0.20	0.25	0.30	0.35	0.40	0.45	0.50	0.55	0.60	0.70	0.80	0.90
(1)	.0283	.0646	.1104	.1676	.2382	.3247	.4301	.5584	.7148	.9068	1.145	1.445	2.348	4.142	9.450
(2)	.2599	.5359	.8318	1.152	1.503	1.891	2.325	2.816	3.382	4.044	4.835	5.801	8.614	14.06	30.01
(3)	.0195	.0422	.0690	.1008	.1392	.1864	.2459	.3230	.4272	.5748	.7962	1.148	2.647	6.142	15.77

TABLE 7.22 : I_0^* ($1/\alpha = .10$, $\mu = .60$)

Q: MODE:	0.05	0.10	0.15	0.20	0.25	0.30	0.35	0.40	0.45	0.50	0.55	0.60	0.70	0.80	0.90
(1)	.0283	.0644	.1100	.1668	.2368	.3225	.4269	.5538	.7085	.8983	1.134	1.430	2.321	4.090	9.323
(2)	.0194	.0481	.0957	.1855	.3566	.6248	.9694	1.377	1.851	2.403	3.058	3.853	6.133	10.47	23.00
(3)	.0280	.0574	.0899	.1270	.1709	.2244	.2922	.3817	.5065	.6915	.9841	1.465	3.378	7.344	18.13

TABLE 7.22 : I_0^* ($1/\alpha = .10$, $\mu = .70$)

Q: MODE:	0.05	0.10	0.15	0.20	0.25	0.30	0.35	0.40	0.45	0.50	0.55	0.60	0.70	0.80	0.90
(1)	.0282	.0642	.1095	.1659	.2353	.3201	.4233	.5487	.7015	.8889	1.121	1.414	2.293	4.038	9.199
(2)	.0221	.0515	.0920	.1508	.2403	.3798	.5912	.8874	1.270	1.741	2.313	3.012	5.018	8.800	19.60
(3)	.0196	.0416	.0671	.0983	.1391	.1989	.3032	.5286	1.021	1.795	2.768	3.934	7.147	12.98	29.32

TABLE 7.22 : I_0^* ($1/\alpha = .10$, $\mu = .80$)

Q: MODE:	0.05	0.10	0.15	0.20	0.25	0.30	0.35	0.40	0.45	0.50	0.55	0.60	0.70	0.80	0.90
(1)	.0282	.0641	.1091	.1649	.2336	.3174	.4193	.5431	.6939	.8788	1.108	1.397	2.265	3.988	9.091
(2)	.0244	.0556	.0967	.1522	.2292	.3390	.4967	.7196	1.022	1.414	1.907	2.522	4.305	7.658	17.14
(3)	.0586	.1599	.3553	.7132	1.232	1.861	2.575	3.374	4.277	5.313	6.524	7.979	12.11	19.92	42.40

TABLE 7.22 : I_0^* ($1/\alpha = .20$, $\mu = .20$)

Q: MODE:	0.05	0.10	0.15	0.20	0.25	0.30	0.35	0.40	0.45	0.50	0.55	0.60	0.70	0.80	0.90
(1)	.0389	.0863	.1436	.2128	.2958	.3955	.5152	.6594	.8340	1.047	1.312	1.644	2.647	4.647	10.59
(2)	.0399	.0893	.1509	.2287	.3284	.4573	.6233	.8441	1.127	1.489	1.940	2.531	4.259	7.573	17.05
(3)	.0907	.2205	.4142	.7031	1.106	1.620	2.233	2.942	3.756	4.697	5.804	7.135	10.92	18.04	38.51

TABLE 7.22 : I_0^* ($1/\alpha = .20$, $\mu = .30$)

Q: MODE:	0.05	0.10	0.15	0.20	0.25	0.30	0.35	0.40	0.45	0.50	0.55	0.60	0.70	0.80	0.90
(1)	.0388	.0860	.1430	.2117	.2941	.3929	.5114	.6541	.8268	1.038	1.299	1.627	2.615	4.582	10.43
(2)	.0360	.0809	.1383	.2131	.3128	.4475	.6300	.8736	1.190	1.592	2.094	2.721	4.547	8.016	17.92
(3)	.0434	.0925	.1506	.2227	.3190	.4600	.6883	1.072	1.663	2.449	3.416	4.587	7.865	13.90	30.94

TABLE 7.22 : I_0^* ($1/\alpha = .20$, $\mu = .40$)

Q: MODE:	0.05	0.10	0.15	0.20	0.25	0.30	0.35	0.40	0.45	0.50	0.55	0.60	0.70	0.80	0.90
(1)	.0388	.0858	.1424	.2105	.2920	.3897	.5068	.6476	.8178	1.025	1.282	1.605	2.575	4.500	10.21
(2)	.0311	.0722	.1301	.2167	.3515	.5556	.8373	1.193	1.623	2.134	2.745	3.489	5.628	9.693	21.42
(3)	.0610	.1240	.1930	.2718	.3648	.4780	.6203	.8056	1.055	1.403	1.895	2.592	4.887	9.411	22.06

TABLE 7.22 : I_0^* ($1/\alpha = .20$, $\mu = .50$)

Q: MODE:	0.05	0.10	0.15	0.20	0.25	0.30	0.35	0.40	0.45	0.50	0.55	0.60	0.70	0.80	0.90
(1)	.0387	.0855	.1417	.2091	.2898	.3862	.5015	.6401	.8074	1.011	1.263	1.579	2.527	4.406	9.967
(2)	.2667	.5500	.8539	1.183	1.544	1.943	2.390	2.897	3.482	4.166	4.985	5.988	8.193	14.59	31.28
(3)	.0380	.0822	.1340	.1953	.2687	.3577	.4679	.6073	.7884	1.031	1.366	1.840	3.504	7.041	16.97

TABLE 7.22 : I_0^* ($1/\alpha = .20$, $\mu = .60$)

Q: MODE:	0.05	0.10	0.15	0.20	0.25	0.30	0.35	0.40	0.45	0.50	0.55	0.60	0.70	0.80	0.90
(1)	.0386	.0852	.1409	.2076	.2872	.3822	.4956	.6316	.7956	.9951	1.241	1.550	2.475	4.305	9.713
(2)	.0388	.0948	.1819	.3222	.5362	.8265	1.185	1.608	2.102	2.682	3.374	4.216	6.649	11.31	24.84
(3)	.0518	.1064	.1665	.2346	.3137	.4082	.5238	.6698	.8602	1.118	1.478	1.994	3.806	7.592	18.24

TABLE 7.22 : I_0^* ($1/\alpha$ = .20 , μ = .70)

Q: MODE:	0.05	0.10	0.15	0.20	0.25	0.30	0.35	0.40	0.45	0.50	0.55	0.60	0.70	0.80	0.90
(1)	.0386	.0848	.1401	.2060	.2845	.3778	.4891	.6222	.7826	.9776	1.218	1.519	2.420	4.202	9.469
(2)	.0442	.1022	.1801	.2872	.4354	.6375	.9039	1.241	1.655	2.156	2.763	3.507	5.655	9.736	21.45
(3)	.0358	.0755	.1208	.1743	.2412	.3314	.4677	.7031	1.135	1.832	2.771	3.939	7.228	13.27	30.32

TABLE 7.22 : I_0^* ($1/\alpha$ = .20 , μ = .80)

Q: MODE:	0.05	0.10	0.15	0.20	0.25	0.30	0.35	0.40	0.45	0.50	0.55	0.60	0.70	0.80	0.90
(1)	.0385	.0845	.1393	.2043	.2815	.3730	.4820	.6122	.7687	.9589	1.193	1.487	2.365	4.105	9.253
(2)	.0486	.1105	.1898	.2926	.4264	.6002	.8233	1.105	1.455	1.883	2.409	3.057	4.931	8.468	18.52
(3)	.1164	.3064	.6174	1.072	1.649	2.321	3.078	3.928	4.891	5.999	7.301	8.868	13.34	21.81	46.29

TABLE 7.22 : I_0^* ($1/\alpha$ = .30 , μ = .20)

Q: MODE:	0.05	0.10	0.15	0.20	0.25	0.30	0.35	0.40	0.45	0.50	0.55	0.60	0.70	0.80	0.90
(1)	.0470	.1030	.1694	.2480	.3411	.4515	.5829	.7399	.9290	1.159	1.443	1.801	2.877	5.024	11.42
(2)	.0550	.1219	.2038	.3046	.4293	.5847	.7787	1.021	1.324	1.701	2.173	2.768	4.529	7.924	17.68
(3)	.1226	.2874	.5137	.8219	1.225	1.726	2.322	3.015	3.815	4.745	5.843	7.167	10.94	18.05	38.52

TABLE 7.22 : I_0^* ($1/\alpha$ = .30 , μ = .30)

Q: MODE:	0.05	0.10	0.15	0.20	0.25	0.30	0.35	0.40	0.45	0.50	0.55	0.60	0.70	0.80	0.90
(1)	.0469	.1026	.1685	.2464	.3384	.4475	.5770	.7318	.9179	1.144	1.423	1.774	2.828	4.926	11.16
(2)	.0494	.1099	.1849	.2791	.3989	.5523	.7492	1.000	1.318	1.714	2.208	2.826	4.638	8.100	18.01
(3)	.0650	.1387	.2251	.3314	.4698	.6628	.9483	1.376	1.980	2.765	3.736	4.921	9.280	14.53	32.28

TABLE 7.22 : I_0^* ($1/\alpha$ = .30 , μ = .40)

Q: MODE:	0.05	0.10	0.15	0.20	0.25	0.30	0.35	0.40	0.45	0.50	0.55	0.60	0.70	0.80	0.90
(1)	.0469	.1022	.1675	.2446	.3354	.4428	.5702	.7221	.9046	1.126	1.399	1.741	2.767	4.803	10.84
(2)	.0425	.0967	.1688	.2688	.4110	.6112	.8803	1.222	1.640	2.142	2.748	3.490	5.632	9.719	21.54
(3)	.0914	.1855	.2879	.4039	.5394	.7021	.9026	1.156	1.484	1.917	2.495	3.268	5.702	10.48	24.05

TABLE 7.22 : I_0^* ($1/\alpha$ = .30 , μ = .50)

Q: MODE:	0.05	0.10	0.15	0.20	0.25	0.30	0.35	0.40	0.45	0.50	0.55	0.60	0.70	0.80	0.90
(1)	.0468	.1018	.1665	.2426	.3321	.4377	.5626	.7112	.8894	1.105	1.371	1.703	2.697	4.663	10.48
(2)	.2739	.5648	.8770	1.215	1.586	1.997	2.458	2.981	3.585	4.293	5.141	6.180	9.220	15.14	32.56
(3)	.0554	.1199	.1953	.2839	.3891	.5155	.6696	.8608	1.103	1.415	1.826	2.379	4.173	7.823	18.11

TABLE 7.22 : I_0^* ($1/\alpha$ = .30 , μ = .60)

Q: MODE:	0.05	0.10	0.15	0.20	0.25	0.30	0.35	0.40	0.45	0.50	0.55	0.60	0.70	0.80	0.90
(1)	.0467	.1013	.1654	.2405	.3285	.4319	.5541	.6990	.8724	1.082	1.339	1.661	2.620	4.513	10.10
(2)	.0580	.1405	.2622	.4411	.6880	1.002	1.381	1.825	2.344	2.955	3.686	4.578	7.167	12.15	26.68
(3)	.0721	.1486	.2324	.3268	.4352	.5624	.7146	.9010	1.135	1.436	1.834	2.372	4.146	7.818	18.35

TABLE 7.22 : I_0^* ($1/\alpha$ = .30 , μ = .70)

Q: MODE:	0.05	0.10	0.15	0.20	0.25	0.30	0.35	0.40	0.45	0.50	0.55	0.60	0.70	0.80	0.90
(1)	.0465	.1009	.1642	.2382	.3246	.4257	.5448	.6857	.8540	1.057	1.306	1.616	2.541	4.362	9.735
(2)	.0662	.1522	.2652	.4141	.6084	.8564	1.165	1.539	1.988	2.526	3.174	3.969	6.268	10.65	23.30
(3)	.0495	.1038	.1649	.2356	.3211	.4308	.5840	.8219	1.221	1.865	2.774	3.943	7.312	13.58	31.34

TABLE 7.22 : I_0^* ($1/\alpha$ = .30 , μ = .80)

Q: MODE:	0.05	0.10	0.15	0.20	0.25	0.30	0.35	0.40	0.45	0.50	0.55	0.60	0.70	0.80	0.90
(1)	.0464	.1004	.1630	.2358	.3204	.4191	.5349	.6716	.8344	1.030	1.270	1.570	2.461	4.219	9.414
(2)	.0728	.1645	.2799	.4244	.6043	.8259	1.096	1.420	1.809	2.275	2.837	3.526	5.506	9.241	19.86
(3)	.1735	.4441	.8454	1.376	2.011	2.734	3.543	4.450	5.479	6.664	8.058	9.740	14.55	23.70	50.19

TABLE 7.22 : I_0^* ($1/\alpha$ = .40 , μ = .20)

Q: MODE:	0.05	0.10	0.15	0.20	0.25	0.30	0.35	0.40	0.45	0.50	0.55	0.60	0.70	0.80	0.90
(1)	.0539	.1173	.1916	.2788	.3811	.5016	.6441	.8136	1.017	1.264	1.568	1.950	3.100	5.394	12.24
(2)	.0678	.1494	.2478	.3667	.5108	.6862	.9001	1.161	1.481	1.875	2.361	2.971	4.775	8.257	18.29
(3)	.1491	.3409	.5905	.9130	1.319	1.813	2.398	3.079	3.869	4.790	5.880	7.198	10.96	18.06	38.52

TABLE 7.22 : I_0^* ($1/\alpha$ = .40 , μ = .30)

Q: MODE:	0.05	0.10	0.15	0.20	0.25	0.30	0.35	0.40	0.45	0.50	0.55	0.60	0.70	0.80	0.90
(1)	.0538	.1168	.1904	.2766	.3775	.4962	.6362	.8026	1.002	1.244	1.541	1.914	3.034	5.263	11.89
(2)	.0608	.1340	.2230	.3321	.4667	.6339	.8417	1.100	1.420	1.816	2.306	2.920	4.723	8.181	18.11
(3)	.0867	.1847	.2992	.4384	.6162	.8548	1.187	1.652	2.275	3.071	4.053	5.259	8.704	15.17	33.64

TABLE 7.22 : I_0^* ($1/\alpha$ = .40 , μ = .40)

Q: MODE:	0.05	0.10	0.15	0.20	0.25	0.30	0.35	0.40	0.45	0.50	0.55	0.60	0.70	0.80	0.90
(1)	.0537	.1163	.1892	.2742	.3735	.4900	.6272	.7898	.9843	1.220	1.509	1.870	2.953	5.099	11.46
(2)	.0520	.1167	.1996	.3090	.4565	.6552	.9164	1.248	1.655	2.150	2.751	3.490	5.635	9.744	21.65
(3)	.1218	.2466	.3817	.5336	.7094	.9178	1.170	1.483	1.876	2.378	3.025	3.868	6.454	11.52	26.02

TABLE 7.22 : I_0^* ($1/\alpha$ = .40 , μ = .50)

Q: MODE:	0.05	0.10	0.15	0.20	0.25	0.30	0.35	0.40	0.45	0.50	0.55	0.60	0.70	0.80	0.90
(1)	.0536	.1157	.1878	.2716	.3692	.4832	.6171	.7755	.9644	1.192	1.471	1.820	2.860	4.913	10.98
(2)	.2813	.5802	.9010	1.249	1.630	2.053	2.528	3.068	3.691	4.424	5.301	6.379	9.535	15.69	33.85
(3)	.0719	.1555	.2528	.3668	.5011	.6611	.8538	1.089	1.381	1.749	2.220	2.833	4.743	8.530	19.21

TABLE 7.22 : I_0^* ($1/\alpha$ = .40 , μ = .60)

Q: MODE:	0.05	0.10	0.15	0.20	0.25	0.30	0.35	0.40	0.45	0.50	0.55	0.60	0.70	0.80	0.90
(1)	.0534	.1151	.1864	.2688	.3645	.4758	.6061	.7597	.9424	1.162	1.431	1.765	2.760	4.716	10.48
(2)	.0772	.1852	.3384	.5501	.8258	1.164	1.566	2.034	2.580	3.224	3.995	4.939	7.687	13.00	28.53
(3)	.0897	.1852	.2898	.4068	.5400	.6943	.8759	1.094	1.360	1.693	2.117	2.672	4.432	8.028	18.46

TABLE 7.22 : I_0^* ($1/\alpha$ = .40 , μ = .70)

Q: MODE:	0.05	0.10	0.15	0.20	0.25	0.30	0.35	0.40	0.45	0.50	0.55	0.60	0.70	0.80	0.90
(1)	.0533	.1145	.1849	.2659	.3594	.4679	.5943	.7427	.9187	1.130	1.387	1.707	2.656	4.517	9.998
(2)	.0881	.2016	.3476	.5340	.7677	1.054	1.399	1.809	2.294	2.869	3.561	4.409	6.864	11.56	25.13
(3)	.0611	.1278	.2019	.2865	.3862	.5099	.6742	.9130	1.291	1.895	2.776	3.948	7.398	13.89	32.38

TABLE 7.22 : I_0^* ($1/\alpha$ = .40 , μ = .80)

Q: MODE:	0.05	0.10	0.15	0.20	0.25	0.30	0.35	0.40	0.45	0.50	0.55	0.60	0.70	0.80	0.90
(1)	.0531	.1139	.1833	.2628	.3541	.4595	.5817	.7248	.8938	1.096	1.342	1.648	2.552	4.329	9.573
(2)	.0970	.2180	.3673	.5495	.7690	1.030	1.338	1.700	2.124	2.626	3.226	3.956	6.048	9.986	21.18
(3)	.2300	.5754	1.054	1.652	2.344	3.120	3.984	4.952	6.049	7.313	8.802	10.60	15.75	25.58	54.08

TABLE 7.22 : I_0^* (1/α = .50 , μ = .20)

Q: MODE:	0.05	0.10	0.15	0.20	0.25	0.30	0.35	0.40	0.45	0.50	0.55	0.60	0.70	0.80	0.90
(1)	.0601	.1301	.2116	.3066	.4176	.5477	.7008	.8825	1.100	1.363	1.688	2.095	3.318	5.761	13.05
(2)	.0788	.1729	.2851	.4190	.5790	.7707	1.001	1.278	1.614	2.022	2.525	3.151	5.000	8.574	18.90
(3)	.1715	.3850	.6528	.9868	1.396	1.887	2.465	3.138	3.919	4.832	5.916	7.228	10.98	18.08	38.53

TABLE 7.22 : I_0^* (1/α = .50 , μ = .30)

Q: MODE:	0.05	0.10	0.15	0.20	0.25	0.30	0.35	0.40	0.45	0.50	0.55	0.60	0.70	0.80	0.90
(1)	.0599	.1294	.2101	.3039	.4131	.5408	.6909	.8686	1.081	1.338	1.653	2.049	3.235	5.595	12.62
(2)	.0705	.1545	.2551	.3761	.5226	.7006	.9175	1.182	1.506	1.903	2.393	3.004	4.803	8.260	18.20
(3)	.1083	.2306	.3728	.5441	.7589	1.039	1.412	1.910	2.556	3.368	4.367	5.598	9.136	15.82	35.01

TABLE 7.22 : I_0^* (1/α = .50 , μ = .40)

Q: MODE:	0.05	0.10	0.15	0.20	0.25	0.30	0.35	0.40	0.45	0.50	0.55	0.60	0.70	0.80	0.90
(1)	.0597	.1287	.2085	.3009	.4081	.5331	.6796	.8527	1.059	1.308	1.613	1.994	3.133	5.390	12.08
(2)	.0602	.1335	.2249	.3417	.4934	.6916	.9474	1.271	1.670	2.158	2.754	3.490	5.639	9.770	21.77
(3)	.1521	.3073	.4744	.6609	.8749	1.126	1.426	1.790	2.240	2.802	3.511	4.418	7.162	12.52	27.97

TABLE 7.22 : I_0^* (1/α = .50 , μ = .50)

Q: MODE:	0.05	0.10	0.15	0.20	0.25	0.30	0.35	0.40	0.45	0.50	0.55	0.60	0.70	0.80	0.90
(1)	.0596	.1280	.2060	.2977	.4027	.5247	.6672	.8349	1.034	1.274	1.567	1.932	3.018	5.159	11.48
(2)	.2890	.5962	.9260	1.284	1.676	2.112	2.601	3.158	3.802	4.558	5.466	6.582	9.856	16.26	35.16
(3)	.0875	.1890	.3069	.4443	.6054	.7957	1.023	1.297	1.631	2.046	2.566	3.231	5.250	9.184	20.27

TABLE 7.22 : I_0^* (1/α = .50 , μ = .60)

Q: MODE:	0.05	0.10	0.15	0.20	0.25	0.30	0.35	0.40	0.45	0.50	0.55	0.60	0.70	0.80	0.90
(1)	.0594	.1273	.2051	.2943	.3969	.5157	.6538	.8157	1.007	1.237	1.517	1.865	2.894	4.914	10.86
(2)	.0963	.2291	.4114	.6526	.9550	1.318	1.743	2.236	2.811	3.490	4.303	5.300	8.209	13.85	30.39
(3)	.1051	.2173	.3401	.4768	.6314	.8086	1.015	1.258	1.550	1.908	2.354	2.924	4.681	8.223	18.56

TABLE 7.22 : I_0^* (1/α = .50 , μ = .70)

Q: MODE:	0.05	0.10	0.15	0.20	0.25	0.30	0.35	0.40	0.45	0.50	0.55	0.60	0.70	0.80	0.90
(1)	.0592	.1266	.2032	.2907	.3908	.5060	.6394	.7951	.9786	1.198	1.464	1.794	2.767	4.669	10.26
(2)	.1100	.2504	.4279	.6487	.9177	1.239	1.617	2.061	2.581	3.195	3.932	4.834	7.447	12.46	26.96
(3)	.0711	.1484	.2336	.3295	.4408	.5752	.7475	.9868	1.349	1.922	2.779	3.952	7.488	14.20	33.43

TABLE 7.22 : I_0^* (1/α = .50 , μ = .80)

Q: MODE:	0.05	0.10	0.15	0.20	0.25	0.30	0.35	0.40	0.45	0.50	0.55	0.60	0.70	0.80	0.90
(1)	.0591	.1258	.2031	.2869	.3844	.4959	.6243	.7734	.9485	1.157	1.409	1.721	2.640	4.437	9.731
(2)	.1210	.2707	.4524	.6692	.9239	1.220	1.562	1.956	2.414	2.950	3.587	4.360	6.566	10.71	22.49
(3)	.2859	.7018	1.251	1.910	2.658	3.488	4.409	5.439	6.605	7.951	9.536	11.45	16.95	27.45	57.97

TABLE 7.22 : I_0^* (1/α = .60 , μ = .20)

Q: MODE:	0.05	0.10	0.15	0.20	0.25	0.30	0.35	0.40	0.45	0.50	0.55	0.60	0.70	0.80	0.90
(1)	.0656	.1417	.2300	.3324	.4515	.5907	.7543	.9478	1.179	1.459	1.803	2.235	3.532	6.123	13.86
(2)	.0884	.1933	.3173	.4639	.6374	.8430	1.087	1.379	1.729	2.152	2.670	3.314	5.211	8.877	19.49
(3)	.1907	.4223	.7050	1.049	1.462	1.952	2.525	3.191	3.965	4.873	5.950	7.256	11.00	18.09	38.53

TABLE 7.22 : I_0^* (1/α = .60 , μ = .30)

Q: MODE:	0.05	0.10	0.15	0.20	0.25	0.30	0.35	0.40	0.45	0.50	0.55	0.60	0.70	0.80	0.90
(1)	.0654	.1409	.2281	.3290	.4460	.5824	.7422	.9309	1.156	1.428	1.761	2.179	3.431	5.922	13.34
(2)	.0790	.1722	.2825	.4136	.5699	.7570	.9818	1.253	1.581	1.980	2.471	3.082	4.879	8.337	18.30
(3)	.1299	.2764	.4460	.6485	.8986	1.216	1.627	2.157	2.827	3.659	4.681	5.939	9.576	16.48	36.40

TABLE 7.22 : I_0^* (1/α = .60 , μ = .40)

Q: MODE:	0.05	0.10	0.15	0.20	0.25	0.30	0.35	0.40	0.45	0.50	0.55	0.60	0.70	0.80	0.90
(1)	.0652	.1401	.2262	.3254	.4400	.5732	.7287	.9118	1.130	1.392	1.713	2.113	3.309	5.676	12.70
(2)	.0672	.1479	.2464	.3691	.5244	.7226	.9747	1.291	1.684	2.165	2.757	3.491	5.642	9.797	21.90
(3)	.1823	.3676	.5660	.7059	1.036	1.327	1.670	2.082	2.583	3.198	3.965	4.934	7.839	13.51	29.92

TABLE 7.22 : I_0^* (1/α = .60 , μ = .50)

Q: MODE:	0.05	0.10	0.15	0.20	0.25	0.30	0.35	0.40	0.45	0.50	0.55	0.60	0.70	0.80	0.90
(1)	.0650	.1393	.2242	.3216	.4336	.5632	.7139	.8908	1.100	1.352	1.658	2.039	3.172	5.399	11.97
(2)	.2971	.6128	.9519	1.320	1.724	2.173	2.677	3.251	3.915	4.697	5.636	6.791	10.18	16.83	36.47
(3)	.1021	.2205	.3576	.5168	.7025	.9205	1.178	1.487	1.859	2.314	2.878	3.589	5.711	9.798	21.30

TABLE 7.22 : I_0^* (1/α = .60 , μ = .60)

Q: MODE:	0.05	0.10	0.15	0.20	0.25	0.30	0.35	0.40	0.45	0.50	0.55	0.60	0.70	0.80	0.90
(1)	.0648	.1384	.2221	.3176	.4268	.5525	.6981	.8680	1.068	1.308	1.599	1.959	3.025	5.108	11.23
(2)	.1153	.2722	.4820	.7505	1.078	1.465	1.914	2.434	3.039	3.753	4.609	5.660	8.733	14.71	32.26
(3)	.1186	.2456	.3845	.5386	.7119	.9091	1.136	1.401	1.715	2.093	2.558	3.142	4.903	8.405	18.66

TABLE 7.22 : I_0^* (1/α = .60 , μ = .70)

Q: MODE:	0.05	0.10	0.15	0.20	0.25	0.30	0.35	0.40	0.45	0.50	0.55	0.60	0.70	0.80	0.90
(1)	.0646	.1375	.2199	.3133	.4196	.5412	.6812	.8439	1.035	1.262	1.537	1.876	2.874	4.817	10.52
(2)	.1318	.2987	.5064	.7593	1.061	1.414	1.824	2.300	2.855	3.509	4.291	5.248	8.021	13.34	28.78
(3)	.0799	.1663	.2610	.3666	.4873	.6304	.8091	1.049	1.400	1.946	2.782	3.957	7.580	14.53	34.49

TABLE 7.22 : I_0^* (1/α = .60 , μ = .80)

Q: MODE:	0.05	0.10	0.15	0.20	0.25	0.30	0.35	0.40	0.45	0.50	0.55	0.60	0.70	0.80	0.90
(1)	.0644	.1366	.2177	.3090	.4121	.5294	.6636	.8186	1.000	1.214	1.473	1.791	2.725	4.543	9.887
(2)	.1450	.3229	.5355	.7845	1.071	1.398	1.771	2.196	2.686	3.255	3.930	4.745	7.066	11.42	23.78
(3)	.3414	.8244	1.439	2.157	2.959	3.844	4.823	5.915	7.152	8.579	10.26	12.30	18.14	29.33	61.85

TABLE 7.22 : I_0^* (1/α = .70 , μ = .20)

Q: MODE:	0.05	0.10	0.15	0.20	0.25	0.30	0.35	0.40	0.45	0.50	0.55	0.60	0.70	0.80	0.90
(1)	.0707	.1525	.2470	.3565	.4835	.6315	.8051	1.010	1.255	1.551	1.915	2.371	3.743	6.482	14.67
(2)	.0969	.2111	.3454	.5031	.6882	.9060	1.163	1.467	1.831	2.268	2.801	3.463	5.408	9.170	20.07
(3)	.2075	.4545	.7497	1.102	1.519	2.009	2.579	3.240	4.009	4.911	5.983	7.284	11.02	18.10	38.54

TABLE 7.22 : I_0^* (1/α = .70 , μ = .30)

Q: MODE:	0.05	0.10	0.15	0.20	0.25	0.30	0.35	0.40	0.45	0.50	0.55	0.60	0.70	0.80	0.90
(1)	.0705	.1516	.2449	.3525	.4770	.6217	.7909	.9904	1.228	1.515	1.866	2.306	3.624	6.246	14.06
(2)	.0865	.1877	.3064	.4460	.6107	.8057	1.038	1.314	1.647	2.049	2.542	3.154	4.951	8.411	18.39
(3)	.1516	.3221	.5187	.7518	1.036	1.389	1.834	2.395	3.091	3.946	4.992	6.283	10.02	17.16	37.81

TABLE 7.22 : I_0^* (1/α = .70 , μ = .40)

Q: MODE:	0.05	0.10	0.15	0.20	0.25	0.30	0.35	0.40	0.45	0.50	0.55	0.60	0.70	0.80	0.90
(1)	.0703	.1507	.2427	.3483	.4700	.6109	.7752	.9681	1.197	1.473	1.809	2.229	3.481	5.959	13.31
(2)	.0734	.1604	.2649	.3926	.5510	.7496	.9989	1.311	1.697	2.173	2.760	3.491	5.646	9.823	22.02
(3)	.2125	.4276	.6565	.9087	1.194	1.522	1.905	2.361	2.908	3.574	4.395	5.425	8.493	14.47	31.86

TABLE 7.22 : I_0^* (1/α = .70 , μ = .50)

Q: MODE:	0.05	0.10	0.15	0.20	0.25	0.30	0.35	0.40	0.45	0.50	0.55	0.60	0.70	0.80	0.90
(1)	.0701	.1497	.2403	.3439	.4626	.5994	.7580	.9437	1.163	1.426	1.746	2.143	3.321	5.637	12.46
(2)	.3054	.6299	.9787	1.357	1.773	2.235	2.755	3.347	4.032	4.839	5.810	7.004	10.52	17.40	37.79
(3)	.1159	.2500	.4051	.5845	.7930	1.036	1.322	1.661	2.067	2.559	3.162	3.915	6.137	10.38	22.30

TABLE 7.22 : I_0^* (1/α = .70 , μ = .60)

Q: MODE:	0.05	0.10	0.15	0.20	0.25	0.30	0.35	0.40	0.45	0.50	0.55	0.60	0.70	0.80	0.90
(1)	.0698	.1487	.2379	.3392	.4547	.5871	.7398	.9175	1.127	1.376	1.677	2.051	3.151	5.298	11.60
(2)	.1343	.3148	.5506	.8450	1.197	1.608	2.082	2.628	3.264	4.014	4.914	6.020	9.258	15.56	34.13
(3)	.1306	.3708	.4240	.5936	.7834	.9981	1.243	1.527	1.860	2.256	2.737	3.334	5.103	8.577	18.77

TABLE 7.22 : I_0^* (1/α = .70 , μ = .70)

Q: MODE:	0.05	0.10	0.15	0.20	0.25	0.30	0.35	0.40	0.45	0.50	0.55	0.60	0.70	0.80	0.90
(1)	.0696	.1477	.2354	.3344	.4465	.5741	.7205	.8899	1.088	1.323	1.606	1.955	2.978	4.963	10.77
(2)	.1535	.3465	.5832	.8664	1.198	1.582	2.023	2.530	3.120	3.813	4.641	5.653	8.587	14.22	30.60
(3)	.0876	.1821	.2850	.3989	.5277	.6781	.8619	1.102	1.444	1.969	2.784	3.962	7.675	14.86	35.57

TABLE 7.22 : I_0^* (1/α = .70 , μ = .80)

Q: MODE:	0.05	0.10	0.15	0.20	0.25	0.30	0.35	0.40	0.45	0.50	0.55	0.60	0.70	0.80	0.90
(1)	.0694	.1466	.2328	.3294	.4379	.5606	.7004	.8611	1.048	1.268	1.533	1.858	2.807	4.646	10.04
(2)	.1690	.3745	.6169	.8960	1.212	1.569	1.970	2.425	2.944	3.547	4.258	5.115	7.551	12.11	25.05
(3)	.3963	.9438	1.620	2.395	3.251	4.191	5.227	6.382	7.691	9.200	10.98	13.14	19.33	31.20	65.74

TABLE 7.22 : I_0^* (1/α = .80 , μ = .20)

Q: MODE:	0.05	0.10	0.15	0.20	0.25	0.30	0.35	0.40	0.45	0.50	0.55	0.60	0.70	0.80	0.90
(1)	.0755	.1627	.2632	.3793	.5138	.6704	.8539	1.071	1.329	1.641	2.025	2.505	3.950	6.838	15.48
(2)	.1043	.2269	.3703	.5377	.7332	.9618	1.230	1.546	1.922	2.373	2.921	3.601	5.595	9.452	20.64
(3)	.2222	.4826	.7888	1.149	1.570	2.060	2.628	3.285	4.050	4.947	6.014	7.312	11.04	18.11	38.55

TABLE 7.22 : I_0^* (1/α = .80 , μ = .30)

Q: MODE:	0.05	0.10	0.15	0.20	0.25	0.30	0.35	0.40	0.45	0.50	0.55	0.60	0.70	0.80	0.90
(1)	.0753	.1616	.2607	.3747	.5064	.6591	.8375	1.048	1.297	1.599	1.968	2.430	3.813	6.567	14.78
(2)	.0931	.2013	.3274	.4745	.6465	.8485	1.087	1.369	1.706	2.112	2.606	3.220	5.020	8.484	18.48
(3)	.1732	.3677	.5912	.8540	1.170	1.557	2.036	2.626	3.349	4.229	5.303	6.627	10.47	17.84	39.22

TABLE 7.22 : I_0^* (1/α = .80 , μ = .40)

Q: MODE:	0.05	0.10	0.15	0.20	0.25	0.30	0.35	0.40	0.45	0.50	0.55	0.60	0.70	0.80	0.90
(1)	.0750	.1605	.2581	.3699	.4984	.6468	.8195	1.022	1.262	1.550	1.903	2.342	3.650	6.238	13.91
(2)	.0789	.1713	.2810	.4130	.5743	.7734	1.021	1.328	1.709	2.180	2.762	3.491	5.649	9.850	22.14
(3)	.2426	.4872	.7460	1.029	1.347	1.711	2.132	2.628	3.219	3.932	4.806	5.896	9.128	15.43	33.79

TABLE 7.22 : I_0^* (1/α = .80 , μ = .50)

Q: MODE:	0.05	0.10	0.15	0.20	0.25	0.30	0.35	0.40	0.45	0.50	0.55	0.60	0.70	0.80	0.90
(1)	.0748	.1594	.2555	.3649	.4899	.6337	.8000	.9942	1.223	1.497	1.830	2.244	3.468	5.870	12.95
(2)	.3139	.6477	1.006	1.396	1.824	2.299	2.835	3.445	4.152	4.985	5.987	7.222	10.86	17.99	39.12
(3)	.1288	.2778	.4496	.6479	.8774	1.144	1.456	1.823	2.260	2.785	3.424	4.217	6.536	10.94	23.28

TABLE 7.22 : I_0^* (1/α = .80 , μ = .60)

Q: MODE:	0.05	0.10	0.15	0.20	0.25	0.30	0.35	0.40	0.45	0.50	0.55	0.60	0.70	0.80	0.90
(1)	.0745	.1583	.2527	.3596	.4810	.6197	.7793	.9646	1.182	1.441	1.753	2.140	3.275	5.486	11.96
(2)	.1532	.3568	.6177	.9367	1.313	1.747	2.246	2.820	3.487	4.274	5.218	6.380	9.783	16.42	36.00
(3)	.1413	.2933	.4594	.6428	.8474	1.078	1.339	1.640	1.989	2.401	2.897	3.506	5.286	8.739	18.87

TABLE 7.22 : I_0^* (1/α = .80 , μ = .70)

Q: MODE:	0.05	0.10	0.15	0.20	0.25	0.30	0.35	0.40	0.45	0.50	0.55	0.60	0.70	0.80	0.90
(1)	.0742	.1571	.2499	.3541	.4717	.6051	.7576	.9335	1.139	1.381	1.673	2.032	3.079	5.105	11.02
(2)	.1752	.3938	.6586	.9707	1.331	1.744	2.214	2.754	3.377	4.109	4.983	6.051	9.146	15.10	32.41
(3)	.0945	.1960	.3062	.4273	.5632	.7198	.9080	1.149	1.483	1.990	2.787	3.967	7.773	15.20	36.65

TABLE 7.22 : I_0^* (1/α = .80 , μ = .80)

Q: MODE:	0.05	0.10	0.15	0.20	0.25	0.30	0.35	0.40	0.45	0.50	0.55	0.60	0.70	0.80	0.90
(1)	.0740	.1559	.2470	.3485	.4620	.5900	.7351	.9013	1.094	1.320	1.591	1.923	2.887	4.748	10.19
(2)	.1928	.4256	.6966	1.004	1.349	1.732	2.161	2.643	3.193	3.827	4.574	5.474	8.025	12.79	26.32
(3)	.4508	1.061	1.797	2.626	3.536	4.530	5.624	6.843	8.223	9.816	11.70	13.97	20.52	33.06	69.62

TABLE 7.22 : I_0^* (1/α = .90 , μ = .20)

Q: MODE:	0.05	0.10	0.15	0.20	0.25	0.30	0.35	0.40	0.45	0.50	0.55	0.60	0.70	0.80	0.90
(1)	.0801	.1722	.2785	.4010	.5429	.7079	.9010	1.129	1.400	1.729	2.132	2.637	4.156	7.192	16.28
(2)	.1110	.2410	.3925	.5685	.7733	1.012	1.290	1.618	2.005	2.469	3.032	3.729	5.773	9.726	21.21
(3)	.2353	.5074	.8232	1.190	1.615	2.106	2.672	3.327	4.088	4.981	6.045	7.338	11.05	18.13	38.55

TABLE 7.22 : I_0^* (1/α = .90 , μ = .30)

Q: MODE:	0.05	0.10	0.15	0.20	0.25	0.30	0.35	0.40	0.45	0.50	0.55	0.60	0.70	0.80	0.90
(1)	.0798	.1711	.2757	.3959	.5344	.6950	.8824	1.103	1.365	1.681	2.067	2.551	4.000	6.885	15.49
(2)	.0989	.2135	.3461	.4997	.6783	.8865	1.130	1.418	1.759	2.169	2.666	3.282	5.085	8.555	18.57
(3)	.1948	.4132	.6632	.9552	1.303	1.722	2.232	2.852	3.602	4.510	5.613	6.973	10.93	18.53	40.65

TABLE 7.22 : I_0^* (1/α = .90 , μ = .40)

Q: MODE:	0.05	0.10	0.15	0.20	0.25	0.30	0.35	0.40	0.45	0.50	0.55	0.60	0.70	0.80	0.90
(1)	.0795	.1698	.2728	.3905	.5254	.6812	.8621	1.074	1.325	1.626	1.994	2.452	3.816	6.514	14.52
(2)	.0838	.1810	.2953	.4311	.5949	.7947	1.041	1.345	1.721	2.186	2.765	3.492	5.653	9.878	22.27
(3)	.2727	.5464	.8344	1.148	1.497	1.894	2.351	2.886	3.518	4.276	5.201	6.351	9.748	16.37	35.71

TABLE 7.22 : I_0^* (1/α = .90 , μ = .50)

Q: MODE:	0.05	0.10	0.15	0.20	0.25	0.30	0.35	0.40	0.45	0.50	0.55	0.60	0.70	0.80	0.90
(1)	.0792	.1686	.2698	.3848	.5159	.6664	.8402	1.043	1.282	1.567	1.913	2.342	3.612	6.101	13.44
(2)	.3228	.6659	1.035	1.435	1.876	2.365	2.917	3.546	4.275	5.134	6.169	7.444	11.20	18.58	40.46
(3)	.1409	.3037	.4913	.7072	.9562	1.245	1.580	1.973	2.439	2.994	3.668	4.499	6.913	11.47	24.25

TABLE 7.22 : I_0^* ($1/\alpha = .90$, $\mu = .60$)

Q: MODE:	0.05	0.10	0.15	0.20	0.25	0.30	0.35	0.40	0.45	0.50	0.55	0.60	0.70	0.80	0.90
(1)	.0789	.1673	.2667	.3789	.5060	.6508	.8172	1.010	1.235	1.503	1.827	2.226	3.396	5.671	12.33
(2)	.1720	.3983	.6834	1.026	1.426	1.884	2.408	3.009	3.708	4.532	5.522	6.739	10.31	17.29	37.88
(3)	.1509	.3135	.4912	.6871	.9050	1.149	1.425	1.741	2.105	2.532	3.041	3.662	5.454	8.894	18.96

TABLE 7.22 : I_0^* ($1/\alpha = .90$, $\mu = .70$)

Q: MODE:	0.05	0.10	0.15	0.20	0.25	0.30	0.35	0.40	0.45	0.50	0.55	0.60	0.70	0.80	0.90
(1)	.0786	.1660	.2636	.3728	.4956	.6346	.7930	.9753	1.187	1.437	1.737	2.105	3.178	5.246	11.27
(2)	.1969	.4408	.7328	1.073	1.461	1.901	2.401	2.971	3.629	4.400	5.319	6.443	9.701	15.97	34.22
(3)	.1006	.2084	.3250	.4526	.5946	.7567	.9487	1.190	1.518	2.010	2.789	3.972	7.874	15.55	37.75

TABLE 7.22 : I_0^* ($1/\alpha = .90$, $\mu = .80$)

Q: MODE:	0.05	0.10	0.15	0.20	0.25	0.30	0.35	0.40	0.45	0.50	0.55	0.60	0.70	0.80	0.90
(1)	.0783	.1647	.2603	.3665	.4849	.6178	.7681	.9396	1.138	1.370	1.647	1.985	2.964	4.848	10.35
(2)	.2166	.4761	.7749	1.110	1.481	1.890	2.345	2.855	3.433	4.099	4.882	5.823	8.489	13.47	27.58
(3)	.5050	1.175	1.969	2.852	3.814	4.863	6.015	7.298	8.751	10.43	12.40	14.80	21.70	34.93	73.50

TABLE 7.23 : I_0^* ($\alpha = .10$, $\mu = .20$)

Q: MODE:	0.05	0.10	0.15	0.20	0.25	0.30	0.35	0.40	0.45	0.50	0.55	0.60	0.70	0.80	0.90
(1)	.0491	.1939	.4300	.7562	1.175	1.695	2.323	3.084	3.992	5.080	6.395	8.002	12.59	21.05	44.49
(2)	.3438	.7707	1.306	1.986	2.866	4.027	5.576	7.638	10.33	13.74	17.95	23.13	37.85	65.12	141.5
(3)	.7833	1.636	2.589	3.681	4.965	6.516	8.442	10.90	14.11	18.38	24.09	31.73	55.70	103.1	240.4

TABLE 7.23 : I_0^* ($\alpha = .10$, $\mu = .30$)

Q: MODE:	0.05	0.10	0.15	0.20	0.25	0.30	0.35	0.40	0.45	0.50	0.55	0.60	0.70	0.80	0.90
(1)	.0729	.2820	.6128	1.061	1.633	2.341	3.206	4.255	5.526	7.073	8.972	11.34	18.31	31.78	70.60
(2)	.3155	.6928	1.146	1.697	2.374	3.219	4.295	5.689	7.528	9.980	13.26	17.64	31.41	58.82	138.7
(3)	.5899	1.212	1.875	2.590	3.372	4.236	5.204	6.305	7.577	9.072	10.87	13.07	19.51	31.93	67.57

TABLE 7.23 : I_0^* ($\alpha = .10$, $\mu = .40$)

Q: MODE:	0.05	0.10	0.15	0.20	0.25	0.30	0.35	0.40	0.45	0.50	0.55	0.60	0.70	0.80	0.90
(1)	.0959	.3610	.7663	1.304	1.984	2.822	3.847	5.094	6.613	8.475	10.78	13.68	22.40	39.63	90.51
(2)	.2807	.6035	.9757	1.407	1.909	2.501	3.204	4.051	5.087	6.376	8.011	10.13	16.78	30.39	70.22
(3)	2.785	5.661	8.663	11.84	15.23	18.90	22.94	27.44	32.54	38.43	45.38	53.80	78.02	124.5	259.9

TABLE 7.23 : I_0^* ($\alpha = .10$, $\mu = .50$)

Q: MODE:	0.05	0.10	0.15	0.20	0.25	0.30	0.35	0.40	0.45	0.50	0.55	0.60	0.70	0.80	0.90
(1)	.1178	.4299	.8893	1.485	2.226	3.134	4.236	5.573	7.201	9.196	11.67	14.79	24.18	42.87	98.37
(2)	.2359	.4967	.7852	1.105	1.461	1.857	2.302	2.803	3.373	4.025	4.780	5.667	8.031	11.96	21.81
(3)	4.828	9.027	12.75	16.34	20.04	24.02	28.44	33.44	39.21	45.98	54.10	64.07	93.26	150.0	317.0

TABLE 7.23 : I_0^* ($\alpha = .10$, $\mu = .60$)

Q: MODE:	0.05	0.10	0.15	0.20	0.25	0.30	0.35	0.40	0.45	0.50	0.55	0.60	0.70	0.80	0.90
(1)	.1383	.4882	.9824	1.605	2.365	3.281	4.379	5.697	7.288	9.221	11.60	14.58	23.46	40.90	92.03
(2)	.1735	.3578	.5539	.7626	.9852	1.223	1.479	1.755	2.060	2.414	2.932	5.014	16.73	38.54	99.85
(3)	3.032	3.857	4.693	5.631	6.707	7.955	9.418	11.15	13.22	15.73	18.79	22.59	33.58	53.90	108.7

TABLE 7.23 : I_0^* ($\alpha = .10$, $\mu = .70$)

Q: MODE:	0.05	0.10	0.15	0.20	0.25	0.30	0.35	0.40	0.45	0.50	0.55	0.60	0.70	0.80	0.90
(1)	.1575	.5360	1.047	1.670	2.409	3.278	4.299	5.502	6.925	8.625	10.68	13.21	20.55	34.44	73.82
(2)	3.072	6.298	9.673	13.26	17.12	21.34	26.03	31.32	37.40	44.52	53.04	63.49	94.11	154.0	331.0
(3)	.3111	.6479	1.013	1.408	1.838	2.306	2.816	3.374	3.987	4.663	5.412	6.250	8.284	11.20	20.34

TABLE 7.23 : I_0^* ($\alpha = .10$, $\mu = .80$)

Q: MODE:	0.05	0.10	0.15	0.20	0.25	0.30	0.35	0.40	0.45	0.50	0.55	0.60	0.70	0.80	0.90
(1)	.1752	.5737	1.086	1.685	2.371	3.152	4.040	5.053	6.217	7.566	9.148	11.04	16.24	25.45	49.85
(2)	3.841	7.591	11.24	14.89	18.65	22.62	26.89	31.61	36.91	43.00	50.17	58.84	83.65	130.9	267.7
(3)	5.506	11.20	17.17	23.50	30.32	37.77	46.02	55.30	65.93	78.33	93.12	111.2	164.1	267.0	570.8

TABLE 7.23 : I^* ($\alpha = .15$, $\mu = .20$)

Q: MODE:	0.05	0.10	0.15	0.20	0.25	0.30	0.35	0.40	0.45	0.50	0.55	0.60	0.70	0.80	0.90
(1)	.0328	.1306	.2920	.5169	.8072	1.168	1.606	2.133	2.765	3.523	4.442	5.570	8.816	14.90	32.05
(2)	.3274	.7293	1.224	1.838	2.605	3.568	4.779	6.295	8.173	10.48	13.30	16.77	26.75	45.51	98.61
(3)	.7484	1.560	2.459	3.477	4.653	6.041	7.713	9.767	12.33	15.58	19.75	25.13	41.65	74.36	169.9

TABLE 7.23 : I_0^* ($\alpha = .15$, $\mu = .30$)

Q: MODE:	0.05	0.10	0.15	0.20	0.25	0.30	0.35	0.40	0.45	0.50	0.55	0.60	0.70	0.80	0.90
(1)	.0489	.1918	.4217	.7357	1.136	1.632	2.236	2.966	3.850	4.924	6.244	7.891	12.76	22.21	49.66
(2)	.3012	.6592	1.086	1.597	2.215	2.968	3.898	5.059	6.524	8.388	10.78	13.88	23.40	42.36	98.01
(3)	.5643	1.159	1.792	2.472	3.213	4.026	4.932	5.954	7.123	8.483	10.09	12.05	17.65	28.19	58.05

TABLE 7.23 : I_0^* ($\alpha = .15$, $\mu = .40$)

Q: MODE:	0.05	0.10	0.15	0.20	0.25	0.30	0.35	0.40	0.45	0.50	0.55	0.60	0.70	0.80	0.90
(1)	.0646	.2482	.5346	.9168	1.398	1.990	2.710	3.582	4.641	5.937	7.540	9.556	15.61	27.62	63.15
(2)	.2683	.5760	.9291	1.336	1.806	2.353	2.995	3.757	4.670	5.780	7.152	8.881	14.08	24.30	53.77
(3)	1.973	4.011	6.141	8.394	10.81	13.43	16.32	19.55	23.22	27.47	32.51	38.64	56.35	90.54	190.7

TABLE 7.23 : I_0^* ($\alpha = .15$, $\mu = .50$)

Q: MODE:	0.05	0.10	0.15	0.20	0.25	0.30	0.35	0.40	0.45	0.50	0.55	0.60	0.70	0.80	0.90
(1)	.0798	.2991	.6290	1.059	1.591	2.239	3.020	3.963	5.106	6.501	8.227	10.40	16.94	29.95	68.60
(2)	.2256	.4748	.7501	1.055	1.392	1.763	2.187	2.657	3.189	3.795	4.494	5.313	7.495	11.19	20.90
(3)	3.273	6.280	9.016	11.65	14.32	17.16	20.28	23.78	27.80	32.51	38.14	45.06	65.33	104.8	221.0

TABLE 7.23 : I_0^* ($\alpha = .15$, $\mu = .60$)

Q: MODE:	0.05	0.10	0.15	0.20	0.25	0.30	0.35	0.40	0.45	0.50	0.55	0.60	0.70	0.80	0.90
(1)	.0944	.3438	.7048	1.162	1.716	2.380	3.170	4.112	5.241	6.609	8.286	10.38	16.60	28.81	64.59
(2)	.1660	.3425	.5305	.7308	.9448	1.174	1.422	1.694	1.999	2.368	2.930	4.427	12.54	28.44	73.65
(3)	2.593	3.461	4.244	5.087	6.030	7.100	8.328	9.749	11.41	13.38	15.73	18.60	26.74	41.64	81.99

TABLE 7.23 : I_0^* ($\alpha = .15$, $\mu = .70$)

Q: MODE:	0.05	0.10	0.15	0.20	0.25	0.30	0.35	0.40	0.45	0.50	0.55	0.60	0.70	0.80	0.90
(1)	.1082	.3822	.7625	1.228	1.779	2.422	3.172	4.049	5.081	6.307	7.783	9.593	14.82	24.70	52.68
(2)	2.102	4.349	6.705	9.208	11.90	14.85	18.13	21.83	26.08	31.05	37.00	44.30	65.71	107.6	231.5
(3)	.2974	.6188	.9665	1.343	1.751	2.194	2.676	3.202	3.778	4.412	5.113	5.897	7.813	10.69	20.30

TABLE 7.23 : I_0^* (α = .15 , μ = .80)

Q: MODE:	0.05	0.10	0.15	0.20	0.25	0.30	0.35	0.40	0.45	0.50	0.55	0.60	0.70	0.80	0.90
(1)	.1212	.4144	.8031	1.262	1.786	2.380	3.052	3.814	4.686	5.692	6.867	8.264	12.10	18.89	36.92
(2)	2.608	5.228	7.801	10.38	13.03	15.83	18.84	22.16	25.88	30.16	35.19	41.26	58.64	91.72	187.4
(3)	3.846	7.824	11.99	16.42	21.19	26.39	32.15	38.64	46.07	54.74	65.09	77.75	114.7	186.8	399.4

TABLE 7.23 : I_0^* (α = .20 , μ = .20)

Q: MODE:	0.05	0.10	0.15	0.20	0.25	0.30	0.35	0.40	0.45	0.50	0.55	0.60	0.70	0.80	0.90
(1)	.0247	.0985	.2214	.3940	.6182	.8977	1.239	1.650	2.143	2.738	3.461	4.352	6.938	11.85	25.88
(2)	.3122	.6909	1.150	1.707	2.384	3.208	4.209	5.422	6.887	8.658	10.81	13.46	21.13	35.67	77.16
(3)	.7165	1.490	2.342	3.296	4.384	5.647	7.138	8.925	11.10	13.78	17.14	21.40	34.30	59.78	134.6

TABLE 7.23 : I_0^* (α = .20 , μ = .30)

Q: MODE:	0.05	0.10	0.15	0.20	0.25	0.30	0.35	0.40	0.45	0.50	0.55	0.60	0.70	0.80	0.90
(1)	.0368	.1455	.3226	.5666	.8793	1.267	1.738	2.309	3.000	3.839	4.871	6.160	9.981	17.44	39.21
(2)	.2880	.6284	1.031	1.508	2.077	2.759	3.583	4.589	5.826	7.363	9.292	11.75	19.20	34.00	77.59
(3)	.5410	1.111	1.717	2.367	3.073	3.846	4.703	5.664	6.759	8.024	9.515	11.31	16.42	25.94	52.85

TABLE 7.23 : I_0^* (α = .20 , μ = .40)

Q: MODE:	0.05	0.10	0.15	0.20	0.25	0.30	0.35	0.40	0.45	0.50	0.55	0.60	0.70	0.80	0.90
(1)	.0488	.1896	.4127	.7132	1.093	1.559	2.124	2.809	3.638	4.651	5.904	7.479	12.21	21.61	49.47
(2)	.2569	.5507	.8868	1.272	1.714	2.225	2.820	3.517	4.342	5.333	6.540	8.038	12.45	20.95	45.28
(3)	1.576	3.205	4.909	6.714	8.651	10.76	13.08	15.69	18.66	22.11	26.21	31.20	45.71	73.81	156.4

TABLE 7.23 : I_0^* (α = .20 , μ = .50)

Q: MODE:	0.05	0.10	0.15	0.20	0.25	0.30	0.35	0.40	0.45	0.50	0.55	0.60	0.70	0.80	0.90
(1)	.0604	.2301	.4903	.8319	1.256	1.770	2.389	3.134	4.033	5.130	6.483	8.184	13.30	23.47	53.70
(2)	.2162	.4549	.7183	1.009	1.331	1.689	2.087	2.533	3.035	3.608	4.267	5.039	7.108	10.68	20.38
(3)	2.476	4.831	7.028	9.151	11.30	13.57	16.04	18.80	21.96	25.65	30.06	35.46	51.29	82.14	173.0

TABLE 7.23 : I_0^* (α = .20 , μ = .60)

Q: MODE:	0.05	0.10	0.15	0.20	0.25	0.30	0.35	0.40	0.45	0.50	0.55	0.60	0.70	0.80	0.90
(1)	.0717	.2665	.5547	.9225	1.369	1.902	2.535	3.286	4.184	5.268	6.593	8.245	13.14	22.74	50.84
(2)	.1591	.3286	.5093	.7022	.9089	1.132	1.374	1.642	1.950	2.335	2.928	4.227	10.60	23.55	60.74
(3)	2.207	3.103	3.850	4.627	5.477	6.428	7.504	8.735	10.16	11.82	13.79	16.17	22.88	35.09	68.24

TABLE 7.23 : I_0^* (α = .20 , μ = .70)

Q: MODE:	0.05	0.10	0.15	0.20	0.25	0.30	0.35	0.40	0.45	0.50	0.55	0.60	0.70	0.80	0.90
(1)	.0825	.2987	.6061	.9860	1.436	1.960	2.569	3.279	4.112	5.098	6.282	7.731	11.91	19.78	42.06
(2)	1.611	3.365	5.212	7.176	9.292	11.61	14.18	17.08	20.41	24.32	28.99	34.72	51.53	84.43	181.7
(3)	.2847	.5920	.9240	1.283	1.672	2.093	2.551	3.050	3.596	4.195	4.860	5.604	7.443	10.32	20.28

TABLE 7.23 : I_0^* (α = .20 , μ = .80)

Q: MODE:	0.05	0.10	0.15	0.20	0.25	0.30	0.35	0.40	0.45	0.50	0.55	0.60	0.70	0.80	0.90
(1)	.0928	.3265	.6449	1.025	1.460	1.953	2.510	3.141	3.860	4.688	5.654	6.802	9.952	15.52	30.36
(2)	1.981	4.020	6.044	8.082	10.18	12.39	14.77	17.38	20.32	23.69	27.65	32.42	46.08	72.08	147.3
(3)	3.018	6.140	9.412	12.89	16.63	20.71	25.24	30.33	36.17	42.97	51.10	61.04	90.07	146.7	313.7

TABLE 7.23 : I_0^* (α = .25 , μ = .20)

Q: MODE:	0.05	0.10	0.15	0.20	0.25	0.30	0.35	0.40	0.45	0.50	0.55	0.60	0.70	0.80	0.90
(1)	.0197	.0791	.1784	.3188	.5023	.7322	1.014	1.355	1.766	2.263	2.869	3.621	5.817	10.04	22.22
(2)	.2980	.6552	1.082	1.591	2.196	2.917	3.775	4.795	6.011	7.468	9.232	11.40	17.71	29.74	64.29
(3)	.6871	1.426	2.235	3.135	4.150	5.314	6.667	8.263	10.17	12.49	15.35	18.94	29.70	50.90	113.3

TABLE 7.23 : I_0^* (α = .25 , μ = .30)

Q: MODE:	0.05	0.10	0.15	0.20	0.25	0.30	0.35	0.40	0.45	0.50	0.55	0.60	0.70	0.80	0.90
(1)	.0295	.1173	.2616	.4621	.7205	1.042	1.433	1.908	2.482	3.181	4.041	5.116	8.314	14.58	32.96
(2)	.2758	.6000	.9806	1.428	1.956	2.580	3.325	4.220	5.303	6.628	8.272	10.34	16.57	28.91	65.30
(3)	.5196	1.067	1.649	2.273	2.949	3.688	4.505	5.419	6.456	7.652	9.056	10.75	15.52	24.41	49.53

TABLE 7.23 : I_0^* (α = .25 , μ = .40)

Q: MODE:	0.05	0.10	0.15	0.20	0.25	0.30	0.35	0.40	0.45	0.50	0.55	0.60	0.70	0.80	0.90
(1)	.0392	.1535	.3370	.5863	.9025	1.291	1.763	2.334	3.025	3.869	4.912	6.223	10.16	18.00	41.27
(2)	.2464	.5276	.8481	1.214	1.632	2.113	2.669	3.315	4.075	4.980	6.071	7.416	11.33	18.79	40.03
(3)	1.345	2.735	4.191	5.734	7.392	9.199	11.20	13.44	16.00	18.97	22.52	26.84	39.44	63.91	136.0

TABLE 7.23 : I_0^* (α = .25 , μ = .50)

Q: MODE:	0.05	0.10	0.15	0.20	0.25	0.30	0.35	0.40	0.45	0.50	0.55	0.60	0.70	0.80	0.90
(1)	.0486	.1872	.4031	.6892	1.045	1.478	1.998	2.622	3.375	4.291	5.421	6.840	11.10	19.57	44.76
(2)	.2075	.4366	.6892	.9681	1.276	1.618	1.998	2.424	2.904	3.450	4.079	4.818	6.812	10.31	20.04
(3)	1.992	3.931	5.778	7.581	9.408	11.33	13.41	15.73	18.38	21.46	25.13	29.64	42.82	68.51	144.2

TABLE 7.23 : I_0^* (α = .25 , μ = .60)

Q: MODE:	0.05	0.10	0.15	0.20	0.25	0.30	0.35	0.40	0.45	0.50	0.55	0.60	0.70	0.80	0.90
(1)	.0578	.2180	.4594	.7703	1.149	1.601	2.137	2.772	3.530	4.442	5.556	6.943	11.05	19.08	42.58
(2)	.1529	.3159	.4900	.6763	.8767	1.094	1.331	1.598	1.910	2.310	2.928	4.132	9.536	20.72	53.11
(3)	1.895	2.789	3.508	4.234	5.017	5.883	6.853	7.954	9.216	10.68	12.41	14.50	20.33	30.93	59.78

TABLE 7.23 : I_0^* (α = .25 , μ = .70)

Q: MODE:	0.05	0.10	0.15	0.20	0.25	0.30	0.35	0.40	0.45	0.50	0.55	0.60	0.70	0.80	0.90
(1)	.0666	.2457	.5056	.8300	1.215	1.665	2.187	2.794	3.505	4.345	5.352	6.583	10.12	16.79	35.66
(2)	1.312	2.769	4.310	5.951	7.720	9.655	11.80	14.23	17.02	20.28	24.19	28.98	43.03	70.54	151.9
(3)	.2730	.5673	.8849	1.228	1.599	2.002	2.438	2.914	3.434	4.007	4.642	5.356	7.142	10.04	20.27

TABLE 7.23 : I_0^* (α = .25 , μ = .80)

Q: MODE:	0.05	0.10	0.15	0.20	0.25	0.30	0.35	0.40	0.45	0.50	0.55	0.60	0.70	0.80	0.90
(1)	.0752	.2702	.5420	.8697	1.247	1.676	2.159	2.708	3.332	4.051	4.888	5.882	8.613	13.45	26.37
(2)	1.599	3.280	4.969	6.677	8.439	10.29	12.29	14.49	16.95	19.77	23.09	27.09	38.52	60.27	123.1
(3)	2.523	5.133	7.868	10.77	13.90	17.32	21.10	25.36	30.24	35.93	42.72	51.04	75.31	122.6	262.3

TABLE 7.23 : I_0^* (α = .30 , μ = .20)

Q: MODE:	0.05	0.10	0.15	0.20	0.25	0.30	0.35	0.40	0.45	0.50	0.55	0.60	0.70	0.80	0.90
(1)	.0165	.0661	.1495	.2680	.4237	.6199	.8613	1.155	1.511	1.943	2.473	3.132	5.074	8.841	19.81
(2)	.2849	.6223	1.020	1.487	2.035	2.677	3.431	4.317	5.366	6.616	8.127	9.986	15.40	25.77	55.71
(3)	.6600	1.368	2.139	2.990	3.944	5.026	6.272	7.724	9.442	11.50	14.02	17.17	26.51	44.90	99.08

TABLE 7.23 : I_0^* (α = .30 , μ = .30)

Q: MODE:	0.05	0.10	0.15	0.20	0.25	0.30	0.35	0.40	0.45	0.50	0.55	0.60	0.70	0.80	0.90
(1)	.0246	.0982	.2202	.3908	.6120	.8879	1.226	1.635	2.132	2.737	3.483	4.417	7.202	12.68	28.81
(2)	.2645	.5738	.9347	1.356	1.848	2.426	3.108	3.919	4.890	6.068	7.516	9.331	14.75	25.46	57.06
(3)	.4998	1.027	1.587	2.188	2.838	3.547	4.331	5.207	6.199	7.341	8.681	10.29	14.83	23.29	47.20

TABLE 7.23 : I_0^* (α = .30 , μ = .40)

Q: MODE:	0.05	0.10	0.15	0.20	0.25	0.30	0.35	0.40	0.45	0.50	0.55	0.60	0.70	0.80	0.90
(1)	.0327	.1290	.2851	.4989	.7716	1.108	1.517	2.011	2.610	3.340	4.244	5.379	8.794	15.59	35.81
(2)	.2367	.5062	.8128	1.161	1.559	2.014	2.537	3.143	3.851	4.690	5.697	6.931	10.50	17.25	36.44
(3)	1.196	2.433	3.728	5.102	6.580	8.191	9.974	11.98	14.27	16.94	20.12	24.01	35.34	57.41	122.3

TABLE 7.23 : I_0^* (α = .30 , μ = .50)

Q: MODE:	0.05	0.10	0.15	0.20	0.25	0.30	0.35	0.40	0.45	0.50	0.55	0.60	0.70	0.80	0.90
(1)	.0406	.1579	.3429	.5902	.8996	1.276	1.729	2.272	2.927	3.723	4.703	5.934	9.628	16.97	38.80
(2)	.1995	.4197	.6626	.9305	1.227	1.555	1.920	2.328	2.789	3.314	3.920	4.634	6.575	10.03	19.79
(3)	1.666	3.316	4.915	6.492	8.096	9.782	11.61	13.63	15.93	18.61	21.81	25.72	37.14	59.40	125.0

TABLE 7.23 : I_0^* (α = .30 , μ = .60)

Q: MODE:	0.05	0.10	0.15	0.20	0.25	0.30	0.35	0.40	0.45	0.50	0.55	0.60	0.70	0.80	0.90
(1)	.0484	.1846	.3930	.6639	.9934	1.392	1.861	2.410	3.080	3.878	4.851	6.060	9.637	16.62	37.06
(2)	.1471	.3042	.4723	.6527	.8474	1.059	1.294	1.559	1.876	2.290	2.927	4.077	8.882	18.90	48.10
(3)	1.650	2.520	3.212	3.898	4.629	5.429	6.321	7.327	8.477	9.807	11.37	13.25	18.50	28.03	54.01

TABLE 7.23 : I_0^* (α = .30 , μ = .70)

Q: MODE:	0.05	0.10	0.15	0.20	0.25	0.30	0.35	0.40	0.45	0.50	0.55	0.60	0.70	0.80	0.90
(1)	.0559	.2090	.4350	.7200	1.060	1.457	1.919	2.456	3.083	3.825	4.713	5.798	8.914	14.78	31.37
(2)	1.110	2.367	3.704	5.130	6.668	8.351	10.22	12.33	14.75	17.59	20.99	25.16	37.37	61.29	132.0
(3)	.2622	.5444	.8488	1.178	1.533	1.918	2.336	2.792	3.290	3.840	4.451	5.141	6.889	9.825	20.26

TABLE 7.23 : I_0^* (α = .30 , μ = .80)

Q: MODE:	0.05	0.10	0.15	0.20	0.25	0.30	0.35	0.40	0.45	0.50	0.55	0.60	0.70	0.80	0.90
(1)	.0632	.2308	.4690	.7594	1.096	1.478	1.911	2.401	2.960	3.603	4.353	5.243	7.690	12.03	23.66
(2)	1.342	2.779	4.239	5.723	7.258	8.877	10.62	12.53	14.68	17.14	20.03	23.51	33.46	52.37	107.0
(3)	2.195	4.464	6.843	9.369	12.09	15.06	18.35	22.05	26.30	31.25	37.15	44.38	65.49	106.6	228.1

TABLE 7.23 : I_0^* (α = .40 , μ = .20)

Q: MODE:	0.05	0.10	0.15	0.20	0.25	0.30	0.35	0.40	0.45	0.50	0.55	0.60	0.70	0.80	0.90
(1)	.0124	.0497	.1129	.2034	.3234	.4762	.6663	.8999	1.186	1.537	1.972	2.519	4.153	7.371	16.03
(2)	.2614	.5637	.9119	1.312	1.772	2.302	2.915	3.628	4.465	5.460	6.660	8.137	12.45	20.78	44.98
(3)	.6116	1.264	1.969	2.740	3.594	4.552	5.638	6.887	8.344	10.07	12.16	14.74	22.34	37.24	81.18

TABLE 7.23 : I_0^* (α = .40 , μ = .30)

Q: MODE:	0.05	0.10	0.15	0.20	0.25	0.30	0.35	0.40	0.45	0.50	0.55	0.60	0.70	0.80	0.90
(1)	.0185	.0742	.1674	.2994	.4723	.6901	.9585	1.286	1.685	2.173	2.777	3.536	5.810	10.31	23.63
(2)	.2444	.5273	.8539	1.231	1.667	2.172	2.761	3.452	4.272	5.255	6.453	7.944	12.37	21.06	46.71
(3)	.4647	.9555	1.478	2.039	2.645	3.308	4.039	4.856	5.781	6.845	8.093	9.592	13.83	21.73	44.14

TABLE 7.23 : I_0^* (α = .40 , μ = .40)

Q: MODE:	0.05	0.10	0.15	0.20	0.25	0.30	0.35	0.40	0.45	0.50	0.55	0.60	0.70	0.80	0.90
(1)	.0246	.0978	.2184	.3859	.6019	.8702	1.198	1.595	2.077	2.667	3.395	4.313	7.074	12.58	28.98
(2)	.2194	.4683	.7502	1.070	1.432	1.845	2.317	2.860	3.493	4.237	5.126	6.210	9.323	15.19	31.81
(3)	1.019	2.073	3.178	4.351	5.613	6.991	8.517	10.23	12.20	14.50	17.23	20.58	30.37	49.44	105.8

TABLE 7.23 : I_0^* (α = .40 , μ = .50)

Q: MODE:	0.05	0.10	0.15	0.20	0.25	0.30	0.35	0.40	0.45	0.50	0.55	0.60	0.70	0.80	0.90
(1)	.0306	.1204	.2648	.4610	.7089	1.012	1.378	1.818	2.349	2.994	3.788	4.784	7.771	13.70	31.34
(2)	.1853	.3898	.6153	.8642	1.139	1.445	1.785	2.166	2.597	3.091	3.663	4.342	6.215	9.635	19.48
(3)	1.256	2.527	3.796	5.070	6.380	7.761	9.255	10.91	12.79	14.97	17.56	20.73	29.97	47.96	101.0

TABLE 7.23 : I_0^* (α = .40 , μ = .60)

Q: MODE:	0.05	0.10	0.15	0.20	0.25	0.30	0.35	0.40	0.45	0.50	0.55	0.60	0.70	0.80	0.90
(1)	.0365	.1415	.3060	.5235	.7924	1.116	1.499	1.955	2.497	3.149	3.944	4.931	7.845	13.53	30.14
(2)	.1368	.2834	.4410	.6111	.7963	1.000	1.229	1.495	1.821	2.260	2.927	4.017	8.141	16.73	41.95
(3)	1.300	2.094	2.733	3.356	4.008	4.715	5.497	6.374	7.371	8.520	9.868	11.48	15.99	24.17	46.58

TABLE 7.23 : I_0^* (α = .40 , μ = .70)

Q: MODE:	0.05	0.10	0.15	0.20	0.25	0.30	0.35	0.40	0.45	0.50	0.55	0.60	0.70	0.80	0.90
(1)	.0423	.1612	.3416	.5733	.8527	1.181	1.564	2.009	2.530	3.145	3.882	4.781	7.364	12.22	25.96
(2)	.8538	1.857	2.938	4.096	5.347	6.716	8.238	9.955	11.93	14.24	17.00	20.39	30.32	49.75	107.2
(3)	.2429	.5036	.7845	1.088	1.416	1.771	2.157	2.579	3.043	3.555	4.130	4.785	6.485	9.499	20.25

TABLE 7.23 : I_0^* (α = .40 , μ = .80)

Q: MODE:	0.05	0.10	0.15	0.20	0.25	0.30	0.35	0.40	0.45	0.50	0.55	0.60	0.70	0.80	0.90
(1)	.0480	.1791	.3715	.6107	.8910	1.212	1.576	1.991	2.464	3.009	3.645	4.402	6.487	10.21	20.23
(2)	1.016	2.138	3.302	4.501	5.748	7.066	8.487	10.05	11.80	13.80	16.16	18.99	27.08	42.46	86.88
(3)	1.786	3.633	5.568	7.624	9.837	12.25	14.93	17.94	21.39	25.42	30.22	36.09	53.25	86.69	185.3

TABLE 7.23 : I_0^* (α = .50 , μ = .20)

Q: MODE:	0.05	0.10	0.15	0.20	0.25	0.30	0.35	0.40	0.45	0.50	0.55	0.60	0.70	0.80	0.90
(1)	.0099	.0398	.0907	.1640	.2619	.3877	.5459	.7425	.9862	1.289	1.668	2.149	3.607	6.510	15.09
(2)	.2410	.5137	.8221	1.171	1.568	2.022	2.543	3.148	3.856	4.698	5.714	6.966	10.64	17.76	38.54
(3)	.5697	1.175	1.825	2.531	3.309	4.173	5.147	6.258	7.545	9.061	10.88	13.12	19.69	32.53	70.35

TABLE 7.23 : I_0^* (α = .50 , μ = .30)

Q: MODE:	0.05	0.10	0.15	0.20	0.25	0.30	0.35	0.40	0.45	0.50	0.55	0.60	0.70	0.80	0.90
(1)	.0148	.0596	.1351	.2430	.3858	.5672	.7926	1.070	1.409	1.827	2.346	3.001	4.974	8.900	20.55
(2)	.2269	.4872	.7852	1.127	1.518	1.970	2.492	3.103	3.822	4.682	5.726	7.020	10.84	18.34	40.44
(3)	.4343	.8941	1.385	1.912	2.484	3.109	3.801	4.574	5.452	6.462	7.649	9.077	13.12	20.68	42.19

TABLE 7.23 : I_0^* (α = .50 , μ = .40)

Q: MODE:	0.05	0.10	0.15	0.20	0.25	0.30	0.35	0.40	0.45	0.50	0.55	0.60	0.70	0.80	0.90
(1)	.0197	.0788	.1772	.3155	.4956	.7212	.9984	1.336	1.747	2.251	2.876	3.662	6.035	10.77	24.89
(2)	.2044	.4355	.6966	.9917	1.326	1.705	2.138	2.637	3.215	3.894	4.705	5.692	8.521	13.84	28.93
(3)	.9207	1.873	2.871	3.930	5.071	6.317	7.697	9.251	11.03	13.11	15.59	18.62	27.49	44.79	95.90

TABLE 7.23 : I_0^* (α = .50 , μ = .50)

Q:	0.05	0.10	0.15	0.20	0.25	0.30	0.35	0.40	0.45	0.50	0.55	0.60	0.70	0.80	0.90
MODE:															
(1)	.0246	.0973	.2161	.3795	.5884	.8457	1.158	1.534	1.989	2.542	3.223	4.078	6.642	11.73	26.85
(2)	.1730	.3639	.5747	.8076	1.066	1.352	1.673	2.033	2.442	2.912	3.462	4.117	5.952	9.362	19.27
(3)	1.008	2.043	3.097	4.175	5.295	6.484	7.774	9.204	10.83	12.71	14.94	17.67	25.62	41.06	86.52

TABLE 7.23 : I_0^* (α = .50 , μ = .60)

Q:	0.05	0.10	0.15	0.20	0.25	0.30	0.35	0.40	0.45	0.50	0.55	0.60	0.70	0.80	0.90
MODE:															
(1)	.0294	.1148	.2512	.4341	.6628	.9394	1.269	1.662	2.130	2.693	3.380	4.232	6.749	11.65	25.97
(2)	.1279	.2654	.4140	.5755	.7527	.9503	1.176	1.443	1.779	2.239	2.926	3.985	7.744	15.50	38.35
(3)	1.068	1.780	2.369	2.939	3.534	4.175	4.882	5.672	6.570	7.603	8.815	10.27	14.32	21.70	41.96

TABLE 7.23 : I_0^* (α = .50 , μ = .70)

Q:	0.05	0.10	0.15	0.20	0.25	0.30	0.35	0.40	0.45	0.50	0.55	0.60	0.70	0.80	0.90
MODE:															
(1)	.0341	.1313	.2821	.4789	.7188	1.003	1.335	1.722	2.177	2.714	3.358	4.144	6.403	10.65	22.68
(2)	.6966	1.545	2.472	3.469	4.548	5.731	7.045	8.529	10.23	12.22	14.61	17.53	26.10	42.84	92.36
(3)	.2262	.4682	.7288	1.010	1.315	1.645	2.006	2.401	2.836	3.321	3.868	4.498	6.172	9.265	20.25

TABLE 7.23 : I_0^* (α = .50 , μ = .80)

Q:	0.05	0.10	0.15	0.20	0.25	0.30	0.35	0.40	0.45	0.50	0.55	0.60	0.70	0.80	0.90
MODE:															
(1)	.0386	.1466	.3087	.5138	.7570	1.037	1.358	1.723	2.141	2.624	3.190	3.863	5.727	9.067	18.12
(2)	.8187	1.743	2.723	3.744	4.813	5.949	7.174	8.522	10.03	11.76	13.80	16.24	23.22	36.48	74.74
(3)	1.542	3.138	4.809	6.584	8.495	10.58	12.89	15.49	18.47	21.94	26.08	31.15	45.94	74.75	159.7

TABLE 7.23 : I_0^* (α = .60 , μ = .20)

Q:	0.05	0.10	0.15	0.20	0.25	0.30	0.35	0.40	0.45	0.50	0.55	0.60	0.70	0.80	0.90
MODE:															
(1)	.0082	.0332	.0758	.1374	.2203	.3275	.4637	.6349	.8497	1.120	1.462	1.902	3.248	5.948	13.94
(2)	.2233	.4710	.7469	1.056	1.406	1.804	2.260	2.790	3.411	4.150	5.044	6.148	9.396	15.72	34.25
(3)	.5331	1.097	1.701	2.354	3.069	3.862	4.751	5.761	6.927	8.296	9.936	11.95	17.84	29.31	63.08

TABLE 7.23 : I_0^* (α = .60 , μ = .30)

Q:	0.05	0.10	0.15	0.20	0.25	0.30	0.35	0.40	0.45	0.50	0.55	0.60	0.70	0.80	0.90
MODE:															
(1)	.0124	.0498	.1133	.2047	.3266	.4828	.6786	.9212	1.221	1.592	2.054	2.641	4.417	7.963	18.51
(2)	.2117	.4524	.7263	1.038	1.395	1.804	2.277	2.828	3.477	4.250	5.188	6.349	9.777	16.48	36.22
(3)	.4077	.8405	1.304	1.803	2.345	2.941	3.601	4.341	5.182	6.153	7.297	8.675	12.59	19.93	40.85

TABLE 7.23 : I_0^* (α = .60 , μ = .40)

Q:	0.05	0.10	0.15	0.20	0.25	0.30	0.35	0.40	0.45	0.50	0.55	0.60	0.70	0.80	0.90
MODE:															
(1)	.0165	.0660	.1491	.2672	.4223	.6182	.8606	1.158	1.521	1.967	2.522	3.222	5.337	9.564	22.17
(2)	.1913	.4070	.6502	.9247	1.235	1.588	1.990	2.453	2.990	3.622	4.376	5.294	7.927	12.89	26.94
(3)	.8592	1.748	2.679	3.667	4.732	5.894	7.182	8.631	10.29	12.23	14.54	17.37	25.63	41.76	89.39

TABLE 7.23 : I_0^* (α = .60 , μ = .50)

Q:	0.05	0.10	0.15	0.20	0.25	0.30	0.35	0.40	0.45	0.50	0.55	0.60	0.70	0.80	0.90
MODE:															
(1)	.0205	.0817	.1826	.3232	.5045	.7298	1.004	1.337	1.740	2.231	2.837	3.597	5.880	10.41	23.86
(2)	.1622	.3413	.5393	.7584	1.002	1.273	1.577	1.920	2.312	2.765	3.298	3.937	5.748	9.161	19.13
(3)	.8414	1.715	2.618	3.555	4.541	5.595	6.744	8.021	9.471	11.15	13.15	15.59	22.68	36.44	76.87

TABLE 7.23 : I_0^* (α = .60 , μ = .60)

Q:	0.05	0.10	0.15	0.20	0.25	0.30	0.35	0.40	0.45	0.50	0.55	0.60	0.70	0.80	0.90
MODE:															
(1)	.0245	.0967	.2132	.3718	.5719	.8158	1.108	1.457	1.874	2.377	2.991	3.753	6.003	10.39	23.18
(2)	.1201	.2497	.3904	.5444	.7150	.9073	1.130	1.400	1.746	2.224	2.926	3.965	7.499	14.72	36.00
(3)	.9053	1.543	2.085	2.612	3.160	3.751	4.401	5.129	5.955	6.907	8.024	9.362	13.11	19.95	38.78

TABLE 7.23 : I_0^* (α = .60 , μ = .70)

Q:	0.05	0.10	0.15	0.20	0.25	0.30	0.35	0.40	0.45	0.50	0.55	0.60	0.70	0.80	0.90
MODE:															
(1)	.0285	.1108	.2406	.4125	.6241	.8763	1.173	1.520	1.929	2.412	2.993	3.702	5.743	9.589	20.48
(2)	.5899	1.332	2.155	3.046	4.012	5.071	6.249	7.577	9.102	10.89	13.02	15.63	23.29	38.25	82.46
(3)	.2115	.4373	.6803	.9428	1.227	1.537	1.876	2.248	2.661	3.123	3.649	4.260	5.920	9.088	20.25

TABLE 7.23 : I_0^* (α = .60 , μ = .80)

Q:	0.05	0.10	0.15	0.20	0.25	0.30	0.35	0.40	0.45	0.50	0.55	0.60	0.70	0.80	0.90
MODE:															
(1)	.0324	.1241	.2645	.4450	.6613	.9126	1.201	1.532	1.911	2.351	2.868	3.484	5.197	8.281	16.68
(2)	.6856	1.474	2.326	3.224	4.172	5.183	6.277	7.481	8.833	10.38	12.20	14.38	20.63	32.46	66.63
(3)	1.382	2.810	4.307	5.897	7.608	9.475	11.54	13.87	16.53	19.64	23.34	27.87	41.08	66.82	142.7

TABLE 7.23 : I_0^* (α = .70 , μ = .20)

Q:	0.05	0.10	0.15	0.20	0.25	0.30	0.35	0.40	0.45	0.50	0.55	0.60	0.70	0.80	0.90
MODE:															
(1)	.0071	.0285	.0651	.1183	.1901	.2838	.4037	.5563	.7501	.9971	1.314	1.724	2.994	5.555	13.14
(2)	.2078	.4341	.6833	.9608	1.273	1.629	2.037	2.511	3.069	3.733	4.540	5.539	8.488	14.25	31.18
(3)	.5009	1.029	1.593	2.201	2.865	3.600	4.422	5.356	6.431	7.692	9.201	11.05	16.45	26.96	57.84

TABLE 7.23 : I_0^* (α = .70 , μ = .30)

Q:	0.05	0.10	0.15	0.20	0.25	0.30	0.35	0.40	0.45	0.50	0.55	0.60	0.70	0.80	0.90
MODE:															
(1)	.0106	.0428	.0976	.1769	.2834	.4210	.5950	.8123	1.083	1.420	1.842	2.381	4.018	7.297	17.05
(2)	.1983	.4221	.6753	.9627	1.290	1.666	2.100	2.605	3.200	3.909	4.769	5.835	8.979	15.13	33.19
(3)	.3842	.7934	1.232	1.707	2.225	2.795	3.429	4.142	4.956	5.897	7.008	8.350	12.17	19.36	39.86

TABLE 7.23 : I_0^* (α = .70 , μ = .40)

Q:	0.05	0.10	0.15	0.20	0.25	0.30	0.35	0.40	0.45	0.50	0.55	0.60	0.70	0.80	0.90
MODE:															
(1)	.0141	.0568	.1288	.2319	.3685	.5424	.7590	1.026	1.354	1.759	2.264	2.902	4.834	8.700	20.23
(2)	.1797	.3820	.6096	.8664	1.157	1.487	1.864	2.299	2.804	3.398	4.109	4.975	7.466	12.16	25.48
(3)	.8178	1.663	2.549	3.490	4.503	5.608	6.832	8.209	9.786	11.63	13.82	16.50	24.35	39.64	84.78

TABLE 7.23 : I_0^* (α = .70 , μ = .50)

Q:	0.05	0.10	0.15	0.20	0.25	0.30	0.35	0.40	0.45	0.50	0.55	0.60	0.70	0.80	0.90
MODE:															
(1)	.0176	.0704	.1583	.2818	.4426	.6438	.8907	1.191	1.556	2.002	2.554	3.247	5.329	9.461	21.72
(2)	.1526	.3214	.5081	.7153	.9459	1.204	1.494	1.824	2.201	2.641	3.160	3.788	5.585	9.007	19.03
(3)	.7222	1.478	2.268	3.100	3.983	4.936	5.980	7.146	8.472	10.01	11.84	14.07	20.55	33.11	69.97

TABLE 7.23 : I_0^* (α = .70 , μ = .60)

Q:	0.05	0.10	0.15	0.20	0.25	0.30	0.35	0.40	0.45	0.50	0.55	0.60	0.70	0.80	0.90
MODE:															
(1)	.0211	.0835	.1854	.3256	.5043	.7235	.9879	1.305	1.685	2.143	2.704	3.401	5.461	9.473	21.17
(2)	.1132	.2358	.3697	.5171	.6818	.8698	1.091	1.363	1.718	2.212	2.926	3.951	7.335	14.18	34.35
(3)	.7848	1.360	1.860	2.348	2.858	3.407	4.013	4.692	5.463	6.354	7.401	8.657	12.18	18.64	36.46

TABLE 7.23 : I_0^* ($\alpha = .70$, $\mu = .70$)

Q:	0.05	0.10	0.15	0.20	0.25	0.30	0.35	0.40	0.45	0.50	0.55	0.60	0.70	0.80	0.90
MODE:															
(1)	.0245	.0959	.2100	.3629	.5530	.7813	1.051	1.369	1.743	2.187	2.721	3.375	5.259	8.814	18.89
(2)	.5124	1.178	1.926	2.740	3.626	4.598	5.678	6.898	8.297	9.932	11.89	14.28	21.29	34.98	75.41
(3)	.1986	.4101	.6376	.8836	1.151	1.442	1.762	2.115	2.509	2.952	3.461	4.058	5.710	8.949	20.24

TABLE 7.23 : I_0^* ($\alpha = .70$, $\mu = .80$)

Q:	0.05	0.10	0.15	0.20	0.25	0.30	0.35	0.40	0.45	0.50	0.55	0.60	0.70	0.80	0.90
MODE:															
(1)	.0278	.1077	.2317	.3933	.5889	.8180	1.083	1.387	1.738	2.146	2.625	3.200	4.802	7.702	15.64
(2)	.5898	1.279	2.033	2.843	3.702	4.622	5.621	6.722	7.959	9.377	11.04	13.04	18.75	29.58	60.83
(3)	1.268	2.578	3.952	5.410	6.979	8.691	10.59	12.72	15.16	18.00	21.39	25.54	37.63	61.17	130.5

TABLE 7.23 : I_0^* ($\alpha = .80$, $\mu = .20$)

Q:	0.05	0.10	0.15	0.20	0.25	0.30	0.35	0.40	0.45	0.50	0.55	0.60	0.70	0.80	0.90
MODE:															
(1)	.0062	.0250	.0571	.1038	.1673	.2505	.3579	.4960	.6737	.9032	1.201	1.591	2.806	5.266	12.54
(2)	.1942	.4022	.6291	.8805	1.163	1.485	1.855	2.287	2.795	3.404	4.144	5.064	7.792	13.14	28.87
(3)	.4723	.9689	1.498	2.067	2.689	3.376	4.144	5.016	6.020	7.198	8.607	10.33	15.37	25.16	53.89

TABLE 7.23 : I_0^* ($\alpha = .80$, $\mu = .30$)

Q:	0.05	0.10	0.15	0.20	0.25	0.30	0.35	0.40	0.45	0.50	0.55	0.60	0.70	0.80	0.90
MODE:															
(1)	.0093	.0375	.0857	.1557	.2505	.3738	.5308	.7286	.9766	1.288	1.681	2.183	3.710	6.000	15.97
(2)	.1864	.3954	.6307	.8972	1.201	1.549	1.950	2.419	2.971	3.631	4.432	5.425	8.356	14.08	30.90
(3)	.3634	.7514	1.169	1.623	2.119	2.668	3.280	3.971	4.762	5.680	6.766	8.081	11.83	18.90	39.09

TABLE 7.23 : I_0^* ($\alpha = .80$, $\mu = .40$)

Q:	0.05	0.10	0.15	0.20	0.25	0.30	0.35	0.40	0.45	0.50	0.55	0.60	0.70	0.80	0.90
MODE:															
(1)	.0124	.0498	.1134	.2049	.3271	.4839	.6807	.9246	1.226	1.600	2.067	2.659	4.455	8.051	18.77
(2)	.1695	.3598	.5738	.8152	1.089	1.400	1.756	2.166	2.645	3.210	3.886	4.713	7.095	11.59	24.37
(3)	.7882	1.603	2.457	3.363	4.339	5.403	6.581	7.906	9.422	11.19	13.30	15.88	23.40	38.07	81.36

TABLE 7.23 : I_0^* ($\alpha = .80$, $\mu = .50$)

Q:	0.05	0.10	0.15	0.20	0.25	0.30	0.35	0.40	0.45	0.50	0.55	0.60	0.70	0.80	0.90
MODE:															
(1)	.0154	.0618	.1397	.2499	.3947	.5772	.8025	1.078	1.414	1.826	2.337	2.979	4.911	8.745	20.11
(2)	.1442	.3037	.4805	.6772	.8967	1.143	1.422	1.739	2.105	2.533	3.043	3.662	5.450	8.884	18.95
(3)	.6326	1.298	2.002	2.749	3.552	4.425	5.388	6.468	7.699	9.131	10.83	12.91	18.94	30.61	64.79

TABLE 7.23 : I_0^* ($\alpha = .80$, $\mu = .60$)

Q:	0.05	0.10	0.15	0.20	0.25	0.30	0.35	0.40	0.45	0.50	0.55	0.60	0.70	0.80	0.90
MODE:															
(1)	.0185	.0734	.1641	.2899	.4517	.6517	.8942	1.186	1.537	1.962	2.483	3.131	5.048	8.781	19.66
(2)	.1070	.2235	.3512	.4928	.6524	.8366	1.057	1.331	1.695	2.202	2.926	3.942	7.218	13.79	33.14
(3)	.6923	1.215	1.677	2.132	2.608	3.123	3.692	4.331	5.059	5.902	6.893	8.086	11.44	17.61	34.67

TABLE 7.23 : I_0^* ($\alpha = .80$, $\mu = .70$)

Q:	0.05	0.10	0.15	0.20	0.25	0.30	0.35	0.40	0.45	0.50	0.55	0.60	0.70	0.80	0.90
MODE:															
(1)	.0215	.0846	.1864	.3244	.4975	.7069	.9558	1.250	1.598	2.012	2.510	3.122	4.887	8.223	17.69
(2)	.4535	1.059	1.751	2.509	3.334	4.241	5.250	6.388	7.693	9.218	11.04	13.27	19.80	32.54	70.13
(3)	.1872	.3860	.5999	.8314	1.083	1.359	1.662	1.999	2.376	2.803	3.298	3.883	5.532	8.837	20.24

TABLE 7.23 : I_0^* ($\alpha = .80$, $\mu = .80$)

Q:	0.05	0.10	0.15	0.20	0.25	0.30	0.35	0.40	0.45	0.50	0.55	0.60	0.70	0.80	0.90
MODE:															
(1)	.0244	.0951	.2063	.3529	.5321	.7434	.9888	1.272	1.601	1.984	2.435	2.978	4.496	7.257	14.84
(2)	.5176	1.130	1.812	2.550	3.341	4.191	5.118	6.141	7.291	8.610	10.16	12.02	17.34	27.41	56.46
(3)	1.183	2.406	3.687	5.047	6.510	8.107	9.874	11.86	14.13	16.78	19.94	23.80	35.05	56.94	121.4

TABLE 7.23 : I_0^* ($\alpha = .90$, $\mu = .20$)

Q:	0.05	0.10	0.15	0.20	0.25	0.30	0.35	0.40	0.45	0.50	0.55	0.60	0.70	0.80	0.90
MODE:															
(1)	.0055	.0222	.0508	.0925	.1494	.2243	.3218	.4483	.6131	.8288	1.112	1.486	2.662	5.045	12.09
(2)	.1821	.3744	.5824	.8123	1.071	1.365	1.705	2.102	2.571	3.135	3.823	4.682	7.239	12.27	27.08
(3)	.4468	.9155	1.414	1.950	2.535	3.181	3.905	4.726	5.673	6.784	8.114	9.745	14.50	23.73	50.80

TABLE 7.23 : I_0^* ($\alpha = .90$, $\mu = .30$)

Q:	0.05	0.10	0.15	0.20	0.25	0.30	0.35	0.40	0.45	0.50	0.55	0.60	0.70	0.80	0.90
MODE:															
(1)	.0083	.0334	.0764	.1392	.2245	.3363	.4798	.6620	.8923	1.183	1.553	2.028	3.485	6.415	15.13
(2)	.1759	.3717	.5915	.8400	1.123	1.447	1.823	2.261	2.779	3.399	4.152	5.088	7.853	13.26	29.10
(3)	.3447	.7139	1.113	1.547	2.025	2.554	3.148	3.821	4.593	5.492	6.559	7.852	11.55	18.54	38.49

TABLE 7.23 : I_0^* ($\alpha = .90$, $\mu = .40$)

Q:	0.05	0.10	0.15	0.20	0.25	0.30	0.35	0.40	0.45	0.50	0.55	0.60	0.70	0.80	0.90
MODE:															
(1)	.0110	.0444	.1013	.1836	.2944	.4374	.6181	.8437	1.124	1.473	1.911	2.467	4.158	7.546	17.64
(2)	.1603	.3400	.5420	.7699	1.028	1.323	1.661	2.051	2.508	3.048	3.697	4.492	6.787	11.13	23.48
(3)	.7662	1.558	2.388	3.269	4.216	5.249	6.393	7.678	9.148	10.86	12.90	15.40	22.69	36.87	78.72

TABLE 7.23 : I_0^* ($\alpha = .90$, $\mu = .50$)

Q:	0.05	0.10	0.15	0.20	0.25	0.30	0.35	0.40	0.45	0.50	0.55	0.60	0.70	0.80	0.90
MODE:															
(1)	.0137	.0551	.1250	.2247	.3565	.5239	.7319	.9874	1.301	1.686	2.164	2.767	4.582	8.185	18.85
(2)	.1366	.2879	.4559	.6431	.8527	1.089	1.357	1.664	2.021	2.439	2.941	3.554	5.337	8.785	18.89
(3)	.5628	1.158	1.792	2.472	3.209	4.017	4.914	5.924	7.081	8.429	10.03	11.99	17.67	28.65	60.75

TABLE 7.23 : I_0^* ($\alpha = .90$, $\mu = .60$)

Q:	0.05	0.10	0.15	0.20	0.25	0.30	0.35	0.40	0.45	0.50	0.55	0.60	0.70	0.80	0.90
MODE:															
(1)	.0164	.0656	.1472	.2615	.4096	.5939	.8188	1.091	1.419	1.817	2.306	2.916	4.720	8.238	18.48
(2)	.1016	.2124	.3346	.4710	.6260	.8069	1.026	1.304	1.675	2.194	2.926	3.934	7.130	13.50	32.20
(3)	.6192	1.097	1.526	1.952	2.399	2.884	3.422	4.027	4.719	5.522	6.470	7.612	10.84	16.78	33.25

TABLE 7.23 : I_0^* ($\alpha = .90$, $\mu = .70$)

Q:	0.05	0.10	0.15	0.20	0.25	0.30	0.35	0.40	0.45	0.50	0.55	0.60	0.70	0.80	0.90
MODE:															
(1)	.0191	.0756	.1676	.2935	.4528	.6468	.8786	1.154	1.480	1.870	2.341	2.919	4.591	7.756	16.74
(2)	.4070	.9654	1.613	2.327	3.106	3.963	4.916	5.991	7.223	8.664	10.38	12.49	18.64	30.64	66.02
(3)	.1769	.3645	.5663	.7849	1.023	1.285	1.574	1.896	2.258	2.672	3.154	3.730	5.378	8.744	20.24

TABLE 7.23 : I_0^* ($\alpha = .90$, $\mu = .80$)

Q:	0.05	0.10	0.15	0.20	0.25	0.30	0.35	0.40	0.45	0.50	0.55	0.60	0.70	0.80	0.90
MODE:															
(1)	.0218	.0852	.1860	.3203	.4860	.6828	.9126	1.179	1.489	1.852	2.281	2.798	4.250	6.902	14.21
(2)	.4612	1.013	1.636	2.317	3.053	3.849	4.718	5.680	6.762	8.005	9.463	11.22	16.22	25.71	53.05
(3)	1.117	2.273	3.483	4.767	6.149	7.655	9.323	11.20	13.34	15.84	18.82	22.45	33.05	53.66	114.4

TABLE 7.23 : I_0^* (α = 1.0 , μ = .20)

Q: MODE:	0.05	0.10	0.15	0.20	0.25	0.30	0.35	0.40	0.45	0.50	0.55	0.60	0.70	0.80	0.90
(1)	.0050	.0200	.0458	.0835	.1349	.2031	.2924	.4095	.5638	.7683	1.040	1.402	2.547	4.870	11.73
(2)	.1714	.3499	.5419	.7535	.9915	1.263	1.577	1.946	2.383	2.910	3.557	4.366	6.787	11.57	25.65
(3)	.4239	.8676	1.339	1.846	2.399	3.010	3.695	4.475	5.374	6.431	7.697	9.250	13.78	22.57	48.31

TABLE 7.23 : I_0^* (α = 1.0 , μ = .30)

Q: MODE:	0.05	0.10	0.15	0.20	0.25	0.30	0.35	0.40	0.45	0.50	0.55	0.60	0.70	0.80	0.90
(1)	.0074	.0301	.0689	.1258	.2035	.3058	.4382	.6077	.8235	1.098	1.450	1.903	3.298	6.109	14.47
(2)	.1664	.3507	.5568	.7895	1.055	1.359	1.712	2.125	2.614	3.201	3.916	4.805	7.438	12.58	27.65
(3)	.3279	.6801	1.062	1.479	1.940	2.453	3.030	3.687	4.444	5.327	6.378	7.656	11.32	18.23	38.00

TABLE 7.23 : I_0^* (α = 1.0 , μ = .40)

Q: MODE:	0.05	0.10	0.15	0.20	0.25	0.30	0.35	0.40	0.45	0.50	0.55	0.60	0.70	0.80	0.90
(1)	.0099	.0400	.0915	.1664	.2677	.3994	.5670	.7775	1.040	1.369	1.783	2.311	3.919	7.142	16.74
(2)	.1521	.3223	.5135	.7294	.9746	1.255	1.577	1.950	2.388	2.908	3.533	4.302	6.528	10.75	22.76
(3)	.7493	1.524	2.335	3.196	4.121	5.131	6.247	7.502	8.936	10.61	12.60	15.03	22.12	35.93	76.63

TABLE 7.23 : I_0^* (α = 1.0 , μ = .50)

Q: MODE:	0.05	0.10	0.15	0.20	0.25	0.30	0.35	0.40	0.45	0.50	0.55	0.60	0.70	0.80	0.90
(1)	.0124	.0498	.1131	.2042	.3253	.4802	.6738	.9130	1.207	1.571	2.023	2.594	4.316	7.735	17.85
(2)	.1298	.2737	.4337	.6125	.8133	1.040	1.299	1.598	1.945	2.356	2.851	3.459	5.241	8.702	18.84
(3)	.5069	1.045	1.622	2.246	2.928	3.682	4.524	5.478	6.574	7.854	9.380	11.24	16.65	27.07	57.52

TABLE 7.23 : I_0^* (α = 1.0 , μ = .60)

Q: MODE:	0.05	0.10	0.15	0.20	0.25	0.30	0.35	0.40	0.45	0.50	0.55	0.60	0.70	0.80	0.90
(1)	.0148	.0592	.1335	.2382	.3750	.5463	.7565	1.012	1.321	1.698	2.161	2.740	4.455	7.799	17.54
(2)	.0966	.2025	.3196	.4513	.6022	.7802	.9983	1.279	1.658	2.188	2.926	3.928	7.062	13.27	31.46
(3)	.5599	1.000	1.399	1.799	2.221	2.680	3.190	3.767	4.429	5.198	6.109	7.210	10.33	16.09	32.10

TABLE 7.23 : I_0^* (α = 1.0 , μ = .70)

Q: MODE:	0.05	0.10	0.15	0.20	0.25	0.30	0.35	0.40	0.45	0.50	0.55	0.60	0.70	0.80	0.90
(1)	.0172	.0684	.1523	.2681	.4159	.5970	.8146	1.074	1.383	1.753	2.201	2.751	4.349	7.377	15.98
(2)	.3694	.8892	1.501	2.180	2.922	3.739	4.648	5.673	6.848	8.221	9.859	11.86	17.72	29.12	62.75
(3)	.1677	.3453	.5362	.7433	.9696	1.219	1.495	1.803	2.153	2.555	3.026	3.593	5.244	8.666	20.24

TABLE 7.23 : I_0^* (α = 1.0 , μ = .80)

Q: MODE:	0.05	0.10	0.15	0.20	0.25	0.30	0.35	0.40	0.45	0.50	0.55	0.60	0.70	0.80	0.90
(1)	.0196	.0771	.1694	.2935	.4478	.6323	.8490	1.102	1.396	1.742	2.153	2.649	4.047	6.612	13.71
(2)	.4159	.9179	1.492	2.127	2.818	3.569	4.392	5.304	6.332	7.513	8.899	10.57	15.33	24.34	50.32
(3)	1.065	2.167	3.320	4.544	5.861	7.297	8.885	10.67	12.71	15.09	17.92	21.38	31.46	51.04	108.7

TABLE 7.23 : I_0^* (1/α = .10 , μ = .20)

Q: MODE:	0.05	0.10	0.15	0.20	0.25	0.30	0.35	0.40	0.45	0.50	0.55	0.60	0.70	0.80	0.90
(1)	.0005	.0020	.0046	.0085	.0140	.0217	.0330	.0512	.0874	.1783	.3821	.6976	1.683	3.584	8.981
(2)	.0262	.0490	.0724	.0984	.1287	.1653	.2109	.2701	.3509	.4684	.6528	.9637	2.345	5.445	14.00
(3)	.0751	.1517	.2340	.3267	.4356	.5692	.7403	.9690	1.287	1.742	2.396	3.314	6.237	11.80	27.17

TABLE 7.23 : I_0^* (1/α = .10 , μ = .30)

Q: MODE:	0.05	0.10	0.15	0.20	0.25	0.30	0.35	0.40	0.45	0.50	0.55	0.60	0.70	0.80	0.90
(1)	.0007	.0030	.0070	.0131	.0219	.0346	.0538	.0854	.1436	.2595	.4679	.7792	1.771	3.710	9.248
(2)	.0279	.0561	.0869	.1224	.1648	.2172	.2843	.3738	.4984	.6805	.9577	1.386	2.967	6.228	15.29
(3)	.0615	.1331	.2200	.3298	.4754	.6774	.9683	1.392	1.993	2.797	3.820	5.089	8.687	15.28	33.70

TABLE 7.23 : I_0^* (1/α = .10 , μ = .40)

Q: MODE:	0.05	0.10	0.15	0.20	0.25	0.30	0.35	0.40	0.45	0.50	0.55	0.60	0.70	0.80	0.90
(1)	.0010	.0041	.0095	.0178	.0300	.0481	.0756	.1202	.1972	.3317	.5485	.8616	1.862	3.834	9.492
(2)	.0269	.0563	.0898	.1293	.1775	.2384	.3181	.4267	.5813	.8098	1.154	1.665	3.402	6.816	16.26
(3)	.6345	1.290	1.974	2.699	3.474	4.316	5.242	6.275	7.449	8.808	10.41	12.36	17.98	28.77	60.25

TABLE 7.23 : I_0^* (1/α = .10 , μ = .50)

Q: MODE:	0.05	0.10	0.15	0.20	0.25	0.30	0.35	0.40	0.45	0.50	0.55	0.60	0.70	0.80	0.90
(1)	.0012	.0051	.0119	.0225	.0382	.0615	.0970	.1531	.2447	.3927	.6168	.9327	1.938	3.926	9.636
(2)	.0237	.0507	.0823	.1207	.1687	.2317	.3183	.4445	.6380	.9425	1.406	2.063	4.093	7.894	18.42
(3)	.0510	.1069	.1709	.2476	.3438	.4715	.6510	.9185	1.333	1.963	2.860	4.042	7.476	13.74	31.05

TABLE 7.23 : I_0^* (1/α = .10 , μ = .60)

Q: MODE:	0.05	0.10	0.15	0.20	0.25	0.30	0.35	0.40	0.45	0.50	0.55	0.60	0.70	0.80	0.90
(1)	.0015	.0061	.0144	.0271	.0462	.0743	.1167	.1821	.2843	.4409	.6692	.9853	1.986	3.964	9.637
(2)	.0181	.0398	.0672	.1048	.1628	.2685	.4877	.8883	1.446	2.123	2.925	3.885	6.576	11.57	25.78
(3)	.0580	.1098	.1627	.2210	.2888	.3707	.4737	.6084	.7928	1.058	1.455	2.067	4.336	8.990	21.62

TABLE 7.23 : I_0^* (1/α = .10 , μ = .70)

Q: MODE:	0.05	0.10	0.15	0.20	0.25	0.30	0.35	0.40	0.45	0.50	0.55	0.60	0.70	0.80	0.90
(1)	.0017	.0071	.0167	.0316	.0537	.0861	.1340	.2062	.3149	.4755	.7038	1.016	2.000	3.942	9.497
(2)	.0412	.1524	.4400	.8574	1.335	1.860	2.441	3.091	3.827	4.678	5.683	6.900	10.40	17.08	36.49
(3)	.0292	.0590	.0913	.1276	.1693	.2188	.2794	.3566	.4602	.6094	.8444	1.253	3.251	7.852	20.23

TABLE 7.23 : I_0^* (1/α = .10 , μ = .80)

Q: MODE:	0.05	0.10	0.15	0.20	0.25	0.30	0.35	0.40	0.45	0.50	0.55	0.60	0.70	0.80	0.90
(1)	.0020	.0081	.0191	.0359	.0606	.0964	.1483	.2244	.3358	.4961	.7199	1.024	1.980	3.866	9.261
(2)	.0423	.0992	.1796	.2989	.4810	.7538	1.134	1.620	2.209	2.906	3.733	4.729	7.546	12.81	27.79
(3)	.6648	1.351	2.069	2.828	3.642	4.525	5.496	6.582	7.816	9.244	10.93	12.98	18.90	30.26	63.44

TABLE 7.23 : I_0^* (1/α = .20 , μ = .20)

Q: MODE:	0.05	0.10	0.15	0.20	0.25	0.30	0.35	0.40	0.45	0.50	0.55	0.60	0.70	0.80	0.90
(1)	.0010	.0040	.0092	.0170	.0279	.0430	.0647	.0981	.1562	.2681	.4715	.7802	1.771	3.713	9.267
(2)	.0497	.0942	.1400	.1907	.2493	.3192	.4050	.5135	.6557	.8497	1.126	1.531	3.013	6.205	15.30
(3)	.1384	.2802	.4315	.5993	.7917	1.019	1.297	1.644	2.088	2.663	3.416	4.400	7.416	13.24	29.67

TABLE 7.23 : I_0^* (1/α = .20 , μ = .30)

Q: MODE:	0.05	0.10	0.15	0.20	0.25	0.30	0.35	0.40	0.45	0.50	0.55	0.60	0.70	0.80	0.90
(1)	.0015	.0061	.0140	.0261	.0433	.0679	.1038	.1584	.2461	.3887	.6095	.9257	1.943	3.965	9.804
(2)	.0521	.1055	.1641	.2309	.3097	.4053	.5241	.6761	.8754	1.143	1.511	2.018	3.696	7.097	16.76
(3)	.1119	.2396	.3895	.5706	.7955	1.082	1.452	1.932	2.552	3.338	4.322	5.548	9.083	15.67	34.21

TABLE 7.23 : I_0^* ($1/\alpha = .20$, $\mu = .40$)

Q: MODE:	0.05	0.10	0.15	0.20	0.25	0.30	0.35	0.40	0.45	0.50	0.55	0.60	0.70	0.80	0.90
(1)	.0020	.0081	.0189	.0353	.0591	.0932	.1429	.2168	.3282	.4943	.7332	1.063	2.112	4.207	10.29
(2)	.0496	.1040	.1657	.2376	.3236	.4288	.5609	.7306	.9537	1.252	1.656	2.203	3.947	7.387	17.07
(3)	.6453	1.312	2.008	2.745	3.535	4.393	5.337	6.393	7.593	8.982	10.63	12.63	18.40	29.51	62.00

TABLE 7.23 : I_0^* ($1/\alpha = .20$, $\mu = .50$)

Q: MODE:	0.05	0.10	0.15	0.20	0.25	0.30	0.35	0.40	0.45	0.50	0.55	0.60	0.70	0.80	0.90
(1)	.0025	.0102	.0237	.0444	.0746	.1177	.1797	.2693	.3984	.5814	.8340	1.174	2.245	4.377	10.56
(2)	.0434	.0925	.1493	.2165	.2981	.4001	.5315	.7059	.9438	1.273	1.728	2.345	4.275	7.999	18.47
(3)	.1019	.2132	.3396	.4886	.6710	.9026	1.207	1.616	2.169	2.907	3.867	5.089	8.658	15.32	34.04

TABLE 7.23 : I_0^* ($1/\alpha = .20$, $\mu = .60$)

Q: MODE:	0.05	0.10	0.15	0.20	0.25	0.30	0.35	0.40	0.45	0.50	0.55	0.60	0.70	0.80	0.90
(1)	.0030	.0122	.0284	.0533	.0893	.1404	.2125	.3138	.4548	.6479	.9075	1.252	2.321	4.439	10.55
(2)	.0330	.0716	.1187	.1795	.2644	.3951	.6130	.9691	1.480	2.132	2.925	3.889	6.624	11.74	26.38
(3)	.1156	.2174	.3200	.4316	.5589	.7094	.8927	1.123	1.419	1.810	2.338	3.058	5.398	10.05	22.94

TABLE 7.23 : I_0^* ($1/\alpha = .20$, $\mu = .70$)

Q: MODE:	0.05	0.10	0.15	0.20	0.25	0.30	0.35	0.40	0.45	0.50	0.55	0.60	0.70	0.80	0.90
(1)	.0035	.0142	.0331	.0618	.1029	.1605	.2400	.3488	.4960	.6927	.9519	1.291	2.335	4.385	10.27
(2)	.0809	.2611	.5946	1.031	1.529	2.081	2.693	3.379	4.160	5.065	6.136	7.437	11.19	18.38	39.36
(3)	.0541	.1097	.1698	.2366	.3127	.4013	.5071	.6371	.8028	1.024	1.336	1.801	3.653	7.964	20.23

TABLE 7.23 : I_0^* ($1/\alpha = .20$, $\mu = .80$)

Q: MODE:	0.05	0.10	0.15	0.20	0.25	0.30	0.35	0.40	0.45	0.50	0.55	0.60	0.70	0.80	0.90
(1)	.0040	.0162	.0375	.0697	.1151	.1774	.2615	.3738	.5220	.7159	.9676	1.294	2.289	4.234	9.801
(2)	.0845	.1962	.3476	.5546	.8331	1.194	1.642	2.178	2.808	3.549	4.427	5.487	8.503	14.17	30.36
(3)	.7065	1.436	2.199	3.007	3.874	4.815	5.852	7.012	8.331	9.861	11.67	13.88	20.24	32.51	68.40

TABLE 7.23 : I_0^* ($1/\alpha = .30$, $\mu = .20$)

Q: MODE:	0.05	0.10	0.15	0.20	0.25	0.30	0.35	0.40	0.45	0.50	0.55	0.60	0.70	0.80	0.90
(1)	.0015	.0060	.0138	.0254	.0416	.0639	.0954	.1421	.2169	.3440	.5525	.8610	1.862	3.846	9.558
(2)	.0709	.1359	.2031	.2773	.3626	.4632	.5849	.7360	.9287	1.181	1.523	1.994	3.587	6.929	16.59
(3)	.1925	.3903	.6007	.8315	1.092	1.393	1.750	2.180	2.709	3.367	4.197	5.253	8.427	14.57	32.11

TABLE 7.23 : I_0^* ($1/\alpha = .30$, $\mu = .30$)

Q: MODE:	0.05	0.10	0.15	0.20	0.25	0.30	0.35	0.40	0.45	0.50	0.55	0.60	0.70	0.80	0.90
(1)	.0022	.0091	.0210	.0389	.0644	.1001	.1509	.2246	.3342	.4974	.7334	1.062	2.114	4.224	10.37
(2)	.0731	.1492	.2328	.3276	.4384	.5707	.7320	.9326	1.187	1.515	1.943	2.511	4.303	7.892	18.20
(3)	.1541	.3272	.5260	.7592	1.038	1.377	1.795	2.311	2.950	3.740	4.716	5.928	9.439	16.04	34.71

TABLE 7.23 : I_0^* ($1/\alpha = .30$, $\mu = .40$)

Q: MODE:	0.05	0.10	0.15	0.20	0.25	0.30	0.35	0.40	0.45	0.50	0.55	0.60	0.70	0.80	0.90
(1)	.0030	.0121	.0282	.0525	.0872	.1361	.2049	.3019	.4388	.6300	.8916	1.243	2.352	4.577	11.09
(2)	.0690	.1450	.2309	.3300	.4467	.5868	.7579	.9704	1.239	1.582	2.026	2.606	4.396	7.903	17.84
(3)	.6565	1.335	2.044	2.794	3.599	4.474	5.437	6.515	7.742	9.164	10.85	12.90	18.83	30.27	63.77

TABLE 7.23 : I_0^* ($1/\alpha = .30$, $\mu = .50$)

Q: MODE:	0.05	0.10	0.15	0.20	0.25	0.30	0.35	0.40	0.45	0.50	0.55	0.60	0.70	0.80	0.90
(1)	.0037	.0152	.0353	.0658	.1094	.1702	.2542	.3696	.5265	.7375	1.017	1.368	2.532	4.816	11.48
(2)	.0601	.1276	.2050	.2951	.4022	.5321	.6932	.8972	1.161	1.506	1.961	2.564	4.436	8.100	18.52
(3)	.1527	.3190	.5062	.7238	.9844	1.305	1.707	2.217	2.867	3.687	4.717	6.006	9.765	16.86	37.00

TABLE 7.23 : I_0^* ($1/\alpha = .30$, $\mu = .60$)

Q: MODE:	0.05	0.10	0.15	0.20	0.25	0.30	0.35	0.40	0.45	0.50	0.55	0.60	0.70	0.80	0.90
(1)	.0045	.0182	.0423	.0786	.1301	.2009	.2969	.4252	.5949	.8170	1.106	1.480	2.628	4.893	11.45
(2)	.0454	.0978	.1599	.2371	.3388	.4825	.6994	1.032	1.510	2.140	2.925	3.894	6.674	11.92	26.98
(3)	.1727	.3229	.4722	.6323	.8122	1.021	1.269	1.571	1.946	2.420	3.028	3.818	6.252	10.99	24.21

TABLE 7.23 : I_0^* ($1/\alpha = .30$, $\mu = .70$)

Q: MODE:	0.05	0.10	0.15	0.20	0.25	0.30	0.35	0.40	0.45	0.50	0.55	0.60	0.70	0.80	0.90
(1)	.0052	.0212	.0490	.0906	.1488	.2274	.3316	.4675	.6430	.8678	1.155	1.523	2.635	4.802	11.02
(2)	.1195	.3552	.7287	1.192	1.715	2.296	2.941	3.667	4.494	5.454	6.593	7.979	11.99	19.70	42.25
(3)	.0754	.1535	.2377	.3308	.4357	.5562	.6974	.8666	1.075	1.342	1.695	2.187	3.965	8.070	20.23

TABLE 7.23 : I_0^* ($1/\alpha = .30$, $\mu = .80$)

Q: MODE:	0.05	0.10	0.15	0.20	0.25	0.30	0.35	0.40	0.45	0.50	0.55	0.60	0.70	0.80	0.90
(1)	.0059	.0241	.0555	.1017	.1652	.2491	.3577	.4961	.6708	.8904	1.167	1.517	2.561	4.577	10.32
(2)	.1265	.2913	.5071	.7860	1.137	1.563	2.067	2.653	3.333	4.125	5.063	6.195	9.419	15.50	32.90
(3)	.7491	1.523	2.333	3.190	4.110	5.111	6.214	7.450	8.857	10.49	12.43	14.78	21.61	34.79	73.39

TABLE 7.23 : I_0^* ($1/\alpha = .40$, $\mu = .20$)

Q: MODE:	0.05	0.10	0.15	0.20	0.25	0.30	0.35	0.40	0.45	0.50	0.55	0.60	0.70	0.80	0.90
(1)	.0020	.0080	.0184	.0338	.0552	.0845	.1253	.1839	.2729	.4129	.6286	.9405	1.955	3.983	9.854
(2)	.0899	.1743	.2620	.3587	.4690	.5982	.7528	.9418	1.178	1.480	1.875	2.403	4.110	7.630	17.89
(3)	.2392	.4858	.7475	1.032	1.350	1.712	2.133	2.631	3.228	3.956	4.855	5.983	9.336	15.82	34.51

TABLE 7.23 : I_0^* ($1/\alpha = .40$, $\mu = .30$)

Q: MODE:	0.05	0.10	0.15	0.20	0.25	0.30	0.35	0.40	0.45	0.50	0.55	0.60	0.70	0.80	0.90
(1)	.0030	.0121	.0279	.0516	.0851	.1315	.1950	.2862	.4143	.5957	.8478	1.191	2.284	4.486	10.94
(2)	.0915	.1881	.2942	.4144	.5536	.7180	.9155	1.157	1.456	1.832	2.311	2.931	4.843	8.638	19.60
(3)	.1899	.4008	.6393	.9134	1.233	1.612	2.064	2.608	3.267	4.067	5.045	6.256	9.765	16.39	35.20

TABLE 7.23 : I_0^* ($1/\alpha = .40$, $\mu = .40$)

Q: MODE:	0.05	0.10	0.15	0.20	0.25	0.30	0.35	0.40	0.45	0.50	0.55	0.60	0.70	0.80	0.90
(1)	.0040	.0162	.0374	.0694	.1147	.1773	.2630	.3799	.5387	.7521	1.036	1.412	2.586	4.946	11.89
(2)	.0858	.1805	.2873	.4098	.5525	.7212	.9236	1.170	1.473	1.851	2.328	2.938	4.785	8.379	18.60
(3)	.6682	1.359	2.081	2.845	3.666	4.557	5.541	6.642	7.896	9.352	11.08	13.18	19.27	31.04	65.56

TABLE 7.23 : I_0^* ($1/\alpha = .40$, $\mu = .50$)

Q: MODE:	0.05	0.10	0.15	0.20	0.25	0.30	0.35	0.40	0.45	0.50	0.55	0.60	0.70	0.80	0.90
(1)	.0050	.0202	.0468	.0867	.1429	.2197	.3230	.4603	.6411	.8769	1.183	1.581	2.806	5.247	12.40
(2)	.0743	.1576	.2522	.3612	.4886	.6402	.8235	1.049	1.331	1.688	2.148	2.744	4.581	8.196	18.57
(3)	.2035	.4242	.6708	.9537	1.286	1.685	2.169	2.762	3.490	4.385	5.485	6.850	10.82	18.36	39.96

TABLE 7.23 : I_0^* ($1/\alpha = .40$, $\mu = .60$)

Q: MODE:	0.05	0.10	0.15	0.20	0.25	0.30	0.35	0.40	0.45	0.50	0.55	0.60	0.70	0.80	0.90
(1)	.0060	.0242	.0559	.1031	.1689	.2574	.3738	.5249	.7191	.9669	1.283	1.688	2.916	5.332	12.34
(2)	.0560	.1197	.1940	.2838	.3974	.5497	.7661	1.083	1.537	2.147	2.925	3.899	6.725	12.10	27.60
(3)	.2295	.4262	.6191	.8235	1.050	1.309	1.611	1.971	2.407	2.943	3.612	4.458	6.989	11.84	25.43

TABLE 7.23 : I_0^* ($1/\alpha = .40$, $\mu = .70$)

Q: MODE:	0.05	0.10	0.15	0.20	0.25	0.30	0.35	0.40	0.45	0.50	0.55	0.60	0.70	0.80	0.90
(1)	.0069	.0281	.0646	.1184	.1920	.2891	.4141	.5726	.7718	1.021	1.334	1.730	2.912	5.201	11.76
(2)	.1571	.4415	.8521	1.344	1.896	2.508	3.188	3.954	4.829	5.846	7.054	8.525	12.79	21.03	45.16
(3)	.0939	.1916	.2969	.4128	.5424	.6896	.8596	1.060	1.301	1.599	1.981	2.491	4.224	8.169	20.23

TABLE 7.23 : I_0^* ($1/\alpha = .40$, $\mu = .80$)

Q: MODE	0.05	0.10	0.15	0.20	0.25	0.30	0.35	0.40	0.45	0.50	0.55	0.60	0.70	0.80	0.90
(1)	.0079	.0319	.0729	.1322	.2118	.3144	.4433	.6032	.7999	1.042	1.340	1.713	2.809	4.900	10.83
(2)	.1683	.3847	.6599	1.002	1.413	1.895	2.450	3.085	3.815	4.662	5.662	6.869	10.31	16.80	35.42
(3)	.7926	1.612	2.469	3.377	4.352	5.412	6.583	7.895	9.390	11.13	13.19	15.70	22.99	37.07	78.40

TABLE 7.23 : I_0^* ($1/\alpha = .50$, $\mu = .20$)

Q: MODE:	0.05	0.10	0.15	0.20	0.25	0.30	0.35	0.40	0.45	0.50	0.55	0.60	0.70	0.80	0.90
(1)	.0025	.0100	.0230	.0422	.0687	.1049	.1544	.2241	.3257	.4775	.7015	1.019	2.050	4.124	10.16
(2)	.1070	.2097	.3170	.4350	.5691	.7251	.9102	1.134	1.409	1.755	2.198	2.776	4.600	8.312	19.19
(3)	.2799	.5694	.8761	1.209	1.576	1.991	2.467	3.022	3.679	4.469	5.433	6.633	10.18	17.03	36.87

TABLE 7.23 : I_0^* ($1/\alpha = .50$, $\mu = .30$)

Q: MODE:	0.05	0.10	0.15	0.20	0.25	0.30	0.35	0.40	0.45	0.50	0.55	0.60	0.70	0.80	0.90
(1)	.0037	.0151	.0348	.0642	.1055	.1620	.2390	.3445	.4894	.6875	.9560	1.316	2.454	4.752	11.52
(2)	.1078	.2227	.3494	.4925	.6574	.8505	1.080	1.356	1.694	2.112	2.636	3.305	5.337	9.348	20.98
(3)	.2208	.4638	.7353	1.043	1.396	1.806	2.286	2.854	3.530	4.343	5.330	6.546	10.07	16.73	35.69

TABLE 7.23 : I_0^* ($1/\alpha = .50$, $\mu = .40$)

Q: MODE:	0.05	0.10	0.15	0.20	0.25	0.30	0.35	0.40	0.45	0.50	0.55	0.60	0.70	0.80	0.90
(1)	.0050	.0202	.0466	.0861	.1415	.2169	.3182	.4531	.6315	.8658	1.172	1.572	2.815	5.314	12.69
(2)	.1004	.2116	.3367	.4795	.6446	.8378	1.066	1.341	1.673	2.080	2.587	3.225	5.134	8.824	19.33
(3)	.6805	1.384	2.119	2.898	3.735	4.645	5.649	6.774	8.056	9.546	11.31	13.47	19.72	31.82	67.37

TABLE 7.23 : I_0^* ($1/\alpha = .50$, $\mu = .50$)

Q: MODE:	0.05	0.10	0.15	0.20	0.25	0.30	0.35	0.40	0.45	0.50	0.55	0.60	0.70	0.80	0.90
(1)	.0062	.0252	.0581	.1072	.1754	.2670	.3877	.5447	.7469	1.006	1.337	1.764	3.070	5.671	13.31
(2)	.0867	.1835	.2928	.4177	.5621	.7313	.9323	1.175	1.471	1.840	2.304	2.899	4.713	8.288	18.61
(3)	.2542	.5289	.8335	1.179	1.578	2.047	2.603	3.268	4.066	5.030	6.201	7.646	11.84	19.85	42.90

TABLE 7.23 : I_0^* ($1/\alpha = .50$, $\mu = .60$)

Q: MODE:	0.05	0.10	0.15	0.20	0.25	0.30	0.35	0.40	0.45	0.50	0.55	0.60	0.70	0.80	0.90
(1)	.0074	.0301	.0693	.1270	.2061	.3106	.4454	.6169	.8330	1.105	1.447	1.881	3.190	5.760	13.22
(2)	.0651	.1384	.2227	.3226	.4455	.6042	.8203	1.127	1.562	2.155	2.925	3.903	6.778	12.28	28.22
(3)	.2857	.5273	.7611	1.006	1.274	1.576	1.925	2.333	2.819	3.406	4.124	5.020	7.651	12.64	26.61

TABLE 7.23 : I_0^* ($1/\alpha = .50$, $\mu = .70$)

Q: MODE:	0.05	0.10	0.15	0.20	0.25	0.30	0.35	0.40	0.45	0.50	0.55	0.60	0.70	0.80	0.90
(1)	.0087	.0350	.0799	.1452	.2331	.3467	.4902	.6686	.8891	1.161	1.499	1.922	3.174	5.586	12.48
(2)	.1939	.5227	.9688	1.491	2.073	2.717	3.434	4.241	5.164	6.239	7.517	9.076	13.60	22.36	48.07
(3)	.1101	.2252	.3490	.4848	.6358	.8058	1.000	1.225	1.492	1.816	2.219	2.742	4.445	8.263	20.23

TABLE 7.23 : I_0^* ($1/\alpha = .50$, $\mu = .80$)

Q: MODE:	0.05	0.10	0.15	0.20	0.25	0.30	0.35	0.40	0.45	0.50	0.55	0.60	0.70	0.80	0.90
(1)	.0099	.0397	.0899	.1615	.2558	.3748	.5216	.7001	.9162	1.178	1.497	1.892	3.040	5.209	11.33
(2)	.2099	.4765	.8075	1.206	1.671	2.204	2.807	3.490	4.270	5.173	6.236	7.519	11.18	18.08	37.93
(3)	.8367	1.702	2.607	3.566	4.597	5.718	6.957	8.346	9.931	11.77	13.96	16.63	24.38	39.38	83.42

TABLE 7.23 : I_0^* ($1/\alpha = .60$, $\mu = .20$)

Q: MODE:	0.05	0.10	0.15	0.20	0.25	0.30	0.35	0.40	0.45	0.50	0.55	0.60	0.70	0.80	0.90
(1)	.0030	.0120	.0276	.0505	.0822	.1250	.1830	.2630	.3762	.5392	.7720	1.097	2.147	4.268	10.46
(2)	.1224	.2424	.3683	.5068	.6635	.8448	1.058	1.314	1.625	2.011	2.498	3.125	5.066	8.981	20.48
(3)	.3157	.6431	.9898	1.364	1.777	2.238	2.763	3.369	4.081	4.928	5.956	7.227	10.96	18.19	39.20

TABLE 7.23 : I_0^* ($1/\alpha = .60$, $\mu = .30$)

Q: MODE:	0.05	0.10	0.15	0.20	0.25	0.30	0.35	0.40	0.45	0.50	0.55	0.60	0.70	0.80	0.90
(1)	.0045	.0181	.0417	.0767	.1256	.1918	.2809	.4004	.5607	.7750	1.060	1.437	2.623	5.020	12.10
(2)	.1222	.2537	.3992	.5633	.7515	.9705	1.229	1.537	1.909	2.364	2.930	3.646	5.798	10.03	22.34
(3)	.2477	.5184	.8180	1.154	1.534	1.970	2.475	3.064	3.758	4.584	5.582	6.807	10.35	17.05	36.16

TABLE 7.23 : I_0^* ($1/\alpha = .60$, $\mu = .40$)

Q: MODE:	0.05	0.10	0.15	0.20	0.25	0.30	0.35	0.40	0.45	0.50	0.55	0.60	0.70	0.80	0.90
(1)	.0060	.0242	.0557	.1026	.1677	.2553	.3711	.5226	.7193	.9735	1.302	1.727	3.041	5.681	13.50
(2)	.1133	.2390	.3804	.5411	.7258	.9403	1.192	1.490	1.847	2.281	2.814	3.480	5.452	9.245	20.04
(3)	.6932	1.410	2.159	2.954	3.807	4.735	5.761	6.910	8.222	9.746	11.56	13.77	20.18	32.62	69.19

TABLE 7.23 : I_0^* ($1/\alpha = .60$, $\mu = .50$)

Q: MODE:	0.05	0.10	0.15	0.20	0.25	0.30	0.35	0.40	0.45	0.50	0.55	0.60	0.70	0.80	0.90
(1)	.0074	.0301	.0694	.1272	.2068	.3124	.4492	.6244	.8466	1.128	1.484	1.940	3.328	6.091	14.23
(2)	.0974	.2060	.3281	.4667	.6255	.8095	1.025	1.282	1.591	1.969	2.440	3.035	4.834	8.377	18.66
(3)	.3049	.6331	.9944	1.400	1.862	2.395	3.016	3.746	4.609	5.638	6.881	8.407	12.84	21.32	45.84

TABLE 7.23 : I_0^* ($1/\alpha = .60$, $\mu = .60$)

Q: MODE:	0.05	0.10	0.15	0.20	0.25	0.30	0.35	0.40	0.45	0.50	0.55	0.60	0.70	0.80	0.90
(1)	.0089	.0360	.0825	.1502	.2420	.3614	.5130	.7031	.9396	1.234	1.601	2.065	3.456	6.179	14.10
(2)	.0730	.1546	.2474	.3556	.4860	.6498	.8661	1.165	1.584	2.162	2.925	3.908	6.832	12.47	28.86
(3)	.3415	.6262	.8981	1.180	1.485	1.826	2.214	2.664	3.193	3.823	4.586	5.527	8.257	13.39	27.75

TABLE 7.23 : I_0^* ($1/\alpha = .60$, $\mu = .70$)

Q: MODE:	0.05	0.10	0.15	0.20	0.25	0.30	0.35	0.40	0.45	0.50	0.55	0.60	0.70	0.80	0.90
(1)	.0104	.0418	.0949	.1711	.2723	.4012	.5614	.7581	.9981	1.291	1.652	2.102	3.424	5.959	13.20
(2)	.2300	.6003	1.081	1.634	2.247	2.924	3.678	4.528	5.500	6.633	7.982	9.629	14.42	23.71	50.99
(3)	.1244	.2548	.3952	.5485	.7182	.9080	1.123	1.370	1.658	2.002	2.423	2.956	4.640	8.351	20.23

TABLE 7.23 : I_0^* (1/α = .60 , μ = .80)

Q:	0.05	0.10	0.15	0.20	0.25	0.30	0.35	0.40	0.45	0.50	0.55	0.60	0.70	0.80	0.90
MODE:															
(1)	.0118	.0473	.1065	.1896	.2975	.4315	.5943	.7898	1.024	1.304	1.643	2.059	3.258	5.506	11.82
(2)	.2514	.5670	.9507	1.401	1.916	2.496	3.145	3.876	4.706	5.665	6.792	8.152	12.03	19.36	40.43
(3)	.8815	1.793	2.747	3.758	4.845	6.028	7.336	8.803	10.48	12.43	14.74	17.57	25.78	41.69	88.46

TABLE 7.23 : I_0^* (1/α = .70 , μ = .20)

Q:	0.05	0.10	0.15	0.20	0.25	0.30	0.35	0.40	0.45	0.50	0.55	0.60	0.70	0.80	0.90
MODE:															
(1)	.0035	.0140	.0322	.0588	.0955	.1448	.2110	.3008	.4250	.5987	.8409	1.174	2.245	4.414	10.77
(2)	.1365	.2725	.4162	.5742	.7525	.9579	1.198	1.484	1.829	2.252	2.781	3.455	5.515	9.640	21.77
(3)	.3474	.7086	1.091	1.504	1.956	2.460	3.029	3.683	4.445	5.347	6.436	7.778	11.71	19.33	41.50

TABLE 7.23 : I_0^* (1/α = .70 , μ = .30)

Q:	0.05	0.10	0.15	0.20	0.25	0.30	0.35	0.40	0.45	0.50	0.55	0.60	0.70	0.80	0.90
MODE:															
(1)	.0052	.0211	.0485	.0891	.1454	.2211	.3216	.4543	.6293	.8591	1.161	1.556	2.792	5.289	12.69
(2)	.1350	.2817	.4443	.6276	.8373	1.080	1.364	1.701	2.105	2.596	3.202	3.962	6.235	10.69	23.69
(3)	.2713	.5662	.8903	1.250	1.655	2.113	2.639	3.247	3.957	4.798	5.808	7.044	10.61	17.36	36.63

TABLE 7.23 : I_0^* (1/α = .70 , μ = .40)

Q:	0.05	0.10	0.15	0.20	0.25	0.30	0.35	0.40	0.45	0.50	0.55	0.60	0.70	0.80	0.90
MODE:															
(1)	.0069	.0281	.0647	.1188	.1933	.2926	.4222	.5893	.8035	1.077	1.427	1.878	3.263	6.047	14.31
(2)	.1246	.2633	.4192	.5959	.7981	1.031	1.303	1.623	2.003	2.460	3.018	3.711	5.747	9.645	20.74
(3)	.7065	1.437	2.201	3.011	3.881	4.829	5.877	7.051	8.393	9.953	11.81	14.07	20.65	33.43	71.03

TABLE 7.23 : I_0^* (1/α = .70 , μ = .50)

Q:	0.05	0.10	0.15	0.20	0.25	0.30	0.35	0.40	0.45	0.50	0.55	0.60	0.70	0.80	0.90
MODE:															
(1)	.0087	.0351	.0805	.1469	.2375	.3562	.5082	.7004	.9416	1.244	1.625	2.110	3.581	6.506	15.13
(2)	.1070	.2260	.3592	.5097	.6810	.8778	1.106	1.375	1.695	2.083	2.559	3.157	4.946	8.463	18.71
(3)	.3555	.7368	1.154	1.616	2.138	2.731	3.412	4.202	5.126	6.220	7.533	9.143	13.81	22.77	48.77

TABLE 7.23 : I_0^* (1/α = .70 , μ = .60)

Q:	0.05	0.10	0.15	0.20	0.25	0.30	0.35	0.40	0.45	0.50	0.55	0.60	0.70	0.80	0.90
MODE:															
(1)	.0104	.0419	.0955	.1730	.2767	.4100	.5774	.7849	1.041	1.356	1.748	2.241	3.713	6.592	14.96
(2)	.0800	.1688	.2689	.3842	.5209	.6890	.9055	1.198	1.605	2.169	2.925	3.913	6.887	12.67	29.50
(3)	.3969	.7229	1.030	1.345	1.684	2.059	2.483	2.969	3.536	4.205	5.008	5.992	8.820	14.11	28.87

TABLE 7.23 : I_0^* (1/α = .70 , μ = .70)

Q:	0.05	0.10	0.15	0.20	0.25	0.30	0.35	0.40	0.45	0.50	0.55	0.60	0.70	0.80	0.90
MODE:															
(1)	.0121	.0485	.1097	.1963	.3101	.4531	.6289	.8424	1.101	1.414	1.798	2.273	3.665	6.324	13.90
(2)	.2656	.6752	1.189	1.773	2.418	3.130	3.922	4.814	5.837	7.029	8.449	10.18	15.24	25.05	53.92
(3)	.1371	.2812	.4362	.6053	.7915	.9987	1.232	1.497	1.803	2.164	2.600	3.143	4.813	8.436	20.23

TABLE 7.23 : I_0^* (1/α = .70 , μ = .80)

Q:	0.05	0.10	0.15	0.20	0.25	0.30	0.35	0.40	0.45	0.50	0.55	0.60	0.70	0.80	0.90
MODE:															
(1)	.0138	.0549	.1227	.2168	.3373	.4851	.6627	.8738	1.124	1.422	1.780	2.217	3.466	5.793	12.30
(2)	.2927	.6563	1.090	1.590	2.152	2.777	3.471	4.248	5.128	6.142	7.334	8.771	12.87	20.62	42.92
(3)	.9268	1.885	2.888	3.952	5.095	6.341	7.719	9.265	11.03	13.09	15.53	18.51	27.19	44.02	93.51

TABLE 7.23 : I_0^* (1/α = .80 , μ = .20)

Q:	0.05	0.10	0.15	0.20	0.25	0.30	0.35	0.40	0.45	0.50	0.55	0.60	0.70	0.80	0.90
MODE:															
(1)	.0040	.0160	.0367	.0670	.1087	.1644	.2385	.3377	.4723	.6565	.9083	1.250	2.345	4.564	11.09
(2)	.1492	.3004	.4610	.6376	.8366	1.065	1.331	1.645	2.022	2.482	3.051	3.770	5.949	10.29	23.06
(3)	.3757	.7672	1.182	1.629	2.118	2.660	3.270	3.968	4.778	5.734	6.882	8.295	12.43	20.43	43.79

TABLE 7.23 : I_0^* (1/α = .80 , μ = .30)

Q:	0.05	0.10	0.15	0.20	0.25	0.30	0.35	0.40	0.45	0.50	0.55	0.60	0.70	0.80	0.90
MODE:															
(1)	.0059	.0241	.0554	.1014	.1650	.2498	.3613	.5067	.6957	.9408	1.259	1.673	2.961	5.561	13.28
(2)	.1466	.3069	.4852	.6863	.9158	1.181	1.489	1.853	2.287	2.811	3.454	4.258	6.651	11.34	25.02
(3)	.2923	.6085	.9541	1.336	1.761	2.240	2.784	3.409	4.136	4.991	6.014	7.263	10.86	17.66	37.10

TABLE 7.23 : I_0^* (1/α = .80 , μ = .40)

Q:	0.05	0.10	0.15	0.20	0.25	0.30	0.35	0.40	0.45	0.50	0.55	0.60	0.70	0.80	0.90
MODE:															
(1)	.0079	.0321	.0737	.1348	.2185	.3290	.4717	.6538	.8847	1.177	1.548	2.025	3.484	6.413	15.12
(2)	.1348	.2851	.4539	.6450	.8629	1.113	1.403	1.742	2.143	2.623	3.204	3.923	6.022	10.03	21.43
(3)	.7203	1.465	2.244	3.071	3.959	4.927	5.996	7.197	8.569	10.17	12.06	14.38	21.14	34.25	72.88

TABLE 7.23 : I_0^* (1/α = .80 , μ = .50)

Q:	0.05	0.10	0.15	0.20	0.25	0.30	0.35	0.40	0.45	0.50	0.55	0.60	0.70	0.80	0.90
MODE:															
(1)	.0099	.0400	.0915	.1663	.2674	.3986	.5651	.7735	1.033	1.356	1.761	2.275	3.829	6.918	16.04
(2)	.1154	.2436	.3868	.5478	.7301	.9382	1.178	1.458	1.788	2.183	2.666	3.266	5.051	8.545	18.75
(3)	.4060	.8400	1.311	1.829	2.407	3.056	3.794	4.641	5.624	6.781	8.165	9.858	14.77	24.21	51.69

TABLE 7.23 : I_0^* (1/α = .80 , μ = .60)

Q:	0.05	0.10	0.15	0.20	0.25	0.30	0.35	0.40	0.45	0.50	0.55	0.60	0.70	0.80	0.90
MODE:															
(1)	.0119	.0477	.1083	.1952	.3103	.4569	.6392	.8633	1.138	1.474	1.890	2.412	3.965	6.999	15.83
(2)	.0862	.1813	.2878	.4093	.5513	.7231	.9401	1.228	1.624	2.175	2.926	3.918	6.944	12.86	30.14
(3)	.4517	.8174	1.158	1.504	1.873	2.278	2.734	3.253	3.854	4.559	5.400	6.424	9.350	14.80	29.97

TABLE 7.23 : I_0^* (1/α = .80 , μ = .70)

Q:	0.05	0.10	0.15	0.20	0.25	0.30	0.35	0.40	0.45	0.50	0.55	0.60	0.70	0.80	0.90
MODE:															
(1)	.0138	.0552	.1241	.2209	.3464	.5028	.6932	.9228	1.199	1.532	1.937	2.438	3.899	6.681	14.60
(2)	.3006	.7481	1.295	1.911	2.588	3.334	4.164	5.101	6.173	7.425	8.918	10.74	16.06	26.41	56.86
(3)	.1484	.3048	.4730	.6561	.8571	1.080	1.329	1.610	1.932	2.309	2.757	3.309	4.970	8.516	20.24

TABLE 7.23 : I_0^* (1/α = .80 , μ = .80)

Q:	0.05	0.10	0.15	0.20	0.25	0.30	0.35	0.40	0.45	0.50	0.55	0.60	0.70	0.80	0.90
MODE:															
(1)	.0157	.0624	.1386	.2431	.3754	.5362	.7276	.9532	1.219	1.534	1.910	2.367	3.666	6.073	12.78
(2)	.3339	.7444	1.227	1.773	2.380	3.048	3.785	4.608	5.538	6.608	7.864	9.379	13.70	21.87	45.39
(3)	.9726	1.978	3.031	4.148	5.349	6.657	8.105	9.730	11.59	13.75	16.32	19.46	28.61	46.35	98.57

TABLE 7.23 : I_0^* (1/α = .90 , μ = .20)

Q:	0.05	0.10	0.15	0.20	0.25	0.30	0.35	0.40	0.45	0.50	0.55	0.60	0.70	0.80	0.90
MODE:															
(1)	.0045	.0180	.0413	.0753	.1219	.1839	.2657	.3739	.5185	.7129	.9747	1.326	2.445	4.716	11.40
(2)	.1608	.3261	.5028	.6973	.9162	1.167	1.457	1.799	2.207	2.700	3.308	4.073	6.373	10.93	24.36
(3)	.4010	.8199	1.264	1.742	2.264	2.842	3.492	4.231	5.087	6.094	7.301	8.785	13.12	21.51	46.06

TABLE 7.23 : L_0^* ($1/\alpha = .90$, $\mu = .30$)

Q: MODE:	0.05	0.10	0.15	0.20	0.25	0.30	0.35	0.40	0.45	0.50	0.55	0.60	0.70	0.80	0.90
(1)	.0067	.0271	.0621	.1137	.1843	.2780	.4001	.5577	.7603	1.020	1.355	1.789	3.130	5.834	13.87
(2)	.1570	.3298	.5226	.7401	.9879	1.273	1.605	1.994	2.456	3.012	3.691	4.538	7.051	11.97	26.34
(3)	.3110	.6462	1.011	1.411	1.855	2.352	2.913	3.555	4.297	5.167	6.204	7.466	11.09	17.95	37.55

TABLE 7.23 : L_0^* ($1/\alpha = .90$, $\mu = .40$)

Q: MODE:	0.05	0.10	0.15	0.20	0.25	0.30	0.35	0.40	0.45	0.50	0.55	0.60	0.70	0.80	0.90
(1)	.0089	.0361	.0826	.1507	.2433	.3646	.5199	.7164	.9636	1.274	1.667	2.169	3.702	6.778	15.93
(2)	.1439	.3046	.4852	.6893	.9214	1.187	1.494	1.851	2.271	2.771	3.375	4.119	6.282	10.40	22.10
(3)	.7345	1.494	2.289	3.132	4.039	5.027	6.120	7.347	8.750	10.38	12.33	14.70	21.63	35.09	74.75

TABLE 7.23 : L_0^* ($1/\alpha = .90$, $\mu = .50$)

Q: MODE:	0.05	0.10	0.15	0.20	0.25	0.30	0.35	0.40	0.45	0.50	0.55	0.60	0.70	0.80	0.90
(1)	.0111	.0449	.1024	.1854	.2967	.4399	.6202	.8442	1.121	1.465	1.894	2.436	4.074	7.328	16.95
(2)	.1230	.2594	.4115	.5818	.7739	.9920	1.242	1.531	1.871	2.274	2.762	3.367	5.149	8.625	18.80
(3)	.4564	.9427	1.467	2.039	2.670	3.373	4.164	5.065	6.105	7.324	8.779	10.56	15.71	25.65	54.61

TABLE 7.23 : L_0^* ($1/\alpha = .90$, $\mu = .60$)

Q: MODE:	0.05	0.10	0.15	0.20	0.25	0.30	0.35	0.40	0.45	0.50	0.55	0.60	0.70	0.80	0.90
(1)	.0133	.0535	.1210	.2169	.3430	.5022	.6988	.9388	1.231	1.588	2.027	2.578	4.212	7.401	16.68
(2)	.0917	.1925	.3046	.4314	.5782	.7532	.9708	1.254	1.642	2.181	2.926	3.923	7.002	13.06	30.80
(3)	.5061	.9096	1.281	1.655	2.051	2.485	2.969	3.518	4.150	4.888	5.765	6.828	9.851	15.46	31.04

TABLE 7.23 : L_0^* ($1/\alpha = .90$, $\mu = .70$)

Q: MODE:	0.05	0.10	0.15	0.20	0.25	0.30	0.35	0.40	0.45	0.50	0.55	0.60	0.70	0.80	0.90
(1)	.0155	.0618	.1383	.2448	.3817	.5507	.7550	1.000	1.292	1.644	2.071	2.597	4.126	7.032	15.29
(2)	.3352	.8193	1.399	2.046	2.756	3.537	4.406	5.387	6.511	7.823	9.388	11.30	16.89	27.76	59.80
(3)	.1586	.3261	.5062	.7018	.9161	1.153	1.416	1.712	2.048	2.438	2.898	3.458	5.113	8.593	20.24

TABLE 7.23 : L_0^* ($1/\alpha = .90$, $\mu = .80$)

Q: MODE:	0.05	0.10	0.15	0.20	0.25	0.30	0.35	0.40	0.45	0.50	0.55	0.60	0.70	0.80	0.90
(1)	.0177	.0698	.1541	.2687	.4122	.5852	.7895	1.029	1.310	1.640	2.034	2.511	3.860	6.345	13.25
(2)	.3750	.8316	1.361	1.952	2.601	3.311	4.092	4.960	5.939	7.064	8.386	9.977	14.52	23.11	47.86
(3)	1.019	2.072	3.175	4.345	5.604	6.976	8.494	10.20	12.15	14.42	17.12	20.42	30.03	48.70	103.6

TABLE 7.24 : L_{q2}^* ($\alpha = .10$, $\mu = .20$)

Q: MODE:	0.05	0.10	0.15	0.20	0.25	0.30	0.35	0.40	0.45	0.50	0.55	0.60	0.70	0.80	0.90
(1)	.0142	.0573	.1299	.2336	.3706	.5442	.7589	1.021	1.339	1.727	2.202	2.794	4.540	7.934	17.85
(2)	.0388	.1587	.3704	.6939	1.159	1.806	2.681	3.827	5.283	7.096	9.333	12.10	20.08	34.82	75.21
(3)	.5917	1.262	2.041	2.970	4.109	5.545	7.402	9.854	13.12	17.45	23.07	30.26	51.44	91.43	205.1

TABLE 7.24 : L_{q2}^* ($\alpha = .10$, $\mu = .30$)

Q: MODE:	0.05	0.10	0.15	0.20	0.25	0.30	0.35	0.40	0.45	0.50	0.55	0.60	0.70	0.80	0.90
(1)	.0209	.0840	.1907	.3434	.5460	.8036	1.124	1.517	1.997	2.585	3.310	4.220	6.925	12.22	27.76
(2)	.0347	.1416	.3289	.6124	1.017	1.583	2.365	3.436	4.886	6.808	9.313	12.55	22.36	41.53	96.95
(3)	.5003	1.048	1.656	2.341	3.127	4.041	5.127	6.440	8.062	10.11	12.76	16.27	27.61	51.46	122.8

TABLE 7.24 : L_{q2}^* ($\alpha = .10$, $\mu = .40$)

Q: MODE:	0.05	0.10	0.15	0.20	0.25	0.30	0.35	0.40	0.45	0.50	0.55	0.60	0.70	0.80	0.90
(1)	.0276	.1110	.2523	.4552	.7252	1.070	1.501	2.033	2.687	3.493	4.497	5.764	9.581	17.17	39.65
(2)	.0300	.1218	.2808	.5172	.8470	1.295	1.898	2.709	3.805	5.289	7.295	10.00	18.63	36.12	87.50
(3)	.3839	.7907	1.225	1.691	2.196	2.746	3.351	4.023	4.777	5.633	6.619	7.777	10.90	16.31	30.96

TABLE 7.24 : L_{q2}^* ($\alpha = .10$, $\mu = .50$)

Q: MODE:	0.05	0.10	0.15	0.20	0.25	0.30	0.35	0.40	0.45	0.50	0.55	0.60	0.70	0.80	0.90
(1)	.0343	.1381	.3143	.5677	.9058	1.339	1.883	2.557	3.390	4.424	5.717	7.360	12.36	22.42	52.53
(2)	.0251	.1014	.2322	.4232	.6833	1.026	1.470	2.042	2.783	3.747	5.017	6.709	12.15	23.26	55.16
(3)	2.779	5.664	8.696	11.92	15.40	19.20	23.40	28.12	33.52	39.81	47.29	56.43	82.99	134.5	286.0

TABLE 7.24 : L_{q2}^* ($\alpha = .10$, $\mu = .60$)

Q: MODE:	0.05	0.10	0.15	0.20	.0.25	0.30	0.35	0.40	0.45	0.50	0.55	0.60	0.70	0.80	0.90
(1)	.0410	.1653	.3764	.6803	1.087	1.609	2.265	3.081	4.093	5.353	6.935	8.953	15.13	27.65	65.38
(2)	.0201	.0810	.1843	.3330	.5313	.7855	1.104	1.498	1.984	2.583	3.326	4.255	6.963	11.87	24.08
(3)	4.195	8.227	12.09	15.94	19.92	24.14	28.75	33.90	39.77	46.61	54.76	64.74	93.85	150.5	317.2

TABLE 7.24 : L_{q2}^* ($\alpha = .10$, $\mu = .70$)

Q: MODE:	0.05	0.10	0.15	0.20	0.25	0.30	0.35	0.40	0.45	0.50	0.55	0.60	0.70	0.80	0.90
(1)	.0478	.1925	.4383	.7926	1.266	1.876	2.642	3.596	4.782	6.259	8.116	10.49	17.77	32.56	77.20
(2)	.0151	.0606	.1373	.2464	.3895	.5692	.7886	1.052	1.364	1.732	2.165	2.674	3.997	5.982	10.07
(3)	4.707	7.661	9.809	11.76	13.75	15.89	18.25	20.92	23.98	27.54	31.75	36.84	51.22	77.79	151.6

TABLE 7.24 : L_{q2}^* ($\alpha = .10$, $\mu = .80$)

Q: MODE:	0.05	0.10	0.15	0.20	0.25	0.30	0.35	0.40	0.45	0.50	0.55	0.60	0.70	0.80	0.90
(1)	.0545	.2196	.5000	.9040	1.444	2.138	3.011	4.097	5.444	7.121	9.228	11.92	20.15	36.83	87.08
(2)	.0101	.0404	.0911	.1627	.2555	.3704	.5081	.6697	.8567	1.071	1.314	1.588	2.252	3.197	19.99
(3)	3.256	3.930	4.475	5.030	5.623	6.265	6.967	7.739	8.591	9.538	10.60	11.80	14.79	19.22	28.83

TABLE 7.24 : L_{q2}^* ($\alpha = .15$, $\mu = .20$)

Q: MODE:	0.05	0.10	0.15	0.20	0.25	0.30	0.35	0.40	0.45	0.50	0.55	0.60	0.70	0.80	0.90
(1)	.0106	.0425	.0965	.1739	.2766	.4073	.5700	.7701	1.015	1.316	1.689	2.159	3.566	6.354	14.61
(2)	.0348	.1413	.3258	.5986	.9745	1.472	2.111	2.916	3.916	5.149	6.668	8.555	14.03	24.25	52.64
(3)	.5648	1.200	1.930	2.785	3.809	5.055	6.599	8.532	10.97	14.04	17.90	22.77	37.11	64.56	143.3

TABLE 7.24 : L_{q2}^* ($\alpha = .15$, $\mu = .30$)

Q: MODE:	0.05	0.10	0.15	0.20	0.25	0.30	0.35	0.40	0.45	0.50	0.55	0.60	0.70	0.80	0.90
(1)	.0150	.0602	.1369	.2468	.3928	.5792	.8116	1.098	1.450	1.882	2.420	3.098	5.134	9.176	21.15
(2)	.0322	.1309	.3020	.5562	.9095	1.385	2.013	2.831	3.888	5.241	6.970	9.185	15.90	29.12	67.61
(3)	.4783	1.000	1.577	2.223	2.955	3.797	4.780	5.945	7.350	9.075	11.23	13.99	22.51	39.80	91.17

TABLE 7.24 : L_{q2}^* ($\alpha = .15$, $\mu = .40$)

Q: MODE:	0.05	0.10	0.15	0.20	0.25	0.30	0.35	0.40	0.45	0.50	0.55	0.60	0.70	0.80	0.90
(1)	.0195	.0786	.1789	.3229	.5148	.7603	1.068	1.448	1.917	2.498	3.222	4.140	6.921	12.49	29.10
(2)	.0282	.1142	.2623	.4800	.7791	1.176	1.696	2.371	3.246	4.380	5.856	7.788	13.81	25.97	61.81
(3)	.3673	.7565	1.171	1.616	2.097	2.620	3.194	3.829	4.539	5.344	6.270	7.356	10.31	15.54	30.21

TABLE 7.24 : I^*_{02} (α = .15 , μ = .50)

Q: MODE:	0.05	0.10	0.15	0.20	0.25	0.30	0.35	0.40	0.45	0.50	0.55	0.60	0.70	0.80	0.90
(1)	.0242	.0973	.2214	.4000	.6385	.9446	1.329	1.806	2.397	3.131	4.052	5.224	8.801	16.02	37.73
(2)	.0238	.0959	.2189	.3975	.6386	.9522	1.352	1.858	2.497	3.306	4.337	5.666	9.750	17.82	40.78
(3)	1.956	3.988	6.124	8.400	10.85	13.54	16.52	19.87	23.71	28.18	33.52	40.05	59.09	96.15	205.4

TABLE 7.24 : I^*_{02} (α = .15 , μ = .60)

Q: MODE:	0.05	0.10	0.15	0.20	0.25	0.30	0.35	0.40	0.45	0.50	0.55	0.60	0.70	0.80	0.90
(1)	.0288	.1160	.2641	.4774	.7628	1.129	1.591	2.165	2.878	3.766	4.882	6.308	10.68	19.56	46.34
(2)	.0191	.0769	.1748	.3151	.5014	.7387	1.034	1.396	1.836	2.373	3.028	3.833	6.116	10.13	20.06
(3)	2.836	5.651	8.380	11.12	13.93	16.90	20.14	23.75	27.85	32.62	38.30	45.25	65.52	105.0	221.1

TABLE 7.24 : I^*_{02} (α = .15 , μ = .70)

Q: MODE:	0.05	0.10	0.15	0.20	0.25	0.30	0.35	0.40	0.45	0.50	0.55	0.60	0.70	0.80	0.90
(1)	.0334	.1347	.3067	.5547	.8865	1.313	1.850	2.519	3.350	4.386	5.689	7.355	12.47	22.87	54.28
(2)	.0144	.0578	.1307	.2342	.3699	.5397	.7463	.9933	1.285	1.627	2.028	2.498	3.712	5.556	9.666
(3)	3.283	5.659	7.451	9.035	10.60	12.23	14.00	15.98	18.21	20.80	23.84	27.50	37.83	56.92	110.1

TABLE 7.24 : I^*_{02} (α = .15 , μ = .80)

Q: MODE:	0.05	0.10	0.15	0.20	0.25	0.30	0.35	0.40	0.45	0.50	0.55	0.60	0.70	0.80	0.90
(1)	.0381	.1534	.3493	.6315	1.009	1.494	2.104	2.863	3.804	4.977	6.450	8.330	14.09	25.76	60.92
(2)	.0096	.0385	.0869	.1551	.2436	.3529	.4839	.6375	.8150	1.018	1.249	1.510	2.147	3.121	15.36
(3)	2.680	3.461	4.003	4.523	5.062	5.637	6.259	6.936	7.680	8.502	9.420	10.46	13.05	17.01	26.25

TABLE 7.24 : I^*_{02} (α = .20 , μ = .20)

Q: MODE:	0.05	0.10	0.15	0.20	0.25	0.30	0.35	0.40	0.45	0.50	0.55	0.60	0.70	0.80	0.90
(1)	.0089	.0360	.0818	.1476	.2352	.3471	.4869	.6597	.8723	1.135	1.462	1.876	3.126	5.623	13.06
(2)	.0308	.1244	.2844	.5168	.8300	1.235	1.745	2.378	3.155	4.108	5.279	6.735	10.97	18.96	41.37
(3)	.5401	1.143	1.829	2.622	3.553	4.659	5.991	7.607	9.582	12.01	15.01	18.75	29.79	51.03	112.4

TABLE 7.24 : I^*_{02} (α = .20 , μ = .30)

Q: MODE:	0.05	0.10	0.15	0.20	0.25	0.30	0.35	0.40	0.45	0.50	0.55	0.60	0.70	0.80	0.90
(1)	.0122	.0489	.1112	.2007	.3198	.4722	.6626	.8982	1.188	1.547	1.994	2.560	4.272	7.697	17.91
(2)	.0298	.1205	.2763	.5047	.8166	1.227	1.755	2.427	3.275	4.344	5.693	7.413	12.62	22.89	52.92
(3)	.4581	.9568	1.506	2.117	2.805	3.589	4.493	5.552	6.808	8.323	10.19	12.53	19.58	33.62	75.08

TABLE 7.24 : I^*_{02} (α = .20 , μ = .40)

Q: MODE:	0.05	0.10	0.15	0.20	0.25	0.30	0.35	0.40	0.45	0.50	0.55	0.60	0.70	0.80	0.90
(1)	.0156	.0620	.1430	.2582	.4120	.6090	.8560	1.162	1.541	2.010	2.597	3.343	5.617	10.18	23.86
(2)	.0265	.1070	.2448	.4458	.7187	1.076	1.535	2.118	2.858	3.797	4.994	6.538	11.29	20.81	48.91
(3)	.3522	.7254	1.123	1.550	2.010	2.511	3.059	3.666	4.344	5.112	5.997	7.039	9.891	15.05	29.79

TABLE 7.24 : I^*_{02} (α = .20 , μ = .50)

Q: MODE:	0.05	0.10	0.15	0.20	0.25	0.30	0.35	0.40	0.45	0.50	0.55	0.60	0.70	0.80	0.90
(1)	.0191	.0771	.1755	.3172	.5066	.7498	1.055	1.435	1.907	2.493	3.228	4.166	7.036	12.85	30.35
(2)	.0225	.0906	.2064	.3738	.5982	.8877	1.253	1.709	2.277	2.984	3.870	4.994	8.370	14.91	33.44
(3)	1.550	3.161	4.856	6.662	8.612	10.75	13.12	15.79	18.85	22.43	26.71	31.94	47.24	77.10	165.3

TABLE 7.24 : I^*_{02} (α = .20 , μ = .60)

Q: MODE:	0.05	0.10	0.15	0.20	0.25	0.30	0.35	0.40	0.45	0.50	0.55	0.60	0.70	0.80	0.90
(1)	.0227	.0915	.2084	.3768	.6021	.8918	1.256	1.711	2.275	2.978	3.862	4.993	8.463	15.52	36.85
(2)	.0182	.0731	.1658	.2985	.4741	.6968	.9722	1.308	1.714	2.204	2.797	3.520	5.544	9.064	17.82
(3)	2.145	4.327	6.481	8.641	10.86	13.22	15.77	18.60	21.83	25.57	30.02	35.46	51.32	82.18	173.1

TABLE 7.24 : I^*_{02} (α = .20 , μ = .70)

Q: MODE:	0.05	0.10	0.15	0.20	0.25	0.30	0.35	0.40	0.45	0.50	0.55	0.60	0.70	0.80	0.90
(1)	.0263	.1059	.2413	.4364	.6974	1.033	1.456	1.982	2.637	3.453	4.480	5.793	9.826	18.03	42.83
(2)	.0137	.0551	.1245	.2230	.3519	.5129	.7085	.9417	1.216	1.538	1.914	2.353	3.492	5.251	9.419
(3)	2.519	4.500	6.057	7.430	8.762	10.14	11.61	13.23	15.06	17.16	19.62	22.58	30.90	46.27	89.17

TABLE 7.24 : I^*_{02} (α = .20 , μ = .80)

Q: MODE:	0.05	0.10	0.15	0.20	0.25	0.30	0.35	0.40	0.45	0.50	0.55	0.60	0.70	0.80	0.90
(1)	.0299	.1204	.2741	.4957	.7918	1.172	1.651	2.247	2.986	3.907	5.064	6.540	11.06	20.23	47.85
(2)	.0092	.0368	.0831	.1482	.2327	.3370	.4620	.6085	.7778	.9715	1.192	1.442	2.060	3.067	13.21
(3)	2.230	3.065	3.610	4.108	4.613	5.144	5.713	6.330	7.004	7.749	8.580	9.521	11.89	15.61	24.72

TABLE 7.24 : I^*_{02} (α = .30 , μ = .20)

Q: MODE:	0.05	0.10	0.15	0.20	0.25	0.30	0.35	0.40	0.45	0.50	0.55	0.60	0.70	0.80	0.90
(1)	.0076	.0307	.0699	.1263	.2014	.2977	.4183	.5679	.7526	.9816	1.268	1.631	2.735	4.951	11.58
(2)	.0240	.0967	.2197	.3961	.6304	.9287	1.300	1.754	2.309	2.986	3.817	4.851	7.878	13.66	30.13
(3)	.4962	1.044	1.656	2.348	3.138	4.050	5.112	6.358	7.834	9.603	11.75	14.41	22.22	37.35	81.47

TABLE 7.24 : I^*_{02} (α = .30 , μ = .30)

Q: MODE:	0.05	0.10	0.15	0.20	0.25	0.30	0.35	0.40	0.45	0.50	0.55	0.60	0.70	0.80	0.90
(1)	.0096	.0385	.0877	.1584	.2527	.3736	.5253	.7136	.9463	1.235	1.597	2.057	3.457	6.279	14.74
(2)	.0252	.1015	.2313	.4188	.6702	.9939	1.402	1.909	2.538	3.317	4.292	5.525	9.241	16.59	38.21
(3)	.4224	.8806	1.382	1.936	2.554	3.249	4.040	4.952	6.016	7.277	8.800	10.68	16.22	27.01	58.62

TABLE 7.24 : I^*_{02} (α = .30 , μ = .40)

Q: MODE:	0.05	0.10	0.15	0.20	0.25	0.30	0.35	0.40	0.45	0.50	0.55	0.60	0.70	0.80	0.90
(1)	.0118	.0477	.1086	.1962	.3133	.4637	.6525	.8873	1.178	1.540	1.994	2.572	4.340	7.918	18.68
(2)	.0232	.0937	.2134	.3863	.6179	.9161	1.292	1.760	2.341	3.064	3.969	5.119	8.600	15.52	35.93
(3)	.3254	.6709	1.040	1.435	1.863	2.328	2.838	3.403	4.036	4.756	5.591	6.582	9.338	14.46	29.35

TABLE 7.24 : I^*_{02} (α = .30 , μ = .50)

Q: MODE:	0.05	0.10	0.15	0.20	0.25	0.30	0.35	0.40	0.45	0.50	0.55	0.60	0.70	0.80	0.90
(1)	.0143	.0574	.1307	.2364	.3777	.5593	.7879	1.073	1.426	1.866	2.420	3.127	5.297	9.707	23.02
(2)	.0201	.0809	.1840	.3320	.5289	.7802	1.094	1.480	1.933	2.534	3.250	4.145	6.778	11.79	25.90
(3)	1.154	2.354	3.617	4.964	6.421	8.017	9.793	11.80	14.10	16.80	20.02	23.98	35.57	58.24	125.3

TABLE 7.24 : I^*_{02} (α = .30 , μ = .60)

Q: MODE:	0.05	0.10	0.15	0.20	0.25	0.30	0.35	0.40	0.45	0.50	0.55	0.60	0.70	0.80	0.90
(1)	.0167	.0674	.1534	.2775	.4435	.6572	.9262	1.262	1.678	2.198	2.853	3.691	6.265	11.51	27.38
(2)	.0164	.0661	.1498	.2692	.4266	.6252	.8693	1.165	1.520	1.945	2.456	3.075	4.793	7.777	15.32
(3)	1.445	2.966	4.510	6.081	7.707	9.428	11.29	13.36	15.71	18.42	21.65	25.59	37.06	59.35	125.0

TABLE 7.24 : I_{02}^* ($\alpha = .30$, $\mu = .70$)

Q: MODE:	0.05	0.10	0.15	0.20	0.25	0.30	0.35	0.40	0.45	0.50	0.55	0.60	0.70	0.80	0.90
(1)	.0192	.0774	.1763	.3189	.5098	.7553	1.064	1.450	1.929	2.526	3.278	4.240	7.195	13.21	31.41
(2)	.0125	.0502	.1135	.2032	.3203	.4664	.6435	.8543	1.102	1.392	1.731	2.129	3.168	4.834	9.127
(3)	1.719	3.201	4.448	5.566	6.644	7.741	8.902	10.17	11.58	13.19	15.08	17.33	23.65	35.31	67.93

TABLE 7.24 : I_{02}^* ($\alpha = .30$, $\mu = .80$)

Q: MODE:	0.05	0.10	0.15	0.20	0.25	0.30	0.35	0.40	0.45	0.50	0.55	0.60	0.70	0.80	0.90
(1)	.0217	.0875	.1993	.3604	.5757	.8524	1.201	1.634	2.171	2.841	3.682	4.755	8.043	14.71	34.79
(2)	.0084	.0338	.0762	.1359	.2133	.3090	.4236	.5582	.7139	.8925	1.097	1.330	1.923	2.996	11.26
(3)	1.636	2.461	3.002	3.472	3.932	4.407	4.911	5.453	6.045	6.697	7.427	8.258	10.39	13.87	22.95

TABLE 7.24 : I_{02}^* ($\alpha = .40$, $\mu = .20$)

Q: MODE:	0.05	0.10	0.15	0.20	0.25	0.30	0.35	0.40	0.45	0.50	0.55	0.60	0.70	0.80	0.90
(1)	.0071	.0287	.0653	.1180	.1882	.2782	.3911	.5310	.7040	.9185	1.187	1.528	2.563	4.645	10.67
(2)	.0193	.0776	.1762	.3173	.5042	.7419	1.037	1.398	1.839	2.377	3.039	3.865	6.298	10.99	24.54
(3)	.4585	.9587	1.510	2.125	2.816	3.598	4.494	5.529	6.739	8.173	9.903	12.03	18.28	30.41	65.93

TABLE 7.24 : I_{02}^* ($\alpha = .40$, $\mu = .30$)

Q: MODE:	0.05	0.10	0.15	0.20	0.25	0.30	0.35	0.40	0.45	0.50	0.55	0.60	0.70	0.80	0.90
(1)	.0084	.0340	.0774	.1398	.2232	.3303	.4646	.6316	.8383	1.095	1.417	1.827	3.079	5.607	13.20
(2)	.0214	.0861	.1958	.3535	.5636	.8324	1.169	1.584	2.096	2.728	3.514	4.506	7.492	13.40	30.84
(3)	.3919	.8158	1.278	1.786	2.350	2.981	3.693	4.508	5.451	6.560	7.890	9.520	14.27	23.43	50.15

TABLE 7.24 : I_{02}^* ($\alpha = .40$, $\mu = .40$)

Q: MODE:	0.05	0.10	0.15	0.20	0.25	0.30	0.35	0.40	0.45	0.50	0.55	0.60	0.70	0.80	0.90
(1)	.0101	.0407	.0926	.1673	.2673	.3957	.5572	.7581	1.007	1.318	1.708	2.205	3.729	6.820	16.14
(2)	.0204	.0823	.1872	.3379	.5387	.7956	1.117	1.514	2.003	2.607	3.358	4.306	7.156	12.79	29.37
(3)	.3026	.6246	.9690	1.340	1.741	2.179	2.661	3.197	3.801	4.492	5.298	6.261	8.979	14.11	29.11

TABLE 7.24 : I_{02}^* ($\alpha = .40$, $\mu = .50$)

Q: MODE:	0.05	0.10	0.15	0.20	0.25	0.30	0.35	0.40	0.45	0.50	0.55	0.60	0.70	0.80	0.90
(1)	.0119	.0480	.1092	.1975	.3157	.4676	.6589	.8973	1.193	1.563	2.028	2.622	4.447	8.161	19.39
(2)	.0181	.0726	.1649	.2970	.4722	.6948	.9712	1.310	1.723	2.226	2.843	3.611	5.853	10.10	22.00
(3)	.9642	1.966	3.022	4.148	5.366	6.703	8.190	9.870	11.80	14.06	16.77	20.10	29.86	48.97	105.6

TABLE 7.24 : I_{02}^* ($\alpha = .40$, $\mu = .60$)

Q: MODE:	0.05	0.10	0.15	0.20	0.25	0.30	0.35	0.40	0.45	0.50	0.55	0.60	0.70	0.80	0.90
(1)	.0138	.0556	.1266	.2290	.3660	.5424	.7646	1.042	1.386	1.816	2.357	3.050	5.181	9.524	22.67
(2)	.0150	.0601	.1361	.2443	.3868	.5663	.7866	1.053	1.372	1.753	2.211	2.765	4.305	7.000	13.93
(3)	1.090	2.264	3.485	4.746	6.063	7.464	8.983	10.67	12.57	14.78	17.40	20.59	29.87	47.90	100.9

TABLE 7.24 : I_{02}^* ($\alpha = .40$, $\mu = .70$)

Q: MODE:	0.05	0.10	0.15	0.20	0.25	0.30	0.35	0.40	0.45	0.50	0.55	0.60	0.70	0.80	0.90
(1)	.0157	.0634	.1443	.2610	.4172	.6181	.8712	1.187	1.579	2.068	2.683	3.470	5.890	10.82	25.71
(2)	.0115	.0460	.1040	.1862	.2935	.4273	.5896	.7829	1.011	1.277	1.590	1.958	2.934	4.554	8.960
(3)	1.304	2.488	3.532	4.491	5.420	6.364	7.359	8.439	9.639	11.00	12.59	14.49	19.81	29.63	57.10

TABLE 7.24 : I_{02}^* ($\alpha = .40$, $\mu = .80$)

Q: MODE:	0.05	0.10	0.15	0.20	0.25	0.30	0.35	0.40	0.45	0.50	0.55	0.60	0.70	0.80	0.90
(1)	.0177	.0712	.1622	.2932	.4684	.6936	.9767	1.329	1.766	2.311	2.995	3.867	6.540	11.96	28.27
(2)	.0078	.0311	.0702	.1253	.1968	.2852	.3913	.5160	.6607	.8274	1.019	1.240	1.818	2.951	10.39
(3)	1.281	2.038	2.559	3.006	3.437	3.876	4.339	4.836	5.377	5.976	6.648	7.418	9.422	12.79	21.94

TABLE 7.24 : I_{02}^* ($\alpha = .50$, $\mu = .20$)

Q: MODE:	0.05	0.10	0.15	0.20	0.25	0.30	0.35	0.40	0.45	0.50	0.55	0.60	0.70	0.80	0.90
(1)	.0069	.0277	.0630	.1138	.1815	.2683	.3771	.5120	.6787	.8853	1.144	1.472	2.469	4.471	10.46
(2)	.0160	.0644	.1462	.2634	.4189	.6169	.8631	1.166	1.535	1.988	2.548	3.249	5.331	9.389	21.20
(3)	.4258	.8858	1.388	1.941	2.556	3.246	4.029	4.926	5.969	7.199	8.678	10.50	15.82	26.18	56.57

TABLE 7.24 : I_{02}^* ($\alpha = .50$, $\mu = .30$)

Q: MODE:	0.05	0.10	0.15	0.20	0.25	0.30	0.35	0.40	0.45	0.50	0.55	0.60	0.70	0.80	0.90
(1)	.0078	.0316	.0719	.1300	.2075	.3070	.4320	.5873	.7796	1.019	1.318	1.700	2.866	5.221	12.29
(2)	.0184	.0741	.1685	.3040	.4843	.7148	1.003	1.358	1.796	2.336	3.007	3.855	6.406	11.46	26.41
(3)	.3654	.7600	1.189	1.660	2.181	2.761	3.415	4.160	5.019	6.027	7.230	8.700	12.96	21.15	44.94

TABLE 7.24 : I_{02}^* ($\alpha = .50$, $\mu = .40$)

Q: MODE:	0.05	0.10	0.15	0.20	0.25	0.30	0.35	0.40	0.45	0.50	0.55	0.60	0.70	0.80	0.90
(1)	.0091	.0367	.0836	.1512	.2415	.3576	.5036	.6853	.9108	1.192	1.545	1.995	3.376	6.178	14.63
(2)	.0181	.0729	.1657	.2988	.4759	.7021	.9843	1.332	1.760	2.287	2.941	3.765	6.234	11.10	25.39
(3)	.2828	.5846	.9084	1.258	1.638	2.054	2.515	3.029	3.612	4.283	5.071	6.020	8.724	13.88	28.96

TABLE 7.24 : I_{02}^* ($\alpha = .50$, $\mu = .50$)

Q: MODE:	0.05	0.10	0.15	0.20	0.25	0.30	0.35	0.40	0.45	0.50	0.55	0.60	0.70	0.80	0.90
(1)	.0106	.0425	.0968	.1752	.2799	.4147	.5844	.7959	1.059	1.387	1.799	2.327	3.948	7.248	17.23
(2)	.0163	.0655	.1487	.2677	.4254	.6256	.8739	1.178	1.547	1.997	2.549	3.234	5.231	9.006	19.59
(3)	.8549	1.743	2.679	3.678	4.759	5.944	7.263	8.755	10.47	12.48	14.88	17.84	26.51	43.49	93.78

TABLE 7.24 : I_{02}^* ($\alpha = .50$, $\mu = .60$)

Q: MODE:	0.05	0.10	0.15	0.20	0.25	0.30	0.35	0.40	0.45	0.50	0.55	0.60	0.70	0.80	0.90
(1)	.0121	.0487	.1109	.2006	.3207	.4751	.6698	.9124	1.214	1.591	2.065	2.672	4.538	8.342	19.86
(2)	.0137	.0548	.1243	.2231	.3533	.5172	.7185	.9618	1.254	1.603	2.022	2.531	3.953	6.466	13.03
(3)	.8750	1.833	2.850	3.917	5.042	6.245	7.553	9.005	10.65	12.55	14.81	17.55	25.53	41.01	86.49

TABLE 7.24 : I_{02}^* ($\alpha = .50$, $\mu = .70$)

Q: MODE:	0.05	0.10	0.15	0.20	0.25	0.30	0.35	0.40	0.45	0.50	0.55	0.60	0.70	0.80	0.90
(1)	.0137	.0551	.1254	.2268	.3624	.5369	.7567	1.031	1.371	1.795	2.330	3.013	5.112	9.385	22.30
(2)	.0106	.0424	.0958	.1715	.2705	.3941	.5442	.7232	.9347	1.183	1.476	1.823	2.755	4.349	8.850
(3)	1.050	2.035	2.936	3.781	4.608	5.450	6.338	7.300	8.369	9.582	10.99	12.68	17.40	26.10	50.50

TABLE 7.24 : I_{02}^* ($\alpha = .50$, $\mu = .80$)

Q: MODE:	0.05	0.10	0.15	0.20	0.25	0.30	0.35	0.40	0.45	0.50	0.55	0.60	0.70	0.80	0.90
(1)	.0153	.0615	.1401	.2533	.4045	.5989	.8434	1.148	1.525	1.995	2.585	3.337	5.642	10.31	24.37
(2)	.0072	.0288	.0651	.1162	.1825	.2647	.3635	.4800	.6156	.7725	.9541	1.166	1.733	2.920	9.904
(3)	1.050	1.732	2.226	2.650	3.057	3.470	3.904	4.369	4.877	5.439	6.073	6.804	8.730	12.04	21.28

TABLE 7.24 : I_{02}^{*} (α = .60 , μ = .20)

Q: MODE:	0.05	0.10	0.15	0.20	0.25	0.30	0.35	0.40	0.45	0.50	0.55	0.60	0.70	0.80	0.90
(1)	.0067	.0271	.0616	.1113	.1775	.2624	.3687	.5006	.6634	.8652	1.117	1.437	2.410	4.361	10.19
(2)	.0136	.0548	.1246	.2247	.3578	.5276	.7394	1.001	1.321	1.716	2.206	2.823	4.672	8.316	18.99
(3)	.3974	.8227	1.283	1.786	2.342	2.962	3.663	4.463	5.390	6.482	7.793	9.405	14.13	23.31	50.31

TABLE 7.24 : I_{02}^{*} (α = .60 , μ = .30)

Q: MODE:	0.05	0.10	0.15	0.20	0.25	0.30	0.35	0.40	0.45	0.50	0.55	0.60	0.70	0.80	0.90
(1)	.0075	.0302	.0686	.1240	.1980	.2929	.4121	.5601	.7435	.9713	1.257	1.621	2.731	4.972	11.70
(2)	.0161	.0647	.1472	.2658	.4237	.6258	.8787	1.191	1.577	2.053	2.647	3.397	5.659	10.15	23.45
(3)	.3423	.7114	1.112	1.552	2.037	2.577	3.185	3.877	4.675	5.609	6.722	8.082	12.01	19.55	41.39

TABLE 7.24 : I_{02}^{*} (α = .60 , μ = .40)

Q: MODE:	0.05	0.10	0.15	0.20	0.25	0.30	0.35	0.40	0.45	0.50	0.55	0.60	0.70	0.80	0.90
(1)	.0085	.0343	.0780	.1411	.2253	.3336	.4698	.6393	.8497	1.112	1.441	1.861	3.148	5.760	13.64
(2)	.0162	.0651	.1480	.2670	.4253	.6277	.8803	1.192	1.575	2.048	2.634	3.372	5.586	9.944	22.72
(3)	.2655	.5497	.8557	1.187	1.549	1.948	2.391	2.889	3.456	4.114	4.890	5.830	8.532	13.71	28.86

TABLE 7.24 : I_{02}^{*} (α = .60 , μ = .50)

Q: MODE:	0.05	0.10	0.15	0.20	0.25	0.30	0.35	0.40	0.45	0.50	0.55	0.60	0.70	0.80	0.90
(1)	.0097	.0390	.0889	.1608	.2570	.3807	.5365	.7306	.9718	1.273	1.651	2.135	3.622	6.648	15.80
(2)	.0148	.0595	.1350	.2431	.3864	.5686	.7946	1.071	1.408	1.819	2.323	2.949	4.777	8.237	17.94
(3)	.7850	1.601	2.460	3.377	4.369	5.457	6.668	8.037	9.610	11.45	13.66	16.37	24.32	39.89	86.00

TABLE 7.24 : I_{02}^{*} (α = .60 , μ = .60)

Q: MODE:	0.05	0.10	0.15	0.20	0.25	0.30	0.35	0.40	0.45	0.50	0.55	0.60	0.70	0.80	0.90
(1)	.0110	.0442	.1007	.1821	.2911	.4313	.6080	.8281	1.102	1.443	1.873	2.424	4.115	7.562	17.99
(2)	.0125	.0503	.1141	.2050	.3247	.4757	.6614	.8862	1.156	1.481	1.871	2.347	3.683	6.072	12.39
(3)	.7311	1.542	2.416	3.347	4.339	5.407	6.573	7.869	9.337	11.04	13.05	15.50	22.62	36.40	76.86

TABLE 7.24 : I_{02}^{*} (α = .60 , μ = .70)

Q: MODE:	0.05	0.10	0.15	0.20	0.25	0.30	0.35	0.40	0.45	0.50	0.55	0.60	0.70	0.80	0.90
(1)	.0123	.0496	.1129	.2043	.3264	.4836	.6814	.9279	1.234	1.616	2.097	2.711	4.598	8.437	20.03
(2)	.0098	.0392	.0887	.1588	.2507	.3656	.5053	.6725	.8706	1.104	1.381	1.711	2.610	4.191	8.773
(3)	.8793	1.722	2.515	3.272	4.022	4.791	5.602	6.482	7.459	8.569	9.859	11.40	15.71	23.68	46.03

TABLE 7.24 : I_{02}^{*} (α = .60 , μ = .80)

Q: MODE:	0.05	0.10	0.15	0.20	0.25	0.30	0.35	0.40	0.45	0.50	0.55	0.60	0.70	0.80	0.90
(1)	.0137	.0551	.1255	.2268	.3623	.5363	.7552	1.027	1.365	1.785	2.313	2.986	5.045	9.216	21.77
(2)	.0067	.0268	.0606	.1082	.1701	.2470	.3395	.4489	.5768	.7255	.8987	1.103	1.663	2.897	9.601
(3)	.8882	1.504	1.967	2.370	2.756	3.148	3.559	4.000	4.482	5.018	5.625	6.328	8.202	11.49	20.81

TABLE 7.24 : I_{02}^{*} (α = .70 , μ = .20)

Q: MODE:	0.05	0.10	0.15	0.20	0.25	0.30	0.35	0.40	0.45	0.50	0.55	0.60	0.70	0.80	0.90
(1)	.0066	.0267	.0607	.1097	.1749	.2585	.3632	.4930	.6532	.8517	1.100	1.414	2.369	4.284	9.999
(2)	.0119	.0477	.1085	.1958	.3121	.4608	.6469	.8772	1.161	1.513	1.952	2.508	4.193	7.545	17.41
(3)	.3723	.7676	1.192	1.654	2.162	2.728	3.365	4.091	4.934	5.925	7.117	8.582	12.88	21.24	45.82

TABLE 7.24 : I_{02}^{*} (α = .70 , μ = .30)

Q: MODE:	0.05	0.10	0.15	0.20	0.25	0.30	0.35	0.40	0.45	0.50	0.55	0.60	0.70	0.80	0.90
(1)	.0073	.0292	.0665	.1201	.1917	.2835	.3988	.5420	.7193	.9394	1.215	1.567	2.638	4.799	11.29
(2)	.0142	.0573	.1304	.2357	.3761	.5562	.7821	1.062	1.409	1.838	2.374	3.054	5.110	9.206	21.33
(3)	.3219	.6687	1.045	1.457	1.913	2.420	2.991	3.641	4.390	5.268	6.315	7.593	11.29	18.36	38.81

TABLE 7.24 : I_{02}^{*} (α = .70 , μ = .40)

Q: MODE:	0.05	0.10	0.15	0.20	0.25	0.30	0.35	0.40	0.45	0.50	0.55	0.60	0.70	0.80	0.90
(1)	.0081	.0326	.0743	.1342	.2144	.3174	.4469	.6081	.8080	1.057	1.370	1.768	2.990	5.467	12.93
(2)	.0146	.0586	.1334	.2408	.3840	.5672	.7964	1.080	1.429	1.860	2.396	3.072	5.101	9.096	20.79
(3)	.2502	.5190	.8094	1.125	1.472	1.856	2.284	2.769	3.324	3.971	4.740	5.676	8.381	13.59	28.79

TABLE 7.24 : I_{02}^{*} (α = .70 , μ = .50)

Q: MODE:	0.05	0.10	0.15	0.20	0.25	0.30	0.35	0.40	0.45	0.50	0.55	0.60	0.70	0.80	0.90
(1)	.0091	.0367	.0835	.1510	.2412	.3573	.5034	.6855	.9116	1.193	1.548	2.001	3.393	6.225	14.78
(2)	.0135	.0543	.1234	.2223	.3536	.5208	.7287	.9837	1.295	1.675	2.143	2.725	4.428	7.659	16.73
(3)	.7370	1.503	2.309	3.170	4.101	5.121	6.257	7.540	9.014	10.74	12.81	15.34	22.78	37.35	80.47

TABLE 7.24 : I_{02}^{*} (α = .70 , μ = .60)

Q: MODE:	0.05	0.10	0.15	0.20	0.25	0.30	0.35	0.40	0.45	0.50	0.55	0.60	0.70	0.80	0.90
(1)	.0102	.0411	.0936	.1692	.2705	.4007	.5647	.7691	1.023	1.340	1.739	2.249	3.817	7.009	16.66
(2)	.0116	.0464	.1053	.1893	.3002	.4403	.6128	.8224	1.075	1.379	1.747	2.196	3.468	5.765	11.91
(3)	.6278	1.331	2.100	2.929	3.823	4.791	5.853	7.037	8.380	9.936	11.78	14.02	20.52	33.10	69.97

TABLE 7.24 : I_{02}^{*} (α = .70 , μ = .70)

Q: MODE:	0.05	0.10	0.15	0.20	0.25	0.30	0.35	0.40	0.45	0.50	0.55	0.60	0.70	0.80	0.90
(1)	.0114	.0458	.1042	.1884	.3011	.4460	.6283	.8555	1.138	1.490	1.932	2.498	4.233	7.763	18.42
(2)	.0091	.0364	.0825	.1478	.2335	.3408	.4717	.6288	.8156	1.037	1.301	1.617	2.491	4.065	8.716
(3)	.7561	1.493	2.200	2.889	3.578	4.288	5.041	5.859	6.769	7.803	9.005	10.44	14.47	21.91	42.78

TABLE 7.24 : I_{02}^{*} (α = .70 , μ = .80)

Q: MODE:	0.05	0.10	0.15	0.20	0.25	0.30	0.35	0.40	0.45	0.50	0.55	0.60	0.70	0.80	0.90
(1)	.0126	.0506	.1151	.2081	.3323	.4919	.6926	.9421	1.252	1.637	2.120	2.736	4.621	8.437	19.91
(2)	.0063	.0251	.0566	.1012	.1592	.2314	.3185	.4217	.5429	.6845	.8507	1.048	1.604	2.879	9.395
(3)	.7694	1.327	1.761	2.143	2.511	2.885	3.277	3.698	4.160	4.676	5.262	5.944	7.782	11.06	20.47

TABLE 7.24 : I_{02}^{*} (α = .80 , μ = .20)

Q: MODE:	0.05	0.10	0.15	0.20	0.25	0.30	0.35	0.40	0.45	0.50	0.55	0.60	0.70	0.80	0.90
(1)	.0066	.0264	.0601	.1085	.1731	.2558	.3593	.4876	.6459	.8420	1.087	1.398	2.340	4.228	9.859
(2)	.0105	.0422	.0960	.1734	.2766	.4089	.5750	.7813	1.037	1.355	1.755	2.265	3.827	6.965	16.24
(3)	.3502	.7192	1.114	1.541	2.009	2.530	3.116	3.786	4.562	5.478	6.579	7.934	11.91	19.66	42.44

TABLE 7.24 : I_{02}^{*} (α = .80 , μ = .30)

Q: MODE:	0.05	0.10	0.15	0.20	0.25	0.30	0.35	0.40	0.45	0.50	0.55	0.60	0.70	0.80	0.90
(1)	.0071	.0285	.0649	.1173	.1872	.2769	.3894	.5291	.7020	.9166	1.186	1.528	2.570	4.673	10.98
(2)	.0127	.0514	.1169	.2115	.3379	.5004	.7048	.9592	1.275	1.667	2.159	2.785	4.687	8.486	19.73
(3)	.3039	.6309	.9859	1.375	1.804	2.284	2.824	3.439	4.150	4.984	5.979	7.193	10.71	17.43	36.84

TABLE 7.24 : I^*_{02} (α = .80 , μ = .40)

Q: MODE:	0.05	0.10	0.15	0.20	0.25	0.30	0.35	0.40	0.45	0.50	0.55	0.60	0.70	0.80	0.90
(1)	.0078	.0314	.0716	.1294	.2066	.3058	.4305	.5856	.7780	1.017	1.318	1.701	2.874	5.252	12.41
(2)	.0132	.0533	.1212	.2190	.3496	.5172	.7273	.9879	1.310	1.708	2.205	2.833	4.721	8.444	19.33
(3)	.2367	.4916	.7682	1.071	1.404	1.774	2.191	2.664	3.210	3.850	4.614	5.548	8.260	13.49	28.73

TABLE 7.24 : I^*_{02} (α = .80 , μ = .50)

Q: MODE:	0.05	0.10	0.15	0.20	0.25	0.30	0.35	0.40	0.45	0.50	0.55	0.60	0.70	0.80	0.90
(1)	.0087	.0349	.0795	.1438	.2298	.3403	.4794	.6526	.8676	1.136	1.473	1.903	3.225	5.911	14.02
(2)	.0124	.0499	.1134	.2045	.3257	.4803	.6729	.9100	1.200	1.556	1.994	2.541	4.148	7.207	15.80
(3)	.7022	1.432	2.200	3.020	3.906	4.877	5.958	7.178	8.580	10.22	12.18	14.59	21.65	35.47	76.35

TABLE 7.24 : I^*_{02} (α = .80 , μ = .60)

Q: MODE:	0.05	0.10	0.15	0.20	0.25	0.30	0.35	0.40	0.45	0.50	0.55	0.60	0.70	0.80	0.90
(1)	.0096	.0388	.0883	.1598	.2553	.3782	.5329	.7257	.9652	1.264	1.640	2.120	3.595	6.597	15.67
(2)	.0107	.0431	.0977	.1758	.2789	.4096	.5710	.7676	1.005	1.293	1.641	2.069	3.290	5.518	11.54
(3)	.5502	1.171	1.858	2.609	3.425	4.317	5.300	6.399	7.648	9.095	10.81	12.90	18.94	30.61	64.79

TABLE 7.24 : I^*_{02} (α = .80 , μ = .70)

Q: MODE:	0.05	0.10	0.15	0.20	0.25	0.30	0.35	0.40	0.45	0.50	0.55	0.60	0.70	0.80	0.90
(1)	.0107	.0429	.0977	.1767	.2823	.4181	.5890	.8018	1.066	1.395	1.810	2.339	3.962	7.260	17.21
(2)	.0085	.0340	.0770	.1381	.2184	.3191	.4424	.5907	.7678	.9787	1.231	1.536	2.390	3.961	8.672
(3)	.6633	1.318	1.957	2.588	3.227	3.890	4.596	5.366	6.223	7.198	8.334	9.690	13.50	20.54	40.32

TABLE 7.24 : I^*_{02} (α = .80 , μ = .80)

Q: MODE:	0.05	0.10	0.15	0.20	0.25	0.30	0.35	0.40	0.45	0.50	0.55	0.60	0.70	0.80	0.90
(1)	.0117	.0472	.1074	.1941	.3100	.4589	.6460	.8786	1.167	1.526	1.976	2.550	4.305	7.854	18.52
(2)	.0059	.0235	.0531	.0950	.1496	.2176	.2999	.3978	.5131	.6485	.8084	1.000	1.552	2.865	9.245
(3)	.6784	1.186	1.593	1.956	2.307	2.665	3.041	3.446	3.891	4.390	4.960	5.626	7.437	10.71	20.20

TABLE 7.24 : I^*_{02} (α = .90 , μ = .20)

Q: MODE:	0.05	0.10	0.15	0.20	0.25	0.30	0.35	0.40	0.45	0.50	0.55	0.60	0.70	0.80	0.90
(1)	.0065	.0262	.0596	.1077	.1717	.2537	.3564	.4836	.6405	.8348	1.077	1.385	2.318	4.185	9.751
(2)	.0094	.0378	.0861	.1555	.2483	.3675	.5175	.7045	.9373	1.229	1.597	2.071	3.537	6.512	15.33
(3)	.3304	.6763	1.044	1.442	1.877	2.360	2.905	3.529	4.253	5.108	6.138	7.408	11.14	18.42	39.80

TABLE 7.24 : I^*_{02} (α = .90 , μ = .30)

Q: MODE:	0.05	0.10	0.15	0.20	0.25	0.30	0.35	0.40	0.45	0.50	0.55	0.60	0.70	0.80	0.90
(1)	.0070	.0280	.0638	.1152	.1839	.2720	.3824	.5195	.6891	.8996	1.163	1.498	2.519	4.576	10.74
(2)	.0115	.0465	.1059	.1916	.3065	.4546	.6415	.8748	1.165	1.528	1.985	2.569	4.350	7.919	18.49
(3)	.2877	.5971	.9331	1.301	1.709	2.164	2.678	3.265	3.944	4.741	5.694	6.859	10.23	16.68	35.28

TABLE 7.24 : I^*_{02} (α = .90 , μ = .40)

Q: MODE:	0.05	0.10	0.15	0.20	0.25	0.30	0.35	0.40	0.45	0.50	0.55	0.60	0.70	0.80	0.90
(1)	.0076	.0306	.0696	.1257	.2008	.2971	.4182	.5688	.7554	.9875	1.279	1.650	2.786	5.086	12.01
(2)	.0121	.0487	.1110	.2007	.3208	.4751	.6693	.9108	1.210	1.582	2.047	2.636	4.415	7.926	18.19
(3)	.2245	.4671	.7313	1.021	1.343	1.702	2.108	2.572	3.110	3.744	4.505	5.439	8.160	13.41	28.69

TABLE 7.24 : I^*_{02} (α = .90 , μ = .50)

Q: MODE:	0.05	0.10	0.15	0.20	0.25	0.30	0.35	0.40	0.45	0.50	0.55	0.60	0.70	0.80	0.90
(1)	.0083	.0336	.0765	.1384	.2211	.3274	.4611	.6276	.8343	1.092	1.415	1.829	3.096	5.670	13.44
(2)	.0114	.0461	.1048	.1893	.3017	.4455	.6252	.8471	1.120	1.454	1.869	2.387	3.918	6.841	15.07
(3)	.6759	1.378	2.118	2.906	3.759	4.693	5.731	6.904	8.250	9.825	11.71	14.02	20.79	34.02	73.17

TABLE 7.24 : I^*_{02} (α = .90 , μ = .60)

Q: MODE:	0.05	0.10	0.15	0.20	0.25	0.30	0.35	0.40	0.45	0.50	0.55	0.60	0.70	0.80	0.90
(1)	.0092	.0370	.0844	.1526	.2438	.3611	.5087	.6925	.9208	1.205	1.563	2.021	3.425	6.280	14.90
(2)	.0100	.0401	.0910	.1639	.2604	.3829	.5346	.7199	.9449	1.218	1.551	1.961	3.140	5.313	11.24
(3)	.4896	1.046	1.667	2.354	3.109	3.939	4.860	5.892	7.067	8.431	10.05	12.01	17.70	28.67	60.77

TABLE 7.24 : I^*_{02} (α = .90 , μ = .70)

Q: MODE:	0.05	0.10	0.15	0.20	0.25	0.30	0.35	0.40	0.45	0.50	0.55	0.60	0.70	0.80	0.90
(1)	.0101	.0407	.0927	.1677	.2679	.3967	.5588	.7605	1.011	1.323	1.715	2.216	3.752	6.870	16.27
(2)	.0079	.0319	.0722	.1296	.2050	.3000	.4165	.5571	.7257	.9274	1.170	1.465	2.302	3.874	8.637
(3)	.5907	1.179	1.762	2.346	2.942	3.566	4.233	4.963	5.778	6.706	7.789	9.083	12.72	19.46	38.39

TABLE 7.24 : I^*_{02} (α = .90 , μ = .80)

Q: MODE:	0.05	0.10	0.15	0.20	0.25	0.30	0.35	0.40	0.45	0.50	0.55	0.60	0.70	0.80	0.90
(1)	.0111	.0446	.1014	.1834	.2928	.4333	.6099	.8295	1.102	1.440	1.865	2.406	4.059	7.402	17.45
(2)	.0055	.0221	.0500	.0895	.1411	.2054	.2834	.3765	.4865	.6165	.7709	.9579	1.507	2.853	9.131
(3)	.6065	1.072	1.454	1.799	2.135	2.478	2.839	3.231	3.662	4.147	4.703	5.356	7.148	10.42	19.98

TABLE 7.24 : I^*_{02} (α = 1.0 , μ = .20)

Q: MODE:	0.05	0.10	0.15	0.20	0.25	0.30	0.35	0.40	0.45	0.50	0.55	0.60	0.70	0.80	0.90
(1)	.0065	.0261	.0593	.1070	.1707	.2521	.3542	.4805	.6363	.8292	1.070	1.375	2.300	4.151	9.666
(2)	.0085	.0342	.0780	.1409	.2252	.3337	.4705	.6416	.8556	1.125	1.468	1.912	3.302	6.149	14.60
(3)	.3127	.6382	.9829	1.354	1.761	2.213	2.724	3.309	3.990	4.796	5.768	6.970	10.51	17.41	37.69

TABLE 7.24 : I^*_{02} (α = 1.0 , μ = .30)

Q: MODE:	0.05	0.10	0.15	0.20	0.25	0.30	0.35	0.40	0.45	0.50	0.55	0.60	0.70	0.80	0.90
(1)	.0069	.0277	.0629	.1137	.1814	.2682	.3771	.5122	.6792	.8864	1.146	1.476	2.479	4.500	10.55
(2)	.0105	.0424	.0967	.1751	.2804	.4165	.5886	.8044	1.074	1.412	1.840	2.389	4.073	7.460	17.49
(3)	.2732	.5668	.8858	1.236	1.624	2.058	2.549	3.111	3.763	4.530	5.449	6.574	9.832	16.06	34.02

TABLE 7.24 : I^*_{02} (α = 1.0 , μ = .40)

Q: MODE:	0.05	0.10	0.15	0.20	0.25	0.30	0.35	0.40	0.45	0.50	0.55	0.60	0.70	0.80	0.90
(1)	.0074	.0299	.0680	.1229	.1963	.2904	.4087	.5557	.7378	.9643	1.248	1.610	2.717	4.956	11.69
(2)	.0111	.0449	.1022	.1851	.2961	.4393	.6199	.8454	1.126	1.475	1.914	2.472	4.161	7.504	17.27
(3)	.2136	.4450	.6981	.9771	1.288	1.637	2.034	2.490	3.022	3.652	4.410	5.345	8.075	13.35	28.66

TABLE 7.24 : I^*_{02} (α = 1.0 , μ = .50)

Q: MODE:	0.05	0.10	0.15	0.20	0.25	0.30	0.35	0.40	0.45	0.50	0.55	0.60	0.70	0.80	0.90
(1)	.0081	.0326	.0742	.1342	.2144	.3174	.4469	.6081	.8082	1.057	1.370	1.770	2.994	5.479	12.97
(2)	.0106	.0428	.0974	.1760	.2809	.4154	.5839	.7926	1.050	1.367	1.761	2.256	3.724	6.539	14.48
(3)	.6555	1.337	2.053	2.818	3.644	4.549	5.554	6.689	7.992	9.515	11.34	13.57	20.10	32.87	70.63

TABLE 7.24 : I_{02}^{*} ($\alpha = 1.0$, $\mu = .60$)

Q: MODE:	0.05	0.10	0.15	0.20	0.25	0.30	0.35	0.40	0.45	0.50	0.55	0.60	0.70	0.80	0.90
(1)	.0089	.0357	.0812	.1469	.2347	.3476	.4896	.6664	.8859	1.159	1.503	1.943	3.290	6.027	14.29
(2)	.0093	.0375	.0852	.1534	.2441	.3593	.5025	.6781	.8920	1.153	1.472	1.867	3.012	5.141	10.99
(3)	.4411	.9445	1.513	2.147	2.851	3.631	4.500	5.478	6.594	7.890	9.427	11.30	16.70	27.12	57.55

TABLE 7.24 : I_{02}^{*} ($\alpha = 1.0$, $\mu = .70$)

Q: MODE:	0.05	0.10	0.15	0.20	0.25	0.30	0.35	0.40	0.45	0.50	0.55	0.60	0.70	0.80	0.90
(1)	.0097	.0390	.0888	.1605	.2565	.3798	.5348	.7278	.9674	1.266	1.641	2.119	3.585	6.560	15.53
(2)	.0075	.0300	.0679	.1219	.1932	.2830	.3935	.5273	.6883	.8819	1.116	1.403	2.226	3.799	8.608
(3)	.5324	1.067	1.603	2.146	2.706	3.296	3.930	4.626	5.406	6.296	7.336	8.580	12.08	18.57	36.82

TABLE 7.24 : I_{02}^{*} ($\alpha = 1.0$, $\mu = .80$)

Q: MODE:	0.05	0.10	0.15	0.20	0.25	0.30	0.35	0.40	0.45	0.50	0.55	0.60	0.70	0.80	0.90
(1)	.0105	.0425	.0967	.1748	.2791	.4130	.5813	.7904	1.050	1.372	1.776	2.291	3.863	7.041	16.58
(2)	.0052	.0209	.0473	.0846	.1335	.1945	.2687	.3574	.4628	.5878	.7373	.9199	1.468	2.843	9.043
(3)	.5484	.9779	1.337	1.665	1.987	2.316	2.666	3.044	3.463	3.936	4.480	5.123	6.900	10.18	19.80

TABLE 7.24 : I_{02}^{*} ($1/\alpha = .10$, $\mu = .20$)

Q: MODE:	0.05	0.10	0.15	0.20	0.25	0.30	0.35	0.40	0.45	0.50	0.55	0.60	0.70	0.80	0.90
(1)	.0062	.0250	.0568	.1025	.1634	.2412	.3384	.4586	.6066	.7893	1.017	1.304	2.171	3.894	8.997
(2)	.0009	.0036	.0082	.0149	.0239	.0358	.0514	.0718	.0993	.1385	.1994	.3083	1.021	3.104	8.920
(3)	.0528	.1028	.1542	.2103	.2747	.3520	.4489	.5761	.7516	1.007	1.393	1.984	4.103	8.358	20.08

TABLE 7.24 : I_{02}^{*} ($1/\alpha = .10$, $\mu = .30$)

Q: MODE:	0.05	0.10	0.15	0.20	0.25	0.30	0.35	0.40	0.45	0.50	0.55	0.60	0.70	0.80	0.90
(1)	.0062	.0251	.0571	.1030	.1642	.2425	.3403	.4612	.6102	.7941	1.023	1.313	2.187	3.926	9.082
(2)	.0012	.0047	.0108	.0197	.0320	.0484	.0703	.0997	.1406	.2002	.2931	.4515	1.254	3.338	9.273
(3)	.0492	.1023	.1621	.2320	.3172	.4250	.5671	.7621	1.039	1.440	2.013	2.807	5.282	9.960	22.90

TABLE 7.24 : I_{02}^{*} ($1/\alpha = .10$, $\mu = .40$)

Q: MODE:	0.05	0.10	0.15	0.20	0.25	0.30	0.35	0.40	0.45	0.50	0.55	0.60	0.70	0.80	0.90
(1)	.0063	.0253	.0575	.1038	.1654	.2443	.3429	.4648	.6151	.8007	1.032	1.325	2.208	3.968	9.190
(2)	.0013	.0053	.0123	.0225	.0367	.0559	.0817	.1168	.1660	.2380	.3492	.5324	1.367	3.429	9.327
(3)	.0401	.0872	.1449	.2191	.3199	.4653	.6863	1.028	1.534	2.220	3.091	4.165	7.199	12.77	28.37

TABLE 7.24 : I_{02}^{*} ($1/\alpha = .10$, $\mu = .50$)

Q: MODE:	0.05	0.10	0.15	0.20	0.25	0.30	0.35	0.40	0.45	0.50	0.55	0.60	0.70	0.80	0.90
(1)	.0063	.0255	.0580	.1048	.1670	.2466	.3462	.4694	.6212	.8089	1.043	1.339	2.233	4.017	9.314
(2)	.0014	.0055	.0127	.0233	.0380	.0579	.0846	.1211	.1722	.2463	.3599	.5437	1.356	3.350	9.058
(3)	.5135	1.046	1.606	2.200	2.839	3.535	4.303	5.163	6.143	7.280	8.627	10.26	15.00	24.10	50.71

TABLE 7.24 : I_{02}^{*} ($1/\alpha = .10$, $\mu = .60$)

Q: MODE:	0.05	0.10	0.15	0.20	0.25	0.30	0.35	0.40	0.45	0.50	0.55	0.60	0.70	0.80	0.90
(1)	.0064	.0258	.0587	.1059	.1689	.2494	.3501	.4748	.6284	.8184	1.055	1.355	2.262	4.070	9.444
(2)	.0013	.0052	.0120	.0220	.0358	.0544	.0792	.1128	.1592	.2260	.3275	.4914	1.238	3.146	8.645
(3)	.0445	.0979	.1654	.2555	.3834	.5761	.8756	1.327	1.950	2.741	3.704	4.872	8.154	14.21	31.29

TABLE 7.24 : I_{02}^{*} ($1/\alpha = .10$, $\mu = .70$)

Q: MODE:	0.05	0.10	0.15	0.20	0.25	0.30	0.35	0.40	0.45	0.50	0.55	0.60	0.70	0.80	0.90
(1)	.0065	.0261	.0594	.1073	.1710	.2526	.3546	.4809	.6365	.8290	1.069	1.373	2.291	4.124	9.568
(2)	.0011	.0045	.0103	.0188	.0305	.0459	.0663	.0933	.1301	.1821	.2605	.3889	1.041	2.907	8.354
(3)	.0539	.1119	.1771	.2537	.3475	.4671	.6258	.8446	1.154	1.596	2.211	3.040	5.554	10.25	23.20

TABLE 7.24 : I_{02}^{*} ($1/\alpha = .10$, $\mu = .80$)

Q: MODE:	0.05	0.10	0.15	0.20	0.25	0.30	0.35	0.40	0.45	0.50	0.55	0.60	0.70	0.80	0.90
(1)	.0066	.0265	.0603	.1088	.1734	.2561	.3596	.4876	.6453	.8403	1.083	1.391	2.321	4.174	9.678
(2)	.0008	.0034	.0077	.0139	.0224	.0333	.0474	.0657	.0897	.1228	.1718	.2540	.8013	2.751	8.389
(3)	.0568	.1086	.1607	.2166	.2799	.3547	.4471	.5660	.7265	.9538	1.290	1.800	3.664	7.522	18.21

TABLE 7.24 : I_{02}^{*} ($1/\alpha = .20$, $\mu = .20$)

Q: MODE:	0.05	0.10	0.15	0.20	0.25	0.30	0.35	0.40	0.45	0.50	0.55	0.60	0.70	0.80	0.90
(1)	.0062	.0251	.0570	.1029	.1641	.2422	.3400	.4608	.6096	.7933	1.022	1.311	2.184	3.921	9.069
(2)	.0018	.0071	.0163	.0296	.0475	.0711	.1017	.1415	.1945	.2679	.3761	.5489	1.360	3.459	9.526
(3)	.0983	.1931	.2906	.3963	.5159	.6562	.8265	1.040	1.316	1.683	2.181	2.863	5.084	9.527	22.12

TABLE 7.24 : I_{02}^{*} ($1/\alpha = .20$, $\mu = .30$)

Q: MODE:	0.05	0.10	0.15	0.20	0.25	0.30	0.35	0.40	0.45	0.50	0.55	0.60	0.70	0.80	0.90
(1)	.0063	.0253	.0576	.1040	.1659	.2449	.3438	.4661	.6169	.8031	1.035	1.329	2.217	3.986	9.239
(2)	.0023	.0093	.0213	.0389	.0630	.0951	.1375	.1938	.2701	.3767	.5318	.7671	1.700	3.873	10.21
(3)	.0904	.1877	.2960	.4204	.5673	.7457	.9679	1.251	1.617	2.095	2.722	3.541	6.033	10.81	24.26

TABLE 7.24 : I_{02}^{*} ($1/\alpha = .20$, $\mu = .40$)

Q: MODE:	0.05	0.10	0.15	0.20	0.25	0.30	0.35	0.40	0.45	0.50	0.55	0.60	0.70	0.80	0.90
(1)	.0064	.0257	.0585	.1056	.1684	.2487	.3492	.4735	.6269	.8166	1.053	1.353	2.259	4.070	9.458
(2)	.0026	.0105	.0241	.0441	.0716	.1086	.1577	.2232	.3123	.4362	.6136	.8745	1.842	4.006	10.29
(3)	.0729	.1569	.2566	.3788	.5339	.7371	1.010	1.380	1.875	2.519	3.337	4.361	7.320	12.84	28.40

TABLE 7.24 : I_{02}^{*} ($1/\alpha = .20$, $\mu = .50$)

Q: MODE:	0.05	0.10	0.15	0.20	0.25	0.30	0.35	0.40	0.45	0.50	0.55	0.60	0.70	0.80	0.90
(1)	.0065	.0262	.0596	.1076	.1715	.2534	.3559	.4828	.6395	.8333	1.075	1.382	2.311	4.170	9.707
(2)	.0026	.0107	.0246	.0451	.0733	.1110	.1611	.2277	.3177	.4418	.6173	.8714	1.795	3.840	9.747
(3)	.5270	1.074	1.648	2.259	2.916	3.632	4.424	5.311	6.324	7.499	8.895	10.59	15.52	25.02	52.85

TABLE 7.24 : I_{02}^{*} ($1/\alpha = .20$, $\mu = .60$)

Q: MODE:	0.05	0.10	0.15	0.20	0.25	0.30	0.35	0.40	0.45	0.50	0.55	0.60	0.70	0.80	0.90
(1)	.0066	.0268	.0609	.1099	.1753	.2590	.3639	.4938	.6541	.8526	1.101	1.415	2.368	4.277	9.967
(2)	.0025	.0101	.0231	.0421	.0683	.1030	.1487	.2090	.2895	.3993	.5533	.7757	1.597	3.472	8.949
(3)	.0889	.1950	.3266	.4961	.7216	1.027	1.439	1.977	2.655	3.484	4.487	5.708	9.186	15.69	34.23

TABLE 7.24 : I_{02}^{*} ($1/\alpha = .20$, $\mu = .70$)

Q: MODE:	0.05	0.10	0.15	0.20	0.25	0.30	0.35	0.40	0.45	0.50	0.55	0.60	0.70	0.80	0.90
(1)	.0068	.0274	.0624	.1126	.1796	.2655	.3729	.5061	.6704	.8739	1.128	1.450	2.427	4.384	10.22
(2)	.0021	.0086	.0196	.0356	.0574	.0860	.1231	.1712	.2345	.3196	.4384	.6119	1.299	3.050	8.384
(3)	.1077	.2225	.3498	.4959	.6688	.8789	1.140	1.470	1.890	2.426	3.110	3.978	6.542	11.37	24.85

TABLE 7.24 : I_{02}^{*} (1/α = .20 , μ = .80)

Q: MODE:	0.05	0.10	0.15	0.20	0.25	0.30	0.35	0.40	0.45	0.50	0.55	0.60	0.70	0.80	0.90
(1)	.0070	.0282	.0641	.1157	.1845	.2725	.3828	.5194	.6879	.8965	1.157	1.487	2.486	4.485	10.43
(2)	.0016	.0063	.0144	.0261	.0417	.0618	.0874	.1198	.1613	.2161	.2920	.4056	.9518	2.763	8.456
(3)	.1131	.2148	.3147	.4192	.5342	.6655	.8205	1.009	1.245	1.550	1.952	2.494	4.270	7.921	18.41

TABLE 7.24 : I_{02}^{*} (1/α = .30 , μ = .20)

Q: MODE:	0.05	0.10	0.15	0.20	0.25	0.30	0.35	0.40	0.45	0.50	0.55	0.60	0.70	0.80	0.90
(1)	.0063	.0252	.0573	.1034	.1648	.2433	.3416	.4630	.6126	.7974	1.028	1.319	2.198	3.949	9.142
(2)	.0026	.0106	.0243	.0441	.0709	.1058	.1510	.2094	.2861	.3902	.5377	.7585	1.651	3.807	10.14
(3)	.1379	.2727	.4118	.5619	.7299	.9239	1.154	1.435	1.784	2.229	2.806	3.562	5.925	10.62	24.12

TABLE 7.24 : I_{02}^{*} (1/α = .30 , μ = .30)

Q: MODE:	0.05	0.10	0.15	0.20	0.25	0.30	0.35	0.40	0.45	0.50	0.55	0.60	0.70	0.80	0.90
(1)	.0064	.0256	.0582	.1051	.1676	.2475	.3475	.4712	.6238	.8124	1.048	1.346	2.247	4.046	9.398
(2)	.0034	.0138	.0316	.0576	.0932	.1403	.2019	.2829	.3905	.5366	.7401	1.031	2.072	4.373	11.14
(3)	.1253	.2601	.4090	.5776	.7729	1.004	1.282	1.624	2.047	2.578	3.249	4.105	6.667	11.59	25.57

TABLE 7.24 : I_{02}^{*} (1/α = .30 , μ = .40)

Q: MODE:	0.05	0.10	0.15	0.20	0.25	0.30	0.35	0.40	0.45	0.50	0.55	0.60	0.70	0.80	0.90
(1)	.0065	.0262	.0595	.1075	.1714	.2532	.3557	.4826	.6392	.8331	1.075	1.382	2.312	4.175	9.728
(2)	.0038	.0154	.0354	.0647	.1049	.1584	.2286	.3210	.4436	.6088	.8351	1.150	2.227	4.524	11.21
(3)	.1004	.2142	.3464	.5037	.6953	.9340	1.237	1.624	2.119	2.747	3.538	4.532	7.433	12.91	28.43

TABLE 7.24 : I_{02}^{*} (1/α = .30 , μ = .50)

Q: MODE:	0.05	0.10	0.15	0.20	0.25	0.30	0.35	0.40	0.45	0.50	0.55	0.60	0.70	0.80	0.90
(1)	.0067	.0269	.0612	.1105	.1762	.2605	.3660	.4968	.6584	.8586	1.109	1.426	2.390	4.325	10.10
(2)	.0039	.0156	.0359	.0655	.1061	.1600	.2305	.3228	.4443	.6063	.8255	1.127	2.138	4.267	10.40
(3)	.5411	1.103	1.693	2.320	2.996	3.734	4.549	5.465	6.511	7.727	9.172	10.93	16.05	25.95	55.01

TABLE 7.24 : I_{02}^{*} (1/α = .30 , μ = .60)

Q: MODE:	0.05	0.10	0.15	0.20	0.25	0.30	0.35	0.40	0.45	0.50	0.55	0.60	0.70	0.80	0.90
(1)	.0069	.0278	.0632	.1141	.1820	.2691	.3782	.5135	.6807	.8879	1.147	1.476	2.477	4.487	10.50
(2)	.0036	.0145	.0332	.0606	.0978	.1468	.2103	.2926	.3998	.5412	.7308	.9903	1.867	3.752	9.239
(3)	.1332	.2911	.4840	.7254	1.032	1.422	1.914	2.520	3.255	4.136	5.196	6.487	10.18	17.16	37.16

TABLE 7.24 : I_{02}^{*} (1/α = .30 , μ = .70)

Q: MODE:	0.05	0.10	0.15	0.20	0.25	0.30	0.35	0.40	0.45	0.50	0.55	0.60	0.70	0.80	0.90
(1)	.0071	.0288	.0654	.1182	.1886	.2787	.3918	.5320	.7052	.9200	1.189	1.530	2.566	4.648	10.87
(2)	.0030	.0122	.0279	.0507	.0814	.1213	.1724	.2376	.3212	.4303	.5758	.7763	1.488	3.176	8.414
(3)	.1613	.3320	.5185	.7281	.9692	1.252	1.589	1.995	2.487	3.087	3.825	4.741	7.402	12.40	26.45

TABLE 7.24 : I_{02}^{*} (1/α = .30 , μ = .80)

Q: MODE:	0.05	0.10	0.15	0.20	0.25	0.30	0.35	0.40	0.45	0.50	0.55	0.60	0.70	0.80	0.90
(1)	.0074	.0299	.0679	.1227	.1957	.2893	.4065	.5518	.7313	.9536	1.231	1.584	2.654	4.798	11.20
(2)	.0022	.0090	.0204	.0367	.0585	.0864	.1215	.1653	.2204	.2910	.3847	.5166	1.061	2.774	8.524
(3)	.1691	.3186	.4622	.6091	.7668	.9421	1.143	1.378	1.661	2.009	2.446	3.008	4.752	8.279	18.59

TABLE 7.24 : I_{02}^{*} (1/α = .40 , μ = .20)

Q: MODE:	0.05	0.10	0.15	0.20	0.25	0.30	0.35	0.40	0.45	0.50	0.55	0.60	0.70	0.80	0.90
(1)	.0063	.0253	.0575	.1039	.1656	.2445	.3432	.4653	.6157	.8016	1.033	1.326	2.212	3.976	9.215
(2)	.0035	.0141	.0322	.0585	.0939	.1400	.1993	.2755	.3747	.5067	.6887	.9495	1.918	4.150	10.76
(3)	.1725	.3433	.5201	.7104	.9216	1.163	1.444	1.781	2.191	2.700	3.343	4.167	6.687	11.67	26.09

TABLE 7.24 : I_{02}^{*} (1/α = .40 , μ = .30)

Q: MODE:	0.05	0.10	0.15	0.20	0.25	0.30	0.35	0.40	0.45	0.50	0.55	0.60	0.70	0.80	0.90
(1)	.0064	.0259	.0588	.1062	.1693	.2501	.3513	.4765	.6310	.8221	1.061	1.363	2.278	4.108	9.559
(2)	.0045	.0182	.0416	.0758	.1225	.1839	.2638	.3676	.5035	.6841	.9284	1.265	2.406	4.849	12.06
(3)	.1554	.3223	.5057	.7114	.9465	1.220	1.543	1.929	2.397	2.971	3.683	4.577	7.229	12.32	26.85

TABLE 7.24 : I_{02}^{*} (1/α = .40 , μ = .40)

Q: MODE:	0.05	0.10	0.15	0.20	0.25	0.30	0.35	0.40	0.45	0.50	0.55	0.60	0.70	0.80	0.90
(1)	.0066	.0266	.0606	.1094	.1746	.2580	.3625	.4920	.6520	.8501	1.098	1.412	2.366	4.281	10.00
(2)	.0050	.0202	.0463	.0844	.1365	.2054	.2950	.4114	.5633	.7632	1.030	1.389	2.564	5.004	12.12
(3)	.1237	.2624	.4209	.6055	.8245	1.089	1.413	1.813	2.312	2.933	3.709	4.682	7.540	12.97	28.47

TABLE 7.24 : I_{02}^{*} (1/α = .40 , μ = .50)

Q: MODE:	0.05	0.10	0.15	0.20	0.25	0.30	0.35	0.40	0.45	0.50	0.55	0.60	0.70	0.80	0.90
(1)	.0069	.0276	.0629	.1135	.1812	.2678	.3765	.5113	.6780	.8847	1.143	1.472	2.472	4.483	10.51
(2)	.0050	.0203	.0464	.0846	.1367	.2052	.2939	.4084	.5565	.7494	1.003	1.342	2.430	4.653	11.03
(3)	.5557	1.133	1.739	2.384	3.080	3.839	4.680	5.625	6.705	7.962	9.459	11.29	16.60	26.90	57.20

TABLE 7.24 : I_{02}^{*} (1/α = .40 , μ = .60)

Q: MODE:	0.05	0.10	0.15	0.20	0.25	0.30	0.35	0.40	0.45	0.50	0.55	0.60	0.70	0.80	0.90
(1)	.0072	.0288	.0655	.1184	.1890	.2794	.3929	.5338	.7080	.9243	1.195	1.539	2.588	4.700	11.03
(2)	.0046	.0187	.0426	.0775	.1246	.1862	.2653	.3663	.4954	.6618	.8789	1.166	2.091	4.000	9.516
(3)	.1774	.3865	.6383	.9456	1.323	1.784	2.342	3.009	3.799	4.737	5.860	7.228	11.15	18.61	40.09

TABLE 7.24 : I_{02}^{*} (1/α = .40 , μ = .70)

Q: MODE:	0.05	0.10	0.15	0.20	0.25	0.30	0.35	0.40	0.45	0.50	0.55	0.60	0.70	0.80	0.90
(1)	.0075	.0301	.0686	.1239	.1977	.2924	.4112	.5585	.7409	.9671	1.250	1.611	2.707	4.915	11.53
(2)	.0039	.0156	.0355	.0642	.1027	.1525	.2156	.2950	.3952	.5228	.6882	.9082	1.641	3.288	8.443
(3)	.2148	.4403	.6833	.9513	1.252	1.596	1.994	2.458	3.006	3.659	4.446	5.411	8.184	13.38	28.01

TABLE 7.24 : I_{02}^{*} (1/α = .40 , μ = .80)

Q: MODE:	0.05	0.10	0.15	0.20	0.25	0.30	0.35	0.40	0.45	0.50	0.55	0.60	0.70	0.80	0.90
(1)	.0078	.0316	.0719	.1299	.2072	.3064	.4307	.5848	.7754	1.012	1.307	1.682	2.823	5.114	11.96
(2)	.0028	.0113	.0256	.0461	.0733	.1079	.1510	.2043	.2704	.3532	.4601	.6044	1.148	2.785	8.594
(3)	.2246	.4199	.6035	.7875	.9809	1.191	1.427	1.697	2.013	2.390	2.851	3.428	5.161	8.607	18.78

TABLE 7.24 : I_{02}^{*} (1/α = .50 , μ = .20)

Q: MODE:	0.05	0.10	0.15	0.20	0.25	0.30	0.35	0.40	0.45	0.50	0.55	0.60	0.70	0.80	0.90
(1)	.0063	.0254	.0578	.1044	.1664	.2457	.3449	.4676	.6189	.8059	1.039	1.334	2.226	4.004	9.289
(2)	.0044	.0176	.0401	.0727	.1165	.1736	.2467	.3400	.4604	.6184	.8315	1.128	2.169	4.489	11.39
(3)	.2030	.4062	.6174	.8442	1.095	1.378	1.705	2.091	2.555	3.120	3.823	4.712	7.396	12.68	28.05

TABLE 7.24 : I^*_{02} ($1/\alpha = .50$, $\mu = .30$)

Q:	0.05	0.10	0.15	0.20	0.25	0.30	0.35	0.40	0.45	0.50	0.55	0.60	0.70	0.80	0.90
MODE:															
(1)	.0065	.0261	.0594	.1073	.1711	.2529	.3552	.4819	.6384	.8320	1.074	1.380	2.310	4.171	9.721
(2)	.0056	.0225	.0514	.0936	.1509	.2260	.3232	.4483	.6101	.8217	1.102	1.480	2.716	5.308	12.97
(3)	.1815	.3763	.5897	.8272	1.096	1.405	1.764	2.189	2.695	3.307	4.057	4.990	7.738	13.01	28.10

TABLE 7.24 : I^*_{02} ($1/\alpha = .50$, $\mu = .40$)

Q:	0.05	0.10	0.15	0.20	0.25	0.30	0.35	0.40	0.45	0.50	0.55	0.60	0.70	0.80	0.90
MODE:															
(1)	.0067	.0271	.0617	.1115	.1779	.2629	.3696	.5018	.6652	.8678	1.121	1.443	2.422	4.390	10.28
(2)	.0061	.0248	.0567	.1033	.1666	.2498	.3574	.4955	.6734	.9038	1.205	1.603	2.871	5.459	13.01
(3)	.1438	.3037	.4840	.6908	.9316	1.216	1.557	1.969	2.473	3.092	3.858	4.817	7.641	13.04	28.50

TABLE 7.24 : I^*_{02} ($1/\alpha = .50$, $\mu = .50$)

Q:	0.05	0.10	0.15	0.20	0.25	0.30	0.35	0.40	0.45	0.50	0.55	0.60	0.70	0.80	0.90
MODE:															
(1)	.0071	.0284	.0646	.1167	.1863	.2755	.3874	.5264	.6983	.9117	1.179	1.519	2.555	4.644	10.91
(2)	.0061	.0247	.0564	.1025	.1651	.2470	.3521	.4863	.6574	.8767	1.160	1.531	2.689	5.010	11.64
(3)	.5710	1.164	1.787	2.451	3.167	3.949	4.815	5.790	6.905	8.205	9.753	11.65	17.16	27.87	59.40

TABLE 7.24 : I^*_{02} ($1/\alpha = .50$, $\mu = .60$)

Q:	0.05	0.10	0.15	0.20	0.25	0.30	0.35	0.40	0.45	0.50	0.55	0.60	0.70	0.80	0.90
MODE:															
(1)	.0074	.0299	.0680	.1229	.1961	.2901	.4081	.5547	.7361	.9615	1.244	1.604	2.701	4.916	11.57
(2)	.0056	.0225	.0512	.0929	.1491	.2220	.3148	.4319	.5796	.7669	1.006	1.317	2.285	4.225	9.782
(3)	.2215	.4812	.7896	1.159	1.600	2.123	2.740	3.463	4.309	5.306	6.495	7.943	12.11	20.05	43.01

TABLE 7.24 : I^*_{02} ($1/\alpha = .50$, $\mu = .70$)

Q:	0.05	0.10	0.15	0.20	0.25	0.30	0.35	0.40	0.45	0.50	0.55	0.60	0.70	0.80	0.90
MODE:															
(1)	.0078	.0315	.0718	.1297	.2071	.3063	.4309	.5857	.7772	1.015	1.313	1.693	2.850	5.184	12.19
(2)	.0046	.0186	.0422	.0763	.1218	.1803	.2538	.3453	.4593	.6021	.7835	1.019	1.769	3.390	8.472
(3)	.2681	.5475	.8446	1.166	1.521	1.917	2.366	2.880	3.475	4.174	5.007	6.020	8.911	14.31	29.53

TABLE 7.24 : I^*_{02} ($1/\alpha = .50$, $\mu = .80$)

Q:	0.05	0.10	0.15	0.20	0.25	0.30	0.35	0.40	0.45	0.50	0.55	0.60	0.70	0.80	0.90
MODE:															
(1)	.0083	.0334	.0759	.1372	.2189	.3237	.4552	.6183	.8200	1.070	1.384	1.782	2.994	5.432	12.73
(2)	.0033	.0133	.0302	.0544	.0863	.1267	.1768	.2381	.3132	.4061	.5233	.6771	1.220	2.796	8.665
(3)	.2796	.5188	.7390	.9554	1.179	1.418	1.682	1.978	2.319	2.720	3.199	3.789	5.520	8.911	18.96

TABLE 7.24 : I^*_{02} ($1/\alpha = .60$, $\mu = .20$)

Q:	0.05	0.10	0.15	0.20	0.25	0.30	0.35	0.40	0.45	0.50	0.55	0.60	0.70	0.80	0.90
MODE:															
(1)	.0063	.0255	.0581	.1049	.1672	.2469	.3466	.4701	.6222	.8103	1.045	1.342	2.240	4.033	9.363
(2)	.0052	.0210	.0478	.0867	.1389	.2067	.2931	.4030	.5436	.7260	.9678	1.297	2.409	4.825	12.03
(3)	.2300	.4625	.7051	.9654	1.252	1.574	1.943	2.374	2.885	3.503	4.263	5.216	8.066	13.66	30.00

TABLE 7.24 : I^*_{02} ($1/\alpha = .60$, $\mu = .30$)

Q:	0.05	0.10	0.15	0.20	0.25	0.30	0.35	0.40	0.45	0.50	0.55	0.60	0.70	0.80	0.90
MODE:															
(1)	.0066	.0264	.0601	.1085	.1730	.2557	.3593	.4876	.6460	.8422	1.087	1.398	2.342	4.235	9.884
(2)	.0066	.0267	.0610	.1109	.1784	.2667	.3803	.5254	.7114	.9514	1.265	1.680	3.008	5.754	13.88
(3)	.2043	.4238	.6633	.9287	1.227	1.566	1.958	2.415	2.955	3.602	4.388	5.361	8.209	13.67	29.32

TABLE 7.24 : I^*_{02} ($1/\alpha = .60$, $\mu = .40$)

Q:	0.05	0.10	0.15	0.20	0.25	0.30	0.35	0.40	0.45	0.50	0.55	0.60	0.70	0.80	0.90
MODE:															
(1)	.0069	.0276	.0629	.1136	.1813	.2680	.3769	.5119	.6789	.8860	1.145	1.475	2.479	4.500	10.56
(2)	.0072	.0292	.0667	.1213	.1952	.2919	.4160	.5742	.7756	1.033	1.366	1.800	3.155	5.893	13.88
(3)	.1613	.3394	.5384	.7639	1.023	1.324	1.679	2.101	2.610	3.229	3.991	4.940	7.736	13.10	28.53

TABLE 7.24 : I^*_{02} ($1/\alpha = .60$, $\mu = .50$)

Q:	0.05	0.10	0.15	0.20	0.25	0.30	0.35	0.40	0.45	0.50	0.55	0.60	0.70	0.80	0.90
MODE:															
(1)	.0072	.0292	.0664	.1200	.1916	.2834	.3987	.5419	.7192	.9394	1.216	1.567	2.640	4.807	11.32
(2)	.0071	.0288	.0657	.1193	.1916	.2857	.4057	.5575	.7491	.9916	1.301	1.700	2.925	5.345	12.23
(3)	.5869	1.196	1.837	2.520	3.256	4.062	4.955	5.960	7.112	8.455	10.06	12.01	17.73	28.84	61.62

TABLE 7.24 : I^*_{02} ($1/\alpha = .60$, $\mu = .60$)

Q:	0.05	0.10	0.15	0.20	0.25	0.30	0.35	0.40	0.45	0.50	0.55	0.60	0.70	0.80	0.90
MODE:															
(1)	.0077	.0310	.0705	.1274	.2035	.3011	.4237	.5761	.7649	1.000	1.294	1.669	2.816	5.134	12.11
(2)	.0064	.0260	.0592	.1071	.1715	.2545	.3595	.4908	.6547	.8600	1.119	1.450	2.457	4.432	10.04
(3)	.2656	.5751	.9384	1.365	1.865	2.446	3.118	3.894	4.795	5.851	7.108	8.630	13.05	21.47	45.92

TABLE 7.24 : I^*_{02} ($1/\alpha = .60$, $\mu = .70$)

Q:	0.05	0.10	0.15	0.20	0.25	0.30	0.35	0.40	0.45	0.50	0.55	0.60	0.70	0.80	0.90
MODE:															
(1)	.0082	.0330	.0751	.1357	.2167	.3206	.4511	.6133	.8142	1.064	1.377	1.776	2.994	5.456	12.85
(2)	.0053	.0213	.0484	.0873	.1390	.2052	.2878	.3898	.5156	.6713	.8661	1.114	1.881	3.484	8.500
(3)	.3212	.6536	1.002	1.374	1.777	2.220	2.713	3.270	3.908	4.649	5.526	6.586	9.598	15.21	31.03

TABLE 7.24 : I^*_{02} ($1/\alpha = .60$, $\mu = .80$)

Q:	0.05	0.10	0.15	0.20	0.25	0.30	0.35	0.40	0.45	0.50	0.55	0.60	0.70	0.80	0.90
MODE:															
(1)	.0087	.0352	.0800	.1445	.2307	.3412	.4799	.6521	.8652	1.130	1.461	1.882	3.166	5.751	13.49
(2)	.0038	.0152	.0344	.0618	.0979	.1434	.1995	.2678	.3506	.4518	.5775	.7389	1.282	2.806	8.738
(3)	.3343	.6153	.8688	1.114	1.363	1.626	1.912	2.231	2.592	3.011	3.506	4.108	5.843	9.194	19.13

TABLE 7.24 : I^*_{02} ($1/\alpha = .70$, $\mu = .20$)

Q:	0.05	0.10	0.15	0.20	0.25	0.30	0.35	0.40	0.45	0.50	0.55	0.60	0.70	0.80	0.90
MODE:															
(1)	.0064	.0257	.0584	.1054	.1680	.2481	.3484	.4726	.6256	.8149	1.051	1.350	2.255	4.062	9.438
(2)	.0060	.0244	.0555	.1005	.1610	.2392	.3387	.4646	.6245	.8300	1.099	1.458	2.640	5.159	12.67
(3)	.2541	.5131	.7845	1.076	1.395	1.753	2.161	2.633	3.190	3.857	4.672	5.687	8.707	14.62	31.94

TABLE 7.24 : I^*_{02} ($1/\alpha = .70$, $\mu = .30$)

Q:	0.05	0.10	0.15	0.20	0.25	0.30	0.35	0.40	0.45	0.50	0.55	0.60	0.70	0.80	0.90
MODE:															
(1)	.0066	.0267	.0608	.1097	.1750	.2587	.3635	.4934	.6539	.8528	1.101	1.417	2.375	4.299	10.05
(2)	.0076	.0308	.0703	.1277	.2051	.3061	.4353	.5993	.8079	1.074	1.419	1.869	3.288	6.191	14.79
(3)	.2245	.4657	.7284	1.018	1.343	1.709	2.129	2.616	3.187	3.866	4.687	5.690	8.648	14.30	30.52

TABLE 7.24 : I^*_{02} ($1/\alpha = .70$, $\mu = .40$)

Q:	0.05	0.10	0.15	0.20	0.25	0.30	0.35	0.40	0.45	0.50	0.55	0.60	0.70	0.80	0.90
MODE:															
(1)	.0070	.0282	.0641	.1158	.1848	.2734	.3844	.5223	.6930	.9048	1.170	1.508	2.537	4.611	10.84
(2)	.0083	.0334	.0762	.1384	.2224	.3317	.4713	.6480	.8711	1.154	1.515	1.982	3.424	6.312	14.74
(3)	.1767	.3706	.5858	.8273	1.102	1.417	1.784	2.216	2.731	3.352	4.110	5.053	7.827	13.17	28.56

TABLE 7.24 : I^*_{02} ($1/\alpha = .70$, $\mu = .50$)

Q: MODE:	0.05	0.10	0.15	0.20	0.25	0.30	0.35	0.40	0.45	0.50	0.55	0.60	0.70	0.80	0.90
(1)	.0074	.0300	.0683	.1234	.1970	.2915	.4103	.5578	.7406	.9679	1.253	1.616	2.727	4.973	11.73
(2)	.0081	.0326	.0744	.1349	.2163	.3216	.4553	.6231	.8331	1.097	1.430	1.855	3.143	5.662	12.81
(3)	.6033	1.230	1.889	2.591	3.349	4.179	5.099	6.136	7.324	8.711	10.37	12.39	18.31	29.84	63.85

TABLE 7.24 : I^*_{02} ($1/\alpha = .70$, $\mu = .60$)

Q: MODE:	0.05	0.10	0.15	0.20	0.25	0.30	0.35	0.40	0.45	0.50	0.55	0.60	0.70	0.80	0.90
(1)	.0080	.0321	.0731	.1322	.2110	.3124	.4397	.5981	.7944	1.039	1.345	1.736	2.932	5.355	12.65
(2)	.0072	.0292	.0665	.1202	.1920	.2842	.4001	.5441	.7224	.9436	1.220	1.569	2.613	4.626	10.29
(3)	.3096	.6684	1.085	1.567	2.121	2.755	3.479	4.308	5.263	6.379	7.705	9.319	13.97	22.89	48.83

TABLE 7.24 : I^*_{02} ($1/\alpha = .70$, $\mu = .70$)

Q: MODE:	0.05	0.10	0.15	0.20	0.25	0.30	0.35	0.40	0.45	0.50	0.55	0.60	0.70	0.80	0.90
(1)	.0086	.0345	.0784	.1418	.2264	.3351	.4716	.6414	.8518	1.114	1.442	1.861	3.140	5.730	13.52
(2)	.0059	.0238	.0540	.0972	.1546	.2276	.3183	.4296	.5658	.7326	.9390	1.198	1.980	3.571	8.528
(3)	.3743	.7585	1.157	1.576	2.023	2.507	3.040	3.636	4.312	5.093	6.012	7.119	10.25	16.08	32.51

TABLE 7.24 : I^*_{02} ($1/\alpha = .70$, $\mu = .80$)

Q: MODE:	0.05	0.10	0.15	0.20	0.25	0.30	0.35	0.40	0.45	0.50	0.55	0.60	0.70	0.80	0.90
(1)	.0092	.0370	.0841	.1520	.2426	.3590	.5050	.6863	.9108	1.190	1.539	1.984	3.339	6.072	14.26
(2)	.0042	.0168	.0381	.0684	.1082	.1583	.2197	.2940	.3835	.4918	.6247	.7926	1.336	2.816	8.812
(3)	.3885	.7094	.9933	1.263	1.534	1.818	2.123	2.460	2.839	3.273	3.782	4.395	6.138	9.460	19.31

TABLE 7.24 : I^*_{02} ($1/\alpha = .80$, $\mu = .20$)

Q: MODE:	0.05	0.10	0.15	0.20	0.25	0.30	0.35	0.40	0.45	0.50	0.55	0.60	0.70	0.80	0.90
(1)	.0064	.0258	.0587	.1059	.1689	.2494	.3503	.4751	.6291	.8196	1.057	1.358	2.270	4.091	9.513
(2)	.0069	.0277	.0631	.1142	.1827	.2712	.3834	.5248	.7034	.9309	1.225	1.614	2.866	5.490	13.31
(3)	.2757	.5589	.8567	1.177	1.527	1.918	2.362	2.874	3.473	4.188	5.056	6.134	9.325	15.57	33.86

TABLE 7.24 : I^*_{02} ($1/\alpha = .80$, $\mu = .30$)

Q: MODE:	0.05	0.10	0.15	0.20	0.25	0.30	0.35	0.40	0.45	0.50	0.55	0.60	0.70	0.80	0.90
(1)	.0067	.0270	.0615	.1110	.1771	.2617	.3679	.4995	.6621	.8637	1.116	1.436	2.409	4.365	10.22
(2)	.0086	.0348	.0793	.1440	.2310	.3441	.4882	.6703	.9002	1.192	1.565	2.049	3.557	6.620	15.69
(3)	.2425	.5031	.7865	1.099	1.446	1.837	2.283	2.797	3.396	4.106	4.960	6.010	9.063	14.90	31.71

TABLE 7.24 : I^*_{02} ($1/\alpha = .80$, $\mu = .40$)

Q: MODE:	0.05	0.10	0.15	0.20	0.25	0.30	0.35	0.40	0.45	0.50	0.55	0.60	0.70	0.80	0.90
(1)	.0071	.0287	.0654	.1181	.1885	.2789	.3923	.5331	.7075	.9241	1.196	1.541	2.596	4.725	11.12
(2)	.0093	.0374	.0853	.1547	.2482	.3695	.5236	.7175	.9608	1.267	1.655	2.154	3.679	6.719	15.59
(3)	.1904	.3983	.6276	.8831	1.171	1.499	1.877	2.318	2.838	3.461	4.219	5.157	7.913	13.23	28.59

TABLE 7.24 : I^*_{02} ($1/\alpha = .80$, $\mu = .50$)

Q: MODE:	0.05	0.10	0.15	0.20	0.25	0.30	0.35	0.40	0.45	0.50	0.55	0.60	0.70	0.80	0.90
(1)	.0077	.0308	.0702	.1269	.2026	.2999	.4222	.5742	.7626	.9970	1.291	1.667	2.815	5.140	12.14
(2)	.0090	.0362	.0826	.1495	.2393	.3551	.5012	.6837	.9107	1.193	1.548	1.998	3.348	5.965	13.38
(3)	.6202	1.264	1.942	2.665	3.445	4.299	5.247	6.316	7.541	8.973	10.68	12.78	18.90	30.84	66.10

TABLE 7.24 : I^*_{02} ($1/\alpha = .80$, $\mu = .60$)

Q: MODE:	0.05	0.10	0.15	0.20	0.25	0.30	0.35	0.40	0.45	0.50	0.55	0.60	0.70	0.80	0.90
(1)	.0083	.0333	.0758	.1370	.2188	.3239	.4560	.6204	.8244	1.078	1.397	1.804	3.050	5.577	13.19
(2)	.0080	.0322	.0732	.1322	.2108	.3114	.4372	.5927	.7840	1.019	1.311	1.677	2.756	4.807	10.53
(3)	.3535	.7610	1.229	1.764	2.370	3.055	3.829	4.708	5.717	6.893	8.289	9.987	14.89	24.31	51.74

TABLE 7.24 : I^*_{02} ($1/\alpha = .80$, $\mu = .70$)

Q: MODE:	0.05	0.10	0.15	0.20	0.25	0.30	0.35	0.40	0.45	0.50	0.55	0.60	0.70	0.80	0.90
(1)	.0089	.0360	.0818	.1479	.2363	.3498	.4924	.6699	.8899	1.164	1.508	1.946	3.288	6.005	14.19
(2)	.0065	.0260	.0590	.1062	.1687	.2478	.3458	.4654	.6107	.7874	1.004	1.273	2.070	3.652	8.555
(3)	.4271	.8625	1.309	1.771	2.259	2.781	3.350	3.982	4.694	5.513	6.473	7.626	10.88	16.93	33.96

TABLE 7.24 : I^*_{02} ($1/\alpha = .80$, $\mu = .80$)

Q: MODE:	0.05	0.10	0.15	0.20	0.25	0.30	0.35	0.40	0.45	0.50	0.55	0.60	0.70	0.80	0.90
(1)	.0096	.0388	.0883	.1595	.2547	.3768	.5302	.7208	.9567	1.250	1.617	2.085	3.513	6.394	15.04
(2)	.0046	.0183	.0415	.0744	.1175	.1716	.2377	.3174	.4128	.5274	.6665	.8398	1.385	2.825	8.887
(3)	.4422	.8012	1.113	1.405	1.695	1.996	2.318	2.670	3.064	3.512	4.034	4.657	6.410	9.712	19.48

TABLE 7.24 : I^*_{02} ($1/\alpha = .90$, $\mu = .20$)

Q: MODE:	0.05	0.10	0.15	0.20	0.25	0.30	0.35	0.40	0.45	0.50	0.55	0.60	0.70	0.80	0.90
(1)	.0064	.0259	.0590	.1065	.1697	.2508	.3522	.4778	.6327	.8243	1.064	1.367	2.285	4.121	9.589
(2)	.0077	.0310	.0706	.1277	.2041	.3027	.4274	.5838	.7804	1.029	1.348	1.765	3.086	5.820	13.96
(3)	.2591	.6004	.9226	1.269	1.649	2.071	2.549	3.098	3.739	4.500	5.421	6.561	9.924	16.50	35.78

TABLE 7.24 : I^*_{02} ($1/\alpha = .90$, $\mu = .30$)

Q: MODE:	0.05	0.10	0.15	0.20	0.25	0.30	0.35	0.40	0.45	0.50	0.55	0.60	0.70	0.80	0.90
(1)	.0068	.0273	.0622	.1123	.1792	.2649	.3724	.5057	.6705	.8749	1.131	1.456	2.444	4.432	10.38
(2)	.0096	.0386	.0881	.1598	.2561	.3809	.5393	.7386	.9888	1.304	1.705	2.222	3.818	7.043	16.59
(3)	.2587	.5366	.8387	1.171	1.539	1.953	2.422	2.961	3.587	4.327	5.213	6.301	9.456	15.49	32.87

TABLE 7.24 : I^*_{02} ($1/\alpha = .90$, $\mu = .40$)

Q: MODE:	0.05	0.10	0.15	0.20	0.25	0.30	0.35	0.40	0.45	0.50	0.55	0.60	0.70	0.80	0.90
(1)	.0073	.0293	.0667	.1205	.1923	.2845	.4004	.5442	.7225	.9439	1.222	1.575	2.656	4.839	11.40
(2)	.0102	.0412	.0940	.1703	.2728	.4053	.5730	.7831	1.046	1.374	1.788	2.316	3.925	7.116	16.43
(3)	.2026	.4229	.6647	.9327	1.232	1.572	1.959	2.408	2.935	3.561	4.318	5.254	7.996	13.29	28.63

TABLE 7.24 : I^*_{02} ($1/\alpha = .90$, $\mu = .50$)

Q: MODE:	0.05	0.10	0.15	0.20	0.25	0.30	0.35	0.40	0.45	0.50	0.55	0.60	0.70	0.80	0.90
(1)	.0079	.0317	.0722	.1305	.2084	.3085	.4344	.5910	.7852	1.027	1.330	1.718	2.904	5.309	12.56
(2)	.0098	.0396	.0902	.1632	.2608	.3862	.5440	.7401	.9827	1.283	1.658	2.131	3.541	6.257	13.93
(3)	.6376	1.300	1.997	2.740	3.543	4.422	5.399	6.500	7.764	9.242	11.01	13.17	19.50	31.85	68.36

TABLE 7.24 : I^*_{02} ($1/\alpha = .90$, $\mu = .60$)

Q: MODE:	0.05	0.10	0.15	0.20	0.25	0.30	0.35	0.40	0.45	0.50	0.55	0.60	0.70	0.80	0.90
(1)	.0086	.0345	.0785	.1419	.2267	.3356	.4727	.6432	.8549	1.119	1.450	1.873	3.169	5.802	13.74
(2)	.0087	.0349	.0794	.1432	.2281	.3363	.4712	.6372	.8402	1.089	1.395	1.776	2.888	4.978	10.76
(3)	.3973	.8531	1.372	1.957	2.613	3.346	4.168	5.097	6.160	7.396	8.863	10.65	15.80	25.71	54.65

TABLE 7.24 : I^*_{02} (1/α = .90 , μ = .70)

Q: MODE:	0.05	0.10	0.15	0.20	0.25	0.30	0.35	0.40	0.45	0.50	0.55	0.60	0.70	0.80	0.90
(1)	.0093	.0375	.0853	.1542	.2463	.3647	.5135	.6987	.9284	1.214	1.574	2.032	3.436	6.282	14.86
(2)	.0070	.0281	.0637	.1144	.1815	.2662	.3707	.4978	.6513	.8369	1.063	1.341	2.151	3.728	8.581
(3)	.4799	.9654	1.457	1.961	2.486	3.043	3.646	4.311	5.058	5.913	6.913	8.112	11.49	17.76	35.40

TABLE 7.24 : I^*_{02} (1/α = .90 , μ = .80)

Q: MODE:	0.05	0.10	0.15	0.20	0.25	0.30	0.35	0.40	0.45	0.50	0.55	0.60	0.70	0.80	0.90
(1)	.0101	.0406	.0925	.1671	.2669	.3949	.5557	.7555	1.003	1.311	1.697	2.188	3.688	6.717	15.81
(2)	.0049	.0197	.0445	.0797	.1259	.1836	.2540	.3384	.4391	.5591	.7038	.8820	1.428	2.834	8.964
(3)	.4955	.8907	1.227	1.538	1.845	2.161	2.498	2.864	3.271	3.732	4.266	4.899	6.663	9.951	19.64

TABLE 7.25 : I^*_{02} (α = .10 , μ = .20)

Q: MODE:	0.05	0.10	0.15	0.20	0.25	0.30	0.35	0.40	0.45	0.50	0.55	0.60	0.70	0.80	0.90
(1)	.0180	.0729	.1675	.3068	.4985	.7548	1.094	1.542	2.140	2.950	4.064	5.622	11.03	23.10	60.72
(2)	.3476	.7210	1.126	1.570	2.061	2.611	3.235	3.951	4.787	5.779	6.981	8.474	12.92	21.76	48.09
(3)	1.439	2.920	4.460	6.074	7.783	9.615	11.60	13.78	16.22	18.99	22.22	26.07	36.97	57.52	117.0

TABLE 7.25 : I^*_{02} (α = .10 , μ = .30)

Q: MODE:	0.05	0.10	0.15	0.20	0.25	0.30	0.35	0.40	0.45	0.50	0.55	0.60	0.70	0.80	0.90
(1)	.0158	.0636	.1453	.2643	.4254	.6360	.9069	1.253	1.697	2.271	3.024	4.031	7.348	14.58	36.98
(2)	.2624	.5384	.8309	1.144	1.482	1.856	2.284	2.810	3.546	4.724	6.508	8.865	15.65	28.17	63.43
(3)	2.661	4.198	5.678	7.389	9.446	11.94	14.94	18.54	22.82	27.93	34.07	41.57	63.12	103.9	220.4

TABLE 7.25 : I^*_{02} (α = .10 , μ = .40)

Q: MODE:	0.05	0.10	0.15	0.20	0.25	0.30	0.35	0.40	0.45	0.50	0.55	0.60	0.70	0.80	0.90
(1)	.0135	.0543	.1237	.2235	.3569	.5281	.7431	1.010	1.341	1.751	2.265	2.915	4.869	8.658	19.26
(2)	1.947	3.996	6.124	8.361	10.75	13.36	16.24	19.47	23.18	27.49	32.62	38.88	57.02	91.98	194.1
(3)	.6236	1.309	2.076	2.949	3.960	5.152	6.586	8.346	10.55	13.37	17.03	21.85	37.07	67.55	155.5

TABLE 7.26 : I^*_{02} (α = .10 , μ = .50)

Q: MODE:	0.05	0.10	0.15	0.20	0.25	0.30	0.35	0.40	0.45	0.50	0.55	0.60	0.70	0.80	0.90
(1)	.0112	.0452	.1024	.1842	.2921	.4286	.5968	.8010	1.047	1.342	1.696	2.125	3.307	5.317	10.33
(2)	2.473	4.729	6.818	8.911	11.13	13.58	16.34	19.51	23.20	27.59	32.88	39.43	58.72	96.46	207.7
(3)	.4606	.9482	1.468	2.026	2.629	3.286	4.007	4.806	5.699	6.708	7.864	9.208	12.76	18.61	33.29

TABLE 7.25 : I^*_{02} (α = .10 , μ = .60)

Q: MODE:	0.05	0.10	0.15	0.20	0.25	0.30	0.35	0.40	0.45	0.50	0.55	0.60	0.70	0.80	0.90
(1)	.0090	.0360	.0815	.1460	.2303	.3357	.4636	.6161	.7957	1.006	1.251	1.538	2.290	3.602	10.89
(2)	2.454	3.804	4.875	5.983	7.227	8.664	10.35	12.33	14.70	17.54	20.99	25.25	37.67	61.30	128.5
(3)	4.145	8.431	12.92	17.69	22.82	28.42	34.62	41.59	49.57	58.88	69.98	83.56	123.2	200.5	428.6

TABLE 7.25 : I^*_{02} (α = .10 , μ = .70)

Q: MODE:	0.05	0.10	0.15	0.20	0.25	0.30	0.35	0.40	0.45	0.50	0.55	0.60	0.70	0.80	0.90
(1)	.0067	.0270	.0609	.1087	.1708	.2477	.3400	.4485	.5743	.7190	.8849	1.076	1.629	7.449	30.18
(2)	1.680	2.151	2.606	3.110	3.678	4.327	5.071	5.930	6.926	8.092	9.468	11.11	15.60	23.18	41.56
(3)	6.334	12.05	17.24	22.19	27.14	32.27	37.77	43.81	50.60	58.43	67.65	78.83	111.0	172.7	351.7

TABLE 7.25 : I^*_{02} (α = .10 , μ = .80)

Q: MODE:	0.05	0.10	0.15	0.20	0.25	0.30	0.35	0.40	0.45	0.50	0.55	0.60	0.70	0.80	0.90
(1)	.0045	.0180	.0405	.0721	.1130	.1632	.2230	.2927	.3727	.4638	.5679	.6960	5.308	18.20	56.38
(2)	.8793	1.090	1.317	1.567	1.844	2.151	2.489	2.863	3.276	3.732	4.238	4.802	6.153	8.020	12.82
(3)	6.449	9.053	10.64	12.03	13.40	14.84	16.40	18.10	20.00	22.12	24.55	27.35	34.73	46.92	76.96

TABLE 7.25 : I^*_{02} (α = .15 , μ = .20)

Q: MODE:	0.05	0.10	0.15	0.20	0.25	0.30	0.35	0.40	0.45	0.50	0.55	0.60	0.70	0.80	0.90
(1)	.0172	.0696	.1596	.2911	.4707	.7078	1.016	1.415	1.934	2.614	3.515	4.729	8.728	17.37	44.08
(2)	.3327	.6901	1.077	1.501	1.968	2.489	3.076	3.746	4.521	5.433	6.527	7.871	11.80	19.41	41.64
(3)	1.067	2.167	3.313	4.521	5.807	7.195	8.713	10.40	12.30	14.48	17.05	20.15	29.04	46.07	95.78

TABLE 7.25 : I^*_{02} (α = .15 , μ = .30)

Q: MODE:	0.05	0.10	0.15	0.20	0.25	0.30	0.35	0.40	0.45	0.50	0.55	0.60	0.70	0.80	0.90
(1)	.0151	.0607	.1386	.2514	.4032	.6002	.8508	1.167	1.565	2.071	2.718	3.560	6.213	11.73	28.41
(2)	.2516	.5178	.8018	1.108	1.445	1.822	2.263	2.808	3.526	4.518	5.874	7.656	12.98	23.15	52.43
(3)	1.929	3.327	4.601	5.969	7.522	9.319	11.42	13.88	16.78	20.24	24.39	29.49	44.27	72.45	153.6

TABLE 7.25 : I^*_{02} (α = .15 , μ = .40)

Q: MODE:	0.05	0.10	0.15	0.20	0.25	0.30	0.35	0.40	0.45	0.50	0.55	0.60	0.70	0.80	0.90
(1)	.0129	.0519	.1180	.2130	.3394	.5007	.7022	.9505	1.255	1.629	2.091	2.668	4.364	7.565	16.40
(2)	1.309	2.717	4.196	5.759	7.432	9.253	11.27	13.53	16.12	19.14	22.73	27.12	39.89	64.60	137.0
(3)	.5952	1.246	1.969	2.783	3.714	4.792	6.061	7.576	9.413	11.67	14.50	18.10	29.01	50.48	112.6

TABLE 7.25 : I^*_{02} (α = .15 , μ = .50)

Q: MODE:	0.05	0.10	0.15	0.20	0.25	0.30	0.35	0.40	0.45	0.50	0.55	0.60	0.70	0.80	0.90
(1)	.0107	.0432	.0978	.1758	.2785	.4079	.5670	.7592	.9897	1.265	1.595	1.992	3.086	4.972	9.901
(2)	1.676	3.290	4.823	6.352	7.956	9.701	11.65	13.87	16.45	19.50	23.18	27.73	41.13	67.38	144.8
(3)	.4405	.9064	1.402	1.934	2.506	3.127	3.807	4.555	5.388	6.326	7.394	8.633	11.90	17.40	31.90

TABLE 7.25 : I^*_{02} (α = .15 , μ = .60)

Q: MODE:	0.05	0.10	0.15	0.20	0.25	0.30	0.35	0.40	0.45	0.50	0.55	0.60	0.70	0.80	0.90
(1)	.0086	.0345	.0779	.1395	.2200	.3203	.4421	.5870	.7576	.9572	1.191	1.465	2.198	3.565	10.00
(2)	1.754	2.949	3.891	4.811	5.799	6.905	8.168	9.633	11.35	13.39	15.86	18.89	27.68	44.41	92.12
(3)	2.907	5.914	9.066	12.41	16.02	19.95	24.32	29.23	34.85	41.42	49.27	58.87	86.95	141.8	303.8

TABLE 7.25 : I^*_{02} (α = .15 , μ = .70)

Q: MODE:	0.05	0.10	0.15	0.20	0.25	0.30	0.35	0.40	0.45	0.50	0.55	0.60	0.70	0.80	0.90
(1)	.0064	.0258	.0582	.1040	.1633	.2368	.3250	.4288	.5494	.6885	.8491	1.037	1.621	5.756	22.43
(2)	1.401	1.907	2.339	2.797	3.303	3.871	4.313	5.243	6.080	7.045	8.172	9.503	13.09	19.13	34.01
(3)	4.299	8.336	12.06	15.61	19.15	22.81	26.71	30.98	35.76	41.26	47.73	55.56	78.09	121.2	246.5

TABLE 7.25 : I^*_{02} (α = .15 , μ = .80)

Q: MODE:	0.05	0.10	0.15	0.20	0.25	0.30	0.35	0.40	0.45	0.50	0.55	0.60	0.70	0.80	0.90
(1)	.0043	.0172	.0387	.0690	.1081	.1562	.2135	.2804	.3575	.4458	.5487	.6834	3.843	12.96	40.06
(2)	.8070	1.018	1.234	1.470	1.730	2.016	2.331	2.678	3.060	3.481	3.948	4.468	5.727	7.546	12.78
(3)	4.684	7.140	8.660	9.909	11.10	12.30	13.58	14.95	16.46	18.14	20.05	22.26	28.06	37.75	62.17

TABLE 7.25 : I^*_{02} (α = .20 , μ = .20)

Q: MODE:	0.05	0.10	0.15	0.20	0.25	0.30	0.35	0.40	0.45	0.50	0.55	0.60	0.70	0.80	0.90
(1)	.0165	.0666	.1524	.2773	.4466	.6683	.9533	1.317	1.782	2.380	3.157	4.180	7.462	14.40	35.68
(2)	.3191	.6622	1.034	1.440	1.887	2.384	2.943	3.578	4.310	5.167	6.190	7.439	11.06	18.00	38.13
(3)	.9034	1.836	2.810	3.839	4.940	6.123	7.443	8.905	10.56	12.48	14.75	17.51	25.45	40.76	85.60

TABLE 7.25 : I^*_{02} (α = .20 , μ = .30)

Q: MODE:	0.05	0.10	0.15	0.20	0.25	0.30	0.35	0.40	0.45	0.50	0.55	0.60	0.70	0.80	0.90
(1)	.0144	.0582	.1325	.2399	.3838	.5695	.8039	1.097	1.463	1.921	2.500	3.243	5.529	10.16	23.99
(2)	.2419	.4996	.7768	1.079	1.415	1.798	2.250	2.806	3.517	4.444	5.650	7.204	11.87	20.91	47.22
(3)	1.493	2.698	3.811	4.971	6.249	7.694	9.350	11.27	13.52	16.18	19.39	23.32	34.76	56.70	120.2

TABLE 7.25 : I^*_{02} (α = .20 , μ = .40)

Q: MODE:	0.05	0.10	0.15	0.20	0.25	0.30	0.35	0.40	0.45	0.50	0.55	0.60	0.70	0.80	0.90
(1)	.0124	.0497	.1130	.2036	.3238	.4769	.6672	.9007	1.185	1.533	1.959	2.488	4.026	6.903	14.83
(2)	.9867	2.066	3.215	4.437	5.750	7.180	8.762	10.54	12.58	14.95	17.79	21.25	31.35	50.94	108.5
(3)	.5690	1.188	1.871	2.634	3.497	4.484	5.627	6.969	8.565	10.49	12.85	15.80	24.56	41.61	91.00

TABLE 7.25 : I^*_{02} (α = .20 , μ = .50)

Q: MODE:	0.05	0.10	0.15	0.20	0.25	0.30	0.35	0.40	0.45	0.50	0.55	0.60	0.70	0.80	0.90
(1)	.0103	.0414	.0937	.1682	.2663	.3897	.5410	.7236	.9420	1.203	1.514	1.890	2.927	4.745	9.653
(2)	1.268	2.531	3.759	4.992	6.279	7.671	9.216	10.97	13.00	15.39	18.27	21.82	32.29	52.81	113.4
(3)	.4221	.8684	1.343	1.850	2.396	2.988	3.632	4.342	5.129	6.013	7.020	8.188	11.29	16.61	31.10

TABLE 7.25 : I^*_{02} (α = .20 , μ = .60)

Q: MODE:	0.05	0.10	0.15	0.20	0.25	0.30	0.35	0.40	0.45	0.50	0.55	0.60	0.70	0.80	0.90
(1)	.0082	.0330	.0747	.1336	.2106	.3067	.4231	.5617	.7250	.9164	1.141	1.407	2.131	3.544	9.618
(2)	1.357	2.395	3.241	4.049	4.896	5.826	6.873	8.073	9.468	11.11	13.09	15.51	22.51	35.82	73.79
(3)	2.294	4.666	7.153	9.796	12.64	15.75	19.20	23.09	27.55	32.76	38.98	46.60	68.91	112.5	241.5

TABLE 7.25 : I^*_{02} (α = .20 , μ = .70)

Q: MODE:	0.05	0.10	0.15	0.20	0.25	0.30	0.35	0.40	0.45	0.50	0.55	0.60	0.70	0.80	0.90
(1)	.0062	.0247	.0558	.0996	.1565	.2270	.3116	.4114	.5276	.6623	.8192	1.006	1.616	5.001	18.65
(2)	1.174	1.695	2.112	2.538	3.001	3.514	4.089	4.738	5.475	6.322	7.303	8.459	11.56	16.81	29.92
(3)	3.260	6.414	9.372	12.21	15.04	17.96	21.06	24.44	28.22	32.56	37.66	43.82	61.53	95.42	193.8

TABLE 7.25 : I^*_{02} (α = .20 , μ = .80)

Q: MODE:	0.05	0.10	0.15	0.20	0.25	0.30	0.35	0.40	0.45	0.50	0.55	0.60	0.70	0.80	0.90
(1)	.0041	.0165	.0371	.0661	.1036	.1498	.2049	.2694	.3440	.4302	.5326	.6743	3.143	10.37	31.94
(2)	.7368	.9503	1.159	1.383	1.629	1.898	2.195	2.520	2.878	3.273	3.712	4.202	5.402	7.207	12.76
(3)	3.661	5.886	7.330	8.495	9.574	10.65	11.77	12.97	14.27	15.73	17.37	19.26	24.26	32.69	54.29

TABLE 7.25 : I^*_{02} (α = .30 , μ = .20)

Q: MODE:	0.05	0.10	0.15	0.20	0.25	0.30	0.35	0.40	0.45	0.50	0.55	0.60	0.70	0.80	0.90
(1)	.0152	.0614	.1400	.2538	.4067	.6047	.8557	1.171	1.567	2.065	2.698	3.515	6.058	11.29	27.18
(2)	.2952	.6136	.9591	1.337	1.752	2.214	2.731	3.318	3.992	4.778	5.712	6.848	10.12	16.35	34.35
(3)	.7709	1.567	2.400	3.283	4.229	5.259	6.395	7.666	9.115	10.80	12.79	15.22	22.27	35.90	75.88

TABLE 7.25 : I^*_{02} (α = .30 , μ = .30)

Q: MODE:	0.05	0.10	0.15	0.20	0.25	0.30	0.35	0.40	0.45	0.50	0.55	0.60	0.70	0.80	0.90
(1)	.0133	.0536	.1219	.2202	.3512	.5190	.7291	.9893	1.310	1.707	2.202	2.827	4.712	8.444	19.41
(2)	.2250	.4685	.7355	1.033	1.371	1.765	2.234	2.805	3.509	4.384	5.476	6.850	10.94	18.91	42.29
(3)	1.020	1.924	2.795	3.696	4.669	5.747	6.963	8.353	9.967	11.87	14.15	16.95	25.12	40.89	86.85

TABLE 7.25 : I^*_{02} (α = .30 , μ = .40)

Q: MODE:	0.05	0.10	0.15	0.20	0.25	0.30	0.35	0.40	0.45	0.50	0.55	0.60	0.70	0.80	0.90
(1)	.0114	.0459	.1041	.1873	.2975	.4372	.6101	.8212	1.077	1.388	1.767	2.236	3.589	6.113	13.11
(2)	.6618	1.403	2.211	3.086	4.035	5.074	6.228	7.529	9.020	10.76	12.84	15.39	22.84	37.35	80.15
(3)	.5224	1.085	1.699	2.376	3.130	3.980	4.947	6.059	7.355	8.888	10.73	13.00	19.60	32.30	69.01

TABLE 7.25 : I^*_{02} (α = .30 , μ = .50)

Q: MODE:	0.05	0.10	0.15	0.20	0.25	0.30	0.35	0.40	0.45	0.50	0.55	0.60	0.70	0.80	0.90
(1)	.0095	.0382	.0864	.1551	.2453	.3588	.4977	.6652	.8656	1.105	1.391	1.738	2.707	4.459	9.376
(2)	.8535	1.737	2.629	3.541	4.498	5.529	6.667	7.951	9.430	11.17	13.26	15.82	23.38	38.19	81.91
(3)	.3896	.8013	1.239	1.706	2.208	2.751	3.342	3.991	4.713	5.523	6.450	7.530	10.44	15.61	30.21

TABLE 7.25 : I^*_{02} (α = .30 , μ = .60)

Q: MODE:	0.05	0.10	0.15	0.20	0.25	0.30	0.35	0.40	0.45	0.50	0.55	0.60	0.70	0.80	0.90
(1)	.0076	.0305	.0689	.1233	.1945	.2832	.3910	.5195	.6715	.8506	1.063	1.317	2.036	3.519	9.275
(2)	.9310	1.733	2.431	3.098	3.785	4.525	5.345	6.272	7.340	8.589	10.08	11.89	17.12	27.03	55.28
(3)	1.689	3.435	5.267	7.215	9.314	11.61	14.16	17.03	20.33	24.18	28.79	34.45	51.02	83.45	179.5

TABLE 7.25 : I^*_{02} (α = .30 , μ = .70)

Q: MODE:	0.05	0.10	0.15	0.20	0.25	0.30	0.35	0.40	0.45	0.50	0.55	0.60	0.70	0.80	0.90
(1)	.0057	.0228	.0515	.0920	.1447	.2100	.2887	.3819	.4912	.6193	.7712	.9578	1.611	4.350	15.00
(2)	.8645	1.364	1.755	2.138	2.543	2.986	3.476	4.025	4.645	5.352	6.171	7.133	9.721	14.14	25.48
(3)	2.201	4.423	6.574	8.672	10.77	12.94	15.23	17.72	20.50	23.69	27.42	31.92	44.83	69.49	141.1

TABLE 7.25 : I^*_{02} (α = .30 , μ = .80)

Q: MODE:	0.05	0.10	0.15	0.20	0.25	0.30	0.35	0.40	0.45	0.50	0.55	0.60	0.70	0.80	0.90
(1)	.0038	.0152	.0343	.0611	.0958	.1387	.1901	.2505	.3210	.4042	.5070	.6617	2.497	7.840	23.88
(2)	.6128	.8307	1.027	1.233	1.457	1.701	1.969	2.263	2.587	2.946	3.345	3.796	4.926	6.745	12.74
(3)	2.539	4.350	5.637	6.681	7.632	8.561	9.513	10.52	11.61	12.82	14.18	15.75	19.92	27.07	45.85

TABLE 7.25 : I^*_{02} (α = .40 , μ = .20)

Q: MODE:	0.05	0.10	0.15	0.20	0.25	0.30	0.35	0.40	0.45	0.50	0.55	0.60	0.70	0.80	0.90
(1)	.0141	.0569	.1297	.2345	.3747	.5551	.7821	1.065	1.416	1.854	2.406	3.110	5.270	9.654	22.85
(2)	.2748	.5725	.8966	1.252	1.643	2.079	2.567	3.121	3.757	4.499	5.380	6.451	9.532	15.40	32.31
(3)	.7200	1.464	2.242	3.067	3.952	4.915	5.978	7.169	8.526	10.10	11.98	14.26	20.87	33.67	71.24

TABLE 7.25 : I^*_{02} (α = .40 , μ = .30)

Q: MODE:	0.05	0.10	0.15	0.20	0.25	0.30	0.35	0.40	0.45	0.50	0.55	0.60	0.70	0.80	0.90
(1)	.0124	.0497	.1131	.2039	.3247	.4789	.6712	.9082	1.199	1.557	2.000	2.558	4.224	7.488	17.02
(2)	.2106	.4428	.7024	.9977	1.340	1.743	2.224	2.804	3.506	4.359	5.404	6.705	10.54	18.02	39.96
(3)	.7733	1.486	2.193	2.930	3.727	4.606	5.593	6.718	8.019	9.549	11.38	13.64	20.21	32.93	70.17

TABLE 7.25 : I^{*}_{02} (α = .40 , μ = .40)

Q: MODE:	0.05	0.10	0.15	0.20	0.25	0.30	0.35	0.40	0.45	0.50	0.55	0.60	0.70	0.80	0.90
(1)	.0106	.0426	.0966	.1738	.2758	.4050	.5648	.7596	.9957	1.282	1.631	2.062	3.309	5.645	12.17
(2)	.4980	1.064	1.695	2.390	3.153	3.997	4.938	6.002	7.225	8.657	10.37	12.47	18.61	30.61	66.04
(3)	.4823	.9965	1.553	2.161	2.833	3.584	4.431	5.397	6.513	7.821	9.384	11.29	16.81	27.36	57.79

TABLE 7.25 : I^{*}_{02} (α = .40 , μ = .50)

Q: MODE:	0.05	0.10	0.15	0.20	0.25	0.30	0.35	0.40	0.45	0.50	0.55	0.60	0.70	0.80	0.90
(1)	.0088	.0354	.0803	.1440	.2279	.3334	.4627	.6189	.8061	1.030	1.300	1.628	2.559	4.282	9.225
(2)	.6432	1.324	2.030	2.766	3.545	4.387	5.317	6.365	7.569	8.981	10.67	12.76	18.87	30.83	66.15
(3)	.3610	.7441	1.150	1.584	2.051	2.556	3.107	3.714	4.389	5.151	6.027	7.056	9.871	14.99	29.73

TABLE 7.25 : I^{*}_{02} (α = .40 , μ = .60)

Q: MODE:	0.05	0.10	0.15	0.20	0.25	0.30	0.35	0.40	0.45	0.50	0.55	0.60	0.70	0.80	0.90
(1)	.0071	.0283	.0640	.1147	.1809	.2637	.3646	.4853	.6288	.7992	1.003	1.251	1.972	3.506	9.118
(2)	.7076	1.354	1.946	2.521	3.112	3.745	4.443	5.227	6.125	7.172	8.416	9.928	14.27	22.48	45.89
(3)	1.393	2.835	4.346	5.954	7.687	9.582	11.69	14.06	16.79	19.97	23.79	28.47	42.18	69.05	148.6

TABLE 7.25 : I^{*}_{02} (α = .40 , μ = .70)

Q: MODE:	0.05	0.10	0.15	0.20	0.25	0.30	0.35	0.40	0.45	0.50	0.55	0.60	0.70	0.80	0.90
(1)	.0053	.0212	.0479	.0856	.1347	.1958	.2697	.3576	.4616	.5850	.7340	.9220	1.608	4.073	13.26
(2)	.6774	1.129	1.492	1.844	2.211	2.608	3.046	3.534	4.084	4.712	5.438	6.293	8.605	12.60	23.03
(3)	1.663	3.390	5.107	6.810	8.529	10.31	12.19	14.24	16.52	19.12	22.17	25.85	36.36	56.42	114.6

TABLE 7.25 : I^{*}_{02} (α = .40 , μ = .80)

Q: MODE:	0.05	0.10	0.15	0.20	0.25	0.30	0.35	0.40	0.45	0.50	0.55	0.60	0.70	0.80	0.90
(1)	.0035	.0141	.0319	.0568	.0892	.1293	.1775	.2346	.3021	.3832	.4872	.6535	2.211	6.614	19.90
(2)	.5154	.7314	.9176	1.110	1.317	1.542	1.789	2.061	2.361	2.694	3.068	3.494	4.586	6.438	12.73
(3)	1.941	3.448	4.593	5.544	6.410	7.252	8.109	9.011	9.984	11.06	12.27	13.68	17.43	23.93	41.32

TABLE 7.25 : I^{*}_{02} (α = .50 , μ = .20)

Q: MODE:	0.05	0.10	0.15	0.20	0.25	0.30	0.35	0.40	0.45	0.50	0.55	0.60	0.70	0.80	0.90
(1)	.0132	.0531	.1208	.2182	.3482	.5149	.7238	.9830	1.303	1.701	2.198	2.831	4.754	8.626	20.22
(2)	.2572	.5371	.8433	1.180	1.552	1.967	2.434	2.964	3.574	4.285	5.132	6.161	9.123	14.76	31.03
(3)	.6945	1.412	2.163	2.958	3.811	4.740	5.764	6.912	8.220	9.739	11.54	13.74	20.10	32.42	68.54

TABLE 7.25 : I^{*}_{02} (α = .50 , μ = .30)

Q: MODE:	0.05	0.10	0.15	0.20	0.25	0.30	0.35	0.40	0.45	0.50	0.55	0.60	0.70	0.80	0.90
(1)	.0115	.0464	.1055	.1901	.3026	.4459	.6244	.8441	1.113	1.443	1.852	2.364	3.890	6.868	15.53
(2)	.1983	.4209	.6750	.9695	1.316	1.727	2.218	2.803	3.504	4.345	5.365	6.627	10.33	17.51	38.61
(3)	.6222	1.209	1.800	2.424	3.101	3.850	4.692	5.652	6.763	8.071	9.640	11.57	17.20	28.13	60.16

TABLE 7.25 : I^{*}_{02} (α = .50 , μ = .40)

Q: MODE:	0.05	0.10	0.15	0.20	0.25	0.30	0.35	0.40	0.45	0.50	0.55	0.60	0.70	0.80	0.90
(1)	.0099	.0397	.0902	.1622	.2575	.3783	.5276	.7099	.9309	1.199	1.527	1.932	3.110	5.328	11.57
(2)	.3993	.8586	1.379	1.962	2.611	3.335	4.149	5.074	6.139	7.388	8.884	10.72	16.10	26.60	57.63
(3)	.4476	.9203	1.428	1.980	2.588	3.264	4.023	4.886	5.880	7.043	8.427	10.11	14.98	24.24	50.92

TABLE 7.25 : I^{*}_{02} (α = .50 , μ = .50)

Q: MODE:	0.05	0.10	0.15	0.20	0.25	0.30	0.35	0.40	0.45	0.50	0.55	0.60	0.70	0.80	0.90
(1)	.0082	.0331	.0750	.1346	.2131	.3121	.4336	.5808	.7579	.9708	1.228	1.544	2.451	4.161	9.130
(2)	.5161	1.070	1.656	2.277	2.942	3.665	4.466	5.369	6.408	7.625	9.083	10.87	16.13	26.40	56.69
(3)	.3377	.6947	1.074	1.481	1.918	2.392	2.911	3.485	4.126	4.854	5.695	6.691	9.452	14.56	29.42

TABLE 7.25 : I^{*}_{02} (α = .50 , μ = .60)

Q: MODE:	0.05	0.10	0.15	0.20	0.25	0.30	0.35	0.40	0.45	0.50	0.55	0.60	0.70	0.80	0.90
(1)	.0066	.0264	.0598	.1072	.1693	.2472	.3423	.4568	.5936	.7573	.9551	1.199	1.926	3.497	9.028
(2)	.5705	1.111	1.623	2.129	2.654	3.215	3.833	4.526	5.319	6.241	7.336	8.666	12.48	19.68	40.19
(3)	1.220	2.483	3.807	5.215	6.734	8.394	10.24	12.32	14.70	17.50	20.84	24.94	36.95	60.48	130.2

TABLE 7.25 : I^{*}_{02} (α = .50 , μ = .70)

Q: MODE:	0.05	0.10	0.15	0.20	0.25	0.30	0.35	0.40	0.45	0.50	0.55	0.60	0.70	0.80	0.90
(1)	.0049	.0198	.0448	.0801	.1261	.1836	.2535	.3371	.4368	.5566	.7038	.8939	1.606	3.925	12.25
(2)	.5552	.9581	1.294	1.619	1.957	2.321	2.722	3.168	3.672	4.247	4.913	5.699	7.838	11.58	21.49
(3)	1.337	2.753	4.193	5.646	7.125	8.663	10.30	12.07	14.05	16.31	18.95	22.13	31.22	48.52	98.64

TABLE 7.25 : I^{*}_{02} (α = .50 , μ = .80)

Q: MODE:	0.05	0.10	0.15	0.20	0.25	0.30	0.35	0.40	0.45	0.50	0.55	0.60	0.70	0.80	0.90
(1)	.0033	.0132	.0298	.0531	.0835	.1212	.1668	.2212	.2861	.3656	.4712	.6477	2.055	5.903	17.54
(2)	.4410	.6497	.8267	1.007	1.201	1.411	1.642	1.896	2.178	2.493	2.849	3.257	4.326	6.216	12.72
(3)	1.571	2.854	3.881	4.755	5.556	6.335	7.128	7.961	8.859	9.851	10.97	12.27	15.77	21.88	38.44

TABLE 7.25 : I^{*}_{02} (α = .60 , μ = .20)

Q: MODE:	0.05	0.10	0.15	0.20	0.25	0.30	0.35	0.40	0.45	0.50	0.55	0.60	0.70	0.80	0.90
(1)	.0124	.0498	.1132	.2044	.3258	.4813	.6760	.9169	1.214	1.582	2.041	2.623	4.385	7.913	18.43
(2)	.2418	.5063	.7970	1.118	1.475	1.873	2.323	2.835	3.425	4.115	4.937	5.937	8.819	14.31	30.14
(3)	.6795	1.381	2.116	2.894	3.728	4.636	5.637	6.758	8.034	9.517	11.28	13.42	19.62	31.61	66.78

TABLE 7.25 : I^{*}_{02} (α = .60 , μ = .30)

Q: MODE:	0.05	0.10	0.15	0.20	0.25	0.30	0.35	0.40	0.45	0.50	0.55	0.60	0.70	0.80	0.90
(1)	.0108	.0435	.0989	.1782	.2837	.4181	.5855	.7915	1.044	1.353	1.736	2.216	3.644	6.478	14.51
(2)	.1875	.4021	.6517	.9463	1.297	1.716	2.213	2.803	3.502	4.335	5.341	6.578	10.19	17.20	37.73
(3)	.5205	1.018	1.525	2.065	2.655	3.310	4.050	4.895	5.877	7.034	8.426	10.14	15.15	24.90	53.49

TABLE 7.25 : I^{*}_{02} (α = .60 , μ = .40)

Q: MODE:	0.05	0.10	0.15	0.20	0.25	0.30	0.35	0.40	0.45	0.50	0.55	0.60	0.70	0.80	0.90
(1)	.0093	.0373	.0846	.1523	.2418	.3555	.4963	.6684	.8777	1.132	1.444	1.831	2.959	5.098	11.15
(2)	.3333	.7199	1.164	1.670	2.240	2.884	3.614	4.446	5.408	6.539	7.894	9.557	14.43	23.95	52.05
(3)	.4172	.8541	1.321	1.826	2.382	2.998	3.690	4.477	5.382	6.441	7.701	9.236	13.66	22.07	46.26

TABLE 7.25 : I^{*}_{02} (α = .60 , μ = .50)

Q: MODE:	0.05	0.10	0.15	0.20	0.25	0.30	0.35	0.40	0.45	0.50	0.55	0.60	0.70	0.80	0.90
(1)	.0077	.0310	.0703	.1264	.2003	.2937	.4088	.5487	.7176	.9217	1.170	1.477	2.367	4.072	9.064
(2)	.4309	.8983	1.400	1.939	2.523	3.162	3.874	4.679	5.606	6.692	7.994	9.593	14.28	23.42	50.37
(3)	.3166	.6516	1.008	1.390	1.803	2.252	2.745	3.292	3.907	4.609	5.425	6.398	9.129	14.25	29.21

TABLE 7.25 : I^*_{02} (α = .60 , μ = .60)

Q: MODE:	0.05	0.10	0.15	0.20	0.25	0.30	0.35	0.40	0.45	0.50	0.55	0.60	0.70	0.80	0.90
(1)	.0062	.0248	.0561	.1007	.1593	.2329	.3232	.4324	.5638	.7222	.9155	1.156	1.890	3.491	8.971
(2)	.4778	.9409	1.392	1.845	2.318	2.827	3.388	4.017	4.737	5.574	6.567	7.773	11.23	17.76	36.33
(3)	1.108	2.255	3.457	4.735	6.113	7.620	9.293	11.18	13.34	15.88	18.91	22.62	33.51	54.82	117.9

TABLE 7.25 : I^*_{02} (α = .60 , μ = .70)

Q: MODE:	0.05	0.10	0.15	0.20	0.25	0.30	0.35	0.40	0.45	0.50	0.55	0.60	0.70	0.80	0.90
(1)	.0046	.0186	.0420	.0752	.1187	.1731	.2394	.3194	.4156	.5325	.6786	.8712	1.605	3.833	11.60
(2)	.4697	.8304	1.141	1.443	1.757	2.095	2.467	2.883	3.351	3.888	4.511	5.249	7.268	10.84	20.41
(3)	1.118	2.320	3.566	4.842	6.155	7.527	8.991	10.58	12.36	14.39	16.75	19.60	27.74	43.21	87.98

TABLE 7.25 : I^*_{02} (α = .60 , μ = .80)

Q: MODE:	0.05	0.10	0.15	0.20	0.25	0.30	0.35	0.40	0.45	0.50	0.55	0.60	0.70	0.80	0.90
(1)	.0031	.0124	.0279	.0499	.0785	.1142	.1575	.2094	.2722	.3506	.4579	.6433	1.960	5.443	15.99
(2)	.3838	.5825	.7507	.9211	1.103	1.301	1.518	1.759	2.026	2.327	2.668	3.064	4.119	6.046	12.72
(3)	1.319	2.435	3.362	4.170	4.918	5.649	6.394	7.176	8.021	8.955	10.01	11.24	14.56	20.42	36.44

TABLE 7.25 : I^*_{02} (α = .70 , μ = .20)

Q: MODE:	0.05	0.10	0.15	0.20	0.25	0.30	0.35	0.40	0.45	0.50	0.55	0.60	0.70	0.80	0.90
(1)	.0166	.0468	.1065	.1923	.3066	.4528	.6358	.8621	1.141	1.486	1.916	2.460	4.105	7.386	17.14
(2)	.2282	.4792	.7564	1.064	1.407	1.792	2.228	2.726	3.301	3.975	4.779	5.758	8.583	13.97	29.49
(3)	.6697	1.361	2.085	2.851	3.673	4.567	5.552	6.655	7.911	9.368	11.10	13.20	19.29	31.06	65.55

TABLE 7.25 : I^*_{02} (α = .70 , μ = .30)

Q: MODE:	0.05	0.10	0.15	0.20	0.25	0.30	0.35	0.40	0.45	0.50	0.55	0.60	0.70	0.80	0.90
(1)	.0102	.0410	.0931	.1679	.2673	.3942	.5524	.7472	.9861	1.279	1.643	2.098	3.454	6.097	13.76
(2)	.1780	.3855	.6315	.9267	1.282	1.706	2.210	2.803	3.501	4.329	5.324	6.545	10.10	16.98	37.11
(3)	.4472	.8784	1.323	1.799	2.321	2.905	3.567	4.327	5.212	6.261	7.525	9.085	13.66	22.58	48.73

TABLE 7.25 : I^*_{02} (α = .70 , μ = .40)

Q: MODE:	0.05	0.10	0.15	0.20	0.25	0.30	0.35	0.40	0.45	0.50	0.55	0.60	0.70	0.80	0.90
(1)	.0087	.0351	.0797	.1435	.2282	.3358	.4694	.6332	.8328	1.076	1.376	1.748	2.839	4.922	10.84
(2)	.2860	.6200	1.008	1.457	1.970	2.556	3.224	3.992	4.882	5.929	7.186	8.729	13.25	22.07	48.09
(3)	.3906	.7963	1.228	1.694	2.206	2.774	3.413	4.139	4.976	5.956	7.125	8.549	12.65	20.46	42.87

TABLE 7.25 : I^*_{02} (α = .70 , μ = .50)

Q: MODE:	0.05	0.10	0.15	0.20	0.25	0.30	0.35	0.40	0.45	0.50	0.55	0.60	0.70	0.80	0.90
(1)	.0073	.0292	.0663	.1192	.1892	.2778	.3874	.5210	.6832	.8802	1.121	1.421	2.300	4.003	9.015
(2)	.3699	.7741	1.213	1.691	2.213	2.790	3.435	4.168	5.014	6.007	7.197	8.659	12.94	21.29	45.84
(3)	.2980	.6135	.9500	1.311	1.703	2.130	2.601	3.126	3.720	4.401	5.199	6.156	8.870	14.01	29.05

TABLE 7.25 : I^*_{02} (α = .70 , μ = .60)

Q: MODE:	0.05	0.10	0.15	0.20	0.25	0.30	0.35	0.40	0.45	0.50	0.55	0.60	0.70	0.80	0.90
(1)	.0058	.0234	.0529	.0950	.1505	.2204	.3066	.4113	.5381	.6922	.8820	1.121	1.861	3.487	8.930
(2)	.4110	.8161	1.218	1.629	2.062	2.529	3.046	3.627	4.292	5.066	5.986	7.103	10.30	16.35	33.55
(3)	1.030	2.095	3.212	4.400	5.680	7.079	8.632	10.38	12.39	14.74	17.55	20.99	31.08	50.81	109.2

TABLE 7.25 : I^*_{02} (α = .70 , μ = .70)

Q: MODE:	0.05	0.10	0.15	0.20	0.25	0.30	0.35	0.40	0.45	0.50	0.55	0.60	0.70	0.80	0.90
(1)	.0044	.0175	.0396	.0709	.1121	.1638	.2271	.3040	.3972	.5118	.6571	.8523	1.604	3.771	11.15
(2)	.4068	.7317	1.019	1.301	1.594	1.912	2.261	2.651	3.093	3.600	4.191	4.892	6.824	10.27	19.60
(3)	.9604	2.006	3.107	4.250	5.438	6.689	8.028	9.489	11.12	12.98	15.15	17.77	25.22	39.38	80.33

TABLE 7.25 : I^*_{02} (α = .70 , μ = .80)

Q: MODE:	0.05	0.10	0.15	0.20	0.25	0.30	0.35	0.40	0.45	0.50	0.55	0.60	0.70	0.80	0.90
(1)	.0029	.0116	.0263	.0471	.0741	.1080	.1492	.1992	.2601	.3376	.4467	.6399	1.897	5.124	14.89
(2)	.3389	.5267	.6865	.8478	1.020	1.207	1.413	1.642	1.898	2.186	2.516	2.902	3.947	5.911	12.71
(3)	1.136	2.123	2.967	3.718	4.421	5.112	5.818	6.561	7.365	8.255	9.266	10.44	13.64	19.32	34.96

TABLE 7.25 : I^*_{02} (α = .80 , μ = .20)

Q: MODE:	0.05	0.10	0.15	0.20	0.25	0.30	0.35	0.40	0.45	0.50	0.55	0.60	0.70	0.80	0.90
(1)	.0110	.0442	.1006	.1817	.2898	.4281	.6013	.8156	1.080	1.406	1.814	2.329	3.884	6.979	16.16
(2)	.2161	.4550	.7203	1.016	1.348	1.721	2.146	2.632	3.196	3.857	4.647	5.611	8.393	13.70	28.99
(3)	.6628	1.347	2.063	2.822	3.635	4.518	5.492	6.582	7.823	9.262	10.97	13.04	19.05	30.65	64.63

TABLE 7.25 : I^*_{02} (α = .80 , μ = .30)

Q: MODE:	0.05	0.10	0.15	0.20	0.25	0.30	0.35	0.40	0.45	0.50	0.55	0.60	0.70	0.80	0.90
(1)	.0096	.0387	.0880	.1588	.2530	.3734	.5238	.7093	.9370	1.217	1.565	2.001	3.301	5.837	13.19
(2)	.1696	.3708	.6139	.9100	1.269	1.699	2.207	2.803	3.501	4.325	5.312	6.521	10.03	16.81	36.66
(3)	.3921	.7725	1.167	1.593	2.062	2.589	3.189	3.882	4.693	5.658	6.826	8.271	12.53	20.83	45.16

TABLE 7.25 : I^*_{02} (α = .80 , μ = .40)

Q: MODE:	0.05	0.10	0.15	0.20	0.25	0.30	0.35	0.40	0.45	0.50	0.55	0.60	0.70	0.80	0.90
(1)	.0082	.0332	.0753	.1358	.2162	.3185	.4460	.6026	.7942	1.029	1.318	1.679	2.742	4.782	10.60
(2)	.2505	.5445	.8899	1.294	1.763	2.304	2.927	3.646	4.483	5.470	6.655	8.110	12.37	20.67	45.13
(3)	.3670	.7454	1.146	1.579	2.054	2.582	3.177	3.854	4.637	5.556	6.653	7.992	11.86	19.21	40.30

TABLE 7.25 : I^*_{02} (α = .80 , μ = .50)

Q: MODE:	0.05	0.10	0.15	0.20	0.25	0.30	0.35	0.40	0.45	0.50	0.55	0.60	0.70	0.80	0.90
(1)	.0069	.0276	.0627	.1129	.1793	.2638	.3686	.4969	.6534	.8444	1.080	1.373	2.244	3.949	8.978
(2)	.3240	.6802	1.071	1.500	1.973	2.501	3.095	3.773	4.556	5.479	6.585	7.945	11.93	19.68	42.45
(3)	.2815	.5798	.8984	1.241	1.614	2.022	2.475	2.981	3.557	4.222	5.006	5.951	8.656	13.82	28.93

TABLE 7.25 : I^*_{02} (α = .80 , μ = .60)

Q: MODE:	0.05	0.10	0.15	0.20	0.25	0.30	0.35	0.40	0.45	0.50	0.55	0.60	0.70	0.80	0.90
(1)	.0055	.0221	.0500	.0900	.1427	.2094	.2919	.3927	.5156	.6661	.8531	1.091	1.838	3.484	8.900
(2)	.3606	.7204	1.083	1.459	1.858	2.292	2.773	3.315	3.938	4.664	5.527	6.576	9.587	15.27	31.44
(3)	.9725	1.978	3.033	4.154	5.362	6.682	8.146	9.797	11.69	13.90	16.55	19.79	29.28	47.83	102.7

TABLE 7.25 : I^*_{02} (α = .80 , μ = .70)

Q: MODE:	0.05	0.10	0.15	0.20	0.25	0.30	0.35	0.40	0.45	0.50	0.55	0.60	0.70	0.80	0.90
(1)	.0041	.0165	.0374	.0672	.1063	.1555	.2162	.2903	.3809	.4936	.6385	.8363	1.603	3.727	10.81
(2)	.3586	.6536	.9199	1.184	1.460	1.759	2.089	2.459	2.879	3.362	3.927	4.600	6.465	9.817	18.97
(3)	.8420	1.767	2.755	3.794	4.885	6.041	7.284	8.645	10.16	11.90	13.93	16.36	23.31	36.50	74.58

TABLE 7.25 : I_{02}^* (α = .80 , μ = .80)

Q: MODE:	0.05	0.10	0.15	0.20	0.25	0.30	0.35	0.40	0.45	0.50	0.55	0.60	0.70	0.80	0.90
(1)	.0027	.0110	.0249	.0445	.0702	.1024	.1419	.1900	.2494	.3261	.4370	.6372	1.852	4.890	14.07
(2)	.3031	.4800	.6317	.7849	.9482	1.126	1.323	1.541	1.787	2.065	2.386	2.763	3.803	5.801	12.71
(3)	.9980	1.882	2.656	3.357	4.021	4.678	5.352	6.063	6.834	7.690	8.664	9.802	12.90	18.45	33.82

TABLE 7.25 : I_{02}^* (α = .90 , μ = .20)

Q: MODE:	0.05	0.10	0.15	0.20	0.25	0.30	0.35	0.40	0.45	0.50	0.55	0.60	0.70	0.80	0.90
(1)	.0104	.0419	.0954	.1723	.2750	.4065	.5713	.7754	1.027	1.339	1.728	2.219	3.704	6.655	15.39
(2)	.2052	.4334	.6880	.9734	1.295	1.639	2.074	2.551	3.105	3.756	4.536	5.487	8.237	13.48	28.60
(3)	.6577	1.337	2.047	2.800	3.606	4.482	5.448	6.528	7.758	9.183	10.87	12.93	18.87	30.34	63.92

TABLE 7.25 : I_{02}^* (α = .90 , μ = .30)

Q: MODE:	0.05	0.10	0.15	0.20	0.25	0.30	0.35	0.40	0.45	0.50	0.55	0.60	0.70	0.80	0.90
(1)	.0091	.0367	.0834	.1507	.2403	.3551	.4987	.6762	.8947	1.164	1.498	1.919	3.175	5.627	12.74
(2)	.1620	.3576	.5982	.8955	1.258	1.693	2.205	2.803	3.500	4.321	5.303	6.502	9.982	16.69	36.31
(3)	.3490	.6894	1.044	1.429	1.855	2.336	2.886	3.524	4.275	5.173	6.265	7.620	11.63	19.46	42.38

TABLE 7.25 : I_{02}^* (α = .90 , μ = .40)

Q: MODE:	0.05	0.10	0.15	0.20	0.25	0.30	0.35	0.40	0.45	0.50	0.55	0.60	0.70	0.80	0.90
(1)	.0078	.0314	.0715	.1290	.2055	.3033	.4253	.5758	.7605	.9874	1.268	1.620	2.660	4.667	10.41
(2)	.2228	.4855	.7968	1.165	1.599	2.105	2.692	3.374	4.171	5.112	6.242	7.630	11.69	19.59	42.84
(3)	.3459	.7003	1.075	1.478	1.922	2.416	2.974	3.611	4.349	5.217	6.257	7.529	11.21	18.21	38.27

TABLE 7.25 . I_{02}^* (α = .90 , μ = .50)

Q: MODE:	0.05	0.10	0.15	0.20	0.25	0.30	0.35	0.40	0.45	0.50	0.55	0.60	0.70	0.80	0.90
(1)	.0065	.0262	.0595	.1072	.1705	.2513	.3519	.4756	.6271	.8131	1.043	1.333	2.198	3.904	8.949
(2)	.2883	.6065	.9583	1.348	1.783	2.270	2.823	3.456	4.191	5.057	6.099	7.379	11.13	18.42	39.80
(3)	.2667	.5496	.8523	1.179	1.535	1.927	2.362	2.853	3.414	4.066	4.838	5.775	8.477	13.67	28.84

TABLE 7.25 : I_{02}^* (α = .90 , μ = .60)

Q: MODE:	0.05	0.10	0.15	0.20	0.25	0.30	0.35	0.40	0.45	0.50	0.55	0.60	0.70	0.80	0.90
(1)	.0052	.0209	.0475	.0854	.1357	.1995	.2788	.3762	.4956	.6430	.8278	1.065	1.819	3.481	8.877
(2)	.3212	.6448	.9754	1.321	1.692	2.097	2.549	3.060	3.648	4.336	5.154	6.150	9.011	14.41	29.78
(3)	.9288	1.889	2.896	3.966	5.119	6.379	7.775	9.349	11.15	13.26	15.78	18.86	27.89	45.53	97.69

TABLE 7.25 : I_{02}^* (α = .90 , μ = .70)

Q: MODE:	0.05	0.10	0.15	0.20	0.25	0.30	0.35	0.40	0.45	0.50	0.55	0.60	0.70	0.80	0.90
(1)	.0039	.0157	.0355	.0638	.1010	.1482	.2065	.2781	.3665	.4775	.6222	.8225	1.603	3.693	10.56
(2)	.3206	.5902	.8380	1.086	1.347	1.630	1.944	2.296	2.698	3.161	3.705	4.356	6.167	9.449	18.47
(3)	.7495	1.580	2.476	3.431	4.443	5.523	6.690	7.971	9.403	11.04	12.95	15.25	21.81	34.24	70.10

TABLE 7.25 : I_{02}^* (α = .90 , μ = .80)

Q: MODE:	0.05	0.10	0.15	0.20	0.25	0.30	0.35	0.40	0.45	0.50	0.55	0.60	0.70	0.80	0.90
(1)	.0026	.0104	.0236	.0423	.0667	.0975	.1354	.1818	.2398	.3159	.4285	.6350	1.818	4.712	13.44
(2)	.2739	.4405	.5847	.7303	.8858	1.056	1.244	1.453	1.690	1.960	2.272	2.643	3.678	5.710	12.71
(3)	.8898	1.690	2.404	3.062	3.691	4.318	4.964	5.649	6.392	7.220	8.165	9.271	12.30	17.75	32.91

TABLE 7.25 : I_{02}^* (α = 1.0 , μ = .20)

Q: MODE:	0.05	0.10	0.15	0.20	0.25	0.30	0.35	0.40	0.45	0.50	0.55	0.60	0.70	0.80	0.90
(1)	.0099	.0398	.0907	.1639	.2618	.3873	.5448	.7403	.9818	1.281	1.654	2.127	3.554	6.388	14.77
(2)	.1955	.4139	.6589	.9350	1.247	1.603	2.009	2.478	3.024	3.668	4.439	5.382	8.106	13.30	28.28
(3)	.6538	1.329	2.035	2.783	3.584	4.454	5.414	6.486	7.707	9.122	10.80	12.84	18.73	30.09	63.36

TABLE 7.25 : I_{02}^* (α = 1.0 , μ = .30)

Q: MODE:	0.05	0.10	0.15	0.20	0.25	0.30	0.35	0.40	0.45	0.50	0.55	0.60	0.70	0.80	0.90
(1)	.0087	.0349	.0794	.1434	.2290	.3388	.4764	.6470	.8575	1.117	1.441	1.849	3.069	5.454	12.37
(2)	.1551	.3457	.5842	.8827	1.249	1.688	2.204	2.802	3.500	4.318	5.296	6.488	9.942	16.59	36.03
(3)	.3144	.6224	.9449	1.295	1.686	2.128	2.636	3.228	3.930	4.773	5.802	7.086	10.90	18.36	40.16

TABLE 7.25 : I_{02}^* (α = 1.0 , μ = .40)

Q: MODE:	0.05	0.10	0.15	0.20	0.25	0.30	0.35	0.40	0.45	0.50	0.55	0.60	0.70	0.80	0.90
(1)	.0074	.0299	.0680	.1228	.1959	.2896	.4069	.5520	.7307	.9511	1.225	1.568	2.591	4.572	10.26
(2)	.2006	.4380	.7216	1.061	1.465	1.942	2.501	3.154	3.919	4.824	5.911	7.247	11.15	18.73	41.02
(3)	.3271	.6602	1.011	1.390	1.806	2.270	2.796	3.399	4.100	4.926	5.919	7.135	10.66	17.39	36.63

TABLE 7.25 : I_{02}^* (α = 1.0 , μ = .50)

Q: MODE:	0.05	0.10	0.15	0.20	0.25	0.30	0.35	0.40	0.45	0.50	0.55	0.60	0.70	0.80	0.90
(1)	.0062	.0249	.0566	.1021	.1627	.2401	.3369	.4565	.6037	.7854	1.012	1.297	2.158	3.868	8.926
(2)	.2596	.5473	.8674	1.225	1.627	2.081	2.599	3.195	3.891	4.712	5.701	6.918	10.48	17.40	37.69
(3)	.2534	.5225	.8108	1.123	1.464	1.841	2.262	2.739	3.287	3.927	4.690	5.621	8.324	13.54	28.76

TABLE 7.25 : I_{02}^* (α = 1.0 , μ = .60)

Q: MODE:	0.05	0.10	0.15	0.20	0.25	0.30	0.35	0.40	0.45	0.50	0.55	0.60	0.70	0.80	0.90
(1)	.0049	.0199	.0452	.0814	.1294	.1907	.2670	.3613	.4777	.6225	.8054	1.042	1.802	3.479	8.859
(2)	.2895	.5835	.8870	1.208	1.554	1.935	2.362	2.847	3.406	4.061	4.843	5.796	8.537	13.72	28.44
(3)	.8945	1.820	2.789	3.819	4.928	6.140	7.483	8.996	10.73	12.75	15.17	18.13	26.79	43.70	93.68

TABLE 7.25 : I_{02}^* (α = 1.0 , μ = .70)

Q: MODE:	0.05	0.10	0.15	0.20	0.25	0.30	0.35	0.40	0.45	0.50	0.55	0.60	0.70	0.80	0.90
(1)	.0037	.0149	.0338	.0607	.0963	.1415	.1977	.2671	.3534	.4630	.6077	.8105	1.602	3.667	10.36
(2)	.2898	.5379	.7694	1.003	1.250	1.519	1.818	2.156	2.542	2.989	3.515	4.146	5.915	9.141	18.05
(3)	.6754	1.428	2.250	3.134	4.081	5.098	6.202	7.418	0.781	10.34	12.16	14.35	20.60	32.42	66.50

TABLE 7.25 : I_{02}^* (α = 1.0 , μ = .80)

Q: MODE:	0.05	0.10	0.15	0.20	0.25	0.30	0.35	0.40	0.45	0.50	0.55	0.60	0.70	0.80	0.90
(1)	.0025	.0099	.0224	.0402	.0636	.0930	.1295	.1744	.2311	.3067	.4210	.6331	1.793	4.572	12.94
(2)	.2497	.4067	.5439	.6827	.8311	.9935	1.174	1.376	1.604	1.866	2.172	2.537	3.570	5.632	12.71
(3)	.8027	1.533	2.197	2.815	3.414	4.014	4.636	5.297	6.018	6.823	7.743	8.824	11.79	17.16	32.17

TABLE 7.25 : I_{02}^* ($1/\alpha$ = .10 , μ = .20)

Q: MODE:	0.05	0.10	0.15	0.20	0.25	0.30	0.35	0.40	0.45	0.50	0.55	0.60	0.70	0.80	0.90
(1)	.0018	.0074	.0173	.0324	.0545	.0864	.1326	.2005	.3012	.4499	.6637	.9615	1.920	3.841	9.374
(2)	.0378	.0866	.1523	.2450	.3811	.5847	.8816	1.287	1.804	2.435	3.195	4.118	6.738	11.63	25.52
(3)	.6268	1.274	1.950	2.665	3.430	4.260	5.173	6.191	7.346	8.682	10.26	12.17	17.68	28.23	58.98

TABLE 7.25 : I_{02}^{*} (1/α = .10 , μ = .30)

Q: / MODE:	0.05	0.10	0.15	0.20	0.25	0.30	0.35	0.40	0.45	0.50	0.55	0.60	0.70	0.80	0.90
(1)	.0016	.0065	.0151	.0284	.0479	.0761	.1173	.1788	.2717	.4120	.6176	.9075	1.847	3.724	9.114
(2)	.0349	.1150	.3305	.6995	1.144	1.640	2.189	2.802	3.496	4.297	5.241	6.381	9.645	15.86	33.86
(3)	.0317	.0637	.0984	.1374	.1829	.2376	.3058	.3949	.5185	.7038	1.009	1.551	3.947	8.924	22.21

TABLE 7.25 : I_{02}^{*} (1/α = .10 , μ = .40)

Q: / MODE:	0.05	0.10	0.15	0.20	0.25	0.30	0.35	0.40	0.45	0.50	0.55	0.60	0.70	0.80	0.90
(1)	.0014	.0055	.0129	.0243	.0409	.0650	.1005	.1540	.2370	.3662	.5619	.8432	1.764	3.604	8.874
(2)	.0202	.0449	.0773	.1243	.2026	.3550	.6546	1.120	1.710	2.406	3.225	4.205	6.959	12.09	26.70
(3)	.0542	.1042	.1561	.2136	.2804	.3613	.4631	.5965	.7796	1.043	1.442	2.060	4.376	9.143	22.11

TABLE 7.25 : I_{02}^{*} (1/α = .10 , μ = .50)

Q: / MODE:	0.05	0.10	0.15	0.20	0.25	0.30	0.35	0.40	0.45	0.50	0.55	0.60	0.70	0.80	0.90
(1)	.0011	.0046	.0107	.0201	.0337	.0535	.0825	.1270	.1980	.3142	.4990	.7736	1.685	3.508	8.725
(2)	.0261	.0560	.0914	.1350	.1910	.2665	.3740	.5358	.7886	1.178	1.738	2.487	4.707	8.834	20.34
(3)	.0463	.0968	.1540	.2212	.3037	.4099	.5541	.7619	1.078	1.571	2.314	3.352	6.500	12.28	28.11

TABLE 7.25 : I_{02}^{*} (1/α = .10 , μ = .60)

Q: / MODE:	0.05	0.10	0.15	0.20	0.25	0.30	0.35	0.40	0.45	0.50	0.55	0.60	0.70	0.80	0.90
(1)	.0009	.0037	.0085	.0159	.0265	.0417	.0641	.0986	.1559	.2570	.4321	.7053	1.625	3.461	8.717
(2)	.0293	.0610	.0969	.1394	.1914	.2576	.3449	.4652	.6378	.8934	1.275	1.828	3.646	7.160	16.87
(3)	.6473	1.316	2.014	2.754	3.546	4.406	5.352	6.409	7.611	9.002	10.65	12.65	18.42	29.51	61.91

TABLE 7.25 : I_{02}^{*} (1/α = .10 , μ = .70)

Q: / MODE:	0.05	0.10	0.15	0.20	0.25	0.30	0.35	0.40	0.45	0.50	0.55	0.60	0.70	0.80	0.90
(1)	.0007	.0027	.0063	.0117	.0194	.0301	.0458	.0700	.1121	.1960	.3650	.6477	1.598	3.478	8.855
(2)	.0300	.0592	.0907	.1265	.1690	.2211	.2874	.3749	.4952	.6683	.9274	1.323	2.783	5.824	14.26
(3)	.0683	.1494	.2504	.3823	.5632	.8223	1.201	1.742	2.472	3.397	4.529	5.903	9.758	16.83	36.65

TABLE 7.25 : I_{02}^{*} (1/α = .10 , μ = .80)

Q: / MODE:	0.05	0.10	0.15	0.20	0.25	0.30	0.35	0.40	0.45	0.50	0.55	0.60	0.70	0.80	0.90
(1)	.0005	.0018	.0042	.0076	.0125	.0191	.0284	.0426	.0685	.1321	.3056	.6149	1.614	3.557	9.097
(2)	.0274	.0499	.0724	.0971	.1256	.1595	.2011	.2539	.3240	.4225	.5713	.8152	1.955	4.769	12.69
(3)	.0818	.1641	.2515	.3491	.4630	.6017	.7777	1.010	1.328	1.774	2.400	3.267	5.995	11.18	25.45

TABLE 7.25 : I_{02}^{*} (1/α = .20 , μ = .20)

Q: / MODE:	0.05	0.10	0.15	0.20	0.25	0.30	0.35	0.40	0.45	0.50	0.55	0.60	0.70	0.80	0.90
(1)	.0033	.0135	.0312	.0579	.0957	.1478	.2188	.3155	.4465	.6235	.8610	1.178	2.178	4.185	10.02
(2)	.0683	.1530	.2609	.4013	.5865	.8313	1.150	1.555	2.057	2.669	3.411	4.321	6.930	11.84	25.84
(3)	.6295	1.279	1.958	2.677	3.445	4.279	5.197	6.220	7.382	8.726	10.31	12.24	17.79	28.43	59.45

TABLE 7.25 : I_{02}^{*} (1/α = .20 , μ = .30)

Q: / MODE:	0.05	0.10	0.15	0.20	0.25	0.30	0.35	0.40	0.45	0.50	0.55	0.60	0.70	0.80	0.90
(1)	.0029	.0118	.0274	.0508	.0841	.1303	.1938	.2812	.4012	.5656	.7889	1.090	2.046	3.964	9.513
(2)	.0605	.1721	.3907	.7334	1.160	1.646	2.190	2.802	3.496	4.299	5.246	6.392	9.676	15.94	34.10
(3)	.0634	.1271	.1959	.2731	.3624	.4688	.5995	.7663	.9886	1.300	1.756	2.444	4.943	10.06	24.20

TABLE 7.25 : I_{02}^{*} (1/α = .20 , μ = .40)

Q: / MODE:	0.05	0.10	0.15	0.20	0.25	0.30	0.35	0.40	0.45	0.50	0.55	0.60	0.70	0.80	0.90
(1)	.0025	.0101	.0234	.0435	.0720	.1117	.1666	.2429	.3495	.4985	.7051	.9882	1.900	3.738	9.041
(2)	.0403	.0895	.1531	.2418	.3765	.5936	.9319	1.401	1.987	2.689	3.525	4.534	7.399	12.79	28.23
(3)	.1014	.1966	.2952	.4038	.5284	.6761	.8568	1.084	1.379	1.772	2.307	3.044	5.472	10.34	23.91

TABLE 7.25 : I_{02}^{*} (1/α = .20 , μ = .50)

Q: / MODE:	0.05	0.10	0.15	0.20	0.25	0.30	0.35	0.40	0.45	0.50	0.55	0.60	0.70	0.80	0.90
(1)	.0021	.0084	.0195	.0361	.0596	.0923	.1378	.2017	.2929	.4244	.6131	.8795	1.760	3.555	8.748
(2)	.0522	.1117	.1817	.2665	.3728	.5102	.6933	.9426	1.284	1.744	2.350	3.132	5.452	9.848	22.30
(3)	.0848	.1766	.2791	.3969	.5366	.7079	.9251	1.210	1.595	2.122	2.842	3.811	6.790	12.44	28.19

TABLE 7.25 : I_{02}^{*} (1/α = .20 , μ = .60)

Q: / MODE:	0.05	0.10	0.15	0.20	0.25	0.30	0.35	0.40	0.45	0.50	0.55	0.60	0.70	0.80	0.90
(1)	.0017	.0067	.0155	.0286	.0471	.0726	.1081	.1585	.2326	.3449	.5163	.7716	1.650	3.463	8.732
(2)	.0586	.1214	.1919	.2738	.3720	.4930	.6457	.8430	1.103	1.448	1.907	2.517	4.402	8.045	18.27
(3)	.6714	1.365	2.090	2.858	3.681	4.576	5.562	6.666	7.922	9.379	11.11	13.21	19.28	31.00	65.34

TABLE 7.25 : I_{02}^{*} (1/α = .20 , μ = .70)

Q: / MODE:	0.05	0.10	0.15	0.20	0.25	0.30	0.35	0.40	0.45	0.50	0.55	0.60	0.70	0.80	0.90
(1)	.0012	.0050	.0115	.0212	.0346	.0529	.0783	.1147	.1702	.2616	.4190	.6764	1.599	3.497	9.011
(2)	.0597	.1173	.1781	.2460	.3249	.4190	.5345	.6795	.8661	1.112	1.441	1.890	3.355	6.326	14.74
(3)	.1364	.2971	.4931	.7390	1.055	1.464	1.990	2.653	3.468	4.454	5.638	7.075	11.14	18.70	40.08

TABLE 7.25 : I_{02}^{*} (1/α = .20 , μ = .80)

Q: / MODE:	0.05	0.10	0.15	0.20	0.25	0.30	0.35	0.40	0.45	0.50	0.55	0.60	0.70	0.80	0.90
(1)	.0008	.0033	.0076	.0139	.0224	.0339	.0494	.0717	.1073	.1758	.3287	.6173	1.631	3.658	9.501
(2)	.0543	.0976	.1401	.1860	.2379	.2984	.3707	.4592	.5709	.7170	.9165	1.203	2.289	4.896	12.70
(3)	.1633	.3257	.4948	.6784	.8848	1.124	1.410	1.758	2.190	2.734	3.425	4.310	6.966	12.04	26.29

TABLE 7.25 : I_{02}^{*} (1/α = .30 , μ = .20)

Q: / MODE:	0.05	0.10	0.15	0.20	0.25	0.30	0.35	0.40	0.45	0.50	0.55	0.60	0.70	0.80	0.90
(1)	.0046	.0186	.0428	.0788	.1287	.1959	.2847	.4012	.5535	.7519	1.011	1.349	2.400	4.505	10.65
(2)	.0935	.2064	.3450	.5172	.7327	1.002	1.337	1.749	2.251	2.857	3.594	4.499	7.108	12.05	26.16
(3)	.6322	1.285	1.967	2.688	3.461	4.299	5.221	6.250	7.419	8.771	10.37	12.31	17.90	28.63	59.93

TABLE 7.25 : I_{02}^{*} (1/α = .30 , μ = .30)

Q: / MODE:	0.05	0.10	0.15	0.20	0.25	0.30	0.35	0.40	0.45	0.50	0.55	0.60	0.70	0.80	0.90
(1)	.0040	.0163	.0375	.0691	.1132	.1727	.2520	.3568	.4951	.6772	.9165	1.232	2.216	4.186	9.899
(2)	.0804	.2117	.4334	.7615	1.174	1.652	2.192	2.802	3.497	4.302	5.252	6.403	9.708	16.02	34.33
(3)	.0950	.1902	.2925	.4069	.5385	.6939	.8826	1.118	1.424	1.832	2.394	3.182	5.822	11.16	26.19

TABLE 7.25 : I_{02}^{*} (1/α = .30 , μ = .40)

Q: / MODE:	0.05	0.10	0.15	0.20	0.25	0.30	0.35	0.40	0.45	0.50	0.55	0.60	0.70	0.80	0.90
(1)	.0034	.0139	.0322	.0592	.0970	.1482	.2167	.3082	.4301	.5930	.8102	1.100	2.017	3.862	9.205
(2)	.0604	.1338	.2275	.3541	.5347	.7979	1.168	1.653	2.250	2.966	3.824	4.866	7.847	13.50	29.79
(3)	.1426	.2786	.4197	.5741	.7492	.9538	1.199	1.499	1.875	2.355	2.977	3.797	6.358	11.41	25.63

TABLE 7.25 : I^*_{02} ($1/\alpha = .30$, $\mu = .50$)

Q: MODE:	0.05	0.10	0.15	0.20	0.25	0.30	0.35	0.40	0.45	0.50	0.55	0.60	0.70	0.80	0.90
(1)	.0029	.0116	.0267	.0492	.0804	.1228	.1797	.2563	.3602	.5019	.6959	.9614	1.827	3.600	8.771
(2)	.0782	.1671	.2708	.3948	.5469	.7375	.9805	1.294	1.697	2.212	2.867	3.696	6.148	10.84	24.24
(3)	.1173	.2436	.3832	.5410	.7240	.9415	1.207	1.538	1.961	2.509	3.226	4.166	7.046	12.60	28.26

TABLE 7.25 : I^*_{02} ($1/\alpha = .30$, $\mu = .60$)

Q: MODE:	0.05	0.10	0.15	0.20	0.25	0.30	0.35	0.40	0.45	0.50	0.55	0.60	0.70	0.80	0.90
(1)	.0023	.0093	.0213	.0391	.0637	.0969	.1417	.2025	.2867	.4059	.5774	.8242	1.674	3.465	8.748
(2)	.0877	.1812	.2849	.4037	.5433	.7109	.9158	1.170	1.490	1.894	2.409	3.068	5.051	8.861	19.62
(3)	.6964	1.416	2.169	2.966	3.822	4.753	5.781	6.932	8.244	9.767	11.58	13.78	20.17	32.53	68.81

TABLE 7.25 : I^*_{02} ($1/\alpha = .30$, $\mu = .70$)

Q: MODE:	0.05	0.10	0.15	0.20	0.25	0.30	0.35	0.40	0.45	0.50	0.55	0.60	0.70	0.80	0.90
(1)	.0017	.0069	.0159	.0290	.0470	.0711	.1035	.1478	.2111	.3067	.4587	.7007	1.599	3.517	9.171
(2)	.0893	.1741	.2622	.3590	.4693	.5981	.7519	.9389	1.170	1.461	1.833	2.316	3.812	6.773	15.20
(3)	.2043	.4432	.7290	1.077	1.502	2.023	2.653	3.406	4.299	5.355	6.611	8.132	12.45	20.51	43.46

TABLE 7.25 : I^*_{02} ($1/\alpha = .30$, $\mu = .80$)

Q: MODE:	0.05	0.10	0.15	0.20	0.25	0.30	0.35	0.40	0.45	0.50	0.55	0.60	0.70	0.80	0.90
(1)	.0011	.0046	.0105	.0190	.0306	.0459	.0661	.0938	.1349	.2055	.3470	.6196	1.649	3.762	9.912
(2)	.0807	.1431	.2034	.2674	.3389	.4208	.5166	.6312	.7711	.9468	1.175	1.481	2.543	5.012	12.70
(3)	.2444	.4847	.7303	.9901	1.273	1.591	1.953	2.376	2.876	3.477	4.210	5.120	7.777	12.82	27.09

TABLE 7.25 : I^*_{02} ($1/\alpha = .40$, $\mu = .20$)

Q: MODE:	0.05	0.10	0.15	0.20	0.25	0.30	0.35	0.40	0.45	0.50	0.55	0.60	0.70	0.80	0.90
(1)	.0057	.0229	.0527	.0963	.1562	.2355	.3385	.4709	.6403	.8570	1.135	1.494	2.600	4.806	11.27
(2)	.1147	.2505	.4130	.6091	.8467	1.135	1.484	1.904	2.409	3.017	3.753	4.658	7.275	12.24	26.48
(3)	.6350	1.291	1.976	2.701	3.477	4.319	5.246	6.281	7.457	8.818	10.43	12.38	18.01	28.83	60.41

TABLE 7.25 : I^*_{02} ($1/\alpha = .40$, $\mu = .30$)

Q: MODE:	0.05	0.10	0.15	0.20	0.25	0.30	0.35	0.40	0.45	0.50	0.55	0.60	0.70	0.80	0.90
(1)	.0050	.0201	.0461	.0845	.1373	.2075	.2991	.4177	.5704	.7672	1.021	1.351	2.368	4.393	10.27
(2)	.0966	.2421	.4666	.7855	1.187	1.658	2.194	2.802	3.497	4.304	5.258	6.415	9.739	16.10	34.57
(3)	.1265	.2529	.3884	.5391	.7115	.9135	1.156	1.454	1.831	2.320	2.968	3.839	6.635	12.23	28.19

TABLE 7.25 : I^*_{02} ($1/\alpha = .40$, $\mu = .40$)

Q: MODE:	0.05	0.10	0.15	0.20	0.25	0.30	0.35	0.40	0.45	0.50	0.55	0.60	0.70	0.80	0.90
(1)	.0042	.0172	.0395	.0724	.1177	.1781	.2572	.3602	.4940	.6681	.8951	1.193	2.121	3.979	9.364
(2)	.0805	.1779	.3007	.4625	.6828	.9840	1.383	1.888	2.503	3.239	4.123	5.201	8.304	14.22	31.36
(3)	.1788	.3517	.5314	.7271	.9472	1.201	1.501	1.860	2.301	2.849	3.541	4.428	7.129	12.39	27.30

TABLE 7.25 : I^*_{02} ($1/\alpha = .40$, $\mu = .50$)

Q: MODE:	0.05	0.10	0.15	0.20	0.25	0.30	0.35	0.40	0.45	0.50	0.55	0.60	0.70	0.80	0.90
(1)	.0035	.0143	.0329	.0602	.0977	.1477	.2135	.2997	.4130	.5627	.7621	1.029	1.886	3.643	8.794
(2)	.1042	.2222	.3588	.5202	.7146	.9522	1.246	1.611	2.066	2.631	3.334	4.216	6.814	11.81	26.18
(3)	.1451	.3008	.4714	.6621	.8796	1.133	1.433	1.798	2.248	2.814	3.533	4.459	7.276	12.75	28.34

TABLE 7.25 : I^*_{02} ($1/\alpha = .40$, $\mu = .60$)

Q: MODE:	0.05	0.10	0.15	0.20	0.25	0.30	0.35	0.40	0.45	0.50	0.55	0.60	0.70	0.80	0.90
(1)	.0028	.0114	.0262	.0479	.0775	.1169	.1688	.2372	.3288	.4534	.6260	.8681	1.696	3.467	8.763
(2)	.1168	.2404	.3760	.5292	.7062	.9145	1.163	1.464	1.830	2.282	2.843	3.548	5.636	9.630	20.94
(3)	.7224	1.469	2.250	3.079	3.968	4.937	6.006	7.206	8.575	10.17	12.06	14.37	21.07	34.07	72.30

TABLE 7.25 : I^*_{02} ($1/\alpha = .40$, $\mu = .70$)

Q: MODE:	0.05	0.10	0.15	0.20	0.25	0.30	0.35	0.40	0.45	0.50	0.55	0.60	0.70	0.80	0.90
(1)	.0021	.0086	.0196	.0356	.0573	.0861	.1238	.1740	.2428	.3415	.4905	.7219	1.600	3.537	9.333
(2)	.1186	.2297	.3433	.4660	.6036	.7616	.9465	1.166	1.432	1.756	2.160	2.670	4.202	7.179	15.64
(3)	.2720	.5878	.9592	1.399	1.919	2.532	3.247	4.079	5.044	6.172	7.507	9.118	13.70	22.28	46.81

TABLE 7.25 : I^*_{02} ($1/\alpha = .40$, $\mu = .80$)

Q: MODE:	0.05	0.10	0.15	0.20	0.25	0.30	0.35	0.40	0.45	0.50	0.55	0.60	0.70	0.80	0.90
(1)	.0014	.0057	.0129	.0234	.0375	.0558	.0797	.1115	.1565	.2282	.3622	.6217	1.668	3.869	10.33
(2)	.1055	.1866	.2625	.3422	.4300	.5295	.6442	.7787	.9396	1.136	1.383	1.704	2.751	5.119	12.70
(3)	.3252	.6414	.9586	1.286	1.635	2.014	2.435	2.912	3.462	4.105	4.873	5.808	8.492	13.54	27.87

TABLE 7.25 : I^*_{02} ($1/\alpha = .50$, $\mu = .20$)

Q: MODE:	0.05	0.10	0.15	0.20	0.25	0.30	0.35	0.40	0.45	0.50	0.55	0.60	0.70	0.80	0.90
(1)	.0066	.0267	.0611	.1114	.1797	.2692	.3841	.5300	.7143	.9472	1.243	1.623	2.783	5.093	11.87
(2)	.1329	.2878	.4698	.6849	.9401	1.243	1.604	2.033	2.544	3.156	3.895	4.803	7.432	12.43	26.79
(3)	.6379	1.297	1.985	2.713	3.493	4.340	5.272	6.313	7.496	8.865	10.48	12.45	18.12	29.03	60.89

TABLE 7.25 : I^*_{02} ($1/\alpha = .50$, $\mu = .30$)

Q: MODE:	0.05	0.10	0.15	0.20	0.25	0.30	0.35	0.40	0.45	0.50	0.55	0.60	0.70	0.80	0.90
(1)	.0058	.0234	.0536	.0977	.1579	.2369	.3388	.4688	.6339	.8435	1.111	1.455	2.506	4.590	10.64
(2)	.1101	.2667	.4939	.8064	1.200	1.664	2.196	2.802	3.498	4.306	5.264	6.427	9.772	16.18	34.81
(3)	.1580	.3154	.4833	.6694	.8812	1.128	1.420	1.776	2.218	2.777	3.500	4.447	7.404	13.28	30.18

TABLE 7.25 : I^*_{02} ($1/\alpha = .50$, $\mu = .40$)

Q: MODE:	0.05	0.10	0.15	0.20	0.25	0.30	0.35	0.40	0.45	0.50	0.55	0.60	0.70	0.80	0.90
(1)	.0050	.0200	.0459	.0837	.1354	.2033	.2911	.4036	.5473	.7310	.9671	1.273	2.213	4.090	9.520
(2)	.1006	.2218	.3730	.5677	.8236	1.159	1.586	2.113	2.749	3.508	4.422	5.538	8.766	14.95	32.94
(3)	.2109	.4172	.6320	.8651	1.126	1.423	1.770	2.181	2.677	3.282	4.034	4.983	7.822	13.31	28.93

TABLE 7.25 : I^*_{02} ($1/\alpha = .50$, $\mu = .50$)

Q: MODE:	0.05	0.10	0.15	0.20	0.25	0.30	0.35	0.40	0.45	0.50	0.55	0.60	0.70	0.80	0.90
(1)	.0041	.0167	.0382	.0696	.1124	.1688	.2417	.3356	.4566	.6132	.8176	1.087	1.941	3.684	8.817
(2)	.1302	.2771	.4458	.6430	.8768	1.157	1.496	1.906	2.407	3.018	3.769	4.705	7.457	12.76	28.11
(3)	.1692	.3502	.5474	.7657	1.012	1.294	1.623	2.014	2.486	3.066	3.791	4.711	7.485	12.89	28.41

TABLE 7.25 : I^*_{02} ($1/\alpha = .50$, $\mu = .60$)

Q: MODE:	0.05	0.10	0.15	0.20	0.25	0.30	0.35	0.40	0.45	0.50	0.55	0.60	0.70	0.80	0.90
(1)	.0033	.0133	.0305	.0554	.0893	.1337	.1914	.2659	.3633	.4923	.6664	.9059	1.716	3.469	8.779
(2)	.1458	.2990	.4654	.6508	.8619	1.106	1.392	1.732	2.139	2.631	3.234	3.983	6.179	10.36	22.24
(3)	.7492	1.524	2.334	3.194	4.119	5.125	6.239	7.488	8.915	10.58	12.55	14.97	21.99	35.64	75.82

TABLE 7.25 : I^*_{02} ($1/\alpha$ = .50 , μ = .70)

Q: MODE:	0.05	0.10	0.15	0.20	0.25	0.30	0.35	0.40	0.45	0.50	0.55	0.60	0.70	0.80	0.90
(1)	.0025	.0100	.0227	.0412	.0662	.0988	.1409	.1957	.2687	.3698	.5170	.7405	1.600	3.557	9.498
(2)	.1477	.2841	.4213	.5674	.7289	.9118	1.123	1.369	1.662	2.014	2.443	2.977	4.549	7.554	16.07
(3)	.3396	.7310	1.184	1.709	2.314	3.007	3.798	4.701	5.736	6.936	8.351	10.06	14.90	24.02	50.13

TABLE 7.25 : I^*_{02} ($1/\alpha$ = .50 , μ = .80)

Q: MODE:	0.05	0.10	0.15	0.20	0.25	0.30	0.35	0.40	0.45	0.50	0.55	0.60	0.70	0.80	0.90
(1)	.0016	.0066	.0151	.0272	.0434	.0643	.0911	.1261	.1740	.2467	.3752	.6238	1.687	3.980	10.75
(2)	.1318	.2281	.3177	.4109	.5127	.6268	.7569	.9076	1.085	1.298	1.559	1.889	2.928	5.219	12.70
(3)	.4057	.7957	1.180	1.569	1.973	2.404	2.872	3.393	3.981	4.659	5.456	6.416	9.138	14.22	28.63

TABLE 7.25 : I^*_{02} ($1/\alpha$ = .60 , μ = .20)

Q: MODE:	0.05	0.10	0.15	0.20	0.25	0.30	0.35	0.40	0.45	0.50	0.55	0.60	0.70	0.80	0.90
(1)	.0074	.0300	.0685	.1245	.2002	.2985	.4238	.5814	.7791	1.027	1.340	1.739	2.954	5.369	12.47
(2)	.1487	.3199	.5182	.7491	1.019	1.335	1.707	2.144	2.662	3.279	4.023	4.936	7.580	12.62	27.09
(3)	.6408	1.303	1.994	2.726	3.510	4.362	5.299	6.346	7.536	8.914	10.54	12.52	18.24	29.24	61.38

TABLE 7.25 : I^*_{02} ($1/\alpha$ = .60 , μ = .30)

Q: MODE:	0.05	0.10	0.15	0.20	0.25	0.30	0.35	0.40	0.45	0.50	0.55	0.60	0.70	0.80	0.90
(1)	.0065	.0262	.0600	.1092	.1757	.2624	.3732	.5130	.6889	.9101	1.190	1.547	2.633	4.776	11.00
(2)	.1217	.2873	.5170	.8249	1.211	1.669	2.197	2.802	3.498	4.309	5.270	6.438	9.805	16.26	35.05
(3)	.1894	.3775	.5774	.7981	1.048	1.337	1.677	2.086	2.586	3.210	4.000	5.018	8.141	14.31	32.17

TABLE 7.25 : I^*_{02} ($1/\alpha$ = .60 , μ = .40)

Q: MODE:	0.05	0.10	0.15	0.20	0.25	0.30	0.35	0.40	0.45	0.50	0.55	0.60	0.70	0.80	0.90
(1)	.0056	.0225	.0514	.0936	.1506	.2250	.3202	.4409	.5931	.7855	1.030	1.344	2.301	4.195	9.673
(2)	.1206	.2654	.4442	.6702	.9590	1.325	1.779	2.330	2.989	3.775	4.720	5.877	9.234	15.69	34.54
(3)	.2394	.4760	.7228	.9901	1.287	1.624	2.014	2.470	3.013	3.670	4.475	5.482	8.460	14.19	30.52

TABLE 7.25 : I^*_{02} ($1/\alpha$ = .60 , μ = .50)

Q: MODE:	0.05	0.10	0.15	0.20	0.25	0.30	0.35	0.40	0.45	0.50	0.55	0.60	0.70	0.80	0.90
(1)	.0046	.0187	.0428	.0778	.1251	.1868	.2658	.3663	.4938	.6564	.8656	1.138	1.990	3.723	8.839
(2)	.1561	.3316	.5319	.7634	1.034	1.354	1.733	2.185	2.728	3.383	4.182	5.174	8.083	13.70	30.03
(3)	.1903	.3934	.6135	.8556	1.126	1.432	1.785	2.198	2.688	3.282	4.013	4.932	7.678	13.03	28.48

TABLE 7.25 : I^*_{02} ($1/\alpha$ = .60 , μ = .60)

Q: MODE:	0.05	0.10	0.15	0.20	0.25	0.30	0.35	0.40	0.45	0.50	0.55	0.60	0.70	0.80	0.90
(1)	.0037	.0150	.0341	.0619	.0994	.1482	.2107	.2903	.3925	.5254	.7012	.9391	1.735	3.471	8.795
(2)	.1747	.3571	.5530	.7687	1.011	1.288	1.607	1.981	2.423	2.952	3.594	4.385	6.691	11.07	23.51
(3)	.7769	1.580	2.421	3.314	4.273	5.320	6.477	7.778	9.264	11.00	13.06	15.58	22.93	37.22	79.36

TABLE 7.25 : I^*_{02} ($1/\alpha$ = .60 , μ = .70)

Q: MODE:	0.05	0.10	0.15	0.20	0.25	0.30	0.35	0.40	0.45	0.50	0.55	0.60	0.70	0.80	0.90
(1)	.0028	.0112	.0255	.0461	.0738	.1096	.1554	.2140	.2905	.3937	.5398	.7572	1.600	3.579	9.665
(2)	.1766	.3373	.4963	.6636	.8462	1.051	1.284	1.553	1.869	2.244	2.696	3.252	4.863	7.905	16.49
(3)	.4071	.8728	1.405	2.009	2.691	3.457	4.317	5.287	6.390	7.663	9.159	10.96	16.08	25.74	53.43

TABLE 7.25 : I^*_{02} ($1/\alpha$ = .60 , μ = .80)

Q: MODE:	0.05	0.10	0.15	0.20	0.25	0.30	0.35	0.40	0.45	0.50	0.55	0.60	0.70	0.80	0.90
(1)	.0019	.0075	.0169	.0305	.0485	.0716	.1009	.1385	.1888	.2622	.3865	.6258	1.707	4.093	11.18
(2)	.1566	.2675	.3692	.4741	.5879	.7145	.8576	1.022	1.212	1.438	1.711	2.050	3.084	5.311	12.70
(3)	.4858	.9476	1.395	1.839	2.292	2.766	3.274	3.830	4.451	5.158	5.983	6.968	9.735	14.86	29.37

TABLE 7.25 : I^*_{02} ($1/\alpha$ = .70 , μ = .20)

Q: MODE:	0.05	0.10	0.15	0.20	0.25	0.30	0.35	0.40	0.45	0.50	0.55	0.60	0.70	0.80	0.90
(1)	.0082	.0329	.0750	.1361	.2182	.3244	.4589	.6272	.8369	1.099	1.428	1.846	3.114	5.635	13.05
(2)	.1625	.3478	.5602	.8046	1.087	1.415	1.796	2.242	2.766	3.389	4.139	5.059	7.721	12.80	27.40
(3)	.6439	1.309	2.004	2.740	3.528	4.384	5.326	6.380	7.577	8.964	10.61	12.60	18.36	29.45	61.87

TABLE 7.25 : I^*_{02} ($1/\alpha$ = .70 , μ = .30)

Q: MODE:	0.05	0.10	0.15	0.20	0.25	0.30	0.35	0.40	0.45	0.50	0.55	0.60	0.70	0.80	0.90
(1)	.0071	.0288	.0657	.1192	.1913	.2848	.4033	.5520	.7376	.9695	1.261	1.632	2.752	4.955	11.35
(2)	.1317	.3048	.5370	.8415	1.221	1.674	2.199	2.802	3.498	4.311	5.276	6.451	9.838	16.34	35.29
(3)	.2208	.4392	.6706	.9250	1.212	1.541	1.926	2.385	2.940	3.622	4.476	5.563	8.855	15.34	34.17

TABLE 7.25 : I^*_{02} ($1/\alpha$ = .70 , μ = .40)

Q: MODE:	0.05	0.10	0.15	0.20	0.25	0.30	0.35	0.40	0.45	0.50	0.55	0.60	0.70	0.80	0.90
(1)	.0061	.0247	.0563	.1022	.1639	.2440	.3457	.4735	.6333	.8335	1.086	1.407	2.381	4.295	9.822
(2)	.1406	.3089	.5146	.7705	1.090	1.485	1.966	2.542	3.226	4.039	5.018	6.218	9.707	16.44	36.15
(3)	.2649	.5290	.8052	1.104	1.434	1.807	2.235	2.732	3.319	4.022	4.878	5.939	9.056	15.03	32.08

TABLE 7.25 : I^*_{02} ($1/\alpha$ = .70 , μ = .50)

Q: MODE:	0.05	0.10	0.15	0.20	0.25	0.30	0.35	0.40	0.45	0.50	0.55	0.60	0.70	0.80	0.90
(1)	.0051	.0205	.0469	.0849	.1362	.2026	.2869	.3929	.5262	.6942	.9080	1.184	2.037	3.761	8.861
(2)	.1821	.3859	.6170	.8817	1.188	1.543	1.960	2.451	3.034	3.732	4.579	5.626	8.697	14.64	31.95
(3)	.2088	.4314	.6716	.9344	1.226	1.553	1.926	2.358	2.865	3.471	4.210	5.129	7.856	13.16	28.55

TABLE 7.25 : I^*_{02} ($1/\alpha$ = .70 , μ = .60)

Q: MODE:	0.05	0.10	0.15	0.20	0.25	0.30	0.35	0.40	0.45	0.50	0.55	0.60	0.70	0.80	0.90
(1)	.0041	.0164	.0374	.0677	.1083	.1608	.2274	.3114	.4178	.5541	.7317	.9687	1.753	3.473	8.811
(2)	.2035	.4145	.6389	.8832	1.154	1.460	1.810	2.215	2.689	3.252	3.930	4.764	7.178	11.76	24.76
(3)	.8053	1.638	2.510	3.436	4.432	5.518	6.721	8.074	9.621	11.42	13.57	16.20	23.88	38.82	82.91

TABLE 7.25 : I^*_{02} ($1/\alpha$ = .70 , μ = .70)

Q: MODE:	0.05	0.10	0.15	0.20	0.25	0.30	0.35	0.40	0.45	0.50	0.55	0.60	0.70	0.80	0.90
(1)	.0031	.0123	.0279	.0504	.0805	.1191	.1680	.2299	.3093	.4144	.5597	.7723	1.601	3.600	9.835
(2)	.2053	.3893	.5686	.7549	.9563	1.180	1.432	1.722	2.057	2.453	2.926	3.502	5.153	8.237	16.89
(3)	.4744	1.013	1.621	2.301	3.054	3.888	4.813	5.847	7.016	8.361	9.939	11.84	17.24	27.43	56.72

TABLE 7.25 : I^*_{02} ($1/\alpha$ = .70 , μ = .80)

Q: MODE:	0.05	0.10	0.15	0.20	0.25	0.30	0.35	0.40	0.45	0.50	0.55	0.60	0.70	0.80	0.90
(1)	.0020	.0082	.0185	.0333	.0529	.0780	.1094	.1493	.2015	.2755	.3965	.6278	1.727	4.209	11.62
(2)	.1807	.3051	.4174	.5324	.6566	.7939	.9482	1.124	1.326	1.562	1.845	2.191	3.223	5.398	12.70
(3)	.5655	1.097	1.604	2.098	2.594	3.105	3.647	4.233	4.883	5.618	6.468	7.478	10.29	15.47	30.09

TABLE 7.25 : I_{02}^{*} (1/α = .80 , μ = .20)

Q: MODE:	0.05	0.10	0.15	0.20	0.25	0.30	0.35	0.40	0.45	0.50	0.55	0.60	0.70	0.80	0.90
(1)	.0088	.0355	.0808	.1464	.2342	.3475	.4903	.6684	.8894	1.164	1.508	1.945	3.267	5.892	13.63
(2)	.1748	.3724	.5970	.8532	1.147	1.485	1.875	2.329	2.860	3.490	4.247	5.173	7.855	12.97	27.69
(3)	.6471	1.315	2.014	2.754	3.546	4.407	5.354	6.414	7.619	9.015	10.67	12.68	18.48	29.66	62.36

TABLE 7.25 : I_{02}^{*} (1/α = .80 , μ = .30)

Q: MODE:	0.05	0.10	0.15	0.20	0.25	0.30	0.35	0.40	0.45	0.50	0.55	0.60	0.70	0.80	0.90
(1)	.0077	.0310	.0708	.1282	.2053	.3047	.4302	.5868	.7813	1.023	1.326	1.709	2.863	5.127	11.70
(2)	.1404	.3201	.5545	.8565	1.231	1.679	2.200	2.802	3.499	4.313	5.283	6.463	9.872	16.42	35.54
(3)	.2520	.5006	.7629	1.050	1.372	1.741	2.169	2.674	3.280	4.019	4.933	6.086	9.550	16.35	36.17

TABLE 7.25 : I_{02}^{*} (1/α = .80 , μ = .40)

Q: MODE:	0.05	0.10	0.15	0.20	0.25	0.30	0.35	0.40	0.45	0.50	0.55	0.60	0.70	0.80	0.90
(1)	.0066	.0266	.0606	.1098	.1758	.2609	.3683	.5024	.6691	.8765	1.136	1.465	2.455	4.391	9.969
(2)	.1607	.3521	.5843	.8689	1.218	1.641	2.148	2.749	3.459	4.302	5.316	6.560	10.18	17.20	37.76
(3)	.2878	.5769	.8801	1.207	1.569	1.975	2.437	2.972	3.600	4.346	5.249	6.364	9.618	15.84	33.62

TABLE 7.25 : I_{02}^{*} (1/α = .80 , μ = .50)

Q: MODE:	0.05	0.10	0.15	0.20	0.25	0.30	0.35	0.40	0.45	0.50	0.55	0.60	0.70	0.80	0.90
(1)	.0055	.0221	.0505	.0913	.1460	.2165	.3054	.4165	.5548	.7278	.9459	1.225	2.080	3.798	8.883
(2)	.2079	.4400	.7013	.9979	1.337	1.727	2.180	2.707	3.328	4.068	4.963	6.067	9.300	15.57	33.87
(3)	.2254	.4651	.7232	1.004	1.314	1.660	2.051	2.499	3.021	3.639	4.386	5.308	8.022	13.29	28.62

TABLE 7.25 : I_{02}^{*} (1/α = .80 , μ = .60)

Q: MODE:	0.05	0.10	0.15	0.20	0.25	0.30	0.35	0.40	0.45	0.50	0.55	0.60	0.70	0.80	0.90
(1)	.0044	.0177	.0403	.0728	.1161	.1719	.2421	.3300	.4401	.5794	.7588	.9954	1.771	3.475	8.827
(2)	.2323	.4714	.7232	.9943	1.292	1.625	2.002	2.436	2.940	3.535	4.249	5.123	7.646	12.42	26.00
(3)	.8344	1.697	2.601	3.561	4.594	5.722	6.971	8.376	9.984	11.86	14.10	16.84	24.84	40.44	86.49

TABLE 7.25 : I_{02}^{*} (1/α = .80 , μ = .70)

Q: MODE:	0.05	0.10	0.15	0.20	0.25	0.30	0.35	0.40	0.45	0.50	0.55	0.60	0.70	0.80	0.90
(1)	.0033	.0133	.0301	.0543	.0864	.1274	.1791	.2438	.3258	.4325	.5774	.7861	1.601	3.622	10.01
(2)	.2337	.4400	.6381	.8416	1.060	1.300	1.570	1.877	2.231	2.645	3.136	3.732	5.422	8.552	17.29
(3)	.5415	1.153	1.834	2.585	3.406	4.304	5.290	6.386	7.621	9.037	10.70	12.69	18.37	29.10	59.99

TABLE 7.25 : I_{02}^{*} (1/α = .80 , μ = .80)

Q: MODE:	0.05	0.10	0.15	0.20	0.25	0.30	0.35	0.40	0.45	0.50	0.55	0.60	0.70	0.80	0.90
(1)	.0022	.0088	.0200	.0359	.0569	.0836	.1169	.1587	.2126	.2871	.4055	.6296	1.748	4.328	12.05
(2)	.2043	.3407	.4624	.5864	.7196	.8663	1.030	1.216	1.428	1.674	1.965	2.318	3.349	5.480	12.70
(3)	.6449	1.245	1.807	2.346	2.880	3.425	3.996	4.610	5.285	6.044	6.919	7.953	10.82	16.06	30.80

TABLE 7.25 : I_{02}^{*} (1/α = .90 , μ = .20)

Q: MODE:	0.05	0.10	0.15	0.20	0.25	0.30	0.35	0.40	0.45	0.50	0.55	0.60	0.70	0.80	0.90
(1)	.0094	.0378	.0860	.1556	.2487	.3684	.5188	.7059	.9374	1.225	1.584	2.039	3.413	6.143	14.20
(2)	.1857	.3943	.6297	.8964	1.200	1.547	1.945	2.407	2.946	3.583	4.346	5.281	7.984	13.14	27.99
(3)	.6504	1.322	2.024	2.768	3.565	4.430	5.384	6.450	7.662	9.068	10.73	12.76	18.60	29.88	62.86

TABLE 7.25 : I_{02}^{*} (1/α = .90, μ = .30)

Q: MODE:	0.05	0.10	0.15	0.20	0.25	0.30	0.35	0.40	0.45	0.50	0.55	0.60	0.70	0.80	0.90
(1)	.0082	.0330	.0753	.1362	.2177	.3226	.4544	.6183	.8211	1.072	1.386	1.781	2.968	5.293	12.04
(2)	.1482	.3336	.5701	.8701	1.241	1.684	2.202	2.802	3.499	4.316	5.289	6.475	9.907	16.51	35.78
(3)	.2833	.5617	.8544	1.174	1.530	1.936	2.405	2.955	3.610	4.402	5.374	6.593	10.23	17.36	38.17

TABLE 7.25 : I_{02}^{*} (1/α = .90, μ = .40)

Q: MODE:	0.05	0.10	0.15	0.20	0.25	0.30	0.35	0.40	0.45	0.50	0.55	0.60	0.70	0.80	0.90
(1)	.0070	.0283	.0645	.1166	.1864	.2760	.3886	.5284	.7014	.9154	1.182	1.519	2.525	4.483	10.11
(2)	.1806	.3951	.6533	.9656	1.343	1.793	2.326	2.953	3.690	4.564	5.614	6.903	10.66	17.96	39.39
(3)	.3084	.6205	.9485	1.302	1.692	2.128	2.624	3.194	3.859	4.646	5.595	6.761	10.15	16.62	35.13

TABLE 7.25 : I_{02}^{*} (1/α = .90, μ = .50)

Q: MODE:	0.05	0.10	0.15	0.20	0.25	0.30	0.35	0.40	0.45	0.50	0.55	0.60	0.70	0.80	0.90
(1)	.0059	.0236	.0537	.0970	.1548	.2289	.3220	.4375	.5805	.7580	.9802	1.263	2.120	3.833	8.904
(2)	.2338	.4938	.7847	1.112	1.483	1.906	2.392	2.955	3.613	4.395	5.336	6.497	9.894	16.49	35.78
(3)	.2401	.4953	.7693	1.067	1.393	1.755	2.162	2.625	3.161	3.790	4.545	5.471	8.177	13.42	28.69

TABLE 7.25 : I_{02}^{*} (1/α = .90, μ = .60)

Q: MODE:	0.05	0.10	0.15	0.20	0.25	0.30	0.35	0.40	0.45	0.50	0.55	0.60	0.70	0.80	0.90
(1)	.0047	.0188	.0428	.0773	.1231	.1818	.2553	.3465	.4599	.6021	.7832	1.020	1.787	3.477	8.843
(2)	.2609	.5278	.8059	1.102	1.425	1.783	2.186	2.646	3.178	3.804	4.552	5.466	8.098	13.08	27.23
(3)	.8641	1.758	2.694	3.689	4.760	5.929	7.225	8.683	10.35	12.30	14.63	17.48	25.81	42.06	90.08

TABLE 7.25 : I_{02}^{*} (1/α = .90, μ = .70)

Q: MODE:	0.05	0.10	0.15	0.20	0.25	0.30	0.35	0.40	0.45	0.50	0.55	0.60	0.70	0.80	0.90
(1)	.0035	.0141	.0320	.0577	.0916	.1349	.1889	.2561	.3404	.4485	.5933	.7988	1.602	3.644	10.18
(2)	.2619	.4895	.7050	.9242	1.158	1.413	1.698	2.021	2.392	2.823	3.332	3.946	5.676	8.853	17.68
(3)	.6085	1.291	2.044	2.862	3.748	4.706	5.753	6.909	8.208	9.695	11.44	13.53	19.49	30.77	63.25

TABLE 7.25 : I_{02}^{*} (1/α = .90, μ = .80)

Q: MODE:	0.05	0.10	0.15	0.20	0.25	0.30	0.35	0.40	0.45	0.50	0.55	0.60	0.70	0.80	0.90
(1)	.0023	.0094	.0213	.0382	.0604	.0886	.1235	.1670	.2223	.2974	.4136	.6314	1.770	4.449	12.50
(2)	.2273	.3746	.5045	.6363	.7776	.9326	1.105	1.299	1.520	1.774	2.073	2.432	3.464	5.558	12.71
(3)	.7240	1.390	2.004	2.585	3.153	3.727	4.325	4.963	5.662	6.445	7.342	8.400	11.31	16.62	31.49

TABLE 7.26: I_1^* (α_2 = .1, μ = .1)

α_1 \ Q:	0.05	0.10	0.15	0.20	0.30	0.40	0.50	0.60	0.70	0.80
0.1	.4966	1.042	1.644	2.311	3.900	6.017	9.139	14.39	24.22	44.81
0.2	.4757	.9975	1.569	2.197	3.656	5.482	7.899	11.39	17.18	28.87
0.3	.4584	.9628	1.516	2.121	3.515	5.227	7.416	10.42	15.09	24.07
0.4	.4437	.9348	1.474	2.065	3.423	5.076	7.156	9.943	14.13	21.87
0.5	.4310	.9115	1.441	2.022	3.357	4.974	6.993	9.659	13.58	20.64
0.6	.4199	.8918	1.414	1.988	3.307	4.901	6.880	9.471	13.22	19.85
0.7	.4101	.8749	1.392	1.960	3.267	4.846	6.798	9.338	12.98	19.32
0.8	.4013	.8601	1.372	1.937	3.236	4.803	6.736	9.239	12.80	18.92
0.9	.3934	.8470	1.355	1.917	3.210	4.768	6.686	9.161	12.66	18.63
1.0	.3862	.8353	1.341	1.900	3.188	4.739	6.646	9.100	12.55	18.40

TABLE 7.26: I_1^* (α_2 = .1, μ = .3)

α_1 \ Q:	0.05	0.10	0.15	0.20	0.30	0.40	0.50	0.60	0.70	0.80
0.1	.5187	1.139	1.881	2.771	5.136	8.634	13.97	22.47	37.13	66.75
0.2	.4938	1.077	1.764	2.572	4.649	7.567	11.79	18.21	28.93	50.22
0.3	.4733	1.029	1.681	2.443	4.381	7.068	10.90	16.66	26.16	44.84
0.4	.4561	.9915	1.618	2.350	4.207	6.771	10.41	15.85	24.77	42.19
0.5	.4413	.9604	1.569	2.280	4.083	6.572	10.10	15.35	23.93	40.61
0.6	.4284	.9342	1.528	2.224	3.990	6.428	9.879	15.01	23.37	39.57
0.7	.4169	.9116	1.494	2.178	3.917	6.319	9.717	14.77	22.97	38.82
0.8	.4068	.8920	1.465	2.140	3.858	6.233	9.593	14.58	22.67	38.27
0.9	.3976	.8746	1.441	2.107	3.809	6.164	9.494	14.43	22.43	37.84
1.0	.3892	.8592	1.419	2.079	3.768	6.107	9.414	14.31	22.25	37.50

TABLE 7.26: I_1^* (α_2 = .1, μ = .5)

α_1 \ Q:	0.05	0.10	0.15	0.20	0.30	0.40	0.50	0.60	0.70	0.80
0.1	.5407	1.234	2.112	3.209	6.237	10.78	17.60	28.20	46.10	81.79
0.2	.5112	1.150	1.936	2.892	5.423	9.053	14.34	22.43	35.98	62.97
0.3	.4873	1.087	1.816	2.693	4.987	8.250	12.99	20.22	32.33	56.49
0.4	.4672	1.038	1.727	2.552	4.706	7.768	12.22	19.02	30.42	53.18
0.5	.4500	.9975	1.657	2.446	4.507	7.441	11.72	18.26	29.25	51.17
0.6	.4351	.9636	1.600	2.362	4.356	7.203	11.36	17.74	28.45	49.82
0.7	.4220	.9346	1.553	2.293	4.237	7.021	11.10	17.35	27.87	48.85
0.8	.4103	.9094	1.512	2.236	4.141	6.877	10.89	17.06	27.42	48.11
0.9	.3998	.8872	1.477	2.187	4.061	6.760	10.73	16.82	27.08	47.54
1.0	.3903	.8674	1.446	2.145	3.993	6.663	10.59	16.63	26.80	47.08

TABLE 7.26: I_1^* (α_2 = .1, μ = .7)

α_1 \ Q:	0.05	0.10	0.15	0.20	0.30	0.40	0.50	0.60	0.70	0.80
0.1	.5624	1.325	2.319	3.571	6.970	11.84	18.80	29.09	45.67	77.33
0.2	.5278	1.215	2.073	3.113	5.802	9.492	14.61	22.07	33.95	56.57
0.3	.4999	1.134	1.909	2.834	5.187	8.386	12.81	19.24	29.51	49.07
0.4	.4768	1.071	1.789	2.639	4.791	7.712	11.76	17.66	27.10	45.12
0.5	.4571	1.021	1.696	2.493	4.508	7.249	11.06	16.63	25.57	42.67
0.6	.4401	.9782	1.621	2.378	4.293	6.907	10.55	15.90	24.51	40.99
0.7	.4251	.9422	1.558	2.283	4.123	6.642	10.17	15.36	23.72	39.76
0.8	.4119	.9109	1.505	2.205	3.984	6.429	9.865	14.93	23.12	38.83
0.9	.4000	.8835	1.459	2.137	3.867	6.254	9.619	14.59	22.64	38.09
1.0	.3892	.8592	1.419	2.079	3.768	6.107	9.414	14.31	22.25	37.50

TABLE 7.26: I_1^* (α_2 = .1, μ = .9)

α_1 \ Q:	0.05	0.10	0.15	0.20	0.30	0.40	0.50	0.60	0.70	0.80
0.1	.5837	1.408	2.486	3.819	7.239	11.75	17.67	25.77	37.92	59.93
0.2	.5433	1.267	2.161	3.212	5.756	8.930	12.92	18.17	25.77	39.04
0.3	.5112	1.167	1.952	2.852	4.984	7.593	10.82	15.02	21.00	31.29
0.4	.4847	1.090	1.800	2.605	4.487	6.769	9.579	13.20	18.33	27.09
0.5	.4624	1.028	1.684	2.420	4.131	6.196	8.732	11.99	16.59	24.40
0.6	.4431	.9773	1.590	2.275	3.860	5.768	8.108	11.11	15.34	22.52
0.7	.4263	.9341	1.513	2.157	3.643	5.432	7.625	10.44	14.40	21.11
0.8	.4114	.8968	1.447	2.058	3.466	5.159	7.237	9.906	13.66	20.01
0.9	.3981	.8642	1.390	1.973	3.316	4.932	6.917	9.467	13.05	19.13
1.0	.3862	.8353	1.341	1.900	3.188	4.739	6.646	9.100	12.55	18.40

TABLE 7.26: I_1^* (α_2 = .2, μ = .1)

α_1 \ Q:	0.05	0.10	0.15	0.20	0.30	0.40	0.50	0.60	0.70	0.80
0.1	.3522	.7417	1.174	1.659	2.846	4.525	7.257	12.30	22.01	42.11
0.2	.3370	.7083	1.117	1.569	2.637	4.024	5.969	9.002	14.34	25.35
0.3	.3242	.6816	1.074	1.506	2.511	3.773	5.451	7.888	11.92	20.03
0.4	.3133	.6596	1.040	1.458	2.424	3.618	5.165	7.330	10.77	17.50
0.5	.3037	.6410	1.012	1.420	2.360	3.512	4.982	6.995	10.11	16.05
0.6	.2953	.6250	.9891	1.389	2.310	3.434	4.855	6.772	9.681	15.12
0.7	.2878	.6110	.9693	1.363	2.271	3.375	4.761	6.613	9.383	14.48
0.8	.2810	.5987	.9521	1.341	2.238	3.327	4.689	6.494	9.164	14.01
0.9	.2748	.5877	.9370	1.323	2.211	3.289	4.631	6.402	8.997	13.66
1.0	.2692	.5778	.9236	1.306	2.188	3.257	4.585	6.328	8.865	13.38

TABLE 7.26: I_1^* (α_2 = .2, μ = .3)

α_1 \ Q:	0.05	0.10	0.15	0.20	0.30	0.40	0.50	0.60	0.70	0.80
0.1	.3634	.7913	1.298	1.902	3.518	5.989	9.972	16.68	28,64	52.98
0.2	.3462	.7490	1.218	1.767	3.170	5.158	8.086	12.65	20.41	36.05
0.3	.3318	.7156	1.159	1.673	2.965	4.748	7.298	11.17	17.63	30.50
0.4	.3195	.6882	1.113	1.603	2.827	4.495	6.852	10.38	16.22	27.76
0.5	.3088	.6653	1.076	1.549	2.725	4.321	6.561	9.896	15.37	26.13
0.6	.2994	.6456	1.045	1.504	2.647	4.192	6.355	9.562	14.80	25.05
0.7	.2910	.6285	1.018	1.467	2.584	4.093	6.202	9.318	14.39	24.28
0.8	.2835	.6134	.9951	1.436	2.532	4.015	6.082	9.132	14.09	23.71
0.9	.2767	.6000	.9750	1.409	2.489	3.950	5.986	8.985	13.85	23.27
1.0	.2705	.5879	.9572	1.385	2.452	3.896	5.908	8.866	13.66	22.91

TABLE 7.26: I_1^* (α_2 = .2, μ = .5)

α_1 \ Q:	0.05	0.10	0.15	0.20	0.30	0.40	0.50	0.60	0.70	0.80
0.1	.3744	.8396	1.417	2.133	4.133	7.253	12.18	20.14	33.84	61.29
0.2	.3550	.7867	1.309	1.938	3.604	6.020	9.596	15.12	24.45	43.06
0.3	.3389	.7455	1.231	1.807	3.301	5.418	8.491	13.19	21.07	36.79
0.4	.3251	.7121	1.170	1.711	3.099	5.045	7.854	12.13	19.29	33.59
0.5	.3133	.6843	1.122	1.636	2.950	4.787	7.432	11.46	18.19	31.64
0.6	.3028	.6606	1.081	1.575	2.836	4.596	7.129	10.98	17.44	30.32
0.7	.2936	.6400	1.047	1.525	2.744	4.447	6.900	10.63	16.89	29.37
0.8	.2853	.6219	1.018	1.482	2.669	4.328	6.720	10.36	16.47	28.66
0.9	.2778	.6058	.9918	1.445	2.605	4.230	6.574	10.15	16.14	28.09
1.0	.2709	.5914	.9689	1.413	2.551	4.147	6.454	9.971	15.87	27.64

TABLE 7.26: I_1^* (α_2 = .2, μ = .7)

α_1 \ Q:	0.05	0.10	0.15	0.20	0.30	0.40	0.50	0.60	0.70	0.80
0.1	.3852	.8853	1.524	2.331	4.584	8.000	13.15	21.09	34.24	59.76
0.2	.3634	.8202	1.383	2.066	3.858	6.383	9.973	15.29	23.90	40.44
0.3	.3453	.7704	1.283	1.893	3.453	5.595	8.587	12.97	20.03	33.56
0.4	.3301	.7304	1.207	1.768	3.183	5.105	7.772	11.67	17.93	29.94
0.5	.3170	.6972	1.146	1.671	2.986	4.763	7.224	10.82	16.59	27.69
0.6	.3055	.6692	1.096	1.593	2.834	4.508	6.825	10.21	15.66	26.15
0.7	.2953	.6449	1.054	1.528	2.712	4.307	6.518	9.754	14.97	25.02
0.8	.2862	.6236	1.017	1.473	2.611	4.145	6.274	9.396	14.43	24.15
0.9	.2780	.6048	.9854	1.426	2.526	4.010	6.074	9.106	14.01	23.47
1.0	.2705	.5879	.9572	1.385	2.452	3.896	5.908	8.866	13.66	22.91

TABLE 7.26: I_1^* ($\alpha_2 = .2$, $\mu = .9$)

Q: α_1	0.05	0.10	0.15	0.20	0.30	0.40	0.50	0.60	0.70	0.80
0.1	.3957	.9273	1.615	2.478	4.805	8.114	12.77	19.56	30.35	50.80
0.2	.3712	.8486	1.437	2.139	3.902	6.212	9.255	13.45	19.82	31.52
0.3	.3512	.7893	1.314	1.923	3.406	5.285	7.700	10.96	15.81	24.58
0.4	.3343	.7423	1.221	1.769	3.077	4.708	6.774	9.529	13.59	20.88
0.5	.3199	.7037	1.148	1.651	2.838	4.302	6.143	8.581	12.16	18.54
0.6	.3073	.6711	1.088	1.557	2.653	3.997	5.678	7.896	11.14	16.91
0.7	.2962	.6431	1.038	1.479	2.505	3.756	5.318	7.372	10.37	15.70
0.8	.2862	.6187	.9944	1.413	2.382	3.560	5.027	6.955	9.763	14.76
0.9	.2773	.5971	.9568	1.356	2.278	3.397	4.788	6.614	9.273	14.00
1.0	.2692	.5778	.9236	1.306	2.188	3.257	4.585	6.328	8.865	13.38

TABLE 7.26: I_1^* ($\alpha_2 = .3$, $\mu = .1$)

Q: α_1	0.05	0.10	0.15	0.20	0.30	0.40	0.50	0.60	0.70	0.80
0.1	.2889	.6113	.9730	1.383	2.411	3.934	6.562	11.61	21.31	41.26
0.2	.2762	.5829	.9237	1.304	2.219	3.447	5.245	8.168	13.42	24.25
0.3	.2655	.5601	.8859	1.247	2.099	3.197	4.705	6.980	10.86	18.73
0.4	.2563	.5410	.8556	1.203	2.015	3.040	4.403	6.376	9.612	16.06
0.5	.2482	.5248	.8306	1.168	1.953	2.932	4.209	6.010	8.886	14.51
0.6	.2411	.5107	.8095	1.139	1.904	2.851	4.072	5.764	8.412	13.51
0.7	.2347	.4984	.7913	1.115	1.865	2.789	3.971	5.588	8.081	12.81
0.8	.2289	.4875	.7754	1.094	1.832	2.739	3.892	5.456	7.837	12.29
0.9	.2237	.4777	.7614	1.076	1.805	2.699	3.830	5.353	7.649	11.90
1.0	.2189	.4688	.7490	1.060	1.781	2.665	3.779	5.270	7.501	11.60

TABLE 7.26: I_1^* ($\alpha_2 = .3$, $\mu = .3$)

Q: α_1	0.05	0.10	0.15	0.20	0.30	0.40	0.50	0.60	0.70	0.80
0.1	.2964	.6451	1.058	1.552	2.887	4.985	8.504	14.65	25.78	48.47
0.2	.2824	.6108	.9934	1.441	2.595	4.254	6.746	10.70	17.54	31.37
0.3	.2707	.5832	.9444	1.363	2.418	3.883	6.002	9.253	14.74	25.73
0.4	.2605	.5605	.9055	1.303	2.295	3.651	5.576	8.481	13.32	22.94
0.5	.2517	.5412	.8736	1.256	2.204	3.488	5.296	7.997	12.46	21.27
0.6	.2439	.5246	.8468	1.217	2.132	3.368	5.096	7.664	11.88	20.17
0.7	.2369	.5100	.8237	1.184	2.075	3.274	4.946	7.420	11.47	19.39
0.8	.2306	.4971	.8037	1.156	2.027	3.199	4.828	7.233	11.15	18.80
0.9	.2249	.4856	.7860	1.132	1.987	3.137	4.734	7.086	10.91	18.35
1.0	.2197	.4752	.7703	1.110	1.952	3.084	4.656	6.966	10.71	17.99

TABLE 7.26: I_1^* ($\alpha_2 = .3$, $\mu = .5$)

Q: α_1	0.05	0.10	0.15	0.20	0.30	0.40	0.50	0.60	0.70	0.80
0.1	.3038	.6777	1.139	1.711	3.323	5.908	10.14	17.19	29.51	54.27
0.2	.2884	.6364	1.056	1.560	2.904	4.881	7.854	12.52	20.44	36.29
0.3	.2754	.6036	.9935	1.456	2.656	4.366	6.867	10.71	17.19	30.13
0.4	.2643	.5768	.9448	1.378	2.487	4.043	6.293	9.726	15.48	25.98
0.5	.2547	.5541	.9051	1.316	2.361	3.817	5.910	9.093	14.42	25.06
0.6	.2462	.5347	.8718	1.265	2.263	3.648	5.634	8.649	13.69	23.77
0.7	.2386	.5177	.8434	1.223	2.184	3.516	5.423	8.318	13.16	22.83
0.8	.2318	.5027	.8187	1.187	2.119	3.409	5.257	8.062	12.75	22.12
0.9	.2256	.4894	.7970	1.155	2.063	3.320	5.122	7.857	12.43	21.57
1.0	.2199	.4773	.7776	1.128	2.015	3.246	5.010	7.689	12.17	21.13

TABLE 7.26: I_1^* ($\alpha_2 = .3$, $\mu = .7$)

Q: α_1	0.05	0.10	0.15	0.20	0.30	0.40	0.50	0.60	0.70	0.80
0.1	.3110	.7082	1.212	1.847	3.651	6.486	10.94	18.03	30.01	53.48
0.2	.2940	.6591	1.107	1.651	3.096	5.176	8.200	12.77	20.25	34.75
0.3	.2798	.6207	1.030	1.518	2.775	4.524	7.000	10.67	16.63	28.14
0.4	.2677	.5894	.9710	1.420	2.557	4.114	6.292	9.499	14.68	24.69
0.5	.2572	.5632	.9229	1.343	2.397	3.826	5.814	8.732	13.44	22.55
0.6	.2480	.5408	.8829	1.280	2.271	3.609	5.464	8.185	12.58	21.07
0.7	.2397	.5213	.8488	1.228	2.170	3.438	5.195	7.771	11.94	19.99
0.8	.2324	.5042	.8192	1.183	2.086	3.299	4.980	7.447	11.44	19.17
0.9	.2257	.4889	.7933	1.144	2.014	3.183	4.804	7.184	11.04	18.52
1.0	.2197	.4752	.7703	1.110	1.952	3.084	4.656	6.966	10.71	17.99

TABLE 7.26: I_1^* ($\alpha_2 = .3$, $\mu = .9$)

Q: α_1	0.05	0.10	0.15	0.20	0.30	0.40	0.50	0.60	0.70	0.80
0.1	.3179	.7361	1.273	1.950	3.824	6.618	10.75	17.05	27.38	47.39
0.2	.2992	.6784	1.145	1.706	3.145	5.097	7.760	11.56	17.50	28.71
0.3	.2837	.6339	1.053	1.544	2.759	4.343	6.435	9.340	13.80	22.07
0.4	.2706	.5979	.9826	1.426	2.499	3.869	5.646	8.079	11.77	18.58
0.5	.2592	.5680	.9261	1.334	2.308	3.534	5.109	7.246	10.47	16.38
0.6	.2492	.5426	.8793	1.260	2.160	3.281	4.712	6.644	9.544	14.86
0.7	.2404	.5206	.8397	1.198	2.039	3.081	4.404	6.184	8.851	13.74
0.8	.2325	.5013	.8055	1.145	1.939	2.918	4.157	5.819	8.307	12.86
0.9	.2254	.4842	.7755	1.100	1.855	2.781	3.952	5.521	7.866	12.17
1.0	.2189	.4688	.7490	1.060	1.781	2.665	3.779	5.270	7.501	11.60

TABLE 7.26: I_1^* ($\alpha_2 = .4$, $\mu = .1$)

Q: α_1	0.05	0.10	0.15	0.20	0.30	0.40	0.50	0.60	0.70	0.80
0.1	.2513	.5343	.8549	1.222	2.162	3.606	6.195	11.26	20.97	40.84
0.2	.2402	.5091	.8103	1.150	1.980	3.126	4.868	7.743	12.97	23.71
0.3	.2308	.4885	.7758	1.097	1.864	2.876	4.303	6.510	10.33	18.10
0.4	.2226	.4713	.7479	1.055	1.783	2.718	3.990	5.875	9.026	15.35
0.5	.2155	.4566	.7247	1.022	1.722	2.608	3.787	5.488	8.259	13.74
0.6	.2091	.4438	.7050	.9947	1.673	2.525	3.644	5.227	7.755	12.69
0.7	.2034	.4326	.6880	.9714	1.634	2.461	3.537	5.038	7.401	11.96
0.8	.1983	.4225	.6732	.9514	1.602	2.410	3.454	4.896	7.138	11.41
0.9	.1936	.4136	.6600	.9339	1.574	2.368	3.387	4.785	6.936	11.00
1.0	.1893	.4054	.6482	.9185	1.550	2.333	3.333	4.696	6.776	10.67

TABLE 7.26: I_1^* ($\alpha_2 = .4$, $\mu = .3$)

Q: α_1	0.05	0.10	0.15	0.20	0.30	0.40	0.50	0.60	0.70	0.80
0.1	.2571	.5603	.9207	1.354	2.537	4.438	7.722	13.60	24.34	46.23
0.2	.2449	.5304	.8642	1.257	2.276	3.762	6.035	9.702	16.09	29.05
0.3	.2347	.5063	.8210	1.187	2.114	3.415	5.316	8.266	13.28	23.36
0.4	.2258	.4862	.7864	1.133	2.001	3.194	4.902	7.498	11.85	20.53
0.5	.2181	.4691	.7578	1.090	1.916	3.040	4.628	7.016	10.98	18.84
0.6	.2112	.4543	.7335	1.054	1.849	2.924	4.432	6.682	10.39	17.72
0.7	.2051	.4413	.7127	1.024	1.795	2.833	4.284	6.438	9.973	16.92
0.8	.1995	.4298	.6944	.9982	1.749	2.760	4.167	6.250	9.655	16.33
0.9	.1945	.4194	.6783	.9757	1.711	2.699	4.073	6.101	9.406	15.86
1.0	.1899	.4101	.6639	.9558	1.678	2.648	3.995	5.980	9.207	15.49

TABLE 7.26: I_1^* ($\alpha_2 = .4$, $\mu = .5$)

Q: α_1	0.05	0.10	0.15	0.20	0.30	0.40	0.50	0.60	0.70	0.80
0.1	.2626	.5850	.9824	1.476	2.878	5.174	9.039	15.63	27.27	50.70
0.2	.2494	.5499	.9119	1.349	2.519	4.262	6.924	11.15	18.37	32.86
0.3	.2383	.5219	.8588	1.259	2.302	3.798	6.006	9.424	15.20	26.75
0.4	.2287	.4987	.8166	1.191	2.151	3.505	5.470	8.479	13.54	23.64
0.5	.2203	.4790	.7820	1.136	2.039	3.298	5.111	7.872	12.50	21.74
0.6	.2129	.4621	.7528	1.092	1.951	3.142	4.850	7.445	11.79	20.46
0.7	.2063	.4472	.7277	1.064	1.879	3.019	4.651	7.127	11.27	19.54
0.8	.2004	.4341	.7058	1.022	1.819	2.920	4.494	6.880	10.87	18.84
0.9	.1950	.4223	.6865	.9936	1.768	2.837	4.365	6.682	10.55	18.29
1.0	.1901	.4117	.6693	.9687	1.724	2.768	4.259	6.519	10.30	17.85

TABLE 7.26: I_1^* (α_2 = .4, μ = .7)

α_1 \ Q:	0.05	0.10	0.15	0.20	0.30	0.40	0.50	0.60	0.70	0.80
0.1	.2680	.6079	1.037	1.580	3.137	5.648	9.715	16.36	27.75	50.22
0.2	.2536	.5672	.9513	1.419	2.673	4.510	7.231	11.40	18.32	31.81
0.3	.2416	.5349	.8873	1.308	2.399	3.936	6.140	9.445	14.85	25.35
0.4	.2312	.5083	.8370	1.224	2.211	3.572	5.495	8.348	12.99	22.00
0.5	.2222	.4860	.7960	1.158	2.070	3.315	5.058	7.632	11.81	19.91
0.6	.2143	.4668	.7615	1.104	1.960	3.121	4.738	7.121	10.99	18.49
0.7	.2072	.4500	.7321	1.058	1.871	2.967	4.491	6.734	10.37	17.44
0.8	.2009	.4352	.7065	1.019	1.797	2.842	4.294	6.430	9.899	16.64
0.9	.1951	.4220	.6840	.9856	1.733	2.737	4.132	6.184	9.519	16.01
1.0	.1899	.4101	.6639	.9558	1.678	2.648	3.995	5.980	9.207	15.49

TABLE 7.26: I_1^* (α_2 = .4, μ = .9)

α_1 \ Q:	0.05	0.10	0.15	0.20	0.30	0.40	0.50	0.60	0.70	0.80
0.1	.2732	.6287	1.083	1.658	3.277	5.773	9.612	15.64	25.77	45.59
0.2	.2576	.5817	.9803	1.462	2.718	4.466	6.918	10.50	16.24	27.21
0.3	.2445	.5449	.9050	1.329	2.393	3.810	5.725	8.444	12.70	20.75
0.4	.2334	.5150	.8464	1.230	2.172	3.395	5.015	7.279	10.78	17.36
0.5	.2238	.4898	.7990	1.153	2.008	3.101	4.531	6.510	9.554	15.25
0.6	.2153	.4683	.7595	1.090	1.879	2.878	4.174	5.957	8.687	13.79
0.7	.2077	.4497	.7258	1.037	1.775	2.702	3.897	5.534	8.037	12.71
0.8	.2010	.4332	.6966	.9920	1.688	2.557	3.674	5.199	7.528	11.88
0.9	.1949	.4186	.6710	.9529	1.614	2.436	3.489	4.925	7.117	11.22
1.0	.1893	.4054	.6482	.9185	1.550	2.333	3.333	4.696	6.776	10.67

TABLE 7.26: I_1^* (α_2 = .5, μ = .1)

α_1 \ Q:	0.05	0.10	0.15	0.20	0.30	0.40	0.50	0.60	0.70	0.80
0.1	.2258	.4822	.7752	1.114	1.998	3.393	5.967	11.05	20.77	40.60
0.2	.2157	.4591	.7339	1.046	1.822	2.917	4.614	7.486	12.70	23.40
0.3	.2072	.4401	.7017	.9959	1.709	2.668	4.049	6.222	10.01	17.73
0.4	.1997	.4242	.6755	.9566	1.630	2.509	3.728	5.566	8.672	14.93
0.5	.1932	.4106	.6536	.9249	1.569	2.397	3.518	5.163	7.877	13.28
0.6	.1874	.3987	.6350	.8984	1.521	2.314	3.370	4.890	7.353	12.20
0.7	.1822	.3882	.6189	.8759	1.482	2.248	3.259	4.692	6.981	11.44
0.8	.1775	.3789	.6047	.8566	1.450	2.196	3.172	4.542	6.705	10.88
0.9	.1732	.3705	.5922	.8396	1.422	2.152	3.102	4.425	6.492	10.45
1.0	.1693	.3629	.5809	.8246	1.398	2.116	3.045	4.331	6.323	10.11

TABLE 7.26: I_1^* (α_2 = .5, μ = .3)

α_1 \ Q:	0.05	0.10	0.15	0.20	0.30	0.40	0.50	0.60	0.70	0.80
0.1	.2304	.5034	.8293	1.223	2.310	4.088	7.231	12.95	23.48	44.90
0.2	.2195	.4765	.7782	1.135	2.068	3.449	5.590	9.087	15.22	27.66
0.3	.2103	.4547	.7388	1.070	1.917	3.116	4.887	7.657	12.39	21.93
0.4	.2023	.4364	.7071	1.020	1.810	2.904	4.480	6.893	10.95	19.08
0.5	.1953	.4208	.6807	.9803	1.729	2.754	4.210	6.411	10.08	17.37
0.6	.1891	.4073	.6583	.9471	1.665	2.641	4.016	6.077	9.488	16.24
0.7	.1835	.3953	.6389	.9189	1.613	2.552	3.869	5.831	9.063	15.43
0.8	.1785	.3847	.6220	.8945	1.570	2.481	3.753	5.642	8.741	14.83
0.9	.1740	.3752	.6069	.8733	1.533	2.421	3.659	5.492	8.489	14.36
1.0	.1698	.3666	.5935	.8545	1.501	2.371	3.582	5.370	8.286	13.99

TABLE 7.26: I_1^* (α_2 = .5, μ = .5)

α_1 \ Q:	0.05	0.10	0.15	0.20	0.30	0.40	0.50	0.60	0.70	0.80
0.1	.2349	.5233	.8794	1.323	2.592	4.705	8.340	14.65	25.89	48.54
0.2	.2231	.4923	.8172	1.210	2.270	3.868	6.337	10.30	17.11	30.78
0.3	.2132	.4673	.7697	1.130	2.073	3.437	5.465	8.625	13.98	24.71
0.4	.2046	.4465	.7318	1.068	1.935	3.163	4.954	7.708	12.35	21.62
0.5	.1971	.4288	.7005	1.019	1.831	2.968	4.611	7.119	11.33	19.74
0.6	.1905	.4135	.6740	.9779	1.749	2.821	4.362	6.705	10.63	18.47
0.7	.1846	.4001	.6511	.9434	1.683	2.706	4.171	6.395	10.12	17.55
0.8	.1792	.3882	.6312	.9137	1.627	2.611	4.019	6.154	9.723	16.85
0.9	.1744	.3775	.6135	.8877	1.579	2.533	3.896	5.960	9.413	16.31
1.0	.1699	.3678	.5978	.8648	1.538	2.466	3.792	5.801	9.160	15.87

TABLE 7.26: I_1^* (α_2 = .5, μ = .7)

α_1 \ Q:	0.05	0.10	0.15	0.20	0.30	0.40	0.50	0.60	0.70	0.80
0.1	.2392	.5417	.9237	1.407	2.806	5.108	8.929	15.30	26.34	48.21
0.2	.2265	.5062	.8491	1.268	2.399	4.082	6.612	10.54	17.11	29.99
0.3	.2159	.4778	.7929	1.170	2.155	3.558	5.593	8.672	13.74	23.64
0.4	.2067	.4544	.7485	1.096	1.986	3.225	4.989	7.626	11.94	20.34
0.5	.1987	.4345	.7120	1.037	1.859	2.988	4.580	6.944	10.80	18.30
0.6	.1916	.4174	.6813	.9883	1.759	2.809	4.280	6.457	10.00	16.91
0.7	.1853	.4024	.6549	.9473	1.678	2.667	4.048	6.089	9.413	15.89
0.8	.1796	.3891	.6319	.9122	1.610	2.551	3.863	5.800	8.954	15.11
0.9	.1745	.3773	.6116	.8815	1.551	2.454	3.710	5.565	8.587	14.49
1.0	.1698	.3666	.5935	.8545	1.501	2.371	3.582	5.370	8.286	13.99

TABLE 7.26: I_1^* (α_2 = .5, μ = .9)

α_1 \ Q:	0.05	0.10	0.15	0.20	0.30	0.40	0.50	0.60	0.70	0.80
0.1	.2433	.5582	.9604	1.470	2.923	5.222	8.865	14.74	24.74	44.47
0.2	.2297	.5178	.8725	1.303	2.439	4.054	6.369	9.820	15.43	26.28
0.3	.2182	.4859	.8074	1.188	2.153	3.462	5.264	7.866	12.01	19.92
0.4	.2084	.4597	.7563	1.102	1.957	3.086	4.605	6.764	10.16	16.60
0.5	.1999	.4376	.7147	1.033	1.810	2.818	4.157	6.039	8.977	14.54
0.6	.1924	.4187	.6798	.9774	1.695	2.615	3.826	5.517	8.146	13.12
0.7	.1857	.4022	.6500	.9305	1.601	2.454	3.569	5.119	7.525	12.08
0.8	.1797	.3876	.6240	.8904	1.523	2.322	3.361	4.804	7.039	11.27
0.9	.1743	.3746	.6012	.8554	1.456	2.211	3.190	4.546	6.648	10.63
1.0	.1693	.3629	.5809	.8246	1.398	2.116	3.045	4.331	6.323	10.11

TABLE 7.26: I_1^* (α_2 = .6, μ = .1)

α_1 \ Q:	0.05	0.10	0.15	0.20	0.30	0.40	0.50	0.60	0.70	0.80
0.1	.2070	.4439	.7170	1.036	1.879	3.244	5.810	10.91	20.64	40.44
0.2	.1977	.4224	.6780	.9708	1.708	2.770	4.447	7.313	12.52	23.19
0.3	.1898	.4047	.6475	.9227	1.598	2.520	3.873	6.027	9.798	17.48
0.4	.1829	.3897	.6226	.8849	1.519	2.361	3.545	5.355	8.436	14.65
0.5	.1769	.3769	.6018	.8542	1.459	2.248	3.331	4.940	7.621	12.98
0.6	.1715	.3657	.5840	.8286	1.412	2.163	3.178	4.658	7.081	11.88
0.7	.1666	.3558	.5685	.8068	1.373	2.097	3.063	4.453	6.697	11.10
0.8	.1623	.3470	.5550	.7879	1.340	2.043	2.974	4.297	6.410	10.52
0.9	.1583	.3390	.5429	.7713	1.313	1.999	2.902	4.175	6.189	10.07
1.0	.1547	.3318	.5321	.7567	1.289	1.962	2.843	4.076	6.013	9.722

TABLE 7.26: I_1^* (α_2 = .6, μ = .3)

α_1 \ Q:	0.05	0.10	0.15	0.20	0.30	0.40	0.50	0.60	0.70	0.80
0.1	.2109	.4619	.7630	1.129	2.148	3.843	6.892	12.52	22.90	44.02
0.2	.2009	.4371	.7159	1.047	1.921	3.229	5.283	8.669	14.63	26.74
0.3	.1924	.4170	.6792	.9864	1.777	2.906	4.591	7.244	11.80	20.98
0.4	.1851	.4001	.6495	.9395	1.674	2.700	4.189	6.481	10.35	18.12
0.5	.1786	.3856	.6248	.9016	1.597	2.554	3.922	5.999	9.473	16.40
0.6	.1729	.3729	.6038	.8701	1.535	2.443	3.729	5.664	8.878	15.25
0.7	.1678	.3618	.5855	.8433	1.485	2.356	3.583	5.417	8.448	14.44
0.8	.1613	.3519	.5695	.8201	1.443	2.285	3.467	5.227	8.123	13.83
0.9	.1589	.3430	.5553	.7998	1.407	2.227	3.374	5.076	7.868	13.35
1.0	.1550	.3350	.5426	.7818	1.376	2.177	3.296	4.953	7.663	12.97

TABLE 7.26: I_1^* (α_2 = .6, μ = .5)

α_1 \ Q:	0.05	0.10	0.15	0.20	0.30	0.40	0.50	0.60	0.70	0.80
0.1	.2146	.4786	.8054	1.213	2.390	4.376	7.854	13.98	24.96	47.09
0.2	.2039	.4505	.7489	1.111	2.094	3.591	5.930	9.711	16.25	29.38
0.3	.1949	.4277	.7055	1.037	1.911	3.184	5.090	8.078	13.16	23.34
0.4	.1870	.4086	.6706	.9802	1.782	2.923	4.598	7.182	11.54	20.27
0.5	.1802	.3923	.6416	.9344	1.604	2.738	4.266	6.606	10.54	18.39
0.6	.1741	.3782	.6171	.8964	1.607	2.598	4.025	6.200	9.845	17.13
0.7	.1686	.3659	.5959	.0641	1.544	2.487	3.840	5.896	9.339	16.22
0.8	.1637	.3548	.5774	.8363	1.491	2.397	3.693	5.660	8.951	15.52
0.9	.1593	.3449	.5609	.8120	1.446	2.321	3.573	5.470	8.643	14.98
1.0	.1552	.3360	.5462	.7905	1.407	2.257	3.472	5.314	8.393	14.54

TABLE 7.26: I_1^* (α_2 = .6, μ = .7)

α_1 \ Q:	0.05	0.10	0.15	0.20	0.30	0.40	0.50	0.60	0.70	0.80
0.1	.2182	.4940	.8425	1.284	2.573	4.727	8.376	14.56	25.37	46.85
0.2	.2068	.4621	.7758	1.160	2.206	3.780	6.178	9.944	16.28	28.76
0.3	.1971	.4365	.7251	1.071	1.983	3.292	5.210	8.136	12.99	22.48
0.4	.1887	.4152	.6847	1.004	1.827	2.981	4.636	7.128	11.22	19.23
0.5	.1814	.3971	.6514	.9499	1.709	2.759	4.247	6.470	10.11	17.22
0.6	.1750	.3815	.6233	.9054	1.617	2.590	3.962	6.001	9.335	15.84
0.7	.1692	.3678	.5991	.8677	1.541	2.457	3.741	5.646	8.760	14.84
0.8	.1640	.3557	.5780	.8352	1.477	2.347	3.564	5.367	8.313	14.07
0.9	.1593	.3448	.5593	.8069	1.423	2.255	3.418	5.141	7.956	13.47
1.0	.1550	.3350	.5426	.7818	1.376	2.177	3.296	4.953	7.663	12.97

TABLE 7.26: I_1^* (α_2 = .6, μ = .9)

α_1 \ Q:	0.05	0.10	0.15	0.20	0.30	0.40	0.50	0.60	0.70	0.80
0.1	.2216	.5076	.8729	1.337	2.673	4.830	8.334	14.09	24.03	43.71
0.2	.2094	.4718	.7954	1.190	2.241	3.761	5.979	9.341	14.87	25.64
0.3	.1990	.4432	.7373	1.087	1.982	3.214	4.936	7.461	11.52	19.35
0.4	.1902	.4197	.6913	1.009	1.803	2.865	4.315	6.403	9.723	16.09
0.5	.1825	.3997	.6537	.9472	1.669	2.616	3.892	5.708	8.577	14.06
0.6	.1757	.3826	.6221	.8964	1.563	2.427	3.579	5.209	7.773	12.67
0.7	.1696	.3676	.5950	.8536	1.477	2.277	3.337	4.829	7.172	11.64
0.8	.1641	.3544	.5714	.8169	1.404	2.154	3.141	4.528	6.703	10.86
0.9	.1592	.3425	.5506	.7849	1.343	2.050	2.979	4.282	6.325	10.23
1.0	.1547	.3318	.5321	.7567	1.289	1.962	2.843	4.076	6.013	9.722

TABLE 7.26: I_1^* (α_2 = .7, μ = .1)

α_1 \ Q:	0.05	0.10	0.15	0.20	0.30	0.40	0.50	0.60	0.70	0.80
0.1	.1924	.4143	.6720	.9753	1.789	3.132	5.697	10.81	20.55	40.32
0.2	.1837	.3940	.6349	.9129	1.622	2.660	4.324	7.189	12.40	23.04
0.3	.1763	.3772	.6057	.8664	1.513	2.410	3.743	5.886	9.648	17.31
0.4	.1699	.3630	.5819	.8298	1.435	2.249	3.411	5.202	8.267	14.46
0.5	.1642	.3509	.5619	.8000	1.376	2.136	3.192	4.778	7.437	12.76
0.6	.1591	.3402	.5447	.7751	1.329	2.050	3.036	4.489	6.885	11.64
0.7	.1546	.3308	.5299	.7538	1.290	1.983	2.918	4.278	6.491	10.85
0.8	.1505	.3224	.5167	.7353	1.258	1.928	2.826	4.117	6.196	10.26
0.9	.1468	.3148	.5051	.7191	1.230	1.883	2.752	3.991	5.968	9.806
1.0	.1433	.3080	.4946	.7048	1.206	1.845	2.691	3.888	5.786	9.446

TABLE 7.26: I_1^* (α_2 = .7, μ = .3)

α_1 \ Q:	0.05	0.10	0.15	0.20	0.30	0.40	0.50	0.60	0.70	0.80
0.1	.1957	.4299	.7123	1.057	2.027	3.661	6.644	12.20	22.49	43.39
0.2	.1005	.4068	.6681	.9796	1.809	3.065	5.057	8.366	14.21	26.08
0.3	.1786	.3879	.6335	.9225	1.671	2.751	4.373	6.945	11.37	20.31
0.4	.1717	.3721	.6055	.8779	1.572	2.548	3.975	6.182	9.922	17.43
0.5	.1657	.3584	.5820	.8417	1.497	2.405	3.710	5.699	9.037	15.70
0.6	.1604	.3466	.5621	.8115	1.437	2.296	3.518	5.364	8.438	14.54
0.7	.1556	.3360	.5447	.7857	1.388	2.210	3.372	5.116	8.005	13.72
0.8	.1512	.3267	.5295	.7635	1.347	2.140	3.257	4.925	7.677	13.11
0.9	.1473	.3183	.5159	.7439	1.312	2.082	3.163	4.773	7.419	12.63
1.0	.1437	.3107	.5038	.7266	1.281	2.033	3.085	4.648	7.212	12.24

TABLE 7.26: I_1^* (α_2 = .7, μ = .5)

α_1 \ Q:	0.05	0.10	0.15	0.20	0.30	0.40	0.50	0.60	0.70	0.80
0.1	.1990	.4444	.7491	1.131	2.239	4.131	7.494	13.49	24.28	46.04
0.2	.1891	.4184	.6968	1.036	1.962	3.385	5.629	9.283	15.62	28.37
0.3	.1807	.3972	.6564	.9670	1.789	2.996	4.815	7.678	12.56	22.36
0.4	.1734	.3795	.6238	.9135	1.667	2.746	4.336	6.798	10.96	19.29
0.5	.1670	.3643	.5967	.8703	1.574	2.567	4.013	6.232	9.965	17.43
0.6	.1614	.3611	.5737	.8345	1.500	2.432	3.778	5.832	9.280	16.17
0.7	.1563	.3395	.5538	.8040	1.440	2.325	3.597	5.534	8.778	15.26
0.8	.1517	.3292	.5363	.7776	1.389	2.237	3.453	5.301	8.393	14.57
0.9	.1476	.3200	.5208	.7546	1.346	2.164	3.336	5.114	8.088	14.03
1.0	.1438	.3116	.5069	.7341	1.308	2.102	3.237	4.959	7.839	13.59

TABLE 7.26: I_1^* (α_2 = .7, μ = .7)

α_1 \ Q:	0.05	0.10	0.15	0.20	0.30	0.40	0.50	0.60	0.70	0.80
0.1	.2020	.4576	.7811	1.192	2.399	4.443	7.964	14.01	24.66	45.86
0.2	.1915	.4284	.7201	1.078	2.061	3.554	5.855	9.502	15.68	27.86
0.3	.1826	.4048	.6734	.9969	1.853	3.094	4.926	7.742	12.43	21.64
0.4	.1749	.3852	.6361	.9342	1.707	2.798	4.375	6.761	10.70	18.42
0.5	.1681	.3684	.6052	.8840	1.596	2.587	4.001	6.122	9.610	16.43
0.6	.1622	.3539	.5791	.8424	1.509	2.427	3.727	5.667	8.849	15.08
0.7	.1568	.3412	.5566	.8071	1.438	2.300	3.514	5.322	8.285	14.09
0.8	.1520	.3299	.5368	.7767	1.378	2.195	3.344	5.051	7.849	13.33
0.9	.1477	.3198	.5194	.7501	1.326	2.108	3.203	4.831	7.499	12.73
1.0	.1437	.3107	.5038	.7266	1.281	2.033	3.085	4.648	7.212	12.24

TABLE 7.26: I_1^* (α_2 = .7, μ = .9)

α_1 \ Q:	0.05	0.10	0.15	0.20	0.30	0.40	0.50	0.60	0.70	0.80
0.1	.2049	.4692	.8070	1.237	2.486	4.537	7.936	13.62	23.50	43.16
0.2	.1937	.4367	.7369	1.104	2.092	3.540	5.687	8.983	14.46	25.17
0.3	.1843	.4106	.6839	1.010	1.853	3.028	4.691	7.159	11.17	18.94
0.4	.1761	.3890	.6418	.9386	1.687	2.699	4.098	6.135	9.404	15.71
0.5	.1690	.3707	.6072	.8816	1.562	2.465	3.694	5.463	8.284	13.71
0.6	.1627	.3549	.5781	.8346	1.463	2.286	3.395	4.981	7.499	12.34
0.7	.1571	.3411	.5530	.7950	1.382	2.144	3.163	4.614	6.913	11.33
0.8	.1521	.3288	.5311	.7609	1.314	2.027	2.977	4.324	6.457	10.56
0.9	.1475	.3178	.5119	.7311	1.256	1.929	2.822	4.087	6.089	9.947
1.0	.1433	.3080	.4946	.7048	1.206	1.845	2.691	3.888	5.786	9.446

TABLE 7.26: I_1^* (α_2 = .8, μ = .1)

α_1 \ Q:	0.05	0.10	0.15	0.20	0.30	0.40	0.50	0.60	0.70	0.80
0.1	.1806	.3905	.6360	.9271	1.718	3.045	5.610	10.74	20.48	40.24
0.2	.1724	.3712	.6004	.8667	1.553	2.574	4.230	7.096	12.30	22.93
0.3	.1654	.3552	.5723	.8216	1.446	2.324	3.644	5.780	9.536	17.17
0.4	.1594	.3416	.5493	.7859	1.369	2.162	3.307	5.087	8.141	14.31
0.5	.1540	.3300	.5300	.7568	1.310	2.048	3.085	4.655	7.299	12.60
0.6	.1492	.3198	.5134	.7324	1.263	1.961	2.926	4.359	6.737	11.47
0.7	.1449	.3108	.4990	.7116	1.224	1.894	2.000	4.143	6.336	10.66
0.8	.1410	.3027	.4862	.6935	1.192	1.838	2.712	3.979	6.034	10.07
0.9	.1375	.2955	.4749	.6776	1.164	1.793	2.636	3.849	5.799	9.604
1.0	.1342	.2889	.4648	.6636	1.140	1.754	2.574	3.744	5.612	9.237

TABLE 7.26: $I_1^* (\alpha_2 = .8, \mu = .3)$

α_1 \ Q:	0.05	0.10	0.15	0.20	0.30	0.40	0.50	0.60	0.70	0.80
0.1	.1836	.4044	.6720	1.000	1.931	3.519	6.453	11.96	22.18	42.91
0.2	.1749	.3826	.6300	.9265	1.722	2.937	4.884	8.136	13.90	25.59
0.3	.1675	.3647	.5972	.8717	1.588	2.629	4.205	6.717	11.05	19.80
0.4	.1610	.3497	.5704	.8290	1.492	2.431	3.810	5.954	9.597	16.91
0.5	.1553	.3367	.5480	.7941	1.419	2.289	3.546	5.471	8.707	15.17
0.6	.1503	.3254	.5288	.7650	1.360	2.181	3.355	5.135	8.105	14.01
0.7	.1458	.3154	.5122	.7401	1.312	2.096	3.210	4.886	7.669	13.18
0.8	.1417	.3065	.4976	.7186	1.272	2.027	3.095	4.694	7.338	12.56
0.9	.1379	.2985	.4846	.6997	1.237	1.970	3.001	4.541	7.079	12.08
1.0	.1345	.2913	.4729	.6829	1.208	1.921	2.923	4.416	6.870	11.69

TABLE 7.26: $I_1^* (\alpha_2 = .8, \mu = .5)$

α_1 \ Q:	0.05	0.10	0.15	0.20	0.30	0.40	0.50	0.60	0.70	0.80
0.1	.1864	.4172	.7045	1.066	2.120	3.941	7.217	13.11	23.77	45.26
0.2	.1772	.3928	.6555	.9767	1.859	3.225	5.397	8.957	15.15	27.61
0.3	.1693	.3729	.6175	.9115	1.694	2.850	4.602	7.373	12.11	21.61
0.4	.1625	.3562	.5867	.8607	1.577	2.608	4.134	6.505	10.52	18.56
0.5	.1565	.3419	.5610	.8197	1.488	2.435	3.818	5.946	9.531	16.70
0.6	.1512	.3295	.5392	.7855	1.417	2.303	3.588	5.552	8.851	15.44
0.7	.1464	.3185	.5202	.7564	1.359	2.190	3.411	5.257	8.353	14.54
0.8	.1421	.3088	.5036	.7312	1.310	2.114	3.270	5.027	7.971	13.85
0.9	.1382	.3000	.4889	.7092	1.268	2.042	3.154	4.842	7.668	13.31
1.0	.1346	.2921	.4756	.6896	1.231	1.982	3.057	4.689	7.421	12.88

TABLE 7.26: $I_1^* (\alpha_2 = .8, \mu = .7)$

α_1 \ Q:	0.05	0.10	0.15	0.20	0.30	0.40	0.50	0.60	0.70	0.80
0.1	.1891	.4288	.7326	1.120	2.263	4.222	7.644	13.59	24.12	45.12
0.2	.1793	.4016	.6760	1.014	1.947	3.378	5.605	9.162	15.21	27.19
0.3	.1710	.3796	.6325	.9380	1.751	2.939	4.706	7.438	12.01	21.00
0.4	.1638	.3612	.5976	.8792	1.613	2.656	4.173	6.480	10.30	17.81
0.5	.1575	.3456	.5686	.8318	1.508	2.454	3.811	5.855	9.227	15.84
0.6	.1519	.3320	.5440	.7926	1.425	2.300	3.545	5.410	8.479	14.50
0.7	.1469	.3200	.5227	.7592	1.357	2.178	3.339	5.074	7.925	13.52
0.8	.1424	.3094	.5041	.7305	1.299	2.077	3.174	4.809	7.495	12.77
0.9	.1383	.2999	.4876	.7053	1.250	1.993	3.038	4.594	7.152	12.17
1.0	.1345	.2913	.4729	.6829	1.208	1.921	2.923	4.416	6.870	11.69

TABLE 7.26: $I_1^* (\alpha_2 = .8, \mu = .9)$

α_1 \ Q:	0.05	0.10	0.15	0.20	0.30	0.40	0.50	0.60	0.70	0.80
0.1	.1917	.4389	.7552	1.159	2.340	4.307	7.624	13.25	23.10	42.74
0.2	.1813	.4088	.6907	1.037	1.974	3.367	5.459	8.706	14.14	24.82
0.3	.1725	.3847	.6416	.9496	1.751	2.881	4.500	6.925	10.89	18.62
0.4	.1649	.3646	.6025	.8830	1.595	2.569	3.929	5.927	9.159	15.42
0.5	.1582	.3475	.5702	.8297	1.477	2.345	3.539	5.273	8.058	13.44
0.6	.1524	.3328	.5430	.7856	1.384	2.175	3.252	4.804	7.289	12.09
0.7	.1471	.3198	.5195	.7484	1.307	2.040	3.028	4.448	6.715	11.10
0.8	.1424	.3084	.4991	.7164	1.243	1.928	2.848	4.166	6.268	10.33
0.9	.1381	.2981	.4809	.6883	1.188	1.834	2.699	3.936	5.909	9.730
1.0	.1342	.2889	.4648	.6636	1.140	1.754	2.574	3.744	5.612	9.237

TABLE 7.26: $I_1^* (\alpha_2 = .9, \mu = .1)$

α_1 \ Q:	0.05	0.10	0.15	0.20	0.30	0.40	0.50	0.60	0.70	0.80
0.1	.1709	.3709	.6064	.8875	1.660	2.976	5.542	10.68	20.42	40.17
0.2	.1631	.3523	.5720	.8288	1.498	2.505	4.155	7.023	12.23	22.85
0.3	.1565	.3370	.5448	.7847	1.392	2.254	3.565	5.697	9.449	17.07
0.4	.1507	.3240	.5225	.7499	1.315	2.092	3.224	4.996	8.042	14.19
0.5	.1456	.3128	.5037	.7214	1.256	1.977	2.999	4.558	7.191	12.47
0.6	.1410	.3030	.4876	.6975	1.210	1.890	2.838	4.257	6.621	11.33
0.7	.1369	.2943	.4736	.6770	1.171	1.821	2.716	4.037	6.213	10.52
0.8	.1332	.2865	.4612	.6593	1.139	1.766	2.620	3.869	5.906	9.915
0.9	.1298	.2795	.4502	.6437	1.111	1.720	2.543	3.736	5.667	9.446
1.0	.1267	.2731	.4403	.6298	1.087	1.680	2.479	3.628	5.476	9.073

TABLE 7.26: $I_1^* (\alpha_2 = .9, \mu = .3)$

α_1 \ Q:	0.05	0.10	0.15	0.20	0.30	0.40	0.50	0.60	0.70	0.80
0.1	.1736	.3834	.6389	.9539	1.854	3.406	6.302	11.77	21.94	42.55
0.2	.1653	.3627	.5988	.8831	1.651	2.835	4.746	7.955	13.65	25.21
0.3	.1583	.3456	.5673	.8304	1.521	2.532	4.073	6.538	10.80	19.41
0.4	.1522	.3312	.5416	.7890	1.427	2.336	3.680	5.775	9.343	16.51
0.5	.1468	.3188	.5200	.7553	1.355	2.196	3.417	5.291	8.449	14.76
0.6	.1420	.3080	.5016	.7270	1.298	2.090	3.226	4.954	7.844	13.60
0.7	.1377	.2985	.4856	.7029	1.251	2.005	3.081	4.705	7.406	12.76
0.8	.1338	.2900	.4714	.6820	1.211	1.937	2.966	4.512	7.073	12.14
0.9	.1302	.2823	.4589	.6636	1.177	1.879	2.872	4.358	6.812	11.65
1.0	.1270	.2753	.4476	.6473	1.148	1.831	2.794	4.232	6.601	11.26

TABLE 7.26: $I_1^* (\alpha_2 = .9, \mu = .5)$

α_1 \ Q:	0.05	0.10	0.15	0.20	0.30	0.40	0.50	0.60	0.70	0.80
0.1	.1761	.3949	.6682	1.013	2.025	3.789	6.995	12.81	23.36	44.64
0.2	.1674	.3718	.6218	.9284	1.775	3.097	5.213	8.699	14.78	27.02
0.3	.1599	.3530	.5856	.8662	1.617	2.733	4.433	7.132	11.76	21.03
0.4	.1535	.3371	.5563	.8178	1.504	2.497	3.973	6.274	10.18	17.99
0.5	.1478	.3235	.5318	.7784	1.418	2.329	3.663	5.721	9.192	16.13
0.6	.1428	.3117	.5109	.7456	1.349	2.200	3.437	5.331	8.516	14.88
0.7	.1383	.3013	.4928	.7177	1.293	2.098	3.262	5.039	8.020	13.97
0.8	.1342	.2920	.4769	.6934	1.245	2.015	3.123	4.811	7.640	13.29
0.9	.1305	.2836	.4628	.6722	1.205	1.945	3.009	4.628	7.338	12.75
1.0	.1271	.2760	.4501	.6533	1.169	1.886	2.914	4.477	7.093	12.32

TABLE 7.26: $I_1^* (\alpha_2 = .9, \mu = .7)$

α_1 \ Q:	0.05	0.10	0.15	0.20	0.30	0.40	0.50	0.60	0.70	0.80
0.1	.1785	.4052	.6932	1.061	2.154	4.044	7.388	13.26	23.69	44.53
0.2	.1693	.3797	.6401	.9621	1.855	3.237	5.405	8.891	14.85	26.65
0.3	.1614	.3589	.5991	.8901	1.669	2.815	4.530	7.197	11.67	20.50
0.4	.1546	.3416	.5660	.8344	1.537	2.542	4.012	6.256	9.988	17.32
0.5	.1487	.3268	.5386	.7894	1.436	2.347	3.659	5.644	8.926	15.37
0.6	.1434	.3139	.5152	.7520	1.357	2.198	3.400	5.207	8.188	14.04
0.7	.1387	.3026	.4950	.7202	1.291	2.080	3.199	4.877	7.641	13.07
0.8	.1344	.2925	.4774	.6927	1.236	1.982	3.038	4.618	7.218	12.33
0.9	.1305	.2835	.4617	.6687	1.189	1.901	2.906	4.407	6.879	11.74
1.0	.1270	.2753	.4476	.6473	1.148	1.831	2.794	4.232	6.601	11.26

TABLE 7.26: $I_1^* (\alpha_2 = .9, \mu = .9)$

α_1 \ Q:	0.05	0.10	0.15	0.20	0.30	0.40	0.50	0.60	0.70	0.80
0.1	.1807	.4141	.7131	1.096	2.222	4.123	7.374	12.95	22.78	42.41
0.2	.1710	.3861	.6531	.9820	1.880	3.228	5.275	8.484	13.89	24.54
0.3	.1627	.3634	.6072	.9003	1.669	2.763	4.346	6.737	10.68	18.38
0.4	.1556	.3445	.5704	.8376	1.521	2.464	3.793	5.761	8.964	15.20
0.5	.1494	.3285	.5400	.7873	1.409	2.249	3.415	5.122	7.880	13.23
0.6	.1438	.3146	.5143	.7457	1.320	2.086	3.136	4.663	7.122	11.89
0.7	.1389	.3024	.4921	.7104	1.247	1.955	2.920	4.316	6.558	10.91
0.8	.1345	.2916	.4728	.6800	1.185	1.848	2.745	4.040	6.120	10.16
0.9	.1304	.2819	.4556	.6534	1.133	1.758	2.601	3.816	5.767	9.560
1.0	.1267	.2731	.4403	.6298	1.087	1.680	2.479	3.628	5.476	9.073

TABLE 7.26: I_1^* ($\alpha_2 = 1.0$, $\mu = .1$)

α_1 \ Q:	0.05	0.10	0.15	0.20	0.30	0.40	0.50	0.60	0.70	0.80
0.1	.1627	.3543	.5814	.8543	1.612	2.918	5.487	10.63	20.38	40.12
0.2	.1553	.3364	.5481	.7970	1.451	2.449	4.095	6.964	12.17	22.78
0.3	.1489	.3216	.5216	.7539	1.346	2.197	3.500	5.630	9.380	16.99
0.4	.1433	.3091	.4999	.7197	1.270	2.035	3.157	4.922	7.963	14.10
0.5	.1384	.2983	.4816	.6917	1.212	1.919	2.930	4.479	7.105	12.37
0.6	.1341	.2888	.4659	.6682	1.165	1.831	2.766	4.174	6.580	11.22
0.7	.1302	.2804	.4522	.6481	1.127	1.762	2.643	3.951	6.114	10.40
0.8	.1266	.2729	.4402	.6306	1.094	1.706	2.646	3.780	5.803	9.793
0.9	.1234	.2661	.4294	.6152	1.067	1.659	2.467	3.644	5.560	9.319
1.0	.1204	.2599	.4197	.6016	1.043	1.619	2.401	3.534	5.365	8.942

TABLE 7.26: I_1^* ($\alpha_2 = 1.0$, $\mu = .3$)

α_1 \ Q:	0.05	0.10	0.15	0.20	0.30	0.40	0.50	0.60	0.70	0.80
0.1	.1651	.3657	.6112	.9152	1.790	3.313	6.179	11.62	21.74	42.26
0.2	.1573	.3459	.5726	.8468	1.593	2.752	4.634	7.809	13.45	24.90
0.3	.1505	.3295	.5423	.7958	1.465	2.452	3.964	6.393	10.60	19.10
0.4	.1447	.3157	.5174	.7556	1.373	2.258	3.573	5.630	9.138	16.19
0.5	.1396	.3038	.4966	.7228	1.302	2.119	3.311	5.146	8.242	14.43
0.6	.1350	.2934	.4787	.6953	1.246	2.014	3.121	4.808	7.634	13.27
0.7	.1309	.2842	.4632	.6718	1.200	1.930	2.975	4.558	7.194	12.43
0.8	.1271	.2760	.4495	.6515	1.161	1.862	2.860	4.364	6.859	11.80
0.9	.1238	.2686	.4374	.6335	1.127	1.805	2.766	4.209	6.596	11.31
1.0	.1206	.2619	.4264	.6176	1.098	1.757	2.688	4.083	6.384	10.92

TABLE 7.26: I_1^* ($\alpha_2 = 1.0$, $\mu = .5$)

α_1 \ Q:	0.05	0.10	0.15	0.20	0.30	0.40	0.50	0.60	0.70	0.80
0.1	.1674	.3761	.6378	.9690	1.947	3.664	6.815	12.57	23.04	44.15
0.2	.1591	.3542	.5936	.8882	1.706	2.992	5.063	8.489	14.48	26.54
0.3	.1520	.3362	.5590	.8286	1.553	2.636	4.295	6.937	11.47	20.57
0.4	.1459	.3211	.5309	.7819	1.444	2.406	3.843	6.086	9.896	17.53
0.5	.1405	.3081	.5073	.7440	1.360	2.241	3.537	5.539	8.917	15.68
0.6	.1357	.2968	.4873	.7123	1.293	2.116	3.314	5.152	8.245	14.43
0.7	.1314	.2868	.4699	.6853	1.238	2.016	3.142	4.863	7.752	13.52
0.8	.1275	.2779	.4545	.6619	1.192	1.934	3.004	4.637	7.373	12.84
0.9	.1240	.2698	.4409	.6413	1.152	1.865	2.892	4.455	7.073	12.30
1.0	.1207	.2626	.4287	.6230	1.118	1.807	2.797	4.304	6.828	11.87

TABLE 7.26: I_1^* ($\alpha_2 = 1.0$, $\mu = .7$)

α_1 \ Q:	0.05	0.10	0.15	0.20	0.30	0.40	0.50	0.60	0.70	0.80
0.1	.1696	.3854	.6604	1.013	2.064	3.899	7.177	12.98	23.35	44.06
0.2	.1608	.3613	.6101	.9188	1.779	3.121	5.241	8.670	14.55	26.22
0.3	.1534	.3416	.5712	.8502	1.601	2.712	4.386	7.001	11.40	20.09
0.4	.1469	.3251	.5397	.7970	1.474	2.448	3.879	6.074	9.733	16.94
0.5	.1413	.3110	.5135	.7540	1.377	2.259	3.535	5.471	8.682	15.00
0.6	.1367	.2997	.4912	.7181	1.300	2.114	3.281	5.042	7.952	13.67
0.7	.1318	.2880	.4719	.6876	1.237	1.999	3.085	4.718	7.412	12.71
0.8	.1277	.2783	.4549	.6613	1.184	1.904	2.928	4.462	6.993	11.97
0.9	.1240	.2697	.4399	.6381	1.138	1.825	2.798	4.255	6.659	11.39
1.0	.1206	.2619	.4264	.6176	1.098	1.757	2.688	4.083	6.384	10.92

TABLE 7.26: I_1^* ($\alpha_2 = 1.0$, $\mu = .9$)

α_1 \ Q:	0.05	0.10	0.15	0.20	0.30	0.40	0.50	0.60	0.70	0.80
0.1	.1716	.3934	.6782	1.044	2.126	3.971	7.168	12.71	22.52	42.14
0.2	.1624	.3670	.6217	.9366	1.802	3.113	5.124	8.302	13.68	24.31
0.3	.1545	.3456	.5784	.8594	1.601	2.666	4.220	6.584	10.50	18.18
0.4	.1478	.3277	.5436	.7999	1.459	2.377	3.681	5.625	8.806	15.02
0.5	.1419	.3125	.5147	.7520	1.352	2.170	3.313	4.998	7.735	13.07
0.6	.1366	.2993	.4903	.7123	1.266	2.012	3.042	4.549	6.987	11.73
0.7	.1320	.2878	.4692	.6786	1.196	1.885	2.831	4.208	6.431	10.76
0.8	.1277	.2775	.4507	.6495	1.137	1.782	2.661	3.938	5.999	10.01
0.9	.1239	.2683	.4343	.6241	1.087	1.694	2.520	3.718	5.652	9.423
1.0	.1204	.2599	.4197	.6016	1.043	1.619	2.401	3.534	5.365	8.942

TABLE 7.26: I_1^* ($1/\alpha_2 = .1$, $\mu = .1$)

α_1 \ Q:	0.05	0.10	0.15	0.20	0.30	0.40	0.50	0.60	0.70	0.80
0.1	.0601	.1520	.2864	.4771	1.116	2.389	5.022	10.26	20.03	39.69
0.2	.0570	.1423	.2646	.4338	.9685	1.913	3.575	6.489	11.71	22.24
0.3	.0542	.1342	.2470	.4005	.8691	1.651	2.937	5.078	8.824	16.35
0.4	.0519	.1274	.2325	.3738	.7957	1.477	2.558	4.309	7.328	13.37
0.5	.0498	.1214	.2202	.3517	.7386	1.350	2.301	3.816	6.401	11.57
0.6	.0479	.1162	.2096	.3330	.6923	1.252	2.112	3.467	5.765	10.35
0.7	.0463	.1116	.2004	.3170	.6539	1.174	1.966	3.205	5.299	9.468
0.8	.0448	.1075	.1923	.3031	.6212	1.109	1.849	3.001	4.942	8.802
0.9	.0434	.1039	.1851	.2908	.5931	1.055	1.752	2.835	4.658	8.279
1.0	.0421	.1005	.1786	.2798	.5684	1.008	1.670	2.698	4.426	7.858

TABLE 7.26: I_1^* ($1/\alpha_2 = .1$, $\mu = .3$)

α_1 \ Q:	0.05	0.10	0.15	0.20	0.30	0.40	0.50	0.60	0.70	0.80
0.1	.0604	.1537	.2910	.4865	1.141	2.438	5.100	10.36	20.17	39.90
0.2	.0572	.1437	.2685	.4417	.9893	1.953	3.638	6.579	11.84	22.45
0.3	.0545	.1355	.2504	.4073	.8870	1.685	2.992	5.161	8.950	16.56
0.4	.0521	.1284	.2354	.3798	.8116	1.507	2.609	4.388	7.450	13.58
0.5	.0499	.1223	.2228	.3570	.7528	1.378	2.348	3.890	6.521	11.77
0.6	.0481	.1170	.2119	.3378	.7052	1.278	2.157	3.539	5.882	10.56
0.7	.0464	.1123	.2024	.3212	.6656	1.198	2.008	3.275	5.415	9.675
0.8	.0448	.1081	.1940	.3068	.6320	1.132	1.889	3.068	5.055	9.008
0.9	.0435	.1043	.1866	.2942	.6030	1.076	1.790	2.901	4.769	8.484
1.0	.0422	.1009	.1799	.2829	.5776	1.028	1.707	2.762	4.536	8.062

TABLE 7.26: I_1^* ($1/\alpha_2 = .1$, $\mu = .5$)

α_1 \ Q:	0.05	0.10	0.16	0.20	0.30	0.40	0.50	0.60	0.70	0.80
0.1	.0607	.1550	.2946	.4939	1.162	2.484	5.178	10.47	20.31	40.10
0.2	.0574	.1449	.2714	.4477	1.005	1.985	3.692	6.660	11.96	22.62
0.3	.0546	.1364	.2520	.4122	.8997	1.710	3.036	5.227	9.048	16.72
0.4	.0522	.1292	.2374	.3837	.8220	1.528	2.644	4.443	7.536	13.73
0.5	.0501	.1229	.2244	.3602	.7614	1.395	2.378	3.938	6.597	11.91
0.6	.0482	.1175	.2132	.3404	.7123	1.292	2.182	3.581	5.952	10.68
0.7	.0464	.1127	.2035	.3234	.6716	1.210	2.030	3.312	5.477	9.791
0.8	.0449	.1084	.1949	.3086	.6369	1.142	1.908	3.101	5.113	9.118
0.9	.0435	.1046	.1872	.2955	.6069	1.084	1.806	2.930	4.822	8.588
1.0	.0422	.1011	.1804	.2839	.5807	1.035	1.721	2.788	4.585	8.161

TABLE 7.26: I_1^* ($1/\alpha_2 = .1$, $\mu = .7$)

α_1 \ Q:	0.05	0.10	0.15	0.20	0.30	0.40	0.50	0.60	0.70	0.80
0.1	.0609	.1561	.2972	.4990	1.176	2.513	5.226	10.52	20.35	40.10
0.2	.0576	.1456	.2733	.4513	1.014	2.002	3.717	6.687	11.97	22.60
0.3	.0548	.1370	.2542	.4147	.9057	1.720	3.049	5.239	9.048	16.68
0.4	.0523	.1296	.2384	.3854	.8257	1.534	2.651	4.446	7.525	13.68
0.5	.0501	.1232	.2250	.3613	.7634	1.397	2.379	3.934	6.577	11.84
0.6	.0482	.1177	.2136	.3409	.7129	1.292	2.179	3.570	5.924	10.61
0.7	.0465	.1128	.2036	.3234	.6710	1.208	2.024	3.297	5.444	9.711
0.8	.0449	.1084	.1948	.3082	.6353	1.138	1.898	3.082	5.073	9.031
0.9	.0435	.1045	.1869	.2948	.6045	1.079	1.795	2.907	4.778	8.495
1.0	.0422	.1009	.1799	.2829	.5776	1.028	1.707	2.762	4.536	8.062

TABLE 7.26: I_1^* $(1/\alpha_2 = .1,\ \mu = .9)$

α_1 \ Q:	0.05	0.10	0.15	0.20	0.30	0.40	0.50	0.60	0.70	0.80
0.1	.0611	.1567	.2986	.5014	1.182	2.521	5.228	10.50	20.27	39.90
0.2	.0578	.1461	.2741	.4523	1.015	2.000	3.704	6.647	11.88	22.40
0.3	.0549	.1372	.2545	.4148	.9040	1.713	3.029	5.193	8.951	16.48
0.4	.0524	.1297	.2383	.3848	.8222	1.523	2.626	4.395	7.425	13.47
0.5	.0502	.1232	.2246	.3601	.7584	1.385	2.352	3.880	6.475	11.64
0.6	.0482	.1176	.2129	.3392	.7069	1.278	2.149	3.514	5.819	10.40
0.7	.0465	.1126	.2027	.3213	.6640	1.192	1.991	3.238	5.337	9.508
0.8	.0449	.1081	.1937	.3058	.6275	1.120	1.865	3.021	4.965	8.827
0.9	.0434	.1041	.1857	.2920	.5960	1.060	1.759	2.845	4.669	8.291
1.0	.0421	.1005	.1786	.2798	.5684	1.008	1.670	2.698	4.426	7.858

TABLE 7.26: I_1^* $(1/\alpha_2 = .2,\ \mu = .1)$

α_1 \ Q:	0.05	0.10	0.15	0.20	0.30	0.40	0.50	0.60	0.70	0.80
0.1	.0795	.1889	.3375	.5387	1.187	2.456	5.075	10.30	20.07	39.74
0.2	.0755	.1778	.3140	.4937	1.039	1.982	3.636	6.542	11.76	22.30
0.3	.0721	.1685	.2951	.4591	.9393	1.722	3.004	5.140	8.885	16.42
0.4	.0691	.1606	.2794	.4314	.8662	1.551	2.631	4.379	7.398	13.45
0.5	.0665	.1538	.2661	.4085	.8094	1.426	2.378	3.892	6.480	11.66
0.6	.0642	.1478	.2547	.3892	.7635	1.331	2.193	3.549	5.851	10.44
0.7	.0621	.1425	.2447	.3726	.7255	1.254	2.051	3.293	5.392	9.572
0.8	.0602	.1377	.2359	.3581	.6932	1.192	1.937	3.093	5.040	8.913
0.9	.0584	.1335	.2280	.3453	.6654	1.139	1.843	2.932	4.761	8.396
1.0	.0568	.1296	.2209	.3339	.6411	1.094	1.765	2.799	4.535	7.980

TABLE 7.26: I_1^* $(1/\alpha_2 = .2,\ \mu = .3)$

α_1 \ Q:	0.05	0.10	0.15	0.20	0.30	0.40	0.50	0.60	0.70	0.80
0.1	.0800	.1918	.3456	.5554	1.234	2.551	5.229	10.50	20.34	40.16
0.2	.0760	.1803	.3208	.5076	1.077	2.058	3.759	6.721	12.02	22.72
0.3	.0725	.1706	.3009	.4710	.9720	1.787	3.112	5.305	9.136	16.84
0.4	.0695	.1624	.2844	.4417	.8948	1.608	2.729	4.534	7.641	13.87
0.5	.0668	.1553	.2704	.4176	.8349	1.479	2.469	4.039	6.717	12.07
0.6	.0644	.1491	.2584	.3972	.7864	1.379	2.279	3.690	6.083	10.86
0.7	.0622	.1436	.2479	.3796	.7462	1.299	2.131	3.429	5.620	9.984
0.8	.0603	.1387	.2387	.3644	.7121	1.233	2.013	3.225	5.264	9.323
0.9	.0585	.1342	.2304	.3509	.6826	1.177	1.916	3.060	4.982	8.804
1.0	.0569	.1302	.2230	.3389	.6569	1.130	1.834	2.924	4.753	8.387

TABLE 7.26: I_1^* $(1/\alpha_2 = .2,\ \mu = .5)$

α_1 \ Q:	0.05	0.10	0.15	0.20	0.30	0.40	0.50	0.60	0.70	0.80
0.1	.0806	.1943	.3522	.5689	1.273	2.638	5.381	10.72	20.62	40.56
0.2	.0764	.1823	.3261	.5183	1.106	2.119	3.863	6.878	12.25	23.07
0.3	.0728	.1723	.3051	.4796	.9953	1.835	3.194	5.432	9.329	17.16
0.4	.0697	.1638	.2879	.4487	.9138	1.647	2.796	4.641	7.811	14.16
0.5	.0670	.1564	.2733	.4233	.8504	1.510	2.526	4.132	6.867	12.34
0.6	.0646	.1500	.2607	.4018	.7993	1.405	2.327	3.771	6.219	11.10
0.7	.0624	.1443	.2497	.3834	.7568	1.321	2.173	3.500	5.743	10.22
0.8	.0604	.1392	.2401	.3673	.7207	1.252	2.049	3.288	5.377	9.540
0.9	.0586	.1346	.2315	.3531	.6896	1.193	1.947	3.116	5.085	9.010
1.0	.0569	.1304	.2237	.3405	.6623	1.143	1.860	2.973	4.847	8.582

TABLE 7.26: I_1^* $(1/\alpha_2 = .2,\ \mu = .7)$

α_1 \ Q:	0.05	0.10	0.15	0.20	0.30	0.40	0.50	0.60	0.70	0.80
0.1	.0810	.1963	.3571	.5786	1.300	2.695	5.474	10.82	20.70	40.55
0.2	.0768	.1838	.3297	.5252	1.124	2.151	3.911	6.930	12.28	23.02
0.3	.0731	.1734	.3078	.4845	1.007	1.854	3.220	5.455	9.327	17.07
0.4	.0700	.1646	.2898	.4521	.9209	1.658	2.808	4.646	7.787	14.05
0.5	.0672	.1570	.2745	.4254	.8544	1.515	2.528	4.122	6.826	12.21
0.6	.0647	.1503	.2615	.4029	.8007	1.405	2.321	3.751	6.163	10.96
0.7	.0624	.1445	.2500	.3836	.7560	1.317	2.161	3.471	5.675	10.05
0.8	.0604	.1392	.2400	.3668	.7182	1.244	2.031	3.251	5.299	9.367
0.9	.0586	.1345	.2310	.3520	.6855	1.183	1.925	3.072	4.998	8.825
1.0	.0569	.1302	.2230	.3389	.6569	1.130	1.834	2.924	4.753	8.387

TABLE 7.26: I_1^* $(1/\alpha_2 = .2,\ \mu = .9)$

α_1 \ Q:	0.05	0.10	0.15	0.20	0.30	0.40	0.50	0.60	0.70	0.80
0.1	.0814	.1977	.3602	.5839	1.311	2.711	5.476	10.77	20.53	40.16
0.2	.0770	.1848	.3316	.5279	1.127	2.148	3.886	6.852	12.09	22.62
0.3	.0733	.1740	.3088	.4854	1.005	1.842	3.181	5.365	9.136	16.68
0.4	.0701	.1650	.2900	.4516	.9153	1.639	2.762	4.547	7.589	13.65
0.5	.0673	.1571	.2742	.4238	.8461	1.492	2.475	4.017	6.624	11.81
0.6	.0647	.1503	.2606	.4004	.7903	1.378	2.264	3.640	5.957	10.56
0.7	.0625	.1442	.2488	.3804	.7440	1.287	2.100	3.356	5.465	9.651
0.8	.0604	.1388	.2384	.3629	.7047	1.212	1.967	3.132	5.086	8.962
0.9	.0585	.1340	.2292	.3476	.6708	1.148	1.858	2.951	4.783	8.419
1.0	.0568	.1296	.2209	.3339	.6411	1.094	1.765	2.799	4.535	7.980

TABLE 7.26: I_1^* $(1/\alpha_2 = .3,\ \mu = .1)$

α_1 \ Q:	0.05	0.10	0.15	0.20	0.30	0.40	0.50	0.60	0.70	0.80
0.1	.0945	.2181	.3795	.5912	1.252	2.520	5.128	10.34	20.11	39.79
0.2	.0899	.2059	.3544	.5444	1.102	2.048	3.696	6.595	11.81	22.36
0.3	.0860	.1957	.3343	.5085	1.002	1.790	3.070	5.202	8.947	16.49
0.4	.0825	.1869	.3176	.4798	.9293	1.621	2.702	4.449	7.469	13.53
0.5	.0795	.1794	.3034	.4561	.8725	1.498	2.453	3.968	6.558	11.74
0.6	.0767	.1727	.2912	.4361	.8267	1.404	2.272	3.631	5.936	10.54
0.7	.0743	.1669	.2805	.4189	.7888	1.329	2.133	3.379	5.483	9.676
0.8	.0721	.1616	.2711	.4039	.7566	1.268	2.022	3.184	5.137	9.023
0.9	.0701	.1568	.2627	.3907	.7290	1.216	1.931	3.027	4.864	8.512
1.0	.0682	.1525	.2551	.3789	.7048	1.172	1.855	2.898	4.642	8.102

TABLE 7.26: I_1^* $(1/\alpha_2 = .3,\ \mu = .3)$

α_1 \ Q:	0.05	0.10	0.15	0.20	0.30	0.40	0.50	0.60	0.70	0.80
0.1	.0953	.2222	.3907	.6143	1.318	2.659	5.356	10.65	20.52	40.42
0.2	.0906	.2093	.3637	.5636	1.156	2.157	3.877	6.862	12.20	22.99
0.3	.0865	.1986	.3422	.5248	1.048	1.883	3.228	5.447	9.321	17.12
0.4	.0830	.1894	.3243	.4939	.9691	1.703	2.845	4.678	7.831	14.16
0.5	.0798	.1815	.3092	.4684	.9078	1.572	2.585	4.185	6.911	12.37
0.6	.0771	.1745	.2962	.4469	.8583	1.472	2.395	3.838	6.282	11.16
0.7	.0746	.1683	.2849	.4284	.8173	1.392	2.249	3.579	5.822	10.29
0.8	.0723	.1628	.2749	.4123	.7825	1.326	2.132	3.377	5.470	9.636
0.9	.0702	.1578	.2659	.3980	.7526	1.270	2.036	3.214	5.192	9.123
1.0	.0683	.1533	.2579	.3854	.7264	1.223	1.955	3.080	4.965	8.709

TABLE 7.26: I_1^* $(1/\alpha_2 = .3,\ \mu = .5)$

α_1 \ Q:	0.05	0.10	0.15	0.20	0.30	0.40	0.50	0.60	0.70	0.80
0.1	.0960	.2258	.3999	.6334	1.374	2.784	5.577	10.96	20.93	41.01
0.2	.0912	.2122	.3711	.5785	1.198	2.245	4.028	7.092	12.54	23.51
0.3	.0870	.2009	.3481	.5368	1.081	1.951	3.346	5.633	9.606	17.59
0.4	.0834	.1913	.3292	.5036	.9957	1.757	2.941	4.834	8.081	14.59
0.5	.0802	.1830	.3132	.4763	.9295	1.617	2.666	4.320	7.134	12.76
0.6	.0773	.1757	.2994	.4533	.8762	1.510	2.464	3.956	6.482	11.53
0.7	.0747	.1692	.2874	.4335	.8320	1.423	2.308	3.683	6.004	10.64
0.8	.0724	.1635	.2768	.4163	.7944	1.352	2.183	3.469	5.636	9.960
0.9	.0703	.1583	.2673	.4011	.7621	1.292	2.079	3.296	5.344	9.429
1.0	.0684	.1536	.2588	.3876	.7338	1.241	1.992	3.153	5.105	9.001

TABLE 7.26: I_1^* $(1/\alpha_2 = .3,\ \mu = .7)$

α_1 \ Q:	0.05	0.10	0.15	0.20	0.30	0.40	0.50	0.60	0.70	0.80
0.1	.0967	.2287	.4071	.6475	1.414	2.867	5.711	11.12	21.05	41.00
0.2	.0917	.2144	.3765	.5886	1.223	2.291	4.097	7.166	12.58	23.43
0.3	.0874	.2026	.3521	.5440	1.097	1.979	3.383	5.665	9.601	17.46
0.4	.0837	.1925	.3320	.5086	1.006	1.773	2.958	4.839	8.043	14.42
0.5	.0804	.1839	.3151	.4795	.9354	1.624	2.669	4.305	7.070	12.57
0.6	.0775	.1763	.3005	.4550	.8784	1.510	2.456	3.926	6.398	11.31
0.7	.0748	.1696	.2879	.4340	.8312	1.418	2.291	3.639	5.903	10.39
0.8	.0725	.1636	.2767	.4158	.7911	1.342	2.158	3.414	5.520	9.700
0.9	.0703	.1582	.2668	.3997	.7566	1.278	2.048	3.232	5.215	9.153
1.0	.0683	.1533	.2579	.3854	.7264	1.223	1.955	3.080	4.965	8.709

TABLE 7.26: I_1^* $(1/\alpha_2 = .3,\ \mu = .9)$

α_1 \ Q:	0.05	0.10	0.15	0.20	0.30	0.40	0.50	0.60	0.70	0.80
0.1	.0973	.2309	.4120	.6559	1.431	2.890	5.713	11.03	20.79	40.41
0.2	.0921	.2160	.3795	.5931	1.228	2.288	4.060	7.050	12.30	22.84
0.3	.0877	.2036	.3538	.5458	1.095	1.962	3.327	5.531	9.316	16.87
0.4	.0839	.1931	.3326	.5084	0.999	1.747	2.891	4.693	7.750	13.83
0.5	.0806	.1841	.3149	.4777	.9245	1.592	2.593	4.150	6.770	11.97
0.6	.0776	.1763	.2997	.4520	.8646	1.472	2.373	3.763	6.092	10.71
0.7	.0749	.1693	.2864	.4299	.8149	1.376	2.203	3.471	5.592	9.793
0.8	.0724	.1631	.2748	.4107	.7728	1.297	2.065	3.241	5.205	9.096
0.9	.0702	.1576	.2644	.3939	.7366	1.230	1.951	3.054	4.896	8.547
1.0	.0682	.1525	.2551	.3789	.7048	1.172	1.855	2.898	4.642	8.102

TABLE 7.26: I_1^* $(1/\alpha_2 = .4,\ \mu = .1)$

α_1 \ Q:	0.05	0.10	0.15	0.20	0.30	0.40	0.50	0.60	0.70	0.80
0.1	.1072	.2432	.4161	.6377	1.312	2.582	5.181	10.38	20.15	39.83
0.2	.1021	.2299	.3895	.5892	1.161	2.111	3.756	6.648	11.86	22.42
0.3	.0977	.2189	.3682	.5521	1.060	1.855	3.135	5.264	9.008	16.56
0.4	.0938	.2094	.3506	.5225	.9870	1.687	2.771	4.518	7.639	13.62
0.5	.0904	.2013	.3357	.4980	.9301	1.566	2.526	4.043	6.636	11.83
0.6	.0874	.1941	.3228	.4774	.8843	1.473	2.348	3.711	6.022	10.64
0.7	.0847	.1877	.3116	.4597	.8464	1.400	2.212	3.464	5.575	9.780
0.8	.0822	.1820	.3017	.4442	.8143	1.339	2.103	3.273	5.234	9.133
0.9	.0800	.1768	.2928	.4305	.7867	1.289	2.015	3.120	4.966	8.628
1.0	.0779	.1722	.2848	.4184	.7626	1.246	1.941	2.995	4.749	8.223

TABLE 7.26: I_1^* $(1/\alpha_2 = .4,\ \mu = .3)$

α_1 \ Q:	0.05	0.10	0.15	0.20	0.30	0.40	0.50	0.60	0.70	0.80
0.1	.1082	.2484	.4301	.6668	1.397	2.762	5.480	10.79	20.69	40.69
0.2	.1029	.2343	.4012	.6133	1.229	2.252	3.992	7.001	12.38	23.26
0.3	.0984	.2225	.3781	.5725	1.118	1.974	3.341	5.586	9.505	17.40
0.4	.0944	.2125	.3591	.5400	1.037	1.792	2.957	4.819	8.020	14.45
0.5	.0909	.2039	.3430	.5133	.9744	1.660	2.607	4.328	7.105	12.66
0.6	.0878	.1963	.3291	.4907	.9239	1.559	2.508	3.983	6.479	11.46
0.7	.0850	.1895	.3170	.4714	.8820	1.479	2.362	3.726	6.023	10.60
0.8	.0825	.1835	.3063	.4545	.8466	1.413	2.245	3.526	5.674	9.948
0.9	.0801	.1781	.2967	.4396	.8161	1.357	2.150	3.365	5.399	9.439
1.0	.0780	.1732	.2882	.4263	.7894	1.310	2.070	3.232	5.175	9.030

TABLE 7.26: I_1^* $(1/\alpha_2 = .4,\ \mu = .5)$

α_1 \ Q:	0.05	0.10	0.15	0.20	0.30	0.40	0.50	0.60	0.70	0.80
0.1	.1092	.2530	.4420	.6912	1.468	2.923	5.767	11.20	21.24	41.46
0.2	.1037	.2379	.4107	.6323	1.282	2.364	4.188	7.302	12.82	23.95
0.3	.0990	.2255	.3857	.5877	1.160	2.061	3.493	5.829	9.880	18.02
0.4	.0949	.2149	.3652	.5523	1.071	1.862	3.081	5.023	8.348	15.01
0.5	.0913	.2058	.3479	.5233	1.002	1.718	2.802	4.504	7.396	13.18
0.6	.0881	.1978	.3331	.4988	.9464	1.607	2.596	4.137	6.742	11.95
0.7	.0852	.1907	.3201	.4778	.9004	1.519	2.438	3.861	6.261	11.05
0.8	.0826	.1844	.3087	.4595	.8615	1.446	2.311	3.645	5.892	10.38
0.9	.0802	.1787	.2985	.4434	.8280	1.385	2.206	3.471	5.599	9.845
1.0	.0781	.1735	.2893	.4290	.7987	1.333	2.118	3.327	5.360	9.410

TABLE 7.26: I_1^* $(1/\alpha_2 = .4,\ \mu = .7)$

α_1 \ Q:	0.05	0.10	0.15	0.20	0.30	0.40	0.50	0.60	0.70	0.80
0.1	.1101	.2568	.4514	.7097	1.519	3.030	5.940	11.40	21.39	41.45
0.2	.1044	.2409	.4176	.6454	1.315	2.424	4.275	7.396	12.87	23.84
0.3	.0996	.2277	.3909	.5971	1.181	2.097	3.540	5.869	9.870	17.85
0.4	.0954	.2166	.3689	.5588	1.084	1.882	3.102	5.028	8.295	14.79
0.5	.0916	.2070	.3505	.5275	1.010	1.727	2.805	4.483	7.310	12.92
0.6	.0883	.1986	.3346	.5011	.9494	1.608	2.585	4.095	6.629	11.65
0.7	.0854	.1912	.3208	.4786	.8996	1.512	2.415	3.803	6.127	10.73
0.8	.0827	.1845	.3087	.4590	.8575	1.433	2.278	3.573	5.739	10.03
0.9	.0802	.1786	.2979	.4417	.8212	1.367	2.165	3.387	5.429	9.478
1.0	.0780	.1732	.2882	.4263	.7894	1.310	2.070	3.232	5.175	9.030

TABLE 7.26: I_1^* $(1/\alpha_2 = .4,\ \mu = .9)$

α_1 \ Q:	0.05	0.10	0.15	0.20	0.30	0.40	0.50	0.60	0.70	0.80
0.1	.1109	.2598	.4580	.7212	1.543	3.062	5.942	11.29	21.05	40.66
0.2	.1050	.2430	.4219	.6518	1.322	2.420	4.227	7.243	12.51	23.06
0.3	.1000	.2292	.3934	.5999	1.179	2.075	3.468	5.693	9.494	17.06
0.4	.0957	.2175	.3700	.5590	1.076	1.849	3.015	4.836	7.908	14.00
0.5	.0919	.2074	.3505	.5256	.9964	1.685	2.706	4.280	6.914	12.13
0.6	.0885	.1986	.3337	.4976	.9325	1.560	2.478	3.883	6.225	10.86
0.7	.0854	.1909	.3192	.4737	.8796	1.459	2.301	3.583	5.716	9.933
0.8	.0827	.1840	.3064	.4529	.8348	1.376	2.159	3.347	5.322	9.229
0.9	.0802	.1778	.2950	.4346	.7963	1.306	2.041	3.155	5.007	8.674
1.0	.0779	.1722	.2848	.4184	.7626	1.246	1.941	2.995	4.749	8.223

TABLE 7.26: I_1^* $(1/\alpha_2 = .5,\ \mu = .1)$

α_1 \ Q:	0.05	0.10	0.15	0.20	0.30	0.40	0.50	0.60	0.70	0.80
0.1	.1184	.2655	.4489	.6801	1.368	2.642	5.233	10.42	20.19	39.88
0.2	.1120	.2513	.4210	.6299	1.216	2.172	3.814	6.701	11.91	22.48
0.3	.1080	.2395	.3987	.5917	1.115	1.917	3.198	5.326	9.070	16.64
0.4	.1038	.2294	.3803	.5611	1.041	1.750	2.838	4.586	7.610	13.70
0.5	.1001	.2207	.3647	.5360	.9836	1.630	2.597	4.117	6.714	11.92
0.6	.0968	.2130	.3512	.5148	.9377	1.539	2.422	3.790	6.107	10.74
0.7	.0939	.2063	.3395	.4966	.8997	1.466	2.288	3.548	5.666	9.884
0.8	.0912	.2002	.3291	.4807	.8676	1.407	2.182	3.361	5.331	9.244
0.9	.0887	.1947	.3198	.4667	.8400	1.357	2.096	3.211	5.067	8.744
1.0	.0866	.1897	.3115	.4542	.8160	1.315	2.024	3.089	4.854	8.344

TABLE 7.26: I_1^* $(1/\alpha_2 = .5,\ \mu = .3)$

α_1 \ Q:	0.05	0.10	0.15	0.20	0.30	0.40	0.50	0.60	0.70	0.80
0.1	.1197	.2718	.4657	.7149	1.470	2.861	5.601	10.93	20.87	40.95
0.2	.1139	.2566	.4350	.6586	1.297	2.342	4.104	7.139	12.56	23.54
0.3	.1089	.2439	.4105	.6159	1.183	2.061	3.450	5.724	9.689	17.69
0.4	.1046	.2331	.3903	.5820	1.100	1.877	3.065	4.958	8.209	14.74
0.5	.1007	.2238	.3733	.5541	1.036	1.744	2.806	4.469	7.297	12.96
0.6	.0973	.2157	.3586	.5306	.9847	1.643	2.617	4.126	6.675	11.76
0.7	.0942	.2084	.3458	.5104	.9421	1.562	2.471	3.870	6.222	10.91
0.8	.0915	.2020	.3345	.4928	.9060	1.495	2.355	3.671	5.876	10.26
0.9	.0889	.1961	.3245	.4773	.8749	1.439	2.260	3.512	5.603	9.754
1.0	.0866	.1908	.3154	.4635	.8478	1.392	2.180	3.380	5.382	9.348

TABLE 7.26: I_1^* $(1/\alpha_2 = .5,\ \mu = .5)$

α_1 \ Q:	0.05	0.10	0.15	0.20	0.30	0.40	0.50	0.60	0.70	0.80
0.1	.1209	.2773	.4801	.7444	1.557	3.056	5.952	11.44	21.55	41.91
0.2	.1149	.2610	.4464	.6816	1.361	2.478	4.343	7.508	13.10	24.38
0.3	.1097	.2475	.4197	.6342	1.233	2.165	3.635	6.021	10.15	18.45
0.4	.1052	.2360	.3977	.5968	1.141	1.961	3.216	5.207	8.612	15.44
0.5	.1012	.2261	.3792	.5660	1.069	1.813	2.932	4.683	7.656	13.60
0.6	.0977	.2175	.3634	.5402	1.012	1.700	2.724	4.313	6.998	12.36
0.7	.0945	.2098	.3495	.5181	.9641	1.610	2.563	4.035	6.516	11.47
0.8	.0917	.2030	.4474	.4988	.9238	1.536	2.434	3.818	6.145	10.79
0.9	.0890	.1968	.3265	.4818	.8891	1.473	2.328	3.642	5.851	10.26
1.0	.0866	.1912	.3167	.4667	.8588	1.419	2.239	3.497	5.610	9.829

TABLE 7.26: I_1^* $(1/\alpha_2 = .5,\ \mu = .7)$

α_1 \ Q:	0.05	0.10	0.15	0.20	0.30	0.40	0.50	0.60	0.70	0.80
0.1	.1220	.2821	.4917	.7672	1.619	3.187	6.161	11.68	21.73	41.89
0.2	.1157	.2646	.4550	.6977	1.401	2.551	4.448	7.619	13.16	24.24
0.3	.1104	.2503	.4260	.6457	1.259	2.209	3.691	6.067	10.13	18.23
0.4	.1057	.2381	.4023	.6047	1.157	1.986	3.241	5.211	8.543	15.16
0.5	.1016	.2276	.3824	.5712	1.079	1.824	2.935	4.656	7.546	13.27
0.6	.0980	.2185	.3653	.5432	1.015	1.700	2.710	4.261	6.857	12.00
0.7	.0947	.2104	.3505	.5191	.9633	1.601	2.535	3.963	6.347	11.07
0.8	.0917	.2032	.3375	.4982	.9191	1.520	2.394	3.729	5.954	10.36
0.9	.0891	.1967	.3258	.4799	.8810	1.451	2.278	3.539	5.639	9.801
1.0	.0866	.1908	.3154	.4635	.8478	1.392	2.180	3.380	5.382	9.348

TABLE 7.26: I_1^* $(1/\alpha_2 = .5,\ \mu = .9)$

α_1 \ Q:	0.05	0.10	0.15	0.20	0.30	0.40	0.50	0.60	0.70	0.80
0.1	.1230	.2859	.5002	.7819	1.650	3.226	6.162	11.54	21.31	40.91
0.2	.1165	.2674	.4605	.7060	1.411	2.546	4.388	7.430	12.71	23.27
0.3	.1109	.2521	.4293	.6496	1.258	2.183	3.603	5.850	9.668	17.25
0.4	.1061	.2393	.4039	.6054	1.148	1.946	3.135	4.975	8.064	14.17
0.5	.1019	.2283	.3826	.5694	1.063	1.775	2.815	4.406	7.056	12.29
0.6	.0981	.2186	.3644	.5392	.9956	1.643	2.580	4.000	6.357	11.01
0.7	.0948	.2102	.3487	.5135	.9396	1.538	2.397	3.693	5.839	10.07
0.8	.0917	.2026	.3348	.4912	.8923	1.451	2.249	3.451	5.439	9.362
0.9	.0890	.1958	.3225	.4716	.8516	1.378	2.127	3.253	5.118	8.800
1.0	.0865	.1897	.3115	.4542	.8160	1.315	2.024	3.089	4.854	8.344

TABLE 7.26: I_1^* $(1/\alpha_2 = .6,\ \mu = .1)$

α_1 \ Q:	0.05	0.10	0.15	0.20	0.30	0.40	0.50	0.60	0.70	0.80
0.1	.1286	.2858	.4789	.7192	1.422	2.700	5.285	10.46	20.22	39.93
0.2	.1226	.2708	.4499	.6675	1.267	2.231	3.872	6.754	11.97	22.54
0.3	.1174	.2583	.4266	.6281	1.165	1.977	3.261	5.387	9.132	16.71
0.4	.1129	.2476	.4074	.5968	1.091	1.811	2.904	4.654	7.680	13.78
0.5	.1089	.2384	.3912	.5710	1.034	1.692	2.666	4.191	6.793	12.01
0.6	.1054	.2303	.3772	.5493	.9877	1.602	2.494	3.869	6.191	10.83
0.7	.1022	.2232	.3650	.5306	.9497	1.530	2.363	3.631	5.756	9.988
0.8	.0993	.2167	.3542	.5143	.9176	1.471	2.259	3.447	5.426	9.354
0.9	.0966	.2109	.3446	.5000	.8900	1.422	2.174	3.301	5.167	8.860
1.0	.0942	.2057	.3359	.4872	.8660	1.381	2.104	3.181	4.958	8.464

TABLE 7.26: I_1^* $(1/\alpha_2 = .6,\ \mu = .3)$

α_1 \ Q:	0.05	0.10	0.15	0.20	0.30	0.40	0.50	0.60	0.70	0.80
0.1	.1301	.2931	.4984	.7595	1.540	2.956	5.721	11.07	21.04	41.21
0.2	.1238	.2769	.4660	.7006	1.362	2.429	4.214	7.275	12.74	23.81
0.3	.1185	.2634	.4402	.6561	1.245	2.144	3.557	5.861	9.872	17.97
0.4	.1138	.2519	.4190	.6208	1.160	1.958	3.171	5.096	8.396	15.03
0.5	.1096	.2420	.4011	.5918	1.095	1.824	2.911	4.608	7.488	13.25
0.6	.1059	.2334	.3857	.5674	1.042	1.722	2.722	4.266	6.869	12.07
0.7	.1026	.2257	.3723	.5465	.9985	1.641	2.577	4.012	6.419	11.21
0.8	.0996	.2188	.3605	.5283	.9618	1.574	2.461	3.814	6.076	10.57
0.9	.0969	.2126	.3499	.5122	.9302	1.518	2.366	3.656	5.805	10.07
1.0	.0944	.2070	.3404	.4979	.9027	1.470	2.287	3.526	5.586	9.665

TABLE 7.26: I_1^* $(1/\alpha_2 = .6,\ \mu = .5)$

α_1 \ Q:	0.05	0.10	0.15	0.20	0.30	0.40	0.50	0.60	0.70	0.80
0.1	.1315	.2997	.5153	.7940	1.641	3.185	6.132	11.67	21.85	42.36
0.2	.1250	.2821	.4794	.7274	1.436	2.587	4.493	7.710	13.38	24.82
0.3	.1194	.2676	.4510	.6774	1.303	2.266	3.773	6.210	10.42	18.88
0.4	.1145	.2553	.4277	.6380	1.207	2.056	3.347	5.388	8.873	15.86
0.5	.1102	.2447	.4081	.6057	1.133	1.905	3.059	4.860	7.912	14.02
0.6	.1064	.2355	.3913	.5786	1.073	1.789	2.848	4.486	7.252	12.78
0.7	.1030	.2273	.3766	.5553	1.024	1.697	2.684	4.206	6.767	11.88
0.8	.0999	.2200	.3637	.5352	.9824	1.621	2.554	3.987	6.395	11.20
0.9	.0970	.2134	.3522	.5174	.9466	1.557	2.446	3.810	6.099	10.67
1.0	.0944	.2074	.3419	.5015	.9154	1.502	2.356	3.664	5.858	10.24

TABLE 7.26: I_1^* $(1/\alpha_2 = .6,\ \mu = .7)$

α_1 \ Q:	0.05	0.10	0.15	0.20	0.30	0.40	0.50	0.60	0.70	0.80
0.1	.1328	.3053	.5292	.8210	1.715	3.338	6.375	11.95	22.06	42.33
0.2	.1260	.2864	.4896	.7465	1.483	2.672	4.615	7.838	13.45	24.65
0.3	.1202	.2709	.4585	.6910	1.334	2.317	3.838	6.262	10.39	18.61
0.4	.1151	.2578	.4331	.6474	1.226	2.085	3.376	5.391	8.787	15.52
0.5	.1107	.2465	.4118	.6119	1.144	1.917	3.061	4.825	7.779	13.62
0.6	.1067	.2367	.3936	.5821	1.078	1.789	2.830	4.423	7.081	12.34
0.7	.1032	.2280	.3778	.5567	1.023	1.687	2.651	4.120	6.565	11.40
0.8	.1000	.2203	.3639	.5346	.9770	1.602	2.507	3.881	6.166	10.69
0.9	.0971	.2133	.3515	.5152	.9373	1.531	2.388	3.687	5.847	10.12
1.0	.0944	.2070	.3404	.4979	.9027	1.470	2.287	3.526	5.586	9.665

TABLE 7.26: I_1^* $(1/\alpha_2 = .6,\ \mu = .9)$

α_1 \ Q:	0.05	0.10	0.15	0.20	0.30	0.40	0.50	0.60	0.70	0.80
0.1	.1340	.3100	.5395	.8390	1.752	3.384	6.375	11.78	21.56	41.16
0.2	.1269	.2898	.4963	.7567	1.496	2.668	4.544	7.613	12.91	23.48
0.3	.1209	.2732	.4626	.6960	1.333	2.287	3.734	6.004	9.840	17.44
0.4	.1156	.2593	.4352	.6485	1.216	2.038	3.250	5.111	8.216	14.35
0.5	.1110	.2473	.4123	.6100	1.126	1.860	2.921	4.530	7.196	12.45
0.6	.1069	.2369	.3927	.5778	1.055	1.722	2.678	4.114	6.486	11.15
0.7	.1033	.2278	.3758	.5504	.9960	1.613	2.489	3.800	5.961	10.21
0.8	.1000	.2196	.3609	.5266	.9462	1.522	2.336	3.552	5.553	9.494
0.9	.0970	.2123	.3477	.5058	.9034	1.446	2.210	3.350	5.227	8.926
1.0	.0942	.2057	.3359	.4872	.8660	1.381	2.104	3.181	4.958	8.464

TABLE 7.26: I_1^* $(1/\alpha_2 = .7,\ \mu = .1)$

α_1 \ Q:	0.05	0.10	0.15	0.20	0.30	0.40	0.50	0.60	0.70	0.80
0.1	.1380	.3046	.5069	.7558	1.472	2.757	5.336	10.51	20.26	39.97
0.2	.1316	.2888	.4766	.7026	1.316	2.288	3.929	6.807	12.02	22.60
0.3	.1261	.2756	.4525	.6622	1.214	2.034	3.322	5.448	9.194	16.78
0.4	.1213	.2645	.4326	.6301	1.139	1.870	2.969	4.722	7.751	13.86
0.5	.1170	.2548	.4158	.6037	1.081	1.752	2.734	4.264	6.871	12.10
0.6	.1133	.2463	.4014	.5815	1.035	1.662	2.564	3.947	6.276	10.93
0.7	.1099	.2388	.3887	.5624	.9970	1.591	2.435	3.712	5.846	10.09
0.8	.1068	.2321	.3776	.5458	.9648	1.533	2.333	3.532	5.521	9.464
0.9	.1040	.2260	.3676	.5312	.9373	1.485	2.250	3.389	5.267	8.975
1.0	.1014	.2205	.3587	.5181	.9133	1.444	2.181	3.272	5.062	8.584

TABLE 7.26: I_1^* ($1/\alpha_2 = .7$, $\mu = .3$)

α_1 \ Q:	0.05	0.10	0.15	0.20	0.30	0.40	0.50	0.60	0.70	0.80
0.1	.1397	.3129	.5290	.8014	1.606	3.049	5.838	11.21	21.22	41.47
0.2	.1330	.2957	.4949	.7400	1.423	2.513	4.322	7.410	12.92	24.08
0.3	.1273	.2814	.4679	.6938	1.303	2.225	3.662	5.996	10.06	18.25
0.4	.1222	.2693	.4457	.6572	1.217	2.037	3.275	5.231	8.583	15.32
0.5	.1178	.2589	.4270	.6271	1.150	1.902	3.014	4.746	7.678	13.55
0.6	.1139	.2497	.4110	.6019	1.096	1.799	2.825	4.404	7.062	12.37
0.7	.1104	.2417	.3970	.5803	1.052	1.717	2.680	4.151	6.615	11.52
0.8	.1072	.2344	.3846	.5615	1.015	1.649	2.564	3.955	6.274	10.88
0.9	.1043	.2279	.3736	.5449	.9826	1.593	2.470	3.797	6.006	10.38
1.0	.1016	.2220	.3637	.5302	.9547	1.545	2.391	3.668	5.788	9.981

TABLE 7.26: I_1^* ($1/\alpha_2 = .7$, $\mu = .5$)

α_1 \ Q:	0.05	0.10	0.15	0.20	0.30	0.40	0.50	0.60	0.70	0.80
0.1	.1413	.3204	.5484	.8408	1.722	3.310	6.308	11.90	22.15	42.81
0.2	.1343	.3017	.5103	.7705	1.508	2.693	4.640	7.909	13.66	25.25
0.3	.1283	.2863	.4802	.7180	1.369	2.363	3.908	6.395	10.68	19.30
0.4	.1231	.2732	.4556	.6767	1.270	2.148	3.475	5.567	9.132	16.28
0.5	.1185	.2620	.4350	.6429	1.193	1.993	3.182	5.033	8.167	14.44
0.6	.1144	.2522	.4173	.6146	1.132	1.874	2.968	4.656	7.503	13.19
0.7	.1108	.2435	.4019	.5903	1.081	1.780	2.803	4.374	7.016	12.30
0.8	.1074	.2358	.3883	.5693	1.038	1.703	2.670	4.153	6.643	11.62
0.9	.1044	.2288	.3763	.5507	1.001	1.638	2.562	3.975	6.346	11.08
1.0	.1016	.2225	.3654	.5342	.9691	1.582	2.470	3.827	6.104	10.65

TABLE 7.26: I_1^* ($1/\alpha_2 = .7$, $\mu = .7$)

α_1 \ Q:	0.05	0.10	0.15	0.20	0.30	0.40	0.50	0.60	0.70	0.80
0.1	.1429	.3270	.5644	.8721	1.806	3.484	6.583	12.21	22.39	42.76
0.2	.1355	.3067	.5221	.7925	1.561	2.789	4.777	8.052	13.73	25.05
0.3	.1293	.2901	.4889	.7337	1.404	2.420	3.980	6.452	10.65	18.98
0.4	.1238	.2761	.4619	.6876	1.292	2.180	3.506	5.566	9.028	15.88
0.5	.1191	.2641	.4393	.6501	1.206	2.007	3.184	4.991	8.009	13.97
0.6	.1148	.2536	.4200	.6187	1.137	1.874	2.947	4.582	7.302	12.67
0.7	.1110	.2443	.4033	.5919	1.080	1.769	2.763	4.273	6.780	11.73
0.8	.1076	.2361	.3886	.5687	1.032	1.681	2.616	4.030	6.376	11.01
0.9	.1044	.2287	.3755	.5483	.9907	1.608	2.494	3.832	6.053	10.44
1.0	.1016	.2220	.3637	.5302	.9547	1.545	2.391	3.668	5.788	9.981

TABLE 7.26: I_1^* ($1/\alpha_2 = .7$, $\mu = .9$)

α_1 \ Q:	0.05	0.10	0.15	0.20	0.30	0.40	0.50	0.60	0.70	0.80
0.1	.1442	.3325	.5766	.8933	1.850	3.538	6.582	12.02	21.80	41.41
0.2	.1366	.3107	.5300	.8047	1.576	2.784	4.696	7.791	13.11	23.69
0.3	.1301	.2928	.4938	.7397	1.404	2.386	3.860	6.154	10.01	17.63
0.4	.1244	.2779	.4644	.6891	1.280	2.127	3.362	5.244	8.367	14.52
0.5	.1195	.2651	.4400	.6482	1.186	1.941	3.023	4.650	7.333	12.60
0.6	.1151	.2539	.4191	.6140	1.111	1.799	2.772	4.226	6.614	11.30
0.7	.1111	.2441	.4011	.5850	1.049	1.685	2.578	3.905	6.080	10.35
0.8	.1076	.2354	.3853	.5598	.9972	1.591	2.421	3.651	5.667	9.625
0.9	.1044	.2276	.3713	.5377	.9524	1.512	2.291	3.444	5.335	9.051
1.0	.1014	.2205	.3587	.5181	.9133	1.444	2.181	3.272	5.062	8.584

TABLE 7.26: I_1^* ($1/\alpha_2 = .8$, $\mu = .1$)

α_1 \ Q:	0.05	0.10	0.15	0.20	0.30	0.40	0.50	0.60	0.70	0.80
0.1	.1467	.3221	.5331	.7903	1.521	2.812	5.387	10.55	20.30	40.02
0.2	.1399	.3056	.5010	.7357	1.363	2.343	3.985	6.859	12.07	22.66
0.3	.1341	.2918	.4768	.6943	1.260	2.090	3.382	5.509	9.256	16.85
0.4	.1291	.2802	.4563	.6615	1.184	1.926	3.033	4.789	7.822	13.94
0.5	.1246	.2701	.4389	.6345	1.127	1.809	2.801	4.336	6.949	12.19
0.6	.1206	.2613	.4240	.6118	1.080	1.720	2.633	4.023	6.360	11.03
0.7	.1170	.2534	.4110	.5924	1.042	1.650	2.506	3.793	5.936	10.20
0.8	.1138	.2464	.3996	.5755	1.010	1.593	2.405	3.616	5.616	9.574
0.9	.1108	.2401	.3893	.5606	.9823	1.545	2.324	3.475	5.365	9.090
1.0	.1081	.2343	.3801	.5473	.9583	1.504	2.256	3.361	5.164	8.704

TABLE 7.26: I_1^* ($1/\alpha_2 = .8$, $\mu = .3$)

α_1 \ Q:	0.05	0.10	0.15	0.20	0.30	0.40	0.50	0.60	0.70	0.80
0.1	.1487	.3315	.5678	.8411	1.670	3.139	5.953	11.35	21.39	41.73
0.2	.1416	.3133	.5222	.7773	1.482	2.595	4.428	7.544	13.10	24.35
0.3	.1355	.2983	.4940	.7294	1.359	2.303	3.765	6.129	10.24	18.53
0.4	.1302	.2856	.4709	.6915	1.271	2.113	3.376	5.366	8.769	15.61
0.5	.1255	.2747	.4514	.6606	1.203	1.976	3.115	4.880	7.867	13.84
0.6	.1213	.2651	.4347	.6345	1.148	1.872	2.926	4.540	7.254	12.67
0.7	.1176	.2566	.4202	.6123	1.103	1.790	2.780	4.288	6.809	11.82
0.8	.1142	.2490	.4074	.5929	1.065	1.723	2.665	4.093	6.471	11.19
0.9	.1111	.2422	.3960	.5758	1.033	1.666	2.571	3.937	6.204	10.69
1.0	.1083	.2360	.3857	.5607	1.004	1.618	2.492	3.809	5.989	10.30

TABLE 7.26: I_1^* ($1/\alpha_2 = .8$, $\mu = .5$)

α_1 \ Q:	0.05	0.10	0.15	0.20	0.30	0.40	0.50	0.60	0.70	0.80
0.1	.1505	.3400	.5796	.8854	1.799	3.431	6.480	12.13	22.45	43.26
0.2	.1431	.3201	.5394	.8115	1.576	2.795	4.784	8.105	13.94	25.68
0.3	.1367	.3038	.5077	.7565	1.433	2.456	4.040	6.578	10.95	19.72
0.4	.1311	.2900	.4819	.7134	1.330	2.236	3.600	5.742	9.389	16.70
0.5	.1263	.2782	.4603	.6781	1.251	2.078	3.303	5.204	8.419	14.85
0.6	.1219	.2678	.4418	.6487	1.187	1.957	3.086	4.824	7.752	13.61
0.7	.1180	.2587	.4257	.6235	1.135	1.861	2.918	4.539	7.263	12.71
0.8	.1145	.2506	.4115	.6016	1.091	1.782	2.784	4.316	6.888	12.02
0.9	.1113	.2432	.3989	.5823	1.053	1.716	2.674	4.137	6.590	11.49
1.0	.1084	.2366	.3876	.5652	1.020	1.659	2.581	3.989	6.347	11.06

TABLE 7.26: I_1^* ($1/\alpha_2 = .8$, $\mu = .7$)

α_1 \ Q:	0.05	0.10	0.15	0.20	0.30	0.40	0.50	0.60	0.70	0.80
0.1	.1523	.3474	.5978	.9208	1.895	3.626	6.786	12.48	22.71	43.20
0.2	.1444	.3259	.5528	.8364	1.636	2.903	4.935	8.262	14.01	25.44
0.3	.1378	.3082	.5176	.7742	1.472	2.520	4.119	6.638	10.90	19.35
0.4	.1320	.2933	.4891	.7257	1.355	2.272	3.634	5.738	9.266	16.23
0.5	.1269	.2806	.4653	.6863	1.265	2.093	3.304	5.154	8.236	14.31
0.6	.1224	.2695	.4449	.6534	1.193	1.957	3.061	4.738	7.521	13.01
0.7	.1183	.2597	.4273	.6253	1.134	1.848	2.873	4.424	6.993	12.06
0.8	.1147	.2509	.4118	.6010	1.085	1.758	2.722	4.176	6.584	11.33
0.9	.1113	.2431	.3980	.5796	1.042	1.683	2.597	3.975	6.257	10.76
1.0	.1083	.2360	.3857	.5607	1.004	1.618	2.492	3.809	5.989	10.30

TABLE 7.26: I_1^* ($1/\alpha_2 = .8$, $\mu = .9$)

α_1 \ Q:	0.05	0.10	0.15	0.20	0.30	0.40	0.50	0.60	0.70	0.80
0.1	.1538	.3538	.6119	.9453	1.945	3.686	6.782	12.26	22.04	41.65
0.2	.1457	.3304	.5620	.8505	1.654	2.897	4.842	7.965	13.30	23.90
0.3	.1387	.3113	.5233	.7813	1.472	2.482	3.983	6.300	10.17	17.81
0.4	.1327	.2954	.4921	.7277	1.342	2.213	3.471	5.373	8.516	14.68
0.5	.1274	.2817	.4661	.6844	1.244	2.020	3.122	4.768	7.469	12.76
0.6	.1227	.2699	.4440	.6483	1.165	1.872	2.864	4.336	6.740	11.45
0.7	.1185	.2595	.4249	.6177	1.100	1.754	2.664	4.008	6.199	10.49
0.8	.1147	.2502	.4082	.5911	1.046	1.657	2.503	3.748	5.779	9.755
0.9	.1112	.2419	.3934	.5679	0.999	1.575	2.369	3.537	5.441	9.176
1.0	.1081	.2343	.3801	.5473	.9583	1.504	2.256	3.361	5.164	8.704

TABLE 7.26: I_1^* $(1/\alpha_2 = .9,\ \mu = .1)$

α_1 \ Q:	0.05	0.10	0.15	0.20	0.30	0.40	0.50	0.60	0.70	0.80
0.1	.1549	.3386	.5578	.8230	1.567	2.866	5.437	10.59	20.34	40.07
0.2	.1478	.3214	.5255	.7671	1.408	2.396	4.040	6.912	12.12	22.72
0.3	.1417	.3071	.4998	.7248	1.304	2.145	3.442	5.569	9.318	16.92
0.4	.1364	.2950	.4786	.6913	1.228	1.981	3.095	4.856	7.893	14.02
0.5	.1317	.2846	.4608	.6638	1.170	1.865	2.866	4.408	7.027	12.28
0.6	.1275	.2754	.4455	.6407	1.123	1.776	2.700	4.099	6.444	11.13
0.7	.1238	.2673	.4321	.6209	1.085	1.707	2.575	3.872	6.025	10.30
0.8	.1204	.2600	.4203	.6037	1.053	1.650	2.476	3.698	5.710	9.683
0.9	.1173	.2534	.4098	.5885	1.025	1.603	2.396	3.561	5.463	9.205
1.0	.1144	.2475	.4004	.5750	1.001	1.563	2.330	3.449	5.265	8.823

TABLE 7.26: I_1^* $(1/\alpha_2 = .9,\ \mu = .3)$

α_1 \ Q:	0.05	0.10	0.15	0.20	0.30	0.40	0.50	0.60	0.70	0.80
0.1	.1571	.3490	.5851	.8789	1.731	3.227	6.067	11.49	21.57	41.99
0.2	.1496	.3300	.5480	.8128	1.538	2.674	4.532	7.677	13.27	24.63
0.3	.1432	.3143	.5187	.7633	1.413	2.378	3.865	6.262	10.42	18.81
0.4	.1376	.3011	.4947	.7243	1.323	2.186	3.475	5.499	8.954	15.90
0.5	.1327	.2896	.4745	.6924	1.253	2.049	3.214	5.014	8.055	14.14
0.6	.1283	.2796	.4573	.6656	1.198	1.944	3.024	4.675	7.445	12.97
0.7	.1244	.2708	.4422	.6427	1.152	1.861	2.879	4.424	7.002	12.13
0.8	.1208	.2629	.4290	.6228	1.114	1.793	2.764	4.229	6.666	11.49
0.9	.1176	.2557	.4171	.6053	1.081	1.737	2.670	4.074	6.401	11.00
1.0	.1146	.2493	.4065	.5897	1.052	1.688	2.591	3.947	6.187	10.61

TABLE 7.26: I_1^* $(1/\alpha_2 = .9,\ \mu = .5)$

α_1 \ Q:	0.05	0.10	0.15	0.20	0.30	0.40	0.50	0.60	0.70	0.80
0.1	.1592	.3585	.6093	.9280	1.874	3.549	6.649	12.35	22.74	43.71
0.2	.1513	.3376	.5671	.8506	1.642	2.895	4.925	8.298	14.21	26.12
0.3	.1446	.3204	.5339	.7933	1.494	2.548	4.169	6.759	11.21	20.15
0.4	.1387	.3059	.5070	.7483	1.388	2.322	3.723	5.915	9.643	17.11
0.5	.1336	.2935	.4844	.7118	1.306	2.161	3.421	5.372	8.669	15.26
0.6	.1290	.2827	.4651	.6812	1.241	2.037	3.201	4.989	7.999	14.02
0.7	.1249	.2731	.4483	.6550	1.188	1.939	3.031	4.702	7.508	13.12
0.8	.1212	.2645	.4335	.6324	1.142	1.859	2.895	4.477	7.131	12.43
0.9	.1178	.2569	.4204	.6124	1.104	1.791	2.784	4.297	6.832	11.90
1.0	.1147	.2499	.4086	.5947	1.070	1.734	2.690	4.147	6.589	11.46

TABLE 7.26: I_1^* $(1/\alpha_2 = .9,\ \mu = .7)$

α_1 \ Q:	0.05	0.10	0.15	0.20	0.30	0.40	0.50	0.60	0.70	0.80
0.1	.1611	.3668	.6298	.9676	1.980	3.764	6.984	12.73	23.03	43.63
0.2	.1528	.3440	.5821	.8784	1.709	3.013	5.090	8.468	14.28	25.83
0.3	.1458	.3253	.5450	.8130	1.537	2.618	4.254	6.821	11.15	19.72
0.4	.1396	.3096	.5150	.7621	1.415	2.361	3.758	5.908	9.501	16.59
0.5	.1343	.2962	.4899	.7208	1.322	2.177	3.420	5.314	8.460	14.66
0.6	.1295	.2845	.4686	.6864	1.248	2.036	3.172	4.891	7.738	13.34
0.7	.1252	.2742	.4501	.6571	1.187	1.924	2.980	4.572	7.203	12.38
0.8	.1214	.2650	.4339	.6318	1.135	1.832	2.826	4.320	6.789	11.65
0.9	.1178	.2567	.4194	.6095	1.091	1.755	2.699	4.116	6.459	11.08
1.0	.1146	.2493	.4065	.5897	1.052	1.688	2.591	3.947	6.187	10.61

TABLE 7.26: I_1^* $(1/\alpha_2 = .9,\ \mu = .9)$

α_1 \ Q:	0.05	0.10	0.15	0.20	0.30	0.40	0.50	0.60	0.70	0.80
0.1	.1629	.3740	.6457	.9954	2.036	3.830	6.978	12.48	22.28	41.90
0.2	.1542	.3491	.5925	.8943	1.729	3.007	4.985	8.135	13.49	24.11
0.3	.1468	.3289	.5514	.8211	1.537	2.575	4.103	6.444	10.34	17.99
0.4	.1404	.3119	.5184	.7645	1.402	2.296	3.577	5.501	8.662	14.85
0.5	.1348	.2975	.4910	.7189	1.299	2.096	3.219	4.884	7.603	12.91
0.6	.1298	.2850	.4677	.6810	1.217	1.943	2.954	4.443	6.864	11.59
0.7	.1254	.2740	.4476	.6488	1.149	1.821	2.749	4.109	6.316	10.62
0.8	.1214	.2642	.4299	.6210	1.092	1.720	2.583	3.844	5.889	9.885
0.9	.1177	.2554	.4143	.5966	1.044	1.635	2.446	3.628	5.547	9.300
1.0	.1144	.2475	.4004	.5750	1.001	1.563	2.330	3.449	5.265	8.823

CHAPTER EIGHT

FREE VIBRATION OF BEAMS WITH DISCONTINUITIES IN CROSS-SECTIONAL PROPERTIES

All beams and shafts in the preceding chapters had uniform cross-sectional properties from one end to the other. In this chapter we look at the lateral vibration of beams that have step changes in the properties of their cross sections. In particular, we are interested in step changes in the mass per unit length ρA and in the bending stiffness EI. Such changes could result from an abrupt change in cross-sectional geometry or in beam material.

For illustrative purposes we consider the lateral vibration of a cantilever beam, as shown in the inset of Case 8.1. The beam cross-sectional properties are shown to be constant over the dimensionless distance μ from the clamped end, and $\gamma = 1 - \mu$ for the remainder of the beam. It should be emphasized again that while property discontinuities are shown schematically in the figures to follow, as geometric discontinuities they could be the result of the joining of beam sections of the same cross-sectional geometry but of different material properties. They could also result from the uniform distribution along part of the beam of a flexible material which would add mass to the beam but leave the stiffness unchanged. The pipe, partially filled with liquid, in Illustrative Problem 8.1 is an example of such a case.

It is found desirable to express the displacement of the beam segments of constant cross-sectional properties by individual functions. Accordingly, we express the displacement of the two segments of the cantilever beam (see inset, Case 8.1) as

$$r_1(\xi) = A_1 \sin \beta_1\xi + B_1 \cos \beta_1\xi + C_1 \sinh \beta_1\xi + D_1 \cosh \beta_1\xi \qquad (8.1)$$

and

$$r_2(\xi) = A_2 \sin \beta_2\xi + B_2 \cos \beta_2\xi + C_2 \sinh \beta_2\xi + D_2 \cosh \beta_2\xi \qquad (8.2)$$

where subscripts 1 and 2 refer to the left-hand and right-hand segments, respectively, and

$$\beta_1^4 = \frac{\rho_1 A_1}{E_1 I_1}\omega^2 L^4 \qquad \text{and} \qquad \beta_2^4 = \frac{\rho_2 A_2}{E_2 I_2}\omega^2 L^4$$

The next step to be taken in obtaining a solution for the frequencies is to enforce the boundary conditions. We note that each displacement function has its origin at an outer end of the beam (inset, Case 8.1). Enforcing the boundary conditions at the beam left end, that is,

$$r_1(\xi) = \frac{dr_1(\xi)}{d\xi} = 0 \bigg|_{\xi=0}$$

it is readily shown that $A_1 = -C_1$ and $B_1 = -D_1$. Enforcing the free end boundary conditions, that is,

$$\frac{d^2 r_2(\xi)}{d\xi^2} = \frac{d^3 r_2(\xi)}{d\xi^3} = 0 \bigg|_{\xi=0}$$

it is shown that $A_2 = C_2$ and $B_2 = D_2$

$$\therefore r_1(\xi) = A_1(\sin\beta_1\xi - \sinh\beta_1\xi) + B_1(\cos\beta_1\xi - \cosh\beta_1\xi) \tag{8.3}$$

and

$$r_2(\xi) = A_2(\sin\beta_2\xi + \sinh\beta_2\xi) + B_2(\cos\beta_2\xi + \cosh\beta_2\xi) \tag{8.4}$$

Introducing

$$\phi = \left(\frac{\rho_2 A_2}{\rho_1 A_1}\right)^{1/4} \qquad \text{and} \qquad \alpha = \left(\frac{E_2 I_2}{E_1 I_1}\right)^{1/4}$$

it is seen from the definition of β_1^4 and β_2^4 that $\beta_2/\beta_1 = \phi/\alpha$. There are four more boundary conditions that can be written. They pertain to the continuity of displacement, slope, bending moment, and shear forces across the interface of the two beam segments. Introducing the parameter $\theta = \phi/\alpha$ so that we may write $\beta_2 = \theta\beta_1$, these four boundary conditions are written:

1. Continuity of displacement, $r_1(\mu) = r_2(\gamma)$.

$$\begin{aligned} &A_1(\sin\beta_1\mu - \sinh\beta_1\mu) + B_1(\cos\beta_1\mu - \cosh\beta_1\mu) \\ &\quad + A_2(-\sin\beta_1\theta\gamma - \sinh\beta_1\theta\gamma) + B_2(-\cos\beta_1\theta\gamma - \cosh\beta_1\theta\gamma) = 0 \end{aligned} \tag{8.5}$$

2. Continuity of slope, $\left.\dfrac{dr_1(\xi)}{d\xi}\right|_{\xi=\mu} = -\left.\dfrac{dr_2(\xi)}{d\xi}\right|_{\xi=\gamma}$.

$$\begin{aligned} &A_1(\cos\beta_1\mu - \cosh\beta_1\mu) + B_1(-\sin\beta_1\mu - \sinh\beta_1\mu) \\ &\quad + A_2\theta(\cos\beta_1\theta\gamma + \cosh\beta_1\theta\gamma) + B_2\theta(-\sin\beta_1\theta\gamma + \sinh\beta_1\theta\gamma) = 0 \end{aligned} \tag{8.6}$$

3. Continuity of bending moment, $\left.\dfrac{d^2 r_1(\xi)}{d\xi^2}\right|_{\xi=\mu} = \alpha^4 \left.\dfrac{d^2 r_2(\xi)}{d\xi^2}\right|_{\xi=\gamma}$.

$$A_1(-\sin\beta_1\mu - \sinh\beta_1\mu) + B_1(-\cos\beta_1\mu - \cosh\beta_1\mu) + A_2\alpha^4\theta^2(\sin\beta_1\theta\gamma - \sinh\beta_1\theta\gamma) + B_2\alpha^4\theta^2(\cos\beta_1\theta\gamma - \cosh\beta_1\theta\gamma) = 0 \tag{8.7}$$

4. Continuity of shear force, $\left.\dfrac{d^3r_1(\xi)}{d\xi^3}\right|_{\xi=\mu} = -\alpha^4 \left.\dfrac{d^3r_2(\xi)}{d\xi^3}\right|_{\xi=\gamma}$.

$$A_1(-\cos\beta_1\mu - \cosh\beta_1\mu) + B_1(\sin\beta_1\mu - \sinh\beta_1\mu) + A_2\alpha^4\theta^3(-\cos\beta_1\theta\gamma + \cosh\beta_1\theta\gamma) + B_2\alpha^4\theta^3(\sin\beta_1\theta\gamma + \sinh\beta_1\theta\gamma) = 0 \tag{8.8}$$

The eigenvalues β_1 are now determined from the coefficient matrix of Eqs. 8.5 through 8.8. The eigenvalues are those values of β_1 that make the determinant of the coefficient matrix vanish.

We next turn our attention to expressions for the modal shapes. By arbitrarily setting one of the constants A_1 or B_1 equal to unity, we convert Eqs. 8.5 through 8.8 into a nonhomogeneous set. Only three equations are required to solve for the other three constants, and we choose to utilize the first three. These three nonhomogeneous equations are provided with each case considered, so that they may be solved for the values B_1, A_2, and B_2 and the modal shape expressions may be obtained. The equations are easily solved using Cramer's rule. It is appreciated that there may arise cases for which the only permissible value of the first constant of the left segment of the beam, A_1 in this case, is zero. This situation manifests itself in the obtaining of infinite or indeterminate values for the other constants. In such a case, since both A_1 and B_1, the second constant associated with the left-hand segment, cannot both equal zero, it is evident that B_1 should have been set equal to one. It is seen that solutions for A_2 and B_2, the remaining constants, can easily be obtained in such cases by neglecting the right-hand side of the nonhomogeneous equations provided, setting B_1 equal to one, and solving for A_2 and B_2 from the first two equations of the new nonhomogeneous set. In the family of beams considered with single discontinuities, there are several cases in which it is possible to eliminate immediately three of the constants appearing in the modal shape expressions by means of the boundary conditions. This is true of Cases 8.5 through 8.9 in which an intermediate pinned (simple) support is located at the discontuity and we can write the boundary conditions

$$r_1(\mu) = r_2(\gamma) = 0$$

In cases such as this the constant associated with the modal shape expression for the left-hand span is set equal to unity, and an expression is provided for evaluating the constant associated with the right-hand span. The analyst therefore is not required to solve a set of nonhomogeneous equations to obtain the modal shapes for these beams.

In the tables provided the three variables of interest are α, ϕ, and μ. The range of variables for which tabulated data are provided was selected after numerous considerations. A careful study of the dependence of the eigenvalue β_1 on the stiffness and mass ratios has indicated that it is advisable to present tables with ϕ and α varying in equal increments. The range selected is .6 to 1.0 in intervals of .1 for ϕ and α, and also for the inverse of these parameters, $1/\phi$ and $1/\alpha$. This means, for example, that stiffness and mass ratios may range from $(.6)^4 = .1296$ to 1. Inverses of the ratios may also vary within the same range. In order to minimize storage requirements it has been decided to prepare the tables on the assumption that the stiffness ratio and the mass ratio for any beam are equal to or greater than 1.0, or equal to or less than 1.0. Combinations in which one parameter is greater than 1.0 and the other is less than 1.0 are expected to be rare in occurrence and are not provided for in the tables. Linear interpolation may be used for analyzing beams with parameter values falling between those for which eigenvalues are tabulated. Studies have indicated that maximum errors encountered through interpolation will not be greater than about 1 percent. The tables are provided with values of μ varying between .15 and .85 in increments of .05. It is expected that this range of μ covers most beams of practical interest. Again, linear interpolation involving μ has generally been found to give frequencies with maximum error not greater than about 1 percent. The tables give values for the eigenvalue β_1 associated with the left-hand span of the beam. It is seen that, based on the earlier expression given for β_1, we may write

$$\omega = \frac{1}{L^2}\sqrt{\frac{E_1 I_1}{\rho_1 A_1}}\,\beta_1^2$$

and, having obtained a value for β_1 from the tables, we may obtain the beam frequency from the expression

$$f = \frac{1}{2\pi L^2}\sqrt{\frac{E_1 I_1}{\rho_1 A_1}}\,\beta_1^2 \tag{8.9}$$

In presenting the data, each case studied is given a case number. The same number is utilized in the corresponding table. Beside the case number is found a small sketch of the beam in question and a short description. This is followed by the modal shape expressions and, where appropriate, the nonhomogeneous equations discussed earlier which must be solved to obtain the constants appearing in the expressions. Data are provided for the first four modal shapes. Where a problem possesses symmetry, it is found that data associated with the parameters $1/\phi$ and $1/\alpha$ are not provided. Use of the tables is demonstrated with illustrative problems.

ILLUSTRATIVE PROBLEM 8.1

A steel pipe of 3 in. i.d., .100 in. wall thickness, and 20 ft, 0 in. in length is clamped at one end and given simple support at the other. The 8-ft section adjacent to the pinned end is filled with water.

(a) Assuming the only effect of the water is to add uniformly distributed mass to this latter section, determine the first mode vibration frequency of the pipe.

The moment of inertia of the pipe cross-sectional area

$$=\frac{\pi(D_o^4-D_i^4)}{64}=\frac{\pi}{64}(3.2^4-3^4)=1.171\text{ in.}^4$$

where D_o and D_i are the outer and inner diameters, respectively. The mass per unit length of the unfilled pipe

$$=\frac{\pi(D_o^2-D_i^2)}{4}\times .283=.276\text{ lb/in.}$$

The mass per unit length of the filled pipe

$$=.276+\frac{\pi D_i^2}{4}\times .036=.530\text{ lb/in.}$$

where the density of steel and water are taken as .283 and .036 lb/in.^{3}, respectively.

The problem in question is that of Case 8.2.

$$\phi=\left(\frac{\rho_2 A_2}{\rho_1 A_1}\right)^{1/4}=\text{fourth root of mass-per-unit-length ratio}$$

$$=\left(\frac{.530}{.276}\right)^{1/4}=1.177$$

$$\therefore \frac{1}{\phi}=.849 \qquad \alpha=\left(\frac{E_2 I_2}{E_1 I_1}\right)^{1/4}=1.0 \qquad \mu=\frac{12}{20}=0.6$$

Referring to Table 8.2, it is seen that, for $1/\phi=.8$, $1/\alpha=1.0$, $\mu=.6$,

$$\beta_1=3.464$$

and for $1/\phi=.9$, $1/\alpha=1.0$, $\mu=.6$,

$$\beta_1=3.727$$

Interpolating for $1/\phi=.849$, we obtain

$$\beta_1=3.593$$

Utilizing Eq. 8.9, we obtain

$$f = \frac{1}{2\pi(20 \times 12)^2} \sqrt{\frac{30.0 \times 10^6 \times 1.171 \times 386}{.276}} \times 3.593^2 = 7.91 \text{ Hz}$$

where the modulus of elasticity for steel is taken as 30×10^6 lb/in.^{2}.

(b) Determine the frequency if both ends of the pipe are clamped. This problem corresponds to Case 8.4. In order to utilize the tables available, it is necessary to select the axes so that the filled portion of the beam constitutes the left-hand segment (inset, Case 8.4). Then $\mu = 8/20 = 0.4$, $\phi = .849$, and $\alpha = 1.0$. Referring to Table 8.4, for $\phi = .8$, $\alpha = 1.0$, $\mu = .4$,

$$\beta_1 = 5.431$$

and for $\phi = .9$, $\alpha = 1.0$, $\mu = .4$,

$$\beta_1 = 5.085$$

Interpolating for $\phi = .849$, we obtain

$$\beta_1 = 5.261$$

Then, referring to part (a) above,

$$f = \frac{5.261^2}{3.593^2} \times 7.91 = 16.96 \text{ Hz}$$

(c) Determine the value β_2 to be used for writing the modal shapes of part (a) above.

$$\beta_2 = \frac{\phi}{\alpha} \beta_1 = \frac{1.177}{1} \times 3.593 = 4.229$$

Part 1. Beams with Single Discontinuities

CASE 8.1. Beam clamped at one end and free at the other.

$$r_1(\xi) = \sin \beta_1\xi - \sinh \beta_1\xi + B_1(\cos \beta_1\xi - \cosh \beta_1\xi)$$

$$r_2(\xi) = A_2(\sin \beta_1\theta\xi + \sinh \beta_1\theta\xi) + B_2(\cos \beta_1\theta\xi + \cosh \beta_1\theta\xi)$$

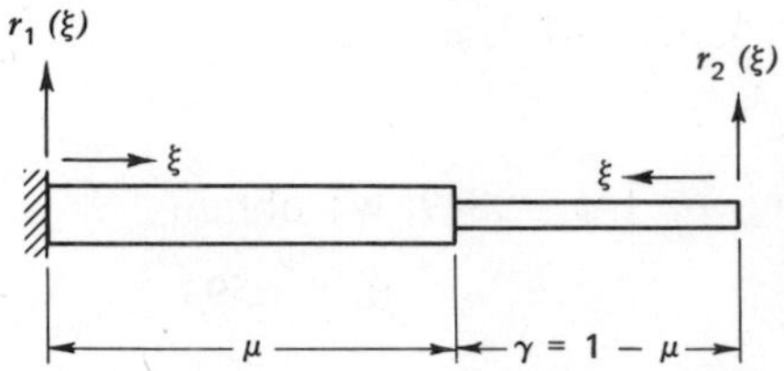

$$B_1(\cos\beta_1\mu - \cosh\beta_1\mu) + A_2(-\sin\beta_1\theta\gamma - \sinh\beta_1\theta\gamma) + B_2(-\cos\beta_1\theta\gamma - \cosh\beta_1\theta\gamma) = \sinh\beta_1\mu - \sin\beta_1\mu$$

$$B_1(-\sin\beta_1\mu - \sinh\beta_1\mu) + A_2\theta(\cos\beta_1\theta\gamma + \cosh\beta_1\theta\gamma) + B_2\theta(-\sin\beta_1\theta\gamma + \sinh\beta_1\theta\gamma) = \cosh\beta_1\mu - \cos\beta_1\mu$$

$$B_1(-\cos\beta_1\mu - \cosh\beta_1\mu) + A_2\alpha^4\theta^2(\sin\beta_1\theta\gamma - \sinh\beta_1\theta\gamma) + B_2\alpha^4\theta^2(\cos\beta_1\theta\gamma - \cosh\beta_1\theta\gamma) = \sin\beta_1\mu + \sinh\beta_1\mu$$

where

$$\alpha = \left(\frac{E_2I_2}{E_1I_1}\right)^{1/4} \qquad \phi = \left(\frac{\rho_2A_2}{\rho_1A_1}\right)^{1/4} \qquad \theta = \frac{\phi}{\alpha}$$

CASE 8.2. Beam clamped at one end, and with simple support at the other.

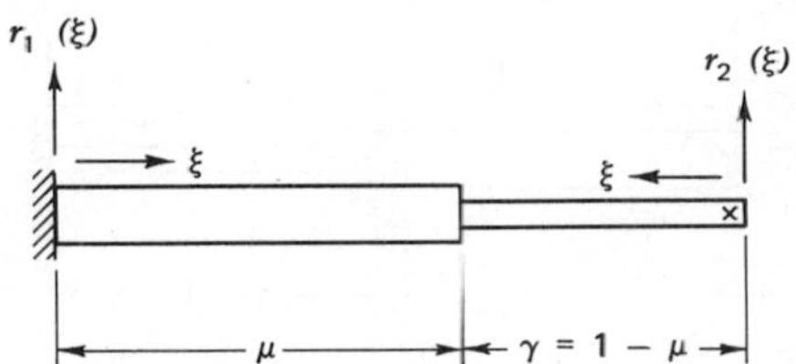

$$r_1(\xi) = \sin\beta_1\xi - \sinh\beta_1\xi + B_1(\cos\beta_1\xi - \cosh\beta_1\xi)$$
$$r_2(\xi) = A_2\sin\beta_1\theta\xi + C_2\sinh\beta_1\theta\xi$$

$$B_1(\cos\beta_1\mu - \cosh\beta_1\mu) + A_2(-\sin\beta_1\theta\gamma) + C_2(-\sinh\beta_1\theta\gamma) = \sinh\beta_1\mu - \sin\beta_1\mu$$

$$B_1(-\sin\beta_1\mu - \sinh\beta_1\mu) + A_2\theta(\cos\beta_1\theta\gamma) + C_2\theta(\cosh\beta_1\theta\gamma) = \cosh\beta_1\mu - \cos\beta_1\mu$$

$$B_1(-\cos\beta_1\mu - \cosh\beta_1\mu) + A_2\alpha^4\theta^2(\sin\beta_1\theta\gamma) + C_2\alpha^4\theta^2(-\sinh\beta_1\theta\gamma) = \sinh\beta_1\mu + \sin\beta_1\mu$$

where

$$\alpha = \left(\frac{E_2I_2}{E_1I_1}\right)^{1/4} \qquad \phi = \left(\frac{\rho_2A_2}{\rho_1A_1}\right)^{1/4} \qquad \theta = \frac{\phi}{\alpha}$$

CASE 8.3. Beam with simple support at each end.

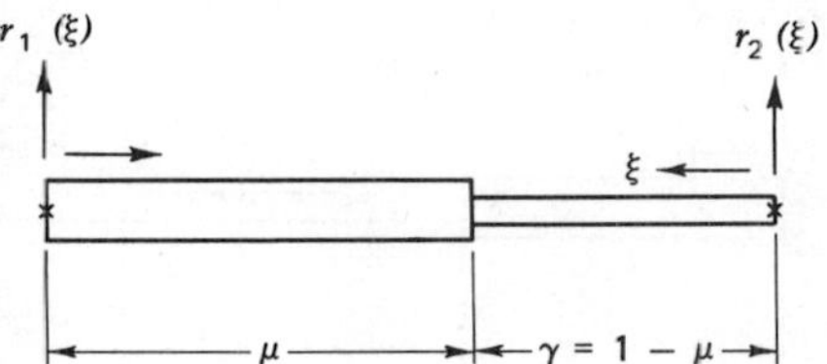

$$r_1(\xi) = \sin \beta_1\xi + C_1 \sinh \beta_1\xi$$
$$r_2(\xi) = A_2 \sin \beta_1\theta\xi + C_2 \sinh \beta_1\theta\xi$$
$$C_1(\sinh \beta_1\mu) + A_2(-\sin \beta_1\theta\gamma) + C_2(-\sinh \beta_1\theta\gamma) = -\sin \beta_1\mu$$
$$C_1(\cosh \beta_1\mu) + A_2\theta(\cos \beta_1\theta\gamma) + C_2\theta(\cosh \beta_1\theta\gamma) = -\cos \beta_1\mu$$
$$C_1(\sinh \beta_1\mu) + A_2\alpha^4\theta^2(\sin \beta_1\theta\gamma) + C_2\alpha^4\theta^2(-\sinh \beta_1\theta\gamma) = \sin \beta_1\mu$$

where

$$\alpha = \left(\frac{E_2I_2}{E_1I_1}\right)^{1/4} \qquad \phi = \left(\frac{\rho_2A_2}{\rho_1A_1}\right)^{1/4} \qquad \theta = \frac{\phi}{\alpha}$$

CASE 8.4. Beam clamped at each end.

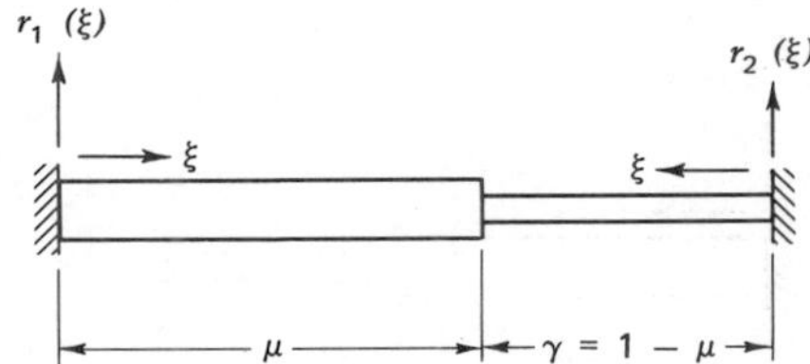

$$r_1(\xi) = \sin \beta_1\xi - \sinh \beta_1\xi + B_1(\cos \beta_1\xi - \cosh \beta_1\xi)$$
$$r_2(\xi) = A_2(\sin \beta_1\theta\xi - \sinh \beta_1\theta\xi) + B_2(\cos \beta_1\theta\xi - \cosh \beta_1\theta\xi)$$
$$B_1(\cos \beta_1\mu - \cosh \beta_1\mu) + A_2(-\sin \beta_1\theta\gamma + \sinh \beta_1\theta\gamma) + B_2(-\cos \beta_1\theta\gamma + \cosh \beta_1\theta\gamma) = \sinh \beta_1\mu - \sin \beta_1\mu$$
$$B_1(-\sin \beta_1\mu - \sinh \beta_1\mu) + A_2\theta(\cos \beta_1\theta\gamma - \cosh \beta_1\theta\gamma) + B_2\theta(-\sin \beta_1\theta\gamma - \sinh \beta_1\theta\gamma) = \cosh \beta_1\mu - \cos \beta_1\mu$$
$$B_1(-\cos \beta_1\mu - \cosh \beta_1\mu) + A_2\alpha^4\theta^2(\sin \beta_1\theta\gamma + \sinh \beta_1\theta\gamma) + B_2\alpha^4\theta^2(\cos \beta_1\theta\gamma + \cosh \beta_1\theta\gamma) = \sin \beta_1\mu + \sinh \beta_1\mu$$

where

$$\alpha = \left(\frac{E_2I_2}{E_1I_1}\right)^{1/4} \qquad \phi = \left(\frac{\rho_2A_2}{\rho_1A_1}\right)^{1/4} \qquad \theta = \frac{\phi}{\alpha}$$

CASE 8.5. Beam clamped at one end, free at the other, and with simple support at the discontinuity.

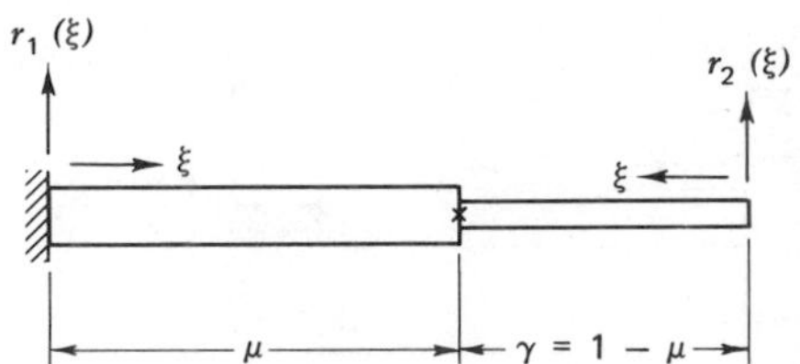

$$r_1(\xi) = \sin \beta_1\xi - \sinh \beta_1\xi + \frac{\sinh \beta_1\mu - \sin \beta_1\mu}{\cos \beta_1\mu - \cosh \beta_1\mu}(\cos \beta_1\xi - \cosh \beta_1\xi)$$

$$r_2(\xi) = \gamma_1\left[\sin \beta_1\theta\xi + \sinh \beta_1\theta\xi - \frac{\sin \beta_1\theta\gamma + \sinh \beta_1\theta\gamma}{\cos \beta_1\theta\gamma + \cosh \beta_1\theta\gamma} \times (\cos \beta_1\theta\xi + \cosh \beta_1\theta\xi)\right]$$

where

$$\gamma_1 = \frac{(\cos \beta_1\mu \cosh \beta_1\mu - 1)(\cos \beta_1\theta\gamma + \cosh \beta_1\theta\gamma)}{\theta(\cos \beta_1\mu - \cosh \beta_1\mu)(\cos \beta_1\theta\gamma \cosh \beta_1\theta\gamma + 1)}$$

$$\alpha = \left(\frac{E_2I_2}{E_1I_1}\right)^{1/4} \qquad \phi = \left(\frac{\rho_2A_2}{\rho_1A_1}\right)^{1/4} \qquad \theta = \frac{\phi}{\alpha}$$

CASE 8.6. Beam clamped at one end, simply supported at the other, and with simple support at the discontinuity.

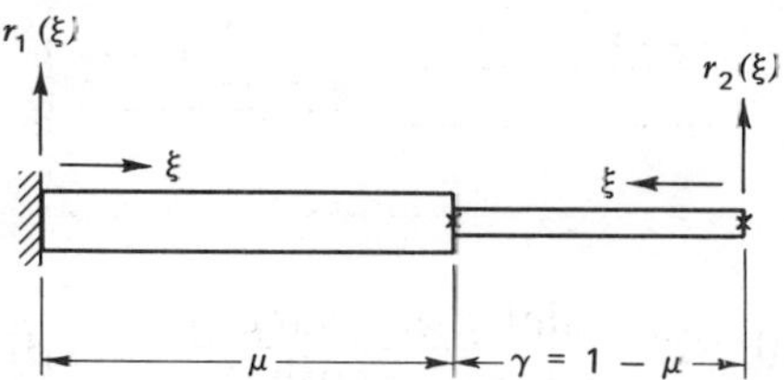

$$r_1(\xi) = \sin \beta_1\xi - \sinh \beta_1\xi + \frac{\sinh \beta_1\mu - \sin \beta_1\mu}{\cos \beta_1\mu - \cosh \beta_1\mu}(\cos \beta_1\xi - \cosh \beta_1\xi)$$

$$r_2(\xi) = \gamma_1\left(\sin \beta_1\theta\xi - \frac{\sin \beta_1\theta\gamma}{\sinh \beta_1\theta\gamma}\sinh \beta_1\theta\xi\right)$$

where

$$\gamma_1 = \frac{2(1 - \cos \beta_1\mu \cosh \beta_1\mu)\sinh \beta_1\theta\gamma}{\theta(\cos \beta_1\mu - \cosh \beta_1\mu)(\sin \beta_1\theta\gamma \cosh \beta_1\theta\gamma - \cos \beta_1\theta\gamma \sinh \beta_1\theta\gamma)}$$

$$\alpha = \left(\frac{E_2I_2}{E_1I_1}\right)^{1/4} \qquad \phi = \left(\frac{\rho_2A_2}{\rho_1A_1}\right)^{1/4} \qquad \theta = \frac{\phi}{\alpha}$$

CASE 8.7. Beam with simple support at each end, and with simple support at the discontinuity.

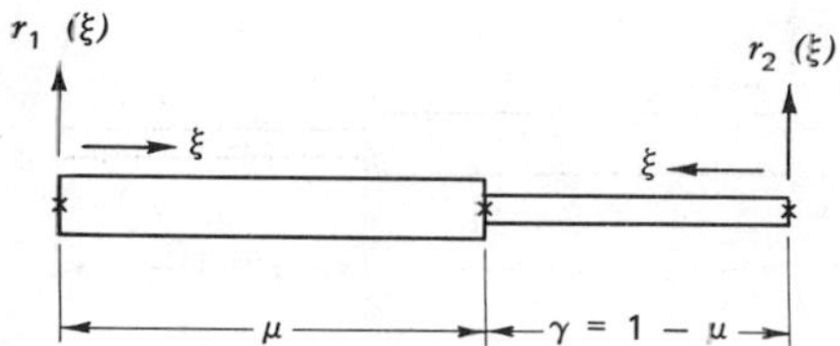

$$r_1(\xi) = \sin \beta_1 \xi - \frac{\sin \beta_1 \mu}{\sinh \beta_1 \mu} \sinh \beta_1 \xi$$

$$r_2(\xi) = \gamma_1 \left(\sin \beta_1 \theta \xi - \frac{\sin \beta_1 \theta \gamma}{\sinh \beta_1 \theta \gamma} \sinh \beta_1 \theta \xi \right)$$

where

$$\gamma_1 = -\frac{(\cos \beta_1 \mu \sinh \beta_1 \mu - \sin \beta_1 \mu \cosh \beta_1 \mu) \sinh \beta_1 \theta \gamma}{\theta \sinh \beta_1 \mu (\cos \beta_1 \theta \gamma \sinh \beta_1 \theta \gamma - \sin \beta_1 \theta \gamma \cosh \beta_1 \theta \gamma)}$$

$$\alpha = \left(\frac{E_2 I_2}{E_1 I_1} \right)^{1/4} \qquad \phi = \left(\frac{\rho_2 A_2}{\rho_1 A_1} \right)^{1/4} \qquad \theta = \frac{\phi}{\alpha}$$

CASE 8.8. Beam clamped at each end, and with simple support at the discontinuity.

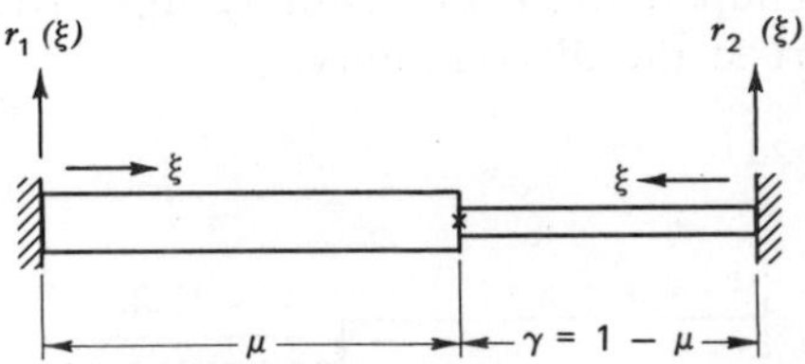

$$r_1(\xi) = \sin \beta_1 \xi - \sinh \beta_1 \xi + \frac{\sinh \beta_1 \mu - \sin \beta_1 \mu}{\cos \beta_1 \mu - \cosh \beta_1 \mu} (\cos \beta_1 \xi - \cosh \beta_1 \xi)$$

$$r_2(\xi) = \gamma_1 \left[\sin \beta_1 \theta \xi - \sinh \beta_1 \theta \xi + \frac{\sinh \beta_1 \theta \gamma - \sin \beta_1 \theta \gamma}{\cos \beta_1 \theta \gamma - \cosh \beta_1 \theta \gamma} \times (\cos \beta_1 \theta \xi - \cosh \beta_1 \theta \xi) \right]$$

where

$$\gamma_1 = \frac{(\cos \beta_1 \mu \cosh \beta_1 \mu - 1)(\cos \beta_1 \theta \gamma - \cosh \beta_1 \theta \gamma)}{\theta (1 - \cos \beta_1 \theta \gamma \cosh \beta_1 \theta \gamma)(\cos \beta_1 \mu - \cosh \beta_1 \mu)}$$

$$\alpha = \left(\frac{E_2 I_2}{E_1 I_1} \right)^{1/4} \qquad \phi = \left(\frac{\rho_2 A_2}{\rho_1 A_1} \right)^{1/4} \qquad \theta = \frac{\phi}{\alpha}$$

CASE 8.9. Beam with simple support at one end, free at the other, and with simple support at the discontinuity.

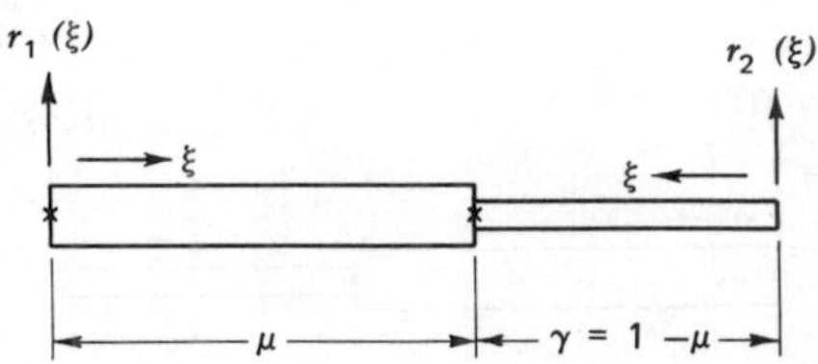

$$r_1(\xi) = \sin \beta_1 \xi - \frac{\sin \beta_1 \mu}{\sinh \beta_1 \mu} \sinh \beta_1 \xi$$

$$r_2(\xi) = \gamma_1 \left[\sin \beta_1 \theta \xi + \sinh \beta_1 \theta \xi - \frac{\sin \beta_1 \theta \gamma + \sinh \beta_1 \theta \gamma}{\cos \beta_1 \theta \gamma + \cosh \beta_1 \theta \gamma} \times (\cos \beta_1 \theta \xi + \cosh \beta_1 \theta \xi) \right]$$

where

$$\gamma_1 = \frac{(\sin \beta_1 \mu \cosh \beta_1 \mu - \cos \beta_1 \mu \sinh \beta_1 \mu)(\cos \beta_1 \theta \gamma + \cosh \beta_1 \theta \gamma)}{2(\cos \beta_1 \theta \gamma \cosh \beta_1 \theta \gamma + 1)}$$

$$\alpha = \left(\frac{E_2 I_2}{E_1 I_1} \right)^{1/4} \qquad \phi = \left(\frac{\rho_2 A_2}{\rho_1 A_1} \right)^{1/4} \qquad \theta = \frac{\phi}{\alpha}$$

Part 2. Beams with Symmetric Discontinuity in Cross-Sectional Properties and Identical Outer End Supports

This family of beams is considered separately because of its particular characteristics. For illustrative purposes we consider a beam with simple support at each end and with a discontinuity in cross-sectional properties distributed symmetrically about the center of the beam as shown in Fig. 8.1.

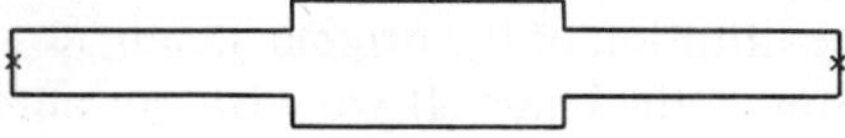

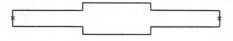

Fig. 8.1

It is noted that the boundary conditions of this beam also present symmetry with respect to the center of the beam. It is known that beams of this type vibrate in modal shapes which are alternately symmetric and antisymmetric with respect to the center of the beam. The first mode is symmetric, the second antisymmetric, the third symmetric, and so on.

It is of utmost importance to observe that the antisymmetric frequencies and modal shapes can be determined by focusing our attention on half the beam only, let us say the left half. We may consider the left half to be severed from the right half and to be given simple support at the end where it is severed. This conceptual simplification is permissible, since for an antisymmetric modal shape the displacement and bending moment must be zero at the center of the beam. For the lowest frequency antisymmetric mode, we may therefore focus our attention on the half-beam as shown in Fig. 8.2. Since the geometry and material properties of the original beam in Fig. 8.1 are known, we also know the geometry and material properties of

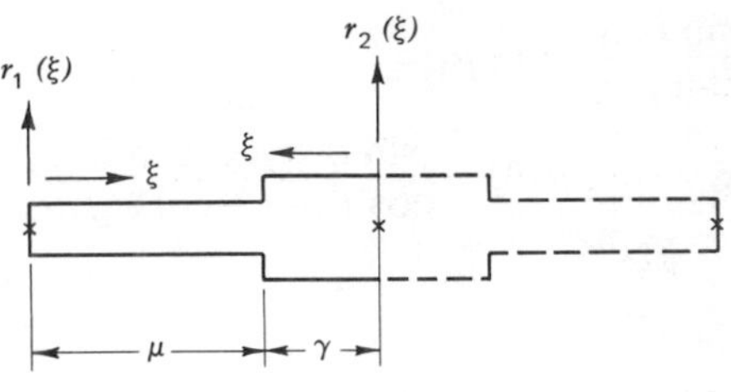

Fig. 8.2

the beam in Fig. 8.2 which has half the length of the original beam. We can easily determine the first and second mode frequencies and modal shapes from the tables in Part 1. The first and second mode frequencies of the half-beam correspond, respectively, to the second and fourth antisymmetric) modes of the full length beam. Since the modal shapes of the left half of the beam can be determined, the modal shapes of the right half can be immediately inferred from the antisymmetry.

For all the beams in Part 2 the information necessary for establishing the antisymmetric (second and fourth) mode frequencies and modal shapes is therefore already available in Part 1. We turn next to the first and third modes of vibration of these beams, that is, the symmetric modes. This time we find that we can take advantage of the symmetry, and that again we need only focus our attention on the left half of the beam. It is appreciated that, for symmetric mode vibration of the original beam, no slope or shear force can exist at the center of the beam. If we wish our analysis to involve the left half of the beam only, we must enforce the condition of zero slope and zero shear at its right-hand end. The half-beam of interest is shown schematically in Fig. 8.3 where the rollers at the right-hand end of the half-beam forbid the existence of slope and shear. The analysis of the half-beam above is conducted in a manner identical to that described for the beams of Part 1. Its first and second modes correspond, respectively, to the first and third (symmetric) modes of the full length beam.

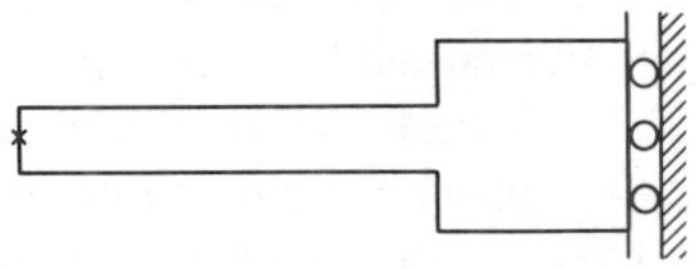

Fig. 8.3

All the beams in Part 2 have been assigned a reference number. Beside each reference number is a sketch of the beam and a brief description, as in Part 1. As indicated above, the antisymmetric modes can be analyzed using existing cases. These are referred to by case number for each beam in Part

2. In addition, information for the analysis of half-beams with zero shear and slope at the inner boundary is given. Tables have numbers corresponding to the case numbers. The objective of the tables is to permit beam vibration analysis, with the length of the middle section of the original beam varying from .15 to .85 of the original beam overall length. The analysis provided for half-beams permits overall beam analysis within these limits for the first four modes. Expressions for the constants used in modal shape expressions are provided or, if there is more than one constant in each expression, nonhomogeneous equations are provided for evaluating the constants. These equations are to be solved in a manner identical to that described for the equations in Part 1. Illustrative problems are worked out to demonstrate the use of the tables.

CASE 8.10. Beam with simple support at each end, and with symmetric discontinuity in cross section.

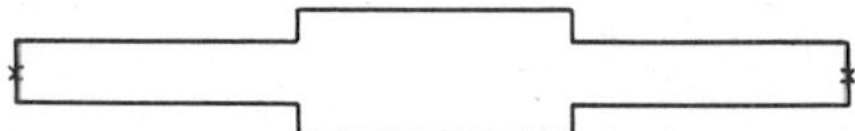

Antisymmetric modes and frequencies (modes 2 and 4). Utilize first and second mode data of Case 8.3.

Symmetric modes and frequencies (modes 1 and 3). Utilize first and second mode data, Table 8.10, and modal shape information given below.

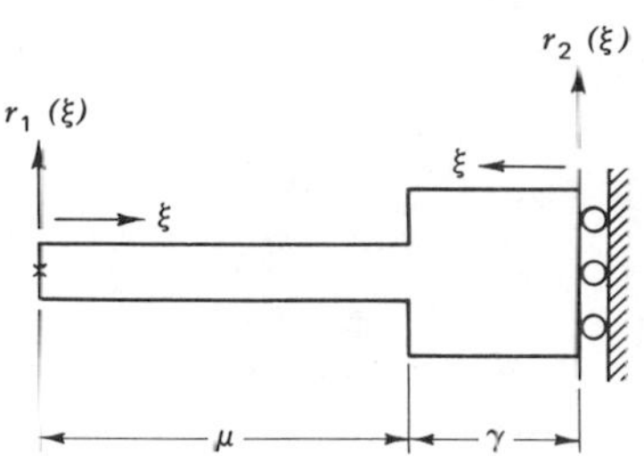

$$r_1(\xi) = \sin \beta_1\xi + C_1 \sinh \beta_1\xi$$

$$r_2(\xi) = B_2 \cos \beta_1\theta\xi + D_2 \cosh \beta_1\theta\xi$$

$$C_1(\sinh \beta_1\mu) + B_2(-\cos \beta_1\theta\gamma) + D_2(-\cosh \beta_1\theta\gamma) = -\sin \beta_1\mu$$

$$C_1(\cosh \beta_1\mu) + B_2\theta(-\sin \beta_1\theta\gamma) + D_2\theta(\sinh \beta_1\theta\gamma) = -\cos \beta_1\mu$$

$$C_1(\sinh \beta_1\mu) + B_2\alpha^4\theta^2(\cos \beta_1\theta\gamma) + D_2\alpha^4\theta^2(-\cosh \beta_1\theta\gamma) = \sin \beta_1\mu$$

$$\alpha = \left(\frac{E_2I_2}{E_1I_1}\right)^{1/4} \qquad \phi = \left(\frac{\rho_2A_2}{\rho_1A_1}\right)^{1/4} \qquad \theta = \frac{\phi}{\alpha}$$

CASE 8.11. Beam clamped at each end, and with symmetric discontinuity in cross section.

Antisymmetric modes and frequencies (modes 2 and 4). Utilize first and second mode data of Case 8.2.

Symmetric modes and frequencies (modes 1 and 3). Utilize first and second mode data, Table 8.11, and modal shape information given below.

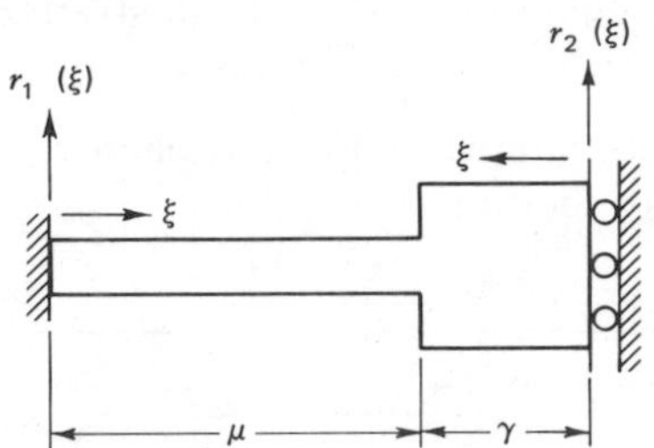

$$r_1(\xi) = \sin \beta_1\xi - \sinh \beta_1\xi + B_1(\cos \beta_1\xi - \cosh \beta_1\xi)$$

$$r_2(\xi) = B_2 \cos \beta_1\theta\xi + D_2 \cosh \beta_1\theta\xi$$

$$B_1(\cos \beta_1\mu - \cosh \beta_1\mu) + B_2(-\cos \beta_1\theta\gamma) + D_2(-\cosh \beta_1\theta\gamma) = \sinh \beta_1\mu - \sin \beta_1\mu$$

$$B_1(-\sin \beta_1\mu - \sinh \beta_1\mu) + B_2\theta(-\sin \beta_1\theta\gamma) + D_2\theta(\sinh \beta_1\theta\gamma) = \cosh \beta_1\mu - \cos \beta_1\mu$$

$$B_1(-\cos \beta_1\mu - \cosh \beta_1\mu) + B_2\alpha^4\theta^2(\cos \beta_1\theta\gamma) + D_2\alpha^4\theta^2(-\cosh \beta_1\theta\gamma) = \sinh \beta_1\mu + \sin \beta_1\mu$$

$$\alpha = \left(\frac{E_2I_2}{E_1I_1}\right)^{1/4} \qquad \phi = \left(\frac{\rho_2A_2}{\rho_1A_1}\right)^{1/4} \qquad \theta = \frac{\phi}{\alpha}$$

CASE 8.12. Beam with simple support at each end, and with simple support at each end of a symmetric discontinuity in cross section.

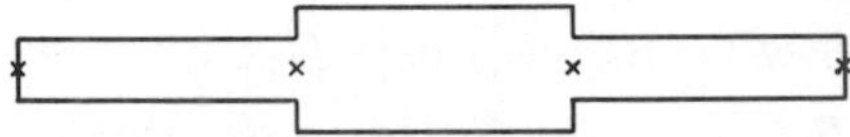

Antisymmetric modes and frequencies (modes 2 and 4). Utilize first and second mode data of Case 8.7.

Symmetric modes and frequencies (modes 1 and 3). Utilize first and second mode data, Table 8.12, and modal shape information given below.

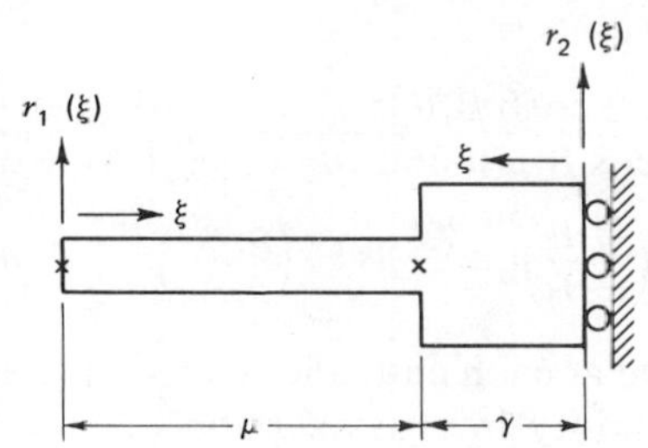

$$r_1(\xi) = \sin \beta_1\xi - \frac{\sin \beta_1\mu}{\sinh \beta_1\mu} \sinh \beta_1\xi$$

$$r_2(\xi) = \gamma_1\left(\cos \beta_1\theta\xi \frac{-\cos \beta_1\theta\gamma}{\cosh \beta_1\theta\gamma} \cosh \beta_1\theta\xi\right)$$

where

$$\gamma_1 = \frac{(\cos \beta_1\mu \sinh \beta_1\mu - \sin \beta_1\mu \cosh \beta_1\mu) \cosh \beta_1\theta\gamma}{\theta \sinh \beta_1\mu (\sin \beta_1\theta\gamma \cosh \beta_1\theta\gamma + \cos \beta_1\theta\gamma \sinh \beta_1\theta\gamma)}$$

$$\alpha = \left(\frac{E_2I_2}{E_1I_1}\right)^{1/4} \qquad \phi = \left(\frac{\rho_2A_2}{\rho_1A_1}\right)^{1/4} \qquad \theta = \frac{\phi}{\alpha}$$

CASE 8.13. Beam clamped at each end, and with simple support at each end of a symmetric discontinuity in cross section.

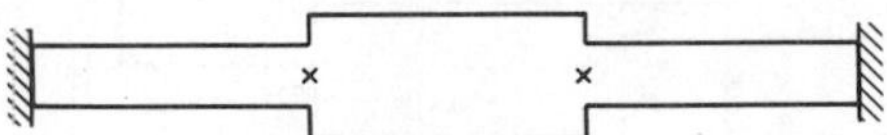

Antisymmetric modes and frequencies (modes 2 and 4). Utilize first and second mode data of Case 8.6.

Symmetric modes and frequencies (modes 1 and 3). Utilize first and second mode data, Table 8.13, and modal shape information given below.

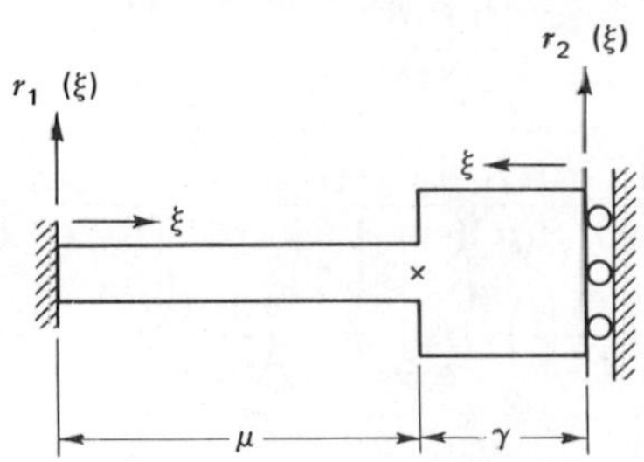

$$r_1(\xi) = \sin \beta_1\xi - \sinh \beta_1\xi + \frac{\sin \beta_1\mu - \sinh \beta_1\mu}{\cosh \beta_1\mu - \cos \beta_1\mu} (\cos \beta_1\xi - \cosh \beta_1\xi)$$

$$r_2(\xi) = \gamma_1\left(\cos \beta_1\theta\xi - \frac{\cos \beta_1\theta\gamma}{\cosh \beta_1\theta\gamma} \cosh \beta_1\theta\xi\right)$$

where

$$\gamma_1 = \frac{2\cosh\beta_1\mu(\cos\beta_1\mu\cosh\beta_1\mu - 1)}{\theta(\cosh\beta_1\mu - \cos\beta_1\mu)(\sin\beta_1\theta\gamma\cosh\beta_1\theta\gamma + \cos\beta_1\theta\gamma\sinh\beta_1\theta\gamma)}$$

$$\alpha = \left(\frac{E_2 I_2}{E_1 I_1}\right)^{1/4} \qquad \phi = \left(\frac{\rho_2 A_2}{\rho_1 A_1}\right)^{1/4} \qquad \theta = \frac{\phi}{\alpha}$$

CASE 8.14. Beam free at each end, and with simple support at each end of a symmetric discontinuity in cross section.

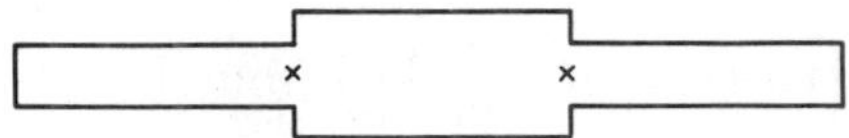

Antisymmetric modes and frequencies (modes 2 and 4). Utilize first and second mode data of Case 8.9.

Symmetric modes and frequencies (modes 1 and 3). Utilize first and second mode data, Table 8.14, and modal shape information given below.

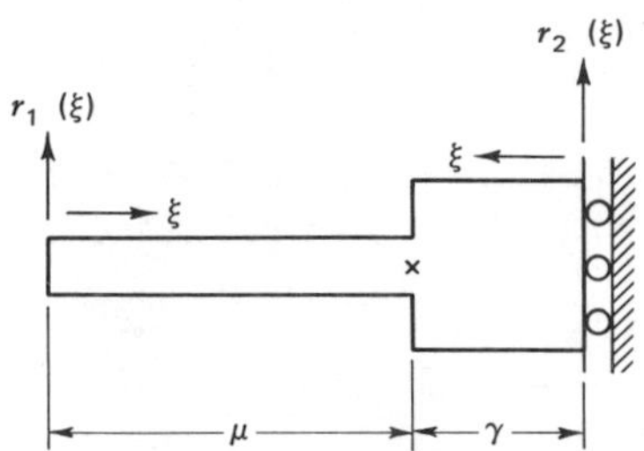

$$r_1(\xi) = \sin\beta_1\xi + \sinh\beta_1\xi - \frac{\sin\beta_1\mu + \sinh\beta_1\mu}{\cos\beta_1\mu + \cosh\beta_1\mu}(\cos\beta_1\xi + \cosh\beta_1\xi)$$

$$r_2(\xi) = \gamma_1\left(\cos\beta_1\theta\xi - \frac{\cos\beta_1\theta\gamma}{\cosh\beta_1\theta\gamma}\cosh\beta_1\theta\xi\right)$$

where

$$\gamma_1 = \frac{[(\cos\beta_1\mu + \cosh\beta_1\mu)^2 - (\sinh^2\beta_1\mu - \sin^2\beta_1\mu)]\cosh\beta_1\theta\gamma}{\theta(\cos\beta_1\mu + \cosh\beta_1\mu)(\sin\beta_1\theta\gamma\cosh\beta_1\theta\gamma + \cos\beta_1\theta\gamma\sinh\beta_1\theta\gamma)}$$

$$\alpha = \left(\frac{E_2 I_2}{E_1 I_1}\right)^{1/4} \qquad \phi = \left(\frac{\rho_2 A_2}{\rho_1 A_1}\right)^{1/4} \qquad \theta = \frac{\phi}{\alpha}$$

ILLUSTRATIVE PROBLEM 8.2

The steel pipe in Illustrative Problem 8.1 is simply supported at each end. A central portion of the pipe of length 8 ft, 0 in. is filled with water.

(a) Determine the first mode frequency.

The problem in question is that of Case 8.10. Referring to Illustrative Problem 8.1, it is seen that

$$\frac{1}{\phi} = .849 \qquad \text{and} \qquad \frac{1}{\alpha} = 1.0$$

Focusing our attention on half the beam (Case 8.10), it is seen that $\mu = 0.6$ and, referring to Table 8.10, we observe that for $1/\phi = 0.8$, $1/\alpha = 1.0$, $\mu = 0.6$

$$\beta_1 = 1.318$$

and for $1/\phi = .9$, $1/\alpha = 1.0$, $\mu = 0.6$,

$$\beta_1 = 1.452$$

Interpolating for $1/\phi = .849$, we obtain,

$$\beta_1 = 1.384$$

Utilizing results of illustrative problem 8.1(a),

$$f = 7.91 \times \frac{1.384^2}{3.593^2} = 1.17 \text{ Hz}$$

(b) Determine the second mode frequency.
Referring to Case 8.10, it is seen that we must utilize the first mode data of Case 8.3. The corresponding problem in Case 8.3 (where advantage is taken of symmetry) has $\mu = (1 - .6) = .4$ and $\phi = .849$. Referring to Table 8.3, we observe that for $\phi = 0.8$, $\alpha = 1.0$, $\mu = 0.4$

$$\beta_1 = 3.576$$

and for $\phi = 0.9$, $\alpha = 1.0$, $\mu = .4$

$$\beta_1 = 3.362$$

Interpolating for $\phi = .849$, we obtain

$$\beta_1 = 3.471$$

$$\therefore f = 7.91 \times \frac{3.471^2}{3.593^2} = 7.38 \text{ Hz}$$

Part 3. Beams with Symmetric Discontinuity in Cross-Sectional Properties and Differing End Supports

Typical of this family of beams is the one shown in the inset for Case 8.15. It is necessary to introduce three displacement functions, as shown in the figure. Although initially there is a total of 12 unknowns in the functions, it is always possible to eliminate 4 of the unknowns by enforcing the boundary conditions at each end of the beam. For beams such as the one in Case 8.15 it is then necessary to write eight homogeneous algebraic equations by enforcing the conditions of continuity of displacement, slope, bending moment, and shear forces at the cross-sectional discontinuities. The eigenvalues β_1 can then be found, as was done in Parts 1 and 2, by obtaining those values of β_1 that make the determinant of the coefficient matrix of the unknowns vanish. For beams where simple support is provided at the cross-sectional discontinuities, four additional constants can be immediately eliminated by enforcing the boundary conditions of zero displacement at the discontinuities. We are then left with four boundary conditions, continuity of slope and bending moment at the discontinuities, from which four homogeneous algebraic equations can be written. The eigenvalues are then obtained by requiring the determinant of the coefficient matrix of the four unknowns in these equations to vanish.

We next focus our attention on the modal shapes. For the beams without simple support at the discontinuities, the first unknown in the left-hand beam segment is arbitrarily set equal to one. This means that the set of eight homogeneous equations now becomes nonhomogeneous. Only the first seven are required to solve for the other seven unknowns which can be obtained using Cramer's rule. These seven equations are provided. It is appreciated that, as in Part 1, there may arise cases for which the only permissible value of the first unknown of the left-hand beam segment is zero. This situation manifests itself in the obtaining of infinite or indeterminate values for the other unknowns. In such cases, since the unknowns associated with the left-hand span cannot be both equal to zero, it is evident that the second unknown should have been set equal to unity. The procedure to be followed, then, is to neglect the right-hand side of the nonhomogeneous equations provided and set the second unknown associated with the left-hand beam segment equal to unity. The set of equations then becomes nonhomogeneous again, and the remaining six unknowns can be obtained by solving the first six equations of the new set. For beams with simple support at the discontinuities, only one unknown is associated with the left-hand beam segment. Since it can never be equal to zero, it is set equal to unity and the set of four algebraic equations are made nonhomogeneous. The first three equations of the set are provided with each case, so that the remaining unknowns appearing in the displacement functions can be obtained.

The objective of the tables is to permit beam vibration analysis, with the length of the middle section of the beam varying from .15 to .85 of the beam overall length. Tables have been prepared with values of μ (Fig. 8.2) varying in such a way as to make vibration analysis possible over this range. Each case is given a number. Tables have numbers corresponding to the cases to which they pertain. Beside the case number are a brief description of the beam in question, expressions for the modal shapes of vibration, and nonhomogeneous equations which permit solution for the constants appearing in the modal shape expressions. Illustrative problems are worked out to demonstrate the use of the tables.

The cases presented are considered those most likely to be encountered in practice. It is noted that data are provided in Case 8.15 for associated values of $1/\phi$ and $1/\alpha$ only, because the only values of α and ϕ likely to be encountered when solving problems of this type are greater than one.

ILLUSTRATIVE PROBLEM 8.3

The steel pipe in Illustrative Problem 8.1 is clamped at one end and simply supported at the other. A central portion of the pipe, of length 6 ft, 0 in., is filled with water. Determine the first mode vibration frequency.

The problem in question is that in Case 8.15. The value of μ is evidently equal to .35. Referring to Illustrative Problem 8.1, it is seen that

$$\frac{1}{\phi} = .849 \qquad \text{and} \qquad \frac{1}{\alpha} = 1.0$$

Utilizing Table 8.15, it is observed that for $1/\phi = .8$, $1/\alpha = 1.0$, $\mu = .35$

$$\beta_1 = 3.373$$

and for $1/\phi = .9$, $1/\alpha = 1.0$, $\mu = .35$

$$\beta_1 = 3.675$$

Interpolating for $1/\phi = .849$, we obtain

$$\beta_1 = 3.521$$

and, utilizing results of Illustrative Problem 8.1(a),

$$f = 7.91 \times \frac{3.521^2}{3.593^2} = 7.60 \text{ Hz}$$

CASE 8.15. Beam clamped at one end, with simple support at the other, and with symmetrical discontinuity in cross section.

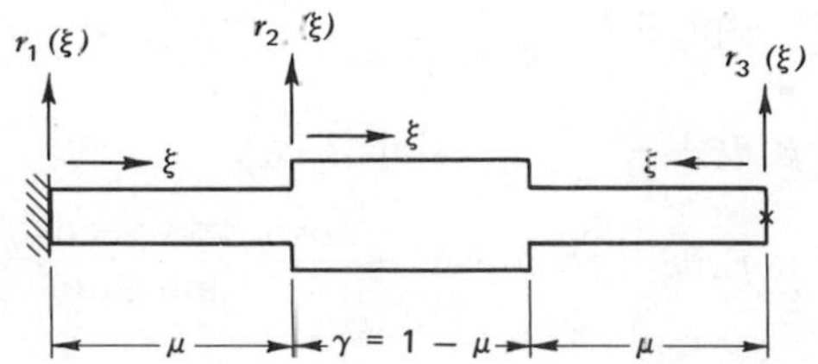

$$r_1(\xi) = \sin\beta_1\xi - \sinh\beta_1\xi + B_1(\cos\beta_1\xi - \cosh\beta_1\xi)$$
$$r_2(\xi) = A_2\sin\beta_1\theta\xi + B_2\cos\beta_1\theta\xi + C_2\sinh\beta_1\theta\xi + D_2\cosh\beta_1\theta\xi$$
$$r_3(\xi) = A_3\sin\beta_1\xi + C_3\sinh\beta_1\xi$$

$$B_1(\cos\beta_1\mu - \cosh\beta_1\mu) + A_2(0) + B_2(-1) + C_2(0) + D_2(-1) = \sinh\beta_1\mu - \sin\beta_1\mu$$

$$B_1(\sin\beta_1\mu + \sinh\beta_1\mu) + A_2(\theta) + B_2(0) + C_2(\theta) + D_2(0) = \cos\beta_1\mu - \cosh\beta_1\mu$$

$$B_1(\cos\beta_1\mu + \cosh\beta_1\mu) + A_2(0) + B_2(-\alpha^4\theta^2) + C_2(0) + D_2(\alpha^4\theta^2) = -\sin\beta_1\mu - \sinh\beta_1\mu$$

$$B_1(\sin\beta_1\mu - \sinh\beta_1\mu) + A_2(\alpha^4\theta^3) + B_2(0) + C_2(-\alpha^4\theta^3) + D_2(0) = \cos\beta_1\mu + \cosh\beta_1\mu$$

$$A_2(\sin\beta_1\theta\gamma) + B_2(\cos\beta_1\theta\gamma) + C_2(\sinh\beta_1\theta\gamma) + D_2(\cosh\beta_1\theta\gamma) + A_3(-\sin\beta_1\mu) + C_3(-\sinh\beta_1\mu) = 0$$

$$A_2(\theta\cos\beta_1\theta\gamma) + B_2(-\theta\sin\beta_1\theta\gamma) + C_2(\theta\cosh\beta_1\theta\gamma) + D_2(\theta\sinh\beta_1\theta\gamma) + A_3(\cos\beta_1\mu) + C_3(\cosh\beta_1\mu) = 0$$

$$A_2(-\alpha^4\theta^2\sin\beta_1\theta\gamma) + B_2(-\alpha^4\theta^2\cos\beta_1\theta\gamma) + C_2(\alpha^4\theta^2\sinh\beta_1\theta\gamma) + D_2(\alpha^4\theta^2\cosh\beta_1\theta\gamma) + A_3(\sin\beta_1\mu) + C_3(-\sinh\beta_1\mu) = 0$$

where

$$\alpha = \left(\frac{E_2I_2}{E_1I_1}\right)^{1/4} \qquad \phi = \left(\frac{\rho_2A_2}{\rho_1A_1}\right)^{1/4} \qquad \theta = \frac{\phi}{\alpha}$$

CASE 8.16. Beam clamped at one end, with simple support at the other, and with simple support at each end of a symmetrical discontinuity in cross section.

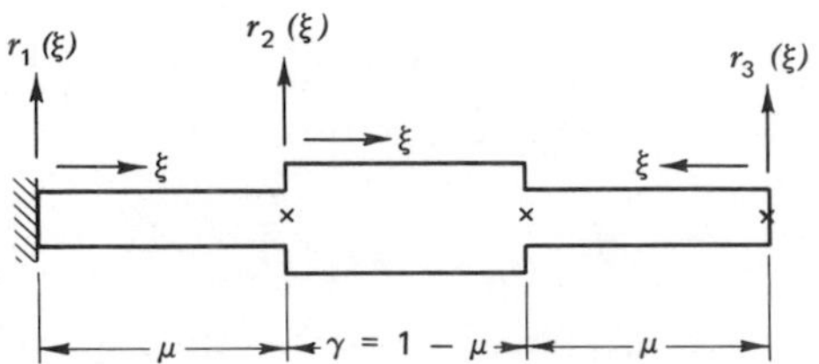

$$r_1(\xi) = \sin\beta_1\xi - \sinh\beta_1\xi + \frac{\sin\beta_1\mu - \sinh\beta_1\mu}{\cosh\beta_1\mu - \cos\beta_1\mu}(\cos\beta_1\xi - \cosh\beta_1\xi)$$

$$r_2(\xi) = A_2\left(\sin\beta_1\theta\xi - \frac{\sin\beta_1\theta\gamma}{\sinh\beta_1\theta\gamma}\sinh\beta_1\theta\xi\right) + B_2\left(\cos\beta_1\theta\xi - \cosh\beta_1\theta\xi + \frac{\cosh\beta_1\theta\gamma - \cos\beta_1\theta\gamma}{\sinh\beta_1\theta\gamma}\sinh\beta_1\theta\xi\right)$$

$$r_3(\xi) = A_3\left(\sin \beta_1\xi - \frac{\sin \beta_1\mu}{\sinh \beta_1\mu} \sinh \beta_1\xi\right)$$

$$A_2\theta\left(\frac{\sin \beta_1\theta\gamma - \sinh \beta_1\theta\gamma}{\sinh \beta_1\theta\gamma}\right) + B_2\theta\frac{\cos \beta_1\theta\gamma - \cosh \beta_1\theta\gamma}{\sinh \beta_1\theta\gamma} + A_3(0) = \frac{2(\cos \beta_1\mu \cosh \beta_1\mu - 1)}{\cos \beta_1\mu - \cosh \beta_1\mu}$$

$$A_2(0) + B_2\alpha^4\theta^2(\cosh \beta_1\mu - \cos \beta_1\mu) + A_3(0) = \sin \beta_1\mu \cosh \beta_1\mu - \sinh \beta_1\mu \cos \beta_1\mu$$

$$A_2\theta\left(\cos \beta_1\theta\gamma - \frac{\sin \beta_1\theta\gamma}{\sinh \beta_1\theta\gamma} \cosh \beta_1\theta\gamma\right) + B_2\theta\frac{1 - \sin \beta_1\theta\gamma \sinh \beta_1\theta\gamma - \cos \beta_1\theta\gamma \cosh \beta_1\theta\gamma}{\sinh \beta_1\theta\gamma} + A_3\left(\cos \beta_1\mu - \frac{\sin \beta_1\mu}{\sinh \beta_1\mu} \cosh \beta_1\mu\right) = 0$$

where

$$\alpha = \left(\frac{E_2I_2}{E_1I_1}\right)^{1/4} \qquad \phi = \left(\frac{\rho_2A_2}{\rho_1A_1}\right)^{1/4} \qquad \theta = \frac{\phi}{\alpha}$$

CASE 8.17. Beam clamped at one end, free at the other, and with simple support at each end of a symmetric discontinuity in cross section.

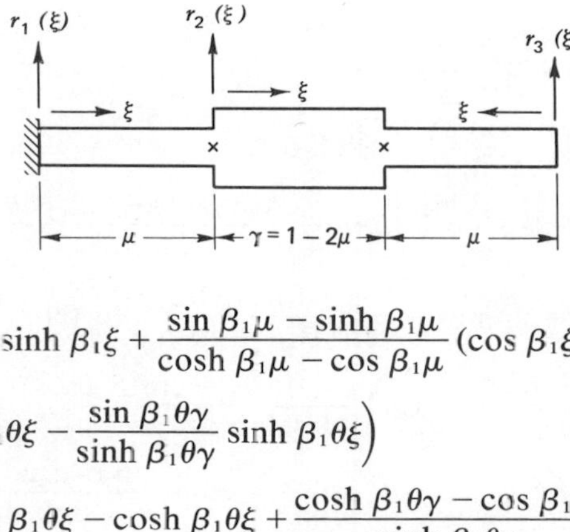

$$r_1(\xi) = \sin \beta_1\xi - \sinh \beta_1\xi + \frac{\sin \beta_1\mu - \sinh \beta_1\mu}{\cosh \beta_1\mu - \cos \beta_1\mu}(\cos \beta_1\xi - \cosh \beta_1\xi)$$

$$r_2(\xi) = A_2\left(\sin \beta_1\theta\xi - \frac{\sin \beta_1\theta\gamma}{\sinh \beta_1\theta\gamma} \sinh \beta_1\theta\xi\right) + B_2\left(\cos \beta_1\theta\xi - \cosh \beta_1\theta\xi + \frac{\cosh \beta_1\theta\gamma - \cos \beta_1\theta\gamma}{\sinh \beta_1\theta\gamma} \sinh \beta_1\theta\xi\right)$$

$$r_3(\xi) = A_3\left[\sin \beta_1\xi + \sinh \beta_1\xi - \frac{\sin \beta_1\mu + \sinh \beta_1\mu}{\cos \beta_1\mu + \cosh \beta_1\mu}(\cos \beta_1\xi + \cosh \beta_1\xi)\right]$$

$$A_2\theta\frac{\sin \beta_1\theta\gamma - \sinh \beta_1\theta\gamma}{\sinh \beta_1\theta\gamma} + B_2\theta\frac{\cos \beta_1\theta\gamma - \cosh \beta_1\theta\gamma}{\sinh \beta_1\theta\gamma} + A_3(0) = \frac{2(\cos \beta_1\mu \cosh \beta_1\mu - 1)}{\cos \beta_1\mu - \cosh \beta_1\mu}$$

$$A_2(0) + B_2\alpha^4\theta^2(\cosh \beta_1\mu - \cos \beta_1\mu) + A_3(0) = \sin \beta_1\mu \cosh \beta_1\mu - \sinh \beta_1\mu \cos \beta_1\mu$$

$$A_2\theta\left(\cos \beta_1\theta\gamma - \frac{\sin \beta_1\theta\gamma}{\sinh \beta_1\theta\gamma} \cosh \beta_1\theta\gamma\right) + B_2\theta \frac{1 - \sin \beta_1\theta\gamma \sinh \beta_1\theta\gamma - \cos \beta_1\theta\gamma \cosh \beta_1\theta\gamma}{\sinh \beta_1\theta\gamma} + A_3 \frac{2(1 + \cos \beta_1\mu \cosh \beta_1\mu)}{\cos \beta_1\mu + \cosh \beta_1\mu} = 0$$

where

$$\alpha = \left(\frac{E_2I_2}{E_1I_1}\right)^{1/4} \qquad \phi = \left(\frac{\rho_2A_2}{\rho_1A_1}\right)^{1/4} \qquad \theta = \frac{\phi}{\alpha}$$

CASE 8.18. Beam with simple support at one end, free at the other, and with simple support at each end of a symmetric discontinuity in cross section.

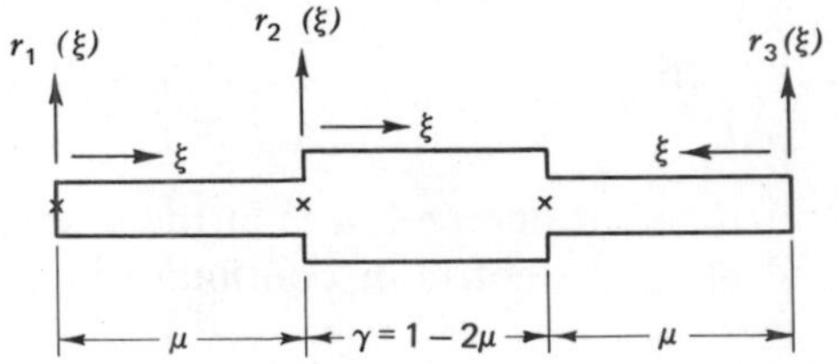

$$r_1(\xi) = \sin \beta_1\xi - \frac{\sin \beta_1\mu}{\sinh \beta_1\mu} \sinh \beta_1\xi$$

$$r_2(\xi) = A_2\left(\sin \beta_1\theta\xi - \frac{\sin \beta_1\theta\gamma}{\sinh \beta_1\theta\gamma} \sinh \beta_1\theta\xi\right) + B_2\left(\cos \beta_1\theta\xi - \cosh \beta_1\theta\xi + \frac{\cosh \beta_1\theta\gamma - \cos \beta_1\theta\gamma}{\sinh \beta_1\theta\gamma} \sinh \beta_1\theta\xi\right)$$

$$r_3(\xi) = A_3\left[\sin \beta_1\xi + \sinh \beta_1\xi - \frac{\sin \beta_1\mu + \sinh \beta_1\mu}{\cos \beta_1\mu + \cosh \beta_1\mu} (\cos \beta_1\xi + \cosh \beta_1\xi)\right]$$

$$A_2\theta \frac{\sinh \beta_1\theta\gamma - \sin \beta_1\theta\gamma}{\sinh \beta_1\theta\gamma} + B_2\theta \frac{\cosh \beta_1\theta\gamma - \cos \beta_1\theta\gamma}{\sinh \beta_1\theta\gamma} + A_3(0) = -\frac{\sin \beta_1\mu \cosh \beta_1\mu}{\sinh \beta_1\mu} + \cos \beta_1\mu$$

$$A_2\theta\left(\cos \beta_1\theta\gamma - \frac{\sin \beta_1\theta\gamma}{\sinh \beta_1\theta\gamma} \cosh \beta_1\theta\gamma\right) + \beta_2\theta\left(\frac{1 - \sin \beta_1\theta\gamma \sinh \beta_1\theta\gamma - \cos \beta_1\theta\gamma \cosh \beta_1\theta\gamma}{\sinh \beta_1\theta\gamma}\right) + A_3 \frac{2(1 + \cos \beta_1\mu \cosh \beta_1\mu)}{\cos \beta_1\mu + \cosh \beta_1\mu} = 0$$

$$B_2 = \frac{\sin \beta_1\mu}{\alpha^4\theta^2}$$

TABLE 8.1 : β_1 (ϕ = 0.6, α = 0.6)

μ MODE:	0.15	0.20	0.25	0.30	0.35	0.40	0.45	0.50	0.55	0.60	0.65	0.70	0.75	0.80	0.85
(1)	2.146	2.248	2.355	2.464	2.571	2.667	2.737	2.764	2.737	2.667	2.571	2.464	2.355	2.248	2.146
(2)	5.294	5.437	5.470	5.322	5.049	4.779	4.593	4.527	4.593	4.779	5.049	5.322	5.470	5.437	5.294
(3)	8.645	8.382	7.829	7.545	7.598	7.837	8.113	8.249	8.113	7.837	7.598	7.545	7.829	8.382	8.645
(4)	11.45	10.75	10.76	11.10	11.40	11.24	10.78	10.55	10.78	11.24	11.40	11.10	10.76	10.75	11.45

TABLE 8.1 : β_1 (ϕ = .6, α = .7)

μ MODE:	0.15	0.20	0.25	0.30	0.35	0.40	0.45	0.50	0.55	0.60	0.65	0.70	0.75	0.80	0.85
(1)	2.451	2.542	2.632	2.716	2.789	2.839	2.856	2.832	2.770	2.680	2.576	2.466	2.355	2.248	2.146
(2)	6.001	6.054	5.953	5.696	5.393	5.144	4.994	4.957	5.027	5.177	5.361	5.504	5.537	5.453	5.296
(3)	9.675	9.164	8.631	8.443	8.521	8.701	8.822	8.742	8.489	8.213	8.038	8.064	8.327	8.642	8.690
(4)	12.69	12.13	12.23	12.47	12.44	12.00	11.59	11.55	11.80	11.[illegible]	11.80	11.44	11.18	11.30	11.76

TABLE 8.1 : β_1 (ϕ = .6, α = .8)

μ MODE:	0.15	0.20	0.25	0.30	0.35	0.40	0.45	0.50	0.55	0.60	0.65	0.70	0.75	0.80	0.85
(1)	2.718	2.786	2.847	2.897	2.930	2.939	2.918	2.865	2.785	2.687	2.578	2.466	2.355	2.248	2.146
(2)	6.631	6.590	6.391	6.079	5.762	5.515	5.365	5.314	5.346	5.434	5.534	5.591	5.566	5.460	5.297
(3)	10.62	9.968	9.446	9.267	9.306	9.394	9.377	9.180	8.885	8.613	8.453	8.460	8.614	8.756	8.710
(4)	13.98	13.48	13.54	13.61	13.31	12.77	12.43	12.42	12.55	12.49	12.17	11.80	11.59	11.67	11.89

TABLE 8.1 : β_1 (ϕ = .6, α = .9)

μ MODE:	0.15	0.20	0.25	0.30	0.35	0.40	0.45	0.50	0.55	0.60	0.65	0.70	0.75	0.80	0.85
(1)	2.943	2.977	3.004	3.019	3.019	2.998	2.952	2.883	2.793	2.690	2.579	2.467	2.355	2.248	2.146
(2)	7.208	7.092	6.828	6.471	6.124	5.850	5.670	5.579	5.561	5.590	5.629	5.635	5.581	5.463	5.297
(3)	11.53	10.77	10.20	9.990	9.981	9.996	9.886	9.613	9.272	8.972	8.777	8.721	8.772	8.812	8.720
(4)	15.25	14.73	14.72	14.64	14.14	13.52	13.17	13.13	13.15	12.94	12.54	12.14	11.90	11.88	11.95

TABLE 8.1 : β_1 (ϕ = .6, α = 1.0)

μ MODE:	0.15	0.20	0.25	0.30	0.35	0.40	0.45	0.50	0.55	0.60	0.65	0.70	0.75	0.80	0.85
(1)	3.124	3.122	3.115	3.100	3.074	3.033	2.972	2.893	2.798	2.692	2.580	2.467	2.356	2.248	2.146
(2)	7.759	7.590	7.264	6.846	6.446	6.126	5.901	5.764	5.701	5.685	5.683	5.660	5.590	5.465	5.298
(3)	12.43	11.51	10.87	10.63	10.60	10.57	10.38	10.02	9.610	9.254	9.005	8.884	8.861	8.843	8.726
(4)	16.44	15.89	15.85	15.60	14.89	14.17	13.81	13.76	13.68	13.35	12.86	12.41	12.10	12.00	11.99

TABLE 8.1 : β_1 (ϕ = .7, α = .6)

μ MODE:	0.15	0.20	0.25	0.30	0.35	0.40	0.45	0.50	0.55	0.60	0.65	0.70	0.75	0.80	0.85
(1)	1.840	1.927	2.019	2.113	2.208	2.298	2.374	2.427	2.448	2.433	2.390	2.328	2.256	2.179	2.101
(2)	4.539	4.674	4.746	4.718	4.596	4.438	4.309	4.249	4.280	4.409	4.623	4.882	5.102	5.189	5.144
(3)	7.449	7.407	7.137	6.912	6.890	7.040	7.276	7.493	7.562	7.458	7.304	7.237	7.398	7.855	8.292
(4)	10.08	9.687	9.585	9.804	10.11	10.27	10.12	9.897	9.935	10.28	10.66	10.69	10.48	10.40	10.88

TABLE 8.1 : β_1 (ϕ = .7, α = .7)

μ MODE:	0.15	0.20	0.25	0.30	0.35	0.40	0.45	0.50	0.55	0.60	0.65	0.70	0.75	0.80	0.85
(1)	2.101	2.179	2.257	2.331	2.399	2.454	2.491	2.504	2.491	2.454	2.399	2.331	2.257	2.179	2.101
(2)	5.150	5.220	5.207	5.103	4.945	4.792	4.686	4.649	4.686	4.792	4.945	5.103	5.207	5.220	5.150
(3)	8.385	8.190	7.886	7.712	7.719	7.842	7.987	8.054	7.987	7.842	7.719	7.712	7.886	8.190	8.385
(4)	11.28	10.90	10.87	11.06	11.21	11.12	10.88	10.75	10.88	11.12	11.21	11.06	10.87	10.90	11.28

TABLE 8.1 : β_1 (ϕ = .7, α = .8)

μ MODE:	0.15	0.20	0.25	0.30	0.35	0.40	0.45	0.50	0.55	0.60	0.65	0.70	0.75	0.80	0.85
(1)	2.330	2.388	2.442	2.489	2.526	2.549	2.556	2.544	2.513	2.465	2.403	2.333	2.257	2.179	2.101
(2)	5.696	5.703	5.630	5.487	5.315	5.159	5.048	4.995	4.999	5.053	5.136	5.215	5.255	5.234	5.152
(3)	9.256	8.973	8.651	8.468	8.440	8.497	8.552	8.527	8.407	8.250	8.130	8.106	8.199	8.352	8.425
(4)	12.49	12.10	12.04	12.14	12.13	11.92	11.66	11.55	11.62	11.71	11.64	11.44	11.27	11.28	11.47

TABLE 8.1 : β_1 (ϕ = .7, α = .9)

μ MODE:	0.15	0.20	0.25	0.30	0.35	0.40	0.45	0.50	0.55	0.60	0.65	0.70	0.75	0.80	0.85
(1)	2.523	2.553	2.578	2.597	2.607	2.606	2.593	2.566	2.524	2.470	2.405	2.334	2.258	2.179	2.101
(2)	6.198	6.160	6.051	5.880	5.683	5.500	5.355	5.262	5.219	5.219	5.246	5.275	5.281	5.242	5.154
(3)	10.11	9.758	9.373	9.135	9.054	9.059	9.061	8.987	8.828	8.633	8.468	8.380	8.382	8.436	8.445
(4)	13.69	13.22	13.10	13.12	13.01	12.70	12.37	12.21	12.21	12.21	12.06	11.81	11.59	11.51	11.57

TABLE 8.1 : β_1 (ϕ = .7, α = 1.0)

μ MODE:	0.15	0.20	0.25	0.30	0.35	0.40	0.45	0.50	0.55	0.60	0.65	0.70	0.75	0.80	0.85
(1)	2.679	2.677	2.675	2.669	2.658	2.641	2.614	2.578	2.530	2.473	2.407	2.334	2.258	2.179	2.101
(2)	6.680	6.614	6.476	6.267	6.023	5.789	5.596	5.454	5.366	5.322	5.310	5.309	5.295	5.246	5.155
(3)	10.97	10.51	10.02	9.715	9.602	9.590	9.564	9.439	9.211	8.947	8.716	8.559	8.489	8.481	8.456
(4)	14.84	14.24	14.09	14.07	13.85	13.39	12.97	12.77	12.74	12.68	12.45	12.12	11.82	11.65	11.62

TABLE 8.1 : β_1 (ϕ = .8, α = .6)

μ MODE:	0.15	0.20	0.25	0.30	0.35	0.40	0.45	0.50	0.55	0.60	0.65	0.70	0.75	0.80	0.85
(1)	1.610	1.686	1.767	1.850	1.934	2.015	2.089	2.149	2.189	2.204	2.198	2.173	2.135	2.089	2.039
(2)	3.973	4.094	4.175	4.193	4.150	4.076	4.010	3.985	4.024	4.137	4.322	4.559	4.794	4.942	4.973
(3)	6.533	6.567	6.460	6.345	6.340	6.460	6.664	6.885	7.033	7.046	6.980	6.950	7.085	7.476	7.962
(4)	8.918	8.753	8.691	8.854	9.130	9.371	9.416	9.316	9.322	9.581	9.984	10.21	10.15	10.11	10.48

TABLE 8.1 : β_1 (ϕ = .8, α = .7)

μ MODE:	0.15	0.20	0.25	0.30	0.35	0.40	0.45	0.50	0.55	0.60	0.65	0.70	0.75	0.80	0.85
(1)	1.838	1.907	1.975	2.041	2.103	2.156	2.199	2.226	2.237	2.231	2.210	2.178	2.137	2.090	2.039
(2)	4.508	4.580	4.599	4.563	4.494	4.422	4.374	4.370	4.417	4.514	4.650	4.801	4.926	4.990	4.984
(3)	7.372	7.308	7.174	7.092	7.114	7.219	7.356	7.460	7.482	7.435	7.385	7.410	7.569	7.852	8.099
(4)	10.04	9.867	9.862	10.01	10.19	10.25	10.17	10.11	10.21	10.43	10.62	10.62	10.54	10.59	10.93

TABLE 8.1 : β_1 (ϕ = .8, α = .8)

μ MODE:	0.15	0.20	0.25	0.30	0.35	0.40	0.45	0.50	0.55	0.60	0.65	0.70	0.75	0.80	0.85
(1)	2.039	2.090	2.138	2.180	2.216	2.244	2.261	2.267	2.261	2.244	2.216	2.180	2.138	2.090	2.039
(2)	4.989	5.012	4.990	4.931	4.854	4.783	4.733	4.716	4.733	4.783	4.854	4.931	4.990	5.012	4.989
(3)	8.160	8.049	7.906	7.817	7.804	7.847	7.903	7.928	7.903	7.847	7.804	7.817	7.906	8.049	8.160
(4)	11.16	10.99	10.95	11.02	11.08	11.04	10.95	10.89	10.95	11.04	11.08	11.02	10.95	10.99	11.16

TABLE 8.1 : β_1 (ϕ = .8, α = .9)

μ MODE:	0.15	0.20	0.25	0.30	0.35	0.40	0.45	0.50	0.55	0.60	0.65	0.70	0.75	0.80	0.85
(1)	2.208	2.234	2.257	2.276	2.289	2.296	2.297	2.289	2.274	2.250	2.219	2.182	2.138	2.090	2.039
(2)	5.432	5.422	5.379	5.308	5.219	5.129	5.050	4.994	4.964	4.960	4.976	5.002	5.023	5.023	4.992
(3)	8.939	8.796	8.615	8.468	8.391	8.373	8.382	8.376	8.332	8.253	8.170	8.117	8.113	8.153	8.191
(4)	12.29	12.05	11.93	11.92	11.92	11.83	11.68	11.56	11.52	11.53	11.51	11.42	11.31	11.25	11.29

TABLE 8.1 : β_1 ($\phi = .8$, $\alpha = 1.0$)

μ MODE:	0.15	0.20	0.25	0.30	0.35	0.40	0.45	0.50	0.55	0.60	0.65	0.70	0.75	0.80	0.85
(1)	2.344	2.343	2.342	2.340	2.336	2.329	2.318	2.302	2.281	2.254	2.221	2.182	2.138	2.090	2.039
(2)	5.858	5.832	5.776	5.686	5.566	5.433	5.307	5.201	5.122	5.072	5.048	5.043	5.042	5.030	4.993
(3)	9.721	9.529	9.264	9.033	8.899	8.854	8.850	8.828	8.746	8.607	8.451	8.319	8.237	8.211	8.208
(4)	13.39	13.02	12.82	12.78	12.75	12.58	12.31	12.09	12.00	12.00	11.94	11.78	11.57	11.41	11.35

TABLE 8.1 : β_1 ($\phi = .9$, $\alpha = .6$)

μ MODE:	0.15	0.20	0.25	0.30	0.35	0.40	0.45	0.50	0.55	0.60	0.65	0.70	0.75	0.80	0.85
(1)	1.431	1.499	1.570	1.644	1.720	1.793	1.862	1.922	1.968	1.998	2.012	2.013	2.003	1.985	1.962
(2)	3.532	3.642	3.721	3.757	3.752	3.727	3.707	3.719	3.780	3.900	4.082	4.314	4.552	4.728	4.805
(3)	5.813	5.875	5.848	5.814	5.851	5.981	6.182	6.408	6.590	6.664	6.654	6.660	6.801	7.177	7.688
(4)	7.969	7.922	7.931	8.035	8.359	8.626	8.766	8.757	8.783	9.014	9.425	9.759	9.817	9.818	10.16

TABLE 8.1 : β_1 ($\phi = .9$, $\alpha = .7$)

μ MODE:	0.15	0.20	0.25	0.30	0.35	0.40	0.45	0.50	0.55	0.60	0.65	0.70	0.75	0.80	0.85
(1)	1.634	1.695	1.756	1.815	1.817	1.921	1.963	1.996	2.017	2.027	2.028	2.020	2.005	1.986	1.962
(2)	4.008	4.077	4.107	4.103	4.080	4.059	4.059	4.092	4.165	4.278	4.420	4.573	4.706	4.791	4.821
(3)	6.568	6.561	6.520	6.518	6.587	6.714	6.866	6.990	7.048	7.046	7.042	7.105	7.290	7.589	7.865
(4)	8.999	8.954	9.017	9.187	9.380	9.495	9.498	9.499	9.624	9.878	10.12	10.20	10.19	10.28	10.65

TABLE 8.1 : β_1 ($\phi = .9$, $\alpha = .8$)

μ MODE:	0.15	0.20	0.25	0.30	0.35	0.40	0.45	0.50	0.55	0.60	0.65	0.70	0.75	0.80	0.85
(1)	1.812	1.858	1.900	1.939	1.973	2.000	2.021	2.035	2.041	2.041	2.035	2.023	2.006	1.986	1.962
(2)	4.437	4.465	4.465	4.446	4.423	4.409	4.412	4.439	4.489	4.559	4.639	4.719	4.782	4.820	4.828
(3)	7.281	7.246	7.211	7.215	7.262	7.336	7.405	7.446	7.450	7.443	7.459	7.527	7.658	7.819	7.946
(4)	10.03	10.00	10.05	10.15	10.23	10.25	10.24	10.27	10.38	10.52	10.60	10.60	10.60	10.71	10.93

TABLE 8.1 : β_1 ($\phi = .9$, $\alpha = .9$)

μ MODE:	0.15	0.20	0.25	0.30	0.35	0.40	0.45	0.50	0.55	0.60	0.65	0.70	0.75	0.80	0.85
(1)	1.962	1.986	2.007	2.025	2.039	2.049	2.055	2.057	2.055	2.049	2.039	2.025	2.007	1.986	1.962
(2)	4.832	4.835	4.822	4.799	4.773	4.750	4.734	4.729	4.734	4.750	4.773	4.799	4.822	4.835	4.832
(3)	7.986	7.943	7.894	7.860	7.848	7.851	7.861	7.865	7.861	7.851	7.848	7.860	7.894	7.943	7.986
(4)	11.07	11.01	10.99	11.00	11.01	11.00	10.98	10.97	10.98	11.00	11.01	11.00	10.99	11.01	11.07

TABLE 8.1 : β_1 ($\phi = .9$, $\alpha = 1.0$)

μ MODE:	0.15	0.20	0.25	0.30	0.35	0.40	0.45	0.50	0.55	0.60	0.65	0.70	0.75	0.80	0.85
(1)	2.084	2.083	2.083	2.082	2.081	2.079	2.075	2.070	2.063	2.053	2.041	2.026	2.008	1.986	1.962
(2)	5.212	5.204	5.187	5.156	5.113	5.061	5.005	4.952	4.908	4.875	4.855	4.846	4.845	4.844	4.835
(3)	8.697	8.635	8.537	8.430	8.348	8.304	8.292	8.291	8.275	8.231	8.164	8.093	8.039	8.012	8.009
(4)	12.11	11.96	11.83	11.79	11.78	11.75	11.66	11.54	11.46	11.44	11.43	11.39	11.30	11.21	11.16

TABLE 8.1 : β_1 ($\phi = 1.0$, $\alpha = .6$)

μ MODE:	0.15	0.20	0.25	0.30	0.35	0.40	0.45	0.50	0.55	0.60	0.65	0.70	0.75	0.80	0.85
(1)	1.288	1.349	1.413	1.480	1.548	1.615	1.679	1.736	1.783	1.820	1.845	1.860	1.869	1.873	1.875
(2)	3.179	3.279	3.354	3.396	3.410	3.411	3.421	3.459	3.540	3.675	3.867	4.108	4.359	4.554	4.655
(3)	5.236	5.305	5.316	5.330	5.403	5.552	5.766	6.010	6.225	6.341	6.361	6.380	6.523	6.907	7.461
(4)	7.192	7.202	7.262	7.445	7.717	8.009	8.211	8.256	8.288	8.503	8.932	9.362	9.511	9.533	9.865

TABLE 8.1 : β_1 ($\phi = 1.0$, $\alpha = .7$)

μ MODE:	0.15	0.20	0.25	0.30	0.35	0.40	0.45	0.50	0.55	0.60	0.65	0.70	0.75	0.80	0.85
(1)	1.471	1.525	1.580	1.634	1.685	1.731	1.771	1.804	1.830	1.849	1.861	1.868	1.872	1.874	1.875
(2)	3.608	3.672	3.706	3.717	3.717	3.726	3.756	3.818	3.917	4.052	4.216	4.387	4.533	4.630	4.676
(3)	5.919	5.937	5.943	5.991	6.101	6.263	6.448	6.607	6.695	6.713	6.722	6.800	7.013	7.356	7.678
(4)	8.136	8.157	8.275	8.482	8.714	8.878	8.920	8.937	9.076	9.371	9.694	9.850	9.864	9.970	10.39

TABLE 8.1 : β_1 ($\phi = 1.0$, $\alpha = .8$)

μ MODE:	0.15	0.20	0.25	0.30	0.35	0.40	0.45	0.50	0.55	0.60	0.65	0.70	0.75	0.80	0.85
(1)	1.631	1.672	1.711	1.746	1.777	1.803	1.825	1.842	1.854	1.863	1.869	1.872	1.874	1.875	1.875
(2)	3.994	4.024	4.033	4.034	4.039	4.058	4.098	4.161	4.246	4.347	4.453	4.549	4.621	4.665	4.686
(3)	6.566	6.569	6.590	6.654	6.760	6.886	6.997	7.062	7.080	7.083	7.119	7.227	7.411	7.623	7.777
(4)	9.082	9.130	9.258	9.423	9.553	9.598	9.604	9.673	9.848	10.07	10.20	10.23	10.25	10.41	10.72

TABLE 8.1 : β_1 ($\phi = 1.0$, $\alpha = .9$)

μ MODE:	0.15	0.20	0.25	0.30	0.35	0.40	0.45	0.50	0.55	0.60	0.65	0.70	0.75	0.80	0.85
(1)	1.766	1.788	1.807	1.823	1.837	1.848	1.856	1.863	1.867	1.871	1.873	1.874	1.875	1.875	1.875
(2)	4.351	4.359	4.360	4.362	4.369	4.387	4.417	4.457	4.505	4.555	4.602	4.640	4.668	4.684	4.692
(3)	7.208	7.211	7.235	7.284	7.348	7.408	7.447	7.461	7.462	7.473	7.513	7.587	7.683	7.771	7.827
(4)	10.04	10.09	10.17	10.25	10.29	10.29	10.31	10.38	10.49	10.58	10.61	10.61	10.65	10.76	10.90

TABLE 8.1 : β_1 ($1/\phi = .6$, $1/\alpha = .6$)

μ MODE:	0.15	0.20	0.25	0.30	0.35	0.40	0.45	0.50	0.55	0.60	0.65	0.70	0.75	0.80	0.85
(1)	1.303	1.250	1.214	1.189	1.172	1.160	1.154	1.152	1.154	1.160	1.172	1.189	1.214	1.250	1.303
(2)	4.237	4.210	4.148	4.066	3.985	3.919	3.878	3.863	3.878	3.919	3.985	4.066	4.148	4.210	4.237
(3)	7.305	7.207	7.203	7.327	7.563	7.873	8.183	8.331	8.183	7.873	7.563	7.327	7.203	7.207	7.305
(4)	10.33	10.41	10.71	11.11	11.43	11.29	10.81	10.57	10.81	11.29	11.43	11.11	10.71	10.41	10.33

TABLE 8.1 : β_1 ($1/\phi = .6$, $1/\alpha = .7$)

μ MODE:	0.15	0.20	0.25	0.30	0.35	0.40	0.45	0.50	0.55	0.60	0.65	0.70	0.75	0.80	0.85
(1)	1.272	1.231	1.201	1.180	1.166	1.156	1.151	1.150	1.153	1.160	1.171	1.189	1.214	1.250	1.303
(2)	3.723	3.730	3.727	3.715	3.702	3.695	3.702	3.728	3.775	3.845	3.934	4.034	4.131	4.203	4.235
(3)	6.438	6.439	6.476	6.583	6.770	7.023	7.310	7.554	7.632	7.516	7.325	7.165	7.098	7.150	7.284
(4)	9.133	9.225	9.452	9.781	10.13	10.31	10.15	9.922	9.955	10.31	10.70	10.71	10.48	10.27	10.26

TABLE 8.1 : β_1 ($1/\phi = .6$, $1/\alpha = .8$)

μ MODE:	0.15	0.20	0.25	0.30	0.35	0.40	0.45	0.50	0.55	0.60	0.65	0.70	0.75	0.80	0.85
(1)	1.231	1.203	1.182	1.167	1.157	1.150	1.147	1.147	1.151	1.159	1.171	1.189	1.214	1.250	1.303
(2)	3.340	3.352	3.368	3.388	3.412	3.446	3.493	3.556	3.638	3.740	3.859	3.987	4.107	4.194	4.233
(3)	5.740	5.787	5.864	5.987	6.166	6.396	6.661	6.919	7.088	7.105	7.023	6.945	6.946	7.063	7.252
(4)	8.156	8.284	8.500	8.794	9.123	9.391	9.447	9.341	9.339	9.599	10.01	10.24	10.17	10.07	10.15

TABLE 8.1 : β_1 ($1/\phi = .6$, $1/\alpha = .9$)

μ MODE:	0.15	0.20	0.25	0.30	0.35	0.40	0.45	0.50	0.55	0.60	0.65	0.70	0.75	0.80	0.85
(1)	1.181	1.168	1.157	1.149	1.144	1.141	1.141	1.143	1.148	1.157	1.170	1.188	1.214	1.250	1.303
(2)	3.048	3.058	3.076	3.105	3.146	3.201	3.272	3.362	3.474	3.607	3.760	3.922	4.071	4.179	4.229
(3)	5.184	5.243	5.338	5.476	5.660	5.889	6.151	6.417	6.626	6.713	6.702	6.684	6.747	6.939	7.204
(4)	7.352	7.501	7.724	8.010	8.333	8.630	8.788	8.781	8.799	9.026	9.442	9.782	9.840	9.817	9.988

TABLE 8.1 : β_1 $(1/\phi = .6,\ 1/\alpha = 1.0)$

μ MODE:	0.15	0.20	0.25	0.30	0.35	0.40	0.45	0.50	0.55	0.60	0.65	0.70	0.75	0.80	0.85
(1)	1.125	1.125	1.126	1.126	1.128	1.130	1.133	1.138	1.145	1.155	1.168	1.187	1.213	1.250	1.303
(2)	2.820	2.827	2.843	2.871	2.914	2.975	3.057	3.163	3.294	3.452	3.636	3.834	4.022	4.159	4.223
(3)	4.739	4.794	4.890	5.032	5.220	5.451	5.719	5.999	6.242	6.376	6.403	6.414	6.515	6.775	7.135
(4)	6.692	6.842	7.066	7.351	7.678	8.000	8.223	8.277	8.304	8.513	8.942	9.378	9.530	9.548	9.782

TABLE 8.1 : β_1 $(1/\phi = .7,\ 1/\alpha = .6)$

μ MODE:	0.15	0.20	0.25	0.30	0.35	0.40	0.45	0.50	0.55	0.60	0.65	0.70	0.75	0.80	0.85
(1)	1.520	1.459	1.417	1.387	1.367	1.353	1.345	1.342	1.343	1.349	1.360	1.377	1.401	1.436	1.484
(2)	4.941	4.904	4.820	4.707	4.589	4.485	4.405	4.350	4.319	4.311	4.319	4.334	4.348	4.351	4.343
(3)	8.498	8.342	8.281	8.359	8.546	8.768	8.905	8.814	8.529	8.194	7.898	7.681	7.555	7.512	7.511
(4)	11.97	11.98	12.22	12.50	12.48	12.03	11.61	11.58	11.85	12.03	11.82	11.41	11.03	10.76	10.66

TABLE 8.1 : β_1 $(1/\phi = .7,\ 1/\alpha = .7)$

μ MODE:	0.15	0.20	0.25	0.30	0.35	0.40	0.45	0.50	0.55	0.60	0.65	0.70	0.75	0.80	0.85
(1)	1.484	1.436	1.401	1.377	1.360	1.348	1.342	1.340	1.342	1.348	1.360	1.377	1.401	1.436	1.484
(2)	4.342	4.347	4.337	4.312	4.278	4.247	4.225	4.218	4.225	4.247	4.278	4.312	4.337	4.347	4.342
(3)	7.498	7.473	7.473	7.541	7.680	7.868	8.045	8.122	8.045	7.868	7.680	7.541	7.473	7.473	7.498
(4)	10.61	10.65	10.82	11.07	11.24	11.15	10.90	10.77	10.90	11.15	11.24	11.07	10.82	10.65	10.61

TABLE 8.1 : β_1 $(1/\phi = .7,\ 1/\alpha = .8)$

μ MODE:	0.15	0.20	0.25	0.30	0.35	0.40	0.45	0.50	0.55	0.60	0.65	0.70	0.75	0.80	0.85
(1)	1.436	1.404	1.379	1.362	1.349	1.341	1.337	1.336	1.340	1.347	1.359	1.376	1.401	1.436	1.484
(2)	3.896	3.907	3.923	3.938	3.954	3.975	4.005	4.045	4.096	4.155	4.219	4.277	4.320	4.341	4.340
(3)	6.689	6.727	6.788	6.887	7.030	7.206	7.384	7.509	7.536	7.477	7.392	7.341	7.350	7.413	7.479
(4)	9.486	9.589	9.767	9.996	10.20	10.27	10.19	10.12	10.22	10.45	10.64	10.64	10.53	10.47	10.53

TABLE 8.1 : β_1 $(1/\phi = .7,\ 1/\alpha = .9)$

μ MODE:	0.15	0.20	0.25	0.30	0.35	0.40	0.45	0.50	0.55	0.60	0.65	0.70	0.75	0.80	0.85
(1)	1.378	1.362	1.350	1.341	1.334	1.331	1.330	1.332	1.337	1.345	1.358	1.376	1.401	1.436	1.484
(2)	3.556	3.565	3.584	3.612	3.651	3.701	3.765	3.842	3.933	4.033	4.136	4.228	4.296	4.331	4.338
(3)	6.043	6.100	6.192	6.320	6.486	6.676	6.867	7.018	7.088	7.087	7.068	7.088	7.179	7.324	7.450
(4)	8.559	8.701	8.906	9.151	9.381	9.512	9.517	9.514	9.634	9.889	10.14	10.22	10.20	10.24	10.42

TABLE 8.1 : β_1 $(1/\phi = .7,\ 1/\alpha = 1.0)$

μ MODE:	0.15	0.20	0.25	0.30	0.35	0.40	0.45	0.50	0.55	0.60	0.65	0.70	0.75	0.80	0.85
(1)	1.313	1.313	1.313	1.314	1.315	1.317	1.321	1.325	1.333	1.343	1.356	1.375	1.400	1.435	1.484
(2)	3.289	3.296	3.312	3.341	3.385	3.446	3.527	3.628	3.748	3.885	4.027	4.160	4.262	4.318	4.334
(3)	5.526	5.581	5.679	5.821	6.004	6.216	6.434	6.617	6.720	6.745	6.750	6.808	6.963	7.200	7.407
(4)	7.794	7.948	8.170	8.438	8.702	8.885	8.935	8.950	9.084	9.378	9.704	9.865	9.879	9.960	10.26

TABLE 8.1 : β_1 $(1/\phi = .8,\ 1/\alpha = .6)$

μ MODE:	0.15	0.20	0.25	0.30	0.35	0.40	0.45	0.50	0.55	0.60	0.65	0.70	0.75	0.80	0.85
(1)	1.737	1.667	1.619	1.585	1.561	1.545	1.535	1.530	1.529	1.533	1.542	1.556	1.576	1.604	1.642
(2)	5.643	5.592	5.476	5.317	5.146	4.987	4.851	4.741	4.657	4.595	4.550	4.517	4.491	4.469	4.453
(3)	9.669	9.417	9.261	9.260	9.365	9.474	9.451	9.226	8.882	8.527	8.221	7.983	7.818	7.715	7.654
(4)	13.53	13.42	13.56	13.65	13.35	12.80	12.45	12.45	12.60	12.52	12.17	11.73	11.33	11.05	10.87

TABLE 8.1 : β_1 $(1/\phi = .8,\ 1/\alpha = .7)$

μ MODE:	0.15	0.20	0.25	0.30	0.35	0.40	0.45	0.50	0.55	0.60	0.65	0.70	0.75	0.80	0.85
(1)	1.696	1.641	1.601	1.573	1.553	1.539	1.531	1.527	1.528	1.533	1.542	1.556	1.576	1.604	1.642
(2)	4.960	4.961	4.938	4.888	4.821	4.749	4.680	4.622	4.577	4.544	4.519	4.500	4.483	4.466	4.452
(3)	8.547	8.472	8.400	8.389	8.449	8.545	8.613	8.582	8.439	8.236	8.035	7.871	7.757	7.688	7.645
(4)	12.04	11.97	12.04	12.16	12.16	11.95	11.68	11.57	11.65	11.74	11.66	11.42	11.16	10.96	10.84

TABLE 8.1 : β_1 $(1/\phi = .8,\ 1/\alpha = .8)$

μ MODE:	0.15	0.20	0.25	0.30	0.35	0.40	0.45	0.50	0.55	0.60	0.65	0.70	0.75	0.80	0.85
(1)	1.642	1.604	1.576	1.556	1.541	1.531	1.526	1.524	1.526	1.531	1.541	1.556	1.576	1.604	1.642
(2)	4.451	4.461	4.471	4.474	4.472	4.467	4.462	4.460	4.462	4.467	4.472	4.474	4.471	4.461	4.451
(3)	7.631	7.646	7.664	7.703	7.774	7.863	7.942	7.974	7.942	7.863	7.774	7.703	7.664	7.646	7.631
(4)	10.79	10.82	10.91	11.03	11.10	11.06	10.96	10.90	10.96	11.06	11.10	11.03	10.91	10.82	10.79

TABLE 8.1 : β_1 $(1/\phi = .8,\ 1/\alpha = .9)$

μ MODE:	0.15	0.20	0.25	0.30	0.35	0.40	0.45	0.50	0.55	0.60	0.65	0.70	0.75	0.80	0.85
(1)	1.575	1.557	1.542	1.532	1.524	1.520	1.518	1.519	1.522	1.529	1.540	1.555	1.576	1.604	1.642
(2)	4.063	4.071	4.087	4.110	4.139	4.174	4.215	4.262	4.312	4.362	4.406	4.437	4.452	4.454	4.449
(3)	6.897	6.943	7.012	7.105	7.215	7.329	7.424	7.476	7.483	7.467	7.459	7.480	7.529	7.583	7.611
(4)	9.747	9.854	9.998	10.15	10.25	10.27	10.25	10.28	10.39	10.53	10.61	10.61	10.59	10.63	10.71

TABLE 8.1 : β_1 $(1/\phi = .8,\ 1/\alpha = 1.0)$

μ MODE:	0.15	0.20	0.25	0.30	0.35	0.40	0.45	0.50	0.55	0.60	0.65	0.70	0.75	0.80	0.85
(1)	1.500	1.500	1.501	1.501	1.502	1.504	1.507	1.512	1.518	1.526	1.538	1.554	1.575	1.604	1.642
(2)	3.758	3.765	3.779	3.804	3.842	3.895	3.963	4.045	4.135	4.230	4.317	4.385	4.427	4.444	4.445
(3)	6.308	6.358	6.444	6.567	6.715	6.869	7.000	7.078	7.102	7.104	7.130	7.212	7.349	7.494	7.582
(4)	8.883	9.019	9.208	9.407	9.556	9.608	9.613	9.679	9.853	10.07	10.21	10.24	10.26	10.38	10.60

TABLE 8.1 : β_1 $(1/\phi = .9,\ 1/\alpha = .6)$

μ MODE:	0.15	0.20	0.25	0.30	0.35	0.40	0.45	0.50	0.55	0.60	0.65	0.70	0.75	0.80	0.85
(1)	1.954	1.875	1.821	1.782	1.755	1.735	1.723	1.715	1.712	1.712	1.716	1.724	1.736	1.751	1.772
(2)	6.343	6.269	6.107	5.882	5.639	5.411	5.210	5.043	4.908	4.801	4.719	4.658	4.615	4.586	4.573
(3)	10.81	10.41	10.12	10.03	10.05	10.07	9.939	9.625	9.226	8.833	8.491	8.214	8.007	7.864	7.776
(4)	14.98	14.73	14.76	14.67	14.16	13.54	13.20	13.17	13.18	12.95	12.50	12.02	11.59	11.25	11.03

TABLE 8.1 : β_1 $(1/\phi = .9,\ 1/\alpha = .7)$

μ MODE:	0.15	0.20	0.25	0.30	0.35	0.40	0.45	0.50	0.55	0.60	0.65	0.70	0.75	0.80	0.85
(1)	1.908	1.846	1.801	1.769	1.746	1.729	1.719	1.712	1.710	1.711	1.716	1.724	1.735	1.751	1.772
(2)	5.577	5.569	5.523	5.436	5.317	5.185	5.057	4.940	4.840	4.759	4.694	4.644	4.608	4.584	4.572
(3)	9.578	9.416	9.230	9.113	9.087	9.112	9.114	9.023	8.830	8.584	8.339	8.126	7.960	7.844	7.769
(4)	13.39	13.16	13.12	13.15	13.03	12.72	12.39	12.23	12.24	12.23	12.06	11.77	11.45	11.19	11.00

TABLE 8.1 : β_1 $(1/\phi = .9,\ 1/\alpha = .8)$

μ MODE:	0.15	0.20	0.25	0.30	0.35	0.40	0.45	0.50	0.55	0.60	0.65	0.70	0.75	0.80	0.85
(1)	1.847	1.804	1.773	1.750	1.732	1.720	1.713	1.708	1.708	1.710	1.715	1.723	1.735	1.751	1.772
(2)	5.005	5.011	5.009	4.991	4.956	4.908	4.852	4.795	4.742	4.696	4.656	4.624	4.598	4.580	4.571
(3)	8.562	8.531	8.471	8.415	8.391	8.401	8.419	8.411	8.352	8.246	8.117	7.992	7.889	7.811	7.759
(4)	12.05	11.96	11.92	11.94	11.94	11.85	11.69	11.57	11.53	11.55	11.53	11.41	11.25	11.09	10.97

TABLE 8.1 : β_1 ($1/\phi = .9$, $1/\alpha = .9$)

μ / MODE:	0.15	0.20	0.25	0.30	0.35	0.40	0.45	0.50	0.55	0.60	0.65	0.70	0.75	0.80	0.85
(1)	1.772	1.751	1.735	1.723	1.714	1.708	1.704	1.703	1.704	1.708	1.714	1.723	1.735	1.751	1.772
(2)	4.569	4.574	4.584	4.594	4.602	4.607	4.611	4.611	4.611	4.607	4.602	4.594	4.584	4.574	4.569
(3)	7.744	7.764	7.784	7.806	7.832	7.859	7.880	7.889	7.880	7.859	7.832	7.806	7.784	7.764	7.744
(4)	10.91	10.94	10.97	11.00	11.02	11.01	10.99	10.98	10.99	11.01	11.02	11.00	10.97	10.94	10.91

TABLE 8.1 : β_1 ($1/\phi = .9$, $1/\alpha = 1.0$)

μ / MODE:	0.15	0.20	0.25	0.30	0.35	0.40	0.45	0.50	0.55	0.60	0.65	0.70	0.75	0.80	0.85
(1)	1.688	1.688	1.688	1.688	1.689	1.690	1.692	1.695	1.699	1.705	1.712	1.722	1.735	1.751	1.772
(2)	4.227	4.231	4.240	4.257	4.282	4.315	4.355	4.401	4.449	4.493	4.528	4.552	4.563	4.566	4.567
(3)	7.085	7.118	7.174	7.250	7.333	7.406	7.454	7.472	7.473	7.481	7.512	7.570	7.641	7.697	7.722
(4)	9.954	10.04	10.15	10.25	10.29	10.30	10.32	10.39	10.49	10.58	10.61	10.61	10.65	10.74	10.83

TABLE 8.1 : β_1 ($1/\phi = 1.0$, $1/\alpha = .6$)

μ / MODE:	0.15	0.20	0.25	0.30	0.35	0.40	0.45	0.50	0.55	0.60	0.65	0.70	0.75	0.80	0.85
(1)	2.171	2.083	2.022	1.979	1.947	1.925	1.908	1.896	1.888	1.883	1.879	1.877	1.876	1.875	1.875
(2)	7.038	6.931	6.703	6.391	6.060	5.754	5.490	5.271	5.095	4.958	4.856	4.784	4.738	4.711	4.699
(3)	11.89	11.29	10.86	10.69	10.67	10.63	10.40	9.999	9.531	9.086	8.700	8.386	8.150	7.989	7.900
(4)	16.30	15.91	15.88	15.63	14.90	14.19	13.84	13.79	13.70	13.33	12.80	12.25	11.78	11.40	11.15

TABLE 8.1 : β_1 ($1/\phi = 1.0$, $1/\alpha = .7$)

μ / MODE:	0.15	0.20	0.25	0.30	0.35	0.40	0.45	0.50	0.55	0.60	0.65	0.70	0.75	0.80	0.85
(1)	2.120	2.051	2.001	1.964	1.938	1.918	1.904	1.894	1.886	1.882	1.879	1.877	1.876	1.875	1.875
(2)	6.191	6.168	6.088	5.943	5.754	5.549	5.355	5.183	5.038	4.923	4.835	4.773	4.732	4.709	4.699
(3)	10.58	10.29	9.947	9.725	9.643	9.636	9.600	9.453	9.191	8.879	8.577	8.316	8.113	7.973	7.894
(4)	14.66	14.23	14.11	14.09	13.86	13.40	12.98	12.79	12.76	12.69	12.43	12.05	11.67	11.35	11.13

TABLE 8.1 : β_1 ($1/\phi = 1.0$, $1/\alpha = .8$)

μ / MODE:	0.15	0.20	0.25	0.30	0.35	0.40	0.45	0.50	0.55	0.60	0.65	0.70	0.75	0.80	0.85
(1)	2.052	2.005	1.969	1.943	1.923	1.908	1.897	1.889	1.884	1.880	1.878	1.876	1.876	1.875	1.875
(2)	5.557	5.555	5.534	5.481	5.395	5.287	5.169	5.056	4.954	4.869	4.803	4.755	4.724	4.706	4.697
(3)	9.477	9.368	9.187	9.016	8.913	8.880	8.877	8.848	8.750	8.586	8.394	8.208	8.056	7.947	7.886
(4)	13.25	12.98	12.82	12.80	12.76	12.59	12.32	12.10	12.02	12.01	11.94	11.76	11.51	11.27	11.10

TABLE 8.1 : β_1 ($1/\phi = 1.0$, $1/\alpha = .9$)

μ / MODE:	0.15	0.20	0.25	0.30	0.35	0.40	0.45	0.50	0.55	0.60	0.65	0.70	0.75	0.80	0.85
(1)	1.969	1.946	1.927	1.913	1.902	1.894	1.888	1.883	1.880	1.878	1.877	1.876	1.875	1.875	1.875
(2)	5.074	5.073	5.070	5.057	5.031	4.992	4.943	4.890	4.839	4.794	4.757	4.730	4.711	4.701	4.696
(3)	8.579	8.552	8.490	8.412	8.347	8.312	8.304	8.302	8.282	8.229	8.148	8.056	7.972	7.909	7.873
(4)	12.03	11.93	11.83	11.79	11.79	11.76	11.66	11.54	11.46	11.44	11.44	11.39	11.28	11.16	11.06

TABLE 8.2 : β_1 ($\phi = .6$, $\alpha = .6$)

μ / MODE:	0.15	0.20	0.25	0.30	0.35	0.40	0.45	0.50	0.55	0.60	0.65	0.70	0.75	0.80	0.85
(1)	4.447	4.598	4.700	4.700	4.564	4.343	4.111	3.908	3.750	3.640	3.582	3.575	3.617	3.698	3.797
(2)	7.843	7.756	7.304	6.891	6.754	6.859	7.098	7.357	7.483	7.377	7.138	6.887	6.690	6.595	6.647
(3)	10.82	10.09	9.908	10.16	10.50	10.61	10.29	9.893	9.778	10.07	10.52	10.62	10.35	10.00	9.770
(4)	13.27	13.09	13.48	13.77	13.37	12.92	13.14	13.66	13.71	13.23	12.92	13.27	13.78	13.56	13.13

TABLE 8.2 : β_1 ($\phi = .6$, $\alpha = .7$)

μ / MODE:	0.15	0.20	0.25	0.30	0.35	0.40	0.45	0.50	0.55	0.60	0.65	0.70	0.75	0.80	0.85
(1)	5.052	5.146	5.155	5.045	4.832	4.582	4.348	4.152	4.000	3.894	3.830	3.804	3.811	3.841	3.878
(2)	8.814	8.488	7.956	7.612	7.538	7.644	7.815	7.925	7.868	7.659	7.400	7.161	6.984	6.894	6.903
(3)	11.98	11.29	11.23	11.45	11.59	11.37	10.93	10.64	10.67	10.93	11.09	10.92	10.58	10.26	10.06
(4)	14.97	14.95	15.23	15.07	14.45	14.30	14.60	14.74	14.33	13.86	13.80	14.15	14.17	13.79	13.38

TABLE 8.2 : β_1 ($\phi = .6$, $\alpha = .8$)

μ / MODE:	0.15	0.20	0.25	0.30	0.35	0.40	0.45	0.50	0.55	0.60	0.65	0.70	0.75	0.80	0.85
(1)	5.588	5.612	5.541	5.360	5.106	4.840	4.597	4.393	4.231	4.108	4.021	3.964	3.933	3.921	3.920
(2)	9.689	9.203	8.641	8.313	8.229	8.287	8.368	8.358	8.205	7.953	7.679	7.434	7.242	7.117	7.059
(3)	13.15	12.50	12.44	12.56	12.48	12.07	11.61	11.37	11.40	11.53	11.48	11.20	10.84	10.52	10.29
(4)	16.69	16.67	16.74	16.22	15.63	15.58	15.73	15.49	14.93	14.54	14.55	14.68	14.47	14.04	13.63

TABLE 8.2 : β_1 ($\phi = .6$, $\alpha = .9$)

μ / MODE:	0.15	0.20	0.25	0.30	0.35	0.40	0.45	0.50	0.55	0.60	0.65	0.70	0.75	0.80	0.85
(1)	6.069	6.034	5.904	5.676	5.390	5.098	4.834	4.607	4.421	4.271	4.155	4.069	4.007	3.967	3.943
(2)	10.53	9.920	9.298	8.936	8.816	8.823	8.837	8.754	8.539	8.247	7.943	7.667	7.438	7.266	7.153
(3)	14.31	13.62	13.53	13.54	13.29	12.75	12.25	11.99	11.98	11.99	11.82	11.48	11.09	10.73	10.45
(4)	18.33	18.26	18.10	17.33	16.74	16.70	16.64	16.14	15.52	15.15	15.12	15.09	14.75	14.28	13.83

TABLE 8.2 : β_1 ($\phi = .6$, $\alpha = 1.0$)

μ / MODE:	0.15	0.20	0.25	0.30	0.35	0.40	0.45	0.50	0.55	0.60	0.65	0.70	0.75	0.80	0.85
(1)	6.516	6.438	6.267	5.996	5.668	5.338	5.039	4.782	4.566	4.389	4.247	4.136	4.053	3.994	3.956
(2)	11.36	10.61	9.887	9.475	9.328	9.309	9.283	9.139	8.862	8.516	8.166	7.348	7.577	7.363	7.209
(3)	15.42	14.66	14.55	14.48	14.04	13.35	12.78	12.51	12.47	12.41	12.15	11.73	11.29	10.88	10.56
(4)	19.87	19.78	19.37	18.33	17.76	17.70	17.44	16.72	16.02	15.65	15.61	15.46	15.02	14.48	13.98

TABLE 8.2 : β_1 ($\phi = .7$, $\alpha = .6$)

μ / MODE:	0.15	0.20	0.25	0.30	0.35	0.40	0.45	0.50	0.55	0.60	0.65	0.70	0.75	0.80	0.85
(1)	3.812	3.946	4.050	4.097	4.066	3.966	3.833	3.703	3.598	3.529	3.502	3.520	3.582	3.680	3.790
(2)	6.743	6.780	6.602	6.348	6.212	6.244	6.408	6.637	6.839	6.907	6.822	6.672	6.544	6.501	6.598
(3)	9.455	9.116	8.898	9.016	9.296	9.553	9.570	9.360	9.204	9.316	9.692	10.03	10.02	9.812	9.658
(4)	11.86	11.59	11.83	12.18	12.27	11.98	11.93	12.30	12.72	12.67	12.39	12.45	13.00	13.19	12.95

TABLE 8.2 : β_1 ($\phi = .7$, $\alpha = .7$)

μ / MODE:	0.15	0.20	0.25	0.30	0.35	0.40	0.45	0.50	0.55	0.60	0.65	0.70	0.75	0.80	0.85
(1)	4.332	4.423	4.464	4.439	4.348	4.218	4.079	3.954	3.857	3.791	3.759	3.758	3.784	3.827	3.873
(2)	7.607	7.504	7.243	7.011	6.917	6.959	7.084	7.220	7.284	7.235	7.105	6.961	6.853	6.817	6.867
(3)	10.58	10.19	10.06	10.18	10.37	10.42	10.25	10.04	9.979	10.14	10.37	10.44	10.29	10.09	9.968
(4)	13.36	13.21	13.43	13.57	13.36	13.12	13.23	13.52	13.55	13.28	13.12	13.31	13.58	13.47	13.22

TABLE 8.2 : β_1 ($\phi = .7$, $\alpha = .8$)

μ / MODE:	0.15	0.20	0.25	0.30	0.35	0.40	0.45	0.50	0.55	0.60	0.65	0.70	0.75	0.80	0.85
(1)	4.795	4.834	4.821	4.750	4.630	4.485	4.339	4.208	4.100	4.017	3.960	3.926	3.911	3.910	3.916
(2)	8.400	8.205	7.907	7.668	7.557	7.557	7.616	7.667	7.652	7.555	7.409	7.256	7.132	7.056	7.033
(3)	11.70	11.28	11.14	11.20	11.26	11.17	10.94	10.73	10.67	10.75	10.83	10.77	10.58	10.37	10.22
(4)	14.89	14.73	14.84	14.78	14.45	14.26	14.34	14.42	14.23	13.93	13.81	13.91	13.96	13.76	13.49

TABLE 8.2 : β_1 ($\phi = .7$, $\alpha = .9$)

μ MODE:	0.15	0.20	0.25	0.30	0.35	0.40	0.45	0.50	0.55	0.60	0.65	0.70	0.75	0.80	0.85
(1)	5.211	5.207	5.159	5.061	4.921	4.757	4.591	4.437	4.304	4.192	4.104	4.036	3.989	3.957	3.939
(2)	9.170	8.909	8.557	8.265	8.102	8.052	8.062	8.066	8.011	7.881	7.704	7.517	7.349	7.218	7.131
(3)	12.82	12.31	12.11	12.10	12.09	11.90	11.59	11.33	11.21	11.21	11.21	11.09	10.86	10.61	10.40
(4)	16.36	16.13	16.15	15.93	15.49	15.25	15.26	15.19	14.87	14.51	14.35	14.36	14.30	14.04	13.72

TABLE 8.2 : β_1 ($\phi = .7$, $\alpha = 1.0$)

μ MODE:	0.15	0.20	0.25	0.30	0.35	0.40	0.45	0.50	0.55	0.60	0.65	0.70	0.75	0.80	0.85
(1)	5.598	5.568	5.498	5.378	5.211	5.016	4.816	4.629	4.462	4.320	4.202	4.108	4.036	3.986	3.953
(2)	9.937	9.601	9.151	8.780	8.566	8.487	8.479	8.461	8.369	8.191	7.961	7.723	7.504	7.324	7.191
(3)	13.90	13.25	12.99	12.97	12.89	12.58	12.15	11.81	11.65	11.63	11.59	11.40	11.10	10.79	10.51
(4)	17.73	17.46	17.42	17.01	16.40	16.14	16.11	15.90	15.43	14.98	14.79	14.77	14.63	14.28	13.89

TABLE 8.2 : β_1 ($\phi = .8$, $\alpha = .6$)

μ MODE:	0.15	0.20	0.25	0.30	0.35	0.40	0.45	0.50	0.55	0.60	0.65	0.70	0.75	0.80	0.85
(1)	3.336	3.455	3.552	3.613	3.623	3.590	3.527	3.460	3.407	3.382	3.391	3.440	3.530	3.651	3.778
(2)	5.908	5.983	5.930	5.811	5.739	5.774	5.909	6.111	6.321	6.457	6.474	6.412	6.353	6.369	6.522
(3)	8.339	8.214	8.095	8.177	8.408	8.674	8.830	8.794	8.707	8.775	9.081	9.472	9.639	9.566	9.498
(4)	10.60	10.45	10.62	10.94	11.18	11.13	11.06	11.30	11.74	11.99	11.90	11.90	12.34	12.76	12.71

TABLE 8.2 : β_1 ($\phi = .8$, $\alpha = .7$)

μ MODE:	0.15	0.20	0.25	0.30	0.35	0.40	0.45	0.50	0.55	0.60	0.65	0.70	0.75	0.80	0.85
(1)	3.792	3.875	3.924	3.932	3.900	3.842	3.775	3.714	3.671	3.652	3.659	3.690	3.743	3.805	3.864
(2)	6.676	6.658	6.550	6.441	6.402	6.449	6.558	6.688	6.785	6.810	6.769	6.708	6.674	6.704	6.812
(3)	9.386	9.210	9.153	9.251	9.423	9.547	9.538	9.452	9.432	9.561	9.791	9.952	9.942	9.853	9.829
(4)	11.98	11.91	12.08	12.27	12.29	12.18	12.24	12.50	12.70	12.66	12.57	12.71	13.02	13.10	12.99

TABLE 8.2 : β_1 ($\phi = .8$, $\alpha = .8$)

μ MODE:	0.15	0.20	0.25	0.30	0.35	0.40	0.45	0.50	0.55	0.60	0.65	0.70	0.75	0.80	0.85
(1)	4.198	4.239	4.248	4.223	4.173	4.107	4.038	3.975	3.925	3.890	3.872	3.869	3.877	3.892	3.909
(2)	7.387	7.314	7.188	7.076	7.022	7.029	7.075	7.127	7.153	7.135	7.082	7.020	6.976	6.966	6.992
(3)	10.42	10.24	10.16	10.20	10.27	10.29	10.22	10.14	10.11	10.18	10.27	10.31	10.25	10.16	10.11
(4)	13.39	13.30	13.38	13.44	13.35	13.26	13.30	13.42	13.43	13.32	13.26	13.34	13.44	13.41	13.29

TABLE 8.2 : β_1 ($\phi = .8$, $\alpha = .9$)

μ MODE:	0.15	0.20	0.25	0.30	0.35	0.40	0.45	0.50	0.55	0.60	0.65	0.70	0.75	0.80	0.85
(1)	4.563	4.571	4.555	4.516	4.455	4.380	4.298	4.218	4.143	4.079	4.027	3.988	3.960	3.943	3.933
(2)	8.080	7.978	7.823	7.669	7.559	7.506	7.497	7.506	7.504	7.468	7.398	7.309	7.219	7.146	7.099
(3)	11.46	11.23	11.07	11.03	11.04	11.01	10.91	10.76	10.67	10.64	10.66	10.64	10.56	10.43	10.31
(4)	14.77	14.58	14.57	14.55	14.40	14.23	14.18	14.19	14.12	13.95	13.81	13.79	13.80	13.72	13.56

TABLE 8.2 : β_1 ($\phi = .8$, $\alpha = 1.0$)

μ MODE:	0.15	0.20	0.25	0.30	0.35	0.40	0.45	0.50	0.55	0.60	0.65	0.70	0.75	0.80	0.85
(1)	4.904	4.892	4.864	4.815	4.741	4.646	4.537	4.425	4.317	4.219	4.135	4.065	4.011	3.972	3.947
(2)	8.773	8.639	8.423	8.192	8.013	7.913	7.878	7.875	7.861	7.800	7.688	7.545	7.396	7.264	7.163
(3)	12.49	12.14	11.88	11.81	11.80	11.72	11.52	11.28	11.10	11.03	11.02	10.98	10.84	10.64	10.44
(4)	16.08	15.77	15.73	15.65	15.35	15.05	14.95	14.94	14.77	14.47	14.24	14.17	14.15	14.01	13.76

TABLE 8.2 : β_1 ($\phi = .9$, $\alpha = .6$)

μ MODE:	0.15	0.20	0.25	0.30	0.35	0.40	0.45	0.50	0.55	0.60	0.65	0.70	0.75	0.80	0.85
(1)	2.966	3.072	3.162	3.224	3.253	3.251	3.229	3.206	3.195	3.207	3.253	3.336	3.459	3.611	3.762
(2)	5.255	5.340	5.342	5.303	5.293	5.357	5.499	5.699	5.915	6.083	6.132	6.144	6.133	6.200	6.418
(3)	7.441	7.414	7.387	7.490	7.711	7.978	8.189	8.249	8.226	8.302	8.588	9.007	9.279	9.300	9.302
(4)	9.518	9.487	9.659	9.964	10.25	10.34	10.32	10.51	10.94	11.33	11.40	11.41	11.81	12.36	12.44

TABLE 8.2 : β_1 ($\phi = .9$, $\alpha = .7$)

μ MODE:	0.15	0.20	0.25	0.30	0.35	0.40	0.45	0.50	0.55	0.60	0.65	0.70	0.75	0.80	0.85
(1)	3.371	3.446	3.496	3.517	3.513	3.494	3.471	3.457	3.459	3.482	3.529	3.598	3.684	3.775	3.852
(2)	5.944	5.960	5.926	5.900	5.923	6.006	6.135	6.277	6.391	6.446	6.446	6.434	6.456	6.553	6.733
(3)	8.398	8.348	8.370	8.500	8.685	8.840	8.894	8.879	8.910	9.069	9.328	9.544	9.603	9.585	9.646
(4)	10.80	10.83	11.02	11.24	11.34	11.32	11.41	11.68	11.96	12.05	12.03	12.20	12.56	12.74	12.73

TABLE 8.2 : β_1 ($\phi = .9$, $\alpha = .8$)

μ MODE:	0.15	0.20	0.25	0.30	0.35	0.40	0.45	0.50	0.55	0.60	0.65	0.70	0.75	0.80	0.85
(1)	3.732	3.772	3.789	3.786	3.770	3.748	3.729	3.718	3.718	3.730	3.756	3.790	3.830	3.868	3.898
(2)	6.583	6.564	6.527	6.509	6.529	6.582	6.653	6.716	6.754	6.761	6.751	6.750	6.776	6.841	6.932
(3)	9.346	9.303	9.327	9.412	9.504	9.552	9.547	9.544	9.598	9.718	9.844	9.907	9.904	9.899	9.956
(4)	12.10	12.13	12.25	12.34	12.34	12.35	12.46	12.62	12.70	12.69	12.72	12.87	13.02	13.05	13.04

TABLE 8.2 : β_1 ($\phi = .9$, $\alpha = .9$)

μ MODE:	0.15	0.20	0.25	0.30	0.35	0.40	0.45	0.50	0.55	0.60	0.65	0.70	0.75	0.80	0.85
(1)	4.058	4.070	4.068	4.056	4.036	4.012	3.988	3.966	3.947	3.933	3.925	3.920	3.920	3.921	3.924
(2)	7.207	7.175	7.131	7.093	7.070	7.064	7.070	7.080	7.085	7.081	7.070	7.056	7.046	7.045	7.052
(3)	10.30	10.24	10.21	10.21	10.22	10.23	10.21	10.19	10.19	10.20	10.22	10.23	10.22	10.20	10.19
(4)	13.39	13.35	13.36	13.37	13.35	13.33	13.34	13.37	13.37	13.35	13.33	13.35	13.37	13.37	13.34

TABLE 8.2 : β_1 ($\phi = .9$, $\alpha = 1.0$)

μ MODE:	0.15	0.20	0.25	0.30	0.35	0.40	0.45	0.50	0.55	0.60	0.65	0.70	0.75	0.80	0.85
(1)	4.362	4.358	4.349	4.333	4.308	4.273	4.231	4.184	4.135	4.087	4.043	4.005	3.975	3.953	3.938
(2)	7.834	7.791	7.716	7.622	7.532	7.466	7.431	7.421	7.420	7.410	7.377	7.320	7.250	7.180	7.123
(3)	11.26	11.13	11.00	10.93	10.91	10.91	10.85	10.75	10.65	10.58	10.57	10.56	10.53	10.44	10.34
(4)	14.63	14.46	14.40	14.40	14.32	14.18	14.08	14.06	14.04	13.93	13.80	13.72	13.71	13.68	13.58

TABLE 8.2 : β_1 ($\phi = 1.0$, $\alpha = .6$)

μ MODE:	0.15	0.20	0.25	0.30	0.35	0.40	0.45	0.50	0.55	0.60	0.65	0.70	0.75	0.80	0.85
(1)	2.669	2.765	2.847	2.908	2.944	2.957	2.959	2.962	2.979	3.020	3.095	3.210	3.368	3.556	3.738
(2)	4.732	4.817	4.844	4.846	4.877	4.970	5.129	5.343	5.578	5.777	5.880	5.899	5.911	6.007	6.282
(3)	6.711	6.729	6.758	6.890	7.120	7.401	7.653	7.774	7.786	7.860	8.140	8.598	8.966	9.052	9.086
(4)	8.614	8.653	8.847	9.159	9.481	9.657	9.674	9.822	10.23	10.72	10.93	10.96	11.32	11.99	12.19

TABLE 8.2 : β_1 ($\phi = 1.0$, $\alpha = .7$)

μ MODE:	0.15	0.20	0.25	0.30	0.35	0.40	0.45	0.50	0.55	0.60	0.65	0.70	0.75	0.80	0.85
(1)	3.034	3.103	3.151	3.177	3.186	3.187	3.190	3.204	3.237	3.294	3.376	3.483	3.608	3.732	3.835
(2)	5.354	5.384	5.387	5.406	5.473	5.594	5.754	5.928	6.073	6.152	6.170	6.175	6.224	6.370	6.626
(3)	7.585	7.594	7.673	7.844	8.063	8.261	8.359	8.370	8.414	8.596	8.909	9.201	9.315	9.324	9.429
(4)	9.790	9.892	10.12	10.39	10.55	10.57	10.66	10.95	11.31	11.51	11.52	11.69	12.13	12.44	12.47

TABLE 8.2 : β_1 (ϕ = 1.0, α = .8)

μ MODE:	0.15	0.20	0.25	0.30	0.35	0.40	0.45	0.50	0.55	0.60	0.65	0.70	0.75	0.80	0.85
(1)	3.359	3.397	3.417	3.424	3.425	3.427	3.436	3.458	3.495	3.547	3.614	3.689	3.766	3.834	3.884
(2)	5.933	5.938	5.946	5.983	6.059	6.165	6.281	6.380	6.439	6.436	6.458	6.479	6.549	6.682	6.851
(3)	8.452	8.481	8.579	8.729	8.873	8.956	8.970	8.987	9.086	9.274	9.475	9.587	9.605	9.628	9.762
(4)	10.99	11.12	11.31	11.45	11.48	11.52	11.68	11.93	12.09	12.12	12.18	12.41	12.67	12.75	12.77

TABLE 8.2 : β_1 (ϕ = 1.0, α = .9)

μ MODE:	0.15	0.20	0.25	0.30	0.35	0.40	0.45	0.50	0.55	0.60	0.65	0.70	0.75	0.80	0.85
(1)	3.653	3.666	3.671	3.672	3.672	3.677	3.687	3.704	3.729	3.761	3.796	3.832	3.865	3.892	3.911
(2)	6.500	6.501	6.512	6.545	6.597	6.656	6.709	6.743	6.756	6.757	6.763	6.788	6.840	6.912	6.987
(3)	9.330	9.362	9.433	9.514	9.569	9.584	9.588	9.622	9.704	9.805	9.879	9.901	9.904	9.937	10.03
(4)	12.18	12.28	12.37	12.41	12.42	12.47	12.59	12.70	12.73	12.74	12.82	12.96	13.03	13.04	13.08

TABLE 8.2 : β_1 (1/ϕ = .6, 1/α = .6)

μ MODE:	0.15	0.20	0.25	0.30	0.35	0.40	0.45	0.50	0.55	0.60	0.65	0.70	0.75	0.80	0.85
(1)	3.461	3.460	3.438	3.399	3.349	3.300	3.262	3.240	3.239	3.263	3.314	3.395	3.505	3.637	3.770
(2)	6.552	6.456	6.399	6.438	6.581	6.812	7.102	7.385	7.524	7.420	7.175	6.916	6.712	6.609	6.655
(3)	9.561	9.576	9.797	10.15	10.52	10.66	10.33	9.921	9.789	10.07	10.52	10.62	10.35	10.00	9.770
(4)	12.72	13.03	13.49	13.81	13.41	12.94	13.14	13.66	13.71	13.23	12.92	13.27	13.78	13.56	13.13

TABLE 8.2 : β_1 (1/ϕ = .6, 1/α = .7)

μ MODE:	0.15	0.20	0.25	0.30	0.35	0.40	0.45	0.50	0.55	0.60	0.65	0.70	0.75	0.80	0.85
(1)	3.055	3.062	3.069	3.071	3.069	3.068	3.072	3.088	3.119	3.169	3.244	3.345	3.472	3.619	3.762
(2)	5.764	5.762	5.774	5.830	5.950	6.135	6.377	6.644	6.866	6.943	6.857	6.702	6.567	6.516	6.605
(3)	8.456	8.511	8.683	8.962	9.298	9.379	9.609	9.392	9.222	9.322	9.692	10.03	10.02	9.811	9.657
(4)	11.20	11.43	11.81	12.20	12.31	12.01	11.94	12.30	12.72	12.67	12.39	12.45	13.00	13.19	12.95

TABLE 8.2 : β_1 (1/ϕ = .6, 1/α = .8)

μ MODE:	0.15	0.20	0.25	0.30	0.35	0.40	0.45	0.50	0.55	0.60	0.65	0.70	0.75	0.80	0.85
(1)	2.759	2.763	2.775	2.791	2.811	2.835	2.868	2.912	2.970	3.047	3.147	3.273	3.424	3.592	3.752
(2)	5.140	5.175	5.229	5.316	5.446	5.622	5.844	6.094	6.331	6.484	6.505	6.440	6.375	6.383	6.530
(3)	7.548	7.650	7.826	8.078	8.384	8.683	8.859	8.824	8.728	8.786	9.083	9.472	9.637	9.564	9.498
(4)	9.996	10.23	10.56	10.94	11.20	11.17	11.08	11.31	11.74	11.99	11.90	11.90	12.34	12.76	12.70

TABLE 8.2 : β_1 (1/ϕ = .6, 1/α = .9)

μ MODE:	0.15	0.20	0.25	0.30	0.35	0.40	0.45	0.50	0.55	0.60	0.65	0.70	0.75	0.80	0.85
(1)	2.535	2.537	2.547	2.564	2.589	2.624	2.670	2.730	2.807	2.906	3.029	3.180	3.358	3.554	3.736
(2)	4.649	4.691	4.761	4.865	5.008	5.191	5.413	5.662	5.910	6.098	6.176	6.169	6.154	6.214	6.425
(3)	6.805	6.928	7.116	7.366	7.663	7.970	8.204	8.275	8.248	8.315	8.592	9.007	9.278	9.299	9.300
(4)	9.009	9.249	9.575	9.945	10.26	10.37	10.34	10.52	10.94	11.33	11.40	11.41	11.81	12.36	12.44

TABLE 8.2 : β_1 (1/ϕ = .6, 1/α = 1.0)

μ MODE:	0.15	0.20	0.25	0.30	0.35	0.40	0.45	0.50	0.55	0.60	0.65	0.70	0.75	0.80	0.85
(1)	2.357	2.361	2.369	2.383	2.407	2.442	2.491	2.558	2.644	2.755	2.895	3.068	3.274	3.502	3.713
(2)	4.259	4.296	4.366	4.473	4.621	4.809	5.036	5.295	5.561	5.781	5.897	5.920	5.930	6.021	6.290
(3)	6.199	6.322	6.512	6.762	7.060	7.379	7.655	7.792	7.806	7.874	8.146	8.599	8.966	9.051	9.085
(4)	8.191	8.431	8.754	9.125	9.478	9.673	9.694	9.833	10.24	10.72	10.93	10.96	11.32	11.99	12.19

TABLE 8.2 : β_1 (1/ϕ = .7, 1/α = .6)

μ MODE:	0.15	0.20	0.25	0.30	0.35	0.40	0.45	0.50	0.55	0.60	0.65	0.70	0.75	0.80	0.85
(1)	4.037	4.033	4.004	3.948	3.878	3.804	3.739	3.689	3.659	3.649	3.662	3.695	3.746	3.806	3.863
(2)	7.629	7.486	7.376	7.366	7.465	7.643	7.843	7.965	7.909	7.695	7.427	7.181	6.998	6.902	6.907
(3)	11.09	11.03	11.20	11.47	11.64	11.41	10.96	10.65	10.67	10.93	11.09	10.92	10.58	10.26	10.06
(4)	14.70	14.94	15.27	15.11	14.48	14.30	14.60	14.74	14.32	13.86	13.80	14.15	14.17	13.79	13.38

TABLE 8.2 : β_1 (1/ϕ = .7, 1/α = .7)

μ MODE:	0.15	0.20	0.25	0.30	0.35	0.40	0.45	0.50	0.55	0.60	0.65	0.70	0.75	0.80	0.85
(1)	3.563	3.571	3.576	3.572	3.562	3.547	3.535	3.530	3.537	3.560	3.598	3.653	3.720	3.793	3.858
(2)	6.717	6.697	6.678	6.699	6.779	6.917	7.088	7.245	7.318	7.268	7.133	6.982	6.867	6.825	6.871
(3)	9.829	9.836	9.957	10.17	10.39	10.45	10.28	10.06	9.986	10.14	10.37	10.44	10.29	10.09	9.967
(4)	12.97	13.15	13.44	13.60	13.39	13.13	13.23	13.52	13.55	13.28	13.12	13.31	13.58	13.47	13.22

TABLE 8.2 : β_1 (1/ϕ = .7, 1/α = .8)

μ MODE:	0.15	0.20	0.25	0.30	0.35	0.40	0.45	0.50	0.55	0.60	0.65	0.70	0.75	0.80	0.85
(1)	3.218	3.223	3.235	3.249	3.266	3.285	3.310	3.342	3.384	3.439	3.508	3.591	3.682	3.773	3.850
(2)	5.992	6.022	6.064	6.132	6.235	6.373	6.537	6.696	6.808	6.837	6.794	6.728	6.688	6.712	6.816
(3)	8.785	8.865	9.009	9.212	9.428	9.571	9.565	9.472	9.442	9.565	9.791	9.951	9.941	9.852	9.829
(4)	11.61	11.80	12.07	12.29	12.31	12.20	12.25	12.50	12.70	12.66	12.57	12.71	13.02	13.10	12.99

TABLE 8.2 : β_1 (1/ϕ = .7, 1/α = .9)

μ MODE:	0.15	0.20	0.25	0.30	0.35	0.40	0.45	0.50	0.55	0.60	0.65	0.70	0.75	0.80	0.85
(1)	2.958	2.959	2.969	2.986	3.011	3.044	3.087	3.143	3.211	3.295	3.394	3.508	3.628	3.744	3.839
(2)	5.420	5.462	5.530	5.626	5.758	5.917	6.096	6.269	6.402	6.465	6.468	6.452	6.470	6.561	6.736
(3)	7.926	8.043	8.219	8.439	8.671	8.851	8.915	8.898	8.922	9.074	9.329	9.544	9.603	9.584	9.646
(4)	10.47	10.70	10.98	11.24	11.36	11.34	11.42	11.68	11.96	12.05	12.03	12.20	12.55	12.74	12.73

TABLE 8.2 : β_1 (1/ϕ = .7, 1/α = 1.0)

μ MODE:	0.15	0.20	0.25	0.30	0.35	0.40	0.45	0.50	0.55	0.60	0.65	0.70	0.75	0.80	0.85
(1)	2.750	2.754	2.762	2.776	2.800	2.836	2.885	2.950	3.034	3.137	3.261	3.404	3.558	3.705	3.823
(2)	4.966	5.004	5.075	5.183	5.329	5.508	5.709	5.911	6.074	6.164	6.186	6.190	6.236	6.377	6.629
(3)	7.221	7.347	7.538	7.779	8.038	8.260	8.373	8.386	8.426	8.602	8.911	9.201	9.315	9.324	9.429
(4)	9.531	9.773	10.08	10.38	10.56	10.59	10.67	10.95	11.32	11.51	11.52	11.69	12.13	12.44	12.47

TABLE 8.2 : β_1 (1/ϕ = .8, 1/α = .6)

μ MODE:	0.15	0.20	0.25	0.30	0.35	0.40	0.45	0.50	0.55	0.60	0.65	0.70	0.75	0.80	0.85
(1)	4.612	4.604	4.562	4.483	4.381	4.271	4.166	4.075	4.002	3.949	3.914	3.896	3.894	3.900	3.911
(2)	8.692	8.475	8.277	8.186	8.213	8.311	8.405	8.397	8.239	7.979	7.698	7.447	7.250	7.121	7.062
(3)	12.55	12.37	12.45	12.59	12.52	12.10	11.63	11.37	11.40	11.53	11.48	11.20	10.84	10.52	10.29
(4)	16.56	16.69	16.78	16.25	15.64	15.58	15.73	15.49	14.93	14.54	14.55	14.68	14.47	14.04	13.63

TABLE 8.2 : β_1 (1/ϕ = .8, 1/α = .7)

μ MODE:	0.15	0.20	0.25	0.30	0.35	0.40	0.45	0.50	0.55	0.60	0.65	0.70	0.75	0.80	0.85
(1)	4.071	4.078	4.078	4.065	4.037	3.999	3.958	3.920	3.889	3.868	3.859	3.862	3.873	3.890	3.907
(2)	7.662	7.608	7.532	7.484	7.492	7.552	7.637	7.697	7.682	7.581	7.428	7.270	7.140	7.061	7.035
(3)	11.17	11.08	11.10	11.21	11.29	11.20	10.96	10.73	10.67	10.75	10.83	10.77	10.58	10.37	10.22
(4)	14.67	14.72	14.87	14.81	14.47	14.26	14.34	14.42	14.23	13.93	13.81	13.91	13.96	13.76	13.49

TABLE 8.2 : β_1 ($1/\phi = .8$, $1/\alpha = .8$)

μ	0.15	0.20	0.25	0.30	0.35	0.40	0.45	0.50	0.55	0.60	0.65	0.70	0.75	0.80	0.85
MODE:															
(1)	3.677	3.681	3.691	3.701	3.709	3.715	3.721	3.728	3.740	3.757	3.781	3.810	3.843	3.874	3.901
(2)	6.839	6.854	6.866	6.887	6.931	6.998	7.078	7.146	7.176	7.156	7.099	7.033	6.984	6.971	6.994
(3)	10.00	10.02	10.09	10.19	10.29	10.31	10.24	10.15	10.12	10.18	10.27	10.30	10.25	10.16	10.11
(4)	13.16	13.25	13.39	13.46	13.37	13.26	13.30	13.42	13.43	13.32	13.26	13.34	13.44	13.41	13.29

TABLE 8.2 : β_1 ($1/\phi = .8$, $1/\alpha = .9$)

μ	0.15	0.20	0.25	0.30	0.35	0.40	0.45	0.50	0.55	0.60	0.65	0.70	0.75	0.80	0.85
MODE:															
(1)	3.379	3.380	3.389	3.404	3.424	3.450	3.482	3.521	3.567	3.620	3.678	3.740	3.800	3.852	3.891
(2)	6.188	6.223	6.276	6.347	6.437	6.540	6.641	6.723	6.769	6.777	6.766	6.761	6.784	6.845	6.934
(3)	9.032	9.121	9.246	9.389	9.510	9.567	9.563	9.555	9.604	9.719	9.843	9.907	9.904	9.898	9.956
(4)	11.91	12.07	12.24	12.35	12.35	12.36	12.46	12.62	12.70	12.69	12.72	12.87	13.02	13.05	13.04

TABLE 8.2 : β_1 ($1/\phi = .8$, $1/\alpha = 1.0$)

μ	0.15	0.20	0.25	0.30	0.35	0.40	0.45	0.50	0.55	0.60	0.65	0.70	0.75	0.80	0.85
MODE:															
(1)	3.143	3.146	3.153	3.166	3.187	3.218	3.261	3.316	3.383	3.464	3.554	3.649	3.742	3.821	3.879
(2)	5.671	5.705	5.768	5.862	5.984	6.124	6.264	6.379	6.446	6.467	6.469	6.488	6.555	6.685	6.852
(3)	8.235	8.347	8.511	8.701	8.869	8.964	8.981	8.996	9.090	9.275	9.476	9.587	9.604	9.627	9.762
(4)	10.85	11.06	11.29	11.46	11.49	11.52	11.68	11.93	12.09	12.12	12.17	12.41	12.67	12.75	12.77

TABLE 8.2 : β_1 ($1/\phi = .9$, $1/\alpha = .6$)

μ	0.15	0.20	0.25	0.30	0.35	0.40	0.45	0.50	0.55	0.60	0.65	0.70	0.75	0.80	0.85
MODE:															
(1)	5.187	5.171	5.109	4.996	4.848	4.687	4.531	4.390	4.268	4.168	4.087	4.026	3.983	3.954	3.938
(2)	9.734	9.404	9.077	8.884	8.830	8.858	8.874	8.786	8.565	8.266	7.956	7.675	7.444	7.269	7.154
(3)	13.93	13.58	13.56	13.58	13.32	12.77	12.25	11.99	11.97	11.99	11.82	11.48	11.08	10.73	10.45
(4)	18.29	18.29	18.13	17.34	16.75	16.69	16.64	16.14	15.52	15.15	15.12	15.09	14.75	14.28	13.83

TABLE 8.2 : β_1 ($1/\phi = .9$, $1/\alpha = .7$)

μ	0.15	0.20	0.25	0.30	0.35	0.40	0.45	0.50	0.55	0.60	0.65	0.70	0.75	0.80	0.85
MODE:															
(1)	4.579	4.582	4.574	4.543	4.488	4.414	4.330	4.245	4.166	4.098	4.041	3.997	3.966	3.946	3.934
(2)	8.596	8.484	8.314	8.164	8.081	8.067	8.087	8.093	8.035	7.899	7.716	7.525	7.354	7.221	7.132
(3)	12.45	12.21	12.11	12.13	12.11	11.92	11.60	11.33	11.21	11.21	11.21	11.09	10.86	10.61	10.40
(4)	16.25	16.15	16.17	15.95	15.50	15.25	15.26	15.19	14.87	14.51	14.35	14.36	14.30	14.04	13.72

TABLE 8.2 : β_1 ($1/\phi = .9$, $1/\alpha = .8$)

μ	0.15	0.20	0.25	0.30	0.35	0.40	0.45	0.50	0.55	0.60	0.65	0.70	0.75	0.80	0.85
MODE:															
(1)	4.136	4.137	4.143	4.144	4.136	4.118	4.092	4.061	4.029	3.999	3.974	3.954	3.941	3.932	3.929
(2)	7.679	7.664	7.618	7.560	7.516	7.501	7.509	7.525	7.522	7.483	7.409	7.316	7.223	7.149	7.100
(3)	11.18	11.10	11.04	11.04	11.06	11.03	10.92	10.77	10.67	10.64	10.66	10.64	10.56	10.43	10.31
(4)	14.64	14.57	14.59	14.57	14.41	14.23	14.18	14.19	14.12	13.95	13.81	13.79	13.80	13.72	13.56

TABLE 8.2 : β_1 ($1/\phi = .9$, $1/\alpha = .9$)

μ	0.15	0.20	0.25	0.30	0.35	0.40	0.45	0.50	0.55	0.60	0.65	0.70	0.75	0.80	0.85
MODE:															
(1)	3.801	3.800	3.806	3.815	3.825	3.835	3.844	3.854	3.863	3.873	3.884	3.894	3.904	3.914	3.920
(2)	6.952	6.970	6.988	7.006	7.026	7.048	7.071	7.089	7.096	7.092	7.078	7.062	7.050	7.047	7.053
(3)	10.12	10.14	10.17	10.21	10.23	10.24	10.22	10.20	10.19	10.20	10.22	10.23	10.22	10.20	10.19
(4)	13.28	13.32	13.36	13.38	13.36	13.33	13.34	13.37	13.37	13.35	13.33	13.35	13.37	13.36	13.34

TABLE 8.2 : β_1 ($1/\phi = .9$, $1/\alpha = 1.0$)

μ	0.15	0.20	0.25	0.30	0.35	0.40	0.45	0.50	0.55	0.60	0.65	0.70	0.75	0.80	0.85
MODE:															
(1)	3.535	3.537	3.542	3.550	3.564	3.584	3.610	3.644	3.683	3.727	3.772	3.816	3.856	3.887	3.909
(2)	6.372	6.395	6.436	6.496	6.568	6.643	6.706	6.746	6.761	6.762	6.768	6.791	6.842	6.913	6.987
(3)	9.234	9.306	9.407	9.507	9.571	9.590	9.593	9.625	9.705	9.806	9.879	9.901	9.904	9.938	10.03
(4)	12.13	12.26	12.37	12.42	12.42	12.48	12.59	12.70	12.73	12.74	12.82	12.95	13.04	13.04	13.08

TABLE 8.2 : β_1 ($1/\phi = 1.0$, $1/\alpha = .6$)

μ	0.15	0.20	0.25	0.30	0.35	0.40	0.45	0.50	0.55	0.60	0.65	0.70	0.75	0.80	0.85
MODE:															
(1)	5.759	5.732	5.640	5.479	5.270	5.045	4.829	4.634	4.464	4.321	4.203	4.109	4.037	3.986	3.953
(2)	10.74	10.25	9.763	9.467	9.357	9.344	9.314	9.164	8.880	8.528	8.174	7.853	7.581	7.365	7.210
(3)	15.19	14.67	14.58	14.51	14.07	13.36	12.79	12.51	12.47	12.41	12.15	11.73	11.29	10.88	10.56
(4)	19.89	19.81	19.40	18.34	17.76	17.70	17.43	16.72	16.02	15.65	15.61	15.46	15.02	14.48	13.98

TABLE 8.2 : β_1 ($1/\phi = 1.0$, $1/\alpha = .7$)

μ	0.15	0.20	0.25	0.30	0.35	0.40	0.45	0.50	0.55	0.60	0.65	0.70	0.75	0.80	0.85
MODE:															
(1)	5.085	5.083	5.061	5.002	4.906	4.780	4.642	4.504	4.375	4.261	4.163	4.084	4.023	3.979	3.950
(2)	9.514	9.307	9.002	8.732	8.568	8.508	8.501	8.481	8.385	8.203	7.970	7.728	7.507	7.325	7.191
(3)	13.66	13.21	13.01	12.99	12.91	12.59	12.16	11.81	11.65	11.63	11.59	11.40	11.10	10.79	10.51
(4)	17.69	17.48	17.44	17.02	16.41	16.14	16.11	15.90	15.43	14.98	14.79	14.77	14.63	14.28	13.89

TABLE 8.2 : β_1 ($1/\phi = 1.0$, $1/\alpha = .8$)

μ	0.15	0.20	0.25	0.30	0.35	0.40	0.45	0.50	0.55	0.60	0.65	0.70	0.75	0.80	0.85
MODE:															
(1)	4.594	4.591	4.589	4.573	4.539	4.484	4.412	4.333	4.252	4.174	4.105	4.047	4.001	3.967	3.945
(2)	8.509	8.442	8.304	8.137	7.998	7.918	7.891	7.888	7.872	7.809	7.695	7.549	7.399	7.266	7.164
(3)	12.32	12.07	11.88	11.82	11.81	11.73	11.53	11.28	11.10	11.03	11.02	10.98	10.84	10.64	10.44
(4)	16.01	15.77	15.74	15.66	15.35	15.05	14.95	14.93	14.77	14.47	14.24	14.17	14.15	14.01	13.76

TABLE 8.2 : β_1 ($1/\phi = 1.0$, $1/\alpha = .9$)

μ	0.15	0.20	0.25	0.30	0.35	0.40	0.45	0.50	0.55	0.60	0.65	0.70	0.75	0.80	0.85
MODE:															
(1)	4.223	4.218	4.218	4.216	4.207	4.191	4.166	4.134	4.098	4.062	4.026	3.995	3.970	3.950	3.937
(2)	7.708	7.694	7.652	7.586	7.517	7.464	7.435	7.427	7.426	7.414	7.380	7.322	7.251	7.181	7.124
(3)	11.17	11.09	10.99	10.93	10.92	10.91	10.86	10.76	10.65	10.58	10.57	10.56	10.53	10.44	10.34
(4)	14.58	14.46	14.41	14.40	14.33	14.18	14.08	14.06	14.03	13.93	13.80	13.72	13.71	13.68	13.58

TABLE 8.3 : β_1 ($\phi = .6$, $\alpha = .6$)

μ	0.15	0.20	0.25	0.30	0.35	0.40	0.45	0.50	0.55	0.60	0.65	0.70	0.75	0.80	0.85
MODE:															
(1)	3.060	2.980	2.890	2.805	2.735	2.683	2.652	2.642	2.652	2.683	2.735	2.805	2.890	2.980	3.060
(2)	5.922	5.819	5.841	5.967	6.166	6.404	6.622	6.718	6.622	6.404	6.166	5.967	5.841	5.819	5.922
(3)	8.962	9.126	9.425	9.738	9.852	9.567	9.165	8.989	9.165	9.567	9.852	9.738	9.425	9.126	8.962
(4)	12.28	12.67	12.99	12.72	12.19	12.24	12.69	13.00	12.69	12.24	12.19	12.72	12.99	12.67	12.28

TABLE 8.3 : β_1 ($\phi = .6$, $\alpha = .7$)

μ	0.15	0.20	0.25	0.30	0.35	0.40	0.45	0.50	0.55	0.60	0.65	0.70	0.75	0.80	0.85
MODE:															
(1)	3.566	3.467	3.352	3.241	3.144	3.067	3.010	2.974	2.957	2.958	2.974	3.003	3.039	3.077	3.110
(2)	6.867	6.705	6.682	6.768	6.918	7.076	7.162	7.103	6.912	6.669	6.439	6.256	6.141	6.104	6.141
(3)	10.34	10.45	10.68	10.83	10.65	10.22	9.893	9.844	10.04	10.27	10.26	9.995	9.671	9.399	9.256
(4)	14.13	14.42	14.41	13.82	13.49	13.69	13.96	13.77	13.28	12.97	13.13	13.43	13.30	12.90	12.54

TABLE 8.3 : β_1 ($\phi = .6, \alpha = .8$)

μ MODE:	0.15	0.20	0.25	0.30	0.35	0.40	0.45	0.50	0.55	0.60	0.65	0.70	0.75	0.80	0.85
(1)	4.069	3.945	3.798	3.650	3.517	3.404	3.313	3.243	3.192	3.157	3.136	3.127	3.125	3.129	3.135
(2)	7.775	7.525	7.430	7.454	7.536	7.606	7.589	7.446	7.213	6.955	6.717	6.522	6.382	6.300	6.269
(3)	11.63	11.66	11.78	11.75	11.38	10.91	10.62	10.58	10.70	10.75	10.58	10.26	9.929	9.653	9.475
(4)	15.86	15.99	15.65	14.98	14.76	14.90	14.88	14.42	13.92	13.71	13.83	13.87	13.57	13.15	12.79

TABLE 8.3 : β_1 ($\phi = .6, \alpha = .9$)

μ MODE:	0.15	0.20	0.25	0.30	0.35	0.40	0.45	0.50	0.55	0.60	0.65	0.70	0.75	0.80	0.85
(1)	4.567	4.410	4.220	4.025	3.845	3.688	3.557	3.448	3.361	3.293	3.241	3.202	3.176	3.158	3.148
(2)	8.632	8.261	8.077	8.036	8.060	8.069	7.986	7.789	7.517	7.229	6.962	6.735	6.556	6.425	6.342
(3)	12.82	12.75	12.78	12.60	12.09	11.57	11.25	11.18	11.21	11.15	10.89	10.53	10.16	9.852	9.620
(4)	17.47	17.45	16.84	16.09	15.89	15.92	15.64	15.03	14.51	14.32	14.36	14.23	13.84	13.39	12.98

TABLE 8.3 : β_1 ($\phi = .6, \alpha = 1.0$)

μ MODE:	0.15	0.20	0.25	0.30	0.35	0.40	0.45	0.50	0.55	0.60	0.65	0.70	0.75	0.80	0.85
(1)	5.057	4.857	4.610	4.356	4.122	3.917	3.744	3.599	3.480	3.384	3.308	3.249	3.206	3.175	3.156
(2)	9.420	8.902	8.636	8.548	8.541	8.512	8.379	8.124	7.800	7.466	7.159	6.892	6.673	6.504	6.386
(3)	13.90	13.78	13.74	13.41	12.74	12.13	11.78	11.68	11.67	11.52	11.18	10.76	10.35	9.994	9.714
(4)	19.01	18.85	17.94	17.10	16.92	16.84	16.32	15.57	15.02	14.83	14.80	14.56	14.09	13.58	13.12

TABLE 8.3 : β_1 ($\phi = .7, \alpha = .6$)

μ MODE:	0.15	0.20	0.25	0.30	0.35	0.40	0.45	0.50	0.55	0.60	0.65	0.70	0.75	0.80	0.85
(1)	2.665	2.638	2.605	2.574	2.550	2.536	2.535	2.549	2.580	2.629	2.695	2.778	2.873	2.971	3.057
(2)	5.264	5.232	5.263	5.363	5.519	5.716	5.924	6.088	6.139	6.065	5.930	5.801	5.728	5.748	5.886
(3)	7.933	8.057	8.289	8.567	8.792	8.806	8.610	8.437	8.479	8.762	9.129	9.282	9.157	8.962	8.864
(4)	10.75	11.06	11.40	11.52	11.26	11.12	11.38	11.81	11.97	11.73	11.56	11.84	12.35	12.36	12.11

TABLE 8.3 : β_1 ($\phi = .7, \alpha = .7$)

μ MODE:	0.15	0.20	0.25	0.30	0.35	0.40	0.45	0.50	0.55	0.60	0.65	0.70	0.75	0.80	0.85
(1)	3.107	3.071	3.026	2.981	2.940	2.908	2.888	2.881	2.888	2.908	2.940	2.981	3.026	3.071	3.107
(2)	6.116	6.047	6.042	6.104	6.216	6.352	6.468	6.516	6.468	6.352	6.216	6.104	6.042	6.047	6.116
(3)	9.178	9.251	9.425	9.603	9.651	9.500	9.286	9.192	9.286	9.500	9.651	9.603	9.425	9.251	9.178
(4)	12.39	12.63	12.80	12.65	12.37	12.38	12.64	12.80	12.64	12.38	12.37	12.65	12.80	12.63	12.39

TABLE 8.3 : β_1 ($\phi = .7, \alpha = .8$)

μ MODE:	0.15	0.20	0.25	0.30	0.35	0.40	0.45	0.50	0.55	0.60	0.65	0.70	0.75	0.80	0.85
(1)	3.546	3.498	3.436	3.368	3.302	3.242	3.192	3.154	3.128	3.112	3.106	3.108	3.115	3.124	3.133
(2)	6.946	6.815	6.745	6.744	6.793	6.858	6.899	6.880	6.791	6.657	6.514	6.390	6.301	6.256	6.250
(3)	10.36	10.34	10.42	10.49	10.42	10.20	9.978	9.881	9.924	10.02	10.03	9.907	9.711	9.528	9.415
(4)	13.93	14.05	14.05	13.76	13.50	13.51	13.62	13.56	13.29	13.06	13.05	13.17	13.13	12.91	12.67

TABLE 8.3 : β_1 ($\phi = .7, \alpha = .9$)

μ MODE:	0.15	0.20	0.25	0.30	0.35	0.40	0.45	0.50	0.55	0.60	0.65	0.70	0.75	0.80	0.85
(1)	3.983	3.918	3.830	3.729	3.627	3.529	3.442	3.367	3.304	3.254	3.215	3.186	3.166	3.154	3.146
(2)	7.742	7.517	7.359	7.285	7.276	7.293	7.293	7.239	7.122	6.960	6.786	6.624	6.490	6.390	6.327
(3)	11.45	11.32	11.33	11.32	11.17	10.87	10.60	10.44	10.42	10.44	10.38	10.21	9.980	9.751	9.574
(4)	15.36	15.38	15.25	14.84	14.51	14.46	14.47	14.26	13.90	13.63	13.56	13.58	13.46	13.18	12.89

TABLE 8.3 : β_1 ($\phi = .7, \alpha = 1.0$)

μ MODE:	0.15	0.20	0.25	0.30	0.35	0.40	0.45	0.50	0.55	0.60	0.65	0.70	0.75	0.80	0.85
(1)	4.414	4.325	4.201	4.057	3.908	3.766	3.637	3.525	3.429	3.350	3.285	3.235	3.197	3.171	3.154
(2)	8.492	8.143	7.886	7.751	7.708	7.704	7.684	7.600	7.441	7.232	7.009	6.800	6.619	6.475	6.373
(3)	12.44	12.22	12.19	12.14	11.88	11.47	11.10	10.90	10.85	10.84	10.73	10.50	10.20	9.914	9.676
(4)	16.70	16.67	16.41	15.80	15.40	15.33	15.26	14.89	14.41	14.08	13.99	13.96	13.76	13.41	13.04

TABLE 8.3 : β_1 ($\phi = .8, \alpha = .6$)

μ MODE:	0.15	0.20	0.25	0.30	0.35	0.40	0.45	0.50	0.55	0.60	0.65	0.70	0.75	0.80	0.85
(1)	2.351	2.347	2.344	2.345	2.352	2.368	2.394	2.432	2.485	2.553	2.638	2.738	2.848	2.959	3.052
(2)	4.702	4.725	4.787	4.892	5.037	5.216	5.409	5.585	5.692	5.705	5.652	5.590	5.573	5.644	5.832
(3)	7.106	7.239	7.447	7.697	7.934	8.063	8.026	7.938	7.964	8.185	8.536	8.809	8.838	8.745	8.725
(4)	9.592	9.865	10.18	10.40	10.38	10.28	10.44	10.82	11.16	11.18	11.07	11.23	11.74	11.99	11.89

TABLE 8.3 : β_1 ($\phi = .8, \alpha = .7$)

μ MODE:	0.15	0.20	0.25	0.30	0.35	0.40	0.45	0.50	0.55	0.60	0.65	0.70	0.75	0.80	0.85
(1)	2.741	2.734	2.726	2.720	2.718	2.723	2.737	2.760	2.793	2.837	2.889	2.947	3.007	3.061	3.103
(2)	5.469	5.474	5.513	5.591	5.700	5.825	5.943	6.020	6.037	6.001	5.944	5.901	5.902	5.963	6.078
(3)	8.238	8.337	8.497	8.669	8.778	8.767	8.683	8.646	8.731	8.923	9.115	9.180	9.122	9.047	9.061
(4)	11.08	11.30	11.49	11.52	11.42	11.42	11.63	11.87	11.92	11.82	11.81	12.04	12.29	12.29	12.19

TABLE 8.3 : β_1 ($\phi = .8, \alpha = .8$)

μ MODE:	0.15	0.20	0.25	0.30	0.35	0.40	0.45	0.50	0.55	0.60	0.65	0.70	0.75	0.80	0.85
(1)	3.129	3.116	3.099	3.080	3.062	3.048	3.039	3.036	3.039	3.048	3.062	3.080	3.099	3.116	3.129
(2)	6.221	6.190	6.183	6.206	6.254	6.312	6.360	6.379	6.360	6.312	6.254	6.206	6.183	6.190	6.221
(3)	9.325	9.349	9.425	9.501	9.517	9.455	9.368	9.329	9.368	9.455	9.517	9.501	9.425	9.349	9.325
(4)	12.49	12.59	12.66	12.60	12.49	12.49	12.60	12.66	12.60	12.49	12.49	12.60	12.66	12.59	12.49

TABLE 8.3 : β_1 ($\phi = .8, \alpha = .9$)

μ MODE:	0.15	0.20	0.25	0.30	0.35	0.40	0.45	0.50	0.55	0.60	0.65	0.70	0.75	0.80	0.85
(1)	3.515	3.492	3.460	3.421	3.378	3.334	3.293	3.255	3.223	3.197	3.177	3.162	3.152	3.146	3.143
(2)	6.951	6.859	6.779	6.731	6.715	6.721	6.731	6.727	6.693	6.631	6.550	6.466	6.393	6.337	6.304
(3)	10.35	10.26	10.25	10.27	10.24	10.14	10.01	9.901	9.857	9.865	9.872	9.826	9.725	9.604	9.503
(4)	13.80	13.79	13.79	13.67	13.48	13.39	13.40	13.39	13.26	13.09	13.00	13.01	13.00	12.90	12.74

TABLE 8.3 : β_1 ($\phi = .8, \alpha = 1.0$)

μ MODE:	0.15	0.20	0.25	0.30	0.35	0.40	0.45	0.50	0.55	0.60	0.65	0.70	0.75	0.80	0.85
(1)	3.897	3.860	3.805	3.736	3.657	3.576	3.496	3.422	3.355	3.298	3.251	3.213	3.185	3.164	3.152
(2)	7.651	7.469	7.297	7.177	7.115	7.097	7.096	7.081	7.027	6.930	6.803	6.666	6.538	6.431	6.353
(3)	11.29	11.09	11.03	11.02	10.96	10.79	10.55	10.35	10.26	10.24	10.23	10.14	9.984	9.793	9.618
(4)	15.01	14.95	14.91	14.67	14.34	14.18	14.16	14.09	13.84	13.57	13.41	13.38	13.34	13.17	12.93

TABLE 8.3 : β_1 ($\phi = .9, \alpha = .6$)

μ MODE:	0.15	0.20	0.25	0.30	0.35	0.40	0.45	0.50	0.55	0.60	0.65	0.70	0.75	0.80	0.85
(1)	2.099	2.106	2.117	2.135	2.161	2.195	2.241	2.299	2.371	2.459	2.563	2.683	2.813	2.940	3.044
(2)	4.228	4.283	4.370	4.490	4.642	4.820	5.011	5.192	5.324	5.379	5.373	5.357	5.385	5.507	5.754
(3)	6.413	6.568	6.776	7.019	7.260	7.431	7.476	7.452	7.500	7.711	8.063	8.397	8.522	8.503	8.547
(4)	8.655	8.924	9.230	9.487	9.571	9.547	9.675	10.02	10.43	10.62	10.59	10.73	11.23	11.63	11.65

TABLE 8.3 : β_1 (ϕ = .9, α = .7)

μ MODE:	0.15	0.20	0.25	0.30	0.35	0.40	0.45	0.50	0.55	0.60	0.65	0.70	0.75	0.80	0.85
(1)	2.447	2.453	2.463	2.478	2.501	2.531	2.570	2.619	2.677	2.745	2.821	2.901	2.979	3.047	3.098
(2)	4.921	4.971	5.048	5.152	5.278	5.414	5.539	5.631	5.672	5.672	5.659	5.666	5.724	5.847	6.021
(3)	7.446	7.585	7.762	7.944	8.077	8.120	8.104	8.121	8.241	8.457	8.685	8.806	8.811	8.803	8.903
(4)	10.02	10.25	10.47	10.56	10.55	10.60	10.81	11.09	11.25	11.25	11.29	11.54	11.86	11.96	11.94

TABLE 8.3 : β_1 (ϕ = .9, α = .8)

μ MODE:	0.15	0.20	0.25	0.30	0.35	0.40	0.45	0.50	0.55	0.60	0.65	0.70	0.75	0.80	0.85
(1)	2.794	2.797	2.802	2.011	2.824	2.842	2.865	2.894	2.927	2.964	3.003	3.041	3.076	3.104	3.125
(2)	5.604	5.634	5.683	5.748	5.823	5.894	5.950	5.979	5.983	5.974	5.969	5.983	6.026	6.097	6.179
(3)	8.447	8.537	8.644	8.734	8.775	8.770	8.762	8.801	8.897	9.017	9.104	9.126	9.112	9.122	9.197
(4)	11.33	11.47	11.56	11.56	11.56	11.64	11.79	11.90	11.91	11.91	11.99	12.15	12.26	12.26	12.26

TABLE 8.3 : β_1 (ϕ = .9, α = .9)

μ MODE:	0.15	0.20	0.25	0.30	0.35	0.40	0.45	0.50	0.55	0.60	0.65	0.70	0.75	0.80	0.85
(1)	3.139	3.136	3.132	3.128	3.124	3.121	3.118	3.118	3.118	3.121	3.124	3.128	3.132	3.136	3.139
(2)	6.270	6.263	6.261	6.266	6.277	6.290	6.301	6.305	6.301	6.290	6.277	6.266	6.261	6.263	6.270
(3)	9.403	9.407	9.425	9.443	9.446	9.432	9.412	9.403	9.412	9.432	9.446	9.443	9.425	9.407	9.403
(4)	12.55	12.57	12.59	12.57	12.55	12.55	12.57	12.59	12.57	12.55	12.55	12.57	12.59	12.57	12.55

TABLE 8.3 : β_1 (ϕ = .9, α = 1.0)

μ MODE:	0.15	0.20	0.25	0.30	0.35	0.40	0.45	0.50	0.55	0.60	0.65	0.70	0.75	0.80	0.85
(1)	3.481	3.469	3.450	3.425	3.395	3.362	3.327	3.292	3.259	3.229	3.204	3.183	3.167	3.155	3.147
(2)	6.914	6.847	6.772	6.708	6.664	6.643	6.638	6.637	6.628	6.600	6.552	6.491	6.426	6.368	6.324
(3)	10.30	10.19	10.14	10.13	10.12	10.08	9.992	9.890	9.814	9.783	9.780	9.766	9.715	9.628	9.536
(4)	13.68	13.62	13.62	13.57	13.44	13.31	13.27	13.27	13.21	13.08	12.97	12.92	12.92	12.88	12.77

TABLE 8.3 : β_1 (ϕ = 1.0, α = .6)

μ MODE:	0.15	0.20	0.25	0.30	0.35	0.40	0.45	0.50	0.55	0.60	0.65	0.70	0.75	0.80	0.85
(1)	1.894	1.905	1.923	1.950	1.985	2.031	2.088	2.160	2.246	2.350	2.473	2.614	2.766	2.914	3.034
(2)	3.832	3.903	4.004	4.135	4.295	4.480	4.680	4.874	5.027	5.107	5.124	5.129	5.181	5.341	5.652
(3)	5.828	5.996	6.211	6.459	6.710	6.912	7.002	7.010	7.066	7.276	7.646	8.043	8.245	8.266	8.343
(4)	7.871	8.145	8.455	8.737	8.883	8.897	9.010	9.342	9.791	10.10	10.15	10.26	10.76	11.31	11.41

TABLE 8.3 : β_1 (ϕ = 1.0, α = .7)

μ MODE:	0.15	0.20	0.25	0.30	0.35	0.40	0.45	0.50	0.55	0.60	0.65	0.70	0.75	0.80	0.85
(1)	2.208	2.220	2.238	2.264	2.300	2.345	2.400	2.467	2.546	2.636	2.736	2.840	2.941	3.027	3.090
(2)	4.461	4.533	4.633	4.760	4.907	5.062	5.209	5.320	5.379	5.393	5.396	5.426	5.520	5.699	5.945
(3)	6.773	6.939	7.141	7.347	7.510	7.585	7.593	7.628	7.771	8.028	8.317	8.497	8.535	8.553	8.707
(4)	9.130	9.387	9.633	9.774	9.792	9.857	10.09	10.43	10.68	10.73	10.78	11.06	11.49	11.67	11.69

TABLE 8.3 : β_1 (ϕ = 1.0, α = .8)

μ MODE:	0.15	0.20	0.25	0.30	0.35	0.40	0.45	0.50	0.55	0.60	0.65	0.70	0.75	0.80	0.85
(1)	2.521	2.532	2.548	2.571	2.601	2.639	2.684	2.737	2.797	2.861	2.926	2.989	3.044	3.088	3.118
(2)	5.082	5.145	5.230	5.333	5.442	5.544	5.622	5.664	5.677	5.678	5.692	5.742	5.838	5.975	6.121
(3)	7.695	7.835	7.987	8.115	8.181	8.190	8.205	8.284	8.440	8.629	8.769	8.817	8.821	8.870	9.031
(4)	10.34	10.54	10.67	10.70	10.72	10.85	11.08	11.27	11.33	11.34	11.47	11.73	11.93	11.96	12.01

TABLE 8.3 : β_1 (ϕ = 1.0, α = .9)

μ MODE:	0.15	0.20	0.25	0.30	0.35	0.40	0.45	0.50	0.55	0.60	0.65	0.70	0.75	0.80	0.85
(1)	2.833	2.839	2.850	2.864	2.883	2.906	2.933	2.962	2.994	3.025	3.056	3.083	3.105	3.122	3.133
(2)	5.691	5.731	5.784	5.842	5.897	5.940	5.965	5.974	5.975	5.979	5.998	6.037	6.095	6.162	6.223
(3)	8.583	8.666	8.743	8.790	8.802	8.804	8.833	8.901	8.993	9.073	9.111	9.117	9.125	9.175	9.267
(4)	11.49	11.59	11.63	11.63	11.67	11.78	11.89	11.94	11.94	11.98	12.09	12.21	12.26	12.26	12.31

TABLE 8.4 : β_1 (ϕ = .6, α = .6)

μ MODE:	0.15	0.20	0.25	0.30	0.35	0.40	0.45	0.50	0.55	0.60	0.65	0.70	0.75	0.80	0.85
(1)	5.337	5.484	5.519	5.364	5.073	4.771	4.520	4.334	4.213	4.151	4.138	4.163	4.205	4.243	4.257
(2)	0.643	8.280	7.831	7.556	7.622	7.872	8.156	8.289	8.139	7.835	7.534	7.306	7.189	7.200	7.302
(3)	11.45	10.75	10.76	11.10	11.39	11.24	10.78	10.56	10.81	11.29	11.43	11.11	10.71	10.41	10.33
(4)	13.94	13.95	14.39	14.50	13.90	13.73	14.20	14.57	14.20	13.74	13.94	14.55	14.41	13.92	13.54

TABLE 8.4 : β_1 (ϕ = .6, α = .7)

μ MODE:	0.15	0.20	0.25	0.30	0.35	0.40	0.45	0.50	0.55	0.60	0.65	0.70	0.75	0.80	0.85
(1)	6.052	6.107	6.001	5.728	5.394	5.088	4.843	4.661	4.536	4.458	4.416	4.397	4.388	4.379	4.365
(2)	9.672	9.164	8.637	8.461	8.552	8.741	8.864	8.771	8.493	8.166	7.879	7.668	7.548	7.508	7.509
(3)	12.69	12.13	12.22	12.47	12.44	12.00	11.60	11.57	11.85	12.03	11.82	11.41	11.03	10.76	10.66
(4)	15.82	15.94	16.18	15.75	15.22	15.34	15.66	15.43	14.88	14.69	15.03	15.12	14.70	14.22	13.88

TABLE 8.4 : β_1 (ϕ = .6, α = .8)

μ MODE:	0.15	0.20	0.25	0.30	0.35	0.40	0.45	0.50	0.55	0.60	0.65	0.70	0.75	0.80	0.85
(1)	6.687	6.644	6.434	6.095	5.732	5.412	5.153	4.952	4.803	4.695	4.618	4.564	4.525	4.497	4.480
(2)	10.61	9.970	9.456	9.291	9.341	9.435	9.411	9.191	8.855	8.509	8.209	7.976	7.814	7.713	7.652
(3)	13.90	13.48	13.54	13.61	13.31	12.78	12.45	12.46	12.60	12.52	12.17	11.73	11.33	11.05	10.87
(4)	17.68	17.76	17.69	16.98	16.59	16.73	16.69	16.15	15.64	15.59	15.75	15.52	15.01	14.51	14.16

TABLE 8.4 : β_1 (ϕ = .6, α = .9)

μ MODE:	0.15	0.20	0.25	0.30	0.35	0.40	0.45	0.50	0.55	0.60	0.65	0.70	0.75	0.80	0.85
(1)	7.265	7.143	6.859	6.466	6.064	5.712	5.421	5.190	5.011	4.874	4.773	4.700	4.650	4.619	4.604
(2)	11.53	10.77	10.22	10.02	10.02	10.03	9.905	9.600	9.208	8.821	8.482	8.209	8.004	7.862	7.775
(3)	15.25	14.73	14.72	14.64	14.14	13.53	13.20	13.17	13.18	12.95	12.50	12.02	11.59	11.25	11.03
(4)	19.44	19.44	19.07	18.19	17.85	17.89	17.54	16.84	16.36	16.32	16.28	15.87	15.30	14.76	14.35

TABLE 8.4 : β_1 (ϕ = .6, α = 1.0)

μ MODE:	0.15	0.20	0.25	0.30	0.35	0.40	0.45	0.50	0.55	0.60	0.65	0.70	0.75	0.80	0.85
(1)	7.816	7.635	7.283	6.824	6.367	5.969	5.641	5.379	5.175	5.020	4.906	4.827	4.777	4.748	4.735
(2)	12.43	11.52	10.89	10.66	10.64	10.59	10.38	9.981	9.519	9.077	8.694	8.383	8.148	7.987	7.898
(3)	16.44	15.89	15.85	15.60	14.89	14.18	13.84	13.80	13.71	13.33	12.80	12.25	11.78	11.40	11.15
(4)	21.12	21.06	20.34	19.29	19.01	18.92	18.27	17.44	16.99	16.93	16.74	16.19	15.53	14.95	14.49

TABLE 8.4 : β_1 (ϕ = .7, α = .6)

μ MODE:	0.15	0.20	0.25	0.30	0.35	0.40	0.45	0.50	0.55	0.60	0.65	0.70	0.75	0.80	0.85
(1)	4.577	4.714	4.790	4.762	4.633	4.459	4.297	4.173	4.098	4.070	4.085	4.131	4.189	4.236	4.255
(2)	7.447	7.405	7.136	6.915	6.900	7.059	7.305	7.528	7.595	7.479	7.295	7.144	7.084	7.143	7.281
(3)	10.08	9.687	9.585	9.803	10.11	10.27	10.12	9.900	9.948	10.31	10.70	10.71	10.48	10.27	10.26
(4)	12.44	12.29	12.61	12.94	12.85	12.57	12.75	13.23	13.42	13.15	13.05	13.54	13.90	13.67	13.40

TABLE 8.4 : β_1 ($\phi = .7, \alpha = .7$)

μ MODE:	0.15	0.20	0.25	0.30	0.35	0.40	0.45	0.50	0.55	0.60	0.65	0.70	0.75	0.80	0.85
(1)	5.193	5.267	5.255	5.145	4.972	4.789	4.632	4.513	4.434	4.391	4.374	4.373	4.377	4.374	4.364
(2)	8.383	8.188	7.887	7.719	7.734	7.867	8.019	8.087	8.011	7.841	7.661	7.527	7.466	7.469	7.496
(3)	11.28	10.90	10.87	11.06	11.21	11.12	10.88	10.76	10.90	11.15	11.24	11.07	10.82	10.65	10.61
(4)	14.07	14.03	14.28	14.32	14.01	13.92	14.17	14.37	14.17	13.92	14.04	14.35	14.30	14.00	13.78

TABLE 8.4 : β_1 ($\phi = .7, \alpha = .8$)

μ MODE:	0.15	0.20	0.25	0.30	0.35	0.40	0.45	0.50	0.55	0.60	0.65	0.70	0.75	0.80	0.85
(1)	5.744	5.752	5.676	5.521	5.325	5.130	4.959	4.821	4.716	4.639	4.585	4.546	4.517	4.494	4.479
(2)	9.254	8.972	8.654	8.478	8.459	8.524	8.582	8.551	8.413	8.217	8.022	7.863	7.753	7.686	7.643
(3)	12.49	12.10	12.04	12.14	12.13	11.92	11.67	11.57	11.65	11.74	11.66	11.43	11.16	10.96	10.84
(4)	15.72	15.65	15.75	15.57	15.23	15.17	15.30	15.23	14.91	14.71	14.80	14.88	14.66	14.33	14.08

TABLE 8.4 : β_1 ($\phi = .7, \alpha = .9$)

μ MODE:	0.15	0.20	0.25	0.30	0.35	0.40	0.45	0.50	0.55	0.60	0.65	0.70	0.75	0.80	0.85
(1)	6.248	6.208	6.092	5.904	5.679	5.452	5.249	5.077	4.938	4.829	4.746	4.686	4.643	4.616	4.604
(2)	10.11	9.757	9.379	9.147	9.074	9.084	9.086	9.000	8.812	8.572	8.331	8.121	7.958	7.842	7.768
(3)	13.69	13.22	13.10	13.12	13.01	12.70	12.38	12.23	12.24	12.23	12.06	11.77	11.45	11.19	11.00
(4)	17.29	17.14	17.13	16.77	16.35	16.25	16.26	16.01	15.60	15.38	15.39	15.32	15.01	14.62	14.29

TABLE 8.4 : β_1 ($\phi = .7, \alpha = 1.0$)

μ MODE:	0.15	0.20	0.25	0.30	0.35	0.40	0.45	0.50	0.55	0.60	0.65	0.70	0.75	0.80	0.85
(1)	6.730	6.660	6.511	6.282	6.009	5.736	5.489	5.281	5.112	4.981	4.883	4.814	4.771	4.746	4.735
(2)	10.97	10.51	10.02	9.727	9.622	9.611	9.579	9.436	9.179	8.872	8.572	8.313	8.111	7.971	7.893
(3)	14.84	14.24	14.09	14.07	13.85	13.39	12.98	12.79	12.77	12.69	12.43	12.05	11.67	11.35	11.13
(4)	18.75	18.58	18.46	17.87	17.34	17.24	17.14	16.69	16.18	15.93	15.90	15.73	15.31	14.84	14.45

TABLE 8.4 : β_1 ($\phi = .8, \alpha = .6$)

μ MODE:	0.15	0.20	0.25	0.30	0.35	0.40	0.45	0.50	0.55	0.60	0.65	0.70	0.75	0.80	0.85
(1)	4.005	4.131	4.214	4.236	4.191	4.110	4.026	3.966	3.941	3.955	4.006	4.082	4.164	4.227	4.252
(2)	6.531	6.565	6.459	6.345	6.344	6.470	6.680	6.909	7.061	7.073	6.995	6.923	6.932	7.056	7.249
(3)	8.918	8.753	8.690	8.853	9.129	9.370	9.415	9.315	9.326	9.595	10.01	10.24	10.17	10.07	10.15
(4)	11.14	11.05	11.29	11.63	11.78	11.67	11.72	12.11	12.52	12.54	12.45	12.75	13.30	13.34	13.21

TABLE 8.4 : β_1 ($\phi = .8, \alpha = .7$)

μ MODE:	0.15	0.20	0.25	0.30	0.35	0.40	0.45	0.50	0.55	0.60	0.65	0.70	0.75	0.80	0.85
(1)	4.546	4.621	4.643	4.607	4.530	4.443	4.367	4.315	4.291	4.291	4.311	4.337	4.359	4.368	4.362
(2)	7.370	7.306	7.173	7.094	7.120	7.232	7.376	7.485	7.507	7.451	7.373	7.327	7.342	7.409	7.477
(3)	10.04	9.866	9.861	10.01	10.19	10.24	10.17	10.11	10.21	10.45	10.64	10.64	10.53	10.47	10.53
(4)	12.64	12.62	12.83	12.99	12.93	12.86	13.04	13.32	13.39	13.28	13.33	13.64	13.81	13.70	13.61

TABLE 8.4 : β_1 ($\phi = .8, \alpha = .8$)

μ MODE:	0.15	0.20	0.25	0.30	0.35	0.40	0.45	0.50	0.55	0.60	0.65	0.70	0.75	0.80	0.85
(1)	5.031	5.056	5.034	4.971	4.883	4.790	4.707	4.639	4.591	4.557	4.535	4.519	4.504	4.489	4.477
(2)	8.158	8.047	7.906	7.820	7.812	7.861	7.923	7.950	7.920	7.846	7.761	7.696	7.659	7.643	7.630
(3)	11.16	10.99	10.95	11.02	11.08	11.04	10.95	10.90	10.96	11.06	11.10	11.03	10.91	10.82	10.79
(4)	14.14	14.10	14.19	14.21	14.09	14.04	14.15	14.23	14.15	14.05	14.10	14.23	14.21	14.07	13.96

TABLE 8.4 : β_1 ($\phi = .8, \alpha = .9$)

μ MODE:	0.15	0.20	0.25	0.30	0.35	0.40	0.45	0.50	0.55	0.60	0.65	0.70	0.75	0.80	0.85
(1)	5.476	5.467	5.421	5.343	5.240	5.127	5.016	4.916	4.830	4.761	4.705	4.664	4.632	4.612	4.602
(2)	8.937	8.795	8.615	8.472	8.400	8.387	8.399	8.391	8.336	8.234	8.109	7.988	7.886	7.809	7.757
(3)	12.29	12.05	11.93	11.92	11.92	11.84	11.69	11.57	11.53	11.55	11.53	11.41	11.25	11.09	10.97
(4)	15.61	15.45	15.46	15.40	15.20	15.07	15.07	15.05	14.89	14.72	14.67	14.69	14.61	14.41	14.21

TABLE 8.4 : β_1 ($\phi = .8, \alpha = 1.0$)

μ MODE:	0.15	0.20	0.25	0.30	0.35	0.40	0.45	0.50	0.55	0.60	0.65	0.70	0.75	0.80	0.85
(1)	5.902	5.874	5.815	5.716	5.583	5.431	5.279	5.139	5.020	4.923	4.848	4.796	4.762	4.743	4.734
(2)	9.720	9.528	9.265	9.037	8.907	8.866	8.862	8.836	8.741	8.580	8.389	8.206	8.053	7.946	7.884
(3)	13.39	13.02	12.81	12.78	12.75	12.58	12.31	12.10	12.02	12.01	11.94	11.76	11.51	11.27	11.10
(4)	16.97	16.73	16.71	16.54	16.18	15.96	15.93	15.83	15.53	15.25	15.16	15.14	14.97	14.67	14.38

TABLE 8.4 : β_1 ($\phi = .9, \alpha = .6$)

μ MODE:	0.15	0.20	0.25	0.30	0.35	0.40	0.45	0.50	0.55	0.60	0.65	0.70	0.75	0.80	0.85
(1)	3.561	3.674	3.757	3.796	3.793	3.765	3.737	3.728	3.750	3.808	3.899	4.013	4.127	4.212	4.248
(2)	5.812	5.873	5.846	5.812	5.852	5.986	6.191	6.423	6.611	6.688	6.676	6.663	6.733	6.931	7.201
(3)	7.969	7.922	7.930	8.095	8.359	8.625	8.765	8.756	8.783	9.020	9.441	9.782	9.841	9.818	9.989
(4)	10.03	10.03	10.26	10.59	10.84	10.85	10.90	11.22	11.69	11.92	11.90	12.13	12.73	12.99	12.96

TABLE 8.4 : β_1 ($\phi = .9, \alpha = .7$)

μ MODE:	0.15	0.20	0.25	0.30	0.35	0.40	0.45	0.50	0.55	0.60	0.65	0.70	0.75	0.80	0.85
(1)	4.042	4.114	4.147	4.145	4.120	4.091	4.075	4.080	4.109	4.159	4.222	4.286	4.334	4.357	4.359
(2)	6.566	6.559	6.518	6.518	6.589	6.721	6.877	7.006	7.067	7.065	7.049	7.075	7.171	7.320	7.448
(3)	8.999	8.954	9.017	9.187	9.380	9.494	9.497	9.499	9.628	9.887	10.14	10.23	10.20	10.24	10.42
(4)	11.40	11.47	11.70	11.90	11.94	11.95	12.13	12.45	12.65	12.64	12.72	13.06	13.34	13.35	13.39

TABLE 8.4 : β_1 ($\phi = .9, \alpha = .8$)

μ MODE:	0.15	0.20	0.25	0.30	0.35	0.40	0.45	0.50	0.55	0.60	0.65	0.70	0.75	0.80	0.85
(1)	4.474	4.505	4.506	4.486	4.458	4.432	4.418	4.416	4.426	4.444	4.464	4.479	4.485	4.481	4.475
(2)	7.279	7.245	7.211	7.215	7.266	7.343	7.417	7.460	7.465	7.452	7.448	7.472	7.525	7.581	7.609
(3)	10.03	10.00	10.05	10.15	10.23	10.25	10.24	10.28	10.39	10.53	10.61	10.61	10.59	10.63	10.71
(4)	12.79	12.85	12.98	13.04	13.03	13.08	13.24	13.38	13.40	13.40	13.52	13.70	13.76	13.74	13.78

TABLE 8.4 : β_1 ($\phi = .9, \alpha = .9$)

μ MODE:	0.15	0.20	0.25	0.30	0.35	0.40	0.45	0.50	0.55	0.60	0.65	0.70	0.75	0.80	0.85
(1)	4.872	4.875	4.862	4.836	4.804	4.770	4.738	4.709	4.685	4.665	4.647	4.631	4.617	4.605	4.601
(2)	7.984	7.941	7.893	7.861	7.850	7.857	7.870	7.876	7.869	7.850	7.825	7.802	7.782	7.762	7.742
(3)	11.07	11.01	10.99	11.00	11.01	11.00	10.98	10.97	10.99	11.01	11.02	11.01	10.97	10.94	10.91
(4)	14.16	14.13	14.15	14.15	14.12	14.12	14.14	14.16	14.14	14.12	14.13	14.16	14.16	14.12	14.07

TABLE 8.4 : β_1 ($\phi = .9, \alpha = 1.0$)

μ MODE:	0.15	0.20	0.25	0.30	0.35	0.40	0.45	0.50	0.55	0.60	0.65	0.70	0.75	0.80	0.85
(1)	5.252	5.243	5.224	5.191	5.144	5.085	5.019	4.953	4.892	4.840	4.798	4.768	4.749	4.737	4.732
(2)	8.695	8.633	8.536	8.430	8.350	8.307	8.296	8.295	8.276	8.224	8.145	8.053	7.970	7.908	7.871
(3)	12.11	11.96	11.83	11.79	11.78	11.75	11.66	11.54	11.46	11.44	11.44	11.39	11.28	11.16	11.06
(4)	15.47	15.31	15.27	15.25	15.14	14.99	14.93	14.92	14.85	14.70	14.60	14.58	14.55	14.44	14.28

TABLE 8.4 : β_1 (ϕ = 1.0, α = .6)

μ MODE:	0.15	0.20	0.25	0.30	0.35	0.40	0.45	0.50	0.55	0.60	0.65	0.70	0.75	0.80	0.85
(1)	3.205	3.308	3.386	3.432	3.448	3.450	3.456	3.482	3.540	3.634	3.765	3.921	4.076	4.191	4.243
(2)	5.234	5.304	5.315	5.329	5.403	5.554	5.771	6.019	6.239	6.358	6.380	6.395	6.501	6.767	7.132
(3)	7.192	7.202	7.262	7.445	7.717	8.009	8.210	8.256	8.288	8.505	8.940	9.378	9.530	9.548	9.782
(4)	9.086	9.153	9.393	9.734	10.04	10.14	10.18	10.46	10.96	11.35	11.41	11.58	12.21	12.65	12.69

TABLE 8.4 : β_1 (ϕ = 1.0, α = .7)

μ MODE:	0.15	0.20	0.25	0.30	0.35	0.40	0.45	0.50	0.55	0.60	0.65	0.70	0.75	0.80	0.85
(1)	3.639	3.705	3.743	3.755	3.756	3.761	3.783	3.829	3.901	3.996	4.107	4.215	4.298	4.343	4.355
(2)	5.918	5.935	5.941	5.990	6.102	6.266	6.453	6.616	6.707	6.728	6.734	6.795	6.955	7.196	7.405
(3)	8.136	8.157	8.275	8.482	8.714	8.877	8.919	8.936	9.077	9.375	9.704	9.865	9.880	9.961	10.26
(4)	10.35	10.48	10.73	11.01	11.12	11.13	11.32	11.68	12.00	12.07	12.14	12.51	12.94	13.02	13.11

TABLE 8.4 : β_1 (ϕ = 1.0, α = .8)

μ MODE:	0.15	0.20	0.25	0.30	0.35	0.40	0.45	0.50	0.55	0.60	0.65	0.70	0.75	0.80	0.85
(1)	4.028	4.060	4.071	4.073	4.075	4.089	4.119	4.166	4.229	4.300	4.368	4.424	4.458	4.470	4.471
(2)	6.564	6.567	6.588	6.654	6.761	6.889	7.002	7.070	7.090	7.092	7.121	7.206	7.345	7.492	7.580
(3)	9.082	9.130	9.258	9.423	9.553	9.597	9.604	9.673	9.850	10.07	10.21	10.24	10.26	10.38	10.60
(4)	11.63	11.78	11.99	12.10	12.11	12.20	12.42	12.66	12.75	12.76	12.94	13.24	13.37	13.39	13.54

TABLE 8.4 : β_1 (ϕ = 1.0, α = .9)

μ MODE:	0.15	0.20	0.25	0.30	0.35	0.40	0.45	0.50	0.55	0.60	0.65	0.70	0.75	0.80	0.85
(1)	4.387	4.396	4.398	4.398	4.404	4.418	4.440	4.471	4.505	4.539	4.568	4.586	4.595	4.597	4.598
(2)	7.206	7.210	7.234	7.283	7.348	7.409	7.450	7.465	7.466	7.475	7.507	7.567	7.639	7.695	7.720
(3)	10.04	10.09	10.17	10.25	10.29	10.29	10.31	10.39	10.49	10.58	10.61	10.61	10.65	10.74	10.83
(4)	12.90	13.01	13.10	13.12	13.14	13.23	13.36	13.43	13.44	13.49	13.62	13.73	13.75	13.78	13.89

TABLE 8.5 : β_1 (ϕ = .6, α = .6)

μ MODE:	0.15	0.20	0.25	0.30	0.35	0.40	0.45	0.50	0.55	0.60	0.65	0.70	0.75	0.80	0.85
(1)	2.194	2.325	2.474	2.643	2.836	3.061	3.323	3.634	4.007	4.455	4.964	5.290	5.147	4.879	4.610
(2)	5.492	5.822	6.194	6.617	7.096	7.632	8.113	7.853	7.250	6.721	6.340	6.401	7.267	8.381	8.254
(3)	9.190	9.742	10.36	11.02	11.14	10.02	9.174	9.397	10.28	11.24	10.90	10.20	9.609	9.555	11.43
(4)	12.86	13.63	14.38	13.32	12.27	13.02	14.07	14.14	13.01	12.23	13.24	14.42	13.68	12.90	12.71

TABLE 8.5 : β_1 (ϕ = .6, α = .7)

μ MODE:	0.15	0.20	0.25	0.30	0.35	0.40	0.45	0.50	0.55	0.60	0.65	0.70	0.75	0.80	0.85
(1)	2.547	2.695	2.861	3.049	3.264	3.512	3.799	4.135	4.529	4.971	5.354	5.386	5.157	4.880	4.610
(2)	6.378	6.750	7.168	7.637	8.156	8.657	8.635	8.004	7.389	6.911	6.713	7.129	8.065	8.578	8.260
(3)	10.68	11.29	11.96	12.45	11.47	10.33	10.07	10.76	11.68	11.74	11.00	10.32	9.856	10.51	11.72
(4)	14.94	15.77	15.77	13.88	14.04	15.01	15.54	14.34	13.29	13.56	14.92	14.63	13.77	13.08	13.85

TABLE 8.5 : β_1 (ϕ = .6, α = .8)

μ MODE:	0.15	0.20	0.25	0.30	0.35	0.40	0.45	0.50	0.55	0.60	0.65	0.70	0.75	0.80	0.85
(1)	2.890	3.050	3.228	3.429	3.657	3.918	4.218	4.562	4.946	5.327	5.529	5.418	5.161	4.880	4.611
(2)	7.243	7.647	8.099	8.598	9.111	9.379	8.878	8.173	7.581	7.192	7.203	7.764	8.538	8.630	8.262
(3)	12.13	12.79	13.45	13.21	11.80	10.92	11.21	12.01	12.54	11.91	11.14	10.50	10.28	11.27	11.79
(4)	16.97	17.74	16.36	15.12	15.83	16.71	15.92	14.60	14.05	15.06	15.60	14.76	13.91	13.40	14.65

TABLE 8.5 : β_1 (ϕ = .6, α = .9)

μ MODE:	0.15	0.20	0.25	0.30	0.35	0.40	0.45	0.50	0.55	0.60	0.65	0.70	0.75	0.80	0.85
(1)	3.219	3.384	3.568	3.775	4.007	4.271	4.570	4.904	5.253	5.541	5.605	5.432	5.163	4.881	4.611
(2)	8.078	8.506	8.980	9.488	9.929	9.817	9.093	8.373	7.818	7.527	7.671	8.235	8.758	8.651	8.264
(3)	13.53	14.23	14.72	13.66	12.22	11.76	12.31	13.04	12.90	12.08	11.30	10.74	10.78	11.71	11.82
(4)	18.93	19.34	17.04	16.65	17.51	17.67	16.20	15.02	15.20	16.22	15.83	14.90	14.09	13.84	14.99

TABLE 8.5 : β_1 (ϕ = .6, α = 1.0)

μ MODE:	0.15	0.20	0.25	0.30	0.35	0.40	0.45	0.50	0.55	0.60	0.65	0.70	0.75	0.80	0.85
(1)	3.530	3.693	3.875	4.079	4.308	4.566	4.853	5.163	5.464	5.663	5.643	5.439	5.164	4.881	4.611
(2)	8.880	9.323	9.811	10.31	10.60	10.14	9.316	8.593	8.073	7.856	8.052	8.548	8.860	8.661	8.264
(3)	14.90	15.59	15.71	14.08	12.75	12.67	13.31	13.76	13.13	12.25	11.49	11.02	11.21	11.93	11.83
(4)	20.83	20.43	17.97	18.18	18.97	18.13	16.52	15.68	16.34	16.86	16.01	15.05	14.30	14.27	15.13

TABLE 8.5 : β_1 (ϕ = .7, α = .6)

μ MODE:	0.15	0.20	0.25	0.30	0.35	0.40	0.45	0.50	0.55	0.60	0.65	0.70	0.75	0.80	0.85
(1)	1.880	1.993	2.120	2.265	2.431	2.624	2.849	3.116	3.438	3.831	4.309	4.830	5.032	4.850	4.602
(2)	4.707	4.990	5.310	5.673	6.087	6.561	7.088	7.489	7.180	6.681	6.255	6.011	6.396	7.560	8.171
(3)	7.877	8.351	8.883	9.477	10.08	9.870	8.973	8.437	8.901	9.851	10.64	10.16	9.565	9.148	10.27
(4)	11.03	11.69	12.41	12.85	11.57	11.28	12.14	13.16	12.91	11.98	11.65	13.11	13.60	12.87	12.29

TABLE 8.5 : β_1 (ϕ = .7, α = .7)

μ MODE:	0.15	0.20	0.25	0.30	0.35	0.40	0.45	0.50	0.55	0.60	0.65	0.70	0.75	0.80	0.85
(1)	2.183	2.310	2.452	2.614	2.798	3.010	3.258	3.549	3.894	4.302	4.757	5.101	5.074	4.854	4.603
(2)	5.467	5.786	6.145	6.551	7.009	7.513	7.964	7.853	7.327	6.832	6.475	6.467	7.128	8.129	8.199
(3)	9.151	9.685	10.28	10.90	11.09	10.14	9.342	9.403	10.18	11.07	10.91	10.27	9.718	9.616	11.12
(4)	12.81	13.55	14.25	13.46	12.39	12.97	13.94	14.14	13.11	12.37	13.13	14.31	13.72	12.99	12.77

TABLE 8.5 : β_1 (ϕ = .7, α = .8)

μ MODE:	0.15	0.20	0.25	0.30	0.35	0.40	0.45	0.50	0.55	0.60	0.65	0.70	0.75	0.80	0.85
(1)	2.477	2.614	2.767	2.939	3.135	3.360	3.619	3.920	4.269	4.662	5.040	5.208	5.089	4.856	4.603
(2)	6.208	6.556	6.946	7.383	7.866	8.347	8.519	8.066	7.500	7.036	6.782	6.970	7.688	8.349	8.209
(3)	10.39	10.98	11.61	12.13	11.55	10.48	10.00	10.47	11.30	11.66	11.07	10.47	9.950	10.23	11.45
(4)	14.56	15.33	15.63	14.02	13.76	14.60	15.30	14.43	13.40	13.30	14.49	14.63	13.85	13.18	13.47

TABLE 8.5 : β_1 (ϕ = .7, α = .9)

μ MODE:	0.15	0.20	0.25	0.30	0.35	0.40	0.45	0.50	0.55	0.60	0.65	0.70	0.75	0.80	0.85
(1)	2.759	2.901	3.059	3.236	3.436	3.664	3.925	4.224	4.560	4.914	5.198	5.254	5.096	4.856	4.603
(2)	6.924	7.293	7.706	8.164	8.650	9.034	8.867	8.273	7.701	7.276	7.117	7.401	8.054	8.433	8.214
(3)	11.60	12.22	12.85	13.04	11.93	10.93	10.83	11.47	12.16	11.93	11.23	10.60	10.24	10.74	11.57
(4)	16.25	17.02	16.44	14.86	15.20	16.07	15.92	14.71	13.89	14.44	15.35	14.01	14.00	13.42	14.03

TABLE 8.5 : β_1 (ϕ = .7, α = 1.0)

μ MODE:	0.15	0.20	0.25	0.30	0.35	0.40	0.45	0.50	0.55	0.60	0.65	0.70	0.75	0.80	0.85
(1)	3.025	3.166	3.322	3.497	3.695	3.919	4.173	4.460	4.773	5.080	5.285	5.278	5.100	4.857	4.603
(2)	7.612	7.995	8.426	8.898	9.367	9.583	9.147	8.485	7.912	7.519	7.420	7.725	8.270	8.473	8.217
(3)	12.78	13.42	14.01	13.66	12.31	11.48	11.67	12.36	12.72	12.14	11.40	10.79	10.54	11.09	11.62
(4)	17.91	18.57	17.08	15.96	16.59	17.24	16.28	15.02	14.60	15.48	15.77	14.97	14.17	13.68	14.37

TABLE 8.5 : β_1 ($\phi = .8, \alpha = .6$)

μ MODE:	0.15	0.20	0.25	0.30	0.35	0.40	0.45	0.50	0.55	0.60	0.65	0.70	0.75	0.80	0.85
(1)	1.645	1.744	1.855	1.982	2.127	2.296	2.493	2.727	3.010	3.357	3.787	4.312	4.789	4.798	4.589
(2)	4.119	4.367	4.646	4.964	5.328	5.748	6.230	6.752	6.994	6.633	6.215	5.889	5.888	6.757	7.960
(3)	6.893	7.308	7.775	8.302	8.886	9.370	8.872	8.170	7.991	8.695	9.735	10.07	9.535	9.052	9.320
(4)	9.650	10.23	10.88	11.55	11.36	10.37	10.69	11.63	12.52	11.90	11.17	11.64	13.25	12.84	12.21

TABLE 8.5 : β_1 ($\phi = .8, \alpha = .7$)

μ MODE:	0.15	0.20	0.25	0.30	0.35	0.40	0.45	0.50	0.55	0.60	0.65	0.70	0.75	0.80	0.85
(1)	1.910	2.021	2.146	2.287	2.448	2.634	2.851	3.107	3.411	3.778	4.215	4.679	4.919	4.810	4.590
(2)	4.784	5.063	5.378	5.734	6.139	6.599	7.096	7.459	7.232	6.780	6.386	6.172	6.463	7.436	8.077
(3)	8.007	8.476	9.000	9.578	10.13	9.953	9.138	8.650	9.029	9.906	10.62	10.22	9.663	9.300	10.20
(4)	11.21	11.86	12.56	12.93	11.80	11.51	12.30	13.24	12.99	12.12	11.86	13.16	13.63	12.95	12.44

TABLE 8.5 : β_1 ($\phi = .8, \alpha = .8$)

μ MODE:	0.15	0.20	0.25	0.30	0.35	0.40	0.45	0.50	0.55	0.60	0.65	0.70	0.75	0.80	0.85
(1)	2.168	2.287	2.421	2.572	2.744	2.940	3.168	3.434	3.747	4.113	4.522	4.877	4.965	4.816	4.591
(2)	5.432	5.736	6.079	6.465	6.900	7.377	7.819	7.853	7.422	6.960	6.612	6.528	6.975	7.840	8.113
(3)	9.096	9.609	10.18	10.77	11.05	10.29	9.514	9.410	10.05	10.89	10.93	10.36	9.837	9.667	10.78
(4)	12.74	13.45	14.12	13.62	12.53	12.90	13.79	14.14	13.24	12.50	13.01	14.18	13.79	13.10	12.82

TABLE 8.5 : β_1 ($\phi = .8, \alpha = .9$)

μ MODE:	0.15	0.20	0.25	0.30	0.35	0.40	0.45	0.50	0.55	0.60	0.65	0.70	0.75	0.80	0.85
(1)	2.415	2.538	2.676	2.831	3.007	3.207	3.437	3.704	4.012	4.361	4.721	4.977	4.986	4.818	4.591
(2)	6.059	6.382	6.746	7.155	7.609	8.079	8.380	8.128	7.621	7.164	6.862	6.873	7.358	8.043	8.128
(3)	10.16	10.70	11.30	11.85	11.62	10.64	10.02	10.24	10.97	11.53	11.14	10.53	10.04	10.05	11.08
(4)	14.23	14.98	15.45	14.19	13.57	14.25	15.04	14.52	13.52	13.15	14.11	14.61	13.94	13.28	13.24

TABLE 8.5 : β_1 ($\phi = .8, \alpha = 1.0$)

μ MODE:	0.15	0.20	0.25	0.30	0.35	0.40	0.45	0.50	0.55	0.60	0.65	0.70	0.75	0.80	0.85
(1)	2.647	2.770	2.907	3.060	3.234	3.431	3.657	3.916	4.211	4.535	4.846	5.030	4.997	4.819	4.591
(2)	6.661	6.997	7.379	7.807	8.274	8.713	8.814	8.371	7.822	7.368	7.100	7.156	7.619	8.145	8.136
(3)	11.18	11.76	12.38	12.78	12.05	11.01	10.60	11.04	11.77	11.91	11.32	10.69	10.25	10.37	11.23
(4)	15.69	16.45	16.44	14.83	14.73	15.52	15.85	14.82	13.87	13.97	15.00	14.85	14.10	13.47	13.58

TABLE 8.5 : β_1 ($\phi = .9, \alpha = .6$)

μ MODE:	0.15	0.20	0.25	0.30	0.35	0.40	0.45	0.50	0.55	0.60	0.65	0.70	0.75	0.80	0.85
(1)	1.462	1.550	1.649	1.762	1.891	2.041	2.216	2.425	2.676	2.986	3.373	3.862	4.430	4.705	4.569
(2)	3.661	3.881	4.130	4.413	4.737	5.112	5.547	6.047	6.541	6.544	6.180	5.837	5.659	6.149	7.511
(3)	6.127	6.496	6.912	7.383	7.916	8.487	8.679	8.076	7.584	7.834	8.746	9.742	9.497	9.012	8.841
(4)	8.578	9.094	9.673	10.31	10.83	10.12	9.688	10.39	11.40	11.78	11.07	10.71	12.17	12.78	12.17

TABLE 8.5 : β_1 ($\phi = .9, \alpha = .7$)

μ MODE:	0.15	0.20	0.25	0.30	0.35	0.40	0.45	0.50	0.55	0.60	0.65	0.70	0.75	0.80	0.85
(1)	1.698	1.797	1.907	2.033	2.176	2.342	2.534	2.762	3.034	3.365	3.768	4.244	4.670	4.740	4.572
(2)	4.252	4.501	4.780	5.098	5.460	5.875	6.345	6.830	7.030	6.719	6.335	6.047	6.063	6.792	7.835
(3)	7.118	7.535	8.003	8.528	9.095	9.494	9.011	8.375	8.251	8.908	9.866	10.11	9.624	9.190	9.468
(4)	9.966	10.55	11.19	11.83	11.54	10.69	11.03	11.93	12.64	12.02	11.37	11.92	13.31	12.91	12.34

TABLE 8.5 : β_1 ($\phi = .9, \alpha = .8$)

μ MODE:	0.15	0.20	0.25	0.30	0.35	0.40	0.45	0.50	0.55	0.60	0.65	0.70	0.75	0.80	0.85
(1)	1.927	2.033	2.152	2.286	2.439	2.614	2.817	3.054	3.336	3.671	4.065	4.487	4.772	4.753	4.573
(2)	4.829	5.099	5.404	5.749	6.141	6.583	7.058	7.426	7.301	6.900	6.529	6.311	6.482	7.250	7.942
(3)	8.086	8.543	9.055	9.615	10.15	10.06	9.323	8.833	9.093	9.887	10.59	10.29	9.779	9.435	10.05
(4)	11.33	11.96	12.64	13.02	12.04	11.68	12.38	13.26	13.08	12.28	11.99	13.13	13.66	13.05	12.58

TABLE 8.5 : β_1 ($\phi = .9, \alpha = .9$)

μ MODE:	0.15	0.20	0.25	0.30	0.35	0.40	0.45	0.50	0.55	0.60	0.65	0.70	0.75	0.80	0.85
(1)	2.146	2.256	2.379	2.517	2.673	2.851	3.057	3.296	3.576	3.903	4.272	4.630	4.820	4.760	4.573
(2)	5.386	5.673	5.998	6.365	6.780	7.238	7.686	7.853	7.530	7.094	6.739	6.581	6.829	7.534	7.986
(3)	9.028	9.519	10.07	10.64	11.01	10.45	9.679	9.417	9.918	10.72	10.96	10.47	9.954	9.707	10.44
(4)	12.66	13.34	14.01	13.78	12.67	12.82	13.65	14.14	13.37	12.62	12.87	14.04	13.86	13.21	12.85

TABLE 8.5 : β_1 ($\phi = .9, \alpha = 1.0$)

μ MODE:	0.15	0.20	0.25	0.30	0.35	0.40	0.45	0.50	0.55	0.60	0.65	0.70	0.75	0.80	0.85
(1)	2.353	2.462	2.584	2.720	2.875	3.051	3.253	3.487	3.759	4.071	4.410	4.713	4.844	4.763	4.573
(2)	5.921	6.221	6.562	6.949	7.384	7.851	8.239	8.185	7.743	7.285	6.939	6.816	7.084	7.700	8.007
(3)	9.943	10.47	11.04	11.61	11.67	10.80	10.07	10.05	10.69	11.38	11.21	10.63	10.13	9.953	10.68
(4)	13.96	14.68	15.27	14.36	13.46	13.96	14.78	14.61	13.64	13.07	13.78	14.57	14.03	13.37	13.11

TABLE 8.5 : β_1 ($\phi = 1.0, \alpha = .6$)

μ MODE:	0.15	0.20	0.25	0.30	0.35	0.40	0.45	0.50	0.55	0.60	0.65	0.70	0.75	0.80	0.85
(1)	1.316	1.395	1.484	1.586	1.702	1.837	1.994	2.182	2.409	2.688	3.039	3.489	4.055	4.547	4.539
(2)	3.295	3.493	3.717	3.972	4.264	4.602	4.996	5.458	5.978	6.341	6.134	5.805	5.561	5.736	6.949
(3)	5.514	5.846	6.221	6.647	7.131	7.676	8.186	7.983	7.449	7.271	7.938	9.028	9.425	8.986	8.666
(4)	7.720	8.185	8.708	9.295	9.915	9.936	9.212	9.430	10.32	11.31	11.00	10.42	11.10	12.63	12.15

TABLE 8.5 : β_1 ($\phi = 1.0, \alpha = .7$)

μ MODE:	0.15	0.20	0.25	0.30	0.35	0.40	0.45	0.50	0.55	0.60	0.65	0.70	0.75	0.80	0.85
(1)	1.528	1.617	1.717	1.830	1.959	2.108	2.281	2.486	2.732	3.031	3.401	3.855	4.357	4.629	4.544
(2)	3.827	4.051	4.302	4.588	4.915	5.292	5.725	6.210	6.645	6.621	6.292	5.983	5.851	6.292	7.444
(3)	6.406	6.781	7.204	7.681	8.213	8.751	8.820	8.254	7.845	8.138	9.014	9.839	9.578	9.136	9.058
(4)	8.970	9.496	10.08	10.72	11.10	10.37	10.10	10.81	11.77	11.90	11.23	11.06	12.48	12.85	12.29

TABLE 8.5 : β_1 ($\phi = 1.0, \alpha = .8$)

μ MODE:	0.15	0.20	0.25	0.30	0.35	0.40	0.45	0.50	0.55	0.60	0.65	0.70	0.75	0.80	0.85
(1)	1.734	1.830	1.937	2.058	2.195	2.353	2.535	2.750	3.005	3.311	3.680	4.109	4.515	4.660	4.547
(2)	4.346	4.589	4.864	5.175	5.531	5.936	6.392	6.857	7.077	6.829	6.474	6.201	6.180	6.734	7.665
(3)	7.278	7.690	8.153	8.671	9.223	9.597	9.181	8.585	8.439	9.019	9.910	10.17	9.734	9.331	9.513
(4)	10.19	10.77	11.41	12.02	11.76	10.97	11.26	12.11	12.74	12.16	11.56	12.07	13.33	13.01	12.48

TABLE 8.5 : β_1 ($\phi = 1.0, \alpha = .9$)

μ MODE:	0.15	0.20	0.25	0.30	0.35	0.40	0.45	0.50	0.55	0.60	0.65	0.70	0.75	0.80	0.85
(1)	1.932	2.031	2.141	2.265	2.406	2.566	2.752	2.968	3.223	3.525	3.881	4.272	4.596	4.675	4.548
(2)	4.847	5.106	5.399	5.731	6.110	6.538	7.003	7.399	7.385	7.031	6.667	6.424	6.474	7.039	7.763
(3)	8.126	8.570	9.069	9.617	10.15	10.18	9.510	8.983	9.110	9.829	10.56	10.39	9.900	9.541	9.884
(4)	11.39	12.02	12.69	13.11	12.28	11.80	12.40	13.26	13.20	12.43	12.08	13.05	13.71	13.16	12.68

TABLE 8.5 : β_1 ($1/\phi$ = .6, $1/\alpha$ = .6)

μ / MODE:	0.15	0.20	0.25	0.30	0.35	0.40	0.45	0.50	0.55	0.60	0.65	0.70	0.75	0.80	0.85
(1)	1.775	1.786	1.811	1.851	1.905	1.974	2.061	2.168	2.301	2.467	2.676	2.942	3.282	3.699	4.097
(2)	4.907	5.145	5.436	5.782	6.189	6.667	7.226	7.853	8.181	7.716	7.189	6.733	6.372	6.164	6.316
(3)	8.512	8.989	9.546	10.18	10.89	11.29	10.35	9.452	9.165	10.00	11.14	11.04	10.39	9.804	9.365
(4)	12.15	12.87	13.67	14.44	13.34	12.27	13.01	14.14	14.06	13.01	12.25	13.30	14.40	13.65	12.93

TABLE 8.5 : β_1 ($1/\phi$ = .6, $1/\alpha$ = .7)

μ / MODE:	0.15	0.20	0.25	0.30	0.35	0.40	0.45	0.50	0.55	0.60	0.65	0.70	0.75	0.80	0.85
(1)	1.645	1.677	1.719	1.772	1.836	1.913	2.007	2.121	2.259	2.430	2.643	2.914	3.260	3.685	4.093
(2)	4.319	4.519	4.762	5.052	5.395	5.800	6.279	6.836	7.414	7.496	7.002	6.660	6.316	6.118	6.279
(3)	7.319	7.792	8.260	8.802	9.419	10.07	10.07	9.285	8.666	8.833	9.814	10.70	10.29	9.740	9.317
(4)	10.50	11.10	11.78	12.54	12.85	11.70	11.41	12.30	13.31	12.86	12.03	11.86	13.51	13.54	12.87

TABLE 8.5 : β_1 ($1/\phi$ = .6, $1/\alpha$ = .8)

μ / MODE:	0.15	0.20	0.25	0.30	0.35	0.40	0.45	0.50	0.55	0.60	0.65	0.70	0.75	0.80	0.85
(1)	1.512	1.559	1.613	1.676	1.750	1.835	1.936	2.056	2.201	2.377	2.596	2.873	3.227	3.663	4.086
(2)	3.870	4.049	4.263	4.516	4.815	5.168	5.586	6.079	6.639	7.075	6.917	6.557	6.237	6.052	6.223
(3)	6.561	6.906	7.311	7.781	8.323	8.926	9.422	9.066	8.455	8.192	8.818	9.902	10.12	9.650	9.249
(4)	9.270	9.786	10.38	11.05	11.71	11.36	10.58	10.96	11.94	12.53	11.87	11.31	12.24	13.33	12.78

TABLE 8.5 : β_1 ($1/\phi$ = .6, $1/\alpha$ = .9)

μ / MODE:	0.15	0.20	0.25	0.30	0.35	0.40	0.45	0.50	0.55	0.60	0.65	0.70	0.75	0.80	0.85
(1)	1.385	1.441	1.503	1.573	1.653	1.745	1.852	1.977	2.127	2.309	2.534	2.818	3.181	3.633	4.076
(2)	3.504	3.674	3.870	4.100	4.369	4.685	5.059	5.502	6.019	6.544	6.686	6.424	6.137	5.965	6.147
(3)	5.906	6.216	6.578	6.996	7.479	8.028	8.595	8.734	8.254	7.866	8.120	9.081	9.838	9.533	9.160
(4)	8.314	8.772	9.301	9.901	10.55	10.85	10.19	10.03	10.81	11.78	11.67	11.09	11.31	12.90	12.67

TABLE 8.5 : β_1 ($1/\phi$ = .6, $1/\alpha$ = 1.0)

μ / MODE:	0.15	0.20	0.25	0.30	0.35	0.40	0.45	0.50	0.55	0.60	0.65	0.70	0.75	0.80	0.85
(1)	1.271	1.330	1.395	1.469	1.553	1.648	1.759	1.888	2.042	2.228	2.458	2.748	3.121	3.591	4.062
(2)	3.197	3.360	3.546	3.759	4.008	4.298	4.640	5.046	5.525	6.055	6.397	6.269	6.019	5.859	6.048
(3)	5.371	5.658	5.988	6.369	6.808	7.311	7.864	8.270	8.029	7.635	7.639	8.412	9.424	9.393	9.056
(4)	7.542	7.960	8.438	8.983	9.508	10.12	9.861	9.443	9.919	10.88	11.38	10.90	10.74	12.23	12.54

TABLE 8.5 : β_1 ($1/\phi$ = .7, $1/\alpha$ = .6)

μ / MODE:	0.15	0.20	0.25	0.30	0.35	0.40	0.45	0.50	0.55	0.60	0.65	0.70	0.75	0.80	0.85
(1)	2.071	2.083	2.113	2.160	2.222	2.303	2.403	2.528	2.682	2.872	3.108	3.400	3.751	4.113	4.326
(2)	5.724	6.002	6.341	6.742	7.213	7.758	8.368	8.795	8.361	7.754	7.222	6.789	6.485	6.418	6.809
(3)	9.930	10.48	11.13	11.85	12.45	11.57	10.45	9.847	10.45	11.51	11.78	11.09	10.42	9.859	9.532
(4)	14.18	15.00	15.87	15.39	13.66	13.98	15.07	15.35	14.15	13.15	13.47	15.01	14.50	13.68	12.98

TABLE 8.5 : β_1 ($1/\phi$ = .7, $1/\alpha$ = .7)

μ / MODE:	0.15	0.20	0.25	0.30	0.35	0.40	0.45	0.50	0.55	0.60	0.65	0.70	0.75	0.80	0.85
(1)	1.919	1.957	2.006	2.067	2.142	2.232	2.341	2.473	2.633	2.829	3.071	3.370	3.730	4.103	4.325
(2)	5.039	5.271	5.555	5.892	6.289	6.754	7.290	7.853	8.037	7.609	7.125	6.715	6.425	6.366	6.775
(3)	8.626	9.088	9.629	10.25	10.91	11.12	10.24	9.447	9.323	10.12	11.10	10.93	10.33	9.793	9.476
(4)	12.25	12.94	13.71	14.34	13.22	12.40	13.11	14.14	13.93	12.96	12.36	13.43	14.28	13.59	12.92

TABLE 8.5 : β_1 ($1/\phi$ = .7, $1/\alpha$ = .8)

μ / MODE:	0.15	0.20	0.25	0.30	0.35	0.40	0.45	0.50	0.55	0.60	0.65	0.70	0.75	0.80	0.85
(1)	1.764	1.819	1.882	1.956	2.041	2.141	2.259	2.398	2.565	2.768	3.018	3.326	3.698	4.087	4.322
(2)	4.515	4.724	4.973	5.268	5.614	6.020	6.494	7.026	7.479	7.386	6.991	6.613	6.340	6.291	6.722
(3)	7.654	8.055	8.524	9.064	9.666	10.21	9.946	9.216	8.765	9.140	10.09	10.64	10.20	9.703	9.397
(4)	10.81	11.41	12.09	12.80	12.75	11.70	11.74	12.63	13.39	12.77	12.03	12.22	13.68	13.46	12.83

TABLE 8.5 : β_1 ($1/\phi$ = .7, $1/\alpha$ = .9)

μ / MODE:	0.15	0.20	0.25	0.30	0.35	0.40	0.45	0.50	0.55	0.60	0.65	0.70	0.75	0.80	0.85
(1)	1.616	1.681	1.753	1.835	1.928	2.036	2.160	2.306	2.480	2.690	2.948	3.266	3.652	4.064	4.318
(2)	4.088	4.286	4.515	4.782	5.094	5.460	5.886	6.377	6.880	7.082	6.822	6.487	6.231	6.191	6.647
(3)	6.890	7.252	7.671	8.133	8.698	9.268	9.495	8.973	8.457	8.480	9.242	10.14	10.03	9.590	9.294
(4)	9.698	10.23	10.84	11.50	11.97	11.29	10.83	11.44	12.38	12.49	11.82	11.52	12.75	13.27	12.73

TABLE 8.5 : β_1 ($1/\phi$ = .7, $1/\alpha$ = 1.0)

μ / MODE:	0.15	0.20	0.25	0.30	0.35	0.40	0.45	0.50	0.55	0.60	0.65	0.70	0.75	0.80	0.85
(1)	1.482	1.551	1.628	1.714	1.811	1.923	2.052	2.202	2.381	2.597	2.861	3.189	3.592	4.032	4.311
(2)	3.730	3.920	4.136	4.385	4.674	5.010	5.403	5.859	6.359	6.726	6.632	6.346	6.108	6.067	6.545
(3)	6.266	6.600	6.984	7.425	7.925	8.473	8.913	8.699	8.210	8.032	8.579	9.538	9.834	9.463	9.173
(4)	8.799	9.284	9.836	10.45	11.04	10.89	10.27	10.53	11.40	12.07	11.61	11.15	11.90	13.02	12.60

TABLE 8.5 : β_1 ($1/\phi$ = .8, $1/\alpha$ = .6)

μ / MODE:	0.15	0.20	0.25	0.30	0.35	0.40	0.45	0.50	0.55	0.60	0.65	0.70	0.75	0.80	0.85
(1)	2.367	2.381	2.415	2.468	2.539	2.631	2.745	2.886	3.059	3.270	3.527	3.830	4.154	4.401	4.447
(2)	6.542	6.858	7.244	7.699	8.228	8.822	9.356	9.098	8.411	7.786	7.268	6.875	6.662	6.771	7.281
(3)	11.35	11.98	12.69	13.43	13.06	11.67	10.69	10.87	11.81	12.52	11.88	11.11	10.45	9.947	9.802
(4)	16.20	17.11	17.73	15.62	14.90	15.87	16.69	15.50	14.25	13.78	15.13	15.43	14.54	13.71	13.07

TABLE 8.5 : β_1 ($1/\phi$ = .8, $1/\alpha$ = .7)

μ / MODE:	0.15	0.20	0.25	0.30	0.35	0.40	0.45	0.50	0.55	0.60	0.65	0.70	0.75	0.80	0.85
(1)	2.194	2.237	2.292	2.362	2.447	2.550	2.674	2.824	3.004	3.223	3.487	3.800	4.136	4.396	4.447
(2)	5.759	6.024	6.346	6.729	7.177	7.691	8.247	8.580	8.208	7.659	7.173	6.797	6.594	6.716	7.255
(3)	9.857	10.38	10.99	11.66	12.16	11.39	10.38	9.883	10.41	11.36	11.61	10.99	10.37	9.876	9.738
(4)	13.99	14.77	15.57	15.13	13.62	13.86	14.84	15.13	14.05	13.14	13.44	14.79	14.39	13.62	13.00

TABLE 8.5 : β_1 ($1/\phi$ = .8, $1/\alpha$ = .8)

μ / MODE:	0.15	0.20	0.25	0.30	0.35	0.40	0.45	0.50	0.55	0.60	0.65	0.70	0.75	0.80	0.85
(1)	2.016	2.078	2.151	2.235	2.333	2.447	2.580	2.739	2.928	3.155	3.430	3.755	4.109	4.386	4.445
(2)	5.160	5.398	5.683	6.017	6.409	6.862	7.372	7.853	7.894	7.483	7.044	6.690	6.499	6.635	7.215
(3)	8.747	9.203	9.733	10.33	10.93	10.94	10.11	9.440	9.485	10.26	11.06	10.81	10.25	9.779	9.646
(4)	12.36	13.03	13.77	14.22	13.09	12.52	13.24	14.14	13.79	12.89	12.48	13.59	14.15	13.51	12.91

TABLE 8.5 : β_1 ($1/\phi$ = .8, $1/\alpha$ = .9)

μ / MODE:	0.15	0.20	0.25	0.30	0.35	0.40	0.45	0.50	0.55	0.60	0.65	0.70	0.75	0.80	0.85
(1)	1.847	1.921	2.004	2.097	2.204	2.326	2.468	2.634	2.832	3.068	3.353	3.694	4.070	4.373	4.444
(2)	4.672	4.898	5.160	5.464	5.817	6.227	6.695	7.188	7.482	7.269	6.889	6.559	6.378	6.524	7.156
(3)	7.874	8.286	8.761	9.299	9.877	10.27	9.808	9.144	8.895	9.420	10.31	10.56	10.10	9.659	9.523
(4)	11.08	11.68	12.36	12.99	12.61	11.74	12.03	12.91	13.36	12.66	12.04	12.56	13.74	13.36	12.79

TABLE 8.5 : β_1 ($1/\phi = .8$, $1/\alpha = 1.0$)

μ MODE:	0.15	0.20	0.25	0.30	0.35	0.40	0.45	0.50	0.55	0.60	0.65	0.70	0.75	0.80	0.85
(1)	1.694	1.773	1.860	1.959	2.070	2.197	2.344	2.516	2.719	2.963	3.258	3.614	4.016	4.353	4.441
(2)	4.263	4.479	4.727	5.011	5.339	5.718	6.154	6.634	7.043	7.031	6.723	6.415	6.236	6.384	7.072
(3)	7.160	7.542	7.979	8.474	9.020	9.523	9.456	8.877	8.494	8.776	9.604	10.24	9.947	9.527	9.374
(4)	10.05	10.61	11.22	11.86	12.04	11.24	11.12	11.86	12.68	12.42	11.77	11.81	13.12	13.20	12.67

TABLE 8.5 : β_1 ($1/\phi = .9$, $1/\alpha = .6$)

μ MODE:	0.15	0.20	0.25	0.30	0.35	0.40	0.45	0.50	0.55	0.60	0.65	0.70	0.75	0.80	0.85
(1)	2.662	2.679	2.717	2.776	2.856	2.959	3.086	3.243	3.432	3.660	3.927	4.218	4.474	4.584	4.512
(2)	7.359	7.714	8.146	8.651	9.227	9.817	9.899	9.180	8.445	7.826	7.335	7.002	6.910	7.180	7.642
(3)	12.76	13.46	14.23	14.66	13.21	11.81	11.36	12.09	13.02	12.76	11.91	11.14	10.51	10.08	10.18
(4)	18.21	19.17	18.38	16.13	16.61	17.67	17.10	15.58	14.50	15.14	16.27	15.50	14.56	13.75	13.21

TABLE 8.5 : β_1 ($1/\phi = .9$, $1/\alpha = .7$)

μ MODE:	0.15	0.20	0.25	0.30	0.35	0.40	0.45	0.50	0.55	0.60	0.65	0.70	0.75	0.80	0.85
(1)	2.468	2.516	2.579	2.657	2.753	2.868	3.007	3.173	3.372	3.609	3.886	4.190	4.461	4.581	4.512
(2)	6.478	6.776	7.137	7.563	8.055	8.599	9.059	8.876	8.275	7.702	7.236	6.917	6.836	7.127	7.627
(3)	11.09	11.67	12.34	13.00	12.75	11.52	10.63	10.74	11.56	12.23	11.73	11.03	10.42	10.00	10.11
(4)	15.73	16.58	17.15	15.42	14.64	15.46	16.25	15.33	14.17	13.69	14.81	15.24	14.44	13.67	13.13

TABLE 8.5 : β_1 ($1/\phi = .9$, $1/\alpha = .8$)

μ MODE:	0.15	0.20	0.25	0.30	0.35	0.40	0.45	0.50	0.55	0.60	0.65	0.70	0.75	0.80	0.85
(1)	2.268	2.338	2.420	2.514	2.624	2.752	2.902	3.078	3.288	3.536	3.827	4.149	4.441	4.576	4.512
(2)	5.804	6.073	6.391	6.765	7.197	7.687	8.186	8.406	8.042	7.540	7.105	6.801	6.731	7.048	7.602
(3)	9.839	10.35	10.93	11.56	11.93	11.19	10.28	9.923	10.45	11.30	11.45	10.87	10.30	9.894	10.01
(4)	13.90	14.64	15.37	14.85	13.55	13.82	14.72	14.91	13.92	13.11	13.49	14.65	14.27	13.55	13.03

TABLE 8.5 : β_1 ($1/\phi = .9$, $1/\alpha = .9$)

μ MODE:	0.15	0.20	0.25	0.30	0.35	0.40	0.45	0.50	0.55	0.60	0.65	0.70	0.75	0.80	0.85
(1)	2.078	2.161	2.254	2.359	2.479	2.617	2.776	2.962	3.181	3.441	3.747	4.091	4.412	4.569	4.511
(2)	5.256	5.510	5.803	6.144	6.537	6.984	7.468	7.853	7.759	7.351	6.949	6.659	6.594	6.937	7.565
(3)	8.858	9.319	9.846	10.43	10.95	10.77	9.971	9.432	9.640	10.41	11.02	10.68	10.16	9.760	9.874
(4)	12.46	13.13	13.84	14.08	12.94	12.64	13.37	14.14	13.64	12.81	12.62	13.74	14.04	13.41	12.89

TABLE 8.5 : β_1 ($1/\phi = .9$, $1/\alpha = 1.0$)

μ MODE:	0.15	0.20	0.25	0.30	0.35	0.40	0.45	0.50	0.55	0.60	0.65	0.70	0.75	0.80	0.85
(1)	1.906	1.994	2.093	2.204	2.329	2.472	2.637	2.829	3.056	3.326	3.646	4.013	4.370	4.559	4.510
(2)	4.796	5.039	5.317	5.635	6.001	6.419	6.883	7.327	7.452	7.154	6.785	6.502	6.431	6.791	7.511
(3)	8.055	8.483	8.970	9.512	10.06	10.26	9.670	9.077	9.035	9.674	10.48	10.48	10.01	9.611	9.698
(4)	11.31	11.92	12.59	13.12	12.46	11.80	12.29	13.15	13.30	12.56	12.08	12.86	13.75	13.27	12.75

TABLE 8.5 : β_1 ($1/\phi = 1.0$, $1/\alpha = .6$)

μ MODE:	0.15	0.20	0.25	0.30	0.35	0.40	0.45	0.50	0.55	0.60	0.65	0.70	0.75	0.80	0.85
(1)	2.958	2.976	3.019	3.084	3.173	3.286	3.426	3.596	3.799	4.037	4.300	4.551	4.707	4.695	4.550
(2)	8.176	8.569	9.044	9.595	10.20	10.62	10.08	9.223	8.480	7.878	7.429	7.179	7.224	7.581	7.880
(3)	14.17	14.94	15.71	15.14	13.31	12.14	12.38	13.29	13.67	12.83	11.94	11.18	10.59	10.28	10.60
(4)	20.21	21.08	18.63	17.39	18.34	18.85	17.20	15.70	15.30	16.54	16.56	15.53	14.59	13.82	13.42

TABLE 8.5 : β_1 ($1/\phi = 1.0$, $1/\alpha = .7$)

μ MODE:	0.15	0.20	0.25	0.30	0.35	0.40	0.45	0.50	0.55	0.60	0.65	0.70	0.75	0.80	0.85
(1)	2.742	2.796	2.865	2.952	3.058	3.186	3.338	3.520	3.734	3.983	4.260	4.528	4.699	4.693	4.549
(2)	7.198	7.528	7.926	8.392	8.917	9.441	9.566	8.987	8.320	7.753	7.324	7.086	7.147	7.537	7.872
(3)	12.32	12.96	13.66	14.09	12.96	11.68	11.19	11.75	12.59	12.54	11.79	11.07	10.49	10.19	10.55
(4)	17.47	18.35	17.93	15.87	16.05	17.00	16.81	15.44	14.40	14.76	15.87	15.35	14.47	13.73	13.33

TABLE 8.5 : β_1 ($1/\phi = 1.0$, $1/\alpha = .8$)

μ MODE:	0.15	0.20	0.25	0.30	0.35	0.40	0.45	0.50	0.55	0.60	0.65	0.70	0.75	0.80	0.85
(1)	2.519	2.598	2.689	2.794	2.915	3.057	3.222	3.416	3.643	3.907	4.202	4.493	4.686	4.691	4.549
(2)	6.449	6.746	7.099	7.509	7.976	8.478	8.857	8.660	8.114	7.593	7.185	6.959	7.035	7.470	7.859
(3)	10.93	11.49	12.12	12.71	12.45	11.34	10.57	10.69	11.44	12.01	11.57	10.92	10.36	10.07	10.45
(4)	15.43	16.22	16.70	15.17	14.48	15.21	15.93	15.13	14.05	13.66	14.65	15.05	14.32	13.60	13.21

TABLE 8.5 : β_1 ($1/\phi = 1.0$, $1/\alpha = .9$)

μ MODE:	0.15	0.20	0.25	0.30	0.35	0.40	0.45	0.50	0.55	0.60	0.65	0.70	0.75	0.80	0.85
(1)	2.309	2.401	2.505	2.621	2.754	2.907	3.083	3.288	3.527	3.806	4.122	4.443	4.667	4.687	4.549
(2)	5.840	6.121	6.447	6.822	7.249	7.722	8.168	8.268	7.880	7.409	7.021	6.800	6.886	7.371	7.840
(3)	9.841	10.35	10.92	11.51	11.76	10.98	10.16	9.957	10.52	11.29	11.31	10.75	10.21	9.912	10.32
(4)	13.85	14.57	15.23	14.58	13.47	13.82	14.67	14.73	13.77	13.07	13.58	14.57	14.14	13.46	13.05

TABLE 8.6 : β_1 ($\phi = .6$, $\alpha = .6$)

μ MODE:	0.15	0.20	0.25	0.30	0.35	0.40	0.45	0.50	0.55	0.60	0.65	0.70	0.75	0.80	0.85
(1)	4.594	4.870	5.181	5.535	5.940	6.402	6.920	7.357	7.122	6.620	6.156	5.753	5.407	5.115	4.874
(2)	8.270	8.767	9.324	9.941	10.50	9.903	8.951	8.350	8.710	9.626	10.52	10.12	9.504	8.950	8.469
(3)	11.95	12.66	13.41	13.17	11.67	11.78	12.72	13.65	12.93	11.96	11.49	12.82	13.55	12.82	12.12
(4)	15.62	16.51	15.92	14.69	15.62	16.71	15.83	14.62	15.58	16.80	15.80	14.80	15.50	16.65	15.78

TABLE 8.6 : β_1 ($\phi = .6$, $\alpha = .7$)

μ MODE:	0.15	0.20	0.25	0.30	0.35	0.40	0.45	0.50	0.55	0.60	0.65	0.70	0.75	0.80	0.85
(1)	5.335	5.646	5.996	6.391	6.836	7.329	7.787	7.751	7.239	6.714	6.252	5.855	5.517	5.234	5.003
(2)	9.606	10.17	10.78	11.40	11.23	10.12	9.275	9.227	9.947	10.85	10.83	10.20	9.579	9.035	8.567
(3)	13.88	14.66	15.18	13.57	12.82	13.58	14.54	14.20	13.09	12.28	12.86	14.14	13.66	12.89	12.20
(4)	18.13	18.86	16.64	16.93	18.02	17.78	16.17	16.38	17.72	17.12	15.95	15.45	17.21	16.74	15.85

TABLE 8.6 : β_1 ($\phi = .6$, $\alpha = .8$)

μ MODE:	0.15	0.20	0.25	0.30	0.35	0.40	0.45	0.50	0.55	0.60	0.65	0.70	0.75	0.80	0.85
(1)	6.057	6.395	6.773	7.197	7.665	8.135	8.347	7.941	7.362	6.830	6.370	5.976	5.642	5.361	5.129
(2)	10.91	11.52	12.16	12.53	11.57	10.43	9.872	10.25	11.05	11.50	10.97	10.29	9.676	9.138	8.679
(3)	15.76	16.55	16.05	14.23	14.36	15.26	15.58	14.42	13.35	13.09	14.19	14.52	13.75	12.98	12.30
(4)	20.57	20.14	18.27	19.15	19.85	18.14	17.22	18.33	18.57	17.30	16.25	16.85	17.71	16.83	15.93

TABLE 8.6 : β_1 ($\phi = .6$, $\alpha = .9$)

μ MODE:	0.15	0.20	0.25	0.30	0.35	0.40	0.45	0.50	0.55	0.60	0.65	0.70	0.75	0.80	0.85
(1)	6.754	7.109	7.507	7.948	8.413	8.791	8.681	8.109	7.502	6.963	6.500	6.105	5.767	5.479	5.236
(2)	12.18	12.81	13.42	13.23	11.89	10.84	10.64	11.21	11.90	11.78	11.09	10.40	9.785	9.249	8.790
(3)	17.59	18.26	16.63	15.31	15.89	16.67	15.97	14.67	13.78	14.14	15.10	14.69	13.86	13.08	12.40
(4)	22.89	20.99	20.22	21.21	20.58	18.67	18.86	19.84	18.84	17.52	16.88	18.07	17.88	16.93	16.03

TABLE 8.6 : β_1 ($\phi = .6$, $\alpha = 1.0$)

μ MODE:	0.15	0.20	0.25	0.30	0.35	0.40	0.45	0.50	0.55	0.60	0.65	0.70	0.75	0.80	0.85
(1)	7.420	7.786	8.196	8.643	9.084	9.307	8.934	8.279	7.650	7.101	6.630	6.226	5.879	5.579	5.320
(2)	13.40	14.05	14.54	13.71	12.25	11.36	11.43	12.05	12.47	11.97	11.22	10.52	9.896	9.356	8.887
(3)	19.36	19.63	17.27	16.58	17.31	17.59	16.26	14.96	14.40	15.13	15.56	14.82	13.97	13.19	12.50
(4)	25.01	22.07	22.17	22.93	21.04	19.63	20.46	20.54	19.08	17.83	17.80	18.75	18.01	17.03	16.13

TABLE 8.6 : β_1 ($\phi = .7$, $\alpha = .6$)

μ MODE:	0.15	0.20	0.25	0.30	0.35	0.40	0.45	0.50	0.55	0.60	0.65	0.70	0.75	0.80	0.85
(1)	3.938	4.174	4.441	4.745	5.093	5.494	5.957	6.471	6.831	6.564	6.138	5.746	5.405	5.114	4.874
(2)	7.089	7.515	7.995	8.536	9.128	9.519	8.861	8.122	7.782	8.345	9.347	9.965	9.479	8.943	8.467
(3)	10.24	10.85	11.53	12.18	11.43	10.48	10.98	11.94	12.66	11.89	11.11	11.22	12.89	12.79	12.11
(4)	13.39	14.18	14.90	13.53	13.47	14.46	15.37	14.31	13.64	14.95	15.68	14.72	14.03	16.07	15.76

TABLE 8.6 : β_1 ($\phi = .7$, $\alpha = .7$)

μ MODE:	0.15	0.20	0.25	0.30	0.35	0.40	0.45	0.50	0.55	0.60	0.65	0.70	0.75	0.80	0.85
(1)	4.573	4.840	5.140	5.479	5.866	6.305	6.789	7.204	7.110	6.675	6.237	5.849	5.515	5.233	5.003
(2)	8.234	8.715	9.252	9.840	10.36	9.960	9.088	8.503	8.699	9.504	10.35	10.13	9.562	9.029	8.566
(3)	11.90	12.58	13.30	13.21	11.86	11.79	12.63	13.51	12.98	12.07	11.60	12.66	13.50	12.87	12.19
(4)	15.55	16.41	16.05	14.76	15.55	16.57	15.92	14.77	15.50	16.68	15.86	14.92	15.38	16.62	15.83

TABLE 8.6 : β_1 ($\phi = .7$, $\alpha = .8$)

μ MODE:	0.15	0.20	0.25	0.30	0.35	0.40	0.45	0.50	0.55	0.60	0.65	0.70	0.75	0.80	0.85
(1)	5.192	5.482	5.807	6.174	6.588	7.044	7.489	7.639	7.279	6.801	6.358	5.972	5.641	5.361	5.129
(2)	9.353	9.878	10.46	11.04	11.15	10.25	9.412	9.142	9.657	10.48	10.77	10.25	9.662	9.134	8.678
(3)	13.52	14.25	14.84	13.72	12.72	13.23	14.12	14.19	13.19	12.38	12.56	13.79	13.67	12.96	12.29
(4)	17.67	18.42	16.77	16.56	17.54	17.75	16.29	16.04	17.26	17.14	16.05	15.39	16.77	16.76	15.92

TABLE 8.6 : β_1 ($\phi = .7$, $\alpha = .9$)

μ MODE:	0.15	0.20	0.25	0.30	0.35	0.40	0.45	0.50	0.55	0.60	0.65	0.70	0.75	0.80	0.85
(1)	5.789	6.095	6.438	6.824	7.253	7.702	8.042	7.916	7.441	6.940	6.491	6.101	5.766	5.479	5.236
(2)	10.44	11.00	11.60	12.08	11.62	10.56	9.845	9.884	10.53	11.18	10.97	10.37	9.774	9.246	8.789
(3)	15.09	15.85	15.94	14.28	13.87	14.61	15.26	14.49	13.45	12.88	13.58	14.36	13.80	13.07	12.40
(4)	19.72	19.96	17.88	18.38	19.27	18.22	16.95	17.60	18.38	17.37	16.31	16.27	17.50	16.88	16.02

TABLE 8.6 : β_1 ($\phi = .7$, $\alpha = 1.0$)

μ MODE:	0.15	0.20	0.25	0.30	0.35	0.40	0.45	0.50	0.55	0.60	0.65	0.70	0.75	0.80	0.85
(1)	6.360	6.676	7.032	7.431	7.866	8.288	8.474	8.143	7.604	7.084	6.624	6.224	5.878	5.579	5.320
(2)	11.49	12.08	12.68	12.94	12.00	10.90	10.35	10.61	11.28	11.62	11.14	10.49	9.889	9.353	8.887
(3)	16.63	17.36	16.68	14.98	15.08	15.87	15.90	14.76	13.75	13.55	14.46	14.65	13.93	13.18	12.50
(4)	21.71	20.98	19.33	20.13	20.42	18.66	18.03	19.03	18.85	17.60	16.67	17.25	17.82	17.00	16.12

TABLE 8.6 : β_1 ($\phi = .8$, $\alpha = .6$)

μ MODE:	0.15	0.20	0.25	0.30	0.35	0.40	0.45	0.50	0.55	0.60	0.65	0.70	0.75	0.80	0.85
(1)	3.445	3.653	3.886	4.152	4.457	4.810	5.220	5.696	6.208	6.413	6.103	5.734	5.400	5.112	4.873
(2)	6.203	6.576	6.997	7.473	8.011	8.579	8.686	8.041	7.483	7.476	8.266	9.347	9.425	8.932	8.464
(3)	8.960	9.499	10.10	10.76	11.09	10.13	9.773	10.51	11.53	11.78	11.04	10.49	11.54	12.69	12.10
(4)	11.72	12.42	13.17	13.20	12.07	12.71	13.74	14.14	13.11	13.23	14.78	14.66	13.83	14.48	15.71

TABLE 8.6 : β_1 ($\phi = .8$, $\alpha = .7$)

μ MODE:	0.15	0.20	0.25	0.30	0.35	0.40	0.45	0.50	0.55	0.60	0.65	0.70	0.75	0.80	0.85
(1)	4.001	4.235	4.497	4.795	5.135	5.525	5.970	6.451	6.784	6.597	6.212	5.840	5.512	5.232	5.003
(2)	7.205	7.627	8.100	8.629	9.194	9.534	8.971	8.289	7.976	8.444	9.359	9.936	9.528	9.020	8.563
(3)	10.41	11.02	11.68	12.27	11.59	10.74	11.15	12.06	12.68	11.98	11.26	11.38	12.85	12.82	12.18
(4)	13.61	14.39	15.04	13.80	13.72	14.66	15.44	14.46	13.90	15.10	15.71	14.82	14.25	16.02	15.80

TABLE 8.6 : β_1 ($\phi = .8$, $\alpha = .8$)

μ MODE:	0.15	0.20	0.25	0.30	0.35	0.40	0.45	0.50	0.55	0.60	0.65	0.70	0.75	0.80	0.85
(1)	4.543	4.797	5.082	5.405	5.771	6.185	6.640	7.053	7.096	6.747	6.339	5.964	5.638	5.360	5.128
(2)	8.185	8.646	9.160	9.720	10.22	10.03	9.243	8.656	8.606	9.359	10.17	10.15	9.628	9.177	8.676
(3)	11.83	12.49	13.17	13.26	12.07	11.79	12.52	13.36	13.05	12.20	11.69	12.48	13.44	12.93	12.28
(4)	15.47	16.29	16.18	14.85	15.47	16.43	16.03	14.92	15.39	16.55	15.95	15.05	15.25	16.57	15.90

TABLE 8.6 : β_1 ($\phi = .8$, $\alpha = .9$)

μ MODE:	0.15	0.20	0.25	0.30	0.35	0.40	0.45	0.50	0.55	0.60	0.65	0.70	0.75	0.80	0.85
(1)	5.065	5.333	5.635	5.976	6.361	6.789	7.227	7.503	7.319	6.900	6.476	6.095	5.764	5.479	5.236
(2)	9.136	9.629	10.17	10.74	11.03	10.38	9.559	9.129	9.419	10.15	10.66	10.30	9.757	9.241	8.788
(3)	13.21	13.92	14.54	13.87	12.73	12.95	13.77	14.14	13.31	12.48	12.36	13.42	13.68	13.05	12.39
(4)	17.29	18.06	16.95	16.29	17.15	17.67	16.42	15.83	16.86	17.16	16.16	15.41	16.34	16.78	16.01

TABLE 8.6 : β_1 ($\phi = .8$, $\alpha = 1.0$)

μ MODE:	0.15	0.20	0.25	0.30	0.35	0.40	0.45	0.50	0.55	0.60	0.65	0.70	0.75	0.80	0.85
(1)	5.565	5.842	6.155	6.511	6.910	7.344	7.745	7.854	7.518	7.055	6.613	6.220	5.877	5.579	5.320
(2)	10.06	10.58	11.15	11.68	11.62	10.70	9.898	9.652	10.11	10.82	10.96	10.45	9.876	9.350	8.886
(3)	14.56	15.30	15.71	14.40	13.57	14.10	14.86	14.55	13.55	12.82	13.11	14.10	13.85	13.16	12.50
(4)	19.06	19.63	17.78	17.80	18.71	18.30	16.89	17.04	18.05	17.45	16.39	15.94	17.16	16.94	16.11

TABLE 8.6 : β_1 ($\phi = .9$, $\alpha = .6$)

μ MODE:	0.15	0.20	0.25	0.30	0.35	0.40	0.45	0.50	0.55	0.60	0.65	0.70	0.75	0.80	0.85
(1)	3.063	3.247	3.454	3.691	3.962	4.276	4.643	5.075	5.574	6.050	6.033	5.715	5.394	5.110	4.873
(2)	5.514	5.846	6.220	6.645	7.129	7.671	8.173	7.951	7.389	7.040	7.444	8.455	9.271	8.912	8.459
(3)	7.965	8.444	8.983	9.586	10.20	9.973	9.205	9.429	10.32	11.30	10.98	10.34	10.47	12.27	12.08
(4)	10.42	11.04	11.73	12.38	11.60	11.39	12.27	13.30	12.99	12.26	13.30	14.49	13.77	13.38	15.55

TABLE 8.6 : β_1 ($\phi = .9$, $\alpha = .7$)

μ MODE:	0.15	0.20	0.25	0.30	0.35	0.40	0.45	0.50	0.55	0.60	0.65	0.70	0.75	0.80	0.85
(1)	3.557	3.764	3.998	4.263	4.566	4.915	5.318	5.778	6.249	6.429	6.168	5.825	5.506	5.230	5.002
(2)	6.405	6.780	7.202	7.677	8.206	8.735	8.777	8.187	7.687	7.708	8.437	9.405	9.461	9.006	8.560
(3)	9.254	9.796	10.40	11.03	11.24	10.36	10.10	10.81	11.76	11.86	11.16	10.72	11.74	12.70	12.17
(4)	12.10	12.01	13.54	13.42	12.48	13.11	14.10	14.26	13.37	13.61	15.02	14.74	13.98	14.75	15.74

TABLE 8.6 : β_1 ($\phi = .9$, $\alpha = .8$)

μ MODE:	0.15	0.20	0.25	0.30	0.35	0.40	0.45	0.50	0.55	0.60	0.65	0.70	0.75	0.80	0.85
(1)	4.038	4.264	4.517	4.805	5.133	5.507	5.932	6.389	6.734	6.645	6.307	5.953	5.634	5.358	5.128
(2)	7.275	7.686	8.147	8.661	9.204	9.551	9.108	8.469	8.134	8.479	9.303	9.907	9.595	9.116	8.674
(3)	10.52	11.11	11.76	12.32	11.78	10.96	11.25	12.10	12.70	12.10	11.42	11.45	12.76	12.87	12.27
(4)	13.76	14.52	15.13	14.09	13.89	14.77	15.49	14.62	14.08	15.15	15.76	14.94	14.40	15.92	15.86

TABLE 8.6 : β_1 (ϕ = .9, α = .9)

μ MODE:	0.15	0.20	0.25	0.30	0.35	0.40	0.45	0.50	0.55	0.60	0.65	0.70	0.75	0.80	0.85
(1)	4.503	4.741	5.009	5.314	5.661	6.054	6.489	6.911	7.079	6.830	6.453	6.087	5.761	5.478	5.235
(2)	8.121	8.562	9.055	9.595	10.11	10.10	9.410	8.799	8.672	9.205	9.998	10.17	9.727	9.233	8.786
(3)	11.75	12.39	13.05	13.31	12.27	11.79	12.39	13.22	13.13	12.34	11.78	12.30	13.36	13.01	12.38
(4)	15.38	16.17	16.31	14.96	15.37	16.30	16.14	15.06	15.27	16.42	16.05	15.18	15.12	16.51	15.98

TABLE 8.6 : β_1 (ϕ = .9, α = 1.0)

μ MODE:	0.15	0.20	0.25	0.30	0.35	0.40	0.45	0.50	0.55	0.60	0.65	0.70	0.75	0.80	0.85
(1)	4.947	5.193	5.473	5.791	6.154	6.563	7.001	7.363	7.354	7.005	6.596	6.214	5.875	5.578	5.320
(2)	8.941	9.408	9.931	10.49	10.90	10.50	9.703	9.154	9.226	9.867	10.52	10.37	9.856	9.345	8.885
(3)	12.95	13.63	14.28	14.00	12.80	12.73	13.47	14.06	13.42	12.58	12.23	13.09	13.68	13.13	12.49
(4)	16.97	17.76	17.13	16.09	16.82	17.55	16.55	15.71	16.52	17.15	16.27	15.45	15.96	16.80	16.10

TABLE 8.6 : β_1 (ϕ = 1.0, α = .6)

μ MODE:	0.15	0.20	0.25	0.30	0.35	0.40	0.45	0.50	0.55	0.60	0.65	0.70	0.75	0.80	0.85
(1)	2.756	2.922	3.109	3.322	3.566	3.849	4.180	4.572	5.036	5.559	5.878	5.682	5.384	5.107	4.872
(2)	4.962	5.261	5.598	5.982	6.419	6.919	7.465	7.750	7.326	6.885	6.876	7.683	8.834	8.876	8.453
(3)	7.168	7.600	8.087	8.635	9.238	9.643	9.027	8.686	9.343	10.35	10.86	10.28	9.915	11.37	12.04
(4)	9.375	9.939	10.57	11.25	11.35	10.55	11.09	12.08	12.78	12.04	12.11	13.72	13.72	13.07	14.93

TABLE 8.6 : β_1 (ϕ = 1.0, α = .7)

μ MODE:	0.15	0.20	0.25	0.30	0.35	0.40	0.45	0.50	0.55	0.60	0.65	0.70	0.75	0.80	0.85
(1)	3.201	3.388	3.598	3.837	4.110	4.425	4.791	5.217	5.697	6.111	6.089	5.802	5.499	5.228	5.001
(2)	5.764	6.102	6.482	6.913	7.399	7.930	8.351	8.083	7.567	7.295	7.710	8.661	9.312	8.983	8.555
(3)	8.329	8.817	9.365	9.967	10.52	10.17	9.523	9.825	10.70	11.50	11.09	10.52	10.81	12.37	12.14
(4)	10.89	11.53	12.23	12.76	11.93	11.90	12.77	13.68	13.16	12.64	13.77	14.60	13.90	13.74	15.60

TABLE 8.6 : β_1 (ϕ = 1.0, α = .8)

μ MODE:	0.15	0.20	0.25	0.30	0.35	0.40	0.45	0.50	0.55	0.60	0.65	0.70	0.75	0.80	0.85
(1)	3.634	3.837	4.066	4.325	4.621	4.961	5.352	5.795	6.244	6.455	6.254	5.936	5.628	5.357	5.128
(2)	6.548	6.918	7.335	7.804	8.322	8.827	8.890	8.359	7.884	7.865	8.508	9.402	9.516	9.100	8.670
(3)	9.465	10.00	10.60	11.22	11.41	10.61	10.34	10.99	11.88	11.96	11.31	10.91	11.81	12.73	12.26
(4)	12.38	13.08	13.80	13.68	12.81	13.38	14.32	14.41	13.55	13.85	15.14	14.86	14.15	14.84	15.79

TABLE 8.6 : β_1 (ϕ = 1.0, α = .9)

μ MODE:	0.15	0.20	0.25	0.30	0.35	0.40	0.45	0.50	0.55	0.60	0.65	0.70	0.75	0.80	0.85
(1)	4.052	4.267	4.509	4.784	5.098	5.458	5.867	6.313	6.690	6.706	6.416	6.075	5.757	5.476	5.235
(2)	7.309	7.707	8.155	8.656	9.190	9.577	9.259	8.644	8.258	8.469	9.214	9.882	9.678	9.222	8.783
(3)	10.57	11.16	11.79	12.36	11.99	11.15	11.30	12.09	12.73	12.23	11.57	11.48	12.65	12.93	12.37
(4)	13.84	14.59	15.23	14.36	14.00	14.82	15.56	14.80	14.20	15.15	15.83	15.07	14.51	15.80	15.94

TABLE 8.6 : β_1 (1/ϕ = .6, 1/α = .6)

μ MODE:	0.15	0.20	0.25	0.30	0.35	0.40	0.45	0.50	0.55	0.60	0.65	0.70	0.75	0.80	0.85
(1)	4.038	4.215	4.438	4.706	5.026	5.405	5.856	6.390	7.000	7.402	7.079	6.632	6.220	5.852	5.524
(2)	7.607	8.024	8.515	9.083	9.731	10.41	10.22	9.343	8.613	8.383	9.201	10.37	10.28	9.711	9.172
(3)	11.24	11.89	12.64	13.44	13.20	11.82	11.69	12.68	13.66	12.94	12.04	11.46	12.77	13.53	12.84
(4)	14.90	15.78	16.70	15.58	14.76	15.81	16.81	15.59	14.60	15.85	16.66	15.62	14.72	15.95	16.48

TABLE 8.6 : β_1 (1/ϕ = .6, 1/α = .7)

μ MODE:	0.15	0.20	0.25	0.30	0.35	0.40	0.45	0.50	0.55	0.60	0.65	0.70	0.75	0.80	0.85
(1)	3.580	3.732	3.918	4.144	4.413	4.734	5.117	5.574	6.115	6.691	6.853	6.526	6.151	5.804	5.492
(2)	6.623	6.970	7.382	7.862	8.417	9.040	9.567	9.159	8.473	7.978	8.185	9.203	10.01	9.621	9,119
(3)	9.721	10.27	10.90	11.61	12.25	11.54	10.66	11.06	12.08	12.66	11.92	11.22	11.40	13.10	12.75
(4)	12.84	13.59	14.41	14.91	13.50	13.71	14.78	15.27	14.18	13.94	15.46	15.48	14.59	14.39	16.28

TABLE 8.6 : β_1 (1/ϕ = .6, 1/α = .8)

μ MODE:	0.15	0.20	0.25	0.30	0.35	0.40	0.45	0.50	0.55	0.60	0.65	0.70	0.75	0.80	0.85
(1)	3.222	3.362	3.530	3.730	3.966	4.247	4.581	4.982	5.462	6.015	6.461	6.369	6.057	5.739	5.446
(2)	5.887	6.190	6.546	6.963	7.446	8.000	8.590	8.808	8.304	7.792	7.631	8.318	9.425	9.484	9.045
(3)	8.590	9.064	9.612	10.24	10.91	11.11	10.30	10.01	10.76	11.78	11.73	11.08	10.74	12.14	12.62
(4)	11.31	11.96	12.69	13.43	13.04	12.33	13.09	14.14	13.93	13.13	13.85	15.11	14.47	13.87	15.71

TABLE 8.6 : β_1 (1/ϕ = .6, 1/α = .9)

μ MODE:	0.15	0.20	0.25	0.30	0.35	0.40	0.45	0.50	0.55	0.60	0.65	0.70	0.75	0.80	0.85
(1)	2.925	3.061	3.219	3.403	3.618	3.872	4.173	4.534	4.967	5.480	6.003	6.160	5.938	5.656	5.387
(2)	5.306	5.580	5.900	6.271	6.702	7.199	7.755	8.233	8.076	7.621	7.334	7.695	8.740	9.290	8.953
(3)	7.710	8.132	8.618	9.175	9.795	10.34	9.994	9.446	9.786	10.73	11.39	10.92	10.47	11.26	12.43
(4)	10.13	10.70	11.35	12.05	12.42	11.62	11.83	12.78	13.49	12.83	12.72	14.17	14.30	13.67	14.76

TABLE 8.6 : β_1 (1/ϕ = .6, 1/α = 1.0)

μ MODE:	0.15	0.20	0.25	0.30	0.35	0.40	0.45	0.50	0.55	0.60	0.65	0.70	0.75	0.80	0.85
(1)	2.671	2.805	2.956	3.130	3.332	3.568	3.846	4.178	4.577	5.054	5.585	5.915	5.804	5.561	5.316
(2)	4.829	5.084	5.378	5.717	6.108	6.562	7.079	7.610	7.778	7.437	7.121	7.255	8.152	9.035	8.846
(3)	6.998	7.383	7.825	8.330	8.900	9.487	9.616	9.105	9.082	9.855	10.81	10.74	10.28	10.61	12.17
(4)	9.177	9.693	10.28	10.93	11.53	11.18	10.92	11.67	12.64	12.57	12.09	13.10	14.05	13.51	13.94

TABLE 8.6 : β_1 (1/ϕ = .7, 1/α = .6)

μ MODE:	0.15	0.20	0.25	0.30	0.35	0.40	0.45	0.50	0.55	0.60	0.65	0.70	0.75	0.80	0.85
(1)	4.711	4.918	5.176	5.489	5.859	6.297	6.810	7.392	7.856	7.603	7.106	6.638	6.221	5.852	5.524
(2)	8.874	9.359	9.928	10.58	11.27	11.35	10.33	9.429	8.945	9.490	10.56	10.90	10.31	9.715	9.172
(3)	13.11	13.87	14.70	15.08	13.39	12.65	13.51	14.56	14.06	13.00	12.18	12.65	14.13	13.58	12.84
(4)	17.37	18.37	18.38	16.28	17.08	18.23	17.25	15.84	16.53	17.80	16.78	15.69	15.41	17.27	16.50

TABLE 8.6 : β_1 (1/ϕ = .7, 1/α = .7)

μ MODE:	0.15	0.20	0.25	0.30	0.35	0.40	0.45	0.50	0.55	0.60	0.65	0.70	0.75	0.80	0.85
(1)	4.177	4.353	4.571	4.833	5.146	5.517	5.955	6.466	7.014	7.268	6.960	6.546	6.156	5.805	5.492
(2)	7.726	8.131	8.608	9.160	9.780	10.36	10.07	9.262	8.622	8.551	9.339	10.32	10.17	9.636	9.121
(3)	11.34	11.97	12.69	13.43	13.03	11.81	11.83	12.76	13.54	12.84	12.00	11.61	12.90	13.42	12.77
(4)	14.98	15.83	16.65	15.47	14.89	15.88	16.67	15.51	14.74	15.94	16.53	15.56	14.77	16.09	16.39

TABLE 8.6 : β_1 (1/ϕ = .7, 1/α = .8)

μ MODE:	0.15	0.20	0.25	0.30	0.35	0.40	0.45	0.50	0.55	0.60	0.65	0.70	0.75	0.80	0.85
(1)	3.759	3.923	4.119	4.351	4.625	4.950	5.334	5.787	6.303	6.757	6.744	6.420	6.068	5.741	5.446
(2)	6.868	7.220	7.634	8.115	8.662	9.250	9.562	9.043	8.424	8.087	8.493	9.464	9.931	9.527	9.051
(3)	10.02	10.57	11.20	11.89	12.31	11.45	10.84	11.39	12.36	12.57	11.84	11.26	11.79	13.09	12.66
(4)	13.20	13.94	14.73	14.84	13.62	14.09	15.11	15.18	14.17	14.33	15.71	15.39	14.57	14.82	16.19

TABLE 8.6 : β_1 $(1/\phi = .7,\ 1/\alpha = .9)$

μ	0.15	0.20	0.25	0.30	0.35	0.40	0.45	0.50	0.55	0.60	0.65	0.70	0.75	0.80	0.85
MODE:															
(1)	3.412	3.571	3.755	3.970	4.220	4.514	4.862	5.272	5.749	6.243	6.467	6.268	5.961	5.661	5.388
(2)	6.190	6.510	6.881	7.310	7.802	8.350	8.845	8.748	8.221	7.814	7.923	8.733	9.568	9.392	8.964
(3)	8.995	9.484	10.04	10.67	11.27	11.06	10.32	10.40	11.24	12.03	11.65	11.05	11.07	12.52	12.54
(4)	11.81	12.47	13.20	13.80	13.00	12.79	13.64	14.48	13.87	13.35	14.42	15.12	14.41	14.10	15.86

TABLE 8.6 : β_1 $(1/\phi = .7,\ 1/\alpha = 1.0)$

μ	0.15	0.20	0.25	0.30	0.35	0.40	0.45	0.50	0.55	0.60	0.65	0.70	0.75	0.80	0.85
MODE:															
(1)	3.117	3.272	3.449	3.652	3.887	4.160	4.482	4.862	5.308	5.801	6.163	6.101	5.844	5.570	5.318
(2)	5.633	5.931	6.273	6.666	7.116	7.625	8.151	8.372	8.003	7.593	7.524	8.153	9.121	9.239	8.867
(3)	8.164	8.612	9.122	9.697	10.30	10.55	9.957	9.706	10.33	11.25	11.41	10.87	10.63	11.85	12.39
(4)	10.71	11.30	11.97	12.64	12.49	11.88	12.47	13.42	13.54	12.86	13.34	14.62	14.24	13.74	15.34

TABLE 8.6 : β_1 $(1/\phi = .8,\ 1/\alpha = .6)$

μ	0.15	0.20	0.25	0.30	0.35	0.40	0.45	0.50	0.55	0.60	0.65	0.70	0.75	0.80	0.85
MODE:															
(1)	5.384	5.620	5.915	6.270	6.689	7.180	7.736	8.260	8.182	7.649	7.116	6.641	6.222	5.852	5.524
(2)	10.14	10.69	11.33	12.04	12.52	11.54	10.41	9.639	9.793	10.70	11.46	10.99	10.33	9.717	9.173
(3)	14.98	15.83	16.66	15.44	13.81	14.24	15.30	15.33	14.13	13.09	12.76	14.14	14.38	13.60	12.84
(4)	19.85	20.85	18.75	18.24	19.41	19.26	17.41	17.16	18.60	18.10	16.83	15.87	17.08	17.44	16.51

TABLE 8.6 : β_1 $(1/\phi = .8,\ 1/\alpha = .7)$

μ	0.15	0.20	0.25	0.30	0.35	0.40	0.45	0.50	0.55	0.60	0.65	0.70	0.75	0.80	0.85
MODE:															
(1)	4.774	4.975	5.223	5.522	5.876	6.293	6.776	7.304	7.667	7.442	6.992	6.554	6.158	5.806	5.492
(2)	8.829	9.289	9.829	10.44	11.07	11.10	10.20	9.380	9.003	9.503	10.45	10.73	10.21	9.642	9.122
(3)	12.96	13.67	14.45	14.74	13.27	12.62	13.36	14.30	13.90	12.92	12.19	12.66	13.93	13.48	12.77
(4)	17.11	18.04	18.03	16.20	16.85	17.90	17.10	15.81	16.37	17.53	16.67	15.65	15.48	17.08	16.41

TABLE 8.6 : β_1 $(1/\phi = .8,\ 1/\alpha = .8)$

μ	0.15	0.20	0.25	0.30	0.35	0.40	0.45	0.50	0.55	0.60	0.65	0.70	0.75	0.80	0.85
MODE:															
(1)	4.296	4.483	4.707	4.971	5.283	5.649	6.077	6.562	7.029	7.138	6.827	6.439	6.073	5.742	5.447
(2)	7.849	8.250	8.719	9.257	9.844	10.30	9.901	9.160	8.633	8.726	9.500	10.27	10.05	9.543	9.053
(3)	11.45	12.07	12.77	13.40	12.84	11.81	11.98	12.86	13.42	12.72	11.96	11.78	13.03	13.30	12.68
(4)	15.08	15.90	16.59	15.33	15.03	15.97	16.53	15.41	14.89	16.05	16.40	15.48	14.84	16.22	16.28

TABLE 8.6 : β_1 $(1/\phi = .8,\ 1/\alpha = .9)$

μ	0.15	0.20	0.25	0.30	0.35	0.40	0.45	0.50	0.55	0.60	0.65	0.70	0.75	0.80	0.85
MODE:															
(1)	3.899	4.081	4.291	4.536	4.821	5.154	5.543	5.991	6.466	6.770	6.631	6.305	5.971	5.663	5.388
(2)	7.074	7.439	7.860	8.343	8.883	9.414	9.489	8.917	8.374	8.226	8.780	9.667	9.843	9.425	8.969
(3)	10.28	10.83	11.46	12.11	12.25	11.35	11.03	11.69	12.56	12.46	11.76	11.32	12.14	13.04	12.57
(4)	13.50	14.24	14.99	14.71	13.79	14.42	15.35	15.06	14.18	14.68	15.83	15.29	14.56	15.22	16.10

TABLE 8.6 : β_1 $(1/\phi = .8,\ 1/\alpha = 1.0)$

μ	0.15	0.20	0.25	0.30	0.35	0.40	0.45	0.50	0.55	0.60	0.65	0.70	0.75	0.80	0.85
MODE:															
(1)	3.562	3.739	3.941	4.173	4.441	4.751	5.114	5.534	6.000	6.398	6.423	6.164	5.860	5.574	5.319
(2)	6.438	6.778	7.167	7.611	8.112	8.642	8.981	8.652	8.140	7.871	8.222	9.082	9.602	9.299	8.876
(3)	9.330	9.839	10.41	11.04	11.49	10.96	10.39	10.77	11.63	12.10	11.56	11.04	11.43	12.70	12.45
(4)	12.23	12.91	13.63	13.95	13.01	13.20	14.10	14.59	13.80	13.65	14.88	15.06	14.35	14.45	15.88

TABLE 8.6 : β_1 $(1/\phi = .9,\ 1/\alpha = .6)$

μ	0.15	0.20	0.25	0.30	0.35	0.40	0.45	0.50	0.55	0.60	0.65	0.70	0.75	0.80	0.85
MODE:															
(1)	6.057	6.321	6.652	7.048	7.513	8.045	8.595	8.779	8.270	7.666	7.121	6.642	6.222	5.852	5.524
(2)	11.41	12.02	12.72	13.42	12.99	11.62	10.54	10.19	10.85	11.78	11.71	11.01	10.33	9.718	9.173
(3)	16.84	17.76	17.99	15.65	14.98	15.91	16.64	15.46	14.19	13.33	13.96	15.08	14.44	13.60	12.84
(4)	22.29	22.60	19.61	20.41	21.36	19.43	18.00	19.09	19.56	18.16	16.91	16.62	18.22	17.47	16.51

TABLE 8.6 : β_1 $(1/\phi = .9,\ 1/\alpha = .7)$

μ	0.15	0.20	0.25	0.30	0.35	0.40	0.45	0.50	0.55	0.60	0.65	0.70	0.75	0.80	0.85
MODE:															
(1)	5.370	5.596	5.875	6.209	6.602	7.058	7.566	8.012	7.961	7.499	7.006	6.557	6.159	5.806	5.492
(2)	9.932	10.45	11.04	11.69	12.11	11.32	10.30	9.615	9.727	10.52	11.20	10.84	10.23	9.646	9.122
(3)	14.57	15.36	16.10	15.17	13.68	13.94	14.88	15.07	14.00	13.03	12.72	13.88	14.21	13.50	12.78
(4)	19.23	20.14	18.49	17.80	18.83	18.95	17.29	16.86	18.12	17.92	16.74	15.85	16.81	17.28	16.42

TABLE 8.6 : β_1 $(1/\phi = .9,\ 1/\alpha = .8)$

μ	0.15	0.20	0.25	0.30	0.35	0.40	0.45	0.50	0.55	0.60	0.65	0.70	0.75	0.80	0.85
MODE:															
(1)	4.833	5.043	5.294	5.590	5.937	6.341	6.800	7.272	7.522	7.278	6.859	6.448	6.075	5.743	5.447
(2)	8.829	9.278	9.799	10.38	10.94	10.87	10.04	9.310	9.064	9.577	10.41	10.58	10.09	9.550	9.054
(3)	12.88	13.57	14.29	14.44	13.11	12.62	13.32	14.14	13.73	12.82	12.20	12.75	13.79	13.36	12.69
(4)	16.95	17.82	17.68	16.13	16.74	17.67	16.91	15.77	16.33	17.33	16.54	15.59	15.59	16.94	16.31

TABLE 8.6 : β_1 $(1/\phi = .9,\ 1/\alpha = .9)$

μ	0.15	0.20	0.25	0.30	0.35	0.40	0.45	0.50	0.55	0.60	0.65	0.70	0.75	0.80	0.85
MODE:															
(1)	4.387	4.591	4.827	5.102	5.420	5.789	6.212	6.672	7.046	7.021	6.693	6.322	5.975	5.665	5.388
(2)	7.958	8.367	8.837	9.366	9.922	10.23	9.730	9.045	8.646	8.896	9.669	10.23	9.929	9.440	8.972
(3)	11.56	12.18	12.85	13.38	12.64	11.80	12.13	12.97	13.31	12.60	11.91	11.95	13.15	13.18	12.58
(4)	15.18	15.98	16.51	15.20	15.16	16.07	16.39	15.30	15.03	16.18	16.27	15.39	14.92	16.34	16.17

TABLE 8.6 : β_1 $(1/\phi = .9,\ 1/\alpha = 1.0)$

μ	0.15	0.20	0.25	0.30	0.35	0.40	0.45	0.50	0.55	0.60	0.65	0.70	0.75	0.80	0.85
MODE:															
(1)	4.007	4.207	4.434	4.694	4.993	5.339	5.737	6.182	6.608	6.734	6.524	6.191	5.860	5.576	5.319
(2)	7.243	7.624	8.059	8.550	9.084	9.542	9.391	8.790	8.329	8.376	9.041	9.815	9.761	9.324	8.881
(3)	10.49	11.06	11.69	12.30	12.15	11.27	11.22	11.95	12.70	12.35	11.67	11.42	12.45	12.99	12.47
(4)	13.76	14.50	15.19	14.56	13.96	14.70	15.53	14.95	14.22	14.99	15.86	15.19	14.55	15.58	16.02

TABLE 8.6 : β_1 $(1/\phi = 1.0,\ 1/\alpha = .6)$

μ	0.15	0.20	0.25	0.30	0.35	0.40	0.45	0.50	0.55	0.60	0.65	0.70	0.75	0.80	0.85
MODE:															
(1)	6.730	7.023	7.388	7.823	8.327	8.877	9.289	8.969	8.304	7.675	7.123	6.643	6.222	5.852	5.524
(2)	12.67	13.35	14.09	14.50	13.13	11.71	10.83	11.05	11.91	12.41	11.78	11.02	10.33	9.719	9.173
(3)	18.70	19.63	18.38	16.12	16.48	17.49	17.02	15.52	14.31	14.00	15.21	15.33	14.45	13.61	12.84
(4)	24.71	23.12	21.39	22.56	21.96	19.63	19.52	20.78	19.71	18.22	17.11	17.98	18.49	17.49	16.52

TABLE 8.6 : β_1 $(1/\phi = 1.0,\ 1/\alpha = .7)$

μ	0.15	0.20	0.25	0.30	0.35	0.40	0.45	0.50	0.55	0.60	0.65	0.70	0.75	0.80	0.85
MODE:															
(1)	5.967	6.218	6.526	6.893	7.322	7.806	8.292	8.472	8.070	7.524	7.013	6.560	6.160	5.806	5.492
(2)	11.03	11.60	12.24	12.86	12.63	11.43	10.45	10.09	10.61	11.43	11.49	10.88	10.23	9.647	9.123
(3)	16.18	17.02	17.35	15.41	14.60	15.35	16.11	15.26	14.08	13.26	13.68	14.75	14.29	13.51	12.78
(4)	21.34	21.80	19.17	19.63	20.59	19.21	17.73	18.44	19.20	18.02	16.83	16.44	17.86	17.34	16.43

TABLE 8.6 : β_1 ($1/\phi = 1.0$, $1/\alpha = .8$)

μ MODE:	0.15	0.20	0.25	0.30	0.35	0.40	0.45	0.50	0.55	0.60	0.65	0.70	0.75	0.80	0.85
(1)	5.370	5.603	5.881	6.208	6.588	7.022	7.489	7.854	7.761	7.335	6.875	6.453	6.077	5.743	5.447
(2)	9.809	10.30	10.87	11.47	11.81	11.09	10.15	9.576	9.732	10.45	11.02	10.68	10.11	9.554	9.055
(3)	14.30	15.05	15.71	14.86	13.55	13.77	14.62	14.81	13.84	12.95	12.73	13.78	14.04	13.38	12.69
(4)	18.81	19.63	18.17	17.52	18.46	18.63	17.13	16.70	17.82	17.72	16.62	15.83	16.73	17.13	16.32

TABLE 8.6 : β_1 ($1/\phi = 1.0$, $1/\alpha = .9$)

μ MODE:	0.15	0.20	0.25	0.30	0.35	0.40	0.45	0.50	0.55	0.60	0.65	0.70	0.75	0.80	0.85
(1)	4.874	5.101	5.363	5.666	6.016	6.417	6.860	7.280	7.413	7.127	6.722	6.331	5.978	5.665	5.389
(2)	8.842	9.293	9.808	10.37	10.86	10.66	9.861	9.223	9.119	9.679	10.43	10.45	9.966	9.447	8.973
(3)	12.84	13.52	14.20	14.18	12.94	12.63	13.32	14.04	13.56	12.70	12.19	12.87	13.71	13.24	12.59
(4)	16.86	17.68	17.36	16.05	16.69	17.53	16.71	15.71	16.34	17.19	16.40	15.52	15.72	16.84	16.20

TABLE 8.7 : β_1 ($\phi = .6$, $\alpha = .6$)

μ MODE:	0.15	0.20	0.25	0.30	0.35	0.40	0.45	0.50	0.55	0.60	0.65	0.70	0.75	0.80	0.85
(1)	4.586	4.858	5.164	5.510	5.899	6.319	6.601	6.283	5.795	5.356	4.979	4.658	4.386	4.160	3.979
(2)	8.255	8.743	9.276	9.733	9.078	8.106	7.545	7.853	8.589	9.474	9.607	9.032	8.475	7.984	7.561
(3)	11.92	12.60	12.57	10.93	10.90	11.68	12.61	12.57	11.57	10.78	11.16	12.63	12.57	11.86	11.20
(4)	15.57	15.71	13.82	14.53	15.54	15.71	14.22	14.14	15.48	15.71	14.63	13.83	15.34	15.71	14.86

TABLE 8.7 : β_1 ($\phi = .6$, $\alpha = .7$)

μ MODE:	0.15	0.20	0.25	0.30	0.35	0.40	0.45	0.50	0.55	0.60	0.65	0.70	0.75	0.80	0.85
(1)	5.318	5.621	5.960	6.337	6.741	7.074	6.917	6.398	5.892	5.454	5.082	4.768	4.504	4.286	4.111
(2)	9.576	10.11	10.64	10.47	9.315	8.462	8.406	8.989	9.781	10.23	9.723	9.109	8.556	8.075	7.666
(3)	13.82	14.42	13.03	11.98	12.57	13.45	13.81	12.76	11.78	11.55	12.67	13.33	12.65	11.93	11.29
(4)	18.00	16.41	15.87	16.82	17.56	16.03	15.15	16.23	16.98	15.87	14.83	15.07	16.52	15.78	14.93

TABLE 8.7 : β_1 ($\phi = .6$, $\alpha = .8$)

μ MODE:	0.15	0.20	0.25	0.30	0.35	0.40	0.45	0.50	0.55	0.60	0.65	0.70	0.75	0.80	0.85
(1)	6.026	6.351	6.710	7.099	7.466	7.550	7.097	6.525	6.013	5.576	5.207	4.895	4.632	4.413	4.233
(2)	10.86	11.41	11.77	10.84	9.641	9.073	9.361	10.03	10.68	10.47	9.828	9.207	8.658	8.184	7.782
(3)	15.65	15.71	13.64	13.41	14.17	14.90	14.16	12.98	12.23	12.73	13.79	13.50	12.74	12.02	11.39
(4)	20.12	17.54	17.96	18.94	18.24	16.55	16.89	18.00	17.30	16.06	15.35	16.56	16.76	15.87	15.02

TABLE 8.7 : β_1 ($\phi = .6$, $\alpha = .9$)

μ MODE:	0.15	0.20	0.25	0.30	0.35	0.40	0.45	0.50	0.55	0.60	0.65	0.70	0.75	0.80	0.85
(1)	6.704	7.039	7.408	7.786	8.054	7.854	7.267	6.671	6.152	5.712	5.339	5.023	4.755	4.527	4.333
(2)	12.09	12.63	12.57	11.19	10.08	9.817	10.26	10.92	11.18	10.63	9.945	9.319	8.770	8.297	7.893
(3)	17.39	16.48	14.57	14.85	15.64	15.71	14.42	13.29	12.99	13.83	14.31	13.63	12.84	12.12	11.50
(4)	21.60	19.22	19.99	20.58	18.68	17.67	18.60	18.85	17.52	16.35	16.33	17.48	16.89	15.97	15.12

TABLE 8.7 : β_1 ($\phi = .6$, $\alpha = 1.0$)

μ MODE:	0.15	0.20	0.25	0.30	0.35	0.40	0.45	0.50	0.55	0.60	0.65	0.70	0.75	0.80	0.85
(1)	7.346	7.686	8.054	8.403	8.526	8.104	7.444	6.827	6.297	5.848	5.466	5.141	4.861	4.619	4.409
(2)	13.27	13.75	13.14	11.57	10.59	10.56	11.08	11.62	11.47	10.78	10.07	9.436	8.883	8.403	7.989
(3)	19.01	17.16	15.71	16.23	16.90	16.13	14.70	13.72	13.86	14.70	14.57	13.76	12.95	12.23	11.60
(4)	22.68	21.03	21.90	21.46	19.24	19.11	20.02	19.21	17.76	16.79	17.39	17.88	17.01	16.07	15.22

TABLE 8.7 : β_1 ($\phi = .7$, $\alpha = .6$)

μ MODE:	0.15	0.20	0.25	0.30	0.35	0.40	0.45	0.50	0.55	0.60	0.65	0.70	0.75	0.80	0.85
(1)	3.930	4.164	4.427	4.725	5.063	5.446	5.850	6.051	5.743	5.338	4.971	4.655	4.385	4.160	3.979
(2)	7.076	7.496	7.963	8.464	8.726	8.011	7.282	6.986	7.439	8.232	9.101	8.976	8.461	7.980	7.560
(3)	10.22	10.82	11.39	10.69	9.687	10.08	10.89	11.77	11.48	10.65	10.12	11.04	12.34	11.83	11.20
(4)	13.36	14.06	12.96	12.54	13.37	14.35	14.06	12.93	13.40	14.79	14.56	13.64	13.51	15.48	14.85

TABLE 8.7 : β_1 ($\phi = .7$, $\alpha = .7$)

μ MODE:	0.15	0.20	0.25	0.30	0.35	0.40	0.45	0.50	0.55	0.60	0.65	0.70	0.75	0.80	0.85
(1)	4.558	4.818	5.110	5.438	5.804	6.191	6.468	6.283	5.855	5.439	5.076	4.766	4.503	4.286	4.111
(2)	8.209	8.676	9.179	9.601	9.144	8.254	7.696	7.853	8.497	9.304	9.563	9.074	8.545	8.071	7.665
(3)	11.86	12.49	12.57	11.14	10.92	11.61	12.47	12.57	11.66	10.92	11.13	12.42	12.57	11.91	11.28
(4)	15.49	15.71	13.96	14.49	15.43	15.71	14.37	14.14	15.34	15.71	14.71	13.96	15.14	15.71	14.92

TABLE 8.7 : β_1 ($\phi = .7$, $\alpha = .8$)

μ MODE:	0.15	0.20	0.25	0.30	0.35	0.40	0.45	0.50	0.55	0.60	0.65	0.70	0.75	0.80	0.85
(1)	5.166	5.444	5.756	6.102	6.474	6.807	6.834	6.447	5.984	5.564	5.202	4.893	4.632	4.413	4.233
(2)	9.309	9.807	10.30	10.38	9.450	8.603	8.334	8.740	9.437	10.01	9.739	9.180	8.649	8.181	7.781
(3)	13.45	14.03	13.17	11.94	12.27	13.05	13.61	12.84	11.90	11.47	12.28	13.17	12.69	12.01	11.39
(4)	17.52	16.57	15.58	16.38	17.20	16.14	15.06	15.80	16.76	15.94	14.94	14.79	16.26	15.82	15.01

TABLE 8.7 : β_1 ($\phi = .7$, $\alpha = .9$)

μ MODE:	0.15	0.20	0.25	0.30	0.35	0.40	0.45	0.50	0.55	0.60	0.65	0.70	0.75	0.80	0.85
(1)	5.746	6.036	6.360	6.714	7.069	7.291	7.088	6.611	6.129	5.702	5.335	5.022	4.754	4.526	4.333
(2)	10.37	10.88	11.31	10.88	9.786	9.053	9.040	9.556	10.21	10.38	9.884	9.299	8.764	8.295	7.892
(3)	14.98	15.34	13.75	13.01	13.57	14.33	14.16	13.09	12.24	12.28	13.25	13.46	12.80	12.11	11.49
(4)	19.41	17.47	17.28	18.17	18.20	16.58	16.30	17.33	17.30	16.16	15.29	15.89	16.67	15.93	15.11

TABLE 8.7 : β_1 ($\phi = .7$, $\alpha = 1.0$)

μ MODE:	0.15	0.20	0.25	0.30	0.35	0.40	0.45	0.50	0.55	0.60	0.65	0.70	0.75	0.80	0.85
(1)	6.298	6.593	6.924	7.278	7.601	7.681	7.312	6.781	6.279	5.840	5.463	5.140	4.860	4.619	4.409
(2)	11.39	11.91	12.18	11.29	10.15	9.558	9.723	10.29	10.81	10.62	10.03	9.422	8.878	8.402	7.988
(3)	16.46	16.36	14.41	14.13	14.80	15.33	14.49	13.36	12.69	13.11	13.94	13.65	12.92	12.22	11.59
(4)	21.07	18.62	18.96	19.78	18.73	17.24	17.66	18.48	17.58	16.40	15.83	16.84	16.89	16.05	15.22

TABLE 8.7 : β_1 ($\phi = .8$, $\alpha = .6$)

μ MODE:	0.15	0.20	0.25	0.30	0.35	0.40	0.45	0.50	0.55	0.60	0.65	0.70	0.75	0.80	0.85
(1)	3.439	3.643	3.874	4.135	4.433	4.774	5.159	5.542	5.613	5.303	4.959	4.650	4.383	4.159	3.979
(2)	6.192	6.560	6.972	7.430	7.892	7.854	7.194	6.664	6.658	7.265	8.147	8.795	8.434	7.972	7.558
(3)	8.945	9.472	10.04	10.32	9.326	8.966	9.575	10.43	11.15	10.59	9.909	9.892	11.38	11.78	11.19
(4)	11.70	12.36	12.57	11.29	11.75	12.63	13.49	12.75	12.07	13.10	14.29	13.58	12.88	14.30	14.81

TABLE 8.7 : β_1 ($\phi = .8$, $\alpha = .7$)

μ MODE:	0.15	0.20	0.25	0.30	0.35	0.40	0.45	0.50	0.55	0.60	0.65	0.70	0.75	0.80	0.85
(1)	3.989	4.216	4.472	4.761	5.087	5.451	5.822	6.013	5.782	5.414	5.066	4.761	4.502	4.285	4.111
(2)	7.184	7.594	8.049	8.521	8.744	8.133	7.461	7.176	7.542	8.270	9.052	8.997	8.526	8.066	7.664
(3)	10.38	10.96	11.48	10.86	9.948	10.25	11.01	11.82	11.55	10.78	10.32	11.12	12.29	11.88	11.28
(4)	13.57	14.22	13.23	12.80	13.58	14.49	14.16	13.15	13.60	14.87	14.62	13.77	13.69	15.43	14.90

TABLE 8.7 : β_1 (ϕ = .8, α = .8)

μ / MODE:	0.15	0.20	0.25	0.30	0.35	0.40	0.45	0.50	0.55	0.60	0.65	0.70	0.75	0.80	0.85
(1)	4.320	4.764	5.039	5.346	5.689	6.050	6.338	6.283	5.931	5.544	5.194	4.890	4.630	4.413	4.233
(2)	8.147	8.590	9.066	9.476	9.221	8.424	7.850	7.853	8.383	9.121	9.506	9.131	8.634	8.176	7.780
(3)	11.77	12.38	12.57	11.36	10.94	11.52	12.32	12.57	11.78	11.06	11.08	12.20	12.57	11.98	11.38
(4)	15.38	15.71	14.13	14.43	15.31	15.71	14.54	14.14	15.18	15.71	14.82	14.09	14.93	15.71	15.00

TABLE 8.7 : β_1 (ϕ = .8, α = .9)

μ / MODE:	0.15	0.20	0.25	0.30	0.35	0.40	0.45	0.50	0.55	0.60	0.65	0.70	0.75	0.80	0.85
(1)	5.028	5.283	5.569	5.889	6.237	6.572	6.729	6.497	6.089	5.687	5.329	5.019	4.753	4.526	4.333
(2)	9.078	9.544	10.02	10.25	9.589	8.758	8.330	8.538	9.138	9.772	9.750	9.264	8.753	8.292	7.891
(3)	13.13	13.71	13.29	12.00	12.03	12.73	13.38	12.92	12.03	11.47	11.95	12.94	12.73	12.10	11.49
(4)	17.14	16.72	15.39	16.03	16.86	16.25	15.08	15.45	16.48	16.02	15.05	14.64	15.92	15.87	15.10

TABLE 8.7 : β_1 (ϕ = .8, α = 1.0)

μ / MODE:	0.15	0.20	0.25	0.30	0.35	0.40	0.45	0.50	0.55	0.60	0.65	0.70	0.75	0.80	0.85
(1)	5.511	5.771	6.066	6.395	6.742	7.036	7.048	6.697	6.249	5.829	5.459	5.138	4.860	4.619	4.409
(2)	9.977	10.46	10.92	10.86	9.936	9.114	8.839	9.181	9.804	10.24	9.939	9.397	8.870	8.400	7.988
(3)	14.45	14.95	13.89	12.78	13.11	13.84	14.09	13.20	12.31	12.00	12.76	13.37	12.88	12.21	11.59
(4)	18.82	17.57	16.77	17.56	18.04	16.68	15.92	16.75	17.24	16.26	15.32	15.39	16.51	16.00	15.21

TABLE 8.7 : β_1 (ϕ = .9, α = .6)

μ / MODE:	0.15	0.20	0.25	0.30	0.35	0.40	0.45	0.50	0.55	0.60	0.65	0.70	0.75	0.80	0.85
(1)	3.057	3.239	3.443	3.676	3.941	4.247	4.599	4.991	5.318	5.236	4.938	4.642	4.380	4.158	3.978
(2)	5.504	5.831	6.199	6.613	7.066	7.431	7.107	6.558	6.241	6.544	7.305	8.256	8.378	7.960	7.555
(3)	7.951	8.423	8.944	9.456	9.158	8.394	8.584	9.313	10.21	10.47	9.843	9.384	10.27	11.63	11.18
(4)	10.40	11.01	11.59	10.87	10.54	11.27	12.18	12.57	11.71	11.78	13.17	13.51	12.77	12.99	14.73

TABLE 8.7 : β_1 (ϕ = .9, α = .7)

μ / MODE:	0.15	0.20	0.25	0.30	0.35	0.40	0.45	0.50	0.55	0.60	0.65	0.70	0.75	0.80	0.85
(1)	3.546	3.748	3.976	4.233	4.526	4.858	5.224	5.567	5.633	5.370	5.050	4.755	4.499	4.284	4.111
(2)	6.386	6.752	7.161	7.608	8.028	7.951	7.351	6.877	6.881	7.436	8.257	8.803	8.491	8.057	7.661
(3)	9.227	9.752	10.30	10.47	9.573	9.284	9.856	10.67	11.24	10.69	10.08	10.15	11.50	11.82	11.27
(4)	12.07	12.72	12.79	11.70	12.13	12.99	13.69	12.91	12.41	13.44	14.38	13.69	13.08	14.48	14.86

TABLE 8.7 : β_1 (ϕ = .9, α = .8)

μ / MODE:	0.15	0.20	0.25	0.30	0.35	0.40	0.45	0.50	0.55	0.60	0.65	0.70	0.75	0.80	0.85
(1)	4.018	4.235	4.480	4.755	5.066	5.411	5.761	5.977	5.836	5.512	5.181	4.885	4.629	4.412	4.233
(2)	7.242	7.639	8.077	8.528	8.768	8.282	7.651	7.333	7.589	8.241	8.975	9.030	8.610	8.169	7.778
(3)	10.47	11.03	11.53	11.06	10.18	10.36	11.06	11.82	11.64	10.92	10.47	11.12	12.23	11.94	11.37
(4)	13.69	14.31	13.52	12.98	13.69	14.55	14.30	13.36	13.70	14.87	14.71	13.91	13.78	15.36	14.97

TABLE 8.7 : β_1 (ϕ = .9, α = .9)

μ / MODE:	0.15	0.20	0.25	0.30	0.35	0.40	0.45	0.50	0.55	0.60	0.65	0.70	0.75	0.80	0.85
(1)	4.470	4.696	4.952	5.241	5.564	5.910	6.218	6.283	6.020	5.662	5.319	5.016	4.752	4.526	4.333
(2)	8.071	8.492	8.949	9.367	9.299	8.599	7.998	7.853	8.258	8.939	9.441	9.200	8.735	8.287	7.890
(3)	11.68	12.26	12.57	11.57	10.98	11.42	12.17	12.57	11.91	11.19	11.04	11.99	12.57	12.07	11.48
(4)	15.28	15.71	14.32	14.36	15.18	15.71	14.70	14.14	15.02	15.71	14.94	14.19	14.71	15.71	15.09

TABLE 8.7 : β_1 (ϕ = .9, α = 1.0)

μ / MODE:	0.15	0.20	0.25	0.30	0.35	0.40	0.45	0.50	0.55	0.60	0.65	0.70	0.75	0.80	0.85
(1)	4.899	5.131	5.396	5.696	6.029	6.371	6.619	6.542	6.198	5.811	5.452	5.135	4.859	4.619	4.409
(2)	8.872	9.316	9.786	10.12	9.721	8.911	8.363	8.375	8.875	9.536	9.752	9.354	8.858	8.396	7.987
(3)	12.86	13.45	13.39	12.11	11.84	12.44	13.16	13.00	12.16	11.51	11.69	12.69	12.78	12.19	11.59
(4)	16.83	16.83	15.30	15.73	16.57	16.36	15.15	15.17	16.20	16.10	15.16	14.57	15.57	15.91	15.20

TABLE 8.7 : β_1 (ϕ = 1.0, α = .6)

μ / MODE:	0.15	0.20	0.25	0.30	0.35	0.40	0.45	0.50	0.55	0.60	0.65	0.70	0.75	0.80	0.85
(1)	2.751	2.915	3.099	3.308	3.548	3.824	4.145	4.515	4.907	5.097	4.904	4.631	4.376	4.157	3.978
(2)	4.954	5.248	5.580	5.955	6.376	6.813	6.940	6.493	6.077	6.048	6.631	7.553	8.241	7.941	7.551
(3)	7.156	7.582	8.057	8.568	8.850	8.195	7.889	8.426	9.251	10.08	9.787	9.247	9.424	11.11	11.15
(4)	9.359	9.911	10.50	10.59	9.760	10.18	11.01	11.91	11.59	11.01	11.96	13.26	12.72	12.28	14.45

TABLE 8.7 : β_1 (ϕ = 1.0, α = .7)

μ / MODE:	0.15	0.20	0.25	0.30	0.35	0.40	0.45	0.50	0.55	0.60	0.65	0.70	0.75	0.80	0.85
(1)	3.191	3.373	3.578	3.810	4.075	4.378	4.722	5.091	5.367	5.294	5.025	4.746	4.496	4.283	4.110
(2)	5.747	6.077	6.448	6.860	7.297	7.585	7.248	6.747	6.494	6.796	7.520	8.367	8.429	8.043	7.658
(3)	8.305	8.781	9.300	9.755	9.370	8.723	8.956	9.672	10.51	10.58	9.985	9.638	10.56	11.67	11.25
(4)	10.86	11.47	11.98	11.20	11.02	11.74	12.63	12.73	11.94	12.25	13.56	13.61	12.92	13.37	14.77

TABLE 8.7 : β_1 (ϕ = 1.0, α = .8)

μ / MODE:	0.15	0.20	0.25	0.30	0.35	0.40	0.45	0.50	0.55	0.60	0.65	0.70	0.75	0.80	0.85
(1)	3.616	3.812	4.032	4.281	4.564	4.884	5.234	5.562	5.666	5.459	5.162	4.878	4.626	4.411	4.233
(2)	6.518	6.877	7.278	7.712	8.111	8.073	7.534	7.080	7.037	7.518	8.279	8.815	8.570	8.159	7.776
(3)	9.425	9.942	10.47	10.64	9.835	9.523	10.04	10.81	11.33	10.83	10.25	10.31	11.52	11.87	11.36
(4)	12.33	12.97	13.04	12.03	12.40	13.23	13.84	13.10	12.67	13.64	14.46	13.82	13.27	14.52	14.92

TABLE 8.7 : β_1 (ϕ = 1.0, α = .9)

μ / MODE:	0.15	0.20	0.25	0.30	0.35	0.40	0.45	0.50	0.55	0.60	0.65	0.70	0.75	0.80	0.85
(1)	4.023	4.227	4.458	4.720	5.017	5.348	5.693	5.949	5.903	5.623	5.305	5.010	4.750	4.525	4.333
(2)	7.264	7.647	8.072	8.515	8.801	8.443	7.834	7.460	7.596	8.173	8.892	9.077	8.708	8.280	7.889
(3)	10.52	11.07	11.57	11.26	10.38	10.42	11.06	11.80	11.75	11.07	10.60	11.06	12.17	12.02	11.48
(4)	13.77	14.39	13.80	13.12	13.75	14.59	14.46	13.54	13.74	14.86	14.82	14.04	13.82	15.29	15.06

TABLE 8.8 : β_1 (ϕ = .6, α = .6)

μ / MODE:	0.15	0.20	0.25	0.30	0.35	0.40	0.45	0.50	0.55	0.60	0.65	0.70	0.75	0.80	0.85
(1)	5.534	5.867	6.242	6.667	7.150	7.688	8.155	7.853	7.236	6.680	6.204	5.799	5.455	5.167	4.931
(2)	9.188	9.740	10.36	11.02	11.14	10.02	9.193	9.460	10.35	11.29	10.89	10.18	9.545	8.989	8.511
(3)	12.86	13.63	14.38	13.32	12.27	13.02	14.06	14.14	13.01	12.27	13.34	14.44	13.67	12.87	12.15
(4)	16.54	17.45	16.09	15.71	16.79	17.52	15.94	15.71	17.19	17.06	15.88	15.75	17.61	16.75	15.82

TABLE 8.8 : β_1 (ϕ = .6, α = .7)

μ / MODE:	0.15	0.20	0.25	0.30	0.35	0.40	0.45	0.50	0.55	0.60	0.65	0.70	0.75	0.80	0.85
(1)	6.427	6.802	7.223	7.695	8.215	8.710	8.643	7.988	7.344	6.789	6.318	5.920	5.583	5.301	5.070
(2)	10.67	11.29	11.96	12.45	11.47	10.34	10.13	10.84	11.75	11.74	10.99	10.27	9.636	9.090	8.626
(3)	14.94	15.77	15.77	13.88	14.04	15.01	15.53	14.35	13.31	13.65	14.99	14.62	13.75	12.95	12.25
(4)	19.19	19.56	17.36	18.18	19.24	17.94	16.84	18.03	18.53	17.23	16.32	17.79	17.83	16.82	15.90

TABLE 8.8 : β_1 ($\phi = .6, \alpha = .8$)

μ MODE:	0.15	0.20	0.25	0.30	0.35	0.40	0.45	0.50	0.55	0.60	0.65	0.70	0.75	0.80	0.85
(1)	7.298	7.705	8.160	8.661	9.171	9.408	8.864	8.130	7.477	6.922	6.453	6.055	5.719	5.434	5.195
(2)	12.12	12.79	13.45	13.21	11.80	10.96	11.28	12.09	12.56	11.90	11.10	10.37	9.746	9.206	8.746
(3)	16.97	17.74	16.36	15.11	15.83	16.71	15.92	14.61	14.11	15.15	15.61	14.74	13.84	13.05	12.36
(4)	21.76	20.49	19.48	20.54	20.38	18.45	18.77	19.89	18.79	17.50	17.66	18.85	17.93	16.91	16.00

TABLE 8.8 : β_1 ($\phi = .6, \alpha = .9$)

μ MODE:	0.15	0.20	0.25	0.30	0.35	0.40	0.45	0.50	0.55	0.60	0.65	0.70	0.75	0.80	0.85
(1)	8.139	8.569	9.046	9.553	9.981	9.817	9.056	8.289	7.626	7.066	6.593	6.189	5.845	5.549	5.295
(2)	13.53	14.22	14.72	13.66	12.24	11.83	12.39	13.10	12.89	12.04	11.22	10.49	9.865	9.323	8.857
(3)	18.93	19.34	17.04	16.65	17.51	17.67	16.21	15.05	15.29	16.28	15.82	14.85	13.95	13.16	12.47
(4)	24.14	21.54	21.66	22.59	20.86	19.63	20.67	20.57	19.04	18.07	19.11	19.13	18.04	17.01	16.10

TABLE 8.8 : β_1 ($\phi = .6, \alpha = 1.0$)

μ MODE:	0.15	0.20	0.25	0.30	0.35	0.40	0.45	0.50	0.55	0.60	0.65	0.70	0.75	0.80	0.85
(1)	8.946	9.391	9.879	10.37	10.63	10.11	9.249	8.455	7.779	7.209	6.724	6.309	5.951	5.641	5.369
(2)	14.89	15.59	15.71	14.08	12.79	12.74	13.39	13.78	13.11	12.18	11.35	10.61	9.979	9.428	8.949
(3)	20.83	20.43	17.97	18.18	18.97	18.13	16.54	15.74	16.44	16.87	15.98	14.97	14.06	13.27	12.57
(4)	26.18	23.08	23.77	23.86	21.42	21.25	22.15	20.88	19.35	19.05	20.07	19.29	18.16	17.12	16.20

TABLE 8.8 : β_1 ($\phi = .7, \alpha = .6$)

μ MODE:	0.15	0.20	0.25	0.30	0.35	0.40	0.45	0.50	0.55	0.60	0.65	0.70	0.75	0.80	0.85
(1)	4.743	5.029	5.350	5.716	6.133	6.611	7.139	7.518	7.176	6.661	6.197	5.796	5.454	5.166	4.931
(2)	7.876	8.349	8.882	9.476	10.08	9.871	8.975	8.465	8.962	9.920	10.67	10.15	9.536	8.986	8.510
(3)	11.03	11.69	12.41	12.85	11.57	11.28	12.13	13.16	12.91	11.98	11.71	13.19	13.61	12.85	12.15
(4)	14.18	15.01	15.51	13.87	14.45	15.53	15.74	14.45	14.89	16.39	15.80	14.85	15.94	16.70	15.81

TABLE 8.8 : β_1 ($\phi = .7, \alpha = .7$)

μ MODE:	0.15	0.20	0.25	0.30	0.35	0.40	0.45	0.50	0.55	0.60	0.65	0.70	0.75	0.80	0.85
(1)	5.509	5.830	6.192	6.600	7.061	7.567	8.006	7.853	7.305	6.774	6.312	5.917	5.582	5.301	5.070
(2)	9.149	9.683	10.28	10.90	11.09	10.15	9.359	9.460	10.25	11.12	10.91	10.25	9.629	9.087	8.625
(3)	12.81	13.55	14.25	13.46	12.39	12.97	13.94	14.14	13.11	12.40	13.22	14.34	13.71	12.94	12.25
(4)	16.47	17.33	16.29	15.71	16.70	17.44	16.07	15.71	17.04	17.09	15.98	15.75	17.47	16.79	15.89

TABLE 8.8 : β_1 ($\phi = .7, \alpha = .8$)

μ MODE:	0.15	0.20	0.25	0.30	0.35	0.40	0.45	0.50	0.55	0.60	0.65	0.70	0.75	0.80	0.85
(1)	6.255	6.605	6.998	7.438	7.921	8.397	8.534	8.046	7.448	6.910	6.448	6.053	5.718	5.434	5.195
(2)	10.39	10.97	11.60	12.13	11.55	10.49	10.05	10.54	11.37	11.67	11.05	10.36	9.741	9.205	8.746
(3)	14.56	15.33	15.63	14.02	13.76	14.59	15.30	14.43	13.41	13.37	14.57	14.62	13.82	13.04	12.36
(4)	18.70	19.28	17.29	17.75	18.76	18.03	16.75	17.56	18.41	17.32	16.35	17.29	17.82	16.89	15.99

TABLE 8.8 : β_1 ($\phi = .7, \alpha = .9$)

μ MODE:	0.15	0.20	0.25	0.30	0.35	0.40	0.45	0.50	0.55	0.60	0.65	0.70	0.75	0.80	0.85
(1)	6.977	7.347	7.762	8.222	8.705	9.070	8.856	8.229	7.604	7.058	6.589	6.188	5.844	5.549	5.295
(2)	11.60	12.22	12.85	13.04	11.93	10.95	10.89	11.54	12.21	11.92	11.18	10.48	9.861	9.322	8.857
(3)	16.25	17.02	16.44	14.86	15.20	16.06	15.92	14.71	13.92	14.52	15.39	14.78	13.93	13.15	12.47
(4)	20.87	20.51	18.86	19.72	20.20	18.49	18.11	19.24	18.83	17.57	17.15	18.45	17.98	17.00	16.10

TABLE 8.8 : β_1 ($\phi = .7, \alpha = 1.0$)

μ MODE:	0.15	0.20	0.25	0.30	0.35	0.40	0.45	0.50	0.55	0.60	0.65	0.70	0.75	0.80	0.85
(1)	7.669	8.053	8.485	8.956	9.417	9.596	9.113	8.411	7.763	7.202	6.721	6.308	5.951	5.641	5.369
(2)	12.78	13.42	14.01	13.66	12.32	11.52	11.74	12.43	12.73	12.10	11.32	10.60	9.976	9.427	8.949
(3)	17.91	18.57	17.08	15.95	16.59	17.24	16.28	15.04	14.67	15.55	15.77	14.93	14.05	13.26	12.57
(4)	22.95	21.44	20.61	21.56	20.88	19.16	19.65	20.34	19.11	17.92	18.22	19.01	18.11	17.11	16.20

TABLE 8.8 : β_1 ($\phi = .8, \alpha = .6$)

μ MODE:	0.15	0.20	0.25	0.30	0.35	0.40	0.45	0.50	0.55	0.60	0.65	0.70	0.75	0.80	0.85
(1)	4.150	4.400	4.682	5.002	5.369	5.791	6.276	6.797	7.006	6.624	6.184	5.791	5.452	5.165	4.931
(2)	6.892	7.306	7.773	8.300	8.884	9.369	8.873	8.176	8.031	8.757	9.798	10.07	9.520	8.981	8.509
(3)	9.650	10.23	10.88	11.55	11.36	10.36	10.69	11.63	12.52	11.91	11.18	11.72	13.29	12.83	12.15
(4)	12.41	13.15	13.92	13.39	12.76	13.65	14.71	14.28	13.43	14.56	15.63	14.74	14.36	16.46	15.79

TABLE 8.8 : β_1 ($\phi = .8, \alpha = .7$)

μ MODE:	0.15	0.20	0.25	0.30	0.35	0.40	0.45	0.50	0.55	0.60	0.65	0.70	0.75	0.80	0.85
(1)	4.820	5.102	5.419	5.777	6.185	6.647	7.144	7.485	7.223	6.748	6.302	5.913	5.580	5.300	5.070
(2)	8.006	8.474	8.998	9.576	10.13	9.953	9.142	8.678	9.088	9.972	10.64	10.20	9.616	9.084	8.624
(3)	11.21	11.86	12.56	12.93	11.80	11.51	12.30	13.24	12.99	12.12	11.91	13.24	13.62	12.92	12.24
(4)	14.41	15.23	15.64	14.20	14.70	15.73	15.83	14.67	15.11	16.45	15.88	15.01	16.04	16.72	15.88

TABLE 8.8 : β_1 ($\phi = .8, \alpha = .8$)

μ MODE:	0.15	0.20	0.25	0.30	0.35	0.40	0.45	0.50	0.55	0.60	0.65	0.70	0.75	0.80	0.85
(1)	5.473	5.780	6.125	6.513	6.950	7.427	7.858	7.853	7.393	6.890	6.440	6.050	5.717	5.433	5.195
(2)	9.095	9.607	10.17	10.77	11.05	10.30	9.528	9.460	10.12	10.94	10.93	10.33	9.732	9.202	8.745
(3)	12.74	13.45	14.12	13.62	12.52	12.90	13.79	14.14	13.24	12.52	13.09	14.22	13.77	10.03	12.36
(4)	16.38	17.21	16.51	15.72	16.60	17.36	16.23	15.71	16.88	17.13	16.11	15.74	17.31	16.85	15.98

TABLE 8.8 : β_1 ($\phi = .8, \alpha = .9$)

μ MODE:	0.15	0.20	0.25	0.30	0.35	0.40	0.45	0.50	0.55	0.60	0.65	0.70	0.75	0.80	0.85
(1)	6.105	6.430	6.795	7.206	7.661	8.125	8.399	8.106	7.565	7.043	6.583	6.186	5.843	5.549	5.295
(2)	10.15	10.70	11.30	11.85	11.62	10.65	10.05	10.30	11.04	11.55	11.11	10.46	9.854	9.320	8.857
(3)	14.23	14.98	15.45	14.19	13.57	14.25	15.04	14.52	13.53	13.20	14.18	14.61	13.90	13.15	12.47
(4)	18.30	18.99	17.37	17.40	18.37	18.12	16.78	17.19	18.21	17.41	16.43	16.88	17.78	16.97	16.09

TABLE 8.8 : β_1 ($\phi = .8, \alpha = 1.0$)

μ MODE:	0.15	0.20	0.25	0.30	0.35	0.40	0.45	0.50	0.55	0.60	0.65	0.70	0.75	0.80	0.85
(1)	6.710	7.048	7.431	7.860	8.324	8.750	8.809	8.328	7.735	7.192	6.717	6.307	5.951	5.640	5.369
(2)	11.18	11.76	12.38	12.78	12.06	11.03	10.65	11.10	11.82	11.90	11.27	10.59	9.971	9.426	8.949
(3)	15.69	16.45	16.44	14.82	14.72	15.52	15.85	14.82	13.89	14.04	15.05	14.83	14.03	13.26	12.57
(4)	20.18	20.42	18.50	19.07	19.86	18.58	17.69	18.64	18.84	17.67	16.91	17.95	18.01	17.09	16.20

TABLE 8.8 : β_1 ($\phi = .9, \alpha = .6$)

μ MODE:	0.15	0.20	0.25	0.30	0.35	0.40	0.45	0.50	0.55	0.60	0.65	0.70	0.75	0.80	0.85
(1)	3.689	3.911	4.162	4.446	4.773	5.150	5.588	6.090	6.575	6.544	6.163	5.783	5.449	5.164	4.930
(2)	6.126	6.495	6.911	7.382	7.914	8.486	8.679	8.078	7.598	7.883	8.808	9.775	9.489	8.973	8.507
(3)	8.578	9.094	9.673	10.31	10.83	10.12	9.686	10.39	11.40	11.78	11.07	10.75	12.25	12.78	12.14
(4)	11.03	11.69	12.42	12.91	11.83	12.19	13.18	13.98	13.10	13.09	14.63	14.67	13.90	15.34	15.77

TABLE 8.8 : β_1 (ϕ = .9, α = .7)

μ MODE:	0.15	0.20	0.25	0.30	0.35	0.40	0.45	0.50	0.55	0.60	0.65	0.70	0.75	0.80	0.85
(1)	4.285	4.535	4.817	5.136	5.501	5.918	6.390	6.871	7.038	6.702	6.286	5.907	5.578	5.299	5.070
(2)	7.116	7.533	8.001	8.526	9.094	9.493	9.013	8.384	8.292	8.968	9.922	10.11	9.596	9.078	8.622
(3)	9.966	10.55	11.19	11.83	11.54	10.68	11.03	11.93	12.64	12.02	11.38	11.99	13.34	12.89	12.24
(4)	12.81	13.56	14.29	13.66	13.20	14.06	15.02	14.43	13.80	14.94	15.71	14.87	14.69	16.49	15.86

TABLE 8.8 : β_1 (ϕ = .9, α = .8)

μ MODE:	0.15	0.20	0.25	0.30	0.35	0.40	0.45	0.50	0.55	0.60	0.65	0.70	0.75	0.80	0.85
(1)	4.865	5.138	5.445	5.792	6.186	6.629	7.102	7.449	7.288	6.858	6.428	6.046	5.715	5.433	5.195
(2)	8.084	8.542	9.053	9.614	10.15	10.06	9.328	8.858	9.148	9.949	10.61	10.28	9.717	9.198	8.744
(3)	11.33	11.97	12.65	13.02	12.04	11.68	12.38	13.26	13.09	12.28	12.04	13.20	13.66	13.01	12.35
(4)	14.57	15.36	15.78	14.49	14.85	15.02	15.95	14.88	15.23	16.47	15.99	15.18	16.03	16.76	15.97

TABLE 8.8 : β_1 (ϕ = .9, α = .9)

μ MODE:	0.15	0.20	0.25	0.30	0.35	0.40	0.45	0.50	0.55	0.60	0.65	0.70	0.75	0.80	0.85
(1)	5.426	5.716	6.042	6.411	6.827	7.284	7.722	7.853	7.495	7.019	6.574	6.183	5.842	5.548	5.295
(2)	9.026	9.518	10.06	10.64	11.01	10.45	9.691	9.460	9.979	10.77	10.95	10.43	9.844	9.317	8.856
(3)	12.66	13.34	14.01	13.78	12.67	12.82	13.64	14.14	13.37	12.64	12.94	14.08	13.84	13.13	12.46
(4)	16.28	17.10	16.73	15.72	16.48	17.29	16.39	15.71	16.71	17.18	16.24	15.73	17.12	16.92	16.09

TABLE 8.8 : β_1(ϕ = .9, α = 1.0)

μ MODE:	0.15	0.20	0.25	0.30	0.35	0.40	0.45	0.50	0.55	0.60	0.65	0.70	0.75	0.80	0.85
(1)	5.965	6.266	6.608	6.996	7.431	7.893	8.259	8.163	7.687	7.175	6.711	6.305	5.950	5.640	5.369
(2)	9.941	10.46	11.04	11.61	11.67	10.81	10.09	10.10	10.75	11.40	11.18	10.57	9.965	9.424	8.948
(3)	13.96	14.68	15.27	14.36	13.46	13.96	14.78	14.61	13.65	13.11	13.85	14.58	13.99	13.25	12.57
(4)	17.97	18.74	17.51	17.12	18.03	18.18	16.85	16.88	17.98	17.51	16.51	16.55	17.72	17.06	16.19

TABLE 8.8 : β_1 (ϕ = 1.0, α = .6)

μ MODE:	0.15	0.20	0.25	0.30	0.35	0.40	0.45	0.50	0.55	0.60	0.65	0.70	0.75	0.80	0.85
(1)	3.320	3.520	3.745	4.002	4.296	4.637	5.034	5.498	6.018	6.356	6.125	5.772	5.445	5.163	4.930
(2)	5.513	5.845	6.220	6.645	7.130	7.674	8.185	7.984	7.453	7.302	7.993	9.085	9.425	8.962	8.504
(3)	7.720	8.185	8.708	9.295	9.915	9.936	9.211	9.428	10.32	11.31	11.01	10.43	11.18	12.65	12.13
(4)	9.926	10.52	11.19	11.87	11.53	11.10	11.90	12.95	12.98	12.27	13.30	14.51	13.81	14.06	15.71

TABLE 8.8 : β_1 (ϕ = 1.0, α = .7)

μ MODE:	0.15	0.20	0.25	0.30	0.35	0.40	0.45	0.50	0.55	0.60	0.65	0.70	0.75	0.80	0.85
(1)	3.856	4.082	4.335	4.623	4.952	5.331	5.766	6.252	6.672	6.615	6.261	5.898	5.575	5.298	5.070
(2)	6.405	6.780	7.203	7.680	8.212	8.750	8.820	8.257	7.863	8.188	9.073	9.861	9.562	9.069	8.620
(3)	8.970	9.496	10.08	10.72	11.10	10.37	10.10	10.81	11.77	11.90	11.23	11.11	12.55	12.84	12.23
(4)	11.53	12.21	12.93	13.22	12.27	12.73	13.70	14.19	13.33	13.61	15.03	14.78	14.13	15.69	15.83

TABLE 8.8 : β_1 (ϕ = 1.0, α = .8)

μ MODE:	0.15	0.20	0.25	0.30	0.35	0.40	0.45	0.50	0.55	0.60	0.65	0.70	0.75	0.80	0.85
(1)	4.379	4.624	4.901	5.214	5.571	5.978	6.434	6.893	7.081	6.803	6.410	6.039	5.713	5.432	5.195
(2)	7.276	7.688	8.152	8.669	9.221	9.597	9.183	8.595	8.477	9.077	9.960	10.17	9.694	9.191	8.743
(3)	10.19	10.77	11.41	12.02	11.76	10.97	11.25	12.11	12.74	12.17	11.58	12.14	13.36	12.98	12.35
(4)	13.11	13.85	14.55	13.97	13.53	14.35	15.22	14.62	14.08	15.17	15.81	15.02	14.90	16.51	15.95

TABLE 8.8 : β_1 (ϕ = 1.0, α = .9)

μ MODE:	0.15	0.20	0.25	0.30	0.35	0.40	0.45	0.50	0.55	0.60	0.65	0.70	0.75	0.80	0.85
(1)	4.884	5.144	5.438	5.772	6.152	6.581	7.042	7.420	7.368	6.982	6.562	6.178	5.841	5.548	5.295
(2)	8.124	8.568	9.067	9.616	10.15	10.18	9.515	9.004	9.161	9.886	10.58	10.36	9.829	9.313	8.855
(3)	11.39	12.02	12.69	13.11	12.28	11.79	12.40	13.26	13.20	12.44	12.12	13.12	13.71	13.11	12.46
(4)	14.66	15.45	15.92	14.74	14.94	15.87	16.09	15.07	15.28	16.47	16.12	15.32	15.96	16.81	16.07

TABLE 8.9 : β_1 (ϕ = .6, α = .6)

μ MODE:	0.15	0.20	0.25	0.30	0.35	0.40	0.45	0.50	0.55	0.60	0.65	0.70	0.75	0.80	0.85
(1)	2.189	2.319	2.465	2.631	2.821	3.040	3.295	3.593	3.940	4.311	4.520	4.389	4.152	3.914	3.692
(2)	5.482	5.807	6.173	6.582	7.028	7.367	7.002	6.407	5.901	5.546	5.549	6.134	7.094	7.656	7.352
(3)	9.173	9.713	10.28	10.37	9.199	8.362	8.555	9.275	10.16	10.40	9.740	9.114	8.655	9.222	10.79
(4)	12.84	13.53	12.77	11.46	12.04	12.94	13.66	12.68	11.68	11.74	13.12	13.44	12.65	11.96	12.32

TABLE 8.9 : β_1 (ϕ = .6, α = .7)

μ MODE:	0.15	0.20	0.25	0.30	0.35	0.40	0.45	0.50	0.55	0.60	0.65	0.70	0.75	0.80	0.85
(1)	2.538	2.682	2.844	3.026	3.233	3.469	3.741	4.048	4.376	4.626	4.609	4.402	4.154	3.914	3.692
(2)	6.358	6.721	7.123	7.559	7.948	7.802	7.154	6.554	6.102	5.911	6.190	6.884	7.676	7.709	7.354
(3)	10.64	11.23	11.68	10.74	9.537	9.254	9.816	10.61	11.13	10.56	9.850	9.271	9.094	10.21	10.90
(4)	14.88	15.28	13.30	13.09	13.91	14.77	14.10	12.89	12.38	13.38	14.27	13.56	12.76	12.20	13.50

TABLE 8.9 : β_1 (ϕ = .6, α = .8)

μ MODE:	0.15	0.20	0.25	0.30	0.35	0.40	0.45	0.50	0.55	0.60	0.65	0.70	0.75	0.80	0.85
(1)	2.874	3.026	3.196	3.387	3.601	3.844	4.115	4.404	4.665	4.767	4.640	4.406	4.155	3.914	3.692
(2)	7.207	7.594	8.017	8.437	8.591	8.037	7.332	6.758	6.392	6.372	6.774	7.442	7.905	7.726	7.355
(3)	12.06	12.65	12.57	11.09	10.14	10.32	11.00	11.69	11.46	10.69	10.00	9.524	9.715	10.82	10.93
(4)	16.83	16.19	14.33	14.77	15.65	15.71	14.35	13.32	13.65	14.74	14.55	13.68	12.92	12.67	14.08

TABLE 8.9 : β_1 (ϕ = .6, α = .9)

μ MODE:	0.15	0.20	0.25	0.30	0.35	0.40	0.45	0.50	0.55	0.60	0.65	0.70	0.75	0.80	0.85
(1)	3.192	3.345	3.516	3.706	3.919	4.155	4.409	4.659	4.834	4.830	4.654	4.409	4.155	3.914	3.692
(2)	8.020	8.422	8.847	9.193	9.006	8.261	7.545	7.008	6.729	6.813	7.238	7.794	7.992	7.734	7.356
(3)	13.43	13.94	13.11	11.53	10.95	11.36	12.04	12.33	11.66	10.86	10.20	9.857	10.27	11.09	10.95
(4)	18.65	16.89	15.71	16.39	17.09	16.09	14.67	14.10	14.90	15.48	14.71	13.83	13.14	13.22	14.27

TABLE 8.9 : β_1 (ϕ = .6, α = 1.0)

μ MODE:	0.15	0.20	0.25	0.30	0.35	0.40	0.45	0.50	0.55	0.60	0.65	0.70	0.75	0.80	0.85
(1)	3.486	3.633	3.797	3.979	4.181	4.400	4.628	4.829	4.930	4.861	4.661	4.410	4.155	3.914	3.692
(2)	8.796	9.205	9.615	9.829	9.335	8.497	7.778	7.274	7.055	7.180	7.576	7.993	8.032	7.738	7.356
(3)	14.75	15.06	13.58	12.06	11.80	12.32	12.91	12.69	11.84	11.04	10.42	10.20	10.68	11.21	10.95
(4)	20.28	17.68	17.14	17.91	18.02	16.42	15.11	15.09	15.96	15.82	14.88	13.99	13.39	13.68	14.35

TABLE 8.9 : β_1 (ϕ = .7, α = .6)

μ MODE:	0.15	0.20	0.25	0.30	0.35	0.40	0.45	0.50	0.55	0.60	0.65	0.70	0.75	0.80	0.85
(1)	1.877	1.988	2.113	2.255	2.418	2.606	2.825	3.084	3.391	3.750	4.118	4.261	4.116	3.901	3.688
(2)	4.699	4.978	5.292	5.647	6.046	6.472	6.713	6.342	5.859	5.458	5.220	5.424	6.198	7.239	7.308
(3)	7.864	8.329	8.843	9.346	9.049	8.131	7.637	8.027	8.792	9.676	9.662	9.081	8.565	8.412	10.021
(4)	11.01	11.64	12.11	10.84	10.44	11.15	12.05	12.48	11.58	10.83	11.40	12.86	12.61	11.91	11.51

TABLE 8.9 : β_1 ($\phi = .7$, $\alpha = .7$)

μ MODE:	0.15	0.20	0.25	0.30	0.35	0.40	0.45	0.50	0.55	0.60	0.65	0.70	0.75	0.80	0.85
(1)	2.176	2.299	2.437	2.594	2.771	2.975	3.211	3.483	3.795	4.122	4.347	4.310	4.122	3.902	3.688
(2)	5.450	5.761	6.110	6.498	6.912	7.234	7.016	6.491	6.018	5.683	5.632	6.061	6.876	7.499	7.317
(3)	9.123	9.638	10.17	10.32	9.337	8.533	8.569	9.197	10.01	10.35	9.794	9.205	8.781	9.143	10.57
(4)	12.77	13.42	12.90	11.61	12.01	12.85	13.54	12.76	11.82	11.75	12.95	13.42	12.71	12.06	12.24

TABLE 8.9 : β_1 ($\phi = .7$, $\alpha = .8$)

μ MODE:	0.15	0.20	0.25	0.30	0.35	0.40	0.45	0.50	0.55	0.60	0.65	0.70	0.75	0.80	0.85
(1)	2.464	2.594	2.740	2.903	3.088	3.299	3.538	3.807	4.096	4.351	4.442	4.328	4.124	3.902	3.688
(2)	6.178	6.512	6.884	7.286	7.663	7.713	7.226	6.673	6.231	5.987	6.088	6.594	7.307	7.576	7.321
(3)	10.35	10.89	11.35	10.82	9.698	9.201	9.563	10.27	10.89	10.58	9.938	9.381	9.117	9.815	10.72
(4)	14.47	14.96	13.47	12.87	13.54	14.36	14.12	13.02	12.35	13.00	14.03	13.60	12.85	12.29	13.01

TABLE 8.9 : β_1 ($\phi = .7$, $\alpha = .9$)

μ MODE	0.15	0.20	0.25	0.30	0.35	0.40	0.45	0.50	0.55	0.60	0.65	0.70	0.75	0.80	0.85
(1)	2.736	2.867	3.014	3.178	3.362	3.569	3.801	4.053	4.302	4.477	4.484	4.337	4.125	3.902	3.688
(2)	6.876	7.225	7.610	8.008	8.284	8.041	7.441	6.884	6.479	6.312	6.490	6.988	7.536	7.607	7.323
(3)	11.53	12.08	12.30	11.22	10.17	9.990	10.50	11.19	11.37	10.77	10.11	9.597	9.502	10.27	10.78
(4)	16.12	16.10	14.23	14.21	14.98	15.46	14.44	13.36	13.20	14.14	14.49	13.75	13.01	12.60	13.51

TABLE 8.9 : β_1 ($\phi = .7$, $\alpha = 1.0$)

μ MODE:	0.15	0.20	0.25	0.30	0.35	0.40	0.45	0.50	0.55	0.60	0.65	0.70	0.75	0.80	0.85
(1)	2.988	3.114	3.255	3.412	3.589	3.786	4.003	4.229	4.434	4.547	4.505	4.341	4.126	3.903	3.688
(2)	7.542	7.902	8.294	8.670	8.794	8.324	7.664	7.106	6.728	6.607	6.807	7.256	7.654	7.622	7.324
(3)	12.67	13.20	13.03	11.64	10.71	10.78	11.36	11.92	11.66	10.95	10.29	9.827	9.845	10.55	10.81
(4)	17.70	16.93	15.19	15.53	16.30	16.07	14.75	13.83	14.14	15.03	14.74	13.91	13.19	12.91	13.77

TABLE 8.9 : β_1 ($\phi = .8$, $\alpha = .6$)

μ MODE:	0.15	0.20	0.25	0.30	0.35	0.40	0.45	0.50	0.55	0.60	0.65	0.70	0.75	0.80	0.85
(1)	1.642	1.739	1.849	1.973	2.116	2.280	2.473	2.700	2.972	3.298	3.678	4.016	4.048	3.882	3.682
(2)	4.112	4.356	4.631	4.943	5.297	5.695	6.099	6.187	5.813	5.417	5.110	5.037	5.532	6.563	7.215
(3)	6.881	7.289	7.746	8.241	8.599	8.026	7.337	7.190	7.752	8.591	9.361	9.037	8.528	8.167	9.037
(4)	9.632	10.20	10.79	10.61	9.573	9.818	10.60	11.50	11.47	10.68	10.31	11.50	12.48	11.89	11.34

TABLE 8.9 : β_1 ($\phi = .8$, $\alpha = .7$)

μ MODE:	0.15	0.20	0.25	0.30	0.35	0.40	0.45	0.50	0.55	0.60	0.65	0.70	0.75	0.80	0.85
(1)	1.904	2.012	2.133	2.269	2.425	2.604	2.811	3.053	3.336	3.659	3.982	4.148	4.067	3.884	3.682
(2)	4.769	5.042	5.348	5.692	6.074	6.470	6.689	6.401	5.964	5.593	5.379	5.522	6.163	7.066	7.250
(3)	7.983	8.439	8.935	9.393	9.140	8.310	7.849	8.154	8.862	9.665	9.688	9.161	8.685	8.564	9.855
(4)	11.18	11.79	12.19	11.08	10.67	11.31	12.16	12.52	11.69	11.02	11.54	12.84	12.65	12.01	11.67

TABLE 8.9 : β_1 ($\phi = .8$, $\alpha = .8$)

μ MODE:	0.15	0.20	0.25	0.30	0.35	0.40	0.45	0.50	0.55	0.60	0.65	0.70	0.75	0.80	0.85
(1)	2.156	2.270	2.397	2.541	2.703	2.888	3.100	3.344	3.620	3.911	4.147	4.199	4.075	3.884	3.682
(2)	5.406	5.699	6.028	6.393	6.782	7.105	7.033	6.594	6.151	5.822	5.711	5.981	6.634	7.285	7.263
(3)	9.055	9.545	10.05	10.27	9.497	8.711	8.586	9.099	9.841	10.28	9.864	9.315	8.905	9.067	10.27
(4)	12.68	13.30	13.03	11.78	11.96	12.73	13.43	12.86	11.97	11.75	12.76	13.40	12.80	12.18	12.18

TABLE 8.9 : β_1 ($\phi = .8$, $\alpha = .9$)

μ MODE:	0.15	0.20	0.25	0.30	0.35	0.40	0.45	0.50	0.55	0.60	0.65	0.70	0.75	0.80	0.85
(1)	2.394	2.509	2.637	2.781	2.943	3.126	3.335	3.570	3.825	4.072	4.232	4.221	4.079	3.885	3.682
(2)	6.017	6.325	6.669	7.046	7.419	7.603	7.300	6.801	6.363	6.073	6.035	6.349	6.942	7.377	7.269
(3)	10.09	10.61	11.07	10.87	9.870	9.224	9.363	9.968	10.63	10.60	10.04	9.493	9.163	9.502	10.45
(4)	14.14	14.67	13.66	12.74	13.23	14.02	14.13	13.15	12.38	12.70	13.73	13.64	12.95	12.38	12.65

TABLE 8.9 : β_1 ($\phi = .8$, $\alpha = 1.0$)

μ MODE:	0.15	0.20	0.25	0.30	0.35	0.40	0.45	0.50	0.55	0.60	0.65	0.70	0.75	0.80	0.85
(1)	2.615	2.725	2.848	2.986	3.142	3.319	3.517	3.737	3.967	4.172	4.278	4.233	4.081	3.885	3.682
(2)	6.600	6.919	7.276	7.660	7.997	8.007	7.547	7.011	6.575	6.311	6.307	6.619	7.130	7.422	7.272
(3)	11.10	11.63	12.00	11.35	10.26	9.794	10.11	10.76	11.21	10.83	10.21	9.677	9.413	9.815	10.53
(4)	15.56	15.88	14.29	13.80	14.45	15.13	14.53	13.45	12.94	13.63	14.35	13.83	13.11	12.59	12.99

TABLE 8.9 : β_1 ($\phi = .9$, $\alpha = .6$)

μ MODE:	0.15	0.20	0.25	0.30	0.35	0.40	0.45	0.50	0.55	0.60	0.65	0.70	0.75	0.80	0.85
(1)	1.460	1.546	1.644	1.754	1.881	2.027	2.198	2.401	2.644	2.939	3.296	3.693	3.928	3.850	3.673
(2)	3.655	3.872	4.117	4.395	4.711	5.073	5.475	5.825	5.732	5.382	5.061	4.867	5.074	5.938	7.011
(3)	6.116	6.480	6.888	7.343	7.811	7.854	7.233	6.776	6.982	7.690	8.608	8.943	8.499	8.089	8.326
(4)	8.563	9.070	9.623	10.06	9.309	8.885	9.463	10.31	11.09	10.61	9.979	10.36	11.94	11.85	11.28

TABLE 8.9 : β_1 ($\phi = .9$, $\alpha = .7$)

μ MODE:	0.15	0.20	0.25	0.30	0.35	0.40	0.45	0.50	0.55	0.60	0.65	0.70	0.75	0.80	0.85
(1)	1.692	1.788	1.896	2.017	2.156	2.315	2.500	2.716	2.972	3.273	3.612	3.909	3.979	3.855	3.674
(2)	4.239	4.482	4.754	5.062	5.410	5.792	6.158	6.227	5.904	5.541	5.266	5.212	5.623	6.510	7.127
(3)	7.097	7.504	7.955	8.431	8.712	8.178	7.556	7.440	7.956	8.750	9.399	9.106	8.636	8.335	9.090
(4)	9.937	10.50	11.05	10.80	9.898	10.14	10.89	11.72	11.57	10.84	10.60	11.72	12.52	11.97	11.49

TABLE 8.9 : β_1 ($\phi = .9$, $\alpha = .8$)

μ MODE:	0.15	0.20	0.25	0.30	0.35	0.40	0.45	0.50	0.55	0.60	0.65	0.70	0.75	0.80	0.85
(1)	1.916	2.018	2.131	2.258	2.403	2.568	2.758	2.978	3.233	3.520	3.813	4.008	3.999	3.858	3.674
(2)	4.805	5.066	5.360	5.690	6.055	6.431	6.669	6.480	6.090	5.739	5.522	5.582	6.069	6.848	7.165
(3)	8.050	8.491	8.969	9.405	9.255	8.508	8.032	8.225	8.863	9.613	9.729	9.263	8.815	8.669	9.630
(4)	11.28	11.87	12.26	11.33	10.84	11.40	12.20	12.57	11.83	11.20	11.59	12.78	12.72	12.12	11.79

TABLE 8.9 : β_1 ($\phi = .9$, $\alpha = .9$)

μ MODE:	0.15	0.20	0.25	0.30	0.35	0.40	0.45	0.50	0.55	0.60	0.65	0.70	0.75	0.80	0.85
(1)	2.128	2.230	2.344	2.472	2.616	2.781	2.968	3.183	3.427	3.689	3.930	4.055	4.009	3.859	3.674
(2)	5.348	5.623	5.933	6.279	6.652	6.991	7.051	6.709	6.291	5.953	5.782	5.903	6.389	7.025	7.182
(3)	8.974	9.442	9.931	10.23	9.666	8.884	8.606	8.988	9.674	10.22	9.947	9.431	9.016	9.005	9.939
(4)	12.58	13.18	13.15	11.95	11.91	12.60	13.32	12.97	12.12	11.76	12.57	13.37	12.89	12.29	12.13

TABLE 8.9 : β_1 ($\phi = .9$, $\alpha = 1.0$)

μ MODE:	0.15	0.20	0.25	0.30	0.35	0.40	0.45	0.50	0.55	0.60	0.65	0.70	0.75	0.80	0.85
(1)	2.324	2.422	2.532	2.655	2.794	2.953	3.133	3.338	3.565	3.801	3.999	4.080	4.014	3.860	3.674
(2)	5.867	6.152	6.476	6.837	7.212	7.490	7.368	6.929	6.488	6.157	6.011	6.151	6.604	7.117	7.190
(3)	9.869	10.36	10.85	10.90	10.04	9.285	9.205	9.709	10.39	10.61	10.14	9.599	9.211	9.279	10.11
(4)	13.86	14.44	13.83	12.69	12.97	13.73	14.11	13.28	12.45	12.46	13.45	13.68	13.05	12.46	12.41

TABLE 8.9 : β_1 (ϕ = 1.0, α = .6)

μ / MODE:	0.15	0.20	0.25	0.30	0.35	0.40	0.45	0.50	0.55	0.60	0.65	0.70	0.75	0.80	0.85
(1)	1.314	1.392	1.479	1.579	1.693	1.824	1.979	2.161	2.381	2.648	2.978	3.375	3.742	3.801	3.661
(2)	3.289	3.485	3.705	3.955	4.242	4.571	4.946	5.347	5.562	5.336	5.028	4.790	4.795	5.434	6.647
(3)	5.505	5.832	6.201	6.615	7.070	7.446	7.143	6.625	6.468	6.976	7.825	8.672	8.463	8.051	7.947
(4)	7.707	8.163	8.673	9.196	9.124	8.400	8.585	9.315	10.21	10.50	9.886	9.634	10.96	11.78	11.25

TABLE 8.9 : β_1 (ϕ = 1.0, α = .7)

μ / MODE:	0.15	0.20	0.25	0.30	0.35	0.40	0.45	0.50	0.55	0.60	0.65	0.70	0.75	0.80	0.85
(1)	1.523	1.609	1.706	1.816	1.940	2.084	2.250	2.446	2.678	2.956	3.283	3.631	3.848	3.814	3.662
(2)	3.815	4.034	4.279	4.557	4.873	5.230	5.613	5.909	5.813	5.496	5.206	5.049	5.244	5.993	6.913
(3)	6.387	6.754	7.165	7.615	8.045	7.997	7.424	7.043	7.268	7.949	8.797	9.008	8.599	8.238	8.520
(4)	8.944	9.456	10.00	10.32	9.578	9.285	9.858	10.68	11.27	10.74	10.20	10.72	12.10	11.93	11.42

TABLE 8.9 : β_1 (ϕ = 1.0, α = .8)

μ / MODE:	0.15	0.20	0.25	0.30	0.35	0.40	0.45	0.50	0.55	0.60	0.65	0.70	0.75	0.80	0.85
(1)	1.725	1.816	1.918	2.033	2.163	2.311	2.483	2.683	2.917	3.189	3.492	3.771	3.890	3.819	3.662
(2)	4.325	4.560	4.825	5.125	5.462	5.830	6.180	6.283	6.019	5.683	5.421	5.345	5.642	6.374	7.007
(3)	7.246	7.645	8.087	8.547	8.816	8.360	7.775	7.624	8.071	8.815	9.427	9.199	8.762	8.482	9.033
(4)	10.15	10.70	11.23	11.02	10.19	10.36	11.07	11.85	11.70	11.01	10.80	11.82	12.57	12.08	11.63

TABLE 8.9 : β_1 (ϕ = 1.0, α = .9)

μ / MODE:	0.15	0.20	0.25	0.30	0.35	0.40	0.45	0.50	0.55	0.60	0.65	0.70	0.75	0.80	0.85
(1)	1.915	2.007	2.110	2.225	2.355	2.503	2.674	2.870	3.097	3.354	3.625	3.844	3.910	3.822	3.662
(2)	4.814	5.061	5.342	5.659	6.011	6.380	6.657	6.572	6.226	5.881	5.642	5.616	5.945	6.605	7.047
(3)	8.077	8.504	8.970	9.408	9.384	8.706	8.185	8.254	8.821	9.551	9.785	9.377	8.939	8.736	9.386
(4)	11.33	11.91	12.34	11.58	10.98	11.43	12.20	12.63	11.99	11.35	11.58	12.71	12.80	12.24	11.87

TABLE 8.9 : β_1 (1/ϕ = .6, 1/α = .6)

μ / MODE:	0.15	0.20	0.25	0.30	0.35	0.40	0.45	0.50	0.55	0.60	0.65	0.70	0.75	0.80	0.85
(1)	1.698	1.698	1.716	1.748	1.794	1.855	1.933	2.030	2.151	2.300	2.486	2.716	2.989	3.271	3.444
(2)	4.853	5.096	5.392	5.740	6.145	6.613	7.124	7.382	6.950	6.442	6.003	5.649	5.409	5.373	5.738
(3)	8.469	8.551	9.504	10.12	10.51	9.624	8.702	8.324	8.931	9.895	10.52	9.967	9.365	8.854	8.517
(4)	12.12	12.82	13.55	12.82	11.49	11.96	12.93	13.65	12.72	11.78	11.64	13.16	13.44	12.69	12.02

TABLE 8.9 : β_1 (1/ϕ = .6, 1/α = .7)

μ / MODE:	0.15	0.20	0.25	0.30	0.35	0.40	0.45	0.50	0.55	0.60	0.65	0.70	0.75	0.80	0.85
(1)	1.591	1.612	1.644	1.687	1.742	1.810	1.894	1.996	2.120	2.274	2.464	2.698	2.977	3.265	3.443
(2)	4.264	4.464	4.709	5.000	5.342	5.741	6.197	6.651	6.706	6.322	5.922	5.587	5.359	5.331	5.709
(3)	7.344	7.743	8.212	8.740	9.207	9.297	8.522	7.894	7.920	8.656	9.628	9.796	9.282	8.796	8.471
(4)	10.46	11.05	11.71	12.12	11.02	10.53	11.22	12.17	12.46	11.63	10.97	11.60	13.03	12.60	11.97

TABLE 8.9 : β_1 (1/ϕ = .6, 1/α = .8)

μ / MODE:	0.15	0.20	0.25	0.30	0.35	0.40	0.45	0.50	0.55	0.60	0.65	0.70	0.75	0.80	0.85
(1)	1.476	1.512	1.557	1.611	1.674	1.750	1.840	1.948	2.078	2.236	2.431	2.671	2.958	3.256	3.441
(2)	3.821	3.997	4.209	4.461	4.757	5.104	5.507	5.950	6.279	6.140	5.807	5.501	5.288	5.270	5.665
(3)	6.510	6.854	7.257	7.719	8.226	8.626	8.281	7.680	7.378	7.789	8.654	9.421	9.156	8.715	8.404
(4)	9.221	9.735	10.32	10.89	10.65	9.817	10.01	10.82	11.69	11.44	10.76	10.63	12.02	12.45	11.89

TABLE 8.9 : β_1 (1/ϕ = .6, 1/α = .9)

μ / MODE:	0.15	0.20	0.25	0.30	0.35	0.40	0.45	0.50	0.55	0.60	0.65	0.70	0.75	0.80	0.85
(1)	1.361	1.409	1.463	1.524	1.595	1.677	1.773	1.887	2.022	2.186	2.386	2.634	2.931	3.242	3.439
(2)	3.466	3.629	3.821	4.046	4.310	4.620	4.981	5.392	5.789	5.892	5.662	5.392	5.196	5.188	5.602
(3)	5.861	6.167	6.524	6.935	7.395	7.854	7.931	7.467	7.079	7.195	7.895	8.815	8.980	8.610	8.318
(4)	8.267	8.721	9.238	9.790	10.07	9.430	9.185	9.778	10.65	11.11	10.59	10.18	11.05	12.19	11.79

TABLE 8.9 : β_1 (1/ϕ = .6, 1/α = 1.0)

μ / MODE:	0.15	0.20	0.25	0.30	0.35	0.40	0.45	0.50	0.55	0.60	0.65	0.70	0.75	0.80	0.85
(1)	1.255	1.308	1.367	1.434	1.510	1.597	1.697	1.815	1.955	2.124	2.330	2.585	2.894	3.224	3.435
(2)	3.169	3.324	3.503	3.712	3.954	4.237	4.567	4.947	5.351	5.602	5.494	5.266	5.087	5.086	5.518
(3)	5.333	5.614	5.939	6.311	6.733	7.183	7.479	7.229	6.849	6.784	7.310	8.201	8.747	8.488	8.214
(4)	7.501	7.912	8.380	8.895	9.337	9.076	8.643	8.978	9.757	10.55	10.39	9.934	10.31	11.78	11.67

TABLE 8.9 : β_1 (1/ϕ = .7, 1/α = .6)

μ / MODE:	0.15	0.20	0.25	0.30	0.35	0.40	0.45	0.50	0.55	0.60	0.65	0.70	0.75	0.80	0.85
(1)	1.981	1.981	2.001	2.039	2.092	2.164	2.254	2.365	2.503	2.671	2.873	3.109	3.353	3.530	3.557
(2)	5.661	5.945	6.287	6.689	7.150	7.655	7.994	7.584	6.993	6.476	6.055	5.746	5.601	5.742	6.253
(3)	9.879	10.43	11.06	11.63	10.90	9.729	9.052	9.449	10.32	11.13	10.68	9.997	9.400	8.931	8.749
(4)	14.13	14.92	15.04	13.09	12.96	13.87	14.77	13.93	12.80	12.19	13.31	14.24	13.49	12.72	12.09

TABLE 8.9 : β_1 (1/ϕ = .7, 1/α = .7)

μ / MODE:	0.15	0.20	0.25	0.30	0.35	0.40	0.45	0.50	0.55	0.60	0.65	0.70	0.75	0.80	0.85
(1)	1.856	1.881	1.918	1.968	2.032	2.111	2.208	2.325	2.468	2.641	2.849	3.091	3.343	3.527	3.557
(2)	4.975	5.208	5.492	5.828	6.219	6.663	7.113	7.236	6.833	6.370	5.975	5.680	5.545	5.697	6.230
(3)	8.567	9.030	9.562	10.13	10.35	9.500	8.686	8.469	9.058	9.948	10.39	9.879	9.321	8.869	8.695
(4)	12.19	12.87	13.50	12.66	11.60	12.07	12.98	13.50	12.62	11.77	11.82	13.20	13.33	12.64	12.03

TABLE 8.9 : β_1 (1/ϕ = .7, 1/α = .8)

μ / MODE:	0.15	0.20	0.25	0.30	0.35	0.40	0.45	0.50	0.55	0.60	0.65	0.70	0.75	0.80	0.85
(1)	1.721	1.764	1.817	1.879	1.953	2.041	2.145	2.270	2.419	2.599	2.814	3.065	3.328	3.522	3.556
(2)	4.458	4.663	4.909	5.200	5.540	5.932	6.359	6.698	6.593	6.223	5.864	5.589	5.466	5.631	6.194
(3)	7.594	7.993	8.455	8.966	9.419	9.168	8.442	7.975	8.198	8.939	9.760	9.695	9.211	8.784	8.617
(4)	10.75	11.34	11.96	12.06	10.99	10.83	11.55	12.41	12.36	11.57	11.10	11.96	13.01	12.52	11.95

TABLE 8.9 : β_1 (1/ϕ = .7, 1/α = .9)

μ / MODE:	0.15	0.20	0.25	0.30	0.35	0.40	0.45	0.50	0.55	0.60	0.65	0.70	0.75	0.80	0.85
(1)	1.588	1.644	1.707	1.778	1.861	1.957	2.068	2.199	2.355	2.542	2.766	3.028	3.306	3.515	3.555
(2)	4.044	4.233	4.456	4.718	5.022	5.374	5.768	6.153	6.282	6.042	5.729	5.475	5.363	5.541	6.141
(3)	6.837	7.192	7.603	8.065	8.535	8.694	8.182	7.673	7.621	8.177	9.014	9.424	9.072	8.676	8.514
(4)	9.642	10.16	10.74	11.17	10.56	10.02	10.46	11.27	11.87	11.36	10.78	11.06	12.39	12.37	11.85

TABLE 8.9 : β_1 (1/ϕ = .7, 1/α = 1.0)

μ / MODE:	0.15	0.20	0.25	0.30	0.35	0.40	0.45	0.50	0.55	0.60	0.65	0.70	0.75	0.80	0.85
(1)	1.464	1.526	1.595	1.673	1.761	1.862	1.979	2.116	2.278	2.472	2.704	2.978	3.275	3.505	3.554
(2)	3.697	3.878	4.087	4.328	4.608	4.932	5.299	5.685	5.942	5.842	5.580	5.345	5.239	5.425	6.069
(3)	6.222	6.548	6.923	7.346	7.799	8.133	7.892	7.418	7.216	7.591	8.361	9.061	8.915	8.554	8.388
(4)	8.750	9.224	9.751	10.25	10.12	9.485	9.626	10.33	11.13	11.11	10.55	10.45	11.63	12.18	11.73

TABLE 8.9 : β_1 ($1/\phi = .8$, $1/\alpha = .6$)

μ MODE:	0.15	0.20	0.25	0.30	0.35	0.40	0.45	0.50	0.55	0.60	0.65	0.70	0.75	0.80	0.85
(1)	2.263	2.264	2.287	2.330	2.391	2.471	2.573	2.698	2.849	3.030	3.237	3.453	3.627	3.685	3.614
(2)	6.469	6.792	7.180	7.630	8.130	8.565	8.315	7.641	7.027	6.522	6.135	5.894	5.876	6.173	6.663
(3)	11.29	11.91	12.57	12.46	11.02	9.947	9.944	10.70	11.59	11.47	10.72	10.03	9.451	9.056	9.102
(4)	16.13	16.92	15.38	13.99	14.72	15.71	15.37	14.01	13.01	13.48	14.80	14.37	13.51	12.75	12.21

TABLE 8.9 : β_1 ($1/\phi = .8$, $1/\alpha = .7$)

μ MODE:	0.15	0.20	0.25	0.30	0.35	0.40	0.45	0.50	0.55	0.60	0.65	0.70	0.75	0.80	0.85
(1)	2.121	2.149	2.192	2.249	2.322	2.411	2.521	2.653	2.811	2.998	3.213	3.438	3.621	3.683	3.614
(2)	5.685	5.950	6.273	6.651	7.082	7.534	7.780	7.419	6.888	6.418	6.050	5.822	5.815	6.130	6.648
(3)	9.788	10.31	10.89	11.35	10.69	9.640	9.083	9.436	10.23	10.91	10.55	9.920	9.370	8.987	9.043
(4)	13.92	14.65	14.69	13.00	12.87	13.70	14.49	13.78	12.74	12.23	13.21	14.05	13.39	12.67	12.14

TABLE 8.9 : β_1 ($1/\phi = .8$, $1/\alpha = .8$)

μ MODE:	0.15	0.20	0.25	0.30	0.35	0.40	0.45	0.50	0.55	0.60	0.65	0.70	0.75	0.80	0.85
(1)	1.967	2.017	2.076	2.147	2.232	2.332	2.450	2.590	2.757	2.952	3.177	3.415	3.610	3.681	3.614
(2)	5.095	5.328	5.608	5.936	6.314	6.730	7.099	7.090	6.698	6.279	5.937	5.723	5.728	6.066	6.626
(3)	8.677	9.128	9.638	10.15	10.17	9.354	8.666	8.615	9.207	10.01	10.26	9.772	9.261	8.892	8.955
(4)	12.28	12.93	13.44	12.48	11.69	12.20	13.05	13.35	12.51	11.77	12.02	13.25	13.21	12.56	12.05

TABLE 8.9 : β_1 ($1/\phi = .8$, $1/\alpha = .9$)

μ MODE:	0.15	0.20	0.25	0.30	0.35	0.40	0.45	0.50	0.55	0.60	0.65	0.70	0.75	0.80	0.85
(1)	1.815	1.878	1.950	2.032	2.127	2.236	2.362	2.511	2.685	2.891	3.128	3.381	3.595	3.678	3.614
(2)	4.621	4.837	5.091	5.387	5.728	6.109	6.493	6.688	6.472	6.114	5.799	5.598	5.612	5.975	6.592
(3)	7.812	8.215	8.674	9.160	9.473	9.022	8.361	8.081	8.453	9.192	9.814	9.588	9.129	8.772	8.835
(4)	11.02	11.60	12.16	11.93	10.99	11.10	11.83	12.57	12.24	11.51	11.27	12.28	12.95	12.43	11.93

TABLE 8.9 : β_1 ($1/\phi = .8$, $1/\alpha = 1.0$)

μ MODE:	0.15	0.20	0.25	0.30	0.35	0.40	0.45	0.50	0.55	0.60	0.65	0.70	0.75	0.80	0.85
(1)	1.674	1.744	1.823	1.912	2.013	2.128	2.261	2.417	2.599	2.814	3.063	3.336	3.574	3.673	3.613
(2)	4.225	4.431	4.670	4.944	5.259	5.614	5.992	6.283	6.229	5.939	5.648	5.455	5.471	5.856	6.542
(3)	7.110	7.481	7.902	8.361	8.764	8.651	8.080	7.698	7.875	8.518	9.279	9.376	8.985	8.637	8.683
(4)	9.997	10.53	11.08	11.28	10.49	10.26	10.84	11.64	11.89	11.27	10.83	11.46	12.57	12.28	11.80

TABLE 8.9 : β_1 ($1/\phi = .9$, $1/\alpha = .6$)

μ MODE:	0.15	0.20	0.25	0.30	0.35	0.40	0.45	0.50	0.55	0.60	0.65	0.70	0.75	0.80	0.85
(1)	2.546	2.547	2.573	2.620	2.688	2.778	2.890	3.026	3.187	3.372	3.567	3.735	3.815	3.774	3.645
(2)	7.277	7.638	8.069	8.558	9.050	9.112	8.409	7.678	7.068	6.588	6.250	6.103	6.219	6.596	6.934
(3)	12.69	13.37	13.91	12.68	11.15	10.52	11.05	11.91	12.29	11.54	10.75	10.06	9.528	9.247	9.536
(4)	18.12	18.41	15.71	15.51	16.46	16.95	15.48	14.13	13.76	14.94	15.32	14.42	13.54	12.81	12.39

TABLE 8.9 : β_1 ($1/\phi = .9$, $1/\alpha = .7$)

μ MODE:	0.15	0.20	0.25	0.30	0.35	0.40	0.45	0.50	0.55	0.60	0.65	0.70	0.75	0.80	0.85
(1)	2.387	2.418	2.465	2.530	2.611	2.711	2.832	2.977	3.146	3.340	3.545	3.723	3.811	3.773	3.645
(2)	6.395	6.692	7.050	7.465	7.916	8.277	8.079	7.492	6.934	6.481	6.160	6.025	6.156	6.560	6.926
(3)	11.01	11.58	12.16	12.09	10.84	9.886	9.840	10.49	11.28	11.27	10.60	9.959	9.442	9.169	9.480
(4)	15.65	16.33	15.10	13.79	14.35	15.25	15.11	13.89	12.96	13.28	14.45	14.22	13.43	12.73	12.31

TABLE 8.9 : β_1 ($1/\phi = .9$, $1/\alpha = .8$)

μ MODE:	0.15	0.20	0.25	0.30	0.35	0.40	0.45	0.50	0.55	0.60	0.65	0.70	0.75	0.80	0.85
(1)	2.213	2.269	2.336	2.416	2.510	2.622	2.754	2.908	3.088	3.293	3.511	3.705	3.805	3.772	3.645
(2)	5.731	5.992	6.304	6.666	7.071	7.466	7.603	7.242	6.759	6.340	6.038	5.915	6.064	6.505	6.915
(3)	9.759	10.26	10.80	11.14	10.46	9.525	9.113	9.483	10.21	10.76	10.40	9.820	9.326	9.059	9.395
(4)	13.81	14.48	14.36	12.89	12.85	13.62	14.28	13.61	12.65	12.29	13.21	13.88	13.28	12.61	12.20

TABLE 8.9 : β_1 ($1/\phi = .9$, $1/\alpha = .9$)

μ MODE:	0.15	0.20	0.25	0.30	0.35	0.40	0.45	0.50	0.55	0.60	0.65	0.70	0.75	0.80	0.85
(1)	2.042	2.113	2.194	2.286	2.392	2.514	2.655	2.820	3.011	3.228	3.464	3.679	3.797	3.770	3.645
(2)	5.199	5.441	5.725	6.053	6.422	6.808	7.083	6.951	6.559	6.176	5.891	5.776	5.940	6.425	6.897
(3)	8.787	9.234	9.728	10.17	9.995	9.197	8.645	8.752	9.365	10.08	10.14	9.657	9.188	8.921	9.274
(4)	12.38	13.01	13.36	12.30	11.78	12.34	13.13	13.21	12.38	11.77	12.21	13.29	13.09	12.48	12.07

TABLE 8.9 : β_1 ($1/\phi = .9$, $1/\alpha = 1.0$)

μ MODE:	0.15	0.20	0.25	0.30	0.35	0.40	0.45	0.50	0.55	0.60	0.65	0.70	0.75	0.80	0.85
(1)	1.883	1.962	2.051	2.151	2.264	2.394	2.543	2.715	2.916	3.147	3.401	3.642	3.784	3.768	3.645
(2)	4.753	4.985	5.252	5.557	5.902	6.275	6.605	6.648	6.353	6.004	5.730	5.614	5.785	6.315	6.871
(3)	7.998	8.412	8.872	9.331	9.468	8.875	8.287	8.197	8.683	9.416	9.823	9.485	9.041	8.763	9.111
(4)	11.24	11.82	12.31	11.78	11.01	11.33	12.08	12.65	12.12	11.44	11.46	12.56	12.88	12.34	11.91

TABLE 8.9 : β_1 ($1/\phi = 1.0$, $1/\alpha = .6$)

μ MODE:	0.15	0.20	0.25	0.30	0.35	0.40	0.45	0.50	0.55	0.60	0.65	0.70	0.75	0.80	0.85
(1)	2.829	2.830	2.858	2.911	2.986	3.083	3.204	3.348	3.514	3.691	3.853	3.950	3.937	3.826	3.663
(2)	8.084	8.482	8.950	9.459	9.818	9.314	8.457	7.715	7.121	6.678	6.410	6.371	6.595	6.956	7.097
(3)	14.09	14.79	14.72	12.80	11.43	11.43	12.18	12.93	12.48	11.58	10.78	10.12	9.640	9.516	9.963
(4)	20.07	18.95	16.53	17.14	18.09	17.25	15.56	14.44	15.01	16.09	15.43	14.44	13.57	12.89	12.66

TABLE 8.9 : β_1 ($1/\phi = 1.0$, $1/\alpha = .7$)

μ MODE:	0.15	0.20	0.25	0.30	0.35	0.40	0.45	0.50	0.55	0.60	0.65	0.70	0.75	0.80	0.85
(1)	2.652	2.686	2.739	2.810	2.900	3.010	3.141	3.295	3.471	3.660	3.834	3.942	3.935	3.826	3.663
(2)	7.105	7.432	7.824	8.265	8.693	8.766	8.198	7.541	6.987	6.566	6.313	6.288	6.534	6.930	7.094
(3)	12.22	12.83	13.30	12.37	11.00	10.38	10.77	11.52	11.96	11.38	10.64	10.01	9.546	9.430	9.920
(4)	17.35	17.68	15.44	15.03	15.84	16.43	15.28	14.02	13.55	14.49	15.05	14.29	13.46	12.80	12.57

TABLE 8.9 : β_1 ($1/\phi = 1.0$, $1/\alpha = .8$)

μ MODE:	0.15	0.20	0.25	0.30	0.35	0.40	0.45	0.50	0.55	0.60	0.65	0.70	0.75	0.80	0.85
(1)	2.459	2.521	2.595	2.684	2.788	2.911	3.055	3.221	3.410	3.614	3.805	3.929	3.932	3.825	3.662
(2)	6.368	6.656	6.998	7.388	7.796	8.077	7.852	7.319	6.814	6.419	6.180	6.170	6.444	6.889	7.088
(3)	10.84	11.38	11.90	11.75	10.63	9.803	9.806	10.40	11.10	11.07	10.46	9.868	9.419	9.307	9.852
(4)	15.32	15.92	14.77	13.64	14.14	14.95	14.85	13.73	12.90	13.21	14.25	14.06	13.32	12.68	12.45

TABLE 8.9 : β_1 ($1/\phi = 1.0$, $1/\alpha = .9$)

μ MODE:	0.15	0.20	0.25	0.30	0.35	0.40	0.45	0.50	0.55	0.60	0.65	0.70	0.75	0.80	0.85
(1)	2.269	2.348	2.438	2.540	2.657	2.792	2.947	3.126	3.329	3.550	3.764	3.911	3.927	3.824	3.662
(2)	5.776	6.044	6.357	6.712	7.097	7.438	7.460	7.072	6.620	6.248	6.018	6.016	6.320	6.828	7.080
(3)	9.761	10.25	10.75	10.98	10.23	9.393	9.132	9.552	10.25	10.65	10.26	9.708	9.268	9.146	9.750
(4)	13.75	14.37	14.06	12.75	12.85	13.59	14.14	13.43	12.54	12.34	13.27	13.75	13.16	12.54	12.28

TABLE 8.10 : β_1 (ϕ = .6, α = .6)

μ MODE:	0.15	0.20	0.25	0.30	0.35	0.40	0.45	0.50	0.55	0.60	0.65	0.70	0.75	0.80	0.85
(1)	1.559	1.544	1.523	1.495	1.464	1.431	1.398	1.367	1.340	1.316	1.298	1.286	1.282	1.287	1.306
(2)	4.497	4.360	4.270	4.244	4.283	4.382	4.531	4.720	4.923	5.091	5.159	5.107	4.977	4.818	4.662

TABLE 8.10 : β_1 (ϕ = .6, α = .7)

μ MODE:	0.15	0.20	0.25	0.30	0.35	0.40	0.45	0.50	0.55	0.60	0.65	0.70	0.75	0.80	0.85
(1)	1.818	1.801	1.774	1.741	1.703	1.662	1.621	1.581	1.545	1.513	1.486	1.465	1.451	1.445	1.450
(2)	5.230	5.049	4.915	4.851	4.856	4.919	5.026	5.156	5.274	5.338	5.317	5.219	5.076	4.921	4.774

TABLE 8.10 : β_1 (ϕ = .6, α = .8)

μ MODE:	0.15	0.20	0.25	0.30	0.35	0.40	0.45	0.50	0.55	0.60	0.65	0.70	0.75	0.80	0.85
(1)	2.077	2.056	2.026	1.984	1.937	1.886	1.834	1.783	1.736	1.692	1.654	1.620	1.593	1.572	1.558
(2)	5.949	5.705	5.508	5.386	5.338	5.350	5.402	5.469	5.520	5.522	5.459	5.342	5.196	5.042	4.897

TABLE 8.10 : β_1 (ϕ = .6, α = .9)

μ MODE:	0.15	0.20	0.25	0.30	0.35	0.40	0.45	0.50	0.55	0.60	0.65	0.70	0.75	0.80	0.85
(1)	2.336	2.311	2.273	2.223	2.165	2.101	2.035	1.970	1.908	1.850	1.797	1.749	1.706	1.669	1.636
(2)	6.645	6.316	6.032	5.837	5.728	5.687	5.691	5.713	5.723	5.692	5.609	5.482	5.329	5.169	5.017

TABLE 8.10 : β_1 (ϕ = .6, α = 1.0)

μ MODE:	0.15	0.20	0.25	0.30	0.35	0.40	0.45	0.50	0.55	0.60	0.65	0.70	0.75	0.80	0.85
(1)	2.594	2.654	2.517	2.456	2.383	2.303	2.220	2.138	2.059	1.985	1.916	1.852	1.793	1.740	1.691
(2)	7.312	6.067	6.481	6.207	6.040	5.957	5.930	5.928	5.917	5.870	5.772	5.631	5.464	5.290	5.121

TABLE 8.10 : β_1 (ϕ = .7, α = .6)

μ MODE:	0.15	0.20	0.25	0.30	0.35	0.40	0.45	0.50	0.55	0.60	0.65	0.70	0.75	0.80	0.85
(1)	1.342	1.337	1.330	1.320	1.308	1.295	1.282	1.270	1.258	1.250	1.244	1.244	1.249	1.263	1.289
(2)	3.967	3.919	3.889	3.892	3.934	4.016	4.138	4.294	4.471	4.643	4.766	4.803	4.760	4.670	4.566

TABLE 8.10 : β_1 (ϕ = .7, α = .7)

μ MODE:	0.15	0.20	0.25	0.30	0.35	0.40	0.45	0.50	0.55	0.60	0.65	0.70	0.75	0.80	0.85
(1)	1.566	1.560	1.550	1.538	1.522	1.505	1.487	1.470	1.453	1.438	1.426	1.418	1.416	1.419	1.432
(2)	4.618	4.548	4.491	4.466	4.478	4.529	4.612	4.717	4.825	4.913	4.952	4.933	4.868	4.776	4.682

TABLE 8.10 : β_1 (ϕ = .7, α = .8)

μ MODE:	0.15	0.20	0.25	0.30	0.35	0.40	0.45	0.50	0.55	0.60	0.65	0.70	0.75	0.80	0.85
(1)	1.789	1.781	1.769	1.753	1.733	1.710	1.686	1.660	1.635	1.611	1.590	1.571	1.556	1.545	1.540
(2)	5.261	5.155	5.055	4.982	4.947	4.947	4.977	5.025	5.075	5.108	5.107	5.066	4.994	4.903	4.810

TABLE 8.10 : β_1 (ϕ = .7, α = .9)

μ MODE:	0.15	0.20	0.25	0.30	0.35	0.40	0.45	0.50	0.55	0.60	0.65	0.70	0.75	0.80	0.85
(1)	2.012	2.002	1.987	1.965	1.939	1.908	1.874	1.838	1.802	1.766	1.731	1.699	1.669	1.642	1.618
(2)	5.890	5.732	5.567	5.429	5.332	5.277	5.257	5.262	5.276	5.281	5.263	5.213	5.135	5.038	4.935

TABLE 8.10 : β_1 (ϕ = .7, α = 1.0)

μ MODE:	0.15	0.20	0.25	0.30	0.35	0.40	0.45	0.50	0.55	0.60	0.65	0.70	0.75	0.80	0.85
(1)	2.234	2.222	2.201	2.174	2.138	2.097	2.050	2.000	1.950	1.899	1.849	1.801	1.756	1.713	1.673
(2)	6.500	6.268	6.017	5.801	5.641	5.538	5.484	5.465	5.463	5.458	5.431	5.371	5.280	5.168	5.046

TABLE 8.10 : β_1 (ϕ = .8, α = .6)

μ MODE:	0.15	0.20	0.25	0.30	0.35	0.40	0.45	0.50	0.55	0.60	0.65	0.70	0.75	0.80	0.85
(1)	1.177	1.176	1.175	1.173	1.171	1.169	1.168	1.168	1.169	1.173	1.179	1.190	1.206	1.229	1.264
(2)	3.523	3.523	3.538	3.573	3.643	3.724	3.842	3.987	4.150	4.315	4.451	4.528	4.537	4.501	4.447

TABLE 8.10 : β_1 (ϕ = .8, α = .7)

μ MODE:	0.15	0.20	0.25	0.30	0.35	0.40	0.45	0.50	0.55	0.60	0.65	0.70	0.75	0.80	0.85
(1)	1.373	1.372	1.370	1.367	1.363	1.360	1.356	1.353	1.352	1.351	1.354	1.359	1.368	1.383	1.406
(2)	4.104	4.094	4.095	4.114	4.156	4.220	4.305	4.404	4.506	4.595	4.653	4.670	4.651	4.611	4.565

TABLE 8.10 : β_1 (ϕ = .8, α = .8)

μ MODE:	0.15	0.20	0.25	0.30	0.35	0.40	0.45	0.50	0.55	0.60	0.65	0.70	0.75	0.80	0.85
(1)	1.569	1.567	1.563	1.559	1.553	1.546	1.539	1.531	1.524	1.517	1.511	1.508	1.506	1.508	1.514
(2)	4.678	4.650	4.626	4.612	4.615	4.635	4.670	4.714	4.759	4.794	4.811	4.806	4.780	4.741	4.698

TABLE 8.10 : β_1 (ϕ = .8, α = .9)

μ MODE:	0.15	0.20	0.25	0.30	0.35	0.40	0.45	0.50	0.55	0.60	0.65	0.70	0.75	0.80	0.85
(1)	1.765	1.761	1.756	1.748	1.738	1.727	1.713	1.698	1.682	1.666	1.649	1.633	1.618	1.604	1.592
(2)	5.243	5.185	5.118	5.053	5.002	4.967	4.951	4.949	4.957	4.965	4.966	4.954	4.925	4.882	4.830

TABLE 8.10 : β_1 (ϕ = .8, α = 1.0)

μ MODE:	0.15	0.20	0.25	0.30	0.35	0.40	0.45	0.50	0.55	0.60	0.65	0.70	0.75	0.80	0.85
(1)	1.960	1.954	1.946	1.934	1.919	1.900	1.877	1.852	1.824	1.795	1.765	1.735	1.705	1.676	1.647
(2)	5.796	5.691	5.562	5.431	5.317	5.231	5.175	5.145	5.135	5.134	5.130	5.113	5.077	5.020	4.948

TABLE 8.10 : β_1 (ϕ = .9, α = .6)

μ MODE:	0.15	0.20	0.25	0.30	0.35	0.40	0.45	0.50	0.55	0.60	0.65	0.70	0.75	0.80	0.85
(1)	1.048	1.048	1.050	1.052	1.055	1.059	1.064	1.071	1.081	1.093	1.108	1.128	1.154	1.187	1.233
(2)	3.158	3.181	3.221	3.280	3.360	3.463	3.589	3.738	3.902	4.068	4.211	4.304	4.339	4.335	4.317

TABLE 8.10 : β_1 (ϕ = .9, α = .7)

μ MODE:	0.15	0.20	0.25	0.30	0.35	0.40	0.45	0.50	0.55	0.60	0.65	0.70	0.75	0.80	0.85
(1)	1.222	1.223	1.224	1.225	1.228	1.231	1.236	1.242	1.250	1.261	1.274	1.290	1.311	1.338	1.372
(2)	3.679	3.699	3.735	3.786	3.856	3.943	4.048	4.156	4.265	4.359	4.423	4.454	4.457	4.446	4.436

TABLE 8.10 : β_1 (ϕ = .9, α = .8)

μ MODE:	0.15	0.20	0.25	0.30	0.35	0.40	0.45	0.50	0.55	0.60	0.65	0.70	0.75	0.80	0.85
(1)	1.396	1.397	1.397	1.398	1.399	1.401	1.403	1.407	1.411	1.417	1.424	1.434	1.446	1.461	1.480
(2)	4.195	4.207	4.228	4.260	4.303	4.356	4.414	4.524	4.542	4.562	4.584	4.590	4.585	4.576	4.572

TABLE 8.10 : β_1 (ϕ = .9, α = .9)

μ MODE:	0.15	0.20	0.25	0.30	0.35	0.40	0.45	0.50	0.55	0.60	0.65	0.70	0.75	0.80	0.85
(1)	1.570	1.570	1.569	1.568	1.567	1.565	1.564	1.562	1.560	1.559	1.557	1.556	1.556	1.556	1.558
(2)	4.705	4.699	4.693	4.690	4.690	4.695	4.703	4.713	4.723	4.731	4.735	4.734	4.728	4.719	4.709

TABLE 8.10 : β_1 (ϕ = .9, α = 1.0)

μ MODE:	0.15	0.20	0.25	0.30	0.35	0.40	0.45	0.50	0.55	0.60	0.65	0.70	0.75	0.80	0.85
(1)	1.744	1.742	1.740	1.736	1.730	1.724	1.716	1.706	1.695	1.683	1.670	1.656	1.642	1.628	1.613
(2)	5.205	5.169	5.121	5.067	5.013	4.967	4.931	4.908	4.897	4.893	4.893	4.891	4.881	4.862	4.833

TABLE 8.10 : β_1 ($\phi = 1.0$, $\alpha = .6$)

μ MODE:	0.15	0.20	0.25	0.30	0.35	0.40	0.45	0.50	0.55	0.60	0.65	0.70	0.75	0.80	0.85
(1)	2.855	2.890	2.942	3.014	3.107	3.221	3.357	3.515	3.690	1.015	1.037	1.063	1.097	1.139	1.194
(2)	4.823	4.940	5.097	5.291	5.510	5.736	5.930	6.042	6.067	3.867	4.022	4.128	4.177	4.187	4.189

TABLE 8.10 : β_1 ($\phi = 1.0$, $\alpha = .7$)

μ MODE:	0.15	0.20	0.25	0.30	0.35	0.40	0.45	0.50	0.55	0.60	0.65	0.70	0.75	0.80	0.85
(1)	1.101	1.102	1.105	1.109	1.114	1.121	1.130	1.141	1.155	1.172	1.192	1.217	1.248	1.285	1.331
(2)	3.327	3.362	3.415	3.486	3.576	3.683	3.804	3.934	4.062	4.171	4.247	4.285	4.296	4.297	4.308

TABLE 8.10 : β_1 ($\phi = 1.0$, $\alpha = .8$)

μ MODE:	0.15	0.20	0.25	0.30	0.35	0.40	0.45	0.50	0.55	0.60	0.65	0.70	0.75	0.80	0.85
(1)	1.258	1.259	1.261	1.265	1.269	1.275	1.283	1.293	1.304	1.318	1.335	1.355	1.378	1.406	1.437
(2)	3.795	3.826	3.872	3.932	4.006	4.089	4.177	4.261	4.332	4.383	4.411	4.421	4.421	4.425	4.444

TABLE 8.10 : β_1 ($\phi = 1.0$, $\alpha = .9$)

μ MODE:	0.15	0.20	0.25	0.30	0.35	0.40	0.45	0.50	0.55	0.60	0.65	0.70	0.75	0.80	0.85
(1)	1.414	1.415	1.417	1.419	1.422	1.426	1.431	1.437	1.444	1.452	1.462	1.473	1.486	1.500	1.515
(2)	4.258	4.278	4.307	4.344	4.387	4.432	4.475	4.511	4.538	4.554	4.560	4.561	4.562	4.568	4.584

TABLE 8.10 : β_1 ($1/\phi = .6$, $1/\alpha = .6$)

μ MODE:	0.15	0.20	0.25	0.30	0.35	0.40	0.45	0.50	0.55	0.60	0.65	0.70	0.75	0.80	0.85
(1)	1.559	1.544	1.523	1.495	1.464	1.431	1.398	1.367	1.340	1.316	1.298	1.286	1.282	1.287	1.306
(2)	4.497	4.360	4.270	4.244	4.283	4.382	4.531	4.720	4.923	5.091	5.159	5.107	4.977	4.818	4.662

TABLE 8.10 : β_1 ($1/\phi = .6$, $1/\alpha = .7$)

μ MODE:	0.15	0.20	0.25	0.30	0.35	0.40	0.45	0.50	0.55	0.60	0.65	0.70	0.75	0.80	0.85
(1)	1.342	1.337	1.330	1.320	1.308	1.295	1.282	1.270	1.258	1.250	1.244	1.244	1.249	1.263	1.289
(2)	3.967	3.919	3.889	3.892	3.934	4.016	4.138	4.294	4.471	4.643	4.766	4.803	4.760	4.670	4.566

TABLE 8.10 : β_1 ($1/\phi = .6$, $1/\alpha = .8$)

μ MODE:	0.15	0.20	0.25	0.30	0.35	0.40	0.45	0.50	0.55	0.60	0.65	0.70	0.75	0.80	0.85
(1)	1.177	1.176	1.175	1.173	1.171	1.169	1.168	1.168	1.169	1.173	1.179	1.190	1.206	1.229	1.264
(2)	3.523	3.523	3.538	3.573	3.634	3.724	3.842	3.987	4.150	4.315	4.451	4.528	4.537	4.501	4.447

TABLE 8.10 : β_1 ($1/\phi = .6$, $1/\alpha = .9$)

μ MODE:	0.15	0.20	0.25	0.30	0.35	0.40	0.45	0.50	0.55	0.60	0.65	0.70	0.75	0.80	0.85
(1)	1.048	1.048	1.050	1.052	1.055	1.059	1.064	1.071	1.081	1.093	1.108	1.128	1.154	1.187	1.233
(2)	3.158	3.181	3.221	3.280	3.360	2.463	3.589	3.738	3.902	4.068	4.211	4.304	4.339	4.335	4.317

TABLE 8.10 : β_1 ($1/\phi = .6$, $1/\alpha = 1.0$)

μ MODE:	0.15	0.20	0.25	0.30	0.35	0.40	0.45	0.50	0.55	0.60	0.65	0.70	0.75	0.80	0.85
(1)	2.855	2.890	2.942	3.014	3.107	3.221	3.357	3.515	3.690	1.015	1.037	1.063	1.097	1.139	1.194
(2)	4.823	4.940	5.097	5.291	5.510	5.736	5.930	6.042	6.067	3.867	4.022	4.128	4.177	4.187	4.189

TABLE 8.10 : β_1 ($1/\phi = .7$, $1/\alpha = .6$)

μ MODE:	0.15	0.20	0.25	0.30	0.35	0.40	0.45	0.50	0.55	0.60	0.65	0.70	0.75	0.80	0.85
(1)	1.818	1.801	1.774	1.741	1.703	1.662	1.621	1.581	1.545	1.513	1.486	1.465	1.451	1.445	1.450
(2)	5.230	5.049	4.915	4.851	4.856	4.919	5.026	5.156	5.274	5.338	5.317	5.219	5.076	4.921	4.774

TABLE 8.10 : β_1 ($1/\phi = .7$, $1/\alpha = .7$)

μ MODE:	0.15	0.20	0.25	0.30	0.35	0.40	0.45	0.50	0.55	0.60	0.65	0.70	0.75	0.80	0.85
(1)	1.566	1.560	1.550	1.538	1.522	1.505	1.487	1.470	1.453	1.438	1.426	1.418	1.416	1.419	1.432
(2)	4.618	4.548	4.491	4.466	4.478	4.529	4.612	4.717	4.825	4.913	4.952	4.933	4.868	4.776	4.682

TABLE 8.10 : β_1 ($1/\phi = .7$, $1/\alpha = .8$)

μ MODE:	0.15	0.20	0.25	0.30	0.35	0.40	0.45	0.50	0.55	0.60	0.65	0.70	0.75	0.80	0.85
(1)	1.373	1.372	1.370	1.367	1.363	1.360	1.356	1.353	1.352	1.351	1.354	1.359	1.368	1.383	1.406
(2)	4.104	4.094	4.095	4.114	4.156	4.220	4.305	4.404	4.506	4.595	4.653	4.670	4.651	4.611	4.565

TABLE 8.10 : β_1 ($1/\phi = .7$, $1/\alpha = .9$)

μ MODE:	0.15	0.20	0.25	0.30	0.35	0.40	0.45	0.50	0.55	0.60	0.65	0.70	0.75	0.80	0.85
(1)	1.222	1.223	1.224	1.225	1.228	1.231	1.236	1.242	1.250	1.261	1.274	1.290	1.311	1.338	1.372
(2)	3.679	3.699	3.735	3.786	3.856	3.943	4.045	4.156	4.265	4.359	4.423	4.454	4.457	4.445	4.436

TABLE 8.10 : β_1 ($1/\phi = .7$, $1/\alpha = 1.0$)

μ MODE:	0.15	0.20	0.25	0.30	0.35	0.40	0.45	0.50	0.55	0.60	0.65	0.70	0.75	0.80	0.85
(1)	1.101	1.102	1.105	1.109	1.114	1.121	1.130	1.141	1.155	1.172	1.192	1.217	1.248	1.285	1.331
(2)	3.327	3.362	3.415	3.486	3.576	3.683	3.804	3.934	4.062	4.171	4.247	4.285	4.296	4.297	4.308

TABLE 8.10 : β_1 ($1/\phi = .8$, $1/\alpha = .6$)

μ MODE:	0.15	0.20	0.25	0.30	0.35	0.40	0.45	0.50	0.55	0.60	0.65	0.70	0.75	0.80	0.85
(1)	2.077	2.056	2.025	1.984	1.937	1.886	1.834	1.783	1.736	1.692	1.654	1.620	1.593	1.572	1.558
(2)	5.949	5.705	5.508	5.386	5.338	5.350	5.402	5.469	5.520	5.522	5.459	5.342	5.196	5.042	4.897

TABLE 8.10 : β_1 ($1/\phi = .8$, $1/\alpha = .7$)

μ MODE:	0.15	0.20	0.25	0.30	0.35	0.40	0.45	0.50	0.55	0.60	0.65	0.70	0.75	0.80	0.85
(1)	1.789	1.781	1.769	1.753	1.733	1.710	1.686	1.660	1.635	1.611	1.590	1.571	1.556	1.545	1.540
(2)	5.261	5.155	5.055	4.982	4.947	4.947	4.977	5.025	5.075	5.108	5.107	5.066	4.994	4.903	4.810

TABLE 8.10 : β_1 ($1/\phi = .8$, $1/\alpha = .8$)

μ MODE:	0.15	0.20	0.25	0.30	0.35	0.40	0.45	0.50	0.55	0.60	0.65	0.70	0.75	0.80	0.85
(1)	1.569	1.567	1.563	1.559	1.553	1.546	1.539	1.531	1.524	1.517	1.511	1.508	1.506	1.508	1.514
(2)	4.678	4.650	4.626	4.612	4.615	4.635	4.670	4.714	4.759	4.794	4.811	4.806	4.780	4.741	4.698

TABLE 8.10 : β_1 ($1/\phi = .8$, $1/\alpha = .9$)

μ MODE:	0.15	0.20	0.25	0.30	0.35	0.40	0.45	0.50	0.55	0.60	0.65	0.70	0.75	0.80	0.85
(1)	1.396	1.397	1.397	1.398	1.399	1.401	1.403	1.407	1.411	1.417	1.424	1.434	1.446	1.461	1.480
(2)	4.195	4.207	4.228	4.260	4.303	4.356	4.414	4.472	4.524	4.562	4.584	4.590	4.585	4.576	4.572

TABLE 8.10 : β_1 ($1/\phi = .8$, $1/\alpha = 1.0$)

μ MODE:	0.15	0.20	0.25	0.30	0.35	0.40	0.45	0.50	0.55	0.60	0.65	0.70	0.75	0.80	0.85
(1)	1.258	1.259	1.261	1.265	1.269	1.275	1.283	1.293	1.304	1.318	1.335	1.355	1.378	1.406	1.437
(2)	3.795	3.826	3.872	3.932	4.006	4.089	4.177	4.261	4.332	4.383	4.411	4.421	4.421	4.425	4.444

TABLE 8.10 : β_1 ($1/\phi = .9$, $1/\alpha = .6$)

μ MODE:	0.15	0.20	0.25	0.30	0.35	0.40	0.45	0.50	0.55	0.60	0.65	0.70	0.75	0.80	0.85
(1)	2.336	2.311	2.273	2.223	2.165	2.101	2.035	1.970	1.908	1.850	1.797	1.749	1.706	1.669	1.636
(2)	6.645	6.316	6.032	5.837	5.728	5.687	5.691	5.713	5.723	5.692	5.609	5.482	5.329	5.169	5.017

TABLE 8.10 : β_1 $(1/\phi = .9,\ 1/\alpha = .7)$

μ	0.15	0.20	0.25	0.30	0.35	0.40	0.45	0.50	0.55	0.60	0.65	0.70	0.75	0.80	0.85
MODE:															
(1)	2.012	2.002	1.987	1.965	1.939	1.908	1.874	1.838	1.802	1.766	1.731	1.699	1.669	1.642	1.618
(2)	5.890	5.732	5.567	5.429	5.332	5.277	5.257	5.262	5.276	5.281	5.263	5.213	5.135	5.038	4.935

TABLE 8.10 : β_1 $(1/\phi = .9,\ 1/\alpha = .8)$

μ	0.15	0.20	0.25	0.30	0.35	0.40	0.45	0.50	0.55	0.60	0.65	0.70	0.75	0.80	0.85
MODE:															
(1)	1.765	1.761	1.756	1.748	1.738	1.727	1.713	1.698	1.682	1.666	1.649	1.633	1.618	1.604	1.592
(2)	5.243	5.185	5.118	5.053	5.002	4.967	4.951	4.949	4.957	4.965	4.966	4.954	4.925	4.882	4.830

TABLE 8.10 : β_1 $(1/\phi = .9,\ 1/\alpha = .9)$

μ	0.15	0.20	0.25	0.30	0.35	0.40	0.45	0.50	0.55	0.60	0.65	0.70	0.75	0.80	0.85
MODE:															
(1)	1.570	1.570	1.569	1.568	1.567	1.565	1.564	1.562	1.560	1.559	1.557	1.556	1.556	1.556	1.558
(2)	4.705	4.699	4.693	4.690	4.690	4.695	4.703	4.713	4.723	4.731	4.735	4.734	4.728	4.719	4.709

TABLE 8.10 : β_1 $(1/\phi = .9,\ 1/\alpha = 1.0)$

μ	0.15	0.20	0.25	0.30	0.35	0.40	0.45	0.50	0.55	0.60	0.65	0.70	0.75	0.80	0.85
MODE:															
(1)	1.414	1.415	1.417	1.419	1.422	1.426	1.431	1.437	1.444	1.452	1.462	1.473	1.486	1.500	1.515
(2)	4.258	4.278	4.307	4.344	4.387	4.432	4.475	4.511	4.538	4.554	4.560	4.561	4.562	4.568	4.584

TABLE 8.10 : β_1 $(1/\phi = 1.0,\ 1/\alpha = .6)$

μ	0.15	0.20	0.25	0.30	0.35	0.40	0.45	0.50	0.55	0.60	0.65	0.70	0.75	0.80	0.85
MODE:															
(1)	2.594	2.564	2.517	2.456	2.383	2.303	2.220	2.138	2.059	1.985	1.916	1.852	1.793	1.740	1.691
(2)	7.312	6.867	6.481	6.207	6.040	5.957	5.930	5.928	5.917	5.870	5.772	5.631	5.464	5.290	5.121

TABLE 8.10 : β_1 $(1/\phi = 1.0,\ 1/\alpha = .7)$

μ	0.15	0.20	0.25	0.30	0.35	0.40	0.45	0.50	0.55	0.60	0.65	0.70	0.75	0.80	0.85
MODE:															
(1)	2.234	2.222	2.201	2.174	2.138	2.097	2.050	2.000	1.950	1.899	1.849	1.801	1.756	1.713	1.673
(2)	6.500	6.268	6.017	5.801	5.641	5.538	5.484	5.465	5.463	5.458	5.431	5.371	5.280	5.168	5.046

TABLE 8.10 : β_1 $(1/\phi = 1.0,\ 1/\alpha = .8)$

μ	0.15	0.20	0.25	0.30	0.35	0.40	0.45	0.50	0.55	0.60	0.65	0.70	0.75	0.80	0.85
MODE:															
(1)	1.960	1.954	1.946	1.934	1.919	1.900	1.877	1.852	1.824	1.796	1.765	1.735	1.705	1.676	1.647
(2)	5.796	5.691	5.562	5.431	5.317	5.231	5.175	5.145	5.135	5.134	5.130	5.113	5.077	5.020	4.948

TABLE 8.10 : β_1 $(1/\phi = 1.0,\ 1/\alpha = .9)$

μ	0.15	0.20	0.25	0.30	0.35	0.40	0.45	0.50	0.55	0.60	0.65	0.70	0.75	0.80	0.85
MODE:															
(1)	1.744	1.742	1.740	1.736	1.730	1.724	1.716	1.706	1.695	1.683	1.670	1.656	1.642	1.628	1.613
(2)	5.205	5.169	5.121	5.067	5.013	4.967	4.931	4.908	4.897	4.893	4.893	4.891	4.881	4.862	4.833

TABLE 8.11 : β_1 $(\phi = .6,\ \alpha = .6)$

μ	0.15	0.20	0.25	0.30	0.35	0.40	0.45	0.50	0.55	0.60	0.65	0.70	0.75	0.80	0.85
MODE:															
(1)	2.697	2.814	2.929	3.035	3.118	3.160	3.146	3.077	2.970	2.847	2.722	2.603	2.495	2.401	2.324
(2)	6.173	6.285	6.189	5.869	5.519	5.277	5.185	5.240	5.410	5.641	5.848	5.924	5.848	5.689	5.508

TABLE 8.11 : β_1 $(\phi = .6,\ \alpha = .7)$

μ	0.15	0.20	0.25	0.30	0.35	0.40	0.45	0.50	0.55	0.60	0.65	0.70	0.75	0.80	0.85
MODE:															
(1)	3.074	3.172	3.257	3.320	3.348	3.331	3.266	3.164	3.043	2.915	2.790	2.674	2.570	2.480	2.407
(2)	6.983	6.956	6.692	6.303	5.969	5.767	5.704	5.757	5.883	6.023	6.103	6.075	5.954	5.788	5.616

TABLE 8.11 : β_1 $(\phi = .6,\ \alpha = .8)$

μ	0.15	0.20	0.25	0.30	0.35	0.40	0.45	0.50	0.55	0.60	0.65	0.70	0.75	0.80	0.85
MODE:															
(1)	3.407	3.471	3.517	3.536	3.520	3.464	3.373	3.256	3.128	2.998	2.873	2.757	2.653	2.561	2.485
(2)	7.705	7.551	7.185	6.763	6.427	6.222	6.142	6.155	6.220	6.281	6.287	6.214	6.077	5.909	5.740

TABLE 8.11 : β_1 $(\phi = .6,\ \alpha = .9)$

μ	0.15	0.20	0.25	0.30	0.35	0.40	0.45	0.50	0.55	0.60	0.65	0.70	0.75	0.80	0.85
MODE:															
(1)	3.693	3.719	3.729	3.714	3.668	3.591	3.485	3.359	3.225	3.091	2.962	2.842	2.733	2.636	2.549
(2)	8.375	8.126	7.683	7.213	6.843	6.606	6.488	6.458	6.475	6.490	6.459	6.363	6.215	6.041	5.864

TABLE 8.11 : β_1 $(\phi = .6,\ \alpha = 1.0)$

μ	0.15	0.20	0.25	0.30	0.35	0.40	0.45	0.50	0.55	0.60	0.65	0.70	0.75	0.80	0.85
MODE:															
(1)	3.939	3.932	3.913	3.876	3.814	3.723	3.606	3.472	3.329	3.187	3.051	2.923	2.805	2.697	2.600
(2)	9.027	8.697	8.165	7.621	7.196	6.916	6.763	6.704	6.696	6.688	6.637	6.523	6.360	6.170	5.977

TABLE 8.11 : β_1 $(\phi = .7,\ \alpha = .6)$

μ	0.15	0.20	0.25	0.30	0.35	0.40	0.45	0.50	0.55	0.60	0.65	0.70	0.75	0.80	0.85
MODE:															
(1)	2.312	2.413	2.512	2.606	2.686	2.744	2.768	2.755	2.709	2.642	2.564	2.484	2.407	2.339	2.282
(2)	5.296	5.420	5.434	5.308	5.112	4.945	4.863	4.883	4.998	5.183	5.388	5.536	5.568	5.501	5.389

TABLE 8.11 : β_1 $(\phi = .7,\ \alpha = .7)$

μ	0.15	0.20	0.25	0.30	0.35	0.40	0.45	0.50	0.55	0.60	0.65	0.70	0.75	0.80	0.85
MODE:															
(1)	2.635	2.720	2.795	2.856	2.896	2.909	2.894	2.851	2.788	2.714	2.634	2.556	2.483	2.419	2.366
(2)	5.998	6.033	5.942	5.753	5.553	5.409	5.347	5.368	5.453	5.570	5.675	5.721	5.692	5.608	5.501

TABLE 8.11 : β_1 $(\phi = .7,\ \alpha = .8)$

μ	0.15	0.20	0.25	0.30	0.35	0.40	0.45	0.50	0.55	0.60	0.65	0.70	0.75	0.80	0.85
MODE:															
(1)	2.920	2.977	3.021	3.048	3.055	3.040	3.002	2.946	2.876	2.799	2.719	2.641	2.568	2.502	2.445
(2)	6.631	6.588	6.434	6.215	6.005	5.852	5.770	5.753	5.785	5.837	5.876	5.875	5.824	5.736	5.631

TABLE 8.11 : β_1 $(\phi = .7,\ \alpha = .9)$

μ	0.15	0.20	0.25	0.30	0.35	0.40	0.45	0.50	0.55	0.60	0.65	0.70	0.75	0.80	0.85
MODE:															
(1)	3.166	3.191	3.206	3.207	3.192	3.161	3.113	3.050	2.976	2.895	2.812	2.730	2.651	2.578	2.512
(2)	7.222	7.128	6.931	6.676	6.429	6.235	6.110	6.048	6.034	6.046	6.053	6.032	5.971	5.876	5.763

TABLE 8.11 : β_1 $(\phi = .7,\ \alpha = 1.0)$

μ	0.15	0.20	0.25	0.30	0.35	0.40	0.45	0.50	0.55	0.60	0.65	0.70	0.75	0.80	0.85
MODE:															
(1)	3.378	3.375	3.367	3.352	3.326	3.287	3.232	3.163	3.083	2.995	2.905	2.814	2.726	2.642	2.564
(2)	7.801	7.670	7.423	7.105	6.796	6.548	6.378	6.282	6.243	6.237	6.232	6.201	6.128	6.017	5.884

TABLE 8.11 : β_1 $(\phi = .8,\ \alpha = .6)$

μ	0.15	0.20	0.25	0.30	0.35	0.40	0.45	0.50	0.55	0.60	0.65	0.70	0.75	0.80	0.85
MODE:															
(1)	2.023	2.111	2.199	2.282	2.356	2.415	2.453	2.465	2.454	2.426	2.386	2.341	2.296	2.256	2.225
(2)	4.636	4.755	4.805	4.770	4.681	4.596	4.559	4.592	4.697	4.861	5.054	5.219	5.303	5.300	5.247

TABLE 8.11 : β_1 $(\phi = .8,\ \alpha = .7)$

μ	0.15	0.20	0.25	0.30	0.35	0.40	0.45	0.50	0.55	0.60	0.65	0.70	0.75	0.80	0.85
MODE:															
(1)	2.306	2.380	2.447	2.503	2.545	2.569	2.574	2.562	2.535	2.499	2.457	2.414	2.373	2.337	2.310
(2)	5.254	5.307	5.286	5.207	5.113	5.046	5.028	5.064	5.145	5.253	5.357	5.425	5.441	5.412	5.362

TABLE 8.11 : β_1 (ϕ = .8, α = .8)

μ MODE:	0.15	0.20	0.25	0.30	0.35	0.40	0.45	0.50	0.55	0.60	0.65	0.70	0.75	0.80	0.85
(1)	2.556	2.606	2.646	2.674	2.689	2.690	2.678	2.655	2.622	2.584	2.542	2.500	2.459	2.422	2.391
(2)	5.813	5.811	5.752	5.657	5.560	5.488	5.451	5.452	5.481	5.525	5.565	5.585	5.577	5.543	5.497

TABLE 8.11 : β_1 (ϕ = .8, α = .9)

μ MODE:	0.15	0.20	0.25	0.30	0.35	0.40	0.45	0.50	0.55	0.60	0.65	0.70	0.75	0.80	0.85
(1)	2.771	2.794	2.809	2.816	2.814	2.803	2.783	2.755	2.720	2.680	2.636	2.590	2.545	2.500	2.459
(2)	6.337	6.304	6.224	6.112	5.990	5.882	5.801	5.752	5.732	5.732	5.740	5.742	5.726	5.689	5.636

TABLE 8.11 : β_1 (ϕ = .8, α = 1.0)

μ MODE:	0.15	0.20	0.25	0.30	0.35	0.40	0.45	0.50	0.55	0.60	0.65	0.70	0.75	0.80	0.85
(1)	2.956	2.955	2.952	2.946	2.935	2.919	2.896	2.865	2.826	2.781	2.731	2.677	2.622	2.567	2.512
(2)	6.851	6.800	6.698	6.548	6.375	6.211	6.079	5.987	5.935	5.915	5.912	5.908	5.887	5.838	5.766

TABLE 8.11 : β_1 (ϕ = .9, α = .6)

μ MODE:	0.15	0.20	0.25	0.30	0.35	0.40	0.45	0.50	0.55	0.60	0.65	0.70	0.75	0.80	0.85
(1)	1.798	1.877	1.955	2.029	2.097	2.154	2.195	2.218	2.225	2.219	2.205	2.187	2.169	2.156	2.152
(2)	4.122	4.232	4.294	4.299	4.270	4.244	4.253	4.315	4.434	4.604	4.800	4.976	5.084	5.113	5.099

TABLE 8.11 : β_1 (ϕ = .9, α = .7)

μ MODE:	0.15	0.20	0.25	0.30	0.35	0.40	0.45	0.50	0.55	0.60	0.65	0.70	0.75	0.80	0.85
(1)	2.050	2.116	2.176	2.227	2.267	2.295	2.309	2.311	2.305	2.292	2.276	2.260	2.246	2.238	2.238
(2)	4.673	4.730	4.738	4.714	4.685	4.679	4.710	4.781	4.885	5.006	5.119	5.197	5.230	5.228	5.215

TABLE 8.11 : β_1 (ϕ = .9, α = .8)

μ MODE:	0.15	0.20	0.25	0.30	0.35	0.40	0.45	0.50	0.55	0.60	0.65	0.70	0.75	0.80	0.85
(1)	2.272	2.316	2.353	2.380	2.398	2.406	2.406	2.400	2.388	2.374	2.359	2.345	2.333	2.324	2.320
(2)	5.172	5.187	5.170	5.140	5.117	5.114	5.135	5.178	5.234	5.291	5.336	5.361	5.366	5.359	5.352

TABLE 8.11 : β_1 (ϕ = .9, α = .9)

μ MODE:	0.15	0.20	0.25	0.30	0.35	0.40	0.45	0.50	0.55	0.60	0.65	0.70	0.75	0.80	0.85
(1)	2.463	2.484	2.499	2.508	2.511	2.510	2.504	2.494	2.482	2.467	2.452	2.436	2.420	2.404	2.390
(2)	5.641	5.634	5.608	5.575	5.541	5.515	5.498	5.493	5.495	5.503	5.510	5.515	5.513	5.506	5.496

TABLE 8.11 : β_1 (ϕ = .9, α = 1.0)

μ MODE:	0.15	0.20	0.25	0.30	0.35	0.40	0.45	0.50	0.55	0.60	0.65	0.70	0.75	0.80	0.85
(1)	2.628	2.627	2.626	2.624	2.621	2.616	2.608	2.597	2.583	2.566	2.546	2.524	2.499	2.473	2.446
(2)	6.102	6.086	6.053	6.000	5.933	5.861	5.793	5.739	5.702	5.682	5.676	5.675	5.672	5.659	5.632

TABLE 8.11 : β_1 (ϕ = 1.0, α = .6)

μ MODE:	0.15	0.20	0.25	0.30	0.35	0.40	0.45	0.50	0.55	0.60	0.65	0.70	0.75	0.80	0.85
(1)	1.618	1.689	1.759	1.827	1.889	1.942	1.983	2.010	2.026	2.033	2.035	2.035	2.037	2.046	2.066
(2)	3.710	3.812	3.876	3.900	3.902	3.911	3.953	4.041	4.182	4.371	4.586	4.783	4.911	4.957	4.961

TABLE 8.11 : β_1 (ϕ = 1.0, α = .7)

μ MODE:	0.15	0.20	0.25	0.30	0.35	0.40	0.45	0.50	0.55	0.60	0.65	0.70	0.75	0.80	0.85
(1)	1.845	1.904	1.958	2.005	2.043	2.071	2.089	2.098	2.102	2.103	2.103	2.105	2.113	2.127	2.153
(2)	4.207	4.263	4.285	4.287	4.294	4.326	4.393	4.499	4.636	4.787	4.926	5.022	5.065	5.073	5.076

TABLE 8.11 : β_1 (ϕ = 1.0, α = .8)

μ MODE:	0.15	0.20	0.25	0.30	0.35	0.40	0.45	0.50	0.55	0.60	0.65	0.70	0.75	0.80	0.85
(1)	2.045	2.085	2.118	2.144	2.162	2.173	2.179	2.181	2.181	2.181	2.183	2.189	2.198	2.214	2.236
(2)	4.657	4.679	4.682	4.685	4.704	4.746	4.815	4.903	5.001	5.090	5.156	5.191	5.201	5.201	5.212

TABLE 8.11 : β_1 (ϕ = 1.0, α = .9)

μ MODE:	0.15	0.20	0.25	0.30	0.35	0.40	0.45	0.50	0.55	0.60	0.65	0.70	0.75	0.80	0.85
(1)	2.217	2.236	2.250	2.259	2.265	2.268	2.269	2.269	2.269	2.270	2.273	2.278	2.285	2.295	2.308
(2)	5.081	5.086	5.086	5.093	5.112	5.144	5.187	5.235	5.280	5.314	5.335	5.342	5.343	5.345	5.357

TABLE 8.11 : β_1 ($1/\phi$ = .6, $1/\alpha$ = .6)

μ MODE:	0.15	0.20	0.25	0.30	0.35	0.40	0.45	0.50	0.55	0.60	0.65	0.70	0.75	0.80	0.85
(1)	1.895	1.889	1.891	1.894	1.893	1.888	1.878	1.865	1.850	1.835	1.824	1.819	1.822	1.839	1.875
(2)	5.025	4.977	4.905	4.845	4.830	4.874	4.984	5.160	5.394	5.658	5.883	5.966	5.889	5.725	5.539

TABLE 8.11 : β_1 ($1/\phi$ = .6, $1/\alpha$ = .7)

μ MODE:	0.15	0.20	0.25	0.30	0.35	0.40	0.45	0.50	0.55	0.60	0.65	0.70	0.75	0.80	0.85
(1)	1.726	1.713	1.710	1.713	1.718	1.723	1.728	1.731	1.734	1.738	1.745	1.755	1.773	1.803	1.850
(2)	4.411	4.418	4.416	4.419	4.442	4.499	4.599	4.747	4.943	5.175	5.407	5.569	5.605	5.536	5.419

TABLE 8.11 : β_1 ($1/\phi$ = .6, $1/\alpha$ = .8)

μ MODE:	0.15	0.20	0.25	0.30	0.35	0.40	0.45	0.50	0.55	0.60	0.65	0.70	0.75	0.80	0.85
(1)	1.604	1.590	1.585	1.585	1.589	1.596	1.604	1.615	1.627	1.642	1.683	1.583	1.714	1.756	1.817
(2)	3.944	3.964	3.991	4.029	4.084	4.164	4.276	4.424	4.611	4.828	5.054	5.240	5.333	5.331	5.276

TABLE 8.11 : β_1 ($1/\phi$ = .6, $1/\alpha$ = .9)

μ MODE:	0.15	0.20	0.25	0.30	0.35	0.40	0.45	0.50	0.55	0.60	0.65	0.70	0.75	0.80	0.85
(1)	1.506	1.497	1.493	1.493	1.495	1.501	1.509	1.521	1.536	1.555	1.579	1.610	1.650	1.703	1.775
(2)	3.586	3.604	3.637	3.687	3.759	3.855	3.981	4.140	4.332	4.554	4.785	4.985	5.106	5.140	5.125

TABLE 8.11 : β_1 ($1/\phi$ = .6, $1/\alpha$ = 1.0)

μ MODE:	0.15	0.20	0.25	0.30	0.35	0.40	0.45	0.50	0.55	0.60	0.65	0.70	0.75	0.80	0.85
(1)	1.419	1.419	1.420	1.422	1.425	1.430	1.437	1.448	1.463	1.482	1.508	1.542	1.587	1.647	1.728
(2)	3.305	3.319	3.348	3.398	3.472	3.575	3.708	3.875	4.078	4.312	4.560	4.782	4.924	4.977	4.983

TABLE 8.11 : β_1 ($1/\phi$ = .7, $1/\alpha$ = .6)

μ MODE:	0.15	0.20	0.25	0.30	0.35	0.40	0.45	0.50	0.55	0.60	0.65	0.70	0.75	0.80	0.85
(1)	2.211	2.204	2.205	2.208	2.206	2.198	2.184	2.165	2.142	2.120	2.100	2.085	2.077	2.081	2.100
(2)	5.858	5.791	5.685	5.586	5.531	5.536	5.607	5.734	5.898	6.056	6.144	6.115	5.990	5.819	5.641

TABLE 8.11 : β_1 ($1/\phi$ = .7, $1/\alpha$ = .7)

μ MODE:	0.15	0.20	0.25	0.30	0.35	0.40	0.45	0.50	0.55	0.60	0.65	0.70	0.75	0.80	0.85
(1)	2.013	1.998	1.995	1.998	2.003	2.007	2.010	2.012	2.012	2.011	2.012	2.015	2.025	2.043	2.074
(2)	5.144	5.145	5.131	5.112	5.109	5.136	5.200	5.304	5.439	5.585	5.704	5.755	5.726	5.637	5.525

TABLE 8.11 : β_1 ($1/\phi$ = .7, $1/\alpha$ = .8)

μ MODE:	0.15	0.20	0.25	0.30	0.35	0.40	0.45	0.50	0.55	0.60	0.65	0.70	0.75	0.80	0.85
(1)	1.871	1.855	1.849	1.849	1.853	1.859	1.868	1.878	1.889	1.901	1.917	1.936	1.961	1.994	2.040
(2)	4.600	4.619	4.643	4.672	4.714	4.776	4.863	4.975	5.108	5.249	5.373	5.451	5.469	5.439	5.385

TABLE 8.11 : β_1 ($1/\phi = .7$, $1/\alpha = .9$)

μ MODE:	0.15	0.20	0.25	0.30	0.35	0.40	0.45	0.50	0.55	0.60	0.65	0.70	0.75	0.80	0.85
(1)	1.757	1.747	1.742	1.741	1.744	1.749	1.758	1.769	1.784	1.802	1.825	1.854	1.890	1.936	1.996
(2)	4.183	4.201	4.234	4.282	4.349	4.438	4.550	4.684	4.835	4.989	5.123	5.214	5.252	5.251	5.235

TABLE 8.11 : β_1 ($1/\phi = .7$, $1/\alpha = 1.0$)

μ MODE:	0.15	0.20	0.25	0.30	0.35	0.40	0.45	0.50	0.55	0.60	0.65	0.70	0.75	0.80	0.85
(1)	1.656	1.656	1.657	1.659	1.662	1.667	1.674	1.685	1.700	1.719	1.745	1.778	1.821	1.876	1.947
(2)	3.854	3.869	3.899	3.950	4.024	4.126	4.254	4.409	4.583	4.762	4.921	5.030	5.080	5.090	5.093

TABLE 8.11 : β_1 ($1/\phi = .8$, $1/\alpha = .6$)

μ MODE:	0.15	0.20	0.25	0.30	0.35	0.40	0.45	0.50	0.55	0.60	0.65	0.70	0.75	0.80	0.85
(1)	2.527	2.518	2.520	2.521	2.517	2.505	2.484	2.455	2.422	2.387	2.354	2.323	2.299	2.283	2.277
(2)	6.687	6.591	6.434	6.272	6.152	6.096	6.104	6.165	6.251	6.321	6.327	6.249	6.107	5.933	5.758

TABLE 8.11 : β_1 ($1/\phi = .8$, $1/\alpha = .7$)

μ MODE:	0.15	0.20	0.25	0.30	0.35	0.40	0.45	0.50	0.55	0.60	0.65	0.70	0.75	0.80	0.85
(1)	2.301	2.283	2.279	2.282	2.286	2.289	2.289	2.285	2.279	2.270	2.260	2.251	2.245	2.245	2.252
(2)	5.874	5.865	5.825	5.769	5.716	5.688	5.693	5.731	5.794	5.862	5.908	5.906	5.852	5.759	5.649

TABLE 8.11 : β_1 ($1/\phi = .8$, $1/\alpha = .8$)

μ MODE:	0.15	0.20	0.25	0.30	0.35	0.40	0.45	0.50	0.55	0.60	0.65	0.70	0.75	0.80	0.85
(1)	2.138	2.120	2.112	2.112	2.115	2.121	2.128	2.135	2.142	2.149	2.157	2.167	2.179	2.195	2.218
(2)	5.254	5.269	5.281	5.290	5.302	5.323	5.358	5.409	5.472	5.536	5.586	5.609	5.600	5.564	5.513

TABLE 8.11 : β_1 ($1/\phi = .8$, $1/\alpha = .9$)

μ MODE:	0.15	0.20	0.25	0.30	0.35	0.40	0.45	0.50	0.55	0.60	0.65	0.70	0.75	0.80	0.85
(1)	2.008	1.996	1.990	1.989	1.991	1.996	2.003	2.013	2.025	2.040	2.058	2.079	2.105	2.136	2.175
(2)	4.778	4.794	4.822	4.859	4.909	4.971	5.046	5.129	5.214	5.290	5.346	5.377	5.384	5.376	5.365

TABLE 8.11 : β_1 ($1/\phi = .8$, $1/\alpha = 1.0$)

μ MODE:	0.15	0.20	0.25	0.30	0.35	0.40	0.45	0.50	0.55	0.60	0.65	0.70	0.75	0.80	0.85
(1)	1.892	1.892	1.893	1.895	1.898	1.902	1.909	1.918	1.931	1.948	1.969	1.997	2.031	2.073	2.125
(2)	4.404	4.417	4.443	4.488	4.553	4.639	4.743	4.860	4.979	5.083	5.159	5.200	5.212	5.213	5.222

TABLE 8.11 : β_1 ($1/\phi = .9$, $1/\alpha = .6$)

μ MODE:	0.15	0.20	0.25	0.30	0.35	0.40	0.45	0.50	0.55	0.60	0.65	0.70	0.75	0.80	0.85
(1)	2.842	2.832	2.833	2.833	2.825	2.805	2.774	2.733	2.684	2.632	2.580	2.530	2.484	2.444	2.410
(2)	7.510	7.369	7.135	6.882	6.677	6.545	6.486	6.484	6.512	6.528	6.493	6.392	6.238	6.059	5.878

TABLE 8.11 : β_1 ($1/\phi = .9$, $1/\alpha = .7$)

μ MODE:	0.15	0.20	0.25	0.30	0.35	0.40	0.45	0.50	0.55	0.60	0.65	0.70	0.75	0.80	0.85
(1)	2.588	2.568	2.563	2.564	2.566	2.566	2.560	2.549	2.531	2.509	2.484	2.457	2.431	2.407	2.387
(2)	6.601	6.572	6.491	6.371	6.246	6.142	6.076	6.049	6.054	6.073	6.081	6.058	5.992	5.893	5.776

TABLE 8.11 : β_1 ($1/\phi = .9$, $1/\alpha = .8$)

μ MODE:	0.15	0.20	0.25	0.30	0.35	0.40	0.45	0.50	0.55	0.60	0.65	0.70	0.75	0.80	0.85
(1)	2.405	2.384	2.375	2.374	2.376	2.379	2.383	2.385	2.384	2.382	2.377	2.371	2.364	2.358	2.353
(2)	5.906	5.912	5.901	5.872	5.829	5.786	5.753	5.737	5.736	5.747	5.759	5.761	5.743	5.703	5.647

TABLE 8.11 : β_1 ($1/\phi = .9$, $1/\alpha = .9$)

μ MODE:	0.15	0.20	0.25	0.30	0.35	0.40	0.45	0.50	0.55	0.60	0.65	0.70	0.75	0.80	0.85
(1)	2.258	2.245	2.238	2.236	2.236	2.240	2.244	2.250	2.257	2.264	2.272	2.280	2.289	2.300	2.312
(2)	5.372	5.382	5.396	5.410	5.424	5.439	5.454	5.472	5.491	5.508	5.520	5.526	5.525	5.516	5.503

TABLE 8.11 : β_1 ($1/\phi = .9$, $1/\alpha = 1.0$)

μ MODE:	0.15	0.20	0.25	0.30	0.35	0.40	0.45	0.50	0.55	0.60	0.65	0.70	0.75	0.80	0.85
(1)	2.129	2.129	2.129	2.130	2.132	2.135	2.139	2.145	2.153	2.164	2.177	2.194	2.213	2.236	2.263
(2)	4.952	4.960	4.978	5.007	5.048	5.100	5.159	5.220	5.274	5.314	5.338	5.348	5.349	5.351	5.362

TABLE 8.11 : β_1 ($1/\phi = 1.0$, $1/\alpha = .6$)

μ MODE:	0.15	0.20	0.25	0.30	0.35	0.40	0.45	0.50	0.55	0.60	0.65	0.70	0.75	0.80	0.85
(1)	3.158	3.146	3.145	3.142	3.128	3.098	3.052	2.993	2.924	2.850	2.774	2.701	2.631	2.566	2.507
(2)	8.324	8.116	7.772	7.404	7.102	6.895	6.780	6.737	6.731	6.721	6.665	6.546	6.378	6.184	5.986

TABLE 8.11 : β_1 ($1/\phi = 1.0$, $1/\alpha = .7$)

μ MODE:	0.15	0.20	0.25	0.30	0.35	0.40	0.45	0.50	0.55	0.60	0.65	0.70	0.75	0.80	0.85
(1)	2.875	2.853	2.846	2.845	2.844	2.838	2.823	2.799	2.766	2.725	2.679	2.630	2.580	2.531	2.484
(2)	7.323	7.262	7.115	6.905	6.686	6.501	6.369	6.294	6.264	6.260	6.254	6.220	6.143	6.029	5.892

TABLE 8.11 : β_1 ($1/\phi = 1.0$, $1/\alpha = .8$)

μ MODE:	0.15	0.20	0.25	0.30	0.35	0.40	0.45	0.50	0.55	0.60	0.65	0.70	0.75	0.80	0.85
(1)	2.673	2.649	2.638	2.634	2.634	2.633	2.630	2.623	2.611	2.594	2.571	2.544	2.515	2.484	2.452
(2)	6.555	6.543	6.495	6.403	6.284	6.162	6.059	5.985	5.943	5.928	5.926	5.922	5.898	5.847	5.772

TABLE 8.11 : β_1 ($1/\phi = 1.0$, $1/\alpha = .9$)

μ MODE:	0.15	0.20	0.25	0.30	0.35	0.40	0.45	0.50	0.55	0.60	0.65	0.70	0.75	0.80	0.85
(1)	2.509	2.494	2.486	2.482	2.480	2.480	2.480	2.479	2.476	2.471	2.463	2.453	2.441	2.427	2.412
(2)	5.963	5.962	5.952	5.925	5.882	5.830	5.777	5.733	5.703	5.687	5.682	5.682	5.678	5.664	5.636

TABLE 8.12 : β_1 ($\phi = .6$, $\alpha = .6$)

μ MODE:	0.15	0.20	0.25	0.30	0.35	0.40	0.45	0.50	0.55	0.60	0.65	0.70	0.75	0.80	0.85
(1)	2.762	2.925	3.110	3.319	3.559	3.835	4.153	4.517	4.889	5.027	4.808	4.521	4.252	4.015	3.814
(2)	6.421	6.801	7.227	7.696	8.129	7.854	7.118	6.493	6.055	6.059	6.652	7.561	8.177	7.854	7.443

TABLE 8.12 : β_1 ($\phi = .6$, $\alpha = .7$)

μ MODE:	0.15	0.20	0.25	0.30	0.35	0.40	0.45	0.50	0.55	0.60	0.65	0.70	0.75	0.80	0.85
(1)	3.202	3.384	3.588	3.819	4.081	4.378	4.711	5.056	5.271	5.149	4.865	4.574	4.310	4.082	3.893
(2)	7.448	7.871	8.334	8.794	8.818	8.041	7.283	6.718	6.490	6.816	7.523	8.302	8.301	7.900	7.493

TABLE 8.12 : β_1 ($\phi = .6$, $\alpha = .8$)

μ MODE:	0.15	0.20	0.25	0.30	0.35	0.40	0.45	0.50	0.55	0.60	0.65	0.70	0.75	0.80	0.85
(1)	3.627	3.821	4.037	4.281	4.554	4.858	5.178	5.444	5.469	5.236	4.934	4.644	4.306	4.166	3.988
(2)	8.442	8.893	9.365	9.681	9.146	8.245	7.512	7.055	7.053	7.520	8.219	8.641	8.378	7.959	7.560

TABLE 8.12 : β_1 ($\phi = .6$, $\alpha = .9$)

μ MODE:	0.15	0.20	0.25	0.30	0.35	0.40	0.45	0.50	0.55	0.60	0.65	0.70	0.75	0.80	0.85
(1)	4.031	4.228	4.450	4.697	4.972	5.268	5.552	5.712	5.607	5.328	5.018	4.729	4.476	4.262	4.089
(2)	9.398	9.862	10.30	10.31	9.418	8.486	7.795	7.459	7.606	8.117	8.708	8.811	8.456	8.032	7.641

TABLE 8.12 : β_1 ($\phi = .6$, $\alpha = 1.0$)

μ MODE:	0.15	0.20	0.25	0.30	0.35	0.40	0.45	0.50	0.55	0.60	0.65	0.70	0.75	0.80	0.85
(1)	4.409	4.604	4.823	5.069	5.338	5.618	5.855	5.919	5.735	5.431	5.115	4.827	4.575	4.361	4.186
(2)	10.31	10.78	11.14	10.77	9.692	8.748	8.103	7.868	8.095	8.603	9.039	8.941	8.543	8.116	7.729

TABLE 8.12 : β_1 ($\phi = .7$, $\alpha = .6$)

μ MODE:	0.15	0.20	0.25	0.30	0.35	0.40	0.45	0.50	0.55	0.60	0.65	0.70	0.75	0.80	0.85
(1)	2.367	2.508	2.666	2.845	3.051	3.288	3.564	3.887	4.257	4.617	4.694	4.488	4.241	4.011	3.812
(2)	5.504	5.830	6.198	6.610	7.058	7.401	7.036	6.445	5.953	5.654	5.847	6.577	7.575	7.788	7.430

TABLE 8.12 : β_1 ($\phi = .7$, $\alpha = .7$)

μ MODE:	0.15	0.20	0.25	0.30	0.35	0.40	0.45	0.50	0.55	0.60	0.65	0.70	0.75	0.80	0.85
(1)	2.745	2.901	3.076	3.274	3.499	3.758	4.053	4.386	4.724	4.909	4.789	4.546	4.300	4.078	3.892
(2)	6.384	6.748	7.154	7.594	7.992	7.854	7.210	6.621	6.203	6.132	6.584	7.381	8.039	7.854	7.482

TABLE 8.12 : β_1 ($\phi = .7$, $\alpha = .8$)

μ MODE:	0.15	0.20	0.25	0.30	0.35	0.40	0.45	0.50	0.55	0.60	0.65	0.70	0.75	0.80	0.85
(1)	3.109	3.275	3.461	3.670	3.908	4.176	4.476	4.792	5.047	5.068	4.873	4.620	4.377	4.163	3.988
(2)	7.237	7.629	8.058	8.488	8.662	8.115	7.415	6.858	6.553	6.681	7.233	7.965	8.220	7.922	7.551

TABLE 8.12 : β_1 ($\phi = .7$, $\alpha = .9$)

μ MODE:	0.15	0.20	0.25	0.30	0.35	0.40	0.45	0.50	0.55	0.60	0.65	0.70	0.75	0.80	0.85
(1)	3.455	3.625	3.815	4.029	4.271	4.541	4.833	5.114	5.274	5.195	4.966	4.709	4.469	4.260	4.089
(2)	8.058	8.467	8.904	9.271	9.110	8.364	7.651	7.138	6.944	7.189	7.763	8.340	8.340	8.001	7.633

TABLE 8.12 : β_1 ($\phi = .7$, $\alpha = 1.0$)

μ MODE:	0.15	0.20	0.25	0.30	0.35	0.40	0.45	0.50	0.55	0.60	0.65	0.70	0.75	0.80	0.85
(1)	3.779	3.946	4.136	4.350	4.591	4.859	5.138	5.379	5.460	5.320	5.071	4.809	4.568	4.359	4.185
(2)	8.844	9.266	9.697	9.948	9.467	8.618	7.900	7.427	7.319	7.628	8.187	8.593	8.452	8.091	7.723

TABLE 8.12 : β_1 ($\phi = .8$, $\alpha = .6$)

μ MODE:	0.15	0.20	0.25	0.30	0.35	0.40	0.45	0.50	0.55	0.60	0.65	0.70	0.75	0.80	0.85
(1)	2.071	2.194	2.332	2.490	2.670	2.878	3.121	3.407	3.744	4.127	4.446	4.424	4.221	4.005	3.810
(2)	4.816	5.102	5.424	5.788	6.196	6.626	6.812	6.390	5.908	5.529	5.402	5.851	6.775	7.613	7.409

TABLE 8.12 : β_1 ($\phi = .8$, $\alpha = .7$)

μ MODE:	0.15	0.20	0.25	0.30	0.35	0.40	0.45	0.50	0.55	0.60	0.65	0.70	0.75	0.80	0.85
(1)	2.402	2.538	2.691	2.865	3.063	3.290	3.552	3.856	4.196	4.522	4.641	4.499	4.283	4.072	3.890
(2)	5.586	5.906	6.264	6.662	7.084	7.386	7.099	6.560	6.100	5.824	5.954	6.580	7.468	7.760	7.464

TABLE 8.12 : β_1 ($\phi = .8$, $\alpha = .8$)

μ MODE:	0.15	0.20	0.25	0.30	0.35	0.40	0.45	0.50	0.55	0.60	0.65	0.70	0.75	0.80	0.85
(1)	2.720	2.866	3.028	3.212	3.421	3.659	3.931	4.235	4.551	4.778	4.763	4.581	4.362	4.158	3.986
(2)	6.333	6.678	7.061	7.476	7.856	7.854	7.324	6.768	6.353	6.207	6.505	7.180	7.868	7.854	7.535

TABLE 8.12 : β_1 ($\phi = .8$, $\alpha = .9$)

μ MODE:	0.15	0.20	0.25	0.30	0.35	0.40	0.45	0.50	0.55	0.60	0.65	0.70	0.75	0.80	0.85
(1)	3.023	3.172	3.338	3.527	3.740	3.983	4.256	4.552	4.828	4.965	4.878	4.676	4.456	4.256	4.087
(2)	7.051	7.414	7.815	8.227	8.496	8.187	7.557	7.003	6.637	6.600	6.984	7.643	8.100	7.948	7.620

TABLE 8.12 : β_1 ($\phi = .8$, $\alpha = 1.0$)

μ MODE:	0.15	0.20	0.25	0.30	0.35	0.40	0.45	0.50	0.55	0.60	0.65	0.70	0.75	0.80	0.85
(1)	3.307	3.453	3.619	3.808	4.024	4.268	4.540	4.824	5.061	5.129	4.998	4.781	4.558	4.356	4.185
(2)	7.741	8.118	8.531	8.925	9.022	8.478	7.792	7.239	6.917	6.958	7.388	8.002	8.273	8.049	7.713

TABLE 8.12 : β_1 ($\phi = .9$, $\alpha = .6$)

μ MODE:	0.15	0.20	0.25	0.30	0.35	0.40	0.45	0.50	0.55	0.60	0.65	0.70	0.75	0.80	0.85
(1)	1.841	1.950	2.073	2.213	2.373	2.558	2.775	3.030	3.335	3.697	4.089	4.300	4.189	3.995	3.808
(2)	4.281	4.535	4.822	5.147	5.515	5.926	6.319	6.283	5.868	5.477	5.218	5.355	6.101	7.194	7.370

TABLE 8.12 : β_1 ($\phi = .9$, $\alpha = .7$)

μ MODE:	0.15	0.20	0.25	0.30	0.35	0.40	0.45	0.50	0.55	0.60	0.65	0.70	0.75	0.80	0.85
(1)	2.135	2.256	2.392	2.547	2.723	2.925	3.160	3.435	3.753	4.105	4.394	4.417	4.257	4.064	3.888
(2)	4.966	5.250	5.569	5.928	6.325	6.724	6.862	6.490	6.043	5.697	5.592	5.975	6.805	7.560	7.435

TABLE 8.12 : β_1 ($\phi = .9$, $\alpha = .8$)

μ MODE:	0.15	0.20	0.25	0.30	0.35	0.40	0.45	0.50	0.55	0.60	0.65	0.70	0.75	0.80	0.85
(1)	2.418	2.547	2.692	2.855	3.041	3.255	3.500	3.782	4.098	4.409	4.582	4.517	4.340	4.151	3.984
(2)	5.629	5.937	6.281	6.663	7.065	7.364	7.182	6.699	6.259	5.977	6.021	6.523	7.321	7.727	7.512

TABLE 8.12 : β_1 ($\phi = .9$, $\alpha = .9$)

μ MODE:	0.15	0.20	0.25	0.30	0.35	0.40	0.45	0.50	0.55	0.60	0.65	0.70	0.75	0.80	0.85
(1)	2.687	2.819	2.968	3.135	3.326	3.545	3.795	4.078	4.382	4.645	4.733	4.623	4.438	4.249	4.086
(2)	6.268	6.593	6.956	7.353	7.734	7.854	7.449	6.921	6.495	6.275	6.423	6.975	7.679	7.854	7.601

TABLE 8.12 : β_1 ($\phi = .9$, $\alpha = 1.0$)

μ MODE:	0.15	0.20	0.25	0.30	0.35	0.40	0.45	0.50	0.55	0.60	0.65	0.70	0.75	0.80	0.85
(1)	2.939	3.070	3.218	3.386	3.579	3.801	4.054	4.337	4.628	4.849	4.878	4.738	4.543	4.351	4.183
(2)	6.882	7.220	7.601	8.007	8.342	8.247	7.698	7.140	6.723	6.554	6.774	7.350	7.947	7.976	7.698

TABLE 8.12 : β_1 ($\phi = 1.0$, $\alpha = .6$)

μ MODE:	0.15	0.20	0.25	0.30	0.35	0.40	0.45	0.50	0.55	0.60	0.65	0.70	0.75	0.80	0.85
(1)	1.657	1.755	1.866	1.992	2.136	2.303	2.498	2.729	3.005	3.339	3.732	4.092	4.135	3.980	3.803
(2)	3.853	4.082	4.340	4.633	4.967	5.346	5.759	6.041	5.812	5.441	5.140	5.064	5.574	6.637	7.298

TABLE 8.12 : β_1 ($\phi = 1.0$, $\alpha = .7$)

μ MODE:	0.15	0.20	0.25	0.30	0.35	0.40	0.45	0.50	0.55	0.60	0.65	0.70	0.75	0.80	0.85
(1)	1.921	2.031	2.153	2.292	2.451	2.633	2.846	3.095	3.389	3.730	4.085	4.284	4.216	4.051	3.884
(2)	4.469	4.725	5.013	5.339	5.704	6.102	6.444	6.374	5.993	5.632	5.412	5.552	6.228	7.194	7.387

TABLE 8.12 : β_1 ($\phi = 1.0$, $\alpha = .8$)

μ MODE:	0.15	0.20	0.25	0.30	0.35	0.40	0.45	0.50	0.55	0.60	0.65	0.70	0.75	0.80	0.85
(1)	2.176	2.292	2.423	2.570	2.738	2.930	3.153	3.412	3.711	4.041	4.330	4.417	4.307	4.140	3.981
(2)	5.067	5.344	5.655	6.005	6.389	6.769	6.923	6.616	6.198	5.864	5.738	6.026	6.752	7.495	7.475

TABLE 8.12 : β_1 ($\phi = 1.0$, $\alpha = .9$)

μ MODE:	0.15	0.20	0.25	0.30	0.35	0.40	0.45	0.50	0.55	0.60	0.65	0.70	0.75	0.80	0.85
(1)	2.418	2.537	2.671	2.822	2.995	3.193	3.421	3.685	3.984	4.295	4.523	4.542	4.410	4.241	4.084
(2)	5.642	5.935	6.266	6.635	7.027	7.349	7.278	6.848	6.414	6.107	6.059	6.433	7.158	7.690	7.572

TABLE 8.12 : β_1 $(1/\phi = .6,\ 1/\alpha = .6)$

μ	0.15	0.20	0.25	0.30	0.35	0.40	0.45	0.50	0.55	0.60	0.65	0.70	0.75	0.80	0.85
MODE:															
(1)	2.264	2.326	2.417	2.534	2.680	2.858	3.076	3.341	3.667	4.067	4.546	4.979	4.976	4.757	4.522
(2)	5.757	6.060	6.421	6.842	7.324	7.854	8.162	7.644	7.027	6.503	6.104	5.997	6.651	7.854	8.100

TABLE 8.12 : β_1 $(1/\phi = .6,\ 1/\alpha = .7)$

μ	0.15	0.20	0.25	0.30	0.35	0.40	0.45	0.50	0.55	0.60	0.65	0.70	0.75	0.80	0.85
MODE:															
(1)	2.058	2.117	2.196	2.296	2.419	2.570	2.754	2.979	3.256	3.599	4.020	4.488	4.743	4.642	4.451
(2)	5.036	5.286	5.586	5.940	6.350	6.816	7.290	7.362	6.895	6.411	6.015	5.787	6.063	7.073	7.885

TABLE 8.12 : β_1 $(1/\phi = .6,\ 1/\alpha = .8)$

μ	0.15	0.20	0.25	0.30	0.35	0.40	0.45	0.50	0.55	0.60	0.65	0.70	0.75	0.80	0.85
MODE:															
(1)	1.883	1.945	2.022	2.116	2.228	2.363	2.527	2.726	2.970	3.272	3.646	4.088	4.458	4.496	4.359
(2)	4.498	4.714	4.974	5.279	5.636	6.046	6.498	6.852	6.691	6.285	5.913	5.660	5.739	6.513	7.568

TABLE 8.12 : β_1 $(1/\phi = .6,\ 1/\alpha = .9)$

μ	0.15	0.20	0.25	0.30	0.35	0.40	0.45	0.50	0.55	0.60	0.65	0.70	0.75	0.80	0.85
MODE:															
(1)	1.726	1.794	1.872	1.964	2.072	2.200	2.351	2.534	2.757	3.032	3.373	3.785	4.189	4.335	4.254
(2)	4.071	4.268	4.501	4.773	5.090	5.457	5.873	6.283	6.403	6.124	5.792	5.541	5.533	6.123	7.219

TABLE 8.12 : β_1 $(1/\phi = .6,\ 1/\alpha = 1.0)$

μ	0.15	0.20	0.25	0.30	0.35	0.40	0.45	0.50	0.55	0.60	0.65	0.70	0.75	0.80	0.85
MODE:															
(1)	1.587	1.658	1.738	1.829	1.935	2.058	2.204	2.377	2.587	2.844	3.163	3.552	3.963	4.180	4.147
(2)	3.717	3.903	4.118	4.367	4.657	4.992	5.375	5.784	6.060	5.936	5.654	5.415	5.366	5.833	6.904

TABLE 8.12 : β_1 $(1/\phi = .7,\ 1/\alpha = .6)$

μ	0.15	0.20	0.25	0.30	0.35	0.40	0.45	0.50	0.55	0.60	0.65	0.70	0.75	0.80	0.85
MODE:															
(1)	2.642	2.714	2.819	2.956	3.125	3.333	3.584	3.889	4.256	4.689	5.117	5.218	5.016	4.764	4.523
(2)	6.716	7.068	7.486	7.966	8.497	8.898	8.424	7.693	7.067	6.576	6.316	6.648	7.597	8.377	8.130

TABLE 8.12 : β_1 $(1/\phi = .7,\ 1/\alpha = .7)$

μ	0.15	0.20	0.25	0.30	0.35	0.40	0.45	0.50	0.55	0.60	0.65	0.70	0.75	0.80	0.85
MODE:															
(1)	2.401	2.469	2.561	2.678	2.822	2.997	3.210	3.469	3.783	4.160	4.582	4.886	4.851	4.662	4.454
(2)	5.875	6.165	6.513	6.919	7.379	7.854	8.011	7.516	6.948	6.473	6.149	6.182	6.851	7.854	7.986

TABLE 8.12 : β_1 $(1/\phi = .7,\ 1/\alpha = .8)$

μ	0.15	0.20	0.25	0.30	0.35	0.40	0.45	0.50	0.55	0.60	0.65	0.70	0.75	0.80	0.85
MODE:															
(1)	2.196	2.269	2.359	2.468	2.599	2.756	2.946	3.175	3.453	3.789	4.179	4.541	4.651	4.536	4.366
(2)	5.247	5.499	5.800	6.152	6.555	6.997	7.366	7.245	6.787	6.348	6.019	5.931	6.371	7.317	7.794

TABLE 8.12 : β_1 $(1/\phi = .7,\ 1/\alpha = .9)$

μ	0.15	0.20	0.25	0.30	0.35	0.40	0.45	0.50	0.55	0.60	0.65	0.70	0.75	0.80	0.85
MODE:															
(1)	2.014	2.093	2.184	2.292	2.417	2.565	2.741	2.953	3.207	3.516	3.879	4.250	4.449	4.402	4.267
(2)	4.749	4.979	5.249	5.563	5.925	6.330	6.733	6.883	6.587	6.199	5.883	5.748	6.044	6.889	7.582

TABLE 8.12 : β_1 $(1/\phi = .7,\ 1/\alpha = 1.0)$

μ	0.15	0.20	0.25	0.30	0.35	0.40	0.45	0.50	0.55	0.60	0.65	0.70	0.75	0.80	0.85
MODE:															
(1)	1.852	1.934	2.027	2.134	2.258	2.401	2.570	2.770	3.011	3.302	3.647	4.018	4.268	4.274	4.166
(2)	4.336	4.553	4.803	5.092	5.424	5.799	6.197	6.481	6.362	6.036	5.737	5.583	5.790	6.552	7.375

TABLE 8.12 : β_1 $(1/\phi = .8,\ 1/\alpha = .6)$

μ	0.15	0.20	0.25	0.30	0.35	0.40	0.45	0.50	0.55	0.60	0.65	0.70	0.75	0.80	0.85
MODE:															
(1)	3.019	3.102	3.221	3.377	3.570	3.805	4.088	4.427	4.822	5.236	5.447	5.295	5.031	4.767	4.524
(2)	7.675	8.074	8.544	9.072	9.568	9.338	8.491	7.732	7.127	6.723	6.761	7.433	8.378	8.513	8.142

TABLE 8.12 : β_1 $(1/\phi = .8,\ 1/\alpha = .7)$

μ	0.15	0.20	0.25	0.30	0.35	0.40	0.45	0.50	0.55	0.60	0.65	0.70	0.75	0.80	0.85
MODE:															
(1)	2.744	2.822	2.927	3.060	3.224	3.422	3.663	3.951	4.293	4.677	5.009	5.062	4.891	4.670	4.456
(2)	6.714	7.044	7.436	7.887	8.367	8.668	8.240	7.578	7.004	6.576	6.414	6.780	7.623	8.221	8.019

TABLE 8.12 : β_1 $(1/\phi = .8,\ 1/\alpha = .8)$

μ	0.15	0.20	0.25	0.30	0.35	0.40	0.45	0.50	0.55	0.60	0.65	0.70	0.75	0.00	0.85
MODE:															
(1)	2.510	2.593	2.696	2.820	2.969	3.148	3.362	3.618	3.923	4.274	4.623	4.801	4.727	4.554	4.369
(2)	5.997	6.283	6.623	7.016	7.451	7.854	7.857	7.369	6.849	6.433	6.208	6.383	7.070	7.854	7.868

TABLE 8.12 : β_1 $(1/\phi = .8,\ 1/\alpha = .9)$

μ	0.15	0.20	0.25	0.30	0.35	0.40	0.45	0.50	0.55	0.60	0.65	0.70	0.75	0.80	0.85
MODE:															
(1)	2.302	2.392	2.496	2.619	2.762	2.930	3.130	3.367	3.648	3.976	4.321	4.560	4.563	4.430	4.273
(2)	5.427	5.689	5.996	6.349	6.745	7.152	7.379	7.115	6.668	6.276	6.032	6.104	6.669	7.500	7.710

TABLE 8.12 : β_1 $(1/\phi = .8,\ 1/\alpha = 1.0)$

μ	0.15	0.20	0.25	0.30	0.35	0.40	0.45	0.50	0.55	0.60	0.65	0.70	0.75	0.80	0.85
MODE:															
(1)	2.116	2.210	2.317	2.439	2.580	2.743	2.935	3.161	3.429	3.742	4.083	4.357	4.415	4.314	4.175
(2)	4.956	5.203	5.487	5.813	6.181	6.575	6.896	6.838	6.476	6.110	5.860	5.872	6.351	7.191	7.561

TABLE 8.12 : β_1 $(1/\phi = .9,\ 1/\alpha = .6)$

μ	0.15	0.20	0.25	0.30	0.35	0.40	0.45	0.50	0.55	0.60	0.65	0.70	0.75	0.80	0.85
MODE:															
(1)	3.397	3.489	3.624	3.798	4.014	4.274	4.586	4.949	5.343	5.638	5.586	5.326	5.038	4.769	4.524
(2)	8.632	9.078	9.593	10.14	10.31	9.457	8.532	7.782	7.231	7.010	7.383	8.197	8.801	8.558	8.148

TABLE 8.12 : β_1 $(1/\phi = .9,\ 1/\alpha = .7)$

μ	0.15	0.20	0.25	0.30	0.35	0.40	0.45	0.50	0.55	0.60	0.65	0.70	0.75	0.80	0.85
MODE:															
(1)	3.087	3.175	3.293	3.441	3.624	3.846	4.110	4.422	4.775	5.115	5.262	5.137	4.910	4.675	4.457
(2)	7.552	7.920	8.353	8.833	9.253	9.065	8.325	7.630	7.083	6.756	6.845	7.446	8.231	8.352	8.034

TABLE 8.12 : β_1 $(1/\phi = .9,\ 1/\alpha = .8)$

μ	0.15	0.20	0.25	0.30	0.35	0.40	0.45	0.50	0.55	0.60	0.65	0.70	0.75	0.80	0.85
MODE:															
(1)	2.824	2.918	3.033	3.172	3.339	3.538	3.775	4.053	4.373	4.706	4.938	4.927	4.762	4.562	4.371
(2)	6.745	7.066	7.443	7.868	8.298	8.484	8.045	7.437	6.920	6.568	6.522	6.948	7.704	8.104	7.900

TABLE 8.12 : β_1 $(1/\phi = .9,\ 1/\alpha = .9)$

μ	0.15	0.20	0.25	0.30	0.35	0.40	0.45	0.50	0.55	0.60	0.65	0.70	0.75	0.80	0.85
MODE:															
(1)	2.590	2.691	2.808	2.946	3.106	3.294	3.516	3.775	4.076	4.397	4.664	4.729	4.615	4.444	4.276
(2)	6.105	6.399	6.739	7.126	7.536	7.854	7.709	7.216	6.737	6.386	6.277	6.587	7.287	7.854	7.763

TABLE 8.12 : β_1 $(1/\phi = .9,\ 1/\alpha = 1.0)$

μ	0.15	0.20	0.25	0.30	0.35	0.40	0.45	0.50	0.55	0.60	0.65	0.70	0.75	0.80	0.85
MODE:															
(1)	2.381	2.486	2.607	2.744	2.901	3.084	3.298	3.547	3.835	4.154	4.442	4.561	4.484	4.333	4.179
(2)	5.575	5.852	6.170	6.529	6.921	7.284	7.352	6.987	6.548	6.201	6.055	6.288	6.944	7.625	7.638

TABLE 8.12 : β_1 ($1/\phi = 1.0$, $1/\alpha = .6$)

μ MODE:	0.15	0.20	0.25	0.30	0.35	0.40	0.45	0.50	0.55	0.60	0.65	0.70	0.75	0.80	0.85
(1)	3.774	3.877	4.025	4.218	4.455	4.739	5.073	5.444	5.784	5.868	5.646	5.341	5.042	4.770	4.524
(2)	9.589	10.08	10.63	11.09	10.61	9.513	8.576	7.857	7.412	7.462	8.055	8.834	8.964	8.579	8.151

TABLE 8.12 : β_1 ($1/\phi = 1.0$, $1/\alpha = .7$)

μ MODE:	0.15	0.20	0.25	0.30	0.35	0.40	0.45	0.50	0.55	0.60	0.65	0.70	0.75	0.80	0.85
(1)	3.430	3.527	3.658	3.822	4.024	4.265	4.550	4.875	5.210	5.435	5.390	5.172	4.919	4.677	4.457
(2)	8.390	8.793	9.260	9.737	9.902	9.216	8.381	7.693	7.207	7.048	7.384	8.081	8.586	8.407	8.042

TABLE 8.12 : β_1 ($1/\phi = 1.0$, $1/\alpha = .8$)

μ MODE:	0.15	0.20	0.25	0.30	0.35	0.40	0.45	0.50	0.55	0.60	0.65	0.70	0.75	0.80	0.85
(1)	3.138	3.242	3.369	3.524	3.708	3.926	4.182	4.476	4.792	5.057	5.125	4.989	4.780	4.567	4.372
(2)	7.494	7.847	8.256	8.697	9.040	8.809	8.134	7.499	7.018	6.782	6.953	7.535	8.174	8.212	7.916

TABLE 8.12 : β_1 ($1/\phi = 1.0$, $1/\alpha = .9$)

μ MODE:	0.15	0.20	0.25	0.30	0.35	0.40	0.45	0.50	0.55	0.60	0.65	0.70	0.75	0.80	0.85
(1)	2.877	2.990	3.120	3.272	3.450	3.657	3.897	4.174	4.478	4.759	4.892	4.817	4.642	4.452	4.277
(2)	6.783	7.107	7.479	7.889	8.272	8.335	7.857	7.284	6.820	6.552	6.629	7.124	7.807	8.022	7.790

TABLE 8.13 : β_1 ($\phi = .6$, $\alpha = .6$)

μ MODE:	0.15	0.20	0.25	0.30	0.35	0.40	0.45	0.50	0.55	0.60	0.65	0.70	0.75	0.80	0.85
(1)	2.767	2.933	3.120	3.334	3.578	3.861	4.192	4.582	5.042	5.547	5.809	5.586	5.273	4.978	4.720
(2)	6.432	6.819	7.255	7.747	8.298	8.847	8.701	7.974	7.331	6.863	6.890	7.703	8.811	8.797	8.354

TABLE 8.13 : β_1 ($\phi = .6$, $\alpha = .7$)

μ MODE:	0.15	0.20	0.25	0.30	0.35	0.40	0.45	0.50	0.55	0.60	0.65	0.70	0.75	0.80	0.85
(1)	3.213	3.400	3.610	3.848	4.120	4.433	4.794	5.211	5.668	6.020	5.945	5.642	5.326	5.038	4.792
(2)	7.471	7.907	8.394	8.935	9.491	9.628	8.889	8.129	7.540	7.291	7.730	8.646	9.198	8.849	8.400

TABLE 8.13 : β_1 ($\phi = .6$, $\alpha = .8$)

μ MODE:	0.15	0.20	0.25	0.30	0.35	0.40	0.45	0.50	0.55	0.60	0.65	0.70	0.75	0.80	0.85
(1)	3.647	3.849	4.075	4.331	4.622	4.953	5.330	5.745	6.134	6.258	6.034	5.711	5.396	5.116	4.882
(2)	8.484	8.958	9.483	10.04	10.45	9.961	9.081	8.342	7.859	7.883	8.498	9.284	9.330	8.907	8.462

TABLE 8.13 : β_1 ($\phi = .6$, $\alpha = .9$)

μ MODE:	0.15	0.20	0.25	0.30	0.35	0.40	0.45	0.50	0.55	0.60	0.65	0.70	0.75	0.80	0.85
(1)	4.063	4.274	4.510	4.776	5.077	5.416	5.793	6.180	6.457	6.411	6.126	5.794	5.482	5.208	4.980
(2)	9.464	9.966	10.51	11.04	11.08	10.22	9.306	8.607	8.259	8.473	9.119	9.634	9.425	8.978	8.539

TABLE 8.13 : β_1 ($\phi = .6$, $\alpha = 1.0$)

μ MODE:	0.15	0.20	0.25	0.30	0.35	0.40	0.45	0.50	0.55	0.60	0.65	0.70	0.75	0.80	0.85
(1)	4.459	4.671	4.910	5.180	5.483	5.823	6.189	6.531	6.696	6.544	6.227	5.889	5.577	5.306	5.079
(2)	10.41	10.93	11.48	11.91	11.52	10.47	9.551	8.900	8.674	8.990	9.592	9.846	9.519	9.060	8.624

TABLE 8.13 : β_1 ($\phi = .7$, $\alpha = .6$)

μ MODE:	0.15	0.20	0.25	0.30	0.35	0.40	0.45	0.50	0.55	0.60	0.65	0.70	0.75	0.80	0.85
(1)	2.372	2.514	2.675	2.857	3.067	3.310	3.595	3.932	4.335	4.814	5.321	5.473	5.244	4.969	4.717
(2)	5.513	5.845	6.219	6.643	7.125	7.664	8.148	7.885	7.286	6.770	6.448	6.757	7.791	8.644	8.334

TABLE 8.13 : β_1 ($\phi = .7$, $\alpha = .7$)

μ MODE:	0.15	0.20	0.25	0.30	0.35	0.40	0.45	0.50	0.55	0.60	0.65	0.70	0.75	0.80	0.85
(1)	2.754	2.914	3.094	3.298	3.532	3.801	4.114	4.480	4.906	5.368	5.675	5.568	5.302	5.030	4.790
(2)	6.404	6.778	7.198	7.670	8.193	8.702	8.686	8.057	7.451	7.007	6.954	7.592	8.586	8.765	8.383

TABLE 8.13 : β_1 ($\phi = .7$, $\alpha = .8$)

μ MODE:	0.15	0.20	0.25	0.30	0.35	0.40	0.45	0.50	0.55	0.60	0.65	0.70	0.75	0.80	0.85
(1)	3.126	3.299	3.493	3.713	3.963	4.249	4.580	4.959	5.378	5.758	5.854	5.652	5.375	5.109	4.880
(2)	7.272	7.680	8.136	8.641	9.164	9.449	8.952	8.251	7.675	7.352	7.552	8.292	8.994	8.846	8.448

TABLE 8.13 : β_1 ($\phi = .7$, $\alpha = .9$)

μ MODE:	0.15	0.20	0.25	0.30	0.35	0.40	0.45	0.50	0.55	0.60	0.65	0.70	0.75	0.80	0.85
(1)	3.483	3.664	3.866	4.095	4.354	4.651	4.989	5.367	5.756	6.027	5.987	5.745	5.464	5.202	4.979
(2)	8.113	8.546	9.028	9.548	10.01	9.917	9.190	8.473	7.942	7.750	8.102	8.822	9.204	8.929	8.527

TABLE 8.13 : β_1 ($\phi = .7$, $\alpha = 1.0$)

μ MODE:	0.15	0.20	0.25	0.30	0.35	0.40	0.45	0.50	0.55	0.60	0.65	0.70	0.75	0.80	0.85
(1)	3.822	4.004	4.209	4.441	4.705	5.006	5.345	5.715	6.064	6.236	6.115	5.849	5.563	5.301	5.077
(2)	8.923	9.375	9.877	10.40	10.72	10.26	9.429	8.707	8.221	8.140	8.568	9.210	9.355	9.020	8.615

TABLE 8.13 : β_1 ($\phi = .8$, $\alpha = .6$)

μ MODE:	0.15	0.20	0.25	0.30	0.35	0.40	0.45	0.50	0.55	0.60	0.65	0.70	0.75	0.80	0.85
(1)	2.075	2.200	2.340	2.500	2.684	2.896	3.146	3.442	3.798	4.230	4.744	5.205	5.188	4.954	4.713
(2)	4.824	5.114	5.442	5.814	6.239	6.725	7.259	7.607	7.228	6.727	6.324	6.221	6.930	8.169	8.296

TABLE 8.13 : β_1 ($\phi = .8$, $\alpha = .7$)

μ MODE:	0.15	0.20	0.25	0.30	0.35	0.40	0.45	0.50	0.55	0.60	0.65	0.70	0.75	0.80	0.85
(1)	2.410	2.550	2.707	2.886	3.090	3.326	3.601	3.925	4.308	4.756	5.218	5.418	5.260	5.017	4.786
(2)	5.603	5.931	6.299	6.716	7.186	7.701	8.135	7.937	7.393	6.910	6.618	6.850	7.746	8.561	8.355

TABLE 8.13 : β_1 ($\phi = .8$, $\alpha = .8$)

μ MODE:	0.15	0.20	0.25	0.30	0.35	0.40	0.45	0.50	0.55	0.60	0.65	0.70	0.75	0.80	0.85
(1)	2.735	2.886	3.057	3.249	3.468	3.720	4.012	4.351	4.745	5.175	5.523	5.544	5.341	5.098	4.877
(2)	6.363	6.721	7.122	7.573	8.069	8.556	8.668	8.161	7.591	7.154	7.017	7.464	8.344	8.719	8.425

TABLE 8.13 : β_1 ($\phi = .8$, $\alpha = .9$)

μ MODE:	0.15	0.20	0.25	0.30	0.35	0.40	0.45	0.50	0.55	0.60	0.65	0.70	0.75	0.80	0.85
(1)	3.047	3.206	3.383	3.583	3.811	4.073	4.374	4.721	5.111	5.503	5.734	5.659	5.435	5.193	4.976
(2)	7.099	7.480	7.907	8.382	8.884	9.260	9.012	8.384	7.814	7.432	7.435	7.983	8.739	8.836	8.508

TABLE 8.13 : β_1 ($\phi = .8$, $\alpha = 1.0$)

μ MODE:	0.15	0.20	0.25	0.30	0.35	0.40	0.45	0.50	0.55	0.60	0.65	0.70	0.75	0.80	0.85
(1)	3.344	3.504	3.683	3.887	4.119	4.386	4.693	5.043	5.425	5.774	5.912	5.779	5.540	5.294	5.076
(2)	7.808	8.207	8.656	9.150	9.637	9.817	9.295	8.606	8.039	7.710	7.818	8.408	9.012	8.948	8.600

TABLE 8.13 : β_1 ($\phi = .9$, $\alpha = .6$)

μ MODE:	0.15	0.20	0.25	0.30	0.35	0.40	0.45	0.50	0.55	0.60	0.65	0.70	0.75	0.80	0.85
(1)	1.845	1.955	2.080	2.222	2.386	2.575	2.796	3.060	3.378	3.767	4.245	4.793	5.077	4.930	4.707
(2)	4.288	4.546	4.837	5.169	5.548	5.984	6.483	7.004	7.105	6.687	6.273	6.005	6.307	7.460	8.219

TABLE 8.13 : β_1 ($\phi = .9$, $\alpha = .7$)

μ	0.15	0.20	0.25	0.30	0.35	0.40	0.45	0.50	0.55	0.60	0.65	0.70	0.75	0.80	0.85
MODE: (1)	2.142	2.266	2.407	2.565	2.747	2.957	3.202	3.491	3.836	4.250	4.728	5.148	5.187	4.997	4.781
(2)	4.981	5.272	5.600	5.971	6.394	6.871	7.377	7.658	7.321	6.856	6.488	6.415	7.033	8.122	8.304

TABLE 8.13 : β_1 ($\phi = .9$, $\alpha = .8$)

μ	0.15	0.20	0.25	0.30	0.35	0.40	0.45	0.50	0.55	0.60	0.65	0.70	0.75	0.80	0.85
MODE: (1)	2.431	2.566	2.717	2.888	3.083	3.307	3.568	3.873	4.233	4.652	5.091	5.355	5.285	5.081	4.872
(2)	5.656	5.974	6.332	6.736	7.191	7.684	8.106	8.009	7.524	7.063	6.769	6.893	7.636	8.459	8.386

TABLE 8.13 : β_1 ($\phi = .9$, $\alpha = .9$)

μ	0.15	0.20	0.25	0.30	0.35	0.40	0.45	0.50	0.55	0.60	0.65	0.70	0.75	0.80	0.85
MODE: (1)	2.709	2.849	3.007	3.185	3.388	3.622	3.892	4.207	4.573	4.984	5.367	5.514	5.390	5.179	4.973
(2)	6.311	6.649	7.032	7.462	7.938	8.420	8.650	8.276	7.737	7.291	7.074	7.336	8.102	8.658	8.477

TABLE 8.13 : β_1 ($\phi = .9$, $\alpha = 1.0$)

μ	0.15	0.20	0.25	0.30	0.35	0.40	0.45	0.50	0.55	0.60	0.65	0.70	0.75	0.80	0.85
MODE: (1)	2.972	3.114	3.274	3.456	3.663	3.902	4.178	4.500	4.870	5.270	5.598	5.662	5.503	5.283	5.073
(2)	6.941	7.297	7.700	8.154	8.646	9.084	9.061	8.517	7.945	7.512	7.363	7.720	8.467	8.817	8.576

TABLE 8.13 : β_1 ($\phi = 1.0$, $\alpha = .6$)

μ	0.15	0.20	0.25	0.30	0.35	0.40	0.45	0.50	0.55	0.60	0.65	0.70	0.75	0.80	0.85
MODE: (1)	1.660	1.760	1.872	2.000	2.147	2.317	2.517	2.754	3.041	3.393	3.832	4.373	4.874	4.890	4.697
(2)	3.859	4.092	4.354	4.652	4.994	5.389	5.846	6.364	6.801	6.628	6.238	5.919	5.918	6.818	8.050

TABLE 8.13 : β_1 ($\phi = 1.0$, $\alpha = .7$)

μ	0.15	0.20	0.25	0.30	0.35	0.40	0.45	0.50	0.55	0.60	0.65	0.70	0.75	0.80	0.85
MODE: (1)	1.928	2.040	2.166	2.309	2.472	2.661	2.882	3.143	3.456	3.835	4.292	4.792	5.064	4.966	4.773
(2)	4.483	4.745	5.040	5.375	5.758	6.195	6.685	7.156	7.192	6.806	6.424	6.203	6.503	7.539	8.214

TABLE 8.13 : β_1 ($\phi = 1.0$, $\alpha = .8$)

μ	0.15	0.20	0.25	0.30	0.35	0.40	0.45	0.50	0.55	0.60	0.65	0.70	0.75	0.80	0.85
MODE: (1)	2.188	2.309	2.445	2.599	2.775	2.977	3.212	3.489	3.818	4.209	4.659	5.076	5.194	5.056	4.866
(2)	5.091	5.377	5.700	6.065	6.480	6.945	7.432	7.713	7.439	7.005	6.652	6.554	7.045	8.026	8.322

TABLE 8.13 : β_1 ($\phi = 1.0$, $\alpha = .9$)

μ	0.15	0.20	0.25	0.30	0.35	0.40	0.45	0.50	0.55	0.60	0.65	0.70	0.75	0.80	0.85
MODE: (1)	2.438	2.565	2.706	2.867	3.050	3.260	3.505	3.792	4.130	4.526	4.958	5.291	5.319	5.159	4.968
(2)	5.680	5.985	6.330	6.721	7.162	7.644	8.080	8.095	7.666	7.213	6.894	6.904	7.496	8.344	8.428

TABLE 8.13 : β_1 ($1/\phi = .6$, $1/\alpha = .6$)

μ	0.15	0.20	0.25	0.30	0.35	0.40	0.45	0.50	0.55	0.60	0.65	0.70	0.75	0.80	0.85
MODE: (1)	2.329	2.392	2.481	2.598	2.742	2.920	3.136	3.402	3.731	4.143	4.664	5.307	5.818	5.710	5.446
(2)	5.809	6.106	6.464	6.884	7.375	7.944	8.581	8.961	8.415	7.785	7.248	6.849	6.929	8.102	8.950

TABLE 8.13 : β_1 ($1/\phi = .6$, $1/\alpha = .7$)

μ	0.15	0.20	0.25	0.30	0.35	0.40	0.45	0.50	0.55	0.60	0.65	0.70	0.75	0.80	0.85
MODE: (1)	2.110	2.175	2.258	2.361	2.486	2.639	2.825	3.052	3.333	3.686	4.134	4.703	5.332	5.535	5.357
(2)	5.092	5.339	5.638	5.991	6.406	6.892	7.453	8.043	8.158	7.666	7.164	6.757	6.595	7.301	8.589

TABLE 8.13 : β_1 ($1/\phi = .6$, $1/\alpha = .8$)

μ	0.15	0.20	0.25	0.30	0.35	0.40	0.45	0.50	0.55	0.60	0.65	0.70	0.75	0.80	0.85
MODE: (1)	1.921	1.991	2.075	2.174	2.291	2.431	2.599	2.803	3.054	3.367	3.764	4.273	4.886	5.301	5.243
(2)	4.548	4.768	5.028	5.336	5.697	6.120	6.615	7.174	7.641	7.478	7.048	6.660	6.435	6.809	8.092

TABLE 8.13 : β_1 ($1/\phi = .6$, $1/\alpha = .9$)

μ	0.15	0.20	0.25	0.30	0.35	0.40	0.45	0.50	0.55	0.60	0.65	0.70	0.75	0.80	0.85
MODE: (1)	1.753	1.828	1.913	2.012	2.126	2.260	2.419	2.609	2.840	3.128	3.491	3.957	4.537	5.050	5.113
(2)	4.112	4.315	4.552	4.829	5.152	5.530	5.974	6.486	7.019	7.203	6.899	6.546	6.307	6.500	7.638

TABLE 8.13 : β_1 ($1/\phi = .6$, $1/\alpha = 1.0$)

μ	0.15	0.20	0.25	0.30	0.35	0.40	0.45	0.50	0.55	0.60	0.65	0.70	0.75	0.80	0.85
MODE: (1)	1.605	1.682	1.760	1.866	1.979	2.109	2.262	2.444	2.664	2.935	3.276	3.713	4.265	4.820	4.981
(2)	3.748	3.941	4.163	4.418	4.714	5.060	5.466	5.939	6.464	6.849	6.723	6.417	6.180	6.276	7.272

TABLE 8.13 : β_1 ($1/\phi = .7$, $1/\alpha = .6$)

μ	0.15	0.20	0.25	0.30	0.35	0.40	0.45	0.50	0.55	0.60	0.65	0.70	0.75	0.80	0.85
MODE: (1)	2.717	2.790	2.895	3.030	3.199	3.405	3.657	3.965	4.343	4.811	5.379	5.935	5.988	5.731	5.449
(2)	6.777	7.123	7.539	8.025	8.587	9.216	9.699	9.195	8.458	7.823	7.326	7.127	7.799	9.013	9.018

TABLE 8.13 : β_1 ($1/\phi = .7$, $1/\alpha = .7$)

μ	0.15	0.20	0.25	0.30	0.35	0.40	0.45	0.50	0.55	0.60	0.65	0.70	0.75	0.80	0.85
MODE: (1)	2.462	2.537	2.634	2.754	2.900	3.078	3.294	3.557	3.881	4.283	4.780	5.350	5.702	5.597	5.366
(2)	5.940	6.228	6.576	6.985	7.463	8.010	8.591	8.812	8.298	7.715	7.226	6.917	7.156	8.236	8.825

TABLE 8.13 : β_1 ($1/\phi = .7$, $1/\alpha = .8$)

μ	0.15	0.20	0.25	0.30	0.35	0.40	0.45	0.50	0.55	0.60	0.65	0.70	0.75	0.80	0.85
MODE: (1)	2.241	2.323	2.421	2.536	2.673	2.836	3.031	3.268	3.557	3.915	4.360	4.892	5.365	5.430	5.262
(2)	5.306	5.562	5.865	6.222	6.638	7.120	7.660	8.141	8.045	7.566	7.109	6.779	6.808	7.637	8.559

TABLE 8.13 : β_1 ($1/\phi = .7$, $1/\alpha = .9$)

μ	0.15	0.20	0.25	0.30	0.35	0.40	0.45	0.50	0.55	0.60	0.65	0.70	0.75	0.80	0.85
MODE: (1)	2.045	2.132	2.232	2.347	2.481	2.637	2.821	3.042	3.310	3.640	4.049	4.548	5.056	5.251	5.146
(2)	4.797	5.034	5.311	5.632	6.005	6.438	6.933	7.448	7.684	7.379	6.969	6.645	6.581	7.210	8.261

TABLE 8.13 : β_1 ($1/\phi = .7$, $1/\alpha = 1.0$)

μ	0.15	0.20	0.25	0.30	0.35	0.40	0.45	0.50	0.55	0.60	0.65	0.70	0.75	0.80	0.85
MODE: (1)	1.873	1.962	2.063	2.177	2.308	2.460	2.638	2.850	3.105	3.417	3.804	4.280	4.799	5.082	5.031
(2)	4.373	4.598	4.856	5.153	5.497	5.895	6.353	6.857	7.256	7.162	6.815	6.503	6.395	6.885	7.977

TABLE 8.13 : β_1 ($1/\phi = .8$, $1/\alpha = .6$)

μ	0.15	0.20	0.25	0.30	0.35	0.40	0.45	0.50	0.55	0.60	0.65	0.70	0.75	0.80	0.85
MODE: (1)	3.105	3.189	3.308	3.463	3.655	3.890	4.176	4.523	4.946	5.453	6.000	6.244	6.039	5.740	5.450
(2)	7.744	8.139	8.611	9.159	9.778	10.36	10.10	9.253	8.495	7.883	7.492	7.708	8.704	9.370	9.040

TABLE 8.13 : β_1 ($1/\phi = .8$, $1/\alpha = .7$)

μ	0.15	0.20	0.25	0.30	0.35	0.40	0.45	0.50	0.55	0.60	0.65	0.70	0.75	0.80	0.85
MODE: (1)	2.814	2.900	3.010	3.147	3.314	3.517	3.762	4.060	4.423	4.864	5.375	5.812	5.838	5.621	5.370
(2)	6.788	7.117	7.512	7.975	8.506	9.084	9.467	9.027	8.354	7.768	7.336	7.251	7.908	8.896	8.893

TABLE 8.13 : β_1 ($1/\phi = .8$, $1/\alpha = .8$)

μ MODE:	0.15	0.20	0.25	0.30	0.35	0.40	0.45	0.50	0.55	0.60	0.65	0.70	0.75	0.80	0.85
(1)	2.561	2.655	2.766	2.898	3.054	3.240	3.462	3.730	4.055	4.451	4.920	5.397	5.599	5.478	5.270
(2)	6.064	6.356	6.701	7.105	7.572	8.096	8.604	8.663	8.161	7.624	7.197	7.006	7.399	8.372	8.710

TABLE 8.13 : β_1 ($1/\phi = .8$, $1/\alpha = .9$)

μ MODE:	0.15	0.20	0.25	0.30	0.35	0.40	0.45	0.50	0.55	0.60	0.65	0.70	0.75	0.80	0.85
(1)	2.337	2.437	2.551	2.683	2.835	3.013	3.223	3.473	3.775	4.142	4.582	5.059	5.364	5.329	5.160
(2)	5.482	5.753	6.068	6.433	6.853	7.332	7.840	8.170	7.920	7.455	7.046	6.817	7.051	7.934	8.515

TABLE 8.13 : β_1 ($1/\phi = .8$, $1/\alpha = 1.0$)

μ MODE:	0.15	0.20	0.25	0.30	0.35	0.40	0.45	0.50	0.55	0.60	0.65	0.70	0.75	0.80	0.85
(1)	2.140	2.242	2.358	2.488	2.638	2.811	3.014	3.255	3.543	3.892	4.314	4.789	5.157	5.191	5.052
(2)	4.998	5.255	5.549	5.887	6.276	6.720	7.209	7.646	7.648	7.272	6.888	6.640	6.775	7.576	8.329

TABLE 8.13 : β_1 ($1/\phi = .9$, $1/\alpha = .6$)

μ MODE:	0.15	0.20	0.25	0.30	0.35	0.40	0.45	0.50	0.55	0.60	0.65	0.70	0.75	0.80	0.85
(1)	3.493	3.588	3.722	3.895	4.110	4.373	4.691	5.075	5.532	6.046	6.436	6.358	6.060	5.744	5.451
(2)	8.712	9.154	9.679	10.28	10.91	11.08	10.20	9.291	8.543	7.991	7.836	8.452	9.425	9.475	9.050

TABLE 8.13 : β_1 ($1/\phi = .9$, $1/\alpha = .7$)

μ MODE:	0.15	0.20	0.25	0.30	0.35	0.40	0.45	0.50	0.55	0.60	0.65	0.70	0.75	0.80	0.85
(1)	3.165	3.262	3.386	3.540	3.728	3.954	4.227	4.557	4.952	5.413	5.873	6.057	5.893	5.632	5.372
(2)	7.636	8.005	8.446	8.957	9.525	10.04	9.850	9.103	8.403	7.848	7.538	7.784	8.645	9.188	8.921

TABLE 8.13 : β_1 ($1/\phi = .9$, $1/\alpha = .8$)

μ MODE:	0.15	0.20	0.25	0.30	0.35	0.40	0.45	0.50	0.55	0.60	0.65	0.70	0.75	0.80	0.85
(1)	2.881	2.987	3.112	3.260	3.436	3.644	3.892	4.189	4.545	4.965	5.419	5.732	5.701	5.500	5.274
(2)	6.822	7.149	7.535	7.984	8.491	9.018	9.283	8.844	8.224	7.693	7.342	7.390	8.066	8.836	8.769

TABLE 8.13 : β_1 ($1/\phi = .9$, $1/\alpha = .9$)

μ MODE:	0.15	0.20	0.25	0.30	0.35	0.40	0.45	0.50	0.55	0.60	0.65	0.70	0.75	0.80	0.85
(1)	2.629	2.742	2.870	3.018	3.189	3.388	3.623	3.902	4.235	4.629	5.071	5.442	5.514	5.365	5.167
(2)	6.168	6.472	6.824	7.230	7.692	8.195	8.619	8.521	8.017	7.521	7.161	7.110	7.642	8.490	8.614

TABLE 8.13 : β_1 ($1/\phi = .9$, $1/\alpha = 1.0$)

μ MODE:	0.15	0.20	0.25	0.30	0.35	0.40	0.45	0.50	0.55	0.60	0.65	0.70	0.75	0.80	0.85
(1)	2.408	2.523	2.652	2.799	2.968	3.162	3.390	3.658	3.977	4.357	4.792	5.199	5.349	5.241	5.063
(2)	5.622	5.911	6.242	6.619	7.049	7.527	7.998	8.158	7.798	7.341	6.980	6.870	7.298	8.179	8.473

TABLE 8.13 : β_1 ($1/\phi = 1.0$, $1/\alpha = .6$)

μ MODE:	0.15	0.20	0.25	0.30	0.35	0.40	0.45	0.50	0.55	0.60	0.65	0.70	0.75	0.80	0.85
(1)	3.881	3.986	4.135	4.327	4.565	4.854	5.203	5.616	6.091	6.545	6.662	6.407	6.071	5.747	5.451
(2)	9.678	10.17	10.74	11.38	11.90	11.32	10.25	9.331	8.616	8.187	8.374	9.195	9.810	9.516	9.055

TABLE 8.13 : β_1 ($1/\phi = 1.0$, $1/\alpha = .7$)

μ MODE:	0.15	0.20	0.25	0.30	0.35	0.40	0.45	0.50	0.55	0.60	0.65	0.70	0.75	0.80	0.85
(1)	3.517	3.625	3.762	3.933	4.140	4.390	4.690	5.047	5.464	5.912	6.224	6.171	5.919	5.638	5.373
(2)	8.484	8.892	9.376	9.927	10.49	10.69	9.984	9.154	8.463	7.975	7.877	8.413	9.224	9.303	8.935

TABLE 8.13 : β_1 ($1/\phi = 1.0$, $1/\alpha = .8$)

μ MODE:	0.15	0.20	0.25	0.30	0.35	0.40	0.45	0.50	0.55	0.60	0.65	0.70	0.75	0.80	0.85
(1)	3.201	3.318	3.458	3.622	3.816	4.046	4.319	4.643	5.022	5.445	5.818	5.915	5.750	5.512	5.276
(2)	7.580	7.942	8.367	8.854	9.384	9.817	9.608	8.927	8.282	7.791	7.581	7.904	8.680	9.055	8.797

TABLE 8.13 : β_1 ($1/\phi = 1.0$, $1/\alpha = .9$)

μ MODE:	0.15	0.20	0.25	0.30	0.35	0.40	0.45	0.50	0.55	0.60	0.65	0.70	0.75	0.80	0.85
(1)	2.921	3.046	3.189	3.353	3.542	3.763	4.022	4.327	4.685	5.092	5.488	5.680	5.587	5.384	5.172
(2)	6.853	7.190	7.579	8.024	8.517	8.999	9.140	8.666	8.081	7.601	7.342	7.532	8.242	8.809	8.660

TABLE 8.14 : β_1 ($\phi = .6$, $\alpha = .6$)

μ MODE:	0.15	0.20	0.25	0.30	0.35	0.40	0.45	0.50	0.55	0.60	0.65	0.70	0.75	0.80	0.85
(1)	1.817	1.873	1.890	1.857	1.788	1.705	1.623	1.549	1.486	1.435	1.396	1.369	1.355	1.356	1.377
(2)	4.709	4.223	3.903	3.791	3.843	4.018	4.288	4.644	5.085	5.582	5.829	5.589	5.269	4.972	4.710

TABLE 8.14 : β_1 ($\phi = .6$, $\alpha = .7$)

μ MODE:	0.15	0.20	0.25	0.30	0.35	0.40	0.45	0.50	0.55	0.60	0.65	0.70	0.75	0.80	0.85
(1)	2.119	2.186	2.204	2.165	2.082	1.983	1.885	1.797	1.721	1.659	1.611	1.576	1.554	1.547	1.559
(2)	5.481	4.898	4.514	4.372	4.422	4.611	4.904	5.283	5.720	6.056	5.954	5.639	5.317	5.026	4.777

TABLE 8.14 : β_1 ($\phi = .6$, $\alpha = .8$)

μ MODE:	0.15	0.20	0.25	0.30	0.35	0.40	0.45	0.50	0.55	0.60	0.65	0.70	0.75	0.80	0.85
(1)	2.422	2.498	2.518	2.470	2.373	2.256	2.140	2.036	1.946	1.872	1.812	1.765	1.733	1.714	1.712
(2)	6.242	5.549	5.091	4.914	4.955	5.149	5.451	5.825	6.189	6.283	6.035	5.701	5.381	5.098	4.860

TABLE 8.14 : β_1 ($\phi = .6$, $\alpha = .9$)

μ MODE:	0.15	0.20	0.25	0.30	0.35	0.40	0.45	0.50	0.55	0.60	0.65	0.70	0.75	0.80	0.85
(1)	2.725	2.809	2.830	2.773	2.659	2.521	2.385	2.263	2.157	2.068	1.994	1.934	1.888	1.854	1.834
(2)	6.986	6.165	5.625	5.406	5.432	5.623	5.919	6.261	6.504	6.423	6.118	5.777	5.461	5.184	4.953

TABLE 8.14 : β_1 ($\phi = .6$, $\alpha = 1.0$)

μ MODE:	0.15	0.20	0.25	0.30	0.35	0.40	0.45	0.50	0.55	0.60	0.65	0.70	0.75	0.80	0.85
(1)	3.028	3.121	3.142	3.073	2.937	2.775	2.617	2.475	2.351	2.245	2.155	2.080	2.018	1.967	1.928
(2)	7.705	6.736	6.105	5.841	5.848	6.031	6.313	6.606	6.729	6.544	6.211	5.866	5.551	5.276	5.047

TABLE 8.14 : β_1 ($\phi = .7$, $\alpha = .6$)

μ MODE:	0.15	0.20	0.25	0.30	0.35	0.40	0.45	0.50	0.55	0.60	0.65	0.70	0.75	0.80	0.85
(1)	1.569	1.640	1.692	1.713	1.697	1.652	1.594	1.533	1.478	1.431	1.394	1.368	1.355	1.356	1.377
(2)	4.332	4.016	3.693	3.504	3.463	3.549	3.741	4.023	4.394	4.856	5.355	5.486	5.243	4.963	4.708

TABLE 8.14 : β_1 ($\phi = .7$, $\alpha = .7$)

μ MODE:	0.15	0.20	0.25	0.30	0.35	0.40	0.45	0.50	0.55	0.60	0.65	0.70	0.75	0.80	0.85
(1)	1.831	1.913	1.974	1.998	1.978	1.923	1.852	1.779	1.712	1.655	1.609	1.574	1.553	1.547	1.559
(2)	5.048	4.665	4.273	4.042	3.983	4.072	4.279	4.584	4.975	5.418	5.704	5.572	5.295	5.019	4.775

TABLE 8.14 : β_1 ($\phi = .7$, $\alpha = .8$)

μ MODE:	0.15	0.20	0.25	0.30	0.35	0.40	0.45	0.50	0.55	0.60	0.65	0.70	0.75	0.80	0.85
(1)	2.092	2.186	2.255	2.281	2.256	2.190	2.105	2.018	1.937	1.867	1.809	1.764	1.732	1.714	1.712
(2)	5.760	5.294	4.824	4.543	4.461	4.546	4.759	5.071	5.453	5.807	5.870	5.648	5.363	5.092	4.858

TABLE 8.14 : β_1 ($\phi = .7$, $\alpha = .9$)

μ MODE:	0.15	0.20	0.25	0.30	0.35	0.40	0.45	0.50	0.55	0.60	0.65	0.70	0.75	0.80	0.85
(1)	2.354	2.459	2.537	2.564	2.531	2.451	2.349	2.245	2.148	2.063	1.992	1.933	1.887	1.854	1.834
(2)	6.463	5.897	5.336	4.997	4.887	4.962	5.174	5.481	5.829	6.065	5.992	5.733	5.445	5.179	4.952

TABLE 8.14 : β_1 ($\phi = .7$, $\alpha = 1.0$)

μ MODE:	0.15	0.20	0.25	0.30	0.35	0.40	0.45	0.50	0.55	0.60	0.65	0.70	0.75	0.80	0.85
(1)	2.615	2.733	2.817	2.845	2.801	2.704	2.581	2.457	2.342	2.241	2.153	2.079	2.017	1.967	1.928
(2)	7.155	6.463	5.797	5.396	5.256	5.320	5.528	5.824	6.128	6.260	6.109	5.829	5.537	5.272	5.046

TABLE 8.14 : β_1 ($\phi = .8$, $\alpha = .6$)

μ MODE:	0.15	0.20	0.25	0.30	0.35	0.40	0.45	0.50	0.55	0.60	0.65	0.70	0.75	0.80	0.85
(1)	1.378	1.450	1.516	1.565	1.589	1.583	1.553	1.510	1.466	1.424	1.391	1.367	1.354	1.355	1.377
(2)	3.940	3.802	3.542	3.328	3.223	3.236	3.356	3.573	3.881	4.285	4.784	5.233	5.193	4.950	4.704

TABLE 8.14 : β_1 ($\phi = .8$, $\alpha = .7$)

μ MODE:	0.15	0.20	0.25	0.30	0.35	0.40	0.45	0.50	0.55	0.60	0.65	0.70	0.75	0.80	0.85
(1)	1.608	1.692	1.768	1.826	1.853	1.844	1.806	1.754	1.699	1.648	1.605	1.573	1.552	1.547	1.559
(2)	4.595	4.423	4.103	3.841	3.708	3.712	3.839	4.072	4.402	4.819	5.263	5.435	5.257	5.007	4.771

TABLE 8.14 : β_1 ($\phi = .8$, $\alpha = .8$)

μ MODE:	0.15	0.20	0.25	0.30	0.35	0.40	0.45	0.50	0.55	0.60	0.65	0.70	0.75	0.80	0.85
(1)	1.838	1.933	2.021	2.086	2.115	2.102	2.055	1.990	1.923	1.860	1.806	1.763	1.731	1.714	1.712
(2)	5.248	5.031	4.640	4.320	4.152	4.141	4.268	4.509	4.844	5.241	5.560	5.550	5.331	5.082	4.855

TABLE 8.14 : β_1 ($\phi = .8$, $\alpha = .9$)

μ MODE:	0.15	0.20	0.25	0.30	0.35	0.40	0.45	0.50	0.55	0.60	0.65	0.70	0.75	0.80	0.85
(1)	2.068	2.175	2.273	2.346	2.376	2.356	2.296	2.217	2.134	2.057	1.989	1.932	1.887	1.854	1.834
(2)	5.897	5.622	5.142	4.755	4.547	4.517	4.639	4.880	5.210	5.563	5.758	5.655	5.418	5.170	4.950

TABLE 8.14 : β_1 ($\phi = .8$, $\alpha = 1.0$)

μ MODE:	0.15	0.20	0.25	0.30	0.35	0.40	0.45	0.50	0.55	0.60	0.65	0.70	0.75	0.80	0.85
(1)	2.297	2.417	2.525	2.606	2.634	2.604	2.528	2.430	2.329	2.235	2.150	2.078	2.017	1.967	1.928
(2)	6.543	6.186	5.598	5.137	4.886	4.836	4.954	5.195	5.515	5.822	5.922	5.765	5.516	5.265	5.044

TABLE 8.14 : β_1 ($\phi = .9$, $\alpha = .6$)

μ MODE:	0.15	0.20	0.25	0.30	0.35	0.40	0.45	0.50	0.55	0.60	0.65	0.70	0.75	0.80	0.85
(1)	1.228	1.296	1.365	1.428	1.476	1.501	1.500	1.479	1.448	1.415	1.386	1.364	1.353	1.355	1.377
(2)	3.576	3.565	3.402	3.205	3.068	3.026	3.084	3.241	3.491	3.839	4.294	4.829	5.092	4.928	4.698

TABLE 8.14 : β_1 ($\phi = .9$, $\alpha = .7$)

μ MODE:	0.15	0.20	0.25	0.30	0.35	0.40	0.45	0.50	0.55	0.60	0.65	0.70	0.75	0.80	0.85
(1)	1.432	1.512	1.592	1.665	1.721	1.749	1.746	1.719	1.680	1.638	1.600	1.570	1.551	1.546	1.559
(2)	4.172	4.152	3.948	3.702	3.530	3.470	3.526	3.693	3.963	4.331	4.784	5.181	5.192	4.989	4.766

TABLE 8.14 : β_1 ($\phi = .9$, $\alpha = .8$)

μ MODE:	0.15	0.20	0.25	0.30	0.35	0.40	0.45	0.50	0.55	0.60	0.65	0.70	0.75	0.80	0.85
(1)	1.637	1.728	1.820	1.903	1.966	1.995	1.988	1.953	1.903	1.850	1.801	1.760	1.731	1.713	1.712
(2)	4.766	4.733	4.474	4.169	3.955	3.870	3.918	4.089	4.366	4.736	5.144	5.376	5.280	5.066	4.851

TABLE 8.14 : β_1 ($\phi = .9$, $\alpha = .9$)

μ MODE:	0.15	0.20	0.25	0.30	0.35	0.40	0.45	0.50	0.55	0.60	0.65	0.70	0.75	0.80	0.85
(1)	1.842	1.945	2.047	2.140	2.209	2.239	2.226	2.178	2.114	2.047	1.984	1.930	1.886	1.854	1.834
(2)	5.360	5.303	4.971	4.596	4.332	4.218	4.255	4.425	4.705	5.064	5.411	5.522	5.377	5.157	4.946

TABLE 8.14 : β_1 ($\phi = .9$, $\alpha = 1.0$)

μ MODE:	0.15	0.20	0.25	0.30	0.35	0.40	0.45	0.50	0.55	0.60	0.65	0.70	0.75	0.80	0.85
(1)	2.046	2.161	2.275	2.377	2.452	2.480	2.455	2.392	2.310	2.226	2.146	2.076	2.016	1.967	1.928
(2)	5.951	5.859	5.431	4.972	4.655	4.512	4.539	4.709	4.992	5.339	5.628	5.658	5.482	5.255	5.042

TABLE 8.14 : β_1 ($\phi = 1.0$, $\alpha = .6$)

μ MODE:	0.15	0.20	0.25	0.30	0.35	0.40	0.45	0.50	0.55	0.60	0.65	0.70	0.75	0.80	0.85
(1)	1.106	1.171	1.238	1.305	1.366	1.412	1.437	1.439	1.425	1.403	1.380	1.361	1.351	1.354	1.376
(2)	3.257	3.318	3.253	3.101	2.959	2.882	2.891	2.994	3.192	3.489	3.894	4.416	4.900	4.892	4.690

TABLE 8.14 : β_1 ($\phi = 1.0$, $\alpha = .7$)

μ MODE:	0.15	0.20	0.25	0.30	0.35	0.40	0.45	0.50	0.55	0.60	0.65	0.70	0.75	0.80	0.85
(1)	1.291	1.366	1.444	1.522	1.593	1.646	1.674	1.674	1.654	1.624	1.593	1.567	1.550	1.546	1.559
(2)	3.799	3.868	3.781	3.587	3.408	3.306	3.305	3.411	3.623	3.940	4.360	4.837	5.079	4.961	4.759

TABLE 8.14 : β_1 ($\phi = 1.0$, $\alpha = .8$)

μ MODE:	0.15	0.20	0.25	0.30	0.35	0.40	0.45	0.50	0.55	0.60	0.65	0.70	0.75	0.80	0.85
(1)	1.475	1.561	1.651	1.740	1.820	1.880	1.908	1.904	1.876	1.836	1.794	1.757	1.729	1.713	1.712
(2)	4.341	4.414	4.295	4.048	3.821	3.688	3.670	3.774	3.993	4.317	4.728	5.113	5.198	5.043	4.845

TABLE 8.14 : β_1 ($\phi = 1.0$, $\alpha = .9$)

μ MODE:	0.15	0.20	0.25	0.30	0.35	0.40	0.45	0.50	0.55	0.60	0.65	0.70	0.75	0.80	0.85
(1)	1.659	1.756	1.857	1.957	2.047	2.112	2.139	2.127	2.087	2.033	1.978	1.927	1.885	1.853	1.834
(2)	4.883	4.956	4.789	4.471	4.189	4.019	3.983	4.082	4.303	4.630	5.019	5.314	5.312	5.139	4.941

TABLE 8.14 : β_1 ($1/\phi = .6$, $1/\alpha = .6$)

μ MODE:	0.15	0.20	0.25	0.30	0.35	0.40	0.45	0.50	0.55	0.60	0.65	0.70	0.75	0.80	0.85
(1)	1.847	1.962	2.090	2.234	2.396	2.574	2.759	2.913	2.947	2.845	2.694	2.541	2.400	2.273	2.158
(2)	5.529	5.838	6.082	5.802	5.147	4.595	4.186	3.944	3.950	4.225	4.695	5.314	5.792	5.669	5.408

TABLE 8.14 : β_1 ($1/\phi = .6$, $1/\alpha = .7$)

μ MODE:	0.15	0.20	0.25	0.30	0.35	0.40	0.45	0.50	0.55	0.60	0.65	0.70	0.75	0.80	0.85
(1)	1.584	1.682	1.792	1.915	2.054	2.208	2.372	2.531	2.638	2.638	2.555	2.442	2.327	2.219	2.121
(2)	4.739	5.007	5.257	5.308	4.923	4.455	4.076	3.818	3.725	3.857	4.205	4.729	5.329	5.501	5.318

TABLE 8.14 : β_1 ($1/\phi = .6$, $1/\alpha = .8$)

μ MODE:	0.15	0.20	0.25	0.30	0.35	0.40	0.45	0.50	0.55	0.60	0.65	0.70	0.75	0.80	0.85
(1)	1.386	1.471	1.568	1.676	1.797	1.932	2.079	2.227	2.350	2.407	2.385	2.316	2.231	2.147	2.069
(2)	4.147	4.383	4.612	4.753	4.608	4.262	3.936	3.696	3.578	3.632	3.886	4.325	4.900	5.279	5.207

TABLE 8.14 : β_1 ($1/\phi = .6$, $1/\alpha = .9$)

μ MODE:	0.15	0.20	0.25	0.30	0.35	0.40	0.45	0.50	0.55	0.60	0.65	0.70	0.75	0.80	0.85
(1)	1.232	1.308	1.394	1.490	1.598	1.718	1.849	1.984	2.107	2.187	2.205	2.173	2.119	2.060	2.003
(2)	3.686	3.896	4.105	4.264	4.247	4.026	3.765	3.557	3.443	3.467	3.663	4.037	4.568	5.042	5.082

TABLE 8.14 : β_1 $(1/\phi = .6,\ 1/\alpha = 1.0)$

μ MODE:	0.15	0.20	0.25	0.30	0.35	0.40	0.45	0.50	0.55	0.60	0.65	0.70	0.75	0.80	0.85
(1)	1.108	1.177	1.254	1.341	1.438	1.546	1.665	1.788	1.905	1.992	2.032	2.027	1.998	1.961	1.926
(2)	3.318	3.507	3.697	3.854	3.896	3.768	3.570	3.400	3.303	3.320	3.486	3.818	4.313	4.826	4.957

TABLE 8.14 : β_1 $(1/\phi = .7,\ 1/\alpha = .6)$

μ MODE:	0.15	0.20	0.25	0.30	0.35	0.40	0.45	0.50	0.55	0.60	0.65	0.70	0.75	0.80	0.85
(1)	2.155	2.287	2.434	2.597	2.774	2.954	3.104	3.142	3.038	2.873	2.702	2.544	2.401	2.273	2.158
(2)	6.434	6.740	6.676	5.917	5.203	4.674	4.340	4.261	4.460	4.858	5.395	5.922	5.949	5.689	5.408

TABLE 8.14 : β_1 $(1/\phi = .7,\ 1/\alpha = .7)$

μ MODE:	0.15	0.20	0.25	0.30	0.35	0.40	0.45	0.50	0.55	0.60	0.65	0.70	0.75	0.80	0.85
(1)	1.847	1.960	2.087	2.226	2.379	2.539	2.688	2.787	2.786	2.698	2.575	2.449	2.329	2.220	2.121
(2)	5.516	5.793	5.933	5.584	5.011	4.533	4.199	4.042	4.107	4.383	4.822	5.359	5.677	5.559	5.326

TABLE 8.14 : β_1 $(1/\phi = .7,\ 1/\alpha = .8)$

μ MODE:	0.15	0.20	0.25	0.30	0.35	0.40	0.45	0.50	0.55	0.60	0.65	0.70	0.75	0.80	0.85
(1)	1.616	1.715	1.826	1.948	2.082	2.224	2.363	2.474	2.523	2.495	2.420	2.328	2.235	2.148	2.069
(2)	4.827	5.074	5.251	5.149	4.753	4.351	4.049	3.881	3.884	4.075	4.434	4.922	5.358	5.398	5.225

TABLE 8.14 : β_1 $(1/\phi = .7,\ 1/\alpha = .9)$

μ MODE:	0.15	0.20	0.25	0.30	0.35	0.40	0.45	0.50	0.55	0.60	0.65	0.70	0.75	0.80	0.85
(1)	1.437	1.525	1.623	1.732	1.851	1.978	2.104	2.214	2.282	2.291	2.253	2.192	2.126	2.061	2.004
(2)	4.290	4.513	4.690	4.696	4.451	4.135	3.876	3.723	3.706	3.851	4.156	4.598	5.066	5.228	5.113

TABLE 8.14 : β_1 $(1/\phi = .7,\ 1/\alpha = 1.0)$

μ MODE:	0.15	0.20	0.25	0.30	0.35	0.40	0.45	0.50	0.55	0.60	0.65	0.70	0.75	0.80	0.85
(1)	1.293	1.372	1.461	1.559	1.666	1.780	1.896	2.000	2.074	2.102	2.089	2.052	2.007	1.964	1.927
(2)	3.862	4.063	4.231	4.281	4.136	3.897	3.685	3.555	3.540	3.665	3.938	4.349	4.825	5.070	5.003

TABLE 8.14 : β_1 $(1/\phi = .8,\ 1/\alpha = .6)$

μ MODE:	0.15	0.20	0.25	0.30	0.35	0.40	0.45	0.50	0.55	0.60	0.65	0.70	0.75	0.80	0.85
(1)	2.462	2.611	2.775	2.951	3.131	3.285	3.344	3.252	3.075	2.885	2.706	2.545	2.402	2.273	2.158
(2)	7.321	7.527	6.910	5.981	5.274	4.803	4.601	4.698	5.018	5.482	6.001	6.212	5.996	5.697	5.409

TABLE 8.14 : β_1 $(1/\phi = .8,\ 1/\alpha = .7)$

μ MODE:	0.15	0.20	0.25	0.30	0.35	0.40	0.45	0.50	0.55	0.60	0.65	0.70	0.75	0.80	0.85
(1)	2.110	2.238	2.379	2.530	2.688	2.835	2.937	2.944	2.857	2.724	2.585	2.452	2.330	2.220	2.121
(2)	6.279	6.518	6.359	5.704	5.088	4.642	4.392	4.368	4.564	4.927	5.399	5.802	5.804	5.581	5.330

TABLE 8.14 : β_1 $(1/\phi = .8,\ 1/\alpha = .8)$

μ MODE:	0.15	0.20	0.25	0.30	0.35	0.40	0.45	0.50	0.55	0.60	0.65	0.70	0.75	0.80	0.85
(1)	1.846	1.958	2.081	2.214	2.353	2.488	2.596	2.644	2.617	2.537	2.437	2.334	2.237	2.149	2.069
(2)	5.495	5.724	5.736	5.344	4.849	4.456	4.213	4.147	4.268	4.555	4.969	5.407	5.577	5.443	5.232

TABLE 8.14 : β_1 $(1/\phi = .8,\ 1/\alpha = .9)$

μ MODE:	0.15	0.20	0.25	0.30	0.35	0.40	0.45	0.50	0.55	0.60	0.65	0.70	0.75	0.80	0.85
(1)	1.641	1.741	1.850	1.969	2.093	2.215	2.319	2.382	2.387	2.345	2.277	2.201	2.129	2.062	2.004
(2)	4.885	5.096	5.168	4.946	4.570	4.242	4.027	3.955	4.043	4.284	4.654	5.089	5.355	5.300	5.126

TABLE 8.14 : β_1 $(1/\phi = .8,\ 1/\alpha = 1.0)$

μ MODE:	0.15	0.20	0.25	0.30	0.35	0.40	0.45	0.50	0.55	0.60	0.65	0.70	0.75	0.80	0.85
(1)	1.477	1.567	1.665	1.772	1.884	1.995	2.093	2.159	2.181	2.162	2.117	2.064	2.012	1.965	1.927
(2)	4.397	4.590	4.682	4.556	4.277	4.009	3.828	3.767	3.844	4.062	4.406	4.835	5.162	5.170	5.023

TABLE 8.14 : β_1 $(1/\phi = .9,\ 1/\alpha = .6)$

μ MODE:	0.15	0.20	0.25	0.30	0.35	0.40	0.45	0.50	0.55	0.60	0.65	0.70	0.75	0.80	0.85
(1)	2.768	2.933	3.110	3.291	3.455	3.544	3.483	3.304	3.093	2.891	2.708	2.546	2.402	2.273	2.158
(2)	8.178	8.078	7.013	6.046	5.378	5.007	4.963	5.188	5.580	6.061	6.418	6.318	6.016	5.702	5.410

TABLE 8.14 : β_1 $(1/\phi = .9,\ 1/\alpha = .7)$

μ MODE:	0.15	0.20	0.25	0.30	0.35	0.40	0.45	0.50	0.55	0.60	0.65	0.70	0.75	0.80	0.85
(1)	2.372	2.514	2.666	2.823	2.973	3.084	3.109	3.031	2.892	2.738	2.589	2.453	2.331	2.220	2.121
(2)	7.020	7.128	6.573	5.787	5.183	4.801	4.663	4.763	5.049	5.456	5.881	6.032	5.855	5.592	5.332

TABLE 8.14 : β_1 $(1/\phi = .9,\ 1/\alpha = .8)$

μ MODE:	0.15	0.20	0.25	0.30	0.35	0.40	0.45	0.50	0.55	0.60	0.65	0.70	0.75	0.80	0.85
(1)	2.076	2.200	2.333	2.471	2.606	2.716	2.771	2.750	2.668	2.559	2.445	2.337	2.238	2.149	2.069
(2)	6.147	6.299	6.039	5.460	4.946	4.598	4.439	4.479	4.693	5.038	5.449	5.724	5.671	5.464	5.236

TABLE 8.14 : β_1 $(1/\phi = .9,\ 1/\alpha = .9)$

μ MODE:	0.15	0.20	0.25	0.30	0.35	0.40	0.45	0.50	0.55	0.60	0.65	0.70	0.75	0.80	0.85
(1)	1.845	1.955	2.074	2.197	2.318	2.422	2.486	2.493	2.448	2.373	2.289	2.206	2.130	2.063	2.004
(2)	5.466	5.624	5.504	5.094	4.675	4.375	4.228	4.248	4.424	4.730	5.121	5.452	5.494	5.334	5.133

TABLE 8.14 : β_1 $(1/\phi = .9,\ 1/\alpha = 1.0)$

μ MODE:	0.15	0.20	0.25	0.30	0.35	0.40	0.45	0.50	0.55	0.60	0.65	0.70	0.75	0.80	0.85
(1)	1.661	1.760	1.866	1.978	2.087	2.184	2.249	2.269	2.245	2.194	2.132	2.070	2.014	1.966	1.927
(2)	4.920	5.073	5.020	4.726	4.389	4.137	4.013	4.033	4.193	4.479	4.859	5.225	5.340	5.216	5.033

TABLE 8.14 : β_1 $(1/\phi = 1.0,\ 1/\alpha = .6)$

μ MODE:	0.15	0.20	0.25	0.30	0.35	0.40	0.45	0.50	0.55	0.60	0.65	0.70	0.75	0.80	0.85
(1)	3.073	3.251	3.436	3.612	3.734	3.722	3.559	3.330	3.102	2.894	2.709	2.546	2.402	2.273	2.158
(2)	8.977	8.370	7.085	6.128	5.529	5.292	5.387	5.695	6.122	6.545	6.629	6.363	6.026	5.704	5.410

TABLE 8.14 : β_1 $(1/\phi = 1.0,\ 1/\alpha = .7)$

μ MODE:	0.15	0.20	0.25	0.30	0.35	0.40	0.45	0.50	0.55	0.60	0.65	0.70	0.75	0.80	0.85
(1)	2.634	2.787	2.946	3.101	3.226	3.277	3.218	3.079	2.911	2.745	2.592	2.454	2.331	2.220	2.121
(2)	7.726	7.566	6.694	5.872	5.310	5.019	4.998	5.193	5.533	5.938	6.214	6.137	5.879	5.597	5.333

TABLE 8.14 : β_1 $(1/\phi = 1.0,\ 1/\alpha = .8)$

μ MODE:	0.15	0.20	0.25	0.30	0.35	0.40	0.45	0.50	0.55	0.60	0.65	0.70	0.75	0.80	0.85
(1)	2.305	2.439	2.579	2.716	2.833	2.901	2.892	2.813	2.697	2.571	2.450	2.339	2.239	2.149	2.069
(2)	6.772	6.764	6.219	5.557	5.065	4.785	4.721	4.852	5.130	5.497	5.830	5.895	5.717	5.475	5.239

TABLE 8.14 : β_1 $(1/\phi = 1.0,\ 1/\alpha = .9)$

μ MODE:	0.15	0.20	0.25	0.30	0.35	0.40	0.45	0.50	0.55	0.60	0.65	0.70	0.75	0.80	0.85
(1)	2.049	2.168	2.292	2.415	2.523	2.593	2.607	2.563	2.483	2.389	2.296	2.209	2.131	2.063	2.004
(2)	6.024	6.070	5.722	5.206	4.791	4.545	4.478	4.582	4.825	5.167	5.520	5.675	5.562	5.351	5.137

TABLE 8.15 : β_1 $(1/\phi = .6, 1/\alpha = .6)$

μ	0.075	0.100	0.125	0.150	0.175	0.200	0.225	0.250	0.275	0.300	0.325	0.350	0.375	0.400	0.425
MODE:															
(1)	3.452	3.424	3.395	3.360	3.319	3.273	3.226	3.180	3.140	3.108	3.087	3.079	3.088	3.120	3.184
(2)	6.505	6.411	6.295	6.179	6.078	6.000	5.949	5.929	5.945	6.001	6.099	6.240	6.419	6.616	6.794
(3)	9.499	9.320	9.181	9.117	9.146	9.283	9.534	9.885	10.30	10.68	10.86	10.79	10.57	10.33	10.11
(4)	12.46	12.29	12.27	12.43	12.76	13.21	13.69	13.98	13.88	13.60	13.36	13.13	12.86	12.67	12.72

TABLE 8.15 : β_1 $(1/\phi = .6, 1/\alpha = .7)$

μ	0.075	0.100	0.125	0.150	0.175	0.200	0.225	0.250	0.275	0.300	0.325	0.350	0.375	0.400	0.425
MODE:															
(1)	3.071	3.048	3.032	3.019	3.007	2.995	2.983	2.973	2.966	2.965	2.970	2.986	3.017	3.067	3.148
(2)	5.714	5.687	5.657	5.629	5.610	5.605	5.620	5.659	5.728	5.829	5.968	6.145	6.355	6.576	6.771
(3)	8.368	8.325	8.302	8.320	8.393	8.534	8.752	9.048	9.406	9.780	10.07	10.19	10.16	10.05	9.933
(4)	11.02	11.00	11.05	11.21	11.48	11.85	12.29	12.68	12.87	12.83	12.72	12.61	12.50	12.45	12.60

TABLE 8.15 : β_1 $(1/\phi = .6, 1/\alpha = .8)$

μ	0.075	0.100	0.125	0.150	0.175	0.200	0.225	0.250	0.275	0.300	0.325	0.350	0.375	0.400	0.425
MODE:															
(1)	2.785	2.767	2.758	2.753	2.753	2.756	2.763	2.773	2.788	2.809	2.837	2.875	2.926	2.997	3.097
(2)	5.109	5.107	5.114	5.131	5.160	5.205	5.269	5.356	5.470	5.617	5.799	6.018	6.266	6.519	6.738
(3)	7.465	7.484	7.529	7.607	7.727	7.899	8.127	8.415	8.754	9.114	9.434	9.640	9.719	9.725	9.706
(4)	9.839	9.901	10.02	10.22	10.49	10.84	11.24	11.64	11.94	12.06	12.08	12.08	12.09	12.17	12.44

TABLE 8.15 : β_1 $(1/\phi = .6, 1/\alpha = .9)$

μ	0.075	0.100	0.125	0.150	0.175	0.200	0.225	0.250	0.275	0.300	0.325	0.350	0.375	0.400	0.425
MODE:															
(1)	2.554	2.546	2.543	2.545	2.551	2.561	2.576	2.597	2.623	2.657	2.700	2.754	2.824	2.913	3.033
(2)	4.637	4.645	4.667	4.703	4.756	4.827	4.921	5.040	5.190	5.375	5.598	5.860	6.152	6.447	6.696
(3)	6.746	6.789	6.863	6.974	7.127	7.326	7.576	7.878	8.226	8.594	8.933	9.179	9.319	9.399	9.458
(4)	8.879	8.979	9.137	9.361	9.653	10.01	10.42	10.83	11.17	11.37	11.48	11.57	11.66	11.84	12.23

TABLE 8.15 : β_1 $(1/\phi = .6, 1/\alpha = 1.0)$

μ	0.075	0.100	0.125	0.150	0.175	0.200	0.225	0.250	0.275	0.300	0.325	0.350	0.375	0.400	0.425
MODE:															
(1)	2.358	2.362	2.367	2.375	2.386	2.402	2.422	2.448	2.480	2.521	2.572	2.636	2.717	2.821	2.959
(2)	4.254	4.272	4.301	4.345	4.406	4.488	4.595	4.732	4.903	5.115	5.372	5.675	6.016	6.357	6.644
(3)	6.165	6.217	6.301	6.423	6.589	6.804	7.071	7.391	7.759	8.153	8.522	8.801	8.977	9.102	9.215
(4)	8.096	8.207	8.378	8.615	8.920	9.289	9.709	10.15	10.53	10.79	10.94	11.08	11.23	11.47	11.98

TABLE 8.15 : β_1 $(1/\phi = .7, 1/\alpha = .6)$

μ	0.075	0.100	0.125	0.150	0.175	0.200	0.225	0.250	0.275	0.300	0.325	0.350	0.375	0.400	0.425
MODE:															
(1)	4.027	3.993	3.958	3.914	3.862	3.802	3.740	3.679	3.623	3.574	3.535	3.509	3.498	3.507	3.540
(2)	7.584	7.463	7.311	7.152	7.007	6.884	6.786	6.716	6.677	6.669	6.692	6.743	6.813	6.889	6.954
(3)	11.06	10.81	10.60	10.47	10.43	10.49	10.65	10.89	11.13	11.29	11.28	11.10	10.86	10.60	10.35
(4)	14.47	14.21	14.09	14.16	14.39	14.70	14.93	14.90	14.61	14.28	13.97	13.65	13.36	13.18	13.16

TABLE 8.15 : β_1 $(1/\phi = .7, 1/\alpha = .7)$

μ	0.075	0.100	0.125	0.150	0.175	0.200	0.225	0.250	0.275	0.300	0.325	0.350	0.375	0.400	0.425
MODE:															
(1)	3.582	3.554	3.535	3.518	3.501	3.482	3.463	3.444	3.427	3.414	3.408	3.410	3.423	3.452	3.503
(2)	6.662	6.625	6.579	6.530	6.484	6.450	6.431	6.432	6.453	6.499	6.567	6.655	6.755	6.853	6.933
(3)	9.748	9.676	9.614	9.585	9.605	9.658	9.827	10.02	10.25	10.44	10.55	10.55	10.46	10.32	10.17
(4)	12.82	12.74	12.73	12.82	13.01	13.26	13.52	13.68	13.66	13.53	13.36	13.20	13.06	13.00	13.06

TABLE 8.15 : β_1 $(1/\phi = .7, 1/\alpha = .8)$

μ	0.075	0.100	0.125	0.150	0.175	0.200	0.225	0.250	0.275	0.300	0.325	0.350	0.375	0.400	0.425
MODE:															
(1)	3.248	3.228	3.215	3.209	3.206	3.207	3.210	3.216	3.226	3.240	3.260	3.288	3.326	3.379	3.451
(2)	5.958	5.951	5.952	5.960	5.977	6.006	6.049	6.109	6.188	6.287	6.405	6.538	6.675	6.802	6.903
(3)	8.699	8.707	8.736	8.789	8.877	9.005	9.173	9.377	9.596	9.801	9.952	10.03	10.03	10.00	9.952
(4)	11.46	11.49	11.58	11.73	11.94	12.19	12.46	12.66	12.77	12.77	12.74	12.70	12.69	12.76	12.93

TABLE 8.15 : β_1 $(1/\phi = .7, 1/\alpha = .9)$

μ	0.075	0.100	0.125	0.150	0.175	0.200	0.225	0.250	0.275	0.300	0.325	0.350	0.375	0.400	0.425
MODE:															
(1)	2.980	2.970	2.965	2.966	2.971	2.980	2.994	3.013	3.038	3.069	3.107	3.155	3.215	3.290	3.385
(2)	5.408	5.414	5.434	5.467	5.516	5.581	5.665	5.769	5.895	6.043	6.211	6.393	6.575	6.738	6.865
(3)	7.863	7.903	7.972	8.074	8.212	8.386	8.596	8.832	9.077	9.302	9.476	9.588	9.649	9.683	9.708
(4)	10.34	10.44	10.58	10.78	11.04	11.32	11.61	11.85	12.01	12.09	12.13	12.18	12.27	12.47	12.77

TABLE 8.15 : β_1 $(1/\phi = .7, 1/\alpha = 1.0)$

μ	0.075	0.100	0.125	0.150	0.175	0.200	0.225	0.250	0.275	0.300	0.325	0.350	0.375	0.400	0.425
MODE:															
(1)	2.751	2.754	2.760	2.768	2.780	2.795	2.815	2.841	2.873	2.913	2.963	3.023	3.098	3.191	3.308
(2)	4.961	4.979	5.009	5.053	5.115	5.196	5.301	5.431	5.590	5.777	5.991	6.223	6.455	6.660	6.817
(3)	7.187	7.240	7.325	7.447	7.610	7.815	8.060	8.335	8.621	8.884	9.091	9.229	9.320	9.394	9.466
(4)	9.433	9.545	9.716	9.949	10.24	10.56	10.90	11.18	11.38	11.50	11.59	11.69	11.84	12.13	12.57

TABLE 8.15 : β_1 $(1/\phi = .8, 1/\alpha = .6)$

μ	0.075	0.100	0.125	0.150	0.175	0.200	0.225	0.250	0.275	0.300	0.325	0.350	0.375	0.400	0.425
MODE:															
(1)	4.602	4.561	4.517	4.463	4.395	4.317	4.235	4.151	4.072	3.999	3.935	3.883	3.843	3.818	3.809
(2)	8.658	8.501	8.297	8.078	7.870	7.682	7.517	7.378	7.265	7.179	7.119	7.082	7.064	7.057	7.056
(3)	12.59	12.26	11.94	11.69	11.54	11.49	11.53	11.63	11.73	11.74	11.63	11.41	11.13	10.85	10.58
(4)	16.44	16.02	15.75	15.67	15.76	15.89	15.90	15.68	15.31	14.90	14.52	14.14	13.82	13.57	13.43

TABLE 8.15 : β_1 $(1/\phi = .8, 1/\alpha = .7)$

μ	0.075	0.100	0.125	0.150	0.175	0.200	0.225	0.250	0.275	0.300	0.325	0.350	0.375	0.400	0.425
MODE:															
(1)	4.093	4.060	4.036	4.013	3.988	3.959	3.928	3.894	3.861	3.830	3.802	3.781	3.767	3.764	3.773
(2)	7.608	7.554	7.482	7.397	7.309	7.225	7.151	7.091	7.045	7.016	7.001	7.001	7.010	7.024	7.037
(3)	11.12	10.99	10.86	10.75	10.68	10.66	10.69	10.76	10.85	10.92	10.93	10.87	10.74	10.58	10.41
(4)	14.59	14.41	14.29	14.25	14.30	14.41	14.50	14.50	14.38	14.18	13.96	13.74	13.55	13.42	13.35

TABLE 8.15 : β_1 $(1/\phi = .8, 1/\alpha = .8)$

μ	0.075	0.100	0.125	0.150	0.175	0.200	0.225	0.250	0.275	0.300	0.325	0.350	0.375	0.400	0.425
MODE:															
(1)	3.712	3.687	3.671	3.661	3.654	3.648	3.644	3.642	3.640	3.641	3.645	3.654	3.668	3.690	3.723
(2)	6.805	6.789	6.776	6.765	6.756	6.751	6.753	6.764	6.784	6.813	6.850	6.893	6.938	6.978	7.010
(3)	9.925	9.909	9.897	9.898	9.919	9.964	10.03	10.12	10.22	10.30	10.35	10.36	10.33	10.27	10.20
(4)	13.05	13.04	13.05	13.10	13.19	13.31	13.42	13.49	13.49	13.44	13.36	13.28	13.22	13.21	13.25

TABLE 8.15 : β_1 $(1/\phi = .8, 1/\alpha = .9)$

μ	0.075	0.100	0.125	0.150	0.175	0.200	0.225	0.250	0.275	0.300	0.325	0.350	0.375	0.400	0.425
MODE:															
(1)	3.405	3.393	3.386	3.384	3.386	3.392	3.402	3.415	3.433	3.455	3.481	3.514	3.553	3.600	3.658
(2)	6.176	6.178	6.190	6.214	6.247	6.292	6.348	6.415	6.492	6.578	6.670	6.762	6.847	6.919	6.974
(3)	8.974	9.001	9.050	9.121	9.214	9.327	9.455	9.588	9.713	9.816	9.888	9.930	9.948	9.955	9.961
(4)	11.79	11.86	11.96	12.10	12.26	12.42	12.57	12.67	12.72	12.74	12.75	12.78	12.85	12.97	13.11

TABLE 8.15 : β_1 $(1/\phi = .8,\ 1/\alpha = 1.0)$

μ MODE:	0.075	0.100	0.125	0.150	0.175	0.200	0.225	0.250	0.275	0.300	0.325	0.350	0.375	0.400	0.425
(1)	3.143	3.146	3.151	3.158	3.169	3.182	3.200	3.222	3.250	3.283	3.324	3.373	3.430	3.499	3.581
(2)	5.667	5.683	5.709	5.748	5.801	5.870	5.957	6.062	6.184	6.320	6.464	6.608	6.739	6.849	6.931
(3)	8.203	8.250	8.325	8.430	8.566	8.730	8.914	9.103	9.277	9.417	9.516	9.581	9.626	9.667	9.720
(4)	10.76	10.86	11.00	11.20	11.42	11.65	11.85	12.00	12.09	12.14	12.20	12.28	12.44	12.67	12.95

TABLE 8.15 : β_1 $(1/\phi = .9,\ 1/\alpha = .6)$

μ MODE:	0.075	0.100	0.125	0.150	0.175	0.200	0.225	0.250	0.275	0.300	0.325	0.350	0.375	0.400	0.425
(1)	5.175	5.127	5.073	5.003	4.915	4.812	4.702	4.589	4.478	4.374	4.278	4.193	4.118	4.055	4.004
(2)	9.723	9.518	9.241	8.939	8.646	8.376	8.131	7.912	7.721	7.558	7.423	7.314	7.231	7.170	7.126
(3)	14.10	13.62	13.16	12.77	12.48	12.29	12.21	12.20	12.20	12.13	11.96	11.70	11.40	11.09	10.80
(4)	18.32	17.68	17.21	16.96	16.90	16.89	16.76	16.42	15.95	15.47	15.00	14.56	14.17	13.85	13.61

TABLE 8.15 : β_1 $(1/\phi = .9,\ 1/\alpha = .7)$

μ MODE:	0.075	0.100	0.125	0.150	0.175	0.200	0.225	0.250	0.275	0.300	0.325	0.350	0.375	0.400	0.425
(1)	4.603	4.564	4.533	4.502	4.465	4.421	4.371	4.316	4.258	4.200	4.144	4.091	4.044	4.003	3.969
(2)	8.548	8.469	8.357	8.217	8.065	7.911	7.765	7.629	7.508	7.402	7.312	7.239	7.182	7.139	7.109
(3)	12.46	12.26	12.03	11.80	11.61	11.46	11.37	11.33	11.33	11.32	11.28	11.18	11.03	10.84	10.64
(4)	16.31	15.98	15.69	15.48	15.38	15.36	15.35	15.26	15.07	14.80	14.50	14.21	13.94	13.71	13.54

TABLE 8.15 : β_1 $(1/\phi = .9,\ 1/\alpha = .8)$

μ MODE:	0.075	0.100	0.125	0.150	0.175	0.200	0.225	0.250	0.275	0.300	0.325	0.350	0.375	0.400	0.425
(1)	4.175	4.145	4.124	4.108	4.094	4.079	4.062	4.044	4.024	4.004	3.983	3.964	3.946	3.931	3.921
(2)	7.647	7.617	7.581	7.537	7.483	7.425	7.366	7.309	7.256	7.210	7.170	7.139	7.114	7.096	7.084
(3)	11.14	11.08	11.00	10.91	10.83	10.77	10.72	10.70	10.70	10.70	10.70	10.68	10.62	10.54	10.44
(4)	14.62	14.51	14.39	14.30	14.25	14.23	14.23	14.23	14.18	14.08	13.94	13.79	13.65	13.54	13.45

TABLE 8.15 : β_1 $(1/\phi = .9,\ 1/\alpha = .9)$

μ MODE:	0.075	0.100	0.125	0.150	0.175	0.200	0.225	0.250	0.275	0.300	0.325	0.350	0.375	0.400	0.425
(1)	3.829	3.814	3.804	3.798	3.795	3.794	3.795	3.798	3.801	3.807	3.813	3.821	3.831	3.843	3.858
(2)	6.942	6.934	6.933	6.935	6.939	6.946	6.954	6.964	6.975	6.989	7.003	7.017	7.030	7.042	7.051
(3)	10.07	10.08	10.08	10.09	10.11	10.13	10.15	10.18	10.21	10.23	10.24	10.25	10.24	10.23	10.21
(4)	13.22	13.23	13.24	13.27	13.30	13.33	13.36	13.38	13.39	13.38	13.36	13.34	13.32	13.32	13.33

TABLE 8.15 : β_1 $(1/\phi = .9,\ 1/\alpha = 1.0)$

μ MODE:	0.075	0.100	0.125	0.150	0.175	0.200	0.225	0.250	0.275	0.300	0.325	0.350	0.375	0.400	0.425
(1)	3.535	3.537	3.540	3.545	3.552	3.561	3.572	3.586	3.603	3.623	3.647	3.675	3.707	3.742	3.782
(2)	6.369	6.380	6.397	6.422	6.456	6.498	6.550	6.610	6.675	6.744	6.812	6.876	6.932	6.977	7.011
(3)	9.212	9.243	9.291	9.356	9.437	9.530	9.625	9.713	9.787	9.842	9.877	9.900	9.918	9.940	9.974
(4)	12.07	12.13	12.22	12.34	12.46	12.57	12.65	12.70	12.73	12.75	12.79	12.85	12.95	13.07	13.19

TABLE 8.15 : β_1 $(1/\phi = 1.0,\ 1/\alpha = .6)$

μ MODE:	0.075	0.100	0.125	0.150	0.175	0.200	0.225	0.250	0.275	0.300	0.325	0.350	0.375	0.400	0.425
(1)	5.748	5.690	5.622	5.531	5.416	5.280	5.133	4.982	4.833	4.692	4.560	4.439	4.328	4.229	4.140
(2)	10.78	10.50	10.13	9.717	9.323	8.961	8.632	8.335	8.070	7.838	7.641	7.478	7.348	7.248	7.175
(3)	15.55	14.89	14.25	13.70	13.25	12.94	12.77	12.68	12.62	12.50	12.29	12.00	11.66	11.31	10.98
(4)	20.09	19.19	18.49	18.08	17.91	17.82	17.58	17.11	16.54	15.96	15.41	14.90	14.44	14.04	13.73

TABLE 8.15 : β_1 $(1/\phi = 1.0,\ 1/\alpha = .7)$

μ MODE:	0.075	0.100	0.125	0.150	0.175	0.200	0.225	0.250	0.275	0.300	0.325	0.350	0.375	0.400	0.425
(1)	5.113	5.066	5.027	4.983	4.929	4.863	4.786	4.701	4.611	4.519	4.429	4.341	4.257	4.179	4.107
(2)	9.482	9.366	9.194	8.976	8.738	8.499	8.271	8.059	7.865	7.690	7.537	7.407	7.302	7.220	7.159
(3)	12.78	13.47	13.08	12.71	12.38	12.11	11.92	11.80	11.74	11.70	11.62	11.49	11.31	11.08	10.84
(4)	17.95	17.42	16.93	16.55	16.32	16.22	16.15	16.00	15.72	15.35	14.96	14.58	14.23	13.92	13.67

TABLE 8.15 : β_1 $(1/\phi = 1.0,\ 1/\alpha = .8)$

μ MODE:	0.075	0.100	0.125	0.150	0.175	0.200	0.225	0.250	0.275	0.300	0.325	0.350	0.375	0.400	0.425
(1)	4.637	4.601	4.574	4.549	4.523	4.492	4.457	4.416	4.370	4.321	4.269	4.216	4.163	4.111	4.060
(2)	8.485	8.432	8.360	8.263	8.145	8.015	7.881	7.749	7.623	7.508	7.404	7.314	7.239	7.180	7.136
(3)	12.33	12.20	12.02	11.81	11.61	11.43	11.28	11.17	11.11	11.07	11.04	10.99	10.91	10.80	10.66
(4)	16.13	15.88	15.60	15.34	15.15	15.04	14.99	14.94	14.85	14.68	14.45	14.21	13.98	13.76	13.59

TABLE 8.15 : β_1 $(1/\phi = 1.0,\ 1/\alpha = .9)$

μ MODE:	0.075	0.100	0.125	0.150	0.175	0.200	0.225	0.250	0.275	0.300	0.325	0.350	0.375	0.400	0.425
(1)	4.253	4.234	4.219	4.206	4.195	4.183	4.169	4.154	4.137	4.118	4.097	4.074	4.050	4.025	4.000
(2)	7.703	7.680	7.654	7.621	7.579	7.529	7.474	7.415	7.356	7.299	7.247	7.200	7.161	7.129	7.106
(3)	11.16	11.12	11.06	10.98	10.89	10.80	10.72	10.66	10.62	10.59	10.58	10.56	10.54	10.50	10.44
(4)	14.61	14.53	14.42	14.30	14.20	14.12	14.08	14.06	14.04	13.99	13.91	13.80	13.68	13.57	13.48

TABLE 8.16 : β_1 $(\phi = .6,\ \alpha = .6)$

μ MODE:	0.075	0.100	0.125	0.150	0.175	0.200	0.225	0.250	0.275	0.300	0.325	0.350	0.375	0.400	0.425
(1)	5.528	5.858	6.230	6.653	7.137	7.696	8.345	9.098	9.916	10.28	9.770	9.171	8.632	8.167	7.781
(2)	9.179	9.728	10.35	11.05	11.84	12.72	13.42	12.76	11.85	11.67	11.94	11.35	10.69	10.10	9.605
(3)	12.85	13.62	14.48	15.44	16.41	15.93	14.75	15.10	14.46	13.52	13.58	15.26	16.58	15.83	15.00
(4)	16.53	17.51	18.60	19.65	18.45	18.07	17.65	16.46	17.30	19.09	19.39	18.25	18.17	17.75	16.83

TABLE 8.16 : β_1 $(\phi = .6,\ \alpha = .7)$

μ MODE:	0.075	0.100	0.125	0.150	0.175	0.200	0.225	0.250	0.275	0.300	0.325	0.350	0.375	0.400	0.425
(1)	6.415	6.784	7.198	7.667	8.200	8.810	9.504	10.26	10.81	10.54	9.925	9.326	8.799	8.351	7.983
(2)	10.65	11.27	11.96	12.73	13.58	14.37	14.13	13.10	12.47	12.62	12.21	11.51	10.86	10.29	9.816
(3)	14.92	15.78	16.73	17.72	17.97	16.51	16.19	15.88	14.82	14.28	15.26	16.88	16.82	15.96	15.15
(4)	19.18	20.28	21.42	21.31	19.77	19.71	18.24	18.24	19.72	20.75	19.67	18.77	18.79	17.90	16.98

TABLE 8.16 : β_1 $(\phi = .6,\ \alpha = .8)$

μ MODE:	0.075	0.100	0.125	0.150	0.175	0.200	0.225	0.250	0.275	0.300	0.325	0.350	0.375	0.400	0.425
(1)	7.274	7.669	8.112	8.611	9.175	9.809	10.50	11.14	11.28	10.75	10.11	9.513	8.994	8.553	8.194
(2)	12.08	12.75	13.48	14.29	15.12	15.47	14.56	13.58	13.25	13.09	12.43	11.71	11.06	10.50	10.04
(3)	16.93	17.85	18.85	19.67	18.72	17.48	17.33	16.32	15.36	15.46	16.71	17.63	17.00	16.12	15.32
(4)	21.77	22.93	23.87	22.28	21.62	20.44	19.38	20.28	21.66	21.25	20.03	19.56	19.04	18.07	17.16

TABLE 8.16 : β_1 $(\phi = .6,\ \alpha = .9)$

μ MODE:	0.075	0.100	0.125	0.150	0.175	0.200	0.225	0.250	0.275	0.300	0.325	0.350	0.375	0.400	0.425
(1)	8.094	8.502	8.960	9.473	10.05	10.68	11.33	11.78	11.59	10.97	10.31	9.718	9.199	8.758	8.394
(2)	13.46	14.15	14.91	15.73	16.42	16.15	15.01	14.15	13.88	13.41	12.66	11.93	11.28	10.72	10.24
(3)	18.86	19.83	20.81	21.09	19.47	18.63	18.01	16.81	16.11	16.62	17.78	17.99	17.18	16.30	15.51
(4)	24.26	25.44	25.67	23.57	22.92	21.22	20.95	22.11	22.76	21.62	20.50	20.10	19.25	18.26	17.36

TABLE 8.16 : β_1 (ϕ = .6, α = 1.0)

μ MODE:	0.075	0.100	0.125	0.150	0.175	0.200	0.225	0.250	0.275	0.300	0.325	0.350	0.375	0.400	0.425
(1)	8.867	9.275	9.733	10.25	10.82	11.43	12.01	12.25	11.87	11.20	10.53	9.927	9.400	8.949	8.570
(2)	14.77	15.47	16.25	17.04	17.48	16.71	15.49	14.72	14.36	13.70	12.90	12.15	11.49	10.91	10.42
(3)	20.72	21.70	22.60	22.11	20.31	19.63	18.55	17.34	16.91	17.60	18.51	18.26	17.38	16.48	15.69
(4)	26.67	27.78	26.96	25.07	23.82	22.21	22.53	23.63	23.37	22.00	21.01	20.44	19.46	18.45	17.54

TABLE 8.16 : β_1 (ϕ = .7, α = .6)

μ MODE:	0.075	0.100	0.125	0.150	0.175	0.200	0.225	0.250	0.275	0.300	0.325	0.350	0.375	0.400	0.425
(1)	4.739	5.021	5.340	5.703	6.118	6.598	7.159	7.818	8.589	9.401	9.598	9.128	8.617	8.162	7.780
(2)	7.868	8.338	8.868	9.470	10.16	10.94	11.80	12.33	11.68	11.01	11.15	11.22	10.66	10.09	9.603
(3)	11.02	11.68	12.42	13.25	14.18	15.01	14.30	13.60	14.02	13.35	12.67	13.35	15.34	15.75	14.98
(4)	14.17	15.01	15.96	17.00	17.71	16.43	16.67	16.02	15.33	16.58	18.40	18.12	17.25	17.59	16.81

TABLE 8.16 : β_1 (ϕ = .7, α = .7)

μ MODE:	0.075	0.100	0.125	0.150	0.175	0.200	0.225	0.250	0.275	0.300	0.325	0.350	0.375	0.400	0.425
(1)	5.499	5.815	6.170	6.572	7.031	7.558	8.166	8.864	9.614	10.10	9.815	9.290	8.786	8.346	7.982
(2)	9.131	9.658	10.25	10.92	11.67	12.49	13.19	12.85	12.05	11.69	11.87	11.43	10.83	10.28	9.814
(3)	12.79	13.53	14.35	15.26	16.17	16.03	14.95	14.98	14.57	13.74	13.63	14.97	16.38	15.90	15.13
(4)	16.44	17.39	18.43	19.41	18.67	18.01	17.77	16.70	17.21	18.80	19.39	18.41	18.06	17.81	16.96

TABLE 8.16 : β_1 (ϕ = .7, α = .8)

μ MODE:	0.075	0.100	0.125	0.150	0.175	0.200	0.225	0.250	0.275	0.300	0.325	0.350	0.375	0.400	0.425
(1)	6.234	6.574	6.954	7.383	7.870	8.425	9.054	9.742	10.36	10.46	10.02	9.480	8.981	8.549	8.192
(2)	10.36	10.93	11.56	12.28	13.06	13.83	14.03	13.26	12.54	12.41	12.23	11.65	11.04	10.50	10.04
(3)	14.51	15.31	16.19	17.12	17.68	16.68	15.97	15.06	15.00	14.35	14.82	16.23	16.76	16.07	15.31
(4)	18.66	19.68	20.75	21.17	19.71	19.54	18.44	17.97	19.10	20.36	19.77	18.84	18.69	18.01	17.15

TABLE 8.16 : β_1 (ϕ = .7, α = .9)

μ MODE:	0.075	0.100	0.125	0.150	0.175	0.200	0.225	0.250	0.275	0.300	0.325	0.350	0.375	0.400	0.425
(1)	6.938	7.288	7.681	8.123	8.624	9.190	9.817	10.46	10.90	10.75	10.24	9.689	9.188	8.754	8.393
(2)	11.64	12.13	12.80	13.54	14.31	14.92	14.59	13.70	13.10	12.94	12.52	11.88	11.26	10.71	10.24
(3)	16.17	17.01	17.93	18.80	18.69	17.42	17.00	16.43	15.49	15.11	15.88	17.06	17.01	16.26	15.50
(4)	20.81	21.88	22.88	22.31	21.09	20.50	19.26	19.47	20.72	21.15	20.14	19.36	19.04	18.21	17.34

TABLE 8.16 : β_1 (ϕ = .7, α = 1.0)

μ MODE:	0.075	0.100	0.125	0.150	0.175	0.200	0.225	0.250	0.275	0.300	0.325	0.350	0.375	0.400	0.425
(1)	7.600	7.950	8.345	8.791	9.295	9.862	10.47	11.05	11.32	11.02	10.47	9.902	9.391	8.946	8.570
(2)	12.66	13.27	13.95	14.71	15.45	15.79	15.10	14.16	13.62	13.35	12.79	12.11	11.48	10.91	10.42
(3)	17.77	18.64	19.57	20.27	19.49	18.25	17.84	16.92	16.01	15.87	16.75	17.61	17.24	16.45	15.68
(4)	22.89	23.98	24.78	23.34	22.44	21.26	20.26	20.91	22.01	21.65	20.52	19.86	19.31	18.41	17.53

TABLE 8.16 : β_1 (ϕ = .8, α = .6)

μ MODE:	0.075	0.100	0.125	0.150	0.175	0.200	0.225	0.250	0.275	0.300	0.325	0.350	0.375	0.400	0.425
(1)	4.146	4.394	4.673	4.990	5.353	5.774	6.266	6.848	7.541	8.351	9.101	9.037	8.591	8.153	7.777
(2)	6.885	7.296	7.760	8.287	8.890	9.583	10.38	11.22	11.45	10.82	10.42	10.85	10.60	10.08	9.598
(3)	9.640	10.22	10.87	11.60	12.43	13.35	13.86	13.05	12.90	13.13	12.45	12.20	13.71	15.47	14.96
(4)	12.39	13.14	13.97	14.90	15.90	15.96	15.13	15.56	14.75	14.77	16.41	17.82	17.06	16.98	16.77

TABLE 8.16 : β_1 (ϕ = .8, α = .7)

μ MODE:	0.075	0.100	0.125	0.150	0.175	0.200	0.225	0.250	0.275	0.300	0.325	0.350	0.375	0.400	0.425
(1)	4.811	5.088	5.399	5.751	6.163	6.616	7.153	7.779	8.500	9.244	9.556	9.221	8.763	8.338	7.979
(2)	7.990	8.451	8.969	9.556	10.22	10.97	11.78	12.31	11.84	11.24	11.20	11.26	10.79	10.27	9.810
(3)	11.19	11.84	12.56	13.38	14.27	15.02	14.52	13.85	14.07	13.53	12.95	13.46	15.19	15.77	15.11
(4)	14.39	15.22	16.15	17.15	17.79	16.78	16.00	16.26	16.66	16.72	18.36	18.23	17.46	17.61	16.94

TABLE 8.16 : β_1 (ϕ = .8, α = .8)

μ MODE:	0.075	0.100	0.125	0.150	0.175	0.200	0.225	0.250	0.275	0.300	0.325	0.350	0.375	0.400	0.425
(1)	5.455	5.752	6.085	6.461	6.888	7.378	7.941	8.584	9.281	9.849	9.836	9.422	8.961	8.542	8.190
(2)	9.063	9.561	10.12	10.75	11.46	12.23	12.93	12.91	12.25	11.78	11.81	11.54	11.01	10.49	10.03
(3)	12.70	13.40	14.10	15.06	15.91	16.09	15.17	14.91	14.70	13.97	13.69	14.65	16.08	15.98	15.29
(4)	16.33	17.24	18.23	19.17	18.88	18.02	17.89	16.95	17.10	18.47	19.35	18.58	18.05	17.89	17.13

TABLE 8.16 : β_1 (ϕ = .8, α = .9)

μ MODE:	0.075	0.100	0.125	0.150	0.175	0.200	0.225	0.250	0.275	0.300	0.325	0.350	0.375	0.400	0.425
(1)	6.070	6.377	6.721	7.110	7.551	8.055	8.628	9.266	9.907	10.28	10.09	9.639	9.170	8.748	8.391
(2)	10.09	10.62	11.20	11.86	12.60	13.35	13.82	13.40	12.69	13.32	12.23	11.80	11.24	10.70	10.24
(3)	14.15	14.89	15.72	16.60	17.32	16.84	15.94	15.79	15.19	14.48	14.51	15.62	16.58	16.18	15.48
(4)	18.21	19.17	20.19	20.89	19.83	19.36	18.63	17.87	18.57	19.87	19.85	18.96	18.59	18.13	17.33

TABLE 8.16 : β_1 (ϕ = .8, α = 1.0)

μ MODE:	0.075	0.100	0.125	0.150	0.175	0.200	0.225	0.250	0.275	0.300	0.325	0.350	0.375	0.400	0.425
(1)	6.650	6.957	7.302	7.695	8.142	8.652	9.229	9.855	10.43	10.64	10.34	9.860	9.377	8.942	8.569
(2)	11.08	11.61	12.22	12.90	13.65	14.35	14.52	13.85	13.13	12.80	12.57	12.05	11.45	10.90	10.42
(3)	15.55	16.32	17.19	18.06	18.48	17.52	16.76	16.46	15.64	15.00	15.26	16.38	16.94	16.39	15.67
(4)	20.03	21.04	22.06	22.24	20.86	20.45	19.32	18.93	19.91	20.88	20.26	19.36	19.00	18.35	17.52

TABLE 8.16 : β_1 (ϕ = .9, α = .6)

μ MODE:	0.075	0.100	0.125	0.150	0.175	0.200	0.225	0.250	0.275	0.300	0.325	0.350	0.375	0.400	0.425
(1)	3.686	3.906	4.154	4.436	4.759	5.133	5.571	6.090	6.713	7.462	8.318	8.825	8.547	8.140	7.773
(2)	6.120	6.485	6.898	7.366	7.904	8.523	9.242	10.06	10.83	10.68	10.15	10.20	10.47	10.05	9.592
(3)	8.569	9.081	9.659	10.32	11.06	11.91	12.78	12.80	12.15	12.51	12.32	11.78	12.43	14.61	14.91
(4)	11.02	11.68	12.42	13.26	14.19	15.06	14.50	14.42	14.49	13.84	14.76	16.69	16.94	16.33	16.69

TABLE 8.16 : β_1 (ϕ = .9, α = .7)

μ MODE:	0.075	0.100	0.125	0.150	0.175	0.200	0.225	0.250	0.275	0.300	0.325	0.350	0.375	0.400	0.425
(1)	4.277	4.523	4.799	5.112	5.470	5.882	6.362	6.925	7.588	8.346	9.034	9.091	8.726	8.326	7.976
(2)	7.102	7.512	7.973	8.496	9.092	9.772	10.54	11.32	11.55	11.04	10.68	10.90	10.71	10.25	9.805
(3)	9.946	10.52	11.17	11.90	12.72	13.59	14.02	13.36	13.18	13.28	12.70	12.54	13.84	15.43	15.07
(4)	12.79	13.53	14.36	15.28	16.22	16.24	15.55	15.78	15.08	15.18	16.69	17.89	17.25	17.11	16.89

TABLE 8.16 : β_1 (ϕ = .9, α = .8)

μ MODE:	0.075	0.100	0.125	0.150	0.175	0.200	0.225	0.250	0.275	0.300	0.325	0.350	0.375	0.400	0.425
(1)	4.849	5.113	5.409	5.743	6.124	6.561	7.067	7.654	8.326	9.033	9.477	9.321	8.928	8.531	8.188
(2)	8.056	8.499	8.997	9.560	10.20	10.92	11.68	12.26	12.01	11.46	11.27	11.32	10.95	10.47	10.03
(3)	11.29	11.91	12.61	13.40	14.25	15.00	14.75	14.08	14.14	13.75	13.20	13.48	14.94	15.78	15.26
(4)	14.52	15.33	16.22	17.19	17.86	17.10	16.91	16.52	15.93	16.73	18.23	18.37	17.66	17.63	17.09

TABLE 8.16 : β_1 (ϕ = .9, α = .9)

μ / MODE:	0.075	0.100	0.125	0.150	0.175	0.200	0.225	0.250	0.275	0.300	0.325	0.350	0.375	0.400	0.425
(1)	5.396	5.669	5.975	6.320	6.714	7.166	7.686	8.282	8.944	9.571	9.814	9.555	9.143	8.739	8.389
(2)	8.972	9.437	9.961	10.55	11.22	11.96	12.68	12.94	12.44	11.91	11.78	11.64	11.19	10.69	10.24
(3)	12.58	13.24	13.99	14.81	15.66	16.09	15.39	14.91	14.82	14.20	13.77	14.35	15.72	16.04	15.45
(4)	16.19	17.05	18.00	18.95	19.05	18.11	17.99	17.19	17.01	18.14	19.25	18.75	18.11	17.96	17.30

TABLE 8.16 : β_1 (ϕ = .9, α = 1.0)

μ / MODE:	0.075	0.100	0.125	0.150	0.175	0.200	0.225	0.250	0.275	0.300	0.325	0.350	0.375	0.400	0.425
(1)	5.911	6.184	6.492	6.841	7.242	7.702	8.231	8.833	9.481	10.03	10.12	9.791	9.354	8.935	8.567
(2)	9.846	10.32	10.87	11.49	12.18	12.93	13.56	13.50	12.86	12.34	12.20	11.93	11.42	10.89	10.42
(3)	13.82	14.52	15.30	16.16	16.96	16.95	16.04	15.71	15.34	14.63	14.33	15.09	16.29	16.28	15.65
(4)	17.81	18.72	19.71	20.57	20.00	19.21	18.80	17.88	18.14	19.38	19.89	19.11	18.54	18.23	17.50

TABLE 8.16 : β_1 (ϕ = 1.0, α = .6)

μ / MODE:	0.075	0.100	0.125	0.150	0.175	0.200	0.225	0.250	0.275	0.300	0.325	0.350	0.375	0.400	0.425
(1)	3.317	3.515	3.738	3.992	4.283	4.620	5.014	5.483	6.046	6.732	7.560	8.389	8.468	8.121	7.768
(2)	5.508	5.837	6.208	6.630	7.114	7.673	8.325	9.086	9.933	10.42	10.03	9.741	10.19	10.01	9.582
(3)	7.712	8.173	8.694	9.285	9.961	10.73	11.60	12.30	11.88	11.69	12.09	11.63	11.60	13.45	14.81
(4)	9.916	10.51	11.18	11.94	12.79	13.73	14.13	13.50	13.97	13.54	13.55	15.23	16.71	16.12	16.51

TABLE 8.16 : β_1 (ϕ = 1.0, α = .7)

μ / MODE:	0.075	0.100	0.125	0.150	0.175	0.200	0.225	0.250	0.275	0.300	0.325	0.350	0.375	0.400	0.425
(1)	3.849	4.070	4.319	4.601	4.923	5.294	5.727	6.237	6.843	7.561	8.351	8.842	8.666	8.308	7.971
(2)	6.392	6.761	7.176	7.647	8.185	8.803	9.512	10.30	10.98	10.87	10.42	10.42	10.58	10.22	9.797
(3)	8.952	9.470	10.05	10.71	11.46	12.29	13.09	13.05	12.53	12.77	12.54	12.12	12.76	14.67	15.01
(4)	11.51	12.18	12.93	13.77	14.69	15.43	14.91	14.87	14.78	14.30	15.24	16.98	17.11	16.62	16.80

TABLE 8.16 : β_1 (ϕ = 1.0, α = .8)

μ / MODE:	0.075	0.100	0.125	0.150	0.175	0.200	0.225	0.250	0.275	0.300	0.325	0.350	0.375	0.400	0.425
(1)	4.364	4.602	4.868	5.169	5.512	5.907	6.365	6.900	7.526	8.240	8.922	9.141	8.878	8.516	8.183
(2)	7.250	7.649	8.098	8.606	9.185	9.844	10.58	11.33	11.66	11.27	10.89	10.97	10.86	10.45	10.02
(3)	10.16	10.72	11.35	12.07	12.87	13.70	14.18	13.66	13.39	13.46	12.97	12.77	13.81	15.34	15.21
(4)	13.07	13.80	14.61	15.52	16.42	16.53	15.89	16.00	15.41	15.45	16.78	17.96	17.47	17.21	17.03

TABLE 8.16 : β_1 (ϕ = 1.0, α = .9)

μ / MODE:	0.075	0.100	0.125	0.150	0.175	0.200	0.225	0.250	0.275	0.300	0.325	0.350	0.375	0.400	0.425
(1)	4.856	5.102	5.377	5.689	6.044	6.452	6.925	7.475	8.110	8.801	9.363	9.413	9.101	8.727	8.386
(2)	8.075	8.494	8.967	9.504	10.11	10.80	11.55	12.19	12.18	11.68	11.36	11.39	11.12	10.67	10.23
(3)	11.32	11.92	12.60	13.36	14.19	14.96	14.96	14.30	14.21	13.96	13.43	13.45	14.65	15.75	15.41
(4)	14.57	15.35	16.23	17.17	17.92	17.41	17.02	16.77	16.14	16.67	18.08	18.50	17.84	17.67	17.26

TABLE 8.16 : β_1 ($1/\phi$ = .6, $1/\alpha$ = .6)

μ / MODE:	0.075	0.100	0.125	0.150	0.175	0.200	0.225	0.250	0.275	0.300	0.325	0.350	0.375	0.400	0.425
(1)	4.593	4.718	4.898	5.132	5.422	5.778	6.212	6.743	7.397	8.209	9.205	10.18	10.11	9.608	9.105
(2)	8.017	8.376	8.824	9.364	10.00	10.76	11.65	12.67	13.44	12.81	12.00	11.58	12.01	11.57	10.97
(3)	11.57	12.17	12.88	13.73	14.70	15.80	16.57	15.49	14.72	15.14	14.33	13.51	13.63	15.92	16.34
(4)	15.17	16.01	16.99	18.11	19.33	19.18	17.86	18.24	17.04	16.55	18.28	19.69	18.69	17.86	18.13

TABLE 8.16 : β_1 ($1/\phi$ = .6, $1/\alpha$ = .7)

μ / MODE:	0.075	0.100	0.125	0.150	0.175	0.200	0.225	0.250	0.275	0.300	0.325	0.350	0.375	0.400	0.425
(1)	4.168	4.291	4.453	4.656	4.906	5.210	5.579	6.032	6.590	7.284	8.151	9.165	9.736	9.441	9.003
(2)	7.104	7.405	7.779	8.231	8.770	9.409	10.17	11.06	12.03	12.38	11.78	11.22	11.34	11.35	10.85
(3)	10.13	10.62	11.22	11.93	12.76	13.71	14.73	14.99	14.10	13.97	14.02	13.31	12.91	14.38	16.02
(4)	13.19	13.89	14.72	15.67	16.75	17.76	17.06	16.63	16.63	15.71	16.19	18.16	18.45	17.60	17.73

TABLE 8.16 : β_1 ($1/\phi$ = .6, $1/\alpha$ = .8)

μ / MODE:	0.075	0.100	0.125	0.150	0.175	0.200	0.225	0.250	0.275	0.300	0.325	0.350	0.375	0.400	0.425
(1)	3.803	3.936	4.097	4.289	4.519	4.795	5.126	5.529	6.024	6.638	7.408	8.347	9.193	9.211	8.864
(2)	6.397	6.672	7.005	7.401	7.871	8.429	9.091	9.875	10.78	11.58	11.46	10.97	10.82	11.05	10.70
(3)	9.046	9.482	10.00	10.61	11.33	12.17	13.11	13.95	13.67	13.15	13.44	13.06	12.59	13.35	15.46
(4)	11.72	12.33	13.04	13.87	14.82	15.85	16.32	15.55	15.78	15.30	14.99	16.45	17.95	17.37	17.20

TABLE 8.16 : β_1 ($1/\phi$ = .6, $1/\alpha$ = .9)

μ / MODE:	0.075	0.100	0.125	0.150	0.175	0.200	0.225	0.250	0.275	0.300	0.325	0.350	0.375	0.400	0.425
(1)	3.479	3.621	3.786	3.976	4.198	4.459	4.770	5.143	5.598	6.160	6.863	7.733	8.644	8.937	8.697
(2)	5.813	6.078	6.389	6.752	7.178	7.680	8.275	8.981	9.808	10.68	11.04	10.70	10.47	10.72	10.52
(3)	8.183	8.583	9.053	9.602	10.24	10.99	11.84	12.75	13.11	12.66	12.71	12.73	12.32	12.70	14.77
(4)	10.57	11.11	11.75	12.48	13.33	14.28	15.15	14.92	14.73	14.84	14.37	15.17	17.04	17.11	16.78

TABLE 8.16 : β_1 ($1/\phi$ = .6, $1/\alpha$ = 1.0)

μ / MODE:	0.075	0.100	0.125	0.150	0.175	0.200	0.225	0.250	0.275	0.300	0.325	0.350	0.375	0.400	0.425
(1)	3.192	3.339	3.506	3.696	3.914	4.167	4.455	4.820	5.250	5.778	6.437	7.258	8.176	8.653	8.515
(2)	5.317	5.576	5.873	6.215	6.612	7.076	7.624	8.273	9.041	9.901	10.53	10.40	10.18	10.40	10.33
(3)	7.465	7.844	8.281	8.785	9.371	10.05	10.84	11.72	12.40	12.23	12.10	12.34	12.05	12.23	14.13
(4)	9.622	10.13	10.71	11.38	12.15	13.03	13.94	14.28	13.94	14.23	13.93	14.25	16.01	16.79	16.46

TABLE 8.16 : β_1 ($1/\phi$ = .7, $1/\alpha$ = .6)

μ / MODE:	0.075	0.100	0.125	0.150	0.175	0.200	0.225	0.250	0.275	0.300	0.325	0.350	0.375	0.400	0.425
(1)	5.359	5.505	5.714	5.986	6.324	6.738	7.241	7.853	8.599	9.496	10.45	10.66	10.17	9.617	9.107
(2)	9.354	9.771	10.29	10.92	11.66	12.52	13.51	14.36	13.86	12.92	12.24	12.44	12.21	11.59	10.97
(3)	13.49	14.19	15.02	15.99	17.08	18.04	17.09	15.91	16.16	15.43	14.46	13.98	15.40	17.02	16.38
(4)	17.69	18.67	19.80	21.05	21.54	19.70	19.83	18.64	17.62	18.84	20.62	19.92	18.81	18.69	18.20

TABLE 8.16 : β_1 ($1/\phi$ = .7, $1/\alpha$ = .7)

μ / MODE:	0.075	0.100	0.125	0.150	0.175	0.200	0.225	0.250	0.275	0.300	0.325	0.350	0.375	0.400	0.425
(1)	4.863	5.007	5.195	5.432	5.722	6.075	6.504	7.026	7.664	8.441	9.346	10.06	9.918	9.467	9.007
(2)	8.288	8.639	9.074	9.599	10.22	10.96	11.81	12.74	13.23	12.63	11.94	11.71	11.86	11.41	10.86
(3)	11.82	12.39	13.09	13.90	14.84	15.86	16.33	15.36	14.84	14.96	14.21	13.56	14.05	16.01	16.17
(4)	15.39	16.20	17.15	18.23	19.32	18.94	17.97	18.04	16.97	16.86	18.48	19.48	18.59	17.97	17.97

TABLE 8.16 : β_1 ($1/\phi$ = .7, $1/\alpha$ = .8)

μ / MODE:	0.075	0.100	0.125	0.150	0.175	0.200	0.225	0.250	0.275	0.300	0.325	0.350	0.375	0.400	0.425
(1)	4.437	4.592	4.779	5.004	5.272	5.592	5.977	6.442	7.009	7.702	8.528	9.354	9.582	9.271	8.873
(2)	7.463	7.784	8.171	8.632	9.177	9.819	10.57	11.42	12.22	12.22	11.65	11.28	11.42	11.19	10.72
(3)	10.55	11.06	11.67	12.37	13.19	14.11	14.98	14.82	14.10	14.17	13.88	13.27	13.29	14.93	15.86
(4)	13.68	14.38	15.20	16.15	17.18	17.85	16.99	16.90	16.50	15.86	16.77	18.46	18.30	17.60	17.66

TABLE 8.16 : β_1 $(1/\phi = .7,\ 1/\alpha = .9)$

MODE: \ μ	0.075	0.100	0.125	0.150	0.175	0.200	0.225	0.250	0.275	0.300	0.325	0.350	0.375	0.400	0.425
(1)	4.059	4.225	4.417	4.639	4.897	5.202	5.562	5.994	6.517	7.153	7.921	8.753	9.209	9.046	8.714
(2)	6.782	7.091	7.453	7.875	8.370	8.949	9.628	10.41	11.24	11.67	11.33	10.95	11.00	10.95	10.55
(3)	9.546	10.01	10.56	11.19	11.93	12.77	13.66	14.10	13.59	13.38	13.45	12.98	12.82	14.10	15.50
(4)	12.33	12.96	13.70	14.54	15.49	16.39	16.28	15.80	15.89	15.30	15.58	17.18	17.90	17.31	17.31

TABLE 8.16 : β_1 $(1/\phi = .7,\ 1/\alpha = 1.0)$

MODE: \ μ	0.075	0.100	0.125	0.150	0.175	0.200	0.225	0.250	0.275	0.300	0.325	0.350	0.375	0.400	0.425
(1)	3.724	3.896	4.090	4.311	4.566	4.861	5.208	5.620	6.115	6.717	7.445	8.267	8.854	8.818	8.542
(2)	6.203	6.505	6.851	7.249	7.711	8.248	8.877	9.607	10.41	11.05	10.98	10.64	10.64	10.70	10.38
(3)	8.709	9.151	9.659	10.24	10.92	11.69	12.54	13.25	13.10	12.77	12.95	12.66	12.45	13.45	15.11
(4)	11.23	11.81	12.49	13.26	14.13	15.05	15.49	15.01	15.14	14.82	14.74	16.06	17.39	17.03	16.97

TABLE 8.16 : β_1 $(1/\phi = .8,\ 1/\alpha = .6)$

MODE: \ μ	0.075	0.100	0.125	0.150	0.175	0.200	0.225	0.250	0.275	0.300	0.325	0.350	0.375	0.400	0.425
(1)	6.124	6.291	6.530	6.840	7.225	7.695	8.264	8.949	9.765	10.67	11.18	10.78	10.19	9.621	9.107
(2)	10.69	11.17	11.76	12.47	13.30	14.25	15.19	15.05	13.99	13.09	12.84	12.88	12.27	11.60	10.98
(3)	15.42	16.21	17.16	18.23	19.33	18.97	17.38	17.12	16.68	15.55	14.71	15.17	17.03	17.23	16.39
(4)	20.22	21.32	22.58	23.72	22.09	21.29	20.52	18.99	19.40	21.15	21.27	20.02	19.11	19.11	18.22

TABLE 0.16 : β_1 $(1/\phi = .8,\ 1/\alpha = .7)$

MODE: \ μ	0.075	0.100	0.125	0.150	0.175	0.200	0.225	0.250	0.275	0.300	0.325	0.350	0.375	0.400	0.425
(1)	5.558	5.722	5.937	6.207	6.538	6.939	7.424	8.010	8.713	9.531	10.30	10.40	9.975	9.477	9.009
(2)	9.472	9.872	10.37	10.96	11.67	12.48	13.38	14.05	13.60	12.78	12.25	12.32	12.01	11.44	10.87
(3)	13.50	14.16	14.95	15.86	16.07	17.65	16.82	16.86	15.92	15.23	14.39	14.18	15.53	16.76	16.21
(4)	17.59	18.51	19.57	20.72	21.07	19.56	19.50	18.44	17.69	18.80	20.30	19.72	18.75	18.58	18.03

TABLE 8.16 : β_1 $(1/\phi = .8,\ 1/\alpha = .8)$

MODE: \ μ	0.075	0.100	0.125	0.150	0.175	0.200	0.225	0.250	0.275	0.300	0.325	0.350	0.375	0.400	0.425
(1)	5.071	5.248	5.462	5.718	6.024	6.388	6.824	7.348	7.977	8.718	9.502	9.925	9.708	9.294	8.877
(2)	8.529	8.896	9.337	9.861	10.48	11.19	12.00	12.81	13.02	12.44	11.88	11.78	11.70	11.24	10.73
(3)	12.06	12.64	13.32	14.12	15.02	15.93	16.08	15.21	14.92	14.78	14.08	13.66	14.49	16.07	15.97
(4)	15.63	16.43	17.35	18.39	19.30	18.69	18.05	17.83	16.92	17.18	18.70	19.27	18.46	18.04	17.80

TABLE 8.16 : β_1 $(1/\phi = .8,\ 1/\alpha = .9)$

MODE: \ μ	0.075	0.100	0.125	0.150	0.175	0.200	0.225	0.250	0.275	0.300	0.325	0.350	0.375	0.400	0.425
(1)	4.639	4.829	5.047	5.301	5.596	5.943	6.352	6.839	7.422	8.112	8.876	9.447	9.419	9.088	8.722
(2)	7.751	8.104	8.517	8.997	9.557	10.21	10.95	11.75	12.29	12.04	11.54	11.35	11.37	11.02	10.57
(3)	10.91	11.44	12.06	12.78	13.60	14.47	15.09	14.64	14.14	14.20	13.73	13.28	13.77	15.36	15.71
(4)	14.09	14.81	15.64	16.58	17.54	17.78	16.98	16.99	16.37	16.12	17.30	18.58	18.14	17.63	17.54

TABLE 8.16 : β_1 $(1/\phi = .8,\ 1/\alpha = 1.0)$

MODE: \ μ	0.075	0.100	0.125	0.150	0.175	0.200	0.225	0.250	0.275	0.300	0.325	0.350	0.375	0.400	0.425
(1)	4.256	4.452	4.674	4.927	5.217	5.554	5.948	6.415	6.970	7.630	8.378	9.027	9.143	8.881	8.555
(2)	7.089	7.434	7.829	8.283	8.807	9.414	10.11	10.88	11.57	11.61	11.20	10.98	11.06	10.81	10.40
(3)	9.953	10.46	11.04	11.70	12.46	13.29	14.06	14.06	13.55	13.57	13.36	12.93	13.21	14.73	15.44
(4)	12.83	13.50	14.26	15.13	16.06	16.72	16.21	16.06	15.83	15.39	16.19	17.70	17.80	17.28	17.28

TABLE 8.16 : β_1 $(1/\phi = .9,\ 1/\alpha = .6)$

MODE: \ μ	0.075	0.100	0.125	0.150	0.175	0.200	0.225	0.250	0.275	0.300	0.325	0.350	0.375	0.400	0.425
(1)	6.889	7.077	7.346	7.693	8.124	8.648	9.276	10.02	10.06	11.57	11.43	10.82	10.20	9.622	9.108
(2)	12.02	12.56	13.22	14.01	14.92	15.89	16.29	15.24	14.12	13.44	13.49	13.01	12.29	11.60	10.98
(3)	17.34	18.23	19.27	20.41	20.99	19.26	18.13	18.03	16.83	15.71	15.35	16.65	17.92	17.29	16.40
(4)	22.74	23.96	25.29	25.16	22.90	22.65	20.79	20.09	21.51	22.61	21.41	20.15	19.74	19.21	18.23

TABLE 8.16 : β_1 $(1/\phi = .9,\ 1/\alpha = .7)$

MODE: \ μ	0.075	0.100	0.125	0.150	0.175	0.200	0.225	0.250	0.275	0.300	0.325	0.350	0.375	0.400	0.425
(1)	6.252	6.437	6.679	6.982	7.352	7.800	8.337	8.977	9.721	10.49	10.85	10.52	10.00	9.483	9.009
(2)	10.66	11.10	11.66	12.32	13.09	13.96	14.76	14.67	13.77	12.99	12.76	12.63	12.07	11.45	10.87
(3)	15.19	15.92	16.00	17.79	18.76	18.62	17.18	16.82	16.38	15.39	14.73	15.26	16.80	16.97	16.23
(4)	19.78	20.80	21.96	22.95	21.71	20.85	20.17	18.87	19.20	20.71	20.95	19.85	19.06	18.88	18.06

TABLE 8.16 : β_1 $(1/\phi = .9,\ 1/\alpha = .8)$

MODE: \ μ	0.075	0.100	0.125	0.150	0.175	0.200	0.225	0.250	0.275	0.300	0.325	0.350	0.375	0.400	0.425
(1)	5.704	5.904	6.144	6.432	6.774	7.182	7.666	8.241	8.913	9.648	10.21	10.16	9.760	9.306	8.880
(2)	9.595	10.01	10.50	11.09	11.77	12.54	13.35	13.81	13.33	12.62	12.24	12.20	11.81	11.26	10.73
(3)	13.57	14.21	14.98	15.85	16.78	17.34	16.53	15.79	15.71	15.02	14.33	14.43	15.75	16.55	16.02
(4)	17.58	18.47	19.48	20.53	20.64	19.40	19.23	18.23	17.78	18.90	20.09	19.50	18.66	18.47	17.85

TABLE 8.16 : β_1 $(1/\phi = .9,\ 1/\alpha = .9)$

MODE: \ μ	0.075	0.100	0.125	0.150	0.175	0.200	0.225	0.250	0.275	0.300	0.325	0.350	0.375	0.400	0.425
(1)	5.219	5.432	5.678	5.963	6.294	6.682	7.138	7.676	8.305	9.011	9.643	9.798	9.505	9.108	8.726
(2)	8.720	9.117	9.579	10.12	10.74	11.45	12.22	12.88	12.82	12.24	11.82	11.78	11.53	11.06	10.57
(3)	12.27	12.87	13.56	14.36	15.22	16.00	15.84	15.07	14.94	14.59	13.96	13.85	14.93	16.09	15.79
(4)	15.85	16.65	17.57	18.56	19.25	18.45	18.07	17.61	16.91	17.51	18.92	19.08	18.32	18.06	17.63

TABLE 8.16 : β_1 $(1/\phi = .9,\ 1/\alpha = 1.0)$

MODE: \ μ	0.075	0.100	0.125	0.150	0.175	0.200	0.225	0.250	0.275	0.300	0.325	0.350	0.375	0.400	0.425
(1)	4.788	5.009	5.259	5.543	5.869	6.246	6.687	7.204	7.810	8.499	9.172	9.465	9.263	8.910	8.561
(2)	7.975	8.363	8.807	9.315	9.900	10.57	11.31	12.03	12.28	11.87	11.45	11.40	11.27	10.85	10.41
(3)	11.20	11.76	12.41	13.15	13.97	14.79	15.09	14.48	14.21	14.12	13.59	13.37	14.26	15.64	15.56
(4)	14.43	15.18	16.03	16.97	17.83	17.63	17.03	16.95	16.27	16.45	17.78	18.59	18.00	17.67	17.41

TABLE 8.16 : β_1 $(1/\phi = 1.0,\ 1/\alpha = .6)$

MODE: \ μ	0.075	0.100	0.125	0.150	0.175	0.200	0.225	0.250	0.275	0.300	0.325	0.350	0.375	0.400	0.425
(1)	7.655	7.863	8.161	8.545	9.020	9.594	10.27	11.05	11.83	12.06	11.52	10.84	10.20	9.623	9.108
(2)	13.36	13.95	14.68	15.54	16.50	17.30	16.68	15.36	14.33	14.02	13.79	13.06	12.30	11.60	10.98
(3)	19.27	20.24	21.37	22.45	21.60	19.63	19.27	18.32	16.96	16.03	16.46	17.98	18.20	17.31	16.40
(4)	25.25	26.50	27.78	25.67	24.61	23.05	21.29	21.88	23.42	22.98	21.51	20.40	20.17	19.25	18.23

TABLE 8.16 : β_1 $(1/\phi = 1.0,\ 1/\alpha = .7)$

MODE: \ μ	0.075	0.100	0.125	0.150	0.175	0.200	0.225	0.250	0.275	0.300	0.325	0.350	0.375	0.400	0.425
(1)	6.947	7.152	7.420	7.755	8.164	8.655	9.239	9.918	10.65	11.21	11.10	10.58	10.01	9.486	9.010
(2)	11.84	12.34	12.95	13.67	14.50	15.35	15.72	14.91	13.93	13.33	13.22	12.76	12.09	11.45	10.87
(3)	16.87	17.69	18.63	19.66	20.21	18.88	17.82	17.59	16.57	15.60	15.39	16.49	17.54	17.05	16.24
(4)	21.97	23.09	24.28	24.38	22.44	22.04	20.51	19.83	20.96	22.04	21.15	20.01	19.54	18.98	18.07

TABLE 8.16 : β_1 ($1/\phi = 1.0$, $1/\alpha = .8$)

μ / MODE:	0.075	0.100	0.125	0.150	0.175	0.200	0.225	0.250	0.275	0.300	0.325	0.350	0.375	0.400	0.425
(1)	6.338	6.560	6.827	7.145	7.524	7.972	8.500	9.115	9.800	10.43	10.61	10.27	9.786	9.312	8.881
(2)	10.66	11.12	11.66	12.30	13.04	13.84	14.50	14.31	13.51	12.87	12.68	12.41	11.86	11.27	10.73
(3)	15.07	15.79	16.62	17.55	18.38	18.07	16.94	16.61	16.08	15.21	14.77	15.46	16.74	16.73	16.04
(4)	19.53	20.50	21.58	22.41	21.29	20.54	19.81	18.74	19.17	20.50	20.64	19.65	19.00	18.67	17.88

TABLE 8.16 : β_1 ($1/\phi = 1.0$, $1/\alpha = .9$)

μ / MODE:	0.075	0.100	0.125	0.150	0.175	0.200	0.225	0.250	0.275	0.300	0.325	0.350	0.375	0.400	0.425
(1)	5.798	6.036	6.309	6.625	6.992	7.419	7.919	8.499	9.153	9.804	10.15	9.958	9.547	9.119	8.728
(2)	9.688	10.13	10.64	11.23	11.91	12.66	13.39	13.62	13.07	12.45	12.21	12.06	11.61	11.08	10.58
(3)	13.64	14.30	15.06	15.91	16.77	17.09	16.25	15.71	15.51	14.81	14.29	14.71	15.99	16.39	15.82
(4)	17.61	18.49	19.48	20.43	20.26	19.24	18.98	18.02	17.89	19.05	19.95	19.28	18.58	18.35	17.67

TABLE 8.17 : β_1 ($\phi = .6$, $\alpha = .6$)

μ / MODE:	0.075	0.100	0.125	0.150	0.175	0.200	0.225	0.250	0.275	0.300	0.325	0.350	0.375	0.400	0.425
(1)	4.540	4.687	4.711	4.575	4.341	4.089	3.859	3.662	3.498	3.365	3.260	3.182	3.131	3.109	3.122
(2)	7.882	7.621	7.277	7.219	7.452	7.889	8.483	9.227	10.13	11.13	11.64	11.17	10.54	9.949	9.429
(3)	10.66	10.33	10.61	11.19	11.94	12.84	13.87	14.74	14.25	13.23	12.30	11.49	10.80	10.21	9.727
(4)	13.47	13.84	14.59	15.54	16.62	17.72	17.41	15.94	14.65	13.70	13.76	15.40	17.62	17.59	16.71

TABLE 8.17 : β_1 ($\phi = .6$, $\alpha = .7$)

μ / MODE:	0.075	0.100	0.125	0.150	0.175	0.200	0.225	0.250	0.275	0.300	0.325	0.350	0.375	0.400	0.425
(1)	5.285	5.451	5.475	5.310	5.029	4.727	4.451	4.212	4.012	3.846	3.711	3.605	3.526	3.474	3.452
(2)	9.174	8.849	8.422	8.331	8.577	9.055	9.702	10.50	11.39	12.08	11.90	11.28	18.40	17.70	9.568
(3)	12.38	11.96	12.27	12.91	13.75	14.70	15.61	15.51	14.47	13.41	12.48	11.69	19.16	18.07	9.979
(4)	15.63	16.03	16.88	17.92	19.02	19.27	17.76	16.23	15.05	14.56	15.45	17.29	21.67	24.57	16.80

TABLE 8.17 : β_1 ($\phi = .6$, $\alpha = .8$)

μ / MODE:	0.075	0.100	0.125	0.150	0.175	0.200	0.225	0.250	0.275	0.300	0.325	0.350	0.375	0.400	0.425
(1)	6.019	6.203	6.222	6.023	5.688	5.330	5.003	4.719	4.478	4.275	4.105	3.966	3.853	3.766	3.705
(2)	10.45	10.04	9.511	9.371	9.616	10.11	10.78	11.57	12.32	12.54	12.07	11.41	10.78	10.21	9.740
(3)	14.06	13.53	13.84	14.54	15.41	16.32	16.72	15.87	14.70	13.64	12.71	11.92	11.26	10.69	10.23
(4)	17.72	18.15	19.07	20.14	20.93	19.91	18.13	16.65	15.68	15.74	16.99	18.57	18.66	17.81	16.93

TABLE 8.17 : β_1 ($\phi = .6$, $\alpha = .9$)

μ / MODE:	0.075	0.100	0.125	0.150	0.175	0.200	0.225	0.250	0.275	0.300	0.325	0.350	0.375	0.400	0.425
(1)	6.740	6.935	6.946	6.706	6.310	5.890	5.506	5.173	4.887	4.644	4.438	4.262	4.114	3.992	3.892
(2)	11.70	11.19	10.53	10.32	10.55	11.05	11.72	12.44	12.96	12.83	12.24	11.57	10.94	10.39	9.932
(3)	15.67	15.01	15.32	16.04	16.91	17.63	17.36	16.19	14.97	13.89	12.97	12.17	11.50	10.92	10.44
(4)	19.73	20.17	21.12	22.13	22.17	20.41	18.57	17.17	16.48	16.93	18.23	19.27	18.85	17.96	17.08

TABLE 8.17 : β_1 ($\phi = .6$, $\alpha = 1.0$)

μ / MODE:	0.075	0.100	0.125	0.150	0.175	0.200	0.225	0.250	0.275	0.300	0.325	0.350	0.375	0.400	0.425
(1)	7.442	7.644	7.641	7.352	6.887	6.399	5.955	5.569	5.238	4.954	4.710	4.500	4.319	4.163	4.029
(2)	12.92	12.27	11.45	11.17	11.37	11.87	12.51	13.14	13.43	13.09	12.44	11.75	11.13	10.58	10.12
(3)	17.21	16.38	16.69	17.42	18.24	18.64	17.85	16.52	15.26	14.16	13.22	12.41	11.72	11.12	10.60
(4)	21.62	22.09	23.05	23.86	23.01	20.90	19.05	17.75	17.32	17.97	19.18	19.69	19.04	18.12	17.25

TABLE 8.17 : β_1 ($\phi = .7$, $\alpha = .6$)

μ / MODE:	0.075	0.100	0.125	0.150	0.175	0.200	0.225	0.250	0.275	0.300	0.325	0.350	0.375	0.400	0.425
(1)	3.918	4.097	4.229	4.260	4.170	4.003	3.816	3.640	3.487	3.359	3.257	3.181	3.130	3.109	3.122
(2)	6.926	6.961	6.743	6.560	6.608	6.887	7.344	7.957	8.730	9.671	10.68	10.96	10.49	9.932	9.424
(3)	9.629	9.290	9.291	9.681	10.29	11.05	11.95	12.97	13.68	13.13	12.27	11.48	10.79	10.21	9.727
(4)	12.02	12.03	12.57	13.35	14.29	15.36	16.32	15.77	14.56	13.52	12.87	13.50	15.58	17.29	16.67

TABLE 8.17 : β_1 ($\phi = .7$, $\alpha = .7$)

μ / MODE:	0.075	0.100	0.125	0.150	0.175	0.200	0.225	0.250	0.275	0.300	0.325	0.350	0.375	0.400	0.425
(1)	4.560	4.765	4.914	4.946	4.834	4.630	4.403	4.188	4.000	3.840	3.709	3.604	3.525	3.473	3.452
(2)	8.062	8.092	7.813	7.573	7.606	7.904	8.403	9.067	9.881	10.79	11.38	11.14	17.17	17.53	9.563
(3)	11.20	10.76	10.74	11.17	11.84	12.68	13.62	14.44	14.22	13.34	12.46	11.68	19.12	18.06	9.979
(4)	13.95	13.93	14.55	15.41	16.43	17.44	17.38	16.10	14.88	13.98	13.88	15.18	20.06	22.31	16.77

TABLE 8.17 : β_1 ($\phi = .7$, $\alpha = .8$)

μ / MODE:	0.075	0.100	0.125	0.150	0.175	0.200	0.225	0.250	0.275	0.300	0.325	0.350	0.375	0.400	0.425
(1)	5.194	5.421	5.584	5.613	5.473	5.226	4.952	4.694	4.465	4.269	4.103	3.964	3.852	3.765	3.705
(2)	9.183	9.198	8.840	8.525	8.527	8.830	9.350	10.03	10.83	11.57	11.72	11.30	10.74	10.20	9.736
(3)	12.74	12.18	12.11	12.57	13.29	14.16	15.02	15.29	14.55	13.58	12.69	11.92	11.25	10.69	10.23
(4)	15.82	15.76	16.43	17.36	18.38	18.92	17.88	16.47	15.31	14.67	15.09	16.58	17.99	17.69	16.90

TABLE 8.17 : β_1 ($\phi = .7$, $\alpha = .9$)

μ / MODE:	0.075	0.100	0.125	0.150	0.175	0.200	0.225	0.250	0.275	0.300	0.325	0.350	0.375	0.400	0.425
(1)	5.815	6.060	6.234	6.254	6.079	5.782	5.455	5.148	4.876	4.639	4.435	4.261	4.114	3.991	3.892
(2)	10.28	10.27	9.807	9.396	9.353	9.649	10.17	10.85	11.59	12.09	11.98	11.48	10.91	10.38	9.929
(3)	14.24	13.51	13.39	13.88	14.62	15.47	16.12	15.83	14.86	13.85	12.95	12.17	11.50	10.92	10.44
(4)	17.61	17.50	18.21	19.17	20.06	19.83	18.34	16.90	15.84	15.47	16.19	17.64	18.41	17.86	17.05

TABLE 8.17 : β_1 ($\phi = .7$, $\alpha = 1.0$)

μ / MODE:	0.075	0.100	0.125	0.150	0.175	0.200	0.225	0.250	0.275	0.300	0.325	0.350	0.375	0.400	0.425
(1)	6.420	6.679	6.858	6.865	6.647	6.290	5.906	5.546	5.228	4.949	4.708	4.499	4.319	4.163	4.029
(2)	11.36	11.30	10.70	10.17	10.08	10.36	10.88	11.54	12.19	12.50	12.22	11.68	11.10	10.57	10.12
(3)	15.70	14.74	14.57	15.08	15.84	16.63	16.96	16.26	15.17	14.13	13.21	12.41	11.72	11.12	10.60
(4)	19.30	19.14	19.90	20.86	21.47	20.50	18.81	17.36	16.40	16.25	17.13	18.42	18.71	18.04	17.23

TABLE 8.17 : β_1 ($\phi = .8$, $\alpha = .6$)

μ / MODE:	0.075	0.100	0.125	0.150	0.175	0.200	0.225	0.250	0.275	0.300	0.325	0.350	0.375	0.400	0.425
(1)	3.439	3.619	3.785	3.906	3.940	3.876	3.750	3.606	3.469	3.350	3.253	3.179	3.129	3.108	3.122
(2)	6.129	6.308	6.288	6.130	6.061	6.195	6.524	7.022	7.681	8.515	9.529	10.44	10.38	9.904	9.416
(3)	8.673	8.560	8.406	8.602	9.066	9.703	10.49	11.43	12.44	12.84	12.20	11.46	10.79	10.21	9.726
(4)	10.96	10.77	11.10	11.73	12.53	13.48	14.54	15.23	14.45	13.44	12.63	12.42	13.85	16.34	16.59

TABLE 8.17 : β_1 ($\phi = .8$, $\alpha = .7$)

μ / MODE:	0.075	0.100	0.125	0.150	0.175	0.200	0.225	0.250	0.275	0.300	0.325	0.350	0.375	0.400	0.425
(1)	4.003	4.209	4.399	4.535	4.569	4.485	4.329	4.151	3.981	3.831	3.704	3.602	3.524	3.473	3.452
(2)	7.134	7.336	7.296	7.085	6.979	7.110	7.465	8.005	8.713	9.576	10.48	10.85	10.52	17.12	9.556
(3)	10.09	9.931	9.715	9.922	10.44	11.14	12.00	12.94	13.59	13.19	12.41	11.66	11.00	18.04	9.978
(4)	12.74	12.47	12.83	13.54	14.43	15.44	16.30	15.87	14.76	13.79	13.21	13.68	15.49	20.24	16.71

TABLE 8.17 : β_1 (ϕ = .8, α = .8)

μ / MODE:	0.075	0.100	0.125	0.150	0.175	0.200	0.225	0.250	0.275	0.300	0.325	0.350	0.375	0.400	0.425
(1)	4.559	4.788	4.998	5.146	5.176	5.067	4.874	4.655	4.447	4.260	4.098	3.962	3.851	3.765	3.704
(2)	8.127	8.345	8.271	7.988	7.828	7.941	8.307	8.869	9.592	10.42	11.08	11.09	10.67	10.18	9.730
(3)	11.49	11.25	10.95	11.16	11.72	12.47	13.34	14.14	14.19	13.48	12.65	11.90	11.25	10.69	10.22
(4)	14.47	14.10	14.49	15.26	16.20	17.15	17.34	16.30	15.15	14.27	14.01	14.92	16.68	17.44	16.85

TABLE 8.17 : β_1 (ϕ = .8, α = .9)

μ / MODE:	0.075	0.100	0.125	0.150	0.175	0.200	0.225	0.250	0.275	0.300	0.325	0.350	0.375	0.400	0.425
(1)	5.104	5.351	5.578	5.735	5.755	5.614	5.375	5.110	4.857	4.630	4.431	4.260	4.113	3.991	3.892
(2)	9.102	9.329	9.202	8.822	8.590	8.675	9.041	9.610	10.32	11.07	11.49	11.31	10.85	10.36	9.924
(3)	12.87	12.51	12.10	12.31	12.89	13.67	14.51	15.03	14.62	13.77	12.92	12.16	11.49	10.92	10.44
(4)	16.14	15.63	16.05	16.87	17.82	18.54	18.00	16.73	15.58	14.82	14.85	15.96	17.47	17.67	17.02

TABLE 8.17 : β_1 (ϕ = .8, α = 1.0)

μ / MODE:	0.075	0.100	0.125	0.150	0.175	0.200	0.225	0.250	0.275	0.300	0.325	0.350	0.375	0.400	0.425
(1)	5.635	5.897	6.135	6.297	6.301	6.119	5.827	5.511	5.211	4.942	4.705	4.498	4.318	4.163	4.029
(2)	10.06	10.28	10.08	9.573	9.256	9.310	9.674	10.25	10.94	11.60	11.83	11.54	11.05	10.56	10.12
(3)	14.21	13.69	13.15	13.36	13.98	14.77	15.53	15.71	15.01	14.07	13.19	12.40	11.72	11.12	10.60
(4)	17.74	17.07	17.53	18.39	19.32	19.63	18.55	17.16	16.03	15.38	15.61	16.81	18.02	17.90	17.20

TABLE 8.17 : β_1 (ϕ = .9, α = .6)

μ / MODE:	0.075	0.100	0.125	0.150	0.175	0.200	0.225	0.250	0.275	0.300	0.325	0.350	0.375	0.400	0.425
(1)	3.062	3.233	3.405	3.563	3.674	3.706	3.656	3.557	3.444	3.337	3.246	3.176	3.128	3.108	3.122
(2)	5.479	5.710	5.832	5.786	5.692	5.717	5.928	6.318	6.875	7.608	8.540	9.622	10.17	9.857	9.404
(3)	7.823	7.923	7.791	7.824	8.147	8.673	9.357	10.19	11.18	12.12	14.88	11.42	10.77	10.20	9.725
(4)	10.02	9.878	9.997	10.48	11.17	12.01	13.00	14.03	14.21	13.36	20.15	11.99	12.60	14.92	16.44

TABLE 8.17 : β_1 (ϕ = .9, α = .7)

μ / MODE:	0.075	0.100	0.125	0.150	0.175	0.200	0.225	0.250	0.275	0.300	0.325	0.350	0.375	0.400	0.425
(1)	3.564	3.760	3.957	4.136	4.261	4.291	4.223	4.097	3.954	3.817	3.697	3.599	3.523	3.473	3.452
(2)	6.378	6.642	6.773	6.698	6.561	6.563	6.782	7.203	7.805	8.583	9.513	10.33	10.38	16.24	9.546
(3)	9.106	9.203	9.011	9.022	9.376	9.961	10.71	11.61	12.63	12.87	12.32	11.63	10.99	18.01	9.977
(4)	11.66	11.44	11.55	12.10	12.87	13.79	14.79	15.33	14.63	13.69	12.95	12.80	14.06	19.11	16.61

TABLE 8.17 : β_1 (ϕ = .9, α = .8)

μ / MODE:	0.075	0.100	0.125	0.150	0.175	0.200	0.225	0.250	0.275	0.300	0.325	0.350	0.375	0.400	0.425
(1)	4.059	4.276	4.495	4.693	4.828	4.852	4.759	4.598	4.419	4.246	4.092	3.960	3.850	3.765	3.704
(2)	7.265	7.556	7.689	7.567	7.368	7.332	7.545	7.982	8.607	9.390	10.24	10.73	10.56	10.15	9.722
(3)	10.37	10.45	10.17	10.14	10.52	11.15	11.94	12.83	13.49	13.28	12.59	11.88	11.24	10.69	10.22
(4)	13.27	12.94	13.03	13.63	14.46	15.42	16.24	16.01	15.02	14.09	13.50	13.77	15.27	16.92	16.77

TABLE 8.17 : β_1 (ϕ = .9, α = .9)

μ / MODE:	0.075	0.100	0.125	0.150	0.175	0.200	0.225	0.250	0.275	0.300	0.325	0.350	0.375	0.400	0.425
(1)	4.545	4.780	5.016	5.229	5.370	5.381	5.256	5.053	4.831	4.618	4.426	4.257	4.113	3.991	3.892
(2)	8.137	8.450	8.573	8.381	8.097	8.010	8.209	8.652	9.285	10.05	10.78	11.04	10.77	10.33	9.917
(3)	11.62	11.65	11.24	11.17	11.57	12.23	13.05	13.86	14.16	13.64	12.87	12.14	11.49	10.92	10.44
(4)	14.84	14.35	14.43	15.07	15.95	16.88	17.30	16.52	15.43	14.54	14.11	14.66	16.20	17.32	16.95

TABLE 8.17 : β_1 (ϕ = .9, α = 1.0)

μ / MODE:	0.075	0.100	0.125	0.150	0.175	0.200	0.225	0.250	0.275	0.300	0.325	0.350	0.375	0.400	0.425
(1)	5.017	5.266	5.516	5.740	5.884	5.876	5.709	5.456	5.187	4.931	4.700	4.496	4.318	4.162	4.029
(2)	8.989	9.319	9.419	9.127	8.738	8.593	8.777	9.228	9.868	10.61	11.23	11.32	10.98	10.54	10.11
(3)	12.84	12.80	12.22	12.10	12.54	13.24	14.06	14.75	14.69	13.98	13.15	12.39	11.71	11.12	10.60
(4)	16.36	15.65	15.73	16.44	17.35	18.19	18.10	16.99	15.84	14.98	14.69	15.43	16.93	17.63	17.15

TABLE 8.17 : β_1 (ϕ = 1.0, α = .6)

μ / MODE:	0.075	0.100	0.125	0.150	0.175	0.200	0.225	0.250	0.275	0.300	0.325	0.350	0.375	0.400	0.425
(1)	2.759	2.918	3.086	3.254	3.403	3.505	3.531	3.489	3.409	3.319	3.237	3.171	3.126	3.107	3.122
(2)	4.947	5.190	5.384	5.457	5.411	5.383	5.494	5.782	6.245	6.886	7.728	8.789	9.774	9.782	9.387
(3)	7.094	7.314	7.309	7.264	7.450	7.867	8.456	9.202	10.11	11.15	11.77	11.36	10.75	10.20	9.723
(4)	9.166	9.192	9.176	9.512	10.09	10.84	11.73	12.76	13.62	13.25	12.46	11.82	11.81	13.63	16.10

TABLE 8.17 : β_1 (ϕ = 1.0, α = .7)

μ / MODE:	0.075	0.100	0.125	0.150	0.175	0.200	0.225	0.250	0.275	0.300	0.325	0.350	0.375	0.400	0.425
(1)	3.211	3.393	3.585	3.777	3.947	4.059	4.082	4.021	3.915	3.798	3.688	3.594	3.521	3.472	3.452
(2)	5.758	6.037	6.255	6.324	6.245	6.185	6.206	6.592	7.092	7.779	8.653	9.629	10.14	9.927	9.531
(3)	8.259	8.503	8.466	8.376	8.570	9.033	9.685	10.50	11.44	12.25	12.17	11.59	10.97	10.43	9.976
(4)	10.67	10.66	10.60	10.98	11.63	12.45	13.41	14.34	14.38	13.59	12.82	12.38	12.99	15.07	16.43

TABLE 8.17 : β_1 (ϕ = 1.0, α = .8)

μ / MODE:	0.075	0.100	0.125	0.150	0.175	0.200	0.225	0.250	0.275	0.300	0.325	0.350	0.375	0.400	0.425
(1)	3.657	3.859	4.072	4.285	4.472	4.592	4.604	4.518	4.379	4.227	4.083	3.956	3.849	3.764	3.704
(2)	6.559	6.869	7.106	7.159	7.027	6.916	6.993	7.304	7.824	8.531	9.388	10.19	10.39	10.10	9.709
(3)	9.408	9.668	9.571	9.415	9.608	10.11	10.81	11.65	12.53	12.92	12.49	11.84	11.23	10.68	10.22
(4)	12.15	12.07	11.95	12.36	13.07	13.95	14.90	15.43	14.85	13.98	13.27	13.08	14.08	16.05	16.64

TABLE 8.17 : β_1 (ϕ = 1.0, α = .9)

μ / MODE:	0.075	0.100	0.125	0.150	0.175	0.200	0.225	0.250	0.275	0.300	0.325	0.350	0.375	0.400	0.425
(1)	4.094	4.313	4.544	4.773	4.974	5.097	5.092	4.972	4.792	4.600	4.418	4.254	4.111	3.991	3.892
(2)	7.346	7.681	7.932	7.952	7.742	7.561	7.606	7.914	8.447	9.160	9.978	10.61	10.63	10.30	9.907
(3)	10.54	10.80	10.61	10.36	10.55	11.09	11.83	12.68	13.40	13.40	12.80	12.12	11.48	10.92	10.44
(4)	13.60	13.41	13.21	13.66	14.42	15.34	16.18	16.17	15.29	14.38	13.75	13.78	14.98	16.69	16.86

TABLE 8.17 : β_1 ($1/\phi$ = .6, $1/\alpha$ = .6)

μ / MODE:	0.075	0.100	0.125	0.150	0.175	0.200	0.225	0.250	0.275	0.300	0.325	0.350	0.375	0.400	0.425
(1)	4.194	4.370	4.585	4.842	5.144	5.492	5.874	6.206	6.203	5.884	5.517	5.174	4.867	4.593	4.348
(2)	7.739	8.135	8.592	9.085	9.414	8.907	8.157	7.654	7.726	8.372	9.359	10.62	11.61	11.37	10.84
(3)	11.34	11.94	12.51	11.94	10.84	10.96	11.76	12.80	14.00	14.77	14.10	13.21	12.39	11.67	11.02
(4)	14.96	15.65	14.59	13.97	14.80	15.90	17.16	17.88	16.76	15.51	14.45	13.66	13.84	16.20	17.90

TABLE 8.17 : β_1 ($1/\phi$ = .6, $1/\alpha$ = .7)

μ / MODE:	0.075	0.100	0.125	0.150	0.175	0.200	0.225	0.250	0.275	0.300	0.325	0.350	0.375	0.400	0.425
(1)	3.709	3.873	4.065	4.290	4.550	4.849	5.183	5.516	5.709	5.599	5.330	5.040	4.768	4.520	4.298
(2)	6.756	7.095	7.484	7.912	8.314	8.350	7.854	7.371	7.207	7.566	8.342	9.432	10.66	11.03	10.67
(3)	9.836	10.35	10.88	11.06	10.29	9.893	10.35	11.19	12.25	13.38	13.66	13.00	12.26	11.57	10.95
(4)	12.93	13.57	13.63	12.62	12.94	13.82	14.92	16.08	16.26	15.27	14.28	13.47	13.16	14.60	17.18

TABLE 8.17 : β_1 $(1/\phi = .6,\ 1/\alpha = .8)$

μ MODE:	0.075	0.100	0.125	0.150	0.175	0.200	0.225	0.250	0.275	0.300	0.325	0.350	0.375	0.400	0.425
(1)	3.317	3.476	3.657	3.863	4.099	4.367	4.665	4.974	5.216	5.251	5.091	4.865	4.636	4.422	4.227
(2)	6.002	6.308	6.653	7.033	7.415	7.644	7.448	7.079	6.875	7.057	7.660	8.599	9.787	10.58	10.45
(3)	8.701	9.153	9.630	9.988	9.713	9.249	9.407	10.06	10.96	12.04	12.90	12.69	12.07	11.44	10.85
(4)	11.41	11.98	12.36	11.79	11.67	12.33	13.26	14.36	15.26	14.91	14.05	13.28	12.83	13.61	16.18

TABLE 8.17 : β_1 $(1/\phi = .6,\ 1/\alpha = .9)$

μ MODE:	0.075	0.100	0.125	0.150	0.175	0.200	0.225	0.250	0.275	0.300	0.325	0.350	0.375	0.400	0.425
(1)	2.990	3.144	3.317	3.512	3.732	3.980	4.254	4.540	4.787	4.893	4.821	4.659	4.476	4.299	4.137
(2)	5.395	5.679	5.995	6.341	6.696	6.973	6.973	6.743	6.575	6.688	7.174	7.998	9.106	10.09	10.19
(3)	7.804	8.214	8.649	9.029	9.047	8.720	8.728	9.217	10.00	10.99	12.01	12.29	11.83	11.27	10.74
(4)	10.21	10.73	11.16	11.01	10.76	11.21	12.01	13.00	14.04	14.37	13.75	13.05	12.58	12.98	15.27

TABLE 8.17 : β_1 $(1/\phi = .6,\ 1/\alpha = 1.0)$

μ MODE:	0.075	0.100	0.125	0.150	0.175	0.200	0.225	0.250	0.275	0.300	0.325	0.350	0.375	0.400	0.425
(1)	2.716	2.863	3.028	3.213	3.421	3.653	3.909	4.177	4.418	4.556	4.544	4.436	4.297	4.156	4.028
(2)	4.893	5.160	5.454	5.774	6.105	6.389	6.489	6.378	6.270	6.370	6.792	7.538	8.579	9.647	9.936
(3)	7.070	7.450	7.851	8.222	8.371	8.203	8.182	8.563	9.248	10.15	11.19	11.81	11.57	11.08	10.60
(4)	9.244	9.721	10.15	10.22	10.03	10.34	11.02	11.92	12.94	13.68	13.40	12.79	12.32	12.53	14.54

TABLE 8.17 : β_1 $(1/\phi = .7,\ 1/\alpha = .6)$

μ MODE:	0.075	0.100	0.125	0.150	0.175	0.200	0.225	0.250	0.275	0.300	0.325	0.350	0.375	0.400	0.425
(1)	4.892	5.096	5.344	5.637	5.973	6.339	6.658	6.672	6.332	5.916	5.526	5.177	4.868	4.593	4.348
(2)	9.024	9.475	9.973	10.38	9.990	9.082	8.411	8.303	8.813	9.670	10.77	11.86	11.93	11.41	10.85
(3)	13.21	13.85	13.81	12.34	11.99	12.67	13.65	14.79	15.69	15.15	14.15	13.23	12.40	11.67	11.02
(4)	17.39	17.39	15.60	16.14	17.20	18.44	19.37	18.32	16.85	15.60	14.61	14.23	15.59	18.03	24.22

TABLE 8.17 : β_1 $(1/\phi = .7,\ 1/\alpha = .7)$

μ MODE:	0.075	0.100	0.125	0.150	0.175	0.200	0.225	0.250	0.275	0.300	0.325	0.350	0.375	0.400	0.425
(1)	4.327	4.516	4.738	4.993	5.283	5.601	5.912	6.089	5.973	5.678	5.353	5.047	4.770	4.521	4.298
(2)	7.878	8.264	8.692	9.106	9.212	8.682	8.086	7.800	8.027	8.672	9.603	10.71	11.38	11.15	10.69
(3)	11.46	12.01	12.39	11.69	10.95	11.21	11.97	12.96	14.03	14.50	13.87	13.04	12.27	11.57	10.95
(4)	15.04	15.53	14.44	14.20	14.99	16.04	17.18	17.58	16.53	15.38	14.41	13.80	14.30	16.47	17.65

TABLE 8.17 : β_1 $(1/\phi = .7,\ 1/\alpha = .8)$

μ MODE:	0.075	0.100	0.125	0.150	0.175	0.200	0.225	0.250	0.275	0.300	0.325	0.350	0.375	0.400	0.425
(1)	3.869	4.053	4.261	4.496	4.759	5.045	5.332	5.547	5.564	5.386	5.135	4.879	4.640	4.423	4.228
(2)	6.998	7.347	7.728	8.111	8.355	8.162	7.724	7.438	7.521	8.006	8.797	9.814	10.72	10.82	10.48
(3)	10.14	10.63	11.05	10.90	10.27	10.24	10.79	11.63	12.63	13.51	13.45	12.79	12.09	11.44	10.86
(4)	13.27	13.80	13.46	12.91	13.41	14.29	15.34	16.27	16.04	15.08	14.17	13.52	13.59	15.27	17.12

TABLE 8.17 : β_1 $(1/\phi = .7,\ 1/\alpha = .9)$

μ MODE:	0.075	0.100	0.125	0.150	0.175	0.200	0.225	0.250	0.275	0.300	0.325	0.350	0.375	0.400	0.425
(1)	3.488	3.666	3.865	4.087	4.332	4.597	4.864	5.082	5.161	5.070	4.887	4.682	4.483	4.301	4.137
(2)	6.290	6.614	6.964	7.319	7.594	7.587	7.315	7.089	7.131	7.520	8.212	9.148	10.12	10.48	10.26
(3)	9.092	9.543	9.954	10.04	9.649	9.521	9.917	10.63	11.54	12.50	12.91	12.49	11.88	11.28	10.74
(4)	11.89	12.39	12.42	11.95	12.22	12.96	13.90	14.91	15.34	14.71	13.89	13.25	13.13	[illegible]	16.52

TABLE 8.17 : β_1 $(1/\phi = .7,\ 1/\alpha = 1.0)$

μ MODE:	0.075	0.100	0.125	0.150	0.175	0.200	0.225	0.250	0.275	0.300	0.325	0.350	0.375	0.400	0.425
(1)	3.168	3.339	3.529	3.739	3.970	4.218	4.469	4.684	4.791	4.755	4.627	4.468	4.307	4.159	4.028
(2)	5.705	6.009	6.336	6.667	6.944	7.027	6.882	6.730	6.777	7.122	7.752	8.631	9.625	10.15	10.03
(3)	8.237	8.656	9.047	9.229	9.028	8.914	9.220	9.842	10.67	11.63	12.32	12.17	11.65	11.10	10.60
(4)	10.76	11.24	11.41	11.13	11.28	11.91	12.76	13.74	14.51	14.29	13.59	12.97	12.76	13.76	15.95

TABLE 8.17 : β_1 $(1/\phi = .8,\ 1/\alpha = .6)$

μ MODE:	0.075	0.100	0.125	0.150	0.175	0.200	0.225	0.250	0.275	0.300	0.325	0.350	0.375	0.400	0.425
(1)	5.590	5.820	6.098	6.419	6.771	7.093	7.171	6.841	6.376	5.930	5.530	5.178	4.868	4.593	4.348
(2)	10.30	10.80	11.28	11.22	10.21	9.299	8.925	9.241	9.966	10.94	12.00	12.45	12.03	11.43	10.85
(3)	15.07	15.58	14.32	13.12	13.51	14.40	15.49	16.51	16.31	15.24	14.18	13.23	12.40	11.67	11.02
(4)	19.73	18.15	17.43	18.37	19.57	20.72	20.13	18.44	16.93	15.72	14.95	15.40	17.41	18.74	25.03

TABLE 8.17 : β_1 $(1/\phi = .8,\ 1/\alpha = .7)$

μ MODE:	0.075	0.100	0.125	0.150	0.175	0.200	0.225	0.250	0.275	0.300	0.325	0.350	0.375	0.400	0.425
(1)	4.944	5.158	5.405	5.686	5.991	6.290	6.478	6.386	6.073	5.710	5.364	5.051	4.771	4.521	4.298
(2)	8.995	9.419	9.852	10.09	9.646	8.913	8.450	8.495	8.999	9.793	10.77	11.61	11.64	11.20	10.69
(3)	13.07	13.60	13.35	12.21	12.06	12.69	13.59	14.62	15.31	14.83	13.93	13.06	12.27	11.57	10.95
(4)	17.10	16.85	15.58	16.07	17.05	18.18	18.91	17.99	16.65	15.49	14.63	14.48	15.81	17.79	17.79

TABLE 8.17 : β_1 $(1/\phi = .8,\ 1/\alpha = .8)$

μ MODE:	0.075	0.100	0.125	0.150	0.175	0.200	0.225	0.250	0.275	0.300	0.325	0.350	0.375	0.400	0.425
(1)	4.420	4.628	4.861	5.119	5.396	5.670	5.878	5.901	5.717	5.441	5.154	4.886	4.642	4.424	4.228
(2)	7.990	8.374	8.766	9.063	8.952	8.457	8.051	7.997	8.351	9.012	9.890	10.80	11.17	10.92	10.50
(3)	11.56	12.05	12.18	11.46	11.10	11.49	12.23	13.16	14.06	14.24	13.61	12.83	12.10	11.44	10.86
(4)	15.10	15.31	14.36	14.44	15.22	16.22	17.21	17.28	16.26	15.20	14.35	13.98	14.79	16.74	17.42

TABLE 8.17 : β_1 $(1/\phi = .8,\ 1/\alpha = .9)$

μ MODE:	0.075	0.100	0.125	0.150	0.175	0.200	0.225	0.250	0.275	0.300	0.325	0.350	0.375	0.400	0.425
(1)	3.985	4.187	4.409	4.652	4.910	5.167	5.374	5.447	5.348	5.146	4.916	4.693	4.487	4.302	4.137
(2)	7.182	7.538	7.902	8.202	8.244	7.950	7.642	7.582	7.862	8.437	9.238	10.15	10.72	10.63	10.28
(3)	10.37	10.83	11.06	10.69	10.35	10.58	11.20	12.04	12.96	13.53	13.22	12.57	11.90	11.29	10.74
(4)	13.52	13.86	13.32	13.22	13.83	14.72	15.70	16.33	15.79	14.86	14.05	13.60	14.09	15.86	17.03

TABLE 8.17 : β_1 $(1/\phi = .8,\ 1/\alpha = 1.0)$

μ MODE:	0.075	0.100	0.125	0.150	0.175	0.200	0.225	0.250	0.275	0.300	0.325	0.350	0.375	0.400	0.425
(1)	3.620	3.813	4.025	4.255	4.499	4.741	4.943	5.039	4.991	4.846	4.665	4.482	4.312	4.161	4.029
(2)	6.514	6.849	7.189	7.482	7.591	7.432	7.219	7.187	7.442	7.969	8.724	9.625	10.32	10.36	10.07
(3)	9.395	9.823	10.10	9.933	9.675	9.844	10.38	11.14	12.04	12.80	12.82	12.29	11.68	11.11	10.60
(4)	12.24	12.61	12.35	12.22	12.72	13.52	14.46	15.30	15.26	14.50	13.73	13.24	13.53	15.15	16.66

TABLE 8.17 : β_1 $(1/\phi = .9,\ 1/\alpha = .6)$

μ MODE:	0.075	0.100	0.125	0.150	0.175	0.200	0.225	0.250	0.275	0.300	0.325	0.350	0.375	0.400	0.425
(1)	6.286	6.541	6.843	7.181	7.511	7.674	7.414	6.908	6.396	5.936	5.533	5.179	4.868	4.593	4.348
(2)	11.58	12.09	12.40	11.59	10.40	9.676	9.703	10.27	11.11	12.11	12.85	12.67	12.07	11.44	10.85
(3)	16.88	16.81	14.79	14.38	15.11	16.12	17.19	17.51	16.48	15.27	14.19	13.24	12.40	11.67	11.02
(4)	21.73	19.14	19.47	20.59	21.83	22.11	20.32	18.52	17.03	15.93	15.65	16.90	18.85	18.95	25.23

TABLE 8.17 : ε_1 $(1/\phi = .9,\ 1/\alpha = .7)$

μ MODE:	0.075	0.100	0.125	0.150	0.175	0.200	0.225	0.250	0.275	0.300	0.325	0.350	0.375	0.400	0.425
(1)	5.559	5.796	6.066	6.360	6.655	6.868	6.830	6.519	6.117	5.725	5.369	5.052	4.771	4.521	4.298
(2)	10.11	10.55	10.92	10.73	9.901	9.212	9.017	9.345	10.01	10.87	11.75	12.09	11.74	11.22	10.70
(3)	14.65	14.98	13.90	13.04	13.39	14.18	15.16	16.02	15.88	14.94	13.96	13.07	12.28	11.57	10.95
(4)	19.03	17.68	17.17	17.99	19.08	20.07	19.64	18.14	16.75	15.65	15.04	15.56	17.29	18.38	17.84

TABLE 8.17 : ε_1 $(1/\phi = .9,\ 1/\alpha = .8)$

μ MODE:	0.075	0.100	0.125	0.150	0.175	0.200	0.225	0.250	0.275	0.300	0.325	0.350	0.375	0.400	0.425
(1)	4.971	5.201	5.454	5.725	5.995	6.211	6.261	6.086	5.788	5.467	5.164	4.889	4.643	4.424	4.228
(2)	8.977	9.381	9.736	9.789	9.303	8.752	8.520	8.717	9.249	10.01	10.87	11.46	11.37	10.96	10.51
(3)	12.96	13.36	12.90	12.11	12.18	12.80	13.65	14.56	15.02	14.51	13.67	12.85	12.11	11.45	10.86
(4)	16.85	16.34	15.58	16.11	17.03	18.05	18.54	17.63	16.39	15.34	14.64	14.76	16.13	17.67	17.54

TABLE 8.17 : ε_1 $(1/\phi = .9,\ 1/\alpha = .9)$

μ MODE:	0.075	0.100	0.125	0.150	0.175	0.200	0.225	0.250	0.275	0.300	0.325	0.350	0.375	0.400	0.425
(1)	4.481	4.704	4.946	5.202	5.455	5.665	5.750	5.657	5.440	5.183	4.931	4.698	4.489	4.302	4.137
(2)	8.069	8.445	8.785	8.923	8.663	8.262	8.068	8.218	8.677	9.364	10.19	10.89	11.01	10.70	10.30
(3)	11.63	12.02	11.88	11.30	11.26	11.75	12.50	13.38	14.10	14.01	13.34	12.60	11.91	11.29	10.74
(4)	15.10	14.99	14.34	14.68	15.47	16.43	17.24	17.00	15.98	14.99	14.28	14.19	15.28	16.98	17.23

TABLE 8.17 : ε_1 $(1/\phi = .9,\ 1/\alpha = 1.0)$

μ MODE:	0.075	0.100	0.125	0.150	0.175	0.200	0.225	0.250	0.275	0.300	0.325	0.350	0.375	0.400	0.425
(1)	4.070	4.284	4.515	4.757	4.997	5.199	5.300	5.254	5.095	4.891	4.683	4.489	4.315	4.162	4.029
(2)	7.318	7.673	7.996	8.168	8.038	7.759	7.624	7.770	8.195	8.843	9.653	10.43	10.69	10.46	10.09
(3)	10.53	10.92	10.92	10.53	10.48	10.90	11.58	12.42	13.24	13.49	13.01	12.34	11.70	11.11	10.60
(4)	13.68	13.74	13.28	13.51	14.21	15.11	16.02	16.30	15.55	14.64	13.92	13.71	14.58	16.36	16.95

TABLE 8.17 : ε_1 $(1/\phi = 1.0,\ 1/\alpha = .6)$

μ MODE:	0.075	0.100	0.125	0.150	0.175	0.200	0.225	0.250	0.275	0.300	0.325	0.350	0.375	0.400	0.425
(1)	6.981	7.257	7.577	7.909	8.151	8.031	7.522	6.940	6.406	5.940	5.534	5.180	4.868	4.593	4.348
(2)	12.84	13.32	13.16	11.80	10.66	10.27	10.61	11.31	12.21	13.09	13.29	12.76	12.09	11.44	10.85
(3)	18.61	17.45	15.57	15.82	16.71	17.77	18.56	17.88	16.54	15.29	14.19	13.24	12.40	11.67	11.02
(4)	22.98	20.74	21.56	22.78	23.79	22.56	20.42	18.60	17.18	16.33	16.74	18.39	19.62	19.03	26.31

TABLE 8.17 : ε_1 $(1/\phi = 1.0,\ 1/\alpha = .7)$

μ MODE:	0.075	0.100	0.125	0.150	0.175	0.200	0.225	0.250	0.275	0.300	0.325	0.350	0.375	0.400	0.425
(1)	6.174	6.431	6.715	7.008	7.250	7.292	7.019	6.584	6.139	5.733	5.372	5.053	4.772	4.521	4.298
(2)	11.21	11.64	11.80	11.09	10.16	9.652	9.741	10.25	11.00	11.85	12.44	12.31	11.79	11.23	10.70
(3)	16.19	15.99	14.46	14.15	14.76	15.66	16.60	16.91	16.08	14.99	13.98	13.07	12.28	11.57	10.95
(4)	20.65	18.71	18.93	19.91	21.00	21.34	19.90	18.24	16.87	15.91	15.73	16.82	18.45	18.61	17.87

TABLE 8.17 : ε_1 $(1/\phi = 1.0,\ 1/\alpha = .8)$

μ MODE:	0.075	0.100	0.125	0.150	0.175	0.200	0.225	0.250	0.275	0.300	0.325	0.350	0.375	0.400	0.425
(1)	5.520	5.770	6.037	6.308	6.538	6.632	6.493	6.182	5.825	5.481	5.169	4.891	4.644	4.424	4.228
(2)	9.955	10.36	10.59	10.27	9.589	9.137	9.131	9.517	10.16	10.94	11.65	11.81	11.46	10.98	10.51
(3)	14.33	14.44	13.47	12.99	13.38	14.13	15.01	15.71	15.48	14.62	13.70	12.86	12.11	11.45	10.86
(4)	18.43	17.23	17.04	17.81	18.81	19.63	19.17	17.79	16.51	15.54	15.13	15.80	17.36	18.11	17.59

TABLE 8.17 : β_1 $(1/\phi = 1.0,\ 1/\alpha = .9)$

μ MODE:	0.075	0.100	0.125	0.150	0.175	0.200	0.225	0.250	0.275	0.300	0.325	0.350	0.375	0.400	0.425
(1)	4.976	5.218	5.474	5.730	5.951	6.064	5.995	5.772	5.488	5.203	4.939	4.701	4.490	4.303	4.137
(2)	8.949	9.325	9.578	9.447	8.982	8.627	8.613	8.942	9.519	10.26	11.01	11.36	11.15	10.73	10.30
(3)	12.86	13.07	12.50	12.05	12.31	12.96	13.78	14.58	14.81	14.21	13.40	12.62	11.91	11.29	10.74
(4)	16.57	15.90	15.59	16.19	17.09	18.02	18.26	17.28	16.11	15.16	14.65	15.05	16.48	17.62	17.32

TABLE 8.18 : β_1 $(\phi = .6,\ \alpha = .6)$

μ MODE:	0.075	0.100	0.125	0.150	0.175	0.200	0.225	0.250	0.275	0.300	0.325	0.350	0.375	0.400	0.425
(1)	4.536	4.681	4.705	4.570	4.337	4.085	3.855	3.657	3.493	3.359	3.254	3.175	3.123	3.100	3.112
(2)	7.876	7.614	7.270	7.207	7.434	7.863	8.444	9.160	9.956	10.29	9.773	9.173	8.634	8.169	7.784
(3)	10.65	10.22	10.69	11.16	11.90	12.76	13.44	12.76	11.86	11.70	11.06	11.33	10.67	10.08	9.573
(4)	13.46	13.82	14.57	15.49	16.44	15.94	14.76	15.13	14.44	13.49	13.56	15.26	16.58	15.83	15.00

TABLE 8.18 : β_1 $(\phi = .6,\ \alpha = .7)$

μ MODE:	0.075	0.100	0.125	0.150	0.175	0.200	0.225	0.250	0.275	0.300	0.325	0.350	0.375	0.400	0.425
(1)	5.276	5.440	5.463	5.300	5.020	4.718	4.442	4.204	4.003	3.837	3.701	3.594	3.514	3.462	3.441
(2)	9.161	8.836	8.406	8.307	8.542	9.002	9.618	10.33	10.85	10.55	9.930	9.330	8.803	8.354	7.987
(3)	12.36	11.94	12.23	12.86	13.65	14.41	14.14	13.11	12.49	12.64	12.20	11.48	10.82	10.25	9.770
(4)	15.61	16.00	16.82	17.77	17.99	16.52	16.22	15.86	14.78	14.24	15.25	16.88	16.82	15.96	15.15

TABLE 8.18 : β_1 $(\phi = .6,\ \alpha = .8)$

μ MODE:	0.075	0.100	0.125	0.150	0.175	0.200	0.225	0.250	0.275	0.300	0.325	0.350	0.375	0.400	0.425
(1)	6.003	6.182	6.201	6.004	5.672	5.316	4.990	4.706	4.464	4.261	4.092	3.952	3.840	3.753	3.694
(2)	10.43	10.02	9.484	9.329	9.554	10.02	10.63	11.22	11.30	10.76	10.12	9.518	8.999	8.558	8.198
(3)	14.03	13.49	13.78	14.44	15.20	15.50	14.57	13.59	13.28	13.09	12.40	11.67	11.01	10.45	9.981
(4)	17.69	18.09	18.95	19.71	18.74	17.50	17.34	16.27	15.30	15.43	16.70	17.63	17.00	16.12	15.32

TABLE 8.18 : β_1 $(\phi = .6,\ \alpha = .9)$

μ MODE:	0.075	0.100	0.125	0.150	0.175	0.200	0.225	0.250	0.275	0.300	0.325	0.350	0.375	0.400	0.425
(1)	6.713	6.901	6.911	6.676	6.286	5.870	5.488	5.155	4.870	4.628	4.422	4.248	4.102	3.981	3.884
(2)	11.66	11.15	10.48	10.26	10.45	10.90	11.46	11.84	11.61	10.98	10.32	9.724	9.204	8.762	8.397
(3)	15.64	14.95	15.22	15.87	16.49	16.18	15.03	14.18	13.90	13.38	12.62	11.88	11.22	10.66	10.18
(4)	19.67	20.07	20.91	21.12	19.49	18.66	17.97	16.73	16.05	16.59	17.77	17.99	17.18	16.30	15.51

TABLE 8.18 : β_1 $(\phi = .6,\ \alpha = 1.0)$

μ MODE:	0.075	0.100	0.125	0.150	0.175	0.200	0.225	0.250	0.275	0.300	0.325	0.350	0.375	0.400	0.425
(1)	7.400	7.592	7.589	7.310	6.855	6.373	5.932	5.549	5.220	4.938	4.696	4.487	4.309	4.155	4.024
(2)	12.86	12.22	11.40	11.08	11.24	11.66	12.13	12.31	11.90	11.21	10.54	9.932	9.404	8.952	8.572
(3)	17.15	16.30	16.55	17.17	17.54	16.73	15.51	14.75	14.35	13.65	12.84	12.09	11.42	10.84	10.35
(4)	21.54	21.94	22.69	22.14	20.34	19.63	18.48	17.26	16.86	17.59	18.50	18.26	17.38	16.48	15.69

TABLE 8.18 : β_1 $(\phi = .7,\ \alpha = .6)$

μ MODE:	0.075	0.100	0.125	0.150	0.175	0.200	0.225	0.250	0.275	0.300	0.325	0.350	0.375	0.400	0.425
(1)	3.914	4.092	4.223	4.254	4.165	3.998	3.811	3.635	3.482	3.354	3.251	3.174	3.122	3.100	3.112
(2)	6.920	6.955	6.736	6.551	6.595	6.866	7.315	7.913	8.649	9.437	9.608	9.131	8.619	8.164	7.782
(3)	9.622	9.281	9.278	9.662	10.26	11.00	11.84	12.35	11.68	11.02	11.18	11.22	10.64	10.07	9.571
(4)	12.01	12.01	12.55	13.32	14.22	15.03	14.31	13.62	14.04	13.32	12.63	13.33	15.34	15.75	14.98

TABLE 8.18 : β_1 ($\phi = .7, \alpha = .7$)

μ MODE:	0.075	0.100	0.125	0.150	0.175	0.200	0.225	0.250	0.275	0.300	0.325	0.350	0.375	0.400	0.425
(1)	4.553	4.755	4.902	4.933	4.823	4.620	4.394	4.179	3.991	3.831	3.698	3.593	3.514	3.462	3.440
(2)	8.050	8.078	7.800	7.555	7.579	7.864	8.344	8.971	9.680	10.12	9.823	9.294	8.789	8.349	7.986
(3)	11.19	10.75	10.71	11.13	11.78	12.56	13.22	12.86	12.06	11.72	11.88	11.41	10.80	10.24	9.768
(4)	13.94	13.90	14.50	15.34	16.21	16.04	14.96	15.00	14.55	13.70	13.59	14.96	16.38	15.90	15.13

TABLE 8.18 : β_1 ($\phi = .7, \alpha = .8$)

μ MODE:	0.075	0.100	0.125	0.150	0.175	0.200	0.225	0.250	0.275	0.300	0.325	0.350	0.375	0.400	0.425
(1)	5.180	5.402	5.562	5.590	5.453	5.209	4.937	4.680	4.452	4.255	4.089	3.951	3.839	3.753	3.694
(2)	9.161	9.174	8.816	8.494	8.481	8.762	9.247	9.858	10.43	10.49	10.03	9.486	8.986	8.553	8.196
(3)	12.72	12.15	12.07	12.50	13.18	13.89	14.05	13.27	12.56	12.44	12.22	11.62	11.00	10.44	9.979
(4)	15.80	15.71	16.35	17.20	17.71	16.69	15.99	15.85	14.95	14.29	14.80	16.22	16.76	16.07	15.31

TABLE 8.18 : β_1 ($\phi = .7, \alpha = .9$)

μ MODE:	0.075	0.100	0.125	0.150	0.175	0.200	0.225	0.250	0.275	0.300	0.325	0.350	0.375	0.400	0.425
(1)	5.791	6.029	6.197	6.218	6.049	5.757	5.435	5.130	4.858	4.622	4.420	4.247	4.101	3.981	3.884
(2)	10.25	10.23	9.771	9.350	9.284	9.546	10.02	10.57	10.95	10.77	10.25	9.695	9.193	8.758	8.396
(3)	14.21	13.47	13.32	13.77	14.43	14.97	14.61	13.72	13.12	12.95	12.49	11.83	11.20	10.65	10.18
(4)	17.57	17.42	18.09	18.87	18.71	17.44	17.02	16.38	15.42	15.06	15.86	17.06	17.01	16.26	15.50

TABLE 8.18 : β_1 ($\phi = .7, \alpha = 1.0$)

μ MODE:	0.075	0.100	0.125	0.150	0.175	0.200	0.225	0.250	0.275	0.300	0.325	0.350	0.375	0.400	0.425
(1)	6.383	6.630	6.803	6.813	6.606	6.259	5.880	5.525	5.208	4.932	4.693	4.487	4.308	4.155	4.024
(2)	11.31	11.25	10.65	10.11	9.985	10.22	10.67	11.16	11.37	11.04	10.47	9.908	9.395	8.949	8.571
(3)	15.64	14.68	14.48	14.93	15.56	15.84	15.11	14.17	13.64	13.34	12.74	12.05	11.41	10.84	10.35
(4)	19.24	19.04	19.71	20.33	19.51	18.27	17.82	16.85	15.93	15.83	16.74	17.61	17.24	16.45	15.68

TABLE 8.18 : β_1 ($\phi = .8, \alpha = .6$)

μ MODE:	0.075	0.100	0.125	0.150	0.175	0.200	0.225	0.250	0.275	0.300	0.325	0.350	0.375	0.400	0.425
(1)	3.436	3.615	3.780	3.899	3.933	3.869	3.744	3.601	3.464	3.345	3.247	3.172	3.121	3.099	3.112
(2)	6.124	6.302	6.281	6.122	6.050	6.178	6.501	6.988	7.626	8.406	9.129	9.043	8.594	8.156	7.780
(3)	8.666	8.553	8.395	8.586	9.042	9.668	10.43	11.25	11.46	10.83	10.44	10.87	10.59	10.05	9.567
(4)	10.95	10.76	11.08	11.70	12.49	13.38	13.88	13.06	12.93	13.12	12.42	12.17	13.70	15.47	14.96

TABLE 8.18 : β_1 ($\phi = .8, \alpha = .7$)

μ MODE:	0.075	0.100	0.125	0.150	0.175	0.200	0.225	0.250	0.275	0.300	0.325	0.350	0.375	0.400	0.425
(1)	3.997	4.200	4.387	4.521	4.555	4.473	4.318	4.141	3.972	3.821	3.694	3.591	3.513	3.461	3.440
(2)	7.123	7.322	7.282	7.070	6.958	7.078	7.419	7.937	8.596	9.302	9.577	9.229	8.767	8.342	7.983
(3)	10.08	9.916	9.695	9.891	10.39	11.07	11.84	12.33	11.85	11.25	11.23	11.26	10.76	10.23	9.765
(4)	12.72	12.45	12.80	13.49	14.33	15.05	14.53	13.88	14.09	13.50	12.90	13.44	15.18	15.77	15.11

TABLE 8.18 : β_1 ($\phi = .8, \alpha = .8$)

μ MODE:	0.075	0.100	0.125	0.150	0.175	0.200	0.225	0.250	0.275	0.300	0.325	0.350	0.375	0.400	0.425
(1)	4.547	4.770	4.976	5.122	5.152	5.047	4.856	4.640	4.432	4.246	4.084	3.949	3.838	3.753	3.694
(2)	8.107	8.320	8.246	7.962	7.793	7.889	8.231	8.753	9.382	9.902	9.854	9.431	8.966	8.547	8.194
(3)	11.47	11.23	10.92	11.11	11.63	12.33	12.98	12.93	12.26	11.80	11.82	11.51	10.96	10.43	9.976
(4)	14.44	14.06	14.43	15.16	15.97	16.11	15.18	14.93	14.67	13.92	13.65	14.64	16.08	15.98	15.29

TABLE 8.18 : β_1 ($\phi = .8, \alpha = .9$)

μ MODE:	0.075	0.100	0.125	0.150	0.175	0.200	0.225	0.250	0.275	0.300	0.325	0.350	0.375	0.400	0.425
(1)	5.083	5.323	5.542	5.696	5.718	5.584	5.351	5.090	4.839	4.613	4.416	4.245	4.101	3.981	3.884
(2)	9.069	9.290	9.165	8.785	8.539	8.599	8.929	9.438	10.00	10.33	10.11	9.647	9.176	8.752	8.394
(3)	12.83	12.48	12.05	12.23	12.77	13.44	13.86	13.41	12.70	12.34	12.22	11.75	11.18	10.64	10.18
(4)	16.11	15.58	15.96	16.71	17.37	16.85	15.96	15.79	15.13	14.41	14.47	15.61	16.58	16.18	15.48

TABLE 8.18 : β_1 ($\phi = .8, \alpha = 1.0$)

μ MODE:	0.075	0.100	0.125	0.150	0.175	0.200	0.225	0.250	0.275	0.300	0.325	0.350	0.375	0.400	0.425
(1)	5.602	5.853	6.082	6.241	6.251	6.080	5.797	5.487	5.191	4.924	4.690	4.485	4.308	4.155	4.024
(2)	10.01	10.23	10.03	9.525	9.190	9.211	9.528	10.02	10.52	10.68	10.36	9.867	9.381	8.945	8.570
(3)	14.16	13.64	13.08	13.25	13.81	14.43	14.55	13.86	13.14	12.81	12.54	11.99	11.39	10.83	10.35
(4)	17.69	16.99	17.41	18.16	18.52	17.54	16.78	16.42	15.56	14.93	15.23	16.37	16.94	16.39	15.67

TABLE 8.18 : β_1 ($\phi = .9, \alpha = .6$)

μ MODE:	0.075	0.100	0.125	0.150	0.175	0.200	0.225	0.250	0.275	0.300	0.325	0.350	0.375	0.400	0.425
(1)	3.060	3.229	3.400	3.556	3.666	3.698	3.649	3.551	3.438	3.332	3.240	3.169	3.120	3.099	3.112
(2)	5.474	5.704	5.824	5.779	5.683	5.704	5.908	6.290	6.833	7.537	8.364	8.840	8.552	8.143	7.776
(3)	7.816	7.915	7.782	7.811	8.127	8.645	9.313	10.11	10.86	10.69	10.16	10.23	10.47	10.03	9.561
(4)	10.02	9.869	9.982	10.46	11.14	11.95	12.81	12.80	12.16	12.54	12.30	11.75	12.41	14.61	14.91

TABLE 8.18 : β_1 ($\phi = .9, \alpha = .7$)

μ MODE:	0.075	0.100	0.125	0.150	0.175	0.200	0.225	0.250	0.275	0.300	0.325	0.350	0.375	0.400	0.425
(1)	3.558	3.751	3.945	4.122	4.246	4.277	4.211	4.086	3.943	3.807	3.687	3.588	3.512	3.461	3.440
(2)	6.368	6.628	6.758	6.683	6.543	6.539	6.745	7.149	7.722	8.428	9.077	9.104	8.732	8.330	7.980
(3)	9.092	9.188	8.994	8.997	9.339	9.904	10.62	11.37	11.57	11.05	10.70	10.92	10.69	10.21	9.760
(4)	11.64	11.42	11.53	12.06	12.80	13.63	14.04	13.37	13.21	13.26	12.66	12.50	13.83	15.42	15.07

TABLE 8.18 : β_1 ($\phi = .9, \alpha = .8$)

μ MODE:	0.075	0.100	0.125	0.150	0.175	0.200	0.225	0.250	0.275	0.300	0.325	0.350	0.375	0.400	0.425
(1)	4.048	4.261	4.475	4.668	4.802	4.828	4.739	4.581	4.403	4.232	4.078	3.946	3.837	3.752	3.694
(2)	7.247	7.533	7.663	7.543	7.339	7.292	7.486	7.895	8.467	9.115	9.513	9.334	8.935	8.536	8.192
(3)	10.35	10.43	10.14	10.10	10.46	11.05	11.76	12.29	12.03	11.47	11.29	11.32	10.91	10.42	9.972
(4)	13.24	12.91	12.99	13.56	14.33	15.04	14.76	14.10	14.15	13.70	13.14	13.45	14.93	15.77	15.26

TABLE 8.18 : β_1 ($\phi = .9, \alpha = .9$)

μ MODE:	0.075	0.100	0.125	0.150	0.175	0.200	0.225	0.250	0.275	0.300	0.325	0.350	0.375	0.400	0.425
(1)	4.526	4.754	4.983	5.190	5.330	5.345	5.227	5.030	4.811	4.600	4.410	4.243	4.100	3.980	3.884
(2)	8.107	8.412	8.534	8.346	8.057	7.953	8.124	8.528	9.084	9.647	9.844	9.566	9.149	8.744	8.392
(3)	11.58	11.62	11.20	11.11	11.48	12.09	12.75	12.97	12.45	11.93	11.79	11.61	11.14	10.63	10.17
(4)	14.80	14.30	14.36	14.97	15.73	16.12	15.41	14.93	14.78	14.13	13.71	14.32	15.71	16.04	15.45

TABLE 8.18 : β_1 ($\phi = .9, \alpha = 1.0$)

μ MODE:	0.075	0.100	0.125	0.150	0.175	0.200	0.225	0.250	0.275	0.300	0.325	0.350	0.375	0.400	0.425
(1)	4.988	5.226	5.467	5.685	5.828	5.829	5.673	5.429	5.165	4.913	4.685	4.483	4.307	4.154	4.024
(2)	8.944	9.265	9.366	9.082	8.688	8.520	8.670	9.070	9.610	10.09	10.15	9.801	9.359	8.938	8.569
(3)	12.79	12.75	12.17	12.02	12.41	13.04	13.61	13.52	12.87	12.35	12.20	11.88	11.36	10.82	10.35
(4)	16.31	15.59	15.64	16.29	17.02	16.97	16.05	15.71	15.28	14.55	14.28	15.07	16.29	16.28	15.65

TABLE 8.18 : β_1 ($\phi = 1.0$, $\alpha = .6$)

μ / MODE:	0.075	0.100	0.125	0.150	0.175	0.200	0.225	0.250	0.275	0.300	0.325	0.350	0.375	0.400	0.425
(1)	2.756	2.914	3.081	3.247	3.396	3.497	3.524	3.482	3.402	3.313	3.231	3.164	3.118	3.098	3.112
(2)	4.942	5.184	5.376	5.449	5.403	5.373	5.478	5.759	6.211	6.832	7.621	8.421	8.475	8.124	7.771
(3)	7.088	7.306	7.301	7.253	7.434	7.843	8.422	9.144	9.969	10.43	10.03	9.756	10.21	9.992	9.552
(4)	9.159	9.183	9.164	9.494	10.07	10.79	11.64	12.32	11.88	11.71	12.09	11.60	11.56	13.44	14.81

TABLE 8.18 : β_1 ($\phi = 1.0$, $\alpha = .7$)

μ / MODE:	0.075	0.100	0.125	0.150	0.175	0.200	0.225	0.250	0.275	0.300	0.325	0.350	0.375	0.400	0.425
(1)	3.206	3.385	3.575	3.764	3.932	4.044	4.067	4.008	3.904	3.787	3.677	3.583	3.510	3.460	3.440
(2)	5.749	6.024	6.240	6.309	6.229	6.164	6.256	6.548	7.027	7.671	8.416	8.867	8.674	8.313	7.975
(3)	8.246	8.487	8.450	8.355	8.539	8.987	9.615	10.36	11.01	10.88	10.43	10.44	10.57	10.18	9.753
(4)	10.65	10.64	10.58	10.94	11.57	12.36	13.13	13.06	12.55	12.79	12.51	12.07	12.74	14.67	15.00

TABLE 8.18 : β_1 ($\phi = 1.0$, $\alpha = .8$)

μ / MODE:	0.075	0.100	0.125	0.150	0.175	0.200	0.225	0.250	0.275	0.300	0.325	0.350	0.375	0.400	0.425
(1)	3.647	3.845	4.053	4.261	4.445	4.565	4.580	4.498	4.362	4.212	4.068	3.942	3.835	3.752	3.694
(2)	6.542	6.846	7.080	7.134	7.001	6.883	6.946	7.234	7.720	8.353	8.983	9.163	8.887	8.522	8.188
(3)	9.386	9.641	9.545	9.381	9.557	10.03	10.69	11.39	11.68	11.28	10.91	10.98	10.83	10.40	9.967
(4)	12.12	12.05	11.91	12.30	12.98	13.76	14.20	13.67	13.41	13.43	12.92	12.72	13.79	15.34	15.20

TABLE 8.18 : β_1 ($\phi = 1.0$, $\alpha = .9$)

μ / MODE:	0.075	0.100	0.125	0.150	0.175	0.200	0.225	0.250	0.275	0.300	0.325	0.350	0.375	0.400	0.425
(1)	4.077	4.290	4.513	4.736	4.933	5.056	5.058	4.945	4.770	4.581	4.401	4.239	4.098	3.980	3.884
(2)	7.318	7.646	7.891	7.915	7.706	7.517	7.541	7.818	8.302	8.908	9.415	9.431	9.109	8.732	8.389
(3)	10.50	10.76	10.57	10.31	10.48	10.98	11.65	12.24	12.19	11.69	11.38	11.38	11.07	10.61	10.17
(4)	13.57	13.37	13.15	13.57	14.29	15.01	14.98	14.32	14.21	13.91	13.36	13.41	14.64	15.75	15.41

TABLE 8.18 : β_1 ($1/\phi = .6$, $1/\alpha = .6$)

μ / MODE:	0.075	0.100	0.125	0.150	0.175	0.200	0.225	0.250	0.275	0.300	0.325	0.350	0.375	0.400	0.425
(1)	4.128	4.302	4.517	4.776	5.080	5.430	5.817	6.162	6.185	5.878	5.514	5.173	4.866	4.592	4.348
(2)	7.678	8.078	8.540	9.038	9.381	8.898	8.148	7.634	7.677	8.299	9.237	10.19	10.11	9.609	9.106
(3)	11.29	11.90	12.47	11.92	10.82	10.92	11.70	12.69	13.44	12.81	12.00	11.58	11.94	11.48	10.89
(4)	14.91	15.61	14.58	13.93	14.75	15.81	16.57	15.49	14.72	15.06	14.23	13.42	13.61	15.92	16.34

TABLE 8.18 : β_1 ($1/\phi = .6$, $1/\alpha = .7$)

μ / MODE:	0.075	0.100	0.125	0.150	0.175	0.200	0.225	0.250	0.275	0.300	0.325	0.350	0.375	0.400	0.425
(1)	3.658	3.814	4.002	4.223	4.481	4.780	5.114	5.464	5.668	5.582	5.322	5.036	4.765	4.519	4.297
(2)	6.698	7.036	7.425	7.855	8.264	8.323	7.840	7.353	7.165	7.491	8.228	9.192	9.740	9.442	9.003
(3)	9.780	10.29	10.83	11.03	10.27	9.853	10.29	11.10	12.04	12.38	11.78	11.22	11.32	11.27	10.77
(4)	12.88	13.52	13.60	12.59	12.89	13.75	14.75	14.99	14.10	13.96	13.93	13.22	12.85	14.37	16.02

TABLE 8.18 : β_1 ($1/\phi = .6$, $1/\alpha = .8$)

μ / MODE:	0.075	0.100	0.125	0.150	0.175	0.200	0.225	0.250	0.275	0.300	0.325	0.350	0.375	0.400	0.425
(1)	3.281	3.431	3.604	3.805	4.036	4.300	4.596	4.906	5.161	5.220	5.075	4.856	4.631	4.419	4.226
(2)	5.955	6.254	6.596	6.974	7.358	7.600	7.425	7.057	6.836	6.985	7.548	8.403	9.207	9.213	8.865
(3)	8.650	9.099	9.577	9.943	9.688	9.216	9.346	9.969	10.82	11.59	11.46	10.97	10.82	10.99	10.62
(4)	11.36	11.93	12.32	11.76	11.62	12.25	13.14	13.96	13.67	13.15	13.39	12.97	12.52	13.34	15.46

TA[illegible] β_1 ($1/\phi = .6$, $1/\alpha = .9$)

μ / MODE:	0.075	[illegible]	[illegible]	[illegible]	0.175	0.20[illegible]	[illegible]	0.250	0.275	0.300	0.325	0.350	0.375	0.400	0.425
(1)	2.966	3.112	3.27[illegible]	[illegible]466	3.679	3.921	4.190	4.475	4.729	4.853	4.799	4.646	4.467	4.293	4.133
(2)	5.360	5.636	5.946	6.288	6.640	6.923	6.940	6.716	6.537	6.621	7.067	7.821	8.675	8.942	8.698
(3)	7.762	8.167	8.598	8.980	9.014	8.688	8.674	9.136	9.875	10.71	11.04	10.70	10.47	10.68	10.45
(4)	10.17	10.68	11.12	10.98	10.72	11.14	11.91	12.78	13.12	12.66	12.70	12.66	12.25	12.67	14.77

TABLE 8.18 : β_1 ($1/\phi = .6$, $1/\alpha = 1.0$)

μ / MODE:	0.075	0.100	0.125	0.150	0.175	0.200	0.225	0.250	0.275	0.300	0.325	0.350	0.375	0.400	0.425
(1)	2.700	2.842	3.000	3.179	3.380	3.606	3.856	4.120	4.364	4.514	4.517	4.419	4.284	4.148	4.022
(2)	4.868	5.127	5.415	5.730	6.056	6.342	6.452	6.348	6.232	6.308	6.693	7.376	8.224	8.664	8.517
(3)	7.039	7.412	7.808	8.177	8.335	8.171	8.133	8.403	9.138	9.946	10.54	10.41	10.18	10.37	10.27
(4)	9.209	9.680	10.11	10.18	9.989	10.27	10.93	11.76	12.42	12.23	12.10	12.29	11.98	12.19	14.13

TABLE 8.18 : β_1 ($1/\phi = .7$, $1/\alpha = .6$)

μ / MODE:	0.075	0.100	0.125	0.150	0.175	0.200	0.225	0.250	0.275	0.300	0.325	0.350	0.375	0.400	0.425
(1)	4.815	5.016	5.265	5.560	5.899	6.269	6.601	6.645	6.322	5.912	5.524	5.176	4.867	4.593	4.348
(2)	8.952	9.408	9.912	10.33	9.973	9.071	8.389	8.250	8.731	9.546	10.46	10.66	10.17	9.617	9.107
(3)	13.15	13.80	13.79	12.32	11.94	12.60	13.53	14.37	13.86	12.92	12.24	12.40	12.13	11.60	10.89
(4)	17.34	17.37	15.57	16.08	17.11	18.05	17.09	15.91	16.10	15.33	14.36	13.92	15.39	17.02	16.38

TABLE 8.18 : β_1 ($1/\phi = .7$, $1/\alpha = .7$)

μ / MODE:	0.075	0.100	0.125	0.150	0.175	0.200	0.225	0.250	0.275	0.300	0.325	0.350	0.375	0.400	0.425
(1)	4.267	4.448	4.663	4.915	5.204	5.523	5.840	6.038	5.949	5.667	5.347	5.043	4.768	4.520	4.297
(2)	7.810	8.195	8.624	9.043	9.172	8.663	8.063	7.754	7.945	8.552	9.390	10.07	9.919	9.467	9.007
(3)	11.39	11.95	12.34	11.67	10.91	11.15	11.87	12.76	13.24	12.64	11.94	11.70	11.79	11.33	10.78
(4)	14.98	15.48	14.41	14.14	14.91	15.88	16.34	15.36	14.84	14.88	14.11	13.48	14.03	16.01	16.17

TABLE 8.18 : β_1 ($1/\phi = .7$, $1/\alpha = .8$)

μ / MODE:	0.075	0.100	0.125	0.150	0.175	0.200	0.225	0.250	0.275	0.300	0.325	0.350	0.375	0.400	0.425
(1)	3.827	4.001	4.200	4.428	4.686	4.969	5.257	5.484	5.525	5.365	5.123	4.872	4.635	4.420	4.226
(2)	6.943	7.285	7.662	8.046	8.301	8.130	7.697	7.395	7.445	7.889	8.610	9.384	9.588	9.272	8.874
(3)	10.08	10.57	11.00	10.86	10.24	10.17	10.70	11.48	12.24	12.22	11.66	11.28	11.38	11.12	10.64
(4)	13.21	13.75	13.43	12.86	13.33	14.16	15.00	14.82	14.10	14.14	13.79	13.19	13.26	14.93	15.86

TABLE 8.18 : β_1 ($1/\phi = .7$, $1/\alpha = .9$)

μ / MODE:	0.075	0.100	0.125	0.150	0.175	0.200	0.225	0.250	0.275	0.300	0.325	0.350	0.375	0.400	0.425
(1)	3.460	3.629	3.820	4.033	4.271	4.530	4.795	5.020	5.116	5.042	4.870	4.670	4.475	4.295	4.134
(2)	6.249	6.564	6.908	7.259	7.538	7.547	7.283	7.047	7.060	7.410	8.040	8.803	9.222	9.049	8.715
(3)	9.044	9.489	9.899	9.998	9.613	9.462	9.830	10.50	11.28	11.68	11.33	10.95	10.98	10.88	10.48
(4)	11.83	12.34	12.38	11.91	12.15	12.85	13.69	14.11	13.59	13.38	13.38	12.90	12.77	14.09	15.50

TABLE 8.18 : β_1 ($1/\phi = .7$, $1/\alpha = 1.0$)

μ / MODE:	0.075	0.100	0.125	0.150	0.175	0.200	0.225	0.250	0.275	0.300	0.325	0.350	0.375	0.400	0.425
(1)	3.149	3.313	3.496	3.699	3.923	4.165	4.412	4.628	4.746	4.725	4.606	4.452	4.296	4.151	4.023
(2)	5.676	5.972	6.291	6.617	6.894	6.986	6.848	6.688	6.713	7.022	7.596	8.338	8.877	8.824	8.545
(3)	8.202	8.612	8.999	9.187	8.992	8.863	9.142	9.728	10.47	11.08	10.98	10.65	10.63	10.64	10.31
(4)	10.72	11.19	11.37	11.09	11.22	11.81	12.60	13.27	13.10	12.78	12.91	12.59	12.39	13.44	15.11

TABLE 8.18 : β_1 ($1/\phi = .8$, $1/\alpha = .6$)

μ MODE:	0.075	0.100	0.125	0.150	0.175	0.200	0.225	0.250	0.275	0.300	0.325	0.350	0.375	0.400	0.425
(1)	5.502	5.729	6.007	6.331	6.688	7.022	7.129	6.826	6.369	5.926	5.528	5.177	4.867	4.593	4.348
(2)	10.22	10.72	11.21	11.19	10.19	9.278	8.873	9.155	9.843	10.70	11.18	10.78	10.19	9.621	9.107
(3)	14.99	15.53	14.31	13.08	13.44	14.29	15.20	15.05	13.99	13.09	12.83	12.80	12.18	11.51	10.89
(4)	19.67	18.13	17.37	18.28	19.34	18.97	17.38	17.09	16.57	15.44	14.62	15.15	17.03	17.23	16.39

TABLE 8.18 : β_1 ($1/\phi = .8$, $1/\alpha = .7$)

μ MODE:	0.075	0.100	0.125	0.150	0.175	0.200	0.225	0.250	0.275	0.300	0.325	0.350	0.375	0.400	0.425
(1)	4.875	5.080	5.321	5.597	5.902	6.206	6.413	6.351	6.056	5.700	5.358	5.047	4.769	4.520	4.297
(2)	8.918	9.340	9.776	10.03	9.618	8.888	8.405	8.412	8.876	9.599	10.32	10.41	9.976	9.478	9.009
(3)	12.99	13.53	13.31	12.18	11.99	12.59	13.42	14.06	13.60	12.79	12.25	12.28	11.93	11.35	10.79
(4)	17.03	16.81	15.52	15.99	16.91	17.66	16.82	15.86	15.86	15.13	14.30	14.13	15.52	16.76	16.21

TABLE 8.18 : β_1 ($1/\phi = .8$, $1/\alpha = .8$)

μ MODE:	0.075	0.100	0.125	0.150	0.175	0.200	0.225	0.250	0.275	0.300	0.325	0.350	0.375	0.400	0.425
(1)	4.372	4.569	4.792	5.042	5.314	5.588	5.806	5.853	5.689	5.425	5.144	4.879	4.637	4.420	4.226
(2)	7.928	8.303	8.693	8.995	8.909	8.425	8.007	7.920	8.234	8.834	9.551	9.939	9.711	9.295	8.878
(3)	11.49	11.98	12.12	11.43	11.04	11.39	12.08	12.84	13.03	12.44	11.88	11.76	11.63	11.16	10.65
(4)	15.03	15.25	14.31	14.36	15.10	15.96	16.09	15.21	14.91	14.70	13.99	13.60	14.48	16.07	15.97

TABLE 8.18 : β_1 ($1/\phi = .8$, $1/\alpha = .9$)

μ MODE:	0.075	0.100	0.125	0.150	0.175	0.200	0.225	0.250	0.275	0.300	0.325	0.350	0.375	0.400	0.425
(1)	3.953	4.144	4.357	4.592	4.843	5.096	5.307	5.395	5.314	5.124	4.901	4.682	4.479	4.296	4.134
(2)	7.135	7.482	7.839	8.139	8.195	7.913	7.598	7.511	7.753	8.274	8.952	9.474	9.426	9.091	8.723
(3)	10.31	10.77	11.01	10.65	10.29	10.49	11.08	11.81	12.31	12.04	11.54	11.34	11.32	10.95	10.49
(4)	13.47	13.80	13.28	13.15	13.73	14.53	15.11	14.65	14.14	14.15	13.65	13.21	13.75	15.36	15.71

TABLE 8.18 : β_1 ($1/\phi = .8$, $1/\alpha = 1.0$)

μ MODE:	0.075	0.100	0.125	0.150	0.175	0.200	0.225	0.250	0.275	0.300	0.325	0.350	0.375	0.400	0.425
(1)	3.598	3.784	3.988	4.210	4.447	4.684	4.886	4.991	4.957	4.822	4.646	4.468	4.301	4.153	4.023
(2)	6.481	6.806	7.139	7.429	7.545	7.394	7.176	7.124	7.345	7.824	8.476	9.069	9.155	8.885	8.557
(3)	9.354	9.774	10.05	9.894	9.625	9.768	10.27	10.96	11.60	11.62	11.21	10.98	11.02	10.74	10.33
(4)	12.20	12.56	12.31	12.16	12.63	13.37	14.09	14.07	13.55	13.55	13.29	12.86	13.18	14.72	15.44

TABLE 8.18 : β_1 ($1/\phi = .9$, $1/\alpha = .6$)

μ MODE:	0.075	0.100	0.125	0.150	0.175	0.200	0.225	0.250	0.275	0.300	0.325	0.350	0.375	0.400	0.425
(1)	6.187	6.439	6.742	7.084	7.423	7.612	7.387	6.897	6.390	5.932	5.530	5.178	4.867	4.593	4.348
(2)	11.49	12.00	12.34	11.57	10.38	9.632	9.617	10.15	10.91	11.58	11.43	10.82	10.20	9.622	9.108
(3)	16.80	16.77	14.76	14.31	15.00	15.92	16.29	15.24	14.12	13.44	13.44	12.92	12.20	11.51	10.89
(4)	21.67	19.10	19.39	20.44	20.99	19.26	18.12	17.94	16.71	15.61	15.31	16.64	17.92	17.29	16.40

TABLE 8.18 : β_1 ($1/\phi = .9$, $1/\alpha = .7$)

μ MODE:	0.075	0.100	0.125	0.150	0.175	0.200	0.225	0.250	0.275	0.300	0.325	0.350	0.375	0.400	0.425
(1)	5.483	5.709	5.971	6.263	6.560	6.787	6.781	6.494	6.103	5.716	5.363	5.049	4.769	4.520	4.297
(2)	10.02	10.46	10.84	10.69	9.873	9.171	8.938	9.224	9.827	10.53	10.86	10.53	10.00	9.483	9.010
(3)	14.57	14.91	13.86	12.98	13.28	14.03	14.79	14.67	13.77	12.99	12.74	12.56	11.98	11.36	10.79
(4)	18.95	17.64	17.09	17.87	18.78	18.52	17.18	16.80	16.28	15.29	14.65	15.25	16.80	16.97	16.23

TABLE 8.18 : β_1 ($1/\phi = .9$, $1/\alpha = .8$)

μ MODE:	0.075	0.100	0.125	0.150	0.175	0.200	0.225	0.250	0.275	0.300	0.325	0.350	0.375	0.400	0.425
(1)	4.917	5.134	5.377	5.641	5.908	6.129	6.200	6.050	5.766	5.453	5.154	4.882	4.639	4.421	4.226
(2)	8.907	9.302	9.658	9.728	9.265	8.709	8.448	8.603	9.082	9.726	10.24	10.17	9.762	9.307	8.880
(3)	12.89	13.28	12.86	12.06	12.09	12.66	13.40	13.83	13.33	12.62	12.23	12.15	11.73	11.18	10.65
(4)	16.77	16.29	15.50	16.00	16.83	17.36	16.53	15.79	15.65	14.93	14.25	14.39	15.74	16.55	16.02

TABLE 8.18 : β_1 ($1/\phi = .9$, $1/\alpha = .9$)

μ MODE:	0.075	0.100	0.125	0.150	0.175	0.200	0.225	0.250	0.275	0.300	0.325	0.350	0.375	0.400	0.425
(1)	4.445	4.657	4.889	5.136	5.383	5.593	5.691	5.616	5.412	5.164	4.917	4.687	4.481	4.297	4.134
(2)	8.017	8.383	8.717	8.863	8.619	8.217	8.002	8.115	8.525	9.120	9.691	9.812	9.510	9.110	8.727
(3)	11.57	11.96	11.83	11.25	11.18	11.63	12.31	12.91	12.83	12.25	11.83	11.76	11.47	10.98	10.50
(4)	15.04	14.94	14.28	14.58	15.31	16.04	15.85	15.08	14.91	14.51	13.88	13.80	14.92	16.09	15.79

TABLE 8.18 : β_1 ($1/\phi = .9$, $1/\alpha = 1.0$)

μ MODE:	0.075	0.100	0.125	0.150	0.175	0.200	0.225	0.250	0.275	0.300	0.325	0.350	0.375	0.400	0.425
(1)	4.046	4.251	4.474	4.708	4.941	5.142	5.249	5.215	5.067	4.870	4.666	4.476	4.304	4.154	4.024
(2)	7.281	7.626	7.942	8.116	7.996	7.717	7.564	7.680	8.061	8.632	9.237	9.488	9.271	8.914	8.563
(3)	10.49	10.87	10.88	10.49	10.41	10.79	11.42	12.09	12.30	11.87	11.46	11.39	11.22	10.79	10.34
(4)	13.63	13.69	13.23	13.43	14.08	14.84	15.11	14.49	14.21	14.07	13.52	13.31	14.24	15.63	15.56

TABLE 8.18 : β_1 ($1/\phi = 1.0$, $1/\alpha = .6$)

μ MODE:	0.075	0.100	0.125	0.150	0.175	0.200	0.225	0.250	0.275	0.300	0.325	0.350	0.375	0.400	0.425
(1)	6.871	7.145	7.466	7.805	8.065	7.985	7.503	6.930	6.401	5.936	5.532	5.178	4.868	4.593	4.348
(2)	12.74	13.23	13.11	11.78	10.63	10.19	10.49	11.14	11.86	12.07	11.52	10.84	10.20	9.624	9.108
(3)	18.53	17.42	15.51	15.72	16.55	17.32	16.69	15.36	14.33	14.00	13.71	12.97	12.21	11.52	10.89
(4)	22.94	20.66	21.44	22.47	21.60	19.63	19.22	18.19	16.84	15.95	16.44	17.98	18.20	17.31	16.40

TABLE 8.18 : β_1 ($1/\phi = 1.0$, $1/\alpha = .7$)

μ MODE:	0.075	0.100	0.125	0.150	0.175	0.200	0.225	0.250	0.275	0.300	0.325	0.350	0.375	0.400	0.425
(1)	6.088	6.334	6.611	6.903	7.153	7.222	6.982	6.564	6.127	5.725	5.366	5.050	4.769	4.520	4.297
(2)	11.11	11.55	11.72	11.05	10.12	9.583	9.626	10.08	10.73	11.24	11.10	10.58	10.01	9.486	9.010
(3)	16.09	15.93	14.41	14.05	14.62	15.40	15.73	14.91	13.93	13.33	13.17	12.67	12.01	11.37	10.79
(4)	20.58	18.65	18.81	19.71	20.22	18.89	17.81	17.51	16.46	15.51	15.35	16.48	17.54	17.05	16.24

TABLE 8.18 : β_1 ($1/\phi = 1.0$, $1/\alpha = .8$)

μ MODE:	0.075	0.100	0.125	0.150	0.175	0.200	0.225	0.250	0.275	0.300	0.325	0.350	0.375	0.400	0.425
(1)	5.460	5.696	5.953	6.216	6.448	5.558	6.444	6.153	5.805	5.468	5.160	4.884	4.639	4.421	4.226
(2)	9.878	10.27	10.51	10.22	9.545	9.071	9.025	9.361	9.918	10.48	10.63	10.27	9.788	9.313	8.881
(3)	14.25	14.37	13.42	12.90	13.25	13.92	14.53	14.31	13.51	12.87	12.65	12.34	11.78	11.19	10.65
(4)	18.35	17.17	16.94	17.65	18.42	18.08	16.94	16.58	15.99	15.12	14.71	15.45	16.74	16.73	16.04

TABLE 8.18 : β_1 ($1/\phi = 1.0$, $1/\alpha = .9$)

μ MODE:	0.075	0.100	0.125	0.150	0.175	0.200	0.225	0.250	0.275	0.300	0.325	0.350	0.375	0.400	0.425
(1)	4.936	5.166	5.411	5.659	5.877	5.998	5.946	5.739	5.465	5.185	4.925	4.691	4.482	4.297	4.134
(2)	8.890	9.258	9.509	9.394	8.937	8.566	8.518	8.802	9.309	9.881	10.18	9.967	9.551	9.121	8.729
(3)	12.79	13.01	12.45	11.98	12.20	12.79	13.44	13.64	13.07	12.46	12.20	12.01	11.54	11.00	10.50
(4)	16.50	15.84	15.50	16.06	16.83	17.11	16.25	15.71	15.45	14.72	14.22	14.68	15.99	16.39	15.82

CHAPTER NINE

MISCELLANEOUS PROBLEMS IN FREE LATERAL VIBRATIONS OF BEAMS

Part 1. The Vibration of Beams Subjected to Axial Compression or Tension

In some practical applications it is necessary to determine the effect of axial compressive or tensile forces on the free vibration frequencies and modal shapes of beams. Examples are the effect of compressive forces on some nuclear reactor fuel pins. The analysis presented here is limited to three practical cases, the first of which is worked out in detail.

Consider the simply supported beam with axial force P applied at each end as shown in Fig. 9.1. The dynamic equilibrium equation is written as

$$\frac{-\partial^2 v(x,t)}{\partial t^2} = \frac{EI}{\rho A}\frac{\partial^4 v(x,t)}{\partial x^4} + \frac{P}{\rho A}\frac{\partial^2 v(x,t)}{\partial x^2} \tag{9.1}$$

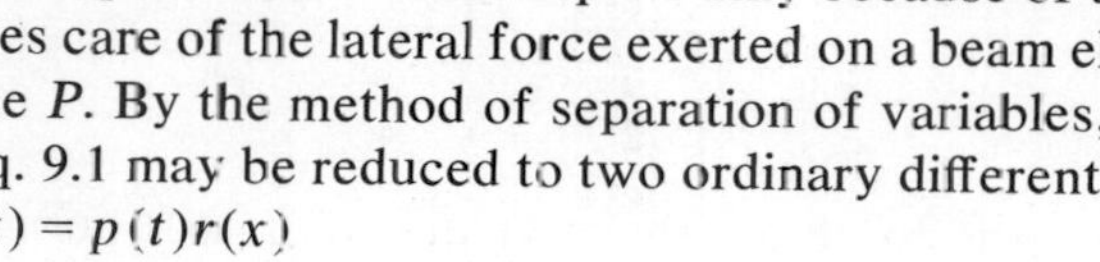

Fig. 9.1

It is noted that Eq. 9.1 differs from Eq. 1.1 only because of the final term. This term takes care of the lateral force exerted on a beam element due to the axial force P. By the method of separation of variables, discussed in Chapter 1, Eq. 9.1 may be reduced to two ordinary differential equations. Letting $v(x, t) = p(t)r(x)$
we have

$$\frac{d^2p(t)}{dt^2} + \omega^2 p(t) = 0 \tag{9.2}$$

and

$$\frac{d^4r(x)}{dx^4} + \frac{P}{EI}\frac{d^2r(x)}{dx^2} - \frac{\rho A\omega^2}{EI}r(x) = 0 \tag{9.3}$$

The solution is then expressed as

$$v(x, t) = r(x)\cos(\omega t - \alpha) \tag{9.4}$$

Before seeking the conditions that permit a nontrivial solution for Eq. 9.3 to exist, it is desirable to express it in dimensionless form. Proceeding as in Chapter 1, and nondimensionalizing the distance along the beam x and displacement $r(x)$ through division by the beam length L, we obtain

$$\frac{d^4 r(\xi)}{d\xi^4} + \frac{PL^2}{EI}\frac{d^2 r(\xi)}{d\xi^2} - \frac{\rho A\omega^2 L^4}{EI} r(\xi) = 0 \tag{9.5}$$

or

$$\frac{d^4 r(\xi)}{d\xi^4} + \pi^2\alpha\frac{d^2 r(\xi)}{d\xi^2} - \beta^4 r(\xi) = 0 \tag{9.6}$$

where

$$\xi = \frac{x}{L} \qquad \beta^4 = \frac{\rho A\omega^2 L^4}{EI}$$

and

$$\alpha = \frac{PL^2}{\pi^2 EI}$$

which is the ratio of axial compression load to the Euler buckling load for a beam with simple support at each end. The general solution for Eq. 9.6 is expressed as

$$r(\xi) = A\sin a\xi + B\cos a\xi + C\sinh b\xi + D\cosh b\xi \tag{9.7}$$

where

$$a = \sqrt{\frac{\pi^2\alpha}{2} + \sqrt{\frac{\pi^4\alpha^2}{4} + \beta^4}} \tag{9.8}$$

and

$$b = \sqrt{\frac{-\pi^2\alpha}{2} + \sqrt{\frac{\pi^4\alpha^2}{4} + \beta^4}} \tag{9.9}$$

The next step is to obtain those values of β that permit a nontrivial solution for Eq. 9.7, subject to the prescribed boundary conditions. With permissible values of β obtained, the beam frequencies are readily established according to the relationship

$$f = \frac{\beta^2}{2\pi L^2}\sqrt{\frac{EI}{\rho A}} \tag{9.10}$$

From the boundary condition

$$r(\xi) = 0|_{\xi=0}$$

we obtain

$$B = -D \tag{9.11}$$

and from the condition

$$\left.\frac{d^2r(\xi)}{d\xi^2} = 0\right|_{\xi=0}$$

we obtain

$$-Ba^2 + Db^2 = -B(a^2+b^2) = 0 \tag{9.12}$$

Since a^2+b^2 cannot equal zero, we require that $B = D = 0$. Utilizing the remaining two boundary conditions, that is,

$$r(\xi) = \left.\frac{d^2r(\xi)}{d\xi^2} = 0\right|_{\xi=1}$$

we obtain the homogeneous algebraic equations

$$A \sin a + C \sinh b = 0 \tag{9.13}$$

and

$$-Aa^2 \sin a + Cb^2 \sinh b = 0 \tag{9.14}$$

The requirement that a nontrivial solution exist for the constants A and C of the above equations (i.e., that the determinant of their coefficient matrix vanish) is expressed by the relationship

$$(a^2+b^2) \sin a \sinh b = 0 \tag{9.15}$$

and, since (a^2+b^2) must not equal zero, we have

$$\sin a = 0 \tag{9.16}$$

that is,

$$a = n\pi$$

where $n = 1, 2, 3$, and so on.

Substituting for a in Eq. 9.8, it is readily shown that

$$\beta_n^4 = n^2\pi^4(n^2 - \alpha) \tag{9.17}$$

It is of course noted that for α equal zero, that is, no applied compression load, Eq. 9.17 gives results pertaining to a beam with simple-simple support as discussed in Chapter 1 (see Table 1.1).

We next focus our attention on the modal shapes associated with each vibration frequency. Referring to Eq. 9.13, it is seen that we may write

$$C = \frac{-A \sin a}{\sinh b} \tag{9.18}$$

and, setting the constant A equal to unity, we obtain for the modal shape expression

$$r(\xi) = \sin a\xi - \frac{\sin a \sinh b\xi}{\sinh b} = \sin a\xi \tag{9.19}$$

since $\sin a = 0$.

It is recalled that α is a measure of how close the compressive axial load approaches the Euler buckling load. It is obvious therefore that α can never take on a value equal to or greater than that associated with buckling for the particular boundary conditions enforced. However, negative values for α represent tensile axial forces, and there is no similar limit on the magnitude of the negative values. It is also noted that tensile axial forces have the effect of increasing the vibration frequency (Eq. 9.17). This is to be expected, since they have the effect of increasing the elastic restoring forces when the beam is disturbed from its neutral configuration.

As stated earlier, beams with three different sets of boundary conditions are considered here. Only in the case discussed above can the eigenvalues β be expressed in a concise analytic form as in Eq. 9.17. For the other two cases the results are expressed for the first six modes in graphical form. In the graphs the ratio of the eigenvalue β, to the known eigenvalue with zero axial loading (Table 1.1), is presented. The range of axial loading covered extends from the beam compressive buckling load to a tensile load equal in magnitude to 10 times the Euler buckling load of a beam with simple-simple support. In presentation of the results, each case is given a reference number. The beam is then illustrated in a small sketch. Information pertaining to the vibration frequencies and modal shapes is supplied.

CASE 9.1. Beam with simple support at each end and subjected to an axial compressive load P.

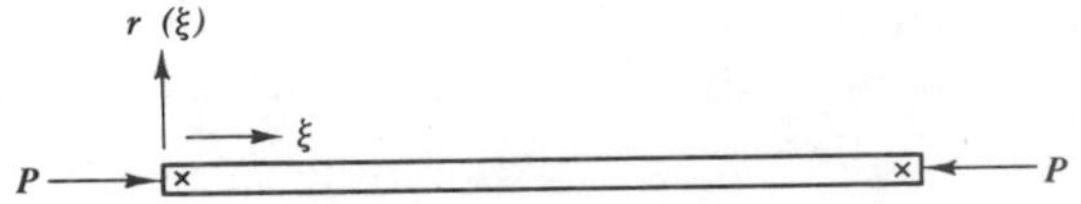

r (ξ)
ξ
P
P

$$\beta_n^4 = n^2\pi^4(n^2 - \alpha)$$

$$r(\xi) = \sin a\xi$$

where

$$\alpha = \frac{PL^2}{\pi^2 EI}$$

$$a = \sqrt{\frac{\pi^2\alpha}{2} + \sqrt{\frac{\pi^4\alpha^2}{4} + \beta^4}}$$

CASE 9.2. Beam clamped at each end and subjected to axial compressive load P.

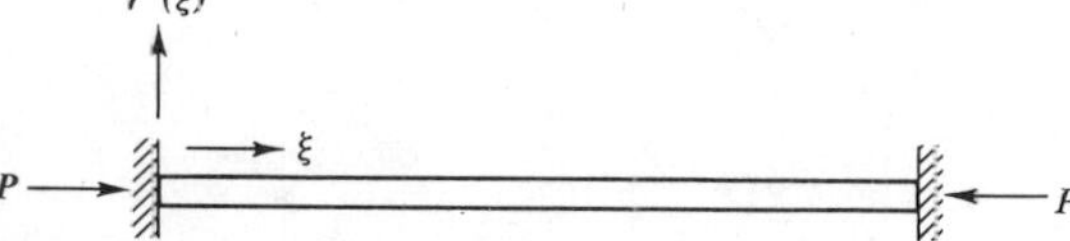

The determinant equation is

$$\cos a \cosh b + \frac{(a^2 - b^2)}{2ab} \sin a \sinh b - 1 = 0$$

where

$$\alpha = \frac{PL^2}{\pi^2 EI} \qquad a = \sqrt{\frac{\pi^2\alpha}{2} + \sqrt{\frac{\pi^4\alpha^2}{4} + \beta^4}}$$

$$b = \sqrt{\frac{-\pi^2\alpha}{2} + \sqrt{\frac{\pi^4\alpha^2}{4} + \beta^4}} \qquad \beta_n \quad \text{(see Fig. 9.2)}$$

$$r(\xi) = \sin a\xi - \frac{a}{b} \sinh b\xi + a \frac{\cos a - \cosh b}{a \sin a + b \sinh b} (\cos a\xi - \cosh b\xi)$$

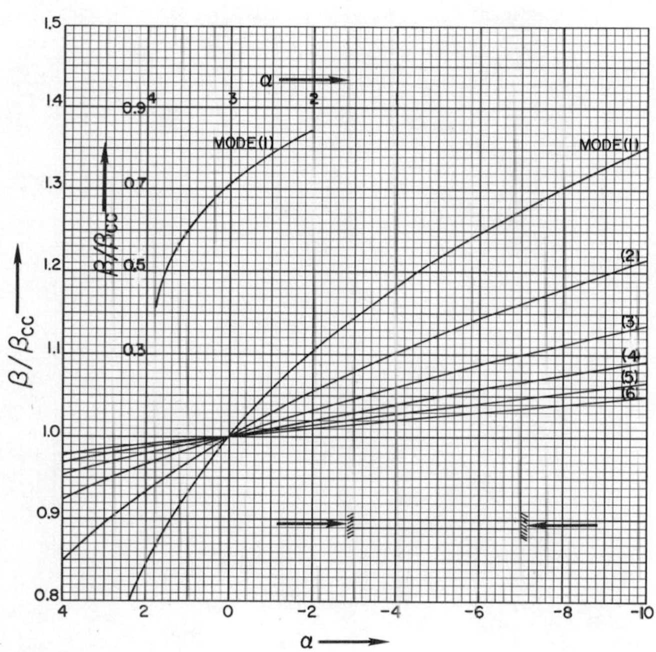

Fig. 9.2

CASE 9.3. Beam clamped at one end, simply supported at the other, and subjected to axial compressive load P.

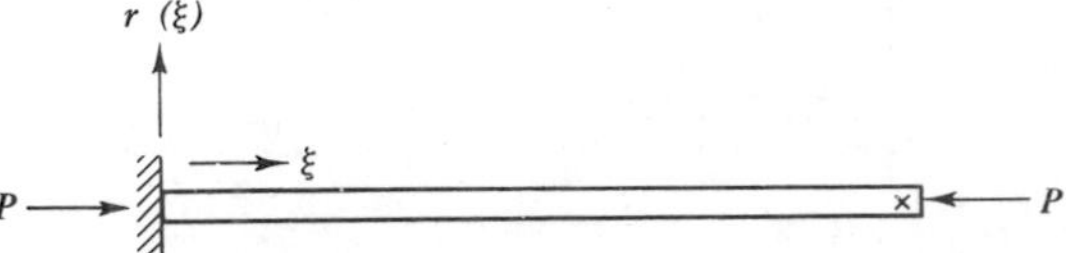

The determinant equation is

$$a \cos a \sinh b - b \sin a \cosh b = 0$$

where

$$\alpha = \frac{PL^2}{\pi^2 EI} \qquad a = \sqrt{\frac{\pi^2\alpha}{2} + \sqrt{\frac{\pi^4\alpha^2}{4} + \beta^4}}$$

$$b = \sqrt{\frac{-\pi^2\alpha}{2} + \sqrt{\frac{\pi^4\alpha^2}{4} + \beta^4}} \qquad \beta_n \quad \text{(see Fig. 9.3)}$$

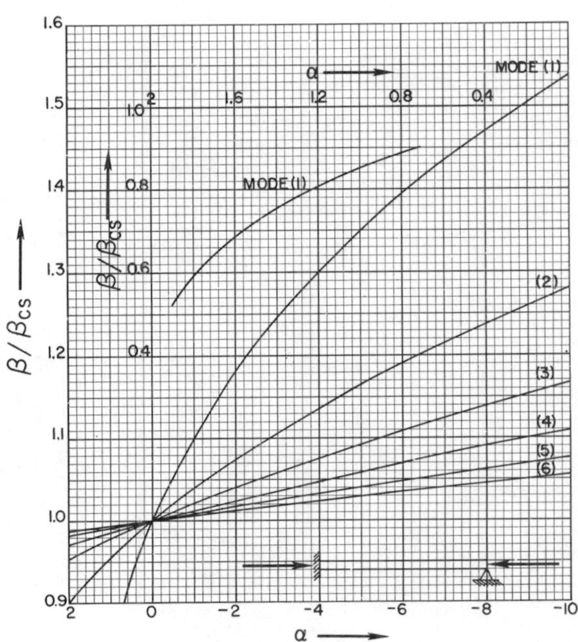

Fig. 9.3

$$r(\xi) = \sin a\xi - \frac{a}{b}\sinh b\xi + \frac{\sin a - \frac{a}{b}\sinh b}{\cosh b - \cos a}(\cos a\xi - \cosh b\xi)$$

Part 2. The Vibration of Tapered Beams

A list of articles related to this subject is presented at the end of the book.

1. *Linear Variation in Cross Section*

This section is concerned with a family of beams in which the shape of the cross section is the same at all points along the beam, however, the transverse dimensions, for example, width and depth, as well as wall thickness in the case of hollow beams, are a linear function of distance along the beam.

Consider such a tapered beam as shown in Fig. 9.4. For the purpose of analysis it is advantageous to select the apex of the taper triangle as the origin.

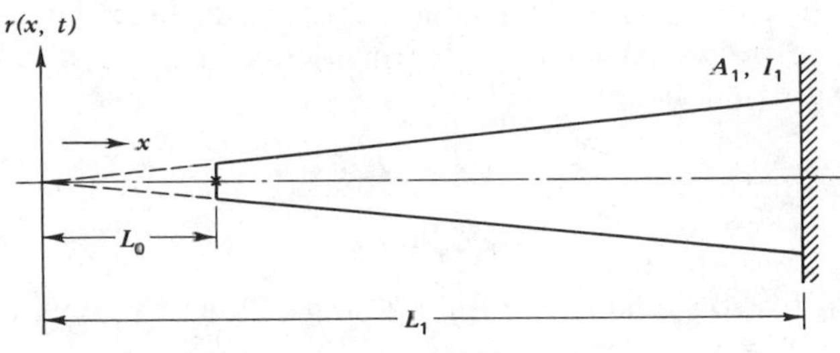

Fig. 9.4

The beam has simple support at one end and is clamped at the other. We denote the cross-sectional area and moment of inertia at the large (clamped) end as A_1 and I_1, respectively.

We look next at the equilibrium equation for a tapered beam. This equation differs only slightly from that for a uniform beam (Eq. 1.1) and, referring to the theory of strength of materials, we find that the equation becomes

$$\frac{\partial^2}{\partial x^2}\left[EI(x)\frac{\partial^2 v(x,t)}{\partial x^2}\right] = -\rho A(x)\frac{\partial^2 v(x,t)}{\partial t^2} \tag{9.20}$$

where x is distance measured along the beam. By the method of separation of variables as utilized in Chapter 1, Eq. 9.20 may be reduced to two ordinary differential equations as follows,

$$\frac{d^2p(t)}{dt^2} + \omega^2 p(t) = 0 \tag{9.21}$$

and

$$\frac{d^2}{dx^2}\left[EI(x)\frac{d^2r(x)}{dx^2}\right] - \rho A(x)\omega^2 r(x) = 0 \tag{9.22}$$

We next introduce ξ (see Fig. 9.5), where ξ equals distance from the origin to any point along the beam, divided by the height of the taper

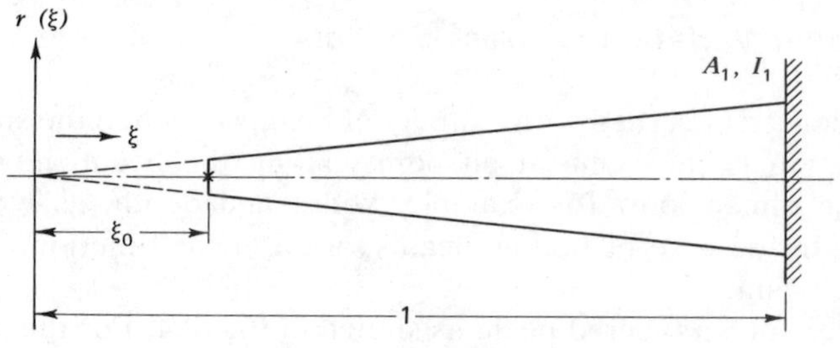

Fig. 9.5

triangle L_1 (see Fig. 9.4), and we nondimensionalize beam lateral displacement with respect to the same reference length L_1. $A(\xi)$ and $I(\xi)$ may now be expressed as

$$A(\xi) = A_1\xi^2 \tag{9.23}$$

and

$$I(\xi) = I_1\xi^4 \tag{9.24}$$

Substituting for $A(\xi)$ and $I(\xi)$ in Eq. 9.22 and dividing by ξ^2, we obtain

$$\frac{1}{\xi^2}\left[\frac{d^2}{d\xi^2}\left(\xi^4\frac{d^2r(\xi)}{d\xi^2}\right)\right] - \beta^4 r(\xi) = 0 \tag{9.25}$$

where

$$\beta^4 = \frac{\rho A_1 L_1^4 \omega^2}{EI_1}$$

and $r(\xi)$ equals beam lateral displacement.

The solution to Eq. 9.25 is expressed as

$$r(\xi) = \xi^{-1}[C_1J_2(Z) + C_2Y_2(Z) + C_3I_2(Z) + C_4K_2(Z)] \tag{9.26}$$

where

$$Z = 2\beta\xi^{1/2}$$

J_2 and Y_2 are Bessel functions of second order and of first and second kind, respectively, and I_2 and K_2 are second order modified Bessel functions of first and second kind, respectively. C_1, C_2, and so on, are constants to be determined.

Enforcement of the boundary conditions of zero slope and bending moment requires utilization of the following expressions

$$\frac{dr(\xi)}{d\xi} = -\beta\xi^{-3/2}[C_1J_3(Z) + C_2Y_3(Z) - C_3I_3(Z) + C_4K_3(Z)] \quad (9.27)$$

and

$$\frac{d^2r(\xi)}{d\xi^2} = \beta^2\xi^{-2}[C_1J_4(Z) + C_2Y_4(Z) + C_3I_4(Z) + C_4K_4(Z)] \quad (9.28)$$

Enforcement of the condition of zero shear at the boundary is expressed as

$$\frac{d}{d\xi}\left(EI(\xi)\frac{d^2r(\xi)}{d\xi^2}\right) = EI_1\frac{d}{d\xi}\left(\xi^4\frac{d^2r(\xi)}{d\xi^2}\right) = 0 \quad (9.29)$$

at the boundary where

$$\frac{d}{d\xi}\left(\xi^4\frac{d^2r(\xi)}{d\xi^2}\right) = \beta^3\xi^{3/2}[C_1J_3(Z) + C_2Y_3(Z) + C_3I_3(Z) - C_4K_3(Z)] \quad (9.30)$$

Referring to Fig. 9.5, it is seen that the appropriate boundary conditions to be applied for the problem in question are

$$r(\xi)\Big|_{\xi=\xi_0} = \frac{d^2r(\xi)}{d\xi^2}\Big|_{\xi=\xi_0} = 0 \qquad \text{and} \qquad \frac{dr(\xi)}{d\xi}\Big|_{\xi=1} = r(\xi)\Big|_{\xi=1} = 0$$

Enforcing these four boundary conditions, we obtain four simultaneous homogeneous algebraic equations for the unknowns C_1, C_2, and so on. The requirement that a nontrivial solution exists for these unknowns is satisfied by requiring that the determinant of their coefficient matrix vanishes. We therefore obtain the determinant equation shown below.

$$\begin{vmatrix} J_2(Z_0) & Y_2(Z_0) & I_2(Z_0) & K_2(Z_0) \\ J_4(Z_0) & Y_4(Z_0) & I_4(Z_0) & K_4(Z_0) \\ J_2(2\beta) & Y_2(2\beta) & I_2(2\beta) & K_2(2\beta) \\ J_3(2\beta) & Y_3(2\beta) & -I_3(2\beta) & K_3(2\beta) \end{vmatrix} = 0 \quad (9.31)$$

where

$$Z_0 = 2\beta\xi_0^{1/2}$$

The problem is solved by finding those values of β, hence ω^2, that satisfy Eq. 9.31. It is seen that $L = L_1(1-\xi_0)$, where L is equal to the actual beam length. At this time it is advantageous to introduce the quantity $\beta^* = \beta(1-\xi_0)$. We can then write

$$\beta^{*4} = \frac{\rho A_1}{EI_1}\omega^2L^4 \quad (9.32)$$

With β^* known, the frequency of vibration of the beam can then be written as

$$f = \frac{\omega}{2\pi} = \frac{\beta^{*2}}{2\pi L^2}\sqrt{\frac{\rho A_1}{EI_1}} \tag{9.33}$$

Values of β^* versus b_0/b_1 are plotted for the first six modes of vibration

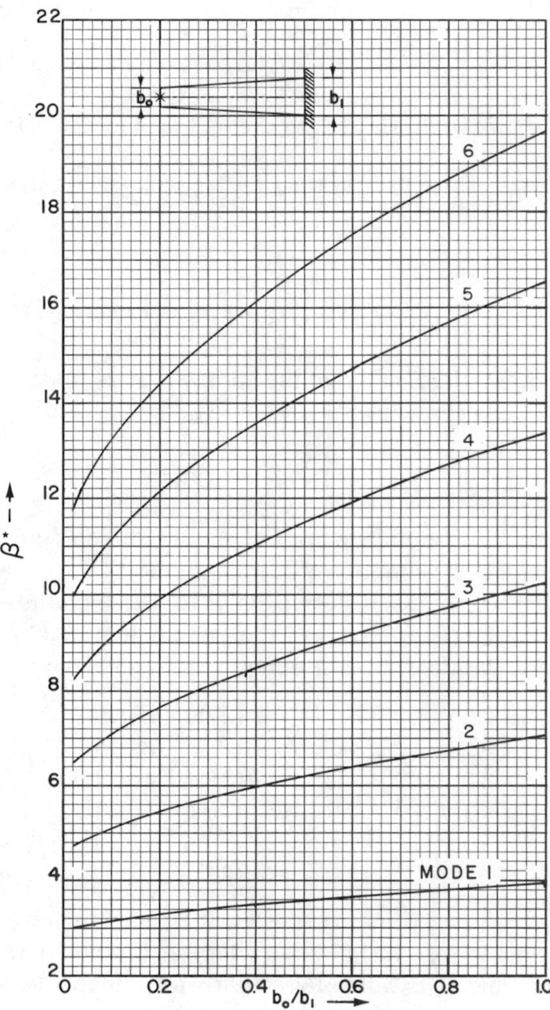

Fig. 9.6 Frequency chart for beam with linear variation in cross section (simple-clamped support).

of various beams in the class under discussion in Figs. 9.6 through 9.11. The ratio $b_0/b_1 = \xi_0$ is the ratio of a transverse dimension at the narrow end to the corresponding transverse dimension at the other end. Each graphical presentation has an inset figure showing the boundary conditions to which the curves apply. Values of b_0/b_1 generally range from .02 to 1.0. It is noted that in the limit as b_0/b_1 approaches unity (a condition of zero taper) the

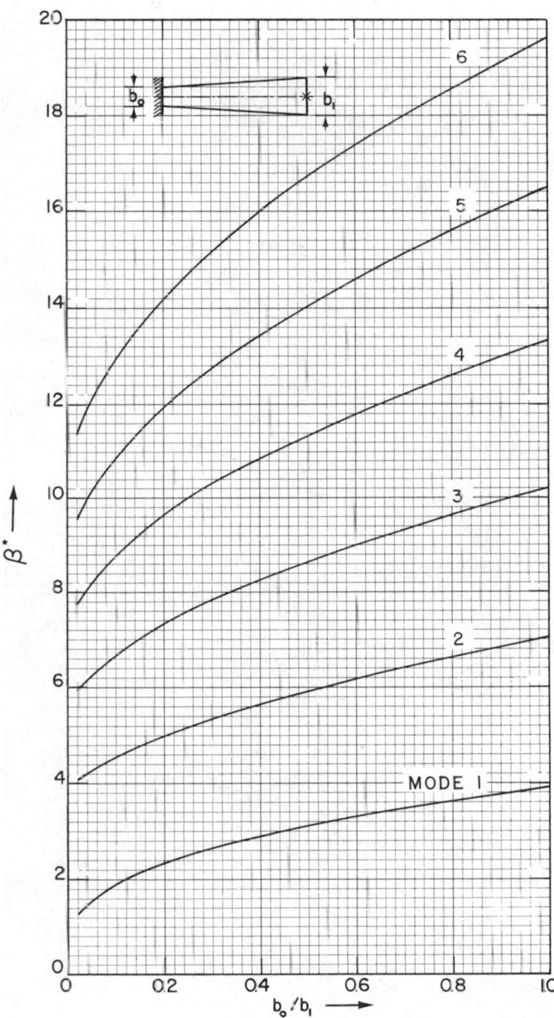

Fig. 9.7 Frequency chart for beam with linear variation in cross section (clamped-simple support).

values of β^* approach those of uniform beams as presented in Table 1.1.

We next turn our attention to the modal shapes associated with the vibration. Once the appropriate value of β^* is obtained from the tables, the value of β is obtained from the relationship $\beta = \beta^*/(1-\xi_0)$. With β established we then write three of the four boundary conditions in which

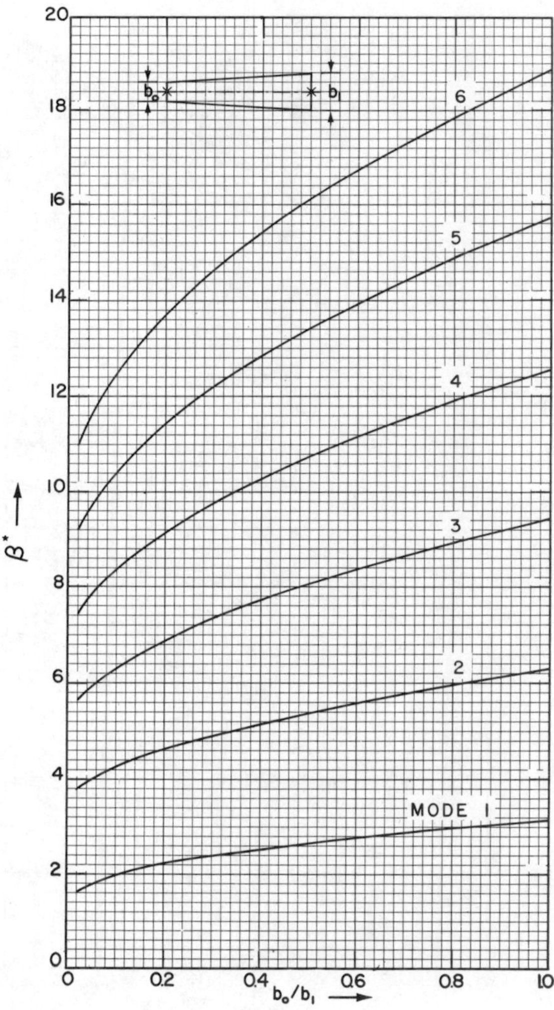

Fig. 9.8 Frequency chart for beam with linear variation in cross section (simple-simple support).

the constants C_1 through C_4 (Eq. 9.26) are the only unknowns. All these equations are already available in the set 9.26 through 9.30. The appropriate value of ξ is either ξ_0 or 1, depending on the end of the beam at which the boundary conditions are enforced. The set of three boundary conditions selected provides three linear homogeneous equations. Following the same

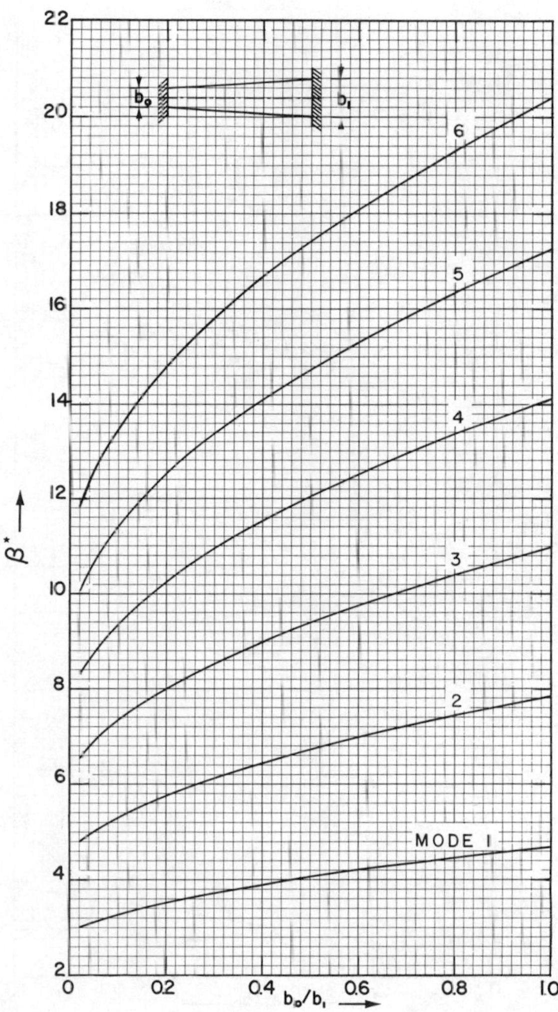

Fig. 9.9 Frequency chart for beam with linear variation in cross section (clamped-clamped support).

practice as in earlier chapters, we can make them nonhomogeneous by setting one of the unknowns, say C_1, equal to unity. The value of the other three unknowns can then be readily established (using Cramer's rule), and the modal shape is obtained by substituting the results in Eq. 9.26. Should a case arise in which C_1 can have only the value zero, any one of the other

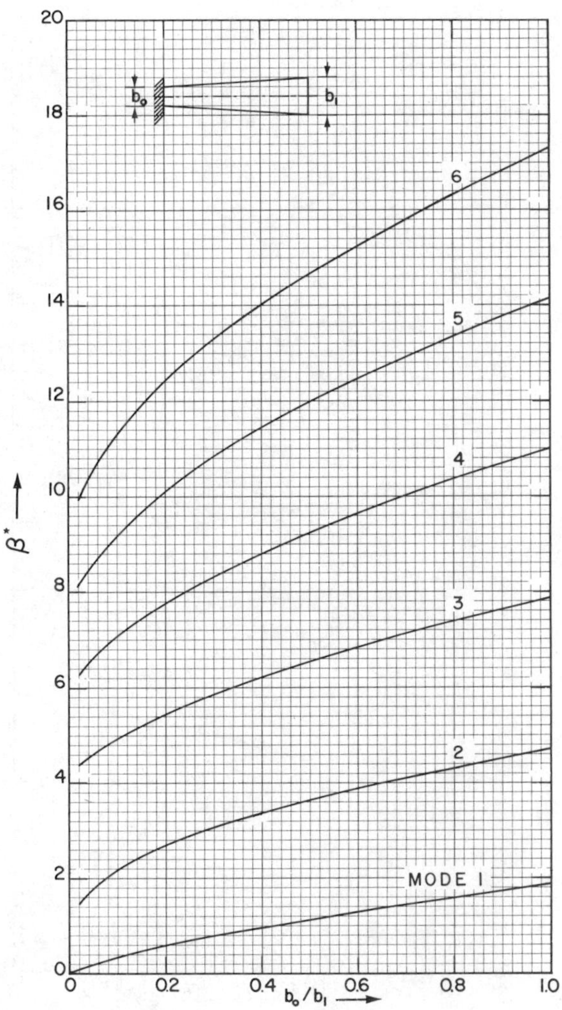

Fig. 9.10 Frequency chart for beam with linear variation in cross section (clamped-free support).

nonzero unknowns could be set equal to unity in order to make the set homogeneous. In view of the fact that only classical boundary conditions are discussed here, and the ease with which they can be applied, the nonhomogeneous set of equations required for assigning values to the unknowns C_1, C_2, and so on, are not provided for each individual case.

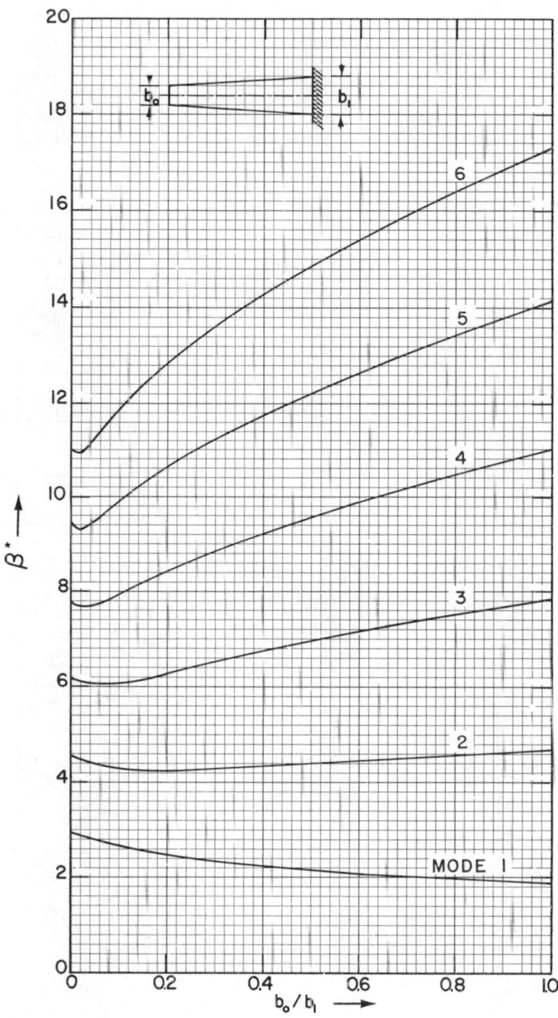

Fig. 9.11 Frequency chart for beam with linear variation in cross section (free-clamped support).

2. *Wedge-Shaped Beams*

This section is concerned with a family of beams that are tapered in such a way that all transverse dimensions of the beam cross section, taken in the plane of vibration, are a linear function of distance along the beam. Dimensions taken in directions normal to the plane of vibration are constant along the beam. Such beams are said to be wedge-shaped.

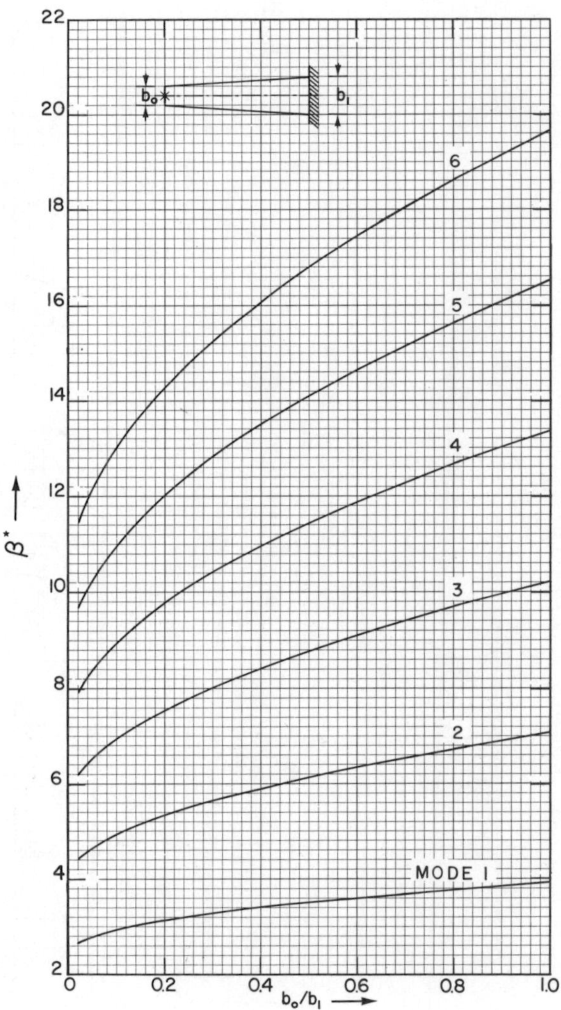

Fig. 9.12 Frequency chart for wedge-shaped beam (simple-clamped support).

Proceeding as in Section 1, it is observed that we may write

$$A(\xi) = A_1\xi \tag{9.34}$$

and

$$I(\xi) = I_1\xi^3 \tag{9.35}$$

Substituting in the differential equation 9.22, we obtain

$$\frac{d^2}{d\xi^2}\left(\xi^3\frac{d^2r(\xi)}{d\xi^2}\right) - \beta^4\xi r(\xi) = 0 \tag{9.36}$$

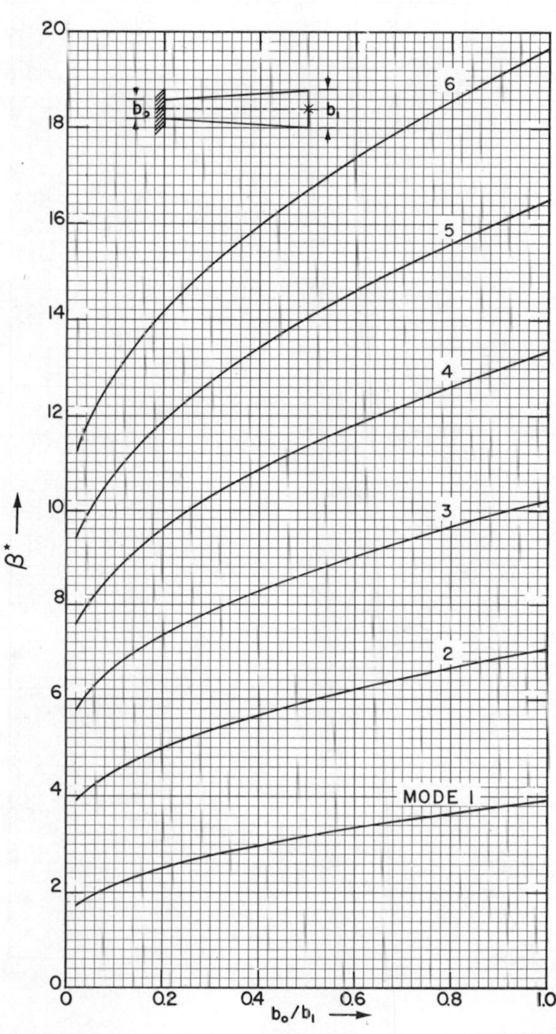

Fig. 9.13 Frequency chart for wedge-shaped beam (clamped-simple support).

for which the solution is

$$r(\xi) = \xi^{-1/2}[C_1 J_1(Z) + C_2 Y_1(Z) + C_3 I_1(Z) + C_4 K_1(Z)] \tag{9.37}$$

where

$$Z = 2\beta\xi^{1/2} \qquad \text{and } \beta^4 = \frac{\rho A_1 \omega^2 L_1^4}{EI_1}$$

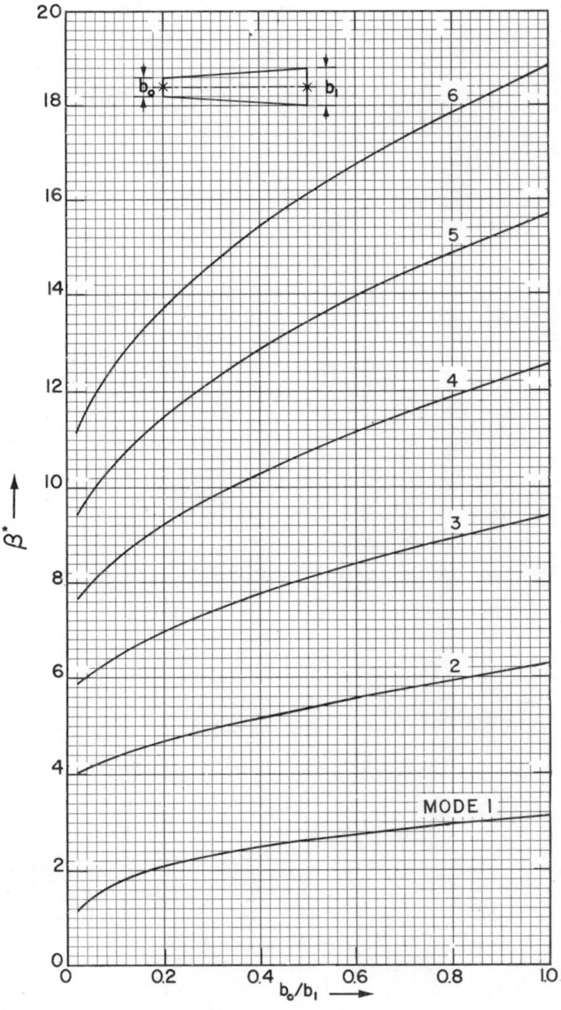

Fig. 9.14 Frequency chart for wedge-shaped beam (simple-simple support).

To apply the boundary conditions, other than the zero shear condition, we require

$$\frac{dr(\xi)}{d\xi} = -\beta\xi^{-1}[C_1J_2(z) + C_2Y_2(z) - C_3I_2(z) + C_4K_2(z)] \tag{9.38}$$

$$\frac{d^2r(\xi)}{d\xi^2} = \beta^2\xi^{-3/2}[C_1J_3(Z) + C_2Y_3(Z) + C_3I_3(Z) + C_4K_3(Z)] \tag{9.39}$$

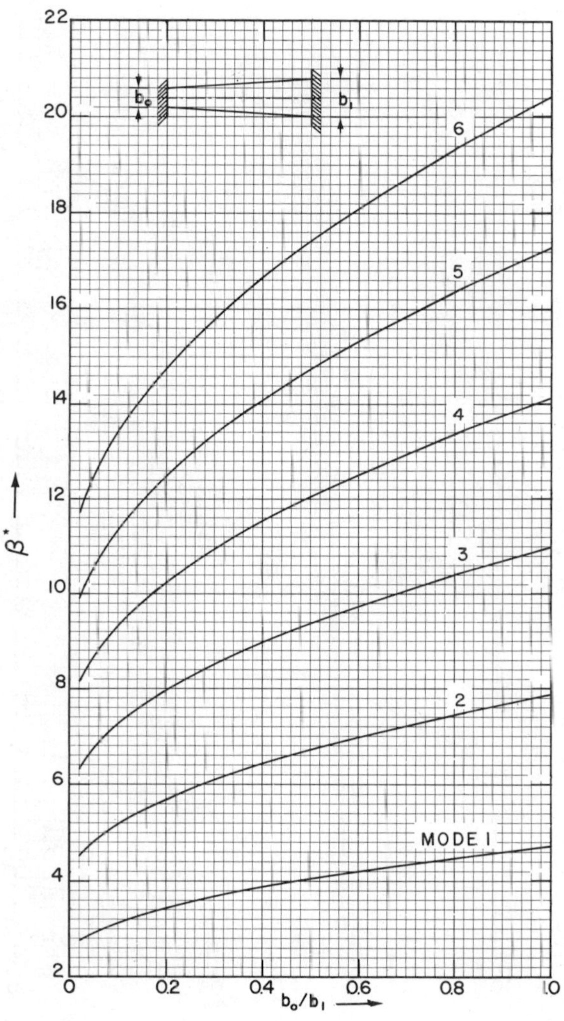

Fig. 9.15 Frequency chart for wedge-shaped beam (clamped-clamped support).

and to apply the zero shear condition we require (see Section 1)

$$\frac{d}{d\xi}\left(\xi^3 \frac{d^2 r(\xi)}{d\xi^2}\right) = \beta^3 \xi [C_1 J_2(Z) + C_2 Y_2(Z) + C_3 I_2(Z) - C_4 K_2(Z)] \quad (9.40)$$

Eigenvalues β^* (see Section 1) for this family of beams are obtained in a

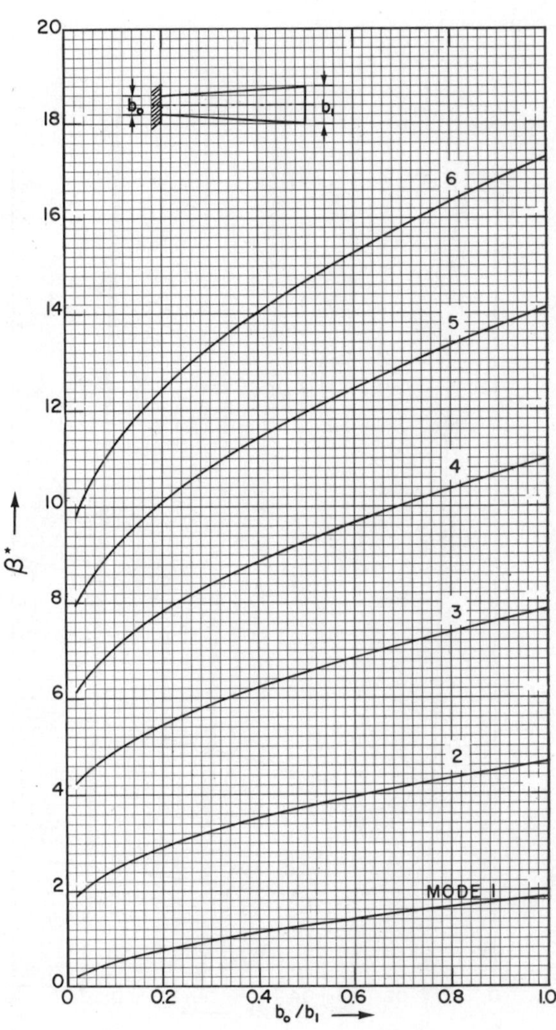

Fig. 9.16 Frequency chart for wedge-shaped beam (clamped-free support).

manner identical to that described in Section 1. Values of β^* versus b_0/b_1 are plotted in Figs. 9.12 through 9.17 for beams with various combinations of boundary conditions. Lateral vibration frequencies are obtained from Eq. 9.33. Modal shapes may be obtained by following the procedure described in Section 1.

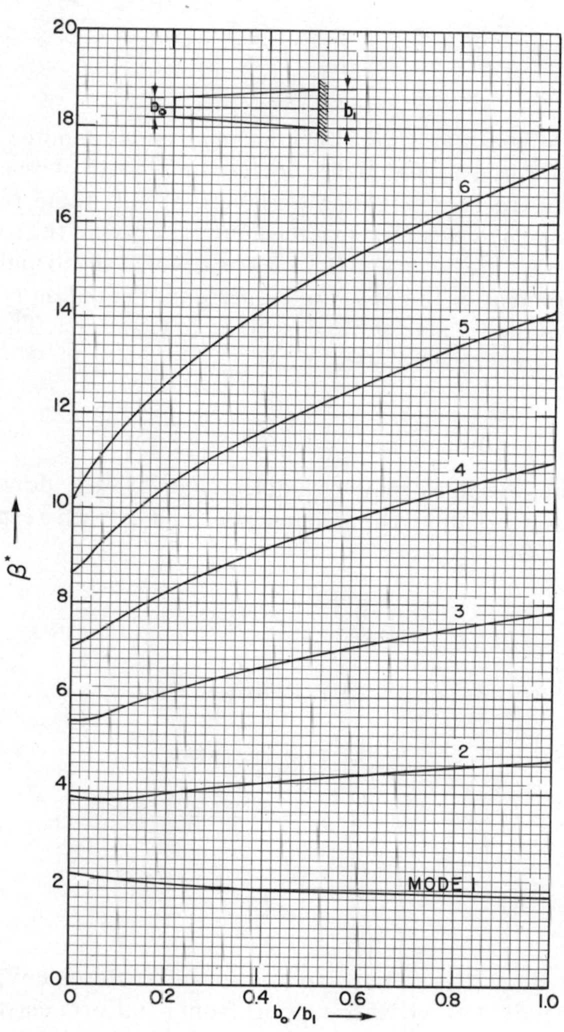

Fig. 9.17 Frequency chart for wedge-shaped beam (free-clamped support).

Part 3. The Vibration of Uniform Beams on Elastic Foundations

The foundation on which a beam rests is said to be an elastic foundation if it exerts a restoring force per unit length of beam proportional to the beam displacement. Such a beam is shown schematically in Fig. 9.18. The

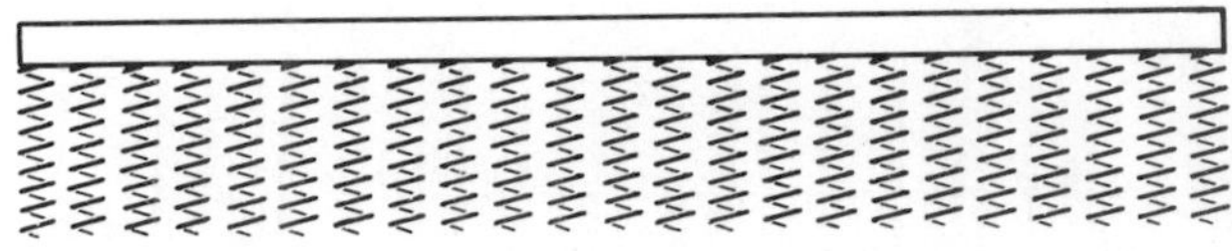

Fig. 9.18

foundation stiffness is uniform along the beam and is denoted as k, where k has the units of force per unit deflection per unit length of beam. Examples of such beams are beams resting on layers of soft foam rubber. In the theoretical model it is assumed that the inertial effects of the foundation are negligible and that the beam and foundation never cease to make contact.

The development of the governing differential equation is achieved by following steps identical to those described in Chapter 1. The equilibrium equation may be written as (Eq. 1.1)

$$\frac{EI}{\rho A}\frac{\partial^4 v(x,t)}{\partial x^4}+\frac{k}{\rho A}v(x,t)=-\frac{\partial^2 v(x,t)}{\partial t^2} \tag{9.41}$$

Separating the variables in the same manner as was done in Eqs. 1.2 through 1.6 and nondimensionalizing $r(x)$ and x with respect to beam overall length L, we obtain

$$\frac{d^4 r(\xi)}{d\xi^4}-\beta^4 r(\xi)=0 \tag{9.42}$$

and

$$p(t)=\cos(\omega t-\alpha) \tag{9.43}$$

where

$$v(x,t)=r(x)p(t)$$

$$\beta^4=\frac{\rho A\omega^2 L^4}{EI}-k^* \tag{9.44}$$

and

$$k^*=\frac{kL^4}{EI}$$

It is seen that Eq. 9.42 is identical to Eq. 1.6, and so the eigenvalue β defined here is independent of whether support from a uniform elastic foundation is provided or not. We therefore have available the eigenvalue β, or can obtain it from the Q factor provided, for all uniform beam vibration problems considered up to this time. The modal shapes for all the uniform beams are also independent of whether or not elastic foundation support is provided.

The only parameter affected by the presence of an elastic foundation is the frequency. Returning to Eq. 9.44, we may write

$$\frac{\rho A \omega^2 L^4}{EI} = \beta^4 + k^* \tag{9.45}$$

whereas, without an elastic foundation we have

$$\frac{\rho A \omega^2 L^4}{EI} = \beta^4 \tag{9.46}$$

Denoting the frequency of vibration of the beam without an elastic foundation as f_c, we may then write

$$\frac{f}{f_c} = \sqrt{1 + \frac{k^*}{\beta^4}} \tag{9.47}$$

or, for a beam on an elastic foundation,

$$f = f_c \sqrt{1 + \frac{k^*}{\beta^4}} \tag{9.48}$$

where β is the eigenvalue for the beam without an elastic foundation. We note, as expected, that as k^* approaches infinity the beam frequency also approaches infinity, and as k^* approaches zero the frequency approaches that of a beam without elastic foundation.

ILLUSTRATIVE PROBLEM 9.1

The aluminum cantilever beam in Illustrative Problem 1.1 is supported on an elastic foundation of stiffness 1.5 lb/in.^{2}.

(a) Determine the fundamental frequency of vibration of this beam if the coil spring is removed from the outer end.

$$f_c = f_{cF} = 17.8 \text{ Hz} \qquad \text{(Illustrative Problem 1.1)}$$

$$\beta = 1.875$$

$$k^* = \frac{kL^4}{EI} = \frac{1.5 \times L^4}{EI} = 11.66$$

$$f = f_c \sqrt{1 + \frac{k^*}{\beta^4}} = 17.8 \times 1.394 = 24.8 \text{ Hz}$$

(b) Determine the fundamental frequency of vibration of the above beam if the coil spring of Illustrative Problem 1.1 ($k = 80.0$ lb/in.) is connected to the outer end.

$$f_c = 27.2 \text{ Hz} \qquad \text{(Illustrative Problem 1.1)}$$

$$\frac{\beta}{\beta_{cF}} = \sqrt{\frac{f}{f_{cF}}}$$

$$\beta = 1.875\sqrt{(27.2/17.8)} = 2.318$$

$$f = f_c\sqrt{1 + \frac{k^*}{\beta^4}} = 27.2 \times 1.185 = 32.2 \text{ Hz}$$

REFERENCES

BOOKS

1. K. N. Tong, *Theory of Mechanical Vibration*, John Wiley and Sons, 1960.
2. S. Timenshenko, *Vibration Problems in Engineering*, 3rd ed., D. Van Nostrand, 1955.
3. J. P. Den Hartog, *Mechanical Vibration*, 4th ed., McGraw-Hill, 1956.
4. W. T. Thomson, *Theory of Vibration with Applications*, Prentice-Hall, 1972.
5. C. M. Harris and C. E. Crede, *Shock and Vibration Handbook*, McGraw-Hill, 1961.
6. R. F. Steidel, Jr., *An Introduction to Mechanical Vibrations*, John Wiley and Sons, 1971.
7. L. Meirovitch, *Analytical Methods in Vibrations*, Macmillan, 1967.

ARTICLES

8. H. D. Conway, E. C. H. Becker, and J. F. Dubil, "Vibration Frequencies of Tapered Bars and Circular Plates," *Journal of Applied Mechanics*, Vol. 31, June 1964, pp. 329–331.
9. H. D. Conway and J. F. Dubil, "Vibration Frequencies of Truncated-Cone and Wedge Beams," *Journal of Applied Mechanics*, Vol. 32, December 1965, pp. 932–934.
10. E. T. Cranch and Alfred A. Adler, "Bending Vibrations of Variable Section Beams," *Journal of Applied Mechanics*, Vol. 23, March 1956, pp. 103–108.

INDEX